Handbook of Digital Techniques for High-Speed Design

Prentice Hall Modern Semiconductor Design Series

James R. Armstrong and F. Gail Gray
VHDL Design Representation and Synthesis

Mark Gordon Arnold
Verilog Digital Computer Design: Algorithms into Hardware

Jayaram Bhasker
A VHDL Primer, Third Edition

Mark D. Birnbaum
Essential Electronic Design Automation (EDA)

Eric Bogatin
Signal Integrity: Simplified

Douglas Brooks
Signal Integrity Issues and Printed Circuit Board Design

Kanad Chakraborty and Pinaki Mazumder
Fault-Tolerance and Reliability Techniques for High-Density Random-Access Memories

Ken Coffman
Real World FPGA Design with Verilog

Alfred Crouch
Design-for-Test for Digital IC's and Embedded Core Systems

Daniel P. Foty
MOSFET Modeling with SPICE: Principles and Practice

Tom Granberg
Handbook of Digital Techniques for High-Speed Design

Nigel Horspool and Peter Gorman
The ASIC Handbook

Howard Johnson and Martin Graham
High-Speed Digital Design: A Handbook of Black Magic

Howard Johnson and Martin Graham
High-Speed Signal Propagation: Advanced Black Magic

William K. Lam
Hardware Design Verification: Simulation and Formal Method-Based Approaches

Pinaki Mazumder and Elizabeth Rudnick
Genetic Algorithms for VLSI Design, Layout, and Test Automation

Farzad Nekoogar and Faranak Nekoogar
From ASICs to SOCs: A Practical Approach

Farzad Nekoogar
Timing Verification of Application-Specific Integrated Circuits (ASICs)

Samir Palnitkar
Design Verification with **e**

David Pellerin and Scott Thibault
Practical FPGA Programming in C

David Pellerin and Douglas Taylor
VHDL Made Easy!

Christopher T. Robertson
Printed Circuit Board Designer's Reference: Basics

Chris Rowen
Engineering the Complex SOC

Frank Scarpino
VHDL and AHDL Digital System Implementation

Wayne Wolf
FPGA-Based System Design

Wayne Wolf
Modern VLSI Design: System-on-Chip Design, Third Edition

Kiat-Seng Yeo, Samir S. Rofail, and Wang-Ling Goh
CMOS/BiCMOS ULSI: Low Voltage, Low Power

Brian Young
Digital Signal Integrity: Modeling and Simulation with Interconnects and Packages

Bob Zeidman
Verilog Designer's Library

Handbook of Digital Techniques for High-Speed Design

Design Examples, Signaling and Memory Technologies, Fiber Optics, Modeling and Simulation to Ensure Signal Integrity

Tom Granberg, Ph.D.

PRENTICE HALL PTR
UPPER SADDLE RIVER, NJ 07458
WWW.PHPTR.COM

Library of Congress Cataloging-in-Publication Data

Granberg, Tom.
Handbook of digital techniques for high-speed design : design examples, signaling and memory technologies, fiber optics, modeling and simulation to ensure signal integrity / Tom Granberg.
p. cm.
Includes bibliographical references and index.
ISBN 0-13-142291-X
1. Signal processing--Digital techniques--Handbooks, manuals, etc. 2. Very high speed integrated circuits--Design and construction--Handbooks, manuals, etc. I. Title.

TK5102.9.G72 2004
621.382'2--dc22

2004046509

Editorial/production supervision: *Vanessa Moore*
Cover design director: *Jerry Votta*
Cover designer: *Jennifer Bergamini*
Manufacturing manager: *Alexis R. Heydt-Long*
Manufacturing buyer: *Maura Zaldivar*
Acquisitions editor: *Bernard Goodwin*
Editorial assistant: *Michelle Vincenti*
Marketing manager: *Dan Depasquale*

Publishing as Prentice Hall PTR
Upper Saddle River, New Jersey 07458

Prentice Hall offers excellent discounts on this book when ordered in quantity for bulk purchases or special sales. For more information, please contact: U.S. Corporate and Government Sales, 1-800-382-3419, corpsales@pearsontechgroup.com. For sales outside of the U.S., please contact: International Sales, 1-317-581-3793, international@pearsontechgroup.com.

Printed in the United States of America, Fifth Printing, August 2006

ISBN 0-13-142291-X

Pearson Education Ltd.
Pearson Education Australia Pty., Limited
Pearson Education South Asia Pte. Ltd.
Pearson Education Asia Ltd.
Pearson Education Canada, Ltd.
Pearson Educación de Mexico, S.A. de C.V.
Pearson Education—Japan
Pearson Malaysia S.D.N. B.H.D.

To
Randy Kriegh,
the world's nicest boss
and a man of high personal integrity.

CONTENTS

Part 2 Signaling Technologies and Devices 49

Preface

How This Book Is Organized

This book was designed from the beginning to be used as both an engineering textbook in colleges and universities and as an in-depth reference for practicing engineers. Chapters in the book are grouped by topic or function. The chapters would likely be most enjoyable if they were not read in numerical sequence but according to the reader's interests and goals, or the goals of the classroom. Chapters 4 through 9 discuss high-speed signaling technologies in so much detail that reading them all sequentially would result in glazed eyes and cure the worst insomniac!

You can gain a good appreciation for the content by taking a few moments to scan the table of contents, which is titled simply "Contents." This will not only give you an understanding of the book's subject matter but also the wide range of topics related to high-speed digital design. For use as a reference, you can, of course, just flip to the topic of interest.

Most readers will find a quick reading of Chapters 1 through 3 worthwhile. These provide a quick introduction to trends in high-speed design, SerDes and bus technologies, and very basic background information on signal integrity. After reading the first three chapters, you can read chapters according to your interest or the goals of the university course.

Chapter 4 starts off the discussion of high-speed signaling technologies with GTL, GTLP, GTL+, and AGTL+. Chapter 5 introduces LVDS. Chapter 6 talks about the differences among BLVDS, LVDM, and M-LVDS. Chapter 7 details HSTL and SSTL. Chapter 8 categorizes and gives organization to the multitude of ECL devices out in the marketplace today. Chapter 9 discusses CML and explains how it differs from ECL.

Chapter 10 discusses high-speed features of today's Altera and Xilinx FPGAs such as 3.125 Gbps RocketIOs. It also discusses HardCopy devices, a shorter time-to-market and less expensive alternative to ASICs for some applications. Chapter 12 discusses high-speed interconnects and cabling. As we'll mention again below, Chapter 23 models and simulates 5-to-10 Gbps board-to-board interconnects.

Those interested in memory technologies should read Chapter 7 on HSTL and SSTL signaling technologies, which are used by memory devices, and then move on to Chapter 13, which provides an overview of memory technologies. After that, you can elect to read about the memory technologies that interest you the most in Chapters 14 through 18, which discuss the various memory technologies in detail.

Chapter 19 introduces the basic theory behind S-parameters and continues on into the more challenging realm of differential and mixed-mode S-parameters. N-port, mixed-mode S-parameters are discussed in the Appendix to Chapter 23. Given that all of today's high-speed serial links are differential in nature, differential and mixed-mode S-parameters have become very relevant topics and are frequently used today in the modeling of high-speed systems.

Chapter 20 discusses TDR and TDT, time-domain reflectometry and time-domain transmission. These techniques are also in frequent use today in both the simulation and testing of high-speed circuit boards.

Three-and-a-half chapters are dedicated to fiber optics due to its increasing importance in high-speed design. Chapter 11 is a broad introduction to fiber optics. Chapter 28 discusses electrical optical circuit boards and Chapter 24 discusses modeling and simulation of fiber-optic transceivers. The second half of Chapter 31 discusses the testing and measurement of optical systems in Section 31.6, which starts on page 880.

Two chapters introduce modeling and simulation and four other chapters provide major, detailed design examples. Chapter 21 introduces modeling with IBIS. Chapter 22 introduces tools from Mentor Graphics Corporation and talks about the flow of the design process. Chapter 23 discusses the advances in design, modeling, simulation, and validation of board-to-board 5-to-10 Gbps interconnects through an actual example. As mentioned before, Chapter 24 discusses modeling and simulation of fiber-optic transceivers, and Chapter 26 is an excellent design example which uses BLVDS SerDes devices. Chapter 27 is dedicated entirely to an example of designing 3.125 Gbaud high-speed serial links using Motorola's WarpLink serial devices. Chapter 25, while not presenting a detailed design example, does discuss every aspect of designing with LVDS.

Four chapters discuss LVDS and Bus LVDS. As stated before, Chapter 5 discusses the LVDS signaling standard and Chapter 6 discusses the BLVDS, LVDM, and M-LVDS signaling standards. Chapter 25 discusses how to design using LVDS and Chapter 26 is an excellent example of a detailed design using BLVDS SerDes devices.

Chapters 29 and 30 discuss the emerging protocols and technology behind RapidIO (Chapter 29) and PCI-Express (Chapter 30). These chapters provide an insight into what is being done to address the issues of faster and faster data rates in embedded systems, and ever-increasing bus speeds and system architecture complexity.

Finally, Chapter 31 is an informative discussion on the high-speed aspects of laboratory and test equipment, including such topics as equivalent-time oscilloscopes.

Throughout this textbook, there are numerous simulation examples and discussions of modeling. The reader is referred to the index for a detailed listing of "simulations" and "models."

This Textbook Was Written with Educational Institutions in Mind

This textbook was prepared from the beginning to be used not only by practicing engineers in the field working with high-speed digital electronics, but by instructors at universities and colleges who wish to teach high-speed digital design fundamentals to their students. High-speed digital design concepts and techniques can no longer be left for students to learn on the job after graduation. Today's clock rates in the hundreds of megahertz for typical applications and in the multigigahertz range for SerDes applications make this knowledge basic and fundamental to any electrical engineering degree.

Prior to this textbook, there has been no suitable textbook available from which to teach these required concepts. The handful of books written on this subject do not make good textbooks for a number of reasons. Two-thirds of these books have no exercises — no homework problems — and leave the extremely time-consuming task of developing these problems to the instructor who could better spend his or her time assisting students. Many of these books have no discussion of the various high-speed signaling technologies such as SSTL (Stub-Series-Terminated Logic), CML (Current-Mode Logic), or proprietary signaling technologies such as Motorola's controlled-impedance WarpLink technology. Today's CML is not a synonym for ECL as was the case two decades ago.

Many of these unsuitable-for-the-classroom books do not provide the basic background material on which high-speed design concepts are based. For example, many of them include no discussion of SerDes devices, double data rate (DDR) and quad data rate (QDR) devices, and no discussion of embedded clocks or echo clocks used by memory devices. They do not provide any discussion of the wide range of high-speed memory devices available today — SigmaRAM, Fast Cycle RAM, Network DRAM, Reduced Latency DRAM, Zero Bus Turnaround SRAM, and so on — at all. These books provide no background in TDR (time-domain reflectometry), which has changed considerably from years ago when it was used to find faults on power cables. The books that do mention it assume the reader is versatile in making TDR measurements. They provide no discussion of S-parameters, let alone the differential and mixed-mode S-parameters in use today, and leave it for the reader to hunt down microwave books or application notes for material on these subjects. Many books do not discuss how jitter is measured and how jitter masks are created and used.

Some of these books are too specialized and leave out the broad scope that this field invloves. Two signal integrity books I have seen are so mathematically rigorous that any course taught using them would be a math course and not an electrical engineering or design course. One is so mathematically rigorous that I believe it would intimidate the great majority of math professors as well, and is just not appropriate for an engineering or digital design course.

Of critical importance today are the EDA (Electronic Design Automation) tools that are used to generate high-speed digital designs. Most books make no mention of these, let alone discuss design constraints and how constraints are implemented and integrated into the design. Today, also, fiber optics is taking on increasing importance in high-speed designs and, as with the many other subjects mentioned above, most other high-speed design and signal integrity books do not touch on this subject.

Enough negativity — these other books do make excellent references for the practicing engineer and allow the reader to research the topics presented in great detail.

University Courses for Which This Book Is Suitable

This book is suitable for use in a one- or two-semester course in high-speed digital design or in a one-semester course in the use of high-speed memory devices. Six of the 31 chapters are dedicated to memory devices. The material could be taught as either a junior-level or senior-level undergraduate course in that there is a minimum of math throughout the book. The material could also be taught at a more accelerated pace at the graduate level.

Solutions Manual Is Available

A 160-page (approximately) solutions manual is available to instructors teaching courses based upon this textbook. The manual includes answers to every exercise listed in the textbook with the exercises listed in the order of the chapters. Each exercise is restated in the manual and the solution is provided immediately below the restated exercise. The solutions manual includes figures and is fully typeset. There is no handwritten material.

Cash for Identifying Textbook Errors

In writing this book, I have made every effort to be as accurate and diligent as possible. However, it seems like small typographical and other errors always creep into all books no matter how diligent the publisher and author have been. In an effort to make this book as error-free as possible, a $5 reward is offered for the identification of each error, no matter how small. To claim this reward, please provide the following:

- Book edition
- Section number where the error occurs
- Section title
- Page number
- Description of the error
- Correction to the error, if possible
- Your name and postal mailing address (to send the check to)

and send this information to ErrorsFor5@comcast.net. In return, I'll send you $5 per error for errors not previously identified.

A list of identified errors is listed at Web site *www.high-speedsolutions.com*. (Please note the dash in the URL.) Please allow several weeks for payment to be made. For practical reasons,

I need to reserve the right to decide what constitutes an error and whether payment should be made, and to terminate this offer at any time. However, please be assured that I am very willing and happy to pay this reward in order to make this book as accurate and error-free as possible.

How This Book Was Prepared

This book was prepared on a Gateway 700SE, Pentium 4, 2 GHz PC with 1,024 MB of PC800 RDRAM and a 80 GB hard drive. Adobe FrameMaker 7.0 was used to edit the text and import the figures, and Adobe Illustrator 10.0 was used to create many of the figures. Adobe Photoshop was used later on by the publisher to adjust the contrast of the figures and to add text where appropriate.

Figure files were kept separate from the FrameMaker text file and most figures were imported by reference, rather than embedded in the document. The FrameMaker text file was 36 MB, and the separate figure files totalled 2,468 MB when submitted to the publisher. The text file included a few embedded figures which made it larger than if it were just pure text, and the figure file size total included both source files and the figure files exported from those source files to other formats. After editing at Prentice Hall, the text file was 250 KB (including embedded figures) and the figure files totalled 264 MB after compression and excluding the source files.

Prentice Hall prefers source files to be in EPS format, which are then exported into high-resolution TIFF format for importing by reference into the FrameMaker document. Prentice Hall supplied electronic template files that allowed the book to be properly formatted for publication right from the beginning. I found Adobe Illustrator and FrameMaker to be excellent tools, and Adobe Technical Support was excellent, but necessary, due to the powerful capability and extreme complexity of these tools. I highly recommend these Adobe products, which is in agreement with Prentice Hall's recommendation to use them as well.

All documents and information that were reviewed for general knowledge or research, or for inclusion in the book, were downloaded from the Internet using Comcast's high-speed Internet service. Comcast proved to be 100 percent reliable and as fast as the Internet connections that I have had at any engineering company. While at the start of the book project, I intentionally took two days to identify and visit the best technical libraries in the area for access to material, I subsequently never set foot in them again. It took approximately 3,500 hours over 25 months to write this over 900-page book which is a testimony to the power of the Internet.

Personal Acknowledgments

In writing this book, there was overwhelming support from a multitude of friends. I would like to thank Lilly Santos for her constant encouragement and morale-boosting humor over the two years it took to write the book. Joelle Henningsen was always so positive and provided me with enlightenment and support during the tedious preparation of numerous drafts. Her daily phone calls kept the walls of the cabin from closing in too tightly. Ursula Walker always made life easier with her warmth and understanding of my commitment to the book, which frequently left little time for social activities. Beverly Petersen was encouraging and supportive and special thanks go to her as well. Nancy Lee was always cheerful and supportive.

Patty Koledo's enthusiasm was genuine and almost overwhelming, a real boost on those sometimes unproductive days. Stephanie Nguyen helped on several occasions by dragging me off to a restaurant and providing a well-needed break. Heidi Hausauer was supportive and encouraging, although her numerous vacations did seem to contrast at times with the austere lifestyle of an author. Heidi Foster needs to be thanked for her constant reminders for me to type the words, "The End" — which, by the way, never made it into the book. I guess they're only put in movie scripts!

I also need to thank Lindsay Van Sant for being so positive and encouraging, and for supplying levity which helped to foster a creative and productive environment. Lindsay is a teacher, and her genuine interest in my success was evidence as to why she is such a good teacher.

Special thanks need to go to Blanca Domingo-Yenes for her great positive attitude and her encouragement and excitement about the book over the many months it took to write it.

I need to thank Barbara Zingalis for influencing me to get a new computer, which made a huge improvement in the book effort. The higher speed of the newer machine and the much larger hard disk drive were of great help when it came to working with large figure files and updating the FrameMaker book file.

Rosemary Steiner, a family friend, was delighted to hear of the book project, and Shirley Glas, and Phyllis and Bill Allen were very supportive through their spontaneous belief in the success of the book.

Steve Florek helped tremendously by reviewing ideas for book content and in gathering a large volume of information on TDR, and in acting as a sounding board for ideas over several lunches and telephone conversations.

Michelle and Ricardo Castro were always so uplifting to be around, while Lesley McIntosh continually created a positive environment by her wonderful good nature and patient ways. David Weise and Valentina Kostenko were continually motivating to be around due to their happy nature and enthusiasm for just about everything.

Denise Jordaine must be highly intuitive as she told me that she thought I was an author upon meeting me for the first time, not knowing what I did for a living. This was before I had even thought about working on the book . . . a little spooky!

Dianne Pittman needs to be acknowledged for her large amounts of encouragement and a very helpful, great positive attitude. Mary Cheryl Walker must also be acknowledged for her astute perceptions, high intellect, and uplifting spirit.

Iris Renner, my Swiss connection, always provided encouragement, although from nine time zones over.

Special thanks to Mike Michaelian, a friend and good family man, who was never reluctant to take on a challenge.

I would also like to congratulate Tom Kung, a fellow engineer, on his recent marriage.

Special thanks go to a squirrel, name unknown, who, on frequent occasions, would jump from redwood tree to redwood tree at 50 feet off the ground, providing pleasant entertainment breaks from the discipline of writing this book. I'd just barely look up and all of a sudden this squirrel would go flying off of one tree and barely land on the other by grabbing the bare tips of a branch and spring back and forth dangling at the end of the branch — a high-wire act that would draw good crowds at Ringling Brothers and Barnum and Baily's Circus.

Molly, Kate, Stan, Sara, and Brighten deserve special mention and hopefully will take no offense to being listed after the squirrel, as none was meant.

And lastly, and most importantly, I need to say thank you to my parents, Jane and Bertil, for teaching me to value education.

Technical Acknowledgments

First and foremost, I would like to thank John Congistre, one of the most talented engineers whom I have come to know in my career, for his numerous rescues and bailouts resolving design problems which otherwise might have impacted project schedules and my career!

Bill Slattery is to be thanked for his patience and good people skills during FPGA and PCB designs and I wish him all the best as a co-founder of Skystream Networks in Sunnyvale, California.

Special thanks go out to all the people at Prentice Hall and their contractors who made the publication of this book possible. I am most grateful to Bernard Goodwin, my acquisition editor, for agreeing to publish the book. Thanks also go to Nick Radhuber for reviewing electronic files and answering technical questions on FrameMaker. Vanessa Moore performed the major editing of figures and text, as well as checked permission approvals throughout the entire text for quoted material and reprinted figures, a very big job! Anne Garcia served as the overall production manager for the publication of the book. Jennifer Bergamini was the talented artist who developed the book cover. Michelle Vincenti helped out tremendously with reviewer comments. Lawrence Hargett performed the copy editing for the book and also prepared the back cover copy.

Several people were engaged by Prentice Hall to review chapters of the book as the chapters were written. Jay Michlin was the major reviewer and proofread the entire text from beginning to end. Thank you, Jay, as that took special commitment and is fully appreciated. Much appreciation also goes out to the other reviewers as well — David Cruzado, Jim Peterson, Sam Shaw, and others. A lot of what went into making this book easier to read and better organized is due to their efforts. Not all of their recommendations were able to be implemented due to time constraints and page limitations, but these recommendations were nonetheless appreciated just as fully.

I need to thank the engineering staff at Mentor Graphics Corporation who enthusiastically supported the development of Chapter 22 on Mentor Graphics' software EDA tools. A tremendous amount of appreciation and praise goes to Matthew Hogan who was able to assist me in writing and pulling together the entire chapter. Matt was fun to work with and is a very smart fellow with a very bright future. Dave Kohlmeier and Joe Curcurio also supported this effort and need to be thanked as well.

John Goldie of National Semiconductor Corporation authored several of the application notes from which material was drawn for inclusion in this handbook. He is also to be thanked for providing the answers to several technical questions which were included in the content of this book as well.

Special thanks to Greg Peters at Agilent Technologies for his success in acquiring the necessary resources to provide information on test equipment, and to Greg Le Cheminant, also at Agilent Technologies, for his authoring of Chapter 31 on test equipment. Two other Agilent professionals, Rich Mills and Lou Eckert, were very helpful in providing information. Lou answered several difficult technical questions on vector network analyzers, which were included in the book to the reader's benefit. In a business environment where some reprint permission approvals took many months, I found Agilent Technologies to be extremely professional and expedient in granting approval, thanks to Danielle Flint. These folks were just a pleasure to work with.

I also need to thank Diane Trevino at National Semiconductor Corporation for assisting me in obtaining copyright permissions on several occasions, and Lucinda Mattera, Associate Chief Editor at Penton Media, Inc., for assisting me in getting reprint permissions for articles in *Electronic Design* magazine. Special thanks go to Yuzo Ishii at NTT Corporation for a very quick response to my permission request from halfway around the world, and to Juergen Schrage at C-LAB for help in obtaining permission approvals.

I would also like to thank several additional people, Charlie Leber for being an all-around great guy to work with, Itsu Wang for helping me review material on Atmel devices, and Actel's Bill Bailey for suggesting that I include information on memory devices. As a result, six chapters on memory devices were included in the book, quite an impact for a single suggestion!

The professionals at Xilinx were extremely helpful — Jeff Weintraub for coordinating everything, Trisa Fleckenstein for legal permissions, Anna Acevedo, a manager for communicating status, and Austin Lesea and Mark Alexander for reviewing material.

Special thanks to Brad Cole of Ansoft Corporation for taking the time to explain Ansoft products and EDA tools. While I was unable to describe Ansoft products in as much detail as I had originally planned due to time and page-count limitations, I was impressed with what I learned about these valuable tools. I will be highly recommending them to engineers I come in contact with in the future for suitable applications.

Two others played critical parts in the development of the book: Garth Sundberg at Maxim Integrated Products who provided detailed technical answers to questions about signaling devices and S-parameters, and Steven Woo at Rambus who supplied needed technical material about RDRAM and XDR DRAM technologies.

Finally, special thanks go out to a multitude of professionals, too numerous to mention, who assisted me in obtaining reprint permissions from over 60 different companies. The information they granted permission to use has resulted in a book which, I hope, the reader will find very factual, hands-on, and more exciting to read.

PART 1

Introduction

CHAPTER 1

Trends in High-Speed Design

Chapter Objectives

- Discuss emerging technologies and industry trends for high-speed digital design.
- Discuss major drivers of PCB technology.
- Discuss how bus architectures have moved from single-segment broadcast to bridged-hierarchy broadcast and on to packet-switched point-to-point architectures.
- Discuss that many high-speed devices — particularly memory devices — are moving to on-die terminations.
- Discuss how some aspects of high-speed digital design are an offshoot from microwave theory.
- Provide a list of educational conferences and forums available to engineers and designers who wish to keep their high-speed digital design skills current.
- Present an overview of this textbook's organization and how the book may best be enjoyed.

1.1 Everything Keeps Getting Faster and Faster!

Everything does keep getting faster. Today the computing speed in a throwaway musical birthday card is greater than the fastest mainframes of a few decades ago. Multigigahertz processors are shipping in mainstream PCs and workstations, and 20 GHz processors are running in labs. Boards have been successfully designed that transmit 12.5 Gbps over a single differential pair of copper traces, and individual devices are now transmitting data at rates up to 40 Gbps over parallel traces.[1] Devices have converged — today, most computing devices communicate and most communication devices compute. "Billion-transistor processors are expected to be produced by the end of the decade and the range of devices that can be manufactured in silicon is expanding. Processing power, measured in millions of instructions per second (MIPS), has risen because of increased transistor counts.

Over a third of a century ago in 1965, just six years after the invention of the integrated circuit, Intel co-founder Gordon Moore observed that the number of transistors on a given piece of silicon would double every couple of years — a profound insight that was dubbed Moore's Law. Today, Moore's Law remains valid. This exponential growth and ever-shrinking transistor size result in ever-increasing performance, speed, and decreased cost. What began as an observation has become a compass and engine, setting the bar for the semiconductor industry and producing an expanding universe of new applications and opportunities at higher and higher speeds.

Moore's Law is not a law of science or nature. It is a principle that describes the unique opportunity for exponential improvements provided by advances in semiconductor technology. The genesis of the law was an article Moore wrote for the 35th anniversary issue of *Electronics* magazine, published in April 1965. Moore had been asked to describe the future of electronics. Integrated circuits at the time were limited to 30 transistors, but Moore's research team was finishing a component with 60 transistors. Balancing innovation and economic factors, Moore extrapolated that the number of devices on a silicon chip could double each year for the next decade. Professor Carver Mead, a colleague at Caltech, later dubbed the prediction Moore's Law, and the name stuck. By 1975, the number of devices on a chip was running slightly better than predicted. Moore, however, adjusted the doubling cycle to 24 months, to compensate for expected increases in the complexity of semiconductors. In the late 1980s, an Intel executive observed that Moore's Law was driving a doubling of computing performance every 18 months. The term Moore's Law is also used to describe the law's results: the continuing exponential growth of digital capability and improved price/performance. In the *Harvard Business Review,*[2] Shona Brown noted that Moore's Law functions as the 'time pacing of technology.'

Silicon's power — and its uniqueness — is that nearly all parameters of microprocessor technology improve as transistor counts climb. For example, speed and performance have climbed even more sharply than the number of transistors (see Figure 1.1). The i486 processor ran at 25 MHz. Intel Pentium 4 processors run at 3+ GHz and rising. The predicted billion-transistor processor will likely run at speeds approaching 20 GHz. To look at it from another perspective, in the early 1990s it took three years to move the i486 from 25 MHz to 50 MHz. Today, engineers are adding frequency at the rate of 25 MHz a week and, in a few years, engineers anticipate adding 25 MHz in a single day.

When Moore first stated his law, the cost of a single transistor was in the neighborhood of \$5. Today, \$5 will buy you 5 million transistors, or roughly 1 million transistors for \$1.

1. These data rates were obtained using RO4350 material with an ε_r of 3.48.
2. Shona L. Brown, Kathleen M. Eisenhardt, "Patching: Restitching Business Portfolios in Dynamic Markets," *Harvard Business Review OnPoint Enhanced Edition*, Feb 1, 2000.

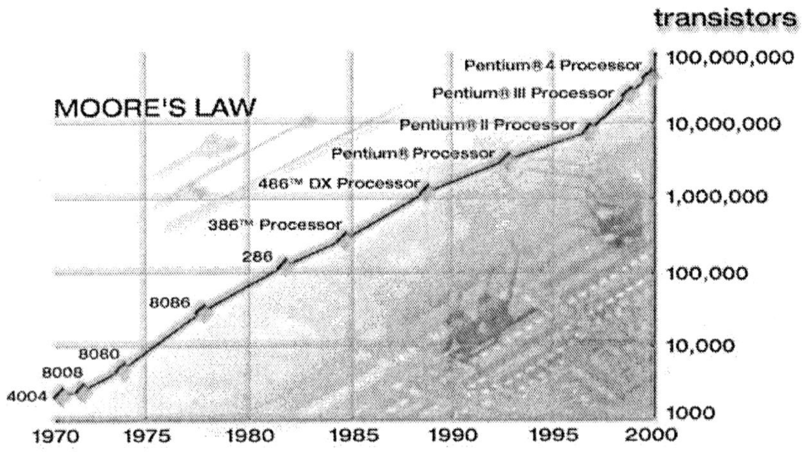

Figure 1.1 The exponential increase in the number of transistors from 2,250 for Intel's 4004 device to 55 million for the Pentium 4. [Reprinted with permission of Intel Corporation. Copyright Intel Corporation 2004.]

As transistors continue to shrink and are packed more tightly onto slivers of silicon running at higher speeds, power consumption and heat can become potential limiting factors for the continuance of Moore's Law. To address this problem of power density, the use of new structures is being explored, such as the tri-gate transistor, and new materials, such as strained silicon, that allow for increased performance while improving power efficiency. Intel's terahertz transistor is perhaps the best example of this. Featuring a depleted substrate transistor structure and a new high-k gate dielectric material, this experimental device can turn on and off a trillion times per second. (It would take you more than 15,000 years to turn a light switch on and off a trillion times.) Intel expects to incorporate elements of this new transistor design into its production line in the second half of this decade.

Today's silicon chips are connected to their packaging via tiny balls of solder, or "bumps," that make the electrical and mechanical connections between the package and the chip. As the frequency of future processors increases exponentially, the performance of the bumps, thickness of the packaging, and number of connection points become concerns. An innovative packaging breakthrough has been developed — bumpless build-up layer (BBUL) — that eliminates these bumps by growing the package around the silicon. This technique reduces the thickness of the package and enables the processor to run at a lower voltage. BBUL packaging technology is just one example of an innovative way to embed powerful computer chips into very small spaces.

The Pentium 4 processor with its newest 0.13-micron technology (now manufactured with 300 mm wafers) packs 55 million transistors onto a piece of silicon the size of your fingernail. In the not-too-distant future, the predicted billion-transistor processor will likely run at speeds approaching 20 GHz."[3] [Int02a]

3. Reprinted by permission of Intel Corporation. Copyright Intel Corporation 2004.

1.2 Emerging Technologies and Industry Trends

1.2.1 Major Drivers of Printed Circuit Board (PCB) Technology[4]

"In terms of PCB technology advancement, three major drivers are occurring — silicon advances, network throughput requirements, and the evolving design chain. Silicon will only become faster and bigger in terms of I/Os, functionality, and power consumption, creating new, high-speed challenges for engineers. Board layer counts continue to increase at the high end — driven by the increase in I/Os on chips. We already see ICs with more than 2,000 I/O pins consuming greater than 100 watts and PCBs with more than 50 layers.

We can expect such complexity and density to result in advances in IC packaging. System-in-packages, such as multichip digital and mixed-signal packages, will help producers get to market faster and help certain techniques, such as embedded components and microvias, become more mainstream. We also will see the adoption of new design methodologies that allow for the concurrent design of the silicon, the package, and the board, as many in the industry begin to realize that package decisions cannot be made independently of the chip and the system.

Optical data transmission, used in datacom and telecom network applications, will increasingly affect PCB design — especially in the conversion between optical and electrical (photons and electrons). Speed is the driver of this market, and as the electrical portion of these designs runs at very high speeds, we are approaching the stage of designing digital signals in the RF domain. Expect terms such as frequency domain, S-parameters, and resonance to become more prevalent in high-speed digital design.

Changes in the way goods are manufactured continue to affect PCB design technologies. Outsourcing continues at a steady clip, as OEMs (original equipment manufacturers) spend more time marketing and less time designing and manufacturing. This new model increases the need for design collaboration between IC makers, OEMs, and EMS (electronic manufacturing services) providers. To be successful, IC companies must aid systems companies so they may efficiently design new, complex chips into their systems. Such support will come in the form of high-speed design kits, reference designs, and intelligent data sheets.

Finally, the communications bandwidth demand will be answered by the market — PCB designers will no longer look for new design challenges. Differential signals, clock data recovery, data transfer protocols, and new types of interconnects will all be explored as we keep pace with the consumers' desire to stay connected and informed." [Pcd02a]

1.2.2 Drivers of Innovation

"It is being predicted that the telecom and personal computer (PC) markets will be the next big drivers of innovation. Practical applications include connecting telecommunications companies with the Internet, miniaturizing PCs, while connecting them to the Internet (utilizing new stan-

4. Reprinted with permission of Printed Circuit Design.

dards in wireless networking), and revolutionizing the PC and telecommunications industries in their entirety (enhancing workstations, mainframes, peripherals, disk drives, switching networks, and networking).

Regardless of such predictions, one thing is certain: PCBs continue to play a dominant role in literally shaping the future of the products they control. While designers respond to shrinking form factors and increased product functionality, high-speed circuitry and mixed-technology designs are already at the mainstream of the telecommunications and PC industries. New technologies, such as optoelectronics, flex PCBs, and microelectronics are rapidly creating new problems and constraints, such as crosstalk and clocked, synchronous timing requirements.

Looking into the crystal ball of PCB design, the future of design tools seems clearer than the industries that embrace such tools. From a high-speed perspective, asynchronous signaling, multilingual modeling, adaptive impedances, programmable I/O buffers, fiber optic interconnects, embedded passive components, and the use of phase-locked- and delay-locked-loop circuitry are adding to the complexity of PCB designs.

Advances in field-programmable gate arrays (FPGAs) and ICs are driving smaller, faster PCBs — PCBs that are only as good as the tools and the designers involved in their creation. A growing need exists to consolidate board functions into a single programmable chip, ultimately driving miniaturization, pin density, and board intelligence. Automatic and interactive routing will play a critical role in enhancing throughput, manufacturability, reliability, and quality.

Changes in PCB technology will occur simultaneously and vertically across a multitude of end-markets. PC boards, communication boards, and even military boards will all feel the impact of changes looming on the horizon. One of the most proactive, business-savvy decisions a forward-thinking company can make is to invest in sophisticated design software and empower designers to master the most basic, to the most advanced, features of that software." [Pcd02a]

1.2.3 I/O Signaling Standards

"Moore's Law has been instrumental in driving the reduction of the end cost of semiconductor products by approximately 200% annually, while higher levels of integration, with lower cost, have served as a principal catalyst in the pervasive 'electronification' of many common items. By employing electronics, it is possible to build a better product, with more features than the previous generation — at a lower cost.

Sustaining Moore's Law has already highlighted the need to adapt to some physical limitations to increase integration at lower cost. A good example is found in the introduction of new I/O signaling standards. These signaling standards, including HSTL, SSTL, LVDS, BLVDS, and CML, result from the need to provide greater switching speeds than the industry standard TTL could support. Surviving mainstream technologies tend to be the ones which are created by standards bodies such as JEDEC.[5]

5. JEDEC refers to the JEDEC Solid State Technology Association, which was once known as the Joint Electron Device Engineering Council. JEDEC is the semiconductor engineering standardization body of the Electronic Industries Alliance (EIA), a trade association that represents all areas of the electronics industry.

As I/O signaling performance moves into the range of gigabits per second, differential-signaling technologies, combined with clock and data recovery (CDR), will be a requirement — noise immunity and clock skew suggest no alternative. This trend is also causing system architects to rethink traditional multidrop, bus-based architectures and shift toward point-to-point serial switched technologies.

Reprogrammable logic technologies, such as FPGAs, are key technology and product requirements, as Moore's Law reduces the cost per gate to reach parity with fixed ASICs. Low cost, field upgradability, and flexibility features will be designed into almost every application imaginable." [Pcd02a]

1.2.4 Web Site as Retailer

"Early in the twentieth century, department stores began to drive many of the small specialty retailers out of business. In the second half of the century, warehouse-type retailers correspondingly displaced both specialty and department stores. The third wave is Internet-supported e-commerce which will displace over the next 25 years many of the specialty, department, and warehouse retailers and bring broad changes in personal services and in industrial marketing. The Web site is the warehouse retailer of the future." [Ele00]

1.2.5 Memories

"As pervasive computing, networking, and graphics-intensive applications continue to require increasing bandwidth, chip makers have driven on-chip clock speeds ever higher to meet this demand. At the same time, system and chip designers have found it progressively more difficult to speed communications from chip to chip, resulting in performance bottlenecks. Innovative solutions are required to break these bottlenecks and speed communications between chips and systems.

One of the bottlenecks is the interface between processors and memory chips. To minimize this bottleneck, memory chips and systems have had to operate at increasingly faster data rates. Memory devices have continually grown into higher and higher densities. For example, RDRAM has densities to 512 Mb with a roadmap to 1 Gb, speed bins of 1333 Mbps with a roadmap to 1600 Mbps, and dual RIMM (Rambus inline memory module) memory configuration modules supporting bandwidths to 5.3 GB/s with a roadmap to support 12.8 GB/s for a quad-channel module."[6] [Ram03a] Games, networking, and communications are the drivers behind DRAM development.

Even with all the increases in speed, density, and high-speed protocols, only "dumb" bits are stored on a memory die. There is no ECC (error correction code) on any memory chip. Parity bits may be stored in memory, but the system external to the chip does the error checking and any

6. Reprinted with permission of Rambus Inc.

error correction. The reason is that an error checking or correction system has to protect the memory data busses as well as the stored memory values, and this cannot be done internal to the chip.

1.2.6 On-Die Terminations

On-die terminations are placing terminations closer to where they are needed and reducing the board space needed for discrete termination resistors. On-the-fly switching of on-die termination resistors can save power and adjust terminations for signals coming from different sources. Memory devices are terminated three ways: (1) external resistors, (2) on-die terminators (resistors) whose values are controlled by programmable values in registers, and (3) on-die termination resistors whose values are controlled by an external resistor. In (3), the external resistor does not terminate the trace or transmission line, but sets the value of the termination resistor, which is on the die. This resistor is usually five or six times larger than the termination resistance it controls.

1.3 Trends in Bus Architecture

"Computer and embedded system development continues to be burdened by divergent requirements. On the one hand, the system performance must continue to increase at a nearly exponential rate while, on the other hand, the system cost must remain constant or even decrease. Several applications, such as those found in networking and telecommunications infrastructure equipment, are also burdened with increasing capability and reliability requirements.

The connections between microprocessors and peripherals have traditionally been composed of a hierarchy of buses (see Figure 1.2). Devices are placed at the appropriate level in the hierarchy according to the performance level they require. Low-performance devices are placed on lower-performance buses, which are bridged to the higher-performance buses so as to not burden the higher-performance devices. Bridging may also be used to address legacy interfaces.

The need for higher levels of bus performance is driven by two key factors. First, the need for higher raw data bandwidth to support higher peripheral device performance requirements, second the need for more system concurrency. The overall system bandwidth requirements have also increased because of the increasing use of DMA (direct memory access), smart processor-based peripherals, and multiprocessing in systems.

Over the past several years the shared multidrop bus has been exploited to its full potential. Many techniques have been applied, such as increasing frequency, widening the interface, pipelining transactions, splitting transactions, and allowing out of order completion. Continuing to work with a bus in this manner creates several design issues. Increasing bus width, for example, reduces the maximum achievable frequency due to skew between signals. More signals will also result in more pins on a device, traces on boards and larger connectors, resulting in a higher product cost and a reduction in the number of interfaces a system or device can provide.

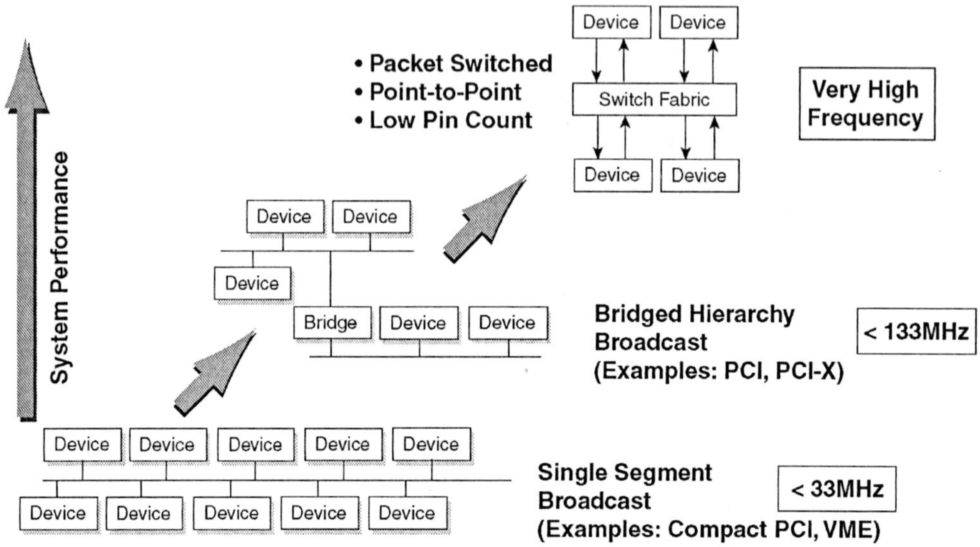

Figure 1.2 Higher system performance levels require adoption of point-to-point switched interconnects. [©RapidIO Trade Association. Reprinted with permission.]

Worsening the situation is the desire to increase the number of devices that can communicate directly with each other. As frequency and width increase, the ability to have more than a few devices attached to a shared bus becomes a difficult design challenge. In many cases, system designers have inserted a hierarchy of bridges to reduce the number of loads on a single bus."[7] [Bou03]

To solve these problems, busses are moving to packet-switched, point-to-point technologies utilizing both mesh and fabric backplane architectures, and new emerging protocols such as RapidIO have been developed. We will touch on the mesh and fabric backplane architectures in the next chapter, Chapter 2 starting on page 17, and discuss them in detail in Chapter 27 starting on page 775. RapidIO is discussed extensively in Chapter 29 starting on page 829.

1.3.1 Moving from Parallel to Serial

"System architecture is moving from parallel I/O standards to serial I/O standards in order to meet the ever-increasing data bandwidth. Serial I/Os provide significant benefits over standard parallel I/O standards for high-bandwidth data transfer. For example, 256 PCB traces are needed to implement a full-duplex, 10 Gbps interface using two 128-bit wide synchronous LVTTL buses running at 78 MHz. The same data bandwidth can be implemented using only 16 PCB traces — 94% less PCB routing area — by using four serial I/Os, each operating at 2.5 Gbps.

7. © RapidIO Trade Association. Reprinted with permission.

Compared with wide parallel buses, the high-speed serial interface uses less board area with less power dissipation." [Int02a]

The trend in backplane driver devices has been to go from TTL to faster versions of TTL (e.g., LVTTL) and on to fast ABT and CMOS devices (e.g., FCT), then on to BTL, ECL, GTL, GTLP, and to LVDS and Bus LVDS. Finally, the trend in devices has moved to serial devices such as CML[8] and some vendor-proprietary serial devices such as controlled-impedance drivers (e.g., Motorola's WarpLink devices). High-speed interfaces connecting chipsets to memory devices have moved to HSTL and SSTL. Throughout the last 40 years, ECL has had a continual use in applications and today is the underlying technology for many high-speed serial links including CML. (See Section 9.7 on page 213 for a discussion of the differences between CML and ECL.) While the latest devices in the trend are serial in nature, the other technologies are still very relevant and may offer the best solution for many applications. For example, using high-speed CML serial links in a fabric, switched backplane may require a more costly backplane layout with several additional layers. It may also require extensive modeling and simulation NRE (non-recurring engineering) where perhaps the system requirements could be met with other, more cost-effective signaling standards or technologies.

As mentioned above, solely increasing the bus width reduces the maximum achievable frequency due to skew between signals. More signals will also result in more pins on a device, more traces on boards, and larger connectors. This in turn results in higher product cost and a reduction in the number of interfaces a system or device can provide.

Serial backplane technologies (e.g., the mesh and fabric switched interconnects presented earlier) allow the connection of any number of line cards using point-to-point, high-speed serial interconnects.

1.3.2 The Power of Tools

One aspect of the latest methods for designing high-speed circuit boards is the absence of significant amounts of high-level mathematical calculations. While most engineers have typically excelled in calculus, differential equations, and higher levels of math, and even completed graduate-level math courses, today's high-speed board-level design process does not use these skills on a daily basis. The EDA (electronic design automation) tools do most of the math for them automatically. On the other hand, if engineers are going to be designing the tool itself, such as a three-dimensional field solver, they will not only use math, but will need to excel in it.

High-speed design today involves considerations as to thermal, microwave, electromagnetic, and transmission line effects, and manufacturing. Looking at industry trends, everything is getting faster and smaller, compounding signal integrity problems. Whether it's CPU clocks, I/Os, memory capacity, storage capacity, gate counts, it's all getting faster, while power supply voltages

8. BTL is backplane transceiver logic; GTL is Gunning Transceiver Logic; GTLP is Gunning Transceiver Logic Plus; LVDS is low-voltage differential signaling; CML is current mode logic; ABT is advanced BiCMOS technology; HSTL is high-speed transceiver logic; SSTL is stub-series terminated logic.

are being reduced, resulting in smaller noise margins. Higher gate counts alone result in power and thermal problems. Today, and continuing in the future, companies must have a strategy to use IP (intellectual property) and reuse building blocks.

1.3.3 ASSPs and ASMs

ASICs (application specific integrated circuits) have been around for a long time, but two newer terms are in frequent use today — ASSPs and ASMs.

"An ASSP (application-specific standard product) is a semiconductor-device integrated-circuit (IC) product that is dedicated to a specific application market and sold to more than one user (and thus, 'standard'). The ASSP is marketed to multiple customers just as a general-purpose product is, but to a smaller number of customers since it is for a specific application. Like an ASIC, the ASSP is for a special application but it is sold to any number of companies. (An ASIC is designed and built to order for a specific company.)

An ASSP generally offers the same performance characteristics and has the same die size as an ASIC. According to a Dataquest study, 17% of all semiconductor products sold recently were ASSPs; 83% were general-purpose with a trend toward more application-specific products."[9] [Wha03]

ASMs — application-specific memories — are similar to ASSPs, but are memory devices dedicated to a specific memory-application market and sold to multiple customers.

1.4 High-Speed Design as an Offshoot from Microwave Theory

To some extent, much of the high-speed digital design knowledge base in use today is nothing new. A fair amount of the theory that is now being applied to high-speed digital designs originated with microwave theory that has been around for years — more accurately — for four decades. Terms such as "skin effect," "crowding effect," and "insertion loss" now being heard in digital design labs are not new to microwave engineers.

What is new is that microwave theory is now being applied within a framework of digital design, in applications where microwave theory has not been applied before. The forums for these new applications include communications, networking, and high-speed backplane applications. What is also new is the wide variety of high-speed digital devices now available in the marketplace, devices that did not exist before. New materials for PCBs are also available, as are sophisticated EDA tools tailored specifically for high-speed digital design. So while the theory has been around for years, the applications, protocols, tools, and devices are new and designers need some knowledge and experience with them in order to be proficient in high-speed design.

9. Definition of "ASSP" copyrighted and reused with permission of whatis.com (*www.whatis.com*) and TechTarget Inc.

1.5 Background Disciplines Needed for High-Speed Design

High-speed digital design is one of the most challenging disciplines in engineering and design today. To be conversant, if not proficient, in the discipline of high-speed design requires an understanding of several different engineering disciplines. These technical disciplines are almost overwhelming. Depending upon the application they may include many, if not all, of the following: PCB design skills, transmission line theory, microwave theory, fiber optic components, digital logic design, skills in using EDA tools, knowledge of FPGAs and CPLDs, S-parameters, TDR and TDR theory, knowledge of memory devices and memory technologies, experience with several different kinds of lab equipment, the ability to model and simulate both single-ended and differential circuits, skill in using SPICE and/or IBIS, an understanding of the 8 to 10 high-speed signaling technologies, high-speed connectors and cabling, and an understanding of emerging protocols and technologies and their relevancy to current and future applications.

The good news (there's always good news) is that, once you acquire this skillset, you are extremely marketable as an engineer or designer!

1.5.1 High-Speed Conferences and Forums

Several conferences and forums are available to engineers and designers who wish to keep their high-speed digital design skills current and for those who wish to get up to speed on high-speed digital design technology altogether. These conferences and forums include:

- DesignCon West and DesignCon East
- DAC (Design Automation Conference)
- Denali MEMCON
- Intel Developer Forum (international locations)
- Rambus Chip 2 Chip Conference/Rambus Developer Forum (international locations)
- Photonics West and Photonics East
- Hot Chips and Hot Interconnects Conference (Stanford University, California)
- PCB Design Conference East and PCB Design Conference West
- Communications Design Conference
- Supercomm
- High-speed digital design seminars (provided by vendors, educational institutions, and individuals)

Information on schedules and content can be found by entering the above conference names at *www.google.com* or other Internet search engine. A list and schedule of networking conferences can be found at *www.ee.unsw.edu.au/~timm/netconf/* and on the Cisco Web site at *www.cisco.com/warp/public/688/events.html.*

1.6 Book Organization

This book was designed from the beginning to be used as both an engineering textbook in colleges and universities and as an in-depth reference for practicing engineers. Chapters in the book are grouped by topic or function. The chapters would likely be most enjoyable if they were not read in numerical sequence but according to the reader's interests and goals, or the goals of the classroom. Chapters 4 through 9 discuss high-speed signaling technologies in so much detail that reading them all sequentially would result in glazed eyes and cure the worst insomniac!

You can gain a good appreciation for the content by taking a few moments to scan the table of contents, which is titled simply "Contents." This will not only give you an understanding of the book's subject matter but also the wide range of topics related to high-speed digital design. For use as a reference, you can, of course, just flip to the topic of interest.

Most readers will find a quick reading of Chapters 1 through 3 worthwhile. These provide a quick introduction to trends in high-speed design, SerDes and bus technologies, and very basic background information on signal integrity. After reading the first three chapters, you can read chapters according to your interest or the goals of the university course.

Chapter 4 starts off the discussion of high-speed signaling technologies with GTL, GTLP, GTL+, and AGTL+. Chapter 5 introduces LVDS. Chapter 6 talks about the differences among BLVDS, LVDM, and M-LVDS. Chapter 7 details HSTL and SSTL. Chapter 8 categorizes and gives organization to the multitude of ECL devices out in the marketplace today. Chapter 9 discusses CML and explains how it differs from ECL.

Chapter 10 discusses high-speed features of today's Altera and Xilinx FPGAs such as 3.125 Gbps RocketIOs. It also discusses HardCopy devices, a shorter time-to-market and less expensive alternative to ASICs for some applications. Chapter 12 discusses high-speed interconnects and cabling. As we'll mention again below, Chapter 23 models and simulates 5-to-10 Gbps board-to-board interconnects.

Those interested in memory technologies should read Chapter 7 on HSTL and SSTL signaling technologies, which are used by memory devices, and then move on to Chapter 13, which provides an overview of memory technologies. After that, you can elect to read about the memory technologies that interest you the most in Chapters 14 through 18, which discuss the various memory technologies in detail.

Chapter 19 introduces the basic theory behind S-parameters and continues on into the more challenging realm of differential and mixed-mode S-parameters. N-port, mixed-mode S-parameters are discussed in the Appendix to Chapter 23. Given that all of today's high-speed serial links are differential in nature, differential and mixed-mode S-parameters have become very relevant topics and are frequently used today in the modeling of high-speed systems.

Chapter 20 discusses TDR and TDT, time-domain reflectometry and time-domain transmission. These techniques are also in frequent use today in both the simulation and testing of high-speed circuit boards.

Three-and-a-half chapters are dedicated to fiber optics due to its increasing importance in high-speed design. Chapter 11 is a broad introduction to fiber optics. Chapter 28 discusses electrical optical circuit boards and Chapter 24 discusses modeling and simulation of fiber-optic transceivers. The second half of Chapter 31 discusses the testing and measurement of optical systems in Section 31.6 which starts on page 880.

Two chapters introduce modeling and simulation and four other chapters provide major, detailed design examples. Chapter 21 introduces modeling with IBIS. Chapter 22 introduces tools from Mentor Graphics Corporation and talks about the flow of the design process. Chapter 23 discusses the advances in design, modeling, simulation, and validation of board-to-board 5-to-10 Gbps interconnects through an actual example. As mentioned before, Chapter 24 discusses modeling and simulation of fiber-optic transceivers, and Chapter 26 is an excellent design example which uses BLVDS SerDes devices. Chapter 27 is dedicated entirely to an example of designing 3.125 Gbaud high-speed serial links using Motorola's WarpLink serial devices. Chapter 25, while not presenting a detailed design example, does discuss every aspect of designing with LVDS.

Four chapters discuss LVDS and Bus LVDS. As stated before, Chapter 5 discusses the LVDS signaling standard and Chapter 6 discusses the BLVDS, LVDM, and M-LVDS signaling standards. Chapter 25 discusses how to design using LVDS and Chapter 26 is an excellent example of a detailed design using BLVDS SerDes devices.

Chapters 29 and 30 discuss the emerging protocols and technology behind RapidIO (Chapter 29) and PCI-Express (Chapter 30). These chapters provide an insight into what is being done to address the issues of faster and faster data rates in embedded systems, and ever-increasing bus speeds and system architecture complexity.

Finally, Chapter 31 is an informative discussion on the high-speed aspects of laboratory and test equipment, including such topics as equivalent-time oscilloscopes.

Throughout this textbook, there are numerous simulation examples and discussions of modeling. The reader is referred to the index for a detailed listing of "simulations" and "models."

1.7 Exercises

1-1. What is the percentage decrease in the cost of an integrated transistor from the time that Moore first stated his law through today?

1-2. What is meant by the statement, "25 MHz a day is being added"?

1-3. How fast are processors running in labs today?

1-4. What is the phenomenon of "convergence" as described in this chapter?

1-5. What is the highest data rate stated in this chapter that has been generated from a single chip?

1-6. (a) What is the highest data rate stated in this chapter at which signals have been sent over copper traces on a PCB? (b) What type of material was used for the PCB?

1-7. What is BBUL technology and what benefit is it being developed to create?

1-8. What three factors are driving PCB technology advancement?

1-9. In what types of applications are optical data transmissions used?

1-10. What technologies generally tend to survive in the competitive marketplace?

CHAPTER 2

ASICs, Backplane Configurations, and SerDes Technology

Chapter Objectives

- Present the current state of ASIC technology and the applicability of ASICs to high-speed designs.
- Introduce the multidrop, multipoint, and point-to-point backplane architectures.
- Discuss switched backplanes and the point-to-point fabric and mesh architectures.
- Introduce the concept of serialization and deserialization (SerDes) and corresponding advantages.
- Discuss the internal architecture of a SerDes device.
- Explore design considerations for using SerDes technology at 5 Gbps.
- Discuss the trade-offs of fiber optic vs. electrical interconnects.
- Clarify some subtle terminology on device families and family categorization.

2.1 Application-Specific Integrated Circuits (ASICs)

A very large number of high-speed designs today use ASICs. ASICs are custom chips developed either in-house or on a proprietary basis, or are purchased from a vendor under an intellectual property agreement (an agreement that lets you use patented technology in exchange for a lot of money). It is very likely that a design will use ASICs rather than off-the-shelf devices (i.e., chips readily available from a manufacturer for any variety of applications). The designer needs to be aware of any ASICs that can be or must be used in her design. ASICs usually adhere to a signaling standard and are members of a logic family, although the signaling standard or logic family can sometimes be proprietary.

At present, IBM Microelectronics is the top world-wide supplier of ASICs. IBM has done 38-million-gate ASICs, has a 65-million-gate ASIC under contract, and has done a 1 GHz ASIC using 0.13-micron technology. A large ASIC today can consume as much as 60 watts of power.

More than 7 Mb of memory can be integrated into an ASIC and the trend is going from using eSRAM to using eDRAM.[1] Plasma TVs, camcorders, and many other consumer products all include embedded DRAM ASICs, and total market sales of these ASICs are 5 to 10 million units a year.

FPGAs are also being embedded into ASICs. Two things are needed when embedding an FPGA (eFPGA): (1) a macro design for the FPGA (the FPGA design) and (2) a higher level of metallurgy. When adding flash memory to an ASIC, the cheapest solution is to add flip-chip flash on top of the ASIC as a MCP (multichip package)." [Nat00]

2.2 Bus Configurations

"Backplanes present special challenges to data transmission systems. This is due to the variety of interconnections (multidrop, multipoint, and switch fabrics) and also due to the close spacing of the loads. There are a number of ways of implementing high-speed backplanes. Each of these ways of implementing a backplane has advantages and disadvantages.

2.2.1 Single-Termination Multidrop

"A single-termination multidrop bus is shown in Figure 2.1 and consists of one driver (D) and multiple receivers (R). This bus is often referred to as a broadcast bus. Note that the driver is restricted to be located at one end of the bus and the other end has a resistive termination (RT). This configuration is useful for data distribution applications and may employ standard devices such as LVDS or Bus LVDS devices.

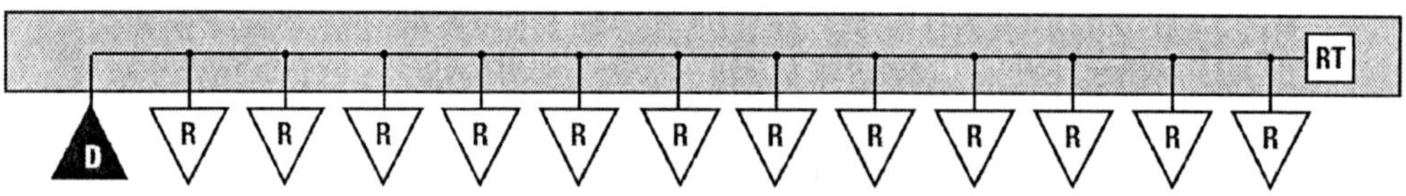

Figure 2.1 Multidrop application with single termination. [Reprinted with permission of National Semiconductor Corporation.]

Single-termination multidrop architecture lends itself to serialization of the bus but without the need for a central switch card or chip. Serialization is the process where multiple CMOS or TTL signals are multiplexed into a single data channel at a higher data rate, transmitted down the backplane, and then demultiplexed back into the original multiple CMOS or TTL signals. During serialization, clock synchronization of the original signals is maintained. Serialization results in less interconnect (fewer connector pins and backplane traces) which in many cases

1. eSRAM is embedded SRAM; eDRAM is embedded DRAM.

reduces the stack up of the backplane to fewer layers. Disadvantages of the single-termination multidrop configuration are the restricted location of the driver to one end of the bus and, if required, the complexity of a return channel (communication path from the loads back to the source)." [Nat00]

2.2.2 Double-Termination Multidrop

"A double-termination multidrop bus consists of one transmitter positioned between multiple receivers and is also referred to as a broadcast bus. The driver can be placed anywhere in the multidrop bus and the bus is terminated at both ends as shown in Figure 2.2

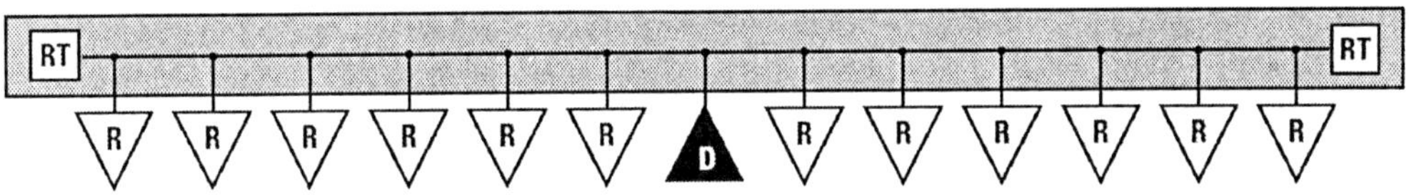

Figure 2.2 Multidrop application with double termination. [Reprinted with permission of National Semiconductor Corporation.]

As mentioned before, this architecture lends itself to serialization of the bus but without the need for a central switch card or chip. As mentioned before, serialization results in less interconnect (fewer connector pins and backplane traces)." [Nat00]

2.2.3 Data Distribution with Point-to-Point Links

"A distribution amplifier can also be used to buffer the signal into multiple copies which then drive independent interconnects to their loads (Figure 2.3). This offers optimized signal quality and the capability to drive long leads (long stubs) to the loads, but at the expense of high interconnect density." [Nat00]

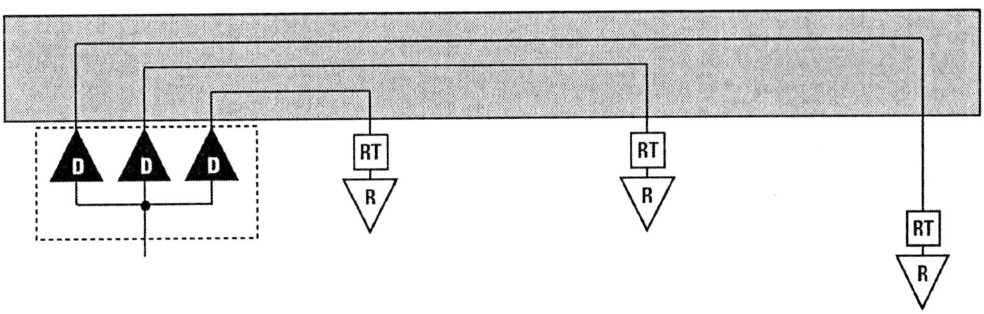

Figure 2.3 Point-to-point data distribution application using a distribution amplifier. [Reprinted with permission of National Semiconductor Corporation.]

2.2.4 Multipoint

"The multipoint bus (Figure 2.4) requires the least amount of interconnect (routing channels and connector pins) while providing bidirectional, half-duplex communication. However, on this type of bus, there can only be one transaction at a time. Thus, a priority and arbitration scheme are typically required. These may be protocol or hardware solutions, depending upon the application." [Nat00]

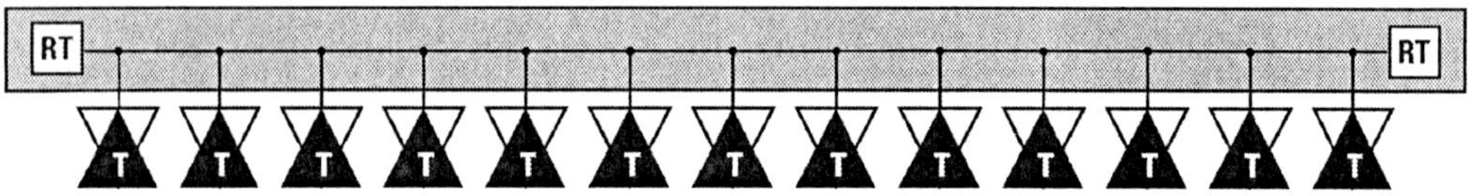

Figure 2.4 Multipoint application. Reprinted with permission of National Semiconductor Corporation.

2.2.5 Switch Matrix

"Switch busses are growing in popularity for systems that require the very highest throughput possible. A switch application is shown in Figure 2.5. Advantages of the switch bus are that simultaneous transactions can occur on the bus at the same time and it has the cleanest electrical signal path of all the bus options (i.e., when compared to multidrop and multipoint) due to the no-stub effect. The disadvantages of this type of scheme are that interconnect density and complexity of the central switching card increases with the number of loads. The switch application, due to its inherent optimized signal quality, is commonly used for links running hundreds of megabits per second into the gigabit per second range. The top speed tends to be limited by the bandwidth of the interconnect."[2] [Nat00]

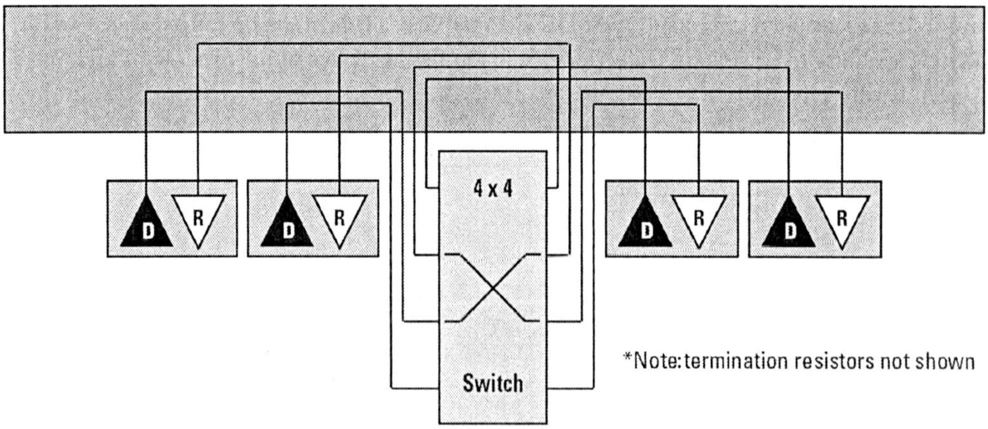

Figure 2.5 Switch application. [Reprinted with permission of National Semiconductor Corporation.]

2. Reprinted with permission of National Semiconductor Corporation.

The major strength of the switched bus is that it allows you to connect all the cards in your backplane using only point-to-point interconnects.

2.2.5.1 Mesh and Fabric Point-to-Point Bus Architectures

Two types of switch backplane bus architectures are the fabric and mesh, as shown in Figure 2.6. The key significance of these architectures is that they allow for the interconnection of multiple line cards using only point-to-point interconnects. These two architectures are discussed in detail in Section 27.2.3.1 which starts on page 779.

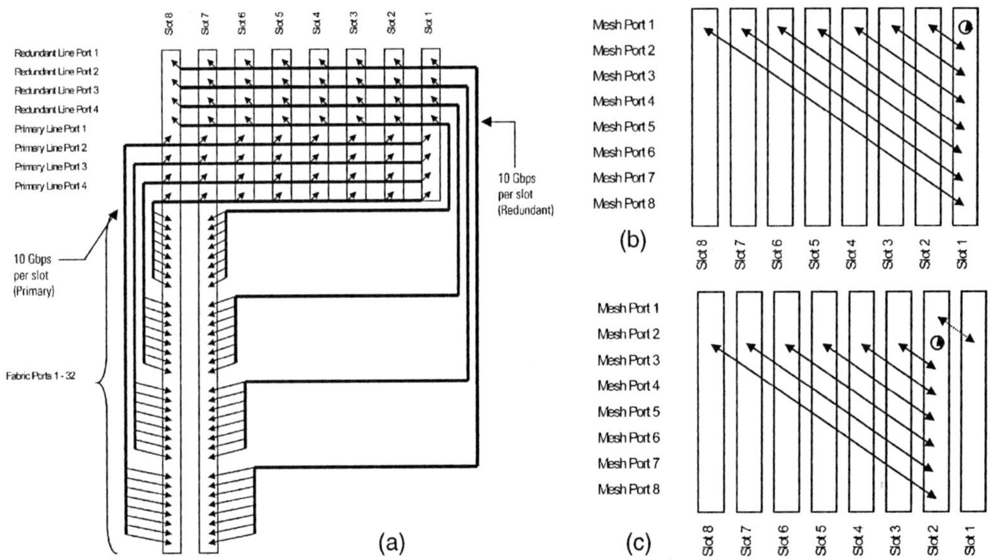

Figure 2.6 Point-to-point switch interconnects: (a) backplane fabric interconnects, (b) backplane mesh interconnects from slot 1, and (c) backplane mesh interconnects from slot 2. [Copyright of Motorola. Used by permission.]

2.3 SerDes Devices

"An important concept is that the maximum data rate of a transceiver is limited to the maximum data rate of the input LVTTL/LVCMOS data stream. This is overcome in devices by using serializers where several bits of LVTTL/LVCMOS parallel data are serialized (multiplexed) into a much higher-speed serial data stream." [Mot03]

2.3.1 SerDes Device Fundamentals

"SerDes devices create a high-speed bus without a lot of connections on a backplane or cables between boxes. The applications for SerDes transceivers are primarily where you want to eliminate interconnections yet maintain a high data throughput.

The 'Ser' stands for serializer. It takes parallel data and serializes it into a serial bit stream as shown in Figure 2.7. The input is typically 8 parallel data bits which are encoded with an optional 8B/10B encoder. This encoding scheme converts the 8-bit data into a 10-bit format that is transmitted over a serial output 'link.' The data rate on the link, in this 1 gigabit per second data input example, is therefore 1.25 gigabaud.

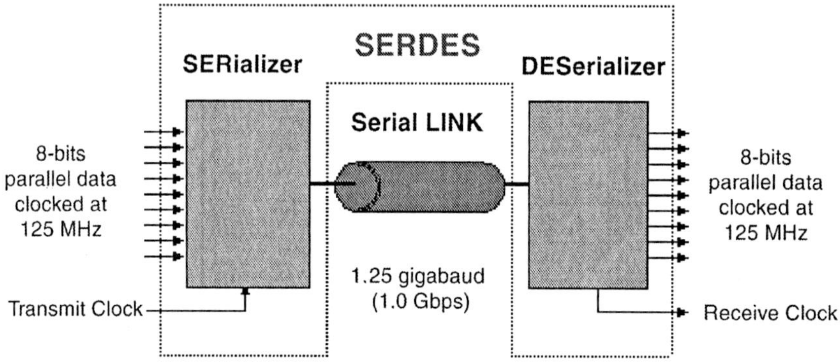

Figure 2.7 SerDes transceiver. [Copyright of Motorola. Used by permission.]

The deserializer, or 'Des,' works in reverse as it takes the serial data, decodes it, and converts it back to a parallel data interface along with a 'recovered' data clock.

A generic functional diagram of a SerDes link is shown in Figure 2.8. The transmitter portion of a SerDes chip has a parallel digital interface, FIFO, 8B/10B encoder (that can be used or bypassed), and serializer. The transmitter output drives a differential signal into, typically, 50 Ω media (100 Ω differential)[3] and may require no external termination or interface components.[4]

The receiver performs the deserializer function. It has a transition-tracking loop that does the data and clock recovery, along with byte alignment, an 8B/10B decoder, word alignment FIFO and digital, parallel-data interface.

This figure is typical for each of the four links contained on quad SerDes devices. There are some additional common functional blocks within each chip. These are the configuration logic, functional test logic and a single-system, phase-locked loop (PLL).

Other important diagnostic features include built-in self-test (BIST) functions and, on most devices, a JTAG[5] test interface. Devices can also be put into a loop-back mode for system testing. In this mode the output data is looped back internally and can test the device even if the links are open or shorted.

3. These impedance values assume that LVDS signaling technology is being used.
4. Termination is built into some receiver chips.
5. JTAG is Joint Test Action Group.

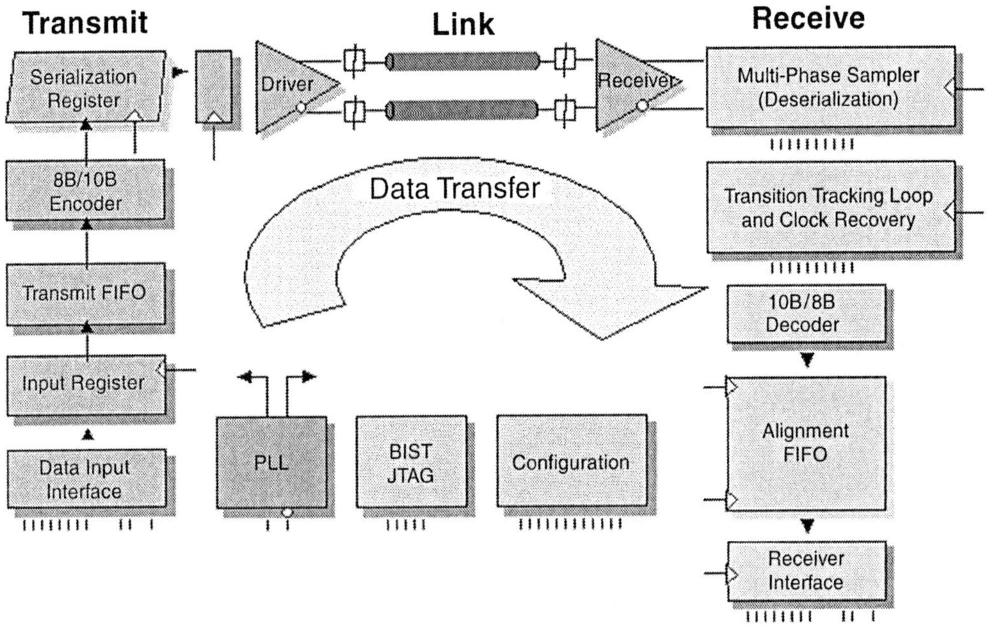

Figure 2.8 SerDes functional diagram. [Copyright of Motorola. Used by permission.]

A block diagram for each SerDes channel which uses the common chip clock is shown in more detail in Figure 2.9. The data input reference clock (chip clock) is nominally in the range of 50 to 156 MHz depending on the configuration and the serial bit stream on the output link may be up to 3.125 gigabaud.

For example, one of the leading vendors of SerDes devices is Motorola. The Motorola SerDes product line is a family of quad, high-speed, full-duplex, serializer/deserializer (SerDes) transceivers and is currently based on two different platforms, the MC92600 quad SerDes with data rates up to 1.25 gigabaud and the MC92610 quad device with data throughput per channel of up to 3.125 gigabaud. They form data interfaces that can be used to transmit parallel data between chips, across a board, through a backplane, or through cabling. The gigabit SerDes transceivers are carefully designed in CMOS to significantly lower power consumption and overall system cost. They are used for point-to-point data transfer operations in high-speed data networking applications such as routers, backplane and access switches, storage area network equipment, and high-speed automatic test equipment. Aggressively packaged as quad devices in fine-pitch BGA packaging, the family offers excellent board density without making unreasonable signal integrity demands of the system logic or backplanes to which they interface."[6] [Mot03]

6. Copyright of Motorola. Used by permission.

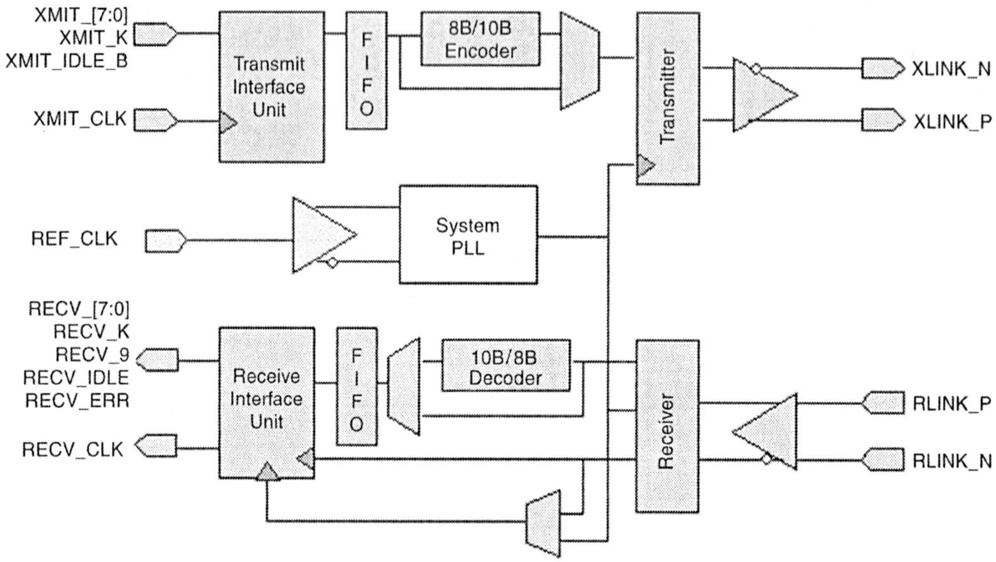

Figure 2.9 SerDes block diagram. [Copyright of Motorola. Used by permission.]

National Semiconductor is also a major supplier of SerDes devices including LVDS (low-voltage differential signaling) and Bus LVDS products. A third and equivalent leading vendor is Maxim Integrated Products, which also provides SerDes BLVDS products.

2.3.2 SerDes at 5 Gbps[7]

"System designs are currently being designed incorporating PCI Express, XAUI, and other interfaces capable of operating in the multi-Gbps range, but these interfaces typically offer data rates topping out at slightly over 3 Gbps. Some designers are now pushing the envelope with goals of getting to the 5 Gbps range.

To break the barrier of 5 Gbps in CMOS technology, 0.13 micron chip technology will be required to be able to easily implement the design. 0.18 micron technology could be made to work, but it would be very challenging to make a robust design that delivers commercially acceptable yields from the fab facility. Outside of CMOS technology, silicon germanium (SiGe), indium phosphide (InP), or gallium arsenide (GaAs) are good alternatives, although not cost-effective for a large integrated device.

At 5 Gbps, it can be assumed that data will have to be encoded, to ensure clean eye patterns for data recovery. The typical encoding method will be 8B/10B encoded data packets. Assuming a simple 10:1 serialization ratio (where the bit rate on the serial portion of the link is

7. Copyright © 2003 by CMP Media LLC, 600 Community Drive, Manhasset, NY 11030. Reprinted from *www.CommsDesign.com* with permission.

10 times the parallel word rate), the link protocol engine and encoder technology need to operate at 500 MHz in order to keep up with a 5 Gbps line speed.

Alternatively, the data path width would need to increase in order to accommodate a lower core clock frequency. The challenge of using a wider data path is processing larger quantities of data per clock cycle, and properly maintaining the data flow. Often, the trade-off is increasing buffer sizes and potentially adding small amounts of RAM to the controller to maintain the protocol stack. While this was a challenge at 2.5 and 3.125 Gbps, the problem only gets worse at 5 Gbps or greater channel speeds.

At data rates in excess of 5 Gbps, cycle time is less than 200 ps. This makes the task of accurately placing the recovered clock edge in the middle of the data eye extremely challenging. (See Section 3.4,“Jitter Measurements Using Eye Patterns,” on page 43 for a basic discussion of eye diagrams and how they are created. Section 26.13,“Interconnect Jitter Margin,” on page 752 provides a more in-depth discussion on the use of eye diagrams.) In most serial standards, eye closure at the input of the data recovery block can be as high as 140 ps, leaving less than 60 ps of open eye. Analog and digital implementations are challenged with maintaining good tracking of the line data clock and keeping the data signal locked to that clock, in order to recover the data.

Much like the serializer, the data path, once deserialized in an 8B/10B-encoded system, will be operating at very high frequencies. For example, 500 MHz clock frequencies are required for a 1:10 system.

Clock recovery at high frequencies puts a large burden on the system's ability to lock the clock to the incoming data stream quickly and reduce overall cycle-to-cycle jitter in order to improve the jitter tolerance of the circuit. At 5 Gbps data rates, minimizing the cycle-to-cycle jitter will be the dominant issue in any type of design.

Noise on the power supply creates an additional issue for designers. This noise increases the likelihood that the recovered clock will have significant amounts of jitter, making low levels of jitter on the data stream even more difficult to obtain. Designers will therefore want to make every effort to ensure that power supply noise is kept to an absolute minimum.

The challenge to reduce common-mode noise has always been present in high-speed differential systems, but with the frequency content of signals pushing into the 10 GHz range, the effect of common-mode noise is even more damaging on the eye opening than with prior systems. By tackling the common-mode noise issue, the designer will directly address the issue of radiated EMI.

Perhaps, more than anything else, pushing serial interfaces into the 5 Gbps range will bring the problems of lossy effects from IC packaging, boards, cables and connectors right to the fore. Designers need to take pains to ensure that (as one example) differential circuits adhere as closely as possible to the differential ideal. In other words, paths for such circuits should be routed together (i.e., the two signal lines should not take different routes, not be of different lengths, and not pass through differing numbers of gates).

Every effort should also be made to minimize impedance discontinuities, as the signal flows though the circuit(s). Maintaining a uniform impedance will improve signal integrity, by minimizing impedance reflections. It will also result in lower radiated EMI, thus helping to meet another design challenge.

In the rarefied realm of 5 Gbps serial interfaces, we find that working examples of such interfaces that exist today are proprietary. Designers will use these to connect function blocks within their own circuits. For these applications, it usually doesn't matter if the interface is not standards-based. Over time, we can expect to see interface standards emerge that can meet these data rates. Such a development will allow designers to more easily combine function blocks from a variety of sources." [Nik03]

2.3.3 SerDes Multibit Signal Encoding

"At present, most SerDes are using one of two types of signaling at data rates greater than 3.125 Gbps: current mode logic (CML is discussed in Chapter 9 starting on page 199) or a type of multibit signal encoding. Where CML signaling sends one bit per symbol, multibit signal encoding packs multiple bits into each symbol by encoding the multiple bits into unique voltage/current values on the lines. At the receiver, signal-processing techniques are used to convert these complex symbols back into a standard binary bit-stream.

The advantage of multibit signaling is that — for a given data transfer rate — a lower symbol rate can be used, compared with what would be needed for CML. The main challenge of multibit signaling is the loss of noise margin, due to the fact that this signaling method does not provide as much of a voltage step between the possible symbol values. CML, being binary, can make use of the full voltage swing available for the channel. Multibit signaling breaks the overall voltage range into several smaller steps, in order to encode the more complex symbols. An additional challenge for multibit signaling is the high frequencies at which the signal processor on the receiver side must operate, in order to resolve the input signal into the correct digital values." [Nik03] To date, the vast majority of devices employ CML signaling rather than multibit signal encoding.

2.4 Electrical Interconnects vs. Fiber Optics

"OEM manufacturers covering datacom, telecom, and high-end computing are setting the development agenda for printed wire boards (PWBs) with performance specifications that can only be met using optical waveguides. At ultra-high bandwidths of greater than 10 Gbps, electrical PWBs suffer badly from electromagnetic interference and large power and cooling requirements.

Embedded optical backplanes are a technology that offers both ultra-high bandwidth and the design freedom of a more distributed architecture. Critically, such backplanes avoid the need for large amounts of fiber at the back of a system which limits system expansion and adds to system maintenance complexity."[8] [Ter03] Chapter 28 starting on page 809 discusses new technology on electrical optical circuit boards (EOCBs) in detail.

8. Reprinted with permission of Terahertz Photonics.

Figure 2.10 illustrates how different technologies are more competitive with one another at different data rates and interconnection distances. In the figure, "UTP" refers to unshielded twisted pair wire, "BB Coax" refers to broadband coaxial cable , and "Free Space" refers to light propagation in free space (air) outside of fiber. Much more will be said about the trade-offs between fiber optic and electrical interconnects in Chapter 11, Fiber-Optic Components, starting on page 263.

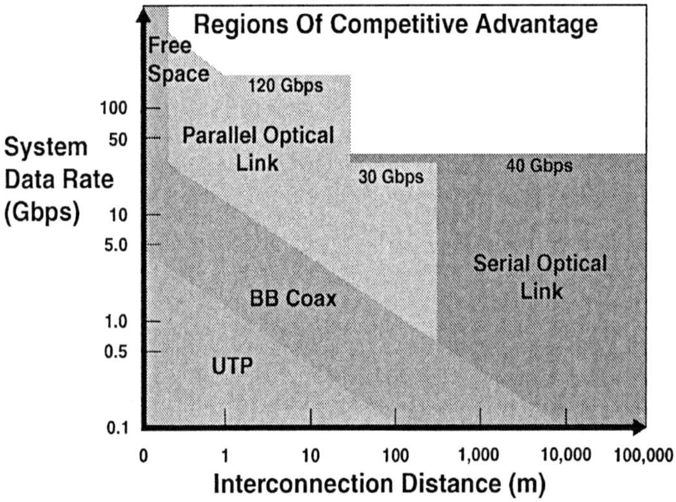

Figure 2.10 The competitive advantages of transmission technologies are a function of data rate and interconnection distance. [Courtesy of Infineon Technologies AG, manufacturer of PAROLI Parallel Optical Links. PAROLI® is a registered trademark of Infineon Technologies AG.]

2.5 Subtleties of Device Families

Chapters 4 through 9 discuss the high-speed signaling standards of GTLP, LVDS, BLVDS, HSTL, SSTL, ECL, and CML.[9] Before examining these signaling standards in detail, a few general words are in order about terminology and categorization in dealing with device families.

2.5.1 Logic vs. Interface Families

In discussing digital integrated circuits, *logic family* generally refers to digital integrated circuit families that are used to send and receive digital signals primarily across an individual circuit card. *Interface family* refers to digital integrated circuit families that are used primarily to transmit

9. GTLP is Gunning Transceiver Logic Plus; LVDS is low-voltage differential signaling; BLVDS is bus low-voltage differential signaling; HSTL is high-speed transceiver logic; SSTL is stub-series terminated logic; ECL is emitter-coupled logic; CML is current-mode logic.

and receive digital signals from one circuit card to another circuit card or to a backplane through a connector, or from a circuit card or backplane through cables to another circuit card or backplane. Interface families may also be used to transmit digital signals over wide backplanes (long trace lengths). Many digital families fit into both categories in that they are used to send and receive digital signals both on and off the card. In these cases, the distinction between whether the family is a logic family or interface family becomes blurred. The important thing to remember is that when you go to a semiconductor manufacturer's Web site and review the digital families available, you need to search under both "logic" and "interface" devices or else you may overlook some of the manufacturer's products.

2.5.2 Murky Device Categories

A lot of devices today no longer fall into nice little neat categories like ECL or CML or HSTL. There has been a huge proliferation of proprietary device drivers, for example, Motorola's Warp-Link controlled-impedance driver. A lot of proprietary ASICs exist that adhere to proprietary signaling standards and that are only available to engineers developing products within their company.

2.5.3 Logic Family vs. Signaling Standard

Another distinction that is worthy of mention is the difference between a logic or interface family and a signaling standard. For example, LVDS is both a family of devices (actually several families) and a signaling standard. A *signaling standard* is a specification that states the requirements and specifications to achieve a transfer of data in a particular way at a particular data rate. The logic or interface family implements this signaling standard in actual devices (integrated circuits).

2.6 *EDN* Magazine's Microprocessor Directory

EDN magazine's "Annual Microprocessor Directory" provides detailed information about microprocessors including company name, device or device family, CPU frequency, bus interface width, instruction width, operating voltages, typical power, power-down modes, and DSP/multiplication hardware support (in bits). It also provides information for each microprocessor on FPUs (floating point units), caching, memory, memory controller, MMU, package selection, timers, serial and parallel I/O, interrupts, ADC/DAC, additional features, and price in quantities of 10,000. The Web version of the "Microprocessor Directory" is more comprehensive.

> *EDN* magazine publishes an "Annual Microprocessor Directory" and an "Annual Programmable-Logic Directory." Both directories are also available online at EDN's Web site at *www.edn.com.*

2.7 Exercises

2-1. (a) Referring to Figure 2.10 on page 27, over what interconnection range of distances are unshielded twisted pair wire, broadband coaxial cable, serial optical links, and parallel optical links all used for interconnects? (b) What range of data rates does this correspond to?

2-2. (a) When using LVDS SerDes devices over electrical interconnects, what is the single-ended characteristic impedance of each of the traces of the differential trace pair? (b) What is the approximate differential impedance of the pair of differential traces?

2-3. What advantage does a switched backplane have over multidrop and multipoint backplanes?

2-4. When searching for high-speed drivers, receivers, and transceivers on vendor Web sites, what two major classifications of parts should you look under?

2-5. (a) An Intel Pentium 4 processor consumes about 100 watts of power. By comparison, how much power does a large ASIC consume? (b) An Intel Pentium 4 processor has 55 million transistors. By comparison, how many transistors are in the larger of the two IBM ASICs mentioned in this chapter? (c) How much power would you guess this larger IBM ASIC consumes?

2-6. What are five design considerations in designing a 5 Gbps SerDes system?

2-7. What are two disadvantages to using the single-termination multidrop configuration?

2-8. What is the difference between a logic family and a signaling standard?

2-9. What is the difference between a multipoint bus and a multidrop bus?

2-10. What is the maximum frequency stated in this chapter that an ASIC device has been operated at?

2-11. What is the difference between gigabaud (Gbaud) and Gbps?

CHAPTER 3

A Few Basics on Signal Integrity

Chapter Objectives

- Introduce the distributed parameter transmission line model and the telegrapher equations.
- Discuss the propagation function γ, attenuation constant α, and phase constant β for transmission lines.
- Present an example of a lattice diagram, show how it is used to illustrate the effects of transmission line reflections, and discuss proper transmission line termination.
- Discuss the high-frequency characteristics of skin effect, crowding effect, return path resistance, and frequency-dependent dielectric loss.
- Define the length of the rising edge and knee frequency; discuss edge rate and frequency content, and single-ended vs. differential signal transmission.
- Discuss assessing signal quality using eye diagrams and bit error ratio (BER) tests.

3.1 Transmission Lines and Termination

3.1.1 Transmission Line Equations

Several years ago, the highest frequencies encountered in digital circuits were below 20 MHz. To be a transmission line or an antenna at these frequencies, a structure had to be about one meter (about 3 feet) long. Practical circuits did not behave like transmission lines or antennas. When clock rates reached 320 MHz and edge rates became a few hundred picoseconds, interconnects as short as 25 mm (about 1 inch) begin to look like transmission lines and antennas. Today, with serial interconnects commonly running at 3.125 Gbps and others at 10–12 Gbps with edge rates down to 30 ps, all chip-to-chip and chip-to-connector interconnects at these rates must be treated as transmission lines. Modeling of the interconnects operating at these high speeds is mandatory.

The main effects of transmission lines is that they delay, disperse, and attenuate signals as the signals propagate down the lines. Transmission lines have traditionally been modeled at high frequencies as a chain of R, L, C, and G linear passive-circuit stages, as shown in Figure 3.1. The R and L represent the series resistance and series self-inductance of the wire per unit length, respectively. The C represents shunt capacitance between the metallic conductor and ground plane. The G represents shunt conductance caused by insulation leakage and, at higher frequencies, energy absorption caused by the dielectric nature of circuit board materials such as FR-4. All four parameters are specified per unit length. This type of model or other more sophisticated models are used whenever a transmission line or trace cannot be treated as lumped R, L, C, and G elements, as in a single R-L-C-G stage. This is the case when the transmission line's physical length is longer than 1/6 of the length of the signal's rising edge (i.e., the rising edge of the signal which will be transmitted on the line). The length of the rising edge is the distance the signal will travel down the line in the time it takes to transition from low to high (as measured by 10% and 90% amplitude values). This type of model is referred to as a distributed model or as the model for a distributed system.

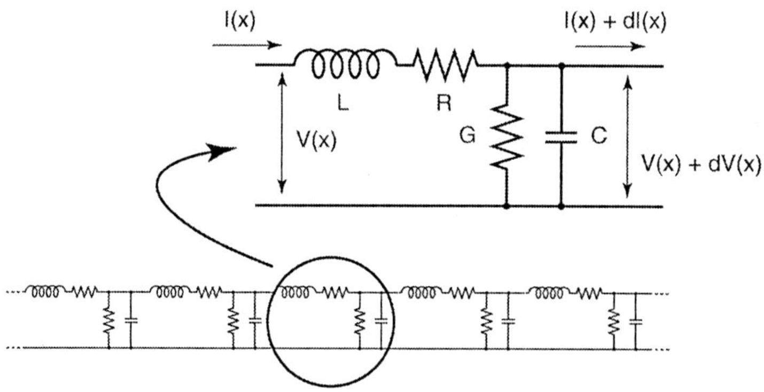

Figure 3.1 RLCG model for a transmission line.

Several important and well-known equations can be derived from the model for a single R, L, C, and G stage of the transmission line. By analyzing the currents flowing into and out of circuit nodes (junctions) and the voltages across circuit elements, the following two circuit equations can be generated:

$$\begin{aligned}(I(x)+dI(x))-I(x) &= (G+j\omega C)V(x)dx \\ (V(x)+dV(x))-V(x) &= (R+j\omega L)I(x)dx\end{aligned} \tag{3.1}$$

In these equations, x is the physical distance down the transmission line to the input of the stage, I(x) is the current flowing into the top conductor of the stage, and I(x) + dI(x) is the current flowing out of the top conductor of the stage, as shown in Figure 3.1. dI(x) represents the effect of current flow through the shunt elements. Similarly, V(x) is the voltage across the input terminals of the stage and V(x) + dV(x) represents the voltage across the output terminals of the stage. dV(x) represents the effect of voltage changes across the series elements as a function of distance x.

In developing these equations, R, L, G, and C are treated as constants while in reality, R, L, G, and C do vary with frequency. The equations assume that a narrow-enough range of frequencies is being used so that R, L, G, and C can be considered to be constant.

If partial derivatives are taken with respect to time t and position x, the above equations can be transformed into the following equations, which are referred to as the *telegrapher's equations*:

$$
\begin{aligned}
-\frac{\partial I}{\partial x} &= GV + C\frac{\partial V}{\partial t} \\
-\frac{\partial V}{\partial x} &= RI + L\frac{\partial I}{\partial t}
\end{aligned}
\tag{3.2}
$$

The solution to the telegrapher's equations can be shown to be

$$
\begin{aligned}
V(x) &= V_0^+ e^{-\gamma x} + V_0^- e^{\gamma x} \\
I(x) &= \frac{V_0^+}{Z_0} e^{-\gamma x} + \frac{V_0^-}{Z_0} e^{\gamma x}
\end{aligned}
\tag{3.3}
$$

where γ is the propagation function of the transmission line given by

$$\gamma = \alpha + j\beta = \sqrt{(R + j\omega L)(G + j\omega C)} \tag{3.4}$$

and where α is the attenuation constant and β is the phase constant. As can be seen, the propagation function, attenuation constant, and phase constant are, in turn, functions of the R, L, G, and C parameters and are dependent upon frequency.

The solution to the telegrapher's equations state the voltage and current as a function of distance down the transmission line. In these equations, V_0^+ represents a voltage wave with phasor value V_0^+ at x = 0 and traveling in the direction of increasing x with phase velocity $v_p = \omega/\beta$. This voltage wave diminishes exponentially in amplitude according to the term $e^{-\alpha x}$ as it travels. In a similar fashion, V_0^- represents a voltage wave with phasor value V_0^- at x = 0 and traveling in the direction of decreasing x with phase velocity of magnitude $|v_p| = \omega/\beta$. This wave also diminishes exponentially in amplitude according to the term $e^{\alpha(-x)}$ as it travels. V_0^+ is the phasor value of the first wave as it leaves the point x = 0 and V_0^- is the phasor value of the second wave as it arrives at x = 0. $V_0^+ + V_0^-$ is the phasor voltage at the input terminals of the line at x = 0.

Another significant equation is that for the characteristic impedance of the transmission line, Z_0. This impedance is derived by taking the ratio of voltage to current and is also a function of R, L, G, and C:

$$Z_0 = \frac{V_0^+}{I_0^+} = -\frac{V_0^-}{I_0^-} = \sqrt{\frac{R + j\omega L}{G + j\omega C}} \tag{3.5}$$

The characteristic impedance, Z_0, of a transmission line is the impedance that a step-impulse signal will see when it is inputted into an infinitely long transmission line. Stated in a more practical context, Z_0 is the impedance a high-frequency signal will initially see when it is applied to a transmission line before the signal has a chance to propagate down the line.

In the above equation for Z_0, at frequencies where $|j\omega L| >> R$ and $|j\omega C| >> G$, the equation for Z_0 reduces to the characteristic impedance equation for lossless transmission lines:

$$Z_0 = \sqrt{\frac{L}{C}} \tag{3.6}$$

In cases where these conditions are not met, the transmission lines in question are classified as "lossy" transmission lines. Under the previous conditions of $|j\omega L| >> R$ and $|j\omega C| >> G$, the propagation function reduces to

$$= 1/\sqrt{LC} \tag{3.7}$$

where v is the phase velocity of the signal on the transmission line.

Z_0 is of critical importance to high-speed design because, as we'll show in the next section, Z_0 is one of the parameters used to determine whether a signal is properly terminated or whether the signal will instead be subject to multiple reflections and ringing, a situation which may make data communication impossible.

3.1.2 Reflection Coefficients, Lattice Diagrams, and Termination

When a signal travels down a transmission line of characteristic impedance Z_0, and the line is terminated at the end with a load Z_L not equal to Z_0, all or part of the signal will reflect back off of the load at the end of the line. In these cases, a reflected signal will travel back up the line toward the signal source where it may cause interference. When such a reflected signal hits the output impedance of the signal driver, it may get reflected again. The signal may continue on for several round trips getting reflected up and down the transmission line and cause ringing and disruption of data signals until the signal finally becomes dissipated.

On the other hand, if the signal hits a load at the end of the transmission line that is exactly equal to the characteristic impedance of the transmission line, the signal will be fully absorbed by the load and no unwanted signal reflection occurs. If a signal does get reflected at the end of the line and travels back up the line to the signal source, and if the signal source has an output impedance that is exactly matched to the characteristic impedance of the transmission line, the reflected signal will be absorbed by the source output impedance and no further reflections will occur.

Whether a signal is reflected when it arrives at a load is determined by the reflection coefficient, ρ, where ρ is equal to

$$\rho = \frac{Z_L - Z_0}{Z_L + Z_0} \tag{3.8}$$

In this equation, Z_L is the load or termination impedance at the end of the line. When trying to determine whether a reflected signal will reflect a second time off of the source impedance, the source impedance is substituted in this equation for Z_L.

The most important concept dealing with this equation is that when Z_L and Z_0 are equal, the reflection coefficient is zero and no reflection occurs. In almost all digital and analog designs, this is usually the desired condition.

Figure 3.2 shows a signal source driving a transmission line and load, which we'll use as an example to explain signal reflection concepts. Let's assume that our signal source generates an internal voltage V_S = 2.5 V, Z_0 = 50 Ω, and Z_L = 75 Ω. Because Z_L does not equal Z_0, a reflection will occur. Let Z_S, the internal signal-source output impedance, be equal to 25 Ω, which will also involve a mismatch to the transmission line's Z_0.

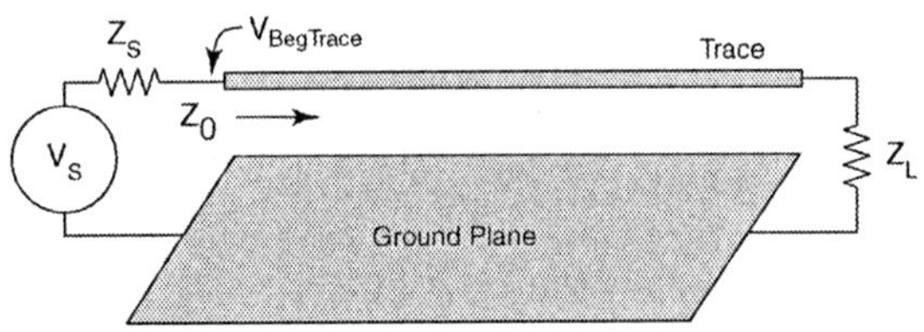

Figure 3.2 A signal source driving a transmission line and load.

There are two reflection coefficients for this circuit, one for the load end and one for the source end. The source reflection coefficient is calculated to be

$$\rho_{source} = \frac{Z_S - Z_0}{Z_S + Z_0} = \frac{25 - 50}{25 + 50} = -0.333 \tag{3.9}$$

and the load-end reflection coefficient is calculated as

$$\rho_{load} = \frac{Z_L - Z_0}{Z_L + Z_0} = \frac{75 - 50}{75 + 50} = 0.20 \tag{3.10}$$

Note that the reflection coefficient can take on both positive and negative values. In our example, when a signal hits the load end of the line, 0.20 of the signal will be reflected back, adding to the incident signal. When this reflected signal hits the source end of the line, the reflected signal will be reflected a second time back down toward the load again. The reflected signal after the second reflection will be 0.333 of the reflected signal after the first reflection.

To determine the voltages seen at one end of the line or the other, the initial voltage wave is added to the reflected voltage wave where the reflected voltage wave may be positive or negative according to the reflection coefficient. The voltage at the load after the first reflection at the load is the sum of the initial signal plus the reflected signal. The voltage at the source, after the signal has had a chance to propagate down the line, get reflected, propagate back up to the source and get reflected a second time, is the sum of the initial signal voltage, the voltage of the reflected signal off of the load end, and the voltage of the reflected signal after the second reflection. Reflected signals are either added to or subtracted from the original signal depending upon whether the reflection coefficient is positive or negative. For example, the load-end reflection coefficient is +0.20 so that the voltage at the load becomes 1.2 (1 + 0.20) times the voltage of the initial signal applied to the line. The voltage at the source end of the transmission line after the second reflection (one load-end reflection plus one source-end reflection) is 1.133 (1 + [0.20] + [0.20][–0.333]) times the voltage of the initial signal applied to the line. The reflected signal from the source is subtracted due to the negative sign in the reflection coefficient (i.e., ρ_{source} is –0.333).

Calculating the total cumulative signal after more than a couple of reflections quickly becomes tedious. A way to make this easier is by use of what's called a "lattice diagram" or "bounce diagram." Figure 3.3 shows the lattice diagram for this example.

As is shown in the diagram, while 2.5 V is generated by the internal source, only 1.667 V initially gets applied to the transmission line. The other 0.833 V gets dropped across the internal impedance of the source, Z_S. This initial voltage value is calculated as a simple voltage divider:

$$V_{Initial} = V_S \frac{Z_0}{Z_S + Z_0} = 2.5 \frac{50}{25 + 50} = 1.667 \text{ V}$$

The 1.667 V signal propagates down the transmission line to the end where it hits load Z_L. With a load-end reflection coefficient of 0.20, 20% of the signal or 0.333 V gets reflected back up the line. Because this reflection coefficient is positive, the reflected voltage adds to the initial voltage yielding a total voltage at the load of exactly 2.000 V. The voltage across the load stays at this value until further reflections make a round trip and reach the load again.

The reflected signal travels back up the line until it hits the source impedance. With a reflection coefficient of –0.333, –0.333 of the reflected 0.333 V signal, or –0.111 V, gets reflected a second time. This time, because the reflection coefficient is negative, the second reflection gets subtracted from the voltage seen at the beginning of the trace. The voltage at the beginning of the trace, $V_{BegTrace}$, is shown in both Figures 3.2 and 3.3 and, after two reflections, is given by

$$V_{BegTrace} = 1.667 \text{ V} + 0.333 \text{ V} - 0.111 \text{ V} = 1.889 \text{ V} \tag{3.11}$$

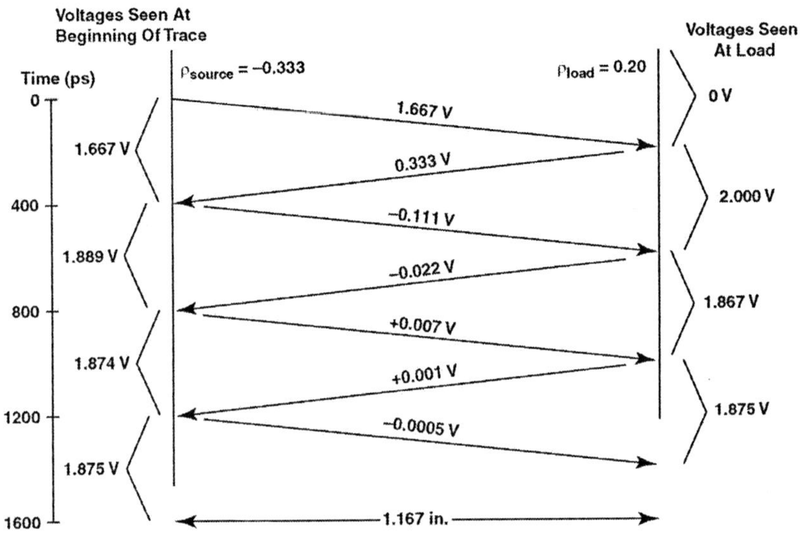

Figure 3.3 Lattice diagram for a trace with a 50 Ω characteristic impedance, 75 Ω load, 25 Ω source impedance, 200 ps flight time, dielectric constant of 4.1, and an internal source voltage of 2.5 V.

The voltage at the input to the transmission line stays at 1.889 V until reflections travel additional round trips and come back to reach the source impedance again. Further reflections and their effects are shown in the lattice diagram of Figure 3.3.

In this example, there are some additional calculations that are important. The velocity v at which the signal propagates down the transmission line is given by

$$v = \frac{c}{\sqrt{\varepsilon_r}} = \frac{3 \times 10^{10}}{\sqrt{4.1}} \text{cm/sec} \ (1 \text{ in}/2.54 \text{ cm}) = 5.83 \times 10^{-3} \text{ in/ps} \qquad (3.12)$$

where c is the speed of light and ε_r is the dielectric constant of the material surrounding the trace (e.g., FR-4).

The propagation delay t_{delay} of the signal as it propagates down the transmission line is given by

$$t_{delay} = 1/v = 1/(5.83 \times 10^{-3} \text{ in/ps}) = 172 \text{ ps/in} \qquad (3.13)$$

In order to make things easier for this example, the total propagation delay of the signal from the source end of the transmission line to the load end, t_D, was assumed to be 200 ps. This corresponds to a transmission line length of

$$\text{Length} = \frac{t_D\, c}{\sqrt{\varepsilon_r}} = \frac{(200\ \text{ps})(3 \times 10^{10}\text{cm/sec})}{\sqrt{4.1}}(1\ \text{in}/2.54\ \text{cm}) = 1.167\ \text{in} \tag{3.14}$$

The above calculation example is provided in a somewhat reversed order. In many situations, you might know what your trace length is from a layout or simulation floorplan, and you may want to calculate the total propagation delay, t_D, based upon ε_r and the trace length and use it in timing analysis. Another name for t_D is the "flight time." Flight time can be calculated from the equivalent equation

$$t_D = \text{Length}\frac{\sqrt{\varepsilon_r}}{c} \tag{3.15}$$

A plot of the voltages at the source and load ends of the transmission line is shown in Figure 3.4. Note that voltage levels at the transmission line ends only change when a signal or reflected signal hits that end. The steady-state or final-voltage levels at both ends is 1.875 V, which can be calculated from a simple voltage divider of the source impedance Z_S and the load impedance Z_L. Z_0 is not a determinant of the final voltage levels.

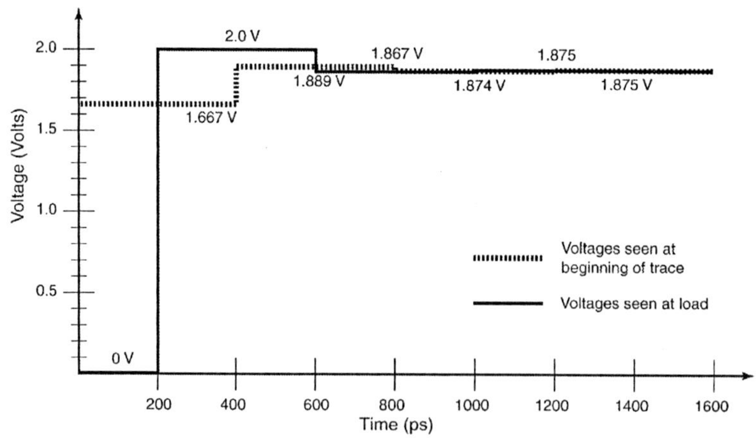

Figure 3.4 Waveforms seen at the beginning of the trace and at the load.

3.2 Important High-Speed Concepts

3.2.1 Rise Time and Edge Rate

Edge rate of a signal is important for several reasons. As we'll show, it is a factor in the determination of whether a transmission line or circuit can be accurately modeled as a lumped-element model when the signal is applied to it or whether the line or circuit has to be modeled as a distributed system. Edge rate is also a factor in the determination of the frequency content of the signal and, in turn, an indication of how far up the frequency spectrum the signal's power lies.

Figure 3.5 illustrates a periodic signal of constant amplitude where the rise time t_r of the signal is measured between the 10% and 90% points of the signal's peak-to-peak amplitude. The rise time is the time it takes for the signal amplitude to go from 10% to 90% of its peak-to-peak voltage value.[1] Rise time is commonly expressed in nanoseconds or picoseconds depending upon how "fast" the signal is.

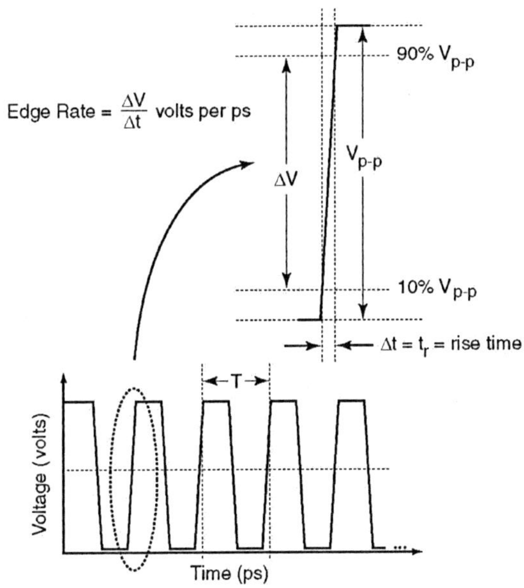

Figure 3.5 Edge rate of a periodic signal of constant amplitude.

Figure 3.5 also illustrates the edge rate, defined to be the slope of the line as measured between the same 10% and 90% points. Edge rate can be expressed in volts per nanosecond, millivolts per nanosecond, millivolts per picosecond, and microvolts per picosecond. When data sheets state minimum or maximum edge rates for a device, they are almost stated in an equivalent way as the propagation delay for the signal to transition from a low-to-high or high-to-low state.

3.2.2 Length of the Rising Edge

As mentioned earlier, the length of a signal's rising edge l_{re} is the distance the signal will travel down a transmission medium in the time it takes to transition from low to high (as measured by 10% and 90% amplitude values). In mathematical terms, the length of the rising edge is given by

1. Rise time has several definitions. In this book, the 10% to 90% definition is used.

$$l_{re} = \frac{t_r c}{\sqrt{\varepsilon_r}} \tag{3.16}$$

where, as before, ε_r is the dielectric constant of the material through which the signal is propagating, c is the speed of light, and t_r is the rise time.

This parameter is used most often to provide general guidance on whether a circuit needs to be modeled as a distributed transmission system or as a lumped-element model. Generally, a trace, transmission line, or circuit can be accurately modeled with lumped parameters if its physical length l_p is much less than 1/6 of the signal's length of the rising edge:

$$l_p \ll l_{re}/6 \tag{3.17}$$

It is advantageous if a circuit is small enough to be modeled using lumped parameters because the modeling is much simpler and less time consuming, and computer simulations run much quicker.

3.2.3 Knee Frequency

In working with signals and circuits, it is often very helpful to know what the frequency content of a signal is, that is, up to what frequency does the signal still contain frequency components with significant power. The "knee frequency" has been defined[2] to provide a rough rule of thumb for this limit of a signal's frequency content and is defined as

$$f_{knee} = \frac{0.5}{t_r}$$

where t_r is the rise time of the signal's edges (either rising or falling).

A common belief is that the frequency content of a signal is determined mainly by its period T, as shown in Figure 3.5. However, whether a periodic signal is analyzed mathematically through Fourier analysis or the frequency spectrum for an actual pulse train is displayed on a spectrum analyzer, it will become apparent that how much of the frequency spectrum a signal occupies is determined by its edge rate and not by its period. This is why the equation for knee frequency is based on rise time t_r but the period T is not in the equation.

3.2.4 Single-Ended vs. Differential Transmission

"In many applications, the use of differential transmission technologies is ruled out by myths alone. When the word differential is spoken, it immediately generates a vision of two pins per signal and gigantic buses. It is true that differential transmission uses two lines per signal, as the

2. See reference [Joh93].

logic state is denoted by the difference voltage, whereas single-ended transmission relies on a voltage level and only one active signal line. However, in designing large single-ended backplane buses, the large return current must be taken into account. To provide a low-impedance path, it is common to assign many ground pins to cable conductors for the cables which connect to the backplane. The ratio of signals to grounds is application dependent but ranges from 3:1 down to 1:1. If the 1:1 ratio is selected, then the 'pins required' for a differential bus and the single-ended bus come close to par.

Differential transmission can even beat the 3:1 applications, and reduce pins required even further. Since differential technologies use small swings to enable high-speed operation, faster signal paths are possible. Combining differential bus driving and a serializer/deserializer function can reduce pin count by as much as 7/8 or 87%, as in the case for 16 LVTTL lines serialized onto a single differential pair. This reduction in bus width requires less PC-board real estate, allows for smaller connectors and smaller interconnect media, and even eases the termination design.

Differential clocks minimize skew, radiate less EMI, and are resistant to common-mode noise.

There are a number of additional benefits of differential technology over single-line technology. Differential technology has

- Lower voltage swings resulting in faster circuits,
- Less crosstalk susceptibility,
- Less susceptibility to current return path discontinuities,
- Reduced electromagnetic interference (EMI), and
- Reduced power consumption where lower supply voltages are possible."[3] [Nat00]

3.2.5 Fast Edge Rate Effects

In the discussion on lattice diagrams and reflection coefficients, it was noted that a signal may reflect back and forth several times off both the load and source ends of a transmission line before the signal's power is completely dissipated. At low frequencies in the kilohertz range or of a few megahertz, these reflections cause ringing at the load, but the ringing dies out before the signal is sampled. The signal has time to propagate down to the load, back to the source, and back to the load again several times before a flip-flop at the load clocks the signal's voltage in. Accurate termination to prevent reflections is not that important.

In high-speed applications using much higher system clock frequencies, the signals have much faster edge rates and data is sampled much more quickly. The flip-flop at the load will sample data so fast that there is no time to allow the signal to bounce back and forth between the load and source and settle before the sample is clocked in by the flip-flop. The circuit must be designed so that the signal travels to the load and has no reflection at all, allowing the sample to

3. Reprinted with permission of National Semiconductor Corporation.

be taken as soon as the signal arrives. In systems that clock data at extremely high rates, there may be several pulses on a trace at the same time where the propagation time down the trace is several times greater than the clock period. In cases like these, no significant level of reflections can be allowed and traces must be very carefully terminated.

Signals with fast edge rates tend to ring more easily and severely in circuits than signals with low edge rates. For this reason, an important design concept is that you normally want to implement your design with the slowest-speed, slowest-edge-rate technology or family of devices that will meet the design specifications. Doing this will save considerable design time and frustration in the prototype development lab.

3.2.6 Parasitics

What tends to make high-speed digital design different from lower-frequency digital design is that, at high clock frequencies, parasitic inductances and capacitances distort and degrade the signals. Every digital signal, in effect, becomes an analog signal whose integrity is compromised. This is because inductive impedance and capacitive impedance are functions of frequency — $j\omega L$ and $1/j\omega C$, respectively. At 25 MHz, the 2 nH inductance of a PCB via is negligible. At 2.5 GHz, the 2 nH via will have an impedance of 31.4 Ω, enough to create a discontinuity in the trace and have a significant enough effect on the signal to require modeling and simulation. Similarly, 10 pf of parasitic capacitance is negligible at 25 MHz, but at 2.5 GHz, it has an impedance of only 6.37 Ω and may attenuate or short out the signal if the capacitance shunts to ground.

Parasitic elements occur everywhere on a circuit board. There's series inductance in the traces that, at high frequencies, become much more significant than the trace series resistance. There's capacitance between traces and their neighboring traces and between adjacent pins on every device. Vias have both inductance and capacitance. The key to establishing good signal integrity in a design becomes how accurately a designer is able to model these parasitics when performing SPICE, IBIS, and other circuit-board-level simulations.

3.3 High-Frequency Effects: Skin Effect, Crowding Effect, Return Path Resistance, and Frequency-Dependent Dielectric Loss

Several additional phenomena occur at high frequencies and are defined briefly as follows:

a. *Skin effect* is the tendency of current to move to the outside edges of a conductor as the frequency of the signal increases, owing to the conductor's self-inductance.
b. *Crowding effect* is the situation where conductors sense each other's presence through the distortion of the electromagnetic field, resulting in the high-frequency current bunching up closer toward the side of the neighboring conductor.
c. *Return path resistance* is the effect of having non-zero impedance in the return path (i.e., the ground plane) of a signal.

d. *Frequency-dependent dielectric loss* is an energy loss caused by PCB materials containing dipole molecules, where very high frequencies cause these molecules to oscillate and absorb energy. This has the effect of attenuating signals as they propagate down a PCB.

3.4 Jitter Measurements Using Eye Patterns

"Signal quality may be measured by a variety of means. Common methods are as follows:

- Measuring the signal waveshape at the load for rise time, amplitude, ringing, and other distortions
- Measuring jitter in an eye pattern
- Determining the margin in an eye pattern around a jitter mask
- Bit error ratio testing (BERT)
- Other means [Nat00]

Measuring jitter in an eye pattern is discussed in this section and BERT is discussed in the following section.

The questions on signal propagation of 'How far?' and 'How fast?' seem simple to answer at first, but after detailed study, their answers become quite complex. These are not simple device parameter specification questions. Rather, they are system-level questions, and to be answered correctly, a number of other parameters besides the switching characteristics of the drivers and receivers must be known. For example, other parameters include the measurement criteria for signal quality that has been selected and also the pulse coding that will be used, for instance, nonreturn to zero (NRZ). Additionally, other system-level components should be known, too. This includes details about cables, connectors, and the printed circuit boards. Since the purpose is to measure signal quality and performance, it should be done in a test fixture that closely matches the end environment — or even better — in the actual application.

Eye pattern measurements are useful in measuring the amount of jitter versus the unit interval to establish data rate versus cable length curves. Eye patterns are therefore a very accurate way to measure the expected signal quality in the end application.

The eye pattern is used to measure the effects of inter-symbol interference on random data being transmitted through a particular medium. The prior data bits affect the transition time of the signal. This is especially true for NRZ data which does not guarantee transitions on the line. For example, in NRZ coding, a transition to high after a long series of lows has a slower rise time than the rise time of a periodic (010101) waveform. This is due to the low pass filter effects of the cable. Figure 3.6 illustrates the superposition of six different data patterns. Overlaid, they form the eye pattern that is the input to the cable. The right-hand side of this figure illustrates the same pattern at the end of the cable. Note the rounding of the formerly sharp transitions and the

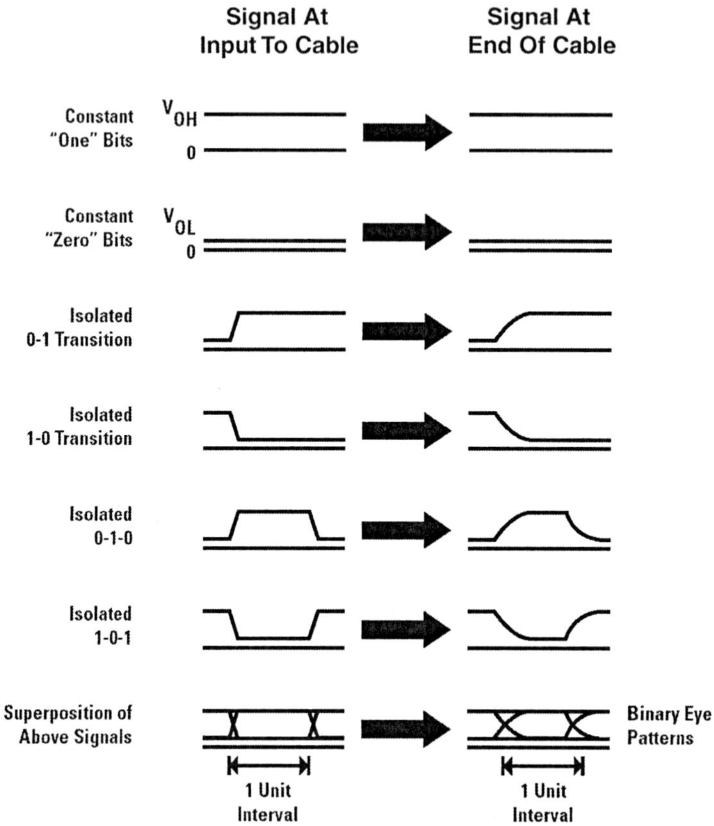

Figure 3.6 Formation of an eye pattern by superposition. [Reprinted with permission of National Semiconductor Corporation.]

decrease in the opening of the eye. Figure 3.7 shows a more detailed view of an eye opening and how the superposition of different data patterns causes the eye opening to be reduced in both the width and height. The template for XAUI (10-Gigabit Attachment Unit Interface) is shown in the figure and is used to determine the signal's integrity at the receiver. The eye opening must stay outside of this template to assure that the receiver will operate properly." [Nat00]

Figure 3.8 illustrates jitter parameters. The threshold-crossing jitter, t_{tcs}, is the jitter measured at the 0 V receiver input voltage level (the vertical center of the waveform). It is measured in time and is the extent to which the signal's 0 V crossing point moves to the left and right horizontally. t_{tcs} is the "minimum" jitter because measuring the jitter at the 0 V crossing results in the smallest jitter values. Explained a little differently, threshold-crossing jitter is the width of the range of signal crossing where they cross optimal receiver thresholds. For a differential receiver, this corresponds to 0 V (differential).

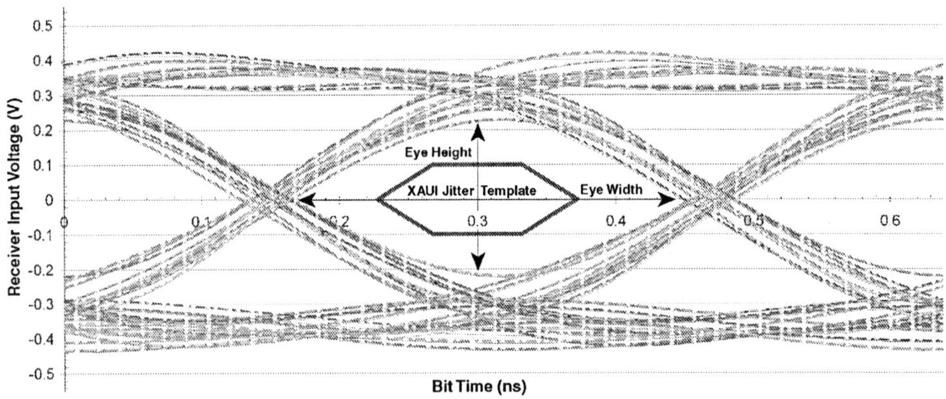

Figure 3.7 Eye diagram with XAUI jitter template. [Copyright of Motorola. Used by permission.]

For a receiver specified to switch between –100 mV and +100 mV, a worst-case jitter measurement can be made by drawing a box between ±100 mV and measuring the jitter between the first and last crossing at ±100 mV. As can be seen from Figure 3.8, the jitter measured at higher or lower voltage levels (relative to 0 V) results in larger jitter values. The jitter measured at the maximum receiver switching thresholds for logic states (e.g., ±100 mV) is taken as the maximum or worst-case jitter.

The unit interval t_{ui} of the eye diagram is the width of the ideal eye opening as measured at 0 V if no jitter were present. A unit of measurement sometimes used with eye diagrams is the milliunit interval, mUI, equal to 1/1000 of the unit interval.

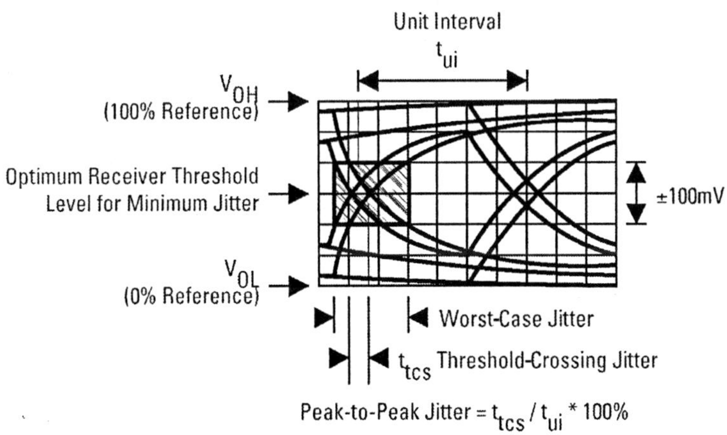

Figure 3.8 NRZ data eye pattern. [Reprinted with permission of National Semiconductor Corporation.]

Peak-to-peak jitter is the jitter measured from the leftmost range of travel to the rightmost range of travel of the signal waveforms and is usually expressed as a percentage of the unit interval. When peak-to-peak jitter is measured at 0 V, it is equivalent to the threshold-crossing jitter but expressed as a percentage. Peak-to-peak jitter is calculated as

$$\text{Peak-to-peak jitter} = \frac{t_{tcs}}{t_{ui}} \times 100\,\% \tag{3.18}$$

An eye diagram can be created on an oscilloscope screen by setting the persistence level of the scope to high. As the scope traces out different data patterns, the eye diagram will be created. With high persistence, the previous traces remain on the scope screen along with the most recent one, thereby creating the eye diagram.

3.5 BER Testing

BER is an acronym that can mean either bit error ratio or bit error rate. The difference between the two is very subtle. Bit error rate is a measure of the number of bit errors over time, such as one error every 37.4 hours. Bit error ratio is a measure of the number of bit errors as a function of the number of bits transmitted, such as one error for every one trillion bits transmitted. While there is this distinction between the two definitions, the subtlety is ignored by most practicing engineers as the application generally determines which definition is being used.

"BER testing (BERT) is another approach to determine signal quality and is frequently used to measure the performance of communications systems. It is performed by transmitting a known digital bit sequence through a system and checking for bit errors at the output of the system. The errors are determined by comparing the known bit sequence applied to the system input to the bit sequence at the system output. The standard equation for a bit error ratio measurement is:

$$\text{Bit Error Ratio} = \text{(Number of Bit Errors)/(Total Number of Bits)} \tag{3.19}$$

Common measurement points are bit error ratios of

$$\leq 1 \times 10^{-12} \Rightarrow \text{One or less errors in 1 trillion bits sent}$$

or

$$\leq 1 \times 10^{-14} \Rightarrow \text{One or less errors in 100 trillion bits sent}$$

Note that BER testing is time intensive. The time length of the test is determined by the data rate and also by the desired performance benchmark. For example, if the data rate is 50 Mbps, and the benchmark is an error ratio of 1×10^{-14} or better, a run time of 2,000,000 seconds is required for a serial channel. 2,000,000 seconds equates to 555.6 hours or 23.15 days!" [Nat00]

A detailed example of BERT is given in Section 25.6, which starts on page 727.

3.6 Exercises

3-1. A periodic signal has a period of length T where the segment of the signal found in the length T is repeated continuously over time. In looking at the eye diagrams in Figures 3.6, 3.7, and 3.8 and reviewing the definition of the eye diagram's unit interval, is the eye diagram's unit interval equal to the period of the waveform, or is it equal to something else? If equal to something else, what else?

3-2. Data sheets for drivers and receivers normally do not specify the edge rate directly. What types of parameters do they specify instead that allow you to determine edge rates?

3-3. It was stated that by going from single-ended to differential technology and using a SerDes device that serializes 16 LVTTL lines onto a single differential pair, that the pin count can be reduced by 87 percent. (a) What do we mean by the pin count? (b) How is this 87 percent calculated? (c) What is one advantage and one disadvantage of doing this?

3-4. (a) What is the propagation function (equation)? (b) What is the real part of the propagation function? (c) Is the real part of the propagation function constant or does it vary with frequency? (d) Given that most signals are composed of various components at different frequencies, what does the real part of the propagation function tell you will happen as the signal propagates down a transmission line?

3-5. Draw a lattice diagram for a circuit similar to the one shown in Figure 3.2 on page 35, but where V_S = 3 V, Z_S = 60 Ω, Z_0 = 100 Ω, Z_L = 45 Ω, and ε_r = 3.9. Indicate all the voltages, including voltages at the beginning of the trace, voltages at the load, and voltages of the reflected waves as was done in the lattice diagram of Figure 3.3 on page 37. Assume that the flight time down the trace is 200 ps. Be sure to show the two reflection coefficient values as well.

3-6. What is the length of the trace in inches for the circuit described in Exercise 3-5 above?

3-7. (a) Repeat Exercise 3-5, but with Z_S = 100 Ω. (b) Repeat Exercise 3-5 but with Z_L = 100 Ω and Z_S = 60 Ω.

3-8. What are eight advantages of differential technology over single-ended technology?

3-9. Define each of the following terms in a single sentence: (a) unit interval, (b) mUI, (c) threshold-crossing jitter, (d) peak-to-peak jitter, (e) worst-case jitter, (f) minimum jitter, and (g) maximum jitter.

3-10. (a) What is the difference between a lossy and lossless transmission line? (b) What is one technique for estimating whether a transmission line will be lossy or lossless?

PART 2

Signaling Technologies and Devices

CHAPTER 4

Gunning Transceiver Logic (GTL, GTLP, GTL+, AGTL+)

Chapter Objectives

- Present the evolution of several backplane signaling technologies and the benefits of the technologies when they first appeared.
- Discuss which technologies have superseded Backplane Transceiver Logic (BTL) and Gunning Transceiver Logic (GTL).
- Discuss Gunning Transceiver Logic Plus (GTLP), its applications, throughput and performance, and backplane design considerations.
- Discuss GTLP device features including bushold, output edge control (OEC), A and B Ports, and live insertion and extraction.
- Discuss Gunning Transceiver Logic+ (GTL+) and Assisted GTL+ (AGTL+).

4.1 Evolution from Backplane Transceiver Logic (BTL)

BTL (Backplane Transceiver Logic) has been superseded by GTLP (Gunning Transceiver Logic Plus), BLVDS (Bus Low Voltage Differential Signaling), and various SerDes (serializer/deserializer) technologies for backplane applications. However, a quick review of BTL is beneficial as it allows a quicker grasp and more complete understanding of this chapter's GTL/GTLP and GTL+/AGTL+ discussions.

BTL evolved from ECL (Emitter-Coupled Logic) in an effort to reduce power consumption while providing near-equivalent performance. [Fai01b] The initial application for BTL was the telecom sector, where BTL met the live insertion and withdrawal requirements. The physical layer of the 5 V V_{CC} FutureBus (IEEE896.2) was also implemented in BTL. [Tex02c]

Several years ago the standard solutions for driving bus lines on backplane systems were TTL or CMOS logic circuits. However TTL and CMOS had high voltage swings of from 3.3 V to 5 V, resulting in high power consumption. They also could not provide correct termination under all load conditions and required large current drive capabilities to enable incident wave switching. [Tex02c] BTL solved many of these types of problems with a reduced voltage swing and an open collector output stage as shown in Figure 4.1. In BTL, the falling signal edge is actively generated by the driver and the rising edge is generated by the passive pullup network. The resistor R in the passive pullup network of Figure 4.1 is matched to the resistance of the loaded backplane to eliminate signal reflections and ensure signal integrity. [Tex02c]

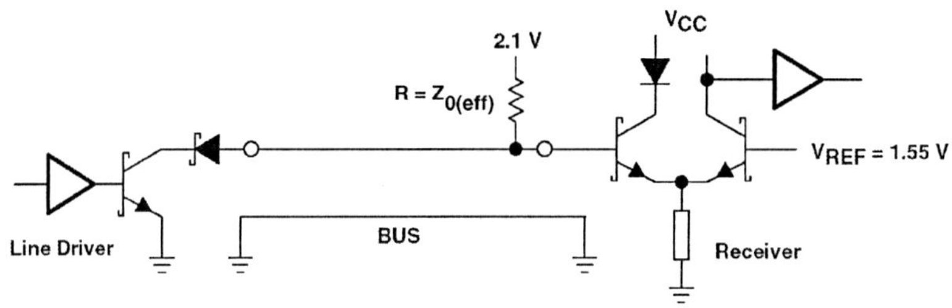

Figure 4.1 Setup of an open collector bus system using BTL devices. [Courtesy Texas Instruments.]

The driver has a low output impedance which allows it to sink a large current and switch a heavily loaded backplane with the incident signal edge. Due to this desire to provide incident wave switching, BTL devices are designed with high current-sinking capability and can sink currents (I_{OL}) of up to 100 mA depending upon the device. [Tex02d] Generally, BTL has a maximum frequency of about 40 MHz.

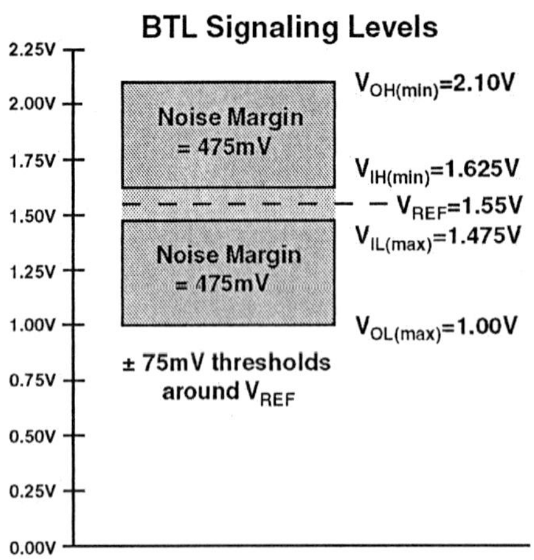

Figure 4.2 BTL signal levels.

BTL provides a reduced voltage swing of 1.1 V, as shown in Figure 4.2. The 2.1 V V_{OH} level is established by tying the pullup resistor R to the 2.1 V termination voltage. The 1.0 V V_{OL} value is the sum of the voltage drop across the driver's decoupling diode and the saturation voltage of the pulldown transistor when the transistor is turned on. The decoupling diode reduces the output capacitance of the driver output stage to less than 5 pF which in turn increases the bus impedance. The receiver switches state when the signaling level reaches 75 mV on either side of the 1.55 V reference voltage, positioned exactly in the middle of the voltage swing. The threshold region ranges from 1.475 V to 1.625 V. BTL generates V_{REF} internally and is set at 1.55 V.

BTL is designed to drive very large loads, such as lines terminated at both ends with 22 Ω resistors for an equivalent load of 11 Ω. BTL generates low switching noise owing to the low output impedance of the driver.

To reduce I_{CC} (collector current) spikes, BTL requires that the fall time be 2 ns or slower resulting in a trapezoidal waveform which reduces system noise. The rise time is not generated by active electronics, but by the pull-up resistor. For safe detection of the logic levels, the receiver is designed with a differential amplifier. BTL was standardized by IEEE Standard 1194.1-1991 which has since been withdrawn.

4.2 Gunning Transceiver Logic (GTL)

GTL (Gunning Transceiver Logic) is specified in standard EIA/JESD8-3 and was originally designed for small buses to be used at the board level. For example, GTL could be used to implement a bus between a processor and its memory modules. GTL has a fast edge rate and reduced voltage output levels from those of BTL, resulting in reduced power consumption, higher speeds, and improved signal integrity. "Because the target application for GTL is a bus on a board rather than a backplane bus, GTL has no requirements for live insertion/withdrawal." [Tex02d] The basic design of a GTL bus is similar to a BTL system and is shown in Figure 4.3. The GTL driver has an open collector output, but the diode in the open collector output that is found in BTL is not present in GTL. This reduces the output voltage low level maximum down to 0.4 V. [Tex02c] Figure 4.4 illustrates the signaling levels for GTL.

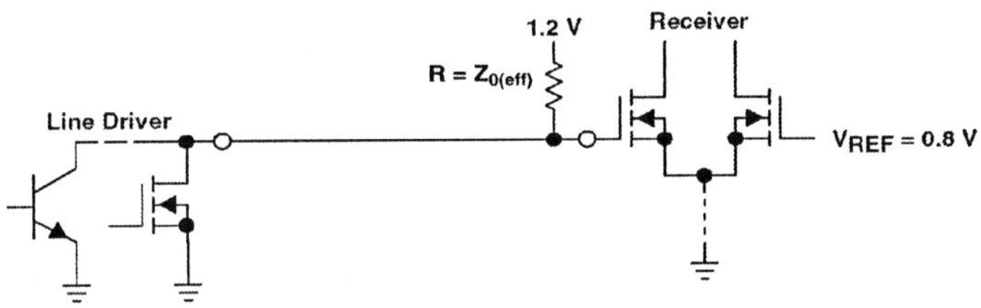

Figure 4.3 Open collector bus system using GTL devices. [Courtesy Texas Instruments.]

The driver's output swing is 0.8 V, from 0.4 V for $V_{OL(max)}$ to 1.2 V for $V_{OH(min)}$, where the low voltage swing results in low switching noise. The $V_{OL(max)}$ is dependant on the device, trace, and effective termination resistance, but is normally 0.4 V or below. The reference voltage V_{REF} is 0.8 V and is centered on the voltage swing. The receiver thresholds for logic states are centered on V_{REF} and are ±50 mV from it. The receiver (as with BTL) is designed as a differential amplifier, guaranteeing stable threshold voltages at the receiver. The termination voltage V_{TT} is 1.2 V as shown in Figure 4.3. In the figure, a pull-up resistor R at the line end is matched to the loaded trace impedance $Z_{0(eff)}$ to avoid line reflections.

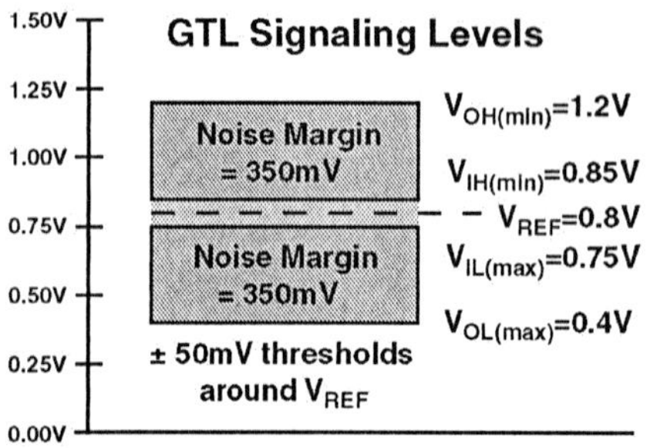

Figure 4.4 GTL has a 0.8 V voltage swing and 50 mV thresholds centered on the reference voltage of 0.8 V.

Typical GTL drivers can normally sink up to 40 mA of current and drive bus loads up to 20 Ω (800 mV ÷ 40 mA = 20 Ω). If the bus line is terminated correctly with a resistor at each end of the bus, the lowest line impedance that can be driven by a GTL driver in the middle of the bus is 40 Ω. The driver effectively sees two 40 Ω resistors in parallel for an equivalent load of 20 Ω. As a result of the 0.4 V $V_{OL(max)}$ and the 40 mA I_{OL}, the maximum power dissipation of one output is 16 mW. It is thus possible to integrate these low-power drivers into ASICs.

Some present GTL devices now provide up to 100 mA of sink current capability, enabling the outputs to drive heavily loaded backplane busses down to 11 Ω using two parallel 22 Ω termination resistors. Some of these devices also allow live insertion and withdrawal although it is not required by the specification. The devices may also include a selectable edge rate control (ERC) circuit to vary rise and fall times so that designers can fine tune their circuits for maximum data throughput as system loading changes dynamically. The edge rate control minimizes bus-settling time. [Tex02d]

4.3 Gunning Transceiver Logic Plus (GTLP)

4.3.1 GTLP General Description and Applications

"Gunning Transceiver Logic Plus (GTLP) devices are high-speed, high-performance backplane transceivers which operate like the GTL family except for two major differences: (1) they have been optimized with slower edge rates for the distributed loads found in multi-slot backplanes, and (2) they support live insertion applications with internal precharge circuitry."[1] [Tex02d] "The primary application for these devices is the logic signal translation between a LVTTL/TTL processor (like the Motorola PowerPC) and a GTLP backplane, such as those used in telecom/networking and other applications that need high-speed transmission of large amounts of information in a multipoint environment. The GTLP family is an improvement over existing LVTTL and BTL logic for higher frequency operation in multipoint applications and a lower cost parallel alternative to high-speed serial solutions in high data rate transfer applications."[2] [Phi02a]

GTLP devices can be used to provide robust 1394[3] link layer controller signals to telecom cards. The 1394 backplane PHY (physical layer), in conjunction with the 1394 link layer and microprocessor, provides absolute control over each card and is able to monitor and shut down

1. Courtesy Texas Instruments.
2. Copyright of Koninklijke Philips Electronics N.V. and Philips Semiconductors. Used by permission.
3. IEEE 1394b-2002, "High-Performance Serial Bus," for gigabit signaling up to 100 meters using glass optical fiber cables for data storage, home network backbone, and other applications.

any card that locks up the system, allowing a safe restart and continued operation. Many optical routers are using GTLP devices in this type of implementation. [Fai02a]

GTLP is a vendor derivative of the JEDEC standard for GTL (JESD8-3). Initially GTLP followed an established specification set by Intel for its microprocessor interface designs in an effort to optimize on-board signal integrity characteristics. However, off-board applications remain more varied in nature and the flexibility of the GTLP output allows it to interface to a number of interface signal levels. The evolution from ECL to BTL to GTLP was an effort to solve power consumption problems in earlier applications while providing near-equivalent performance. ECL is a pure bipolar design and BTL designs are BiCMOS. GTLP was the first all-CMOS solution and provides more power savings than ECL or BTL. [Fai01b]

"GTLP is a single-ended, open-drain (open-collector) I/O technology. GTLP's I/O characteristics and output control circuitry make it possible for GTLP to operate with loads of over 30 cards. Figure 4.5 shows a simplified partial schematic of a typical GTLP device. GTLP devices are commonly designed with two 'ports,' an LVTTL/TTL I/O which is referred to as the A Port and a GTLP I/O which is referred to as the B Port. Both ports are bidirectional as the term 'I/O' indicates. On many devices, A Port I/Os and the control inputs operate with LVTTL signal levels, may be 5.5 V tolerant, and can be CMOS transceiver designs. GTLP transceivers translate TTL/LVTTL signals to GTLP levels for the backplane environment and receive GTLP signals to translate back to TTL/LVTTL signal levels. GTLP's open drain output allows the designer to adjust the pull up termination in the system for better impedance matching and thereby optimize the performance of the backplane." [Fai02b]

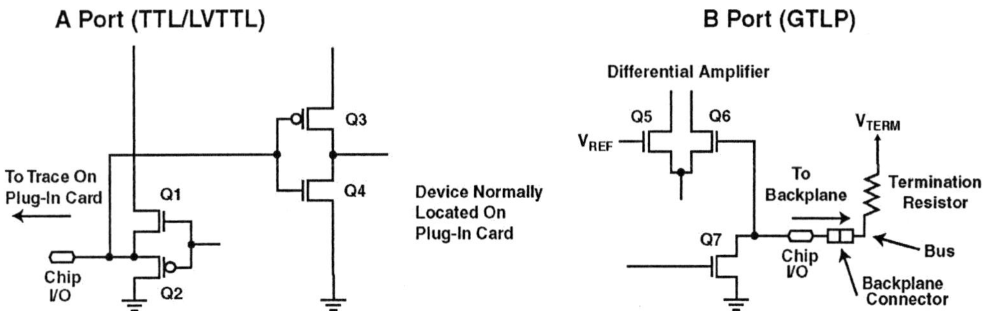

Figure 4.5 A and B Ports of a GTLP device. [Reprinted with permission of Fairchild Semiconductor Corporation.]

The benefit of "open drain" technology is that the output is either sinking current or in high impedance state (3-STATE) but never sources current. GTLP devices range from 1 to 36 bits wide with package options that include US8, SOIC, SSOP, TSSOP, TVSOP, BGA, VFBGA, and LFBGA. [Fai02b] GTLP offers up to four times the performance of TTL devices in backplane upgrade applications. [Tex02c]

GTLP also features a unique edge rate control characteristic that ensures smoother transitions. By tuning the backplane with proper termination resistors and reference voltages, GTLP can act as a drop-in electrical solution for BTL. [Fai01b] GTLP devices have a completely different set of characteristics for the LVTTL driver's A Port interface from that of the GTLP B Port's interface such as medium drive levels, bushold and over-voltage tolerance. [Fai01b]

4.3.2 GTLP Throughput and Performance

When used at the recommended voltage levels, GTLP can operate at frequencies ranging from 60–180 MHz+ depending upon the number of backplane cards, as shown in Figure 4.6. [Fai01b] Most GTLP multipoint applications operate in the 60–80 MHz range. The throughput rates for GTLP devices are in the order of 200 Mbps to 3.6 Gbps, depending upon the device (1–36 bits). [Fai02a] GTLP can operate at its intended voltage levels, but it can also operate at BTL voltage levels. Using GTLP at BTL voltage levels results in a maximum frequency similar to BTL of about 40 MHz.

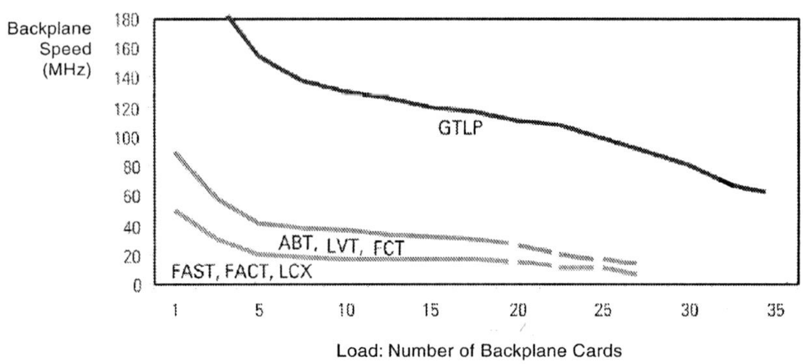

Figure 4.6 GTLP performance compared to other technologies. [Reprinted with permission of Fairchild Semiconductor Corporation.]

4.3.3 GTLP Signaling Levels, Noise Margins, and Current Drive

Under nominal conditions, GTLP has a 0.95 V voltage swing from 0.55 V to 1.5 V with a termination voltage V_{TT} of 1.5 V. The reference voltage V_{REF} is 1.0 V with switching thresholds of ±50 mV around V_{REF}. GTLP signaling levels are shown in Figure 4.7 along with those for GTL and BTL. It can be seen from the figure that the upper and lower noise margins for GTLP are 450 mV and 400 mV, respectively, which are slightly greater than those for GTL but slightly less than those of BTL. The combination of a 50 mV threshold margin coupled with a wider V_{OL} noise margin means that GTLP can operate within BTL signal level specifications and interface to BTL.

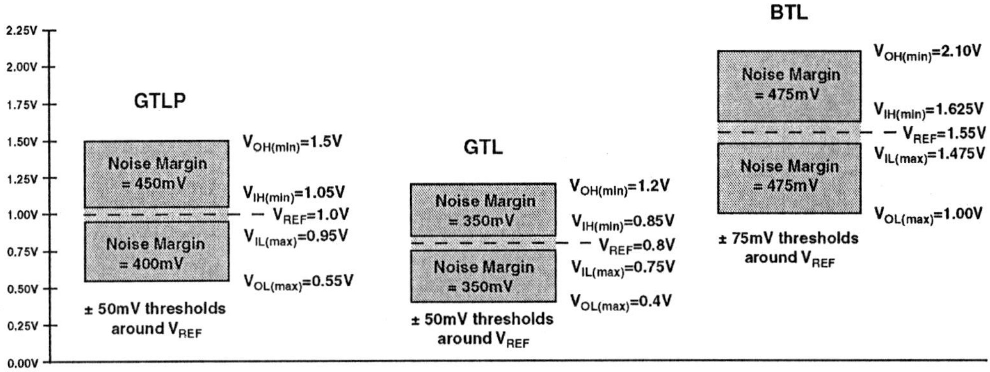

Figure 4.7 Signaling level comparison between GTLP, GTL, and BTL.

GTLP devices also offer comparable noise margins to that of its predecessors (ABT, LVT, and FCT) as shown in Figure 4.8. The designer must take into consideration the amount of noise seen by the receiver and must select a device able to handle the amount of noise generated on the backplane. Noise which the receiver usually contends with is from effects of ground bounce, EMI, and cross-talk. GTLP receivers are able to reject much of this noise due to the input noise margin and adjustable V_{REF}. [Fai02a]

With a reduced output swing of 0.95 V or less, GTLP has many advantages over TTL technologies. The most obvious is much faster transitions which allow for higher backplane speeds. Also, since the device does not have to transition as far, the transition can be much less steep than a 0 V to 3 V transition. This significantly lowers $\Delta V/\Delta t$ and results in much lower noise levels from ground bounce, ringing, cross-talk, and EMI.

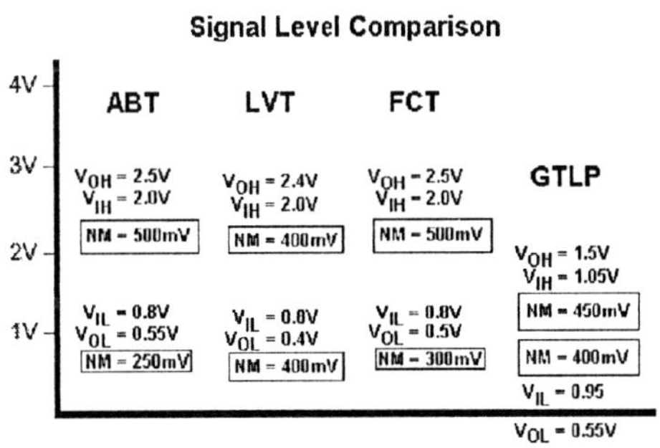

Figure 4.8 GTLP noise margin is similar to noise margins of ABT, LVT, and FCT. [Reprinted with permission of Fairchild Semiconductor Corporation.]

GTLP devices allow the adjustment of three key variables: termination resistance (R_{TT}), termination voltage (V_{TT}), and reference voltage (V_{REF}). V_{TT} and R_{TT} fine-tune a termination scheme, which contributes to factors like power consumption and noise margin. V_{REF} allows the voltage output signal to swing above and below a fixed point, typically 1 V. [Fai02a]

GTLP devices have varying amounts of sink-current capability, $I_{OL(min)}$, (30 mA, 40 mA, 50 mA, and 100 mA) to allow the designer flexibility in matching the device to backplane length, slot spacing and termination resistor values. A 50 mA device can drive lines in point-to-point configurations down to 19 Ω.

The lowest termination resistor that can be driven by the driver in the middle of a bus is 38 Ω (effectively the driver's load is 38 Ω || 38 Ω = 19 Ω). A 100 mA device can drive loads of 9.5 Ω (0.95 V ÷ 100 mA) so that the minimum termination resistor for bus configuration is 19 Ω.

"GTLP requires a termination voltage (V_{TT}) that can be regulated directly from V_{CC}. GTLP uses a reference voltage pin (V_{REF}) which is an input voltage used to reference an incoming GTLP or BTL signal. As a starting point for prototype designs, V_{REF} should be generated from V_{TT} with the use of a simple voltage-divider circuit, resistors R and 2R connected in series between V_{TT} and ground with V_{REF} tapped from the top of the 2R resistor." [Phi02a] By having V_{REF} track with V_{TT}, if V_{TT} were to decrease, V_{REF} would decrease proportionately and maintain roughly equivalent upper and lower noise margins. If instead V_{TT} were to decrease and V_{REF} did not track with V_{TT} but remained at the same voltage, the upper noise margin could become significantly reduced and result in signaling errors. As a separate consideration, by allowing the designer to vary V_{REF} by slightly changing the R and 2R resistor values, noise margins can be controlled in noisy environments. If the ground bus has large amounts of noise, V_{REF} can be set higher to allow more lower noise margin at the cost of reduced upper noise margin. If the V_{TT} power bus is noisy, V_{REF} can be decreased to allow more upper noise margin at the trade-off of reduced lower noise margin. The adjustable V_{REF} allows the design engineer to make engineering trade-offs.

Many GTLP devices have 5 V tolerant LVTTL inputs and outputs, can operate in "wired-OR" configurations, have reduced termination power consumption over Thevenin terminations, and are simple to implement. The GTLP B Port backplane outputs are better suited for backplane applications than GTL, FB or FBL devices since they have a slower edge rate that reduces ringing. The key to improved backplane operation using GTLP is matching the termination pull-up resistors (R_{TT}) with the loaded backplane trace impedance $Z_{0(eff)}$ to ensure incident wave switching and excellent signal integrity.

4.3.4 GTLP Device Features

4.3.4.1 Live Insertion and Extraction

GTLP allows insertion and extraction of a board in a backplane without powering down the host system or suspending an active backplane. All GTLP devices incorporate power-up/power-down 3-STATE circuitry. Some devices provide output pins with precharge circuitry. Precharge places the voltage on the output pins at midswing levels during insertion or extraction. Setting the voltage at midswing during insertion minimizes the capacitive loading effects on the signals of an active backplane. [Fai02a]

4.3.4.2 Controlled Edge Rates

GTLP devices incorporate output edge control (OEC) circuitry to address one of the most prevalent problems with high-speed devices: output switching noise. Overshoot, undershoot, and ground bounce all fall under the category of output switching noise. Edge control incorporates wave-shaping techniques that optimize GTLP devices for driving backplanes. The main focus is to control the output level transition. This minimizes switching noise and electromagnetic interface, and reduces signal-settling time. Figure 4.9 illustrates the areas of the output transition (low-to-high and high-to-low) that are addressed by the GTLP output control circuitry. [Fai02a] The ability to vary rise and fall times compensates for various loads. The output edge rate con-

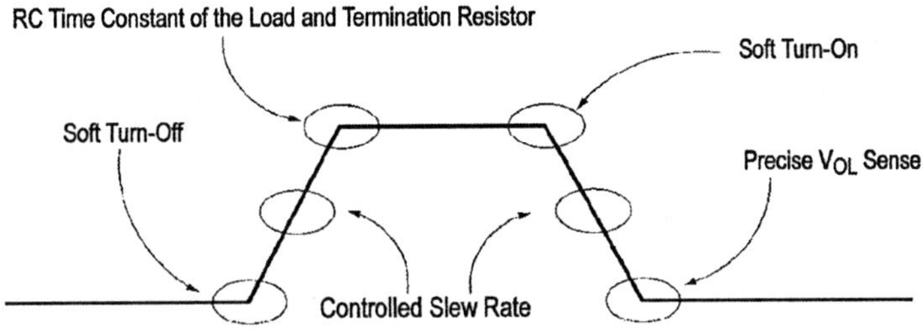

Figure 4.9 GTLP transition waveform. [Reprinted with permission of Fairchild Semiconductor Corporation.]

trol feature coupled with an adequate termination scheme allows the system designer to fine tune the output edges for optimum performance. Proper termination of the system is important for optimizing performance. The backplane impedance will vary from system to system depending on stub lengths, number of slots, slot pitch, and types of connectors used. Figure 4.10 shows how the controlled edge rates of GTLP reduce both the noise amplitude and high-frequency content when compared to TTL.

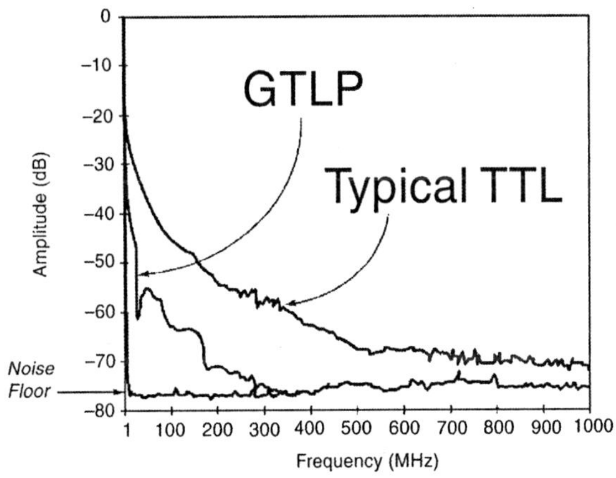

Figure 4.10 Controlled edge rates reduce noise and EMI. [Reprinted with permission of Fairchild Semiconductor Corporation.]

4.3.4.3 Bushold (A Port)

"The bushold feature designed into the A Ports of many GTLP and other devices is an innovative approach to solving the problem of floating inputs. This circuitry maintains a valid logic state on an unloaded input and eliminates the need for external conditioning components for floating or undriven inputs. Bushold circuitry is designed to tolerate floating input conditions. If non-bushold devices were used, these floating input conditions would lead to increased leakage, oscillations, and compromises in system data integrity. Another feature of bushold circuitry is the ability to hold an undriven data-bus line in a valid logic state. Through the bus, the bushold device can condition the other inputs on the bus.

Bushold circuitry works by using a low drive inverter in the device input stage that provides feedback to the input of the device and the bus as shown in Figure 4.11. When the signal driving the input is removed, the inverter will maintain the last received valid signal level on the device input and bus line until it is overdriven by the next incoming signal.

Two types of bushold inverter specifications are provided (see Figure 4.12). $I_{I(HOLD)}$ is the bushold input minimum drive. This is the minimum amount of current the circuit is capable of supplying. $I_{I(OD)}$ is the bushold input over-drive current to change state. This is the minimum amount of current that is necessary to overcome the bushold circuit and cause the input to change states. Table 4.1 lists typical values for $I_{I(HOLD)}$ and $I_{I(OD)}$.

Devices with bushold are designed for use in systems where devices or data-bus lines will be left in an undriven condition. Non-bushold inputs must be maintained in a valid logic state. Leaving an input floating will allow it to oscillate and cause invalid data on the output. False data problems can lead to bus contention, high current draw, and device or system damage. System data buses must be maintained in a valid logic state for the same reasons that individual device inputs must be maintained in a valid logic state. Data integrity problems and device or

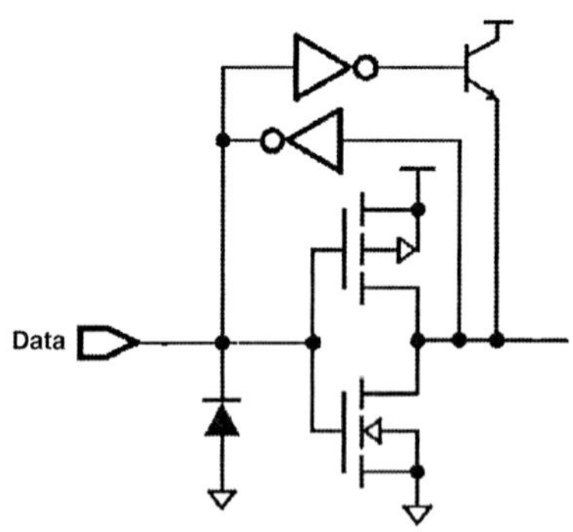

Figure 4.11 Simplified schematic diagram of bushold circuitry. [Reprinted with permission of Fairchild Semiconductor Corporation.]

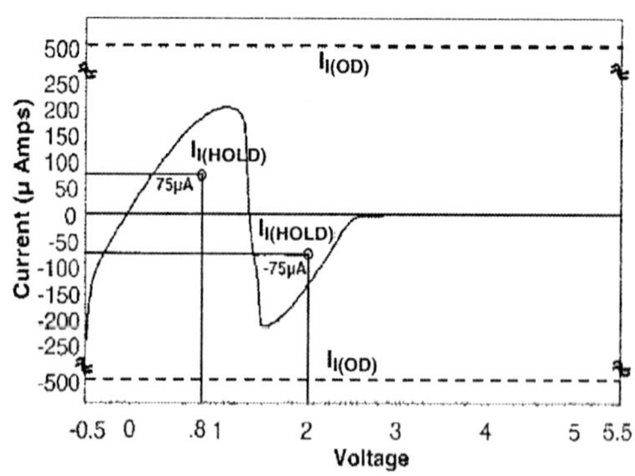

Figure 4.12 Voltage-in versus current-in sweep of bushold device (74LVTHxxx). [Reprinted with permission of Fairchild Semiconductor Corporation.]

system damage can occur if a bus is allowed to float. With a data bus however, the number of device inputs on the bus multiplies the problem.

Table 4.1 Typical Bushold Specifications

Symbol	VCC(V)	Conditions	Minimum	Units
$I_{I(HOLD)}$	3.0	V_{IN} = 0.8 V	75	µA
	3.0	V_{IN} = 2.0 V	–75	µA
$I_{I(OD)}$	3.6	(Note 1)	500	µA
	3.6	(Note 2)	–500	µA

Note 1: An external driver must source at least the specified current to switch from low to high.
Note 2: An external driver must sink at least the specified current to switch from high to low.
[Reprinted with permission of Fairchild Semiconductor Corporation.]

By holding the device or bus at the last received valid logic high or low, devices with bushold can also serve a pseudo memory function. In contrast, a conditioning resistor will always pull the bus to the same state. When designing a data bus using devices with bushold, consideration must be given to the bushold drive and current requirements. With $I_{I(HOLD)}$ specified at ±75 µA, a single device with bushold will hold a large data bus, with power dissipation held at a low level. Calculations for the I_{IN} (Input current) of all device inputs on the bus must be taken into consideration. I_{IN} is typically in the range of 1–5 µA for CMOS devices. With $I_{I(OD)}$ specified at ±500 µA, the data bus will still switch easily. CMOS devices typically have drives in the ±24 mA range or greater.

Bushold circuitry on all system or data-bus inputs can create problems. On a data bus with multiple receiver drops, bushold circuitry should be used on the minimum number of devices that are needed to hold the data bus in a valid state. The $I_{I(OD)}$ specification is additive for every bushold circuit on the data bus. For example, using a device with an $I_{I(OD)}$ specification of ±500 µA, if 10 inputs on a data bus have bushold circuitry, it will take at least 5 mA to overcome and switch the data bus. Using a device with an $I_{I(HOLD)}$ specification of ±75 µA, 750 µA of drive will be available to hold this theoretical 10-input bus. This much drive is not required on an all-CMOS, input bus and unnecessarily adds to the power budget needed for the system."[4] [Fai99]

4.3.5 GTLP Backplane Design Considerations

The GTLP backplane is an open-drain bus almost identical to the GTL bus. The ideal GTLP backplane architecture for promoting increased throughput and incident wave switching is one that has low-capacitive loading through low-driver/receiver capacitance, short stub lengths, and controlled-impedance, high-frequency connectors. Under these conditions, the GTLP family of devices can demonstrate incident wave switching and very high data throughput.

4. Reprinted with permission of Fairchild Semiconductor Corporation.

It is important to pick a termination resistor which matches the backplane impedance for best signal integrity and which is within the current sink capacity of the driver. Impedance is a function of natural trace impedance (Z_0), stub length, connector impedance, device impedance, and card spacing. Closer spacing reduces the effective impedance and requires a smaller termination resistor as shown in Figure 4.13. [Tex02c] Figure 4.14 shows how proper matching improves the waveform shape.

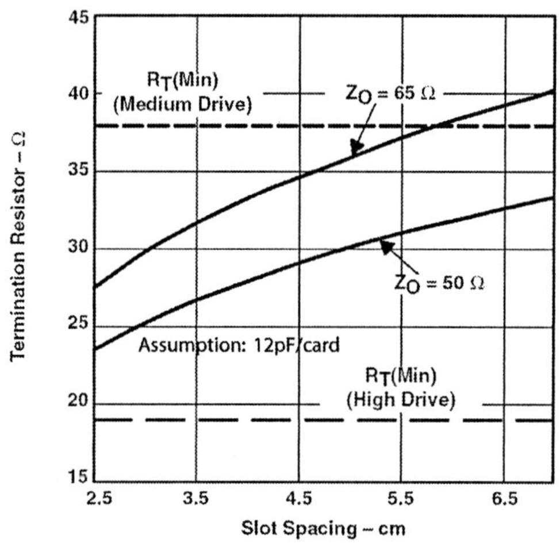

Figure 4.13 R_T vs. slot spacing with GTLP medium and high drive devices. [Courtesy Texas Instruments.]

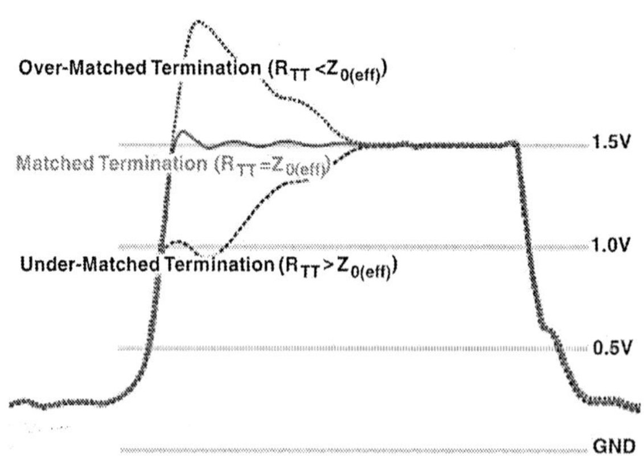

Figure 4.14 Matched vs. over-matched and under-matched terminations. [Copyright of Koninklijke Philips Electronics N.V. and Philips Semiconductors. Used by permission.]

"GTLP backplane construction is simple to implement and uses a pullup resistor on both ends of the backplane trace instead of the typical Thevenin terminations used for totem pole outputs. When all GTLP outputs connected to that trace are turned off there is no current drain, unlike the Thevenin termination where current is continuously flowing through the resistors. This reduces power consumption up to 50%. Additionally, since GTLP has open drain outputs, there can be multiple devices turned on for wired-OR operation without the danger of bus contention." [Phi02a]

"The GTLP reduced output swing (<1 V), reduced input threshold levels and slower edge rates allow higher backplane frequencies. GTLP devices are functionally and footprint equivalent to common ABT, FCT, LVT, ALVT and FB+ devices[5] but provided an alternative to more complex serial technologies in high data throughput applications. High drive devices can be used in place of medium drive devices even with larger termination resistor values (e.g., 75 Ω) without concern although since the medium drive devices are less costly the best drive match for the application would provide the optimum solution.

Texas Instruments provides a GTLP Evaluation Module (EVM) and has measured the unloaded and loaded characteristic impedances, propagation delays, and capacitances per unit length for it. Loaded trace impedance is a function of natural trace impedance (Z_0), stub length, connector impedance, device impedance and card spacing. The GTLP EVM is constructed like a typical open-drain telecom backplane that would look like the double-ended pull-up terminated trace as shown in Figure 4.15.

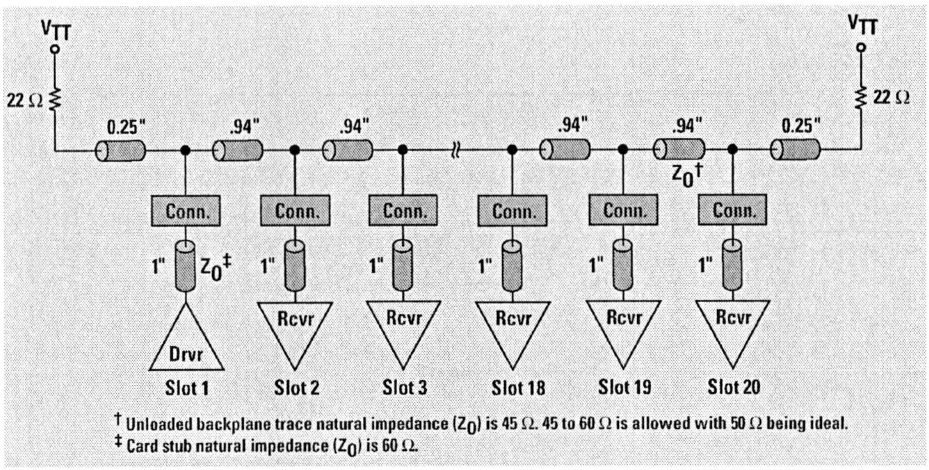

Figure 4.15 Typical GTLP backplane. [Courtesy Texas Instruments.]

5. ABT (Advanced BiCMOS Technology), FCT (Fast CMOS Technology, LVT (Low-Voltage BiCMOS Technology), ALVT (Advanced Low-Voltage BiCMOS Technology), FB+ (Futurebus+) where FB+ is fully compatible with BTL (Backplane Transceiver Logic).

The bus is pulled high to the termination voltage (V_{TT} = 1.5 V) through the termination resistance (R_{TT} = 22 Ω) when the GTLP driver's open drain output stage is off and pulled low when GTLP driver's open drain output stage is on. The advantage of the open drain backplane is that there is no bus contention, it is simple to implement and there is less power consumption than in Thevenin terminated backplanes. Actual range of measurements from the GTLP EVM is shown in Table 4.2. The table clearly shows that in a heavily loaded backplane the termination resistance should easily go as low as 21 Ω, but is limited to 22 Ω by the GTLP high drive maximum I_{OL} recommendations for the devices used on the EVM. Increasing the natural trace impedance can change the termination resistor to a value that is within the capacity of the high-drive devices and better approximate what the medium-drive devices can handle. The down side is that the backplane time of flight will now be about 58% slower. Typical card capacitance will be between 10 and 18 pF depending on device C_{IO} (input/output capacitance) and stub length. (The stub length is the most critical and should always be less than 1 inch.)" [Tex02d]

Table 4.2 Measured Range of Characteristics for the GTLP EVM

Unloaded (Natural) Trace			Trace and 20 Empty Connectors at 0.94"		Fully Loaded with 20 Cards at 0.94"		Card C_t
Z_0 (Ω)	T_{PD} (ps/in)	C_0 (pF/in)	Z_0 (Ω)	t_{PD} (ps/in)	Z_0 (Ω)	t_{PD} (ps/in)	(pF)
47	164	3.49	39	197	21	371	11.6
94	211	2.24	72	274	34	585	12.1

[Courtesy Texas Instruments]

4.3.6 GTLP Power Consumption

Many factors, both internal and external, affect the power consumption of a device: (1) number of outputs simultaneously switching, (2) operating frequency of the system, (3) junction temperature, and (4) capacitive load. The output power of both GTLP and BTL are supplied by an external output voltage supply (V_{TT}). When calculating the static high power consumption of the device, the device power is minimal during a static high event. This is because the output is in the high impedance state, and essentially zero current flows. When considering static low drive current, the output power is at the worst case point. GTLP uses less power than its predecessor BTL does. The power requirement is increased when considering that two 8-bit BTL devices are needed for one 16-bit GTLP device. [Fai01b]

The dynamic power consumption of BTL and GTLP are approximately the same given the same operating conditions, and when GTLP is used at its recommended operating conditions (1.0 V for V_{REF}) it uses much less power. GTLP offers twice the number of bits per device than the BTL technology does making the power consumption per bit lower for GTLP. [Fai01b]

4.4 Intel's AGTL+ and GTL+

Assisted Gunning Transceiver Logic+ (AGTL+) is an Intel® derivative of the JESD8-3 GTL standard and is used as the system bus signaling technology for Pentium® 4 and Xeon™ processors, some of the Pentium III and Celeron® processors, and for various Intel chipsets. "'System Bus' refers to the interface between the processor and system core logic (also known as the chipset components). The system bus is a multiprocessing interface to processors, memory, and I/O."[6] [Int02b] AGTL+ has superseded GTL+, an earlier Intel derivative of GTL. GTL+ has been used as the system bus signaling technology for Pentium II processors, some of the Pentium III and Celeron processors, and their chipsets. Figure 4.16 shows the signaling voltage levels for AGTL+, GTL+ and GTL.

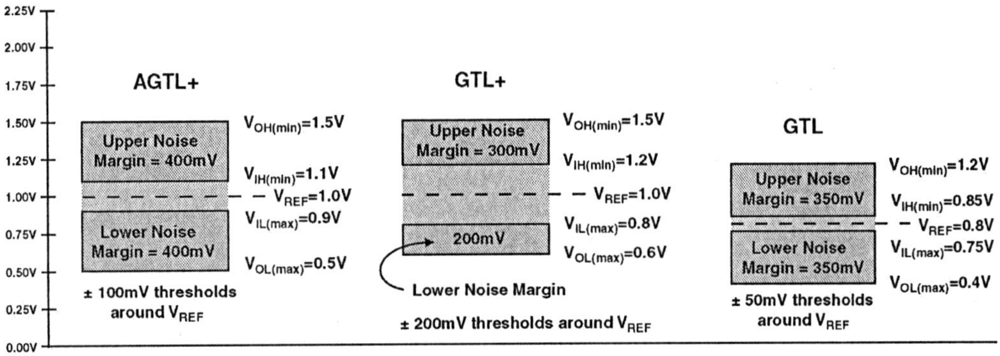

Figure 4.16 Signaling level comparison among JESD8-3 GTL and Intel's GTL+ and AGTL+.

AGTL+ is an incident wave switching, open-drain signaling standard. It has a 1 V swing from 0.5 V to 1.5 V with the reference voltage V_{REF} set to 2/3 of the termination voltage V_{TT}. When V_{TT} is nominally 1.5 V, V_{REF} is 1.0 V. The switching thresholds are ±100 mV around V_{REF}. The AGTL+ bus is terminated at both ends with a 50 Ω R_{TT} resistor. It should be mentioned that AGTL+ is used for both motherboard and on-chip buses as well as for several different groups of signals other than data. Because of these differing applications, there are variations to the typical voltage and termination values presented here for "signal groups."

GTL+ has a 0.9 V swing from 0.6 V to 1.5 V with V_{REF} also set to 1 V and V_{TT} of 1.5 V. A major difference is the ±200 mV switching thresholds centered on V_{REF}. GTL+ is also incident wave switching (logic state determined when the switching signal first arrives rather than having to wait for signal reflections to die down). The specification for GTL+ [Int97] specifies minimum and maximum edge rates of 0.3 V/ns and 0.8 V/ns, respectively.

6. Reprinted by permission of Intel Corporation, Copyright Intel Corporation 2002.

AGTL+ and GTL+ are not compatible as they have different voltage switching thresholds. AGTL+ is used with Intel's newer NetBurst™ microarchitecture system bus for Pentium 4 processors whereas GTL+ was used for the earlier P6 processor family bus. It is also important to note that Intel's GTL+ as shown in Figure 4.16 is not the same as GTLP which is shown in Figure 4.7 on page 57. Figure 4.17 illustrates an AGTL+ system bus with Pentium 4 processors and core logic.

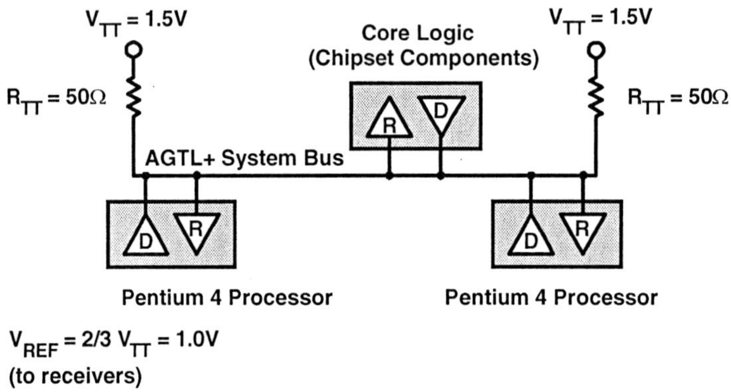

Figure 4.17 AGTL+ system bus.

A useful term when discussing AGTL+ is that of "bus agent." A bus agent is a component or group of components that, when combined, represents a single load on the AGTL+ bus. Because of the speed improvements to data and address buses, signal integrity and platform design methods have become more critical than with previous processor families. Some AGTL+ devices include on-die termination while others do not; these must be terminated on the system board as shown in Figure 4.17.

"The Intel NetBurst microarchitecture system bus uses Source-Synchronous Transfer (SST) of address and data to improve performance by transferring data four times per bus clock (4X data transfer rate as in AGP 4X). Along with the 4X data bus, the address bus can deliver addressees two times per bus clock and is referred to as a "double-clocked" or 2X address bus. Working together, the 4X data bus and 2X address bus provide a data bus bandwidth (data transfer rate) of up to 4.3 GB/s." [Int02b]

The AGTL+ signaling specification is found in reference [Int00] and includes many additional requirements to those of signaling levels. For example, the specification provides tolerances on several parameters, and provides receiver setup and hold times, and overshoot, undershoot and rising and falling edge ringback voltage requirements. The specification also states minimum and maximum edge rates of 0.3 V/ns and 0.8 V/ns, respectively. Tektronix and Agilent Technologies both provide logic analyzer interfaces (LAIs) for use in debugging Pentium 4 processors.

4.5 GTLP/GTL/GTL+/AGTL+ Summary

Table 4.3 Summary of Gunning Transceiver Logic (GTLP, GTL, GTL+, AGTL+)

GTLP/GTL/GTL+ (Gunning Transceiver Logic)

GTLP

Applications: Replacement for BTL
Driving distributed-load backplanes at high speeds
Backplanes with loads up to 30 cards
Multidrop and multipoint buses
Logic signal translation between a LVTTL/TTL processor (e.g., Motorola or IBM PowerPC) and a GTLP backplane such as those used in telecom/networking applications

Where Not to Use:
Where required clock speeds exceed 200 MHz
Where required to drive differential media

Device Voltages: 5 V, 3.3 V

Additional Required Voltages: V_{REF} = 1.00 V, V_{TT} = 1.5 V

Switching Thresholds: ±50 mV from V_{REF}

Driver Voltage Swing: Max of 1 V swing from 0.5 V to 1.5 V

Drive Current: 30 mA into 25 Ω equivalent load, 40 mA into 20 Ω, 50 mA into 19 Ω, or 100 mA into 12 Ω (I_{OL} sink current), device dependent

Process: CMOS

Technology: Single-ended, open-drain outputs

Standard: Philips Semiconductor, Fairchild Semiconductor, Texas Instruments derivative of JESD8-3

Power Consumption: Medium; e.g., 563 mW (typical) for 16-bit GTLP to LVTTL/TTL bus transceiver device, plus another 462 mW worst case (all outputs low) power consumption by termination resistors

Transmission Medium Characteristic Impedance: 50 Ω

Loaded Effective Backplane Characteristic Impedance, $Z_{0(eff)}$: 21–40 Ω depending on natural trace impedance, card spacing, stub length, and device pin capacitance

Termination: Two R_{TT} termination resistors in parallel, one at each end of the bus. Each resistor value should be equal to $Z_{0(eff)}$. Resistive load on GTLP output is then $Z_{0(eff)}/2$

Hot Pluggable: Yes

Noise Margin: 450 mV upper noise margin, 400 mV lower noise margin

Maximum Drive Distance: 60 inches (152 cm) and greater over FR-4

A Port: LVTTL/TTL logic levels

B Port: GTLP signal levels

Maximum Data Rate: Depends upon number of backplane cards (see Figure 4.6 on page 56):

Backplane Speed (MHz):	>180	155	130	120	110	100	85	65
Number of Backplane Cards:	1	5	10	15	20	25	30	35

Generally 60–80 MHz operation in multipoint applications
When GTLP is used at BTL voltage levels, maximum frequency is about 40 MHz

Table 4.3 Summary of Gunning Transceiver Logic (GTLP, GTL, GTL+, AGTL+) (Continued)

AGTL+
Applications: On-chip and motherboard buses. System buses using Intel Pentium 4 Processors, Intel Xeon Processor, some Intel Pentium III and Celeron Processors, Mobile Intel Pentium 4 Processor-M; Intel E7500, E7501, 850, 845, 845E, 845GV, 840, 815, 815E, 810 and 810E2 chipsets
Standard: Intel derivative of JESD8-3. See reference [Int00] for specification.
Termination Voltage V_{TT}: 1.5 V ± 9%, 1.5 V ± 3% during system bus idle
Reference Voltage V_{REF}: 2/3 VTT ± 2% (typically 1.0 V)
Voltage Swing: 1 V ranging from 0.5 V to 1.5 V
Voltage Threshold Levels: ±100 mV centered on VREF
Drive Current: 40 mA into 25 Ω load (I_{OL} sink current)
Motherboard Trace Characteristic Impedance: 50 Ω ± 10%
Termination Resistor R_{TT}: 50 Ω ± 10% at each end of bus
Bus Frequency: 90–100.20 MHz (data can be transferred 4 times per bus clock [Int02b])
Hot Pluggable: No
Noise Margin: 400 mV lower noise margin, 400 mV upper noise margin (nominal)

GTL (Superseded by GTLP)
Applications: Small buses to be used at the board level such as a processor bus to its memory modules.
Standard: JESD8-3
Hot Pluggable: No, with exception that a few GTL devices are hot pluggable. Check data sheet.
Process: BiCMOS
Reference Voltage V_{REF}: 0.8 V
Termination Voltage V_{TERM}: 1.2 V
Voltage Swing: 0.8 V ranging from 0.4 V to 1.2 V
Voltage Threshold Levels: ±50 mV centered on V_{REF}
Transmission Medium Characteristic Impedance: 50 Ω
Termination Resistor R_{TT}: 50 Ω resistor at each end of bus
Noise Margin: 350 mV

GTL+ (Superseded by AGTL+)
Applications: System buses using Intel Pentium II processor, some Pentium III and Celeron processors, 82440FX PCIset, 440MX chipset, 440BX AGPset, and 440GX AGPset
Standard: Intel derivative of JESD8-3. See reference [Int97] for specification.
Termination Voltage V_{TT}: 1.5 V ± 9%, 1.5 V ± 3% during system bus idle
Reference Voltage V_{REF}: 2/3 V_{TT} ± 2% (typically 1.0 V)
Voltage Swing: 0.9 V ranging from 0.6 V to 1.5 V
Voltage Threshold Levels: ±200 mV centered on V_{REF}
Drive Current: 32 mA into 28 Ω load (I_{OL} sink current)
Motherboard Trace Characteristic Impedance: 65 Ω ± 20%
Termination Resistor R_{TT}: 56 Ω ± 5% at each end of bus
Bus Max Frequency: 100 MHz
Hot Pluggable: No
Noise Margin: 200 mV lower noise margin, 300 mV upper noise margin (nominal)

4.6 Exercises

4-1. Which technologies among BTL, GTL, and GTLP are hot pluggable?

4-2. Which among BTL, GTL, GTLP, GTL+, and AGTL+ are single-ended signaling technologies and which are differential signaling technologies?

4-3. Compare the voltage switching range and threshold voltages of BTL with GTLP by drawing a side-by-side voltage diagram for each. Which technology has the larger voltage swing? Which one has the smaller switching thresholds? What is the reference voltage for each one? What voltage would you tie termination pullup resistors to in order to properly terminate a backplane bus for BTL? For GTLP?

4-4. About how much power does a 16-bit GTLP to LVTTL/TTL bus transceiver device consume? State your answer in milliwatts and use a typical value (not maximum or minimum).

4-5. Define and explain "bushold."

4-6. Define "OEC" and explain what it is.

4-7. State the historical order in which the various technologies used to drive backplanes evolved. What were the advantages of moving on to each new technology when it first appeared? Include AGTL+, GTL+, GTL, BTL, CMOS, TTL, ECL, and GTLP.

4-8. What is the main difference between the intended applications for GTL and GTLP?

4-9. Refer to Figure 4.5, A and B Ports of a GTLP Device, on page 55. The figure shows the basic I/O structure for both the A Port and B Port of a GTLP device. In a single paragraph of 200 words or less, explain how the B Port works and how it is able to perform bidirectional data transfers. Similarly, explain in an additional 200 words or less how the A Port works and how it performs bidirectional data transfers.

4-10. Draw a schematic for a working circuit where data can be sent from a BTL transceiver to a GTLP transceiver and from the same GTLP transceiver to the BTL transceiver. Access the Internet and locate two devices, a BTL transceiver and a GTLP transceiver, for use in your design. Print out the first three pages of each data sheet. Do not include in your design any clock or control circuits. Assume backplane traces have a loaded characteristic impedance of 50 Ω. State any assumptions you make. Turn in your schematic along with the data sheet pages and be prepared to discuss your design in class. What is one of the most important things to consider when trying to ensure that GTLP and BTL transceivers interoperate properly? What is the maximum data rate that your circuit can transfer data?

4-11. How is the reference voltage V_{REF} generated for BTL and GTLP? What is the advantage of each method?

4-12. How are the termination voltage V_{TT} and reference voltage V_{REF} usually generated for use by a GTLP circuit?

4-13. What are three differences between GTLP and GTL+? Are they generally compatible with each other? Are they roughly the same signaling standard?

4-14. Log onto the Internet and access reference [Int00], "Pentium III Xeon Processor at 500 and 550 MHz," on Intel's Web site. Find the section, "AGTL+ System Bus Specifications." What is this section's definition of "ringback"? What is the minimum amplitude of ringback? What is the duration of squarewave ringback?

4-15. Assume you are designing a backplane bus and plan to have 20 plug-in cards. Your unloaded backplane trace characteristic impedance is 45 Ω and your plug-in cards are spaced 2 inches apart and have 12 pF of capacitance each. What value of termination resistors will you need to use based upon the information in Figure 4.14, Matched vs. over-matched and under-matched termination, on page 62? If you change your card spacing to $1^3/_8$ inch, what is the new value of your termination resistor? Round your answers to the nearest whole number. State any assumptions you make.

CHAPTER 5

Low Voltage Differential Signaling (LVDS)

Chapter Objectives

- Introduce the LVDS signaling standards.
- Discuss LVDS families of devices.
- Discuss LVDS device features and specifications.
- Identify applications for which LVDS devices are well suited.

5.1 Introduction to LVDS[1]

LVDS stands for low voltage differential signaling. It is a signaling standard that has been implemented with several families of integrated circuits. As a signaling technology, LVDS allows single channel data transmission at hundreds and even thousands of megabits per second (Mbps). Its low swing (small voltage transitions) and its current-mode driver outputs create low noise and provide very low power consumption across frequency. [Nat00]

LVDS drivers are capable of transmitting a signal over cable up to 10–15 meters (32–49 feet). The distance that can be driven varies inversely with the transmission rate, as the greater the distance, the lower the data rate has to be to maintain adequate signal integrity. Also, greater distances can be achieved while maintaining signal integrity by transmitting over higher quality cables or interconnects. LVDS devices usually require controlled-impedance circuit-board traces, connectors, and cables to maintain signal integrity at high data rates. High quality interconnects that are commonly used have an impedance precisely matching the rest of the transmission media, connectors, differential cable impedances, and termination resistors, while creating minimal distortion in the transmitted signal. LVDS is a differential signaling

1. All text and figures in this chapter are reprinted with permission of National Semiconductor Corporation except where indicated otherwise.

technology, meaning that each driver requires two traces (or conductors) in addition to the ground plane. One trace carries the transmitted signal (true signal) while the second trace carries the return current (inverted signal). LVDS communicates data using a very low voltage swing of about 350 mV differentially over the two traces or a balanced cable. A balanced cable is one which has the same electrical characteristics on the outgoing and return current conductors, for example, twisted pair wire, twin-coaxial cable, and so on.

In the commercial marketplace, several families of LVDS devices are available. These families of devices include line drivers, receivers, and transceivers for high-speed point-to-point transmission, as well as Bus LVDS devices for transmission up and down buses from one or several drivers to multiple receivers. For Bus LVDS, only one driver may transmit at a time on any given pair of traces. LVDS families also include Scan LVDS with JTAG (Joint Test Action Group) capabilities, Channel Link Serializers/Deserializers, Flat Panel Display Links (FPD-Links), LVDS Display Interfaces (LDIs), Flat Panel Display Timing Controllers, Special Functions, and Digital Crosspoint Switches. A summary of these families is presented in Section 5.1.9, "Low Voltage Differential Signaling (LVDS) Families," on page 90.

Consumers are demanding more realistic, visual information in the office and in the home. This is driving the need to move video, 3D graphics, and photo-realistic image data from camera to PCs and printers through LAN (local area network), phone, and satellite systems to home set top boxes and digital VCRs. Solutions exist today to move this high-speed digital data both very short and very long distances, on a printed circuit board and across fiber or satellite networks. Moving this data from board-to-board or box-to-box, however, requires an extremely high-performance solution that consumes a minimum of power, generates little noise (must meet increasingly stringent FCC/CISPR EMI requirements — Federal Communications Commission/ International Special Committee on Radio Interference Electromagnetic Interference requirements), is relatively immune to noise, and is inexpensive. Unfortunately, existing solutions are a compromise of these four basic ingredients: performance, power, noise, and cost.

The high-speed and low power/noise/cost benefits of LVDS broaden the scope of LVDS applications far beyond those for traditional technologies. Table 5.1 provides some examples of LVDS applications.

Table 5.1 Scope of LVDS Applications

PC/Computing	Telecom/Datacom	Consumer/Commercial
Flat Panel Displays	Switches	Home/commercial video links
Monitor Link	Add/drop multiplexers	Set-top boxes
SCI Processor Interconnect	Hubs	In-flight entertainment
Printer Engine Links	Routers	Game displays/controls
Digital Copiers		
System Clustering (Box-to-Box and Rack-to-Rack)		
Multimedia Peripheral Links		

5.1.1 How LVDS Works

LVDS outputs consist of a current source (nominal 3.5 mA) which drives a differential pair line, as shown in Figure 5.1(a). The basic receiver has high dc input impedance, so the majority of the driver current flows across the 100 Ω termination resistor, generating about 350 mV across the resistor and receiver inputs (3.5 mA × 100 Ω = 350 mV). When the driver switches, it changes the direction of current flow across the resistor, thereby creating a valid "one" or "zero" logic state.

Part (a) of Figure 5.1 shows the driver with its true output high and its inverted output low. Current flows down through the termination resistor generating a positive voltage across the resistor and across the receiver's differential inputs. The return current flows through the inverting output of the receiver and on to ground. This output state corresponds to a valid "one" or "high" logic level.

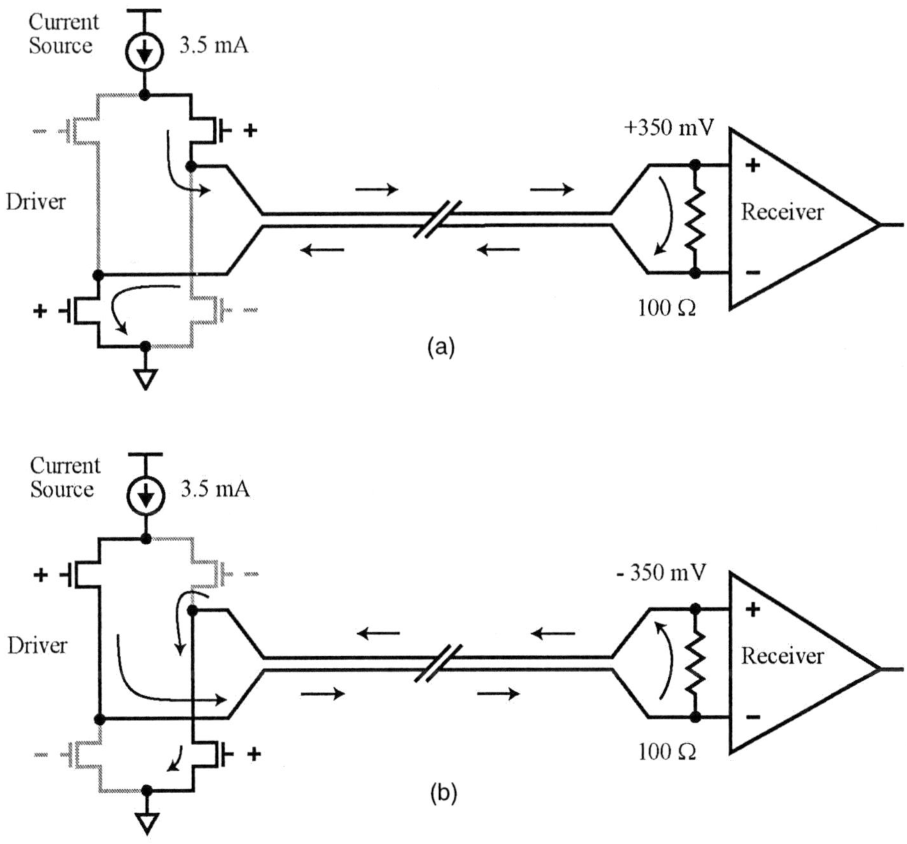

Figure 5.1 Simplified diagram of an LVDS driver and receiver connected via 100 Ω differential impedance medium. (a) Current flowing from the driver's true output down through the 100 Ω termination resistor, and (b) current flowing from the driver's inverted output up through the 100 Ω termination resistor.

Part (b) of Figure 5.1 shows the driver now with its true output low and its inverted output high. Current flows in the reverse direction, up through the termination resistor generating a negative voltage with respect to the receiver's differential inputs. The return current flows back to the true output and on to ground. This output state corresponds to a valid "zero" or "low" logic level.

Figure 5.2 parts (a) and (b) show even more clearly the voltages generated across the termination resistor in an LVDS circuit. This figure assumes that the voltages at the driver and termination resistor (receiver) are the same, an ideal condition which would only be met for very short traces. In other words, the voltages and parameters (V_{OH}, V_{OL}, V_{OD}, and V_{OS}) specified in the figure of the termination resistor are more correctly stated as the voltages and parameters at the output of the driver, not at the termination resistor (receiver). (You'll find V_{OH}, V_{OL}, V_{OD}, and V_{OS} listed on the data sheet for the driver.) Over a long cable or trace, the voltages at the receiver would be distorted from what is shown in this figure. However, the figure is very helpful in generating an understanding of how LVDS drivers and receivers work. As a matter of fact, it turns out that the voltages generated at the driver outputs depend entirely on the value of termination resistance used in the circuit and on the loading of the line by other drivers and receivers which may be present. More will be said about this later on.

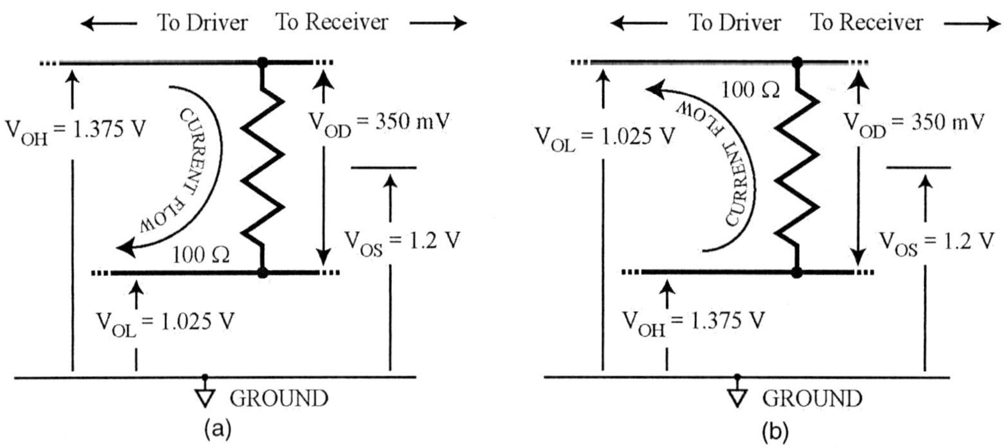

Figure 5.2 Voltages generated across the LVDS terminating resistor assuming no losses or distortion over the interconnect between the driver and terminating resistor at the receiver. (a) Current flowing from the driver's true output to its inverted output, and (b) current flowing from the driver's inverted output to its true output.

Figure 5.2(a) shows the voltages generated as current flows down through the 100 Ω resistor. LVDS typically generates about 350 mV across the termination resistor by a current of 3.5 mA. This is the differential output voltage, or V_{OD}, and is stated as "differential" because it's the voltage appearing across the plus and minus receiver inputs and not a voltage referenced to ground. The low voltage side, V_{OL}, is usually at about 1.025 volts, and the high voltage side, V_{OH}, is usually at about 1.375 volts. Both V_{OL} and V_{OH} are said to be "single-ended" voltages because they are measured with respect to ground. As the driver changes from a "one" to a "zero" output state, the current in the termination resistor reverses direction and the polarity of voltage across the resistor changes from positive to negative. At the instantaneous moment that the current is zero as it changes direction, the voltage across the resistor will be zero while, at exactly that same instant, all points on the resistor will be at the offset voltage, V_{OS}. At the instant of zero current, every point on the resistor will instantaneously be held at V_{OS} by the driver circuitry as no current will be flowing through the resistor. V_{OS} is usually halfway between V_{OL} and V_{OH}. The rise time, the time it takes for the true output to go from V_{OL} plus 10% to within 90% of V_{OH}, is typically about 300 to 500 ps (picoseconds, 10^{-12} second), so the voltage across the resistor doesn't stay at V_{OS} for very long. Figure 5.2(b) shows an output "low" or "zero" state, where current flows in the opposite direction, up through the resistor. The bottom of the resistor is now at the high voltage, V_{OH}, and the top of the resistor is now at the low voltage or V_{OL}. V_{OS} stays the same, but V_{OD} has changed polarity and is now negative.

Figure 5.3 shows both the single-ended and differential LVDS waveforms generated when switching from high to low and back to high. The "A" single-ended waveform is the voltage appearing at the non-inverting (true) output of the driver and the "B" single-ended waveform is the voltage appearing at the inverting driver output. Both waveforms move between V_{OL} of 1.025 V and V_{OH} of 1.375 V for a total change of 350 mV. The offset voltage, V_{OS}, is 1.2 V. Examining the differential waveform obtained by subtracting the "B" waveform voltage from the "A" waveform voltage, the differential waveform can be seen to have twice the amplitude of the single-ended waveforms, or 700 mV peak-to-peak. The differential waveform is centered around ground potential, 0 V, and goes both positive and negative, whereas the single-ended waveforms are centered around the offset voltage, V_{OS}, at 1.2 V, but are always at a positive voltage.

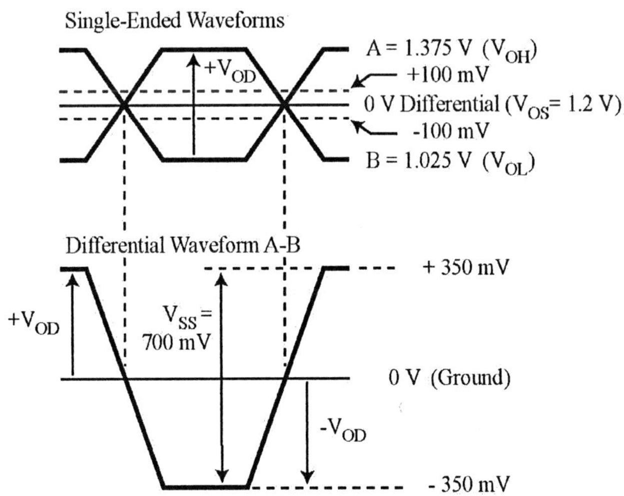

Figure 5.3 Single-ended and differential waveforms (not to scale).

A key parameter depicted in Figure 5.3 is the receiver input differential voltage at which an LVDS receiver output will switch from high to low or low to high (receiver input threshold). The receiver ideally switches state at the crossing point of the two single-ended waveforms. This occurs as the single-ended waveforms pass through the offset voltage, V_{OS}, of 1.2 V. This is also where the differential voltage passes through 0 V. Actual LVDS devices don't change state precisely at the point where the single-ended voltages cross each other at V_{OS}. As a practical matter, they change state within a range of about ±100 mV around V_{OS} as depicted in the figure. As the receiver's non-inverting input waveform "A" passes through V_{OS} in going from V_{OL} to V_{OH}, the receiver's output may not change state unless "A" reaches a differential voltage of up to +100 mV. For example, if the driver's V_{OD} only reaches +50 mV and then drops back negative, the receiver's output may not change state.

Figure 5.4 illustrates that LVDS can tolerate 1 V of difference between the common-mode voltage of the driver and the common-mode voltage of the receiver. This is useful in situations where there may be up to 1 V of difference between the driver's ground and the receiver's ground due to resistive voltage drops over long backplane or cable distances, or due to ground potential variations from one chassis to another.

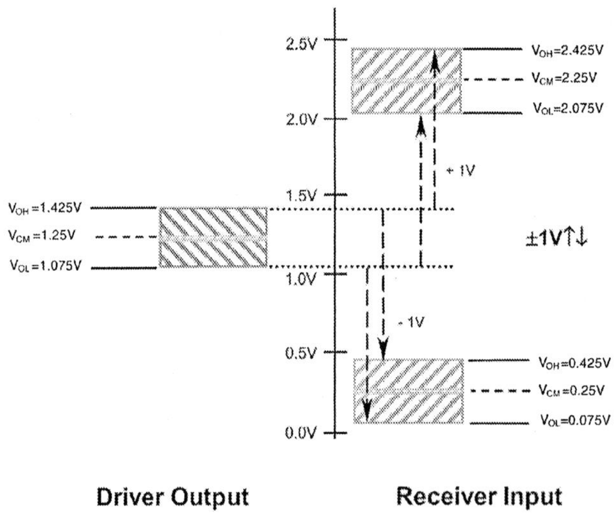

* This figure uses an offset voltage of 1.25 V instead of 1.2 V. There are small variations among vendors for LVDS parameters.

Figure 5.4 LVDS receivers can tolerate ±1 V of common-mode voltage difference between the driver and receiver.* [Reprinted with permission of Fairchild Semiconductor Corporation.]

LVDS current-mode, low-swing outputs mean that LVDS can drive at high-speeds (up to several hundred or even thousands of Mbps over short distances). If high-speed differential design techniques are used, signal noise and electromagnetic interference (EMI) can also be reduced with LVDS because of several factors:

- The low output voltage swing (~350 mV)
- Relatively slow edge rates, dV/dt ~ 0.300 V / 0.3 ns = 1 V/ns

- Differential (odd mode operation) so magnetic fields tend to cancel
- "Soft" output corner transitions
- Minimum I_{CC} spikes owing to low current-mode operation

LVDS can be designed using CMOS processes, allowing LVDS to be integrated with standard digital blocks. LVDS can be used in commercial, industrial, and even military temperature ranges and operate from power supplies down to 2 V. LVDS uses common copper PCB traces and readily available cables and connectors as transmission media, unlike fiber optics.

At present, the major limitations of LVDS are its point-to-point nature (as opposed to multipoint — see "Bus LVDS (BLVDS)" on page 108) and short transmission distance (15 meters) where other technologies must be used to exceed these limitations.

5.1.2 Why Low Swing Differential?

The differential data transmission method used in LVDS is less susceptible to common-mode noise than single-ended schemes. Differential transmission uses two wires with opposite current/voltage swings instead of the one wire used in single-ended methods to convey data information. The advantage of the differential approach is that, if noise is coupled onto the two wires as common-mode (the noise appears on both lines equally), the noise is rejected by the receiver which looks at only the difference between the two signals. The differential signals also tend to radiate less noise than single-ended signals due to the canceling of magnetic fields. And, the current-mode driver is not prone to ringing and switching spikes, further reducing noise.

Because differential technologies such as LVDS reduce concerns about noise, they can use lower signal voltage swings. This advantage is crucial, because it is impossible to raise data rates and lower power consumption without using low voltage swings. The low swing nature of the driver means data can be quickly switched. Since the driver is also current-mode, very low — almost flat — power consumption across frequency is obtained. Switching spikes in the driver are quite small, so that supply current I_{CC} does not increase exponentially as switching frequency is increased. Also, the power consumed by the load (3.5 mA × 350 mV = 1.2 mW) is very small in magnitude.

5.1.3 The LVDS and M-LVDS Standards

The LVDS family of devices is currently specified by the standard[2]

ANSI/TIA/EIA-644-A-2001
(American National Standards Institute/Telecommunications Industry Association/ Electronic Industries Alliance-644-A-2001)
Electrical Characteristics of Low Voltage Differential Signaling (LVDS) Interface Circuits

2. An earlier LVDS standard, IEEE1596.3, has been withdrawn with no superseding document. (IEEE is Institute of Electrical and Electronics Engineers.)

and M-LVDS (Multipoint LVDS), a separate family of devices, is specified by:

ANSI/TIA/EIA-899-2002
Electrical Characteristics of Multipoint-Low-Voltage Differential Signaling (M-LVDS) Interface Circuits for Multipoint Data Interchange

Multipoint LVDS (M-LVDS) is an enhanced version of LVDS which provides higher current and voltage output than LVDS. The M-LVDS standard is discussed in Section 6.4.1 on page 117 and is used in bus applications having both multiple drivers and multiple receivers. LVDS is used primarily in point-to-point applications. Another enhanced version of LVDS is Bus LVDS or BLVDS which also provides higher current output than LVDS for multipoint bus applications. BLVDS adheres to an internal National Semiconductor standard and is not identical to M-LVDS. BLVDS is discussed in Section 6.2 on page 108.

5.1.3.1 The TIA/EIA-644-A Standard[3]

The generic LVDS standard ANSI/TIA/EIA-644-A-2001 is often simply referred to as the TIA/EIA-644-A standard. It originated in the TIA Data Transmission Interface committee TR30.2. The TIA/EIA-644-A standard defines driver output and receiver input characteristics; thus it is an electrical-only standard. It does not include functional specifications, protocols, or even complete cable characteristics since these are application dependent. TIA/EIA-644-A is intended to be referenced by other standards that specify the complete interface (connectors, protocol, etc.). This allows it to be easily adopted into many applications.

The TIA/EIA-644-A standard provides a theoretical maximum data signaling rate of 1.923 Gbps by assuming a lossless interconnection medium and maximum rise and fall times of 260 ps each which do not exceed 0.5 of the unit interval (tui, reciprocal of signaling frequency) at the load (1/520 ps = 1.923 Gbps). It also provides a recommended maximum data rate of 655 Mbps based on a more realistic set of assumptions. These assumptions are 260 ps maximum rise and fall times degraded by 500 ps by the interconnecting media (typical of about 5 meters of cable) to become 760 ps at the receiver. The sum of the rise and fall times is then 1520 ps which corresponds to 655 Mbps. These maximum and recommended signaling data rates are per differential pair. Employing a parallel bus structure (4-, 8-, 16-, 32-, 64-, etc. bus width) can easily extend the obtainable equivalent bit rate into the multi-Gbps range. This allows the referencing standard to specify the maximum data rate required depending upon bus width, required signal quality, and media length or type. The standard also covers minimum media specifications, failsafe operation of the receiver under fault conditions, termination internal to and external to the receiver,

3. Reproduced under written permission from the Telecommunications Industry Association.

and other configuration issues such as multiple receiver operation. The initial TIA/EIA-644 standard was approved in November 1995.[4] In 2001, an "A" revision of the specification was created to include additional information about multiple receiver operation (multidrop buses) with up to 32 parallel-connected receivers (receivers with no internal termination).

Several of the parameters of the TIA/EIA-644-A standard are summarized below in Table 5.2. A few of these parameters may need some definition. ΔV_{OD} is the difference in the

Table 5.2 TIA/EIA-644-A Standard for LVDS

Parameter	Description	Minimum[b]	Maximum	Units
—	Data Rate (Theoretical / Recommended)[a]		1923 / 655	Mbps
Z_0	Characteristic Impedance of Unloaded Balanced Differential Media	90	132	Ω
Z_T	Termination Impedance	90	132	Ω
V_{OD}	Driver Differential Output Voltage	247	454	mV
V_{OS}	Driver Offset Voltage	1.125	1.375	V
ΔV_{OD}	|Driver Change to V_{OD}|		50	mV
ΔV_{OS}	|Driver Change to V_{OS}| [c]		50	mV
I_{SA}, I_{SB}	Driver Short-Circuit Current to Ground		24	mA
I_{SAB}	|Driver Short-Circuit Current, Shorted Terminals|		12	mA
t_r/t_f	Rise/Fall Times Measured at Load (Measured at Receiver)		Greater of 0.26 ns or 50% of tui [d]	ns
I_{IN}	|Receiver Input Current|		20	μA
V_{TH}	Receiver Threshold Voltage		±100	mV
V_{IN}	Receiver Input Voltage Range	0	2.4	V
$\|I_A - I_B\|$	|Receiver Input Current Balance|		6	μA
V_{ID}	Receiver Differential Input Voltage	100	600	mV
—	Interconnect Media Dc Loop Resistance		135	W
—	Voltage Drop Across Interconnect Media		150	mV

a. Actual datasheet specifications for LVDS devices may be significantly better than what is specified by the standard in this table.

b. Data rates are per differential pair and measured at the load. Theoretical maximum is obtained assuming rise and fall times each of 260 ps. Recommended maximum data rate is obtained assuming rise and fall times each of 760 ps, where 500 ps of the 760 ps is due to media distortion. Maximum data signaling rate varies with the amount of media distortion and is therefore application dependent. Drivers and receivers meeting this standard may operate at lower data signaling rates.

c. The steady-state magnitude of the difference of V_{OS} for one binary state and V_{OS}* for the opposite binary state shall be 50 mV or less.

d. tui is unit interval length (i.e., bit width).

[Reproduced under written permission from Telecommunications Industry Association.]

4. National Semiconductor held the editor position for this standard and chairs the subcommittee responsible for electrical TIA interface standards.

differential output voltage magnitudes between one logic state and the opposite logic state. Similarly, ΔV_{OS} is the difference in the offset voltages between one logic state and the opposite logic state. Most of the minimum and maximum values specified in the standard are determined with specific test circuits detailed in the standard. The actual standard should be referred to for the details of these test circuits and for the specification of other parameters which are not listed in Table 5.2.

In the interest of promoting a wider standard, no specific process technology, interconnect medium, or power supply voltages are defined by TIA/EIA-644-A. This means that LVDS can be implemented in BiCMOS, CMOS, GaAs, or other applicable technologies, migrate from 5 V to 3.3 V to sub-3 V supplies, and transmit over PCB traces or cable, thereby serving a broad range of applications in many industry segments. The LVDS interface circuit is not intended for direct inter-operation with other interface electrical technologies such as TIA/EIA-422, TIA/EIA-485, TIA/EIA-612, ITU-T Recommendation V.11, emitter coupled logic (ECL), or PECL.

5.1.4 Appearance of Laboratory LVDS Waveforms

Figure 5.5 National Semiconductor LVDS evaluation board used to demonstrate LVDS waveforms.

An easy way to get familiar with what LVDS waveforms actually look like is to examine sample waveforms recorded in the laboratory from powered-up LVDS evaluation PCBs. Figure 5.5 shows an evaluation PCB available from National Semiconductor. LVDS evaluation PCBs are able to demonstrate line driving capability of LVDS technology across short PCB interconnects and across variable lengths of RJ45 cable. The schematic for the second of four test channels (test circuits) of this evaluation PCB is shown in Figure 5.6. The #2 test channel connects driver #2 to receiver #2 via a PCB trace. The quad LVDS drivers and receivers on the evaluation board are National Semiconductor's DS90LV047A/048A. In the figure, the differential signal is received at connector J6 on pin 1 and pin 3. Pin 1 on connector J6 (J6-1) is connected to the negative input (inverting signal) on the receiver and pin 3 is connected to the positive input (true signal) on the receiver. Single-ended signals are measured from each signal (true or inverting signal) with respect to ground, e.g., from J6-1 to ground or J6-3 to ground. The receiver ideally switches at the crossing point of the two signals known as the offset voltage, V_{OS}. LVDS signals have a V_{OD} (differential output voltage) specification of 250 mV to 450 mV with a typical V_{OS} of 1.2 V. The differential waveform is constructed by subtracting the J6-1 (inverting) signal from the J6-3 (true) signal, that is, V_{OD} = voltage on (J6-3) minus the voltage on (J6-1). The V_{OD} magnitude is either positive or negative, so the differential voltage swing (V_{SS}) is twice the V_{OD} magnitude.

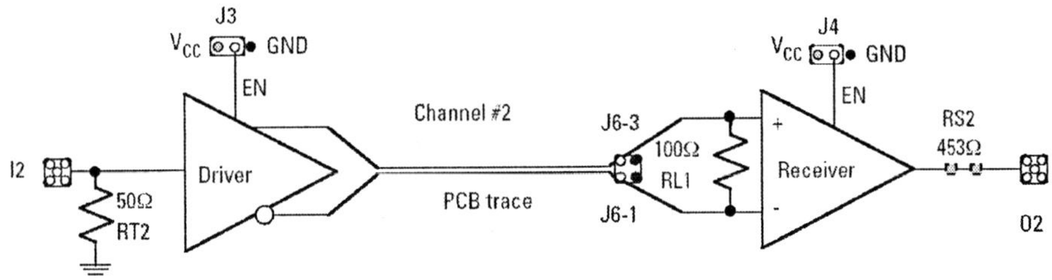

Figure 5.6 Channel #2 test circuit on National Semiconductor's generic LVDS evaluation kit LVDS4748EVK.

The LVDS interconnect signal on Channel #2 can be measured at the receiver inputs (test points J6-1 and J6-3) on the evaluation board. Due to the short interconnect path via the PCB, little distortion to the waveform is caused by the interconnect, as shown in Figure 5.7. In this figure (Figure 5.7) and the next six figures displaying waveforms (Figures 5.10 through 5.12 and Figures 5.14 through 5.16), the overlapping top two waveforms are the single-ended voltages (voltage measured between the receiver's true input and ground and between the receiver's inverted input and ground), the middle waveform is the calculated differential output voltage from the two single-ended signals, and the bottom waveform is the output TTL voltage from the receiver. In Figure 5.7, note that the data rate is 100 Mbps and the differential waveform (V_{DIFF} = (D_{OUT2+}) – (D_{OUT2-})) shows fast transition times with little distortion.

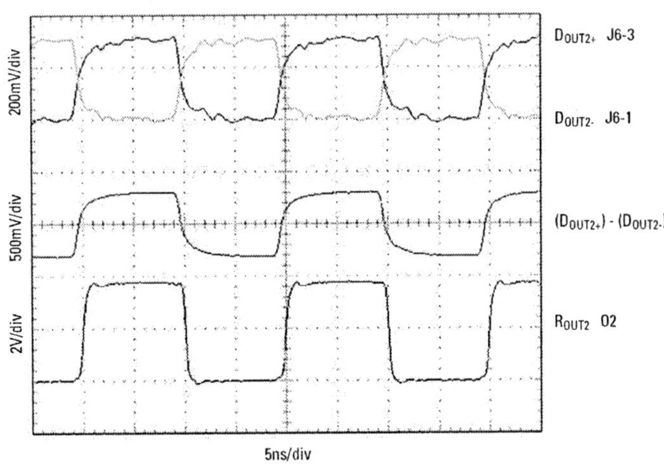

Figure 5.7 LVDS Channel #2 waveforms — PCB interconnect.

5.1.4.1 More Discussion of the Evaluation Board

The LVDS evaluation board is used to measure LVDS signaling performance over different media, i.e., how much the signal degrades over various lengths of traces on the board or various cable lengths between the driver and receiver. Individual LVDS channels can be evaluated over a PCB trace, RJ45 connector and CAT5 UTP cable (category 5 unshielded twisted pair cable), or custom transmission medium. Although LVDS quad drivers and receivers are used on the board, they can represent the standard LVDS I/O characteristics of most of National's LVDS devices. We can also use the same 3 V drivers and receivers on the evaluation board (DS90LV047A/048A) to represent the LVDS I/O characteristics of 3 V Channel Link and FPD-Link devices.

Test Channel #2 connects Driver #2 to Receiver #2 via a pure PCB interconnect. A test point interface for the LVDS signal is provided at test points J6-1 and J6-3 (refer back to Figure 5.6). The driver input signal (I2) is terminated with a 50 Ω termination resistor (RT2) on the bottom side of the PCB. The receiver output signal may be probed at test point O2. A PCB option for a series 453 Ω resistor (RS2) is also provided in case 50 Ω probes are employed on the receiver output signal. The default setting is for use with high impedance probes and the series 453 Ω resistor is not installed. A direct probe connection is possible with a TEK P6247 differential probe high impedance probe (>1 GHz bandwidth) on the LVDS signals at test points J6-1 and J6-3. This channel may be used for analyzing the LVDS signal without the bandwidth limiting effects of a cable interconnect.

The schematic for the third channel (and fourth channel) of the evaluation PCB is shown in Figure 5.8. This test channel connects Driver #3 to Receiver #3 via a cable interconnect rather than a PCB trace as in Channel #2. A test point interface is provided at the receiver input side of the cable at connector J7 and LVDS signals are probed via these test points on J7. The driver input signal (I3) is terminated with a 50 Ω termination resistor (RT3) on the bottom side of the PCB. LVDS signals are probed via test points on J7. The receiver output signal may be probed at test point O3. A PCB option for a series 453 Ω resistor (RS3) is also provided as was done for Channel #2 in case 50 Ω probes are employed on the receiver output signal. The default setting is without the series 453 Ω resistor installed. In the default setting, high-impedance probes are to be used. A differential probe connection is possible with a TEK P6247 differential probe (>1 GHz bandwidth) on the LVDS signals at test point J7-1 and J7-3.

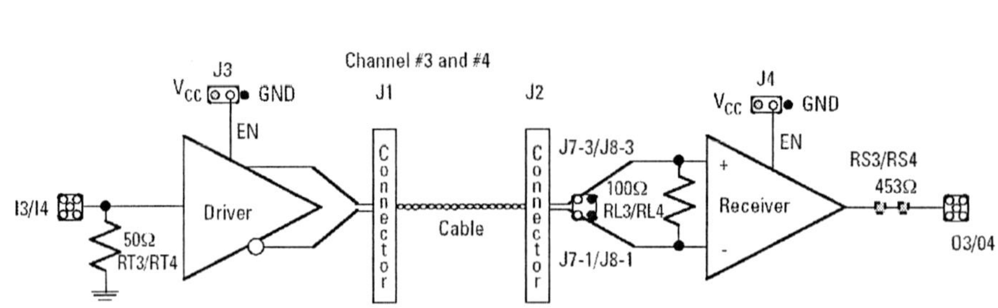

Figure 5.8 Schematic for Channel #3 of LVDS evaluation board.

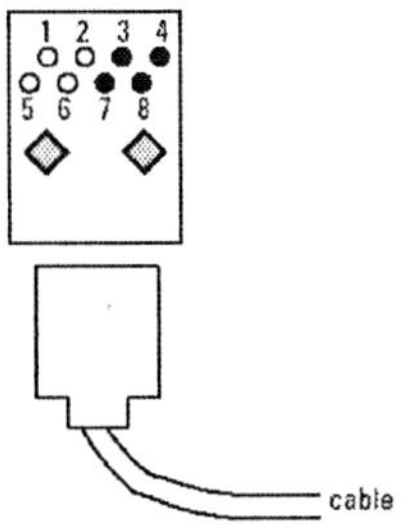

Figure 5.9 RJ45 connector used to connect cable on Channel #3 of the evaluation board.

The evaluation PCB has been designed to directly accommodate a CAT5, four twisted pair (8-pin) RJ45 cable. The pinout, connector, and cable electrical/mechanical characteristics are defined in the Ethernet standard and the cable is widely available. The connector is 8 position, with 0.10 inch centers and the pairs are pinned out up and down. For example, pair 1 is on pins 1 and 5, not pins 1 and 2 (see Figure 5.9). The two unused pairs are connected to ground. Other custom-made cables may also be used where they have been made with RJ45 connectors to fit the mating connector on the board.

Waveform distortion occurs due to the characteristics of the cable. Depending upon the cable length and quality, the transition time of the signal at the end of the cable will be slower than the signal at the driver's outputs. This effect can be measured by taking rise and fall measurements and increasing the cable length. A ratio of transition time to unit interval (minimum bit width) is a common gauge of signal quality. Depending upon the application, ratios of 30% to 50% are common. These measurements tend to be more conservative than jitter measurements. The waveforms acquired with an RJ45 cable of 1 meter, 5 meters, and 10 meters in length are shown in Figures 5.10, 5.11, and 5.12 respectively. Note the additional transition time slowing due to the cable's filter effects on the 5 meter and 10 meter test cases.

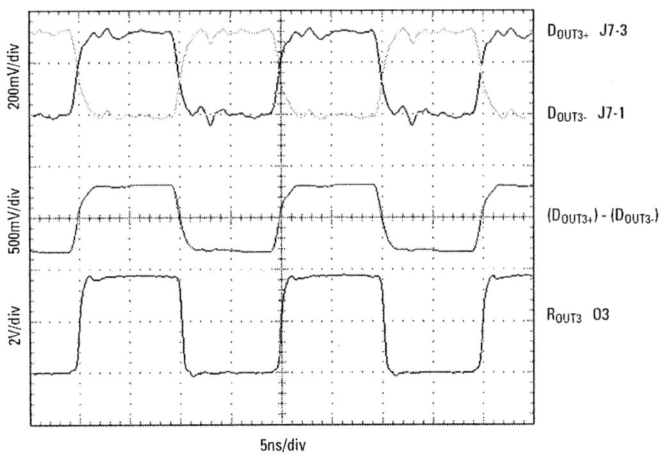

Figure 5.10 LVDS Channel #3 waveforms — 1 m cable interconnect.

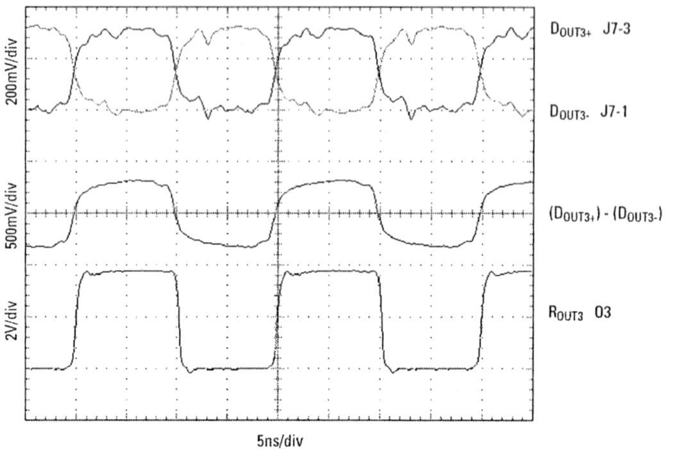

Figure 5.11 LVDS Channel #3 waveforms — 5 m cable interconnect.

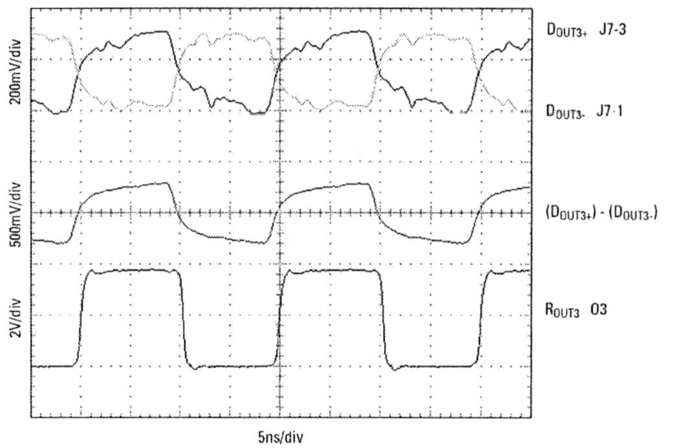

Figure 5.12 LVDS Channel #3 waveforms — 10 m cable interconnect.

5.1.4.2 Common-Mode Noise

Figure 5.13 shows Channel #1B of the LVDS evaluation board. This channel comprises a single isolated LVDS receiver with test points. (Channel #1A, a single isolated LVDS driver, and Channel #4, which is identical to Channel #3, are not discussed.) Termination options on the Channel #1B receiver inputs accommodate either two separate 50 Ω terminations (RT5 and RT6) (each line to ground) or a 100 Ω resistor connected across the inputs (differentially). The first option allows for a standard signal generator interface. Input signals are connected at test points I5 (RIN–) and I6 (RIN+). A PCB option for a series 453 Ω resistor (RS1) is also provided in case 50 Ω probes are employed on the receiver output signal. The default setting is with two separate 50 Ω terminations (RT5 and RT6) and without the series 453 Ω resistor (RS1) for use of high impedance probes. The receiver output signal may be probed at test point O1.

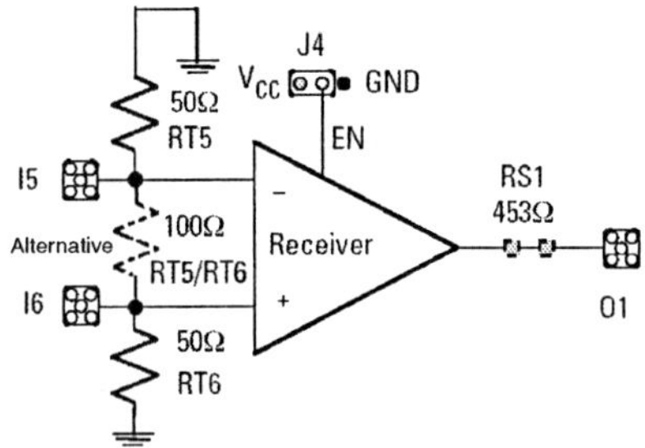

Figure 5.13 Channel #1B circuit of LVDS evaluation board.

When the receiver (DS90LV048A) is enabled, a small amount of common-mode noise is passed from the output of the receiver back to the inputs as shown in Figure 5.14. This noise shows up on the single-ended waveforms as ripples, but does not impact the differential waveform that carries the data. More simply, unwanted noise generated at the receiver output feeds back and gets coupled onto both the true and inverting inputs of the receiver. Because the same fed-back noise is on both the true (+) and inverting (–) inputs, it is called common-mode noise. It cancels itself out and does not show up in the output. A design improvement was made to the

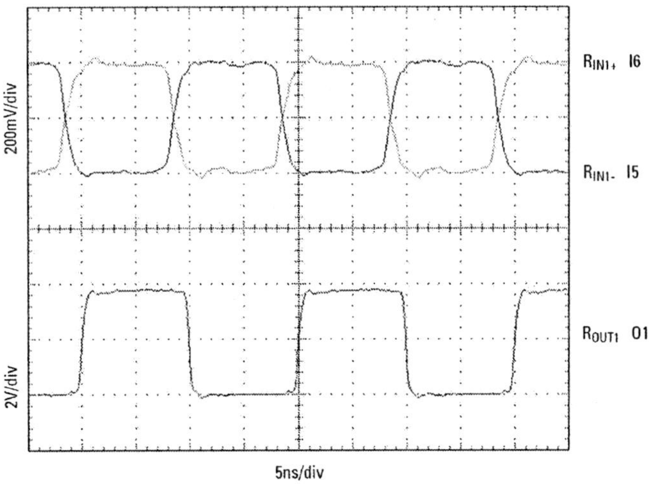

Figure 5.14 LVDS Channel #1B waveforms. A small amount of common-mode noise coupled from output back to input.

DS90LV048A to reduce the magnitude of the noise coupled back to the inputs, reducing the feedback by 30% compared with prior devices. This noise will not be observed if the receiver device is disabled by setting J4 (in Figure 5.13) to "OFF" as shown in Figure 5.15.

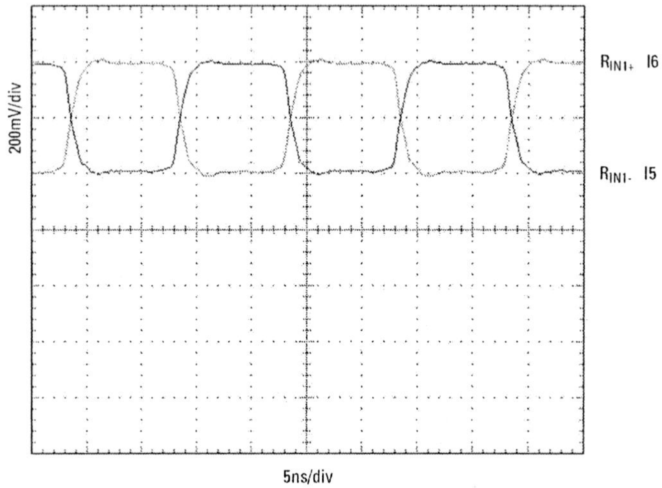

Figure 5.15 LVDS Channel #1B waveforms — output disabled.

5.1.4.3 Probing of High-Speed LVDS Signals

Probe specifications for measuring LVDS signals are unique due to the low drive level of LVDS (3.5 mA typical). Either a high impedance probe (100 kΩ or greater) or the TEK P6247 differential probe (>1 GHz bandwidth) or equivalent probe must be used. The capacitive loading of the probe should be kept in the low pF range, and the bandwidth of the probe should be at least 1 GHz (4 GHz preferred) to accurately acquire the waveform under measurement.

One system that meets the requirements of LVDS particularly well is a TEK TDS 684B Digital Real Time scope (>1 GHz bandwidth) and TEK P6247 differential probe heads. These probes offer 200 kΩ, 1 pF loading and a bandwidth of 1 GHz. This test equipment was used to acquire the waveforms just shown.

The TEK P6247 differential probes may be used to measure the differential LVDS signal or each signal of the differential pair single-ended. This test equipment was used to acquire the waveforms differentially as well as single-endedly with the differential signal calculated by (DOUT+) – (DOUT–), shown in Figure 5.16. You can see that both of the differential signals look identical. The method in which you acquire the single-ended signals is important (such as matching probe types and lengths) if you intend to calculate the differential signal from the two single-ended signals.

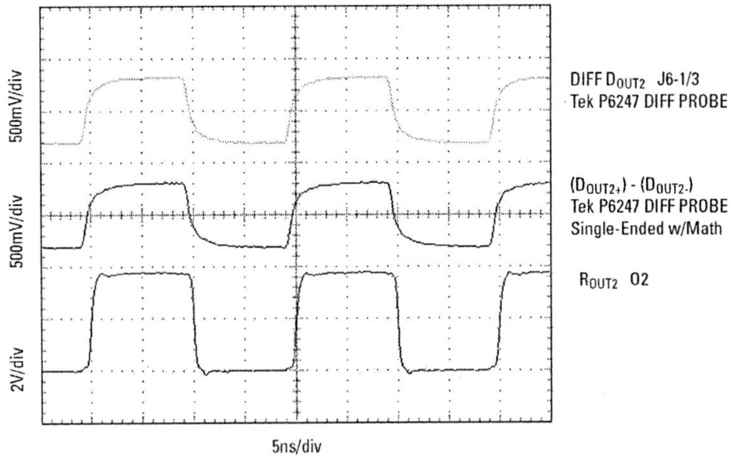

Figure 5.16 LVDS Channel #2 waveforms — differential and calculated differential from single-ended waveform.

LVDS waveforms may also be measured with high impedance probes such as common SD14 probe heads. These probes offer 100 kΩ, 0.4 pF loading, a bandwidth of 4 GHz, and connect to a TEK 11801B scope (50 GHz bandwidth). Probes with standard 50 Ω loading should not be used on LVDS lines since they will load them too heavily. 50 Ω probes may be used on the receiver output signal in conjunction with the 453 Ω series resistor option discussed on page 82. Note that the scope waveform is an attenuated signal (50 Ω / (450 Ω + 50 Ω) or 1/10) of the output signal. The receiver output was loaded with 500 Ω to ground.

5.1.5 Easy Termination

Whether the LVDS transmission medium consists of a cable or controlled impedance traces on a printed circuit board, the transmission medium must be terminated to its characteristic differential impedance to complete the current loop and terminate the high-speed (edge rates) signals. If the medium is not properly terminated, signals reflect from the end of the cable or trace and may interfere with succeeding signals. Proper termination also reduces unwanted electromagnetic emissions and provides the optimum signal quality.

To prevent reflections, LVDS requires a terminating resistor that is matched to the actual cable or PCB traces differential impedance. Commonly 100 Ω media and terminations are employed. This resistor completes the current loop and properly terminates the signal (refer back to Figure 5.1 on page 73). This resistor is placed across the differential signal lines as close as possible to the receiver input.

The simplicity of the LVDS termination scheme makes it easy to implement in most applications. ECL and PECL can require more complex termination than the one-resistor solution for

LVDS. PECL drivers commonly require 220 Ω pull-down resistors from each driver output, along with a 100 Ω resistor across the receiver input.

5.1.6 Maximum Switching Speed

Maximum switching speed of an LVDS interface is a complex question, and its answer depends upon several factors. These factors are the performance of the line driver (edge rate) and receiver, the bandwidth of the media, and the required signal quality for the application.

Since the driver outputs are very fast, the limitation on speed is commonly restricted by

- How fast TTL data can be delivered to the driver — in the case of simple PHY devices that translate a TTL/CMOS signal to LVDS (i.e., DS90LV047A)
- Bandwidth performance of the selected media (cable) — type and length dependent

In the case of LVDS drivers, like the DS90LV047A used on the evaluation board, its speed is limited by how fast the TTL data can be delivered to the driver. Channel Link devices capitalize on the speed mismatch between TTL and LVDS by serializing the TTL data into a narrower LVDS data stream — more will be said about this later.

5.1.7 Saving Power

LVDS technology saves power in several important ways. The power dissipated by the load (the 100 Ω termination resistor) is a mere 1.2 mW. In comparison, an RS-422 driver typically delivers 3 V across a 100 Ω termination, for 90 mW power consumption — 75 times more than LVDS.

LVDS devices are implemented in CMOS processes, which provide low static power consumption. The circuit design of the drivers and receiver require roughly one-tenth the power supply current of PECL/ECL devices (quad device comparison).

Aside from the power dissipated in the load and static I_{CC} current, LVDS also lowers system power through its current-mode driver design. This design greatly reduces the frequency component of I_{CC}. The I_{CC} versus frequency plot for LVDS is virtually flat between 10 MHz and 100 MHz for the quad devices (DS90C031/2), and the total supply current is less than 50 mA for both the driver and receiver at 100 MHz. Compare this with TTL/CMOS transceivers whose dynamic power consumption increases exponentially with frequency.

5.1.8 LVDS Configurations

LVDS drivers and receivers are commonly used in a point-to-point configurations as shown in Figure 5.17. However, other topologies/configurations are also possible.

The bidirectional half-duplex configuration shown in the top half of Figure 5.18 allows bidirectional communication over a single twisted pair. Data can flow in only one direction at a time. The requirement for two terminating resistors reduces the signal (and thus the differential noise margin), so this configuration should be considered only where noise is low and transmis-

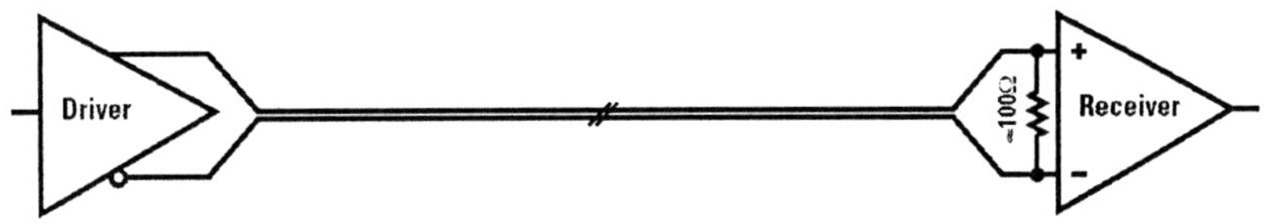

Figure 5.17 Point-to-point configuration.

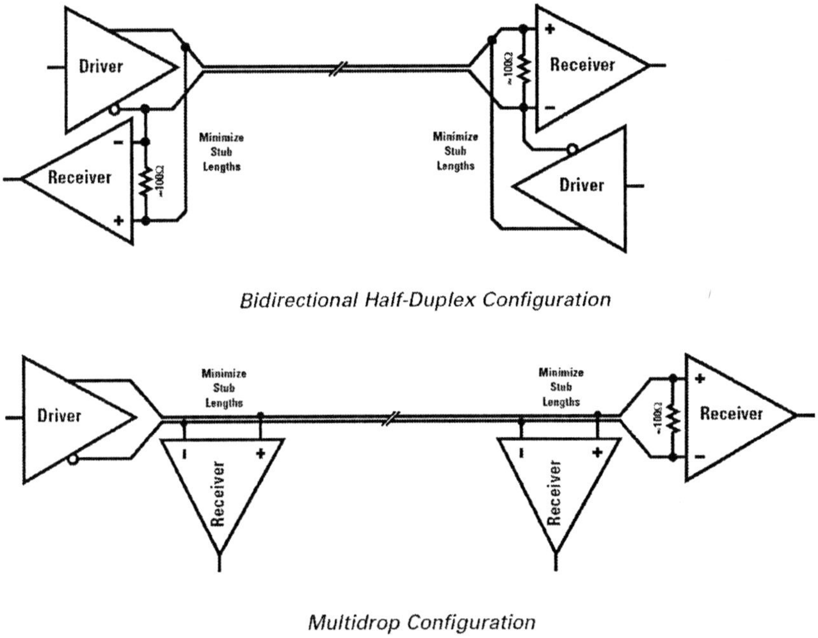

Figure 5.18 LVDS bidirectional half-duplex and multidrop configurations.

sion distance is short (<10 m). This configuration would be best implemented with Bus LVDS devices discussed in "Bus LVDS (BLVDS)" on page 108. Bus LVDS devices are designed for double termination loads and provide full LVDS compatible voltage levels.

A multidrop configuration connects multiple receivers to a driver as shown in the bottom half of Figure 5.18. This configuration is useful in data distribution applications. In multidrop configurations, stub lengths must be as short as possible (<12 mm), although acceptable stub lengths are always dependent upon the application.

Dedicated point-to-point links provide the best signal quality due to the clear path they provide. LVDS has many advantages that make it likely to become the next famous data transmission standard for data rates from dc to hundreds of Mbps and short haul distances in the tens of meters. In this role, LVDS will far exceed the 20 Kbps to 30 Mbps rates of the common RS-232, RS-422, and RS-485 standards.

5.1.9 Low Voltage Differential Signaling (LVDS) Families

LVDS provides high-speed data transmission while consuming little power (megabits or gigabits per second at milliwatts). It also rejects noise, is robust (reliable and suited for many different applications), and is ideal for interconnects of a few inches up to 15 meters in length. It provides an ideal interface for chip-to-chip, card-to-card, shelf-to-shelf, rack-to-rack, or box-to-box communication.

LVDS products are categorized among several families of devices, mostly based upon how the devices are used and their application. Table 5.3 lists these LVDS families.

Table 5.3 LVDS Families

Line Drivers, Receivers, and Transceivers
Bus LVDS
Channel Link Serializers/Deserializers
Flat Panel Display Link (FPD-Link)
LVDS Display Interface (LDI)
Flat Panel Display Timing Controllers
LVDS Special Functions
Digital Crosspoint Switches
Scan LVDS

LVDS line drivers and receivers are used to convey information over PCB trace or cable in situations where only a few channels of information are being transmitted or where the data to be transmitted is already serialized. These devices are also used for small buses and for point-to-point (one driver transmitting to one receiver) and multidrop configurations (one driver transmitting to multiple receivers). Line drivers and receivers are also referred to as simple PHYs (physical layer devices, so named because of their use in the physical layer of networking applications).

Bus LVDS (BLVDS) is an extension of the LVDS line drivers and receivers family. It is specifically designed for multipoint applications (multiple drivers and multiple receivers) where the bus is resistively terminated at both ends. It may also be used in heavily loaded backplanes or multidrop configurations where the effective impedance is lower than 100 Ω. In this case, the drivers may see a load in the 30–50 Ω range. Bus LVDS drivers provide about 10 mA of output current so that they provide the same LVDS-level voltage swings with these heavier termination loads (smaller resistance). Transceivers and repeaters are currently available in this product family.

LVDS Channel link serializers/deserializers are used to transmit wide TTL buses. Channel link will serialize the data, potentially saving money on cables and connectors and helping to avoid complex skew problems associated with a completely parallel solution. A "10-bit" serializer and deserializer family of devices is available (much wider devices are also available) that embeds and recovers the clock from a single serial stream. This chipset (a matched transmitter device and separate receiver device designed to be used together) also provides a high level of integration thereby reducing complexity and overhead to link layer ASICs when used in networking applications. Clock recovery and "Random Lock" digital blocks are integrated with the core interface line driving and receiving functions. The deserializer can also be hot-plugged into a live data bus and does not require PLL (phase locked loop) training.

Whether the architecture of a design is parallel, serial, or serialized depends upon the application. Small busses typically use the simple PHY parts in a serial or parallel approach. However, if the bus is wide, then serialization may make the most sense. Serialization provides a smaller interconnect and reduces cable and connector size and cost.

FPD-link (flat panel display link) is used to convey graphics data from a PC or notebook motherboard to flat panel displays. This family has been extended with the LVDS Display Interface (LDI) chipset that provides higher resolution support and long cable drive enhancements. The LDI chipset is ideal for desktop monitor applications and also industrial display applications. The FPD-link receiver function is also integrated into the timing controller devices to provide a small single-chip solution for TFT (thin film transistor) panels. Special chips have been developed for the control timing of flat panel displays resulting in another family or type of devices.

Special functions are also being developed using BLVDS/LVDS technology. These families provide additional functionality over the simple PHY devices. Currently a special low-skew clock transceiver with 6 CMOS outputs and a repeater/MUX with selectable drive levels is available.

For routing of high-speed point-to-point buses, digital crosspoint switches may be used. Crosspoint switches are also very useful in applications with redundant backup interconnects for fault tolerance. A 2 × 2 crosspoint switch (2 inputs and 2 outputs) that operates above 800 Mbps and generates extremely low jitter is available.

The last family of LVDS devices listed in Table 5.3 is Scan LVDS. These devices include IEEE Standard 1149.1 circuitry for computerized JTAG testing (Joint Test Action Group) of assembled PCBs. JTAG devices are typically used only on products which will be manufactured in large quantities. The additional development time and expense (both additional hardware and software non-recurring engineering) usually do not warrant use of these devices in low-volume products. A lifetime run of 30 prototypes would be considered low volume whereas production runs of 1,500 per month might be considered high volume.

5.1.10 LVDS as a Low-Cost Design Solution

LVDS can provide low-cost design solutions in several important ways:

- Most LVDS solutions are inexpensive CMOS implementations as compared with custom solutions on elaborate processes.

- High performance can be achieved using low cost, off-the-shelf CAT3 (category 3) cable and connectors, and/or FR4 material (typical PCB material).
- LVDS consumes very little power, so power supplies, fans, etc. can be reduced or eliminated.
- LVDS is a low noise producing, noise tolerant technology. Power supply and EMI noise headaches are greatly minimized.
- LVDS transceivers are relatively inexpensive and can also be integrated around digital cores providing a higher level of integration.
- Since LVDS can move data so much faster than TTL, multiple TTL signals can be serialized or multiplexed into a single LVDS channel, reducing board, connector, and cable costs.

In fact, in some applications, the PCB, cable, and connector cost savings greatly overshadow any additional silicon costs. Smaller PCBs, cables, and connectors also result in a much more ergonomic (user friendly) system.

5.1.11 Example of the Wide Range of LVDS Solutions

A large number of LVDS products are available from several vendors. For example, National Semiconductor offers LVDS technology in several forms through over 75 products, for example, National's 5 V DS90C031/DS90C032 and 3 V DS90LV047A/DS90LV048A quad line driver/receiver devices implement LVDS technology in discrete packages for general-purpose use. This family of basic line drivers and receivers also contains singles, duals, and quad footprints.

For the specialized task of connecting laptop and notebook computers to their high-resolution LCD screens, Flat Panel Display Link (FPD-Link) and LVDS Display Interface (LDI) devices are offered. These parts provide a high bandwidth, low power, small size, low power interface enabling XGA/SXGA/UXGA and beyond displays for notebook and monitor applications.

Another more generalized use of LVDS is in the Channel Link family, which can take 21-, 28-, or 48-bits of TTL data and convert it to 3, 4, or 8 channels of LVDS data plus an LVDS clock. These devices provide fast data pipes (up to 5.4 Gbps throughput) and are well suited for high-speed network hubs or router applications or anywhere a low cost, high-speed link is needed. Their serializing nature provides an overall savings to system cost as cable and connector physical size and cost are greatly reduced (see Figure 5.19).

Bus LVDS is an extension of the LVDS line drivers and receivers family. Bus LVDS devices are specifically designed for multipoint applications where the bus is terminated at both ends. They may also be used in heavily loaded backplanes where the effective impedance is lower than 100 Ω. In this case, the drivers may see a load in the 30–50 Ω range. Bus LVDS drivers provide about 10 mA of output current so that they provide LVDS voltage level swings with heavier termination loads. Transceivers and repeaters are currently available in this product family. A 10-bit serializer and deserializer family of devices is available that embeds and recovers

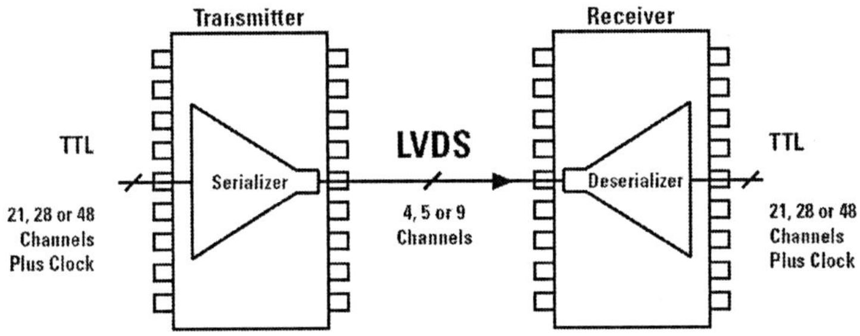

Figure 5.19 National Semiconductor's Channel Link chipsets convert a TTL bus into a compact LVDS data stream and back to TTL.

clock from a single serial stream. This chip set also provides a high level of integration with on-chip clock recovery circuitry. Certain deserializers provide a random data lock capability (an industry first introduced by National Semiconductor). The deserializer can be hot-plugged into a live data bus and does not require PLL (phase-locked loop) training.

Special functions are also being developed using LVDS technology. This family provides additional functionality over the simple PHY devices. Currently a special low-skew clock transceiver with 6 CMOS outputs (DS92CK16) and a line of crosspoint switches are available.

LVDS technology solutions eliminate the trade-offs in speed, power, noise, and cost for high-performance data transmission applications. In doing so, LVDS not only achieves great benefits in existing applications, but opens the door to many new ones.

5.2 Comparison of LVDS to Other Signaling Technologies Using Design Examples

Table 5.4 compares basic LVDS signaling levels with those of PECL and shows that LVDS has half the voltage swing of PECL. LVDS swings are 1/10 of RS-422 and traditional TTL/CMOS levels. (Note that the bottom half of Table 5.4 compares specific LVDS devices.) Another voltage characteristic of LVDS is that the drivers and receivers do not depend on a specific power supply, such as 5 V. Therefore, LVDS has an easy migration path to lower supply voltages such as 3.3 V or even 2.5 V, while still maintaining the same signaling levels and performance. In contrast, technologies such as ECL or PECL have a greater dependence on the supply voltage, which make it difficult to migrate systems utilizing these technologies to lower supply voltages. A comparison of different logic and optical families is provided in Table 5.5. A plus sign (+) indicates that the devices provide the listed advantage. A minus sign (–) indicates that it does not.

Table 5.4 A Quick Comparison Between Differential Signaling Technologies

Parameter	RS-422	PECL	LVDS
Differential Driver Output Voltage	±2 to ±5 V	±600–1000 mV	±250–450 mV
Receiver Input Threshold	±200 mV	±200–300 mV	±100 mV
Data Rate	<30 Mbps	>400 Mbps	>400 Mbps
Parameter	**RS-422**	**PECL**	**LVDS[a]**
Supply Current Quad Driver (no load, static)	60 mA (max)	32–65 mA (max)	8.0 mA
Supply Current Quad Receiver (no load, static)	23 mA (max)	40 mA (max)	15 mA (max)
Propagation Delay of Driver	11 ns (max)	4.5 ns (max)	1.7 ns (max)
Propagation Delay of Receiver	30 ns (max)	7.0 ns (max)	2.7 ns (max)
Pulse Skew (Driver or Receiver)	N/A	500 ps (max)	400 ps (max)

a. LVDS devices noted are DS90LV047A/048A.

Table 5.5 Advantages of Logic and Optical Families

Advantages	LVDS	PECL	Optics	RS-422	GTL	TTL
Data Rate up to 1 Gbps	+	+	+	–	–	–
Very Low Skew	+	+	+	–	+	–
Low Dynamic Power	+	–	+	–		–
Cost-Effective	+	–	–	+	+	+
Low Noise/EMI	+	+	+	–	–	–
Single Power Supply/Reference	+	–	+	+	–	+
Migration Path to Low Voltage	+	–	+	–	+	+
Simple Termination	+	–	–	+	–	+
Wide Common-Mode Range	–	+	+	+	–	–
Process Independent	+	–	+	+	+	+
Allows Integration with Digital	+	–	–	–	+	+
Cable Breakage/Splicing Issues	+	+	–	+	+	+
Long-Distance Transmission	–	+	+	+	–	–
Industrial Temp/Voltage Range	+	+	+	+	+	+

5.2.1 LVDS Drivers and Receivers

The most basic LVDS devices are the driver and receiver. These translate TTL to LVDS and back to TTL (Figure 5.20). Drivers and receivers transmit high-speed data across distances up to 15 meters with very low power, noise, and cost.

Table 5.6 summarizes that only LVDS can deliver the high speed, ultra-low power, and low cost without compromise for a point-to-point application of 15 meters or less. PECL and ECL are expensive and consume too much power. TTL and CMOS are cheap, but they're noisy and they burn a lot of power at high speeds. Fiber optics are expensive and have cables and connectors that are hard to manage.

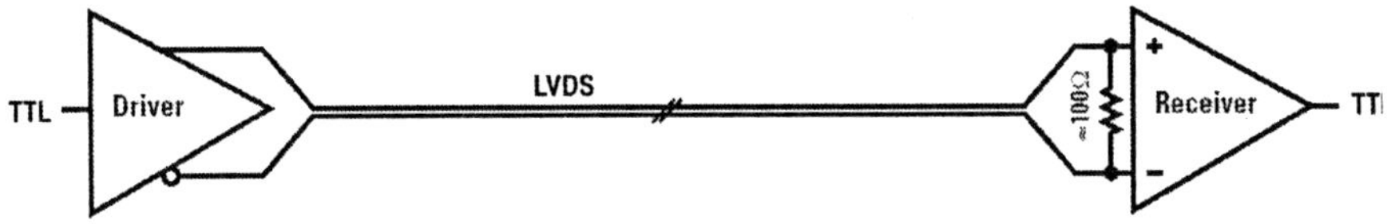

Figure 5.20 Some LVDS drivers and receivers convert TTL to LVDS and back to TTL.

Table 5.6 Speed, Power, Cost Comparison for Logic and Optical Families

Advantages	LVDS	PECL	Optics	RS-422	GTL	TTL
Output Voltage Swing	±350 mV	±800 mV	n/a	±2 V	1.2 V	2.4 V
Receiver Threshold	±100 mV	±200 mV	n/a	±200 mV	100 mV	1.2 V
Speed (Mbps)	>400	>400	>1000	<30	<200	<100
Dynamic Power	Low	High	Low	Low	High	High
Noise	Low	Low	Low	Low	Med	High
Cost	Low	High	High	Low	Low	Low

Figure 5.21 plots driver supply current versus frequency for +5 V LVDS (DS90C031), +3.3 V LVDS (DS90LV031A), RS-422 (26C31), and PECL (41LG) devices. The current shown is per quad device, not per driver. Observe that all of the devices draw fairly constant current over several frequency decades, with the exception of RS-422 which draws increasing current after 1 MHz. Figure 5.22 shows a similar plot of supply current versus frequency, but this time for the corresponding receivers matched to the drivers plotted in Figure 5.21. The plots reflect that LVDS devices consume the least amount of current (power), but that all the receiver devices consume increasing amounts of current at frequencies above about 20 MHz.

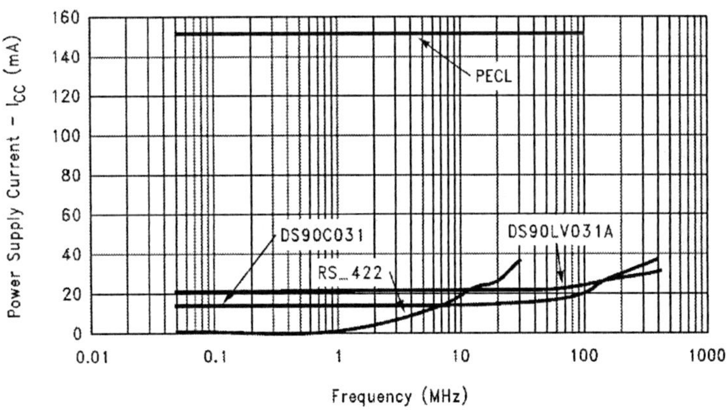

Figure 5.21 Driver supply current I_{CC} vs. frequency for 5 V LVDSDS (DS90C031), 3.3 V LVDS (DS90LV031A), PECL (41LG), and RS-422 (26C31) devices.

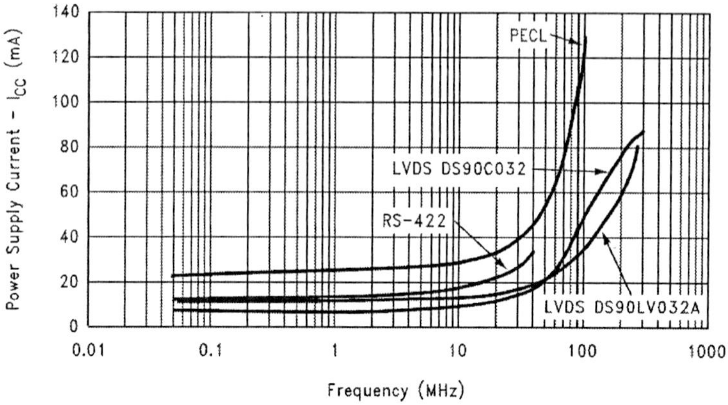

Figure 5.22 Receiver supply current I_{CC} vs. frequency for 5 V LVDS (DS90C032), 3.3 V LVDS (DS90LV032A), PECL (41LF), and RS-422 (26C32) devices.

5.2.2 100 Mbps Serial Interconnect

LVDS drivers and receivers are generally used to create serial or pseudo-serial point-to-point interconnects from 1 Mbps to greater than 400 Mbps per channel (i.e., per differential pair of traces) as illustrated in Figure 5.23. Tables 5.7 and 5.8 summarize the total performance and cost advantages of using LVDS over PECL or TTL for a serial 100 Mbps, 1 meter point-to-point link. Significantly higher data rates can be achieved for LVDS and PECL.

In Table 5.7, plus signs (+) indicate a favorable rating for the associated parameter. The more plus signs, the more favorable the rating. The minus signs (–) indicate the reverse — the more minus signs, the more unfavorable the rating.

The costs detailed in Table 5.8 are "relative costs," meaning that only costs that vary among LVDS, TTL, and PECL are listed. Costs that are the same for all three technologies are not shown. The total cost for the design project would therefore be higher than the total relative system costs calculated in the table. The total subsystem costs listed are per system rather than per board.

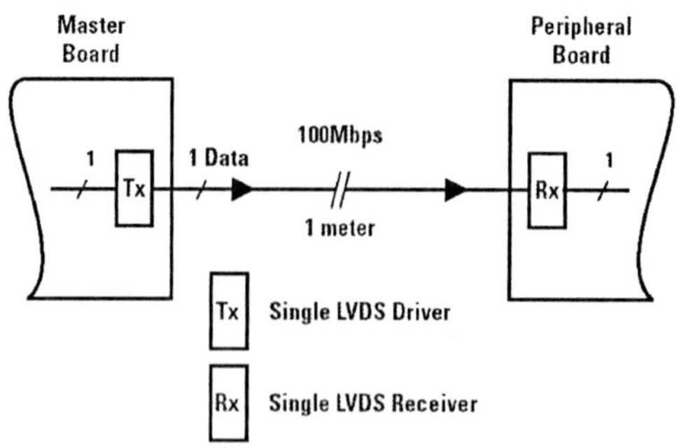

Figure 5.23 100 Mbps board-to-board link.

LVDS provides a high-speed link with minimal noise, power, and cost. LVDS also creates an easy migration path to higher speeds, lower supply voltages, and higher integration that the others do not.

Table 5.7 Performance Estimate for 100 Mbps Serial Bit Stream

Characteristic	Parameter	LVDS	TTL	PECL
Speed	Application Data Rate (Mbps)	100	100	100
	Max Capability per Channel (Mbps)	400	100	400
Power Consumption	Dynamic (mA, @50 MHz)	Low	High	Medium
	Static (mA)	8	10	48
Noise	Low EMI	+++	---	++
	Low Bounce	+++	---	++

Table 5.8 Cost Estimate for 100 Mbps Serial Bit Stream

Subsystem	Parameter	LVDS	TTL	PECL
General	Single-Ended or Differential	Differential	Single-Ended	Differential
	TTL Bus Width	1	1	1
	TTL Bus Speed (MHz)	50	50	50
	# Master Boards	1	1	1
	# Slave Boards	1	1	1
Transceivers	Description	DS90LV017A/018A	74LVT125	10ELT20/21
	# Drivers/Board (Master Board)	1	1	1
	# Rec/Board (Peripheral Board)	1	1	1
	Unit Cost	$0.70	$0.55	$2.00
	Silicon Cost per System	$1.40	$1.10	$4.00
Termination	Voltage	None	None	None
	# Termination Regulators	0	0	0
	Unit Cost	$0	$0	$0
	# Termination Resistors	1	2	2
	Unit Cost	$0.05	$0.05	$0.05
	# Termination Capacitors	0	0	0
	Unit Cost	$0	$0	$0
	Total Termination Cost per System	$0.05	$0.10	$0.10
Transmission Medium	Cable Type	2 Pair CAT3	2 Pair CAT3	2 Pair CAT3
	Distance	1 m	1 m	1 m
	# Conductors	2	2	2
	# Cables	1	1	1
	Connector Type	4-pin Wire to Board	4-pin Wire to Board	4-pin Wire to Board
	Unit Cable + Connector Assembly Cost	$2.00	$2.00	$2.00
	Total Media Cost Per System	$2.00	$2.00	$2.00
Total Relative System Cost		$3.45	$3.20	$6.10

5.2.3 LVDS Channel Link Serializers

The speed of the LVDS line drivers and receivers is limited by how fast the TTL signals can be switched. This is because TTL is used to input data into the LVDS transmitter and is also used to receive data from LVDS receivers. To provide a workaround for this limitation, a family of Channel Link serializers and deserializers was introduced. Instead of using one LVDS channel for every TTL channel, the Channel Link devices send multiple TTL channels through every LVDS channel, thereby matching the speed of LVDS to that of TTL as depicted in Figure 5.24.

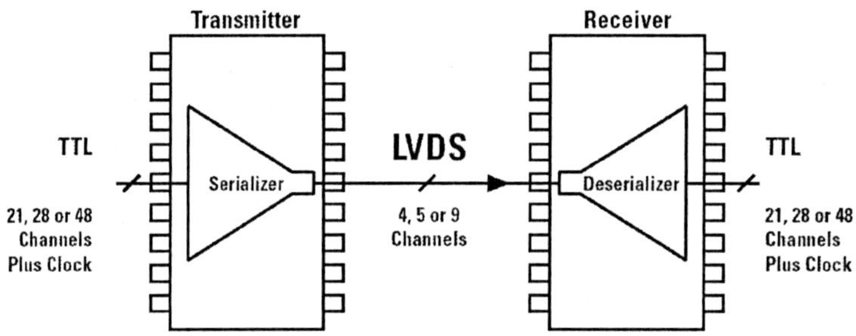

Figure 5.24 Channel Link serializers/deserializers can dramatically reduce the size (and cost) of cables and connectors.

Using fewer channels to convey data also means power and noise can be lower. The biggest advantage, however, is the significant reduction of cable and connector size. Since cables and connectors are usually quite expensive compared with silicon, dramatic cost savings can be achieved. Channel Link chipsets reduce cable size by up to 80%, reducing cable costs by as much as 50%. Plus, smaller cables are more flexible and user-friendly.

LVDS Channel Link serializer/deserializer devices take the inherent high speed, low power, noise, and cost advantages of LVDS and capitalize on the slow speed of TTL to generate significant benefits. For a small increase in silicon cost, Channel Link products can dramatically reduce total system costs and improve total system performance. Therefore, the total system should be evaluated if the true advantages are to be quantified. The following sections summarize the cost and performance benefits of using Channel Link devices.

5.2.4 1 Gbps 16-Bit Interconnect

Channel Link serializers/deserializers take the benefits of LVDS (high speed and low power, noise, and cost) and add serialization to further reduce cable, connector, and PCB size and cost. Channel Link is a great solution for high-speed data bus extension when the overhead of protocols is not desired. The example in Tables 5.9 and 5.10 compares the total performance and cost of moving a 16-bit 66 MHz bus across 1 meter of cable using the 3 V 66 MHz 21-bit DS90CR215/216 Channel Link devices versus other solutions. Driving TTL signals over 1 meter of distance may be very risky due to the limited tolerance to noise (<400 mV) and also transmission line problems generated by the TTL driver. In Table 5.10, all total subsystem costs and the total relative system cost are on a per system basis rather than per board. As in the previous example, the total system cost is a relative cost, meaning that system costs that are the same for all technologies are not included in this total cost estimate.

Table 5.9 Performance Estimate for 16-Bit Cable Interconnect

Characteristic	Parameter	Channel Link	TTL	GTL	ECL	Fibre Channel
Speed	Application Data Rate (Mbps)	1056	1056	1056	1056	1056
	Max Capability per Channel (Mbps)	462	100	150	800	800
Power Consumption	Dynamic (mA, @66 MHz)	180	300	500	300	?
	Static (mA, Outputs Disabled)	0.02 (Power Down)	1	50	50	135
Noise	Low EMI	+++	– – –	– –	+	+++
	Low Bounce	+++	– – –	– –	++	+++
Ergonomics	Compact System Size	+++	– –	– –	– – –	+++
	Compact Transmission Medium Size	+++	–	+	+	+++
	Low Weight	+++	–	–	–	+++

Table 5.10 Cost Estimate for 16-Bit Cable Interconnect

Subsystem	Parameter	Channel Link	TTL	GTL	ECL	Fibre Channel
General	Single-Ended or Differential	Differential	Single-Ended	Single-Ended	Differential	Differential ECL
	TTL Bus Width	16	16	16	16	16
	TTL Bus Speed (MHz)	66	66	66	66	66
	Multiplexed Scheme?	Yes	No	No	No	Yes
	# Master Boards	1	1	1	1	1
	# Slave Boards	1	1	1	1	1
Transceivers	Description	3 V 21:4 Channel Link	ALVT 16-Bit	GTL 18-Bit	9-Bit Translators	Fibre Channel
	# Drivers/Board (Master Bd)	1	1	1	2	1
	# Rec/Board (Peripheral Bd)	1	1	1	2	1
	Unit Cost	$ 3.70	$ 2.40	$ 3.25	$ 5.00	$ 20.00
	Silicon Cost per System	$ 7.40	$ 4.80	$ 6.50	$ 20.00	$ 40.00

Table 5.10 Cost Estimate for 16-Bit Cable Interconnect (Continued)

Subsystem	Parameter	Channel Link	TTL	GTL	ECL	Fibre Channel
PC Board	Layers	4	12	12	12	12
	Size (Normalized)	1.15	1	1.15	15.63	11.97
	Total Add'l PCB Cost/Sys	$0	$15.00	$15.00	$15.00	$15.00
Termination	Voltage	None	None	1.5 V	2.1 V	3.0 V
	# Termination Regulators	0	0	1	1	1
	Unit Cost	$0	$0	$1.00	$1.00	$1.00
	# Termination Resistors	10	16	16	16	32
	Unit Cost	$0.05	$0.05	$0.05	$0.05	$0.05
	# Termination Capacitors	0	0	0	0	0
	Unit Cost	$0	$0	$0	$0	$0
	Total Termination Cost/Sys	$0.50	$0.80	$1.80	$1.80	$2.60
Transmission Medium	Cable Type	SCSI2 CAT3	Shielded Flat	Shielded Flat	SCSI2 CAT3	CAT5
	Distance	2 m	2 m	2 m	2 m	2 m
	# Data+Clock Conductors	8	17	17	34	2
	#Power+Ground Conductors	4	10	10	15	2
	# Cables	1	1	1	1	1
	Connector Type	0.050-20	D-37	D-37	0.05 D-50	DB-9
	Unit Cable+Connector Assembly Cost	$20.00	$30.00	$30.00	$30.00	$15.00
	Total Media Cost/System	$20.00	$30.00	$30.00	$30.00	$15.00
Power Supply	Special Supply Voltages	0	0	1.5 V	2.1 V	3.0 V
	Pwr Sply Size (Normalized)	1	1.3	1.2	1.2	1.2
	Total Add'l Pwr Supply Cost	$0.00	$5.00	$5.00	$5.00	$5.00
Total Relative System Cost		$27.90	$55.60	$58.30	$71.80	$77.60

5.2.5 1.4 Gbps 56-Bit Backplane

In some large datacom and telecom systems, it is necessary to construct a very large, high-speed backplane. There is generally an inverse relationship between the size of a backplane and its maximum speed. In other words, if you try to make a backplane too large, the heavy loading will severely hamper backplane speed and power and noise will become a big problem. Therefore, connecting or extending smaller backplanes via a high-speed cable interconnect is often the only solution. The previous examples illustrated how Channel Link may be used to accomplish this over cable. The cost benefits of using Channel Link to shrink cable and connector costs are clear. What would happen, however, if Channel Link was used to form or extend a backplane using a PCB as the medium? An example of a 1.4 Gbps, 56-bit backplane is outlined in Figure 5.25. This backplane example, enumerated in Tables 5.11 and 5.12, show how Channel Link can reduce the size and number of layers of the printed circuit board transmission medium in the same way as Channel Link reduces the size and cost of cables.

Table 5.11 provides a comparison performance estimate for Channel Link, TTL, GTL/BTL, ECL, and Fibre Channel. It shows that Channel Link consumes the least power of all the technologies while exhibiting no performance liabilities.

Table 5.12 provides a cost estimate for the 56-bit backplane using the various technologies. As before, the total relative system cost is a relative cost and does not include costs common to all solutions. In Table 5.12, all total subsystem costs are listed on a per system basis except for the total silicon cost, total termination cost, and total connector costs, which are listed per board. The per system costs must be divided by 10 to calculate the equivalent per-board cost before figuring totals per board.

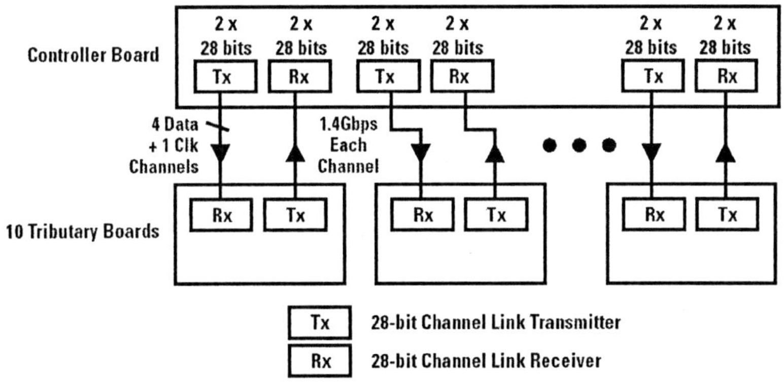

Figure 5.25 1.4 Gbps backplane using point-to-point channel links.

Table 5.11 Performance Estimate for 56-Bit Backplane

Characteristic	Parameter	Channel Link	TTL	GTL/ BTL	ECL	Fibre Channel
Speed	Application Data Rate (Mbps)	1400	1400	1400	1400	1400
	Max Capability per Channel (Mbps)	462	100	150	800	800
Power Consumption (Loaded Tx/ Rx's Only)	Dynamic (mA, @50 MHz)	2600	10,000	6000	16,000	?
	Static (mA, Outputs Disabled)	0.4 (Pwr On)	40	1840	3402	3818
Noise	Low EMI	+++	− − −	+	+	++
	Low Bounce	+++	− − −	+	+	++
Ergonomics	Compact System Size	++	+	+	+	++
	Compact Transmission Medium Size	++	−	+	+	++
	Fans?	No	No	No	Yes	Yes
	Low Weight	+++	−	−	−	+++

Table 5.12 Cost Estimate for 56-Bit Backplane

Subsystem	Parameter	Channel Link	TTL	GTL/BTL	ECL	Fibre Channel
General	Single-Ended or Differential	Differential	Single-Ended	Single-Ended	Single-Ended[a]	Differential
	TTL Bus Width	56	56	56	56	56
	TTL Bus Speed (MHz)	50	50	50	50	50
	Multiplexed Scheme?	Yes	No	No	No	Yes
	Number Tributary Boards	10	10	10	10	10
	Number Channels in Link	10	56	56	56	14
	Number Conductors (Data)	20	56	56	56	28
	Number Conductors (CLK)	1	1	1	1	1
Transceivers	Description	28:5 Channel Link	LVT 16-Bit	GTL 18-Bit	9-Bit	Fibre Channel
	# Transceivers/Board (Trib Board)	4	4	4	14	14
	# Transceivers/Board (Ctrlr Board)	4	4	4	14	14
	Unit Cost	$3.70	$2.40	$3.25	$5.00	$20.00
	Total Silicon Cost per Board	$29.60	$19.20	$26.00	$140.00	$560.00
PC Board	Layers	12	26	26	26	12
	Size (Normalized)	1.15	1	1.15	15.63	11.97
	Total Add'l PCB Cost/Sys	$0	$100.00	$100.00	$150.00	$50.00
Termination	Voltage	None	None	1.5 V	2.1 V	3.0 V
	# Termination Regulators	0	0	14	14	14
	Unit Cost	$0	$0	$1.00	$1.00	$1.00
	# Termination Resistors	10	224	128	128	14
	Unit Cost	$0.05	$0.05	$0.05	$0.05	$0.05
	# Termination Capacitors	0	0	0	0	0
	Unit Cost	$0	$0	$0	$0	$0
	Total Term Cost per Board	$0.50	$11.20	$20.40	$20.40	$14.70
Transmission Medium	Type	PCB Trace Backplane	PCB Trace Backplane	PCB Trace Backplane	PCB Trace Backplane	PCB Trace Backplane
	Distance	<1 m	<1 m	<1 m	<1 m	<1 m
	Layers	12	26	26	26	12
	Size (Normalized)	1	1	1	1	1
	Number Media	1	1	1	1	1
	Additional Media Cost	$0	$200.00	$200.00	$200.00	$0
	Total Add'l Trans. Media Cost per System	$0	$200.00	$200.00	$200.00	$0

Table 5.12 Cost Estimate for 56-Bit Backplane (Continued)

Subsystem	Parameter	Channel Link	TTL	GTL/BTL	ECL	Fibre Channel
Connectors	Connector Type	Header	VME	VME	VME	Header
	Number Pins (Data + CLK)	21	57	57	57	29
	Number Pins (Power/GND)	5	38	38	7	7
	Total Connector Pins	26	96	96	64	36
	Number Connector Pairs	1	1	1	1	1
	Cost of Pair	$3.00	$10.00	$10.00	$8.00	$3.75
	Connector Cost per Board	$3.00	$10.00	$10.00	$8.00	$3.75
Power Supply	Special Supply Voltages	0	0	1.5 V	2.1 V	3.0 V
	Pwr Sply Size (Normalized)	1	1.5	1.5	1.7	1.4
	Total Add'l Power Supply Cost per System	$0	$50.00	$50.00	$60.00	$40.00
Total Relative Cost per Board		$33.10	$75.40	$91.40	$209.40	$587.45
Total Relative System Cost		$331.00	$754.00	$914.00	$2,094.00	$5,874.50

a. ECL can be implemented as either single-ended or differential. For this comparison, a single-ended implementation was used.

5.3 Summary of LVDS Features and Applications

Table 5.13 LVDS Summary

LVDS (Low Voltage Differential Signaling) Summary

Applications:

Use where —

- Data signaling rate is too great for unbalanced (single-ended) operation.
- Data signaling rate exceeds the capability of TIA/EIA-422-B, 485-A, and 612 differential interfaces (10 Mbps, 20 Mbps, and 52 Mbps, respectively)
- Balanced interconnecting media subject to up to ±1 V of common-mode noise
- Necessary to minimize electromagnetic emissions and interference

Use for —

- Point-to-Point (on boards, backplanes, cables)
- Singly-Terminated Multipoint
- Singly-Terminated Backplanes
- Driving Cables
- Where high-speed at low power is needed ("megabits @ milliwatts")

Where Not to Use:

- Multipoint (has dual terminations)
- Multidrop with Dual Terminations
- Driving Low Impedance Lines (<<90 Ω differential impedance)
- When driving unbalanced transmission media

Table 5.13 LVDS Summary (Continued)

Device Voltages: +5 V, +3.3 V
Drive Current: +3.5 mA
Differential Voltage Swing: ±300 mV
Technology: Differential, usually CMOS
Supply Current: <20 mA per quad device below 40 MHz
Power Consumption: <100 mW per quad device below 40 MHz at 5 V
<66 mW per quad device below 40 MHz at 3.3 V
Standard: TIA/EIA-644-A
Termination: Single 100 Ω resistor at far bus end
Hot-Pluggable: Yes
Maximum Drive Distance: 15 meters (varies with data rate, quality of transmission medium)
Maximum data rate: 1.923 Gbps (Theoretical), 655 Mbps (Recommended), per differential pair
Maximum Data Rate Over Interconnect: Per channel, over CAT3 cable, 28AWG TP, controlled impedance traces to AMP Amplite 50 series connectors, 20% jitter (minimum, 0 V differential), see "Eye Pattern Test Results and Data Points" on page 725:

1 meter	400 Mbps
2 meters	391 Mbps
3 meters	370 Mbps
5 meters	295 Mbps
10 meters	180 Mbps

(Divide above values in approximately half for 20% jitter measured at ±100 mV from offset.)
Data rate to 4 GHz for custom ASIC drivers using LVDS threshold voltages at greater than LVDS frequencies over short, controlled-impedance copper.

5.4 Exercises

5-1. Is LVDS a signaling technology or a family of integrated circuits? What is the difference between the two?

5-2. Assume you are working on a new LVDS bus design that will be able to be operated at either the recommended maximum data rate or the theoretical maximum data signaling rate per differential pair. Further assume your bus will have a bit width of 4, 8, 16, 32, or 64 bits where each bit is implemented in a differential pair of traces. Create a table specifying the total overall signaling rate for each combination of signaling rate and bus width.

5-3. For a differential interconnect between an LVDS driver and an LVDS receiver, what circuit elements in the driver, receiver, or load establish the offset voltage at the receiver? Ignore interconnect losses for purposes of answering this question.

5-4. What is the difference between the offset voltage and the common-mode voltage for an LVDS device?

5-5. Does LVDS have to be used with a balanced interconnect medium? Why? What does the term "balanced" refer to?

5-6. Draw two signaling waveforms, one to represent the theoretical maximum signaling rate of 1.923 GHz and the other to represent the recommended maximum data rate of 655 MHz. Label the rise and fall times for each of the two figures. Label the total period for each signal. What is the difference in assumptions between the theoretical and recommended maximum data rates for LVDS?

5-7. A high-speed digital design engineer boasts of designing LVDS circuits with a data rate of 4 Gbps per differential pair. Since the LVDS standard has a theoretical maximum data rate of only 1.923 Gbps, why do you suppose the engineer can make this statement?

5-8. An LVDS circuit is located next to an extremely noisy 220 V ac motor that induces as much as 619 mV of common-mode noise onto each line of the differential pair when the motor starts up. Can the LVDS circuit operate under such noisy conditions? Why? State any assumptions that you need to make. Does it matter whether the induced noise is positive or negative?

5-9. The "A" revision to the TIA/EIA-644 standard includes additional information about multiple receiver operation. How many parallel-connected receivers does it show can be connected in a multidrop bus configuration?

5-10. Define the ΔV_{OD} and ΔV_{OS} parameters. What are their maximum values as specified in the TIA/EIA-644-A standard?

5-11. Compare the top two waveforms in Figures 5.7, 5.10, 5.11, and 5.12. These two waveforms are the true and inverted single-ended signals of the differential pair as seen at the receiver. What effect do you see upon the signals as the transmission medium changes from a PCB trace to a 1 meter cable, a 5 meter cable, and then to a 10 meter cable?

5-12. Why should probes with 50 Ω loading not be used on LVDS circuits?

5-13. Draw examples of (a) point-to-point, (b) bidirectional half-duplex, and (c) multidrop configurations. Can the half-duplex configuration transmit data in both directions at the same time? How could you transmit data in both directions using only multidrop configurations?

5-14. List two advantages that each of these logic families has over most of the others: LVDS, PECL, Optics, RS-422, GTL, and TTL.

5-15. Refer to Table 5.10, "Cost Estimate for 16-Bit Cable Interconnect," on page 99. What two items listed in the table are responsible for the greatest system cost savings from using LVDS Channel Link technology versus other technologies?

CHAPTER 6

Bus LVDS (BLVDS), LVDS Multipoint (LVDM), and Multipoint LVDS (M-LVDS)

Chapter Objectives

- Introduce the BLVDS, LVDM, and M-LVDS families of devices.
- Show how these enhanced families of devices differ from LVDS.
- Discuss the data transfer rates possible for each family, including Bus LVDS with varying numbers of backplane slots.
- Discuss several electrical parameters specified in the TIA/EIA-899 standard and the difference between Type 1 and Type 2 M-LVDS receivers.

6.1 Justification for Enhanced Versions of LVDS

Many high-speed applications require more load than just the ability to transmit from one driver to a receiver or from one driver to multiple receivers terminated with a single resistor. Multiple drivers and multiple receivers on one interconnect (a multipoint configuration) are an efficient and common bus application (see Figure 6.1). This requires double termination of the interconnect to properly terminate the signal at both ends of the bus. LVDS drivers are not intended for double termination loads. As a result, enhanced families of devices were invented to bring the advantages of LVDS to bus multipoint applications. Multipoint configurations offer an efficient balance between interconnect density and throughput as higher connector density often demands slower bus speeds.

Increased current-drive versions of LVDS for bus and cabling applications fall into three main families of devices, Bus LVDS (BLVDS[1]), LVDS Multipoint (LVDM), and Multipoint

1. Perfectionists may be dismayed to find out that in technical literature, BLVDS and LVDM are written without a hyphen while M-LVDS is written with a hyphen. This is, of course, inconsistent, but that's the way it is.

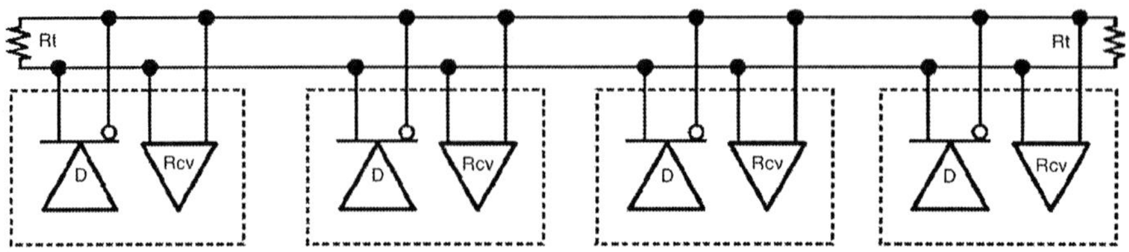

Figure 6.1 Multipoint application with several drivers (D) and receivers (Rcv), and two resistive terminations (Rt). [Reprinted with permission of National Semiconductor Corporation.]

LVDS (M-LVDS). As before, LVDS stands for Low Voltage Differential Signaling. Bus LVDS is standardized by proprietary standards of the vendors, primarily National Semiconductor Corporation and Maxim Integrated Products. LVDS Multipoint is a Texas Instruments proprietary standard and Multipoint LVDS is a publicly released standard, TIA/EIA-899.[2] The names can be easily confused as LVDS Multipoint (the Texas Instruments standard) is different from Multipoint LVDS (the TIA/EIA-899 standard). It may be easiest to compartmentalize the three families in your mind as BLVDS, LVDM, and M-LVDS. The reason for their being three or more standards is that individual companies created competitive, state-of-the-art enhancements to LVDS before the agreed-upon standard, TIA/EIA-899 was developed. Table 6.1 lists several BLVDS and LVDM devices currently available.

Table 6.1 Sampling of Bus LVDS and LVDS Multipoint Devices Currently Available

10-Bit Serializer	16-Channel Transceiver
10-Bit Deserializer	Registered Transceiver
8-Channel 10:1 Serializer	Full-Duplex Transceiver
Six 1:10 Deserializers	Half-Duplex Transceiver
16-Bit SerDes	Quad Bus Driver
1:6 Clock Buffer/Bus Transceiver	Repeater
Single Transceiver	2-Channel MUXed Repeater
Dual Transceiver	Evaluation Kits for Select Devices
Quad Transceiver	Random Lock Deserializer with Embedded Clock Recovery
9-Channel Transceiver	

6.2 Bus LVDS (BLVDS)[3]

"BLVDS is an extension of the LVDS line drivers and receivers family and similarly employs differential technology. It is designed specifically for multipoint applications where the bus is terminated at both ends. Bus LVDS drivers generate about 10 mA of output current (versus the

2. TIA is Telecommunications Industry Association; EIA is Electronics Industries Alliance.
3. This section reprinted with permission of National Semiconductor Corporation.

3.5 mA for LVDS) so that they provide LVDS voltage swings with heavier termination loads. BLVDS conforms to most other LVDS specifications such as the 1.2 V differential offset voltage. Bus LVDS devices may be used in heavily loaded backplanes where the effective impedance is lower than 100 Ω and often as low as 27–50 Ω. Ten mA of driver current into a 30 Ω backplane will provide a differential output voltage of 300 mV, large enough to be adequately detected in a noisy environment. BLVDS adheres to an internal National Semiconductor standard and is not identical to M-LVDS.

For multidrop (with the driver located in the middle of two or more receivers) and multipoint applications, the bus normally requires termination at both ends of the transmission media. If the bus was terminated with 100 Ω at both ends, a driver in the middle would see an equivalent load of 50 Ω. Depending upon the load spacing, the effective impedance of the bus may drop further, and for signal quality reasons, the terminations at each end of the bus may be as low as 60 Ω. In this case, a driver would see both terminations (resistors) in parallel, a 30 Ω load. Standard LVDS drivers (TIA/EIA-644A) have only 3.5 mA of loop current. If these were used to drive the doubly terminated bus with a termination load of 30 Ω, they would only generate a 105 mV differential voltage on the bus. This small differential voltage is not sufficient when noise margins, reflections, over-drive and signal quality are taken into account. In fact, even doubling the drive is not enough for a heavily loaded backplane application. Bus LVDS addresses the issue of driving a low impedance interconnect by boosting its driver current to about 10 mA. This means that a 300 mV differential signal is maintained into a load as low as 30 Ω. LVDS-like signaling with all of its benefits is obtained in doubly terminated bus applications." [Nat00]

Transceivers and repeaters are currently available in the BLVDS product family. A 10-bit serializer and deserializer family of devices is also available that embeds and recovers the clock from a single serial stream. This chipset provides a high level of integration reducing complexity and overhead to link layer ASICs. Clock recovery and "random lock" digital blocks are integrated with the core interface line driving and receiving functions. The deserializer[4] can also be hot-plugged into a live data bus and does not require PLL (phased-locked loop) training.

Special functions are also being developed using BLVDS and LVDS technologies. Currently a special low-skew clock transceiver with 6 CMOS outputs and a repeater/mux with selectable drive levels are available.[5]

6.2.1 System Benefits of Bus LVDS

"There are numerous system benefits to using Bus LVDS over other technologies that have historically been used for bus interconnects. Many of these advantages are discussed next, but can be summed up with the phrase 'gigabits @ milliwatts,' that is, the advantage of a very high-speed data transfer with very low power consumption.

4. Part DS92LV1212/1224 of National Semiconductor Corporation.
5. Parts DS92CK16 and DS92LV222A, respectively.

In addition to Bus LVDS providing a very low cost solution for many applications, there are also other system savings in using Bus LVDS:

- Low power — Its CMOS design helps reduce the cost of system power supplies and cooling, enabling fan-free applications.
- Simple passive terminations — A bus can be implemented using only discrete resistors for termination. Other high-speed bus technologies such at GTL, BTL, and ECL require active termination devices and/or oddball supply rails (+1.5 V, +2.1 V for example) which add to system cost and power distribution complexity.
- Serialization — Available Bus LVDS products include serializers and deserializers designed for bus applications which reduce system interconnect and connector size.
- Low noise — The low noise characteristic of Bus LVDS (due to the differential signalling) helps to limit EMI within a system which in turn reduces system cycle times and system costs." [Nat00]

6.2.2 High-Speed Capability

"Bus LVDS is capable of driving multidrop or multipoint buses at high speeds. For example, Bus LVDS can drive (per differential channel):

- at 155–200 Mbps across a 20 slot multipoint FR-4 backplane,[6]
- at 400 Mbps across a 10 slot multidrop FR-4 backplane, or
- at 1280 Mbps for point-to-point links.[7]

Previously these data rates were only achievable with the more costly high-speed ECL products. ECL devices present a voltage translation challenge to and from common TTL devices and also present power supply and termination problems. Other single-ended technologies are limited to sub 100 MHz applications and present tough termination and power dissipation problems." [Nat00]

6.2.3 Low Power

"Bus LVDS switches an interconnect with only 10 mA of loop current. This is much less than other high-performance bus technologies that use large amounts of current (as much as 80 mA in the case of BTL and 40+ mA for GTL) to switch a bus with an incident wave. Typically, the load power for Bus LVDS is only 2.5 mW. The current-mode drivers tend to offer low power dissipation even at high data signaling rates. With low voltage swings required, Bus LVDS supply rails may be less than 5 V, 3.3 V, or even 2.5 V. These reasons make Bus LVDS components extremely low power." [Nat00]

6. To support these speed claims, see references such as [Say02a] and [Deh02] ,which analyze 20 and 21 slot backplanes, respectively, at 200 Mbps.
7. See Section 26.1 on page 733 for a discussion of the DS92LV16, which provides 2.56 Gbps full-duplex.

6.2.4 Low Swing, Low Noise, and Low EMI

"Bus LVDS uses a low-swing, differential signal. This small, balanced signal transition generates small EMI and the current-mode driver limits spiking into the supply rails. These reasons make the LVDS driver capable of running at better than 10 times the frequency of TTL at lower EMI levels.

CMOS and TTL technologies use a large voltage swing and often large currents in order to switch a bus. (See Figure 6.2.) This switching can cause ringing and undershoot which can be a large contributor to system EMI. BTL addressed this noise issue by reducing its output swing to just 1.1 V. However, it still uses large amounts of current to switch the bus with an incident wave (80 mA typical) and still uses the single-ended approach with limited noise margins and a complex termination scheme. As we have seen, Bus LVDS uses only 10 mA to switch the bus with an incident wave, is low swing (300 mV typical), and is differential. "LVDS and Lower EMI" on page 707 shows how differential signaling can help substantially reduce system EMI. The small swing also provides the high-speed capability while consuming minimal power." [Nat00]

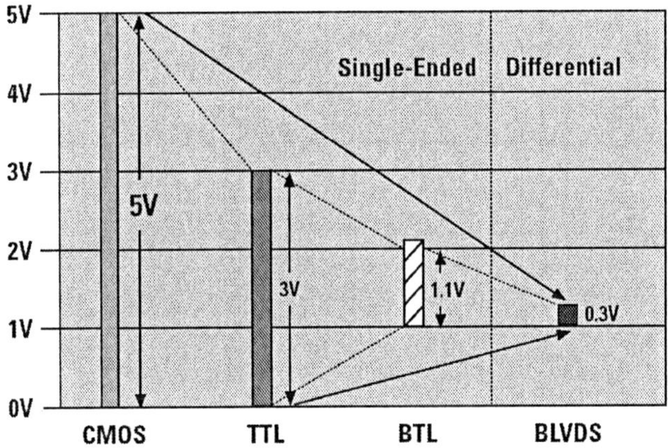

Figure 6.2 Comparison of voltage swings for various backplane technologies. [Reprinted with permission of National Semiconductor Corporation.]

6.2.5 Low System Cost

"Most if not all Bus LVDS products are implemented in a core CMOS technology that allows for low manufacturing cost and the ability to integrate more functionality onto one piece of silicon. By putting the coding, clock recovery, and other phy/link[8] layer digital functions into the inter-

8. Physical layer and data link layer, referring to the two lower layers of the 7-layer Open Systems Interconnection (OSI) protocol model for computer communications architecture.

face device, ASIC[9] complexity and risk are greatly reduced. Immense system savings can be obtained through the use of the serializer and deserializer chipsets. Connectors, cable size, and cost may be reduced. In most cases, these savings more than compensate for the interface silicon cost." [Nat00]

A summary of Bus LVDS is presented in Table 6.2 on page 115.

6.2.6 Bus Failsafe Biasing

"Failsafe biasing may be required if a known state is required on the bus when either of the following conditions occur:

- All drivers are in TRI-STATE® — common in multipoint applications.
- All drivers are removed or powered off.

When these conditions occur, additional biasing (beyond the internal failsafe biasing of the receivers) may be provided with failsafe biasing (power) termination, as shown in Figure 6.3.

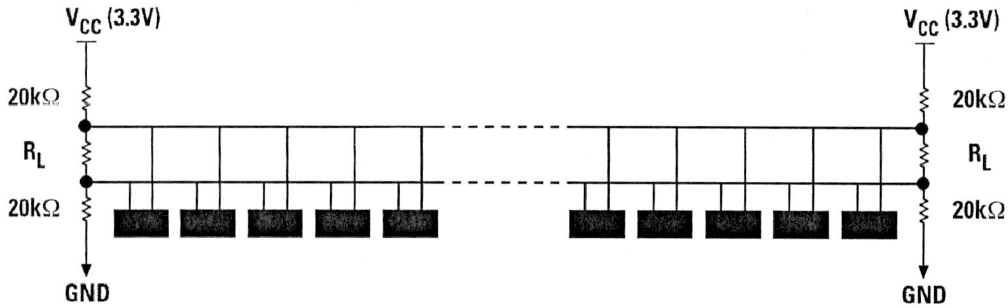

Figure 6.3 Multipoint bus with failsafe bias.

In selecting failsafe resistor values, the following concepts should be considered:

- The magnitude of the resistors should be 1 to 2 orders of magnitude higher than the termination resistor. This prevents excessive loading on the driver and waveform distortion.
- The midpoint of the failsafe bias should be close to the offset voltage of the driver (+1.25 V) to prevent a large common-mode shift from occurring between active and TRI-STATE (passive) bus conditions.
- The pull-up and pull-down resistors should be used at both ends of the bus for quickest response.

9. Application-specific integrated circuit.

- Signal quality is reduced when drivers are TRI-STATED or powered down and external failsafe resistors have been added to create known bus logic states during these conditions. The signal quality is reduced compared to when the drivers are powered up and actually driving." [Nat00]

See "Choosing External Failsafe Resistors" on page 721 for equations used to determine failsafe biasing resistor values.

6.2.7 Hot Plugging (Live Insertion)

"Live insertion, or hot swapping, is of particular importance to the telecommunications marketplace. In these applications, it is critical that maintenance, upgrading, and repair be performed without shutting down the entire system or causing disruption to the traffic on the backplane. BLVDS's wide common-mode range of ±1 V plays a key role in supporting this function. Upon insertion of a card into a live backplane, the occurrence of abnormalities are common on both single-ended signals of the differential pair and differential data is not impacted, as shown in Figure 6.4. This allows for safe hot swapping of cards, making the systems robust and reliable.

Lab testing done in an LVDS interface lab[10] has shown zero-bit errors while plugging in or removing cards from an active bus (BERT test). During the test, up to four cards were inserted at once without an error. This is due to the fact that the differential lines equally load the active line on contact and any glitch seen is a common-mode modulation that is ignored by the receivers.

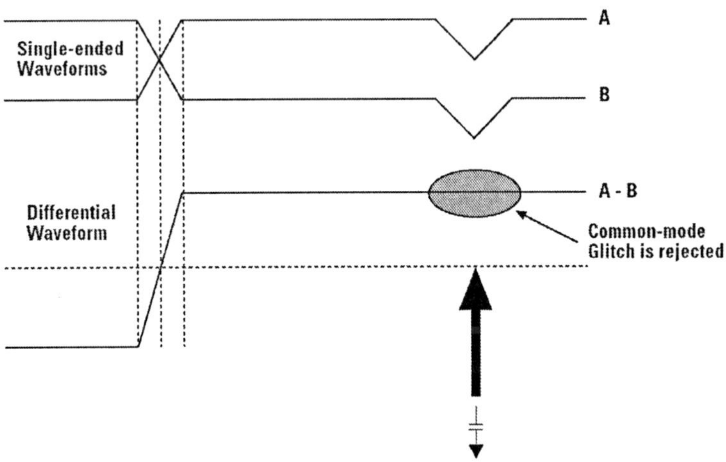

Figure 6.4 Live insertion waveforms.

10. At National Semiconductor Corporation.

Standard power sequencing is still recommended to ensure proper biasing of the devices (substrate). For insertion, the following pin contact sequence should be guaranteed by hardware design: (1) ground, (2) power, and (3) I/O (input/output) pins. For removal, the reverse order is recommended (3-2-1). This sequence can be supported a number of ways. Staggered power pin connectors may be employed (available from multiple sources and even compatible with many matrix connector styles). Multiple connectors are also commonly used. A DIN connector for the I/O, and jack-like connectors for power and ground is one approach. Yet another option uses card edge contact on the rails to establish a GND bias when the card is first inserted into the card rail." [Nat00]

6.3 LVDS Multipoint (LVDM)[11]

"LVDM is LVDS with doubled driver output current and provides the same advantages as LVDS. LVDM stands for LVDS Multipoint and is an internal Texas Instruments standard. In multipoint applications where the bus or other medium is terminated at both ends doubling the load on a transmitter, LVDM can supply LVDS voltage levels and speeds. LVDM allows bidirectional data transfer and the attachment of several drivers, receivers, and/or transceivers.

In a bidirectional transfer application (multipoint application), each termination resistor must still match the characteristic impedance of the loaded backplane or transmission medium. In the case of LVDS, the characteristic impedance of the transmission medium is 100 Ω. In the multipoint application, each end of the bus is terminated with 100 Ω. The parallel combination of these two resistors will equal half the impedance (50 Ω) or twice the load. To ensure that the same voltage levels that would appear across an LVDS 100 Ω load are generated, LVDM devices provide twice the current across the doubled load (50 Ω) and achieve the same LVDS voltage levels.

LVDM can be used as a replacement for TIA/EIA-485 interconnections, especially for high speed, low power, and/or low EMI interfaces. Like LVDS, the common-mode input range is between 0 V and 2.4 V, which limits the line length and the applicability in an electrically noisy environment. It is well suited for backplane applications and cabled interfaces. In addition, LVDM devices are low cost and some provide the same pinout as existing TIA/EIA-422 and TIA/EIA-485 parts. Some 3.3 V LVDM devices provide data rates exceeding 630 Mbps. Texas Instruments uses LVDS and LVDM in its part designations to distinguish between the output-current drive specifications of its LVDS and LVDM line drivers." [Ali00]

A summary of LVDM is presented in Table 6.3 on page 116.

11. This section courtesy of Texas Instruments.

Table 6.2 Bus LVDS (BLVDS) Summary

BLVDS Summary

Applications:

Use where —

- Data signaling rate is too great for unbalanced (single-ended) operation
- Data signaling rate exceeds the capability of TIA/EIA-422, 485, and 612 differential interfaces (10 Mbps, 20 Mbps, and 52 Mbps, respectively)
- Balanced interconnecting media is subject to up to ±1 V of common-mode noise
- Necessary to minimize electromagnetic emissions and interference

Use for —

- Backplanes, multipoint buses, point-to-point interconnects, driving cables
- Multidrop buses with dual terminations
- Driving low impedance loads (27–50 Ω)

Where Not to Use:

- Where multiple simultaneous transactions on the same bus are required
- Where transmitting over double termination interconnects greater than 15 meters
- Where LVDS will suffice and low power consumption is a critical design constraint
- When using unbalanced transmission media.

Power Supply Voltages: +5 V or +3.3 V

Drive Current: 10 mA (approximate, varies with loading)

Differential Output Voltage: ±300 mV into 30 Ω load, ±500 mV (typical) into 100 Ω, 10 pF load

Technology: Differential

Process: CMOS

Power Consumption: Medium. 62 mW (typ) per 200 MHz transceiver, 420 mW (typ) per channel for 2.5 Gbps SerDes device

Standard: Internal vendor standards (National Semiconductor, Maxim), which are enhancements of TIA/EIA-644A (LVDS)

Transmission Medium Impedance: 100 Ω (differential impedance)

Termination: Resistive termination at both ends of bus matched to bus differential impedance

Hot Pluggable: Yes

Maximum Drive Distance: 15+ meters (varies with data rate, quality of transmission medium)

Data Rates:

- 155–200 Mbps across 20 slot multipoint FR-4 backplane (See [Say02a], [Deh02])
- 400 Mbps across 10 slot multidrop FR-4 backplane
- 800 Mbps for point-to-point links
- 2500 Mbps per channel for SerDes device

Table 6.3 LVDS Multipoint (LVDM) Summary

LVDM Summary

Applications:

Use where —

- Backplane or cable loads are 50 Ω or greater
- Data signaling rate is too great for unbalanced (single-ended) operation
- Data signaling rate exceeds the capability of TIA/EIA-422, 485, and 612 differential interfaces (10 Mbps, 20 Mbps, and 52 Mbps, respectively)
- Balanced interconnecting media is subject to up to ±1 V of common-mode noise
- Necessary to minimize electromagnetic emissions and interference

Use for —

- Backplanes, multipoint buses, point-to-point interconnects, driving cables
- Multidrop buses with dual terminations

Where Not to Use:

- Where multiple simultaneous transactions on the same bus are required
- Where transmitting over double termination interconnects greater than 15 meters
- Where LVDS will suffice and low power consumption is a critical design constraint
- When using unbalanced transmission media
- When driven loads are much less than 50 Ω

Power Supply Voltages: +5 V or +3.3 V

Drive Current: 7 mA (approximate, varies with loading, device)

Differential Output Voltage: 250 mV (min) to 450 mV (max) across 50 Ω load (approximate)

Technology: Differential

Process: CMOS

Power Consumption: Medium. 50 mW (typ) driver at 200 MHz, 60 mW (typ) receiver at 200 MHz

Standard: Texas Instruments enhancement of TIA/EIA-644A

Transmission Medium Impedance: 100 Ω (differential impedance before termination and loading)

Termination: 100 Ω resistor at each end of bus

Hot Pluggable: Yes

Maximum Drive Distance: 15+ meters (varies with data rate, quality of transmission medium)

Data Rates: 630 Mbps and greater (See TI's CML device for 3.125 Gbps, TLK3101)

[Courtesy of Texas Instruments]

6.4 Multipoint LVDS (M-LVDS)

6.4.1 The TIA/EIA-899 Standard

Multipoint Low-Voltage Differential Signaling (M-LVDS) is defined by the TIA/EIA-899 standard. It specifies low-voltage differential signaling drivers, receivers, and transceivers for data interchange across half-duplex and multipoint data bus structures. M-LVDS is capable of operating at signaling rates up to 500 Mbps. Devices may be designed for signaling rates less than 500 Mbps, 100 Mbps for example, when economically or technically required for an application.

Driver output characteristics are independent of power supply voltages and may be designed for standard 5 V, 3.3 V, or lower power supplies. The requirements of this standard may be implemented with any integrated circuit technology, such as BiCMOS, CMOS, or GaAs technology. The low-voltage swing (565 mV typical) reduces power dissipation and the potential for radiated emissions. Differential signaling provides multiple benefits over single-ended signaling, notably common-mode rejection and magnetic field cancellation.

The logic function of the driver and the receiver and the communication protocol are not defined by this standard as they are application dependent. The actual interconnecting media is also not specified and may be twisted-pair cable, twinaxial cable (parallel pair), flat-ribbon cable, or printed-circuit board (PCB) traces. Optional receiver input characteristics are specified to allow idle-line failsafe or wired-or signaling at slower signaling rates. (Type 1 vs. Type 2 receivers are discussed later in this section.)

Failsafe is a feature which guarantees that a receiver output will be in a known state (high or low) when the receiver input bus is floating, idle (no active drivers), or experiencing fault conditions (open or short). "Failsafe is a highly desirable feature when transceivers are used with asynchronous controllers such as UARTs." [Nat98]

A wired-AND function is created by tying together the outputs of several open-drain gates with a single pull-up resistor tied to V_{CC}. "An AND function is obtained, since the wired output is high if and only if all of the individual gate outputs are high (actually, open): any output going low is sufficient to pull the wired output low." [Wak01] "'Wired-AND' and 'wired-OR' describe the same thing: it's wired-AND for positive-true logic and wired-OR for negative-true logic. These connections are used in primarily two situations. Most ECL devices have emitter follower outputs and can be wired-OR'ed painlessly, and shared lines in computer buses such as interrupts are sometimes wired-OR'ed." [Hor89]

"It is intended that this standard be referenced by other standards that specify the complete interface (e.g., connector, pin assignments, function, etc.) for applications where the electrical characteristics of a multipoint low-voltage differential interface circuit are required. The referencing standard(s) also specify other characteristics of the interface (such as signal quality, protocol, bus structure, and timing) essential for proper operation across the interface.

The M-LVDS bus consists of balanced interconnecting media terminated at the ends by transmission-line termination resistors. There may be as many as 32 M-LVDS circuits (circuit cards) connected to the media (backplane) at the interchange points. An interchange point is where the interface circuit (plug-in card) plugs into the transmission medium or backplane.

The M-LVDS interface circuit will normally be utilized on data, timing, or control circuits where the signaling rate is up to a maximum limit of 500 Mbps. This limit is determined by the driver output transition times, the media characteristics, and the distance between the driver and the load.

The maximum data rate of 500 Mbps is derived from a calculation of signal transition time at the load assuming a lossless balanced interconnecting media. The recommended signal transition time at the load should not exceed 0.5 of the time unit interval (tui). This should keep jitter from intersymbol interference to less than 5% of the unit interval. This standard specifies that the transition time of the driver into a test load be 1 ns or slower. Therefore, the fastest driver transition time, a lossless balanced interconnecting media, and the 0.5 tui restriction yield a minimum unit interval of 2 ns or 500 Mbps maximum signaling rate. The actual minimum unit interval and the maximum signaling rate are to be specified by the referencing standard while staying within the limits of this standard. 500 Mbps is the maximum signaling rate for a single line (differential pair). Employing a parallel multipoint line structure (4-, 8-, 16-, 32-, 64-, etc. bus width) can easily extend the obtainable data transfer rate into the Gbps range.

The signaling rate of a line is the number of transitions (voltage or frequency changes) that are made per second expressed in the units bits per second. It may be different from the equipment's data transfer rate, which employs the same units. For example, with 8b/10b encoding where 8 bits of data are coded into 10 transmitted bits, the signaling rate may be 500 Mbps while the data transfer rate is only 400 Mbps. In the following discussion, unless otherwise specified, the transition time is the 10% to 90% rise or fall time of a binary signal. The time unit interval (tui) is the mathematical inverse of the signaling rate.

A maximum length of the balanced media is not specified directly but is a function of signal degradation and noise coupling. These, in turn, are functions of the signaling rate, the balanced media characteristics, and the application environment. The ground plane of an M-LVDS transmitter (driver) can vary from the ground plane of an M-LVDS receiver by up to ±1 V. It is anticipated that keeping the ground plane voltage differential of a transmitter and receiver to less than 1 V will be difficult at distances greater than 100 m.

The M-LVDS interface circuit consists of any combination of drivers (D), receivers (R), or transceivers (T) up to a total of 32, a balanced interconnecting medium, and termination.

Table 6.4 lists a few of the driver and receiver parameters and their values under the TIA/EIA-899 standard.

The driver may not be damaged by continuous operation in any short-circuit output condition including a differential voltage of zero volts (short circuit). The outputs of a driver that are neither high nor low are to appear as high impedances (off) relative to the load.

Table 6.4 TIA/EIA-899 Standard for M-LVDS

Parameter[a]	Description	Minimum	Maximum	Units
—	Data Rate (per Differential Pair)		500	Mbps
Z_0[b]	Differential Characteristic Impedance of Unloaded Balanced Interconnect Media	90	110	Ω
Z_T	Termination Impedance (at Each End)	90	110	Ω
V_{OD}	|Driver Differential Output Voltage| (into 48.44 Ω)	480	650	mV
V_{OS}	Driver Offset Voltage (into 49.8 Ω)	0.3	2.1	V
ΔV_{OD}	|Driver Change to V_{OD}|		50	mV
ΔV_{OS}	|Driver Change to V_{OS}|		50	mV
I_{SA}, I_{SB}	Driver Short-Circuit Current to Ground		43	mA
V_{OUT}	Driver Output Voltage Range (into 1.62 KΩ)	0	2.4	V
t_r/t_f	Rise/Fall Times at Load Assuming Lossless Balanced Media		Greater of 1.0 ns or 50% of minimum tui[c]	ns
I_{IN}	|Receiver Input Current|		20[d]	μA
V_{TH}	Receiver Threshold Voltage (Type 1 Receiver)		±50[e]	mV
$\|I_A - I_B\|$	|Receiver Differential Current Balance|		4	μA
—	Interconnect Media One-Way dc Resistance		50	Ω
—	Interconnect Media Signal Attenuation (at 1/tui Hz)		6	dB

a. Actual datasheet specifications for M-LVDS devices may be significantly better than what is specified by the standard in this table.

b. The TIA/EIA-899 standard uses Z_0 to symbolize this parameter although Z_{DIFF} would be a more accurate choice.

c. tui is time unit interval (i.e., the bit width), which is the reciprocal of the signaling frequency.

d. 20 μA over input voltage range of 0 V to 2.4 V with other input at 1.2 V. 32 μA over input voltage range of −1.4 V to 3.8 V with other input at 1.2 V.

e. The TIA/EIA-899 standard specifies ±50 mV but recommends that the referencing standard specify ±100 mV. Referenced to the offset voltage.

[Reproduced under written permission from Telecommunications Industry Association.]

The minimum unit interval should be specified in the referencing standard. In actual practice, tui should allow some time margin for data sampling and be longer than 2 ns. Two types of differential receivers are specified by TIA-EIA-899 and shown in Figure 6.5. Type 1 receivers include no provisions for failsafe and have their differential input voltage thresholds symmetrical about zero volts. Type 2 receivers have their differential input voltage thresholds offset from zero volts to detect the absence of voltage difference. Type 1 receivers maximize the differential noise margin and are intended for the highest signaling rates. Type 2 receivers are intended for control signals and slower signaling rates. The logic function of the receiver is beyond the scope of the standard and therefore is not defined. Receivers may or may not include the transmission line matching impedance.

The difference between the minimum driver differential output voltage (480 mV) and the maximum receiver differential input threshold is the differential signal noise margin (430 mV for Type 1 receivers and 330 mV for Type 2 receivers).

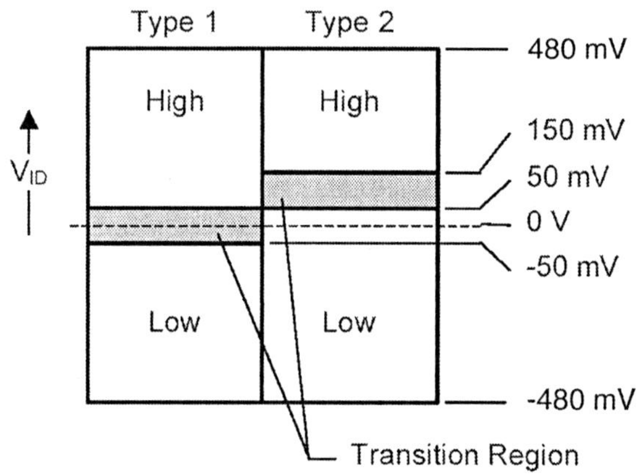

Figure 6.5 Differential input voltage threshold requirements for Type 1 and Type 2 receivers specified by TIA/EIA-899. [Reproduced under written permission from Telecommunications Industry Association.]

The standard specifies several impedances: the media characteristic impedance, a minimum characteristic impedance for a fully loaded M-LVDS line segment (backplane with plug-in cards), and the minimum input impedance of an M-LVDS circuit (plug-in card). The media could be either a cable feeding an M-LVDS backplane or the differential traces on the backplane itself. The media characteristic impedance is specified as 100 Ω ± 10%, or 90 Ω to 110 Ω. Thus the driver typically sees a load of 50 Ω (100 Ω ll 100 Ω) before consideration of transmission line effects.

The standard specifies the minimum input characteristic impedance to M-LVDS circuits as being given by the equation

$$Z_{IN} \geq \frac{tui}{2.5\pi d C_0} \ \Omega \qquad (6.1)$$

where

Z_{IN} = minimum input characteristic impedance to an M-LVDS circuit,

tui = time unit interval,

d = minimum distance between interchanges (e.g., plug-in cards, M-LVDS transceivers, etc.),

and

C_0 = capacitance per unit length (farads/meter) of the media that the M-LVDS circuit is plugged into, before loading.

For example, signaling at 100 Mbps over twisted-pair cable with a distributed capacitance of 40 pF/m and 1 meter between M-LVDS interchanges, the input impedance requirement would be

$$Z_{IN} > \frac{10^{-8}}{2.5 \times 3.14 \times 1 \times 40 \times 10^{-12}} \ \Omega \qquad (6.2)$$
$$Z_{IN} > 32 \ \Omega$$

Z_{IN} is to be measured at the input to the M-LVDS circuit at any frequency below 1/tui Hz.

Annex D to the standard also allows us to calculate the maximum input capacitance C_L to each M-LVDS circuit as

$$C_L \leq 1.25dC_0 \quad \text{farad} \qquad (6.3)$$

where d is, as before, the minimum distance between interchanges (e.g., plug-in cards) and C_0 is the capacitance per unit length (farads/meter) of the media. The above two equations assume that the entire input impedance Z_{IN} is due to capacitance C_L. For the above example, the input capacitance C_L to each plug-in card has to be less than 50 picofarads.

The characteristic impedance for a fully loaded M-LVDS line segment is stated as having to be 60 Ω or greater. The characteristic impedance for a fully loaded M-LVDS line segment is the differential impedance you see looking into the differential media (cable or differential traces) with a full load of M-LVDS circuits (plug-in cards or M-LVDS drivers, receivers, or transceivers) connected to or loading it. In determining this value, the standard states that at the impedance discontinuity where an M-LVDS circuit plugs into the interconnect medium (the backplane), the ratio of transmitted to incident voltage from the interconnect medium to the M-LVDS circuit must be 0.80. Using the 90 Ω minimum characteristic impedance of the transmission medium and this 0.80 transmittance ratio, the standard calculates in its Annex D that the plug-in card must have a characteristic impedance greater than 60 Ω. Given this 60 Ω plug-in card input impedance, it derives an equation for the minimum input capacitance.

Compliance to the minimum requirements of this standard will not ensure interoperability with other differential interface standards such as TIA/EIA-422, TIA/EIA-485, or their ITU-T/ISO equivalents. Interoperation with these interfaces may be possible with proper signal conditioning or use of circuits that exceed the minimum requirements of this standard. Generally, the signal levels of these standards are higher than those of M-LVDS circuits and must be attenuated to levels within the allowed range of M-LVDS operation. Interoperability requires detection of the difference signal and the minimum receiver input differential-voltage threshold is 50 mV for a Type-1 M-LVDS receiver, 100 mV for a TIA/EIA-644 receiver, and 200 mV for a TIA/EIA-422 or -485 receiver. M-LVDS and LVDS will interoperate with certain interchange circuits and signaling rates."[12] [Tia02]

12. This section reproduced under written permission from Telecommunications Industry Association.

"M-LVDS is faster than many other signaling standards. While the TIA/EIA-899 standard specifies a 500-Mbps maximum line speed, limitations such as driver edge rates and stub lengths restrict some M-LVDS buses to about 200 to 300 Mbps. Noise margins or slower edge rates limit other high-performance multipoint-bus technologies to even lower rates. Reduced-swing single-ended technologies, such as BTL and GTL, tend to operate at around 50 Mbps with 100 Mbps as a maximum rate for a 20-node multipoint bus." [Gin02]

Please see the M-LVDS summary in Table 6.5.

Table 6.5 Multipoint LVDS (M-LVDS) Summary Specified in the TIA/EIA-899 Standard

TIA/EIA-899 (M-LVDS) Summary

Applications:

Use where —

- Multiple bus loads are being driven
- Data rates up to 500 Mbps are required
- Data signaling rate is too great for unbalanced (single-ended) operation
- Data signaling rate exceeds the capability of TIA/EIA-422-B, 485-A, and 612 differential interfaces (10 Mbps, 20 Mbps, and 52 Mbps, respectively)
- System power constraints do not allow use of a TIA/EIA-485 interface
- Balanced interconnecting media subject to up to ±1 V of common-mode noise
- Necessary to minimize electromagnetic emissions and interference

Use for —

- Backplanes, multipoint, multidrop with dual terminations
- Driving low impedance lines (loaded impedance of 50 Ω or greater)
- Driving cables, point-to-point

Where Not To Use:

- Where multiple simultaneous transactions on the same bus are required
- Where transmitting over double termination interconnects greater than 15 meters
- Where LVDS will suffice and low power consumption is a critical design constraint
- When using unbalanced transmission media
- When driving low impedance lines much less than 50 Ω when loaded

Power Supply Voltages: No requirement

Driver Differential Output Voltage: ±480 mV to 650 mV across 50 Ω load

Minimum Output Current: 9.6 mA into 50 Ω load

Driver Offset Voltage: ±0.3 V to 2.1 V

Technology: Differential

Process: No requirement — can be BiCMOS, CMOS, or GaAs, but usually CMOS

Power Consumption: Medium. Device dependent.

Standard: TIA/EIA-899

Transmission Medium Characteristic Impedance (Unloaded): 100 Ω ± 10% from 10 MHz to the application signaling rate

Minimum Impedance To Any Plug-in Card: $tui/(2.5\pi dC_0)$ Ω

Termination: Resistive termination at both ends of bus matched to bus differential impedance

Hot Pluggable: Yes

Maximum Noise Rejection: ±1 V common-mode noise

Table 6.5 Multipoint LVDS (M-LVDS) Summary Specified in the TIA/EIA-899 Standard (Continued)

Receiver Types: Type 1 has differential input voltage thresholds centered on offset voltage; Type 2 has differential input voltage thresholds centered on (offset voltage +100 mV).
Maximum Drive Distance: 100 meters due to requirement that grounds of transmitter and receiver cannot differ by more than 1 V. Practical limit is lower, 15+ meters varying with data rate and quality of transmission medium.
Data Rates: 500 Mbps or less depending upon the application

6.5 Selecting BLVDS, BLVM, and M-LVDS Devices

BLVDS, BLVM, and M-LVDS devices do not follow the same standard and it is often difficult and time consuming to identify the best devices for use in a specific application. We examine three actual devices and list a few important device parameters.

The devices to be discussed are not exact equivalents, but were chosen to illustrate the difficulty in selecting devices for a specific application. As a practical matter, it's probably not possible to compare equivalent devices among all vendors given the limited product lines and varying standards. The three devices that we examine parameters on are National's DS92LV040A (4-Channel Bus LVDS Transceiver), Texas Instruments' SN65LVDM179 (High-Speed Differential Line Driver and Receiver), and Maxim's MAX9157EGJ (Quad Bus LVDS Transceiver). The parameters we look at are the magnitude of the differential output voltage $|V_{OD}|$, the rise time t_r or t_{TLH}, the fall time t_f or t_{THL}, and the offset voltage V_{OS} or $V_{OC(SS)}$ for the driver portion of the devices. We also examine the high-side switching voltage threshold V_{TH} or V_{IT+} and the low-side switching voltage threshold V_{TL} or V_{IT-} for the receiver portion of the devices. The devices and their parameters are listed in Table 6.6.

One important parameter is the differential output voltage of the driver under loading, as this determines whether the device is capable of driving the load in your application. Referring to Figure 6.6, it is seen that the differential output voltage varies tremendously with the output load, as you would expect. What makes devices hard to compare is that each vendor provides data under different test conditions. Looking at Table 6.6, we see that one vendor and the TIA/EIA-899 standard provide the output voltage specification under a load of 50 Ω (or almost 50 Ω in the case of the standard) and others provide it at 27 Ω. Without having full load-curve information such as in Figure 6.6, comparing the output voltage specification is difficult. Devices tested at 50 Ω are usually intended for applications with 50 Ω loads or greater and devices tested at 27 Ω are usually intended for lower-impedance loads of 27 Ω or greater.

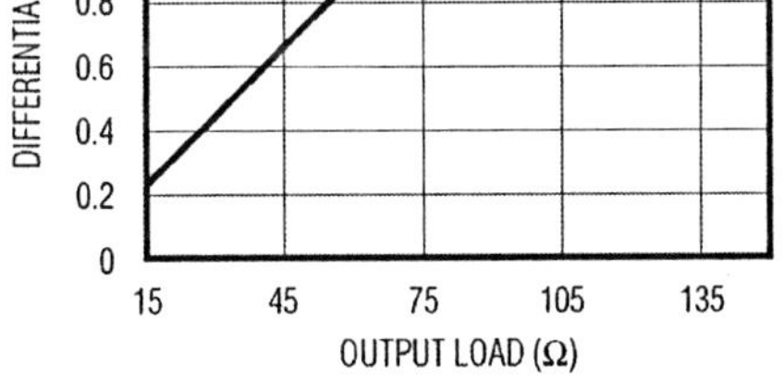

Figure 6.6 Differential output voltage vs. output load for a Bus LVDS transceiver (MAX9157). [Reprinted with permission of Maxim Integrated Products.]

Table 6.6 TIA/EIA-899 (M-LVDS), BLVDS, and LVDM Vendors Specify Their Devices Under Different Loading and Test Conditions

Device	Standard or Vendor	Parameter	Test Loading	Minimum	Nominal	Maximum	Units
Driver[a]							
	TIA/EIA-899	$\|V_{OD}\|$	$R_L = 48.44\ \Omega$	480		650	mV
DS92LV040A	National	$\|V_{OD}\|$	$R_L = 27\ \Omega$	200	300	460	mV
SN65LVDM179	TI	$\|V_{OD}\|$	$R_L = 50\ \Omega$	247	340	454	mV
MAX9157EGJ	Maxim	$\|V_{OD}\|$	$R_L = 27\ \Omega$	250	405	460	mV
	TIA/EIA-899	t_r	$R_L = 49.8\ \Omega$, C_L unspecified	1.0		0.5 tui[b]	ns
DS92LV040A	National	t_{TLH}	$R_L = 27\ \Omega$, $C_L = 10$ pF	0.4	0.75	1.3	ns
SN65LVDM179	TI	t_r	$R_L = 50\ \Omega$, $C_L = 10$ pF		0.6	1.0	ns
MAX9157EGJ	Maxim	t_{TLH}	$R_L = 27\ \Omega$, $C_L = 10$ pF	0.6	1.13	1.4	ns
	TIA/EIA-899	t_f	$R_L = 49.8\ \Omega$, C_L unspecified	1.0		0.5 tui	ns
DS92LV040A	National	t_{THL}	$R_L = 27\ \Omega$, $C_L = 10$ pF	0.4	0.75	1.3	ns
SN65LVDM179	TI	t_f	$R_L = 50\ \Omega$, $C_L = 10$ pF		0.6	1.0	ns
MAX9157EGJ	Maxim	t_{THL}	$R_L = 27\ \Omega$, $C_L = 10$ pF	0.6	1.16	1.4	ns
	TIA/EIA-899	V_{OS}	$R_L = 49.8\ \Omega$	0.3		2.1	V
DS92LV040A	National	V_{OS}	$R_L = 27\ \Omega$	1.1	1.3	1.5	V
SN65LVDM179	TI	$V_{OC(SS)}$	$R_L = 50\ \Omega$	1.125	1.2	1.375	V
MAX9157EGJ	Maxim	V_{OS}	$R_L = 27\ \Omega$	1.185	1.302	1.435	V
Receiver							
	TIA/EIA-899	V_{TH}	(Type 1) R_L unspecified			50	mV
DS92LV040A	National	V_{TH}	$R_L = 27\ \Omega$		–40	0	mV
SN65LVDM179	TI	V_{IT+}	$R_L = 50\ \Omega$			50	mV
MAX9157EGJ	Maxim	V_{TH}	$R_L = 27\ \Omega$		26	100	mV
	TIA/EIA-899	V_{TL}	(Type 1) R_L unspecified	–50			mV
DS92LV040A	National	V_{TL}	$R_L = 27\ \Omega$	–70	–40		mV
SN65LVDM179	TI	V_{IT-}	$R_L = 50\ \Omega$	–50			mV
MAX9157EGJ	Maxim	V_{TL}	$R_L = 27\ \Omega$	–100	–26		mV

a. TIA/EIA-899 is the standard, SN65LVDM179 is an LVDM device, and DS92LV040A and MAX9157EGJ are BLVDS devices.

b. tui (time, unit interval) is the mathematical inverse of the signaling rate.

[Parameters in this table are reprinted with permission. Reproduced under written permission from Telecommunications Industry Association. Courtesy of Texas Instruments. Reprinted with permission of National Semiconductor. Reprinted with permission of Maxim Integrated Products.]

Table 6.6 also reveals some other interesting points. Several parameters of all three devices are different from the TIA/EIA-899 standard. This is not surprising, as the vendor standards were developed earlier than the TIA/EIA-899 standard. Also, some of the devices differ in their offset voltages by about 100 mV. For most applications, this won't make any difference. But in applications where you drive a line with a device from one vendor and receive it with a device from another vendor (such as a new design that has to interface to old legacy equipment), the offset variance might cost you 100 mV in common-mode noise immunity. In almost all applications, you'll be better off using devices from a single vendor to ensure compatibility.

Selecting devices to use is not easy given the multiple and varying standards. The basic approach of working from a well-defined system specification and matching particular device specifications to your requirements is still the best approach to help you find devices that are the best for your design.

6.6 Exercises

6-1. For what type of applications do you need a driver or transmitter with more current drive than LVDS can provide?

6-2. What words do the following three acronyms stand for: BLVDS, LVDM, and M-LVDS?

6-3. How many backplane cards can Bus LVDS drive at 200 MHz?

6-4. Explain the difference between a Type 1 and Type 2 receiver under multipoint LVDS.

6-5. Assume you are designing a backplane where the loaded backplane impedance of each differential trace is 50 Ω. Further assume that there are four devices available in the marketplace for your design and that they are the four devices listed in Table 6.6 on page 124. (Assume that the specifications for the TIA/EIA-899 standard represent one of the four devices.) (a) For each device, calculate the minimum and maximum load current magnitudes by taking the minimum and maximum $|V_{OD}|$ and dividing each by the test load specified in the table. List your results in a table specifying the device, type of device (e.g., BLVDS, LVDM, etc.), and the minimum and maximum load currents for that device. (b) If you further assume that LVDS, BLVDS, LVDM, and M-LVDS output drivers are constant current in nature over all load conditions, which among LVDS, BLVDS, LVDM, and M-LVDS do you feel would deliver the most drive current for driving larger loads (smaller load impedances)? Base your answer on the calculations you made in (a) and assume that these devices are representative of their entire device family. (c) Given that actual devices are not perfectly constant current in nature over all load conditions, and given that device vendors may only specify parameters in data sheets for one load condition, how might you select the highest drive current device for your 50 Ω design load? Assume that obtaining the highest drive current is a design requirement.

6-6. Log on to the Internet and pull up the PDF data sheet for the DS92LV8028 8-channel 10:1 serializer. Does this device have BLVDS or M-LVDS outputs? Does the serial link data rate of each channel exceed the maximum data rate specified in TIA-EIA-899? Identify three parameters in the data sheet that seem to violate parameter values or test conditions from the summary of M-LVDS shown in Table 6.4 on page 119.

6-7. Why do you think the TIA/EIA-899 standard limits the signaling rate to 500 Mbps when there are devices of similar families already in the marketplace that go faster?

6-8. What are three advantages of BLVDS over BTL and GTL?

6-9. What transmission medium impedance would you use for (a) BLVDS, (b) LVDM, and (c) M-LVDS? In answering, provide only the unloaded, unterminated differential impedance of the transmission medium (the backplane or cable).

6-10. What loaded transmission medium differential impedance is a driver likely to see when driving (a) BLVDS, (b) LVDM, and (c) M-LVDS backplanes?

6-11. What is the difference between balanced and unbalanced transmission media? Why don't you want to use unbalanced transmission media with LVDS, BLVDS, LVDM, or M-LVDS?

6-12. Under the M-LVDS standard, if you decrease the space between backplane plug-in cards by half, how does this affect the minimum input impedance of the plug-in cards allowed by the standard? Assume in your answer that all other factors remain unchanged.

6-13. What advantage does a Type 2 receiver under the TIA/EIA-899 standard have over the Type 1 receiver? In other words, what benefit is there to a higher switching threshold?

6-14. For each of the four device families, LVDS, BLVDS, LVDM, and M-LVDS, provide (a) two examples of applications for which they are suited, and (b) two examples of applications for which they are not suited. Make all 16 examples unique, that is, do not repeat any example.

6-15. Go to the Web site for Maxim Integrated Products and pull up and print out the data sheet for the MAX9157EHJ Quad Bus LVDS Transceiver. (a) Draw the schematic for four half-duplex communication channels. (A full-duplex channel can transmit and receive in both directions simultaneously whereas the half-duplex channel can only transmit in one direction at a time.) Use two of these chips total for all four channels. Each channel is point-to-point only and goes from one chip to the other chip. (This chip and other BLVDS chips are primarily intended for multipoint applications. However, to keep the schematic manageable for this exercise, only a point-to-point link is requested.) Show all power and ground connections, connections between the two chips, and all other pin connections except for LVTTL/LVCMOS data and control inputs. Show LVTTL/LVCMOS data and control inputs to the two chips as coming from outside your design. Show power bypass capacitors. Include device pin numbers. (b) How far could you separate the chips in an actual design and still expect it to work? (c) What's the fastest data rate in Mbps that your circuit can operate at if you assume your loaded backplane impedance is 27 Ω? (d) What temperature range can this chip operate over? (e) What technology process is used to make the chip?

6-16. In Table 6.6 on page 124, why isn't a capacitive load specified in the test conditions for the parameters, $|V_{OD}|$, V_{OS}, V_{TH}, and V_{TL}?

CHAPTER 7

High-Speed Transceiver Logic (HSTL) and Stub-Series Terminated Logic (SSTL)

Chapter Objectives

- Introduce the HSTL and SSTL logic families and discuss their intended applications.
- Discuss the four classes of HSTL output buffers, their current drive capabilities, and their applications.
- Discuss the differences among SSTL_3, SSTL_2, and SSTL_18, the two classes of output buffers, and which ones provide both single-ended and differential drivers.
- Explain why both ac and dc signaling levels (ac and dc voltage thresholds) are provided for HSTL and SSTL.

The HSTL (High-Speed Transceiver Logic) and SSTL (stub-series terminated logic) families of devices are grouped together in one chapter because both are primarily intended for applications that interface with main memory chips. They are also commonly used for the parallel data inputs to and the parallel data outputs from SerDes devices (serializer/deserializer devices). (Current mode logic (CML), Low Voltage Differential Signalling (LVDS), or Bus LVDS (BLVDS) are often used for the high-speed serial output of these same SerDes devices.)

7.1 High-Speed Transceiver Logic (HSTL)

7.1.1 The HSTL Standard

In computing, "slow memory access times have traditionally hindered fast processor operations. In the mid-frequency range (between 100 MHz and 180 MHz), the I/O interface options for all single-ended signals are HSTL, GTL/GTL+, SSTL, and LVTTL. Beyond 180 MHz, HSTL is one of the few single-ended I/O interfaces available. HSTL commonly operates at frequencies in excess of 200 Mhz and is usually implemented in CMOS and BiCMOS devices although there is no requirement to be implemented in any technology. With HSTL speeds, faster I/O interfaces significantly improve overall system performance. HSTL is the I/O interface of choice for high-speed memory applications and is ideal for driving address buses to multiple memory banks."[1] [Xil01]

HSTL devices accept a minimal differential input swing from 0.65 V to 0.85 V (nominally) with the outputs driving LVTTL levels. The HSTL output logic switching range is nominally 0 V to 1.5 V. HSTL circuits may require as many as four voltages to operate. These are:

- the main HSTL chip power supply voltage V_{DD},
- a second and different voltage V_{DDQ} supplying the output buffer of the same chip,
- the reference voltage V_{REF} input into the differential receiver,
- and the V_{TT} voltage feeding pull-up resistors in the HSDL circuit external to the chip.

These are shown in Figure 7.1. In the figure, V_{DD} is shown as 3.3 V, although HSTL allows for any integrated circuit supply voltage to be applied as long as the supply voltage for the output drivers V_{DDQ} does not exceed it. Also, V_{TT} and V_{REF} are shown above as being the same voltage whereas they may be different voltages depending upon the implemented circuit. In most cases the use of a separate output supply voltage source for V_{DDQ} is expected, although other implementations that meet the parametric requirements of the HSTL standard are allowed.

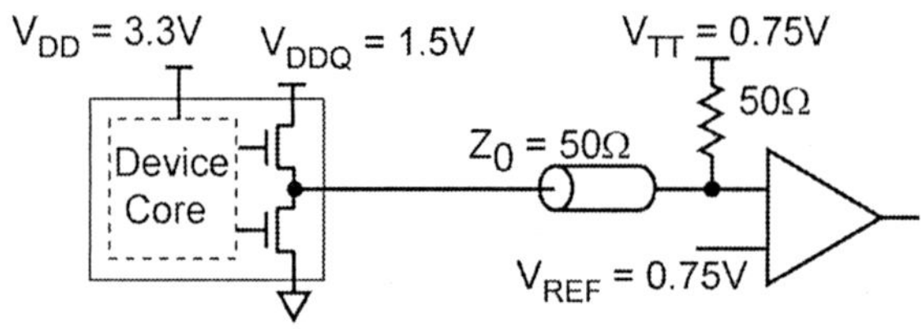

Figure 7.1 An HSTL Class I symmetrically parallel terminated circuit.

The standard governing HSTL is EIA/JESD8-6: "High Speed Transceiver Logic (HSTL). A 1.5 V Output Buffer Supply Voltage Based Interface Standard for Digital Integrated Circuits." Single-ended and differential input buffer specifications as well as various push-pull output driver specifications are incorporated in this standard. The standard is intended to provide for interoperability between all types of HSTL-compliant digital integrated circuits.

1. Material based on or adapted from text owned by Xilinx, Inc. courtesy of Xilinx, Inc. © Xilinx, Inc. 2000–2003. All rights reserved.

The output specifications are divided into four classes (Class I through Class IV) depending upon output drive requirements. A given integrated circuit need not be equipped with all four classes of output drivers, but each must support at least one to claim HSTL output compliance. The full input reference level (V_{REF}) range specified is required on each integrated circuit in order to allow any HSTL integrated circuit to receive signals from any HSTL output driver.

HSTL is a technology-independent interface standard for digital integrated circuits. "Although not intended to be 'sub-optimum,' this standard places a higher priority on flexibility and being technology independent, even supply voltage independent, than on being tailored to a particular semiconductor technology."[2] [Jes95] It was developed for voltage scalable and technology independent I/O structures. The I/O structures required by this standard are differential amplifier inputs (with one input internally tied to a user-supplied input reference voltage for single ended inputs), and outputs using output power supply inputs (V_{DDQ}) that may differ from those operating the device itself.

"To this end, this standard has been developed with voltage scalable (V_{DD} can take on any value) and technology independent I/O structures in mind, namely differential amplifier inputs (with one input internally tied to a user-supplied input reference voltage, V_{REF}, for single-ended inputs) and outputs that use output power supply inputs (V_{DDQ}) that may be different from those that operate the integrated circuit (V_{DD})." [Jes95]

7.1.2 Supply Voltages and Logic Levels

There is no specific device V_{DD} supply voltage requirement for HSTL compliance. Table 7.1 lists the allowed range of voltages for V_{DD}, V_{DDQ}, and V_{REF}.

Table 7.1 HSTL Device and Output Supply Voltages and Input Reference Voltages

Symbol	Parameter	Minimum	Nominal	Maximum	Units
V_{DD}	Device Supply Voltage	—	—	—	V
V_{DDQ}	Output Supply Voltage	1.4	1.5	1.6	V
V_{REF}	Input Reference Voltage	0.68	0.75	0.90	V

[Reprinted with permission of JEDEC.]

V_{REF} of the receiving device should track the variations in the dc value of the V_{DDQ} of the sending device for best noise margins. The value of V_{REF} is to be selected by the user to provide optimum noise margin for the user's application. Peak-to-peak ac noise on V_{REF} may not exceed 2% V_{REF}(dc). For the maximum V_{REF} of 0.9 V (see Table 7.1), this is 18 mV (0.02 × 0.9 V). Although the JEDEC specification states the maximum value of V_{DDQ} as 1.6 V, several chip

2. Reprinted with permission of JEDEC

manufacturers with HSTL I/Os require a V_{DDQ} of 1.8 V. This results in V_{REF} and V_{TT} values having to track this higher V_{DDQ} value. Values are referenced to V_{SS}, which is ground.

The HSTL standard defines input and output specifications and ac test conditions. The HSTL logic levels and their relative positions are depicted in Figure 7.2. "The standard defines both ac and dc input signal values. Making this distinction is important for the design of the high gain, differential receivers that are required. The ac values are chosen to indicate the levels at which the receiver must meet its timing specifications. The dc values are chosen such that the final logic state is unambiguously defined. That is, once the receiver input has crossed this value, the receiver will change to and maintain the new logic state. The reason for this approach is that many input waveforms will include a certain amount of 'ringing.' The system designer can be sure that the device will switch state a certain amount of time after the input has crossed the ac threshold and not switch back as long as the input stays beyond the dc threshold." [Jes02b]

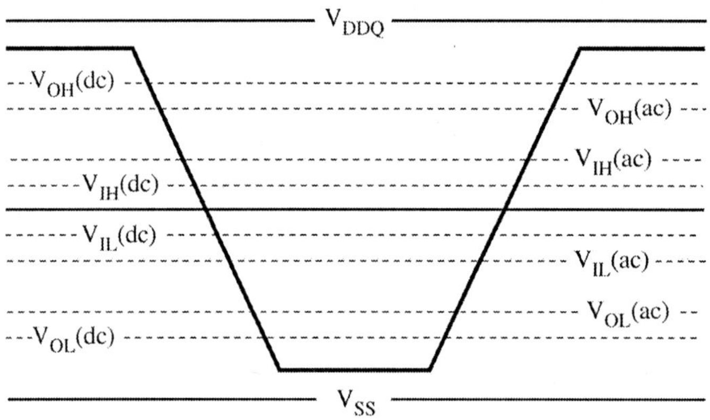

Figure 7.2 HSTL I/O levels for single-ended inputs and outputs. [Reprinted with permission of JEDEC.]

Table 7.2 lists the single-ended input logic levels specified in the standard. The ac specifications are 100 mV outside of the dc specifications.

Table 7.2 HSTL Single-Ended Input Logic Levels

Symbol	Parameter	Minimum	Maximum	Units
$V_{IH}(dc)$	dc Input Logic High	$V_{REF} + 0.10$	$V_{DDQ} + 0.3$	V
$V_{IL}(dc)$	dc Input Logic Low	–0.30	$V_{REF} - 0.10$	V
$V_{IH}(ac)$	ac Input Logic High	$V_{REF} + 0.20$		V
$V_{IL}(ac)$	ac Input Logic Low		$V_{REF} - 0.20$	V

[Reprinted with permission of JEDEC.]

HSTL also allows for differential inputs. The parameters associated with differential inputs are illustrated in Figure 7.3 and the values associated with these parameters are listed in Table 7.3. V_{TR} is the voltage value of the "true" signal and V_{CP} is the voltage value of the "complement" signal. V_{DIF} is the difference or differential signal, which is $V_{TR} - V_{CP}$. The crossing point is the voltage at which V_{TR} and V_{CP} are momentarily equal as the device changes logic state and is denoted as V_X. V_{CM}(dc) is the dc common-mode input voltage and specifies the

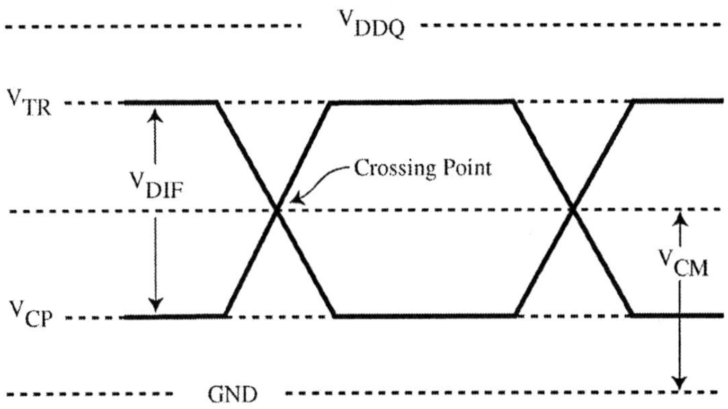

Figure 7.3 HSTL differential input parameters. [Reprinted with permission of JEDEC.]

maximum allowable range of $V_{TR} - ((V_{TR} - V_{CP}) / 2)$. For symmetrical loads, the dc common-mode input voltage and the crossing point voltage should be approximately equal and is sometimes referred to as the "offset." V_{IN}(dc) specifies the maximum allowable dc excursion of each differential input. V_{DIF}(ac) specifies the minimum input differential voltage ($V_{TR} - V_{CP}$) required for switching. Note in Table 7.3 that the minimum ac differential input voltage of 400 mV is 200 mV larger than the minimum dc differential input voltage of 200 mV. A driving high-speed differential input signal must have a differential voltage of at least 400 mV.

Table 7.3 HSTL Differential Input Logic Levels

Symbol	Parameter	Mininum	Maximum	Units
V_{IN}(dc)	dc Input Signal Voltage	–0.30	V_{DDQ} + 0.30	V
V_{DIF}(dc)	dc Differential Input Voltage	0.20	V_{DDQ} + 0.60	V
V_{CM}(dc)	dc Common-Mode Input Voltage	0.68	0.90	V
V_{DIF}(ac)	ac Differential Input Voltage	0.40	V_{DDQ} + 0.60	V
V_X	Differential Crosspoint Voltage	0.68	0.90	V

[Reprinted with permission of JEDEC.]

So far we've discussed input logic levels for both single-ended and differential inputs. The HSTL standard also defines output logic levels, as shown in Table 7.4. The standard specifies four classes of output buffers but the output logic level requirements are the same for all [four classes.[3]

Table 7.4 Output Logic Levels for All Classes of HSTL Output Buffers

Symbol	Parameter	Minimum	Maximum	Units
V_{OH}(dc)	dc Output Logic High	$V_{DDQ} - 0.40$		V
V_{OL}(dc)	dc Output Logic Low		0.4	V
V_{OH}(ac)	ac Output Logic High	$V_{DDQ} - 0.50$[a]		
V_{OL}(ac)	ac Output Logic Low		0.50	

a. V_{OH}(ac) is n/a for Class III outputs as the output voltage is expected to settle very near V_{DDQ} owing to the pull-up termination resistor.

[Reprinted with permission of JEDEC.]

Within this standard, it is the relationship of the sending integrated circuit's V_{DDQ} and the receiving integrated circuit's V_{REF} that determines noise margins. However, in the case of V_{IH} (max.) (i.e., input overdrive) it is the V_{DDQ} of the receiving integrated circuit that is referenced.

Noise margins can be calculated from the specified logic levels provided in Tables 7.2 through 7.4 and are illustrated in Figure 7.4. In calculating noise margins, it is assumed that guaranteed logic states are based upon the dc logic levels specified in the HSTL standard and that the ac levels specified have built-in noise margins of 100 mV at each high and low logic level. That is, the ac specifications already include 100 mV of noise margin for each single-ended logic level and 200 mV of noise margin for a differential input. From Table 7.2 and Table 7.4, we can calculate the single-ended noise margin as V_{OH}(dc) – V_{IH}(dc) or 250 mV assuming minimum values and V_{REF} = 0.75 V. The resulting noise margin is the same on the low side, as shown in Figure 7.4(a). If VREF is increased to the maximum level allowed, 0.90 V, we see that the noise margin for the low logic level increases to 400 mV but the noise margin for the high logic level decreases to only 100 mV, as shown in Figure 7.4(b). When HSTL devices are used in an actual application, the value of V_{REF} is selected by the user to provide optimum noise margins for the specific application. Stated differently, asymmetrical loading can be used to provide maximum immunity to noise for the user's application.

For differential inputs, up to 650 mV of common-mode noise can be rejected depending upon the specific device. The HSTL standard states no requirements for common-mode noise rejection on differential inputs. Given the minimum value of the dc differential input voltage of 200 mV for a valid logic state and assuming that the specific HSTL device can operate with inputs all the way down to 0 V and all the way up to 1.5 V, we see that 650 mV of noise can be

3. Except for Class III V_{OH}(ac). See footnote to Table 7.4.

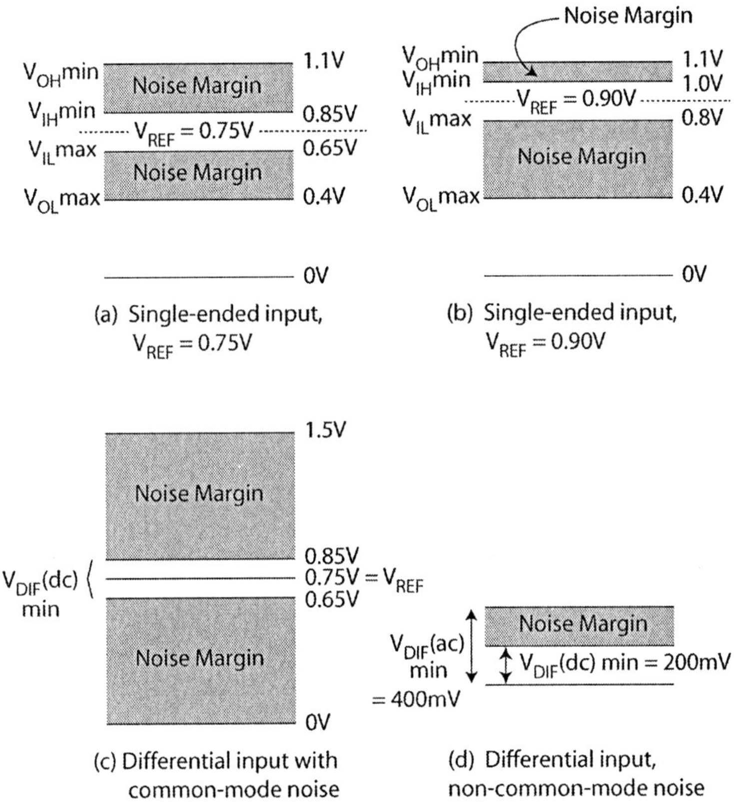

Figure 7.4 HSTL noise margins for different types of noise and V_{REF} levels.

rejected. This is shown in Figure 7.4(c) where we also assume V_{REF} is 0.75 V and that the device is designed to reject this amount of common-mode noise.

When working with differential inputs, most noise coupled onto the inputs will be common-mode; that is, the noise will appear equally on both the true and the complement inputs. This is because differential traces are routed with very small separation between them and any noise is usually picked up by both of them equally. Where noise is coupled onto only one input in a non-common-mode fashion, at least 200 mV of noise rejection is obtainable as shown in Figure 7.4(d). This is because, given that an ac differential signal must have a differential input of at least 400 mV and that a differential logic state is guaranteed for as long as the differential input voltage is at least 200 mV, it would take at least 200 mV of noise on one of the two inputs to cause the differential voltage to fall below 200 mV.

The standard specifies testing of HSTL devices with a test signal having a 1.0 V swing, 0.75 V for V_{REF}, and that test signal edge rates of ±1.0 V/ns or less be maintained in the 20% to 80% range of the input signal swing. The HSTL standard places no restriction on the maximum frequency.

7.1.3 Classes of HSTL Output Buffers

It was mentioned briefly in Section 7.1.2 that there are four classes of HSTL output buffers. These differ primarily in their output current drive capability and the circuit configuration that they will be tested in. These four classes and their supported terminations are listed in Table 7.5. The table lists both series (Class II) and parallel (Class I, III, and IV) terminated loads.

Table 7.5 Classes of HSTL Output Buffers Specified in the EIA/JESD8-6 Standard

Class	Supported Terminations	Source Current I_{OH} (mA, min)[a]	Sink Current I_{OL} (mA, max)
I	Unterminated and Lumped Capacitive Loads	8	–8
	Symmetrically Parallel Terminated Loads	8	–8
II	Externally Source Series Terminated Loads	16	–16
	Symmetrically Double Parallel Terminated Loads	16	–16
III	Asymmetrically Parallel Terminated Loads	8	–24
IV	Asymmetrically Double Parallel Terminated Loads	8	–48

a. The convention for current used in the EIA/JESD8-6 standard is opposite to the normal convention. In this standard, source current out of the device is taken as positive and sink current into the device is taken as negative.

The HSTL standard provides "load models" (shown in Figure 7.5) for each class of output buffers. "Although the load models may accurately represent many actual applications, the models are actually intended to represent an automatic test equipment (ATE) environment typical of semiconductor integrated circuit test operations. The object of these properly terminated test loads is high fidelity testing. Such testing allows successful test correlation between test sites. These loads are intended to produce a clear, repeatable test result that can be easily correlated to simulation data and can be accurately derated to other use environments"[4] [Jes95] (i.e., applications).

HSTL Class I output buffers have three types of loads: push-pull output buffers for unterminated loads and capacitive loads, and push-pull output buffers for symmetrically parallel terminated loads where $V_{TT} = V_{DDQ}/2$. HSTL Class II output buffer models include externally source series terminated loads and symmetrically double parallel terminated loads. HSTL Class III output buffers are push-pull output buffers for asymmetrically parallel terminated loads ($V_{TT} = V_{DDQ}$). HSTL Class IV output buffers are push-pull output buffers for asymmetrically double parallel terminated loads where $V_{TT} = V_{DDQ}$. The V_{TT} pull-up voltage is typically at 0.75 V for Class I, $V_{DDQ}/2$ for Class II, and V_{DDQ} for Classes III and IV.

4. Reprinted with permission of JEDEC

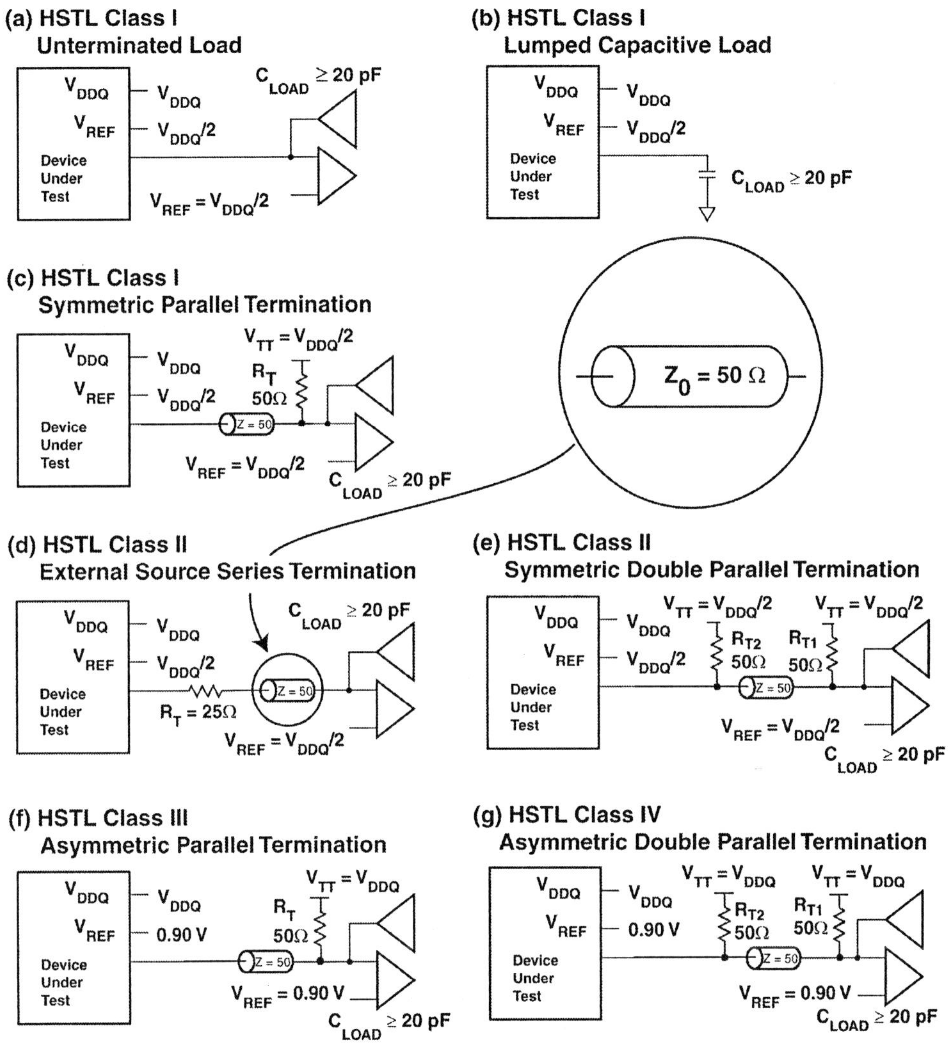

Figure 7.5 Classes of HSTL output buffers. [Reprinted with permission of JEDEC.]

You may notice that all of the buffer output classes and load models are single-ended. The HSTL standard makes no mention of differential outputs, although devices are manufactured in industry that contain HSTL differential outputs.

The speed-limiting factors in high-speed digital I/O circuits are the typical transmission line effects (ringing, reflections, crosstalk, and EMI). Transmission line reflections are the greatest constraint. Controlling reflection requires impedance matching using parallel or series terminations. For Class II applications, design engineers may use IBIS models to compute the source impedance values.

7.1.4 FPGAs with HSTL I/Os

Xilinx Virtex FPGAs provide integrated HSTL I/Os for memory-intensive designs. "The Virtex series of FPGAs feature a 'SelectI/O+' technology which integrates support for 20 single-ended and differential I/O standards. HSTL is one of the single-ended I/O interfaces supported by every Virtex device, eliminating the need for external level translators to interface with high-speed memories and reducing overall system design complexity and cost."[5] [Xil01]

Xilinx Virtex devices support all the push-pull output buffers for parallel terminated loads (Class I, III, and IV.) It is not recommended to use Virtex output buffers for unterminated loads because of signal integrity issues, specifically ringing, which affects overall performance by slowing down the outputs. Using the SelectI/O+ technology, the Virtex series delivers up to 804 single-ended I/Os capable of supporting the HSTL standard. With every I/O capable of supporting HSTL, maximum board lay-out flexibility is provided. By reducing overall system design complexity and cost, direct interfacing to high performance memory devices is easily engineered. Slew rate control is not provided for the HSTL I/Os.

Table 7.6 HSTL Classes Supported by the Xilinx Virtex Series of Devices

Standard	Virtex	Virtex-E	Virtex-II	Virtex-II Pro
HSTL Class I	yes	yes	yes	yes
HSTL Class II	no	no	yes	yes
HSTL Class III	yes	yes	yes	yes
HSTL Class IV	yes	yes	yes	yes

[Table based on or adapted from figures and text owned by Xilinx, Inc. Courtesy of Xilinx, Inc.]

Figure 7.6 provides an example of how HSTL signals can be inputted into a Class I receiver inside an FPGA and outputted by the same FPGA as a Class III output buffer. Figure 7.7 provides a similar example of the FPGA driving a Class IV circuit.

Figure 7.8 shows laboratory test results obtained from an FPGA test circuit similar to that of Figures 7.7 and 7.6. A pulse generator was used to drive the FPGA's HSTL receiver I/O and an FPGA HSTL I/O was used to drive an HSTL load. As shown in the figure, an FPGA can generate a substantial output swing from a very small input swing. The signals marked as "Input Signal" and "Output Signal (Z)" in Figures 7.6 and 7.7 show the connection points for test signals.

5. Material based on or adapted from text owned by Xilinx, Inc. courtesy of Xilinx, Inc. © Xilinx, Inc. 2000–2003. All rights reserved.

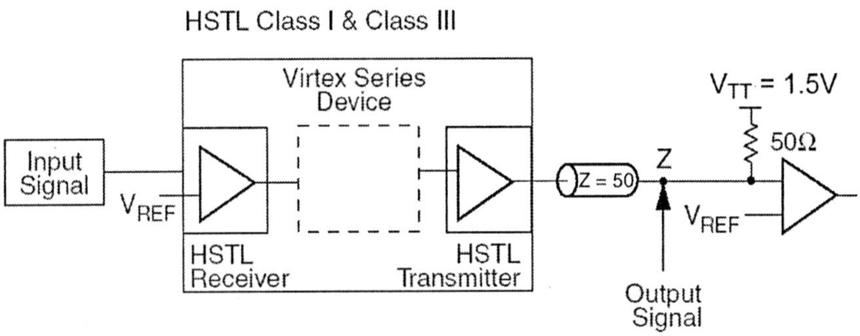

Figure 7.6 HSTL Class I receiver and HSTL Class III transmitter implemented in a Virtex FPGA.

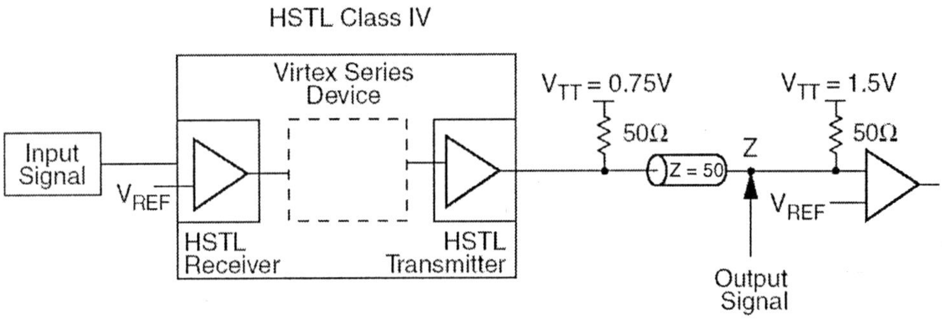

Figure 7.7 Virtex FPGA with HSTL I/Os feeding an HSTL Class IV circuit.

In summary, HSTL compliance does not specify device supply voltages, making it a process independent I/O standard. The lower voltage-level swing associated with this standard makes high-speed HSTL I/O solutions possible for any core voltage-level device. A given circuit need not have all four classes of output drivers, but each circuit must have at least one of the four classes to claim HSTL output compliance. The HSTL nominal logic switching range is 0.0 V to 1.5 V, resulting in faster outputs with reduced power dissipation and minimized EMI concerns. HSTL gives system designers enhanced flexibility in optimizing system performance with an adjustable trip-point reference voltage (V_{REF}).

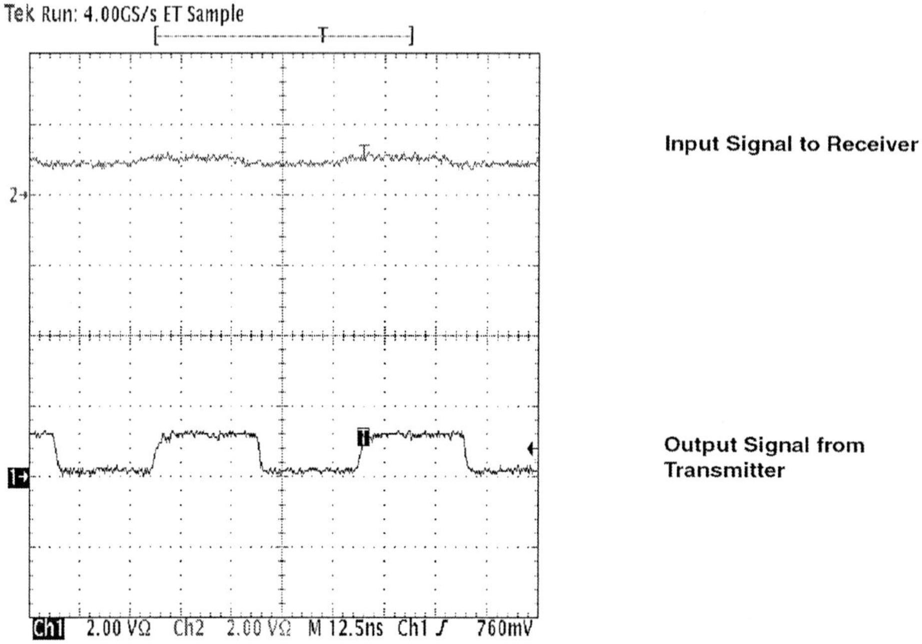

Figure 7.8 HSTL signal waveforms. [Figure based on or adapted from figures owned by Xilinx, Inc. Courtesy of Xilinx, Inc. © Xilinx, Inc. 2000–2003. All rights reserved.]

7.1.5 HSTL Summary

Table 7.7 Summary of HSTL (High-Speed Transceiver Logic) Technology

HSTL (High-Speed Transceiver Logic) Summary
Applications: HSTL is I/O interface for most SRAM memory devices QDR and QDRII SRAM memory chip I/Os Driving high-speed memory chips or banks of memory Where need highest single-ended speed possible Notes: HSTL is one of the few single-ended interfaces available beyond 180 MHz Xilinx Virtex-II Pro FPGA I/Os support HSTL Classes I, II, III, and IV **Where Not to Use:** Where driving differential media **Required Voltages:** Any V_{DD}, 1.5 V nominally for output buffer V_{DDQ}, V_{REF} of 0.68–0.9 V, V_{TT} typically 0.75–1.5 V for transmission medium pull-up

Table 7.7 Summary of HSTL (High-Speed Transceiver Logic) Technology (Continued)

Output Buffer Classes and Minimum Drive Current: (source mA, sink mA, follows opposite current convention)

Class I	Unterminated and lumped capacitive loads	8 mA	–8 mA
	Symmetrically parallel terminated loads	8 mA	–8 mA
Class II	Externally source series terminated loads	16 mA	–16 mA
	Symmetrically double parallel terminated loads	16 mA	–16 mA
Class III	Asymmetrically parallel terminated loads	8 mA	–24 mA
Class IV	Asymmetrically double parallel terminated loads	8 mA	–48 mA

AC Differential Input Voltage Range: 400 mV to 1.5 V nominally
Single-Ended Voltage Swing: 0–1.5 V nominally
Technology: Differential and single-ended inputs, single-ended outputs
Process: Any allowed, but usually CMOS or BiCMOS
Impedance of Transmission Medium: 50 Ω
Supply Current: Depends upon value of V_{DD} and class of output buffer used
Power Consumption: Depends upon value of V_{DD} and class of output buffer used
Standard: EIA/JESD8-6
Output Termination: Series or parallel
Hot Pluggable: No, unless specified in data sheet
Minimum Noise Rejection: 250 mV (p–p, single-ended), 200 mV (peak, non-common-mode, differential), up to 650 mV (common-mode, differential).
Drive Distance: 20 inches (50 cm) or more, depends on frequency and quality of transmission medium.
Maximum Data Rate: 400 Mbps per output. Standard specifies no maximum rate.
Notes: Standard specifies maximum peak-to-peak noise on V_{REF} of 2% (18 mV)

7.2 Stub-Series Terminated Logic (SSTL)

SSTL interface devices are intended for high-speed memory applications, are particularly intended for interfacing with both single and double data rate (DDR) SDRAMs, and support operating frequencies of 333 MHz and higher. SSTL is optimized for main memory applications that have long stubs off of the motherboard bus due to trace routing of the DIMM (dual inline memory module). These long stubs are isolated from the buses by external stub resistors. SSTL supports memories made by Hyundai, NEC, Micron, Samsung, Siemens, and Toshiba. Applications include 3D graphics cards, plasma and LCD displays, DTV interfaces, and set-top boxes.

The need to standardize memory interfaces resulted in JEDEC defining three SSTL interface standards:

- EIA/JESD8-8 Stub-Series Terminated Logic for 3.3 Volts (SSTL_3),
- EIA/JESD8-9B Stub-Series Terminated Logic for 2.5 Volts (SSTL_2), and
- EIA/JESD8-15 Stub-Series Terminated Logic for 1.8 Volts (SSTL_18).

"The SSTL_2 and SSTL_18 signaling level standards were developed to ensure that data signals can meet the higher data throughput speeds needed for newer memory systems. The primary benefits are their ability to drive many stub terminated loads and provide reduced signal voltage swings, increased slew rates, compatibility with the newer memory products, and reduced EMI/RFI. Not all physical parts are unique; the output buffer is essentially the standard P-channel/N-channel CMOS buffer operating at the lower 2.5 V level. The key difference of the SSTL circuits is the receiver and its need of a reference voltage signal, V_{REF}. The input receiver used with SSTL is built from a differential pair common source amplifier. One input to the differential amplifier is the incoming data signal and the other input is the reference voltage. The reference voltage is shared between all memory devices and the memory controller. This type of receiver has the advantage of rejecting common-mode noise while providing better gain, higher bandwidth, and a reduced threshold offset due to proximity and size of the transistor differential pair as compared to the standard TTL type of receiver. This results in improved signal swing sensitivity and reliability, and also higher operating speeds, but requires a higher level of design quality. Many of the signal lines must be very near the same length, attention must be paid to signal line routing, and the reference and termination voltage lines must be kept as noise-free as possible." [Ibm01]

7.2.1 SSTL_3

"Stub-series terminated logic for 3.3 V (SSTL_3), EIA/JESD8-8, is an interface standard for digital integrated circuits designed to operate in the switching range of 0 to 3.3 V. The standard is particularly intended to improve operation in situations where buses must be isolated from relatively large stubs (e.g., memory controllers interfacing to DDR SDRAMs).[6] External resistors provide this isolation and also reduce the on-chip power dissipation of the drivers. The standard provides input and output specifications and ac test conditions. In a typical application, receiver input thresholds of ±200 mV are established around a reference voltage V_{REF}[7] of 1.5 V. A 'high' input logic level is triggered above 1.7 V and a 'low' input logic level is triggered below 1.3 V. The receivers are differential with one input tied to V_{REF} and the logic signal transmitted to the other receiver input. One or more external termination resistors are required that are tied to a termination voltage, V_{TT}. In a typical application, V_{TT} is equal to V_{REF} where both are 1.5 V. A typical output buffer to receiver circuit is shown in Figure 7.9. (In the figure, R_S is a series termination resistor, and the transmission medium, which occurs between R_S and R_T, is not shown.)

While switching thresholds are typically at 200 mV from the reference voltage, there is a requirement that the SSTL_3 driver must create a 400 mV deviation from the reference voltage at the receiver input as a result of the driver V_{OUT} attaining its maximum low or its minimum high value.

6. DDR SDRAM is double data rate synchronous dynamic random access memory.
7. For this discussion and in agreement with convention, V_{DD} is the main device supply voltage, V_{DDQ} is the output buffer supply voltage, V_{SS} is ground, and V_{TT} is the termination voltage that termination resistors are tied to.

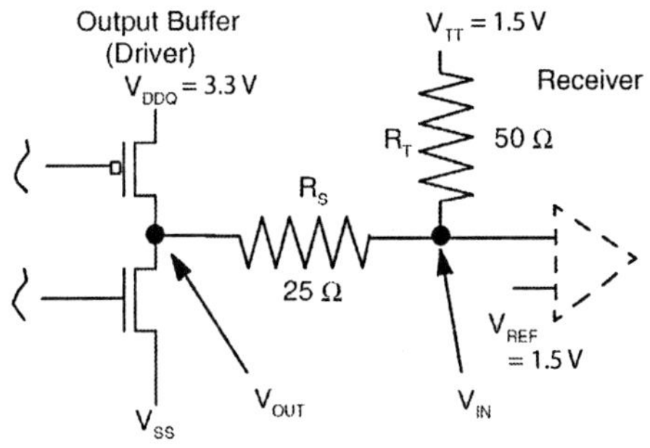

Figure 7.9 Typical output buffer (driver) to receiver circuit. [Reprinted with permission of JEDEC.]

SSTL_3 has two classes of drivers. SSTL_3 Class I drivers provide a sink and source capability of about ±8 mA and usually are intended for driving 50 Ω loads. SSTL_3 Class II drivers provide about ±16 mA of sink and source current and are intended for 25 Ω loads. In addition to the 25 Ω or 50 Ω loads, the specifications for both classes of drivers assume a 25 Ω series termination resistor as shown in Figure 7.9.

The standard specifies no requirement for V_{DD} and separate V_{DD} and V_{DDQ} supply voltages may be used. While in many cases V_{DD} and V_{DDQ} have the same voltage level, the V_{DDQ} value may not exceed that of V_{DD}. For noise rejection reasons, the V_{DD} and V_{DDQ} supplies may be routed separately in the system interconnect.

All of the above values are for a typical application and provided to give you a quick grasp of this technology. The actual standard provides quite a bit of flexibility to meet design requirements. For example, V_{REF} can deviate from 1.5 V to improve noise margins and load impedances vary with the application." [Jes96]

7.2.1.1 Supply Voltage and Logic Input Levels

"Table 7.8 lists voltages required for operation of SSTL_3 devices. Note that V_{TT} is set to V_{REF} and V_{REF} is nominally 1.5 V. There is no specific device V_{DD} supply voltage requirement for SSTL_3 compliance. However under all conditions V_{DDQ} must be less than or equal to V_{DD}. The value of V_{REF} may be selected by the user to provide optimum noise margin in the system. Typically the value of V_{REF} is expected to be about $0.45 \times V_{DDQ}$ of the transmitting device and V_{REF} is expected to track variations in V_{DDQ}. Peak-to-peak ac noise on V_{REF} may not exceed ±2% of the dc value of V_{REF}. V_{TT} of the transmitting device must track the V_{REF} of the receiving device. The full input reference level (V_{REF}) range specified is required for each IC in order to allow any SSTL_3 IC to receive signals from any SSTL_3 output driver.

In typical applications V_{TT} and V_{REF} track as a ratio of V_{DDQ}. In some standards this ratio equals 0.5. If this value were to be applied to SSTL_3, the maximum value of V_{TT} and thus V_{REF} would be 1.8 V (0.5 × 3.6). In order to be more closely compatible with LVTTL, the nominal ratio was selected to be 0.45. This leads to a typical V_{TT} and V_{REF} value of about 1.5 V if V_{DDQ} = 3.3 V. This value is close to the internal reference voltage in current LVTTL designs.

Table 7.8 SSTL_3 Device and Output Supply Voltages and Input Reference and Termination Voltages

Symbol	Parameter	Minimum	Nominal	Maximum	Units
V_{DD}	Device Supply Voltage	V_{DDQ}	—	n/a	V
V_{DDQ}	Output Supply Voltage	3.0	3.3	3.6	V
V_{REF}	Input Reference Voltage	1.3	1.5	1.7	V
V_{TT}	Termination Voltage	$V_{REF} - 0.05$	V_{REF}	$V_{REF} + 0.05$	V

[Reprinted with permission of JEDEC.]

In a manner similar to HSTL, the SSTL_3 standard defines both ac and dc input signal values. Remember that dc levels establish the switching thresholds and that ac levels are provided to indicate levels at which the receiver must meet its timing specifications. The relationship of the different levels is shown in Figure 7.10. An example of ringing is illustrated by the dashed waveform.

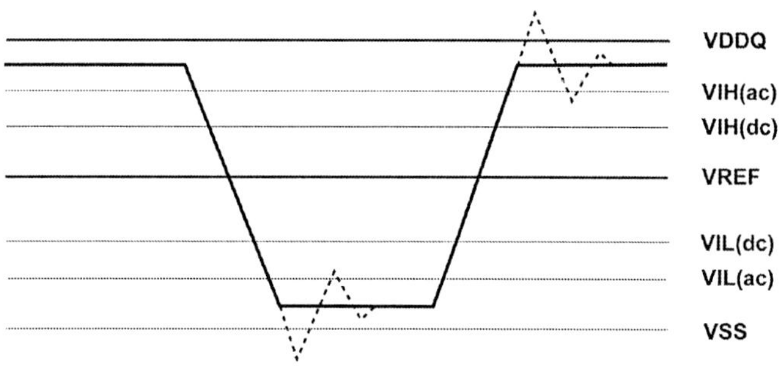

Figure 7.10 SSTL_3 input voltage levels.

Table 7.9 shows the dc and ac receiver input logic levels. For SSTL_3, the relationship of V_{DDQ} of the driving device and V_{REF} of the receiving device is what determines noise margins. In the case of $V_{IH(max)}$, the case of input overdrive,[8] it is the V_{DD} of the receiving device that is referenced. In the case where a device is implemented that supports SSTL_3 inputs but has no SSTL_3 outputs and therefore no V_{DDQ} supply voltage connection (e.g., a translator), inputs must tolerate input overdrive to 3.0 V (high corner V_{DDQ} + 300 mV)." [Jes96]

8. Overdrive is the amount by which the signal voltage exceeds the receiver's supply voltage.

Table 7.9 SSTL_3 dc and ac Input Logic Levels

Symbol	Parameter	Minimum	Maximum	Units
V_{IH}(dc)	dc Input Logic High	V_{REF} + 0.20	V_{DDQ} + 0.3	V
V_{IL}(dc)	dc Input Logic Low	–0.30	V_{REF} – 0.20	V
V_{IH}(ac)	ac Input Logic High	V_{REF} + 0.40		V
V_{IL}(ac)	ac Input Logic Low		V_{REF} – 0.40	V

[Reprinted with permission of JEDEC.]

7.2.1.2 SSTL_3 Output Buffers

"This specification sets minimum requirements for output buffers in such a way that when they are applied within the range of power supply voltages specified in SSTL_3 and are used in conjunction with SSTL_3 input receivers then the input receiver specifications can be met or exceeded. The specifications are quite different from traditional specifications, where minimum values for VOH and maximum values for VOL are set that apply to the entire supply range. In SSTL_3, the input voltage provided to the receiver depends on the driver as well as on the termination voltage and termination resistors. Figure 7.9 is repeated here again as Figure 7.11 and shows the typical dc environment that the output buffer is presented with.

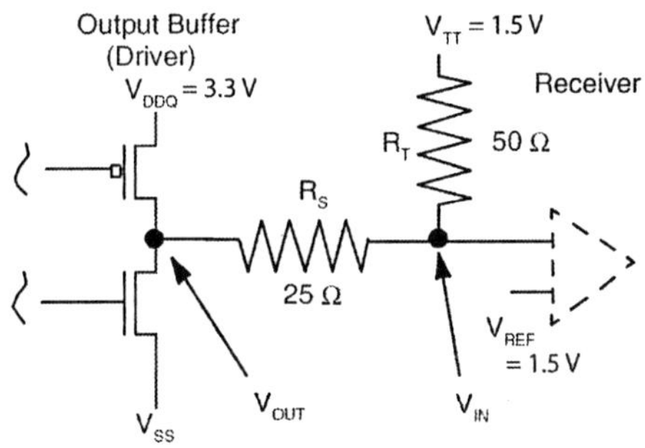

Figure 7.11 Typical output buffer (driver) environment. [Reprinted with permission of JEDEC.]

Of particular interest here are the values V_{OUT} and V_{IN}. These values depend not only on the current drive capabilities of the buffer, but also on the values of V_{DDQ} and V_{TT} (V_{REF} is equal to V_{TT}). The important condition is that V_{IN} be at least 400 mV above or below V_{TT} as a result of V_{OUT} attaining its maximum low or its minimum high value. As will be seen later, the two cases of interest for SSTL_3 are where the series resistor R_S equals 25 Ω and the termination resistor R_T equals 50 Ω (for Class I) or 25 Ω (for Class II). V_{TT} is specified as being equal to 0.45 × V_{DDQ}.

In order to meet the 400 mV minimum requirement for V_{IN}, a minimum of 8 mA must be developed across R_T if R_T equals 50 Ω (Class I) or 16 mA in case R_T equals 25 Ω (Class II). The driver specification now must guarantee that these values of V_{IN} are obtained in the worst-case conditions.

Table 7.10 shows the minimum current requirements of ±8 mA for Class I output buffers and ±16 mA for Class II output buffers. For testing the I_{OH}(dc) requirement of actual devices, V_{DDQ} should be at 3.0 V, and V_{OUT} should be at V_{DDQ} – 1.1 V or 1.9 V. For testing the I_{OL}(dc) requirement, V_{DDQ} should be at 3.0 V but V_{OUT} should be at 0.7 V. For both parameters, V_{REF} is equal to V_{TT} and both are 1.3 V. In other words, the minimum output currents are specified at the minimum required V_{DDQ}, the worst-case condition.

Table 7.10 Output dc Current for Class I and Class II Drivers

Symbol	Parameter	Minimum	Maximum	Units
Class I Output Buffer				
I_{OH}(dc)	Output Source dc Current	–8		mA
I_{OL}(dc)	Output Sink dc Current	8		mA
Class II Output Buffer				
I_{OH}(dc)	Output Source dc Current	–16		mA
I_{OL}(dc)	Output Sink dc Current	16		mA

[Reprinted with permission of JEDEC.]

Figures 7.12 and 7.13 show examples of SSTL_3 Class I and II output buffer circuits, respectively, for symmetrically parallel terminated output loads with a series resistor. The loading is what determines the classification. For example, Figure 7.12 could just as easily be for a Class II output buffer if R_T were 25 Ω instead of 50 Ω. In many applications where interconnections are short, there is no need for any termination at all. Such applications with no R_S and no R_T can be served by either a Class I or II type buffer and an SSTL_3 receiver. Noise margins and possibly speed will be improved over LVTTL.

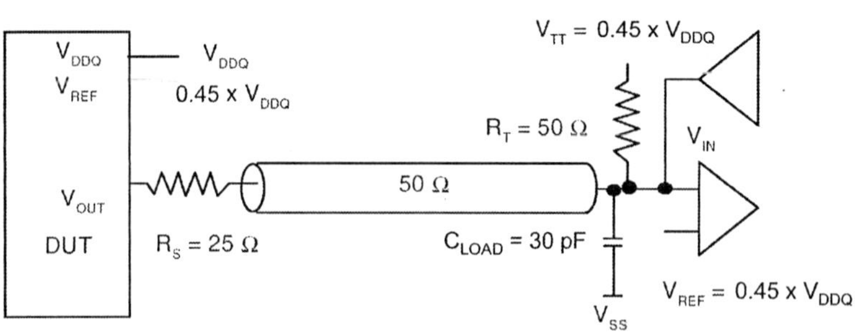

Figure 7.12 An example of SSTL_3 Class I symmetrically parallel terminated output load, with series resistor. [Reprinted with permission of JEDEC.]

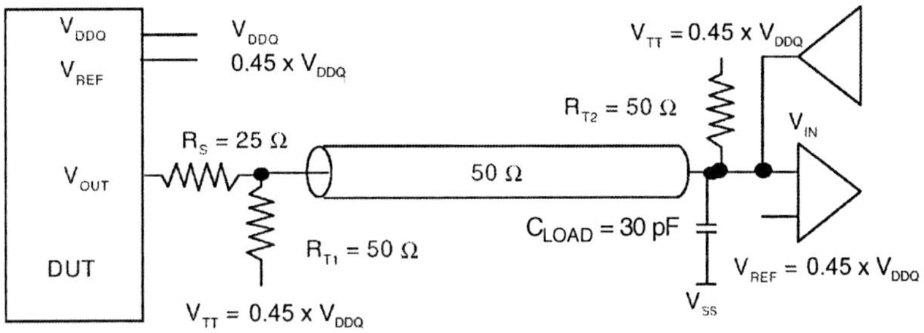

Figure 7.13 An example of SSTL_3 Class II symmetrically double parallel terminated output load, with series resistor. [Reprinted with permission of JEDEC.]

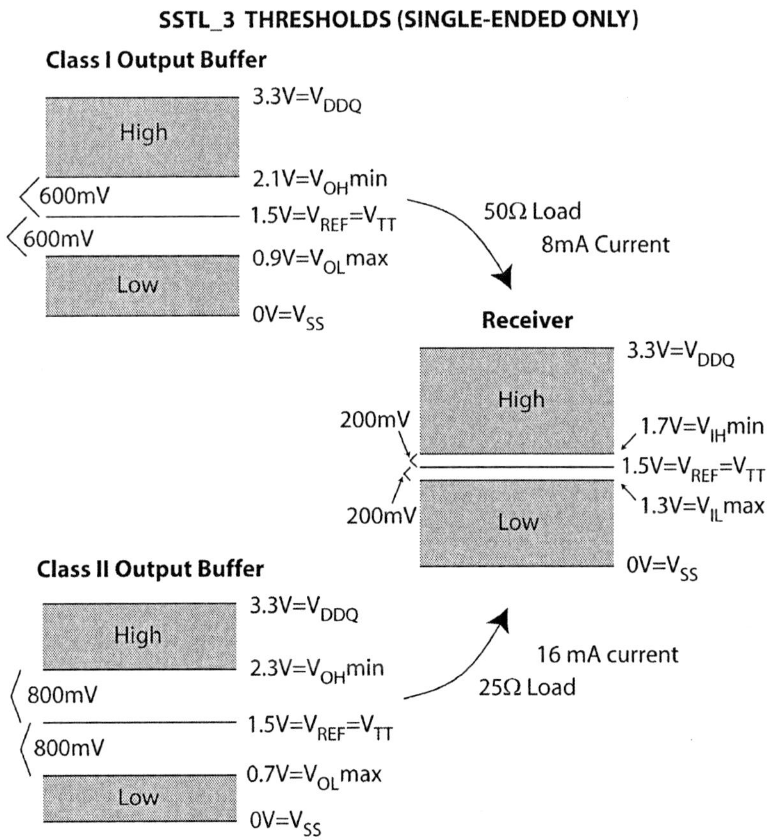

Figure 7.14 SSTL_3 thresholds under nominal conditions.

In other applications, the system designer may wish to terminate only at the signal source rather than at the end of the transmission line. One advantage of this approach is that there is no need for a V_{TT} power supply. This can again be implemented with either a Class I or II type buffer and SSTL_3 receiver. For a series termination only application, a Class II type buffer might be preferred since it comes closer, in conjunction with the series resistor, to matching the characteristic impedance of the transmission line, that is, the Class II type buffer has a lower 'on' resistance than the Class I type buffer, more closely matching the series combination of R_S and the characteristic impedance of the transmission line.

Figure 7.14 illustrates the thresholds for SSTL_3 drivers and receivers and in particular the increased current drive for Class II output buffers (drivers)." [Jes96]

7.2.2 SSTL_2

SSTL_2 is very similar to SSTL_3 but differs from it primarily in two ways. The most obvious difference is that all parameter voltage levels are shifted down to accommodate a nominal V_{DDQ} of 2.5 V and switching range of 0 to 2.5 V. The other difference is that SSTL_2 provides additional specifications for differential inputs and differential outputs. SSTL_2 devices can be either single-ended or differential whereas SSTL_3 devices are only single-ended.

7.2.2.1 SSTL_2 for Single-Ended Inputs and Outputs

"Where V_{REF} was 0.45 of V_{DDQ} for SSTL_3, V_{REF} is now 0.5 of V_{DDQ}, or nominally 1.25 V. As before, peak-to-peak ac noise on V_{REF} may not exceed ±2% of V_{REF}(dc). Minimum output buffer source and sink currents are very slightly different as well, being ±8.1 mA for Class I and ±16.2 mA for Class II drivers. Tables 7.11 and 7.12 show the ranges of some of these parameters for SSTL_2.

Table 7.11 SSTL_2 Device and Output Supply Voltages and Input Reference and Termination Voltages

Symbol	Parameter	Minimum	Nominal	Maximum	Units
V_{DD}	Device Supply Voltage	V_{DDQ}	—	n/a	V
V_{DDQ}	Output Supply Voltage	2.3	2.5	2.7	V
V_{REF}(dc)	Input Reference Voltage	1.13	1.25	1.38	V
V_{TT}	Termination Voltage	$V_{REF} - 0.04$	V_{REF}	$V_{REF} + 0.04$	V

[Reprinted with permission of JEDEC.]

From Table 7.12 it can be seen that the receiver input high and low logic thresholds are 150 mV above and below V_{REF}, respectively. The ac input specifications require that receivers must function properly when supplied with a test signal having a minimum slew rate of 1.0 V/ns and a maximum peak-to-peak swing of 1.5 V where the slew rate is measured at ±310 mV around V_{REF}.

Table 7.12 SSTL_2 dc and ac Input Logic Levels

Symbol	Parameter	Minimum	Maximum	Units
$V_{IH}(dc)$	dc Input Logic High	$V_{REF} + 0.15$	$V_{DDQ} + 0.3$	V
$V_{IL}(dc)$	dc Input Logic Low	–0.3	$V_{REF} - 0.15$	V
$V_{IH}(ac)$	ac Input Logic High	$V_{REF} + 0.31$		V
$V_{IL}(ac)$	ac Input Logic Low		$V_{REF} - 0.31$	V

[Reprinted with permission of JEDEC.]

For SSTL_2, the input voltage to the receiver V_{IN} must be at least 405 mV above or below V_{TT} as a result of V_{OUT} of the output buffer attaining its minimum high or maximum low value, respectively. As before, the input voltage provided to the receiver depends on the driver as well as on the termination voltage and termination resistors. SSTL_2 allows for V_{TT} to deviate from V_{REF} by 40 mV. At these maximum deviations, the requirement for the input voltage to the receiver V_{IN} becomes a minimum 365 mV above or below V_{TT} rather than the 405 mV. The minimum source and sink currents of –16.2 mA[9] and 16.2 mA, respectively, are measured under the worst-case conditions allowed by Table 7.11.

As for SSTL_3, SSTL_2 may be used without any termination at all in design applications where interconnections are short. For applications with longer transmission lines that will only be terminated at one end, such as address drivers on a memory board, either Class I or Class II drivers and an SSTL_2 receiver may be used with an appropriate termination resistor value. Some applications may be implemented with only series termination, which has the advantage of not needing a V_{TT} power source. In this case, a Class II buffer may be preferred since its lower output impedance comes closer, in conjunction with the series resistor, to matching the characteristic impedance of the transmission line. Finally, SSTL_2 may be used in applications where the bus is terminated at both ends. In these double termination cases, if drivers are connected directly onto the bus so that there are no stubs present, series resistors may be eliminated entirely. These applications can be implemented using a Class I or Class II driver and SSTL_2 receiver, although a Class II driver would dissipate more power due to its larger current drive and might require special cooling." [Jes02b]

9. Because of the negative current value for devices sourcing current, this is technically a maximum value.

7.2.2.2 SSTL_2 for Differential Inputs and Outputs

"Differential specifications were added to SSTL_2 as a result of the desire to accommodate differential signals in SSTL_2-based systems. This is particularly relevant to DDR DRAM clock signals, but the specification is intended to cover all differential signals in the SSTL_2 category. SSTL_2 has only Class I for differential output buffers, providing the same minimum ±8.1 mA source and sink current capability as for SSTL_2 Class I single-ended output buffers. Each of the two complementary buffer outputs must be capable of sourcing or sinking this amount of current.

Table 7.13 indicates that the minimum differential voltage magnitude must be at least 300 mV in order for the receiver to receive it properly. It also indicates that the cross point, the voltage level where the true and complementary signals cross each other as they change state, can vary by as much as 200 mV. An illustration of the cross point voltage V_X and differential voltage swing V_{Swing} are shown in Figure 7.15. In the figure, the abbreviation 'TR' represents the true buffer output and 'CP' represents the complementary buffer output.

Table 7.13 SSTL_2 Differential dc and ac Input Logic Levels

Symbol	Parameter	Minimum	Maximum	Units
V_{IH}(dc)	dc Input Signal Voltage	–0.3	V_{DDQ} + 0.3	V
V_{Swing}(dc)	dc Differential Input Voltage	0.3	V_{DDQ} + 0.6	V
V_{Swing}(ac)	ac Differential Input Voltage	0.62	V_{DDQ} + 0.6	V
V_x(ac)	ac Differential Crosspoint Voltage	(0.5 × V_{DDQ}) – 200 mV	(0.5 × V_{DDQ}) + 200 mV	V

[Reprinted with permission of JEDEC.]

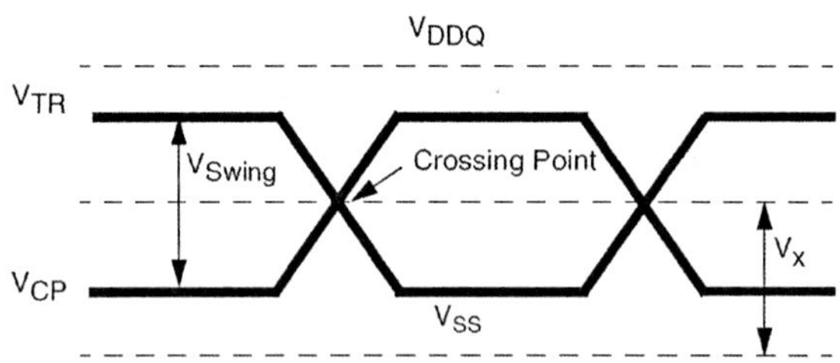

Figure 7.15 Illustration of the crossing point and voltage swing for differential SSTL_2. [Reprinted with permission of JEDEC.]

Figure 7.16 provides an example of SSTL_2 Class I differential signaling using one single termination and one series resistor for each differential line. Each true and complementary signal is separately tied to V_{TT}. The values of I_{OH}(dc) and I_{OL}(dc) have to abide by the Class I specification. Figure 7.17 provides a similar example but using one direct termination resistor and two series resistors. Note that there is no connection to V_{TT} in this configuration and that the source and sink currents are the same current. Note that the termination resistor is 100 Ω rather than 50 Ω." [Jes02b]

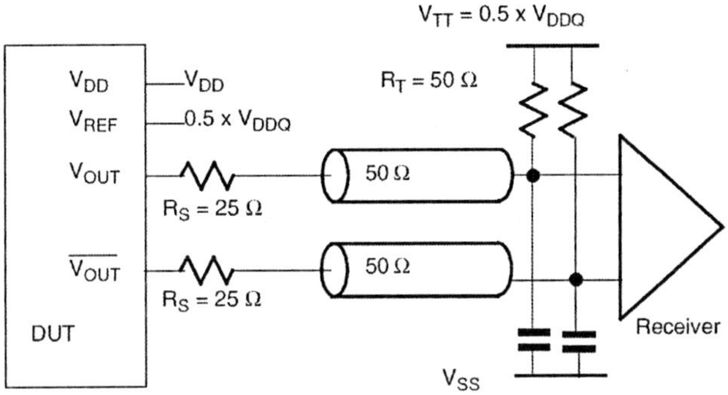

Figure 7.16 Example of SSTL_2 Class I differential signaling using a single termination and a series resistor on each differential line. [Reprinted with permission of JEDEC.]

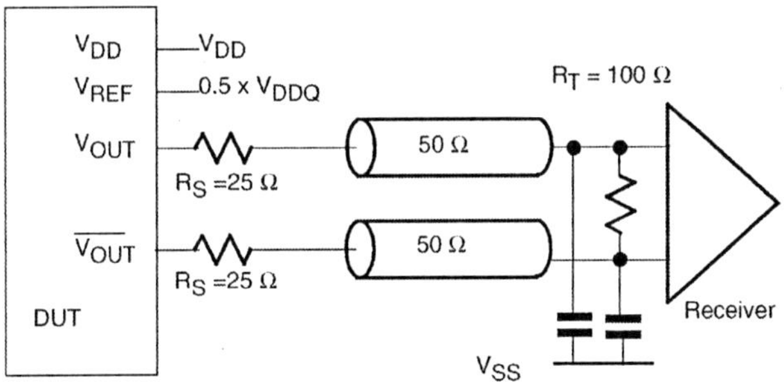

Figure 7.17 Example of SSTL_2 Class I differential signaling using direct termination resistor and two series resistors. [Reprinted with permission of JEDEC.]

7.2.2.3 Illustration of SSTL_2 Thresholds

Figure 7.18 illustrates both driver and receiver SSTL_2 thresholds for Class I and Class II single-ended signaling and for Class I differential signaling.

7.2.2.4 Comparison of SSTL_2 with LVTTL[10]

"Although SSTL_2 is applicable for many different applications, it is particularly optimized for the main memory environment, which has long stubs off of the motherboard bus due to the DIMM routing traces. Figure 7.19 shows simplified driver and receiver topologies used in both LVTTL and SSTL_2. The output buffers for LVTTL and SSTL_2 are very similar and only differ in transistor size and supply voltage, V_{DDQ}. The reduced signal voltage levels of SSTL_2 allow the use of a 2.5 V V_{DDQ}, which is compatible with the migration to lower DRAM supply voltages. It is possible to design a programmable output buffer that can be LVTTL or SSTL_2 compatible, as determined by the level applied to V_{DDQ}.

10. Reprinted with permission of Fairchild Semiconductor Corporation.

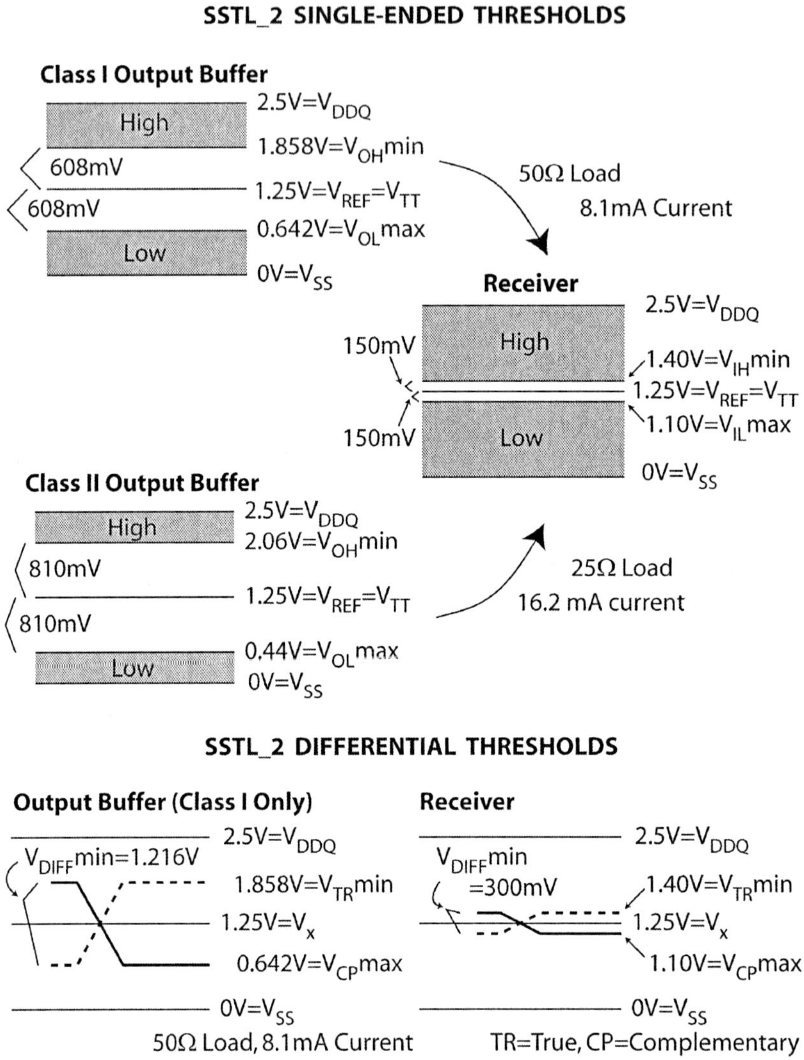

Figure 7.18 SSTL_2 single-ended and differential thresholds for nominal conditions.

The LVTTL input receiver is typically an N-channel/P-channel stacked gate, as shown in Figure 7.19. The advantages of this topology are its simplicity and its very low power dissipation while receiving a steady state logic low or logic high signal. A fundamental disadvantage of the stacked receiver is that the threshold voltage matching of the top and bottom transistors are very poor over voltage, temperature, and process variation. The unpredictability of the switching threshold of this receiver, along with its relatively poor voltage gain, requires a larger input signal swing for reliable switching.

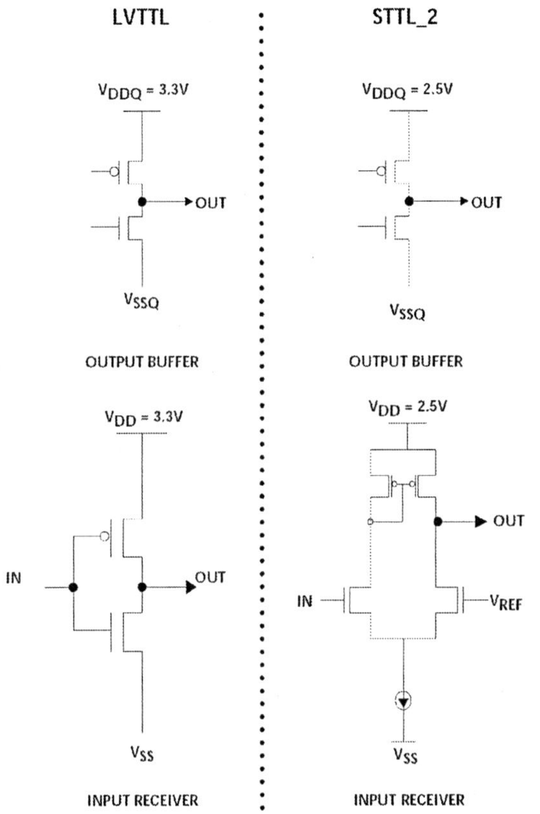

Figure 7.19 Comparison of LVTTL and SSTL_2 drivers and receivers. [Reprinted with permission of Fairchild Semiconductor Corporation.]

The SSTL_2 input receiver is typically a differential pair common source amplifier. This receiver provides better gain and bandwidth, and the variation in threshold voltage is much tighter, since the threshold voltage offset is determined by identical size and technology transistors in a differential pair configuration. The result is that smaller input signal swings can be used reliably. Many variations and enhancements to this input receiver topology are in use today.

Both LVTTL and SSTL input receivers can be turned off while the device is idle to minimize power dissipation, via disabling transistors not shown in the figure. However, the SSTL_2 input receiver will draw dc current while receiving steady state logic low or logic high inputs. To minimize system power dissipation, the DDR SDRAM receivers are turned off when not receiving active inputs, as controlled by the device protocol. With the exception of the CLOCK signal, all inputs are single-ended, and therefore, the second input to the receiver is the reference voltage, V_{REF}. It is important that V_{REF} stay symmetrically positioned between the logic high and logic low input levels over variations in process, voltage, and temperature so timing uncertainty will be minimized.

V_{REF} is defined as 50% of V_{DDQ}, as V_{DDQ} can vary from 2.3 V to 2.7 V. In configurations, V_{REF} will be generated with 1% accuracy. The termination voltage, V_{TT}, is defined as being within 40 mV of V_{REF}. The goal for the V_{REF} and V_{TT} circuits is to generate the V_{REF} and V_{TT} voltages that can track the midpoint of $V_{DDQ} - V_{SSQ}$ over environmental variations, and to be symmetric with respect to V_{OH} and V_{OL}. The V_{REF} and V_{TT} voltages must also track each other. These requirements can easily be met with a low-cost implementation using a switching regulator.

A minimum termination resistance of 23 Ω to V_{TT} can be used and still comply with the minimum output voltages and output currents of the SSTL_2 specification.

Two choices for implementing the parallel termination are shown in Figure 7.20. In Figure 7.20(a), the bus is terminated at both ends with a 50 Ω resistor, for a combined parallel resistance of 25 Ω. In Figure 7.20(b), the bus is terminated at the far end from the controller with a single 25 Ω resistor. It is strongly recommended that the single resistor termination scheme in Figure 7.20(b) be used for best performance. The benefits of this approach include reduced cost, simpler signal routing, reduced reflections, and better signal bandwidth and settling.

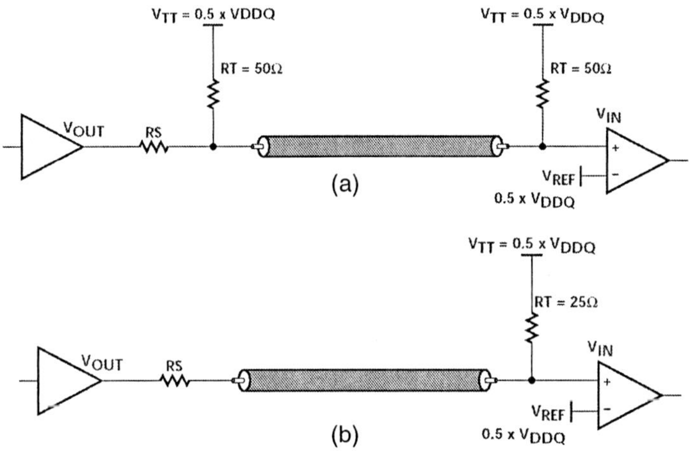

Figure 7.20 SSTL_2 terminations: (a) double-terminated output, and (b) single-terminated output. [Reprinted with permission of Fairchild Semiconductor Corporation.]

The loaded characteristic impedance of this bus environment is on the order of 28 Ω, when the effects of the distributed capacitance of the DIMM modules and components are considered for a fully populated system. So, the 25 Ω resistor will provide a good match for the loaded impedance (considering board and device manufacturer tolerances), and the reflections will be terminated within a single round trip delay on the bus. In the dual termination approach, both ends of the bus are improperly terminated, and therefore, several round trip delays are required before the amplitude of reflections is diminished. The single termination scheme will reduce reflections on the bus, which will provide faster signal rise and fall times, and it will reduce the signal settling time, which will in turn reduce timing jitter due to intersymbol interference (ISI). Finally, the single termination resistor should be placed at the end of the bus farthest from the controller. This placement optimizes the signal integrity for signals transmitted from the controller to the DRAMs. This placement approach takes advantage of the fact that the signals transmitted in this direction must be reliably received by all DRAMs, while signals transmitted from the DRAMs only need to be received in one location — the controller. There are also more signals transmitted in this direction (command/address are unidirectional signals, while the data signals are bidirectional).

Series resistors are incorporated in the SSTL_2 signaling topology for main memory applications (refer to Figure 7.21). 22 Ω stub resistors are included on the data lines of the DDR DIMMs near the connector contacts to improve the signal integrity of the system. The stub resistors are very effective in dissipating the energy of the reflected waves travelling up and down the 30 mm module traces, and they also help to isolate the stubs from the main memory bus. These resistors improve the signal settling times and transition times. A similar approach is used on PC100 DIMMs with 10 Ω resistors on the data lines. The SSTL_2 stub resistors also provide a voltage divider with the parallel termination resistor to reduce the signal voltage swings on the bus. The reduced voltage swings provide headroom to V_{DDQ} and V_{SSQ} for signal overshoot, and therefore, maximum device input voltages are not threatened with this termination scheme.

The SSTL_2 series resistors also minimize on-chip I/O power dissipation, since some of the I/O power is moved off-chip and dissipated in the stub resistor. The nominal I/O power dissipation for this high-speed signaling scheme is about 19 mW/pin, and only 7.5 mW is dissipated on the device. This low I/O power allows DDR SDRAMs to be used without requiring airflow or heat spreaders to cool the DRAMs." [Fai00]

7.2.2.5 SSTL_2 Design Example — DDR SDRAM Memory Subsystem[11]

"Figure 7.21 shows the block diagram of a double data rate (DDR) SDRAM main memory subsystem. There are variations to this diagram, based upon the number of modules supported by the system, whether non-buffered and registered modules are both supported by the system, and the designer's choice of clock topologies. In some applications, only one command/address bus is required. The controller chip is the PC north bridge, and it provides communication and control to the PCI adapter bus, the Accelerated Graphics Port (AGP) bus, the microprocessor host bus, and to the main memory.

Command, address, and data signals are routed to the dual inline memory modules (DIMM). The data signals include series resistors on the DIMMs and on the motherboard near the first DIMM socket, and they are parallel terminated to V_{TT} after the last socket. The data signals are double data rate, but the command/address signals only transition on positive clock edges at a single data rate. The DDR SDRAM input specifications allow the command and address signals to be received as either SSTL_2 or 2.5 V LVCMOS signals. The use of termination resistors on the command/address signals is not necessary when LVCMOS drivers are used; however, the signal integrity and timing margins are improved by using SSTL_2 signaling.

Double data rate (DDR) SDRAM is defined by JEDEC standard JESD79.[12] Enhancements to the SDRAM definition were made to increase the device data rate. These new features included transmitting data on both positive and negative edges of the clock, reducing device input capacitance, adding on-chip delay locked loops (DLLs) to reduce access time uncertainty, adding data strobes to improve data capture reliability, and incorporating SSTL_2 signaling techniques.

11. This section reprinted with permission of Fairchild Semiconductor Corporation.
12. JESD79-2, Double Data Rate (DDR) SDRAM Specification, Release 2, May 2002.

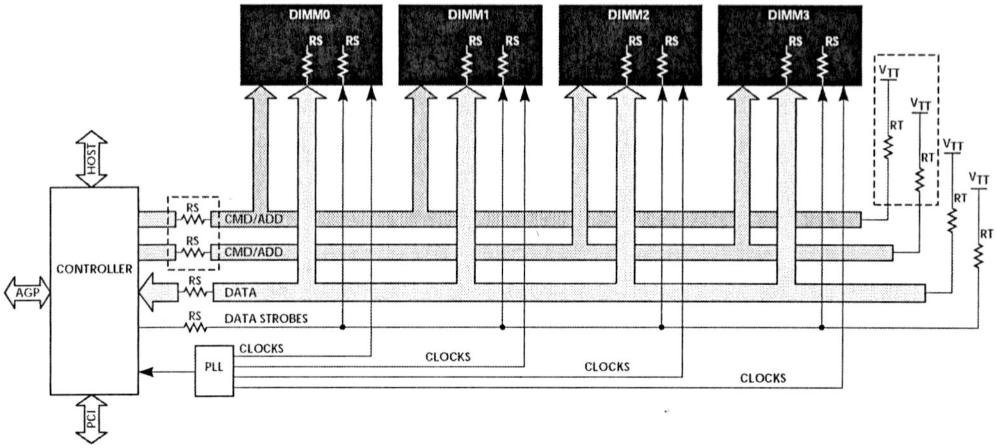

Figure 7.21 DDR SDRAM PC memory subsystem. [Reprinted with permission of Fairchild Semiconductor Corporation.]

The DDR SDRAM motherboard design is very similar to SDRAM motherboard designs with the exception of the SSTL_2 signaling. The use of high-speed, low voltage signaling, such as SSTL_2, requires proper termination voltage and reference voltage design. Some attention must be paid to the generation and location of the termination voltage (V_{TT}) and reference voltage (V_{REF}) circuits, and to the placement and routing to the series and parallel termination resistors." [Fai00]

7.2.3 SSTL_18[13]

"In the continuing challenge to increase speeds and decrease power consumption, SSTL has advanced further with the introduction of the SSTL_18 specification designed for a switching range of 0 to 1.8 V. While SSTL_18 is very similar to the other higher voltage SSTL devices, it does have some distinct differences that stand out. Where SSTL_2 and SSTL_3 output 8.1 mA and 8 mA, respectively, for Class I loads, and 16.2 mA and 16 mA for Class II loads, respectively, SSTL_18 as specified in JESD8-15 makes no mention of classes and specifies only a single current magnitude of 13.4 mA into a 25 Ω load (see Figure 7.22). Similar to SSTL_2, SSTL_18 covers both single-ended and differential devices. As you would anticipate, V_{REF} and V_{TT} are $0.5 \times V_{DDQ}$, or 0.9 V in the nominal SSTL_18 case (see Table 7.14).

Where SSTL_3 switched at ±200 mV and SSTL_2 switched at ±150 mV around their V_{REF} values of 1.5 V and 1.25 V, respectively, SSTL_18 switches at ±125 mV with respect to its V_{REF} of 0.9 V — for single-ended operations. For differential operations, SSTL_18 requires a minimum differential voltage of 250 mV in order to switch. As before, the standard defines both

13. This section reprinted with permission of JEDEC.

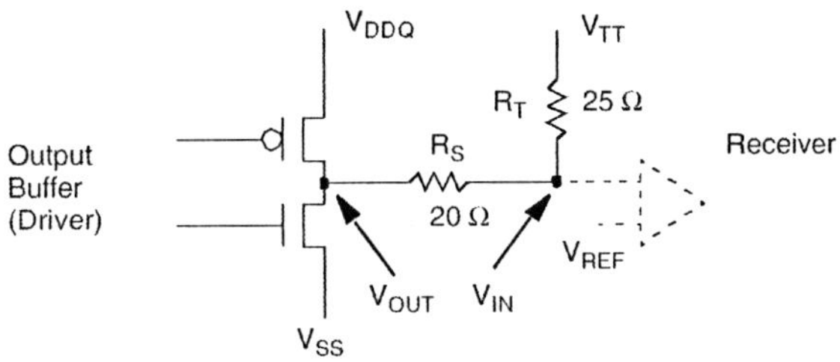

Figure 7.22 Typical output buffer (driver) environment for SSTL_18. [Reprinted with permission of JEDEC.]

Table 7.14 SSTL_18 Driver Output Supply, Reference, and Termination Voltages

Symbol	Parameter	Minimum	Nominal	Maximum	Units
V_{DDQ}	Output Supply Voltage	1.7	1.8	1.9	V
V_{REF}(dc)	Input Reference Voltage	833	900	969	mV
V_{TT}	Termination Voltage	$V_{REF} - 40$	V_{REF}	$V_{REF} + 40$	mV

[Reprinted with permission of JEDEC.]

ac and dc input signal values, which are listed in Table 7.15 and illustrated in Figure 7.23. The specification of ac and dc input signal values is intended to provide predictable receiver timing in the presence of input waveform 'ringing.' An example of ringing is illustrated by the waveform. In the figure, t_S is the minimum time interval within which a valid signal must meet V_{IH}(ac) or V_{IL}(ac) after a transition to supply enough overdrive to the receiver to ensure it switches within its timing specification. t_S is device specific, and would typically be about 15% of the minimum bit time of the system for which the SSTL_18 device is designed.[14]

Table 7.15 SSTL_18 dc and ac Input Logic Levels (Single-Ended)

Symbol	Parameter	Minimum	Maximum	Units
V_{IH}(dc)	dc Input Logic High	$V_{REF} + 125$	$V_{DDQ} + 300$	mV
V_{IL}(dc)	dc Input Logic Low	–300	$V_{REF} - 125$	mV
V_{IH}(ac)	ac Input Logic High	$V_{REF} + 250$		mV
V_{IL}(ac)	ac Input Logic Low		$V_{REF} - 250$	mV

[Reprinted with permission of JEDEC.]

14. t_S values for DDR-II (double data rate II) are provided in the JEDEC JESD90 specification. JESD90 refers to 1.8 V CMOS but not SSTL_18.

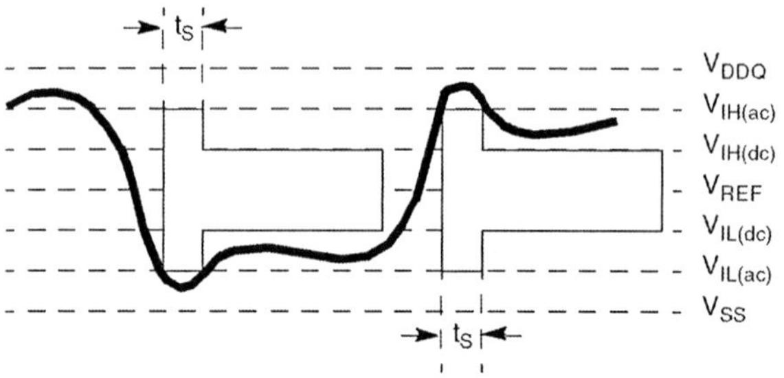

Figure 7.23 SSTL_18 input voltage levels. [Reprinted with permission of JEDEC.]

SSTL_18 has two other distinct differences from the other SSTL families. First, the external series resistor R_S is now specified at 20 Ω for the nominal case, as shown in Figure 7.22, rather than 25 Ω. Secondly, in contrast to the other families, output buffers have a maximum output resistance 'R_{ON}' of 21 Ω at the minimum V_{DDQ}. SSTL_2 and SSTL_3 were specified differently by requiring that V_{IN} be a specified value (e.g., 400 mV, 405 mV) above or below V_{TT} when V_{OUT} attained its maximum low or its minimum high value. As before, peak-to-peak noise on V_{REF} may not exceed ±2% of V_{REF}(dc).

Differential input switching levels are specified in Table 7.16. Differential signaling in SSTL_18 based systems is particularly relevant to DDRII DRAM clocks and data strobes, but is intended for broad use in many other applications as well.

Table 7.16 SSTL_18 Differential dc and ac Input Logic Levels

Symbol	Parameter	Minimum	Maximum	Units
V_{IN}(dc)	dc Input Signal Voltage	–0.3	V_{DDQ} + 0.3	V
V_{ID}(dc)	dc Differential Input Voltage	0.25	V_{DDQ} + 0.6	V
V_{ID}(ac)	ac Differential Input Voltage	0.5	V_{DDQ} + 0.6	V
V_{IX}(ac)	ac Differential Crosspoint Voltage	(0.5 × V_{DDQ}) – 0.175	(0.5 × V_{DDQ}) + 0.175	V

[Reprinted with permission of JEDEC.]

Differential outputs may be created using either true differential drivers or by combining pairs of SSTL_18-compliant, single-ended drivers driven from true and complementary signals. As a result, differential output parameters are assumed to be identical to those for single-ended outputs, except for the ac differential cross point voltage, which can vary ±125 mV from $V_{DDQ}/2$.

Figure 7.24 summarizes SSTL_18 single-ended and differential thresholds for nominal conditions. While JESD8-15 does not specify different classes within the standard, some vendors do specify different classes for SSTL_18 based on their own internal proprietary standards. For example, Xilinx specifies a Class I and Class II for SSTL_18 where Class II is JESD8-15, the standard as discussed above, but Class I is a proprietary class that provides a sink and source current magnitude of 6.7 mA instead of 13.4 mA." [Jes02a]

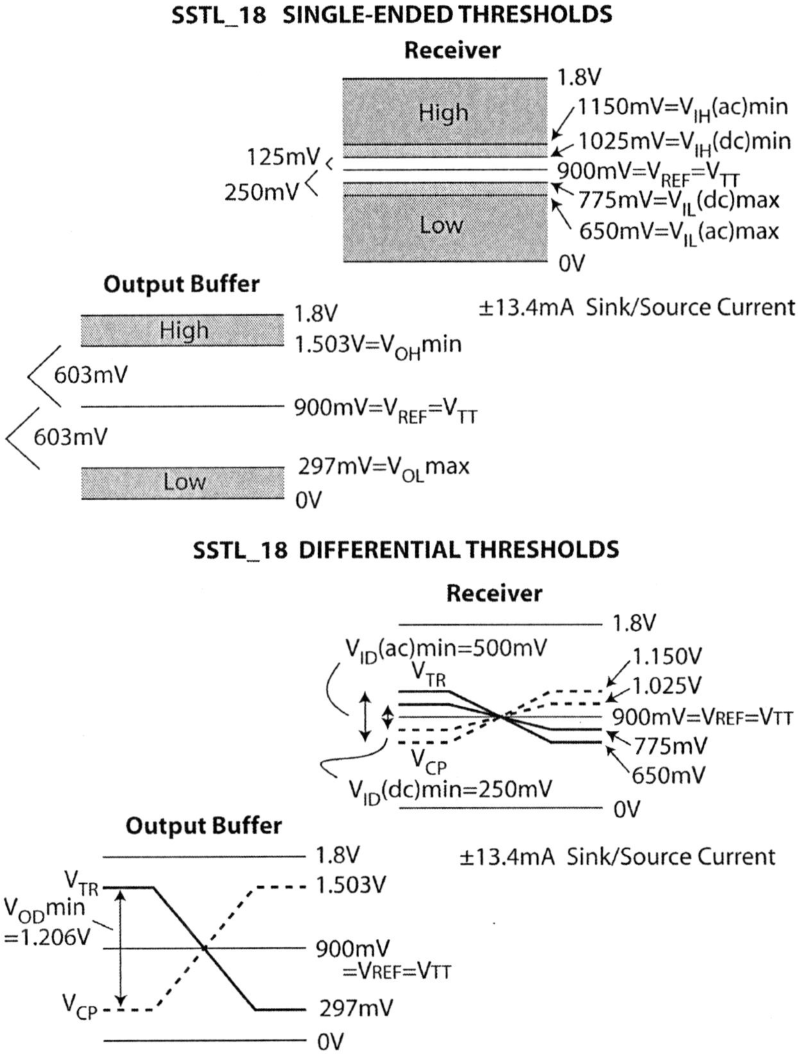

Figure 7.24 SSTL_18 single-ended and differential thresholds for nominal conditions.

7.2.4 Summary of SSTL

Table 7.17 Stub-Series Terminated Logic (SSTL) Summary

Stub-Series Terminated Logic (SSTL) Summary

Applications:

Interconnects of memory controllers to main memory chips
SSTL_18 is used for memory interconnect to DDR2 SDRAM
SSTL_2 is used for memory interconnect to DDR SDRAM
Main memory systems, parallel inputs to and parallel outputs from SerDes devices
Interfacing to DIMMs, SIMMs
Interfacing to DDR SDRAM
Interfacing to multiple loads connected by long stubs to the main bus

Where Not to Use:

Where noise on V_{REF} exceeds ±2% of dc value of V_{REF}
Where more than 8 mA of drive current is needed in differential mode
Where more than 16 mA of drive current is needed in single-ended mode
Where loads are less than 25Ω

Device Supply Voltage: Any
Process: Usually CMOS
Termination/Loading: 50 Ω for Class I drivers, 25 Ω for Class II drivers
Hot-Pluggable: No, unless specified in data sheet
Maximum Drive Distance: 10 in (25.4 cm) or more; keep as short as possible; function of edge rates and ability to control trace impedances
Maximum Noise On V_{REF}: ±2% of the dc value of V_{REF}

SSTL_3 (Single-Ended Only, 2 Classes of Drivers)

Standard: JESD8-8
Source/Sink Current: 8 mA for Class I drivers (50 Ω load), 16 mA for Class II drivers (25 Ω load)
Output Buffer Supply Voltage: 3.3 V nominal, cannot exceed device supply voltage
Additional Voltage Requirements: 1.5 V for V_{TT}, 1.5 V for V_{REF} (0.45 × V_{DDQ})
Receiver Switching Threshold Around V_{REF}: ±200 mV (dc)
Maximum Data Rate: 400 Mbps per output. Standard specifies no maximum rate

SSTL_2 (Single-Ended Class I and II Drivers, Differential Class I Drivers)

Standard: JESD8-9B
Source/Sink Current: 8.1 mA for Class I drivers (50 Ω load), 16.2 mA for Class II drivers (25 Ω load). Same current specifications apply to single-ended and differential currents
Output Buffer Supply Voltage: 2.5 V nominal, cannot exceed device supply voltage
Additional Voltage Requirements: 1.25 V for V_{TT}, 1.25 V for V_{REF}
Receiver Minimum Single-Ended Switching Thresholds Centered on V_{REF}: ±150 mV (dc)
Receiver Minimum Differential Switching Voltage: ±300 mV (dc)
Maximum Data Rate: 400 Mbps per output. Standard specifies no maximum rate

Table 7.17 Stub-Series Terminated Logic (SSTL) Summary (Continued)

SSTL_18 (Single-Ended and Differential, No Classes)
Standard: JESD8-15
Source/Sink Current: 13.4 mA into 25 Ω load (No classes. Only one current level specified. Same current specifications apply to single-ended and differential currents)
Output Buffer Supply Voltage: 1.8 V nominal, cannot exceed device supply voltage
Additional Voltage Requirements: 0.9 V for V_{TT}, 0.9 V for V_{REF}
Receiver Minimum Single-Ended Switching Thresholds Centered on V_{REF}: ±125 mV (dc)
Receiver Minimum Differential Switching Voltage: ±250 mV (dc)
Maximum Data Rate: 400 Mbps per output. Standard specifies no maximum rate.
Maximum Noise Rejection: 41 mV single-ended noise

7.3 Exercises

7-1. Compare the output buffer supply voltage levels (V_{DDQ}) and the receiver input reference voltage levels for HSTL, SSTL_3, SSTL_2, and SSTL_18. Which of these four families has the lowest reference voltage and output buffer supply voltage? Which one has the highest?

7-2. Which among the HSTL, SSTL_3, SSTL_2, and SSTL_18 families come in both differential and single-ended devices? In what applications would you prefer to use single-ended devices and in what applications would you prefer differential devices? State your reason as to why.

7-3. List the output buffer current drive for HSTL and all three of the SSTL families. List the current drive for each class of output buffer as well where there's more than one. Which family of devices seems to have a current drive capability that is the most different from the others? Which two classes of what family sink more current than any of the others?

7-4. Refer to Figure 7.13 on page 147, the example of SSTL_3 Class II, the symmetrically double parallel terminated output load with a series resistor. For what applications would a termination at each end of the transmission line be desired?

7-5. Refer to Figure 7.17 on page 151. Why do you suppose the termination resistor in this differential application is 100 Ω rather than the usual 50 Ω value used in single-ended terminations?

7-6. Why are both ac and dc voltage thresholds specified for HSTL and SSTL devices? What's the significance of the dc switching level? What's the significance of the ac switching level?

7-7. What are the major differences on how receiver switching threshold levels are determined when comparing HSTL and SSTL devices with typical CMOS and TTL devices?

7-8. What is unique about HSTL and SSTL devices that make them well suited for driving transmission lines/backplanes that have a lot of long stubs going to memory chips?

7-9. Among SSTL_3, SSTL_2, and SSTL_18 output buffers, which one would be designed to have the smallest "on" resistances in the output stage?

7-10. Log onto the Internet, access some vendor Web sites, and identify three devices that have HSTL interfaces and three more devices that have SSTL_18 interfaces. List the vendor and part number for each device and state the maximum data rate that the device can handle. Print out the one page of each device's data sheet that shows this data rate and turn that in to your professor with your list of six devices. The goal of this exercise is to give you a feel for the speeds of devices currently on the market. If you have trouble finding the above devices, substitute SSTL_2 devices for some of the six.

7-11. Log onto the Internet, access some SSTL and HSTL chip vendor Web sites, and locate an application note or white paper that provides technical information about using SSTL or HSTL devices in a design. Pick out two design rules or guidelines you feel are important and discuss each briefly in a single paragraph of 300 words or less (two paragraphs total). Explain the type of design they apply to and why they are important to consider when doing a design.

7-12. Given that HSTL and SSTL_18 have driver supply voltages and reference voltages that are both very small and differ by no more than 300 mV from HSTL to SSTL_18, for what applications might you prefer to use HSTL devices and for what applications might you prefer to use SSTL_18 devices?

7-13. Show how current and voltage vary for both Class I and Class II SSTL_3 output buffers and at the receiver input as a function of the output buffer voltage V_{DDQ}. Make calculations for three different values of V_{DDQ}, 3.0, 3.3, and 3.6 V. Refer to Figure 7.11 on page 145 but use correct values in your calculations for V_{TT}, V_{DDQ}, and V_{REF} as appropriate. Do this by reproducing the following table, filling in the missing values, and answering the question, "Where do the most stringent requirements for the output driver transistors occur?"

Conditions	**Units**	**Class I**	**Class I**	**Class I**	**Class II**	**Class II**	**Class II**
V_{DDQ}	V	3.0	3.3	3.6	3.0	3.3	3.6
V_{TT}	V						
V_{REF}	V						
Termination Resistor R_T	Ω	50	50	50	25	25	25
Series Resistor R_S	Ω	25	25	25	25	25	25
Delta V_{IN}	V						

Conditions	Units	Class I	Class I	Class I	Class II	Class II	Class II
Output High Driver							
Voltage at V_{IN}	V						
Voltage at V_{OUT}	V						
Pull-up Source-Drain Voltage	V						
Output Current	A	–0.008					
On Resistance	Ω			162.50			
Output Low Driver							
Voltage at V_{IN}	V						
Voltage at V_{OUT}	V						
Output Current	A						0.016
On Resistance	Ω				31.25		

7-14. Log onto the Internet, access one or more chip vendor Web site(s), and identify an HSTL driver (e.g., memory controller chip) and compatible memory chip. Draw a detailed schematic showing all connections between the two chips and include at least one 8-bit bus. Any required additional devices such as PLLs, clock circuits, and so forth need not be shown in the schematic, only the connections to them. Print out and turn in to your professor the four or five key pages of the electrical specifications from each of the two chips data sheets when you submit your schematic. Your design should be totally complete so that when dropped into an encompassing design, you would be ready to go to design simulation and board layout. How fast can your circuit transmit data in Gbps? Make this exercise as easy as possible by selecting the simplest and easiest devices you can find to complete this assignment.

7-15. Refer to Figure 7.2 on page 132. Photocopy the diagram or draw it again and specify actual voltage levels (in volts) for the 10 parameters shown in the figure. State any assumptions you make, such as the values of V_{DDQ} and V_{REF} you are using.

CHAPTER 8

Emitter Coupled Logic (ECL, PECL, LVPECL, ECLinPS Lite and Plus, SiGe, ECL Pro, GigaPro and GigaComm)

Chapter Objectives

- Discuss the basic operation of an ECL family device.
- Present the various ECL families and clarify the meaning of the multitude of family acronyms associated with ECL.
- Discuss the characteristics of each ECL family member and component nomenclature.
- Discuss the 10K and 100K standards and their output signaling levels.
- Identify the five kinds of ECL family outputs.
- Discuss characteristics, advantages, and disadvantages of single-ended and differential ECL interfaces.
- Present the difference between Reduced Swing ECL and Low Voltage ECL.
- Analyze how CML and ECL differ.

8.1 A Fast Technology — Edge Rates of 20 ps at 12 Gbps!

Today's ECL (emitter coupled logic) devices are fast! The latest families of SiGe ECL devices have edge rates as fast as 20 ps and can operate at data rates of 12 Gbps per differential trace. They can be used for clock and data distribution, backplane transmission, multiplexing, translation (voltage level-shifting and interfacing to other logic families), and state machines. They are used in system applications that include high-speed test equipment, optical networking equipment, ultra high-speed terabit routers, network attached storage devices, OC192 SONET, 10-Gigabit Ethernet, enterprise computing servers, and high-performance workstations. [Ons03a]

ECL devices offer shorter propagation delays and higher toggle rates than any non-ECL type of logic.

"Emitter coupled logic[1] is a nonsaturating form of digital logic that eliminates transistor storage time as a speed-limiting characteristic, permitting very high-speed operation. 'Emitter coupled' refers to the manner in which the emitters of the input differential amplifier within the integrated circuit are connected. The differential amplifier provides high impedance inputs and voltage gain within the circuit. Emitter follower outputs restore the logic levels and provide low output impedance for good line driving and high fanout." [Ons88]

The acronym "ECL" has been used in the past with two different meanings. One meaning refers to the whole range of devices using ECL-related technologies and the other meaning has referred instead to only the families of ECL devices that operate from a –4.2 V to –5.7 V V_{EE} power supply (V_{CC} = 0 V). The convention in use today by ON Semiconductor is for ECL to refer to all emitter-coupled bipolar technologies in general and for NECL (negative emitter coupled logic) to refer to the device families operating from –4.2 V to –5.7 V V_{EE} power supplies.

This book adheres to a slightly different convention since some device vendors still use "ECL" rather than "NECL" to refer to negatively powered devices. In this textbook, the terms "ECL families" or "ECL family" are used to refer to the broad range of emitter-coupled technologies including ECL, NECL, PECL, RSECL, LVPECL, ECLinPS, ECLinPS Lite™, ECLinPS Plus™, CML, ECL Pro™, ECLinPS Pro, Super-300K™, and SiGe devices — plus others. We use "ECL" to refer only to those device families powered from a –4.2 V to –5.7 V V_{EE} power supply. We also use the term "PECL" (positive emitter coupled logic, pronounced "peckle") to refer to devices which are powered from "positive" V_{CC} voltages (V_{EE} = 0 V). Similarly, we use "NECL" (negative emitter coupled logic) to refer to devices which are powered from "negative" V_{EE} voltages (V_{CC} = 0 V). PECL and NECL are really descriptive terms indicating how devices in the ECL families are powered rather than being names of actual family members. If you're not familiar with all of the families of devices listed just above, we discuss their characteristics in detail later in this chapter.

When discussing V_{EE} and V_{CC} and positive and negative power supplies, a convenient term used frequently in the ECL design community is the "power span" or "span." This is always a positive value and refers to the voltage difference, V_{CC}–V_{EE}, e.g., "3.3 V power span."

CML (current-mode logic) is a member of the ECL families and is discussed briefly in this chapter and again in more detail in Chapter 9 starting on page 199. While a member of the ECL families, it tends to be referred to and marketed separately by chip manufacturers. CML is commonly used as the high-speed output stage of SerDes (serializer/deserializer) devices. Twenty years ago, ECL and CML were synonyms for the same technology (ECL), but today's meaning for CML is now different. Today's CML is powered from positive power supply voltages and

1. In this chapter, terms such as "emitter coupled logic," "Low Voltage ECLinPS," and "Reduced Swing ECL" are not hyphenated as proper punctuation might require. The reason is to follow vendor technical publication conventions that don't use a hyphen, possibly with the intent of keeping headings and titles uncluttered.

has a differential output voltage swing magnitude $|V_{OD}|$ of 800 mV, which is much closer to the V_{CC} power rail than the other ECL families; thus, CML uses different signaling voltage levels from other ECL family members. Differential output voltage swing magnitudes of 400 mV are also common for CML. Another lesser-used name for current-mode logic being used by some chip manufacturers today is "common-mode logic," which has the same acronym, CML.

Devices from the ECL families are far less integrated when compared with most other families of integrated circuits. Today's increasingly larger and powerful computer chips, FPGAs, and SOC[2] devices can have millions of gates. By contrast, high-speed ECL family devices have a much lower level of integration and may contain only a few gates.

"Commercial ECL families aren't nearly as popular as CMOS and LVTTL, mainly because they consume much more power. ECL families have a poor speed-power product, do not provide a high level of integration, have fast edge rates requiring designing for transmission-line effects in most applications, and are not directly compatible with TTL and CMOS. Nevertheless, ECL families still find their place as logic and interface technologies in very high-speed communications gear, including fiber-optic transceiver interfaces for Gigabit Ethernet and Asynchronous Transfer Mode (ATM) networks." [Wak01] ECL families are usually implemented with bipolar, BiCMOS, or SiGe process technologies. Many ECL family devices are used in military applications due to their ability to operate over military temperature ranges. They are also used primarily in point-to-point architectures but can be used in multidrop applications.

8.1.1 The ECL Families

Table 8.1 lists the various members of the ECL families. As can be seen from the table, there are a large number ECL family members available today. Coming to grips with the characteristics of each member can be time consuming and challenging if not frustrating when you're under project deadline pressures.

The table is broken out into three categories: (1) families used for new design, (2) ECL descriptive or general terms, and (3) end-of-life families. Each of the devices listed is described in detail in Section 8.6,"The ECL Families and Their Characteristics," beginning on page 184.

"Some ECL family devices are designed for clock distribution and provide fan-out drive capabilities ranging from 2 to 24. They can interface to any number of high-speed signaling standards such as CML, HSTL, LVDS, PECL, and LVPECL. Other ECL family devices perform signal translation allowing a clean differential PECL clock to be transmitted across a backplane and be converted to CMOS/TTL levels for local distribution on target boards." [Ons02a]

8.1.2 ECL Vendor Products

Table 8.2 lists a sampling of ECL family products from various vendors. Product lines change quickly over time and the table is intended to provide only a starting point for gathering device specifications from the various vendors. Arizona Microtek offers components that are pin and

2. SOC is system-on-a-chip.

Table 8.1 The ECL Families

Families used for new design:		
SuperLite™ (CML)	GigaComm™ (SiGe LVECL)	GigaPro™ (SiGe CML)
Super-300K ECL	300 Series ECL	9300 and 9400 Series
Low Voltage ECLinPS Plus (10K/100K)	Low Voltage ECLinPS Lite (10K/100K)	Low Voltage ECLinPS (10K/100K)
ECLinPS Plus (10K/100K) [Same as ECL Pro, ECLinPS Pro]	ECLinPS Lite (10K/100K) [Same as ECL Lite]	ECLinPS (10K/100K)
ECL (100K)	100H ECL	10H ECL (MECL 10KH)
ECL descriptive or general terms:		
PECL	Pseudo ECL	LVPECL
RSECL	RSPECL	RSNECL
LVNECL	NECL	LVECL
End of life families not recommended for new design:		
All MECL 10K[a]		

a. Families adhering to the 10K standard are not end-of-life (EOL).

Table 8.2 A Product Line Sampling of ECL Family Vendors

Vendor	**Product Line Sampling**
ON Semiconductor	100K 300 Series ECL, MECL 10K and 10H, 10/100 ECLinPS, 10/100 ECLinPS Lite, 10/100[a] ECLinPS Plus, GigaComm (SiGe), Reduced Swing ECL, 100K Low Voltage ECLinPS Lite and Plus
Maxim Integrated Products	9300, 9400 Series ECL
Semtech	100K ECLinPS, 100K ECLinPS Lite (ECL Lite) and Plus, Low Voltage ECLinPS
Motorola	100ES Clock Drivers (SiGe:C)
National Semiconductor	100K 300 Series ECL
Fairchild Semiconductor	100K 300 Series ECL, 100K ECLinPS Lite, 100K Low Voltage ECLinPS Lite
Austin Semiconductor	10K, 10H
Micrel	10/100 H (Clock Drivers/Translators), SuperLite (CML), GigaPro (CML), 10/100 ECLinPS Lite, 100K ECLinPS Pro (ECL Pro)[b], Super-300K (100K S300 ECL)
Arizona Microtech	10/100 ECLinPS, 10/100 Low Voltage ECLinPS, 10/100 ECLinPS Lite, 10/100 ECLinPS Lite Translators, 10/100 ECLinPS Plus, 10/100 Low Voltage ECLinPS Lite

a. 10/100 is an abbreviation referring to both 10K and 100K signaling-level standards.

b. The ECLinPS Pro, ECL Pro, and ECLinPS Plus are the same family of devices.

function compatible and direct replacements for ON Semiconductor's ECLinPS, ECLinPS Lite, and ECLinPS Plus products (i.e., they are a "second source").

"Motorola has introduced new products based upon SiGe:C (silicon germanium carbon) process technology. The addition of carbon provides reduced noise levels and improved manufacturing latitude, and engineering samples have demonstrated a reduction in the propagation delay of more than 50% as when compared to the most advanced bipolar processes. The improved noise figures translate into lower clock jitter and improved system timing." [Mot01a]

8.1.3 Comparison of Several ECL Family Members

Table 8.3 lists nine ECL family members and provides several device features for each. The list goes from highest toggle frequency of greater than 12 GHz for the SiGE GigaComm devices to the lowest at 125 MHz for the now end-of-life 10K devices. The formats for the part number designations are listed in the "Family" column below the family name. Note that all of the part number designations (except SiGe) have either a "10" or "100" prefix, indicating whether the part complies with the 10K or 100K standard for dc signaling levels.

8.1.3.1 Power Consumption of ECL Family Devices

The ECL families have fairly high power consumption. By referring to Table 8.3 and multiplying supply current per gate by the corresponding power source voltage, power consumption for the several ECL family members listed can be compared. These values range from 120 mW per gate for Low Voltage ECLinPS Plus at 5 V to a low of 59.4 mW per gate for Low Voltage ECLinPS and Low Voltage ECLinPS Lite. The supply currents in Table 8.3 are only representative and actual supply currents and power consumption vary considerably with the specific device, its function, and the voltage of the power source. An MC100LVEP111 2.5 V/3.3 V 1:10 differential ECL/PECL/HSTL clock driver (a Low Voltage ECLinPS Plus device) draws the same 100 mA typical supply current whether it's powered from 2.5 V or 3.3 V. At 2.5 V, it consumes 2.5 V × 0.1A or 250 mW of power, but at 3.3 V it consumes 330 mW of power. For the 100304 device, a 300 Series ECL quint AND/NAND gate, typical power consumption is 250 mW per device or 50 mW per gate.

In calculating total power consumption by an ECL circuit, it's important to remember that significant power is consumed by pulldown or pullup termination resistors in addition to the ECL device itself. If 50 mA of current is passed through a 50 Ω termination resistor when an output is high, the power consumed is 125 mW or 1/8 W. When the output is low, the output voltage is often very close to the termination voltage (V_{CC}–2 V for 3.3 V PECL) and very little power is consumed in the termination resistor in the low logic state. With a 50% duty cycle, the power consumed in the resistor is cut in half, or about 1/16 W. For the differential output pair requiring two resistors, the pair will consume 1/8 W. In calculating total power for a circuit, the device design should be examined to see whether the device current which you are using to calculate device power consumption also includes the current passing through termination resistors.

An advantage of emitter coupled logic is that, while power consumption is fairly high, the current drawn from the power source is uniform and continuous during switching. This results in much less noise being coupled onto power buses with switching than for comparable devices such as LVTTL.

Table 8.3 Comparison of Several ECL Family Members' Parameters for ON Semiconductor Devices [Ons02a]

Family	Supply Voltage (V)	Supply Current Per Gate (mA)	Toggle Freq (GHz)	Prop Delay (ps)[a]	Output Rise/Fall Times (ps)	Operating Temp Range (°C)	Package Type	Wafer/Die
GigaComm (SiGe) NBSGxxx	2.5/3.3	25	>12	150	40	–40 to +85	FCBGA b	Yes
Low Voltage ECLinPS Plus MC10/100LVEPxxx	2.5/3.3	24	>4	220	130	–40 to +85	SOIC TSSOP TQFP	Yes
ECLinPS Plus MC10/100EPxxx	3.3/5.0	24	>4	220	130	–40 to +85	SOIC TSSOP TQFP	Yes
Low Voltage ECLinPS Lite MC10/100LVELxxx	3.3	18	2.5	275	200	–40 to +85	SOIC TSSOP	Yes
ECLinPS Lite MC10/100ELxxx	5.0	18	2.0	275	200	–40 to +85	SOIC TSSOP	Yes
Low Voltage ECLinPS MC10/100LVExxx	3.3	18	1.1	300	375	–40 to +85	PLCC TQFP	Yes
ECLinPS MC10/100Exxx	5.0	18	1.1	300	375	0 to +85	PLCC TQFP	Yes
MECL 10/100H MC10/100Hxxx	5.2	17	0.350	1000	1000	0 to +75	DIP PLCC	Yes
MECL 10K [c] MC10xxx	5.2	17	0.125	2000	2000	–30 to +85	DIP PLCC	Yes

a. Differential line receiver / driver as baseline.

b. 4 × 4 / 16 × 16 flip-chip ball grid array.

c. MECL 10K is not recommended for new designs.

[This table is reprinted with permission of ON Semiconductor.]

8.2 Basic Device Operation

Figure 8.1 shows an ECL driver connected to an ECL receiver by a transmission media terminated with two termination resistors R_T connected to a termination voltage of V_{TT}. The figure shows the driver with the typical emitter follower output structure and both the driver and receiver are shown with the same power span of V_{CC}–V_{EE}. Dots in the figure signify that an actual device contains connections to additional transistors, resistors, and diodes. While the figure shows the R_t termination resistors at the center of the interconnect between the driver and receiver, the termination resistors are normally placed at the end of interconnect at the receiver input.

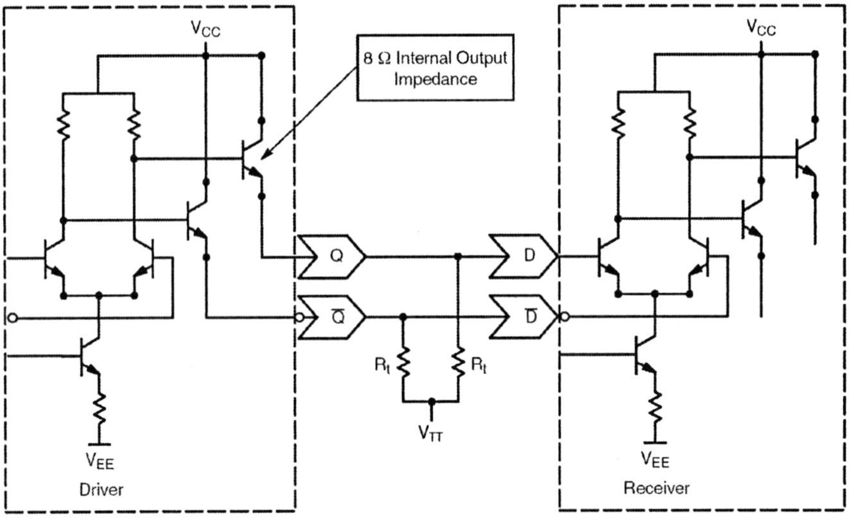

Figure 8.1 Typical emitter follower output structure, termination, and input structure for ECL devices. [Reprinted with permission of ON Semiconductor.]

The driver's output stages are emitter follower circuits. "They provide level shifting from the differential amplifier to ECL output levels and provide a low output impedance for driving transmission lines. Most ECL devices have open emitter follower outputs. The outputs are left open for use with transmission lines, and since the line termination provides an output load, internal pulldown resistors would waste power." [Ons88]

The output emitter follower transistors operate in their active regions (they don't go into saturation) with dc current flowing at all times. This increases switching speeds and helps maintain fast turn-off times. ECL outputs are commonly terminated with an R_T of 50 Ω connected to V_{TT} = (V_{CC}–2 V). The ECL output impedance is low, typically on the order of 4–8 Ω, which provides superior driving capability. When ECL outputs drive a transmission line, the ECL low output impedance generates a mismatch in back termination and can result in high-frequency signal aberrations if the transmission line at the receiver end is not perfectly terminated.

ECL inputs are current switching differential amplifiers with high input impedances. In order to provide operating headroom, the common-mode voltage is around (V_{CC}–1.3 V). High-frequency ECL communication products[3] have two types of ECL input circuit configurations. One configuration has on-chip biasing where inputs have 1 KΩ pullup resistors tied to V_{CC}–1.3 V (not shown in Figure 8.1). The other configuration is without on-chip biasing. In the latter case, proper dc biasing is required to be provided externally.

Pulldown resistors (as opposed to pullup resistors) are also commonly used on ECL inputs. Input pulldown resistors drain off the input transistor leakage current and hold unused inputs at a fixed low-state logic level. Most designs leave unused ECL inputs with pulldown resistors open and do not tie them to ground or power. [Ons88] When working with ECL devices, careful attention must be paid to power supply decoupling in order to keep both ground and the power supply rail noise free.

"ECL devices have fairly constant power supply requirements which are independent of operating frequency. Parts specified to drive transmission lines will drive coaxial cables over distances limited only by the bandwidth of the cable. Unlike totem pole outputs, the emitter followers do not generate a large current spike when switching logic states, so power buses stay comparatively noise free. The low-current switching in signal paths, relatively small voltage swing (typically 800 mV), and low output impedances cut down crosstalk and noise." [Ons88] "ECL's emitter follower outputs have very low impedance in either state, and it is difficult to couple noise from an external source into a signal line driven by such a low-impedance output. ECL gates do not need as much noise margin as CMOS and LVTTL families." [Wak01]

8.3 The Two Major ECL Standards — 10K and 100K

Most ECL family devices adhere to one of two standards, the 10K standard or the 100K standard. The terms "10K" and "100K" once referred to the original 10K and 100K families of devices, which had different electrical characteristics. Today 10K and 100K can still refer to these two original families, but are much more likely to refer to a wide assortment of newer devices. For these newer devices, the terms 10K and 100K specify whether the devices adhere to the "10K" or the "100K" input and output dc electrical characteristics (i.e., the signaling levels) of the two original families — but not other parameters of the original families such as propagation delays and edge rates. Today's devices may vary considerably in what power span they operate from, what their propagation delays and rise times are, and how fast they can toggle, but their signaling levels will either conform to 10K levels or 100K levels. By the way, ECL devices that don't adhere to 10K or 100K standards are CML, Reduced Swing ECL, and adjustable/selectable/variable output devices.

The dc input and output signaling levels for the 10K and 100K standards are shown in Table 8.4 for the case where V_{CC} = 0 V. This table is for Micrel Semiconductor's ECLinPS and ECLinPS Lite devices, where V_{CC} is 0 V and V_{EE} is negative. Where V_{CC} for any 10K or 100K

3. Examples of high-frequency ECL communication products are Maxim's MAX3885 with on-chip biasing and the MAX3867 and MAX3675 without on-chip biasing.

Table 8.4 10K and 100K Series Standard dc Specifications for Micrel ECLinPS and ECLinPS Lite Families [Mic97]

10K Series dc Electrical Characteristics[1]

$V_{EE} = V_{EE}$ (Min.) to V_{EE} (Max.); V_{CC} = GND [4]

Symbol	Parameter	$T_A = -40°C$		$T_A = 0°C$		$T_A = +25°C$		$T_A = +85°C$		Unit
		Min.	Max.	Min.	Max.	Min.	Max.	Min.	Max.	
V_{OH}	Output HIGH Voltage	–1080	–890	–1020	–840	–980	–810	–910	–720	mV
V_{OL}	Output LOW Voltage	–1950	–1650	–1950	–1630	–1950	–1630	–1950	–1595	mV
V_{IH}	Input HIGH Voltage	–1230	–890	–1170	–840	–1130	–810	–1060	–720	mV
V_{IL}	Input LOW Voltage	–1950	–1500	–1950	–1480	–1950	–1480	–1950	–1445	mV
I_{IL}	Input LOW Current	0.5	—	0.5	—	0.5	—	0.3	—	μA

100K Series Dc Electrical Characteristics[2]

$V_{EE} = V_{EE}$ (Min.) to V_{EE} (Max.); V_{CC} = GND [4]

Symbol	Parameter	$T_A = -40°C$			$T_A = 0°C$ to $+85°C$			Unit	Conditions	
		Min.	Typ.	Max.	Min.	Typ.	Max.			
V_{OH}	Output HIGH Voltage	–1085	–1005	–880	–1025	–955	–880	mV	$V_{IN} = V_{IH}(Max)$ or $V_{IL}(Min)$	Loading with 50Ω to –2.0 V
V_{OL}	Output LOW Voltage	–1830	–1695	–1555	–1810	–1705	–1620			
V_{OHA}	Output HIGH Voltage	–1095	—	—	–1035	—	—	mV	$V_{IN} = V_{IH}(Min)$ or $V_{IL}(Max)$	
V_{OLA}	Output LOW Voltage	—	—	–1555	—	—	–1610			
V_{IH}	Input HIGH Voltage[5]	–1165	—	–880	–1165	—	–880	mV		
V_{IL}	Input LOW Voltage[6]	–1810	—	–1475	–1810	—	–1475	mV		
I_{IL}	Input LOW Current	0.5	—	—	0.5	—	—	μA	$V_{IN} = V_{IL}(Min)$	

Absolute Maximum Ratings[3]

Symbol	Parameter		Rating	Unit
V_{EE}	Power Supply (V_{CC} = 0 V)		–8.0 to 0	V
V_I	Input Voltage (V_{CC} = 0 V)		0 to –6.0	V
I_{OUT}	Output Current:	Continuous	50	mA
		Surge	100	
T_A	Operating Temperature Range		–40 to +85	°C
V_{EE} [4]	Operating Range		–5.7 to –4.2	V
T_{store}	Storage Temperature		–65 to +150	°C

(1) 10EL circuits are designed to meet the dc specifications shown in the table after thermal equilibrium has been established. The circuit is in a test socket or mounted on a printed circuit board and transverse airflow greater than 500lfpm is maintained. Outputs are terminated through a 50Ω resistor to –2.0 V except where otherwise specified on the individual data sheets.

(2) This table replaces the three tables traditionally seen in ECL 100K data books. The same dc parameter values at V_{EE} = –4.5 V now apply across the full V_{EE} range of –4.2 V to –5.5 V. Outputs are terminated through a 50Ω resistor to –2.0 V except where otherwise specified on the individual data sheets. 100K circuits are designed to meet the dc specifications shown in the table where transverse airflow greater than 500lfpm is maintained.

(3) Absolute maximum rating, beyond which device life may be impaired unless otherwise specified on the individual data sheet.

(4) Parametric values specified at:

5 volt Power Supply Range	100EL Series	–4.2 V to –5.5 V.
	10EL Series:	–4.75 V to –5.5 V.
3 volt Power Supply Range	10/100EL Series:	–3.0 V to –3.8 V.

(5) Guaranteed HIGH Signal for all Inputs.

(6) Guaranteed LOW Signal for all Inputs.

device is at any other voltage, this table can still be used by adding the value of V_{CC} to each table value. The 10K and 100K output voltages are always referenced to V_{CC}. Outputs are always the same voltage away from V_{CC} no matter whether V_{CC} is at ground, 3.3 V, or 5 V and no matter whether the device is being used as PECL or NECL. If you take any 10K or 100K device designed to operate with V_{CC} at either 3.3 V or 5 V (V_{EE} = 0 V), the differential output swing stays the same as you decrease V_{CC} from 5 V to 3.3 V. The signaling levels simply shift down so that they remain the same voltage away from V_{CC}. Outputs are not affected by changes in V_{EE} as long as V_{EE} is maintained within the specified range of voltages for proper operation. The 10K and 100K standards can vary slightly as you go from one vendor to another, but for the most part, the 10K and 100K standards agree fairly closely among vendors.

A feature of these standards is that the specified signaling voltage levels vary with temperature. Referring back to Table 8.4, the 10K standard is actually a set of four tables, one for each of four ambient operating temperatures, –40°C, 0°C, +25°C, and +85°C. The 100K standard is shown as two tables, one for –40°C and one for the temperature range from 0°C to +85°C. Another feature of the tables is that the voltage swings and output signaling levels don't vary much between 10K and 100K. There's no huge difference in voltage levels and differential output swings between the two standards. 10K devices are not temperature compensated and 100K devices are.

The 10K devices have part numbers with a prefix of "10." Examples are 10xxx or 10LVELxxx where "x" is a one digit number. The 100K devices are numbered similarly with a prefix of "100" such as in 100xxx or 100LVELxxx. The "10" or "100" prefixes identifying the standard don't imply any particular power span.

8.3.1 ECL Output Load Drive Characteristics

"Both of the 10K and 100K standards display similar highly compatible output amplitude swings of about 800 mV_{pp} over a wide range of operating conditions and loads. This is due to drivers exhibiting an internal output impedance ranging from 4–8 Ω when in both high and low logic levels (see Figure 8.2).

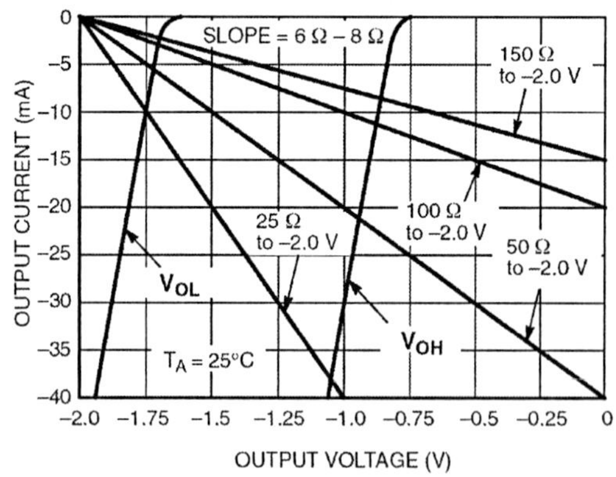

Figure 8.2 Outputs vs. load drive characteristics. [Sho02a] [Reprinted with permission of ON Semiconductor.]

For loads of 35 Ω or less, outputs may need to be 'ganged' (wire 'ANDed'), or have specialized 25 Ω driver circuits deployed. These specialized drivers ensure reduced power dissipation and improve long term reliability. Both standards display similar rise and fall times, propagation delays, and toggle frequencies for a given family of devices." [Sho02a]

8.3.2 The "10" and "100" Prefixes — Both Family and Standard

A "10" or "100" prefix in a device part number can specify the part as an original 10K or 100K part, or it can signify that the device adheres to the "10K" or "100K" dc logic signaling level standard. The difference can be important; as in the first definition, the device will in most cases operate off of a 5.2 V power span. In the second case where the signaling standard is being indicated, the power span is not known until the rest of the part number is examined. For example, a 100304 device is a quint AND/NAND gate operating with a –4.2 V to –5.7 V V_{EE} (V_{CC} = 0 V). A 100E310L (MC100LVE310) operates off of a 3.3 V ± 10% power span. For both the 100304 and the 100E310L, the "100" prefix is indicating that the device adheres to the 100K dc electrical standard.

8.3.3 Five Kinds of ECL Family Outputs

Devices in the ECL families have outputs that generally conform to one of five dc electrical signaling standards:

- 10K dc signaling levels (~800 mV)
- 100K dc signaling levels (~800 mV)
- CML dc signaling levels (~800 mV, but also 400 mV)
- Reduced swing output levels (~400, but also 200, 600 mV)
- Vendor-specific variable, adjustable, or selectable output levels (~0 to 800 mV)

The output voltage values listed in parentheses are peak-to-peak differential signal magnitudes; that is, the maximum voltage magnitude measured from one differential output to the other. Most of today's 10K and 100K signaling levels are specified with outputs terminated with 50 Ω to V_{CC}–2 V. CML and Reduced Swing ECL vendor standards specify multiple voltage output levels corresponding to multiple values of output loading. All five of these output signaling levels are specified by an assortment of vendor standards.

8.4 Single-Ended and Differential Signaling

"Differential signals are used in most ECL 'interfacing' and 'clock distribution' applications because of their low skew and high noise immunity. They are 'low skew' because the timing of a 0-to-1 or 1-to-0 transition does not depend critically on voltage thresholds, which may change with temperature or between devices. Instead, the timing depends only on when the voltages cross over relative to each other. Similarly, the 'relative' definition of 0 and 1 provides outstanding noise immunity, since noise created by variations in the power supply or coupled from external sources tends to be a common-mode signal that affects both differential signals similarly, leaving the difference value unchanged. Single-ended signals are used in most ECL 'logic' applications to avoid the obvious expense of doubling the number of signal lines." [Wak01]

8.4.1 Standard ECL Interface: Differential Driver and Receiver

A typical ECL circuit interface may be defined as a differential driver device sending a paired set of complementary signals — true and invert — over a pair of standard, controlled impedance lines to an ECL differential receiver device. The true and invert signals provide base drives to the two base inputs. Proper differential levels are specified as V_{pp} and V_{IHCMR} (defined in Section 8.4.3.1 on page 181). When an input is interconnected as a differential signal, the dc single-ended parameters of V_{IL} and V_{IH} do not apply. Terminations are required to preserve optimum signal integrity, as shown in Figure 8.3.

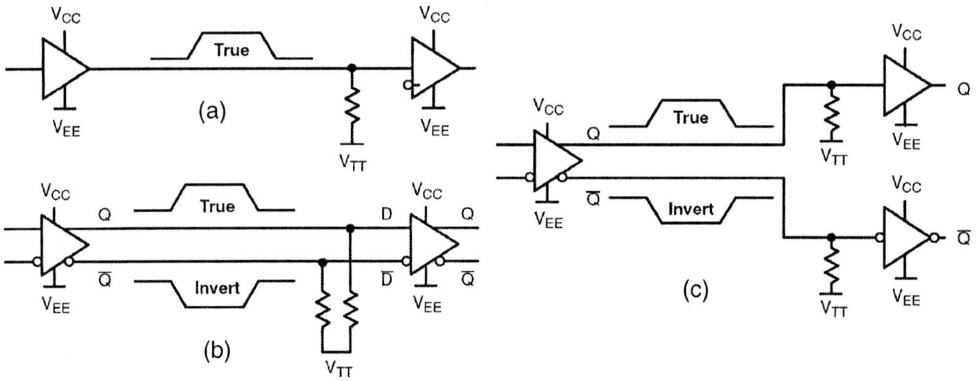

Figure 8.3 Standard (a) single-ended ECL interconnect, (b) differential ECL interconnect, and (c) differential driver with independent single-ended receivers. [Reprinted with permission of ON Semiconductor.]

8.4.1.1 Advantages and Disadvantages of Single-Ended and Differential Interconnects

"Advantages of single-ended (SE) interconnects are decreased board real estate routing and reduced system power demand. Disadvantages may include higher jitter, phase error, and duty cycle skew, high noise sensitivity, critically narrow interface windows, poor receiver sensitivity, and higher EMI emission. Differential interconnect advantages are just the opposite and may include high common-mode noise rejection (low noise sensitivity), wide signal interface windows, high receiver sensitivity, and low EMI emission. Differential interconnect disadvantages may include increased board routing real estate and increased system power demand."[4] [Sho02a]

8.4.2 Single-Ended Interface

Signals may be driven as full differential lines or as an SE line interconnection. The SE interconnection may be seen as a special variation of the typical differential interface using only one

4. This section reprinted with permission of ON Semiconductor.

driver source trace line. This single trace line drives an input pin (transistor base) of the receiver, as shown in Figure 8.3(a). Although a receiver may present only a single, dedicated SE input pin instead of a differential input pair of pins, such a receiver still would have a differential structure with the unavailable input controlled by internal circuitry.

Single-ended receiver input levels are specified in data sheets' dc characteristics block as V_{IH} and V_{IL} parameters. Each temperature has a minimum and maximum limit pair to V_{IH} and V_{IL} parameters, thus defining the single-ended, peak-to-peak input swing, $V_{pp(SE)}$. The $V_{pp(SE)}$ ranges from 595 mV to 890 mV, depending on the temperature and family. The $v_{pp(SE)}$ limits constitute the receiver device's input single-ended sensitivity. Both output lines of the typical differential output may drive two independent single-ended receivers separately (see Figure 8.3(c)).

8.4.2.1 V_{BB} Reference

For a standard differential receiver with two input pins, D and $\overline{D}$, and used in a SE application, only one of these two inputs is suitably selected to receive the signal while the non-driven input must be biased to a dc reference voltage, V_{BB} (see Figure 8.4(a)).

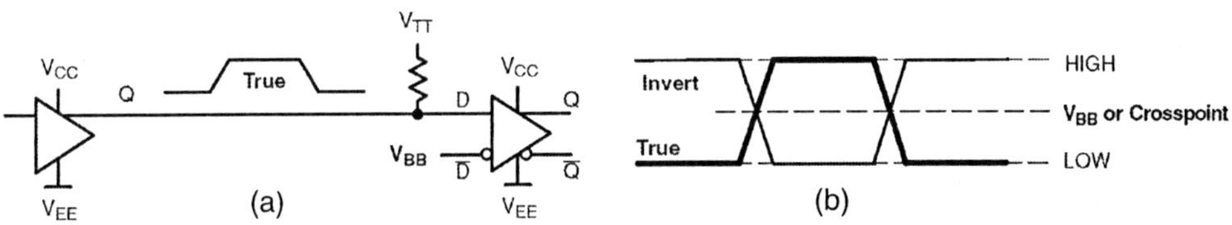

Figure 8.4 (a) Standard single-ended receiver with V_{BB}, and (b) V_{BB} crosspoint voltage. [Sho02a] [Reprinted with permission of ON Semiconductor.]

The V_{BB} value is designed to be maintained midway (50%) between the HIGH and LOW levels of the received signal, that is, at the crosspoint voltage of a differential signal pair, to preserve the duty cycle and signal integrity (see Figure 8.4(b)).

If V_{BB} shifts, due to drift or noise, above the input signal's 50% crosspoint, the device output signal will shift the duty cycle away from a pure 50% point to a decreased, narrower pulse width (see Figure 8.5(a)).

If V_{BB} shifts below the input signal's 50% crosspoint, the device output signal will shift the duty cycle away from a pure 50% point to an increased, wider pulse width (see Figure 8.5(b)). Obviously, any error voltage present on the V_{BB} reference level injects jitter directly into the signal.

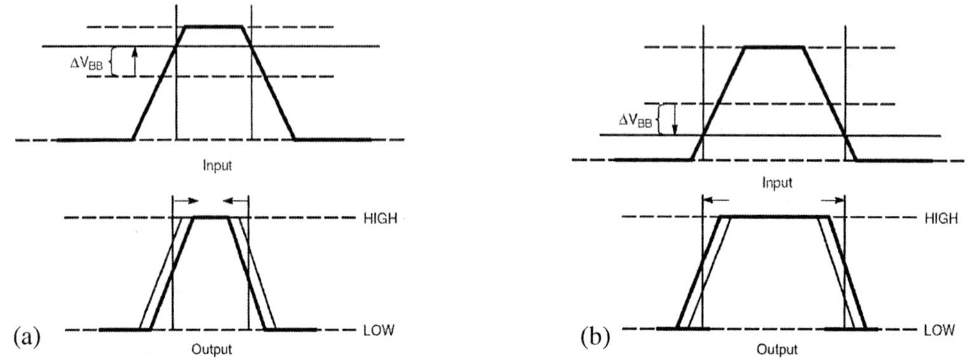

Figure 8.5 Output shift (a) narrower and (b) wider. [Reprinted with permission of ON Semiconductor.]

8.4.2.2 The Voltage Reference Source V_{BB}

A V_{BB} reference voltage output source pin may be available on the receiver device. When present, V_{BB} is an internally generated voltage supply and available only to that device's inputs. Current demand on the V_{BB} pin should be limited to 0.5 mA. Bypass (0.01 μF) V_{BB} to the quietest plane, usually V_{CC}, since noise on V_{BB} will inject jitter and corrupt the duty cycle. The V_{BB} voltage is derived from referencing the V_{CC} supply and will track changes in V_{CC} 100% or 1:1. If V_{CC} shifts 1 mV, then V_{BB} also changes 1 mV. Changes in V_{EE} also affect the V_{BB} voltage and will track at the rate of 0% to 20%, typically 5%. If V_{EE} shifts 100 mV, then V_{BB} follows with a 0 mV to 20 mV shift of the same polarity, typically 5 mV.

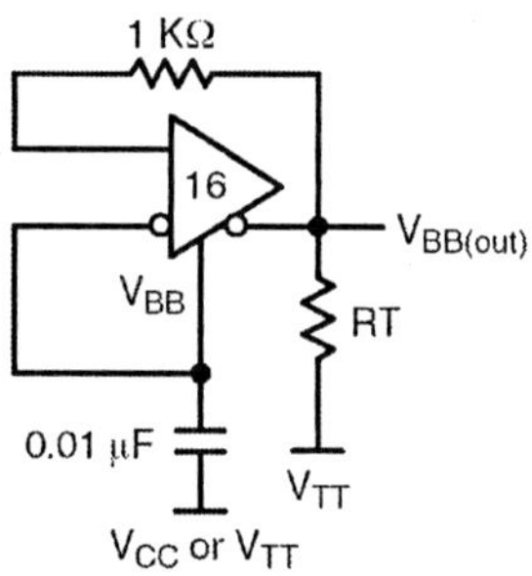

Figure 8.6 V_{BB} voltage reference generator. [Reprinted with permission of ON Semiconductor.]

A V_{BB} reference voltage may be generated off-device and supplied to the input pins. Ripple content must be kept as low as possible on V_{CC} since it transfers to the signal as jitter and phase error. A V_{BB} voltage reference level may be supplied from a V_{BB} generator, as shown in Figure 8.6. Any of the "16" type buffers are recommended to produce a high current gain V_{BB} buffer. For example, devices of type E416, EL16, LVEL16, EP16, LVEP16, EL17, LVEL17, and so forth have a V_{BB} pin available. A 1 KΩ resistor may be needed in the feedback path to stabilize higher gain buffers.

Depending on system requirements, V_{BB} may be generated by a dedicated supply, a "16" type buffer, or by using a bypassed resistor voltage divider.

8.4.2.3 Dedicated Single-Ended Input Structure

A device may have a dedicated single-ended input, having only one of the internal differential base inputs pinned out of the package, available to be driven by a signal. Internal circuitry connects a V_{BB} voltage reference to the other internal, non-driven input base node of the differential buffer gate. This internal, fixed reference voltage, V_{BB}, is maintained at the midpoint between

V_{IL} and V_{IH} for dedicated single-ended inputs. The internal V_{BB} is derived from referencing the V_{CC} supply and tracks changes 1:1 in this supply. Noise and drift in V_{CC} will inject jitter and phase noise directly into the signal.

8.4.2.4 Single-Ended Interface Between 10 and 100 Standards

Single-ended (SE) line signal interconnects require analysis of both the driver output levels, V_{OH} and V_{OL}, across temperature and across the receiver input voltage level limits, V_{IH} and V_{IL}, to determine complete interface compatibility. Although 100 Series standard devices incorporate a temperature compensation network in the output driver, some variation may still be observed. Variation of the driver output levels, V_{OH} and V_{OL}, across temperature is typically present in 10 Series devices.

8.4.2.5 Voltage Transfer Curves

Device series voltage transfer curves characterize the input and output behavior function across temperature. Voltage transfer curves for the 10E,[5] 100E, 10K, and 10H Series are shown in Figures 8.7(a) through (d), respectively. Changes in technology refinements to the 10K Series led

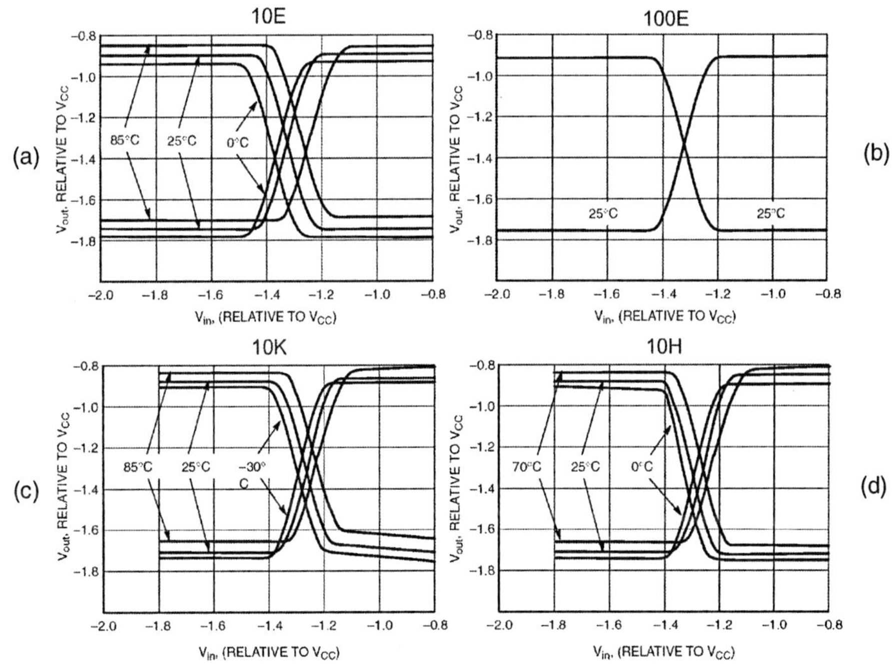

Figure 8.7 V_{in} vs. V_{out} transfer curves for (a) 10E Series, (b) 100E Series, (c) 10K Series, and (d) 10H Series. [Sho02a] [Reprinted with permission of ON Semiconductor.]

5. "10E" is the 10K standard ECLinPS family. "100E" is the 100K standard ECLinPS family. "10K" and "10H" were earlier ECL families.

to the 10H Series with better performance in V_{IH} and V_{OL} as V_{in} approached V_{CC}. The 10E Series is similar to the 10H Series. Temperature compensation allowed the development of the 100 Series.

In some ECL data sheets, values for the "unfamiliar" parameters of V_{ILA}, V_{IHA}, V_{OHA}, and V_{OLA} are listed and often shown in transfer curve plots and diagrams. These are simply minimum and maximum voltage threshold levels for the logic states. V_{OHA} is the minimum value of V_{OH}, V_{OLA} is the maximum value of V_{OL}, V_{IHA} is the minimum value of V_{IH}, and V_{ILA} is the maximum value of V_{IL}. Note that the letter "A" can signify either a maximum or minimum value.

8.4.3 Differential Interface

A standard differential interconnect driver signal will be received as signal swing. Historically, standard ECL driver signal swing may range from 750–1040 mV depending on the family, although 800 mV is typical. Newer devices may offer RSECL (Reduced Swing ECL) or Variable Output Swing (NBSG16VS). Receiver sensitivity is specified by data sheets as the input swing voltage peak-to-peak (V_{pp}). Proper output operation is displayed as the typical amplitude through the entire range of input swing, from minimum to maximum as shown in Figure 8.8 (input voltage swing peak-to-peak, V_{pp}). Input swings greater than the specification limit maximum may cause degraded frequency performance and increased t_{pd} input (input propagation delay). Input swings less than the specification minimum will cause diminished output amplitude owing to the device voltage gain, and low enough input amplitude will result in a loss of output signal. All waveforms are measured with single-ended probes with reference to ground (not as a differential probe value). Operation in the small signal level range less than V_{pp} minimum displays a characteristic gain and may of course be operated as a limited linear amplifier in this input swing range.

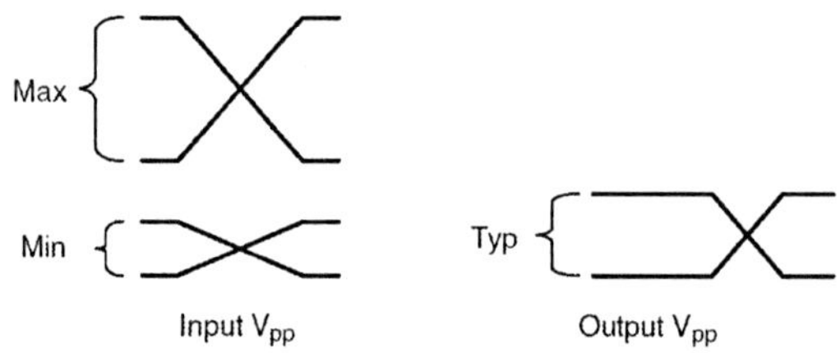

Figure 8.8 Input peak-to-peak swing voltage, V_{pp}. [Reprinted with permission of ON Semiconductor.]

Noise common to both differential lines and within the input operating range will be rejected and ignored by the receiver. The transfer threshold point is determined by the crosspoint of the differential signal. A voltage shift in input operating range of the transfer point has no voltage or timing effect on the signal, therefore preserving integrity. A receiver's tolerance of common-mode interference is illustrated in Figure 8.9.

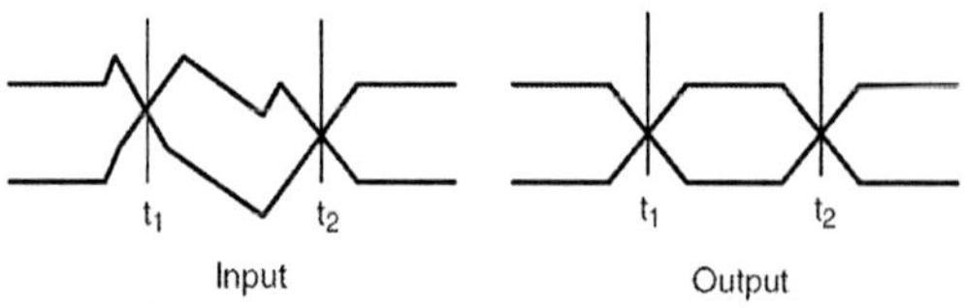

Figure 8.9 Differential input high noise immunity. [Reprinted with permission of ON Semiconductor.]

8.4.3.1 V_{IHCMR}

Each input signal to a differential pair receiver will display a V_{in} high voltage (V_{IH}) level and a V_{in} low voltage (V_{IL}). Proper operation is achieved when the V_{in} high voltage (V_{IH}) level falls within spec limits, V_{IHCMR} (voltage input high common-mode range) minimum to maximum, as represented in Figure 8.10.

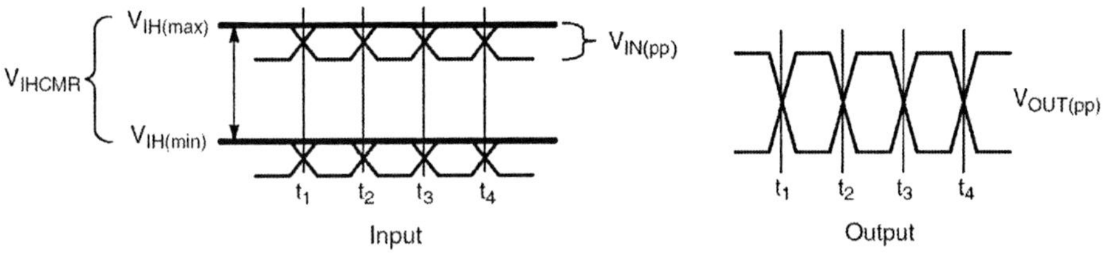

Figure 8.10 V_{IH} common-mode range, V_{IHCMR}. [Reprinted with permission of ON Semiconductor.]

8.4.3.2 Differential Interface Between 10 and 100 Standards

When interfacing differentially, the two basic standards are completely, directly compatible over all operational conditions. This results from receivers of both standards exhibiting wide V_{IH} common-mode range and fine minimum input sensitivity, V_{pp}. Output temperature variations associated with 10 Series devices are well within these receiver input characteristic limitations.

8.4.3.3 ECL Noise Margins

Noise margin quantifies the susceptibility of a driver and receiver interface to any non-signal voltage levels and therefore the risk of false switching. Two measurements — noise margin high and noise margin low — describe the false switching risk across temperature as follows:

$$\text{Noise Margin}_{(High)} = V_{OH(min)} - V_{IH(min)}$$

$$\text{Noise Margin}_{(Low)} = V_{IL(max)} - V_{OL(max)}$$

An MC10EP16DT, operating in LVPECL mode with 3.3 V on V_{CC} and 0 V on V_{EE}, interfaced to an MC10EP16DT receiver, single ended, has a noise margin at the specification ambient temperature as shown in Table 8.5. Notice the safety margin levels are positive for noise margin high, indicating the driver exceeds the receiver's requirement for a minimum and the delta is positive. For noise margin low, the driver's voltage must be below the receiver's maximum voltage.

Table 8.5 Noise Margins for MC10EP16DT Driver Interfaced to MC10EP16DT Receiver

10 to 10 Noise Margin HIGH	Temp	$V_{OH(min)} - V_{IH(min)}$	Delta (mV)
	–40°C	2165 – 2090	75
	25°C	2230 – 2155	75
	85°C	2290 – 2215	75
10 to 10 Noise Margin LOW	**Temp**	$V_{IL(max)} - V_{OL(max)}$	**Delta (mV)**
	–40°C	1690 – 1615	40
	25°C	1755 – 1680	75
	85°C	1810 – 1740	70

When a 10 Series device drives a 100 Series device single ended, the noise margins become a risk factor requiring careful evaluation, as indicated in Table 8.6.

Table 8.6 Noise Margins for an MC10EP16DT Driver Interfaced to an MC100EP16DT Receiver

10 to 100 Noise Margin HIGH	Temp	$V_{OH(min)} - V_{IH(min)}$	Delta (mV)
	–40°C	2165 – 2075	90
	25°C	2230 – 2075	155
	85°C	2290 – 2075	215
10 to 100 Noise Margin LOW	**Temp**	$V_{IL(max)} - V_{OL(max)}$	**Delta (mV)**
	–40°C	1675 – 1615	60
	25°C	1675 – 1680	–5
	85°C	1675 – 1740	–65

When a 100 Series device drives a 10 Series device single-ended, the noise margins are very robust and immunity is optimized (see Table 8.7).

Table 8.7 Noise Margins for an MC100EP16DT Driver Interfaced to an MC10EP16DT Receiver

100 to 10 Noise Margin HIGH	Temp	$V_{OH(min)} - V_{IH(min)}$	Delta (mV)
	–40°C	2405 – 2090	315
	25°C	2405 – 2155	250
	85°C	2405 – 2215	190

Table 8.7 Noise Margins for an MC100EP16DT Driver Interfaced to an MC10EP16DT Receiver (Continued)

100 to 10 Noise Margin LOW	Temp	$V_{IL(max)} - V_{OL(max)}$	Delta (mV)
	–40°C	1690 – 1605	85
	25°C	1755 – 1605	150
	85°C	1810 – 1605	205

8.5 Component Nomenclature

Figure 8.11 lists the part numbering system used by Micrel Semiconductor and many of the ECL families.[6] Of particular interest are the family codes, device types, and function/voltage options.

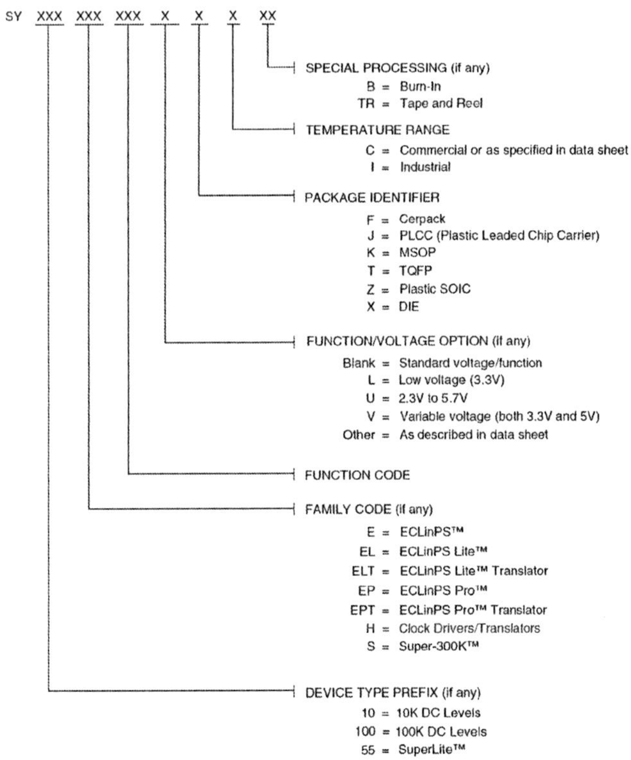

Figure 8.11 The Micrel ECLinPS families part markings.

6. Most part numbers begin with the company abbreviation. NB is used for circuits introduced by ON Semiconductor and MC is used for Motorola circuits that are now owned by ON Semiconductor. MAX is used for Maxim Integrated Products, SK for Semtech, AZ for Arizona Microtek, AS for Austin Semiconductor, SY for Micrel, NS for National Semiconductor, and F for Fairchild Semiconductor.

Family codes are listed for five ECLinPS family variations as well as a family of clock drivers/translators and the Super-300K family. Super-300K is a lower-power, faster family of 300 Series ECL devices. Three device type prefixes (dc signaling level standards) are listed, 10K, 100K, and SuperLite, a CML family. With Micrel, low voltage and special power families of devices are indicated with an "L," "U," or "V" suffix rather than by having additional family designations. In contrast, ON Semiconductor identifies Low Voltage ECL families with additional family designations as shown in Table 8.8. As another variation, Motorola uses a designation of "ES" (e.g., MC100ES6111) where the "E" represents ECL compatibility and the "S" indicates SiGe:C (silicon germanium carbon) process technology.

Table 8.8 ON Semiconductor Family Codes

Family	Code	Family	Code
ECLinPS	E	Low Voltage ECLinPS Lite Translator	LVELT
Low Voltage ECLinPS	LVE	ECLinPS PLus	EP
ECLinPS Lite	EL	ECLinPS Plus Translator	EPT
ECLinPS Lite Translator	ELT	Low Voltage ECLinPS Plus	LVEP
Low Voltage ECLinPS Lite	LVEL	GigaComm	N/A

[Reprinted with permission of ON Semiconductor.]

8.6 The ECL Families and Their Characteristics

8.6.1 A Little MECL History

ECL development went from the original MECL[7] I family (introduced in 1962) to MECL II (introduced in 1966), MECL III (MC1600 series, 1968), MECL 10,000 or 10K (10100, 10200, 10500, 10600, and 10800 Series introduced from 1971 to 1976), and on to the MECL 10H000 family or "10H" for short (introduced in 1981). The 10H000 family was also referred to as the "MECL 10KH" family in the MECL System Design Handbook [Ons88]. As each new ECL family appeared, clock rates were faster and edge transition times were shorter. The MECL I, II, and III families are obsolete (parts not available) and the 10K family is "end-of-life" (parts available but should not be used for most new designs, with some exceptions).

8.6.2 10K

The 10K family was introduced to provide a family of devices that was easier to use than earlier families and had a 2 ns propagation delay and a slowed edge rate of 3.5 ns (10–90%). The slow edge speed permitted use of wire wrap and standard printed circuit lines, although the outputs

7. MECL is Motorola emitter coupled logic.

were still designed to drive transmission lines. The original 10K family was specified to operate from a –5.2 V ± 10% power supply. [Ons88] 10K devices can deliver up to 50 mA of current per output, are designed to operate with V_{CC} = 0 V and V_{EE} = –5.2 V, and are not temperature compensated.

8.6.3 10H

"Because of technological advances in processing as well as market demands for even higher performance devices, Motorola introduced 10H devices. The 10H family provided propagation delays of 1 ns with edge speeds slowed to 1.8 ns (10–90%). These speeds were attained with no increase in power over the 10K family and were due to both advanced circuit design and process technologies which allowed smaller device geometries, reduced parasitic capacitances, and therefore greater bandwidth. Many of the 10H devices are pinout/functional duplications of the 10K family. Generated noise was reduced by relatively slow rise and fall times. For both the 10K and 10H families of devices, the edge speeds are equal to or greater than the propagation delays. ECL 10K and 10H devices have maximum permissible output currents of 50 mA." [Ons88] 10H devices were originally specified to operate from a –5.2 V ± 5% power supply and offered less variation in the output voltage level with changes in the power supply voltage. They do not have temperature compensated outputs.

The 10K and 10H differences are related to V_{OL} isolation from V_{IL} and V_{IH}. In the 10K Series, the V_{OL} would fall, tracking the inverting input as it moved toward V_{CC}. The H Series has compensated input voltage isolation.

8.6.3.1 Dual Meaning of 10H Prefix

The 10H prefix has evolved to have two entirely different meanings. It can refer to the early 10H devices (see Section 8.6.3), which were enhancements to 10K devices, or it can refer to more recent devices which are clock drivers/translators that also use the 10H designation. With either meaning, the device conforms to 10K dc logic signaling levels.

8.6.4 100K

Following the 10K family, the 100K family of devices was introduced. The 100K family had several major differences from the 10K family when it was introduced: reduced power-supply voltage with V_{EE} = –4.5 V rather than –5.2 V, different logic levels as a consequence of the different supply voltage, shorter propagation delays (typically 0.75–1 ns), shorter transition times (typically 0.7–1 ns), and higher power consumption (typically 40 mW+ per gate). "Members of the ECL 100K family have 6-digit part numbers of the form "100xxx" (e.g., 100101, 100117, 100170), but in general their functions are different from those of 10K parts with similar numbers." [Wak01] Today's 100K devices have voltage compensation, allowing them to operate over a wide range of power supply voltages (including –4.5 V and –5.2 V). The 100K devices are also temperature compensated while 10K devices are not.

Modern 100K devices adhere to the 100K dc signaling levels and have temperature-compensated outputs which the 10K devices do not have. "Temperature-compensated outputs" means that the output voltages do not change (by very much) as a function of temperature. The reduced power supply voltage is no longer an advantage as many of today's devices will operate at power spans that include both 4.5 V and 5.2 V (excluding low voltage devices).

So, today when we talk about 100K parts, we're really referring to modern parts that adhere to the 100K standard for dc signal voltage specifications and have temperature-compensated outputs. However, with respect to most other parameters of the original 100K parts when they first arrived in the marketplace, today's parameters are significantly improved. Discussing 100K devices can become confusing because the term can refer to either the original 100K devices when they were first introduced or they can refer to today's devices that adhere to the dc voltage specifications of the 100K standard. Unless otherwise noted, "100K devices" will refer to the modern devices. These devices have a "100" in the part number prefix such as "NB100EP223" (the original 100K family has a "100" in the part number as well).

The major significance of the 10K and 100K families is that they established two standards for dc signaling levels that are still used today by the majority of ECL devices. Put another way, they established the V_{OH}, V_{OL}, V_{IH}, V_{IL} and other dc parameters still in use by today's devices. We refer to these devices as "10K parts" and "100K parts."

The 10K parts are "end-of-life" and should not be used for new designs. There are exceptions to this rule. If your design must interface with legacy designs that do use 10K parts, you may want to use 10K parts to create a compatible interface. However, "when interfacing differentially, the two basic standards are completely, directly compatible over almost all operational conditions." [Sho02a] The only likely use of 10K parts is to interface with legacy systems which are single-ended or to improve single-ended noise margins by using a 100K driver and 10K receiver (see Table 8.7 on page 182).

8.6.5 100H

8.6.5.1 100H Used as Designation for Clock Drivers/Translators

100H devices are not an enhancement to 100K device technology in the way that 10H is to 10K technology. Instead, the "H" is a designation used to indicate that the device is a clock driver/translator. The "100" of the 100H prefix still indicates that the device adheres to the 100K standard for dc signaling levels.

8.6.5.2 Caution: 10H and 100H Devices with "L" Suffix May Use Other Power Options

10H and 100H devices are generally 5.2 V parts — except when they're not! Micrel adds a suffix of "V" (3.3 V or 5 V), "L" (3.3 V only), or "U" (2.5 V, 3.3 V) to indicate other power options. For example, Micrel's SY100H641L 1:9 PECL-to-TTL clock driver operates from a 3.3 V power supply and the data sheet states that it's compatible with 100K logic levels. This can be a little confusing as this is only the case when V_{CC} = 3.3 V. If you drive this device by another 100K device operating from V_{CC} = 5.2 V, the logic levels will be incompatible and your design won't work.

8.6.5.3 Micrel's 10H and 100H

Micrel's 10H and 100H components operate from a 3.3 V power span and are not equivalent to MECL 10H parts that are designed to operate from a 5.2 V ± 5% power span.

8.6.6 ECL, PECL, Psuedo ECL, NECL, LVECL, LVPECL, and LVNECL

ECL was described previously on page 166 to have two meanings. ECL can refer to all of the several device families using bipolar emitter coupled logic technology (the first meaning), or it can refer to families of devices that are powered from a negative V_{EE} power source usually at –4.2 V to –5.7 V with V_{CC} = 0 V (the second meaning). For this discussion of ECL, we're using the second meaning.

PECL was defined earlier as describing devices powered from a positive V_{CC} power supply with V_{EE} = 0 V. "Psuedo ECL" is just another name for PECL. A refinement is in hand, in that PECL (and Psuedo ECL) really refers only to devices powered from a V_{CC} = +5 V nominally (ranging from about +4.2 V to +5.7 V). On the other hand, "Low Voltage PECL" or "LVPECL" refers to emitter coupled logic devices powered from V_{CC} = +3.3 V nominally (ranging from about 3 V to 3.8 V). Similarly, NECL refers to devices powered from –5.2 V nominally and "Low Voltage NECL" or "LVNECL" refers to devices powered from –3.3 V nominally. Low Voltage PECL and Low Voltage NECL can also refer to devices with a power span of 2.5 V and other lower voltages. LVECL (Low Voltage ECL) includes both LVNECL and LVPECL.

ECL, PECL, NECL, Low Voltage PECL, and Low Voltage NECL devices all generate the same 800 mV nominal differential output voltage and they all have their output voltages referenced to the V_{CC} power rail, no matter how the device is powered. The general rule for these devices (there's always exceptions) is that the single-ended outputs swing approximately from V_{CC}–900 mV to V_{CC}–1700 mV, no matter what V_{CC} or V_{EE} is. Whether or not you're using low voltage devices, and PECL or NECL, the output voltages are still referenced to V_{CC} and the differential output (~800 mV) is still the same magnitude (approximately). Most ECL offers 50 mA of output current.

8.6.6.1 300 Series ECL

The 300 Series ECL provides low power operation, 2,000 V ESD protection, has a voltage-compensated operating range of –4.2 to –5.7 V, and is available in industrial temperature grades. It is a socket replacement for the 100 Series (1001xx) and offers inherent total dose radiation tolerance. For the low-power quint AND/NAND gate, the 100304, transition times are between 300 ps and 1.60 ns (min, max) and propagation delays range from 400 ps to 1.8 ns (min, max) at 25°C. Typical power consumption is about 250 mW/pkg or 50 mW/gate.

The 300 Series ECL has part numbers of the form 1003xx (e.g., 100304) where xx are digits that identify the device function. The 3xx portion of the part number indicates that it's a 300 Series part. As the series has a prefix of "100," it adheres to the 100K standard for dc electrical characteristics.

8.6.6.2 Super-300K ECL

The Super-300K family is a higher-speed, lower-power version of the 300 Series family. Super-300K devices have part numbers of the 100S3xx format. It operates from V_{EE} power ranging from –4.2 V to –5.5 V. For the SY100S304 quint AND/NAND gate, transition times are between 300 ps and 900 ps (min, max) and propagation delays range from 300 ps to 1050 ps (min, max) at 25°C (PLCC package). Typical power consumption is about 200 mW/pkg or 40 mW/gate.

8.6.6.3 9300 and 9400 Series ECL/PECL

The 9300 and 9400 Series ECL and PECL components are 100K devices supplied by Maxim Integrated Products. For the most part, these devices are clock and data drivers, receivers, translators (from one device family to another), multiplexors, comparators, phase-frequency detectors, and differential D flip-flops. The 9300 and 9400 Series have at least one input or output that is ECL, Low Voltage ECL, PECL, or Low Voltage PECL. The 9300 Series has switching frequencies ranging from 400 MHz to 3 GHz, maximum random jitter from 0.8 ps to 10.0 ps, maximum deterministic jitter from 45 ps to 140 ps, and maximum propagation delays from 230 ps to 1360 ps, depending upon the device and its function. The 9400 Series devices all have a switching frequency of 3 GHz, maximum random jitter ranging from 0.8 ps to 2.7 ps, maximum deterministic jitter from 45 ps to 80 ps, and maximum propagation delays from 437 ps to 730 ps, again depending upon the device and its function. Most of the 9300 Series devices are second source to the ON Semiconductor product line and sometimes have enhanced features such as a wider power supply range, lower power, or higher maximum frequency. A few of the 9400 Series devices are second source to Semtech, again with improved performance. Most of Maxim's newly introduced products are proprietary.

8.6.6.4 ON Semiconductor's GigaComm Family (SiGe)

GigaComm devices have data bandwidths in excess of 12 Gbps and output edge rates as low as 30 ps. This allows for tight skew performance, improving system jitter levels to less than 0.2 ps. This family uses the SiGe (Silicon Germanium) process technology and is offered in advanced packages including the flip-chip ball grid array (FCBGA, Figure 8.12) and quad flat no-lead (QFN) that optimize device performance with minimal board space.

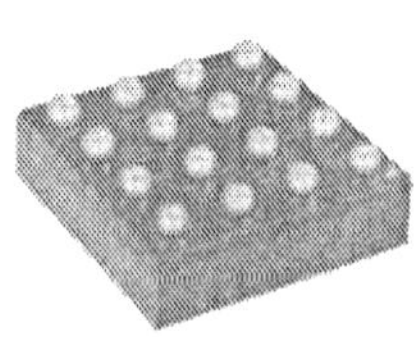

Figure 8.12 Flip-chip ball grid array, 16 pin. [Reprinted with permission of ON Semiconductor.]

GigaComm devices are generally LVNECL and LVPECL devices operating from power spans of 3.3 V or 2.5 V. GigaComm devices also provide several output options that can allow the designer to improve switching speeds. These include a Reduced Swing ECL (RSECL) output of 400 mV, a programmable voltage swing of from 100 mV to full swing (600 mV), and a selectable voltage from 0 mV, 200 mV,

400 mV, and 600 mV full swing. GigaComm inputs accept NECL, PECL, LVTTL, LVCMOS, CML, or LVDS. [Ons03a]

8.6.6.5 Hot Swapping PECL Risk: Powered Driver and Unpowered Receiver

"Hot swapping presents a potential risk to unpowered or powered-down positive mode ECL device inputs driven by a typical PECL level signal. When a receiver V_{CC} is off or powered down, the V_{CC} power supply typically appears as a low impedance source at 0 V capable of sinking considerable current. Typical PECL input signal levels present positive voltages that forward bias the input ESD protection diode structure and the input base collector junction. Potentially lethal current paths may develop through forwarded junctions to V_{CC}.

PECL powered-down receiver risk may be managed in several ways:

1. Physical sequencing — the supplies for V_{EE} (ground) and V_{CC} (power) may be physically connected prior to signal lines by altering the daughterboard edge connection geometry, making V_{EE} and V_{CC} connectors protrude and engage or sequence first. V_{EE} connectors could even be sequenced prior to V_{CC}.
2. Switching — a relay (or analog switch) could be used to open or close the V_{CC} supply insuring the power supply line is opened when powered off.
3. Capacitive coupling — dc isolation of potentially damaging current." [Ons02b]

8.6.7 ECLinPS and Low Voltage ECLinPS

ECLinPS stands for "ECL in Pico Seconds" and is often pronounced as "eclipse." ECLinPS devices are also referred to as "E Series" devices in that they are designated by an "E" in the part number (e.g., MC100E101FN). ECLinPS devices generally have toggle frequencies of about 1.1 GHz, propagation delays of 300 ps, rise and fall times of 375 ps, and come in PLCC and TQFP packages. These devices offer speed and edge rate improvements over the original 100K devices, which had toggle rates of about 350 MHz, propagation delays of from 750 ps to 1000 ps, rise and fall times of 700 ps to 1000 ps, and came in DIP and PLCC packages.

The E series devices generally come in 28-pin packages and have been available for several years. Many ECLinPS products interface directly to CML outputs as CML is popular for interfacing to ASICs and fiber optic modules. Low Voltage ECLinPS devices operate from a 3.3 V power span while still maintaining output signaling levels consistent with the 10K or 100K standards. For ON Semiconductor devices only, ECLinPS will operate from 3.3 V and Low Voltage ECLinPS will operate from 2.5 V or 3.3 V.

8.6.8 ECLinPS Lite, Low Voltage ECLinPS Lite, and ECL Lite

"ECLinPS Lite is a family of single, essential logic primitives — which includes gates, muxes, and flops (see Figure 8.13). They come with part number prefixes of 10EL or 100EL. The '10' or '100' indicates the standard, 'E' indicates 'ECLinPS' and 'L' stands for 'Lite.' These devices have 250 ps typical package gate delays, 2 GHz+ flip-flop toggle frequencies, and come in 8-,

16-, and 20-pin SOIC packages, 8- and 10-pin TSSOP packages, and in die form. They are fully compatible with 10K/100K industry standards, have flow-through pinouts[8] and can operate over the industrial temperature range of from –40°C to +85°C. 10EL devices use a V_{EE} of –5.2 V and 100EL devices use a V_{EE} of –4.5 V (typical) when V_{CC} is at ground. Many devices have an extended V_{EE} range for both 10EL and 100EL devices."[9] [Mic98a] Semtech uses the term "ECL Lite" to refer to its line of 100K ECLinPS Lite devices. ECLinPS Lite has been available in the marketplace for several device generations.

"ECLinPS Lite, being a single gate family, offers better ac specs and tighter skews than the ECLinPS family. At the same time, the family provides increased isolation over multiple gate logic.[10] Its small package size makes it ideal for pin cards and other applications that require density without the expense of signal integrity. The ECLinPS Lite family provides state-of-the-art bipolar process speeds. The 8-pin SOIC package shaves over 50% from the propagation delay associated with the ECLinPS family's 28-pin PLCC package. In addition, special design techniques have been applied to decrease output transition times by nearly 50%, resulting in nearly a 100% improvement in bandwidth and toggle frequencies over a standard ECLinPS device. Like the ECLinPS family, most ECLinPS Lite devices are offered in 10EL and 100EL versions for compatibility with the two existing ECL standards. As with the ECLinPS family, the ac performances of the two versions (10K, 100K) of ECLinPS Lite devices are identical.

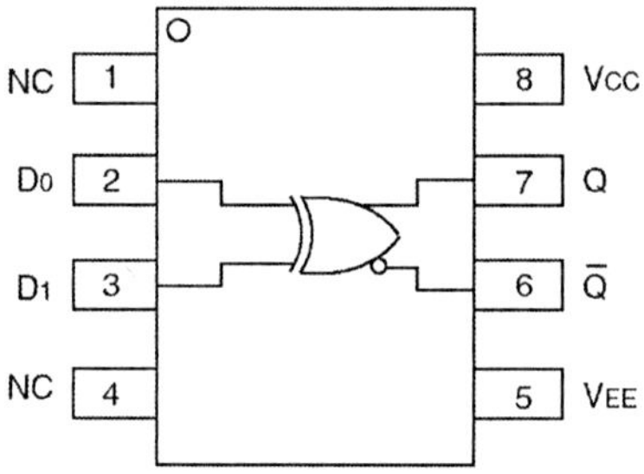

Figure 8.13 An SY100EL07 ECLinPS Lite device (SOIC top view). [Mic03a] [Copyright © 2003. Micrel, Inc. Reprinted with permission of copyright owner. All other rights reserved.]

The V_{EE} power supply range for both ECLinPS Lite versions of some manufacturers has been extended beyond the levels defined in the original ECL standard specifications. The lower end of the 100EL V_{EE} limit has been decreased to –5.5 V to allow for interfacing with three level series gated 100K arrays. In addition, the upper end of the 10EL V_{EE} limit has been increased to –4.75 V to allow for use in PECL designs which use a 5 V ± 5% supply. The –4.75 V V_{EE} for 10EL devices does not always hold for temperatures less than +25°C (see individual data sheets). Therefore, for PECL designs whose environments can fall below +25°C, it is recommended that the 100EL devices be used to allow the designer to choose functions from the entire family. The transfer curves, switching waveforms, and parameter definitions are identical to those of the ECLinPS family. In addition, the application of ECLinPS Lite devices in a system follows the same rules as previous ECL families.

8. "Flow-through pinouts" means that the input pin and corresponding output pin are on opposite sides of the chip and directly across from each other.
9. Copyright © 2003, Micrel, Inc. Reprinted with permission of copyright owner. All other rights reserved.
10. Increased isolation means that noise from a gate in an ECLinPS Lite device introduces less noise on another gate located in a different IC than if both gates were in a multigate IC.

Minimizing skew is the most important aspect of clock distribution and differential interconnect should be used when these devices are used to implement clock circuits. Because all of the inputs of ECLinPS Lite devices have internal input pulldown resistors, the inputs will pull down to V_{EE} when left open." [Mic98a]

When the LVEL (Low Voltage ECLinPS Lite) series was released, it introduced 3.3 V power span devices, a dissipation reduction from the 5 V power span device series: 10H, 100H, 10E, 100E, 10EL, 100EL.

8.6.9 ECLinPS Plus, ECL Pro, ECLinPS Pro, and Low Voltage ECLinPS Plus

ECLinPS Plus can operate off of either 3.3 V or 5 V power spans (and sometimes 2.5 V), comes in small packages, and operates at higher frequencies than the other ECLinPS families. ECLinPS Plus devices use the letters "EP" in their part numbers to indicate that they are "eclipse plus" devices. Device speeds vary with their function. The SY10/100EP33V ÷4 divider has a guaranteed maximum frequency greater than 4 GHz, guaranteed propagation delays less than 460 ps, and 3.3 V and 5 V power supply options. Where EP devices interface to other technologies, they can be relatively slow. For example, the SY100EPT21L differential LVPECL-to-LVTTL translator operates off of a 275 MHz clock bit stream, has a 1.9 ns typical propagation delay, and operates off of a 3.3 V power span. But in general, EP devices will have a toggle frequency greater than 4 GHz, propagation delays of 220 ps, and output rise and fall times of 200 ps for differential drivers and receivers. For comparison to the other ECLinPS families, refer back to Table 8.3 on page 170.

In the preceding two examples of Micrel devices, the part number suffix "V" was used to indicate operation at both 3.3 V and 5 V and "L" to indicate low-voltage 3.3 V operation. Micrel uses the suffix "U" to indicate operation over the range of 2.3 V to 5.7 V. By contrast, ON Semiconductor's EP devices may operate from 3.3 V or 5 V and their Low Voltage ECLinPS Plus parts may operate from 3.3 V or 2.5 V, so it pays to check the data sheet (as always!) for the specific part.

Micrel's ECL Pro and ECLinPS Pro are two trademark names for the same family of devices and are the same family of devices as ON Semiconductor's ECLinPS Plus. The "Pro" and "Plus" devices are replacements for one another when both vendors produce the same part, although the full product line varies somewhat among each vendor.

8.6.10 Reduced Swing ECL (RSECL, RSPECL, RSNECL) and Variable Outputs

Reduced Swing ECL, Reduced Swing NECL and Reduced Swing PECL generally provide output signals with about a 400 mV peak-to-peak output magnitude. This is approximately half of the signal magnitude under the 10K and 100K standards. The advantage of reduced output swings is that their rise/fall transition times are significantly faster than the outputs from larger-swing devices. For example, ON Semiconductor's reduced-swing NBSG16 GigaComm device has a maximum toggle frequency exceeding 12 GHz.

Figure 8.14 illustrates the differential output voltage magnitude from an example RSECL device. In the figure, the magnitude starts out at about 430 mV and then falls off with increasing frequency. This particular reduced swing device can be used at the higher frequencies up to more than 12 GHz when the receiver does not need the full amplitude to properly receive the signal.

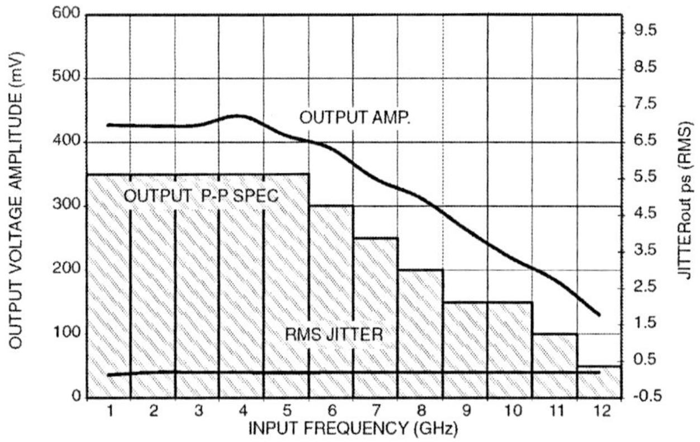

Figure 8.14 Output voltage amplitude and RMS jitter vs. input frequency at ambient temperature (typical) for a RSECL device (NBSG11). [Ons03b] [Reprinted with permission of ON Semiconductor.]

Some driver and receiver devices offer a variable output swing where the output signal's magnitude is controlled by an input pin, V_{CTRL}. For example, Figure 8.15 shows how the differential output voltage swing for the MC100EP16VS differential receiver/driver changes with the input voltage on the V_{CTRL} input pin. The curve in the figure assumes a loading of 50 Ω terminated to V_{CC}–2.0 V. V_{OH} does not change with V_{CTRL}, but V_{OL} does. V_{CTRL} is referenced to V_{CC}.

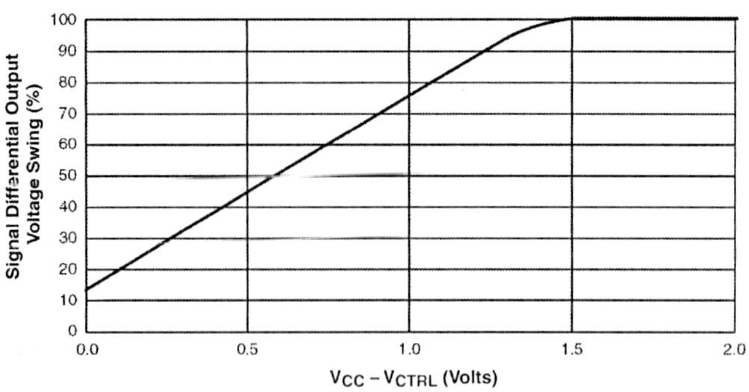

Figure 8.15 Variable differential output voltage swing for an MC100EP16VS device. [Ons02c] [Reprinted with permission of ON Semiconductor.]

8.6.10.1 Reduced-Swing ECL vs. Low Voltage ECL

Low Voltage ECL refers to ECL devices that can operate from lower voltage power supplies, usually with a power span of 3.3 V (sometimes 2.5 V). Reduced Swing ECL refers to ECL devices that have a reduced peak-to-peak output swing magnitude of about 400 mV, down from the approximately 800 mV of regular ECL. These two terms are totally different in that Low Voltage ECL is referring to lower voltage power to the device whereas Reduced Swing ECL is referring a lower signal output voltage swing. Where Low Voltage ECL devices are powered from lower-voltage power supplies, the devices peak-to-peak output swing magnitude still remains approximately 800 mV. Reduced Swing Negative ECL, RSNECL, is Reduced Swing NECL powered from a negative power supply. Reduced Swing Positive ECL, RSPECL, is Reduced Swing PECL powered from a positive power source. RSECL, RSPECL, and RSNECL can never conform to either of the 10K or 100K standards as their output magnitudes are only 400 mV, but Low Voltage ECL generally does conform to either 10K or 100K.

8.7 Summary of the ECL Families

Table 8.9 Summary of the Emitter Coupled Logic (ECL) Families

Emitter Coupled Logic (ECL) Families

General Characteristics of ECL Families

Board-Level Applications:
Clock and data distribution, backplane transmission, multiplexing, translation (voltage level-shifting and interfacing to other logic families), and state machines

System Applications:
High-speed test equipment, optical networking equipment, ultra high-speed terabit routers, network attached storage devices, OC192 SONET, 10-Gigabit Ethernet, enterprise computing servers, and high-performance workstations

Differential Uses: Clock distribution and interfacing

Single-Ended Uses: Logic circuits

Where Not to Use: Circuits requiring low power, low-speed circuits, in designs where you don't want to treat traces and interconnects as transmission lines, designs needing high integration levels

V_{CC} And V_{EE} Device Voltages: Power spans from 2.5 V to 5.2 V. Actual voltage levels ranging from –5.2 V to +5 V

Additional Required Voltages: $V_{TT} = V_{CC}–2$ V (commonly)

Main ECL Standards: 10K and 100K (Both with roughly 800 mV output swings from approximately V_{CC}–1700 mV to V_{CC}–900 mV)

Kinds of ECL Family Outputs: 10K, 100K, CML, reduced swing, vendor-specific variable, adjustable, or selectable

Drive Current: 50 mA or less

Process: CMOS, BiCMOS, SiGE, and SiGE:C

Power Consumption: High, typically 40 mW to 50 mW per gate, maximum of 120 mW/gate, frequently 200 mW to 330 mW per package

Table 8.9 Summary of the Emitter Coupled Logic (ECL) Families (Continued)

Level of Integration: Low
Packages: FCBGA, QFN, MSOP, TSSOP, SOIC, TSSOP, CERPACK, TQFP, PLCC, DIP
Termination: 50 Ω each output
Transmission Medium Characteristic Impedance: 50 Ω (single-ended value)
Loaded Effective Backplane Characteristic Impedance, $Z_{0(eff)}$: 25 Ω to 50 Ω (single-ended values). For loads of 35 Ω or less, outputs may need to be "ganged" (wire "ANDed") or have specialized 25 Ω driver circuits used
Hot Pluggable: Generally no, but can be hot plugged when circuit is designed properly (see Section 8.6.6.5 on page 189)
Noise Margin: 40 mV to 215 mV (single-ended values), depends upon the family of devices connected
Maximum Drive Distance: Limited by bandwidth of interconnect
Maximum Data Rate: 200 MHz to 12 GHz. Varies with device family, function, package, interconnect bandwidth
Edge Rate: 20 ps to 2000 ps
Propagation Delay: 150 ps to 2000 ps
Operating Temperature Range: –40°C to +85°C

Super-300K ECL

Vendor: Micrel Semiconductor
Description: Higher-speed, lower-power version of 300 Series ECL devices
Power Span: 4.2 V to 5.5 V
Edge Rates: 300 ps to 900 ps (min, max, 25°C)
Propagation Delays: 300 ps to 1050 ps (min, max, 25°C)
Power Consumption: 200 mW/pkg or 40 mW/gate
Packages: 24-pin CERPACK, 28-pin PLCC
Function and Pinout Compatibility: National and Signetics F100K
Part Numbers: SY100S3xx

ECLinPS Plus (ECL Pro, ECLinPS Pro)

Description: Fastest ECLinPS family
Power Span: 3.3 V or 5 V (low-voltage versions at 2.5 V and 3.3 V)
Edge Rates: 130 ps to 200 ps (25°C)
Propagation Delays: 220 ps to 460 ps (25°C)
Power Consumption: 86 mW/pkg (typical for 3.3 V 4 GHz ÷ 4 PECL/LVPECL divider)
Maximum Toggle Frequency: 275 MHz to 4 GHz (varies with device function)
Packages: SOIC, TSSOP, TQFP
Part Numbers: 10EPxxx, 100EPxxx

Table 8.9 Summary of the Emitter Coupled Logic (ECL) Families (Continued)

ECLinPS Lite (ECL Lite)
Description: Single-gate family of logic, muxes, and flip-flops
Power Span: 4.5 V to 5.2 V
V_{EE}: –5.2 for 10EL devices, –4.5 V for 100EL devices (when V_{CC} = 0 V), some devices with extended ranges
Edge Rates: 200 ps
Propagation Delays: 275 ps
Power Consumption: 70 mW/pkg
Maximum Toggle Frequency: 1.25 GHz to 2.5 GHz
Packages: 8-, 16-, and 20-pin SOIC, 8- and 10-pin TSSOP, die form
Part Numbers: 10ELxxx, 100ELxxx
GigaComm (SiGe)
Vendor: ON Semiconductor
Description: High-performance SiGe logic family
Process Technology: SiGe
Power Span: 2.5 V or 3.3 V
Edge Rates: 30 ps to 60 ps
Propagation Delays: 300 ps
Power Consumption: 213 mW/pkg (NBSG111 1:10 clock/data driver at 2.5 V)
Maximum Toggle Frequency: 6 GHz to 12 GHz
Packages: FCBGA-49, QFN
Part Numbers: NBSGxxx
Device Outputs: RSECL available
CML
Please see Table 9.10 on page 226 for a detailed summary of CML.

8.8 Exercises

8-1. Which way do inputs pull when they are left floating on a high-frequency ECL communication product such as the MAX3885? Which way do they pull when inputs are left floating on an ECLinPS Lite device?

8-2. What are the two ECL standards to which most ECL devices conform and how do the two standards differ? What part of these original standards are present day devices designed to?

8-3. What does "flow-through pinouts" mean?

8-4. What does it mean when we say that ECLinPS Lite is a single-gate family?

8-5. What ECL family members are end-of-life (EOL) and should not be used for new designs? Name an exception for when you might still want to use these EOL parts.

8-6. What are three designations for the ECL 10H000 family of devices?

8-7. The 10H family provided propagation delays of 1 ns with edge speeds slowed to 1.8 ns (10–90%) when they were first introduced into the marketplace. Are today's 10H devices even faster still? Log onto the Internet and locate at least two 10H parts on a semiconductor Web site. What are their edge rates (10–90%) and propagation delays? Print out the first four pages of each 10H part's datasheet and be prepared to discuss them in class.

8-8. What is meant by a "16" type buffer?

8-9. What is the difference between common-mode logic, current-mode logic, and emitter coupled logic?

8-10. What is the difference, if any, between Low Voltage ECL and Reduced Swing ECL?

8-11. Define "voltage compensation" and "temperature compensation" as they apply to ECL family devices.

8-12. Do all 10H and 100H devices operate from a 5.2 V power span?

8-13. What are the differences between the ECL Pro, ECLinPS Pro, and ECLinPS Plus families?

8-14. (a) By switching from PECL to Low Voltage PECL devices, by approximately how much does the nominal 800 mV differential output voltage magnitude decrease? (b) In switching from PECL to Reduced Swing PECL, by how much does the nominal 800 mV differential output voltage decrease?

8-15. Table 8.10 describes the dc electrical characteristics of a variable output swing device where a control pin voltage, V_{CTRL}, controls the output voltages of the device. The table is for the case where the device is used as PECL with $V_{CC} = 3.3$ V and $V_{EE} = 0$ V. For a temperature of 25°C and assuming this device follows the plot of Figure 8.15 on page 192, (a) what are the values of V_{OH} and V_{OL} that the signal swings between when $V_{CTRL} = V_{CC}$? (b) What are the values of V_{OH} and V_{OL} that the signal swings between when $V_{CTRL} = V_{BB}$? Use typical values in calculating your answers. (c) What is the total differential output swing peak-to-peak magnitude for your answer in (a)? (d) What is the total differential output swing peak-to-peak magnitude for your answer in (b)? For all of your answers in (a) through (d), calculate and use a typical value of V_{OH} by averaging the minimum and maximum values. (e) How does the graph in Figure 8.15 appear to be slightly in error based upon your answers in (c) and (d)? Why do you suppose the graph and your answers don't match up exactly?

Table 8.10 dc Electrical Characteristics for the MC100EP16VS 3.3 V/5 V ECL Differential Receiver/Driver with Variable Output Swing [Ons02c]

Dc Characteristics, PECL V_{CC} = 3.3 V, V_{EE} = 0 V (3)										
Symbol	**Characteristic**	**−40°C**			**25°C**			**85°C**			**Unit**
		Min.	**Typ.**	**Max.**	**Min.**	**Typ.**	**Max.**	**Min.**	**Typ.**	**Max.**	
I_{EE}	Power Supply Current	30	36	42	31	38	44	32	40	48	mA
V_{OH}	Output HIGH Voltage (Max Swing) (4) $V_{CC} \geq V_{CTRL} \geq V_{EE}$	2155		2405	2155		2405	2155		2405	mV
V_{OL}	Output LOW Voltage (Max Swing) (4) $V_{CTRL} \leq V_{BB}$	1355	1490	1605	1355	1520	1605	1355	1520	1605	mV
	$V_{CC} \geq V_{CTRL} > V_{BB}$		See Fig. 8.15			See Fig. 8.15			See Fig. 8.15		
	$V_{CTRL} = V_{CC}$ (Min Swing)	2105	2230	2355	2095	2220	2345	2065	2190	2315	
V_{IH}	D, $\bar{D}$ Input HIGH Voltage (Single-Ended)	2075		2420	2075		2420	2075		2420	mV
V_{IL}	D, $\bar{D}$ Input LOW Voltage (Single-Ended)	1490		1675	1490		1675	1490		1675	mV
V_{BB}	Output Voltage Reference	1805	1905	2005	1805	1905	2005	1805	1905	2005	mV
V_{CTRL}	Input Voltage (V_{CTRL})	V_{EE}		V_{CC}	V_{EE}		V_{CC}	V_{EE}		V_{CC}	mV
V_{IHCMR}	Input HIGH Voltage Common-Mode Range (Differential) (5)	2.0		2.9	2.0		2.9	2.0		2.9	V
I_{IH}	Input HIGH Current			150			150			150	μA
I_{IL}	Input LOW Current D / $\bar{D}$	0.5 / −150			0.5 / −150			0.5 / −150			μA

NOTE: EP circuits are designed to meet the dc specifications shown in the above table after thermal equilibrium has been established. The circuit is in a test socket or mounted on a printed circuit board and transverse airflow greater than 500 lfpm is maintained.

(3) Input and output parameters vary 1:1 with V_{CC}. V_{EE} can vary +0.3 V to −2.2 V.

(4) All l oading with 50 Ω to V_{CC} − 2.0 volts. V_{OH} does not change with V_{CTRL}. V_{OL} changes with V_{CTRL}. V_{CTRL} is referenced to V_{CC}.

(5) V_{IHCMR} min varies 1:1 with V_{EE}. V_{IHCMR} max varies 1:1 with V_{CC}. The V_{IHCMR} range is referenced to the most positive side of the differential input signal.

8-16. Which of the following six termination techniques illustrated in figures (a) through (f) below would you recommend as being proper terminations? Explain why you would or would not recommend each technique.

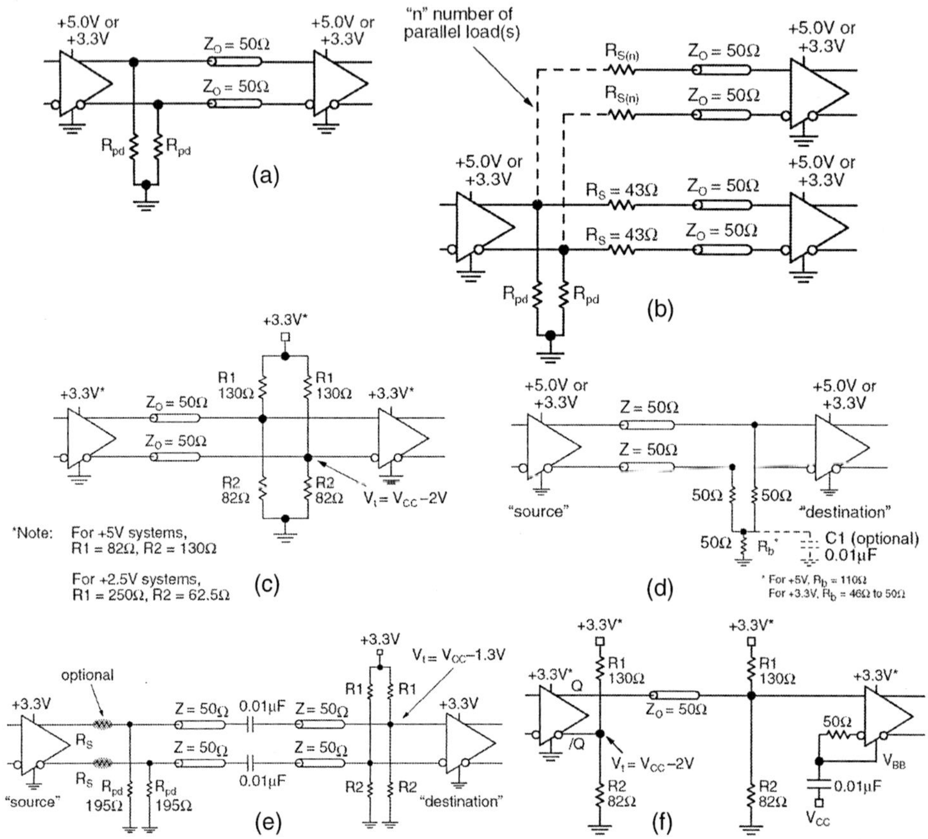

CHAPTER 9

Current-Mode Logic (CML)

Chapter Objectives

- Demonstrate that CML devices are manufactured to several different vendor standards.
- Present typical designs of CML output buffer and receiver input stages.
- Discuss how CML differs from other families of devices.
- Quantize CML differential output and input voltages and their common-mode offsets.
- Discuss design considerations when using ac and dc coupling of the driver to the receiver.
- Discuss pre-emphasis, de-emphasis, transmit and receive equalization, and adaptive equalization.
- Provide examples that show how key parameters vary among device manufacturers.

9.1 CML Overview

CML (current-mode logic) is ECL (emitter coupled logic)[1] technology that is being implemented in many of the newest high-speed devices. It is a de facto[2] standard used extensively industry wide for networking physical layer and SerDes (serializer/deserializer) devices. [Cyp02] CML serial signaling rates typically range from 1 Gbaud[3] to over 10 Gbps and higher, and the data rate that CML can support depends upon the manufacturing process technology for the drivers and receivers. Where with many digital logic families the entire device is constructed using the same

1. See Chapter 8 on page 161 for a discussion of ECL.
2. "De facto" means that it has come to exist without official recognition or legal establishment.
3. Gbaud is the Gbps data rate increased 25% after applying 8B/10B encoding.

type of logic throughout the device, CML is usually implemented only in the output buffer (driver) or input stages of a device. Another name for current-mode logic being used by some chip manufacturers today is "common-mode logic," which has the same acronym, CML. Current-mode logic and common-mode logic are two names for the same technology.

CML devices adhere to no one standard and instead are designed to a wide variety of vendor-specific standards. The XAUI (10 Gigabit Attachment Unit Interface) IEEE standard offers specifications for serial 3.125 Gbaud data links that design engineers may elect to follow when designing CML devices and systems, although XAUI does not define CML. "While not explicitly standardized, CML implantations can meet the requirements of Clause 47 of the IEEE 802.3ae standard that defines the physical layer of the XAUI interface. Most implementations of XAUI SerDes parts feature a CML interface on the high-speed side."[4] [Gol03a] Later on in this chapter we'll examine the XAUI standard and several vendor standards.

CML drivers and receivers may be ac coupled, if encoding is used, or dc coupled as shown in Figure 9.1. Ac coupling can be used for systems in which the receiver and transmitter are at different supply voltages. For ac coupling, the coupling capacitor should be large enough to avoid excessive low-frequency droop when the data signal contains long strings of consecutive identical digits. Careful review of datasheets is recommended to determine whether a driver and receiver can interoperate, especially in dc-coupled applications.

For dc-coupled applications, since one side of the driver output is pulled to the power supply rail, both the driver and receiver should be powered from the same supply potential. Ac coupling provides common-mode tolerance, fault protection, and power supply independence. [Gol03] Ac coupling also offers the ability to interface with other families of differential devices that may have different common-mode (offset) voltages and different differential voltage magnitudes.

Table 9.1 lists key specifications for a typical vendor's implementation of CML. We'll call these specifications "reference" CML specifications for later use in comparing actual device specifications with the ones in this table. The source of this table's information is a simple consensus of specifications from actual vendor data sheets. In the table, V_{TR} and V_{CP} specify the single-ended true and complement voltages of the driver output. $|V_{OD}|$ is the driver's differential output voltage magnitude, V_{OS} is the driver's offset voltage or common-mode voltage, and R_T is the value of the termination resistors used at the receiver. $|V_{ID(min)}|$ is the receiver's differential input threshold voltage magnitude which, at 100 mV, represents a very small threshold for triggering logic states. The receiver's allowable range of input voltage (the voltage on a single input terminal referenced to ground), $V_{IN(min)}$ to $V_{IN(max)}$, rarely ranges from less than –0.3 V to greater than V_{CC} + 0.3 V to avoid damage to the device, but most device specifications restrict it further. Some of the parameter values listed in the table are illustrated in Figure 9.2. In the figure, the differential signal voltage offset seen at the receiver will vary with the type of coupling used between the driver and receiver (i.e., ac or dc), the termination voltage, power supply volt-

4. Copyright © 2003 by CMP Media LLC, 600 Community Drive, Manhasset, NY 11030. Reprinted from *www.PlanetAnalog.com* with permission.

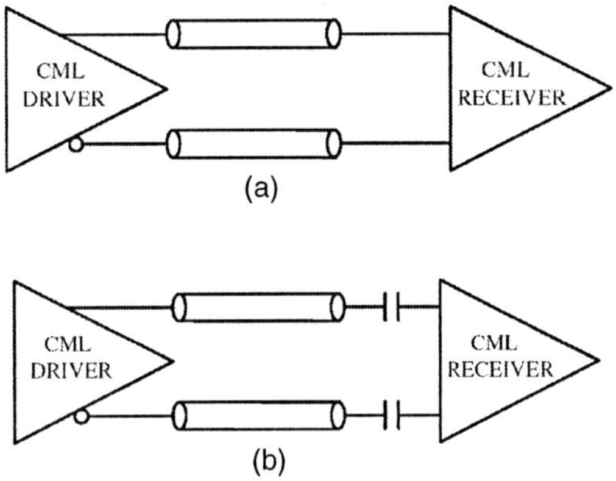

Figure 9.1 (a) Dc coupling between CML and CML, and (b) ac coupling between CML and CML. [Reprinted with permission of Maxim Integrated Products.]

ages, intentional reductions in driver output voltage swings designed to reduce crosstalk, and signal attenuation and losses.

Table 9.1 Reference CML Specifications for an 800 mV Driver Differential Output Example; a 400 mV Driver Output Is Also Common

Parameter	Value
Driver	
V_{TR}	V_{CC}
V_{CP}	V_{CC}–0.8 V
$\lvert V_{OD}\rvert$	800 mV
V_{OS}	V_{CC}–0.4 V
R_T	50 Ω
Receiver	
$\lvert V_{ID(min)}\rvert$	100 mV
$V_{IN(max)}$	V_{CC}+0.3 V, Limited
$V_{IN(min)}$	–0.3 V, Limited

The terminology in dealing with differential voltages in vendor data sheets can get a little confusing. In Figure 9.2, when the data value is "1," the driver's differential output voltage from V_{CP} to V_{TR} is V_{OD}, and V_{OD} is positive. When the data value is "0," the voltage from V_{TR} to V_{CP} is V_{OD}, but now V_{OD} is negative. In the "1" state, V_{OD} is +800 mV and in the "0" state, V_{OD} is –800 mV. The magnitude of V_{OD}, $\lvert V_{OD}\rvert$, is 800 mV in either state, whereas the peak-to-peak voltage value of V_{OD}, $V_{OD(p-p)}$, is 1,600 mV because V_{OD} ranges from negative 800 mV to positive 800 mV. The differential outputs of the driver traverse the full range from +800 mV to

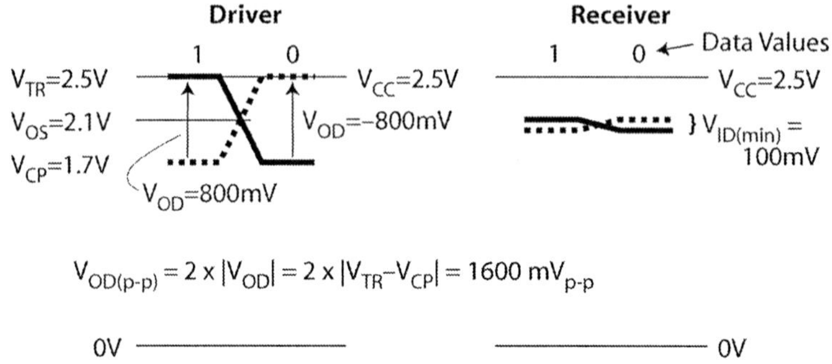

Figure 9.2 Switching levels for a CML 800 mV differential output example.

–800 mV for a total peak-to-peak differential voltage range of 1,600 mV. What can get confusing in looking at data sheets is that it is not always clear whether the vendor is specifying the value of $|V_{OD}|$ or $V_{OD(p\text{-}p)}$. Three ways to tell which parameter is being specified are as follows: (1) In specifying V_{OD} the data sheets use "mV" for the units whereas in specifying $V_{OD(p\text{-}p)}$, the data sheets use "$mV_{p\text{-}p}$" for the units (but there can be typos or exceptions). (2) The data sheet sometimes provides a description of the parameter, such as $V_{OD} = |V_{TR}-V_{CP}|$ or $V_{OD(p\text{-}p)}$ is "peak-to-peak output voltage." Or (3), the data sheet may provide an illustration to clarify the parameters.

9.2 CML Output Structure

A typical CML driver output stage is shown in Figure 9.3 and consists of a common-emitter differential transistor pair with 50 Ω collector resistors. Some devices have the 50 Ω collector resistors built into the device and some devices use external resistors for the collector resistors. The key feature of CML is the driver's constant-current sink which sinks the same current regardless of the load placed upon the outputs or the values of pullups or termination resistors used (within limits).

"The output voltage swing is generated by switching the tail current (I_{BIAS}) through the output transistors. Switching a tail current of 16 mA across a 50 Ω resistor will create a differential signal swing of 800 mV (1,600 mVp-p)." [Cyp02] The CML output transistors are permanently biased in the active region, resulting in the ability for much faster switching than technologies that use saturated transistor operation, such as CMOS or most LVDS implementations.

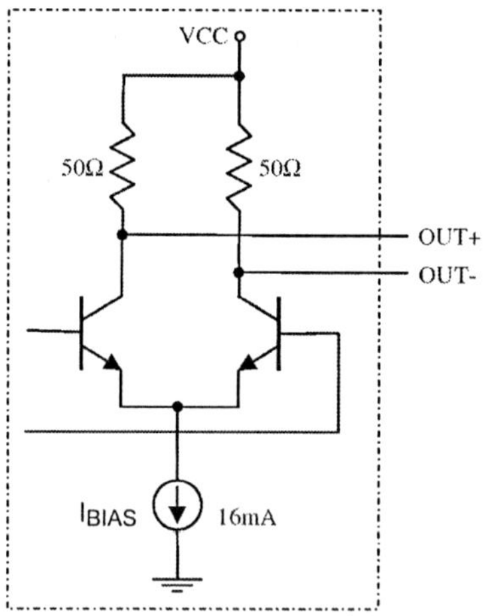

Figure 9.3 CML output structure.
[Reprinted with permission of Maxim Integrated Products.]

"The 50 Ω collector resistors also supply source termination when driving 50 Ω transmission lines to help increase the signal integrity performance of systems using CML outputs. This source termination prevents any reflections due to imperfect impedance matching between the transmission line and the differential load termination being propagated back down the line. This helps to limit any 'ringing' effect seen on the transmission line without requiring any additional external components.

CML output structures are capable of rise and fall times much smaller than those of LVDS while still preserving quality signal transmission. The constant current sink used in the CML output structure creates less switching noise so output rise and fall times of less than 100 ps are possible. CML output structures are currently used almost exclusively for serial transmission above 1 Gbps."[5] [Cyp02]

9.3 CML Input Structure

"The CML input structure has several features that support high-speed operation. As shown in Figure 9.4, the CML input structure has a 50 Ω input impedance for easy termination. The input transistors are emitter followers that drive a differential-pair amplifier." [Max00] In the figure, the termination resistors are internal to the chip and, for most applications, would not need any external termination. Other devices do not have these internal termination resistors and allow external termination resistors to be used for greater flexibility in circuit design.

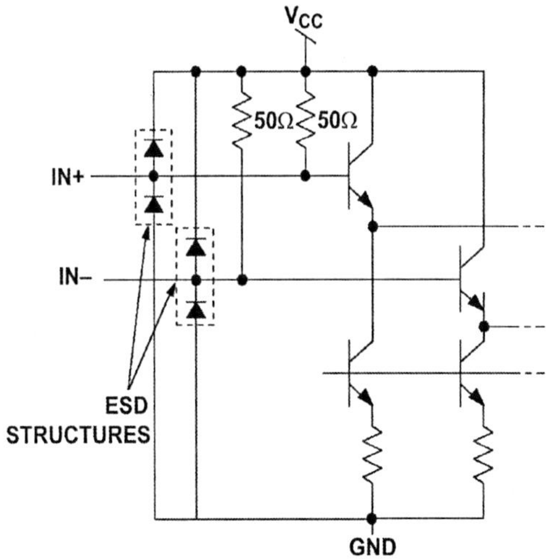

Figure 9.4 CML input circuit configuration. [Reprinted with permission of Maxim Integrated Products.]

5. Reprinted with permission of Cypress Semiconductor Corporation.

9.4 AC- and DC-Coupled CML Circuits

"Today CML has become very popular due to its simplicity and speed, especially for multigigabit SerDes applications." [Gol03a] The main limitation of CML is that it can only be used in point-to-point bus configurations, as shown in Figure 9.5. However, for ultra high-speed applications (>1 Gbps), a restriction to point-to-point is already required. Some CML devices do not require any external resistors as matched source and load termination resistors are provided internally in their driver and receiver stages. Other CML devices do require external driver pullup resistors or termination resistors or both. CML provides an 800 mV driver differential output voltage swing in some implementations, 400 mV in others, and many devices offer the ability to reduce the output swing where the differential output voltage can be adjusted downward to a few hundred millivolts to minimize crosstalk. The 50 Ω collector resistors in the driver as shown in Figure 9.5 also serve as back termination that minimizes the back reflection from the transmission line or receiver reducing high-frequency aberrations. Driver sink current magnitudes are typically about 16 mA for a differential driver output voltage swing of 800 mV across a 50 Ω collector resistor.

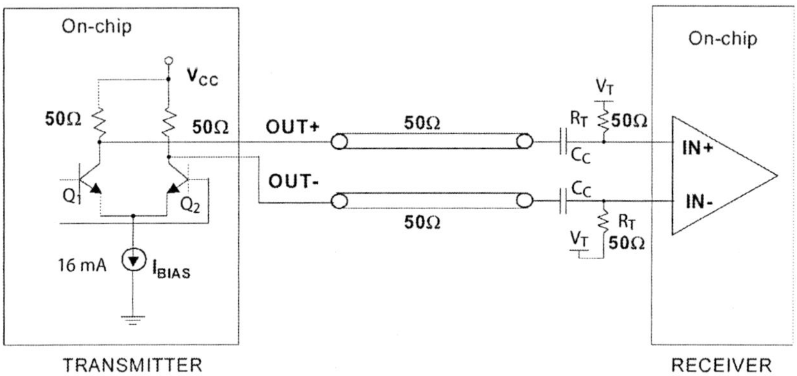

Figure 9.5 Ac-coupled CML circuit with two termination resistors pulled up to V_T.

In Figure 9.5, the coupling capacitors C_C prevent dc current from flowing from V_T through Q_1 or Q_2. In operation, Q1 is off and Q2 is on, Q1 is on and Q2 is off, or they are in transition to these conditions. When a transistor is off, its collector is pulled up to V_{CC}. When a transistor is on, 16 mA of current is pulled through its collector resistor, dropping 800 mV across the resistor. A differential voltage magnitude of 800 mV is developed across the two driver output terminals. At the receiver, V_T establishes and becomes the common-mode or offset voltage at the receiver. The 800 mV differential signal created by the driver is ac coupled to the receiver and superimposed upon the V_T offset voltage. The V_T voltage needs to be less than (V_{CC} minus half the differential output voltage).

Figure 9.6 illustrates a similar CML circuit but now with dc coupling and termination voltages at the receiver of V_{CC}. As before, when a transistor is off, its collector gets pulled up to the VCC power supply rail. When a transistor is on, 16 mA of current flows through the transistor as before, but now the current flows through two resistors, the collector resistor and the termination resistor. The parallel combination of the 50 Ω pullup and the 50 Ω termination resistor is 25 Ω. The resulting voltage drop is only half as much as the above situation, or 400 mV, created by 16 mA of current drawn through the 25 Ω equivalent resistance.

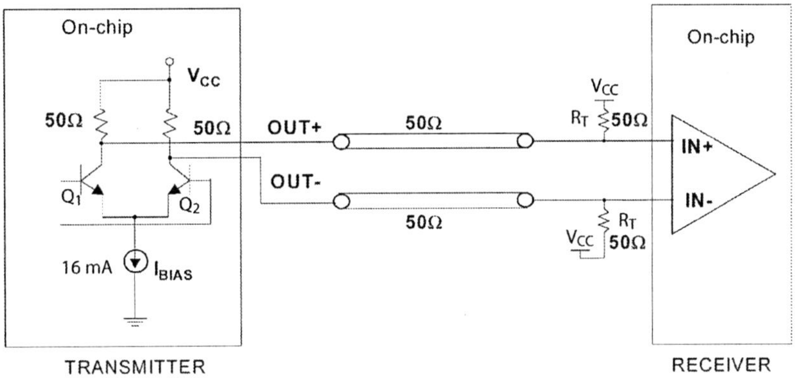

Figure 9.6 Dc-coupled CML circuit with termination resistors pulled up to V_{CC}.

The single-ended output waveform for the ac-coupled case of Figure 9.5 is shown in Figure 9.7(a). The figure assumes $V_T = (V_{CC} - 0.2\text{ V})$. The single-ended output voltage for the dc-coupled case of Figure 9.6 is shown in Figure 9.7(b).

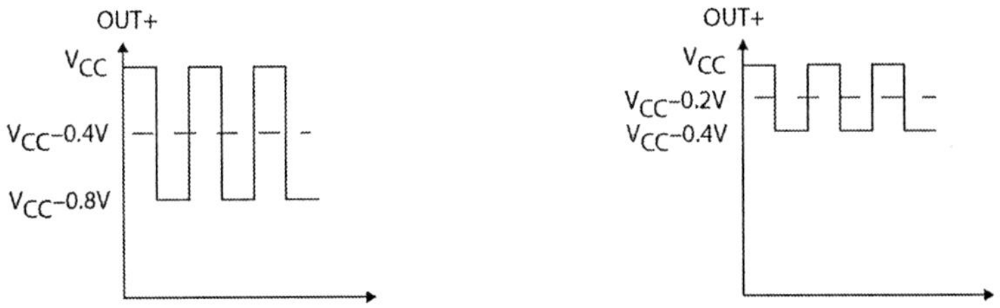

Figure 9.7 CML output waveforms for ac and dc coupling. (a) Ac-coupled, loaded with internal 50 Ω collector resistor, and (b) dc-coupled, loaded with 25 Ω equivalent, internal 50 Ω collector resistor in parallel with 50 Ω termination resistor. [Reprinted with permission of Maxim Integrated Products.]

Figure 9.8 shows another dc-coupled CML circuit, with not only the collector resistors built into the driver but now with termination resistors built into the receiver. As before, the 50 Ω collector resistors of the driver match the 50 Ω single-ended impedances of the transmission line and absorb any back reflections from the receiver. The 50 Ω base pullup resistors of the receiver match the line impedance and properly terminate the line. As before, if a driver transistor is off, its collector will again be pulled up to V_{CC} which in turn will forward bias the base of the transistor to which it is connected. When a driver transistor is on, 16 mA is drawn through its connected 50 Ω base pullup resistor of the receiver, driving down the voltage on the base of its connected receiver transistor and biasing that transistor off. The base pullup resistors in the receiver may be mounted external to the receiver or they may be designed into and part of the receiver chip, as shown.

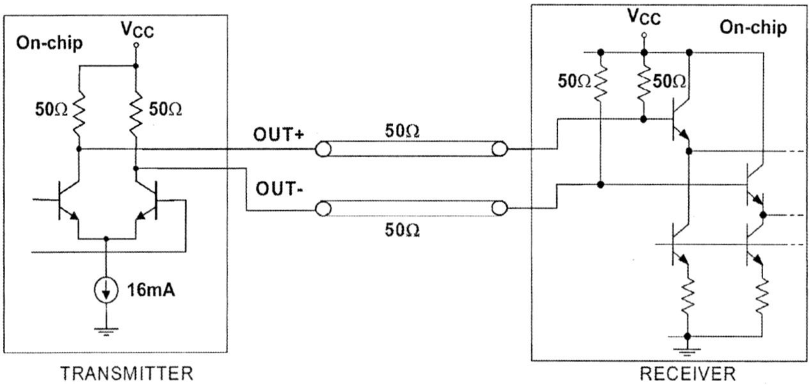

Figure 9.8 Dc-coupled CML circuit with all resistors internal to the chips.

Figure 9.9 shows another dc-coupled case, but now the driver stage has open-collector outputs and the pullup resistors must be supplied externally to the chip. In this configuration, the R_T termination resistors act as the pullup resistors. When a transistor is off, both its output terminal and the receiver input pin to which it is tied will be pulled up to V_{CC}. When a transistor is on, the transistor will sink 16 mA and generate 800 mV across the 50 Ω termination resistor.

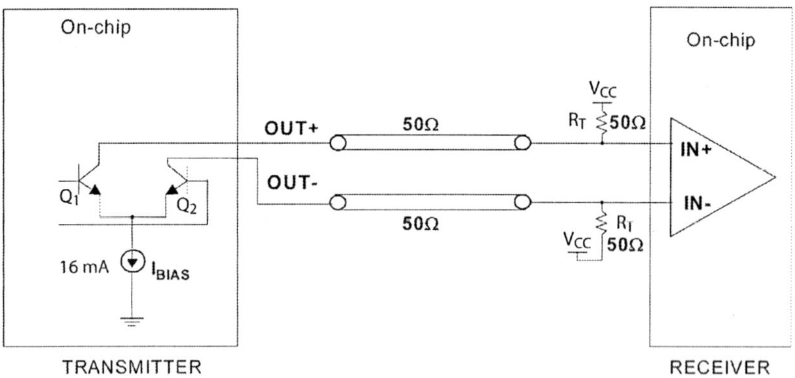

Figure 9.9 Dc-coupled CML circuit with open-collector driver and termination resistors pulled up to V_{CC}.

9.5 XAUI Interface Standard[6]

An overview of the XAUI interface standard is presented here for the reason that many CML components have been designed for use in XAUI applications and adhere to its specifications. The XAUI standard does not define CML, but given the absence of any industry-wide public-domain CML standard, the XAUI standard offers the closest guidelines presently available which design engineers may elect to follow when designing CML devices and systems.

"XAUI stands for '10 Gigabit Attachment Unit Interface' and is specified in Clause[7] 47 of IEEE Standard 802.3ae.[8] The XAUI is primarily intended as a point-to-point interface of up to 50 cm between integrated circuits using controlled impedance traces on low-cost printed circuit boards (PCBs). The recommended differential characteristic impedance of circuit board trace pairs is 100 Ω ± 10% from 100 MHz to 2.5 GHz. The recommended impedance of any connectors, such as ones used between circuit board subsystems, is 100 Ω ± 30%.

Under the XAUI interface, the 10 Gbps data stream is converted into four lanes where each lane is 8B/10B encoded at a nominal rate of 3.125 Gbaud. Each lane is differential with a low voltage swing and the driver and receiver may operate on independent clocks. The majority of the specification deals with the requirements for an individual lane operating at 3.125 Gbaud.

The XAUI is a low-swing, ac-coupled differential interface. Ac coupling allows for interoperability between components operating from different supply voltages. Low swing differential signaling provides noise immunity and improved electromagnetic interference (EMI). Differential signal swings (driver output voltage limits) are discussed below and depend upon several factors such as transmit equalization and transmission line losses.

The XAUI signal paths are point-to-point connections. Each path corresponds to an XAUI lane and is composed of two complementary signals forming a balanced differential pair. There are four differential paths in each direction for a total of 8 pairs, or 16 connections. The signal paths are intended to operate up to approximately 50 cm (20 in.) over controlled impedance traces on standard FR-4 printed circuit boards (PCBs).

The XAUI driver characteristics are summarized in Table 9.2. The XAUI baud is 3.125 Gbaud ± 100 ppm. The corresponding baud period is nominally 320 ps. The load is 100 Ω ± 5% differential measured at 2.5 GHz for these characteristics, unless otherwise noted.

The differential output return loss minimum (see the definition of return loss on page 209) varies from –3.01 dB at 3.125 GHz to –10 dB at 625 MHz. The reference impedance for differential return loss measurements is 100 Ω. Driver differential output amplitude must be less than 1600 $mV_{p\text{-}p}$ including any transmit equalization. (Transmit equalization, a synonym for pre-emphasis, is discussed in Section 9.6.1 on page 211.) Dc referenced logic levels are not defined

6. Quoted material in Section 9.5 is Copyright Apr 2002 IEEE. All rights reserved.
7. A "clause" number of an IEEE document is analogous to a chapter number of a book.
8. IEEE 802 is the IEEE standard, Local and Metropolitan Area Networks (the Ethernet standard). Part 3 of IEEE 802 (IEEE 802.3) is Carrier Sense Multiple Access with Collision Detection (CSMA/CD) Access Method and Physical Layer Specifications.

Table 9.2 XAUI Driver Characteristics

Parameter	Value	Units
Baud Rate Tolerance	3.125 GBd ± 100 ppm	GBd ppm
Unit Interval Nominal	320	ps
Differential Amplitude Maximum	1600	$mV_{p\text{-}p}$
Absolute Output Voltage Limits		
Maximum	2.3	V
Minimum	–0.4	V
Differential Output Return Loss Minimum	–10 for 312.5 MHz < Freq(f) < 625 MHz –10 + 10log(f/625) for 625 MHz ≤ Freq(f) < 3.125 GHz	dB
Output Jitter		
Near-End Maximums		
Total Jitter	±0.175 Peak from the Mean	UI
Deterministic Jitter	±0.085 Peak from the Mean	UI
Far-End Maximums		
Total Jitter	±0.275 Peak from the Mean	UI
Deterministic Jitter	±0.185 Peak from the Mean	UI

since the receiver is ac coupled. Absolute driver output voltage is between –0.4 V and 2.3 V with respect to ground. Figure 9.10 illustrates the absolute driver output voltage limits and defines differential peak-to-peak amplitude. In the figure, SL*i*<P> and SL*i*<N> are the positive and negative sides of the differential signal pair for XAUI Lane *i* (*i* = 0, 1, 2, 3).[9]

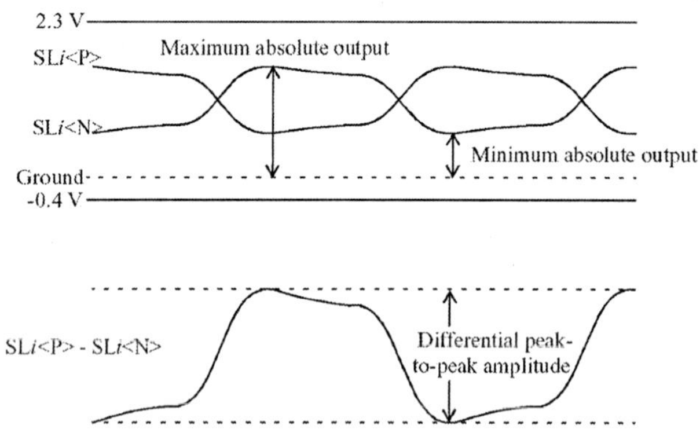

Figure 9.10 Driver output voltage limits and definitions.

9. Quoted material in Section 9.5 is Copyright Apr 2002 IEEE. All rights reserved.

"The concept of 'return loss' is a simple and straightforward one. At any specific point on a uniform transmission line, the power carried by the reflected wave (traveling from the terminal load toward the signal source) will be less than the power carried by the incident wave (traveling from the signal source toward the terminal load) when the magnitude of the reflection coefficient at the terminal load end of the line is less than unity, or when there is attenuation between the specific point and the terminal load.

The return loss in decibels at a point on a transmission circuit is defined as the total loss or attenuation in decibels which the incident wave power at the point would have to experience to be reduced to the reflected wave power at the point. If the voltage reflection coefficient at the terminal load end of a line is ρ, and the line has an attenuation factor α nepers/m, the return loss at a point distant d meters from the terminal load end of the line will be given by

$$\text{return loss} = 10 \log_{10}|\rho|^2 + 8.686 \text{ X } 2\alpha d \text{ decibels."}$$ [Chi68]

Differential transition times between 60 and 130 ps are recommended, as measured between the 20% and 80% levels. Shorter transitions may result in excessive high frequency components and increase EMI and crosstalk. The upper recommended limit of 130 ps corresponds to a sine wave at the half baud.[10]

The driver needs to satisfy either the near-end eye template and jitter requirements, or the far-end eye template and jitter requirements. The eye templates and template intervals are given in Figure 9.11 and Table 9.3. (For a detailed discussion of jitter and eye templates, see Section 26.13 on page 752.) The jitter requirements at the near end are for a maximum total jitter

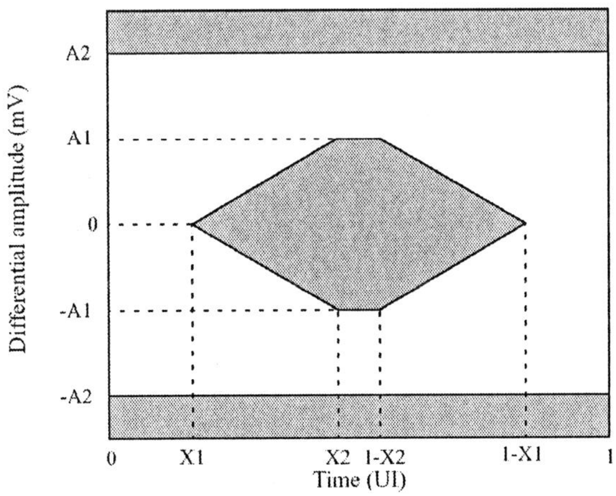

Figure 9.11 Driver jitter template.

10. Quoted material in Section 9.5 is Copyright Apr 2002 IEEE. All rights reserved.

of ±0.175 UI (unit interval) peak from the mean and a maximum deterministic component of ±0.085 UI peak from the mean. The far-end requirements are for a maximum total jitter of ±0.275 UI peak from the mean and a maximum deterministic component of ±0.185 UI peak from the mean. Note that these values assume symmetrical jitter distributions about the mean. If a distribution is not symmetrical, its peak-to-peak total jitter value must be less than these total jitter values to claim compliance to the template requirements. The maximum random jitter is equal to the maximum total jitter minus the actual deterministic jitter. The amount of data represented in the data eye must be adequate to ensure a bit error ratio[11] (BER) of less than 10^{-12}. Jitter is measured with ac coupling and at 0 V differential.

Table 9.3 Driver Template Intervals

Symbol	Near-End Value	Far-End Value	Units
X1	0.175	0.275	UI
X2	0.390	0.400	UI
A1	400	100	mV
A2	800	800	mV

XAUI receivers have to accept differential input signal amplitudes produced by compliant transmitters connected without attenuation to the receiver. Note that these amplitudes may be larger than the 1,600 $mV_{p\text{-}p}$ maximum of Table 9.2 on page 208 owing to actual driver and receiver input impedances. The minimum input amplitude is defined by the far-end driver template and the actual receiver input impedance. Note that the far-end driver template is defined using a well-controlled load impedance. The minimum signal amplitude into an actual receiver may vary from the minimum template height because of the actual receiver input impedance. Since the XAUI receiver is ac coupled to the XAUI, the absolute voltage levels with respect to the receiver ground are dependent on the receiver implementation. Table 9.4 lists XAUI receiver specifications.

The XAUI receiver must be ac coupled to the XAUI to allow for maximum interoperability between various 10 Gbps components (i.e., other families of devices could drive it). Ac coupling is considered to be part of the receiver under the standard unless explicitly stated otherwise. It should be noted that there may be various methods for ac coupling in actual implementations."[12] [Ins02]

11. The terms "bit error ratio" and "bit error rate" can be used interchangeably.
12. Quoted material in Section 9.5 is Copyright Apr 2002 IEEE. All rights reserved.

Table 9.4 Receiver Characteristics

Parameter	Value	Units
Baud Rate Tolerance	3.125 ±100	GBd ppm
Unit Interval (UI) Nominal	320	ps
Receiver Coupling	ac	
Return Loss[a] Differential Common mode	 10 6	 dB dB
Jitter Amplitude Tolerance	0.65	UI_{p-p}

a. Relative to 100 Ω differential and 25 Ω common mode.

9.6 CML Design Considerations

9.6.1 Pre-Emphasis, De-Emphasis, Transmit Equalization, and Receive Equalization

Pre-emphasis and equalization are a means to reduce distortion of signals caused by the effects of transmission media. Transmission media generally have a low-pass frequency response. As a high-frequency signal propagates down a transmission line, high-frequency components of the signal tend to get attenuated more than low-frequency components of the signal. The result is that by the time the signal reaches the receiver at the end of the transmission line (or backplane), the signal is severely distorted. A means of improving the signal quality at the receiver is to transmit the high-frequency components with a larger amplitude than the low-frequency components, so that by the time all the components reach the receiver, the amplitudes of the components are in about the same relative proportion as in the original signal (before the amplitudes were adjusted for transmission).

For example, if low-frequency signal components attenuate 33% but high-frequency components attenuate 45% by the time they reach the receiver, then increasing the magnitude of the high-frequency components by 20% at the driver will allow both low-frequency and high-frequency components to reach the receiver with approximately 33% attenuation (high-frequencies: [1+0.2][1–0.45] = 0.66 or 34% attenuation). Their relative amplitudes will be maintained and distortion will be greatly reduced at the receiver.

One method by which higher frequencies are amplified more prior to transmission is called "pre-emphasis" and sometimes "transmit equalization." Pre-emphasis has the greatest benefit when a signal is driving longer lengths of coax or when its traversing across a large backplane. Rather than run the original signal through filters to identify the high-frequency analog components and then amplify them in a complicated scheme, a simpler approach is to incrementally increase the voltage of the first digital bit in a data stream that occurs immediately after a data transition. This is illustrated in Figure 9.12. In the figure, $V_{OD(p)}$ is the differential driver output voltage following a transition in the data stream and $V_{OD(d)}$ is the differential output voltage when there has been no transition in the data stream. $V_{OD(p)}$ is the differential output voltage with pre-emphasis and $V_{OD(d)}$ is the differential output voltage with de-emphasis. De-emphasis is the opposite of pre-emphasis in that it attenuates or reduces the amplitude of the lower frequency components in the signal before they are transmitted. $V_{OD(pp\_p)}$ and $V_{OD(pp\_d)}$ are the total peak-to-peak differential output voltages with pre-emphasis and de-emphasis, respectively. $V_{OD(pp\_p)}$ and $V_{OD(pp\_d)}$ are sometimes specified in data sheets where $V_{OD(p)}$ and $V_{OD(d)}$ are not provided and $V_{OD(p)}$ and $V_{OD(d)}$ have to be determined by halving the values of $V_{OD(pp\_p)}$ and $V_{OD(pp\_d)}$, as shown in the figure. Also, the voltages in the figure are differential voltages. At the crossing point of 0 V, the differential voltage is 0 V but the single-ended voltages at the driver's terminals are both equal to the offset (common-mode voltage).

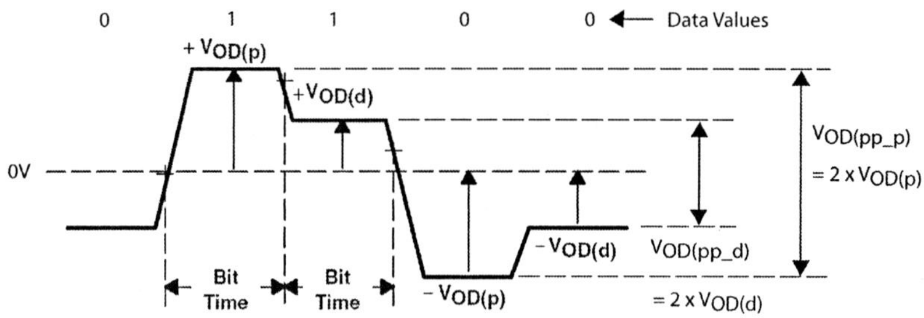

Figure 9.12 Driver differential output voltages vs. time showing pre-emphasis and de-emphasis. [Courtesy Texas Instruments.]

Figure 9.13 shows the same output voltages as in Figure 9.12 but now from a single-ended perspective where both the true and complementary driver outputs are plotted with reference to ground.

Amplification of the higher frequencies of a signal can also be done at the receiver to "open up the eye" of the received signal. When this is done, the process is usually referred to as simply "equalization" but sometimes as "receive equalization" or "post equalization." This is done at the load end with special amplifier stages and or equalizer (analog or digital) circuits in the front end of the receiver device.

For some devices, the amount of pre-emphasis can be selected by the designer, as shown in Table 9.5 for the TLK3101 SerDes transceiver.[13] The level of pre-emphasis is controlled by a PREM (pre-emphasis control) input pin on the device.

13. The Texas Instruments' TLK3101 SerDes transceiver is not a CML device but has a VML (voltage mode logic) serial driver that is described in the data sheet as a pseudo ECL driver (PECL).

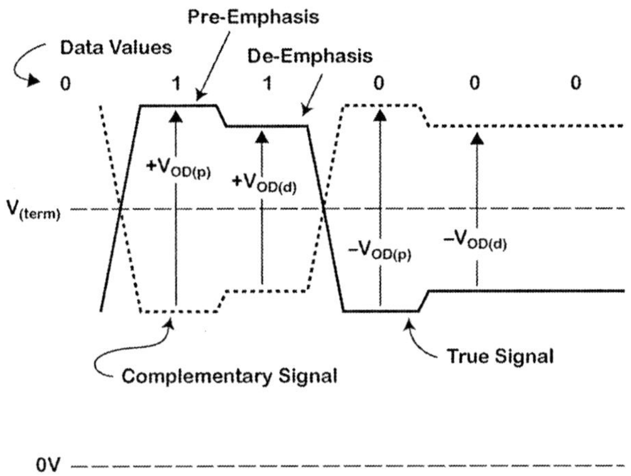

Figure 9.13 Driver output voltages plotted as single-ended and referenced to ground illustrating both pre-emphasis and de-emphasis.

Table 9.5 Pre-Emphasis Levels for the TLK3101 Output Driver Are Controlled by the PREM Input Pin. [Tex01]

PREM	Pre-Emphasis Level (%) ($V_{OD(p)} / V_{OD(d)} - 1$)
0	5%
1	20%

[Courtesy Texas Instruments]

9.6.2 AC Coupling Requires 8B/10B Encoding or DC-Balanced Signal

If a signal contains a dc component that is required for accurate data reception, ac coupling between the driver and receiver will eliminate the dc component and make the data unrecoverable. A benefit of 8B/10B encoding is that it has no dc component and is frequently used in ac coupling schemes. 8B/10B encoding guarantees a minimum number of transitions per frame and dc balance of the data. A disadvantage of 8B/10B encoding is that it requires 25% overhead. Encoding 8 data bits requires 10 signaling bits. Where signal protocols allow long run lengths of consecutive 1s or 0s, ac coupling is not possible. When considering whether a signal contains a dc component, pre-emphasis and equalization should be taken into account. The majority of CML circuits are ac coupled. [Gol03a]

9.7 How CML and ECL Differ

Figure 9.14(a) shows an ECL circuit with a 50 Ω transmission medium (single-ended impedance value), a termination circuit, and a differential receiver. Figures 9.14(b) and (c) show CML circuits, with (b) having pullup resistors in the output structure and (c) having open-collector outputs. All circuits are shown with far-end termination but no source termination external to the device.

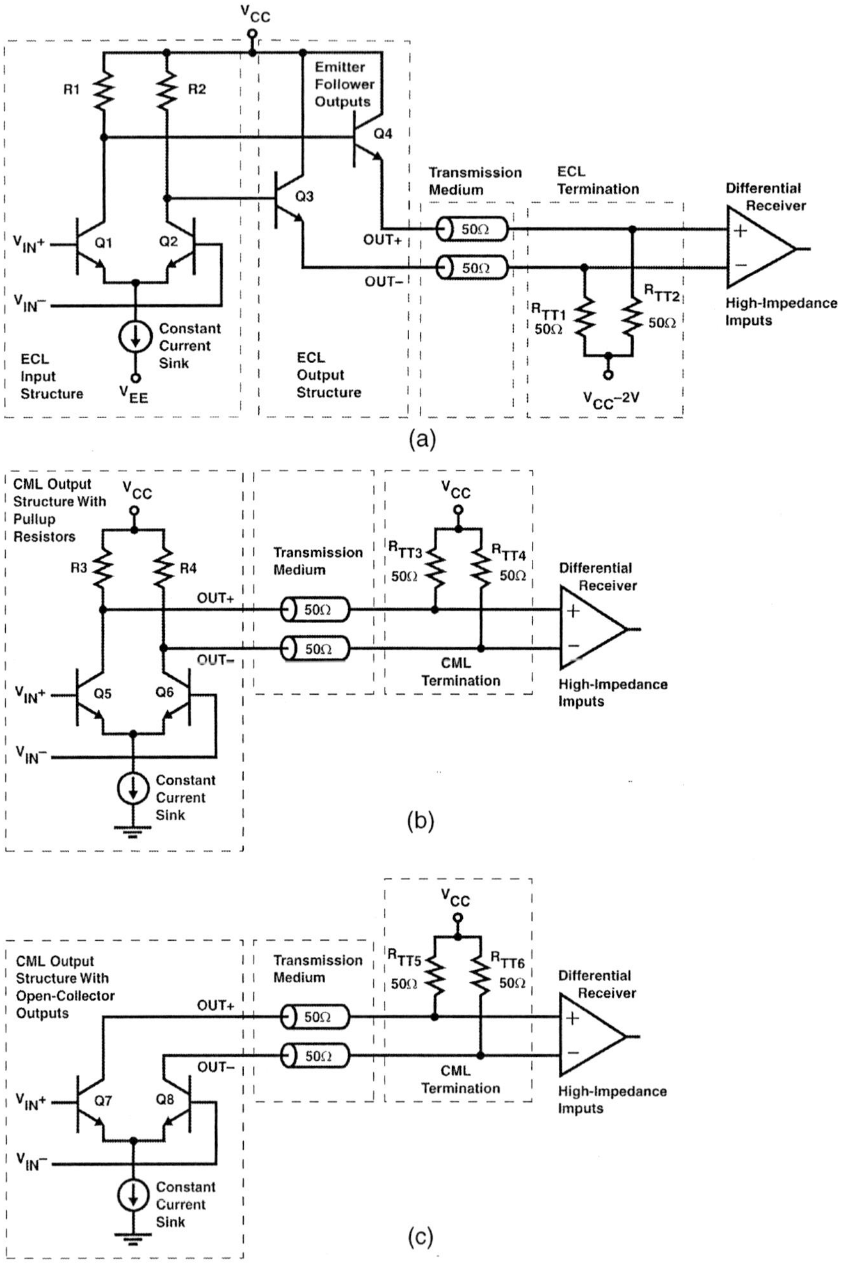

Figure 9.14 CML devices have higher output crosspoint voltages and higher output impedances compared with 10 Series and 100 Series ECL devices. (a) ECL circuit with emitter follower outputs, (b) CML circuit with pullup resistors in the output structure, and (c) CML circuit with an open-collector output structure.

The main difference between ECL and CML is that ECL contains emitter follower circuits in the output stage and CML does not. The effect is to cause the dc voltage level (common-mode dc voltage) of the ECL outputs to be lower than the CML outputs by approximately one diode voltage drop (~700 mV) plus another 100 mV to 200 mV. The diode drop corresponds to the base emitter voltage drop of transistors Q3 and Q4. In Figure 9.14(a), if V_{IN}+ is high and V_{IN}– is low, Q1 is on (but not saturated) and Q2 is off. V_{CC} is at a higher voltage than V_{EE} in relative terms. With Q1 on, current flows through resistor R1 and a voltage drop of about 800 mV occurs across it. The voltage at the bottom of R1 is V_{CC}–800 mV and this voltage is applied to the base of transistor Q4. The output from Q4 on its emitter is an additional diode drop lower, or V_{CC}–800 mV–$V_{diode\ drop}$ or about V_{CC}–1500 mV. With Q2 off, almost no current flows through resistor R2. Only a small base current flows through R2 to the base of Q3. There is very little voltage drop across R2 and the voltage output from Q3 is approximately one diode voltage drop below V_{CC}. Q4's output (OUT+) is low and Q3's output (OUT–) is high. In actual devices that conform to the 100K standard, both of the Q3 and Q4 outputs are lower still. The output from Q3 is typically about V_{CC}–955 mV and the output from Q4 is about V_{CC}–1705 mV (0°C to +85°C) for this logic state (Q1 on and Q2 off).

Figure 9.14(b) shows a circuit similar to the ECL one shown before, but now for a CML device with pullup resistors in the output stage. When V_{IN}+ is high and V_{IN}– is low, Q5 is on and Q6 is off. With this circuit design, when Q6 is off, R4 and the R_{TT4} pullup resistors force OUT– to the V_{CC} rail. Q5 is on and now sinks current from both R3 and its attached pullup termination resistor R_{TT3}. Voltage drops across both R3 and R_{TT3}, and OUT+ is low. How low OUT+ goes depends upon the values of R3 and R_{TT3}; that is, how low it goes is device and application specific. This in turn means that it depends upon the transmission medium's characteristic impedance, how the transmission medium is terminated, and the value of the selected termination impedances. The circuit in Figure 9.14(b) could also be terminated by removing pullup resistors R_{TT3} and R_{TT4} and adding a 100 Ω termination resistor in shunt between the true and complementary inputs at the receiver. The benefit of the shunt termination is that the differential output voltage swing is not reduced by adding the termination.

Figure 9.14(c) illustrates that CML devices are also manufactured with open collector outputs. With open-collector outputs, parasitic capacitances associated with the device's internal pullup resistors are no longer relevant as the pullup resistors are eliminated. The low-state output voltage varies with the value of the termination resistors selected.

We've seen that one effect of the emitter follower output stages of ECL is that ECL output voltages are several hundred millivolts below the V_{CC} rail. For ECL conforming to the 100K standard at 25°C, its outputs swing approximately 750 mV between V_{CC}–955 mV and V_{CC}–1,705 mV. The output swing will vary with temperature and with the standard the device conforms to (10K or 100K), as shown in Table 8.4 on page 173. A second effect of the ECL emitter follower output stages is that the output impedance of emitter follower driver outputs is very low, typically about 4–8 Ω.

Table 9.6 lists the dc electrical characteristics for a CML device. The table lists a typical CML output impedance of 100 Ω, which is considerably higher than the ECL output impedance of 4–8 Ω. The table also shows that CML differential output voltages vary from 200 mV to 800 mV depending upon the loading. Also notice that V_{OH} is almost at the V_{CC} rail, only 10 mV below it and V_{OL} is only 825 mV below V_{CC} (typically). These CML dc levels which are very close to the V_{CC} rail contrast the ECL dc levels, which are often 945 mV or more lower.

Table 9.6 CML DC Electrical Characteristics for Micrel SuperLite™ Device (SY55852U[a])

V_{CC} = 2.3 V to 5.7 V; GND = 0 V; T_A = –40°C to +85°C[b]						
Symbol	**Parameter**	**Min**	**Typ**	**Max**	**Unit**	**Condition**
V_{ID}	Differential Input Voltage	100	—	—	mV	
V_{IH}	Input HIGH Voltage[c]	1.6	—	V_{CC}	V	
V_{IL}	Input LOW Voltage[c]	1.5	—	V_{CC}–0.1	V	
V_{OH}	Output HIGH Voltage	V_{CC}–0.020	V_{CC}–0.010	V_{CC}	V	No Load
V_{OL}	Output LOW Voltage	V_{CC}–0.97	V_{CC}–0.825	V_{CC}–0.660	V	No Load
V_{OS}	Output Voltage Swing[d]	0.660	0.800 0.400 0.200	0.950	V	No Load 100 Ω Environment[e] 50 Ω Environment[f]
R_{DRIVE}	Output Source Impedance	80	100	120	W	

a. Specification for packaged product only.

b. Equilibrium temperature.

c. Inputs must be biased to logic LOW or HIGH when V_{CC} is less than 3.0 V.

d. Actual voltage levels and differential swing will depend on customer termination scheme. Typically, a 400 mV swing is available in the 100 Ω environment and a 200 mV swing in the 50 Ω environment. Refer to Figure 9.15 on page 217.

e. See Figure 9.15(c).

f. See Figures 9.15(a) and (b).

Figures 9.15(a), (b), and (c) illustrate proper CML device terminations for both 50 Ω and 100 Ω characteristic impedance media using the SY55852U SuperLite device. Because of the internal (to the device) 100 Ω pullup resistors in the output stage of the device, the device's outputs are considered to be "source-terminated 100 Ω outputs." Figure 9.15(a) shows 50 Ω media terminated with a 100 Ω shunt impedance. The differential impedance of the two 50 Ω (single-ended characteristic impedance values) transmission lines is 100 Ω, which is exactly matched by the 100 Ω shunt resistor. No significant signal reflections should occur at the far end of the transmission line owing to this exact impedance matching. The device is also terminated at the source

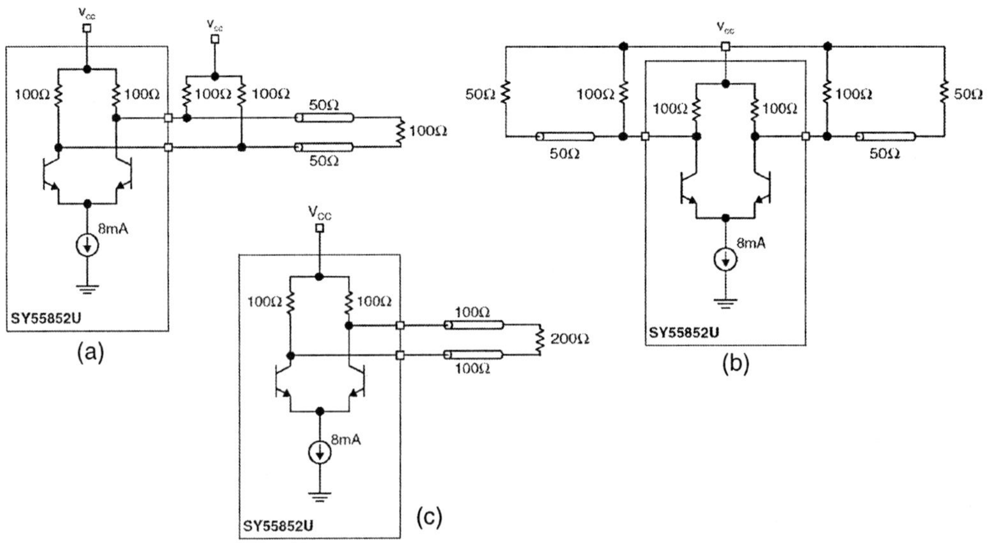

Figure 9.15 Source-terminated 100 Ω output impedance CML differential drivers. (a) Differentially terminated (50 Ω load CML output), (b) Individually terminated (50 Ω load CML output), and (c) 100 Ω load CML output.

with the addition of the two 100 Ω external resistors pulled up to V_{CC}. The parallel combination of the internal 100 Ω pullup resistor with the external 100 Ω pullup resistor for each trace is 50 Ω, which exactly matches the transmission line's single-ended characteristic impedance. No signal reflections should occur at the near end of the transmission line either. Not shown in the figure is the receiver, which is connected across the shunt resistor.

The transmission lines shown in Figures 9.15(b) and (c) are also properly terminated at both their near and far ends, at the source and at the load. As shown in Table 9.6, terminations affect the loading, which in turn affects the differential output voltage swing seen at the load. In referring back to the circuits in Figure 9.14 on page 214, we see that these are only terminated at the far end of the transmission media, that is, at the receiver. They have no source termination.

The characteristic impedances for the transmission lines shown in Figure 9.15 are single-ended values and measured between the transmission line and ground. Differential characteristic impedances for differential traces are approximately double the single-ended values for widely separated true and complement differential traces. As the separation of the differential traces becomes less, capacitive and inductive coupling between the true and complement differential traces increases, and the differential characteristic impedance begins to decrease from the "double-the-single-ended-value." The differential characteristic impedance is measured between the true trace and the complementary trace.

Another difference between ECL and CML is that ECL outputs are pulled down to a voltage less than V_{CC}, often to V_{CC}–2 V or all the way down to V_{EE}, whereas CML outputs are pulled up to V_{CC}. Put differently, ECL uses pulldown resistors and CML uses pullup resistors.

Frequently, ECL devices are used as both PECL or NECL. They may be employed in circuits using either positive or negative power sources. In contrast, CML is almost always connected to a positive voltage power source. CML is often used as the output stage of SerDes devices that also contain LVTTL or HSTL inputs. Operating the SerDes device from a negative power source would destroy the logic levels for the LVTTL or HSTL inputs.

9.8 SuperLite CML and GigaPro™ CML

Micrel Semiconductor's SuperLite CML logic family operates over a wide power span from 2.3 V to 5.7 V and comes in tiny MSOP, TQFP, and QSOP packages. SuperLite devices have guaranteed toggle speeds of 2.5 GHz and beyond and are intended for OC-48 communication systems, telecom, Gigabit Ethernet, fibre channel, SONET, and fiber optic applications. The product line includes D flip-flops, D latches, 2 × 2 protection crosspoint switches, and CML/PECL/LVPECL-to-LVDS translators.

GigaPro CML devices are also offered by Micrel. These devices use a 0.6 micron SiGe fabrication process resulting in communication data rates exceeding 10 Gbps and meet the low jitter and skew requirements of SONET OC-48 to OC-192 systems, 10/100 Gigabit Ethernet, and 2X fiber channel systems. As an example, the SY58016 differential CML buffer has a –3 dB bandwidth greater than 7 GHz, 45 ps maximum rise and fall times, less than 1.5 ps RMS jitter, and internal 50 Ω termination. The SY58016 comes in a 16-pin (3 × 3 mm) MLF package or die form. Operation is guaranteed from a 3.3 V ± 10% power supply over the –40°C to +85°C temperature range.

9.9 Vendor-Specific CML Examples

9.9.1 Texas Instruments' SN65CML100

"The SN65CML100 is a 1.5 Gbps LVDS/LVPECL/CML-to-CML translator/repeater designed to support various high-speed network routing applications. The driver output is compatible with current-mode logic (CML) levels and directly drives 50 Ω or 25 Ω loads connected to 1.8 V, 2.5 V, or 3.3 V nominal supplies. The chip's main power is 3.3 V. The capability for direct connection to loads eliminates the need for coupling capacitors. The receiver input is compatible with LVDS (TIA/EIA–644), LVPECL, and CML signaling levels. The receiver tolerates a wide common-mode voltage range, and may also be directly coupled to the signal source. The internal data path from input to output is fully differential for low noise generation and low pulse-width distortion." [Tex02a]

Table 9.7 illustrates the direct relationship between the termination resistors connected to the output stage of the CML driver and the generated output voltage. The SN65CML100 uses external pullup resistors, R_T, as shown in Figure 9.16. The circuit configuration is similar to

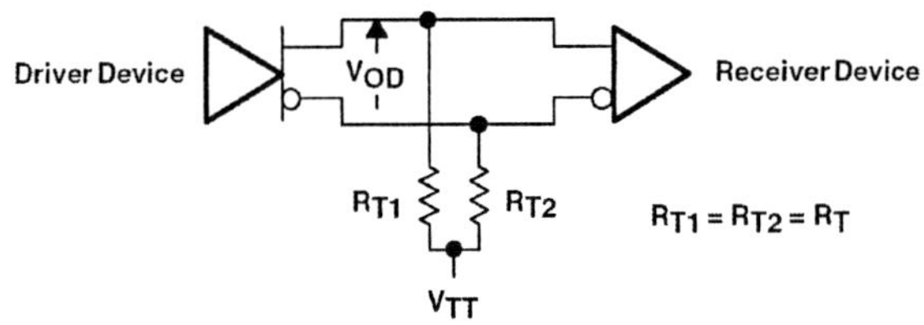

Figure 9.16 Typical termination for output driver of SN65CML100. [Courtesy Texas Instruments.]

Figure 9.9 on page 206, except that V_T replaces V_{CC} for the termination voltage. The important concept demonstrated by the table is that, as the termination resistors R_T change from 50 Ω to 25 Ω, the differential output voltage changes proportionately and is reduced by half. This is due to the constant current sink in the output driver stage that pulls the same current through the termination resistors no matter what their value (within limits). Also, in comparing this table with our CML "reference" table, Table 9.1 on page 201, the typical values of Table 9.7 for a V_{TT} of 3.3 V and R_T of 50 Ω correspond to the reference values of Table 9.1 fairly well.

Table 9.7 Output Electrical Characteristics of the SN65CML100 Translator/Repeater

Symbol	Parameter	Test Conditions	Min	Typ	Max	Unit
V_{OH}	Output High Voltage[a]	$R_T = 50\Omega$ $V_{TT} = 3$ V to 3.6 V or $V_{TT} = 2.5$ V ± 5%	V_{TT}–60	V_{TT}–10	V_{TT}	mV
V_{OL}	Output Low Voltage[a]		V_{TT}–1100	V_{TT}–800	V_{TT}–640	mV
$\lvert V_{OD}\rvert$	Differential Output Voltage Magnitude		640	780	1000	mV
V_{OH}	Output High Voltage[b]	$R_T = 25\Omega$ $V_{TT} = 3$ V to 3.6 V or $V_{TT} = 2.5$ V ± 5%	V_{TT}–60	V_{TT}–10	V_{TT}	mV
V_{OL}	Output Low Voltage[b]		V_{TT}–550	V_{TT}–400	V_{TT}–320	mV
$\lvert V_{OD}\rvert$	Differential Output Voltage Magnitude		320	390	500	mV
V_{OH}	Output High Voltage[a]	$R_T = 50\Omega$ $V_{TT} = 1.8$ V ± 5%	V_{TT}–170	V_{TT}–10	V_{TT}	mV
V_{OL}	Output Low Voltage[a]		V_{TT}–1100	V_{TT}–800	V_{TT}–640	mV
$\lvert V_{OD}\rvert$	Differential Output Voltage Magnitude		570	780	1000	mV
V_{OH}	Output High Voltage[b]	$R_T = 25\Omega$ $V_{TT} = 1.8$ V ± 5%	V_{TT}–85	V_{TT}–10	V_{TT}	mV
V_{OL}	Output Low Voltage[b]		V_{TT}–500	V_{TT}–400	V_{TT}–320	mV
$\lvert V_{OD}\rvert$	Differential Output Voltage Magnitude		285	390	500	mV

a. Outputs are terminated through 50 Ω resistors to V_{TT}; CML level specifications are referenced to V_{TT} and track 1:1 with variation of V_{TT}.

b. Outputs are terminated through 25 Ω resistors to V_{TT}; CML level specifications are referenced to V_{TT} and track 1:1 with variation of V_{TT}.

[Courtesy Texas Instruments.]

9.9.2 Texas Instruments' TLK2501 1.5 to 2.5 Gbps Transceiver

The TLK2501 is a multigigabit transceiver used in ultra high-speed bidirectional point-to-point data transmission systems. The TLK2501 supports an effective CML serial interface speed of 1.5 Gbps to 2.5 Gbps, providing up to 2 Gbps of data bandwidth. The primary application of this chip is to provide very high-speed I/O data channels for point-to-point baseband data transmission over controlled impedance media of approximately 50 Ω (single-ended value). The transmission media can be printed circuit board, copper cable, or fiber-optic cable. The maximum rate and distance of data transfer is dependent upon the attenuation characteristics of the media and the noise coupling to the environment. The TLK2501 is a SerDes (serializer/deserializer) device and performs parallel-to-serial and serial-to-parallel data conversion. It is hot-pluggable, operates off of a 2.5 V power supply, provides on-chip 8B/10B encoding/decoding, comma alignment, and link synchronization. It has a programmable voltage output swing on the serial output, receiver differential input thresholds of 200 mV minimum, and consumes typical power of 360 mW.

"The high-speed data output driver consists of a current-mode logic (CML) differential pair that can be optimized for a particular transmission line impedance and length. The line can be dc-coupled or ac-coupled."[14] [Tex02b] Termination details are shown in Figure 9.17 for the dc-coupled case and Figure 9.18 for the ac-coupled case.

In Figure 9.17 (for dc coupling), TXP and TXN represent the transmitter's positive and negative output terminals of the serial link, and RXP and RXN represent the positive and negative input terminals of the receiver, respectively.[15] $V_{(term)}$ is equal to V_{DD}, which is 2.5 V. The TLK2501 CML serial output driver stage does not have resistor pullups as those shown in Figure 9.3 on page 202, but instead gives the circuit designer the flexibility of using external resistors. The termination resistors R_t in Figure 9.17 act as pullups in the dc-coupled case. When the left transistor in the figure is "off," no current flows through TXP and $V_{(term)}$ pulls RXP to the 2.5 V power supply rail. When the left transistor is "off," the right transistor is "on" and current flows from $V_{(term)}$ through the bottom R_t resistor, the transmission line, terminal TXN, down through the "on" transistor and finally down through the constant-current source to ground. In the case where R_t is 50 Ω and the pre-emphasis current is 21 mA, a voltage drop of 1,050 mV occurs across R_t for a 1,050 mV differential output voltage as measured from RXN to RXP. This agrees exactly with the nominal data sheet value for VOD(p) listed in Table 9.8 on page 223. For the present, we're ignoring any transmission line effects which may occur. If the transmission line has a single-ended characteristic impedance Z_0 of 50 Ω, Z_0 and R_t match and no reflection of the signal occurs at the receiver.

14. Courtesy Texas Instruments.
15. The differential output terminals might be more accurately labeled as "true" and "complementary" as neither output ever goes negative. For purposes of being consistent with vendor data sheets, this book will use the labeling in those data sheets such as TXP and TXN (transmit positive and negative).

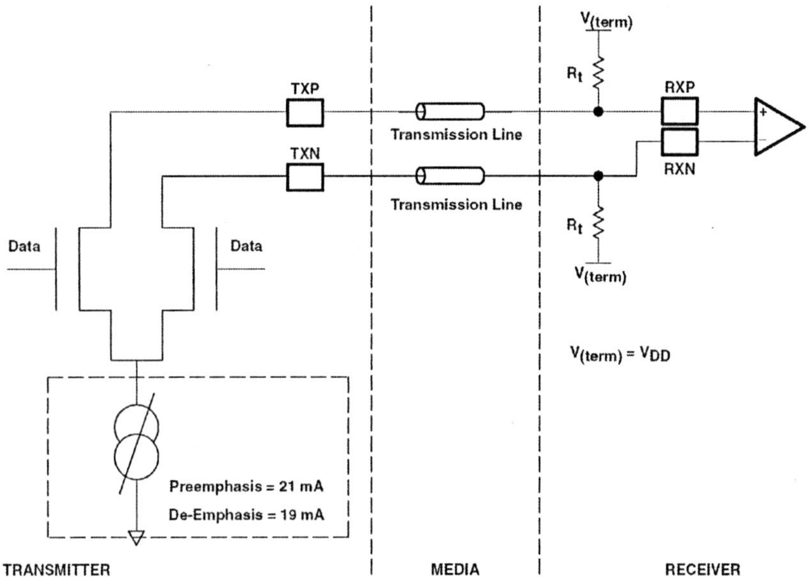

Figure 9.17 High-speed I/O dc-coupled mode using the TLK2501. [Courtesy Texas Instruments.]

In Figure 9.18 for the ac-coupled case, two R_t pullup resistors are located at the driver outputs and two more R_t pullup resistors, separated by the transmission lines and coupling capacitors, are located at the receiver. The resistors located at the driver are tied to V_{DD} (2.5 V) and the ones at the receiver are tied to the termination voltage $V_{(term)}$. For the dc case, $V_{(term)}$ was tied to 2.5 V, but for the ac case, $V_{(term)}$ is tied to 2.0 V as determined by the 200 Ω / 820 Ω voltage divider shown in the figure. The value of $V_{(term)}$ becomes the dc offset (common-mode) voltage at the receiver as was illustrated in Figure 9.12 on page 212.

The differential output voltage of the CML driver is controlled by a reference resistor connected between pin 51 of the device and V_{DD}. Figure 9.19 plots the value of the needed reference resistor as a function of desired differential output voltage for both 50 Ω and 75 Ω transmission media. In the figure, "VODP" is the differential output voltage with pre-emphasis applied and "VODD" is the differential output voltage with de-emphasis applied. See Section 9.6.1 on page 211 for a detailed discussion of pre-emphasis and de-emphasis. Figure 9.19 demonstrates that CML cannot be identified by any characteristic differential output voltage as CML outputs can be programmed to take on a wide range of voltages.

Table 9.8 lists the CML driver and receiver characteristics for the TLK2501 SerDes transceiver. The values listed in the table are for a termination resistor R_t = 50 Ω and reference resistor RREF = 200 Ω. In comparing the table's nominal values with those of Table 9.1 for the reference case, we see that the differential output voltage is higher, 1,260 mV with pre-emphasis and 1140 mV with de-emphasis versus 800 mV for the reference case. We can, of course, make the differential output voltage equal 800 mV by increasing the value of RREF in line with Figure 9.19. The reference receiver has a 100 mV differential threshold whereas the TLK2501 has a 200 mV minimum input voltage requirement. [Tex02b]

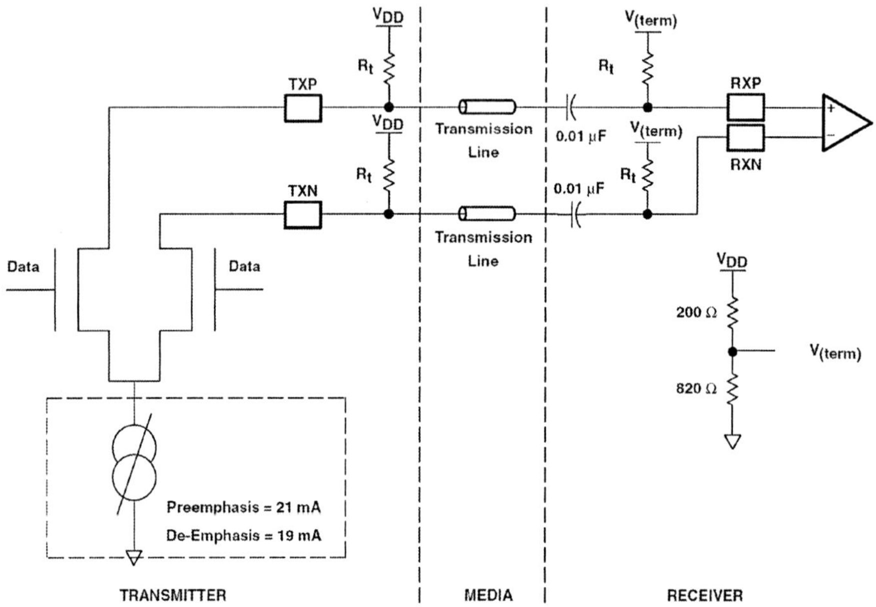

Figure 9.18 High-speed I/O ac-coupled mode using the TLK2501. [Courtesy Texas Instruments.]

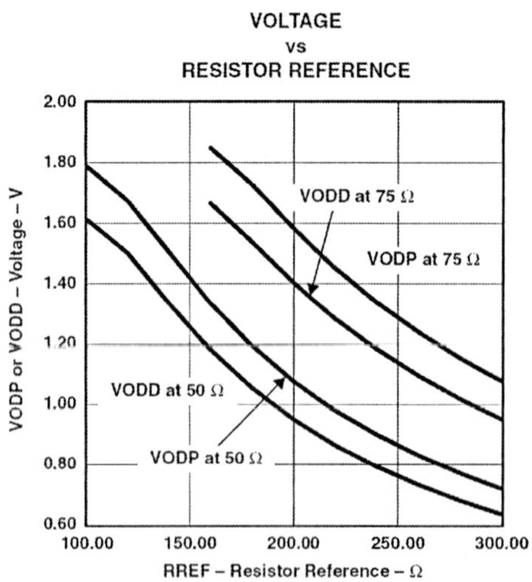

Figure 9.19 Differential transmitter output voltage is controlled by a reference resistor.

Table 9.8 TLK2501 CML Transmitter and Receiver Characteristics

Symbol	Parameter	Min	Nom	Max	Units
$V_{OD(p)}$	Differential Output Voltage with Pre-Emphasis	840	1050	1260	$mV_{p\text{-}p}$
$V_{OD(d)}$	Differential Output Voltage with De-Emphasis	760	950	1140	$mV_{p\text{-}p}$
$V_{(term)}$	Transmit Termination Voltage Range, dc-Coupled			V_{DD}	mV
	Transmit Termination Voltage Range, ac-Coupled	1500		$V_{DD}-V_{ID}/2$	mV
V_{ID}	Differential Receiver Input Voltage Requirement	200			mV
$V_{(cmr)}$	Receiver Common-Mode Voltage Range	1500		$V_{DD}-V_{ID}/2$	mV

[Courtesy Texas Instruments.]

9.9.3 Maxim's MAX3800 3.2 Gbps Adaptive Equalizer and Cable Driver

"The MAX3800 is a +3.3 V adaptive cable equalizer and cable driver implemented together on a single chip. It is designed for coaxial and twin-axial cable point-to-point communications applications, as shown in Figure 9.20. In the figure, each controlled-impedance transmission line shown contains two conductors. The driver's terminals are labeled as DIN and DOUT and both are differential (two conductors) CML stages. The equalizer terminals are labeled as EIN and EOUT and, as with the driver, are differential CML stages. In addition, the driver has an adjustable differential output amplitude and the equalizer includes a loss-of-signal (LOS) output and a cable integrity monitor (CIM) output. The adaptive cable equalizer is capable of equalizing differential or single-ended signals at data rates up to 3.2 Gbps. It automatically adjusts to attenua-

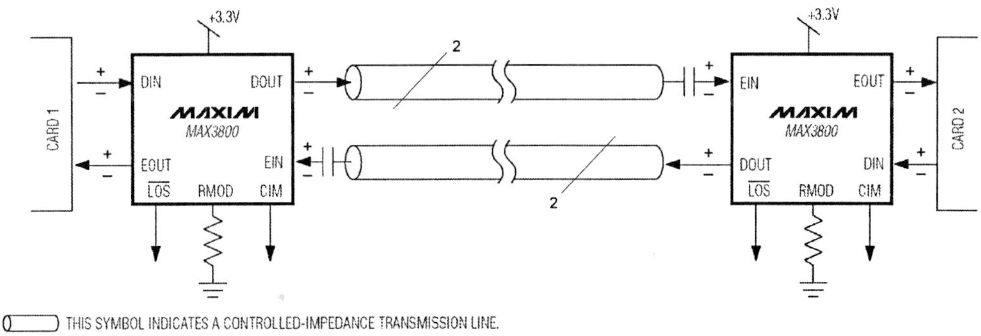

Figure 9.20 Full-duplex CML cable driving circuit employing two MAX3800 chips. [Reprinted with permission of Maxim Integrated Products.]

tion caused by skin-effect losses of up to 30 dB at 1.6 GHz. The equalizer effectively extends the usable length of copper cable in high-frequency interconnect applications.

The MAX3800 is available in a 32-pin TQFP package with exposed pad and consumes only 200 mW at +3.3 V. The driver can be disconnected from the power supply when it is not needed, resulting in a 40% reduction in supply current. Applications for the MAX3800 include high-speed links in communications and data systems, backplane and interconnect applications, and SDH/SONET transmission equipment."[16] [Max01a]

The MAX3800 has several features that differ from other CML devices. Some of these features are evident from Table 9.9, which lists the dc electrical characteristics for the driver output stage and the equalizer input stage. The MAX3800 is designed to drive cables and has a slightly higher output impedance of 62.5 Ω versus the CML "reference case" impedance of 50 Ω, as shown in Figure 9.21. As does the TI TLK2501, the MAX3800 has an adjustable differential output voltage that can range from 400 mV to 1,300 mV, as shown in Figure 9.22. The output voltage is controlled by connecting a resistor between the RMOD pin and ground, where the resistor size determines the output voltage.

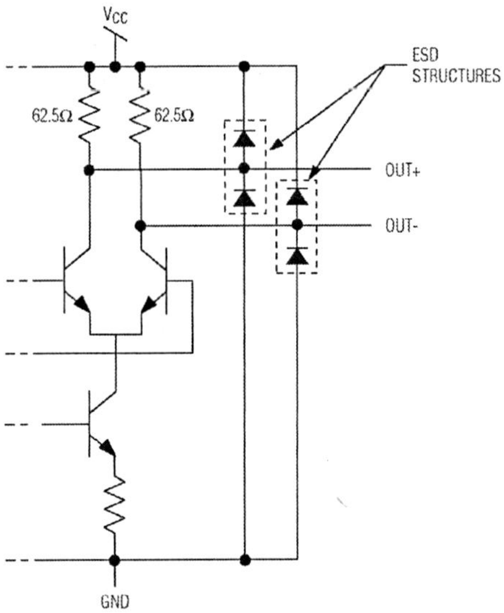

Figure 9.21 CML driver output stage of the MAX3800 using 62.5 Ω pullup resistors.
[Reprinted with permission of Maxim Integrated Products.]

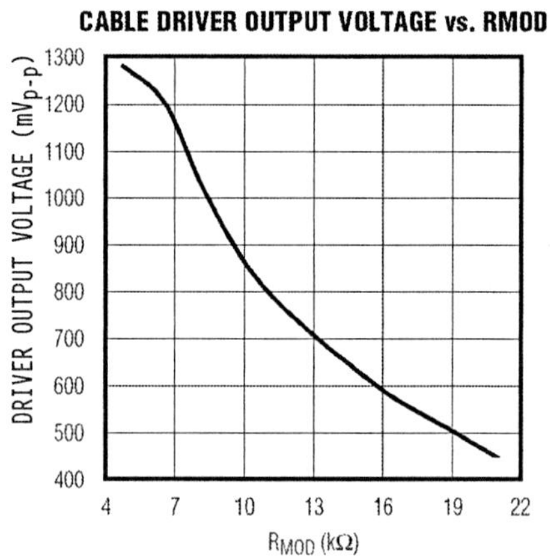

Figure 9.22 The MAX3800 has an adjustable differential output amplitude.

16. Reprinted with permission of Maxim Integrated Products.

Table 9.9 dc Electrical Characteristics of the MAX3800 CML Cable Driver and Equalizer

Parameter	Conditions	Min	Typ	Max	Units
Cable Driver Output Specifications					
Output Voltage (Differential)	$R_{MOD} = 10k\Omega$	750	870	1000	$mV_{p\text{-}p}$
	$R_{MOD} = 20k\Omega$	400	450	550	$mV_{p\text{-}p}$
Output Impedance	Single-Ended	50	62.5	75	Ω
Cable Equalizer Input Specifications					
Minimum Cable Input (Differential)	3.2 Gbps, 30 dB Cable Loss at 1.6 GHz		650	700	$mV_{p\text{-}p}$
Maximum Cable Input (Differential)			1100		$mV_{p\text{-}p}$
Input Impedance		45	55	65	Ω

[Reprinted with permission of Maxim Integrated Products.]

Another feature of the MAX3800 is that it is not a SerDes device. In referring to Figure 9.20, it is seen that the input to the high-speed driver is fed serially as well. This requires an input data rate of 3.2 Gbps if the cable data rate is to be 3.2 Gbps. No serialization or deserialization takes place.

9.9.3.1 Adaptive Equalization

"The MAX3800 adaptive cable equalizer (shown in Figure 9.23) accepts differential CML input data at rates up to 3.2 Gbps and is capable of equalizing differential or single-ended signals. It automatically adjusts to attenuation levels of up to 30 dB at 1.6 GHz (due to skin-effect losses in copper cable). The equalizer consists of a CML input buffer, a loss-of-signal detector, a flat response amplifier, a skin-effect compensation amplifier, a current-steering network, a dual power-detector feedback loop, an output limiting amplifier, and a CML output buffer." [Max01a]

In Figure 9.23, DIN, EIN, DOUT, and EOUT are differential signals requiring two pins per signal for a total of eight pins.

"The shape of the power spectrum of a random bit stream can be described by the square of the sinc function, where sinc f = (sin πf) / πf where f is frequency. For sufficiently long bit patterns (non-random bit streams), $\text{sinc}^2(f)$ is a good approximation. From the shape of the $\text{sinc}^2(f)$ function, we can estimate the ratio of the power densities at any two frequencies. The MAX3800 adaptive equalizer employs this principle by incorporating a feedback loop that continuously monitors the power at two frequencies and dynamically adjusts the equalizer to maintain the correct power ratio." [Max01a]

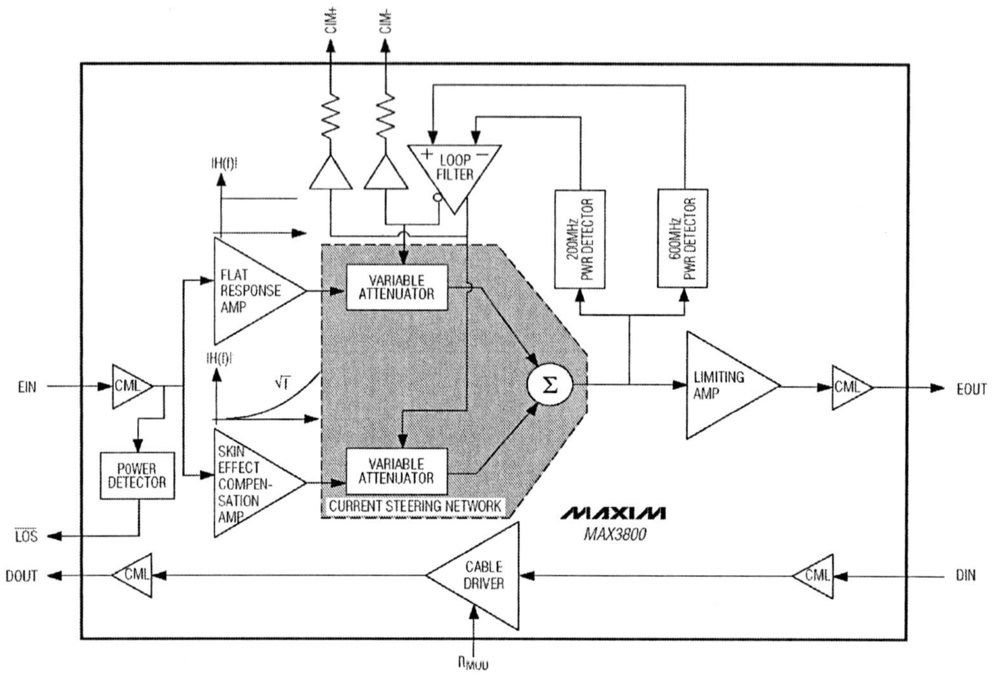

Figure 9.23 Adaptive cable equalizer consists of CML input buffer, loss-of-signal detector, flat response amplifier, skin-effect compensation amplifier, current-steering network, dual power-detector feedback loop, output-limiting amplifier, and a CML output buffer. [Max01a] [Reprinted with permission of Maxim Integrated Products.]

9.10 Summary of Current-Mode Logic (CML)

Table 9.10 Summary of Current-Mode Logic (CML)

CML (Current-Mode Logic) Summary	
Applications:	
Output/input stages of SerDes transceivers	Point-to-point configurations
SDH/SONET Transmission Equipment	Cable interconnects
High-speed backplane interconnects	High-speed serial links
ac coupling to other differential technologies (requires 8B/10B or other encoding)	
Where Not To Use:	
Multipoint or multidrop applications	
For any load whose size, high or low, is inconsistent with the constant current nature of the driver; i.e., for loads much less than 25 Ω or much greater than 100 Ω	
Device Voltages: 3.3 V, 2.5 V, 1.8 V	
Additional Power Sources: V_T termination voltage of 3.3, 2.5, or 1.8 V required for some devices	

Table 9.10 Summary of Current-Mode Logic (CML) (Continued)

Drive Current: 16 mA to 21 mA constant sink current

Driver Differential Output Voltage Magnitude: Vendor specific, commonly 800 mV into 50Ω load, 400 mV into 25Ω load; on some devices, can be tailored to be any voltage using reference resistor; can be as high as 1390 mV with 50Ω load and V_{CC} of 3.3 V

Transmission Media Impedance: Commonly 50 Ω for backplanes, 62.5 Ω for cables (single-ended impedance values), 100 Ω differential impedance

Process: Bipolar, CMOS

Technology: Differential

Power Consumption: Medium, 200 to 420 mW per serial channel depending upon device and function.

Standard: Vendor specific; some compliance to IEEE 802.3ae Clause 47 for CSMA/CD, which defines the XAUI interface

Termination: Vendor/device specific. Both on-chip and external termination. Commonly 50 Ω termination resistor tied to termination voltage on each side of differential pair. Some devices allow termination resistor connected between traces at receiver. Varies with ac vs. dc coupling.

Hot Pluggable: Normally yes, but check data sheet

Maximum Noise Rejection: 800 mV min. differential output detected at receiver having 100 mV differential threshold can give 350 mV single-ended noise immunity over short distances. High common-mode noise rejection due to differential nature of system. Many CML devices have adjustable differential output voltages which allow designer flexibility to design noise margins vs. crosstalk.

Maximum Drive Distance:

- 60 inches (152 cm) of FR-4 backplane trace plus two connectors [Ele02a], [Ins02]
- 115 ft of 50 Ω Gore 89 cable at 2.5 Gbps [Max01a]
- 100 ft of 75 Ω RG179 cable at 2.5 Gbps [Max01a]
- 100 ft of Belden 9207 cable at 2.5 Gbps [Max01a]
- 50 ft of Madison #14887 shielded twisted-pair cable at 2.5 Gbps [Max01a]
- Maximum rate and distance of data transfer is dependent upon the attenuation characteristics of the media and the noise coupling to the environment.

Maximum Data Rate: Common data rates are 2.5 Gbps or 3.125 Gbaud. 10 GHz or higher

9.11 Exercises

9-1. Why was the XAUI interface standard discussed in this chapter on CML?

9-2. Why would you guess that the XAUI standard specifies a maximum differential output voltage for the driver? Wouldn't it make more sense to specify a minimum differential output voltage for the driver to ensure a certain amount of noise margin relative to the receiver?

9-3. Under what situations might you want to lower the CML driver's differential output voltage from the nominal 800 mV?

9-4. The TLK2501 data sheet recommends using coupling capacitors with values from 1 to 10 nF when ac coupling the driver to the receiver. What impedance magnitude will be seen by a 2.5 GHz signal at these two capacitance values? In determining your answer for this exercise, consider only the capacitance and ignore parasitic resistance and inductance.

9-5. When ac coupling a driver to a receiver, a capacitor is inserted into both sides of the differential transmission line. This has the effect of removing the dc component of the differential signal. If an 800 mV differential signal normally ranged from 1,700 mV to 2,500 mV and the dc component were removed, you would be tempted to think that the signal would now be centered on 0 V and that the signal would swing from –400 mV to +400 mV, still totalling the 800 mV total swing. This, of course, would cause major distortion as the receiver can only receive input signals between ground and its power supply voltage (and usually a smaller input voltage range than that); that is, the bottom negative 400 mV would get chopped off. Why doesn't this happen and how does the capacitor really work to isolate the driver and receiver?

9-6. Name at least two ways that the specifications for the TLK2501 CML device differ from the CML reference values listed in Table 9.1.

9-7. Figure 9.19 on page 222 shows the differential output voltage of the TLK2501 CML driver as a function of the reference resistor value. In the figure, there are two pairs of curves, one pair for 50 Ω transmission media and one pair for 75 Ω transmission media. How do you program the TLK2501 or adjust the serial transmission circuit to operate properly with the 50 Ω transmission media versus the 75 Ω transmission media (i.e., how do you shift from one pair of curves to the other)?

9-8. Refer to Figure 9.12 on page 212 showing the CML differential output voltages. Is there ever a time when the differential output voltage between the actual device's terminals will be $V_{OD(pp\_p)} = 2 \times V_{OD(p)}$? Under what circumstances will it reach this voltage?

9-9. Refer to Table 9.9 on page 225. Why should you not use a value of 20 KΩ for the value of R_{MOD} if you're planning to use one MAX3800 to drive another MAX3800 as shown in Figure 9.20?

9-10. Explain pre-emphasis, de-emphasis, transmit equalization, and adaptive equalization. What are the differences between them?

9-11. What are the advantages and disadvantages of using a CML driver and receiver pair that have internal pullup and termination resistors? What are the advantages and disadvantages of using a CML driver and receiver that requires external pullup and termination resistors?

9-12. Under what circumstances would you want to use dc coupling in your CML circuit design? When would you want to use ac coupling?

9-13. Using two Texas Instruments TLK2501 CML transceivers for both the driver and receiver, design and create the schematic for an ac-coupled full-duplex serial data link. Access the TLK2501 data sheet on the Internet to determine the device pin assignments and any special application information. Show external component interconnections (resistors, coupling capacitors) for both the driver and receiver. Do not draw any external integrated circuits such as clock circuits.

9-14. What's the difference between Gbps and Gbaud when used in connection with 8B/10B encoding?

9-15. Access the Internet and search for CML devices. What's the fastest CML device you can find in terms of actual data rate? Limit your search to 20 minutes. Print out the first page of the device's data sheet and turn it in to your instructor. Be prepared to discuss what the device does and applications that it's suited for. (Hint: Look at the Web sites for National Semiconductor, Maxim Integrated Products, Texas Instruments, Cypress Semiconductor, Motorola, Silicon Image, and any others that your search reveals.)

CHAPTER 10

FPGAs — 11.1 Gbps RocketIOs and HardCopy Devices

Chapter Objectives

- Discuss industry trends in Field Programmable Gate Arrays (FPGAs).
- Discuss Altera's transceiver-based FPGAs that support I/O data rates to 3.125 Gbps.
- Identify which FPGA family offers the greatest data bandwidth among Altera FPGAs.
- Discuss Altera's alternative to ASIC development known as HardCopy®.
- Introduce the Virtex-4 11.1 Gbps RocketIO™ and discuss the Virtex-II Pro 3.125 Gbps RocketIO transceiver.
- Describe the Virtex-II Pro PowerPC™ 405 processor core.
- Discuss SelectIOs and their ability to support 18 or more signaling standards.
- Explain the features of digitally controlled impedance, digital clock managers, and double-data-rate (DDR) I/Os.
- Discuss IBIS and SPICE models for Xilinx devices and intellectual property (IP) cores.

10.1 Industry Trends[1]

"Programmable-logic devices are the fastest growing segment of the logic-device family, for two fundamental reasons. Their ever-increasing gate count per device gathers up functions that you might otherwise spread over a number of discrete-logic and memory chips. This approach improves end-system size, power consumption, performance, reliability, and cost. Equally important, you can in seconds or minutes configure and, in many cases, reconfigure these devices at your workstation or in the system-assembly line. This capability provides powerful

1. This section © Reed Business Information, a division of Reed Elsevier Inc.

flexibility to react to last-minute design changes, to prototype ideas before implementation, and to meet time-to-market deadlines driven by both customer needs and competitive pressures.

The term 'logic devices' subdivides into discrete logic, simple and complex PLDs, FPGAs, and standard and custom-cell ASICs. FPGAs, SPLDs and PALs, and CPLDs[2] are all programmable-logic devices, although their internal architecture implementations differ. I use the term 'programmable-logic device' to refer to the entire range of products and specify a category via its abbreviation.

Compared with ASICs, programmable-logic devices have shorter leadtimes, lower upfront NRE (non-recurring engineering) charges, and no minimum-order quantities, all of which simplify inventory management. As per-gate cost decreases and the number of gates per component increases, programmable-logic devices are making significant inroads into traditional gate-array and, to a smaller extent, standard cell-ASIC territory. System designers and manufacturers are only beginning to explore and exploit in-system reprogrammability, either to correct errors and upgrade functions once the end system is in users' hands or to implement reconfigurable computing — that is, using a fixed number of logic gates to time-division-multiplex multiple functions.

Just as your company uses programmable logic's flexibility to differentiate itself from your competition, semiconductor vendors have developed unique PLDs and FPGAs to address specific intersections of performance, power, integration, and cost targets. This diversity is perhaps the most complex challenge you face, because, in many cases, you must analyze each programmable-logic architecture in detail before selecting one that meets your needs. The market leaders are increasingly driving de facto industry standardization, thus simplifying the selection task.

Highly complex programmable-logic architectures rely extensively on design-automation software to produce optimum results for end-system parameters. Prioritizing these parameters depends on the application. Often, for example, designs targeting low power, high performance, or minimal gate count significantly differ from each other. Ideal design-automation software:

- isolates you from the internal device-architecture details,
- enables you to prioritize your design goals and optimizes the software's operation based on this priority ordering,
- efficiently uses silicon resources,
- requires little to no manual intervention,
- quickly compiles and recompiles a design, and
- minimizes or eliminates timing and pinout changes between compilations.

The technical superiority of a programmable-logic vendor's silicon products and the comprehensiveness of the vendor's documentation are not the only determinants of the vendor's success or failure. Equally important are the depth and the breadth of the company's internally developed and third-party-provided software-tool support.

2. SPLDs are simple programmable logic devices, PALs are programmable array logic [devices], and CPLDs are complex programmable logic devices.

Burgeoning amounts of on-chip RAM and single-die ASIC/programmable-logic hybrid devices, along with predictable Moore's Law integration trends, are contributing to the explosion of effective gate counts. These factors are finally making a reality of the long-held vision of systems on chips. To exploit silicon capability in a time frame that still meets time-to-market requirements, many designers are turning from traditional low-level state-machine and schematic-entry synthesis to high-level languages, such as VHDL and Verilog, and even to traditional software languages, such as C. These new language approaches provide the additional benefit of enabling design reuse. Yet, just as with high-level versus assembly-language software development, high-level logic design decreases development time but produces lower performance and less efficient gate usage.

Another technique that has become more popular over the last few years involves leveraging the already-completed designs, or IP (intellectual property), of another company instead of designing your own circuits. The convergence of accelerating silicon gate count, increasing system functions and standardization, and decreasing time to market is driving this approach. Perhaps the biggest IP hurdles still to overcome are legal rather than technical, although robust test and verification suites and core interoperability among vendors and silicon architectures are important.

Although "pure" programmable-logic devices comprise a dominant percentage of industry unit shipments, this fast-moving product category is evolving in several intriguing directions. Vendors historically known as programmable-logic suppliers are adding ASIC-plus-programmable-logic hybrid devices to their portfolios in attempting to optimally address the complicated power-versus-performance-versus-cost balancing act. ASIC vendors and foundries, too, along with standard product suppliers such as communications-processor manufacturers, are developing hybrid-device capabilities. And companies are now offering not only "soft" logic cores that implement specific functions in programmable logic, but also "hard" embedded programmable-logic cores that can implement any function you desire." [Dip02a]

10.2 Altera FPGAs and CPLDs[3]

"From simple glue logic, to complex SOPC (system-on-a-programmable-chip) solutions, to flexible, low-cost ASIC alternatives, Altera's device solutions include:

- High-density FPGAs — the best possible solutions for high-bandwidth systems.
- Mask-programmed devices — for ASIC applications.
- Low-cost/high-volume FPGAs — low-cost FPGAs for price-sensitive, volume-driven systems.
- FPGAs with embedded high-speed transceivers — a low risk path to high-speed transceiver applications.

3. Text, tables, and figures in this section are reprinted with permission of Altera Corporation. Altera is a trademark and service mark of Altera Corporation in the United States and other countries. Altera products are the intellectual property of Altera Corporation and are protected by copyright laws and one or more U.S. and foreign patents and patent applications.

- FPGAs with embedded processors — cost-effective access to embedded processor technologies.
- CPLDs — offering non-volatile, instant-on solutions for a wide variety of digital applications.
- Configuration devices — fast remote system upgrades and in-system programmability (ISP) for all Altera FPGAs.

Figure 10.1 illustrates this broad range of Altera products along with the corresponding product family names.

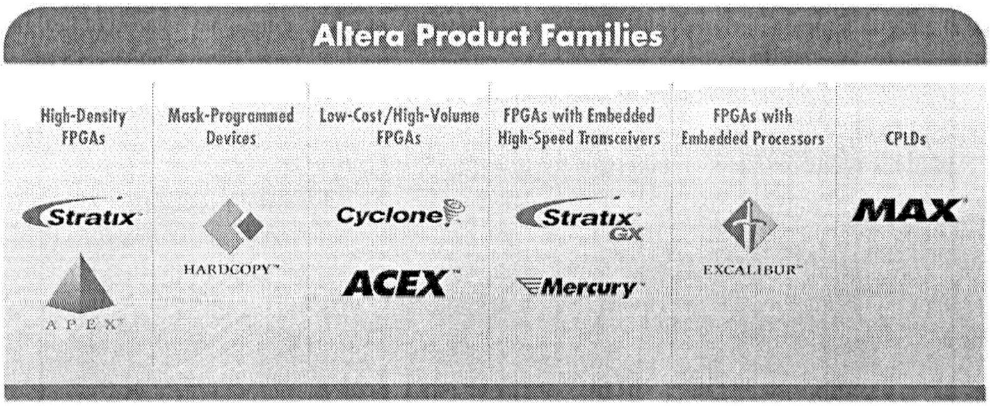

Figure 10.1 Altera product families.

Altera's design tools include the easy-to-use Quartus® II design software and a variety of IP (intellectual property) MegaCore® and Altera Megafunction Partners Program (AMPP℠) megafunctions. These cores are optimized, customizable functions that are ready to drop into your design. Additionally, you can pair FPGAs with the Nios® embedded soft processor, a configurable, general-purpose RISC processor with memory and peripherals that can be tailored to the requirements of virtually any system.

All Altera devices use advanced CMOS processes. The newest product families — including Stratix®, Stratix GX, and Cyclone™ FPGAs — are manufactured on an all-layer-copper 0.13-µm process. In addition, Altera devices are available in a broad range of package options." [Alt03d]

10.2.1 Altera FPGAs with Embedded High-Speed Transceivers

"Ultra high-speed applications need reliable data transfer technology that gets information from source to destination in picoseconds. Altera offers Stratix GX and Mercury™ devices, FPGAs with embedded high-speed transceiver technology that support fast I/O signaling rates and consume minimal power. Using Altera's embedded high-speed transceiver FPGAs, designers can also implement specific interface protocols such as 10 Gigabit Ethernet XAUI or customized functions that require data rates up to 3.125 Gbps. Altera's transceiver-based FPGAs bring the benefits of programmable logic and the capabilities of fast, reliable I/O technology to both new and existing high-speed applications, including bridging applications, switch fabrics, traffic management functions, wireless and high-definition television (HDTV) broadcast applications. These applications include:

- 10 Gigabit Ethernet XAUI and Fibre Channel backplane systems
- SONET/SDH backplane applications
- SD and HD SDI video products" [Alt03d]

10.2.1.1 Stratix GX FPGAs with up to 20 Channels of 3.1825 Gbps SerDes

"The Stratix GX device family is a powerful fusion of 3.125 Gbps transceiver technology with fast FPGA architecture. Including customer-defined features for high-speed design implementation, Stratix GX devices give system architects — from communications to high-end consumer electronics to mass storage systems — a low-risk path to 3.125 Gbps transceiver applications.

Built on a 1.5 V, 0.13 μm, all-layer-copper SRAM process, Stratix GX devices are available in densities ranging from 10,570 to 41,250 LEs (logic elements) and with up to 3.3 Mb of RAM. Based on the Stratix architecture, Stratix GX FPGAs feature dedicated gigabit transceiver blocks with encoded clock-data recovery (CDR) technology to receive and transmit high-speed serial data streams. Each Stratix GX transceiver block has four independent, full-duplex channels that contain dedicated circuitry for speed-critical and logic-intensive physical layer functions. The 3.125 Gbps transceiver blocks also include 8B/10B encoder/decoder circuitry that provides support for several high-speed protocols — including the 10 Gigabit Ethernet (XAUI and XSBI), SONET/SDH, Gigabit Ethernet, InfiniBand, 1G and 2G Fibre Channels, Serial RapidIO, SFI-5, SFI-4, SPI-5, POS-PHY Level 4 (SPI-4 Phase 2), HyperTransport, RapidIO, PCI Express, SMPTE 292M, and UTOPIA IV standards. Figure 10.2 shows the Stratix GX transceiver channel architecture.

To complement the transceiver blocks, Stratix GX devices include a full suite of source-synchronous differential I/O channels with integrated dynamic phase alignment (DPA). There are up to 45 receiver and 45 transmitter high-speed differential I/O channels, each running at up to 1 Gbps. These channels support the LVDS, LVPECL, 3.3 V PCML, and HyperTransport I/O standards.

Stratix GX devices offer up to 3.3 Mb of embedded RAM with the TriMatrix memory structure (see Figure 10.6 on page 242), up to 14 DSP blocks, a complete clock management solution with a hierarchical clock structure and up to eight PLLs, as well as on-chip Terminator™ technology to reduce reflection, improve signal integrity, and simplify board layout. Other features include 40-inch FR4 backplane drive strength capability, low power consumption, and an optimal channel-to-logic ratio. Table 10.1 summarizes the feature benefits of the Stratix GX." [Alt03d]

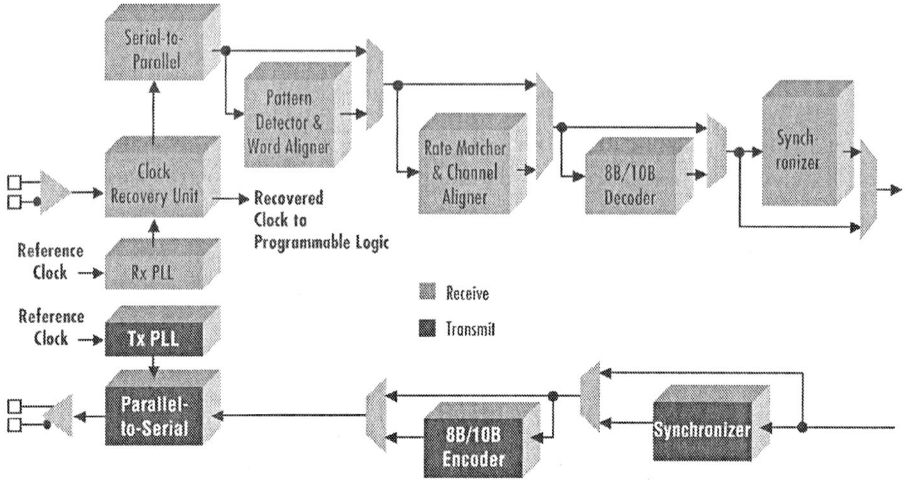

Figure 10.2 Stratix GX transceiver channel architecture.

Table 10.1 Stratix GX Highlights

Feature	Benefit
3.125 Gbps Transceiver Blocks	Embedded transceiver blocks including 8B/10B encoder/decoder circuitry provide support for high-speed applications such as 10 Gigabit Ethernet XAUI, Gigabit Ethernet, InfiniBand, 1G and 2G Fibre Channels, Serial RapidIO, SONET/SDH, SFI-5, SPI-5, PCI Express, and SMPTE 292M standards.
40-Inch Drive-Strength Capability	Transceiver capabilities, embedded programmable pre-emphasis, and embedded receiver equalization combine to enable a 40-inch signal drive capability across an FR4 backplane and two backplane connectors.
Low Power Consumption	450 mW power consumption per 4-channel transceiver block simplifies board design.
Optimal Channel-to-Logic Ratio	Device densities ensure sufficient logic for system implementation.
Source-Synchronous Differential I/O Signaling with DPA1	Support for high-speed I/O standards and high-speed interfaces such as 10 Gigabit Ethernet XSBI, SFI-4, SPI-4 Phase 2 (POS-PHY Level 4), HyperTransport, RapidIO, and CSIX streaming standards at up to 1 Gbps to complement transceiver bandwidth.
High-Performance Stratix Device-Based Architecture	Highly optimized FPGA architecture provides innovative routing architecture for block-based design and maximum system performance. Features such as TriMatrix memory, DSP blocks, Terminator technology, and clock management circuitry enable full system implementation.

10.2.1.2 Mercury FPGAs with up to 45 Gbps of Bandwidth

"Altera's Mercury FPGAs address a wide range of serial backplane, chip-to-chip, and line-side applications. Mercury devices integrate CDR (clock-data recovery) technology and a performance-optimized FPGA architecture into a single device. Data rates of up to 1.25 Gbps per channel and a total CDR bandwidth of up to 45 Gbps provide solutions for today's demanding communications requirements.

High-speed I/O functionality and CDR circuitry are key elements in high-performance communications links. The Mercury CDR circuitry provides the clock management required for complex serial backplane applications and is suitable for use in systems using Gigabit Ethernet, RapidIO, and POS-PHY L4 protocols. These protocols are supported with up to 18 high-speed differential channels.

Mercury devices complement this CDR capability with a robust PLD architecture to support the high bandwidth generated by the CDR circuitry. The Mercury architecture includes over 100 channels of Flexible-LVDS support on regular I/O pins, a fully flexible advanced I/O standard support solution, dedicated circuitry to interface with high-speed memories, and up to six PLLs. In addition, Mercury devices include high-performance logic optimized for speed. This logic performance is enabled using a prioritized interconnect structure, dedicated multiplier circuitry, and quad-port ESBs (embedded system blocks) on a 0.15 µm, 1.8 V all-layer-copper process. Table 10.2 describes some Mercury device highlights. New designs requiring high-speed design implementation can use Stratix GX FPGAs." [Alt03d]

Table 10.2 Mercury Highlights

Feature	Benefit
CDR Circuitry	Data rates of up to 1.25 Gbps with LVDS, LVPECL, and PCML support.
Advanced I/O Standard Support	Supports PCI, PCI-X, HSTL, GTL+, SSTL-2/-3, AGP[a], CTT[b], and LVTTL standards.
External Memory Interface	Built-in support for external ZBT, DDR, and QDR RAM.
Flexible LVDS Support	Provides 624-Mbps LVDS support on up to 100 regular I/O pins.
Enhanced PLLs	Provides up to six PLLs with support for ClockLock, ClockBoost, and ClockShift features.
Quad-Port ESBs	Supports quad-port RAM, dual-port RAM, ROM, CAM, and FIFO buffers.
Prioritized Interconnect Structure	Reduces routing delays for speed-critical signals to increase performance.
Distributed Multiplier Circuitry	Provides 100-MHz, 16 × 16 non-pipelined, push-button performance.
Flip-Chip Technology	Increased I/O performance and I/O count, in-board space-saving Fine-Line BGA packages.
Advanced 0.15 µm, 1.8 V Copper Process	Improved performance over traditional aluminum process technologies.

a. AGP is advanced graphics port.

b. CTT is center tap terminated.

10.2.2 Altera HardCopy Devices

"Altera HardCopy® devices provide a low-cost, time-saving alternative to ASICs, guarantee first-time success, and are a complete prototype-to-volume-production solution for high-density designs. Engineers have the option to target either an FPGA or a high-volume, mask-programmed design at the beginning of the design process using the same architecture, software, and intellectual property (IP). Designers can completely bypass the pain and expense of ASIC development by using an Altera FPGA to prototype and verify in-system, and then seamlessly migrate that FPGA-proven design to a HardCopy device for high-volume production. HardCopy devices not only have the ability to help manage rising costs, risks, and market uncertainty, they also can boost performance an average of 50% and lower power consumption up to 40% compared to the original FPGA." [Alt03c] Figure 10.3 reflects the lower development costs, faster time-to-market, and guaranteed first-silicon success of HardCopy devices relative to ASIC development.

	Development Costs	Time-to-Market	First-Silicon Success
ALTERA® HardCopy™	• Low NRE • Low Mask Costs • $2,000 Design Tools	• 18 to 20 Weeks	• Guaranteed with FPGA-Proven Design
ASIC	• Sizable Engineering • High Mask Costs • $300,000+ Design Tools	• 52+ Weeks	• Potential Expensive & Time-Consuming Re-Spins • New Architecture • New Design

Figure 10.3 Comparison of Altera HardCopy development vs. ASIC development.

"Altera's HardCopy devices preserve the architecture of the FPGA with the programmability removed. As a result, the die size of a HardCopy device is 60 to 70% smaller than the equivalent FPGA. This enables Altera to provide a lower-cost, higher-performance, and lower-power mask-programmed device. HardCopy devices are built on the same process technology and include all of the same embedded features as their FPGA counterparts, enabling seamless migration of the FPGA designs.

HardCopy devices are mask-programmed devices created from a direct mapping of the FPGA architecture. To generate HardCopy devices, Altera uses common base arrays across multiple designs for a particular density and implements customer-specific design information with the top metal layers (the top two layers for HardCopy Stratix devices, and the top three layers for HardCopy APEX 20KE and HardCopy APEX 20KC devices)." [Alt03b]

10.2.2.1 Elimination of ASIC Risk

"In today's market, high-volume applications require a quick, easy-to-use, and low-cost solution that provides enough performance and flexibility to tackle the toughest product development challenges. ASICs, once considered the ideal solution for high-volume, system-on-a-chip (SOC) applications, have become less advantageous and more problematic with their lengthy design and development times, numerous risks, and escalating non-recurring engineering (NRE), mask, and tool costs.

Today's SOC designs demand increasingly complex embedded functions, IP, and logic, plus the flexibility to keep up with rapidly evolving technology and standards. Developing state-of-the-art designs on ASICs requires an enormous budget and months of engineering resources, a hefty investment that necessitates success with the first silicon produced. When an ASIC re-spin is necessary, companies not only have to bear additional development and engineering costs, they also lose time-to-market advantages and prospective market opportunities.

Altera builds HardCopy devices based on customer-provided, in-system-verified FPGA design files and constraints. From these source files, Altera preserves the FPGA-proven netlist, and ensures consistency with the original designs. This seamless migration process (see Figure 10.4) minimizes risk, guarantees first-time success, and accelerates time-to-market. With an easy migration process and minimal customer involvement, Altera HardCopy devices can deliver guaranteed fully operational prototypes and production units in the shortest time possible (see Figure 10.5)." [Alt03c]

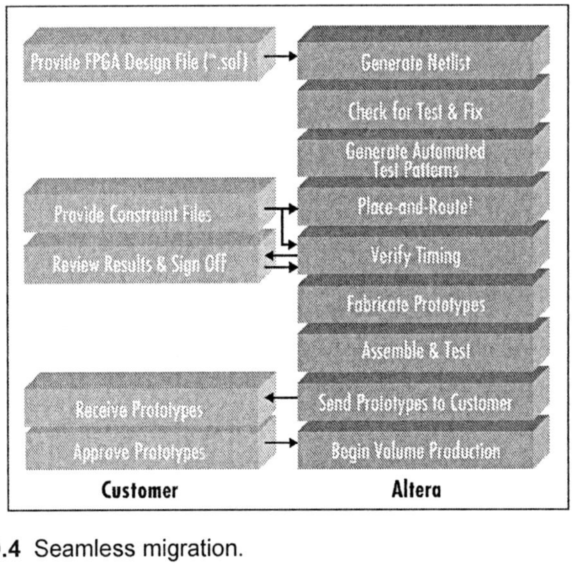

Note: If placement constraints are provided, then Altera will place as per these constraints and only route the design.

Figure 10.4 Seamless migration.

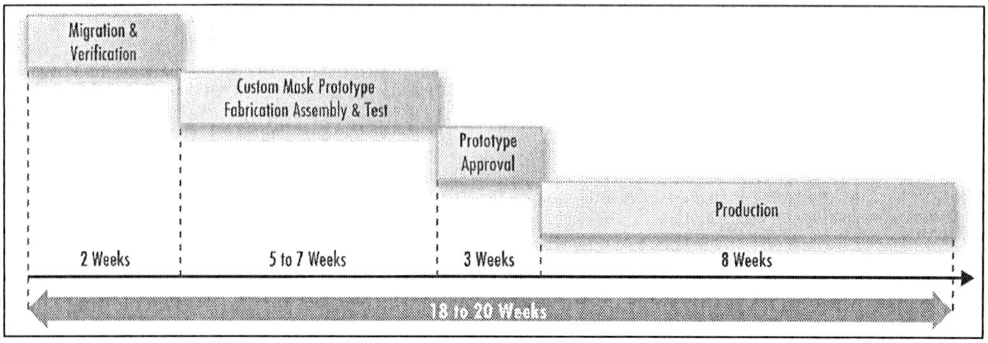

Figure 10.5 Implementation timeline.

10.2.2.2 HardCopy Devices Designed with Quartus II Software

"Altera's Quartus II software offers a unified flow for designing FPGAs and HardCopy devices. Designers using the Quartus II software (version 3.0 and later) can target a HardCopy Stratix device upfront, using similar design flow, architecture, and tools as for a Stratix FPGA. The option to estimate post-migration performance and power consumption and optimize the HardCopy device gives system architects the ability to design their systems for maximum throughput. The Quartus II design software supports all HardCopy Stratix, HardCopy APEX 20KC, and HardCopy APEX 20KE devices." [Alt03c]

10.2.2.3 HardCopy Stratix and APEX Devices

"The HardCopy device family includes HardCopy Stratix, HardCopy APEX 20KE, and HardCopy APEX 20KC devices. Each of the HardCopy devices employ the same high-density architecture as the corresponding FPGA and are available in space-saving FineLine BGA® packages. HardCopy devices support powerful system-level device features, such as:

- Packaging: HardCopy devices are available in the same packages with the same pin-outs as the corresponding FPGA, maintaining pin compatibility.
- Built-in testability: All HardCopy devices include embedded test circuits. Altera automatically generates production test vectors with high fault coverage and does not require any customer-generated test vectors.
- IP: Migration of Altera MegaCores and Megafunctions."[4] [Alt03c]

"HardCopy Stratix devices, Altera's second-generation HardCopy devices, are low-cost, high-performance mask-programmed devices with the same architecture as the high-density Stratix FPGAs. The combination of Stratix FPGAs for in-system prototype and design verifi-

4. The Altera MegaCore functions and Megafunctions are reusable blocks of intellectual property that you can customize and use in a design, allowing you to concentrate on adding proprietary value. Using Megafunctions reduces system implementation and verification times.

cation, HardCopy Stratix devices for high-volume production, and the Quartus II design software (version 3.0 and later) provide a complete and powerful alternative to ASIC design and development.

HardCopy Stratix devices are architecturally equivalent and have the same features as the corresponding Stratix FPGA. The HardCopy Stratix devices use the same base arrays across multiple designs for a given density and are customized using the top two metal layers. Offering pin-to-pin compatibility using the same package as the corresponding Stratix FPGA, HardCopy Stratix devices are drop-in replacements to the FPGA device. Designers can use the Quartus II software to design HardCopy Stratix devices to obtain an average of 50% higher performance and 40% lower power consumption than can be achieved in the corresponding Stratix FPGAs, before migrating their design to the corresponding HardCopy Stratix devices.

HardCopy Stratix devices are manufactured on the same 1.5 V, 0.13 μm all-layer-copper metal fabrication process (up to eight layers of metal) as the Stratix FPGAs. HardCopy Stratix devices:

- Support high-speed networking and communications bus standards including RapidIO, UTOPIA IV, CSIX, HyperTransport™ technology, 10 Gigabit Ethernet XSBI, SPI-4 Phase 2 (POS-PHY Level 4), and SFI-4
- Provide on-chip termination for differential and single-ended I/O pins with impedance matching through Terminator technology
- Support high-speed external memory, including zero bus turnaround (ZBT) SRAM, quad data rate (QDR and QDRII) SRAM, double data rate (DDR) SDRAM, DDR fast cycle RAM (FCRAM), and single data rate (SDR) SDRAM" [Alt03a]

"Altera's HardCopy Stratix devices retain the same rich features as Stratix FPGAs, including the hierarchical clock structure, TriMatrix™ memory, and highly optimized embedded digital signal processing (DSP) blocks. HardCopy Stratix devices support for the wide range of high-speed networking and communication interfaces listed above allows designers to connect to high-speed memory devices such as quad data rate (QDR) and zero bus turnaround (ZBT) SRAMs. The HardCopy Stratix devices are well-suited for high-performance, high-volume applications in the networking, wireless communication, high-end consumer electronics, industrial, test, and medical markets.

HardCopy APEX 20KE and HardCopy APEX 20KC devices represent the first-generation HardCopy devices introduced by Altera. HardCopy APEX 20KE and HardCopy APEX 20KC devices maintain the equivalent APEX™ 20KC and APEX 20KE FPGA architecture and their features." [Alt03c]

10.2.3 High-Density FPGAs

"High-density FPGAs — Stratix and APEX devices — were developed to address the increasingly complex bandwidth requirements of SOPC applications in communications, industrial, automotive, electronic data processing, and digital consumer end markets. Altera's high-density FPGA families range from 1,200 to 79,040 logic elements (LEs) and provide the innovative architectures and rich feature sets needed to reach your next-generation performance and bandwidth targets. Applications include high-end routers and switches, IC testers, and medical imaging equipment." [Alt03d]

10.2.3.1 Stratix FPGAs

"Stratix devices are based on a 1.5 V, 0.13 μm, all-layer-copper SRAM process, with densities ranging from 10,570 to 79,040 LEs. Stratix devices feature the TriMatrix memory structure, an array of three different-sized memory blocks, each optimized to target a different class of applications. The integration of the 512 Kb M-RAM blocks with hundreds of smaller M512 and M4K blocks provides a unique solution to applications requiring large amounts of on-chip memory bits or high-memory bandwidth (see Figure 10.6). Offering up to 7 Mb of RAM, Stratix devices feature the highest memory-to-logic ratio and the highest memory bandwidth in the industry. All memory blocks include extra parity bits for error control, embedded shift register functionality, mixed width mode, and mixed clock mode support.

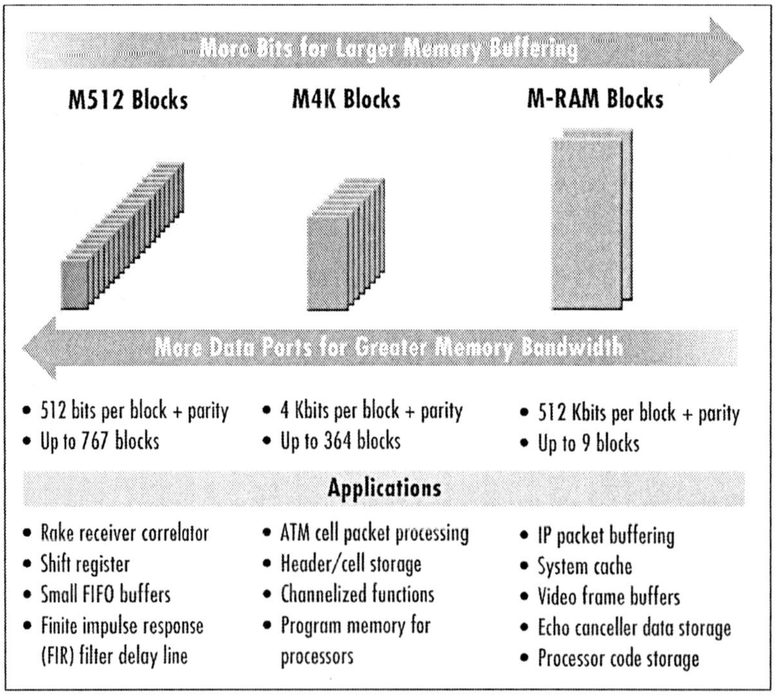

Figure 10.6 TriMatrix memory structure.

Ideal for the wireless communication, telecommunication, video, and image processing markets, Stratix devices offer up to 22 DSP blocks (high-performance embedded arithmetic units) that eliminate the performance bottlenecks in arithmetic applications. Comprised of multiply and accumulate circuitry, the DSP blocks provide predictable performance and significant resource savings for complex applications that require high data throughput. These DSP blocks can implement a variety of typical DSP functions, such as finite impulse response (FIR) filters, fast Fourier transform (FFT) functions, correlators, and encryption/decryption functions, as well as mathematically intensive functions. DSP blocks in Stratix devices can run at over 333 MHz to provide data throughput performance of up to 2.4 GMACS[5] per DSP block. The 22 DSP blocks in the largest Stratix device can provide a combined throughput that is more than 10 times the data throughput available from leading digital signal processors today.

Supporting a variety of single-ended and differential I/O standards, Stratix devices offer designers high aggregate device bandwidth with access to up to 152 high-speed differential I/O channels. Each of these I/O channels includes dedicated serializer/deserializer (SerDes) circuitry for high-speed standards. Stratix devices support standards such as LVDS, LVPECL, PCML, differential SSTL, differential HSTL, LVTTL, LVCMOS, SSTL, HSTL, PCI-X, CTT, AGP, GTL, GTL+, 10 Gigabit Ethernet (XSBI), SFI-4, POS-PHY Level 4, HyperTransport, RapidIO, and UTOPIA IV standards. This comprehensive support for high-speed I/O interfaces and high-bandwidth protocols makes Stratix devices an ideal solution for complete system integration.

Stratix devices are based on innovative high-performance architecture. With as many as 12 PLLs, up to 40 system clocks for system-level clocking management, support for many differential and single-ended I/O electrical standards, high-speed interfaces, on-chip termination, and remote system upgrade capabilities, Stratix devices bring new levels of system integration to SOPC designs." [Alt03d]

10.2.3.2 APEX FPGAs

"With densities of 1,200 to 67,200 LEs and data transfer rates up to 1 Gbps per channel, APEX FPGAs offer system-level integration on a single device. Designed to be 64-bit, 66 MHz PCI-compliant, APEX devices feature the innovative MultiCore™ architecture which combines and enhances the strengths of prior Altera device architectures.

The APEX devices consist of four families: APEX II, APEX 20KC, APEX 20KE, and APEX 20K families. Featuring all-layer-copper interconnects, APEX II and APEX 20KC FPGAs offer the highest APEX performance. APEX II devices also support a host of I/O protocols, such as UTOPIA IV, the RapidIO standard, CSIX, and POS-PHY Level 4 to support complete SOPC solutions. APEX 20KE FPGAs offer tremendous breadth in density, I/O capability, and package options at cost-effective price points.

5. GMACS is giga multiply-accumulates per second.

The APEX MultiCore architecture features embedded system blocks (ESBs) that can implement several types of memory structures, including dual-port RAM, content addressable memory (CAM), first-in first-out (FIFO) buffers, and ROM. The direct implementation of CAM in embedded silicon provides significant performance improvements for any pattern-matching or data-searching applications.

APEX devices meet system performance demands and offer lower supply voltages by supporting multiple I/O interfacing standards, including LVTTL, LVPECL, LVCMOS, GTL+, SSTL-2/-3, AGP, HSTL, CTT, HyperTransport technology, and LVDS with performance up to 1 Gbps. True-LVDS™ support has been specifically incorporated as a key high-performance I/O interface solution, and is supported via the use of dedicated serial-to-parallel conversion circuitry and LVDS-optimized PLLs.

To increase system clock rates, APEX devices feature up to four PLLs with output frequencies up to 335 MHz. These PLLs can provide improvements within an APEX device or in the interactions of an APEX device with other system components. They can also serve as the clock management circuitry for an entire board-level system, providing improvements for unrelated system components." [Alt03d]

10.2.4 Low-Cost/High-Volume FPGAs

"Altera's low-cost/high-volume Cyclone and ACEX® FPGA families provide the reprogrammability of an FPGA at extremely low cost. These FPGAs deliver mid-density SOPC solutions to applications and provide an optimized feature set and abundant on-chip resources specifically tuned for high-volume applications. Applications include digital set-top boxes, DVD player/recorder systems, plasma displays, automotive telematics, and industrial automation equipment." [Alt03d]

10.2.4.1 Cyclone FPGAs

"Cyclone devices are built on a cost-optimized, all-copper 1.5 V SRAM process, and offer powerful functionality at low cost. With up to 20,060 LEs and 288 Kb of on-chip RAM, Cyclone devices can integrate many complex functions. Cyclone devices offer multiple full-featured PLLs to manage board-level clock networks and dedicated I/O interfaces for interfacing with industry-standard external memory devices. Altera's Nios embedded processor and a full IP portfolio are also available for development with Cyclone devices. In fact, with a Nios processor in a Cyclone device, an embedded RISC processor costs less than U.S.$2.00 for high-volume applications. Cyclone family devices are priced at less than $1.50 per 1,000 LEs.

To fully realize the value of Cyclone devices in SOPC designs, Altera developed a new cost-optimized serial configuration device family (see Section 10.2.7 on page 247). Also, to reduce overall development costs, Cyclone devices are fully supported in the free Quartus II Web-edition design software available on the Altera Web site." [Alt03d]

10.2.4.2 ACEX FPGAs

"The ACEX FPGA family provides value and performance for cost-sensitive, volume-driven applications. Capable of interfacing with 5.0 V, 3.3 V, and 2.5 V I/O circuitries, these devices are ideal for the communications and consumer marketplaces.

The ACEX family ranges from 576 to 4,992 LEs. Operating at 2.5 V, ACEX devices are fully 64-bit, 66 MHz PCI compliant and feature embedded dual-port RAM. ACEX devices support PLL circuitry and can drive two separate ClockLock™ and ClockBoost™ circuitry-generated signals for extensive clock management capability.

New mid-density designs requiring the lowest cost should utilize Cyclone devices, the lowest-price FPGAs." [Alt03d]

10.2.5 Altera FPGAs with Embedded Processors

"Combining a high-density programmable architecture and a high-performance hard processor core, Altera's Excalibur™ devices give designers cost-effective access to advanced processor technologies. Excalibur devices span the performance domain up to 200 MHz, providing high-speed processing power with FPGA flexibility. Applications include industrial control, factory automation, medical imaging, and wireless basestations." [Alt03d]

10.2.5.1 Excalibur Devices

"Embedding hard IP into an FPGA and allowing engineers to focus on the delivery of complex system solutions, Excalibur devices combine an industry-standard ARM922T™ hard processor, along with memory and peripherals, with an APEX device-like programmable logic architecture.

Excalibur devices feature an ARM processor-based embedded subsystem, or stripe, which also contains dual-port and single-port SRAM memory, memory interfaces, and peripherals. The integrated SDRAM controller supports SDR or DDR SDRAM at up to 133 MHz or 266 MHz, respectively, while the Excalibur device supports up to 512 MB of SDRAM. Excalibur devices also provide an expansion bus interface for interfacing to flash memory, external SRAM, and memory-mapped external peripherals. The expansion bus interface can support up to four external devices, each up to 32 MB. A high-performance interrupt controller is also provided.

The Excalibur's ARM922T processor is equivalent to 210 Dhrystone MIPS when running at 200 MHz. Excalibur devices also feature data and instruction caches, associated memory management units (MMUs), and on-chip debug and trace capabilities. Bus interfacing to the processor is also provided using the industry-standard AMBA™ high-performance bus (AHB) architecture, which connects to the stripe memory interfaces and peripherals. The embedded processor stripe interfaces to the FPGA using both master- and slave-dedicated AHB bus bridges, and the dual-port SRAM applications interfaces." [Alt03d]

10.2.6 Altera CPLDs

"Altera CPLDs feature instant-on non-volatility for simple logic integration for a wide variety of digital applications. Altera CPLDs contain deterministic routing that guarantees fast, fixed, and predictable pin-to-pin (propagation) delays. These general-purpose devices range in density from 32 to 512 macrocells. A comparison of MAX® device voltages and propagation delays is shown in Figure 10.7. Applications include consumer electronics, automotive telematics, storage servers, and 3G basestations." [Alt03d]

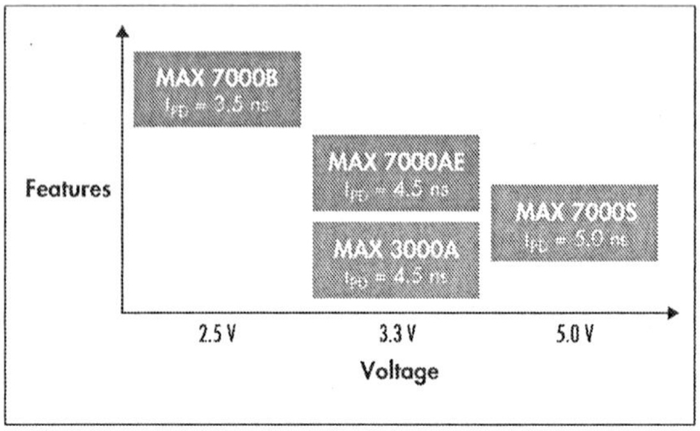

Figure 10.7 Comparison of MAX device propagation delays and voltages.

10.2.6.1 MAX 3000 CPLDs

"The cost-optimized MAX 3000 devices are the ideal choice for high-volume systems. Featuring the packages, densities, and propagation delays customers need, MAX 3000 devices provide designers with exceptional performance at the lowest price per macrocell among MAX devices. The 3.3 V MAX 3000 devices have an enhanced ISP (in-system programming) feature set and range in density from 32 to 512 macrocells." [Alt03d]

10.2.6.2 MAX 7000 CPLDs

"Fabricated on a sophisticated CMOS EEPROM process, the MAX 7000 family is a fast, easy-to-use programmable logic solution. Because of their flexibility and simple programming requirements, MAX 7000 devices are a popular CPLD architecture.

MAX 7000 devices provide a rich feature set that includes global clocking, fast input registers, and programmable slew-rate control. High-speed global clocks coupled with 3.5 ns propagation delays and fast clock-to-out times create superior system performance with high-speed, device-to-device communication. The programmable slew-rate control reduces system noise by slowing the switching time of outputs that are not speed-critical. Additionally, a programmable powersaving feature allows for 50% or greater power reduction.

MAX 7000 devices are well-suited for mixed-voltage environments, offering 2.5 V, 3.3 V, or 5.0 V core supply operation and multivolt I/O operation to interface with 1.8 V, 2.5 V, 3.3 V, and 5.0 V devices." [Alt03d]

10.2.6.3 MAX 7000AE CPLDs

"The popular MAX 7000AE devices are the industry-standard 3.3 V CPLD family. MAX 7000AE devices include an enhanced ISP algorithm for faster programming times, an ISP done-bit to ensure complete programming, and a pull-up resistor on the I/O pins during ISP." [Alt03d]

10.2.6.4 MAX 7000B CPLDs

"MAX 7000B devices feature a wide array of supported I/O standards in an instant-on CPLD. Maintaining pin compatibility with MAX 7000S and MAX 7000AE devices, MAX 7000B devices offer designers a seamless path to lower voltages, faster propagation times, and compatibility with new I/O standards. MAX 7000B devices offer propagation delays as fast as 3.5 ns and counter frequencies of over 300 MHz. Altera offers high-performance MAX 7000B devices that feature 2.5 V ISP and support the following advanced I/O standards: LVCMOS, LVTTL, GTL+, SSTL-2, and SSTL-3." [Alt03d]

10.2.6.5 MAX 7000S CPLDs

"MAX 7000S devices feature 5.0 V core supply operation and are ideal for system-level integration. With 5.0 V ISP, these devices are pin-, function-, and programming-file-compatible with all MAX 7000 family members." [Alt03d]

10.2.7 Configuration Devices

"The enhanced configuration devices — EPC4, EPC8, and EPC16 — are based on a flash process and utilize a patented data compression technology to effectively double the amount of configuration memory, providing up to 30 Mb in a package that is less than 1 square centimeter. A single EPC16 device is sufficient to configure one Stratix EP1S80 device or up to seven APEX EP20K400C devices.

Altera's configuration devices enable remote system upgrades, provide ISP, allow unused memory to be used as general-purpose memory, and dramatically reduce configuration times.

Configuration devices have been relatively expensive compared to low-cost FPGAs, costing between 30% to 50% of the FPGA cost. While the price of a low-cost FPGA may have been low enough to address most price-sensitive applications, the cost of an FPGA combined with a configuration device has not been low, resulting in programmable logic being locked out of these price-sensitive markets. Altera's new serial configuration devices were built to address this issue. Costing on average 10% of the corresponding FPGA, serial configuration devices and Cyclone devices are a perfect solution for cost-sensitive designs. The serial configuration device family provides memory densities at 1 Mb and 4 Mb, and are offered in an 8-pin small-outline package.

Manufactured on either an EPROM or a flash process, these feature-rich devices offer ISP, dynamic configuration, external flash interface, a programmable clock, parallel programming, fast programming times, and small form-factor packages.

All of Altera's SRAM-based FPGA devices interface seamlessly with the configuration devices, allowing for a smooth configuration without the need for an external controller. Through a MultiVolt interface, Altera configuration devices support operating voltages from 1.8 V to 5.0 V." [Alt03d]

10.3 Xilinx FPGAs and CPLDs[6]

Xilinx offers a complete product line of FPGAs and CPLDs including the following major families of devices:

FPGAs		CPLDs
Virtex-4	Virtex	CoolRunner-II
Virtex-II Pro	Spartan-IIE	CoolRunner XPLA3
Virtex-II		XC9500

10.3.1 Virtex FPGAs

The Virtex™ FPGA offered one million system gates when introduced. Subsequently, Xilinx introduced the Virtex-II FPGA architecture containing high-performance logic that defined the Platform FPGA. Virtex Series FPGAs offer high clock speeds, memory bandwidth, I/O performance, and logic density.

Two years after the introduction of the Virtex-II, Xilinx introduced the Virtex-II Pro, which added embedded PowerPC processors and 3.125 Gbps serial transceivers as standard FPGA features.

Xilinx subsequently introduced the Virtex-4 FPGA with 11.1 Gbps RocketIO transceivers and 1+ Gbps SelectIOs. These data rates are per channel, i.e., per pair of differential traces.

10.3.2 Spartan FPGAs

Spartan™ FPGAs are used in consumer applications and are targeted as replacements for gate arrays and for ASSPs (application-specific standard products). They allow the design of competitive consumer product platforms.

The Spartan-IIE FPGA offers programmable support for up to 19 I/O standards (including LVDS), intellectual property (including DSP and processor cores), on-chip block RAM, and digital DLLs (delay-locked loops) for both chip-level and board-level clock management. Spartan-IIE devices provide additional value by integrating LVDS translation, secondary processing, PLLs, FIFOs, and system bus drivers.

10.3.3 CPLDs

Currently, Xilinx offers CPLD products in three categories: CoolRunner-II, CoolRunner XPLA3, and XC9500.

6. Figures and text in Section 10.3 and its subsections are based on or adapted from figures and text owned by Xilinx, Inc., courtesy of Xilinx, Inc., © Xilinx, Inc. 2000–2004. All rights reserved.

10.3.3.1 CoolRunner CPLDs

CoolRunner™ CPLDs combine ultra-low power with high speed, high density, and high I/O counts in a single device. The CoolRunner-II CPLD combines high performance, low power and low cost with a 100% digital core, up to 333 MHz performance, and less than 100 μA of standby current.

The CoolRunner XPLA3 family of CPLDs is targeted for low-power applications that include portable, handheld, and power-sensitive systems. The XPLA3 family offers true pin-to-pin speeds of 5.0 ns, while simultaneously delivering power that is less than 100 μA (standby) without the need for special power down techniques that negatively affect device performance.

CoolRunner XPLA3 The CoolRunner XPLA3 product families offer low power in an all new market segment for CPLDs — portable electronics. With standby current in the low micro amps and low operational power consumption, these parts fit power sensitive applications such as battery powered applications.

CoolRunner-II The new CoolRunner-II CPLD product family provides performance with low power consumption. CoolRunner-II features a 100% digital core, over 300 MHz performance, and less than 100 μA standby current, and can be used in new system designs. CoolRunner CPLDs deliver ultra-low power consumption and true pin-to-pin speeds of 5.0 ns.

10.3.3.2 XC9500

The high-performance, low-cost XC9500™ families of CPLDs are targeted for systems that require rapid design development, long system life, and field upgrade capability. They are in-system programmable (ISP) and deliver high speeds while giving you the flexibility of enhanced pin-locking architecture along with IEEE Std. 1149.1 JTAG boundary scan support.

10.3.4 More About the Virtex-II Pro FPGA

The Virtex-II Pro family consists of 10 members (see Table 10.3), each with 4 to 24 RocketIO multigigabit transceivers based on the Mindspeed SkyRail™ technology. Each RocketIO channel can provide a data rate of from 622 Mbps to 3.125 Gbps (baud rate). Each channel can also propagate a signal 40 inches (102 cm) at 3.125 Gbps over FR4. Further, there are built-in optional 8B/10B, pre-emphasis (amplification of higher frequencies to reduce distortion), and channel bonding (synchronizing) capabilities. Each of the larger Virtex-II Pro devices incorporates one to four PowerPC 405 processor cores, each capable of more than 300 MHz clock frequency and 420 Dhrystone MIPS.

Features common in the Virtex-II/-II Pro Series include SystemIO™ connectivity solutions, digitally controlled impedance (DCI) technology, clocking solutions, Active Interconnect™ routing architecture, and bitstream encryption. The upward compatibility of the Virtex series of products improves engineering productivity, performance, design longevity, and continuing cost reduction.

Xilinx offers documentation and technical support. A Virtex™-II Pro Developer's Kit is available, which includes a PowerPC 405 user manual for information on programming the PowerPC. A RocketIO user guide is also available.

Table 10.3 Virtex-II Pro Family of Ten Members

Feature/Product	XC 2VP2	XC 2VP4	XC 2VP7	XC 2VP20	XC 2VP30	XC 2VP40	XC 2VP50	XC 2VP70	XC 2VP100	XC 2VP125
Logic Cells	3,168	6,768	11,088	20,880	30,816	43,632	53,136	74,448	99,216	125,136
BRAM (Kbits)	216	504	792	1,584	2,448	3,456	4,176	5,904	7,992	10,008
18x18 Multipliers	12	28	44	88	136	192	232	328	444	556
Digital Clock Management Blocks	4	4	4	8	8	8	8	8	12	12
Config (Mb)	1.31	3.01	4.49	8.21	11.36	15.56	19.02	25.6	33.65	42.78
PowerPC Processors	0	1	1	2	2	2	2	2	2	4
Max Available Multi-Gigabit Transceivers	4	4	8	8	8	12	16	20	20	24
Max Available User I/O	204	348	396	564	644	804	852	996	1164	1200

10.3.5 Virtex-II Pro RocketIO Multigigabit Transceiver

The RocketIO multigigabit transceivers are embedded SerDes (Serializer/Deserializer) devices that enable multi-gigabit serial connectivity between buses, backplanes, and subsystems. Each RocketIO block supports a full-duplex baud rate of 622 Mbps to 3.125 Gbps. This provides up to 75 Gbps aggregate baud rate per device. The RocketIO blocks incorporate built-in user-configurable support for 8B/10B encode/decode, output pre-emphasis, channel bonding, and comma detect. RocketIO blocks may be used in conjunction with the SystemIO soft IP cores to support emerging interface standards and protocols, such as XAUI (for 10 Gigabit Ethernet) and Gigabit Ethernet.

The RocketIO transceiver is based on Mindspeed's SkyRail technology. Figure 10.8 depicts an overall block diagram of the transceiver. Up to 24 transceiver modules are available on a single Virtex-II Pro FPGA, depending on the part being used.

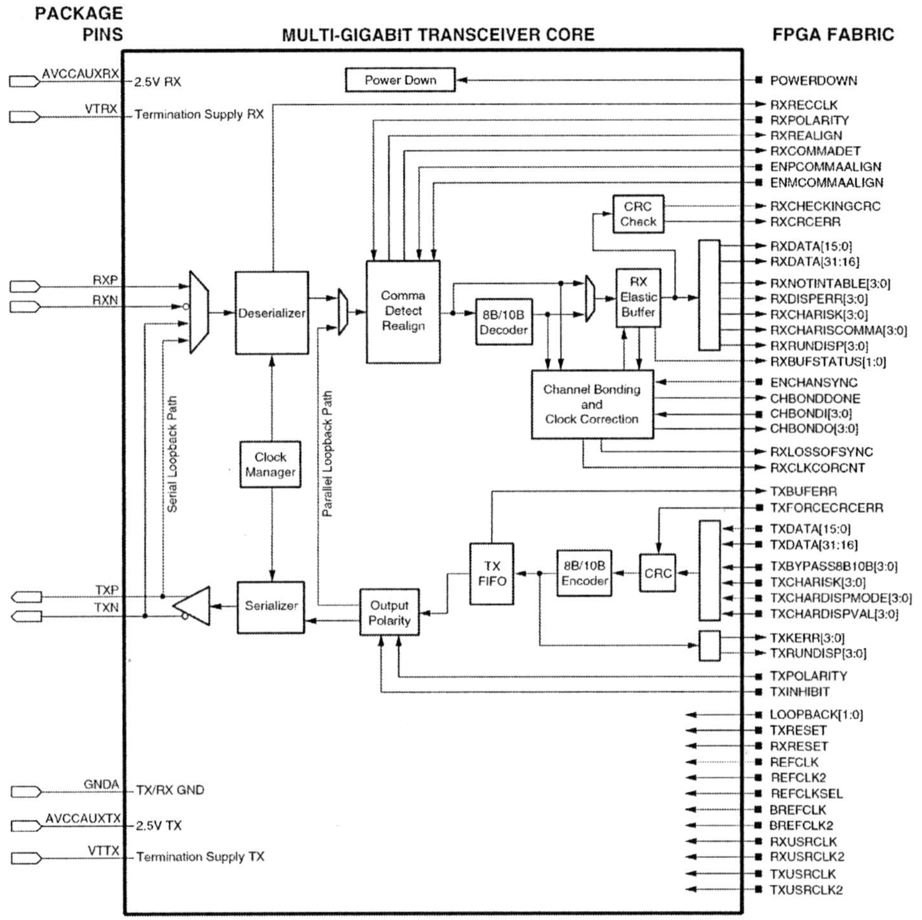

Figure 10.8 RocketIO transceiver block diagram.

10.3.6 The Virtex-II Pro PowerPC 405 Processor Core

The PPC405x3 is a 32-bit implementation of the PowerPC embedded-environment architecture that is derived from the PowerPC architecture. Specifically, the PPC405x3 is an embedded PowerPC 405D5 processor core (PPC405D5). The term "processor block" is used throughout this section to refer to the combination of PPC405D5 core, on-chip memory logic (OCM), and the gasket logic and interface.

The PowerPC architecture provides a software model that ensures compatibility between implementations of the PowerPC family of microprocessors. The PowerPC architecture defines parameters that guarantee compatible processor implementations at the application-program level, allowing broad flexibility in the development of derivative PowerPC implementations that meet specific market requirements.

The PowerPC architecture requires that all PowerPC implementations adhere to the UISA (user instruction set architecture), offering compatibility among all PowerPC application programs. However, different versions of the VEA (virtual environment architecture) and OEA (operating environment architecture) are permitted.[7]

Embedded applications written for the PPC405x3 are compatible with other PowerPC implementations. Privileged software generally is not compatible. The migration of privileged software from the PowerPC architecture to the PPC405x3 is in many cases straightforward because of the simplifications made by the PowerPC embedded-environment architecture.

10.3.6.1 PPC405x3 Hardware Organization

As shown in Figure 10.9, the PPC405x3 processor contains the following elements:

- A 5-stage pipeline consisting of fetch, decode, execute, writeback, and load writeback stages
- A virtual-memory-management unit that supports multiple page sizes and a variety of storage-protection attributes and access-control options
- Separate instruction-cache and data-cache units
- Debug support, including a JTAG interface
- Three programmable timers

The Virtex-II Pro Platform FPGA Developer's Kit provides full detail on the MMU (memory management unit), CPU (central processing unit), PLB (processor local bus), OCM (on-chip memory), MAC (multiply-accumulate unit), ALU (arithmetic-logic unit), TLB (translation look-aside buffer), and other blocks shown in the diagram.

The processor block (PPC405x3, OCM controllers, and gasket logic) interfaces support the attachment of cores and user logic and provide input/output (I/O) signals that can be grouped as follows:

- Clock and power management interface
- CPU control interface
- Reset interface

7. "The UISA, VEA, and OEA are the three levels, or programming environments, of the PowerPC architecture. The UISA defines the architecture level to which user-level software should conform. The UISA defines the base user-level instruction set, user-level registers, data types, memory conventions, and the memory and programming models seen by application programmers. The VEA, which is the smallest component of the PowerPC architecture, defines additional user-level functionality that falls outside typical user-level software requirements. The VEA describes the memory model for an environment in which multiple processors or other devices can access external memory, and defines aspects of the cache model and cache control instructions from a user-level perspective. The OEA defines supervisor-level resources typically required by an operating system. The OEA defines the memory management model, supervisor-level registers, and exception model. Implementations that conform to the OEA also conform to the UISA and VEA." [Mot02b] Copyright of Motorola. Used by permission.

- Instruction-side processor local bus interface
- Data-side processor local bus interface
- Device-control register interface
- External interrupt controller interface
- JTAG interface
- Debug interface
- Trace interface
- On-chip memory controller interface

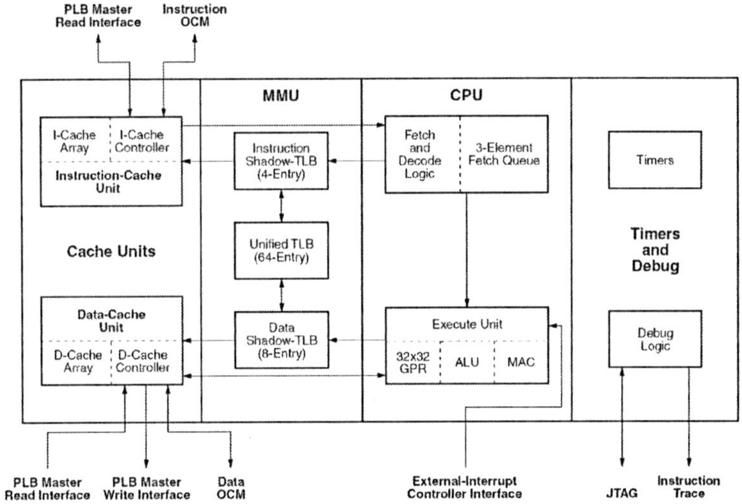

Figure 10.9 PowerPC 405 organization.

10.3.7 Applications of the Virtex-II Pro

10.3.7.1 Data Pipes

For larger bandwidth requirements, multiple RocketIO blocks may be "bonded" (synchronized) together to form larger-data-rate pipes. For instance, in an XAUI interface, four serial channels are channel-bonded to form a single 10 Gbps data-rate pipe. Four RocketIO transceivers, employing 16 PCB traces (4 differential pairs transmit, 4 differential pairs receive), can be used to support a full-duplex 10 Gbps channel by way of the RocketIO channel-bonding feature. This is equivalent to 256 traces of typical LVTTL buses, or 68 traces of a high-speed, source-synchronous parallel LVDS bus. It allows a PCB trace reduction of up to 16X over conventional parallel buses, resulting in significant reductions in PCB complexity and EMI system noise. The RocketIO technology fulfills higher bandwidth system requirements than previously possible, with cost savings coming from faster time-to-market, reduced printed circuit board (PCB) complexity, and lower component count.

10.3.7.2 Reducing PCB Complexity

System architecture is moving from parallel I/O standards to serial I/O standards to meet the increasing data bandwidth. RocketIO blocks provide significant benefits over standard parallel I/O standards for high-bandwidth data transfer. For example, 256 PCB traces are needed to implement a full-duplex 10 Gbps interface using two 128-bit wide synchronous LVTTL buses running at 78 MHz. The same data bandwidth can be implemented using only 16 PCB traces — 94% less PCB routing area — by using four RocketIO blocks. Compared with wide parallel buses, the high-speed serial interface uses less board area with less power dissipation.

10.3.8 Support of Communications Standards

The transceiver module is designed to operate at any serial bit rate in the range of 622 Mbps to 3.125 Gbps per channel, including the specific data rates used by the communications standards listed in Table 10.4.

Table 10.4 Serial Standards That May Be Implemented with Virtex-II Pro Devices

Serial Standard	Data Rate per Channel (Gbps)	Baud Rate per Channel (Gbps)	RefClk (MHz)
InfiniBand	2.0	2.5	125
1 Gb Ethernet 1000BASE-CX/SX/LX	1.0	1.25	62.5
10 Gb Ethernet XAUI	2.5	3.125	156.25
Fibre Channel	0.85 1.7 2.5	1.06 2.12 3.1875[a]	53.125 106.25 159.375
Serial RapidIO	2.5	3.125	156.25
PCI Express (3GIO)	2.0	2.5	125

a. Bit rate is possible with the following topology specification: Maximum 6 inch FR4 and one Molex 74441 connector. See RocketIO user guide.

10.3.8.1 System-on-a-Chip (SOC) Designs

The Virtex-II Pro supports high-performance system-on-a-chip designs, previously the exclusive domain of custom ASICs. The Virtex-II Pro family marks the first paradigm change from programmable logic to programmable systems, with implications for system architectures in networking applications, deeply embedded systems, and digital signal processing systems. It allows custom user-defined system architectures to be synthesized, next-generation connectivity standards to be bridged, and complex hardware and software systems to be codeveloped with in-system debug at system speeds.

10.3.8.2 Network Processing

The Virtex-II Pro family supports network processing where low latency is required, such as storage area networks, wireless infrastructure, and voice-over-IP networks. The digital convergence phenomenon drives the need for packet routing based on type and priority. For example, live voice and video data packets require significantly lower latency than data file packets. New data networking applications must now handle higher bandwidth traffic as well as more complex types of prioritized packets.

In many applications, Virtex-II Pro devices function as specialized network processors (NPs). Using the Virtex-II Pro architecture, the most common packets may be quickly read and routed using FPGA logic, without incurring the lengthy software run-time needed by NPs. The FPGA logic interrupts the PowerPC processor core only when processor instructions are needed for special packet types. For example, packets may be stored into a 16 KB dual-port memory area accessible by both the FPGA logic and the PowerPC 405 on-chip memory (OCM) port, allowing rapid change of control and packet disposition. By using the FPGA logic to process the most common packet types while the processor core handles the more specialized ones as a slave to the logic, the Virtex-II Pro architecture can provide higher overall performance than NPs, as well as more sophisticated processing capabilities than FPGA logic alone.

10.3.8.3 Protocol Bridges

Protocol bridges for tying together disparate data stream formats are well-suited for the Virtex-II Pro solution. New interface standards and protocols include 3GIO, Infiniband, Gigabit Ethernet, XAUI/10 Gigabit Ethernet, RapidIO, and HyperTransport. These must interface with one another, as well as to other standards such as PCI, Fibre Channel, POS Phy Level 4, Flexbus 4, and others. This presents a challenge to system developers because of changing standards, scarcity of off-the-shelf interface components, and the inflexibility of available solutions. System designers have had to assemble their own blend of FPGAs, discrete physical transceivers, and discrete communications processors to solve their complex system challenges. Even newer "programmable ASSPs" (application-specific standard products) with built-in serial transceivers fall short, because they frequently require companion FPGAs to supplement their logic capacity. The Virtex-II Pro solution, using the Xilinx SystemIO™ capability to fully integrate silicon, software, and IP capabilities, provides pre-engineered protocol bridge solutions available for fast time-to-market and low development cost.

10.3.9 Other Features of Virtex-II Pro Devices

10.3.9.1 Global Clock Networks

Virtex-II Pro devices support high-frequency designs and thus require low-skew advanced clock distribution. With device density up to 10 million system gates, numerous global clocks are necessary in most designs. Therefore, to provide a uniform and portable solution (soft-IP), all Virtex-II Pro devices from XC2VP2 to XC2VP125 have 16 global clock buffers and support 16 global clock domains. Up to 8 of these clocks can be used in any quadrant of the device by the synchronous logic elements (that is, registers, 18Kb block RAM, pipeline multipliers) and the IOBs. The software tools place and route these global clocks automatically.

10.3.9.2 Single-Ended SelectIO™-Ultra Resources

Virtex-II Pro SelectIO-Ultra blocks can support up to 18 single-ended I/O standards. Supporting such a variety of I/O standards allows interconnection to almost any digital circuit with positive power supply voltages. The SelectIO-Ultra resource includes a robust set of features including programmable control of output drive strength, slew rate, and input delay and hold time. The supported standards are LVTTL, LVCMOS33, LVCMOS25, LVCMOS18, LVCMOS15, PCI33_3, PCI66_3, PCI-X, GTLP, GTL, HSTL I, HSTL II, HSTL III, HSTL IV, SSTL2 I, SSTL2 II, SSTL18 I, and SSTL 18 II.

10.3.9.3 LVDS I/O

Virtex-II Pro I/Os are designed to comply with the EIA/TIA electrical specifications for LVDS. With the addition of an LVDS current-mode driver in the IOBs, which eliminates the need for external source termination in point-to-point applications, and with the choice of an extended mode, Virtex-II Pro devices provide the most flexible solution for doing an LVDS design in an FPGA.

10.3.9.4 LVPECL I/O

Low Voltage Positive Emitter-Coupled Logic (LVPECL) is a popular and powerful high-speed interface in many system applications. Virtex-II Pro I/Os are designed to comply with the EIA/TIA electrical specifications for 2.5 V LVPECL.

10.3.9.5 Block SelectRAM™ Memory

Virtex-II Pro devices feature a large number of 18 Kb block SelectRAM memories. The block SelectRAM memory is a True Dual-Port™ RAM, offering fast, discrete, and large blocks of memory in the device. The memory is organized in columns, and the total amount of block SelectRAM memory depends on the size of the Virtex-II Pro device. The 18 Kb blocks are cascadable to enable a deeper and wider memory implementation.

10.3.9.6 Distributed SelectRAM Memory

In addition to 18 Kb SelectRAM blocks, Virtex-II Pro devices feature distributed SelectRAM modules. Each function generator or LUT of a CLB[8] resource can implement a 16 × 1-bit synchronous RAM resource. Distributed SelectRAM memory writes synchronously and reads asynchronously. However, a synchronous read can be implemented using the register that is available in the same slice. This 16 × 1-bit RAM is cascadable for a deeper and/or wider memory implementation, with a small timing penalty incurred through specialized logic resources. Distributed SelectRAM modules up to a size of 128 × 1 are available as primitives.

8. LUT is look-up table. CLB is configurable logic block.

10.3.9.7 Bitstream Encryption

Virtex-II Pro devices have an on-chip decryptor that can be enabled to make the configuration bitstream (and thus the whole logic design) secure. The user can encrypt the bitstream in the Xilinx software, and the Virtex-II Pro chip then performs the reverse operation, decrypting the incoming bitstream, and internally recreating the intended configuration.

This method provides a very high degree of design security. Without knowledge of the encryption/decryption key or keys, potential pirates cannot use the externally intercepted bitstream to analyze, or even to clone the design. System manufacturers can be sure that their Virtex-II Pro implemented designs cannot be copied and reverse engineered. The methods available to encrypt the data are Data Encryption Standard (DES) and triple DES.

10.3.9.8 Loopback

To facilitate testing without having the need to either apply patterns or measure data at GHz rates, two programmable loopback features are available. One option, serial loopback, places the gigabit transceiver into a state where transmit data is directly fed back to the receiver. An important point to note is that the feedback path is at the output pads of the transmitter. This tests the entirety of the transmitter and receiver. The second loopback path is a parallel path that checks the digital circuitry. When the parallel option is enabled, the serial loopback path is disabled.

10.3.9.9 Digital Clock Managers (DCMs)

Virtex-II Pro devices have 4 to 12 DCMs, and each DCM provides a wide range of powerful clock management features:

- **Clock Deskew:** The DCM contains a delay-locked loop (DLL) that can completely eliminate clock distribution delays, hence deskewing the DCM's output clocks with respect to the input clock. The DLL contains delay elements (individual small buffers) and control logic. The incoming clock drives a chain of delay elements and the output of each delay element represents a version of the incoming clock delayed by a different amount to minimize the clock skew.

- **Frequency Synthesis:** One output provides a doubled frequency and another output (CLKDV) provides a frequency that is a specified fraction of the input frequency (÷1.5, ÷2, ÷2.5, and so forth, up to ÷15 and ÷16.) An additional two outputs provide an output frequency that is derived from the input clock by simultaneous frequency division and multiplication. The user can specify any integer multiplier (M) and divisor (D) within the range specified in the DCM Timing Parameters section of the Virtex-II Pro Data Sheet.

- **Phase Shifting:** Three outputs drive the same frequency as CLK0 but are delayed by 1/4, 1/2, and 3/4 of a clock period. An additional control optionally shifts all 9 clock outputs by a fixed fraction of the clock period (defined during configuration, and described in multiples of the clock period divided by 256).

10.3.9.10 Digitally Controlled Impedance (DCI)

DCI adjusts the output impedance or input termination to accurately match the characteristic impedance of the transmission line. DCI actively adjusts the impedance of the I/O to equal an external reference resistance. This compensates for changes in I/O impedance due to process variation. It also continuously adjusts the impedance of the I/O to compensate for variations of temperature and supply voltage fluctuations. In the case of controlled impedance drivers, DCI controls the driver impedance to match two reference resistors, or optionally, to match half the value of these reference resistors. DCI eliminates the need for external series termination resistors. DCI can configure output drivers to be the following types:

1. Controlled impedance driver (source termination)
2. Controlled impedance driver with half impedance (source termination)

It can also configure inputs to have the following types of on-chip terminations:

1. Input termination to VCCO (single termination)
2. Input termination to VCCO/2 (split termination, Thevenin equivalent)

For bidirectional operation, both ends of the line can be permanently DCI terminated:

1. Driver with termination to VCCO (single termination)
2. Driver with termination to VCCO/2 (split termination, Thevenin equivalent)

Alternatively, bidirectional point-to-point lines can use controlled-impedance drivers (with three-state buffers) on both ends.

10.3.9.11 Double-Data-Rate (DDR) I/O

Virtex-II Pro devices have dedicated registers in a single IOB (input output block) to implement input, output, and output with three-state control Double-Data-Rate (DDR) registers. Input and output DDR is accomplished with the use of two registers in the IOB. A single clock triggers one register on a low to high transition and a second register on a high to low transition. Output DDR with three-state requires the use of four registers in the IOB clocked in a similar fashion. Since the introduction of DLLs, Xilinx devices can generate low-skew clock signals that are 180 degrees out of phase, with a 50/50 duty cycle. These clocks reach the DDR registers in the IOB via dedicated routing resources.

10.3.10 IBIS and SPICE Models for Xilinx Devices

Analog simulation is a necessary part of PCB and system design. Xilinx supports simulation of its devices through IBIS I/O models that are available on its Web site (*www.xilinx.com*). SPICE models of all Xilinx devices are also available from its Web site. No license agreement is required for IBIS models.

10.3.11 Xilinx Intellectual Property (IP) Cores

IP Cores are defined as very complex, pretested, system-level functions that are used in logic designs to dramatically shorten development time. The IP Core benefits are:

- Faster time-to-market
- Simplifies the development process
- Minimal design risk
- Reduced software compile time
- Reduced verification time
- Predictable performance/functionality

IP Cores are similar to vendor-provided soft macros in that they simplify the design specification step by removing the designer from gate-level details of commonly used functions. IP Cores differ from soft macros in that they are generally much larger system-level functions such as PCI bus interface, DSP filter, PCMCIA interface, and so on. They are extensively tested (and hence rarely free of charge) to offload the designer from having to verify the IP Core functions him or herself.

IP Cores are available for Virtex-II Pro, Virtex-II, Virtex/E, Spartan-II/E, Spartan/XL, CPLD, and XC4000 devices. The Xilinx website lists 462 available cores and specifies which devices support which cores, as the cores cannot be programmed into all devices. Table 10.5 lists some of the available cores.

Table 10.5 Summary of the 462 Intellectual Property Cores Available from Xilinx

Processor Solutions	**System Connectivity**
Microblaze	10/100 Ethernet MAC
Picoblaze	1 Gigabit Ethernet MAC
PowerPC in Virtex-II Pro	10 Gigabit Ethernet MAC
Embedded Development Kit	PCI Express Solutions
Processor IP	PCI 32-bit/64-bit Cardbus
Crossbar Switch	PCI 32-bit/64-bit Cores
IIC	PCI-X 64-bit 66/133 MHz
UART 16450/550	RapidIO
	RapidIO with PowerPC
DSP Solutions	RapidIO Endpoint Core PowerPC
FEC Solutions	HyperTransport
Turbo Product Code	POS-PHY Level 3 Cores
Additive White Gaussian Noise	POS-PHY Level 4 Cores
	Flexbus 4 10Gbps Cores
Communications Solution	PL4 to Flexbus 4 Bridge
HDLC Controller Cores	

10.4 Exercises

10-1. Let's assume that a requirement for your printed circuit board design is to use clock and data rates that exceed those possible for an FPGA by 40 percent. Let's further assume that you plan to build a prototype board using an FPGA. You also plan to use a HardCopy device in place of the FPGA for the high-volume production, which will come after the prototype board checks out. How can you know ahead of time that your higher-speed, high-volume design using the HardCopy device will work when the prototype design using the FPGA can't run at these clock and data rates?

10-2. (a) Is block SelectRAM memory a type of single-port memory or dual-port memory? (b) What is the range of densities (number of bits) of SelectRAM available in the Virtex-II Pro family of devices?

10-3. What is "loopback" and what does it allow you to do?

10-4. Why would you bother to spend another 18 to 20 weeks and custom order and implement a HardCopy device in a design when it comes in the same package and takes up just as much space on the printed circuit board as its corresponding FPGA?

10-5. What are IP Cores and what are the advantages to using them?

10-6. What is the difference in cost between the typical software required to design an ASIC and the Quartus software required to design a HardCopy device?

10-7. What are five actions that will be required of the designer in order to migrate a design from an FPGA to a HardCopy device?

10-8. (a) What is a DCM? (b) How many are there on a Virtex-II device? (c) Name and explain three major functions that the DCM performs.

10-9. If your goal in a design was to maximize bandwidth, would you want to use high-density FPGAs or FPGAs with high-speed transceivers?

10-10. (a) What is the purpose behind using encryption with an FPGA? (b) Does the FPGA perform the encryption, decryption, or both?

10-11. (a) What are the 18 single-ended IOs supported by Virtex-II Pro devices? (b) What are two differential IOs supported by these devices?

10-12. Which Altera devices require the use of a configuration device to load their configuration programs if the configuration program is not loaded through ISP?

10-13. What is the maximum number of user IO pins available with the Virtex-II Pro family of devices?

10-14. How fast can the Virtex-II Pro PowerPC 405 processor cores operate? Express your answer in both clock frequency and Dhrystone MIPS.

10-15. What are the two fundamental reasons why programmable-logic devices are the fastest growing segment of the logic-device family?

CHAPTER 11

Fiber-Optic Components

Chapter Objectives

- Discuss the types of applications where fiber-optic components are used.
- Compare copper and fiber-optic transmission media.
- List high-speed optical protocols and the transmission distances corresponding to each.
- Discuss using fiber optics in the short haul.
- Characterize the X-modules — Xenpak, Xpak, X2, and XFP — and the PAROLI®2 optical link modules.
- Discuss dense wave division multiplexing (DWDM).
- Discuss current trends in the application of fiber optics.
- Present various types of fiber-optic cables and connectors.
- Provide information on laser safety and list fiber-optic vendors.

11.1 Getting On Board with Optics[1]

11.1.1 The Rationale for Optical Interconnects

"Why do designers use optics to interconnect their boards? Today, these interconnects can be as simple as a clock distribution between boards in a system, or as complex as an optical backplane (see Figure 11.1). Optics on board will become more visible with wider use of standards such as 10 Gigabit Ethernet and FibreChannel. There are two driving requirements for optics. Histori-

1. The author of "Getting On Board with Optics" is Lynne Green except where referenced otherwise. Reprinted with permission of Cadence Design Systems Inc.

cally, the driving requirement has been higher data rates. Even with differential signaling, copper runs out of speed somewhere between 1 and 10 Gbps for interconnects between boards.

Optical interconnects, on the other hand, are already being used to carry over 10 THz on existing interconnects, even over distances of a mile or more. Recently, the fact that optical fiber does not produce crosstalk and EMI has become a significant driving factor.

Figure 11.1 Molex ParaLink® optical link with 12 channels at 3.2 Gpbs each. [Reprinted with permission of Molex Fiber Optics.]

Many vendors are helping users move to optical interconnects. New products are introduced almost weekly. Optical backplane connectors allow high-density interconnects, comparable to the best copper connectors, but without the crosstalk normally associated with close spacing. And you do not need to do your own optical interface design; you can now purchase interface modules, such as 10 Gigabit Ethernet, in a single package that is ready to mount on a printed circuit board.

Of course, there is an added cost when using fiber optics. The main cost is associated with the optical modules that convert between electrical signals on the board and optical signals on the fiber, leading to use of fiber as a serial channel. Integrated circuits (a serializer/encoder at the driver and a deserializer/decoder and clock recovery chip at the receiver) may be included inside the optical modules, or separate integrated circuits may be used. With today's technology, the cost of optics has become competitive with the cost of copper at data rates above 1 GHz for many applications, and the crossover data rate is still decreasing." [Cad02a]

11.1.2 Optics in the Physical Design

"In many ways, optical components and modules are similar to their electrical interconnect cousins. The parts have footprints and electrical pins. They come with recommended placement and routing rules, including rules for terminating high-speed signal traces. Because the laser diode inside the optical module switches high currents (typically several mA), power supply distribution and decoupling capacitor selection become critical. Routing rules address power and ground trace widths and distances to power and ground vias, to further reduce power distribution parasitics. There may also be spacing rules on routing, since the fast edge rates lead to increased crosstalk.

There are also placement and routing rules specific to the optical side of the parts. To allow the optical connector on the end of the fiber to attach correctly, the optical 'pin' or port may need to extend over the edge of the board. If the optical port is using free space instead of a fiber, then alignment becomes critical, and both position and orientation will need to be carefully controlled. Finally, the laser diode and its control circuits generate heat, particularly if a Peltier cooler is used to regulate the diode temperature, and layout for thermal design needs to be addressed.

Once production begins, new considerations enter in. First, using laser diodes and optical modules may require review of soldering and reflow processing, since thermal stresses can reduce operating life. Second, and far more important, is the control of dust particles shed during various processing steps. The optical port has been processed to be optically 'flat,' or smooth to about 1 micron. Any small bit of dust can easily be many times bigger than this. When a fiber is connected to the optical port during testing, this dust can be ground against the surface, and the resulting scratch reduces the signal amplitude (light) coupled into a fiber. Even if the dust does not scratch the surface, it can block some of the light, reducing the signal amplitude. Both scratches and blockages often act like a laser quality problem. Problems associated with dust can be reduced by covering the optical port to protect it from dust, and using 'canned air' to remove dust prior to connecting a fiber to the module." [Cad02a]

11.1.3 Modeling Optical Interconnects

"Optical interconnects have been modeled using a variety of approaches. To simulate the electrical signal on the board, a signal integrity simulator such as SPECCTRAQuest from Cadence Design Systems can be used in conjunction with electrical models for the optical diodes. Optical link analysis, in the form of power budget and risetime budget constraint checks, can be done with a spreadsheet. The optical and electrical domains can be simulated together using AMS Designer, using electrical topologies exported from a tool such as SPECCTRAQuest, and optical models written in Verilog-A." [Cad02a]

11.2 Comparison of Copper and Fiber Transmission Media

"In nearly all ITS[2] systems, transmission cables provide the backbone network linking roadside sensors, displays, and monitors with central management and control facilities. The three primary terrestrial-communication media currently used are twisted-pair copper wire, coaxial cable, and optical fiber. The major differences among these three media are the amount of information they carry, the speed at which they can carry it, how susceptible they are to such external forces such as electromagnetic or radio interference (EMI/RFI) or harsh weather conditions, and how far they can carry information.

Evaluated in terms of capacity, speed, and distance, optical fiber is moving to the forefront of transmission media choices. For example, while most twisted copper pair is limited to megahertz bandwidth capability, fiber-based communications networks provide gigahertz bandwidth. This capacity is critical for advanced multimedia communications services that require the bandwidth that optical fiber delivers.

Link lengths are another important factor. Coaxial cable cannot carry signals beyond several thousand feet before amplification is required. In comparison, single-mode optical fiber supports link distances of up to 90 km without any degradation in its bandwidth capacity. Reduc-

2. ITS is Information Technology Systems.

ing the number of electronic components simplifies the network by lowering attenuation and reducing maintenance by limiting the number of components that can fail.

Compared with copper media, optical fiber also maintains a lower noise floor. As a result, optical communication systems exhibit a much lower bit error rate (BER) while operating at much higher data rates. In the gigabit per second (Gbps) range, copper twisted-pair communications can exhibit a BER of 10^{-6}. At 2.5 Gbps, fiber-optic networks typically exhibit a bit error rate of 10^{-9} and are tested at 10^{-11}. This makes communications of ITS data both faster and more reliable over optical systems.

Reliability is another reason why system planners may choose optical cabling. Interference from EMI/RFI, crosstalk, and lightning susceptibility are all factors that can affect a network's performance. Because of their dielectric nature, optical fibers are immune to both EMI and RFI. Copper cables, on the other hand, can act as antennas, receiving EMI/RFI from sources such as electric motors, high-voltage transmission lines, vehicle engines, radio or TV broadcast equipment. These signals, when superimposed on a data stream, may make it difficult for the receiver to differentiate between valid data and EMI/RFI-induced noise. This is especially true for long spans that incorporate components from multiple vendors.

Crosstalk occurs when unwanted signals are coupled between 'transmit' and 'receive' copper pairs. It is especially troublesome when it occurs near the transmit source. Corrupted data caused by crosstalk can make it difficult for the receiver to differentiate between normal and coupled signals. This phenomenon is common with twisted-copper-pair cables, but rarely occurs with fiber.

Lightning can be a problem when it strikes copper cable, because energy may be coupled into the conductor and can propagate in both directions. This can damage the cable and attached electronics if they are not protected.

Since optical-fiber cables do not use metallic conductors for transmission, all-dielectric, optical-fiber communications are less susceptible to electrical surges even if struck directly by lightning. It has been shown that armored loose-tube optical cables can withstand two to four times greater induced voltage from lightning than copper-pair cables.

Many people incorrectly perceive that it's easier to install and maintain copper cable than optical transmission media. In fact, many technicians prefer to install optical cable because fiber is strong and durable, optical cable has better bend performance and flexibility, and the cable's smaller size and lighter weight make it easier to handle. Additionally, connection and termination procedures for fiber are well defined and easy to learn.

Fiber-based systems also are easier to test than copper networks. Technicians typically use just two pieces of optical-test equipment; optical time-domain reflectometers (OTDRs) and optical-power meters to characterize system performance. Fiber's bandwidth capacity supports sophisticated real-time tests, not only for fault detection, but for prevention of fault propagation as well. These capabilities also can help reduce the maintenance costs associated with fiber-based systems.

Harsh environments begin to take their toll the moment system components are deployed. Severe weather, moisture ingress, corrosion, and fatigue are all known to damage communication infrastructure. Given these factors, the typical usable life of most copper systems has been shown to be 15–20 years. Conversely, optical fiber's immunity to adverse conditions makes it possible to project its useful life out to 40 years or more. The oldest fiber systems in operation now are 20 years old and continue to meet installation performance requirements."[3] [Wei03]

11.3 Application Space for High-Speed Optical Data Link Modules

"Table 11.1 shows the universe of communications protocols, data rates, and link distances for optical communications. The link distance categories roughly correspond to those used by various optical communications standards bodies. The vast majority of optical communications applications are included in the table. An 'X' entered into a cell in the table means that there is an applicable protocol and/or transmitter/receiver product for that distance and data rate combination.

The cells of the table that are shaded show the application space covered by the market for high-speed optical data link modules that are used in enterprise or premises data communica-

Table 11.1 Application Space for High-Speed Optical Data Link Modules

Data Rate/Protocol	Category/Distance				
	Very Short Reach (<300 m)	Short Reach–1 (0.3–2 km)	Short Reach–2 (2–10 km)	Intermediate Reach (10–40 km)	Long Reach (>40 km)
100 Mbps FDDI	X	X	X		
10/100 Mbps Ethernet	X	X	X		
Gigabit Ethernet	X	X	X	X	
10 Gigabit Ethernet	X	X	X	X	
1–2 Gbps Fibre Channel	X	X	X		
10 Gbps Fibre Channel	X	X	X		
1 Gbps iSCSI	X	X	X		
10 Gbps iSCSI	X	X	X		
2.5 Gbps InfiniBand	X	X	X		
10 Gbps InfiniBand	X				
30 Gbps InfiniBand	X				
155 Mbps SONET (OC-3)	X	X	X	X	X
622 Mbps SONET (OC-12)	X	X	X	X	X
2.5 Gbps SONET (OC-48)		X	X	X	X
10 Gbps SONET (OC-192)	X		X	X	X
40 Gbps SONET (OC-768)	X				X
Proprietary Parallel Interconnect, 10–40 Gbps	X				

[Reprinted with permission of Strategies Unlimited Inc.]

3. Reprinted with permission of Corning Inc.

tions applications. The term 'high-speed' used here refers to data rates of 1 Gbps and above. Enterprise and premises applications include networks or point-to-point data links that are confined to a single building or campus. They are usually specified as having link distances of 2 km or less.

The data link modules corresponding to the shaded cells of the table are usually supplied in the form of transceivers. These are transmitters (850 nm VCSELs or 1,310 nm Fabry-Perot lasers) and receivers (detectors) combined in a single package with fiber-optic receptacles (multimode and single-mode, respectively). A laser driver and a transimpedance amplifier are usually also included in the package. An exception is parallel optical interconnects in which multichannel transmitter and receiver packages are most often supplied separately, although there are some exceptions (e.g., a four-channel transceiver). An emerging packaging format for higher data rates (10 Gbps) is widely termed a transponder. This product incorporates the normal transceiver components, plus a clock and data recovery chip, and a mux/demux chipset that multiplexes lower data rate streams up to 10 Gbps and vice versa.

The types of applications corresponding to the shaded cells of Table 11.1 include:

- Local area networks (LANs)
- Storage area networks (SANs)
- Server clusters (also known as server area networks)
- Very-short-reach optical interconnects for high-performance telecommunications equipment
- Optical backplanes for high-performance switches and routers

In the first three applications, transceivers are used in switch ports and in various types of adapter cards (e.g., NICs, HBAs)[4] that control the I/O functions on a computer, typically a server. In the case of SANs, transceivers are also used in interfaces that connect storage arrays to the network. In the fourth application, transceivers (or transmit/receive pairs) are used on interface cards or line cards that provide high-speed interconnects between communications equipment chassis. In the last application, parallel optical transmit/receive pairs are used to interconnect internal elements of high-capacity (terabit-class) switches or routers.

Longer distance data communication and telecommunication applications, such as metropolitan-area networks (MANs) or wide-area networks (WANs), fall outside the shaded cells of the table. These generally involve link distances greater than 2 km."[5] [Gii02]

11.4 Using Fiber for the Short Haul

"Despite excess capacity in the long haul, the last few years have been dark days for telecomm- and datacomm-equipment vendors. But innovative fiber-optic technologies for short-haul appli-

4. NIC is network interface card. HBA is host-bus adapter.
5. Reprinted with permission of Strategies Unlimited Inc.

cations are breaking multiple bottlenecks and lighting the way to greater channel densities and a brighter outlook.

Since about five years ago when data traffic outstripped voice, COs (central offices) have required increasingly large numbers of short-haul links to route deaggregated traffic through switching equipment. The I/O transceivers servicing those links must take a minimum of board-edge space, dissipate little power, and carry signals over the OC-3 to OC-192 range. They also must offer low signal attenuation and low distortion. These requirements, particularly in the regions of 2.5 and 10 Gbps, align nicely with those from the datacomm segment, which is looking beyond the throughputs afforded by 1 Gbps and 2 Gbps links for SAN and LAN (local-area network) backbones. This alignment of performance requirements is causing a peak in demand for fiber-optic short-haul and VSR (very-short-reach) transceivers. It is also causing some transceiver vendors to push into the MAN from the short-haul side to fill the gap for intermediate-reach links.

Though fiber-optic devices for long-haul services continue to evolve, fiber's push into LAN and SAN applications has not derived from incremental advances in long-haul technologies. On the contrary, to capture the cost-sensitive datacomm market, fiber-optic-interface technologies depart significantly from those for long haul and even from earlier generation short-reach devices. Low-cost, low-power, small-footprint fiber-optic I/O derives largely from advances in semiconductor processing and in electronic and fiber-optic assembly and packaging.

With port-density demands ever on the rise and newer generation fiber-optic transceivers dissipating less power, the fiber termination becomes a key limiting factor to the number of channels you can fit onto a card edge. Advances in packaging and fiber-termination technologies have given rise to multifiber-I/O modules. These modules not only greatly increase I/O port density, but also allow end customers to connect links carrying many tens of gigabits per second with one fiber-optic ribbon cable.

Perhaps the biggest boon to short-haul fiber optics are the connectorized transmitter and receiver modules (Figure 11.2). Equipment manufacturers don't want high-density I/O cards with many thousands of dollars worth of fiber-optic modules in their finished-goods inventory. Such an arrangement is not only expensive, but also inflexible. It prevents the manufacturer from configuring a board after assembly for channel counts or speeds, which adds even more inventory cost. Soldered fiber-optic modules can also raise the equipment-acquisition cost to the end user by making a coarse-grained I/O channel.

Figure 11.2 Picolight pluggable modules. [Reprinted with permission of Picolight Inc.]

Connectorized BGA/LGA (ball grid array/land grid array) modules solve this problem by allowing manufacturers to 'provision' the fiber-optic I/O to order. The I/O connectors also allow manufacturers to test boards before committing expensive transceivers to them, which can reduce the yielded cost of the finished boards. Boards with connectorized I/O also give users the flexibility of expanding their systems while protecting their base investment. A network-equipment customer might, for example, purchase a 16-module board with only 4 modules in place. As the network expands, the customer may purchase additional modules and install them in the field.

Picolight was one of the first module makers to offer connectorized multifiber modules. The company's Magnus family includes the PL-TCP-00-Sxx series of 12 fiber transmitters based on 850 nm VCSEL arrays and the PL-RCP-00-Sxx family of mating receivers. Individual models operate with data rates to 1.6, 2.5, or 2.7 Gbps for an aggregate bandwidth as high as 32 Gbps in less than an inch of board-edge space. The module's footprint is only twice that of industry-standard, small-form-factor modules that operate single 1 Gbps strands. The 12 channels operate asynchronously for VSR rack-to-rack and switch-to-switch connections.

Picolight's Magnus modules use a standard MTP optical interface with 50/125 or 62.5/125 micron multimode fiber. Their operating range is either 300 or 600 m, depending on the type of fiber you use. The parallel optical interconnect complies with IEEE 802.3 1000-BaseSX. A 12-channel transmitter/receiver pair dissipates 2 W.

Picolight's first foray into long wavelength was the 1310 nm PL-XPL-00-L13 pluggable transceiver modules for links as long as 20 km. The SFP (small-form-factor-pluggable) transceiver operates at 1.06 or 1.25 Gbps for Gigabit Ethernet 1000BaseLX and Fibre Channel 100-SMLC-L service over SMF. The module is the first in Picolight's Extensus family, which targets low-cost, point-to-point applications, such as EFM (Ethernet in the First Mile).

Provable interoperability at the module level is a large concern for equipment makers. It is also a key concern for end users who expect some assurance that equipment they buy to expand their networks will work properly with the existing equipment base.

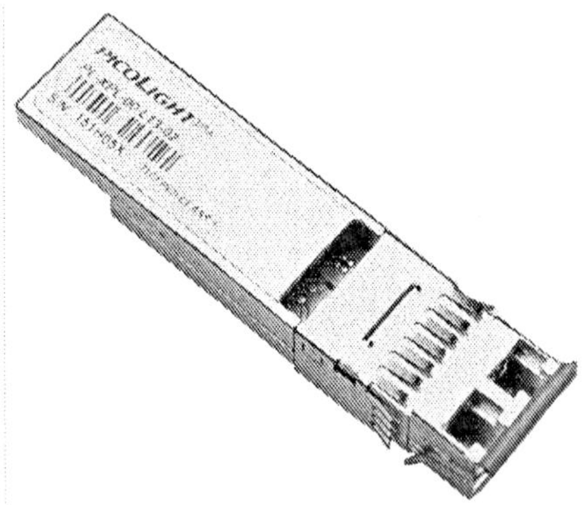

Figure 11.3 Picolight's pluggable transceiver module PL-XPL-00-L13. [Reprinted with permission of Picolight Inc.]

The Xenpak MSA (multisource agreement) is boosting 10 Gbps Ethernet. The agreement defines 850 nm serial, 1,310 nm serial, 1,310 nm WWDM (Wide-Wavelength Division Multiplexing), and 1,550 nm serial pluggable transceivers. Agilent's 10GbaseLR Xenpak transceiver provides an XAUI (10-Gigabit Attachment Unit Interface) electrical interface and meets the proposed IEEE 802.3ae 10 Gbps Ethernet standard. The uncooled, directly modulated, 1,310 nm serial transceiver has a range of 10 km. Agilent's transceiver includes a complete physical layer from the fiber to the four-channel, 3.125 Gbps XAUI interface, including both 8B/10B and 64B/66B encoding and decoding.

Aduro supports the Xenpak MSA by offering Optical Subassemblies (OSAs) that go inside the LX4 Xenpak.[6] Aduro's customers build and sell the LX4 Xenpak transceiver using the LX4-Tx (transmitter OSA) and the LX4-Rx (receiver OSA). The LX4 uses four long-wavelength

6. Aduro Inc. bought out Blaze Network Products.

CWDM (coarse-wavelength-division multiplexing) DFB[7] lasers and is based on a 4 × 3.125 Gbps design. The LX4 OSAs have an operating temperature range from 0–70°C and the optical wavelength spacing is set at 24.5 nm between channels. The optical spacing, in conjunction with the four corresponding CWDM band pass filters, ensures enough 'elbow room' for proper laser operation over recommended temperature. The LX4 is capable of running on both SMF and MMF at the appropriate link lengths. The specification for LX4 over SMF is 10 km, identical to the 10GBASE-LR specification. Over 62.5 m MMF, the LX4 is specified to run 300 m.

Other module suppliers that have signed on to the Xenpak MSA include Agere, Mitsubishi Electric, Pine Photonics Communications, Optillion, and Tyco. IBM has announced that it will also support the Xenpak MSA.

Figure 11.4 Aduro's earlier XGSX4 3.1875 GHz transceiver. [Reprinted with permission of Aduro Inc.]

For those who don't need to drive SMF, Aduro provides OSAs to support 10 Gbps Ethernet, OC-192, and 10 Gbps FC (Fibre Channel). In the past, Aduro built CWDM transceivers around 850 nm Emcore VCSEL arrays. These transceivers contained four oxide VCSELs, which emitted at 778, 800, 825, and 850 nm. Aduro's present transmitter OSAs can drive 100 m of 62.5/125 micron MMF or 300 m of 50/125 micron MMF at 2.488 Gbps. The OSAs can operate to 3.1875 Gbps on 20 percent shorter runs, have channel spacing of 24.5 nm, and have an operating termperature range of 0–70°C."[8] [Isr02]

11.4.1 User Beware

"Beyond safety, fiber-optic module specifiers need to remember a couple of other issues as they develop their designs. Interoperability among modules of similar capability is not a given. Module vendors have been busy demonstrating and documenting the interoperability of their wares with each other at trade-show exhibitions as well as in more traditional applications development labs. Check which modules your design needs to be compatible with before choosing a particular model for your design.

Modules that interoperate are not necessarily interchangeable in your design. For interchangeability, look at which modules share an MSA. MSAs cover, at a minimum, the module's physical dimensions, the mounting mechanisms, pinout, fiber-optic connector, and key operating specifications. Familiarize yourself with the text of the MSA to see what specifications it covers. Ironically, while MSAs ensure mechanical and electrical interchangeability, they do not guarantee equivalent performance. Indeed, some MSAs cover multiple mutually incompatible optical interfaces. A careful analysis of the MSA specifications and those parameters that the drawing omits helps you identify which modules will work with your design.

7. DFB is distributed feedback.
8. © Reed Business Information, a division of Reed Elsevier Inc.

Though fiber-optic cables present no EMI-radiation problem, the electrical end of a fiber-optic module can radiate. In this regard, not all modules are created equal. Some include EMI collars that engage the perimeter of your panel cutout. Some module packages inherently provide better self-shielding. As modules get smaller and packing densities get tighter, low-crosstalk PC-board layout becomes more challenging.

Compared with the purely electronic-component manufacturers with which you are familiar, fiber-optic-module vendors play their cards much closer to their vests. Although this trait is within their rights, it means that you need to allow sufficient extra time in your design cycle to gather basic information. For example, most vendors do not make public their full data sheets. You may need to sign a nondisclosure agreement to get a peek. If so, getting nondisclosure-agreement approval within your own organization will take far more time than the usual half-dozen clicks to download a spec sheet from a component vendor's Web site. Vendors often similarly guard pricing information. Several vendors even omit part numbers in their publicly disclosed information." [Isr02]

11.5 The 10 Gbps X-Modules

"X-modules are 10 Gbps fiber-optic transponders[9] which come in four main module types — Xenpak, Xpak, X2, and XFP. These are front-panel, hot-pluggable fiber-optic modules that are mounted on line cards (PC cards) and can be replaced without disturbing the line card or any of its other ports.

Early and predecessor modules were developed by manufacturers under MSAs. These MSAs included the SFF (small-form factor), SFP (small-form-factor-pluggable), 200PIN, and 300PIN MSAs. These MSAs succeeded in demonstrating the advantages of modularity and reduced the barriers to market acceptance of photonic technologies and the time to market for vendors providing them. MSAs also provided a model for multiple module makers and component vendors to converge on electrical-, optical-, and mechanical-implementation specifications beyond the scope of the standards organizations.

X-modules come with a modular I/O structure. Modular I/O structures bring a host of benefits to the development, manufacture, and maintenance of datacom and telecom ports:

- Modularizing the port largely isolates the design, materials, manufacturing, and test issues linked to the line card's highest-frequency digital circuits from those involving the rest of the product.
- Vendors can populate line cards and adapter boards with the most expensive functions as a post-assembly process, minimizing the yielded cost of their products and reducing the inventory carrying costs.
- End users can populate dense cards as their needs evolve rather than tie up capital in unused capacity.

9. A transponder is a communications device that simultaneously receives and rebroadcasts a signal.

- Interoperability issues are isolated to the modular level, allowing system vendors, their board suppliers, and their customers to agree on and specify proven, stock I/Os.
- New suppliers can enter the market with a predefined, market-accepted qualification regime, fostering competition and securing the supply chain to a greater extent than can strictly proprietary designs." [Isr03]

Table 11.2 lists the module dimensions and minimum pitch of the X-modules.

Table 11.2 Module Dimensions and Minimum Pitch of the X-Modules

MSA	Width (in.)	Length (in.)	Height (in.)	Pitch (in.)
Xenpak	1.42	4.76	0.685	1.87
Xpak (short)	1.42	2.98	0.386	1.65
Xpak (tall)	1.42	2.98	0.876	1.65
X2 (low)	1.42	3.00	0.433	1.73
X2 (mid)	1.42	3.00	0.527	1.73
X2 (high)	1.42	3.00	0.900	1.73
XFP	0.722	2.80	0.335	0.925

[Table reprinted with permission of Strategies Unlimited Inc.]

11.5.1 Xenpak

"Xenpak is an established MSA. The Xenpak MSA was released two years ago and represented an important step in transponder evolution by providing the means for module replacement without disturbing the line card or any of its other ports. A front-panel, hot-pluggable module requires several electrical, electromechanical, and mechanical features that board-mounted modules need not possess.

An example Xenpak architecture is shown in Figure 11.5. For this example, the Xenpak architecture provides an XAUI interface to the media-access controller. The serializer/deserializer distributes the 10 Gbps payload plus the forward-error-correction overhead across the interface's four lanes, reducing the signaling rate to 3.125 Gbps per line.

Though conceptually simple and attractive, a front-panel, hot-swappable transponder module that operates at 10 Gbps requires design inputs from several engineering disciplines. Solid mechanical, thermal, photonic, RF, and power-management engineering must all combine efficiently to address module attachment, EMI, dissipation, and interconnection-integrity requirements.

The electromechanical elements of the design include an EMI shield, its conductive seal, and a dense, multiservice connector that carries power, control, and RF signals. The EMI shield's breakable connection to ground must exhibit a low impedance at the signaling frequency; if it does not, the connection impairs the shield's effectiveness. A gasket compressed between the module's flange and the front panel during module installation establishes the Xen-

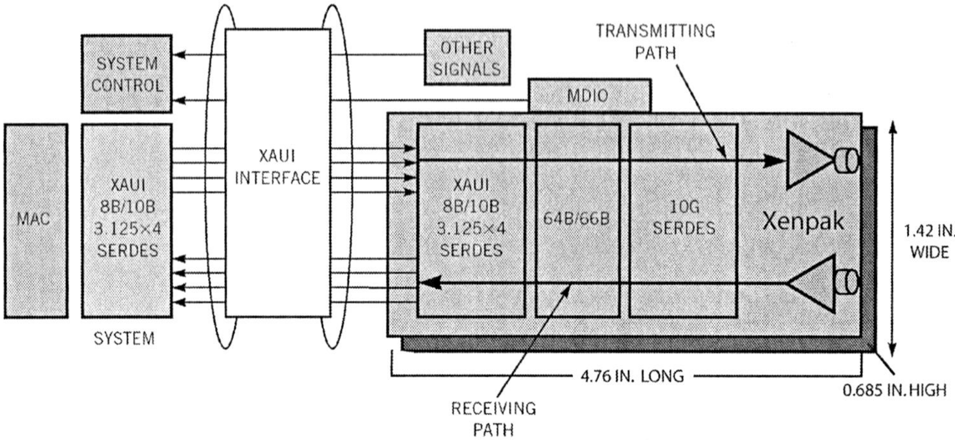

Figure 11.5 Example architecture of a 10 Gbps (12.5 Gbaud) Xenpak module. [Copyright © 2003 Agilent Technologies Inc. Reproduced with permission of Agilent Technologies Inc.]

pak front-shield ground. The module uses a 70-pin connector, 16 pins of which form four differential signal paths in each direction that typically carry signals as fast as 3.125 Gbps.

Key mechanical attributes of the Xenpak MSA include the module guide, the retaining mechanism, and case designs. The guide and the retaining mechanism allow end users to quickly swap modules and establish the appropriate contact positioning and force; they also preclude the possibility of damage to either the module or the line card at the hands of an overexuberant technician. The case design sets the limits on module density and thermal load — critical issues that are always in tension, because, at the board level, they are the physical and thermodynamic reflections of channel count, channel speed, and link length. Implementation technologies link together the physical, thermodynamic, and photonic parameters.

Unlike the 300PIN ('300-pin') MSA that implements the 16-bit, parallel OIF (optical Internetworking forum) SFI-4 interface standard, the Xenpak architecture suits the IEEE 802.3ae standard with its system-side, 4-bit XAUI (10 Gbps attachment-unit interface). IEEE 802.3ae defines physical-layer interfaces for 10 Gbps LAN and OC-192 WAN payloads and seven PMD (physical-media-dependent) interfaces (see Table 11.3). Beyond the raw system-side-bus interface, comprising power, control, status, and data lines, the electrical interface also provides hot-swap protection across its width. Simple management schemes for power-line protection are well-established for a range of supply voltages and currents. Control-line, status-line, and data-line-protection methods, however, are more application specific and need to address signal bandwidths, amplitudes, and polarities.

As attractive as the Xenpak form factor is compared with modules that are not front-panel hot-swappable, the Xenpak MSA does not meet the dimensional requirements of some important market segments. Though many services have converged on data rates at or near 10 Gbps, other parameters, such as the link length and its corresponding power dissipation, cannot converge.

The price the Xenpak form factor pays for a small-footprint module that can dissipate 10W is a tall package that straddles the PC board and, as a result, requires a board cutout. The cutout has been the source of much criticism of the Xenpak form factor because it adds to the PC board manufacturing cost and eliminates precious routing space. Additionally, though the functional segmentation that Xenpak offers has proved successful, it is by no means the only possible arrangement." [Isr03]

Table 11.3 IEEE 802.3ae and Physical-Media-Dependent Interfaces[a]

Device	Features			Media (nm)				Supported Physical Media-Dependent Distances						
	PCS			1550	1310	1310	850	SMF (km)	50 MMF (m)			62.5 MMF (m)		
	8B/10B	64B/66B	WIS	Serial	Serial	WWDM	Serial		400 MHz	500 MHz	2 GHz	160 MHz	200 MHz	500 MHz
10GBASE-LR		✔			✔			10						
10GGBASE-LX4	✔					✔		10	240	300				300
10GBASE-LW		✔			✔			10						
10GBASE-EW		✔	✔	✔				40						
10GBASE-ER		✔		✔				40						
10GBASE-SR		✔					✔		69	86	300	28	35	
10GBASE-SW		✔	✔				✔		69	86	300	28	35	

a. PCS = physical-coding sublayer. WIS = WAN-interface sublayer. WWDM = Wide-Wavelength Division Multiplexing.

11.5.1.1 Xenpak MSA

"An example optical transceiver standard is the Xenpak MSA. These provide a pluggable form-factor with an SC Duplex optical receptacle, as illustrated in Figure 11.6. These modules utilize the XAUI interface on the system (electrical) side, while the external (optical) interface can be any of the various line standards defined in 10 Gigabit Ethernet. These include 10GBase-SR/W, 10GBase-LR/W, 10GBase-ER/W, and 10GBase-LX4." [Irw02]

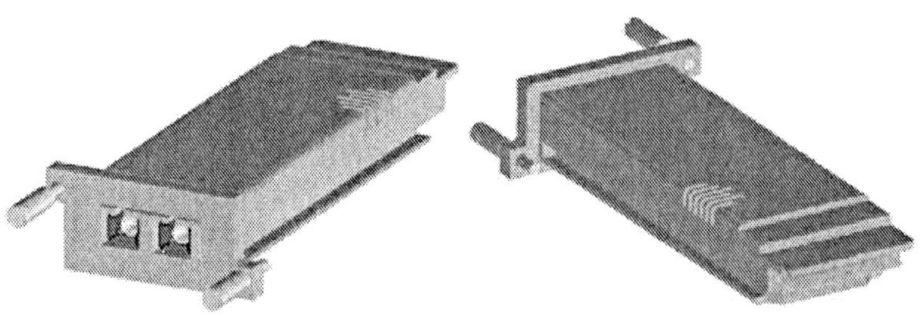

Figure 11.6 Xenpak MSA optical module assembly.

11.5.2 Xpak

"Last year brought definitions of two new Xenpak-like transponder MSAs — Xpak and X2 — that boast smaller package dimensions for high-density line cards, PCI adapters, and other smallformat interface cards. Infineon, Intel, and Picolight founded the Xpak MSA, which some 25 companies have since joined. (For a list of these companies, see the Xpak Web site at *www.xpak.org*.) This module definition is strongly similar to its older brother, Xenpak, in that it segments the modularized functions of the I/O channel at the physical-layer/media-access-controller interface through a four-lane (10 Gbps attachment-unit-interface) connection. The Xpak MSA uses the Xenpak pinout, except that it includes a pair of clock signals dropped from later versions of the Xenpak standard. The additional clock pins provide for operation at 10GFC (10 Gbps Fibre Channel), 10GE (10 Gbps Ethernet), and OIF SFI4- P2 (Optical Internetworking Forum SerDes/Framer Interface). The MSA provides a second pinout for the SFI4-P2 connection.

The Xpak module's dimensions demand less of the I/O card's air rights than does Xenpak, eliminating the board cutout and allowing a standard top-side mount on a PCI card (see Figure 11.7). Two versions of the Xpak module exist (see [Xpa02a]). Both occupy a 1.42-in × 2.98-in. maximum footprint — the length measured from the faceplate. The low-profile version rises to a maximum of 0.386 in. above the board surface to fit within the standard PCI-card-component-height limit. This low-profile form factor can dissipate 4 W with standard mounting in PCI NIC (network-interface card) or InfiniBand HBA (host-bus adapter) and TCA (target-channel-adapter) applications. For longer reach applications outside a small-card environment, a tall version of the Xpak module provides for additional cooling fins to a maximum overall height of 0.876 in. The module footprint and its associated module-holder design address board-edge and mid-board mounting for telecommunications applications (see [Xpa02b]). Installations that require many channels, as is the case in the SAN (storage-area network) and switching-center segments, can take advantage of double-sided belly-to-belly mounting arrangements. You can mount as many as 20 devices on a 17 in. PC board using such a layout.

The Xpak retaining mechanism comprises a rotating wire bale, similar to the SFP's mechanism, that interlocks with the module holder. When you connect fiber-optic cables to the Xpak module, the cable's connectors prevent the wire bale from moving to the unlatched position. Disconnect the cables, and this simple latching mechanism allows a module swap in less time than the Xenpak retainer that uses two captive thumbscrews. The Xpak latch provides a specified amount of compression to a conductive gasket implementing one part of the breakaway EMI shield. A second module-guide mounting option provides an EMI seal for systems with painted faceplates.

With the variety of available fiber-optic module outlines, mounting options, packing densities, and cooling strategies, OEM engineers need to determine the thermal performance of their I/O designs. You may simulate the thermal behavior, but, at some point in the design cycle, you need to reconcile your results with the fact that, from your customers' perspectives, your models and their experiences are not interchangeable. Your simulations ought to correlate to physical mea-

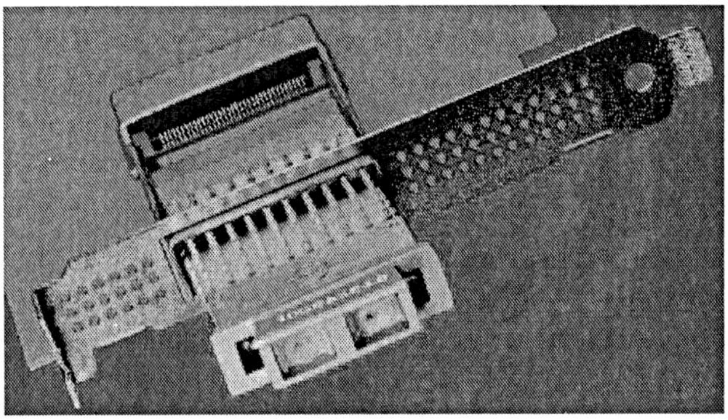

Figure 11.7 The low-profile Xpak module provides the same functions as the larger Xenpak device for 10-Gbps services requiring lower power dissipation.

surements. Appendix A1 of the Xpak MSA contains a useful introductory guide to making controlled thermal performance measurements. It provides dimensions and mounting details for a simple test chamber, recommended temperature- and airflow-measurement locations, and module operating conditions. As the document's authors point out, this appendix can also guide discussions between module makers and OEMs about thermal performance and requirements." [Isr03]

11.5.3 X2 Module

"On the heels of the Xpak MSA announcement, a separate consortium, the X2 MSA group, announced its own module agreement (see [X2m02a]). Eight companies — Agere Systems, Agilent Technologies, JDS Uniphase, Mitsubishi Electric, NEC, Opnext, Optillion, and Tyco — formed the X2 group. (Since signing the X2 MSA, Agere has sold its optoelectronic business to TriQuint.)

X2 also uses the Xenpak electrical interface, though with a few exceptions. X2 provides a 4-bit port-address space — one bit narrower than Xenpak and Xpak. X2 also reduces the number of power pins and makes common the chassis and the electrical ground. The MSA also reserves Xenpak's four vendor-specific pins. On the optical end, the X2 MSA supports 10GE, OC192 SONET, 10GFC, and others (see [X2m02b]).

As dimensionally similar as X2 and Xpak modules may be, they are mechanically incompatible, primarily due to different approaches to the guide mechanism. (See Table 11.2 on page 273.) Xpak uses different guides for its two module heights. Agilent claims that X2 has an advantage by using a single guide design for its three defined module heights (see [Agi02b]). Beyond the obvious reduction of guide-part inventory, the claimed advantage may be real, but only in applications that meet two criteria: The assembly must use a mid-board mounting arrangement; otherwise, the module position is height-specific due to the panel cutout. The application must benefit from mixing module types, ostensibly as a group from instance to instance of the product. Perhaps an underappreciated benefit of Xpak's approach is that it prevents end users from mixing module heights within a single instance of a multimodule product. Such a mix would have airflow implications, the thermal consequence of which the end customer's system operators may not find obvious.

Agilent also claims that X2 has better EMI and thermal performance. X2 uses the Xenpak EMI gasket-and-flange design and thus theoretically inherits the larger module's EMI-seal performance. The Xpak design uses a more modest seal geometry, the adequacy or lack of which the MSA groups should demonstrate, not infer. EMI performance can be challenging to assess. Small differences in the evaluation method, which includes setup geometries, operating conditions, and evaluation criteria, can lead to significant differences in the results. EMI requirements are largely a function of the end application, and many applications have adopted EMI standards that specify the evaluation method. EMI performance claims and comparisons need to either reference a standard or document the evaluation method. This situation is true in the same sense that comparisons between track-and-field athletes need to specify an event.

Though differences in the two modules' case features are clearly visible and suggest dissimilarities in thermal performance, the extent of those performance dissimilarities is not apparent to the unaided eye. Here again, claims of superiority by a vendor may well be valid under application constraints, but these constraints remain unidentified, unspecified, and uncharacterized. Until the vendors identify the operating conditions and evaluation criteria and verify measurement results, consider these excellent topics for you to evaluate as you consider any high-speed, modular I/O." [Isr03]

11.5.4 XFP Module

"A fourth MSA group, XFP (10 Gbps small-form-factor-pluggable module), offers a departure from the Xenpak architecture and its four-lane interface. XFP is a full-speed, single-lane serial alternative to Xenpak and its derivatives using an XFI (10 Gbps serial interface) connection. With no SerDes (serializer/ deserializer) in the module, XFP is smaller and cheaper, and it uses less power. But it also does less of the job and leaves you with PC-board traces operating at four times the speed of the other MSAs. If port-packing density is an issue in the applications you design for, XFP is hard to beat. The dimensions provide for 16 modules on one side of a 17-inch wide PC board. The footprint also provides for belly-to-belly mounting of a second course to the same board.

The trade-offs between XFP and either of the smaller-than-Xenpak alternatives are numerous and need careful consideration. Remember that you need to compare the total channel requirement for a given parameter to compare XFP with either Xpak or X2. You need to assess your product-development costs, bill of materials, power budget, manufacturing costs, and sustaining engineering costs for a channel from end to end, not just for the module, because the XFP segments the channel differently from the other MSAs. You cannot calculate the values one form factor provides outside the context of your project and your organization's resources.

If you go back to the benefits of modularizing the I/O, the XFP form factor does force you and your customers to pay for the SerDes function a board at a time, not with single-channel granularity. On the other side of that issue, with XFP, you can take advantage of some of the multichannel SerDes chips and technologies available from companies specializing in high-speed communication on FR4. This flexibility mitigates some of the costs and may provide an overall channel-density win that none of the other current form factors can match.

Single-lane XFI is more compact to lay out than four-lane XAUI, and maintaining equal lengths and corner counts in your layout is less complicated if you're dealing with differential trace pairs and not clusters of eight. Layouts on FR4 are more challenging at 10 Gbps than they are at 2.5 Gbps, so you should be sure that your team has the necessary engineering resources to generate a high-quality, high-speed board. The XFP-module vendors and the SerDes providers will no doubt have a good deal of application-support information. The end result, however, will be your team's responsibility.

The XFI interface is protocol-agnostic and operates at 9.95 to 10.7 Gbps. XFP implements a two-wire I2C-serial monitor-and-control capability similar to that of Fibre Channel and OC-x SFP modules. Additional control lines provide reset, module detection, interrupt, module-not-ready, and similar signals.

Interestingly, virtually all of the Xpak and X2 MSA group members are also members of XFP. The perception may be that the overall 10 Gbps fiber-optic I/O market will segment into little-Xen XAUI and the littler XFP/XFI form factors. Alternatively, Xpak and X2 may be poised to re-enact the infamous VHS/Betamax battle. A market-adoption delay for Xpak and X2 could give XFP a footing despite the additional OEM engineering load that form factor brings by allowing module makers and SerDes vendors the time to show I/O designers that 10 Gbps on FR4 is manageable." [Isr03]

11.6 PAROLI 2 Parallel Optical Link Modules and Backplane Optical Interconnects

"Infineon Technologies AG has developed second-generation Parallel Optical Link Modules (PAROLI 2) that are designed to speed PCB manufacturing. (See Figure 11.8.) The new version features a ball grid array connector that allows the module to be easily plugged onto a PCB. As a result, users can speed production by running a fully automated PCB line and leaving assembly of the optical module to a later stage.

Infineon's 850 nm VCSEL-based PAROLI 2 modules are products for datacom and telecom applications. Their simple socketed attachment to the PCB makes for easy manufacturing. PAROLI 2 offers data rates up to 3.125 Gbps per channel with 12 channels, for aggregate speed of 37 Gbps over a distance up to 300 m and beyond.

PAROLI 2 modules provide 12 parallel optical channels through an industry standard MPO optical interface. This product supports both LVDS and CML electrical signaling standards through a 100-pin standard connector. An optical shielding collar allows the use of this product in even the most demanding EMI-sensitive environments.

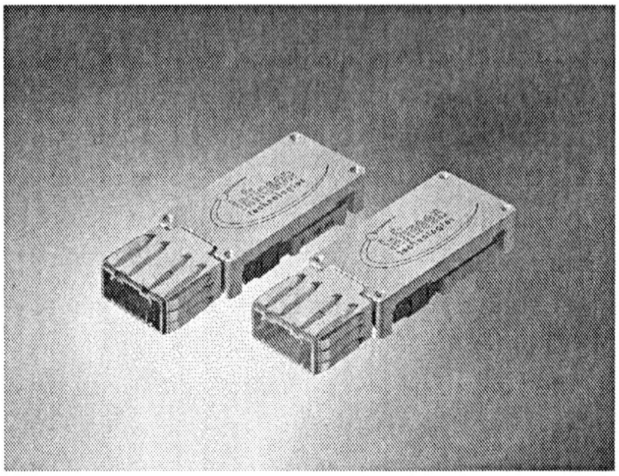

Figure 11.8 Infineon's second-generation Parallel Optical Link (PAROLI 2) transmitter and receiver modules can support either LVDS or CML interfaces to an aggregate data rate of 37 Gbps. [Courtesy of Infineon Technologies AG, manufacturer of PAROLI Parallel Optical Links. PAROLI® is a registered trademark of Infineon Technologies AG.]

The PAROLI 2 link is a very economical solution for high-bandwidth, short-reach applications in which limited space and manufacturing ease are key requirements.

Additional features are:

- Pluggable BGA 10 × 10 electrical socket (FCI megArray 100-pin pluggable BGA electrical connector)
- Industry standard MPO optical connector
- Data rate up to 12 × 3.125 Gbps
- 3.3 V power supply
- Power consumption: typical 1.2 W for the transmitter, 0.8 W for the receiver
- Asynchronous data transfer
- Link length (depends on fiber and data rate): 330 m (@1.25 Gbaud with 62.5 μm/160 MHz×km), 220 m (@2.5 Gbaud with 50 μm / 500 MHz×km)
- Operating temperature of 0–80°C (T_{case})
- ac-coupled module
- Integrated EMI shielding
- SONET and Gigabit Ethernet mask compliant
- IEC Class 1M laser compliant
- Ultra-low profile

"In the feature specifications above, the unit MHz×km is a bandwidth × distance product. For an optical fiber, under specified launching and cabling conditions, and at a specified wavelength, this unit is a figure of merit equal to the product of the 3 dB bandwidth of the optical signal and the fiber's length. The bandwidth × distance product is usually stated in megahertz × kilometer (MHz×km) or gigahertz × kilometer (GHz × km). It is a useful figure of merit for predicting the effective fiber bandwidth for other lengths, but should only be considered as a rough guide if the actual link length for a certain data rate has to be calculated.

The actual link length will depend on many additional parameters than just the modal bandwidth. A few of them are the transmitter and receiver rise/fall times, the spectral width (for chromatic dispersion), the available power budget, the transmitter extinction ratio, and so on. These are usually all compiled into the so called 'IEEE 802.3z link model,' which is an Excel spreadsheet which calculates the expected link length for a certain data rate and fiber on a worst-case basis.

The feature specifications listed above were taken from PAROLI 2 data sheets and are calculated values using this theoretical model. Since these calculated values use the theoretical model and take into account the many aditional parameters, the two link lengths specified above differ from what you would calculate from taking one of the link lengths and scaling it by the new bandwidth or data rate to obtain the other link length."[10] See reference [Pon01] for a detailed discussion of bandwidths, data rates, and link lengths.

The PAROLI 2 modules can be used for terabit/pentabit core and metro routers, intelligent optical switches, and digital cross-connects. Within these applications, it can be used for card-to-card, shelf-to-shelf, rack-to-rack (VSR), and system-to-system interconnects (see Figure 11.9). It can also be used for OC-192 very short reach interconnects (OIF/VSR-1) and for future OC-768 very short reach interconnects (OIF/VSR-5) at 3.32 Gbps per channel. Other applications include use as Infiniband 12-channel optical modules, proprietary data links for storage-area networks, use for server clustering and supercomputing, use for proprietary interconnects for optical test equipment, and for customized interconnect design in aerospace.

Infineon has also developed a system (see Figure 11.10) for the design of high-speed backplane optical interconnects. It is intended to reduce the design effort for high-end systems and reduce the complexity in manufacturing and system maintenance. The system combines the advantages of parallel high-speed transmission over multimode fibers, like low loss and long transmission length with the blind mate feature onto the backplane, the flexibility of the fiber routing, and the robustness of a field-usable multifiber connector.

The blind mate function allows board plugging and unplugging from the backplane without needing access to the rear side optical cables. There is no need for backplane access after system installation and no risk of wrongly connected ports as the cabling always stays in place. Devices can be mounted on only one side or mounted belly to belly for increased port density."[11] [Inf03b]

10. Courtesy of Dr. Thomas Knuth, Infineon Technologies AG.
11. Courtesy of Infineon Technologies AG, manufacturer of PAROLI Parallel Optical Links. PAROLI® is a registered trademark of Infineon Technologies AG.

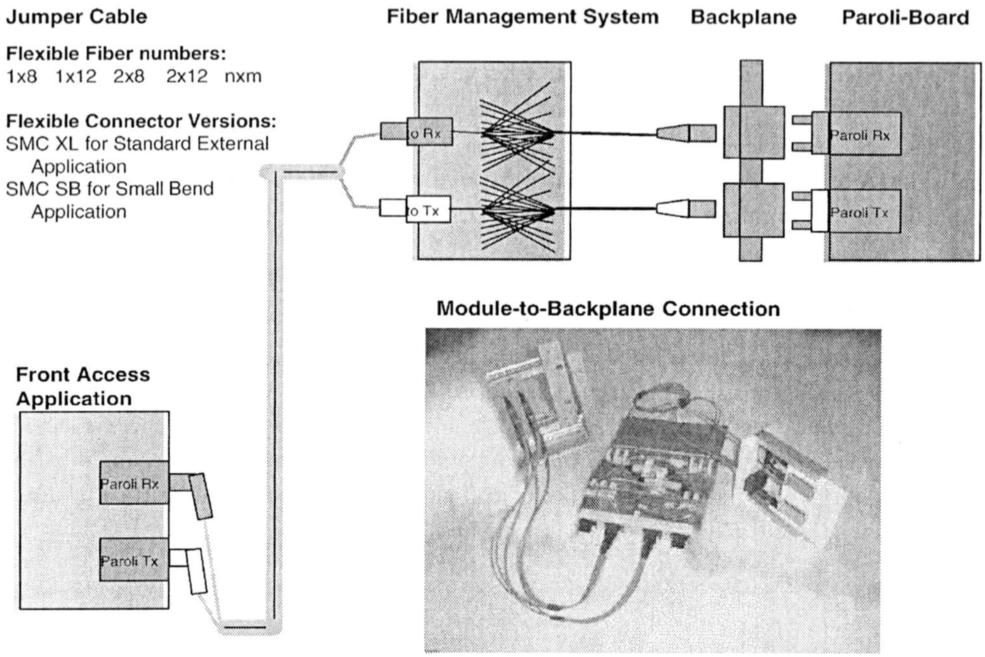

Figure 11.9 PAROLI rack-to-rack connection and module-to-backplane connection (lower-right inset). [Courtesy of Infineon Technologies AG, manufacturer of PAROLI" Parallel Optical Links. PAROLI" is a registered trademark of Infineon Technologies AG.]

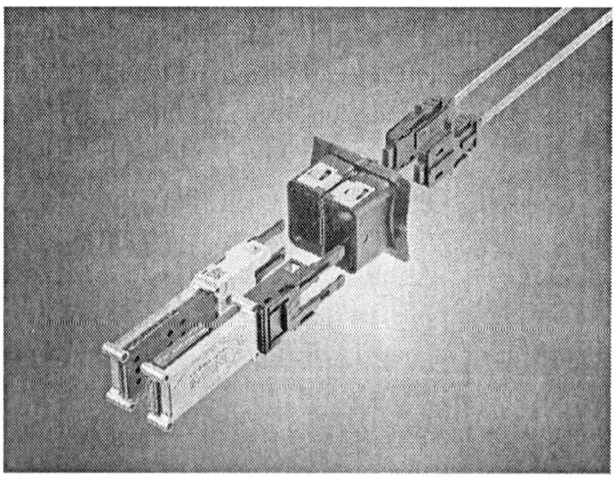

Figure 11.10 Infineon Technologies AG high-speed backplane optical interconnects. [Courtesy of Infineon Technologies AG, manufacturer of PAROLI" Parallel Optical Links. PAROLI" is a registered trademark of Infineon Technologies AG.]

11.7 Dense-Wavelength-Division Multiplexing (DWDM)

"DWDM was the term historically given to wave division multiplexing (WDM) when significantly more channels were being added to systems. Since WDM is increasingly more dense all the time, both WDM and DWDM are now used synonymously."[12] [Com03]

"DWDM and WDM work by combining and transmitting multiple signals simultaneously at different wavelengths on the same fiber. In effect, one fiber is transformed into multiple virtual fibers. So, if you were to multiplex eight OC-48 signals into one fiber, you would increase the carrying capacity of that fiber from 2.5 Gbps to 20 Gbps. Currently, because of DWDM, single fibers have been able to transmit data at speeds up to 400 Gbps.

A key advantage to DWDM is that it's protocol- and bit-rate independent. DWDM-based networks can transmit data in IP, ATM, SONET/SDH, and Ethernet, and handle bit rates between 100 Mbps and 2.5 Gbps. Therefore, DWDM-based networks can carry different types of traffic at different speeds over an optical channel."[13] [Web03]

11.8 Trends in the Application of Fiber Optics

"Many optical component and semiconductor manufacturers are cranking out innovative new products for the fiber-optic marketplace. One fiber-optic market showing promise is metro networks. Sonet/SDH dominates this sector but is being challenged by Ethernet. During the dot-com boom, long-haul and metro networks were overbuilt, and there's still an enormous amount of dark fiber. Some metro networks are being built, although the emphasis is on lower costs and equipment that's easier to provision. 10 Gigabit Ethernet (10GE) just may capture this segment.

Sonet/SDH was originally optimized for voice, although today data is equal to, or even exceeds, voice on some networks. Its fixed time-division multiplexing (TDM) nature makes it inefficient because of the difficulty in scaling and adapting to the nature of bursty data traffic. Considerable effort is under way on a number of fronts to make optical Ethernet a service-bearing technology like Sonet/SDH. Some of the architectural alternatives include Ethernet over Sonet (EoS), Switched Ethernet over Fiber, and Ethernet over WDM (wavelength-division multiplexing). For more information, contact the Metro Ethernet Forum (MEF).

Growth in Ethernet local-area networks (LANs) continues, but with the newer optical versions lagging. Optical options exist in both 1GE and 10GE. The copper CAT5 and CAT6 versions of 1GE have all but won this segment. So why rewire with fiber if plain-old existing unshielded twisted pair works just fine? Most of the newer PCs and laptops have a 10/100/1,000-Mbps RJ-45 optical port, and 1GE switches are growing in number in enterprise networks.

More activity is expected in the 10GE optical space as business recovers with investment in 10GE optical LAN backbones. Lots of equipment and components for this market are now

12. Reproduced with permission from Computer Desktop Encyclopedia. Copyright © 1981–2003. The Computer Language Co., Inc. (*www.computerlanguage.com*).

13. Copyright 2003 Jupitermedia Corporation. All rights reserved. *www.webopedia.com.*

available, and competition is heating up. Thus, the outlook looks strong for optical transceiver modules, switches, and routers in this market.

Another sector showing growth right now is storage area networks (SANs). This may be the fastest growing segment for fiber as more companies install Fibre Channel and iSCSI fiber connections to their servers. (iSCSI is Internet Small Computer Systems Interface and is pronounced, 'I-skuzzy.')

For the future, fiber-to-the-home (FTTH) seems to have the most potential. Unless the traditional carriers get off their duffs, the cable companies are going to steal the broadband market from them. While DSL (Digital Subscriber Line, as in xDSL) has shown some growth, it's just a stopgap that will ultimately have to yield to fiber. That's because DSL over the plain-old telephone-service local loop can't hack the higher speeds required by movies-on-demand, gaming, and other multimedia services. Some of the smaller independent carriers already install passive optical networks (PONs) in the last mile. However, this enormous segment won't go anywhere until the main carriers decide to do it on their own, or unless they're directly and immediately threatened by the cable companies." [Fre03]

11.8.1 Exciting Optical Devices

"Intel has an extensive and growing line of optical transceiver modules and a wide range of chips that address optical networking. One of Intel's optical achievements was its tunable laser modules earlier in the year. No one product makes DWDM more practical than a tunable laser. Instead of buying one laser for each of dozens of different wavelengths, designers can now buy just one tunable laser that can be set to any of the wavelengths in the popular C-band (1,550 nm range). Intel's laser module has an output of 20 mW and is voltage tunable over a 50 GHz range. Besides lowering equipment and maintenance costs, it permits new architectures in which data can be switched dynamically on-the-fly from wavelength to wavelength. Intel is a member of the recently formed Tunable Laser Multisource Agreement (MSA) alliance.

Another company with an extensive portfolio of optical transport chips is PMC-Sierra. Typical of its newest products are the PM8377 and PM8369 Fibre Channel (FC) intelligent port bypass controllers. The PM8377 supports four channels, while the PM8369 supports 18 channels. These chips are designed for 4.25 Gbps enterprise-class storage arrays and SAN applications. They also support the recent Fibre Channel Industry Association (FCIA) vote to extend 4.25 Gbps FC into the SAN fabric.

A companion chip is the PM8358 QuadPHY 10GX, a 4-channel, fully redundant 3.2 Gbps transceiver offering high-speed serial I/O. This SerDes product supports a broad range of optical-system applications that require multigigabit serial interconnect, including Gigabit Ethernet, Fibre Channel, 10GE, 10 Gbps FC, OC-48, OC-192 Sonet/SDH, and InfiniBand interfaces. The chip features bidirectional XGMII-to-XAUI interfaces and includes clock recovery, clock synthesis, byte alignment, trunking, and 8B/10B encode/decode logic.

This year, Scintera Networks' SCN3142 and SCN5028 were two of the top optical-fiber-related chips announced. Scintera Networks is a startup just emerging from stealth mode to announce its Electronic Dispersion Compensation Engines (EDCE). These specialized DSP chips implement filtering and equalization techniques, permitting 10 Gbps data to be transmitted over legacy fiber at greater distances. Scintera's Advanced Signal Processing Platform (ASPP) provides automatic and adaptive dispersion compensation that can be incorporated into almost any standard laser transceiver module.

The SCN3142 device, designed for the enterprise market, can transmit 10 Gbps serial data at a range of up to 300 meters in legacy multimode fiber (MMF). This is a far better solution than the 4-wavelength, coarse-wavelength-division multiplexing (CWDM) solution offered by the 802.3ae LX4 standard. The chip is totally adaptive and protocol agnostic, and it requires no training sequence. It syncs and locks in within tens of microseconds. Moreover, it works with Ethernet or Fibre Channel.

The SCN5028 uses the same ASPP techniques but is optimized for the metro optical market. It compensates for nonlinear impairments and dispersion when transmitting data at 10 Gbps on metro single-mode fiber (SMF). It can extend the typical 80 km connection to over 140 km. The device compensates for both chromatic and polarization dispersion while using the automatic and adaptive EDCE techniques.

Some examples of all-optical components are those from TeraXion and Lightconnect. TeraXion recently announced the availability of its C-band fixed and tunable dispersion compensators. Dispersion, the stretching of a light pulse due to aberrations in the fiber, is a major problem at very high data rates (>10 Gbps). Chromatic dispersion seriously limits transmission distances at 10 and 40 Gbps with multiple wavelengths unless it can be corrected. The TeraXion dispersion compensators use fiber Bragg gratings to make the correction. A tunable version is available, permitting on-the-fly adjustment of the connection for data-rate changes to 40 Gbps or dynamic reconfiguration. The products also come in versions for L-band.

One survivor of the optical shakeout over these past few years is Lightconnect. It has demonstrated that innovation and a clearly superior product will get you through the hard times. The company's newest product is the Fast VOA 5000, the industry's smallest variable optical attenuator. It uses MEMS (micro electromechanical systems) technology with a moving semiconductor structure that modifies the input light to produce an attenuation effect. The attenuation is fully variable from 0.8 to 40 dB with a 0 to 5 V control signal. The device is fast and offers superior attenuation over previous generations of MEMS, liquid crystal, or thermal attenuators. VOAs are used extensively in optical telecom systems to equalize the power levels at different points in the network, as well as correct for differing power levels in multiple-wavelength DWDM systems.

The product can be used for receiver and channel protection, on/off switching, multiplexing, and demultiplexing, as well as in optical add-drop multiplexers (OADMs). Lightconnect also makes dynamic optical equalizers for multiple-wavelength DWDM systems using the same MEMS technology.

One of the fastest growing segments of optical networking involves the aforementioned SANs. They connect servers and other storage devices via a unified network optimized for storage applications, especially database access. The dominant SAN interface to this point has been

Fibre Channel. This optical system is complex and expensive with some speed and distance limitations. However, the industry is addressing them. Nevertheless, a competitor has grown quickly and is beginning to cut into FC sales — iSCSI or Internet Small Computer Systems Interface. This fast serial version of SCSI uses TCP/IP to transmit storage data over fast optical Ethernet network connections. It's less expensive than FC and has higher speeds that scale better. Needless to say, sizable growth is anticipated in this sector.

One of the companies making iSCSI practical and better than ever is Silverback Systems. Its recently announced iSNAP 2100 is a fast IP storage network access processor (Figure 11.11). It is designed for use in servers, iSCSI storage devices, NAS (network attached storage) appliances, blade servers, and multiprotocol storage routers. The 2100 accelerates access to IP storage by offloading both networking and upper-layer protocols like iSCSI and RDMA and then mapping them to the iSNAP architecture. What results is superior throughput. Besides the 2100 processor chip, Silverback offers a full line of supporting firmware and software to adapt the product to the specific storage network application." [Fre03]

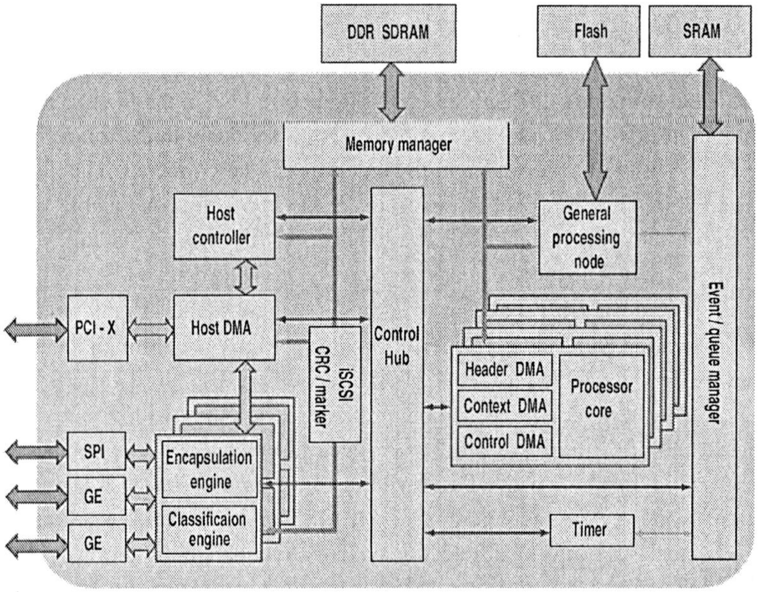

Figure 11.11 The Silverback 2100 iSNAP iSCSI offload processor connects to two Gigabit Ethernet (GE) optical or copper interfaces. [Reprinted with permission of Silverback Systems Inc.]

11.8.2 PON Devices

"Another highly promising area of the fiber-optical business encompasses passive optical networks (PONs). These relatively short-range networks are designed primarily for metro networks and particularly for the FTTH or fiber-to-the-business (FTTB) markets. Their benefit is low cost, because no intermediate active electronic elements are needed between the source and destination. Passive optical splitters/couplers are used to divide a single fiber bandwidth into 16 or 32 channels.

The past few years have seen precious little activity in PONs. But some of the regional Bell operating companies (RBOCs) are beginning to experiment with and invest in PONs, especially for new homes. As the industry works itself out of the usual chicken and egg problem, costs will decrease. No doubt, then, we can look forward to an all-PON telephone system in the distant future that will also provide high-speed Internet access as well as video. Considerable PON activity is taking place in other countries, especially Japan and Korea, who are leading the pack, and to a lesser degree in China and Europe.

Multiple types of PONs are being built in the United States. One of the standards vying for attention is the APON, or PON based upon ATM (asynchronous transfer mode). RBOCs would choose this technology because it fits with their existing ATM switching networks. It can achieve a speed of 155 Mbps uplink and 622 Mbps downlink using relatively inexpensive components. The standard for APONs is defined by the ITU-T as well as the Full Service Access Network (FSAN) Group, a forum of companies working toward a PON future.

Another emerging standard is the IEEE's Ethernet in the First Mile (EFM) 802.3ah protocol. Known as EPON or Ethernet PON, this standard defines both copper and optical Ethernet to the home.

GPONs or Gigabit PONs are PONs that operate at 1 Gbps or higher. The FSAN Group has set standards in coordination with the ITU-T for data rates of 1.25 and 2.5 Gbps.

One of the companies addressing the PON market is OpticalZonu. It makes a line of optical transceivers for use as central-office optical line terminators (OLTs) and customer premise optical network terminals (ONTs). These are protocol agnostic, so they can be used in APONs and EPONs. Some use a two-wavelength (2λ) full-duplex scheme on 1,310 and 1,490 nm.

Another company going after the PON market is Alloptic Inc. It has developed a complete line of central-office and customer premise equipment meeting the EPON standards. Alloptic's system brings fiber to the home, business, or multiunit dwelling. These systems use a symmetrical 1 Gbps upstream/downstream format up to a range of about 20 km or more, depending upon the type of fiber used." [Fre03]

11.9 Optical Cable Applications

"Optical cable solutions can provide the highest combination of data rate and transmission distance. Solutions are now available over a wide range of cost/performance trade-offs, from very short reach, short-wave, multimode fiber to long-haul, long-wave, single-mode fiber supporting distances of over 40 km. In addition to single-channel serial implementations, other options include aggregating multiple channels via parallel optical ribbon cable or WDM.

Some examples of high-speed serial interconnect standards supporting optical cabling include Fibre Channel, Gigabit Ethernet (1000Base-SX/LX), 10 Gigabit Ethernet, InfiniBand and the OIF Very Short Reach (VSR) standards. In addition, there are several optical transceiver form-factor definitions available, including Gigabit Interface Converter (GBIC), Small Form-Factor (SFF), and Small Form-factor Pluggable (SFP, Xenpak, and XGP, to name a few). PAROLI and MPO are common parallel optical ribbon cable and connector solutions." [Irw02]

11.10 Optical Internetworking Forum (OIF)

"Several standards have been defined within the Optical Internetworking Forum (OIF) to support Very Short Reach (VSR) optical communication channels. The following four OIF implementation agreements support the OC-192 (~10 Gbps) generation of SONET equipment:

- VSR4-01.0: Aggregated link of 12 fibers operating at 1.244 Gbps per fiber using 8B/10B encoding. Utilizes multi-mode, parallel optical ribbon cable with MTP™ (MPO) connector (each direction) and shortwave (850 nm) optoelectronics, supporting distances up to 300 m.
- VSR4-02.0: Single fiber operating at 9.953 Gbps (OC-192). Utilizes single-mode fiber and longwave (1,310 nm) optoelectronics, supporting distances up to 600 m.
- VSR4-03.0: Aggregated link of four fibers operating at 2.5 Gbps per fiber. Utilizes multimode, parallel optical ribbon cable with MTP (MPO) connector (one 12-fiber ribbon with 4 fibers each direction and middle four unused) and shortwave (850 nm) optoelectronics, supporting distances up to 300 m.
- VSR4-04.0: Single fiber operating at 9.953 Gbps (OC-192). Utilizes multimode fiber and shortwave (850 nm) optoelectronics, supporting distances to 85–300 m." [Irw02]

11.11 Fiber-Optic Connectors

"The ideal connector for fiber cabling is a fixed splice, where a precision machine operated by a skilled mechanic fuses two fibers permanently together with heat. Obviously, this technique lacks flexibility, and so various connectors have been developed to allow non-permanent connections. In all cases, the job of a fiber-optic connector is to bring two minuscule glass cores in contact with one another with a great deal of precision. The tiny diameter of single-mode fiber, in particular, makes it imperative that the two cores align perfectly to ensure the light is contained. The SMA connector is a screw-on type, not widely used in new installations. The ST was designed by AT&T and Lucent and is a 'bayonet' push-and-turn connection. The SC connector is that specified by TIA 568A wiring standard; the action is push-pull. All of these connectors are available on prefabricated patch cords as well as in kits for field termination."[14] [Bir01]

11.11.1 Small-Form Factor Connectors

"Recently, several new 'small-form factor' (SFF) fiber-optic connectors have entered the fray. Each of these takes up the same space as a twisted pair RJ-45, and is about half the size of traditional connectors. The most common of these are the Volition VF-45, the MT-RJ, and the LC.

14. © 2004 101communications LLC. By Randy Bird. Reprinted with permission from CertCities.com.

While these are still fighting for market share, at this point the VF-45 is mostly seen in multimode fiber-to-the-desktop (FTTD) runs, the MT-RJ in FTTD and multimode building backbone connections, and the LC in single-mode backbone and WAN applications.

Regardless of the connector used and the manufacturer's claims, connectors are always a trouble spot in fiber-optic networks. Because of the small size of the mating fibers, the slightest contamination can result in blocking a fair amount of light and causing a bad, or even worse, intermittent connection. Fiber-optic connectors should always be kept with the manufacturer's protective dust cap in place until the moment of mating, and excessive connection/reconnection discouraged. If dirty fiber is suspected, it should be cleaned with alcohol and a lint-free tissue. A good, clean fiber connection might have 0.5 dB of loss; a poor one, 2 dB or more. These losses are roughly equivalent to those seen in one to several miles of unspliced fiber-optic cable, so you can see how important the connector quality is to the overall system." [Bir01]

11.11.2 InfiniBand Connectors

In addition to backplane and copper cable options, the InfiniBand standard supports 1x, 4x, and 12x optical interconnect options at 2.5 Gbps per fiber. Although the optical module form-factor is not specified, the connector technology is defined. The 1x standard is based on an LC duplex connector, while the 4x and 12x standards utilize the 12-fiber MPO connector, as illustrated in Figure 11.12. The fiber connections can be seen as dark row of holes between the alignment holes on the cable and alignment pins on the socket. The 4x channel requires 8 fibers (4 each direction), leaving 4 of the 12 channels unused. The 12x channel requires two such cables, each carrying 12 fibers in each direction. [Irw02]

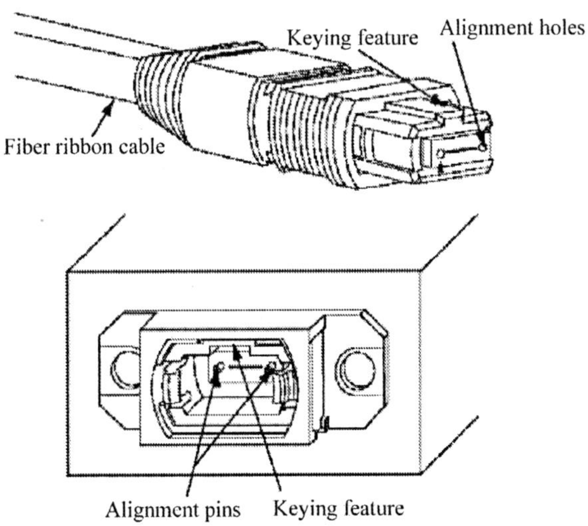

Figure 11.12 InfiniBand™ MPO fiber-optic connector (4x, 12x).

11.12 Laser Safety

"If you're not familiar with eye safety measures and procedures regarding fiber-optic transmitters, you should be before you start to work with them. Module makers go to considerable lengths to comply with established standards. However, if you don't know what the standards specify and under what conditions, the phrase 'eye safe' can lose its meaning.

IEC (International Electrotechnical Commission) 60825-1 specifies laser classifications, exposure levels and times, and tests (see [Iec02]). Of the laser classifications that IEC 60825-1 defines, two are most commonly associated with fiber-optic transmitter modules for short-haul service. Class 1 lasers are unconditionally safe, including cases in which you view them with instruments. Class 1M lasers are safe when you view them without instruments (see [Kol02]). Both classifications assume that the modules are running within their specified operating conditions.

The International Committee on Nonionizing Radiation Exposures has defined biologically safe exposure limits from which IEC 60825-1 derives MPEs (maximum permissible exposures). The MPEs, in turn, form the basis for calculating the AEL (accessible emission limits) for a given laser class and wavelength at an exposure duration of 100 sec. The calculated AELs essentially place an upper bound on the transmitter's output power, depending on which safety class the manufacturer is targeting.

Your eyes are more sensitive to short-wavelength radiation than to long. So, in addition to getting better fiber attenuation and dispersion performance, transmitters output more power from a 1,300 nm VCSEL than an 850 nm device within a given eye-safety classification.

The most recent amendment to IEC 60825-1, which the IEC added last year, uses new MPEs and larger resulting AELs that increase the Class 1 power for 840 nm devices on 50 mm MMF to 1.24 dBm: 2.46 dB, or nearly a factor of two, over previous limits. The same device operating to Class 1M has an AEL of 9.43 dBm — nearly a factor of 12 over previous Class 1 limits. These changes allow additional transmit power, which in turn reduces the demand on receiver sensitivity. But your eyes haven't changed, so the other result is that the safety margin a given class offers is now smaller.

Suppliers and customers of components, modules, and equipment are scattered across the globe. It does little good to have one standard in force in the European Union, another in the Americas, and yet a third in, say, Japan. The need to converge, or 'harmonize,' standards from disparate geographical regions and governing bodies is a familiar one. In this case, the job may be easier by virtue of an unusually close association. IEC Technical Committee 76 was responsible for amending 60825-1. Its chairman is also a key laser safety official at the U.S. FDA (Food and Drug Administration). The FDA Center for Devices and Radiological Health's Standard 21 CFR (Code of Federal Regulations) 1040 is the regulatory document in the United States that most closely corresponds to IEC 60825. ANSI (American National Standards Institute) standard Z136 covers safe use of lasers.

You can purchase a consolidated version of 60825-1, which includes the two amendments currently in force, for about $100 directly from the IEC. A combined set of Z136.1 and Z136.2,

'Safe Use of Lasers' and 'Safe Use of Optical Fiber Communications,' is available from ANSI for $144. Additional transmitter-specific safety information is available from the IEEE, SPIE, and from module vendors." [Isr02]

11.13 Vendors and Organizations for Fiber-Optic Components

Table 11.4 lists various vendors and organizations dealing with fiber-optic components.

Table 11.4 Vendors and Organizations for Fiber-Optic Components

Company	URL[a]
300PIN MSA	http://300pinmsa.org
Aduro Inc.	www.aduroinc.com
Agere Systems	www.agere.com
Agilent Technologies	www.agilent.com
Alloptic Inc.	www.alloptic.com
CDT Optical Systems Division	www.cdtfiber.com
Corning	www.corning.com/opticalfiber
EFMA (Ethernet in the First Mile Alliance)	www.efmalliance.com
Emcore	www.emcore.com
Fibercore Limited	www.fibercore.com
Fibre Channel Industry Association	www.fibrechannel.com
Gore Electronics	www.goreelectronics.com
IBM	www.ibm.com/chips/products/fiberoptic
IEEE	www.ieee802.org
IEEE 802.3ae Task Force	http://grouper.ieee.org/group/802/3/ae
Infineon	www.infineon.com
InfiniBand	www.infinibandta.org
Inphi	www.inphi-corp.com
Intel	www.intel.com
iSuppliCorp.	www.issuppli.com
JDS Uniphase	www.jdsu.com
Lightconnect Inc.	www.lightconnect.com
Lucent Technologies	www.lucnet.com
Maxim Integrated Products	www.maxim-ic.com
Metro Ethernet Forum	www.metroethernetforum.org
Mitsubishi Electric	www.mitsubishielectric.com
Molex	www.molex.com
NEC	www.nec.com
OIF (Optical Internetworking Forum)	www.oiforum.com
Opnext/Hitachi	www.opnext.com
OpticalZonu	www.opticalzonu.com
Optillion	www.optillion.com
Passive Optical Network Forum	www.ponforum.org

Table 11.4 Vendors and Organizations for Fiber-Optic Components (Continued)

Company	URL[a]
Picolight	www.picolight.com
PMC-Sierra	www.pmc-sierra.com
Scintera Networks	www.scinteranetworks.com
SiGe Semiconductor	www.sige.com
Silverback Systems	www.silverbacksystems.com
TeraConnect	www.teraconnect.com
TeraXion Inc.	www.teraxion.com
TriQuint	www.triquint.com
Tunable Laser Multisource Agreement	www.tunablelasermsa.com
Tyco Electronics	www.tycoelectronics.com
X2 MSA	www.x2msa.org
XFP MSA	www.xfpmsa.org
Xpak MSA	www.xpak.org
Zarlink Semiconductor	www.zarlink.com

a. Only Inphi, Maxim Integrated Products, and PMC-Sierra have a dash in their URL.

11.14 Exercises

11-1. What is most likely the biggest disadvantage of the Xenpak module when compared with Xpak, X2, and XFP modules?

11-2. Given that the PAROLI 2 parallel optical link module provides 12 channels, each capable of carrying data at 3.125 Gbps, what applications is this device most suited for?

11-3. What is meant by "passive WDM"? What is meant by "dense WDM"?

11-4. What are four trends in the application of fiber optics today?

11-5. Does Molex make only connectors or does it also make fiber-optic transmitters and receivers?

11-6. What is a transponder? What components comprise the transponder?

11-7. How are optical interconnects modeled and simulated?

11-8. What are MSAs? Do they guarantee that modules from different vendors will have equivalent performance?

11-9. What are VSR transceivers and in what types of applications are they used?

11-10. What are the four kinds of X-modules? Provide a description of about 100 words for each module.

11-11. Why will DSL ultimately be replaced by fiber?

11-12. What are the two most common pieces of equipment used by technicians to check out fiber-optic cable connections?

CHAPTER 12

High-Speed Interconnects and Cabling

Chapter Objectives

- Describe SiliconPipe's 12.8 GHz memory channel, 20 GHz chip-to-chip channel, and 40 GHz backplane interconnect technologies.
- Briefly describe the limiting factors of conventional backplanes.
- Explain the basic theory behind SiliconPipe's high-speed technologies.
- Discuss assignment of signals to connector pins in order to minimize crosstalk and skew.
- Discuss Teradyne's VHDM high-speed differential connector, connectors for XAUI, Tyco's and AMP's Z-PACK HM-Zd connector, and the InfiniBand connector.
- Present cables and connectors for the LVDS signaling standard and discuss cable ground and shield connections.

12.1 SiliconPipe's 12.8 GHz to 40 GHz Interconnect Solutions[1]

"Current multigigabit interconnects are complex and costly, and the performance is limited by connectors, vias, crosstalk, ESD, dielectric loss, and skew. To overcome these existing fundamental challenges, developers need new solutions that are holistic in nature in order to solve the channel problem from die to die. That includes solutions for chip packages, chip-to-chip interconnects, memory channels, and backplanes, yet at a cost equal to or even lower than those of today's solutions.

SiliconPipe's technologies and concepts create that reliable and affordable bridge for 5 to 40 gigabit interconnects. SiliconPipe's holistic view of the interconnect at several gigabits/second

1. Except as indicated, this section is reprinted with permission of SiliconPipe.

has led its engineers to focus on the physical aspects of the channel, including connection technologies, material, termination methods, and dielectric. The team has reviewed and refined manufacturing technology methods as well as silicon design (especially for the memory technology) to create a line of technologies that are the embodiment of the company's highly improved electronic interconnection structures." [Sil04] SiliconPipe markets the intellectual property for these technologies where OEM vendors manufacture the actual products (like Rambus does).

12.1.1 Yosemite — 40 Gbps Backplane Channel Technology

"The Yosemite technology, also known as ChannelPlane, focuses on high-speed backplane solutions with the expressly defined objective of delivering higher performance (40 Gbps/channel) at the cost of today's 2.5 Gbps products. Although, admittedly, there is no silicon generally available today to support the 40 Gbps speed made possible with SiliconPipe's technology, the competitive cost structures allow the use of these higher performance backplanes for today's lower-speed solutions. These lower-speed solutions can also take advantage of the increased budget in signal improvement for further reliability, upgrade paths, and additional benefits.

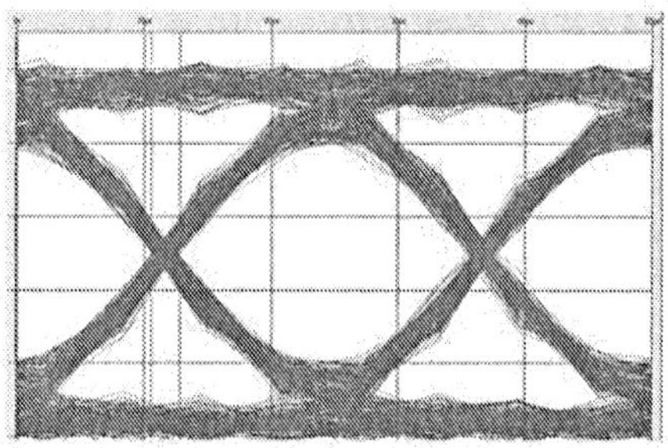

Figure 12.1 Eye diagram for 40 Gbps signal transmitted 30 inches through two interconnects and a backplane. [Eye diagram courtesy of Agilent and TDA Systems.]

The eye diagram shown in Figure 12.1 is for a 40 Gbps signal transmitted through two interconnects and over a backplane totalling 30 inches. Agilent Technologies took the data and TDA Systems extracted and generated the eye diagram.

Advantages of the Yosemite technologies are as follows:

- Up to 40 Gbps/channel (differential pair) passive copper backplane
- Up to 40 inch backplane length
- Low crosstalk and no signal compensation required for NRZ signaling
- Significant reduction in backplane layers
- High reliability and yield
- Complementary interconnect technology allows for chip-to-backplane-to-chip interconnects at 15–20 Gbps with GBX, VHDM, or SIP 1000 connectors, and 40 Gbps with a special HPSP[2] alignment system and packaging improvements
- Cost is competitive

2. HPSP is High-Performance Signal Path.

There are several performance benefits. For example, a 40 Gbps/channel backplane (24 inches) can be manufactured at 3 Gbps/channel cost. Similarly, a 'perfect channel' (32 inches) can be created between line cards across a backplane and allow for low-power communication by 10 Gbps SerDes devices (<100 mW). With this technology, core structures have the same supply voltage and processor geometry as I/O structures. The technology offers a low-skew infrastructure ('clock and data') of over 10 inches, allowing for meshed architectures and low-cost implementation.

Supply chain benefits include that the backplane is structured as a commodity item allowing many companies to bid for the task, there is high yield on backplanes, and the use of low-power SerDes reduces cost of integration and requires less chips.

Yosemite has a standardized architecture. There is one backplane type at low cost for most applications, the channel is scalable without the need for upgrades, and the channel works with all currently proposed standards (OIF, 5–10G backplane, SerDes, etc.). There are shorter design cycles, less signal integrity work, and less system validation. A reference board is available for testing to OEMs, VARs, or system integrators." [Sil04]

"An implementation of ChannelPlane, the CP-OC768, is already being used in a patent-pending Yosemite-HPSP (High-Performance Signal Path) trace and via replacement system."[3] [All03]

12.1.2 Sequoia — 20 Gbps Chip-to-Chip Channel Technologies

"The Sequoia product line offers three distinct technologies to allow for 20 Gbps/channel interconnects at low cost — the X3-Channel, the ESD-Channel, and the Flex-Channel.

The X3-Channel is a cost-effective packaging technology for electrical performances of 20 Gbps per pin, with largely improved thermal conductance. The technology can be used as a standalone packaging option or in a custom solution. The ESD-Channel is a packaging option that adds an ESD-Channel with capacitances of 0.2 pf or less. The Flex-Channel technology enables the link between packages over distances up to 15 inches, with minimal skew at speeds of 20 Gbps per channel. These result in benefits such as 15 Gbps/channel packaging at 0.13-μ CMOS structures, 20 Gbps at 90 nm CMOS structures have improved thermal distribution, low ESD capacitance (below 0.2 pF per pin), and high-density packaging options.

There are supply chain benefits as well. The package is structured as a commodity item allowing many companies to bid for the task. There is a one-time license structure with no royalty. The technology uses standard manufacturing techniques, has a standardized architecture, is applicable to various die sizes and performances, has a flexible design, and has mechanical and electrical compatibility with existing packaging." [Sil04]

3. Reprinted with permission from Electronic Design, October 13, 2003. Copyright 2003. Penton Media Inc.

12.1.3 Grand Canyon — 12.8 Gbps Memory Channel Technologies

12.1.3.1 ChaniPlexer — High-Speed Memory Channel

"ChaniPlexer is the first implementation of SiliconPipe's Grand Canyon memory channel technologies. ChaniPlexer is a high-performance interconnection technology that enables standard DIMM modules to achieve effective data throughput at rates from 3.2 to 12.8 Gbps/channel. ChaniPlexer technology provides performance well beyond the performance offered by DDRI, DDRII, QBM,[4] and RDRAM, at costs that are in line with today's standard DDR solutions. It is applicable for PCs, graphics systems, servers, and supercomputers.

ChaniPlexer-based systems have eight times the throughput of any existing DDRI or DDRII memory solution at equal or better latency.

There is support for point-to-point channels for DDRI, DDRII, or QBM signaling, data transfer speeds from 3.2 Gb/channel to 12.8 Gb/channel, and with no extra PLL needed. ChaniPlexer uses low-power I/O and is compatible with the memory controller signaling standards of LVDS, CML, and SSTL_2. ChaniPlexer also offers high-density packaging with up to 32 DIMMs per controller.

There are also several supply chain benefits of using ChaniPlexer. The technology employs a simple channel, a simplified host board, and standard DIMMs. There are standard manufacturing techniques, lower signal integrity cost, standardized architecture, short implementation cycle, architecture is expandable, and the design is flexible. There is also mechanical and electrical compatible with existing DIMMs." [Sil04]

12.1.3.2 SeriPlexer — Next-Gen Memory Channel Technology

"SeriPlexer is a set of technologies ranging from interconnects to signaling technologies and from ESD improvements to new topologies mainly geared for the markets of servers, workstations, graphic systems, and super computers. It supports speeds of 12.8 Gbps and beyond." [Sil04]

12.1.4 Limitations of Conventional Backplanes

"Copper backplane interconnects, struggling to meet the needs of high-speed backplane communications, are currently stuck at about 10 Gbps. In fact, most systems work at just 3 Gbps. To go any higher requires more complex and costly optical interconnects.

High-speed backplanes built today are limited in bandwidth by the materials and design of the system's interconnects. At high frequencies, attenuation and reflections degrade the signal. These discontinuities in the signal path limit the transmission of high frequencies and increase signal rise times. This, in turn, causes intersymbol interference, deterministic jitter, and collapse of the eye diagram.

Conventional backplane connectors use a 'prong and sleeve' approach: One cable connector that enters into the backplane contains the male prong pins, and the other comes out of the

4. QBM is Quad Band Memory, a type of DDR RAM.

backplane with female sleeve jackets, into which the pins slide. This approach degrades the signal because it introduces interconnect losses that consist of resistive losses and a lack of uniform impedance, as well as channel losses. Channel losses include dielectric, conductor, and fringing losses, which combine to act as performance bottlenecks (see Figure 12.2).

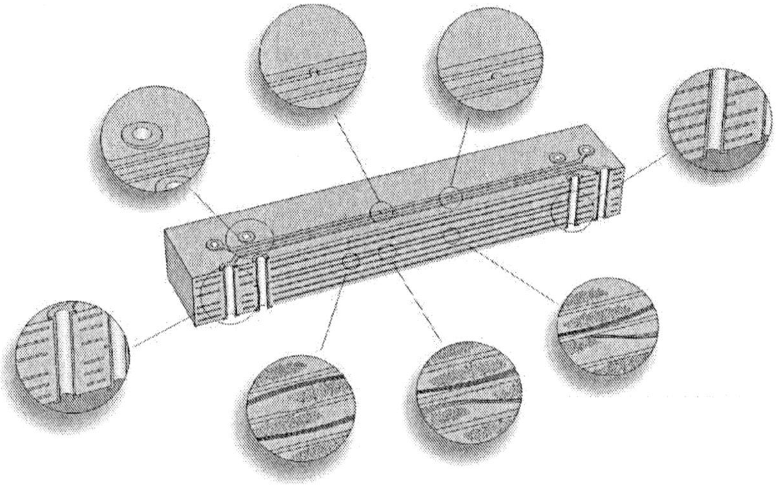

Common problem areas, clockwise from left:
1) Capacitance and reflections in through-hole
2) Localized crosstalk
3) Localized changes in conductor width
4) Localized changes in conductor spacing
5) Reflections due to via stub
6) Non-uniform dielectric
7) Surface treatment thickness non-uniformity
8) Localized changes in foil thickness

Other potential problems:
- Skew caused by lack of routing space
- Dielectric loss
- Conductor loss
- Non-homogeneous dielectrics
- Plated-through hole defects (voids and cracks)
- Back drilling errors
- Wrong hole size

Figure 12.2 Performance-limiting characteristics of conventional printed circuit backplanes. [Sil04] [Reprinted with permission of SiliconPipe.]

Designers can use certain techniques to control fundamental signal-transmission limitations in conventional backplanes. These include the use of low-loss laminates, stub back-drilling, and active signal-conditioning techniques. But these come at a price that increases the cost of backplanes and line cards." [All03]

"To achieve a system solution using copper media, not only is the dielectric of the material crucial but so is the performance of the interconnections. Uniform impedance must be maintained from the line-card silicon, through the line card's connector, the high-density backplane, the switch-card connector and, finally, through to the switch-card silicon." [Sil03]

12.1.5 How SiliconPipe Technology Works

12.1.5.1 Backplanes

"Instead of relying on a conventional backplane connector scheme, a backplane system was created that allows the coaxial cable conductors on one side of the backplane to mate 'head on' with existing conventional backplane connectors (Figure 12.3).

SiliconPipe's ChannelPlane technnology mates coaxial cable conductors on one side of the backplane "head on" with existing cable connectors. This eliminates many signal-degrading factors when transmitting at multigigabit speeds.

Figure 12.3 Head-to-head mating of coaxial cables. [All03] [Reprinted with permission of SiliconPipe.]

The conventional male and female sockets are eliminated with a sexless interface at the end of the cable. The only performance-affecting part is a compliance interface. This interface is designed to be less than one-quarter of a wavelength thick, with controlled impedances and consideration for fringing, effectively eliminating performance-limiting discontinuities when using conventional male/female backplane connectors.

Because no media discontinuities exist in the ChannelPlane system, there are no vias or stubs that cause degradation of the signal. All signals are routed straight from the surface of the ChannelPlane, known as the XL array, keeping impedance mismatches well within the range of 100 Ω (±3%) under all conditions.

Standard coaxial cables and mating connectors are used. A W.L. Gore coaxial cable is used on one side of the backplane and an SIP 1000 Winchester connector is used to mate with the backplane on the other side. Therefore, many standard high-speed connectors from industry leaders like Teradyne, Winchester Electronics, and Tyco Electronics can also be used to get the same performance. Up to 95% of such connectors are 2 mm types that are limited to transmission speeds of 5 Gbps when used in conventional backplanes.

The 184-pin Winchester connector handles 46 differential-pair signals. The XL array connector design has greater than twice the density of the Winchester connector. Because the cables mate head-on with the connector, there's no need for routing channels, as with all other connectors. This allows for 160 channels of signals, each operating at 40 Gbps, in a 1 in. square area." [All03]

"The ChannelPlane system consists of two major components: a conventional FR4 backplane containing power, ground, and low-frequency signals plus an integrated second part, the Omnibus system, which carries all high-speed links over high-performance cables. The ChannelPlane system can be expanded to become a uniform connection from IC packages on one line card, through the backplane, to IC packages on another line card, providing optimal system-level performance. Package connections that eliminate discontinuities in line cards are made close to the silicon die, using uniform low-loss interconnects. Discontinuities from vias, multilayer PCB traces, and IC substrates are eliminated.

The ChannelPlane system can be used as an alternative to optical backplanes for OC-768 systems for distances up to 40 inches. Additionally, the system enables silicon suppliers to use standard low-power 0.07 μ complementary metal-oxide semiconductor (CMOS) silicon to drive backplane channels. Transceiver vendors can provide low-power, high-performance CMOS solutions without the need for receive equalization, multilevel PAM[5] signaling or multitap pre-emphasis. Conventional backplane solutions demand higher-level voltage swings, eliminating the ability to use smaller-geometry CMOS.

Manufacturing costs are similar to conventional high-density backplanes, resulting in 40 Gbps performance at the cost of 3 Gbps. Today's 3 Gbps systems can be upgraded with new line cards as the market demands higher performance." [Sil03]

"A related technology, CleanChannel, can be used for 10 Gbps SerDes with a power requirement of less than 100 mW. The new physical channels allow board designs to run high-speed signals, generally using existing techniques such as common-mode logic, in current power envelopes of 60 to 100 mW.

Today, a router might use a 50- or 60-layer PCB costing as much as $20,000 for the backplane and requiring sophisticated counterboring techniques that only five or six companies in the world can handle. It's likely that such boards can be replaced by ChannelPlane technology at a fraction of the cost and, instead of being limited to five or six vendors, any typical vendor could make the backplane board.

To adopt the SiliconPipe technique, OEMs lay out just power and ground signals on a 10- to 15-layer PC board, have a third party create the cable and backplane connector assembly based on the OEM's Gerber files, and then have the cables assembled onto the board. This requires only standard manufacturing techniques; there are no tricks or specialty processes required."[6] [Mer03]

5. PAM is Pulse Amplitude Modulation.
6. Copyright © 2003 by CMP Media LLC, 600 Community Drive, Manhasset, NY 11030. Reprinted from EETIMES with permission.

12.1.5.2 Memory Interconnects

"The ability to move a minimum of 3.2 Gbps between a controller and memory is achieved using packages with existing driver/receiver technology and without the use of SerDes devices. A substantial array of technologies is needed, however, including differential receivers, CML transmitters, and a proprietary shielding design that runs all the way from the pad ring on one die to the pad ring of the next.

A direct connection between the memory controller and a simple multiplexer chip is made, which in turn is connected to a bank of DIMMs. The design supports up to 32 DIMMs at 3.2 Gbps per pin, while considerably simplifying motherboard layout and yielding precisely predictable performance.

Modifications are made to the drivers and package of the memory controller. The circuit board technology is also modified and the multiplexer chip needs to be designed, but no changes are required for the DIMMs themselves. This is considerably a smaller effort than a JEDEC SerDes approach in development.

By incorporating this technology in the DRAM chips as well, a new kind of DIMM is produced that can talk directly to the memory controller at up to 24 Gbps per pin. The connection provides bus widths up to 128 bits, with no SerDes and with a simple, differential, source-synchronous interconnect scheme." [Wil03]

12.2 High-Speed Connectors

Connectors are a complex topic and are the subject of many heated debates at standards committee meetings. There are two basic types: standard matrix-based connectors (e.g., 3 rows of 32-pins) and special connectors. The special connectors may be optimized for differential signals or may use elaborate techniques to clamp to the PCB. The elaborate connectors tend to avoid via structures and thus offer the highest bandwidth. They are also very application specific and tend to be rather expensive. More common is the use of standard connectors for a mix of differential, power, ground, and single-ended connections (see Figure 12.4). The figure shows a variety of pinout recommendations for single-ended and differential options.

For differential lines in a matrix (single-ended) connector, adjacent pins in a row should be used as the electrical lengths are the same. The row with the shorter path provides a better path

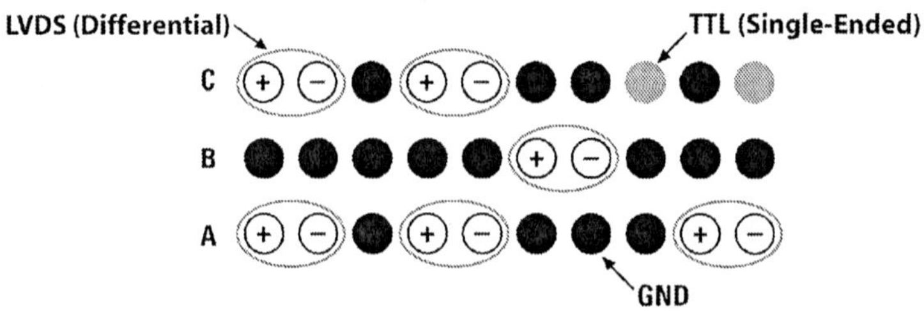

Figure 12.4 Typical connector pinouts. [Reprinted with permission of National Semiconductor Corporation.]

over the longer row also. Ground signal assignment can be used to isolate large-swing (TTL) signals from the small-swing lines, as shown above.[7] [Nat00]

12.2.1 Teradyne VHDM-HSD Connector Example

The Teradyne VHDM-HSD™ connector (high-speed differential) is a differentially optimized high-density connector for applications into the gigabit per second range. It is offered in 8-row (38 real differential pairs per linear inch, 22 mm minimum slot pitch), 6-row (25 real differential pairs per linear inch, 18 mm minimum slot pitch), and 5-row (25 real differential pairs per linear inch, 15 mm minimum slot pitch) configuration. The 8-row connector can coexist with the basic 8-row VHDM single-ended connector and its corresponding power and ground contacts. Shielding is provided between wafers, providing excellent isolation of signal contacts. The backplane layout is also improved over earlier generations and the new footprint now supports wider traces (10 mils) to be routed through the pin field, easing backplane design. The backplane side of the connector accepts either the single-ended or differential versions of VHDM. Skew has been minimized within the pair and the pairs are also routed together (see Figures 12.5 and 12.6). This is an example of a high-throughput, differentially optimized connector that provides excellent signal quality. [Nat00]

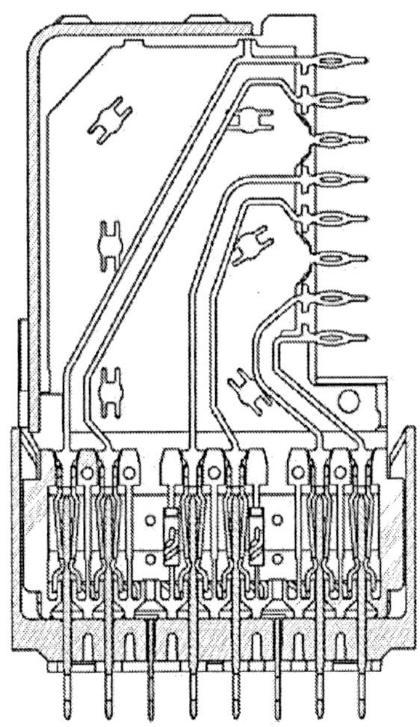

Figure 12.5 Cross-section of Teradyne VHDM-HSD™ connector.

7. Reprinted with permission of National Semiconductor Corporation.

12.2.2 XAUI Connectors

"XAUI is defined in the 10 Gigabit Ethernet standard for extending the useful reach of the XGMII interface between the link and physical layer. It consists of four channel bonded 3.125 Gbps interfaces with an aggregate bandwidth of 12.5 Gbps (10 Gbps useful payload with 8B/10B encoding overhead). The standard defines chip-to-chip connectivity up to 20 inches of FR4 PCB trace. XAUI can also be used in backplane applications.

In addition to chip-to-chip applications, the XAUI can support communication to mezzanine cards or across a backplane. Two examples of high-speed connectors supporting XAUI (and other) backplane links include the Teradyne VHDM-HSD™ and the Tyco/AMP Z-PACK HM-Zd. These are illustrated in Figure 12.6 and Figure 12.7, respectively. For more information on these products, go to *www.teradyne.com* and *www.amp.com*." [Irw02]

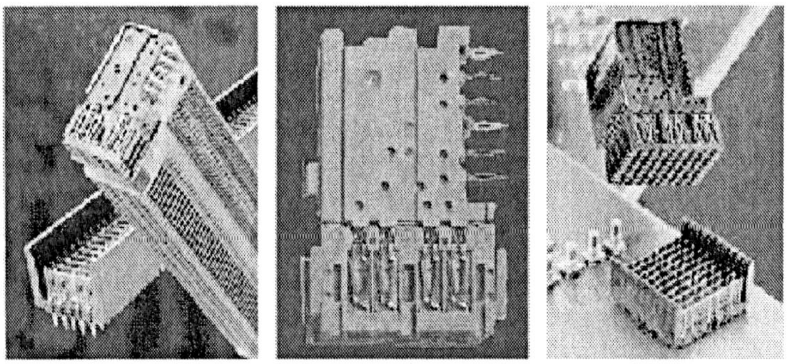

Figure 12.6 Teradyne VHDM-HSD backplane connector. [© Copyright Teradyne, Inc. — All rights reserved. Copied with permission of the copyright owner.]

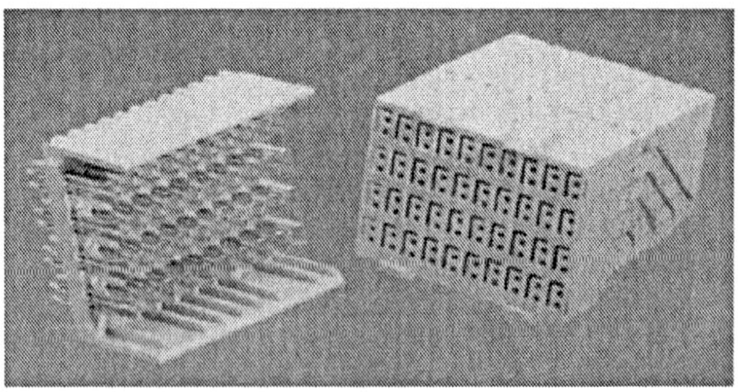

Figure 12.7 Tyco/AMP Z-PACK HM-Zd backplane connector. [Reprinted with permission of Tyco Electronics.]

12.2.3 InfiniBand Connector

The backplane connector specified by the InfiniBand™ standard is shown in Figure 12.8. It is based on the SpeedPac technology from Tyco/AMP and supports 1x, 4x, and 12x serial channel configurations. The connector consists of a card edge "paddle" (with paddle guard) and backplane socket combination.

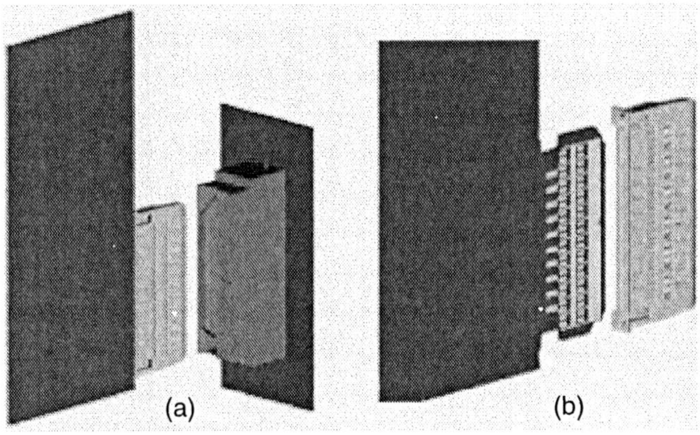

Figure 12.8 InfiniBand (a) backplane connector and board paddle and (b) paddle guard assembly based on SpeedPac technology. [Reprinted with permission of Tyco Electronics.]

12.3 High-Speed Cabling

12.3.1 Copper Cable Applications

Copper cabling can be used to interconnect system components over short to medium distances (generally up to about 20 m) at data rates up to about 3.125 Gbps. These cables are generally based upon shielded twisted pair (STP), coaxial, twin-axial, or other similar technologies. Copper cabling has the advantage of utilizing familiar electrical signaling technology rather than more exotic (and generally more expensive) optical technology. There is a trade-off between data rate and distance, with greater distances coming at the expense of lower data rates. Some type of galvanic isolation (blocking capacitors, transformers, etc.) is typically required. While single-channel, dual-simplex configurations (single differential pair in each direction) are the most common, multiple-channel configurations can also be used.

Some examples of high-speed serial interconnect standards supporting copper cabling include Fibre Channel, Gigabit Ethernet (1000Base-CX), and InfiniBand.

12.3.2 InfiniBand Cabling

An example of a high-speed cabled solution can be found in the InfiniBand standard. This standard supports communication at 2.5 Gbps per differential signal pair over 1, 4, and 12 channels in each direction. Examples of the connector technology adopted by this standard are shown in Figure 12.9 and Figure 12.10. The 1x connector is based upon the HSSDC2 technology from Tyco/AMP.

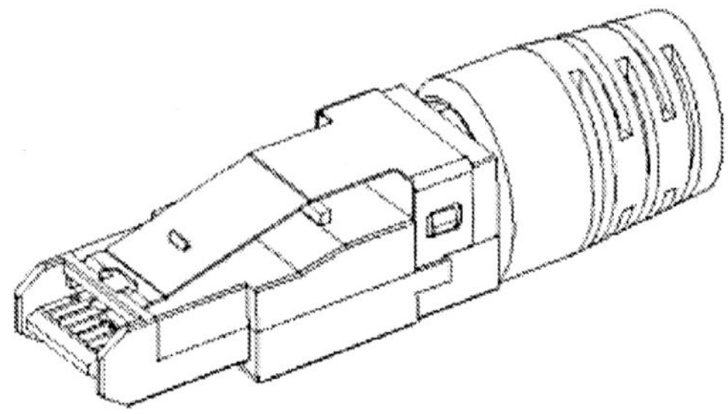

Figure 12.9 InfiniBand connector example (1x) (HSSDC2 technology). [Reprinted with permission of Tyco Electronics.]

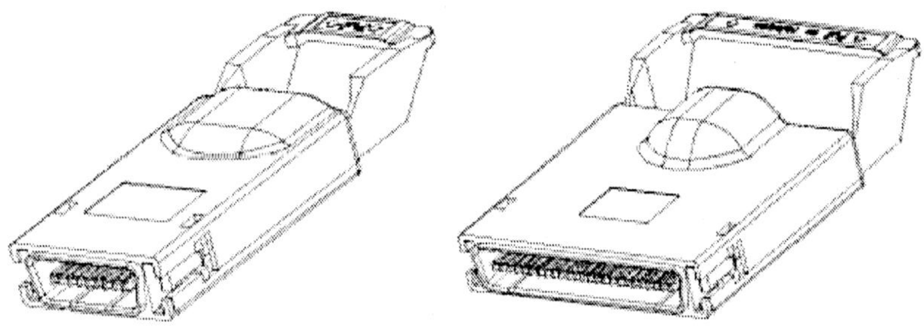

Figure 12.10 InfiniBand connector examples (4x, 12x) (HSSDC2 technology). [Reprinted with permission of Tyco Electronics.]

12.4 Cables and Connectors for LVDS

12.4.1 General Comments on Cables and Connectors

When choosing cables and connectors for LVDS it is important to remember:

1. Use controlled-impedance media. The cables and connectors you use should have a differential impedance of about 100 Ω. They should not introduce major impedance discontinuities that cause signal reflections.
2. Balanced cables (twisted pair) are usually better than unbalanced cables (ribbon cable, multi-conductor) for noise reduction and signal quality. Balanced cables tend to generate less EMI owing to field canceling effects and also tend to pick up electromagnetic radiation as common-mode (not differential-mode) noise, which is rejected by the receiver.
3. For cable distances of less than 0.5 meter, most cables can be made to work effectively. For distances from half a meter to 10 meters, CAT3 (Category 3) twisted-pair cable

works well and is readily available and relatively inexpensive. Other types of cable may also be used as required by a specific application. This includes twin-ax cables built from separate-pair and ribbon-style constructions, which are then coiled.

12.4.2 Cabling Suggestions

As described in Section 12.4.1, try to use balanced cables (twisted pair, twin-ax, or flex circuit with closely coupled differential traces). LVDS was intended to be used on a wide variety of media. The exact media is not specified in the LVDS standard, as it is intended to be specified in the referencing standard that specifies the complete interface. The referencing standard includes the media, data rate, length, connector, function, and pin assignments. In some applications that are very short (less than 0.3 m), ribbon cable or flex circuit may be acceptable. In box-to-box applications, a twisted pair or twin-ax cable would be a better option because of robustness, shielding, and balance. Whatever cable you do choose, following the suggestions below will help you achieve optimal results.

12.4.2.1 Twisted Pair

Twisted-pair cables (see Figure 12.11) provide a good, low cost solution with good balance, are flexible, and capable of medium to long runs depending upon the application skew budget. They are offered with an overall shield or with shields around each pair, as well as an overall shield. Design points to consider:

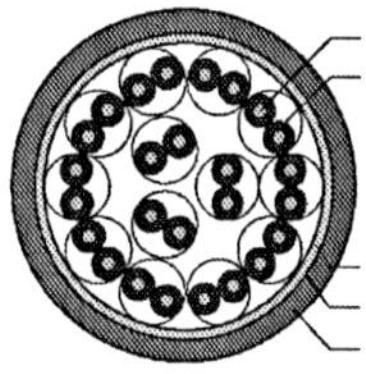

Figure 12.11 Drawing of twisted-pair cable (cross-section). [Reprinted with permission of National Semiconductor Corporation.]

- Twisted pair is a good choice for LVDS. CAT3 cable is good for runs up to about 10 m, while CAT5 has been used for longer runs.
- For the lowest skew, group skew-dependent pairs together (in the same ring to minimize skew between pairs).
- Ground and/or terminate unused conductors (do not float).

12.4.2.2 Twin-Ax Cables

Twin-ax cables (see Figure 12.12) are flexible and have shields around each pair for isolation and have low skew. Since they are not twisted, they tend to have very low skew within a pair and between pairs. These cables are for long runs and have been commonly deployed in Channel Link and FPD-Link applications. Points to consider:

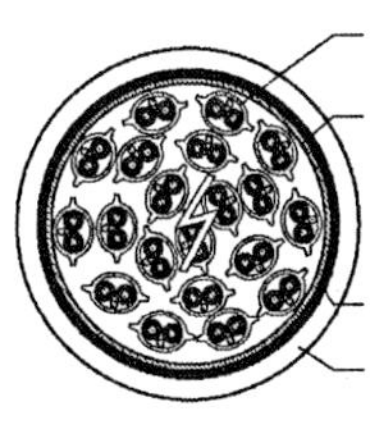

Figure 12.12 Drawing of individually shielded parallel pair twin-ax cable (cross section). [Reprinted with permission of National Semiconductor Corporation.]

- Drain wires per pair may be connected together in the connector header to reduce pin count.
- Ground and/or terminate unused conductors.

12.4.2.3 Flex Circuit

Flex circuit (see Figure 12.13) is a good choice for very short runs, but it is difficult to shield. It can be used as an interconnect between boards within a system.

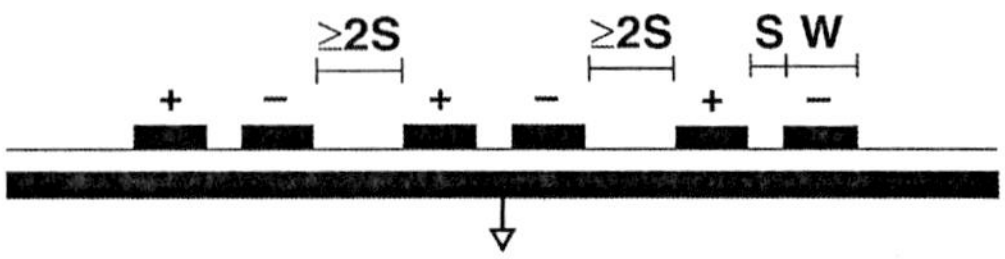

Figure 12.13 Cross-section of a flex circuit. [Reprinted with permission of National Semiconductor Corporation.]

Points to consider:

- Closely couple the members of differential pairs (S < W). Do not run signal pairs near the edges of the cable, as these are not balanced.
- Use a ground plane to establish the impedance.

12.4.2.4 Ribbon Cable

Ribbon cable is cheap and is easy to use and shield. Ribbon cable is not well suited for high-speed differential signaling (good coupling is difficult to achieve), but it is adequate for very short runs.

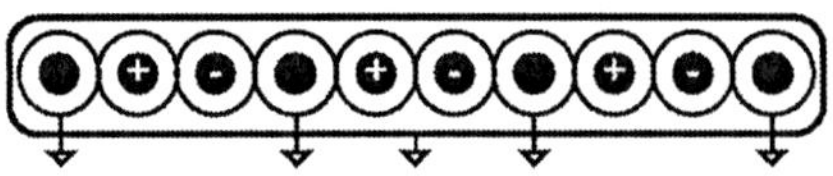

Figure 12.14 Cross-section of flat ribbon cable. [Reprinted with permission of National Semiconductor Corporation.]

Here are some design considerations:

- If ribbon cable must be used, separate the pairs with ground wires as shown in Figure 12.14. Do not run signal pairs at the edges of the ribbon cable.
- Use shielded cable if possible. Shielded flat cable is available.

12.4.3 Connectors

Connectors are also application dependent and depend upon the cable system being used, the number of pins, the need for shielding and other mechanical footprint concerns. Standard connectors have been used at low to medium data rates, and optimized low-skew connectors have been developed for medium to high-speed applications.

Figure 12.15 illustrates two common connector pinout arrangements. In the top connector, each conductor within a pair will be approximately the same length as it runs from the connector face, turns 90°, and runs into the PC board. The top connector will offer low skew among the two conductors within each pair, but will have higher skew between conductors in the top row and conductors in the row immediately beneath it. This is because the length the conductor travels from the top row to reach the PC board is longer than the length the bottom row of conductors takes to reach the PC board. The bottom connector will have higher skew within each pair of conductors as one conductor of each pair is in the top row and one conductor is in the bottom row.

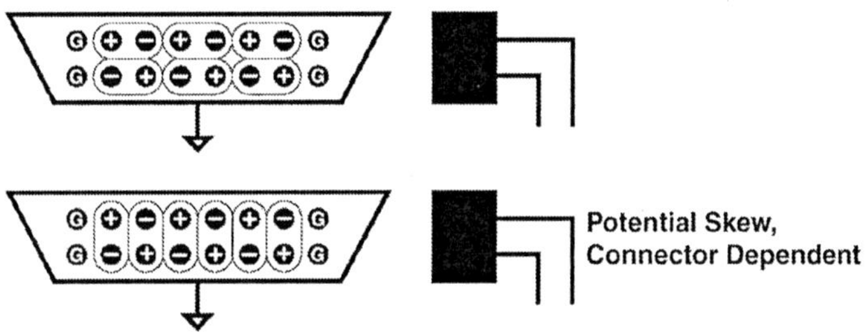

Figure 12.15 Typical connector pinouts. The top connector offers lower skew within each conductor pair than does the bottom connector. [Reprinted with permission of National Semiconductor Corporation.]

Here are some design considerations when connectors are being used:

- Choose low-skew, impedance-matching connectors if possible.
- Group members of each pair together. Pins of a pair should be close together (adjacent), not separated from each other. This is done to maintain balance, and to help ensure that external noise, if picked up, will be common-mode and not differential in nature.
- Some connectors have different length leads for different pins. Group pairs on same-length leads. Consult the connector manufacturer for the orientation of pins that yield the lowest skew and crosstalk for your particular connector. Shorter pin lengths tend to be better than long ones. Select connectors with minimal pin lengths if possible.
- Assign ground pins between signal pin pairs where possible and convenient. Be sure to use ground pins to separate TTL or CMOS signals from LVDS signals.
- Ground the end pins or outside pins on the connector edge. Do not use end pins for high-speed signals, if possible, as they offer less balance.

- Ground and/or terminate unused pins. Many different connector options exist. One such cable-connector system that has been used for LVDS with great results is the 3M "High-speed MDR Digital Data Transmission System."[8] The connector is offered in a surface mount option that has very small skew between all the pins. Different cable types are also supported.

12.4.4 Cable Ground and Shield Connections

In many systems, cable shielding is required for EMC compliance. Although LVDS provides benefits of low EMI when used properly, shielding is still usually a good idea especially for box-to-box applications. Together, cable shielding and ground return wires help reduce EMI. The shielding contains the EMI and the ground return wire (the pair shield or drain wire in some cables) and provides a small loop area return path for common-mode currents. Typically one or more pairs are assigned to ground (circuit common). Using one or more pair reduces the DCR (dc resistance) of the path by the parallel connection of the conductors. This provides a known, very-low-impedance return path for common-mode currents.

In most applications the grounding system will be common to both the receiver and the driver, as shown in Figure 12.16. The cable shield is connected at one end with a dc connection to the common ground (frame ground). Avoid "pig-tail" (high inductance) ground wiring from the cable. The other end of the shield is typically connected with a capacitor or network of a capacitor and a resistor, as shown in the figure. This prevents or reduces dc current flow in the shield. In the case where connectors are involved that penetrate the system's enclosure, the cable shield must have a circumferential contact to the connector's conductive backshell to provide an effective shield and must make good contact. [Nat00]

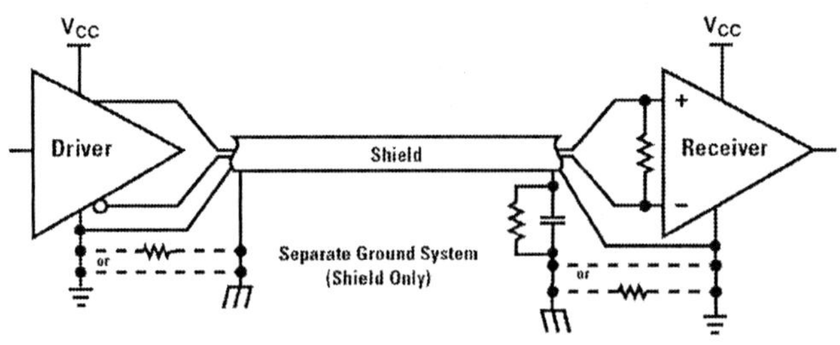

Figure 12.16 Typical grounding scheme. [Reprinted with permission of National Semiconductor Corporation.]

8. This cable system is featured on National Semiconductor's channel-link (48-bit) and LDI evaluation kits.

12.5 Exercises

12-1. In a five-row Teradyne VHDM differential connector, how many differential pairs of signals can be connected? How many with an eight-row connector?

12-2. For up to what distance does the XAUI standard define chip-to-chip connectivity for FR4 PCB traces?

12-3. How many pairs of differential signals are typically found in an InfiniBand connector?

12-4. Should ribbon cable be used in high-speed applications? Why or why not? What are two design guidelines for using ribbon cable with differential signals?

12-5. For a matrix connector, what are general guidelines for assigning differential and single-ended signals, and power and grounds, among the connector pins?

12-6. What are eight performance-limiting characteristics of conventional backplanes?

12-7. How do backplane connectors using the "prong and sleeve" approach cause signal degradation?

12-8. For a theoretically ideal channel of 32 inches in length running between line cards across a backplane, how much power is required to operate a single serial communicaiton link using 10 Gbps SerDes devices using SiliconPipe technology?

12-9. What are seven advantages of the Yosemite technology over the technology of conventional backplanes?

12-10. In proving out the Yosemite backplane technology, an eye diagram was generated at 40 GHz after the signal had traveled 30 in. and through two backplane interconnects. In generating this eye diagram, what was the risetime of the 40 GHz signal used?

PART 3

High-Speed Memory and Memory Interfaces

CHAPTER 13

Memory Device Overview and Memory Signaling Technologies

Chapter Objectives

- Provide an overview of memory devices and identify those that are high speed.
- Provide a quick review of basic principles of memory devices.
- Define memory device banks and memory module ranks.
- Identify those signaling technologies that are used with memory devices.
- Discuss design considerations in the use of memory including power up and initialization, slew rate control, drive capability, and termination.
- Discuss memory nomenclature and speed bins.
- Provide a quick reference look-up table listing and describing most types of memory.

13.1 Overview and Trends

"Memory performance is becoming a key concern of computing system designers across many market segments. From PC applications surrounding multimedia (gaming, streaming media, and video/photo editing) to networking systems routing hundreds of gigabytes per second of packets, high-performance memory systems are a must. Many competing memory technologies strive to meet these performance needs. This chapter will introduce several of these technologies so that they may be discussed in complete detail in subsequent chapters. In the process of analysis, it is critical to look at the issues that surround all memory designs: cost, real-world performance, power, and scalability." [Ram01b] Scalability is the ability to increase (or decrease) the amount of memory available to a system with minimal impact to the system design. Figure 13.1 provides an overview of the different memory technologies available today, where the higher-speed technologies are shown with a black background. The figure indicates that memory technologies have expanded continuously in terms of speed, the number of them, and complexity. The continuing trend toward increasing complexity is easy to demonstrate — for example, the data sheet for Samsung Semiconductor's K4T5104 512 Mb DDR2 SDRAM chip is 80 pages long!

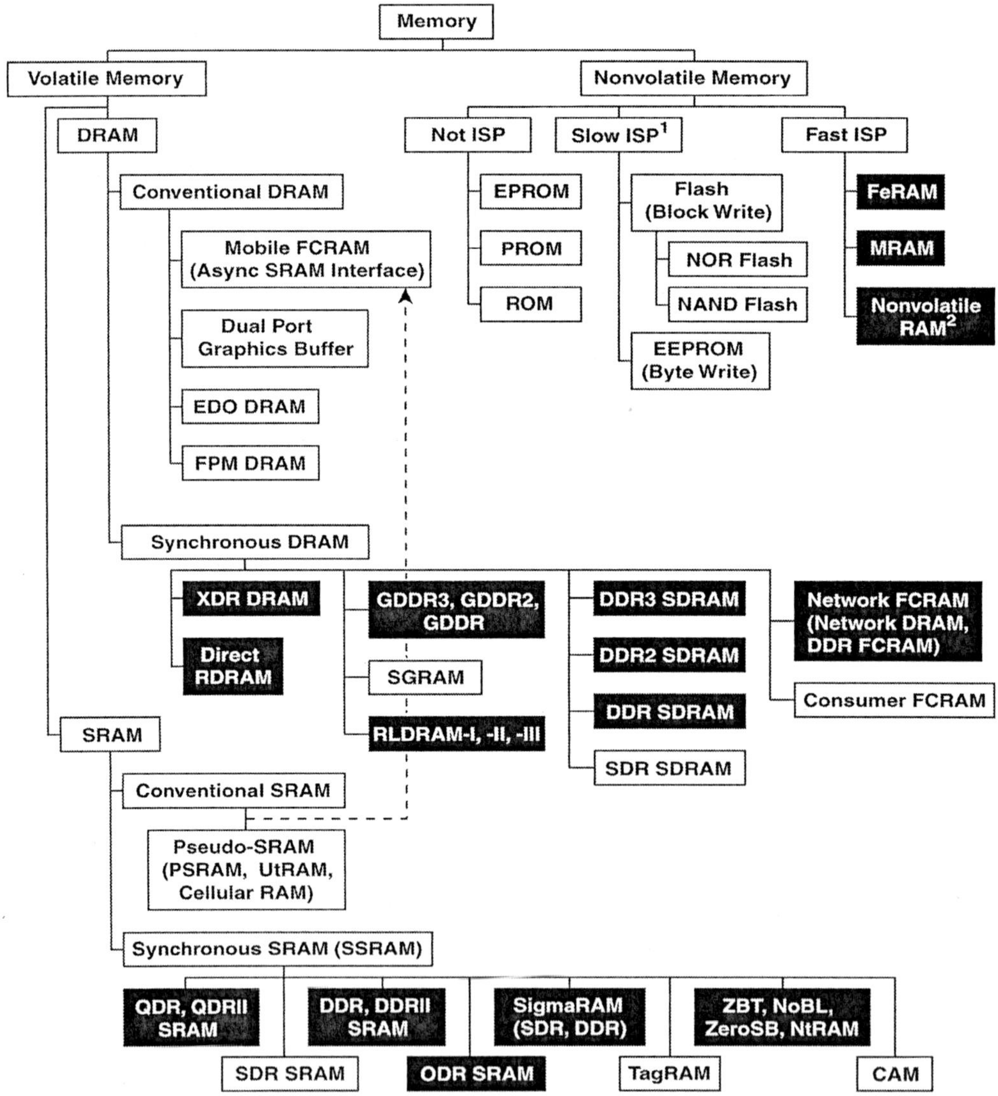

[1] ISP = In-System Programmable

[2] SRAM w/10-year battery enclosed in same package

Figure 13.1 Memory device overview with higher-speed devices shown with black background.

Many high-speed memory devices today are synchronous in nature, transfer data on both edges of the memory clock, and are referred to as DDR (double data rate) devices. The high speeds required in memory system designs have pushed memory manufacturers to focus mostly on synchronous devices to maximize the available window for sampling data. The high speeds have also resulted in the introduction of new device features to enhance device performance and signal integrity. These features include the addition of on-die termination (ODT) and off-chip driver (OCD) impedance adjustments. Often the ODT is programmable, where the termination impedance is determined by loading a register with a binary value corresponding to the desired impedance. ODT has progressed so far that it is also used for power management. Termination impedances are disconnected when reads and writes are not being performed in order to save power. The OCD feature allows the designer to select the I/O output driver impedance to match printed circuit board trace impedances and maximize the signal integrity. Another feature added to enhance the performance of memory devices is that of echo clocks. Echo clocks are generated coincident with the data to help register data downstream.

There are now two generations of DDR technology in use by SDRAM memories (DDR-I and DDR-II/DDR2), and a third generation, DDR3, is likely to be released publicly by the time this book is published. Graphics DDR SDRAM (Synchronous Dynamic RAM) is now in its third generation of DDR technology, GDDR3. "Aggregate speeds achieved with low-cost DDR-I SDRAM technology range from 1.6 Gbps (DDR-200), to 2.7 Gbps (DDR-333), up to 3.2 Gbps (DDR-400)." [Say03] DDR2 SDRAMs, as well as many other types of devices, require complicated, detailed power sequencing followed by initialization sequences after power up. Incidentally, the older technologies of FPM (fast page mode) DRAM and EDO (extended data out) DRAM technologies will continue to have low levels of commercial use for another couple of years.

"Direct Rambus® DRAMs now transfer data at a rate of 1,333 Mbps per pin with a roadmap to 1,600 Mbps per pin and have densities to 512 Mb with a roadmap to 1 Gb. Dual Rambus In-Line Memory Modules (RIMMs™) support bandwidths to 5.3 GB/s and have a roadmap to support 12.8 GB/s for a quad-channel module." [Ram03a]

Rambus's new technology, XDR DRAM (Xtreme Data Rate DRAM) is designed to transfer data at a rate of 3.2 Gbps per pin where a single chip with a 16-bit data bus will provide 6.4 GB/s of bandwidth. XDR uses a 400 MHz memory clock and transfers data 8 times per clock period using its Octal Data Rate (ODR) technology.

Networking applications have put pressure on memory chip designers to introduce new technologies such as Reduced Latency DRAM (RLDRAM™) and Fast Cycle RAM (FCRAM™). RLDRAM has four defining characteristics: high density, high bandwidth, reduced cycle time, and SRAM-like access. Network FCRAM introduced a new DRAM core to enhance performance where the core is refreshed through a hidden precharge. The refresh is "hidden" as it is controlled by the FCRAM internally and does not need external control. The major features of FCRAM products are short t_{RC} (row cycle time), high effective bandwidth, low power consumption, and flexible external interfaces. Row cycle time is the delay before you may access the same region (the bank) of a chip again. FCRAM devices come with one of three types of interfaces: a DDR synchronous interface used in networking, an asynchronous SRAM interface for mobile phone applications, and a synchronous SDR SDRAM interface for use in consumer applications.

Still other devices are continuing to be used in networking applications. ZBT (Zero Bus Turnaround) SRAM, NoBL (No Bus Latency) SRAM, ZeroSB (Zero Speed Bump) SRAM, and NtRAM (No turnaround RAM) are equivalent SRAM devices that have been designed to eliminate dead clock cycles between read-to-write and write-to-read cycle transitions.

Nonvolatile memory has recently been enhanced with two new technologies — ferroelectric RAM (FeRAM) and magnetoresistive RAM (MRAM). FeRAM could well replace flash in the near future. The big difference with FeRAM is its speed. A 32 Mb FeRAM is now available with an access time of 50 nanoseconds and a cycle time of 75 nanoseconds. This is slower than SDRAMs but much faster than flash, where a program time of 200 microseconds is considered good.

By comparison, MRAM may still be a few years away from commercial use. "The basic attractions of MRAM are its speed and nonvolatility. MRAM can reach theoretical write times down to 2.3 nanoseconds. That's more than 1,000 times faster than the fastest nonvolatile flash and is 20 times faster than FeRAM. MRAM access times are as fast as 3 ns, or 20 times faster than DRAM. And since such reads require only 2 milliamps, MRAM consumes less than 1/100 the energy of DRAM. MRAM is also resistant to radiation, which is one of the reasons it is considered a prime replacement for SRAM."[1] [Cle03] FeRAM devices have entered the sampling stages of development, and while some industry experts feel that commercial use of MRAM is still years away, Motorola is expected to sample MRAM devices soon. Cypress has a preliminary data sheet out on a 32 K × 8 MRAM with 70 ns read and write access times.

The SigmaRAM family consists of devices featuring both common and separate I/O busses. Common buses allow both read and write data to be transferred over the same bus whereas separate I/O bus devices have one bus for read data and a second bus for write data. Each bus architecture optimizes a specific sequence of read and write commands. In addition, the family includes single and double data rate operation and supports late and double late write protocols in various combinations. SigmaRAM also provides wide bus widths, from x18 to x72, and high clock rates (up to 333 MHz).

"Quad data rate (QDR) devices have separate input and output ports that operate independently and simultaneously. Each port is equipped with a DDR interface in order to maximize data throughput. Two ports operating in a DDR manner provide four pieces of data being transferred during each clock cycle, thereby leading to the name, quad data rate. The separate ports completely eliminate the possibility of data contention between the SRAM and memory controller, a concern with conventional, common I/O devices. Separate port-selects allow for easy depth

1. Copyright © 2003 by CMP Media LLC, 600 Community Drive, Manhasset, NY 11030. Reprinted from EETIMES with permission.

Denali MemCon

"Denali MemCon is the industry's only conference series for exchanging information about the technology and economics of semiconductor memory. System designers gain insights into new product introductions and technology roadmaps for the latest semiconductor memory devices — directly from the leading memory vendors. An applications track provides other vendors with a forum for presenting case studies of design tools, semiconductor IP, and advanced design techniques for using leading-edge memory devices. Denali Software, Inc. offers electronic design automation (EDA) software and intellectual property (IP) design cores for complex chip interfaces such as DDR memory and PCI Express architectures." [Den03a]

and width expansion. Variable impedance output buffers provide flexibility in drive strength in order to match the impedance seen by the device."[2] [Cyp02c]

As memory speeds increased, the "price to pay was the need for more sophisticated controllers to generate the appropriate read and write commands and the use of length-limited buses with series and powered parallel end-point terminations to decrease reflections and improve signal integrity. As memory speeds and signal efficiency continue to improve, engineers need to be skillful in the design and implementation of memory technologies to ensure efficient, cost-effective products." [Say03]

In this and the following five chapters that cover memory devices and technologies, we discuss these technologies in much greater detail. In this chapter, we start with a short refresher on memory devices and their features, and then move on to discuss memory signaling technologies and design considerations in the use of memory. At the end of this chapter, we provide a detailed six-page table listing each type of memory and its description.

13.2 A Quick Review of Memory Basics[3]

13.2.1 Read/Write Memory[4]

"The name *read/write memory (RWM)* is given to memory arrays in which we can store and retrieve information at any time. Most of the RWMs used in digital systems are *random-access memories (RAMs)*, which means that the time it takes to read or write a bit of memory is independent of the bit's location in RAM. From this point of view, ROMs (read-only memories) are also random-access memories, but the name 'RAM' is generally used only for read/write random-access memories.

In a *static RAM (SRAM)* ('S-ram'), once a word is written at a location, it remains stored as long as power is applied to the chip, unless the same location is written again. In a *dynamic*

2. Reprinted with permission of Cypress Semiconductor Corporation.
3. Engineers or knowledgeable students may wish to skip this section.
4. Sections 13.2.1 through 13.2.3.3, Section 13.2.5, and Figures 13.2 through 13.11 are reprinted by permission of Pearson Education, Inc., Upper Saddle River, NJ. WAKERLY, JOHN F., *DIGITAL DESIGN: PRINCIPLES AND PRACTICES*, 3rd Edition, © 2003.

RAM (DRAM) ('D-ram'), the data stored at each location must be refreshed periodically by reading it and then writing it back again, or else it disappears.

Most RAMs lose their memory when power is removed; they are a form of *volatile memory.* Some RAMs retain their memory even when power is removed; they are called *nonvolatile memory.* An example of a nonvolatile RAM is a CMOS static memory in an extra-large package that includes a lithium battery with a 10-year lifetime." [Wak01] Other types of nonvolatile memory include flash, FeRAM, MRAM, and EEPROM.

13.2.2 Static RAM

13.2.2.1 Static RAM Inputs and Outputs

"The inputs and outputs of a simple $2^n \times$ b-bit static RAM are shown in Figure 13.2. The memory locations in a static RAM behave like D latches, rather than edge-triggered D flip-flops. This means that whenever the WE (write enable) input is asserted, the latches for the selected memory location are 'open' or 'transparent,' and input data flows into and through the latch. The actual value stored is whatever is present when the latch closes. Static RAM normally has just two defined access operations, a 'read' where an address is placed on the address inputs while CS (chip select) and OE (output enable) are asserted, and a 'write' where an address is placed on the address inputs and a data word is placed on DIN (data input) followed by the assertion of CS and WE (write enable).

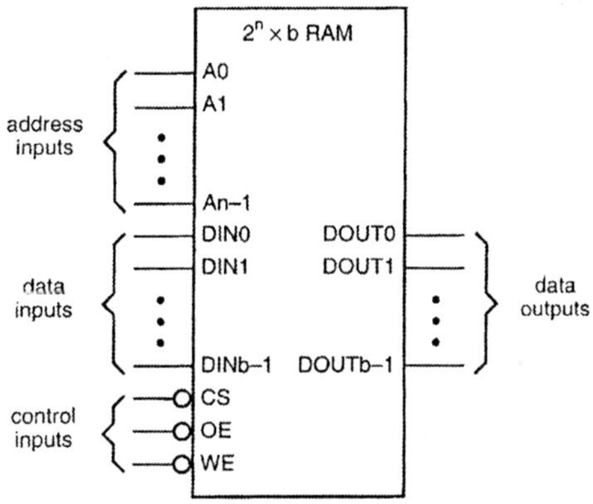

Figure 13.2 Basic structure of a $2^n \times$ b RAM.

A certain amount of care is needed when accessing SRAM, because it is possible to inadvertently 'clobber' one or more locations while writing to a selected one, if the SRAM's timing requirements are not met. The following subsection explains the actual timing behavior and requirements.

Several important aspects of SRAM behavior are:

- During read operations, the output data is a combinational function of the address inputs, as in a ROM. No harm is done by changing the address lines while the output data bus is enabled. The access time for read operations is specified from the time that the last address input becomes stable.
- During write operations, the input data is stored in latches. This means that the data must meet certain setup and hold tines with respect to the trailing edge of the latch enable signal. That is, the input data at a latch's D input need not be stable at the moment WE is asserted; it must only be stable a certain time before WE is negated.

- During write operations, the address inputs must be stable for a certain setup time before WE is asserted internally and for a hold time after WE is negated. Otherwise, data may be 'sprayed' all over the array because of glitches that may appear on the internal memory cell select lines when the address inputs are changed.
- A write cycle begins when both CS and WE are asserted and ends when either is negated. Setup and hold times for address and data are specified with respect to these events." [Wak01]

13.2.2.2 Static RAM Timing

"Figure 13.3 shows the timing parameters that are typically specified for read operations in a static RAM; they are described below:

t_{AA} — Access time from address. Assuming that OE and CS are already asserted, or will be soon enough not to make a difference, this is how long it takes to get stable output data after a change in address. When designers talk about a '70 ns SRAM,' they're usually referring to this number.

t_{ACS} — Access time from chip select. Assuming that the address and OE are already stable, or will be soon enough not to make a difference, this is how long it takes to get stable output data after CS is asserted. Often this parameter is identical to t_{AA}, but sometimes it's longer in SRAMs with a 'powerdown' mode and shorter in SRAMs without one.

t_{OE} — Output-enable time. This is how long it takes for the three-state output buffers to leave the high-impedance state when OE and CS are both asserted. This parameter is normally less than t_{ACS}, so it is possible for the RAM to start accessing data internally before OE is asserted; this feature is used to achieve fast access times while avoiding 'bus fighting' in many applications.

t_{OZ} — Output-disable time. This is how long it takes for the three-state output buffers to enter the high-impedance state after OE _L or CS_L is negated.

t_{OH} — Output-hold time. This parameter specifies how long the output data remains valid after a change in the address inputs.

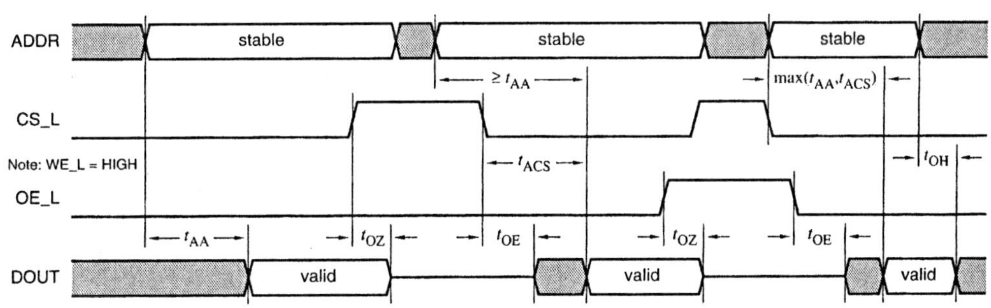

Figure 13.3 Timing parameters for read operations in a static RAM.

When they're not being written, SRAMs can be used just like ROMs. The same is not true for DRAMs, as we see later.

Timing parameters for write operations are shown in Figure 13.4 and are described below:

t_{AS} — Address setup time before write. All of the address inputs must be stable at this time before both CS and WE are asserted. Otherwise, the data stored at unpredictable locations may be corrupted.

t_{AH} — Address hold time after write. Analogous to t_{AS}, all address inputs must be held stable until this time after CS or WE is negated.

t_{CSW} — Chip-select setup before end of write. CS must be asserted at least this long before the end of the write cycle in order to select a cell.

t_{WP} — Write pulse width. WE must be asserted at least this long to reliably latch data into the selected cell.

t_{DS} — Data setup time before end of write. All of the data inputs must be stable at this time before the write cycle ends. Otherwise, the data may not be latched.

t_{DH} — Data hold time after end of write. Analogous to t_{DS}, all data inputs must be held stable until this time after the write cycle ends.

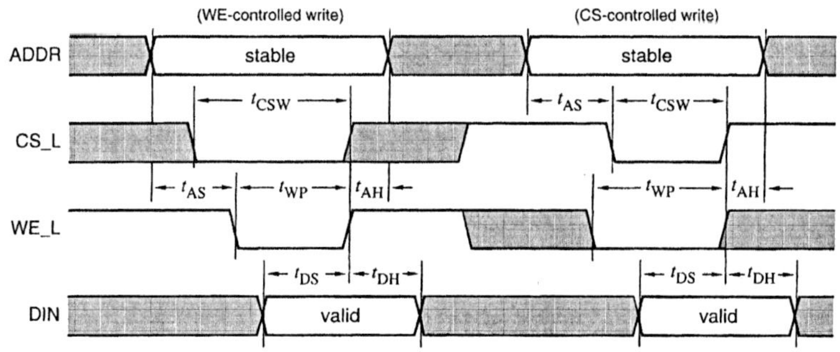

Figure 13.4 Timing parameters for write operations in a static RAM.

Manufacturers of SRAMs specify two write cycle types, WE-controlled and CS-controlled, as shown in Figure 13.4. The only difference between these cycles is whether WE or CS is the last to be asserted and the first to be negated when enabling the SRAM's internal write operation.

The write timing requirements of SRAMs could be relaxed somewhat if, instead of using latches, the cells contained edge-triggered D flip-flops with a common clock input and enable inputs controlled by SEL and WR. However, this just isn't done, because it would at least double the chip area of each cell, since a D flip-flop is built from two latches. Thus, the logic designer is left to reconcile the SRAM's latch-type timing with the edge-triggered register and state machine timing used elsewhere in a system." [Wak01]

13.2.2.3 Synchronous SRAM

"A variety of SRAM, called synchronous SRAM (SSRAM) ('S-S-ram'), still uses latches internally but has a clocked interface for control, address, and data. As shown in Figure 13.5, internal edge-triggered registers AREG and CREG are placed on the signal paths for address and control. As a result, an operation that is set up before the rising edge of the clock is performed internally during a subsequent clock period. Register INREG captures the input data for write operations and, depending on whether the device has 'pipelined' or 'flow-through' outputs, register OUTREG is or is not provided to hold the output data from a read operation, respectively.

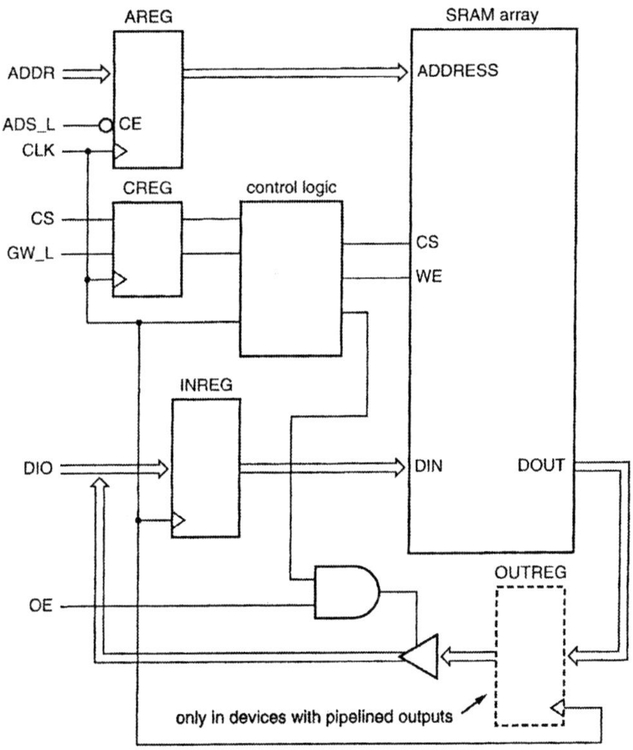

Figure 13.5 Internal structure of a synchronous SRAM.

The first variety of SSRAM to be introduced was the late-write SSRAM with flow-through outputs. For a read operation, the control and address inputs are sampled at the rising edge of the clock, and the internal address register AREG is loaded only if ADS_L is asserted. During the next clock period, the internal SRAM array is accessed and read data is delivered to the device's DIO data bus pins. The device also supports a burst mode, in which data is read from a sequence of addresses. In this mode AREG behaves as a counter, eliminating the need to apply a new address at each cycle. (The control signals that support burst mode are not shown in Figure 13.5.)

For a write operation, the write data is stored temporarily in an on-chip register INREG, which is sampled one clock tick after the address register is loaded. Therefore, ADS_L must be inhibited for at least one tick after loading the address, so that the address in AREG is still valid when the write takes place. The write takes place during the clock period following the edge on which the 'global write' control signal GW_L is asserted. As with reading, the device has a burst mode where a sequence of addresses can be written without supplying a new address.

Note that the 'late-write' protocol makes it impossible to write to two different, nonsequential addresses in successive clock periods; the SRAM array is idle for one clock period between writes (except in burst mode). From the point of view of internal chip capabilities, this behavior is not necessary. However, the late-write protocol was designed this way to match the bus protocols of microprocessors that use these SSRAMs in their cache subsystems.

A late-write SSRAM with pipelined outputs is like the previous version, except that a register OUTREG is placed between the SRAM array output and the device output for read operations. This delays the read output data at the device pins until the beginning of the next clock period, but it also provides the benefit that the data is now valid for almost the entire clock period. The write cycle behaves the same as with flow-through outputs. Compared with flow-through outputs, pipelined outputs provide much better setup time for the device receiving the read data, and therefore may allow operation at higher clock frequencies.

As shown in Figure 13.5, conventional SSRAMs share the same pins for both input data and output data. During a given clock period the data I/O pins can be used for reading or writing but not both. If you study the pattern of data bus and SRAM array use in both styles of late-write SSRAM, you'll find cases where it's not possible to initiate a read one clock cycle after initiating a write or vice versa, due to resource conflict. Thus, late write SSRAMs suffer a turnaround penalty, a clock period in which the internal SRAM array must be idle when a read is followed by a write or vice versa.

The turnaround penalty is eliminated in so called zero-bus-turn-around (ZBT) SSRAMs. The type of operation (read or write) is selected by a control signal R/W that sampled at the same clock edge as the address. Regardless of whether the operation is a read or a write, the DIO bus is used during the next clock period to transfer the read or write data. As a result, there is no data-bus-usage conflict, as long as OE is controlled properly to avoid bus fighting between successive cycles. However, if a write is followed by a read, both operations would like to use the SRAM array during the same clock period. To avoid this resource conflict, the write operation is deferred until the next available SRAM cycle. This opportunity occurs when either another write operation or no operation is initiated on the address and control lines.

Although a ZBT SSRAM can access the internal SRAM array on every clock cycle, this performance improvement is not without a price. While a write operation is pending, the write address and related information must be stored in another register, WAREG, since AREG is reused by other operations; this costs chip area. More significantly for some applications, a write operation may be deferred indefinitely if it is immediately followed by a continuous series of read operations. This anomaly may require tricky controller design to detect the case where one of these read operations attempts to access the address that was just written, since the value stored in the SRAM array is 'stale'!

A ZBT SSRAM with pipelined outputs adds OUTREG to the read data path but is otherwise similar to the previous device. In this device, both reads and writes use the DIO bus during the second clock period following the clock edge, in which the operation was initiated. As in the previous device, writes to the internal SRAM array are deferred until an available cycle, so that reads can take precedence. As implied by the timing, two levels of internal registers are needed for write address and data, since up to two writes may be deferred while a sequence of reads occurs.

Synchronous SRAM access protocols are very beneficial in high-speed systems. For example, address, control, and write inputs can be applied with more or less conventional setup and hold times with respect to the system clock, and read data on pipelined output pins is available for almost a complete clock cycle. Most importantly, the designer does not have to worry about the tricky circuits and timing paths that are otherwise needed to enable conventional SRAM latch-style operation.

Note that among the four styles of SSRAM that we described, no single one is the 'best.' The best SSRAM is the one that best fits the bus protocol and other requirements of the system in which it is used." [Wak01]

13.2.3 Dynamic RAM (DRAM)

"The basic memory cell in an SRAM, a D latch, requires four gates in a discrete design, and four to six transistors in a custom-designed SRAM LSI chip. In order to build RAMS with higher density (more bits per chip), chip designers invented memory cells that use as little as one transistor per bit." [Wak01]

13.2.3.1 Dynamic RAM Structure

"It is not possible to build a bistable element with just one transistor. Instead, the memory cells in a dynamic RAM (DRAM) store information on a tiny capacitor accessed through a MOS transistor. Figure 13.6 shows the storage cell for one bit of a DRAM, which is accessed by setting the word line to a HIGH voltage. To store a 1, a HIGH voltage is placed on the bit line, which charges the capacitor through the 'on' transistor. To store a 0, LOW voltage on the bit line discharges the capacitor.

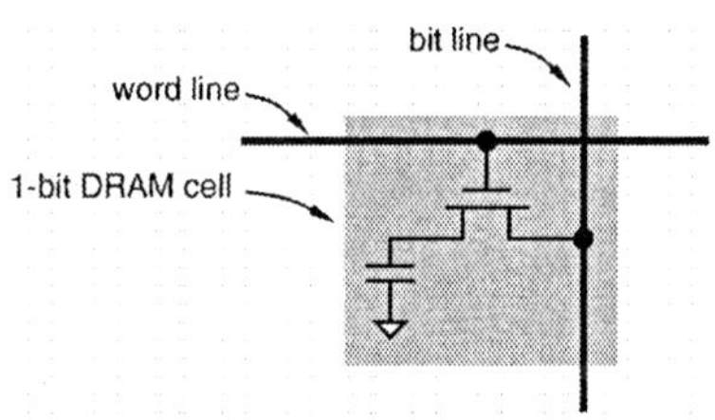

Figure 13.6 Storage cell for one bit in a DRAM.

To read a DRAM cell, the bit line is first precharged to a voltage halfway between HIGH and LOW, and then the word line is set HIGH. Depending on whether the capacitor voltage is HIGH or LOW, the precharged bit line is pulled slightly higher or slightly lower. A sense amplifier detects this small change and recovers a 1 or 0 accordingly. Note that reading a cell destroys the original voltage stored on the capacitor, so that the recovered data must be written back into the cell after reading.

The capacitor in a DRAM cell has a very small capacitance, but the MOS transistor that accesses it has a very high impedance. Therefore, it takes a relatively long time (many milliseconds) for a HIGH voltage to discharge to the point that it looks more like a LOW voltage. In the meantime, the capacitor stores one bit of information.

DRAM-based memory systems use refresh cycles to update every memory cell periodically, once every four milliseconds in early DRAMS. This involves sequentially reading the somewhat degraded contents of each cell into a D latch and writing back a nice solid LOW or HIGH value from the latch. Figure 13.7 illustrates the electrical state of a cell after a write and a sequence of refresh operations.

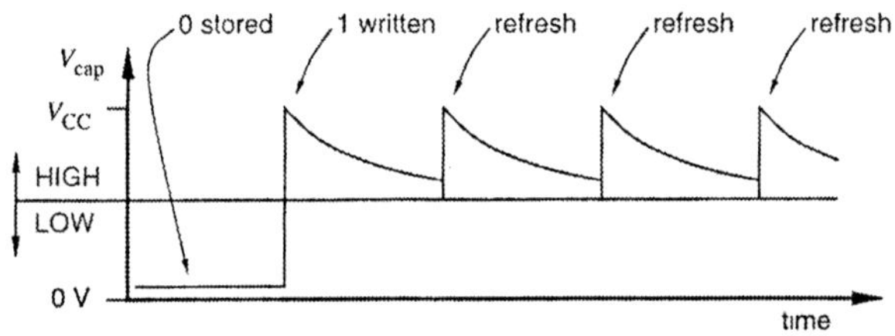

Figure 13.7 Voltage stored in a DRAM cell after writing and refresh operations.

DRAMs are organized using two dimensional arrays, and a single operation refreshes an entire row of the array. Early DRAM arrays had 256 rows, requiring 256 refresh operations every four milliseconds, or one about every 15.6 μsec. Newer arrays have 4,096 rows but need to be refreshed only once every 64 ms, which still works out to one row per 15.6 μsec. A refresh operation typically takes less than 100 ns, so the DRAM is available for useful read and write operations over 99% of the time.

Figure 13.8 is a block diagram of the internal structure of a 64K × 1 DRAM. The logical array has 64K × 1 bits, but the physical array is square, containing 256 × 256 bits. Although the memory has 64K locations, the chip has only eight multiplexed address inputs. A complete 16-bit address is presented to the chip in two steps controlled by two signals RAS_L (row address strobe) and CAS_L (column address strobe). Multiplexing the address inputs saves pins, important for compact design of memory systems, and also fits quite naturally with the two-step DRAM access methods that we describe shortly.

DRAM operation is much more complicated than SRAM operation. Taking advantage of the multiple arrays in larger, high-speed DRAMS, a modern DRAM controller can perform several operations in parallel for example, completing a write operation in one array while initiating a read operation in another. This increases the effective throughput of the memory." [Wak01]

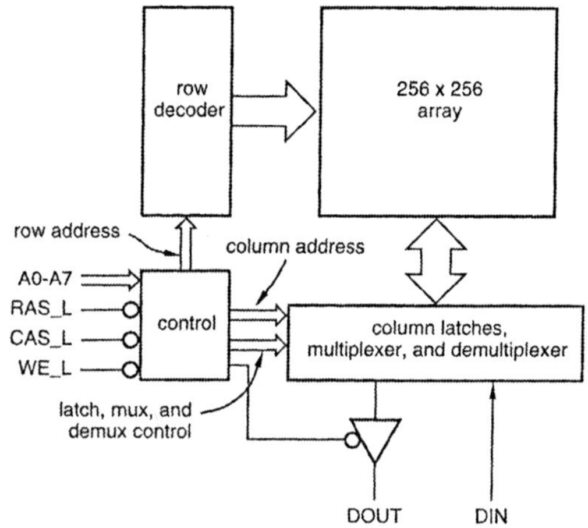

Figure 13.8 Internal structure of a 64 × 1 DRAM.

13.2.3.2 Dynamic RAM Timing

"There are many different timing scenarios for different DRAM types and operations. In this section we describe the most common cycles for conventional DRAM and relate them to the internal structure of the device. The most striking aspect of conventional DRAM timing is that there is no clock. Instead, DRAM operations are initiated and completed on both the rising and falling edges of RAS_L and CAS_L. This is very tricky timing indeed, but the industry has somehow lived with it for over 25 years!

The timing for a RAS-only refresh cycle is shown in Figure 13.9. This RAS-only refresh cycle is used to refresh a row of memory without actually reading or writing any data at the external pins of the DRAM chip. The cycle begins when a row address is applied to the multiplexed address inputs and RAS_L is asserted. The DRAM stores the row address in an internal row-address register on the falling edge of RAS_L and reads the selectcd row of the memory array into an on-chip row latch. When RAS_L is negated, the contents of the row are written back from the row latch.

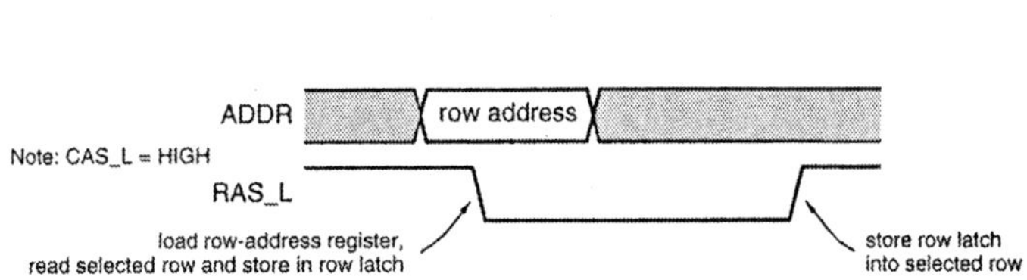

Figure 13.9 DRAM RAS-only refresh-cycle timing.

A read cycle, shown in Figure 13.10, begins like a refresh cycle, where a selected row is read into the row latch. Next, a column address is applied to the multiplexed address inputs and is stored in an on-chip column address register on the falling edge of CAS_L. The column address is used to select one bit of the just-read row, which is made available on the DRAM's DOUT pin. A three-state pin, DOUT is output-enabled as long as CAS_L is asserted. In the meantime, the entire row is written back into the array as soon as RAS_L is negated.

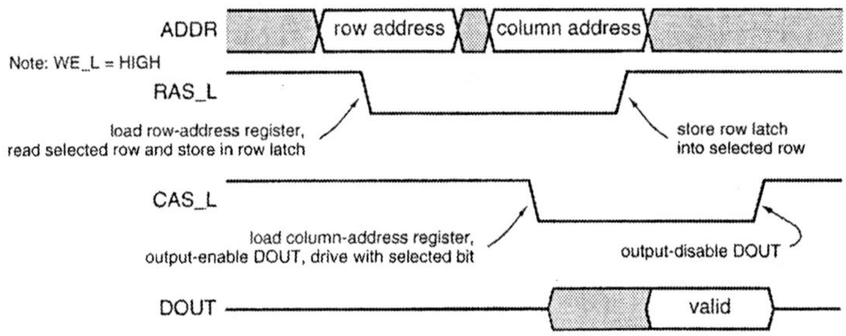

Figure 13.10 DRAM read-cycle timing.

A write cycle, shown in Figure 13.11, also begins like a refresh or read cycle. However, WE_L (write enable) must be asserted before CAS_L is asserted to select a write cycle. One thing this accomplishes is to disable DOUT for the rest of the cycle, even though CAS_L will be asserted subsequently. Once the selected row is read into the row latch, WE _L also forces the input bit on DIN to be merged into the row latch, in the bit position selected by the column address. When the row is subsequently written back into the array on the rising edge of RAS_L, it contains a new value in the selected column.

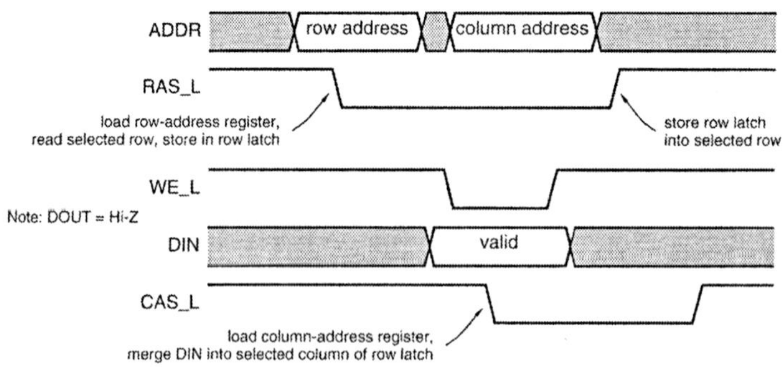

Figure 13.11 DRAM write-cycle timing.

Several other types of cycles, not shown in the figure, are available in typical DRAMS:

- A CAS-before-RAS refresh cycle performs a refresh operation without requiring a row address to be supplied by an external counter. Instead, the DRAM chip uses an internal row address counter. If CAS_L is asserted before RAS_L is asserted, the DRAM refreshes the row selected by the counter and then increments the counter. This simplifies DRAM system design, eliminating the external refresh counter and reducing from three (row, column, refresh) to two the number of multiplexed sources that must drive the DRAM address inputs.
- A read-modify-write cycle begins like a read cycle and allows data to be read on DOUT when CAS_L is asserted. However, WE_L may be asserted subsequently to write new data back at the same location.
- A page-mode read cycle allows up to an entire row ('page') of data to be read without repeating the entire RAS-CAS cycle. Since the entire row is already stored in the row latch, this cycle simply requires CAS_L to be pulsed LOW multiple times while RAS_L remains asserted continuously. A new column address is provided and a new bit becomes available on DOUT on each falling edge of CAS_L. This cycle provides much faster memory access during a sequence of reads to consecutive or otherwise nearby addresses; such sequences often occur in microprocessors when they're fetching instructions or filling caches.
- Similarly, a page-mode write cycle allows multiple bits of a row to be written with a single RAS_L and multiple CAS_L cycles.

In extended data out (EDO) DRAM devices, CAS_L no longer controls the output enable during read cycles; instead, a separate OE_L control input is provided. This is important for high-speed page-mode read cycles, as it provides valid output data for a longer period of time. Instead of turning off between successive CAS_L pulses, the output is left on continuously (as controlled by OE_L), and page mode read data is valid from the beginning of one CAS_L pulse to the beginning of the next.

You may have noticed that the valid intervals of row and column addresses in Figure 13.10 are skewed to the right with respect to the falling edges of RAS_L and CAS_L, which load the addresses into on-chip edge-triggered registers. DRAMs typically have a short setup-time requirement for the address inputs (often 0 ns) but a relatively large, positive hold-time requirement (7–20 ns). This can cause some problems in the design of an otherwise fully synchronous system that uses a single clock and flip-flops with zero hold times, and is often solved by distasteful methods involving tapped delay lines or using the opposite edge of the clock. This is one of the reasons that many designers consider DRAM system design a tricky, black art." [Wak01]

13.2.3.3 Synchronous DRAMs

"The RAS/CAS edge-based access protocol of conventional DRAMs is not only tricky, it's difficult to make it go fast and still meet timing margins when interfacing with the rest of a system. Synchronous DRAMs (SDRAMs) were introduced that use a more conventional clocked interface.

Synchronous DRAMs retain the multiplexed addressing style of conventional DRAMs — row and column addresses are delivered in two steps. However, an SDRAM's control signals as well as its address inputs are sampled only on the rising edge of a common clock signal (CLK). In addition, an SDRAM has a clock enable signal (CKE) so that inputs are ignored if CKE is not asserted. For write operations, data is sampled on the clock edge, and for reads, output data is delivered on the clock edge. Just like a conventional DRAM, an SDRAM requires multiple internal steps to perform an operation, and this means multiple clock ticks externally. Internally, SDRAMs have multiple DRAM banks, typically four, and they can have multiple operations pending simultaneously.

At each clock tick, an SDRAM's control signals RAS_L, CAS_L, and WE_L are interpreted as a 'command word,' as opposed to having individual significance. At the same time, the high-order address bits are interpreted as a 'bank select' to indicate which bank a command applies to. For example, an intelligent SDRAM controller can use four ticks to initiate read operations in four different banks, and then come back and read the results one per clock tick as they complete.

Internal timing in an SDRAM is derived from the clock input CLK. Typically, a RAS signal to the internal array will be asserted immediately after the clock tick when a read or write command is given. To meet internal timing requirements, the chip generates an internal CAS signal at a later clock tick. How many ticks later is a function of the CLK frequency and the speed grade of the SDRAM chip itself. To accommodate differing requirements, the RAS to CAS delay, t_{RCD}, is programmable. This and several other important operating parameters must be downloaded into an SDRAM at initialization. Downloading is fairly easy; the 'load parameters' command is recognized when RAS_L, CAS_L, and WE_L control signals are all asserted simultaneously, and the parameters themselves are transferred on the address lines." [Wak01]

"SDRAM has specifications different from those of conventional DRAM allowing it to realize higher-speed operation. SDRAM has package pin configuration, control signal names, and a number of signals different from those of conventional DRAM as well."[5] [Elp01a]

13.2.3.4 FPM (Fast Page Mode) DRAM

"In the early years of the PC, FPM (fast page mode) DRAM was usually the most common form of DRAM available for personal computers. The page mode process allowed for the accessing of information by providing a constant RAS (row address strobe) signal while at the same time, allowing the CAS (column address strobe) signal to maneuver to different locations. This process acquires data from columns sharing the same row, but without the need to constantly specify the row location in each access. By acquiring rows of information, or pages, in this manner, the operation resulted in a lower memory access time and a reduction in power consumption, an overall improvement over its predecessors." [Elp01a]

5. Reprinted with permission of Elpida Memory, Inc.

13.2.3.5 EDO DRAM (Extended Data Out DRAM)

"EDO DRAM is a still faster version of FPM DRAM. If the data output (read cycle) of FPM DRAM is made higher speed, the data output time becomes shorter. Since EDO DRAM is provided with extended output functions, the data output time does not become short even in higher speed. Thus, wider range for the timing can be allowed in the data output side, and as a result a speed higher than that of FPM DRAM can be realized.

In addition, since EDO DRAM and FPM DRAM are compatible DRAM that have packages with the same pin configuration, EDO DRAM can easily replace FPM DRAM. EDO DRAM is provided with extended output functions, but it has a problem of operating frequency similarly to FPM DRAM if higher speed is demanded. Generally, EDO DRAM can be synchronized only up to the clock of approximately 75 MHz." [Elp01a]

FPM and EDO DRAM are expected to be used at low volume levels in commercial applications for about another two years.

13.2.3.6 RLDRAM (Reduced Latency DRAM)

"Reduced latency DRAM (RLDRAM) memory is a new type of high-performance memory. It combines the performance-critical features that networking and cache applications need — such as high density (256 Mb), high bandwidth (2.4 GB/s), and fast SRAM-like random access. The key to the RLDRAM device's performance is its innovative circuit design. This design minimizes the time between the beginning of the access cycle and the availability of the first data. Plus, RLDRAM memory offers a latency half of the 45 ns for standard DDR SDRAMs."[6] [Mic03f]

13.2.4 Special Application Memory

13.2.4.1 Video RAM (VRAM)

Video RAM (VRAM) "is the type of memory used in a display adapter. Video RAM is designed with dual ports so that it can simultaneously refresh the screen while text and images are drawn in memory." [Com03]

13.2.4.2 Dual-Port Graphics Buffer

"A dual-port graphics buffer is a memory dedicated for images and configured with two ports: a serial port and a RAM port. The serial port performs a read-out of data that is sent to the display unit thereby displaying an image, and the RAM port mainly performs write-in of data to an image memory (i.e., it stores a drawing in memory). Stated differently, the memory receives data through the RAM port and sends the data out to be displayed through the serial port. Although the RAM port is basically configured with DRAM, the operation mode for drawing, which is not provided to DRAM, is added in order to simplify designing of graphics systems." [Elp01a]

6. Used with permission by Micron Technology, Inc. © 2004 Micron Technology, Inc., All rights reserved.

13.2.4.3 SGRAM (Synchronous Graphics RAM)

"Synchronous Graphics RAM (SGRAM) is clock-synchronized random access memory that is used for video memory. It is relatively low-cost video memory. It uses a masked write, which enables selected data to be modified in a single operation rather as a sequence of read, update, and write operations. It also uses block write, which allows data for background or foreground image fills to be handled more efficiently. SGRAM is single-ported. Its special features are what make it a moderately fast form of video memory. The Matrox Mystique is an example of a video card that uses SGRAM." [Wha03]

"SGRAM is a type of SDRAM technology, but includes enhanced graphics features for use with display adapters. Its block write and mask write functions allow the frame buffer to be cleared faster and selected pixels to be modified faster." [Com03]

SGRAM comes in both SDR and DDR versions. DDR SGRAM is also known as GDDR SDRAM (graphics double data rate SDRAM). GDDR SDRAM is discussed in more detail in Chapter 15, Section 15.1, "Graphics Double Data Rate SDRAM (GDDR, GDDR2, GDDR3)," beginning on page 391.

13.2.5 Read-Only Memory (ROM)

"Most of the early integrated-circuit ROMs were mask-programmable ROMs (or, simply, mask ROMs). A mask ROM is programmed by the pattern of connections and no-connections in one of the masks used in the IC manufacturing process. To program or write information into the ROM, the customer gives the manufacturer a listing of the desired ROM contents, using a floppy disk or other transfer medium. The manufacturer uses this information to create one or more customized masks to manufacture ROMs with the required pattern. ROM manufacturers impose a mask charge of several thousand dollars for the 'customized' aspects of mask-ROM production. Because of mask charges and the four-week delay typically required to obtain programmed chips, mask ROMs are normally used today only in very high-volume applications.

A programmable read-only memory (PROM) is similar to a mask ROM, except that data values may be programmed in just a few minutes using a PROM programmer. A PROM chip is manufactured with all of its diodes or transistors 'connected.' This corresponds to having all bits at a particular value, typically 1. The PROM programmer can be used to set desired bits to the opposite value. In bipolar PROMs, this is done by vaporizing tiny fusible links inside the PROM corresponding to each bit. A link is vaporized by selecting it using the PROM's address and data lines, and then applying a high-voltage pulse (10–30 V) to the device through a special input pin.

An erasable programmable read-only memory (EPROM) is programmable like a PROM, but it can also be 'erased' to the all-1s state by exposing it to ultraviolet light. An EPROM has a floating-gate MOS transistor at every bit location. Each transistor has two gates. The 'floating' gate is not connected and is surrounded by extremely high-impedance insulating material. To program an EPROM, the programmer applies a high voltage to the non-floating gate at each bit location where a 0 is to be stored. This causes a temporary breakdown in the insulating material and allows a negative charge to accumulate on the floating gate. When the high voltage is removed, the negative charge remains. During subsequent read operations, the negative charge prevents the MOS transistor from turning on when it is selected.

EPROM manufacturers 'guarantee' that a properly programmed bit will retain 70% of its charge for at least 10 years, even if the part is stored at 125°C, so EPROMs definitely fall into the category of 'nonvolatile memory.' However, they can also be erased. The insulating material surrounding the floating gate becomes slightly conductive if it is exposed to ultraviolet light with a certain wavelength. Thus, EPROMs can be erased by exposing the chips to ultraviolet light, typically for 5–20 minutes. An EPROM chip is normally housed in a package with a quartz lid through which the chip may be exposed to the erasing light.

An electrically erasable programmable read-only memory (EEPROM) is like an EPROM, except that individual stored bits may be erased electrically. The floating gates in an EEPROM are surrounded by a much thinner insulating layer and can be erased by applying a voltage of the opposite polarity as the charging voltage to the nonfloating gate. Large EEPROMs (1 Mb or larger) allow erasing only in fixed-size blocks, typically 128–512 Kb (16–64 KB) at a time. These memories are usually called *flash EPROMs* or flash memories, because erasing occurs 'in a flash.' While their underlying technology is the same, industry treats flash memory and EEPROMs as two different markets, as discussed in Section 13.2.6.

Programming or 'writing' an EEPROM location takes much longer than reading it, so an EEPROM is no substitute for the read/write memories discussed above. Also, because the insulating layer is so thin, it can be worn out by repeated programming operations. As a result, EEPROMs can be reprogrammed only a limited number of times, as few as 10,000 times per location. Therefore, EEPROMs are typically used for storing data that must be preserved when the equipment is not powered, but which doesn't change very often, such as the default configuration data for a computer." [Wak01]

13.2.6 Flash Memory

Flash memory is similar to EEPROM in that it is electrically erasable and programmable and retains its memory contents when power is removed (i.e., flash is nonvolatile memory). "The main difference between flash memory and traditional EEPROM is that flash memory requires data to be written or erased in blocks and EEPROM requires data to be written or erased one byte at a time." [Web03] Because flash memory performs read and write operations a block at a time, it is generally faster than EEPROM and intended for different applications than EEPROM.

13.2.7 ECC (Error-Correcting Code) Memory

"Error-Correcting Code (ECC) memory is a memory system that tests for and corrects errors automatically on the fly, very often without the operating system being aware of it, let alone the user. When writing the data into memory, ECC circuitry generates checksums from the binary sequences in the bytes and stores them in an additional seven bits of memory for 32-bit data paths or eight bits for 64-bit paths. When data are retrieved from memory, the checksum is recomputed to determine if any of the data bits have been corrupted. Such systems can typically detect and automatically correct errors of one bit per word and can detect, but not correct, most errors greater than one bit."[7] [Com03]

ECC is always implemented at the system level (or at least above the physical layer) and ECC features of memory chips simply provide data storage and paths for checksum bits or similar information. The reason for this is that any check of a memory system has to include the buses and interfaces which transfer data between the controller and the actual memory chips.

13.2.8 Banks and Ranks

13.2.8.1 Banks

"In modern DRAM and other devices, the entire memory array of individual bit cells is divided into sub-arrays called banks. Banks serve a dual purpose. They allow a device's memory core to be repaired of manufacturing defects by mapping in redundant core components (word lines, bit lines, individual memory cells) and they increase performance of a single device by facilitating concurrent operations.

Memory core banks are connected to the external device interface by electronic circuits called sense amplifiers. Sense amplifiers retrieve a value stored in an individual memory core cell and periodically refresh the value stored in the cell (in the case of DRAM). Sense amplifiers consume much more area than memory cells and they are shared across all of the data (column) lines within an individual bank to save space on the DRAM die.

The longest portion of a memory read operation is the time to access the core and read the data at an address into the sense amplifiers. Modern computing systems can issue memory operations faster than they can be processed by a memory system. Bank conflict occurs when multiple operations are active and two operations request an address that lies in the same memory bank. Only one column address within a bank can be accessed at a time since the sense amplifiers are shared among the columns (datalines) of the bank. Having more banks in a DRAM device increases performance at the expense of added die area (and cost) due to the additional sense amplifiers and redundant bank elements that must be supplied for each bank." [Ram01d]

13.2.8.2 Ranks

"Rank" is a term that is descriptive of a multichip memory module such as a DIMM (dual in-line memory module). A rank is a group of memory chips on a memory module that get selected simultaneously and that together usually form a data word width that is identical to the width of the memory data bus. Some memory modules have only one rank where a single chip-select input (e.g., CS#) selects every memory chip on the module. These are referred to as "single-rank modules." Other modules have two ranks where there are two chip-select inputs to the memory module (e.g., CS1#, CS2#). These are referred to as "double-rank modules." For double-rank modules, only one rank or the other is selected at a time during a read or write operation. The

7. Reproduced with permission from Computer Desktop Encyclopedia.

chip-select inputs are controlled by the memory controller, which is most often located on the motherboard.

While many double-rank modules have two chip-select inputs, the number of chip selects is not necessarily an indication of the number of ranks in a memory module. Sometimes a single-rank module will have two chip selects and a double-rank module may have four chip selects. Doubling of chip-select inputs may occur in a module in order to reduce the loading on any one chip-select line, or may occur because individual memory chips on the module have two chip-select inputs where both inputs must be asserted to activate the chip. Similarly, whether a memory module has chips on both sides of the module or just on one side is not an indication of whether the module is single-rank or double-rank. The best way to identify the number of ranks is to refer to the manufacturer's specification sheet for the module. The term "module rank density" refers to the total density (total number of megabits of storage) contained in all the chips forming a rank.

Ranks may also cross memory module boundaries and be formed from chips on more than one memory module. For example, rank#1 of a memory system, where the memory system comprises four memory modules, may be formed with the first two chips from each of the four modules, with a total of eight chips. In this case, a single chip-select line may be routed to all four modules and to the chip selects of the first two chips of each module.

13.2.9 Memory Nomenclature

The designations used for memory chips and their modules can often become a little confusing even though they're fairly straightforward. This is because the designations include numbers representing the data rates and it's not always obvious whether the numbers are referring to the data rate in Mbps or MB/s, or referring to the clock frequency. It's also not obvious whether the numbers are referring to data rates per pin, per device, or for the whole module.

In general, designations for chips use a PCxxxx or DDRxxxx format where xxxx represents the data rate *in Mbps per pin*. (The number of digits, i.e., the number of x's, will vary with the data rate value.) In the chip notation, PC indicates that the chip is used in personal computer type applications and DDR indicates that the chip uses a double data rate interface. The xxxx value is always the data rate per pin in mega*bits* per second, even for DDR devices where data is clocked twice during a clock cycle. For DDR devices, the memory clock frequency is half of the xxxx value, but in MHz. The "PC" prefix is also used for memory modules, as discussed next.

The designations for memory modules are in the format of DIMMyyyy, PCyyyy, or RIMMyyyy, where yyyy represents the data rate *in MB/s per module*, i.e., the data rate in mega*bytes* per pin times the total number of data pins. DIMM is Dual In-line Memory Module and RIMM is Rambus In-line Memory Module. For both chip and module designations, the data rates indicated are peak rates with implied pipelining, not effective rates. Effective rates are almost always lower than peak rates due to lost clock cycles from switching from write operations to read operations and from reads to writes, and other factors (see Section 17.1.3, "Bandwidth and Latency," on page 456). Let's look at a few designation examples.

- PC133 SDRAM outputs data at 133 Mbps per pin. It uses a 133 MHz clock.
- DDR266 is SDRAM which outputs data at 266 Mbps per pin. It uses a 133 MHz clock. While RDRAM® uses a dual-edged clocking system, it's not referred to as DDR and a DDR designation always refers to SDRAM (or SRAM).
- PC800 by itself indicates that data will be output at 800 Mbps per data pin and that the chip uses a 400 MHz clock. But you'll need additional information to determine if it's SDRAM or RDRAM. Because 800 is not a very large number (not 4 digits), you can guess that this is a chip designation and not a module designation.
- PC1066 refers to a memory chip that runs off of a 533 MHz clock and that outputs data using both edges of the clock at a data rate of 1,066 (2 × 533) Mbps. At 16 bits per channel, the chip will output (16 bits × 1,066 Mbps) × (1 byte/8 bits) = 2.132 GB/s per chip.
- PC3200 refers to a memory module containing eight 200 MHz DDR400 x8 (8-bit) chips for a total module data rate of 400 Mbps × 8 bits ÷ 8 bits/byte × 8 chips = 3,200 MB/s.
- DIMM3200 is an SDRAM memory module that generates a peak data rate of 3,200 MB/s per module. Without the module specification sheet, you don't know the chip configuration, but one possible configuration would be a module with four DDR400 chips. At 16 data pins per chip, the total module data rate would be 400 Mbps per pin × 16 pins × 1 byte/8bits = 3,200 MB/s or 3.2 GB/s.
- RIMM4200 is a Rambus In-line Memory Module having a performance level of 4,200 MB/s or 4.2 GB/s. It uses one PC1066 chip for each of the two 16-bit channels, 2 chips total, thereby creating a 32-bit module. 1,066 Mbps per pin, times 32 data pins total for both chips, divided by 8 bits/byte = 4.2 GB/s. While the total peak data rate is precisely 4264 MB/s, the module is referred to as a "4200."
- RIMM9600 is a 64-bit wide memory module which uses 4 PC1200 RDRAM chips and has a performance level of 9.6 GB/s.

13.2.9.1 Speed Bin

Memory chips often come with a designation that is referred to as a "speed bin." For example, an SDRAM Samsung chip might be specified as DDR2-533 5-4-4 where the numbers "5-4-4" refer to the speed bin. The speed bin order is CL–t_{RCD}–t_{RP} (the 5-4-4). CL is the CAS (column address strobe) latency, which is the number of memory clock cycles (rounded up) from the time a valid column address first appears at the column address inputs to when valid data first appears at the outputs. t_{RCD} is the RAS# (Row Address Strobe) to CAS#[8] (column address strobe) delay. t_{RP} is the row precharge time, which is the time required to precharge memory cells before new data may be written to them; that is, the time to switch between memory banks. When t_{RCD} and t_{RP} are used as speed bin numbers, they are converted to clock cycles by taking the t_{RCD} and t_{RP} time delays in ns, dividing them by the memory clock period in ns, and rounding up to integers.

8. The "#" symbol after signal names indicates that they are active "low."

13.3 Memory Signaling Technologies

The mainstream signaling technologies used to transfer data between memory chips and their memory controllers are SSTL (stub-series terminated logic), HSTL (high-speed transceiver logic), LVCMOS (low-voltage CMOS), and RSL (Rambus signaling level). In addition to these, there are other high-speed proprietary technologies such as DRSL (differential Rambus signaling levels), QRSL (quad Rambus signaling levels), ODR (octal data rate), and others.

As a rule of thumb, SDR (single data rate) SDRAM tends to use LVTTL, DDR (double data rate) SDRAM tends to use SSTL_2, DDR-II SDRAM tends to use SSTL_18, SRAM tends to use HSTL, and RDRAM uses RSL for their control and data signaling interfaces.

13.3.1 SSTL_18, SSTL_2, and HSTL (Class II)

SSTL was discussed earlier in Chapter 7, Section 7.2, "Stub-Series Terminated Logic (SSTL)," beginning on page 141. Please refer to that section for detailed information on SSTL. Today, SDRAM devices are using SSTL_18 and SSTL_2. SSTL_3 is close to end-of-life (EOL) for most memory applications because of its larger signaling swing and voltages.

HSTL is used with many SRAM (static RAM) devices. When used, device inputs are HSTL and most device outputs are HSTL Class I or II. HSTL was discussed earlier in Chapter 7, Section 7.1, "High-Speed Transceiver Logic (HSTL)," beginning on page 129. Please refer to that section for detailed information on HSTL.

13.3.2 RSL (Rambus Signaling Level)

"Rambus introduced its proprietary signaling technology, Rambus Signaling Level (RSL) at a 500 MHz data rate to address high-bandwidth chip interconnect bottlenecks. RSL now supports 1,333 MHz operation with a roadmap to 1,600 MHz.

There are many key elements to RSL technology:

- Low voltage swing (800 mV p-p) (see Figure 13.12(a))
- Current mode output drivers
- Automatic output current control
- Controlled impedance design, typically 28 to 40 Ω, 30 mA maximum drive
- Low parasitic packages
- Double data rate (transfers data on both edges of clock)
- Bidirectional bus, multiple data bits outstanding on the channel at once
- Pseudo-differential signals,[9] only one pin per signal with common V_{ref} for all input samplers
- Uniform high-speed connection, not a mesh topology
- Advanced driver and receiver circuits on-chip, including the use of DLLs

9. The pseudo-differential signal is a single-ended signal referenced to V_{ref}.

- Common differential clock travels with data and loops around in the controller
- Terminated at one end

The result of the combination of all these techniques is a signaling technology that transfers data at 1,333 MHz with headroom to support higher frequencies in the future.

Figure 13.12(b) shows an eye diagram of a RSL driver, showing high-quality signaling."[10] [Ram01a]

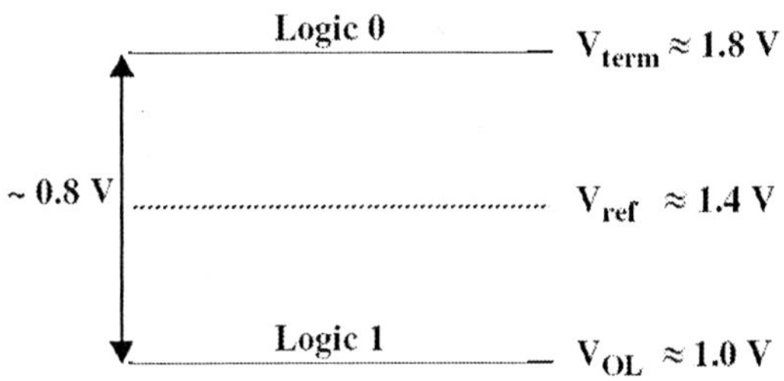

(a)

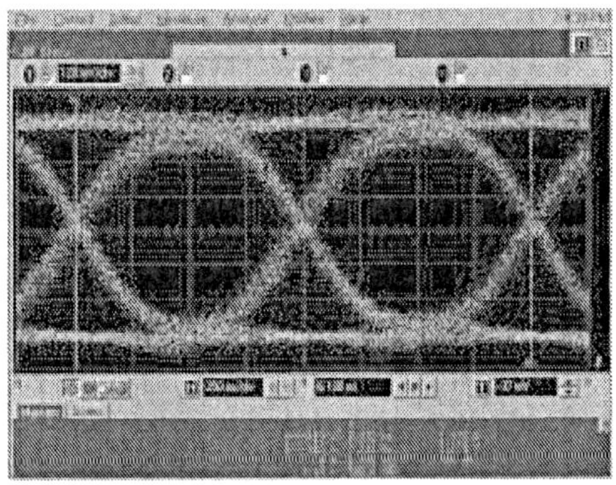

(b)

Figure 13.12 (a) RSL signal levels, and (b) 1,066 MHz RSL eye diagram at the driver. [Reprinted with permission of Rambus Inc.]

13.3.3 DRSL (Differential Rambus Signaling Levels)

"DRSL (Differential Rambus Signaling Levels) is a bidirectional differential signaling technology offering a high-performance, low-power, and cost-effective solution for getting bandwidth on and off chip. DRSL uses an ultra-low 200 mV signal swing from 1.0 to 1.2 V as shown in Figure 13.13. DRSL enables low power operation and scalability for future lower voltage." [Ram03c]

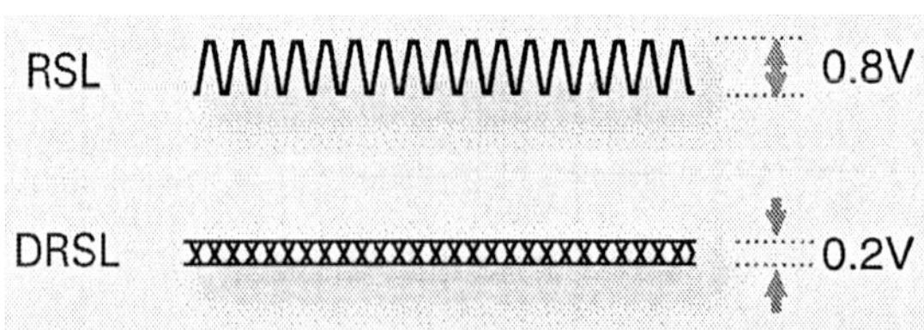

Figure 13.13 DRSL is ultra-low-voltage signaling. [Reprinted with permission of Rambus, Inc.]

10. Reprinted with permission of Rambus Inc.

13.3.4 QRSL (Quad Rambus Signaling Levels)

"QRSL is a new signaling technology, fundamentally different from RSL. Almost all chip interconnect technology today transfers data using binary information, representing 0s and 1s with two voltage levels. QRSL doubles the data rate by using four voltages to represent two bits of information, as shown in Figure 13.14. This multilevel signaling allows higher data bandwidth, twice that of RSL (>2 Gbps) while not increasing the frequency of operation of the channel. Part (b) of the figure shows an eye diagram of a QRSL driver.

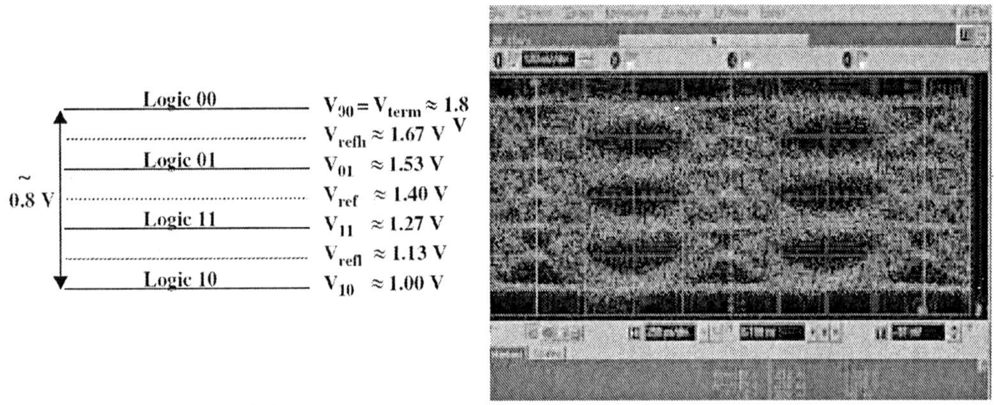

(a) (b)

Figure 13.14 (a) QRSL signal levels, and (b) >2 Gbps QRSL eye diagram. [Reprinted with permission of Rambus Inc.]

Major features of QRSL are:

- Low voltage swing (800 mV p-p, ~267 mV per step)
- Current mode output drivers
- Automatic output current control (for three drive levels)
- Controlled impedance design, typically 40 Ω systems
- Low parasitic packages
- Double data rate and multilevel signaling, providing four bits of information per clock cycle
- Gray-coded logic levels
- Bidirectional channel
- Pseudo-differential receivers, with three V_{ref} inputs
- New driver and receiver circuits, including integrating receivers
- Common differential clock, which travels with the data, and is the same frequency as RSL
- Terminated at one end" [Ram01a]

13.3.5 Octal Data Rate (ODR)

With Octal Data Rate (ODR) technology, data is sampled eight times per memory clock cycle. ODR is discussed in detail in Section 18.2.3 on page 478.

13.4 Design Considerations in Use of Memory

13.4.1 Power Up and Initialization

Most memory devices, particularly DDR2 SDRAMs, must be powered up and initialized in a predefined manner. Operational procedures other than those specified may result in undefined operation. As an example, the following power-up and initialization sequence for Samsung Semiconductor's K4T5104 512 Mb DDR2 SDRAM chip is provided[11]:

1. Apply power and attempt to maintain CKE below $0.2^*V_{DD}Q$ and ODT[12] at a low state (all other inputs may be undefined).
   — V_{DD}, $V_{DD}L$, and $V_{DD}Q$ are driven from a single power converter output, *and*
   — V_{TT} is limited to 0.95 V max, *and*
   — V_{ref} tracks $V_{DD}Q/2$.
   or
   — Apply V_{DD} before or at the same time as $V_{DD}L$.
   — Apply $V_{DD}L$ before or at the same time as $V_{DD}Q$.
   — Apply $V_{DD}Q$ before or at the same time as V_{TT} and V_{ref}.
   At least one of these two sets of conditions must be met.
2. Start clock and maintain stable condition.
3. For the minimum of 200 μs after stable power and clock (CK, $\overline{CK}$), then apply NOP or deselect and take CKE high.
4. Wait minimum of 400 ns then issue precharge all command. NOP or deselect applied during 400 ns period.
5. Issue EMRS(2) command. (To issue EMRS(2) command, provide "Low" to BA0, "High" to BA1.)
6. Issue EMRS(3) command. (To issue EMRS(3) command, provide "High" to BA0 and BA1.)
7. Issue EMRS to enable DLL. (To issue "DLL Enable" command, provide "Low" to A0, "High" to BA0 and "Low" to BA1 and A12.)

11. In this sequence, CKE is clock enable, $V_{DD}Q$ is the DQ power supply voltage for data outputs, $V_{DD}L$ is the DLL (delay locked-loop) power supply voltage, V_{DD} is the IC power supply voltage, V_{TT} is the termination power supply voltage, V_{ref} is the reference voltage, CK is the true signal of the differential clock, $\overline{CK}$ is the complementary signal of the differential clock, NOP is no operation, EMRS is extended mode register set command, and BA0–BA2 are bank address inputs that define to which bank an active, read, write, or precharge command is being applied.
12. To guarantee ODT off, V_{ref} must be valid and a low level must be applied to the ODT pin.

8. Issue a Mode Register Set command for "DLL reset." (To issue DLL reset command, provide "High" to A8 and "Low" to BA0-1.)
9. Issue precharge all command.
10. Issue two or more auto-refresh commands.
11. Issue a mode register set command with low to A8 to initialize device operation (i.e., to program operating parameters without resetting the DLL).
12. At least 200 clocks after step 8, execute OCD Calibration (off-chip driver impedance adjustment). If OCD calibration is not used, EMRS OCD Default command (A9 = A8 = A7 = 1) followed by EMRS OCD. Calibration Mode Exit command (A9 = A8 = A7 = 0) must be issued with other operating parameters of EMRS.
13. The DDR2 SDRAM is now ready for normal operation.

The initialization sequence after power-up is illustrated in Figure 13.15.[13] [Sam03a]

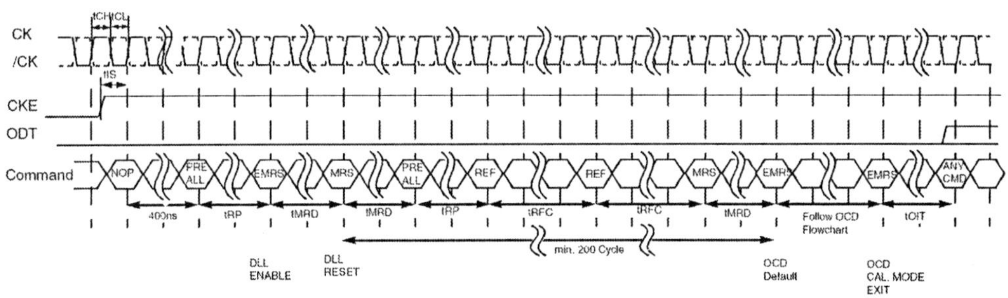

Figure 13.15 Initialization sequence after power up for the Samsung K4T5104 512 Mb M-die DDR2 SDRAM. [Reprinted with permission of Samsung Electronics.]

13.4.2 Slew Rate Control

For the Samsung 512 Mb DDR2 SDRAM, "Slew rate control can be achieved by adjustment of the OCD setting. Recommended SSTL_18 minimum slew rate is ~1 V/ns with a recommended maximum slew rate of ~6 V/ns." [Say03]

13.4.3 Drive Capability

"It is recommended that sufficient drive capability be available to simplify design and avoid 'signal integrity'-induced threshold or timing errors. This involves properly adjusting the ODT and OCD values to achieve optimum amplitude and timing margin performance. SPICE-based simu-

13. These materials are the copyright of Samsung Semiconductor, Inc. (Samsung) with permission granted for publication in this text subject to these materials being "AS IS" and Samsung shall not be responsible for reliance, use, or damages.

lation will materially assist determining proper signal integrity. However, clarification regarding the presence and state of the ODT termination should be sought from the memory and controller suppliers. Programmable characteristics of driver I/O transistors should be balanced to maintain duty cycle symmetry." [Say03]

13.4.4 On-Die Termination (ODT)

"Significant improvements in Signal Integrity for 400 and 500 MHz operation are possible with the use of on-die termination. The JEDEC committee for DDR-II SDRAM ran a significant number of SPICE simulations to determine empirical relationships of ODT values for optimum signal integrity for 400 and 500 MHz DDR-II operation. The flexibility provided by the programmable ODT feature allows for various memory configurations and loadings including number of modules (RDIMM = 1 or 2) and number of ranks (R) per module (R = 0, 1, or 2) for either of the two module locations and whether a particular device or controller is driving or receiving. The choice of Thevenin termination values are R_{TT} = (75 Ω, 150 Ω, and → ∞). The programmable ODT termination feature is included in all DQ, DM, DQS, DQS#, RDQS# pins on DDR-II SDRAM devices.[14] The ODT termination can be altered on-the-fly according to the driver location. This allows for optimum signal fidelity for each drive/receive scenario.

The ODT termination is located between the I/O pin and the DDR-II SDRAM input buffer and is shown schematically in Figure 13.16." [Say03] Specific recommendations for ODT settings are shown in Table 13.1.

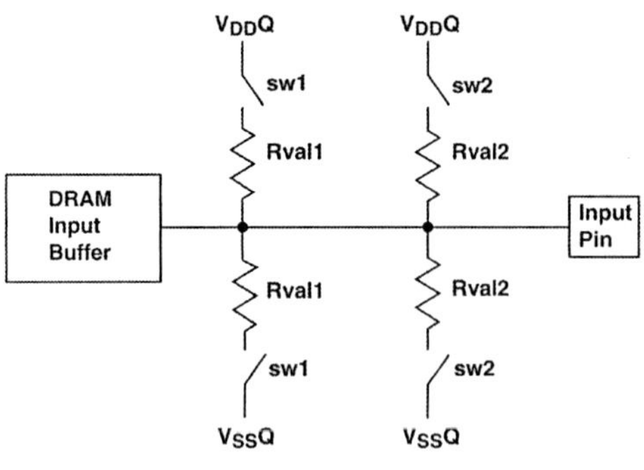

Switch sw1 or sw2 is enable by ODT pin.
Selection between sw1 and sw2 is determined by "Rtt(nominal)" in EMRS.
Termination included on all DQ, DM, DQS#, RDQS and RDQS# pins.
Target Rtt (ohms) = (Rval1)/2 or (Rval2)/2.

Figure 13.16 Functional Representation of ODT. [Jed03b] [Reprinted with permission of JEDEC.]

14. The "#" symbol indicates the signal complement. DQs are data inputs and outputs (bidirectional bus), DM is an input mask signal for write data, and DQS, DQS#, and RDQS# are data strobes.

Table 13.1 Termination Recommendations for ODT Settings vs. Operation and Memory Configurations [Say03]

Termination Matrix for Writes to DRAM						
			Target DQ ODT Resistance R_{TT}			
			Module in Slot 1		Module in Slot 2	
Configuration	Write To	Controller	Rank 1	Rank 2	Rank 1	Rank 2
2R/2R	Slot 1	infinite	infinite	infinite	75 Ω	infinite[a]
	Slot 2	infinite	75 Ω	infinite[a]	infinite	infinite
2R/1R	Slot 1	infinite	infinite	infinite	75 Ω	unpopulated
	Slot 2	infinite	75 Ω	infinite[a]	infinite	unpopulated
1R/2R	Slot 1	infinite	infinite	unpopulated	75 Ω	infinite[a]
	Slot 2	infinite	75 Ω	unpopulated	infinite	infinite
1R/1R	Slot 1	infinite	infinite	unpopulated	75 Ω	unpopulated
	Slot 2	infinite	75 Ω	unpopulated	infinite	unpopulated
2R/Empty	Slot 1	infinite	150 Ω	infinite	unpopulated	unpopulated
Empty/2R	Slot 2	infinite	unpopulated	unpopulated	150 Ω	infinite
1R/Empty	Slot 1	infinite	150 Ω	unpopulated	unpopulated	unpopulated
Empty/1R	Slot 2	infinite	unpopulated	unpopulated	150 Ω	unpopulated
Termination Matrix for Reads from DRAM						
			Target DQ ODT Resistance R_{TT}			
			Module in Slot 1		Module in Slot 2	
Configuration	Read From	Controller	Rank 1	Rank 2	Rank 1	Rank 2
2R/2R	Slot 1	150 Ω	infinite	infinite	75 Ω	infinite[a]
	Slot 2	150 Ω	75 Ω	infinite[a]	infinite	infinite
2R/1R	Slot 1	150 Ω	infinite	infinite	75 Ω	unpopulated
	Slot 2	150 Ω	75 Ω	infinite[a]	infinite	unpopulated
1R/2R	Slot 1	150 Ω	infinite	unpopulated	75 Ω	infinite[a]
	Slot 2	150 Ω	75 Ω	unpopulated	infinite	infinite
1R/1R	Slot 1	150 Ω	infinite	unpopulated	75 Ω	unpopulated
	Slot 2	150 Ω	75 Ω	unpopulated	infinite	unpopulated
2R/Empty	Slot 1	75 Ω	infinite	infinite	unpopulated	unpopulated
Empty/2R	Slot 2	75 Ω	unpopulated	unpopulated	infinite	infinite
1R/Empty	Slot 1	75 Ω	infinite	unpopulated	unpopulated	unpopulated
Empty/1R	Slot 2	75 Ω	unpopulated	unpopulated	infinite	unpopulated

a. Alternatively, the controller may use Rank 2 for termination instead of Rank 1.

13.4.4.1 Modifying Terminations "On-the-Fly"

ODT allows termination values to be changed at computer speeds. Termination resistors may be switched out when reads or writes are not being performed to save power. Termination values can also be switched as controller-to-memory trace lengths change.

13.4.5 Memory Power and Ground Planes

"For DDR-II SDRAM devices such as Samsung's 512 Mb device, special care must be given to the design of memory power and ground planes. DDR-II SDRAM uses SSTL_18, a reduced-swing logic family (ac thresholds, +/–0.25 about V_{REF}). This requires special care with the control of conducted EMC onto the 1.8 V power plane. A locally regulated, isolated, filtered power plane for the memory controller and memory chips (DIMMs), and continuous ground planes are recommended." [Say03]

13.4.6 Memory Controllers

The major concept dealing with memory controllers is that it's usually to the advantage of system developers to buy the controller for DDR2 and other high-speed memories and to instead spend engineering resources on unique product features and architecture. Controllers are usually purchased as intellectual property from companies such as Denali for ASIC cell-based controller designs or from FPGA vendors such as Xilinx and Altera for controller designs implemented in an FPGA or CPLD, such as those used with RLDRAM I and II. Intellectual property generally is valued for two reasons: it has proven, reliable architecture and it lowers risk.

13.5 Summary of Memory Devices and Terminology

Table 13.2 Summary of Memory Devices and Terminology

Memory or Term	Description
1T SRAM	One Transistor SRAM — Pseudo SRAM that requires only one transistor per memory cell. See Pseudo-SRAM.
4i	Four Independent Bank Core — Bank architecture of RDRAM devices.
6T SRAM	Six Transistor SRAM — Conventional SRAM where each cell is made from six transistors, a structure that allows data to be retained without requiring refreshing of the data contents in order to retain their values as in DRAM.
Bank	Bank — The portion of a memory device read out by a single sense amplifier. The cells of a memory device are divided into banks where each bank is read out through a single sense amplifier.
CAM	Content-Addressable Memory — "Includes comparison logic with each bit of storage. A data value is broadcast to all words of storage and compared with the values there. Words that match are flagged in some way. Subsequent operations can then work on flagged words, e.g., read them out one at a time or write to certain bit positions in all of them." [Fol03] CAM is usually synchronous SRAM.
Cellular RAM	Cellular Random Access Memory — A new, multi-generation family of low-power Pseudo-Static RAM (see PSRAM) for wireless handsets.
CompactFlash	A flash memory format invented by SanDisk Corporation that uses a 50-pin PC Card/ATA interface and supports both 3.3 V and 5 V operation.

Table 13.2 Summary of Memory Devices and Terminology (Continued)

Memory or Term	Description
Consumer FCRAM	Consumer Fast Cycle RAM — An FCRAM core with an SDR (single data rate) interface used in consumer applications including digital cameras, battery-driven devices, car navigation systems, etc.
DDR FCRAM	Network FCRAM (Fast Cycle RAM) — Synchronous DRAM that uses an FCRAM memory core and DDR (double data rate) I/Os where data is transferred on both edges of the clock.
DDR SGRAM	Double Data Rate Synchronous Graphics Random Access Memory — Next-generation high-speed SGRAM that adopts the dual-edge system.
DDR SRAM	Double Data Rate Synchronous RAM — SSRAM that transfers data on both edges of the memory clock. Same as DDR SSRAM.
DDR SSRAM	Double Data Rate Synchronous Static RAM — SSRAM that transfers data on both edges of the memory clock. Same as DDR SRAM.
DDR2 SDRAM	Double Data Rate 2 Synchronous Dynamic Random Access Memory — Second-generation DDR SDRAM that features ODT (on-die termination) and programmable OCD (off-chip driver) not found in DDR. DDR2 operates at SSTL_18 (1.8 V) signaling levels vs. SSTL_2 (2.5 V) signaling levels for DDR. Data still transferred on both edges of the clock.
DDR-II SDRAM	Same as DDR2. JEDEC specification uses "DDR2" notation and industry tends to use "DDR-II" notation.
Direct RDRAM	Direct Rambus Dynamic Random Access Memory — Same as DRDRAM® and RDRAM. High bandwidth per pin DRAM memory licensed to chip manufacturers by Rambus. Used with Intel 850, 850E, and 860 chipsets and SiS R659 North Bridge chipset. Offered with either 16- or 18-bit data buses (called channels) and has separate buses for row and column addresses. Data rates per pin of 800 Mbps to 1,333 Mbps with roadmap to 1,600 Mbps per pin, densities from 128 Mb to 512 Mb with roadmap to 1 Gb.
DRDRAM	Direct Rambus Dynamic Random Access Memory — Same as Direct RDRAM and RDRAM.
DRSL	Differential Rambus Signaling Levels — DRSL is a bidirectional, differential signaling technology offering a high-performance, low-power, and cost-effective solution for getting bandwidth on and off chip. DRSL uses an ultra-low 200 mV signal swing from 1.0 to 1.2 V. DRSL enables low-power operation and scalability for future lower voltage.
Dual-Port RAM	Dual-Port RAM provides simultaneous read and write access to any cell in its memory array from either of its two ports. In addition, the two ports may operate at independent clock speeds allowing complete decoupling of the devices being interfaced. Asynchronous multiprocessor systems require a means to transmit data between two independently running processors. Dual-port memory provides a common memory accessible to both processors that can be used to share and transmit data and system status between the two processors.

Table 13.2 Summary of Memory Devices and Terminology (Continued)

Memory or Term	Description
EDO DRAM	Extended Data Out DRAM — In EDO DRAM, "the column address strobe no longer controls the output enable during read cycles as in conventional DRAM. Instead, a separate output enable control input allows valid output data for a longer period of time. Rather than turning off between successive column address strobe pulses, the output is left on continuously (as controlled by the output enable), and page mode read data is valid from the beginning of one column address strobe pulse to the beginning of the next." [Wak01]
eDRAM	Embedded DRAM — DRAM that is embedded in an ASIC with the biggest benefit being increased memory bandwidth.
EEPROM	Electrically Erasable Programmable Read-Only Memory — Nonvolatile memory that can be programmed and erased electrically in its circuit. It can be erased by applying a voltage of the opposite polarity as the charging voltage to the non-floating gate of each cell.
eFlash	Embedded flash memory — Flash memory that is embedded in an ASIC.
EPROM	Electrically Programmable Read-Only Memory — Erased with ultraviolet light.
eSRAM	Embedded SRAM — SRAM that is embedded in an ASIC.
FCRAM	Fast Cycle RAM — Memory devices utilizing a nontraditional DRAM core structure designed to create a low, row cycle time (t_{RC}). t_{RC} is the delay before you may access the same bank of a chip again.
FeRAM	Ferroelectric RAM — "A form of nonvolatile memory using ferroelectric crystals. When an electric field is applied to the crystal, the central atom moves toward the field. If the electric field is removed, the central atom stays in the same place. FeRAMs currently have densities of 32 Mb, access times of 50 nanoseconds and cycle times of 75 nanoseconds, slower than SDRAMs but much faster than flash, where a program time of 200 microseconds is considered good." [Cle03]
Flash Memory	Flash memory is similar to EEPROM in that it is electrically erasable/programmable and retains its memory contents when power is removed (nonvolatile memory). The main difference between flash memory and traditional EEPROM is that flash memory allows data to be written or erased in blocks and EEPROM requires data to be written or erased one byte at a time. This results in their being used in different applications.
Flash PROM	Another name for flash memory. This terminology is a misnomer that originated when flash memory started to replace PROMs, such as those used for BIOS chips.
Flash RAM	Another name for flash memory. See flash memory. This terminology is a misnomer that originated when flash memory started to replace RAM devices.
FlexPhase	FlexPhase is a Rambus XDR (Xtreme Data Rate) technology that creates precise on-chip alignment of data with the clock, eliminating the need for PCB trace-length matching and strict PCB timing constraints. FlexPhase circuits skew individual data bit output timing from the ASIC to the XDR DRAM to present a synchronized parallel data word to the addressed XDR DRAM. FlexPhase I/O circuits can compensate for systematic timing errors with a resolution as fine as 2.5 ps.

Table 13.2 Summary of Memory Devices and Terminology (Continued)

Memory or Term	Description
FPM DRAM	Fast Page Mode DRAM — "Fast page mode or FPM DRAM is slightly faster than conventional DRAM. While standard DRAM requires that a row and column be sent for each access, FPM works by sending the row address just once for many accesses to memory in locations near each other, improving access time. FPM DRAM itself is an improved version of its predecessor, page mode DRAM, although both are very rarely seen now." [Pcg03]
GDDR2 SDRAM (GDDR-II SDRAM)	Graphics Double Data Rate 2 SDRAM (Same as GDDR-II SDRAM) — Second-generation graphics memory with densities up to 256 Mb, chip bandwidths to 12.8 GB/s, data rates per pin from 900 Mbps to 3.2 Gbps, x32 bus widths, SSTL_18 signaling, on-die termination (ODT), and off-chip driver calibration (OCD).
GDDR3 SDRAM	Graphics Double Data Rate SDRAM 3. Third-generation GDDR SDRAM. Operates at higher data rates while consuming approximately half the power of GDDR2 SDRAM.
Graphics Memory	General category of memories used for graphic applications; includes all generations of SGRAM and GDDR SDRAM devices.
MCP Memory	Multi-Chip Package — A single package enclosing two or more memory die, e.g., Samsung's MCP stacks a 64 Mb NAND-type flash memory chip and an 8 Mb SRAM chip together for 3G mobile phone applications.
Memory Stick	A flash memory card from Sony designed for digital storage on cameras, camcorders, and other handheld devices.
Mobile DRAM	DRAM optimized for mobile applications by using lower voltages for power savings, packing in small-form factor BGA packages, and providing high densities, low cost-per-bit, and fast random access times.
Mobile FCRAM	Mobile Fast Cycle RAM — FCRAM core used in mobile phone applications with an SRAM-like interface.
MRAM	Magnetoresistive RAM — A nonvolatile memory technology that uses magnetic, thin-film elements on a silicon substrate. "MRAM's main features are its speed and nonvolatility. MRAM can reach theoretical write times down to 2.3 ns, more than 1,000 times faster than the fastest nonvolatile flash and 20 times faster than either DRAM or FeRAM. Reads require only 2 milliamps, less than 1/100 the energy of DRAM. MRAM is resistant to radiation." [Cle03]
NB DRAM	Nibble Mode DRAM — An older, obsolete technology. (Nibble is 4 bits.)
Network DRAM NetDRAM	DRAM used for networking applications that is 100% compatible with Network FCRAM. High density with short row cycle times and short write-to-read turn around times. See FCRAM.
Network FCRAM	Network Fast Cycle RAM — Synchronous DRAM that uses an FCRAM memory core and DDR (double data rate) I/Os. Same as DDR FCRAM. Used in networking applications.
NoBL SRAM	NoBL (No Bus Latency) SRAM (Cypress). Designed to eliminate dead bus cycles during back-to-back read/write and back-to-back write/read cycles by providing 100% bus utilization. Similar to NtRAM (No turnaround RAM) (Samsung, Toshiba), ZeroSB (Zero Speed Bump) SRAM (NEC), and Zero Bus Turnaround SRAM (IDT, Micron).

Table 13.2 Summary of Memory Devices and Terminology (Continued)

Memory or Term	Description
Nonvolatile	Describes any device that retains its memory contents when power is turned off.
Nonvolatile RAM NVRAM	Nonvolatile RAM — Traditionally, this has been SRAM made nonvolatile through powering with a long-life (e.g., 10-year) battery enclosed in one package. Newer references to NVRAM refer to flash or other nonvolatile technologies embedded in an MCP (multi-chip package) or ASIC.
NtRAM	NtRAM (No turnaround RAM) (Samsung, Toshiba). Designed to eliminate dead bus cycles during back-to-back read/write and back-to-back write/read cycles by providing 100% bus utilization. Similar to NoBL (No Bus Latency) SRAM (Cypress), ZeroSB (Zero Speed Bump) SRAM (NEC), and Zero Bus Turnaround SRAM (IDT, Micron).
ODR SRAM	Octal Data Rate SRAM — SRAM where data is transferred 8 times per clock period.
PBSRAM	Pipeline Burst SRAM — SSRAM that adopts the pipeline system, i.e., pipelined SSRAM.
PROM	Programmable Read-Only Memory — One-time programmable, nonvolatile memory device.
PSRAM Pseudo-SRAM	Pseudo-Static RAM — Asynchronous SRAM that has a memory cell size of one transistor versus the six-transistor cell size that traditional low power SRAM has. Specific features are SRAM-like write/read timing, TTL input/output compatibility, standby mode or deep powerdown standby mode, very high density, and relatively high speed for low power consumption.
QDR SRAM	Quad Data Rate SRAM — Has separate input and output data buses, separate and concurrent read and write ports that operate at DDR rates. Optimized for systems with short-term-balanced read and write operations, i.e., where #reads/#writes < 3:1 for 2-word burst, < 2:1 for 4-word burst.
QRSL	Quad Rambus Signaling Levels — QRSL doubles the data rate by using four voltages to represent two bits of information. This multilevel signaling allows higher data bandwidth, twice that of RSL (>2 Gbps per pin) while not increasing the frequency of operation of the channel.
Quad SerDes	Quad serializer/deserializer interface — A Rambus high-speed quadruple serial point-to-point interface technology designed to connect two chips together across a backplane through connectors.
Rank	Rank — a group of memory chips on a memory module that get selected simultaneously and that together usually form a data word width that is identical to the width of the memory data bus.
RDRAM	Rambus Dynamic Random Access Memory — Same as Direct RDRAM and DRDRAM. See Direct RDRAM
RLDRAM	Reduced Latency DRAM — DRAM that offers high density, high bandwidth, reduced cycle time, and SRAM-like access.
RLDRAM II	Reduced Latency DRAM, second generation. DRAM that offers high density, high bandwidth, reduced cycle time, and SRAM-like access.
ROM	Read-Only Memory

Table 13.2 Summary of Memory Devices and Terminology (Continued)

Memory or Term	Description
RSL	Rambus Signaling Level — A pseudo-differential signaling technology (one pin per signal with a common non-zero reference voltage) having an 800 mV p-p signal swing, current mode output drivers, and double data rate data transfers.
SC DRAM	Static Column-Mode DRAM — An older, obsolete technology.
SDR SDRAM	Single Data Rate SDRAM — Same as SDRAM.
SDR SRAM	Single Data Rate SRAM. Includes NoBL, ZeroSB, NtRAM, and ZBT devices.
SDRAM	Synchronous Dynamic Random Access Memory — RAM memory that is synchronized to a memory clock through the addition of clocked registers and that needs to have its cells refreshed periodically to maintain the memory contents.
SGRAM	Synchronous Graphics RAM — Dedicated for images and basically configured with SDRAM. An operation mode for drawing, which is not provided to SDRAM, is added in order to simplify designing graphics systems. Superseded by GDDR devices.
SigmaRAM (ΣRAM)	SigmaRAM (ΣRAM) is a family of SRAM products jointly defined by the SigmaRAM Consortium. Features include no bus turnaround time, both common and separate I/O busses, single and double data rate operation, late- and double late-write protocols in various combinations, wide bus widths from x18 to x72, and high clock rates (up to 333 MHz).
SmartMedia	An ultra-compact flash memory format developed by Toshiba. SmartMedia cards are as thin as a credit card and can be plugged into a SmartMedia socket or into a standard Type II PC Card slot with a PC Card adapter.
SSRAM	Synchronous Static RAM (Synchronous SRAM)
SyncBurst SRAM	A synchronous SRAM technology supporting both flow-through and pipelined architectures with1–4 Mb, 3.3 V output supply $V_{DD}Q$, and x18, x32, x36 bus width options. Clock cycle timing to 150 MHz.
SynchLink DRAM (SLDRAM)	Synchronous Link DRAM — "A type of memory developed by the now-defunct SynchLink Consortium. SLDRAM was intended to be an enhanced version of SDRAM that used a multiplexed bus to transfer data to and from chips rather than fixed pin settings." [Web03]
TagRAM	TagRAM — A synchronous static RAM (SSRAM) used in networking applications for temporary storage of tags. TagRAM integrates SRAM cells with advanced synchronous peripheral circuitry and a comparator for tag compare operations.
UtRAM	Unit Transistor RAM — Pseudo SRAM. See Pseudo SRAM.
Video RAM	See VRAM.
VMRAM	Vertical Magnetic RAM — Nonvolatile memory. MRAM that has washer-shaped magnetic elements where the magnetic direction is clockwise or counter clockwise vs. the linear (north or south) magnetic orientation found in most other MRAM designs.
Volatile	Descriptive of any device that loses its contents when power is turned off.

Table 13.2 Summary of Memory Devices and Terminology (Continued)

Memory or Term	Description
VRAM	Video RAM — VRAM "is the type of memory used in a display adapter. Video RAM is designed with dual ports so that it can simultaneously refresh the screen while text and images are drawn in memory." [Com03]
XDR DRAM	Xtreme Data Rate DRAM — New Rambus technology. A single, 2-byte wide, 3.2 GHz XDR DRAM component provides up to 6.4 GB/s bandwidth.
ZBT SRAM	Zero Bus Turnaround SRAM (IDT, Micron). Designed to eliminate dead bus cycles during back-to-back read/write and back-to-back write/read cycles by providing 100% bus utilization. Similar to NoBL (No Bus Latency) SRAM (Cypress), ZeroSB (Zero Speed Bump) SRAM (NEC), and NtRAM (No turn-around RAM) (Samsung, Toshiba).
ZeroSB SRAM	Zero Speed Bump SRAM. NEC's version of NoBL (No Bus Latency) SRAM (Cypress), Zero Bus Turnaround SRAM (IDT, Micron), and NtRAM (No turn-around RAM) (Samsung, Toshiba). Designed to eliminate dead bus cycles during back-to-back read/write and back-to-back write/read cycles by providing 100% bus utilization.

13.6 Exercises

13-1. What is the difference between nonvolatile RAM and flash memory?

13-2. What is the difference between PBSRAM and SSRAM?

13-3. (a) What are some of the features of DRDRAM? (b) What are some of the features of RLDRAM?

13-4. What is the speed bin and what do the speed bin numbers represent?

13-5. What is the maximum data output rate per bus bit from (a) PC333 SDRAM, and (b) PC333 RDRAM? (c) What is the data output rate per bus bit for a 64-bit wide RIMM9600?

13-6. What signaling technologies are typically used by (a) SDR SDRAM, (b) DDR SDRAM, (c) DDR-II SDRAM, and (d) DRDRAM?

13-7. What are eDRAM, eSRAM, and eFlash?

13-8. Do memory chips of any kind include totally functional error correction within the chip itself? Provide an example or explain your answer.

13-9. Refer to Figure 13.16 on page 342. How many different ODT (on-die termination) values could be selected at different times based upon the programming of the ODT feature?

13-10. How are banks selected in an SDRAM device?

13-11. What is CAS latency?

13-12. State the three major differences between RSL and DRSL.

13-13. What do chip designers usually do to make a conventional DRAM design into a synchronous DRAM design?

13-14. What's the difference between a "bank" and a "rank"?

13-15. What percentage of a DRAM's overall operation time do refresh operations typically take? That is, how much overhead is there in constantly having to refresh DRAM cores?

CHAPTER 14

Double Data Rate SDRAM (DDR, DDR2) and SPICE Simulation

Chapter Objectives

- Present the features and specifications of double data rate (DDR) SDRAM.
- Explain the feature differences between single data rate (SDR) SDRAM and DDR SDRAM.
- Contrast DDR SDRAM features with those of DDR2 SDRAM.
- Discuss on-die termination (ODT) and off-chip driver (OCD) calibration.
- Present the signal integrity results from simulating a DDR2 SDRAM memory system using SPICE (Simulation Program with Integrated Circuit Emphasis) and IBIS (I/O Buffer Information Specification) models.

14.1 DDR (Double Data Rate) SDRAM

"DDR SDRAM is synchronous DRAM that realizes high-speed data transfer while adhering to the specifications of SDRAM as much as possible. Although DDR SDRAM does not have complete compatibility with SDRAM, DDR SDRAM can be used with simple specification changes to SDRAM controllers since DDR SDRAM has basically the same package pin configuration and control method as SDRAM.

While SDRAM adopts the single-edge system, which performs controls at a single edge of the basic clock, DDR SDRAM adopts the dual-edge system. As for the controller-to-memory device interface, DDR SDRAM adopts SSTL_2 signaling in order to realize speeds higher than that of SDRAM. DDR SDRAM transfers data at rates of 200, 266, 333, and 400 Mbps corresponding to memory clock speeds of 100, 133, 166, and 200 MHz."[1] [Elp01a]

1. Reprinted with permission of Elpida Memory.

DDR is specified in JEDEC Standard JESD79C, "Double Data Rate (DDR) SDRAM Specification."[2] DDR is sometimes referred to as DDR-I, which has been used to distinguish it from DDR-II (DDR2) once DDR-II came into existence.

14.1.1 Differences Between SDRAM and DDR SDRAM

"DDR SDRAM (double data rate synchronous DRAM) is a type of DRAM that realizes twice the data transfer rate of conventional SDRAM. (Hereafter in this section, conventional SDRAM is referred to as SDR SDRAM (single data rate synchronous DRAM) in contrast with DDR SDRAM).

This section explains the differences between DDR SDRAM and SDR SDRAM in the following areas: functions and specifications, commands, and operation timing." [Elp02a]

14.1.1.1 Differences in Functions and Specifications

"DDR SDRAM is a type of SDRAM that inherits technologies from SDR SDRAM and realizes faster operation and lower power consumption. It shares many common aspects with SDR SDRAM, which enables easy transition to DDR SDRAM.

Table 14.1 lists the differences in functions and specifications between DDR SDRAM and SDR SDRAM.

Table 14.1 Differences in Functions and Specifications Between DDR and SDR SDRAM

Item	DDR SDRAM	SDR SDRAM
Data Transfer Frequency	Twice the Operation Frequency	Same as the Operation Frequency
Data Rate	$2/t_{CK}$[a]	$1/t_{CK}$
Clock Input	Differential Clock	Single Clock
Data Strobe Signal (DQS)	Essential	Not Supported
Interface	SSTL_2	LVTTL
Supply Voltage	2.5 V	3.3 V
/CAS Read Latency	2, 2.5	2, 3
/CAS Write Latency	1	0
Burst Length	2, 4, 8	1, 2, 4, 8, Full-Page (256)[b]
Burst Sequence	Sequential/Interleave	Sequential/Interleave
Use of DLL	Essential	Optional
Data Mask	Write Mask Only	Write Mask/Read Mask

a. t_{CK} is clock cycle time.

b. Full-page (256) burst of SDR SDRAM is an option.

[Reprinted with permission of Elpida Memory.]

2. JESD79C is dated March 2003 and is a revision of JESD79B.

Data Transfer Frequency, Data Rate DDR SDRAM achieves a data transfer rate that is twice the clock frequency by employing 2-bit prefetch architecture. Using the read cycle as an example, 2n bits of data are transferred from the memory cell array to the I/O buffer every clock (CK) as shown in Figure 14.1. Data transferred to the I/O buffer is output n bits at a time every half clock (both rising and falling edges of the clock). As the internal bus width is twice the external bus width, DDR SDRAM achieves a data output rate that is twice the data rate of the internal bus. Because data is accessed in 2-bit pairs, only burst lengths of 2, 4, and 8 are supported for DDR SDRAM.

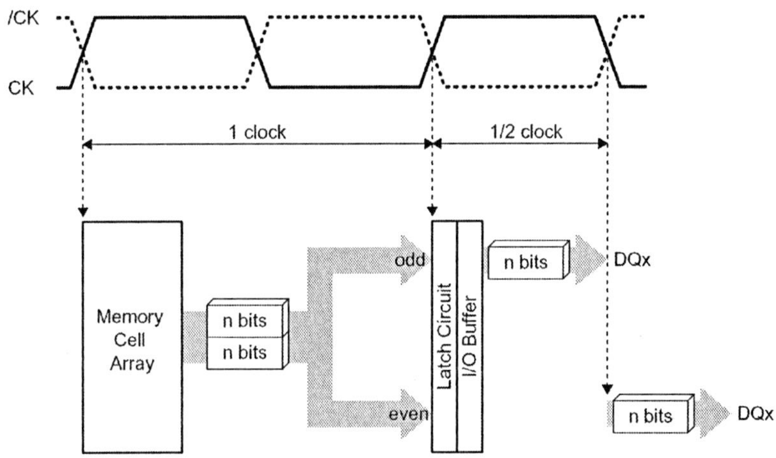

Figure 14.1 2-bit prefetch architecture. [Reprinted with permission of Elpida Memory, Inc.]

Clock Input Due to the influence of various factors, the high-level period and low-level period of the clock input may not be the same (the duty ratio may not be 50%) as shown in Figure 14.2.

In the case of SDR SDRAM, for which data input/output is only synchronized with the rising edges of the clock, there is some margin in the timing. However, in the case of DDR SDRAM, for which data input/output is synchronized with both the rising and falling edges of the clock, it is difficult to accurately control the data input/output timing based on the conventional single clock. Therefore, DDR SDRAM adopts a differential clock scheme that enables accurate memory control. Employing a differential clock scheme enables DDR SDRAM to support a higher clock frequency and limit the negative influence of noise and other factors.

/CK is an input signal that has the same clock period but the reverse phase of CK. The high-level period and low-level period can be made equal by using the intersection of CK and /CK as an input reference level. In the case of DDR SDRAM, data input/output is synchronized with both the rising and falling edges of the data strobe signal, which has the same period as clock input CK.

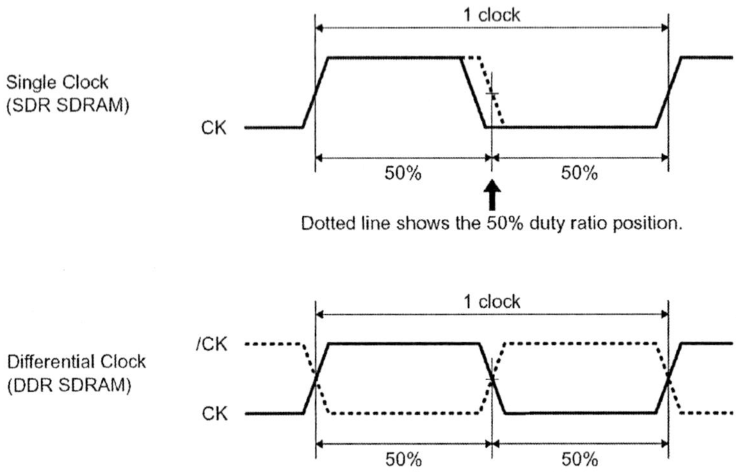

Figure 14.2 Clock input. [Reprinted with permission of Elpida Memory, Inc.]

Data Strobe Signal (DQS) Similarly to SDR SDRAM, DDR SDRAM is controlled by command input at the rising edge of the clock (CK), but the data input/output timing differs from that of SDR SDRAM. To achieve high-speed data transfer, DDR SDRAM adopts a data strobe signal (DQS). DQS is output from the device and received by the receiver, which adjusts the data (DQ) capture timing using DQS.

Since DDR SDRAM performs data input/output at twice the frequency of the external clock, the valid data window is narrower than for SDR SDRAM. If the wiring length between the memory and the controller is different, the time required for data to reach the receiver (flight time) is different. This makes it difficult for the receiver to determine the data acceptance timing.

DDR SDRAM employs a DQS to notify the receiver of the data transfer timing. DQS is a bidirectional strobe signal and functions as the basic operating clock for DQ during read/write operations.

Data is edge-aligned to DQS for read data and center-aligned for write data as shown in Figure 14.3. This means that when controller receives read data from DDR SDRAM, it will internally delay the received strobe to the center of the received data window.

Interface DDR SDRAM employs the JEDEC-compliant SSTL_2 (stub-series terminated logic for 2.5 V) interface to eliminate the signal degradation caused by noise and reflection produced as a result of a high operating frequency. SSTL_2 is a low-voltage (2.5 V), small-amplitude and high-speed interface that reduces the effect of reflection by connecting series resistance between the signal branch point from the bus (stub) and the memory.

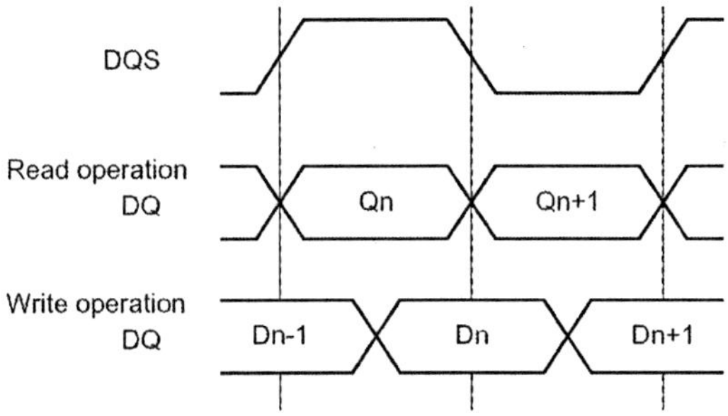

Figure 14.3 Relationship between data strobe signal and data input/output. [Reprinted with permission of Elpida Memory, Inc.]

1. **Stub Resistance** A stub resistance of approximately 25 Ω (22 Ω is generally used for DIMMs (Dual In-Line Memory Modules)) is connected in series to the output pin (V_{OUT}), providing impedance matching between the transmission line and device output.
2. **Termination Voltage** The V_{TT} (termination voltage) line is terminated with a resistance of approximately 25 Ω (27 Ω is generally used for DIMMs). This termination suppresses signal reflection in the transmission line and also reduces voltage spikes, enabling high-speed data transmission.
3. **Reference Voltage** The SSTL_2 interface is symmetrical with respect to high-level and low-level output. V_{REF} is used as a reference voltage to detect high and low levels.
4. **Interface Specification** DDR SDRAM generally uses the SSTL_2 interface with termination at one end as shown in Figure 14.4. SSTL_2 interface specifications are listed in Table 14.2.

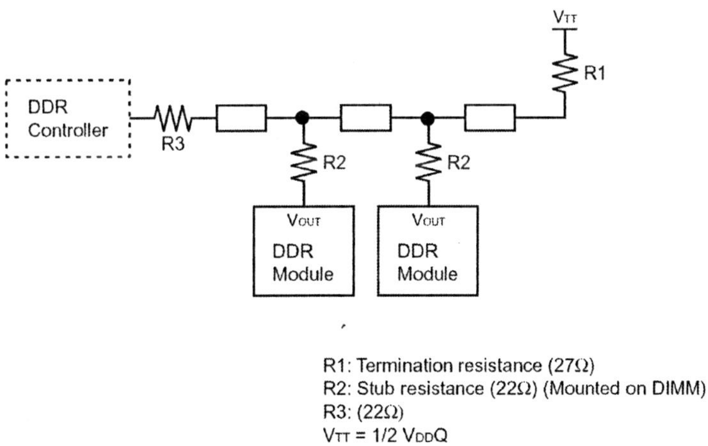

Figure 14.4 SSTL_2 interface (case of DIMM). [Reprinted with permission of Elpida Memory, Inc.]

Table 14.2 SSTL_2 Interface Specifications[a]

Item	Symbol	Minimum	Typical	Maximum	Unit
Power Supply	V_{DD}	V_{DDQ}		n/a	V
DQ Power Supply	V_{DDQ}	2.3	2.5	2.7	V
SSTL_2 Reference Voltage	V_{REF}	1.13	1.25	1.38	V
Termination Voltage	V_{TT}	$V_{REF} - 0.04$	V_{REF}	$V_{REF} + 0.04$	V
High-Level Input Voltage (dc Specifications)	V_{IH} (dc)	$V_{REF} + 0.15$		$V_{DDQ} + 0.30$	V
Low-Level Input Voltage (dc Specifications)	V_{IL} (dc)	–0.30		$V_{REF} - 0.15$	V
High-Level Input Voltage (ac Specifications)	V_{IH} (ac)	$V_{REF} + 0.31$			V
Low-Level Input Voltage (ac Specifications)	V_{IL} (ac)			$V_{REF} - 0.31$	V

a. These specifications may differ depending on the product. For details, refer to the corresponding data sheet of each DDR SDRAM.

[Reprinted with permission of Elpida Memory.]

Power Supply Compared to the 3.3 V power supply of SDR SDRAM, the power supply of DDR SDRAM is 2.5 V. This reduction in power supply voltage reduces the power consumption of the DDR SDRAM circuits.

/CAS Read Latency, /CAS Write Latency, Burst Length, and Burst Sequence Similarly to SDR SDRAM, DDR SDRAM has mode register set commands for the latency, burst length, and burst sequence.

- Programming the mode register — The mode register is used to define various operating parameters of DDR SDRAM, such as the latency mode, burst sequence (wrap type (WT)) and burst length. The extended mode register is used to enable/disable DLL.

  The addresses (A0 through Ax) and the bank addresses (BA0, BA1) are used as input data to set the mode register and the extended mode register. Once the parameters are set, these registers retain the stored information until they are reprogrammed or the power is turned off. To program the mode register, the following steps are taken (see Figure 14.5):

  1. Execute the precharge all banks command (PALL) to precharge all banks. After a time delay of t_{RP} specified in the data sheet, all banks become idle.
  2. Execute the mode register set command (MRS) and extended mode register set command (EMRS) to program the register.

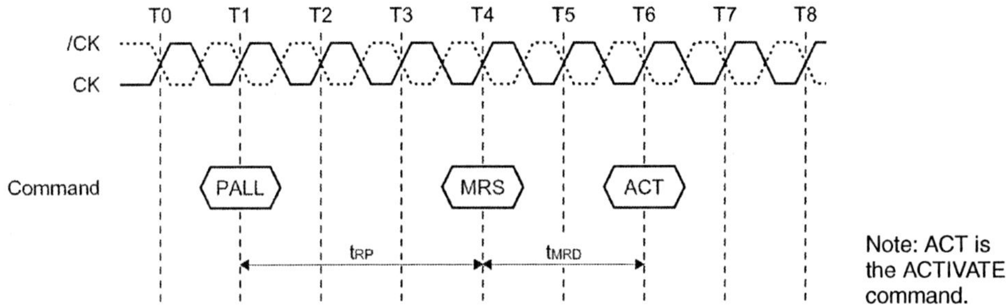

Figure 14.5 Mode register/extended mode register set cycle. [Reprinted with permission of Elpida Memory, Inc.]

- Parameters — The mode register has the following five functions:

1. A0 through A2: Burst length
2. A3: Burst sequence
3. A4 to A6: /CAS latency
4. A7and A9 through Ax: Option
5. A8: DLL reset

The extended mode register has the following two functions:

6. A0: DLL enable/disable
7. A1 through A11: Option

Use of DLL DDR SDRAM is provided with a DLL (Delay-Locked Loop) circuit. The DLL circuit is designed to realize a fast access time and high operation frequencies by controlling and adjusting the time lag between the external clock and internal clock. Refer to the block diagram in Figure 14.6.

Data Mask DDR SDRAM uses a DQ write mask enable signal (DM) which masks write data. SDR SDRAM can mask both read and write data, but the read mask is not supported by DDR SDRAM." [Elp02a]

14.1.1.2 Differences in Commands

"Similarly to SDR SDRAM, DDR SDRAM is controlled by commands, but the commands supported by DDR SDRAM differ from those supported by SDR SDRAM (see Table 14.3).

Clock Suspend In the case of SDR SDRAM, operation can be suspended by making the clock enable signal (CKE) low during a read or write operation, but in the case of DDR SDRAM, clock suspend is not supported.

Full-Page Burst The burst lengths of DDR SDRAM are 2, 4, and 8. Full-page burst is not supported.

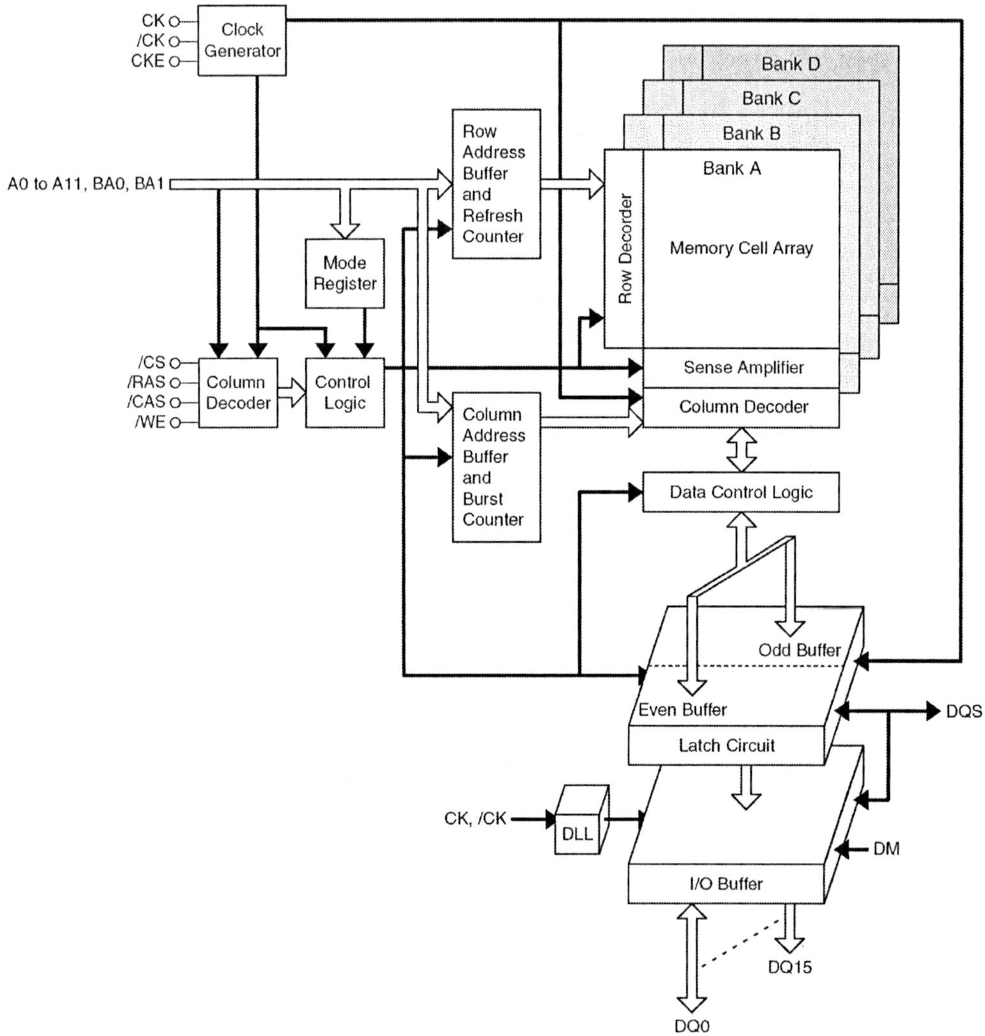

Figure 14.6 Block diagram of 128 Mb DDR SDRAM (EDD1216ALTA). [Reprinted with permission of Elpida Memory, Inc.]

Table 14.3 Differences in Commands Between DDR and SDR SDRAM

Item	DDR SDRAM	SDR SDRAM
Clock Suspend	Invalid	Valid
Full-Page Burst	Invalid	Valid
Burst Stop	Valid Only for Read Operation	Valid
Single Write After Burst Read	Invalid	Valid

[Reprinted with permission of Elpida Memory.]

Burst Stop In the case of DDR SDRAM, data output is suspended by the burst stop command during a burst read operation, but the burst stop command is not available for a write operation. Instead write data can be masked using the DQ write mask enable signal (DM) during a burst write operation, but read data cannot be masked during a burst read operation.

Single Write After Burst Read DDR SDRAM does not support single write after burst read. This is because, whereas command input is synchronized only with the rising edge of the clock, data input/output is synchronized with both the rising and falling edges and thus the operation frequency is twice as high." [Elp02a]

14.1.1.3 Differences in Operation Timing

"Similarly to SDR SDRAM, DDR SDRAM is controlled by inputting commands at the rising edge of the clock (CK). However, the data input/output timing of DDR SDRAM differs from that of SDR SDRAM. DDR SDRAM employs a differential clock (CK, /CK) and data strobe signal (DQS) to realize high-speed data transfer. DQS is synchronized with CK, and data input/output (DQ) is synchronized with both the rising and falling edges of DQS.

The examples in Figures 14.7 through 14.10 show the relationship between the clock input of DDR SDRAM/SDR SDRAM, their control signals (commands), and the data input/output timing." [Elp02a]

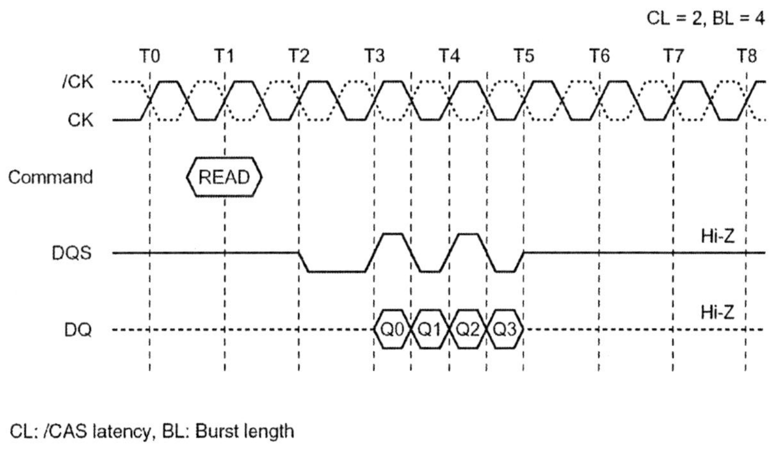

Figure 14.7 DDR SDRAM read cycle timing. [Reprinted with permission of Elpida Memory, Inc.]

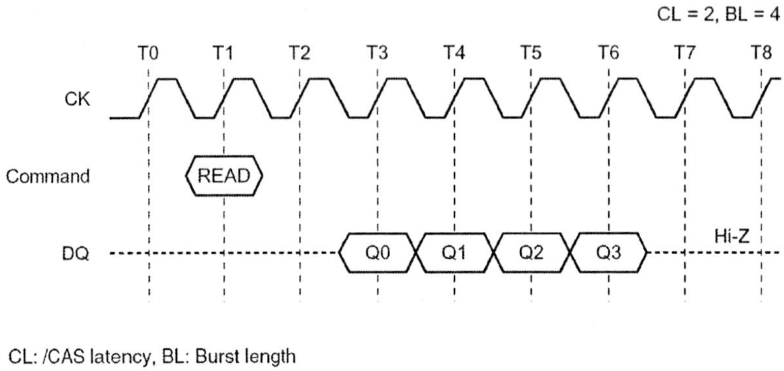

Figure 14.8 SDR SDRAM read cycle timing. [Reprinted with permission of Elpida Memory, Inc.]

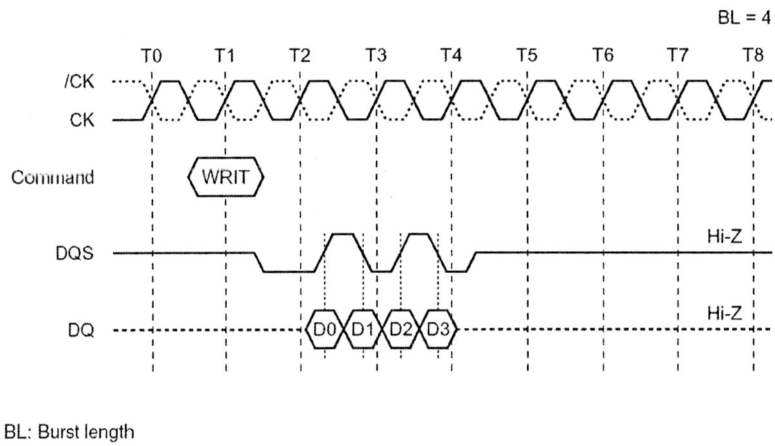

Figure 14.9 DDR SDRAM write cycle timing. [Reprinted with permission of Elpida Memory, Inc.]

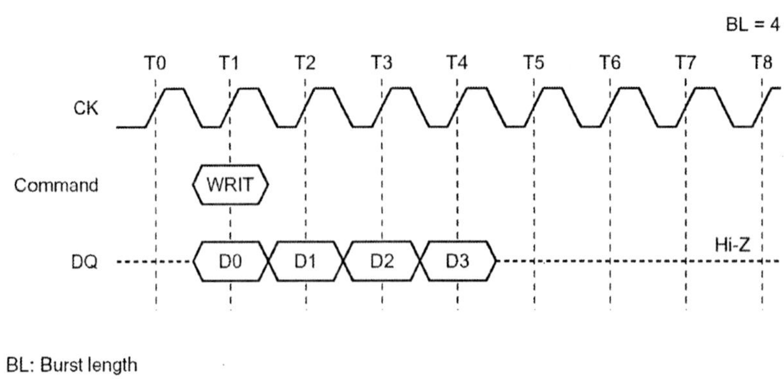

Figure 14.10 SDR SDRAM write cycle timing. [Reprinted with permission of Elpida Memory, Inc.]

14.2 DDR2 (Double Data Rate 2, DDR-II) SDRAM[3]

DDR2 SDRAM introduces features and functions that go beyond the DDR SDRAM specification and enable DDR2 to operate at data rates of 400 Mbps, 533 Mbps, and 667 Mbps with the potential of extending to 800 Mbps and above. The advancements of DDR2 are largely the result of changes in DRAM architecture and signaling, as well as additions to the mode register for lower power and improved command/data bandwidth. This section discusses the various changes and the resulting features and benefits of DDR2 technology.

DRAM architecture changes enable twice the bandwidth without increasing the demand on the DRAM core and while keeping power low. A summary of the functionality changes is shown in Table 14.4.

Table 14.4 Comparison of DDR and DDR2 SDRAM Features and Options

Feature/Option	DDR	DDR2
Data Transfer Rate	200, 266, 333, 400 Mbps	400, 533, 667, 800 Mbps
Clock Frequency	100, 133, 166, 200 MHz	200, 266, 333, 400 MHz
Prefetch Size	2-bit	4-bit
Burst Length	2 / 4 / 8	4 / 8
Data Strobe	Single Data Strobe	Differential Data Strobe
CAS Latency (CL)	2, 2.5, 3 Clocks	3, 4, 5 Clocks
Additive Latency (AL)	No	0, 1, 2, 3, 4 Clocks
Read Latency	CL	AL + CL
Write Latency	Fixed	Read Latency – 1 Clock
I/O Width	x4 / x8 / x16	x4 / x8 / x16
Data Strobes	Bidirectional Strobe (Single-Ended)	Bidirectional Strobe (Single-Ended or Differential) with RDQS
I/O Interface	SSTL_2	SSTL_18
Operating Voltage	2.5 V	1.8 V
I/O Voltage	2.5 V	1.8 V
Densities	64 Mb to 1 Gb	256 Mb to 4 Gb
Internal Banks	4	4 and 8
Power Consumption (400 Mbps)	IDD1: 527 mW (max.)	IDD1: 247 mW (max.)
Package	TSOP(II) and FBGA	FBGA only (lead free (Su-Ag-Cu))
Output Calibration	None	OCD
On-Die Termination	None	Selectable

3. The text, tables, and figures in Section 14.2 were authored by Jeff Janzen of Micron Technologies, Inc., except where indicated. Used with permission by Micron Technology, Inc. © 2004 Micron Technology, Inc., All rights reserved. Source is reference [Mic03c].

Table 14.4 Comparison of DDR and DDR2 SDRAM Features and Options (Continued)

Feature/Option	DDR	DDR2
Command Set		Same as DDR
Basic Device Timing		
New Functions		ODT, OCD Calibration, Posted CAS, AL[a]

a. "ODT is on-die termination, the termination function on DRAM chips for the DQ lines. OCD is off-chip driver, the strength control function of a pull-up/pull-down driver. Posted CAS is the method of improving bus utilization. Allowing RAS/CAS to be issued back to back before t_{RCD} eliminates command bus collisions. AL is additive latency. AL is the internal CAS execution delay relative to external CAS. t_{RCD} = (RAS–CAS command interval) + AL." [Elp03a]

Modifications to the DRAM architecture include shortened row lengths for reduced activation power, burst lengths of four and eight for improved data bandwidth capability, and the addition of eight banks in 1 Gb chip densities and above. New signaling features include ODT (on-die termination) and OCD (off-chip driver). ODT provides improved signal quality owing to better system termination on the data signals, and OCD calibration provides the option of tightening the variance of the pull-up and pull-down output driver at 18 Ω nominal. Modifications were also made to the mode register and extended mode register, including CAS latency, additive latency, and programmable data strobes.

Double Data Rate 2 (DDR2) is specified by JEDEC standard JESD79-2, "DDR2 SDRAM Specification." While this specification consistently refers to this technology as DDR2 throughout the specification, industry and chip manufacturers tend to use the notation "DDR-II" for this technology.

14.2.1 DRAM Architecture Changes

14.2.1.1 DDR2 Array Definition

Changes to the DRAM architecture include shortened page size for reduced activation power, burst lengths of four and eight for improved data bandwidth capability, and the addition of eight banks in 1 Gb densities and above.

14.2.1.2 DDR2 Page Size

Page size is important to the power consumption of a DRAM device. Page size is defined as the minimum number of column locations that are accessed with a single ACTIVATE command. This is calculated as the number of column locations times the number of DQs on the DRAM. For example, 512 Mb, x8 DDR2 SDRAM has 1,024 column locations, so the page size is 1,024 columns times 8 DQs, which equals 8,192 bits. Dividing the 8,192 bits by a word length of 8 equals 1,024 bytes or 1 KB.

Each time an ACTIVATE command is given, all bits in the page are read by the sense amplifiers and restored to the correct value. Because this process is a major contributor to the active power, a device with a shorter page size has a significantly lower operating current. The page size of the 512 Mb DDR2 device is half that of a standard DDR SDRAM. Therefore, the current consumed by the ACTIVATE commands is greatly reduced, resulting in improved IDD0, IDD1, and IDD7 specifications. (See Table 14.5 for a comparison of DDR and DDR2 page size and number of banks.)

Table 14.5 DRAM Architectures

	Density	DDR	DDR2
256 Mb	Page Size	1 KB	1 KB
	Banks	4	4
512 Mb	Page Size	2 KB	1 KB (x4, x8), 2 KB (x16)
	Banks	4	4
1 Gb	Page Size	2 KB	1 KB (x4, x8), 2 KB (x16)
	Banks	4	8
2 Gb	Page Size	–	1 KB (x4, x8), 2 KB (x16)
	Banks	–	8

With the higher-density DDR2 SDRAM (512 Mb, 1 Gb, and 2 Gb), the page size increases on the x16 device. The x4 and x8 devices continue to operate at lower activate currents with a 1 KB page size, similar to the 256 Mb DDR2 SDRAM. However, the x16 DDR2 page size increases to 2 KB. While power consumption increases on the x16 DDR2 SDRAM, system power is typically no higher than that of the x8 device. A 64-bit bus requires eight x8 DRAM devices to support the bus width, while only four x16 DRAM devices are required. Thus, the activate current of a x16 component may be somewhat higher, but the overall system power consumption is still lower.

14.2.1.3 DDR2 Bank Accesses

Another change from DDR to DDR2 (as shown in Table 14.5) is the number of banks on the larger-density DDR2 devices. A typical four-bank access, which is applicable to the 256 Mb and 512 Mb DDR2 SDRAM, is shown in Figure 14.11. Bank accesses for DDR2 devices are similar to DDR, except ACTIVATE command spacing is 7.5 ns for 1 KB and 10 ns for 2 KB page size, due to the increased power consumption of the larger page size. When a bank is precharged, it may be activated again after t_{RC} is met, following the original ACTIVATE command.

The larger-density 1 Gb and 2 Gb DDR2 devices use eight banks, which increases flexibility in DRAM accesses. However, it also increases the power that must be delivered to the DRAM. To keep the power delivery from adversely affecting the functionality of the DRAM and the memory subsystem, an additional restriction (t_{FAW}) is applied to bank ACTIVATE commands, as shown in Figure 14.12. ACTIVATE-to-ACTIVATE spacing remains t_{RRD}, as in the four-bank case; however, there is an additional restriction within any t_{FAW} window where no more than four banks may be activated. This requires a short amount of extra time prior to activating the fifth bank.

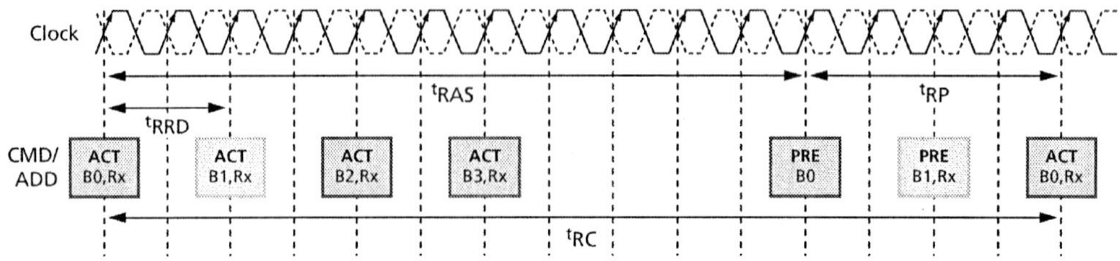

Figure 14.11 ACTIVATE for four-bank DDR or DDR2 devices. [Used with permission by Micron Technology, Inc. © 2004 Micron Technology, Inc. All rights reserved.]

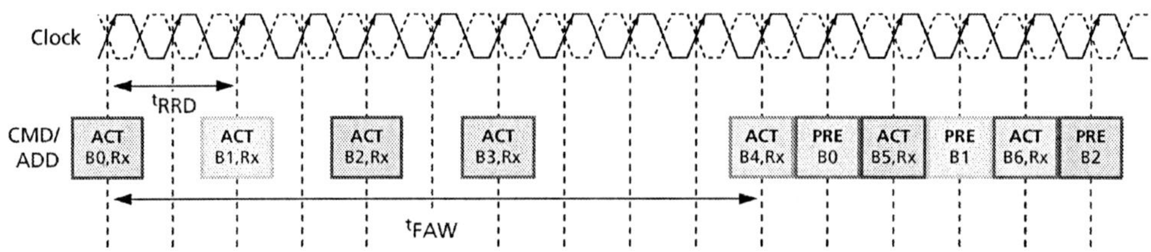

Note: *tRRD is 7.5ns on 1KB page size and 10ns on 2KB page size (two clock minimum).*
tFAW is 37.5ns on 1KB page size and 50ns on 2KB page size.

Figure 14.12 ACTIVATE for eight-bank DDR or DDR2 devices. [Used with permission by Micron Technology, Inc. © 2004 Micron Technology, Inc. All rights reserved.]

14.2.2 On-Die Termination

On-die termination (ODT) has been added to the DDR2 data signals to improve signal integrity in the system. Figure 14.13 shows a functional diagram of the on-die termination. The termination value of R_{TT} is the Thevenin equivalent of the resistors that terminate the DQ inputs to VssQ and $V_{DD}Q$. An ODT pin is added to the DRAM so the system can turn the termination on and off as needed. In a simple system with one DRAM load per DQ signal, the termination is turned on for WRITEs and disabled for READs. Figure 14.14 shows the ODT trace routing on a typical PC system with two memory slots.

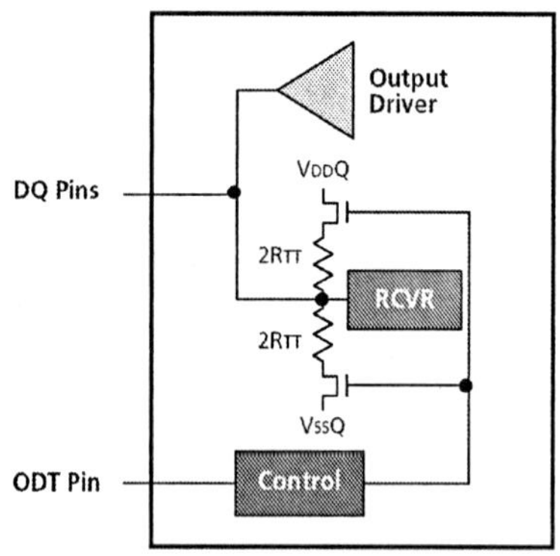

Figure 14.13 Functional diagram for on-die termination. [Used with permission by Micron Technology, Inc. © 2004 Micron Technology, Inc. All rights reserved.]

Each slot contains two ranks of memory (front and back). A separate ODT signal is routed to each rank of memory to provide flexibility in power management and the best system performance. The routing of ODT signals is similar to the routing of CS# signals.

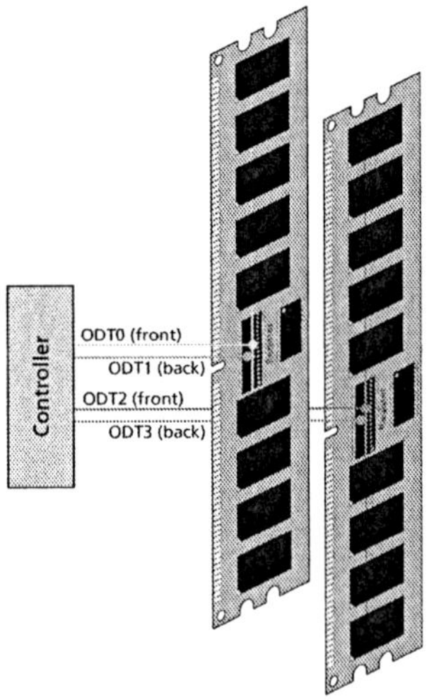

Figure 14.14 Typical 2-slot ODT routing. [Used with permission by Micron Technology, Inc. © 2004 Micron Technology, Inc. All rights reserved.]

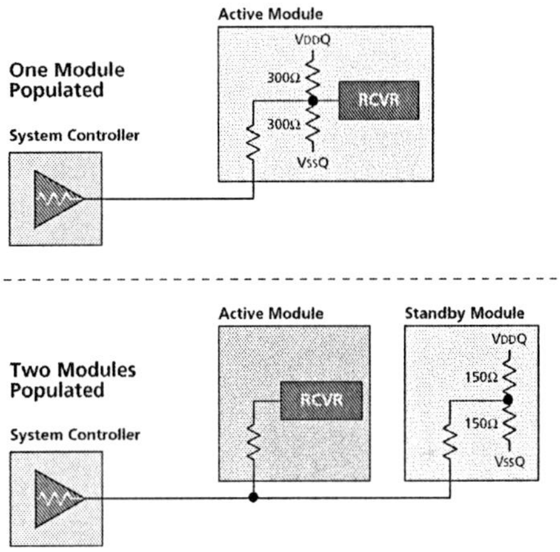

Figure 14.15 Typical ODT configuration for SDRAM writes. [Used with permission by Micron Technology, Inc. © 2004 Micron Technology, Inc. All rights reserved.]

14.2.2.1 ODT WRITEs

Figure 14.15 above shows a typical termination scheme for write data in a two-module system. If only one module is loaded in the system, a 150 Ω equivalent termination (300 Ω pull-up/300 Ω pull-down) must be enabled on the module during the WRITE. If two modules are loaded into the system, the termination on the module not receiving the write data should be enabled at 75 Ω equivalent termination resistance (150 Ω pull-up/150 Ω pull-down). This termination scheme provides optimal trade-offs between signal integrity and voltage swing.

The equivalent comparison for write data at 533 MT/s (mega transfers per second) is shown in Figure 14.16. Neither bus has significant overshoot. However, the voltage margin on the SSTL bus is not sufficiently outside the DRAM input voltage requirements. Thus, when the input slew rate starts slowing down in the two-slot system, the data eye that the DRAM must capture is significantly reduced. The ODT system shows increased voltage margin, and the data eye remains significantly larger, even under more heavily loaded conditions.

14.2.2.2 ODT READs

Figure 14.17 shows a typical termination scheme for read data in a two-module system. For READs, on-die termination must be included on the controller. If only one module is loaded in the system, a 75 Ω termination should be enabled on the system controller. If two modules are loaded, the controller termination should be increased to 150 Ω and the unused module should terminate at 75 Ω. This configuration provides good voltage margin at the controller. Figure 14.18 shows a comparison of a DDR2 DQ bus for a typical SSTL-terminated, two-module system using ODT termination versus a non-ODT system reading data at 533 MT/s.

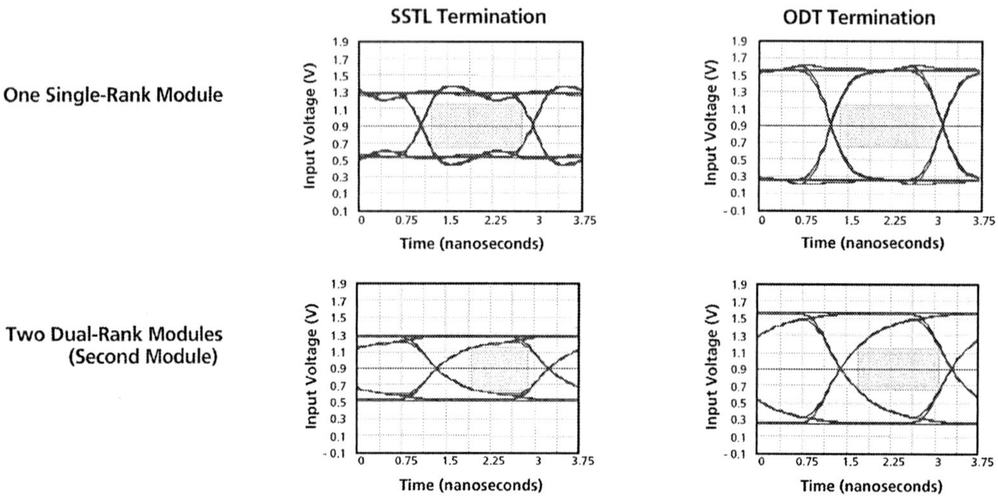

Figure 14.16 SSTL vs. ODT termination comparison for WRITEs. [Used with permission by Micron Technology, Inc. © 2004 Micron Technology, Inc. All rights reserved.]

The most obvious difference between the two buses is the reduction in the overshoot on the data bus. The overshoot reduction significantly improves the voltage margin and reduces inter-symbol interference (ISI). The second major improvement using DDR2 ODT is the increase in the slew rate in the two-module case. The SSTL bus starts "laying over" when more loads are added, which detracts from the timing margin in the system. Both of these improvements dramatically increase the size of the data eye that the system controller must capture to read data from the DRAM.

Overall, ODT implementation provides improved signal integrity, data eyes, and voltage margin for DDR2 to enable systems to attain higher data rates. At the same time, it is an option and can be disabled for systems that require a more traditional SSTL type of bus.

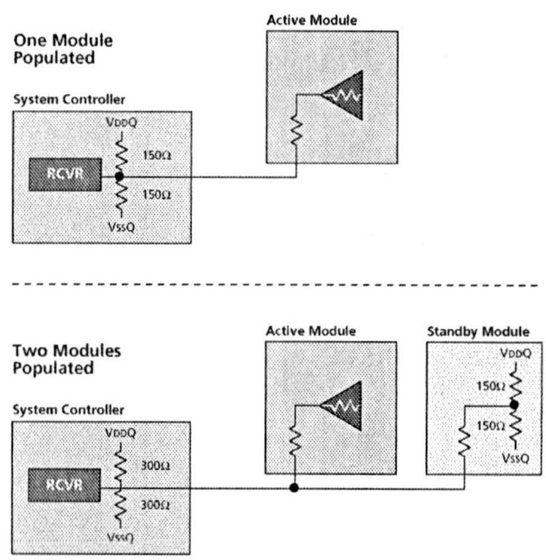

Figure 14.17 Typical ODT configuration for DRAM READs. [Used with permission by Micron Technology, Inc. © 2004 Micron Technology, Inc. All rights reserved.]

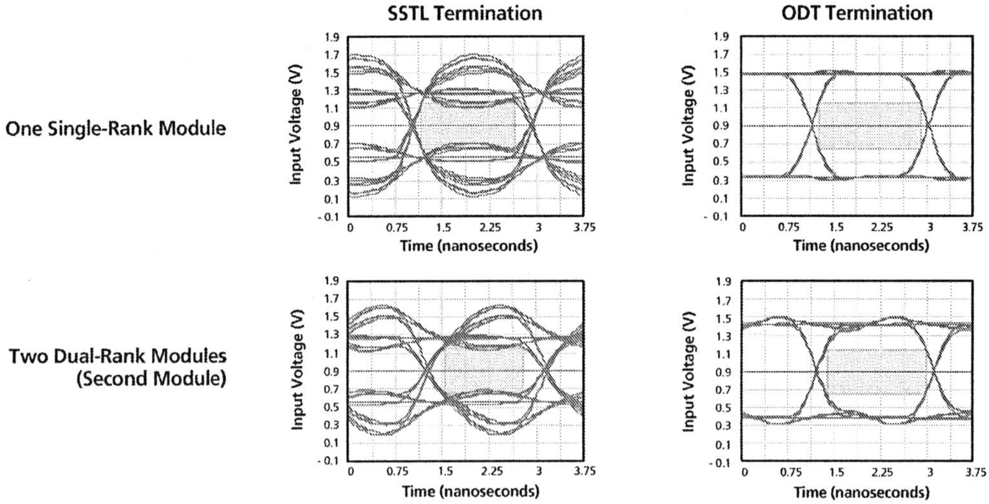

Figure 14.18 SSTL vs. ODT termination comparison for READs. [Used with permission by Micron Technology, Inc. © 2004 Micron Technology, Inc. All rights reserved.]

14.2.2.3 Termination Values

As illustrated in Figures 14.15 and 14.17, on-die termination can have different values, depending on the system configuration. This is controlled via the extended mode register (EMR). (See Figure 14.19.) If extended mode bits 2 and 6 are both "0," the on-die termination is disabled on the DRAM. Two additional options include selecting either an R_{TT} equivalent value of 75 Ω

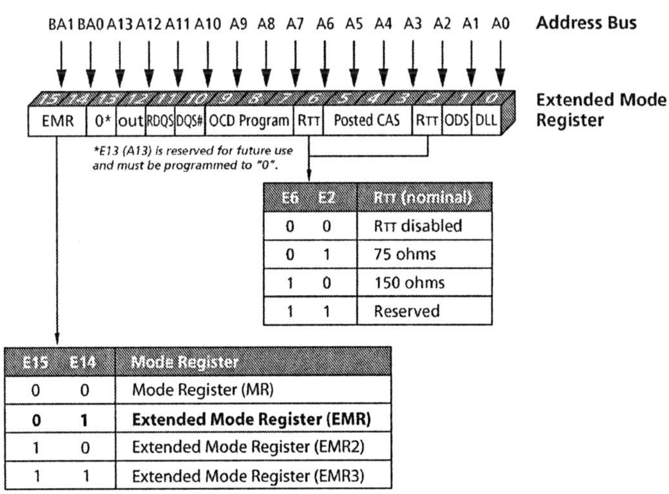

Figure 14.19 Extended mode register control for on-die termination. [Used with permission by Micron Technology, Inc. © 2004 Micron Technology, Inc. All rights reserved.]

(150 Ω pull-up and 150 Ω pull-down) or an R_{TT} equivalent of 150 Ω (300 Ω pull-up and 300 Ω pull-down). The proper termination value is selected to optimize signal integrity at minimum power, as shown in Figures 14.16 and 14.18.

14.2.2.4 On-Die Termination Control in Active Mode

After the proper value of termination is selected, the system controller must switch the termination on and off at the proper times. While the DRAM is in active or fast exit power-down mode, the termination turns on two clock cycles after the ODT signal is registered high, as shown in Figure 14.20. For module 2, the ODT signal is latched high on the same clock cycle as the READ command to module 1. Two clock cycles later, the on-die termination on the module 2 DRAM turns on. To schedule on-die termination, ODT should be transitioned high three clocks before the data is on the bus to ensure it is on prior to the preamble.

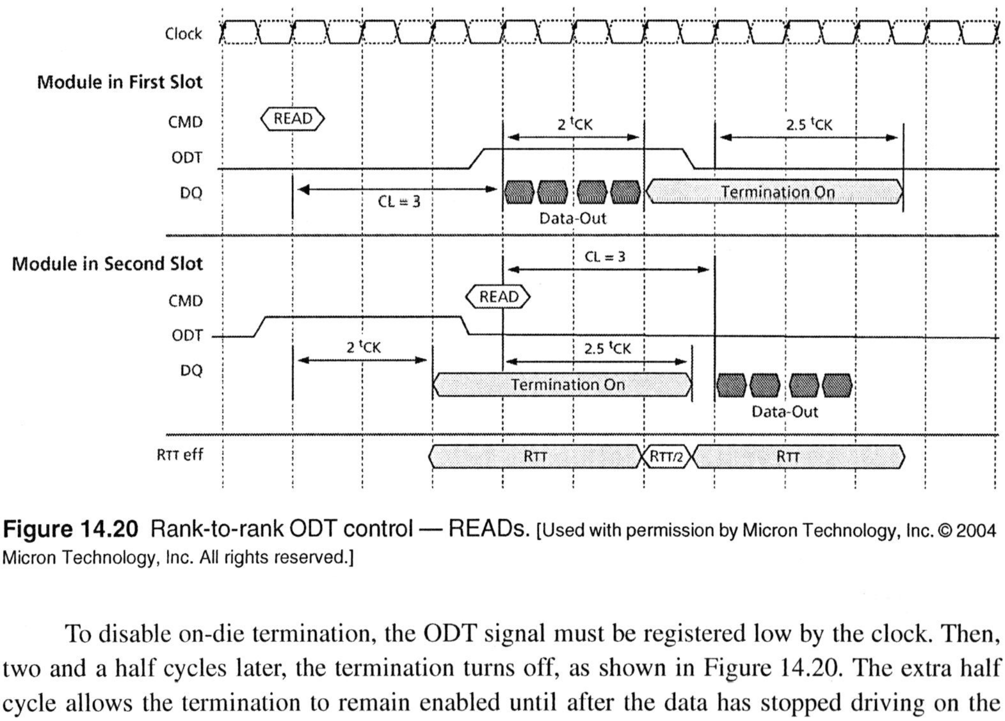

Figure 14.20 Rank-to-rank ODT control — READs. [Used with permission by Micron Technology, Inc. © 2004 Micron Technology, Inc. All rights reserved.]

To disable on-die termination, the ODT signal must be registered low by the clock. Then, two and a half cycles later, the termination turns off, as shown in Figure 14.20. The extra half cycle allows the termination to remain enabled until after the data has stopped driving on the bus. Thus, the ODT pin can be registered low two clock cycles prior to the end of the data burst on the bus.

14.2.2.5 On-Die Termination Control in Power-Down

Asynchronous operations are also associated with the ODT pin when the DRAM is in slow-exit active power-down or precharge power-down mode. During this operation, the turn on and off time is an analog delay off the registration of the ODT signal. The timing isn't as critical since data can't be read from or written to a rank that is in power-down mode.

Thus, the system only has to enable the termination to form a window around the data from the other ranks. See DDR2-specific device data sheets for additional information on specific ODT timings and restrictions.

14.2.2.6 On-Die Termination Control in Self-Refresh

During self-refresh, the DRAM disables on-die termination and ignores the ODT pin, regardless of the values programmed into the EMR. A memory subsystem that is in self-refresh mode cannot read or write data, so disabling on-die termination does not affect signal quality, and it allows the DRAM to minimize power consumption.

14.2.2.7 Off-Chip Driver (OCD) Calibration

DDR2 also contains an off-chip driver (OCD) calibration to reduce the mismatch between pull-up and pull-down output driver characteristics, which can improve the system timing margin. Without OCD calibration, the DRAM has a nominal output driver strength of 18 Ω +30% and a pull-up and pull-down mismatch of up to 4 Ω. Using OCD calibration, a system can reduce the pull-up and pull-down mismatch and target the output driver at 18 Ω to optimize the signal integrity. OCD calibration is performed by the LOAD MODE command using the EMR, which enables the DRAM to be in OCD drive mode with the outputs driving high or low, or the DRAM can be in OCD adjust mode. A system that is not using OCD calibration can enable a default output drive strength of 18 Ω + 30% by selecting the OCD default setting (see Figure 14.21).

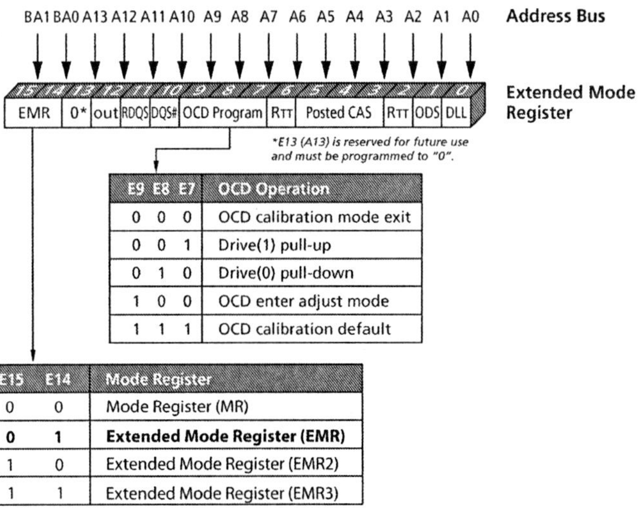

E9	E8	E7	OCD Operation
0	0	0	OCD calibration mode exit
0	0	1	Drive(1) pull-up
0	1	0	Drive(0) pull-down
1	0	0	OCD enter adjust mode
1	1	1	OCD calibration default

E15	E14	Mode Register
0	0	Mode Register (MR)
0	**1**	**Extended Mode Register (EMR)**
1	0	Extended Mode Register (EMR2)
1	1	Extended Mode Register (EMR3)

Figure 14.21 Extended mode register control for OCD calibration. [Used with permission by Micron Technology, Inc. © 2004 Micron Technology, Inc. All rights reserved.]

For a system to utilize OCD calibration, a LOAD MODE command must be issued to the DRAM to enter drive mode by driving a high or a low, as shown in Figure 14.22. When OCD drive mode is entered, the outputs drive the desired high or low against the system termination until an OCD exit mode command is issued.

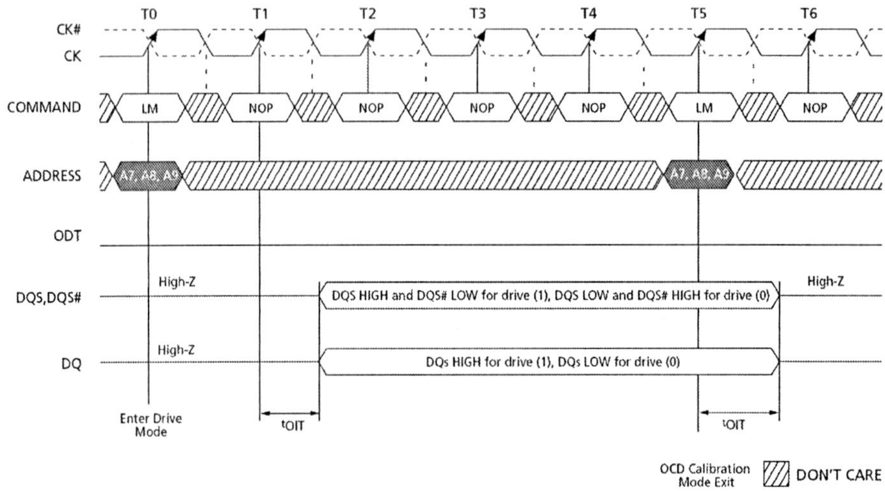

Figure 14.22 OCD drive mode. [Used with permission by Micron Technology, Inc. © 2004 Micron Technology, Inc. All rights reserved.]

While the DRAM is in OCD drive mode, the system controller must measure the voltage it receives against a known reference. If the voltage and termination values of a system are known, the DRAM output drive strength for pull-up and pull-down can be calculated separately. If the pull-up and/or pull-down drive strength is not at 18 Ω, the system controller can enter OCD adjust mode and increase or decrease the drive strength of the DRAM, using the sequence shown in Figure 14.23.

During OCD adjust mode, the memory controller issues a WRITE with a burst length of four to DQ0–DQ3 to adjust the pull-up or pull-down. To maintain commonality across different DRAM widths (x4, x8, x16), the adjust data is always entered on DQ0–DQ3. Refer to the DDR2 data sheet for the specific device for details on the available adjust mode op-codes.

Using OCD calibration, it is possible for the DRAM to set the output drive strength at 18 Ω ± 1.5 Ω. However, in actual systems, the system error of the reference voltages, termination, and comparators must be added to this variation, which will decrease the output drive accuracy in most systems. The OCD calibration is only intended to help match the output driver pull-up and pull-down to 18 Ω at nominal temperature and voltage conditions. Using this feature to move the drive strength higher or lower than 18 Ω is not supported. OCD is only supported for the full-strength output driver option and is not supported in half-strength mode. More specific details regarding OCD adjust mode can be found in DDR2 device data sheets.

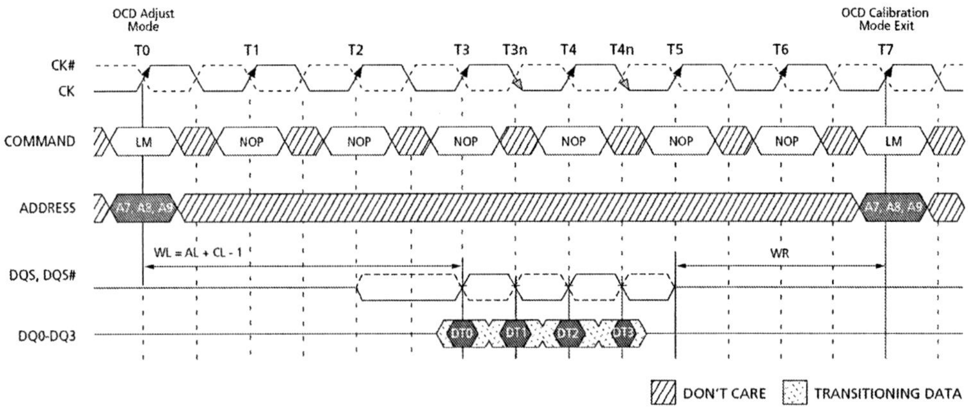

Figure 14.23 OCD adjust mode. [Used with permission by Micron Technology, Inc. © 2004 Micron Technology, Inc. All rights reserved.]

14.2.3 Additional Mode Register Changes

14.2.3.1 Mode Register (MR)

Some of the additional features and functionality that have been added to DDR2 require changes in the mode register and the extended mode register. Figure 14.24 highlights the changes made in the mode register for DDR2. These include support for burst lengths of 4 and 8 only, a modification of the sequential interleave mode, and extension of the CAS latencies supported. In addition, a low-power definition for active power-down mode and a write recovery parameter have been added.

14.2.3.2 CAS Latency

DDR2 operates at higher clock frequencies than DDR SDRAM. Figure 14.25 illustrates the difference between a DDR READ at a 266 MHz data rate and a DDR2 READ at a 533 MHz data rate. The latency of each device is the same, approximately 15 ns. For the DDR SDRAM, this is equivalent to two clocks; for DDR2, four clocks.

As clock rates increase, the number of latencies supported must also increase. Table 14.6 shows the latencies supported for various clock rates and CAS latencies. The DRAM latency is calculated as t_{CK} * CAS latency supported. DDR2 is targeted at latencies of 12–20 ns. If the DRAM latency is beyond this range, it is unlikely that a system will utilize that latency due to lost system performance. Likewise, it is unlikely the DRAM will meet latencies below 12 ns due to DRAM core performance. To support DDR2's higher clock rates, the CAS latency programmability has been set at three to five clocks. Half-clock latencies are not supported.

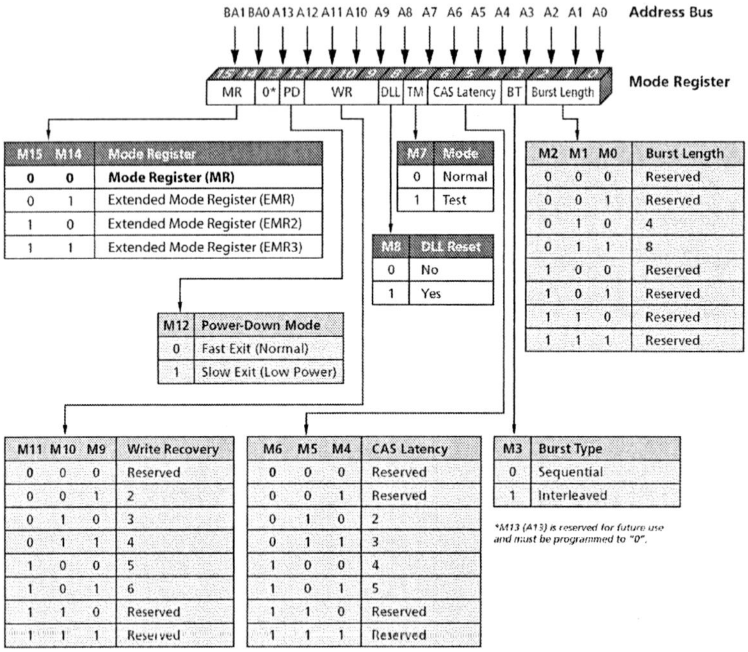

Figure 14.24 Mode register changes for DDR2. [Used with permission by Micron Technology, Inc. © 2004 Micron Technology, Inc. All rights reserved.]

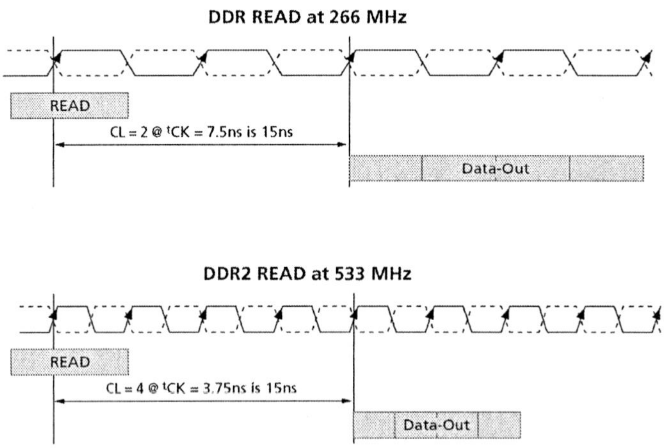

Figure 14.25 DDR vs. DDR2 READ. [Used with permission by Micron Technology, Inc. © 2004 Micron Technology, Inc. All rights reserved.]

Table 14.6 DDR2 READ Latencies

		READ Latency (Clocks)		
	^tCK \ RL	3	4	5
DDR2-400	5 ns	15 ns	20 ns	25 ns
DDR2-533	3.75 ns	11.25 ns	15 ns	18.75 ns
DDR2-667	3 ns	9 ns	12 ns	15 ns
DDR2-800	2.5 ns	7.5 ns	10 ns	12.5 ns

[Used with permission by Micron Technology, Inc. © 2004 Micron Technology, Inc. All rights reserved.]

(shaded) Unlikely operation

14.2.3.3 WRITE Latency

DDR2 also includes a modification to the WRITE latency specification. Figure 14.26 illustrates the differences between DDR and DDR2 SDRAM. DDR SDRAM has a WRITE latency of one clock, which means that approximately one clock after a WRITE command, the DRAM starts clocking in data. However, given the increase in clock rates, this time is very short, making it difficult for the DRAM to prepare to capture data. DDR2 enables the WRITE latency to track the CAS latency minus one clock.

Figure 14.26 shows a 533 MHz DDR2 device with CAS latency set to four clocks. The WRITE latency is set to CAS latency – 1 clock, which is three clocks.

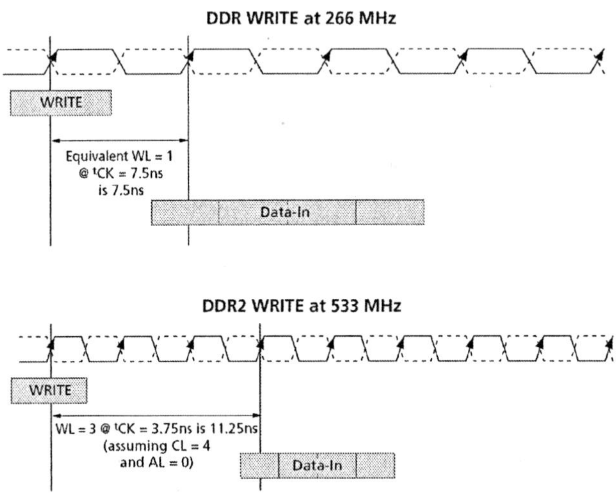

Figure 14.26 DDR vs. DDR2 WRITE. [Used with permission by Micron Technology, Inc. © 2004 Micron Technology, Inc. All rights reserved.]

The modification in WRITE latency allows more time for DRAM input buffers to support higher clock frequencies. Additionally, tracking WRITE latency with CAS latency simplifies system read-to-write timings.

14.2.3.4 Sequential Interleave

To support the high data rates of DDR2, the DRAM must have the flexibility to prefetch four bits of data from the core instead of only two. Consequently, support for a burst length of two has been dropped. Prefetching four bits, however, makes it difficult to support the traditional sequential interleave mode for DRAM burst lengths of eight.

The sequential burst orders for DDR2 are shown in Table 14.7. For burst lengths of four, there is no change from the DDR specification. However, for burst lengths of eight, a new burst type called "sequential nibble" is implemented. During sequential nibble operation, the DRAM breaks the eight-bit burst into two separate four-bit nibbles. The sequential interleave is executed on the nibble with the starting column location, followed by a similar sequential interleave on the other nibble using the starting location that is equal to the lower bits of the starting column.

Using the nibble sequential burst, a system requiring sequential interleave can take advantage of the eight-bit burst and still use any starting column within the burst.

The interleave burst ordering for DDR2 is the same as DDR.

Table 14.7 DDR2 Sequential Burst Ordering

Burst Length	Starting Column (A2, A1, A0)	Order of Accesses Within a Burst Burst Type = Sequential	
		DDR	DDR2
4	000	0,1,2,3	0,1,2,3
	001	1,2,3,0	1,2,3,0
	010	2,3,0,1	2,3,0,1
	011	3,0,1,2	3,0,1,2
8	000	0,1,2,3,4,5,6,7	0,1,2,3,4,5,6,7
	001	1,2,3,4,5,6,7,0	1,2,3,0,5,6,7,4
	010	2,3,4,5,6,7,0,1	2,3,0,1,6,7,4,5
	011	3,4,5,6,7,0,1,2	3,0,1,2,7,4,5,6
	100	4,5,6,7,0,1,2,3	4,5,6,7,0,1,2,3
	101	5,6,7,0,1,2,3,4	5,6,7,4,1,2,3,0
	110	6,7,0,1,2,3,4,5	6,7,4,5,2,3,0,1
	111	7,0,1,2,3,4,5,6	7,4,5,6,3,0,1,2

[Used with permission by Micron Technology, Inc., © 2004 Micron Technology, Inc., All rights reserved.]

14.2.3.5 Active Power-Down Mode

Active power-down mode enables a system designer to optimize power and performance. In this mode, the transition timing between power-down exit and the next command is dependent upon the power-down mode bit.

If the power-down mode is set to fast exit mode, there is a two-clock delay from active power-down exit to any command (including READ). Alternatively, the power-down mode bit may be set to slow exit mode, which extends the power-down exit delay to six clocks, minus the additive latency to a READ command. Active power-down slow exit is the same as the precharge power-down time during all operations of the DRAM. The power-down mode bit is ignored in precharge power-down. See Table 14.8.

Table 14.8 DDR2 Low-Power-Mode Options

	MR (bit 12)	PD Mode	Delay from CKE High to Command
Active Power-Down	0	Fast Exit	2 clocks to READ command
	1	Slow Exit	2 clocks to any non-READ command (Additive latency – 6 clocks) to READ
Precharge Power-Down	Ignored		2 clocks to any non-READ command (Additive latency – 6 clocks) to READ

14.2.3.6 Write Recovery

Write recovery (WR) must be correctly entered into the mode register to ensure proper DRAM operation. Write recovery is measured from the end of write data to the time when the internal auto precharge operation starts (see Figure 14.27). To correctly set the MR, WR is calculated as t_{WR}/t_{CK}, rounded up to the next integer.

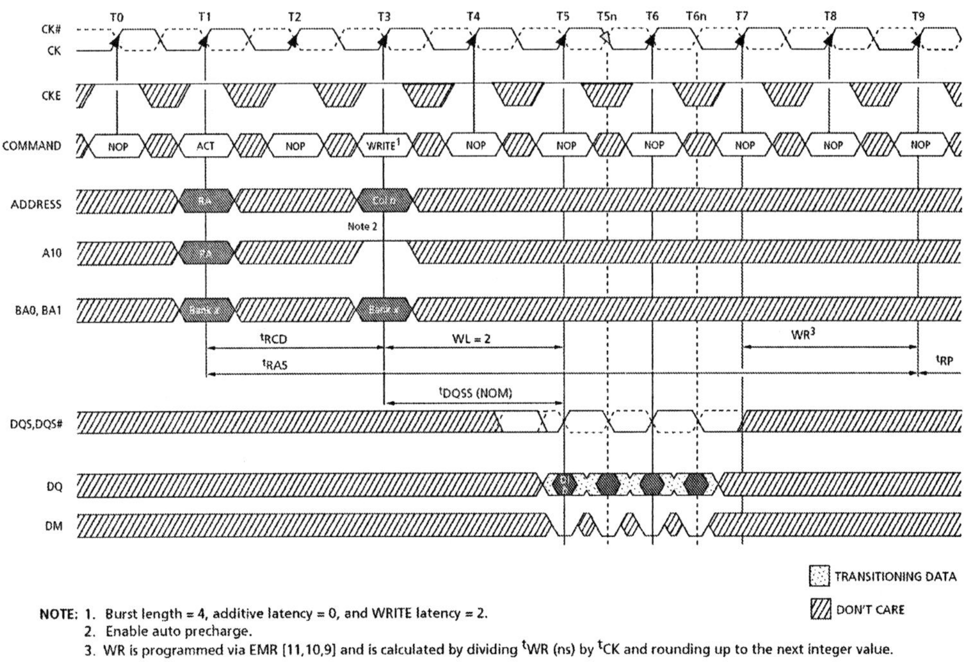

Figure 14.27 DDR2 write-to-precharge timing. [Used with permission by Micron Technology, Inc. © 2004 Micron Technology, Inc. All rights reserved.]

14.2.3.7 Extended Mode Register (EMR)

Additional enhancements have been made to the EMR, as shown in Figure 14.28. ODT and OCD calibration have been discussed previously. Other enhancements include additive latency, output disable, and DQ strobe configurability.

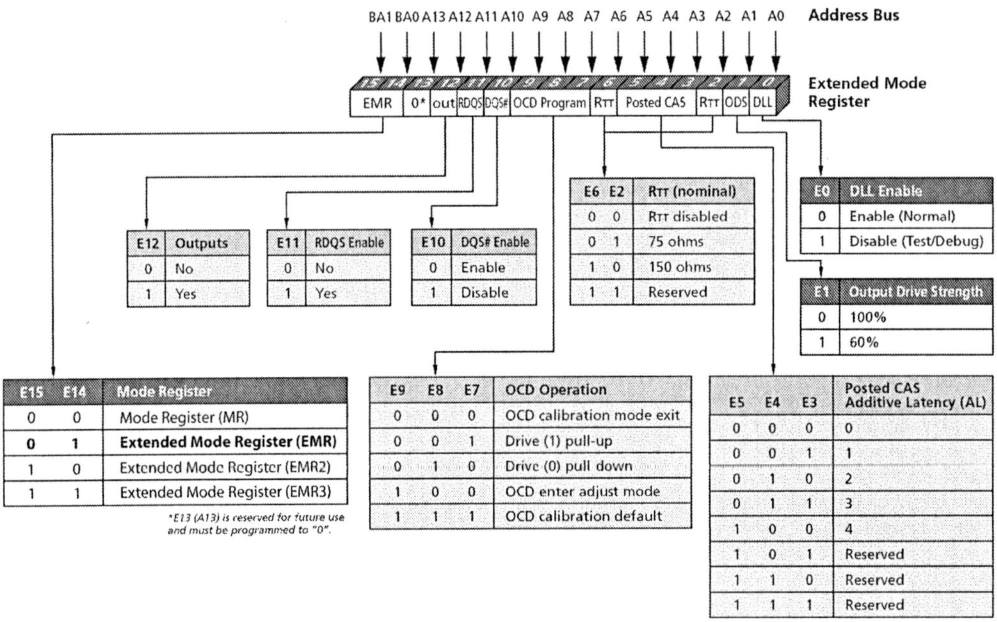

Figure 14.28 Extended mode register. [Used with permission by Micron Technology, Inc. © 2004 Micron Technology, Inc. All rights reserved.]

14.2.3.8 Additive Latency

Figure 14.29 shows an example of DDR2 bank interleave READs. During these accesses, a bank is opened with an ACTIVATE command. After t_{RCD} is met, a READ with auto precharge command is issued and data is read out of the DRAM. As seen in Figure 14.29, there are times when a command cannot be issued at the optimal location because there is another command already there. For example, ACT (to bank 2) must be delayed by one clock because the RD_AP (to bank 0) is already in that spot. The result is a gap in the output data.

To improve system scheduling, DDR2 includes a mode register for additive latency. Using additive latency, the DRAM captures READ or WRITE commands but waits to execute them until the additive latency time expires.

Figure 14.30 illustrates DRAM operation with two different additive latencies. The top part of the figure shows an operation similar to DDR SDRAM using AL = 0 clocks. The bottom part of the figure shows AL = t_{RCD} − 1 clock. In this configuration, the ACTIVATE and READ commands can be issued back-to-back. The DRAM holds the READ command for the additive latency time and then executes it on the same clock as the top diagram.

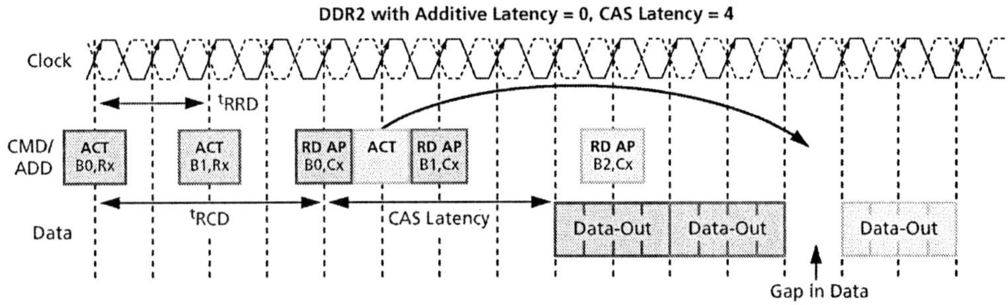

Figure 14.29 DDR2 READs with no additive latency. [Used with permission by Micron Technology, Inc. © 2004 Micron Technology, Inc. All rights reserved.]

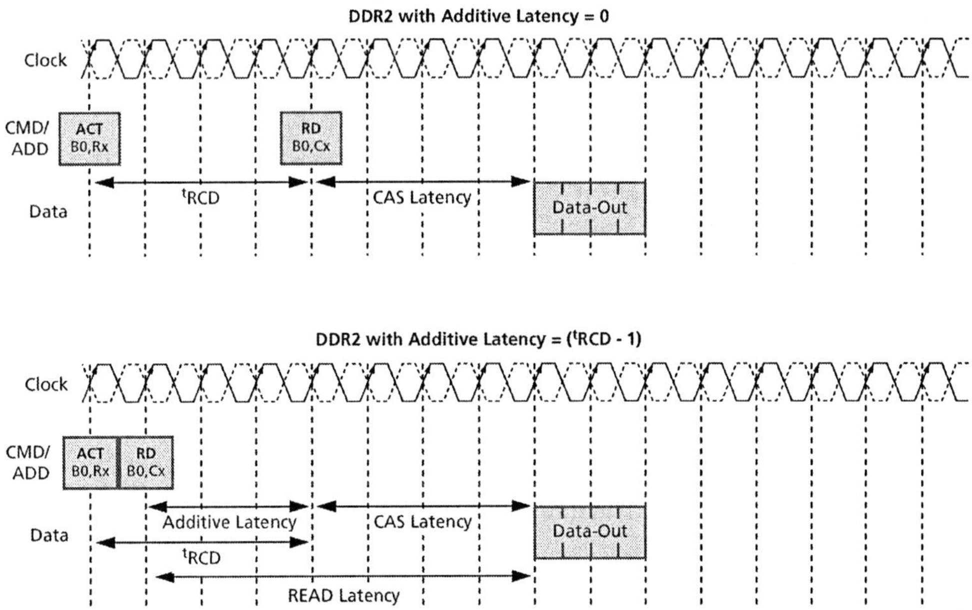

Figure 14.30 DDR2 with additive latency. [Used with permission by Micron Technology, Inc. © 2004 Micron Technology, Inc. All rights reserved.]

With additive latency set to t_{RCD} – 1 clock, a simplified read system timing can be attained. Figure 14.31 assumes t_{RCD} = 4 clocks. Thus, additive latency is set to three clocks. This allows the READ command to directly follow the ACTIVATE command. With a burst length of four, which requires two clocks to burst data, the two clocks for the back-to-back ACT and RD_AP commands match perfectly, enabling data to be continuously output from the DRAM without gaps. The simplicity of system bus scheduling is expected to be particularly popular in applications that have high page misses.

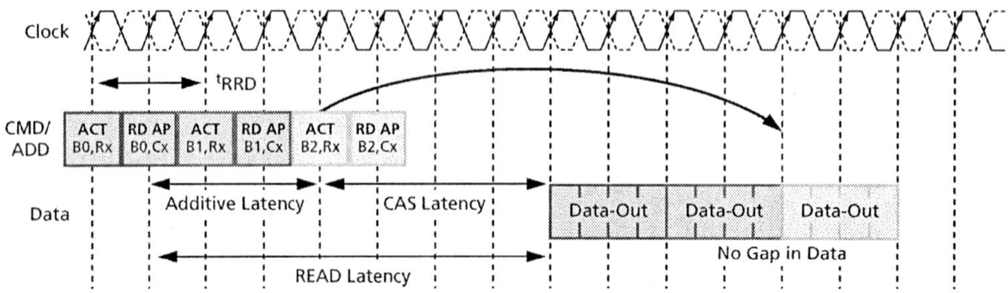

Figure 14.31 DDR2 READs using additive latency. [Used with permission by Micron Technology, Inc. © 2004 Micron Technology, Inc. All rights reserved.]

14.2.3.9 DQS Configuration

The extended mode register also allows configuration of the DQ strobes (DQS) for multiple system configurations. Figure 14.32 highlights the areas of the extended mode register that configure the strobes. Bit 10 switches the strobes between differential and single-ended mode, and bit 11 can be used to enable the redundant strobe on the x8 device.

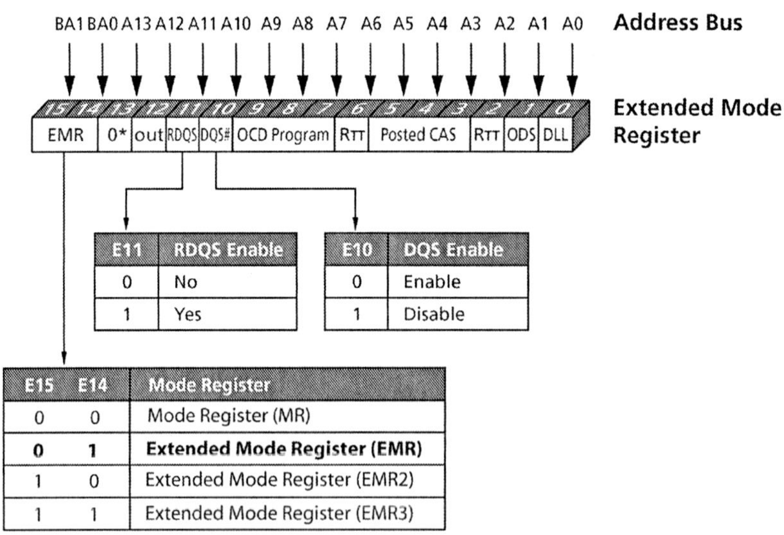

Figure 14.32 Extended mode register for DQS configuration. [Used with permission by Micron Technology, Inc. © 2004 Micron Technology, Inc. All rights reserved.]

14.2.3.10 Single/Differential Data Strobe

DDR2 supports single-ended strobes in the same manner as DDR SDRAM, as shown in Figure 14.33. DDR2 SDRAM also supports differential strobes. If differential strobes are enabled, DQS operates the same as in single-ended mode with the addition of DQS# operating as the complement in differential signals. For READs the DRAM will output both signals, as shown in Figure 14.34. For WRITEs, the system must provide both input signals.

The use of differential strobes is system-dependent and may enable improved system timings due to reduced crosstalk and less simultaneous switching noise on the strobe output drivers.

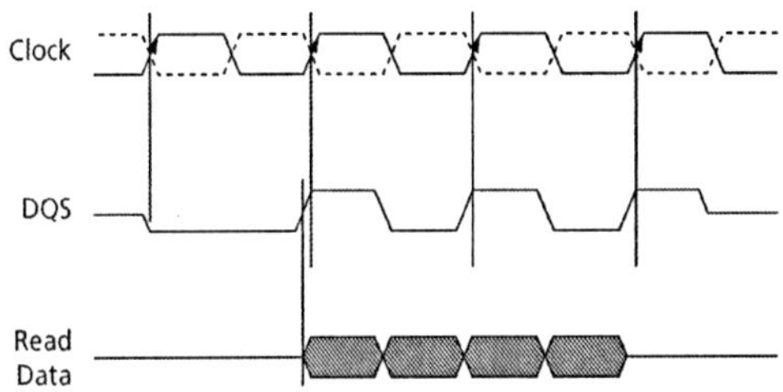

Figure 14.33 DDR2 Single-ended strobe. [Used with permission by Micron Technology, Inc. © 2004 Micron Technology, Inc. All rights reserved.]

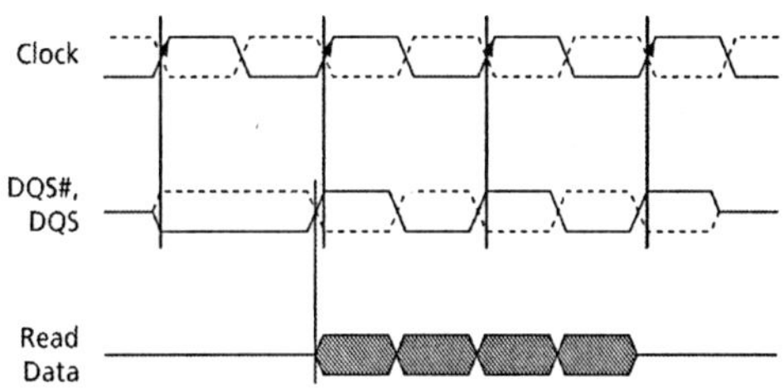

Figure 14.34 DDR2 Differential strobe. [Used with permission by Micron Technology, Inc. © 2004 Micron Technology, Inc. All rights reserved.]

14.2.3.11 Redundant DQS (RDQS)

The RDQS option is only available on the x8 DRAM and is required when mixing x4 and x8 DRAM devices in the same system. An example of this is shown in Figure 14.35. Because the x8 DRAM has only one DQS input for the 8-bit byte and the x4 has a DQS for each 4-bit nibble, the loading of the DQS lines is different when the devices are mixed in the same system.

RDQS reduces this effect. If devices are mixed in the system, the signal routing from the second x4 DRAM is routed to the RDQS pin on the x8 DRAM. During WRITEs, the loading and termination on the RDQS is the same as the DQS signals, so all signaling is matched for the x4 DRAMs. For READs, the x8 DRAM drives the RDQS output the same way it does the DQS. Thus, the controller can use RDQS to capture read data on a nibble device and the x8 DRAM emulates the functionality of the x4.

Systems mixing x4 and x8 devices can utilize RDQS signals to improve timing. RDQS functionality is only an option on the x8 DRAM. Additionally, if RDQS is enabled, the data mask functionality of the x8 DRAM is disabled.

14.2.3.12 Output Disable

For DDR SDRAM, it can be difficult to measure I_{DD} values in system environments. All I_{DD} values are specified with $I_{OUT} = 0$ mA, which is challenging to accomplish in the system. Thus, DDR2 has an option to disable the output drivers from the EMR. If the output disable option is set, the last stage of the output driver is disabled. WRITEs are completed normally and READs execute except for the final output stage. Thus, I_{DD} measurements within a system environment can be made.

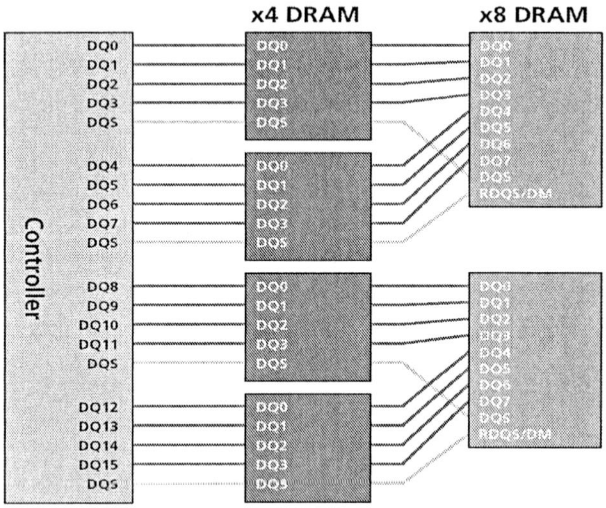

Figure 14.35 x4 and x8 devices in same system. [Used with permission by Micron Technology, Inc. © 2004 Micron Technology, Inc. All rights reserved.]

14.3 SPICE/IBIS Simulation of DDR-II SDRAM[4]

14.3.1 SPICE/IBIS Simulations of DQ and DQS Pins

Simulations were performed for the DQ and DQS pins of Samsung's 512 Mb M-die DDR2 SDRAM. "This 512 Mb DDR2 SDRAM chip is organized as 32 Mb × 4 I/O × 4 banks, as 16 Mb × 8 I/O × 4 banks, or as 8 Mb × 16 I/O × 4 banks. This synchronous device achieves high-speed, double-data-rate transfer rates of up to 533 Mbps/pin (DDR2-533) for general applications. The chip is designed to comply with the following key DDR2 SDRAM features: (1) posted CAS with additive latency, (2) write latency = read latency – 1, (3) off-chip driver (OCD) impedance adjustment, and (4) on-die termination. All of the control and address inputs are synchronized with a pair of externally supplied differential clocks. Inputs are latched at the cross point of differential clocks (CK rising and CK falling). All I/Os are synchronized with a

4. Except where indicated, the source of text, tables, and figures contained in Section 14.3 is a white paper titled, "DDR-II SDRAM Technology and Design Trends," written by Edward P. Sayre, Edward P. Sayre, III, and Michael A. Baxter of North East Systems Associates, Inc. (NESA), reference [Say03]. Reprinted with permission of North East Systems Associates, Inc. (NESA).

pair of bidirectional strobes (DQS and DQS) in a source synchronous fashion. A 14-bit address bus is used to convey row, column, and bank address information in a RAS/CAS multiplexing style. For example, the 512 Mb (x4) device receives 14/11/2 addressing. The 512 Mb DDR2 devices operate with a single 1.8 V ± 0.1 V power supply and 1.8 V ± 0.1 V VDDQ. The 512 Mb DDR2 devices are available in 60 ball FBGAs (x4/8) and in 84 ball FBGAs (x16). Refresh and self-refresh operations of 8192 refresh cycles per 64 ms are supported. (Refresh period 7.8 μs)"[5] [Sam03a]

The specific ODT termination values have been determined by the JEDEC 42.3 Memory Device Committee members from SPICE simulations of specific parts. Unfortunately, users of DDR-II memory devices only have IBIS models to draw on to make a determination of the ODT register settings. The ODT terminations shown in Chapter 13, Figure 13.16 on page 342, are incorporated into the overall model. The end-to-end schematic model is shown in Figure 14.36.

Note that in the SPICE/IBIS model, the ODT terminations form a split Thevenin termination with V_{REF} formed by the voltage divider and the net termination value being the parallel combination of the two resistors of a selected termination (Rval1, Rval2) making up the divider. Thus, where the notation used is $R_{TT}1 = 75\ \Omega$, the resistor termination values are Rval1 = 150 Ω

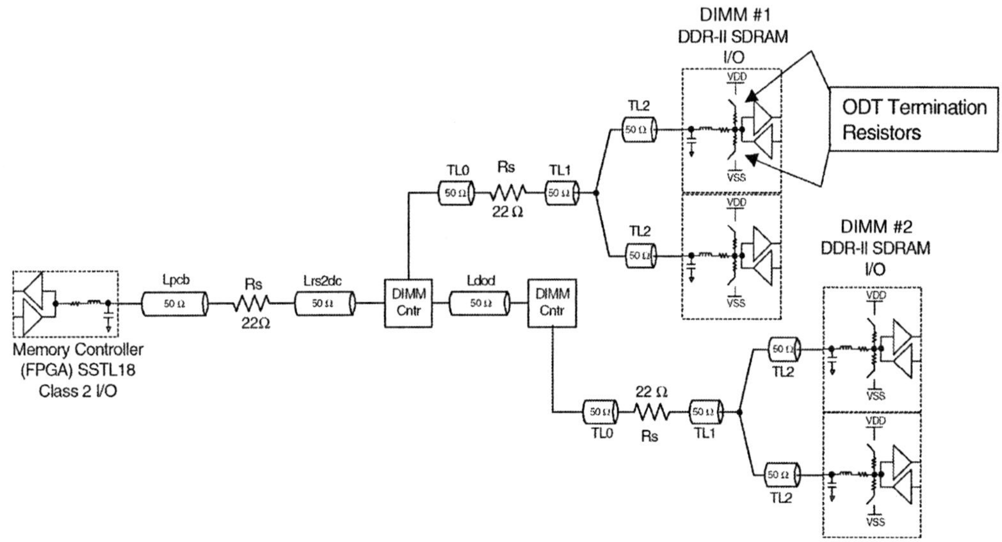

Tline segments	Lpcb	Lrs2dc	Ldod	TL0	TL1	TL2
Lengths (mm)	60–90	10–15	10–15	3.6–6.1	4.8–11	3.3–9.4

Figure 14.36 SSTL_18 DDR-II READ/WRITE DQ, DQS path from/to DIMM. [Reprinted with permission of North East Systems Associates, Inc. (NESA).]

5. Reprinted with permission of Samsung Semiconductor, Inc.

and similarly for $R_{TT}2 = 150\ \Omega$ the resistors are Rval2 = 300 Ω. The following 400 Mbps (200 MHz clock) SPICE/IBIS cases have been developed and run using Samsung IBIS behavioral models for the DDR-II memories and a transistor-level Vertex-II SSTL_18 Class II transistor model. The simulations were run using Synopsys' HSPICE mixed-model simulation capability. The SPICE case variable list and an abridged matrix are shown in Tables 14.9 and 14.10.

Table 14.9 DDR-II SPICE Variable Table

Variables	Range
Interconnect	
On-Chip Termination (OCT)	75 Ω, 150 Ω, open
Read — Driver Location	DIMM1, DIMM2
Write — Receiver Location	DIMM1, DIMM2
Constants	**Value**
Device	
DDR Driver Type	SSTL_18
DDR Receiver Type	SSTL_18
FPGA Driver Type	SSTL_18, Class II
FPGA Rccciver Type	SSTL_18, Class II
Data Rate	533 Mbps[a]
Clock Rate	267 MHz[b]
Write — Driver Location	FPGA
Temperature	25°C
Process	Typical — TT
Voltage	Nominal
Interconnect	
Controller Board Path Lengths	Nominal
Net Structure Routing for Data	Fig 6 of JEDEC Standard #8-15
DIMM Path Lengths	Nominal
DIMM Series Termination	22 Ω ± 5%
Transmission Line Type	Lossless
DIMM Connector	
Molex 0.050" 71243-0001 SLM	Single line model
FPGA Package	
Xilinx package — single line	Provided with SSTL_18 model
DDR Package	
60 ball FPGA — single line	Provided with DDR-II model

a. These are the maximum rates for the memory device. Simulations for DQ and DQS signals were run for a data rate of 400 Mbps and simulations for the differential clock were run at 533 Mbps.

[Reprinted with permission of North East Systems Associates, Inc. (NESA).]

Figure 14.37 shows the DIMM #1 Memory DQ and DQS READ waveforms including the output at the DDR-II memory driver, at the Xilinx FPGA controller receiver and the receiver output. Note that the transmission line, the driver output, and the receiver and ODT input terminations have the effect of morphing the waveform into a more symmetric shape. The receiver output is nonetheless somewhat asymmetric because it is offset from the V_{REF} level.

Table 14.10 Abridged DDR-II SPICE Matrix

Data	Run	OCT value (Ω)	Read Driver Location
Write FPGA Tx/ DDR Rx	1	0	–
	2	75	–
	3	150	–
Read DDR Tx/ FPGA Rx	4	0	DIMM1
	5	75	DIMM1
	6	150	DIMM1
	7	0	DIMM2
	8	75	DIMM2
	9	150	DIMM2

[Reprinted with permission of North East Systems Associates, Inc. (NESA).]

Figure 14.38 shows the DIMM #2 Memory DQ and DQS READ waveforms including the output at the DDR-II memory driver, at the Xilinx FPGA controller receiver and the receiver output. Note that the noise margins are substantially less in this case than the waveforms originating at DIMM #1. The FPGA receiver output shows some asymmetry similar to that shown in the DIMM #1 results.

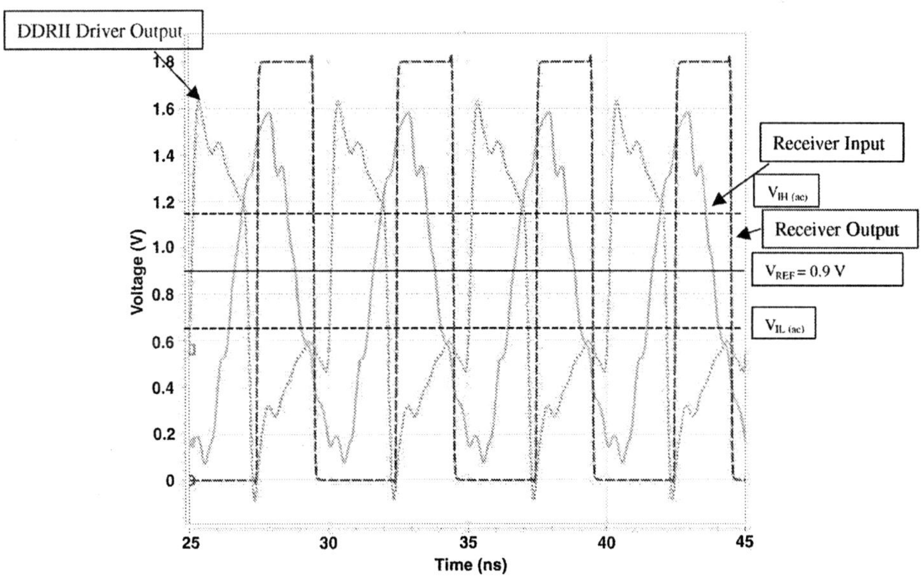

Figure 14.37 DIMM #1 memory DQ and DQS READ waveforms at FPGA receiver and receiver output. [Reprinted with permission of North East Systems Associates, Inc. (NESA).]

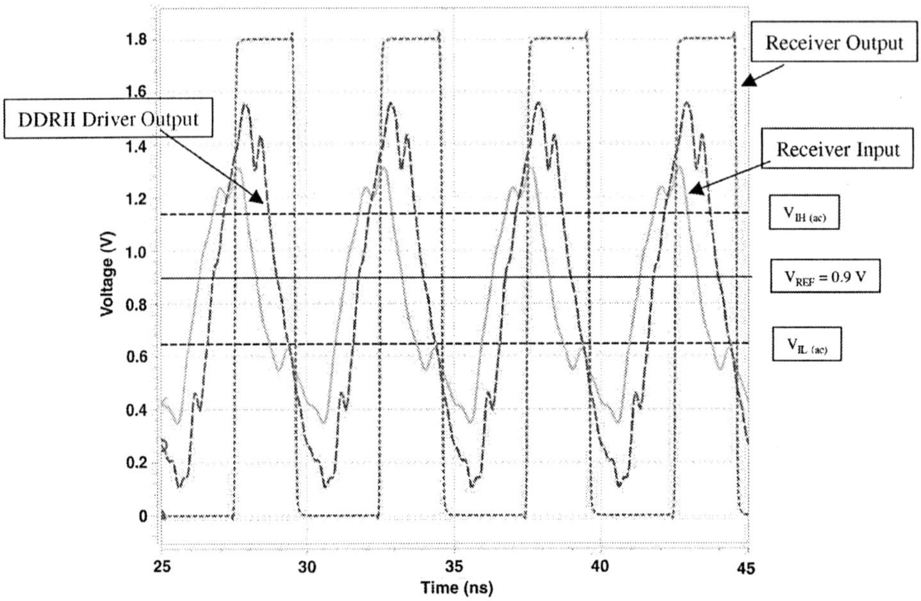

Figure 14.38 DIMM #2 memory DQ and DQS READ waveforms at FPGA receiver and receiver output. [Reprinted with permission of North East Systems Associates, Inc. (NESA).]

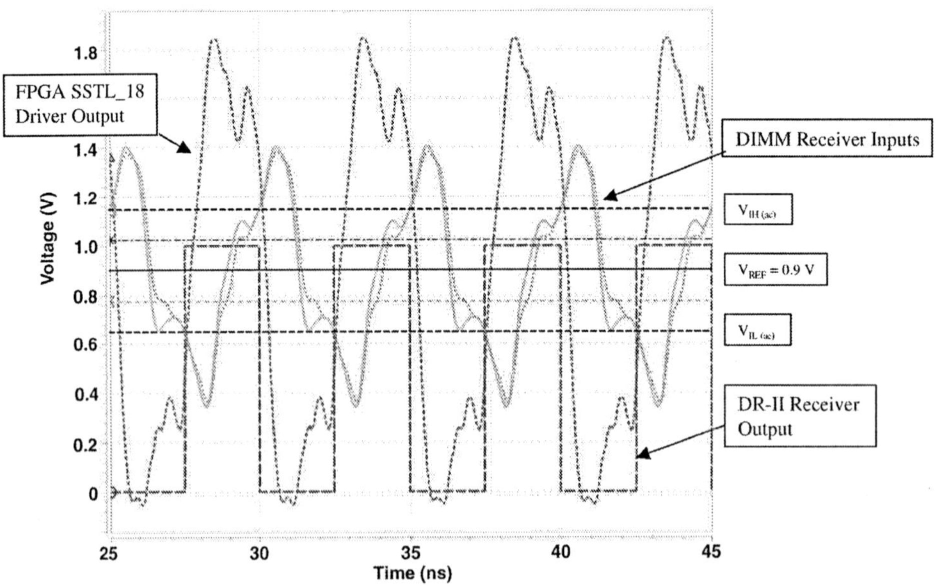

Figure 14.39 Controller initiated DQ and DQS WRITE waveforms at DDR-II receiver and receiver output. [Reprinted with permission of North East Systems Associates, Inc. (NESA).]

The triangular shape of the receiver input waveforms shows that capacitive loading is occurring near the driver. It is clear that operation at a bit rate much higher than 400 Mbps would be questionable, especially READs from the farthest DIMM. Under WRITE mode, there is a single driver with multiple receivers, which is a more advantageous signal integrity environment for a two DIMM memory system. Figure 14.39 shows the added margin at the DIMM memory receivers.

14.3.2 Clock Signal Integrity

DDR-II read/write transfers are always DQS timed, where DDR-I can be either DQS timed or centrally clocked. The clock differential input in DDR-II memories is only used as a reference and edge placement plays no role in the READ or WRITE DQS-based timing. The DDR-I design with centralized DQS clocking is applicable when the controller chip(s) do not have the appropriate I/O cells. An example of the differential clock distribution is shown in Figure 14.40. This ensured that the same transistor process would be used for all I/O functions. The multiload, parallel line-to-line, end-point-terminated DDR-II clock topology is nearly identical to that used for DDR-I memories aside from the new logic thresholds. The resistive termination value is 120 Ω and it should be located as close as possible after the second DIMM memory. The multiload DDR-II clock topology is shown in Figure 14.40.

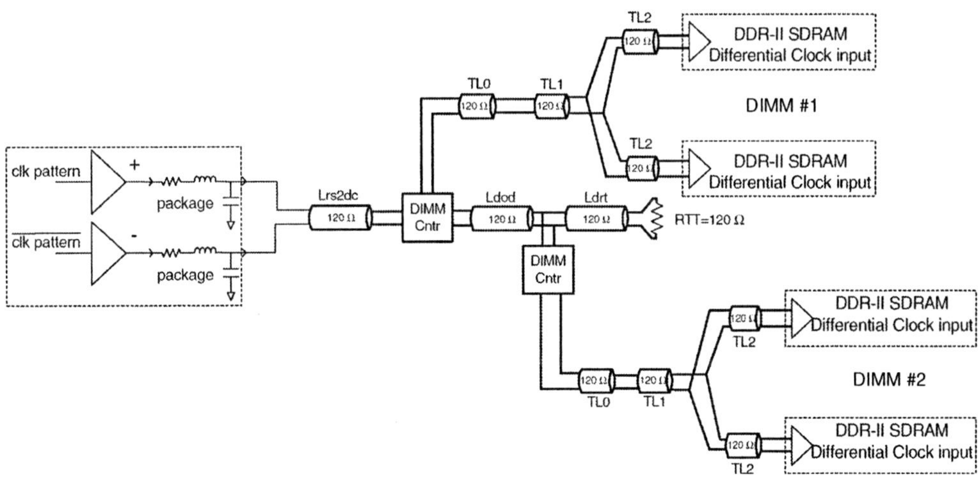

Tline segments	Lrs2dc	Ldod	Ldrt	TL0	TL1	TL2
Lengths (mm)	70–115	10–15	10–15	3.6–6.1	4.8–11	3.3–9.4

Figure 14.40 DDR-II multiload differential clock topology. [Reprinted with permission of North East Systems Associates, Inc. (NESA).]

The clock is issued by the controller and feeds the receivers located in each memory rank of each DIMM card. There is a significant stub load shunting the clock line at each DIMM location. In the simulation outputs that follow, the waveform shape at the receiver is triangular, indicating similar stub-loading effects to that previously seen in the single-ended DQ and DQS signals (see Figure 14.39).

The SPICE simulation outputs are shown in Figure 14.41 for a 267 MHz (513 Mbps per pin data) clock driver. In the top panel, the clock stimulus and differential output waveforms are shown. The excellent signal fidelity and the short delay between clock input to output is visible (since this is a transistor-level model of the driver). In the middle panel, the differential clock signal at the first DIMM receiver pair is shown. The waveform shows evidence of capacitive loading and reflection effects. In the bottom panel, the differential clock signal at the second DIMM receiver pair is shown. Again, the waveform shows evidence of capacitive loading and reflection effects but at a somewhat lesser rise time due the capacitive shunt loads on the clock bus. In both cases, the symmetry around the differential switch point is very good, indicating that the internal clock waveforms will have equally good symmetry.

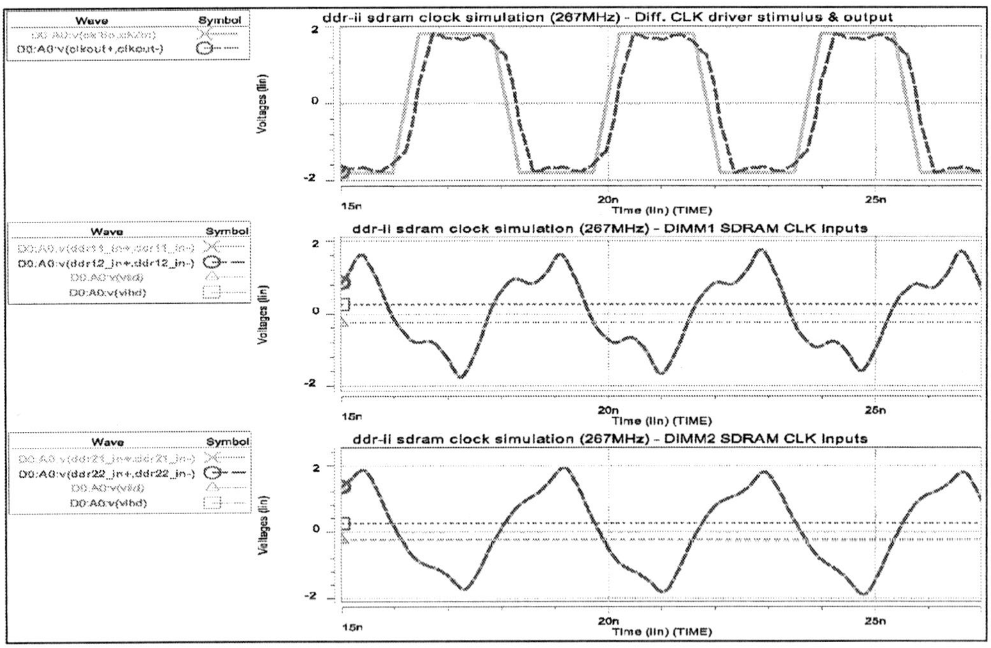

Figure 14.41 SPICE simulation outputs for differential clocks: (top) clock stimulus and differential output, (middle) differential clock at DIMM #1, and (bottom) differential clock at DIMM #2. [Reprinted with permission of North East Systems Associates, Inc. (NESA).]

14.4 Exercises

14-1. As clock rates increase for DDR2 SDRAM devices, what usually happens to the read latency as measured in clock cycles?

14-2. Create a single table comparing the features of SDR, DDR, and DDR-II SDRAM by combining Table 14.1 on page 354 and Table 14.4 on page 363. Where a table item cannot be determined from information contained in this chapter, leave the item blank. Your table will still provide a good comparison between the three generations of devices.

14-3. For the DDR-II SDRAM simulation described in this chapter, what kind of device was used for the memory controller?

14-4. What is the frequency of the memory clock for a DDR SDRAM that transfers data at the rate of 400 Mbps per pin?

14-5. (a) What is the signaling technology used for SDR SDRAM? (b) for DDR SDRAM? (c) for DDR-II SDRAM? (d) for DDR2 SDRAM?

14-6. How does the use of a differential clock in DDR and DDR2 SDRAM help reduce the negative effects of a memory clock that does not have a 50 percent duty cycle?

14-7. Assume you are designing a memory system that uses DDR-II SDRAM devices and you plan to use the ODT (on-die termination) capability of your devices to maximize signal integrity. Since these devices use SSTL for the signaling technology and SSTL normally requires a pull-up termination resistor R_{TT} at the end of the bus on each trace common to all of the memory devices, how do you determine the value of R_{TT} to use?

14-8. SDR SDRAM features both a write and a read mask. Does DDR SDRAM also provide both a write and a read mask capability?

14-9. (a) Which bits in the extended mode register control the values of the ODT for a DDR2 SDRAM device? (b) What values do they take on to program an R_{TT} of 150 Ω?

14-10. For DDR2 SDRAM, what are the two programmable values of the output driver strength?

14-11. (a) For the simulation of the DDR-II SDRAM memory system shown in Figure 14.36 on page 383, which parts of the memory system were represented by IBIS models and which parts were represented by SPICE models? (b) Why was SPICE not used to simulate everything? (c) Before you can run a simulation using IBIS and SPICE models, what do you need to know about your EDA (electronic design automation) tool (software)?

14-12. What are the two most obvious differences in operational read timing between DDR SDRAM and SDR SDRAM as shown in Figure 14.7 and Figure 14.8 on page 362?

14-13. Figure 14.35 on page 382 shows mixing x4 and x8 devices in the same memory system. Why would you ever want to do this?

14-14. For SSTL_2, if V_{REF} is 2.5 V, what is the minimum dc input voltage level for a high logic state and what is the maximum dc input voltage level for a low logic state?

14-15. (a) How many banks does DDR2 memory have? (b) What determines the number of banks in a DDR2 device?

CHAPTER 15

GDDR3, ZBT, FCRAM, SigmaRAM, RLDRAM, DDR SRAM, Flash, FeRAM, and MRAM

Chapter Objectives

- Compare the features of GDDR, GDDR2, and GDDR3 graphics memories.
- Discuss the features of DDR and DDR2 SRAM and show that their signaling technologies are different from those of DDR and DDR2 SDRAM.
- Show that ZBT, NoBL, ZeroSB, and NtRAM SRAM are equivalents.
- Compare Network Fast Cycle RAM to Mobile and Consumer FCRAM.
- Present the SigmaRAM (ΣRAM) family of memory devices.
- Introduce reduced latency DRAM (RLDRAM) and compare the features of RLDRAM-I and RLDRAM-II.
- Discuss densities, access times, and other features of flash memory and when to use NOR versus NAND flash.
- Describe FeRAM and MRAM and how they may soon replace flash memory.

15.1 Graphics Double Data Rate SDRAM (GDDR, GDDR2, GDDR3)

"Graphics memories are at the pinnacle of DRAM performance technology. In graphics applications, two factors determine the quality of the cinematic experience: how much memory is available and its bandwidth. A larger DRAM can store more graphics information that translates into richer textures, more subtle shading, more colors, and larger screen sizes. It isn't uncommon today to see a personal computer shipped with 35 MB of graphics memory, but upgraded with an add-on card to as much as 256 MB. Game consoles, basing their consumer appeal on the quality of the graphics, already ship with more memory than PCs of just a few years ago.

How quickly data can be moved in and out of the graphics memory is critical to smooth graphics or high-speed communications."[1] [Sam03c] At data rates up to 3.2 Gbps per pin, graphics double-data-rate memories are the speediest DRAMs available.

Graphic DDR memories use double-data-rate I/Os and have evolved into three generations, from GDDR and GDDR2 to GDDR3. While the first-generation GDDR devices came in either 16- or 32-bit bus widths, GDDR2 and GDDR3 devices are supplied with 32-bit data bus widths. Table 15.1 compares the features of these three generations of devices.

Table 15.1 Feature Comparison Among GDDR, GDDR2, and GDDR3 SDRAM

Feature	GDDR3	GDDR2	GDDR
Generation	3rd	2nd	1st
Density	256 Mb	128 Mb, 256 Mb	128 Mb, 256 Mb
Bandwidth per Chip	To 6.4 GB/s	To 12.8 GB/s	1.2 to 3.2 GB/s
Data Rate per Pin, (2 × Memory Clock)	800 Mbps to 1.6 Gbps	900 Mbps to 1.6 Gbps, 3.2 Gbps	333–900 Mbps
Memory Clock (MHz)	400, 450, 500, 550 600–800 MHz	450, 500, 550, 600, 700, 800, 1,600 MHz	166, 200, 250, 275, 300, 350, 400, 450 MHz
Bus Width	x32	x32	x16, x32
VDD/VDDQ (V)	1.8/1.8	2.5/1.8 1.8/1.8	2.5/2.5 1.8/1.8
Interface	SSTL_18	SSTL_18	SSTL_2, SSTL_18
On-Die Termination (ODT)	Yes	Yes	No
Off-Chip Driver Calibration (OCD)	Yes	Yes	No
Power Consumption	1/2 of GDDR2 power	1–3 W	1–3 W
Package	FBGA	FBGA	FBGA, TSOP
Banks	4	4	4

"Early graphics devices were synchronous memories dedicated for images and were basically configured with SDRAM. The operation mode for drawing, which is not provided to SDRAM, was added in order to simplify the design of graphics systems. These devices included enhanced graphics features for use with display adapters and had block write and mask write functions allowing the frame buffer to be cleared faster and selected pixels to be modified faster."[2] [Elp01a]

1. Reprinted with permission of Samsung Semiconductor, Inc.
2. Reprinted with permission of Elpida Memory.

In some cases today, vendors are simply taking fast DDR2 SDRAM and marketing it as "Graphics SDRAM" based solely on the density, bus width, and speed of the device, knowing that graphics applications have the greatest speed requirements.

GDDR2 and GDDR3 devices incorporate new technical features such as on-die termination (ODT), off-chip driver calibration (OCD), and posted CAS to boost performance by 50% when compared with first-generation GDDR SDRAM.

One of the main features that distinguishes GDDR3 devices from those of GDDR and GDDR2 is the lower power consumption by GDDR3. Power consumption can vary considerably depending upon the application and whether the device has one bank active or is in burst mode with all banks active, as shown in Table 15.2. In the table, the version (dashed number) is the part number speed suffix. The speed suffix is the period of the memory clock in nanoseconds times 10, e.g., a "-12" speed part has a clock frequency of 800 MHz and a clock period of 1.25 ns. In this example, the $10 \times 1.25 = 12.5$ speed suffix is rounded to 12.

Table 15.2 Power Consumption of 256 Mb GDDR SDRAM Device (Samsung K4D551638D-TC) as Characterized by Various Parameters of Current [Sam03b]

Parameter	Symbol	Test Condition	Version					Unit	Note
			-33	-36	-40	-50	-60		
Operating Current (One Bank Active)	I_{CC1}	Burst Length = 2 $t_{RC} \geq t_{RC}(min)$ $I_{OL} = 0$ mA, $t_{CC} = t_{CC}(min)$	230	220	210	145	125	mA	1
Precharge Standby Current in Power-Down Mode	$I_{CC2}P$	$CKE \leq V_{IL}(max)$, $t_{CC} = t_{CC}(min)$		70		4	3	mA	
Precharge Standby Current in Non Power-Down Mode	$I_{CC2}N$	$CKE \geq V_{IH}(min)$, $CS \geq V_{IH}(min)$, $t_{CC} = t_{CC}(min)$	100	90	80	30	25	mA	
Active Standby Current Power-Down Mode	$I_{CC3}P$	$CKE \leq V_{IL}(max)$, $t_{CC} = t_{CC}(min)$	80	75	70	55	35	mA	
Active Standby Current in Non Power-Down Mode	$I_{CC3}N$	$CKE \geq V_{IH}(min)$, $CS \geq V_{IH}(min)$, $t_{CC} = t_{CC}(min)$	150	140	130	75	55	mA	
Operating Current (Burst Mode)	I_{CC4}	$t_{RC} \geq t_{RFC}(min)$ Page Burst, All Banks Activated.	450	430	410	250	200	mA	
Refresh Current	I_{CC5}	$t_{RC} \geq t_{RFC}(min)$	390	380	370	200	180	mA	2
Self Refresh Current	I_{CC6}	$CKE \leq 0.2$ V		4		3		mA	

Notes: 1. Measured with outputs open.

2. Refresh period is 32 ms for -TC33/36/40 (4K/32 ms); Refresh period is 64 ms for -TC50/60 (8K/64 ms)

[Reprinted with permission of Samsung Electronics.]

15.1.1 GDDR3 (Graphics DDR 3)

"As PC graphics cards' 3D accelerators have grown more and more powerful (ATI Technologies' Radeon 9700 Pro chip has as many transistors — 110 million — as two Intel Pentium 4s), manufacturers have moved from generic memory to faster VRAM (Video RAM), SGRAM, and on to DDR memory. Feeling that still faster memory was needed, ATI collaborated with DRAM vendors to complete a specification for third-generation, double-data-rate memory for graphics, called GDDR3."[3] [Jup02a]

"ATI's partners in the GDDR3 venture include Micron, Infineon, Elpida Memory (USA), and Hynix Semiconductor America. Expanding on the work of the JEDEC industry consortium in designing DDR-II for the mainstream memory market, GDDR3 emphasizes high-speed signaling optimized for graphics processing." [Jup02a] Initial GDDR3 devices operated at 500 MHz (1 Gbps data rate per pin) and later devices operated at 800 MHz (1.6 Gbps data rate per pin) with a roadmap for future devices to go even higher. With a 32-bit width, the 800 MHz device offered 6.4 GB/s of total bandwidth per chip. The GDDR3 standard is proprietary to the venture and has been created prior to the availability of an industry standard.

The intended big advantages behind the introduction of GDDR3 was increased bandwidth and less power consumption. The increase in bandwidth was made possible by using on-die termination (ODT) for improved signal quality, although GDDR2 also uses on-die termination. Because vendors of GDDR2 devices have continued to offer increasing speeds, GDDR3 and GDDR2 devices are still highly competitive with one another in terms of bandwidth.

15.1.1.1 GDDR3 Device Operation

"The I/O interface for GDDR3 is a pseudo-open-drain logic (PODL 1.8), which drives a simple 40 Ω push-pull driver into a 60 Ω transmission line, with an equivalent 60 Ω on-die pull-up resistor to $V_{DD}Q$. Unlike a pure open drain scheme, source termination is achieved on low-to-high transitions using the PMOS device of the push driver. Dc power is minimized because current is only consumed when a logic low is driven onto the bus."[4] [Mic03e]

"GDDR3 utilizes an external precision resistor (RQ) tied to V_{SS} to calibrate the driver to a known value of impedance. This eliminates most of the process variation that can be introduced during manufacturing. Initially, the driver functions as a standard push-pull driver with multiple legs to tune the impedance and adjust the on-die termination.

The calibration process requires the external resistor (RQ) to be six times the desired driver impedance. The PMOS device is first calibrated to the resistor tied to ground. This calibrates one PMOS leg to 240 Ω. The pull-up stage for this output buffer consists of six legs, which is the nominal 40 Ω equivalent pull-up driver. After the PMOS leg is calibrated to the external resistor, the NMOS leg is calibrated to the PMOS leg in a similar fashion." [Mic03f]

3. Reprinted with permission of Jupitermedia Corporation.
4. Used with permission by Micron Technology, Inc., © 2004 Micron Technology, Inc., All rights reserved.

15.1.1.2 GDDR3 Power Consumption Calculations

Data pattern and address pattern are the two most important factors when calculating the current draw due to ODT (on-die termination), with the worst case being when all zeros are driven on the data and address buses. Slew rate and frequency are two other factors that must be accounted for in the current calculation. As the slew rate increases, the amount of time spent at the worst-case power levels also increases. The frequency relationship is opposite the slew rate in terms of current. As the frequency decreases, the amount of time spent at the worst-case power levels increases (assuming a constant slew rate). The ODT value selected will also affect the amount of current consumed. The total system power will not change based on a selected ODT value. Rather, the amount of power dissipated by each chip will change depending on load conditions. For a more thorough discussion of ODT power calculations, see reference [Mic03e].

15.1.2 Terminology: GDDR SDRAM vs. DDR SGRAM

Graphics memories today generally fall under two general classifications:

- Specialty graphics memory
- Commodity-based graphics memory

Specialty graphics memory devices are graphic memory SDRAM devices intended only for graphics applications. Commodity-based graphics devices are high-speed SDRAM devices that are not only intended to be used in graphics applications, but in any other application requiring high-speed SDRAM as well. Specialty graphics memories tend to be lower in density and higher in cost per bit than commodity-based graphics devices. Specialty graphics memories also tend to come in wider data bus widths than commodity-based graphics memories, for example, x32 versus x16, and they are more likely to have special features intended only for graphics applications such as a data masking capability for write data. Both specialty and commodity-based memories can come in SDR and DDR versions.

Today, the newer, increasingly faster graphics memories are falling more and more into the commodity-based category. These newer memories are typically SDRAM devices that offer high-speed DDR I/Os and high density, but whose only graphics feature is a wider data bus. Put differently, much of today's graphics memory is simply just high-speed SDRAM. For example, Infineon uses the same "type" designation in its part numbers for DDR SDRAM as it does for its graphics RAM; it makes no distinction between the two. Infineon also offers some DDR SDRAM devices with a data mask for write data feature, but makes no mention of graphics applications in its data sheets for these devices. Most newer graphics memory devices are 32-bits wide, although 16-bit graphics devices are readily available for appropriate market segments.

DDR graphics memory is referred to by two general names: DDR SGRAM (DDR synchronous graphics RAM) and GDDR SDRAM (graphics DDR SDRAM). Different manufacturers tend to use one name or the other, but no specific device features or characteristics are attached to one name but not the other. There may be significant differences between parts as specified in their data sheets, but the name used is not an indication of these differences. Like most SDRAM, they use SSTL_2 or SSTL_18 as the signaling technology. As mentioned above, GDDR3 is a newer generation of devices than GDDR2, which itself is the generation after GDDR. Similarly, DDR2 SGRAM is a newer generation of devices following DDR SGRAM, which followed SDR (single data rate) SGRAM.

15.2 ZBT, NoBL, ZeroSB, and NtRAM SRAM

ZBT (Zero Bus Turnaround), NoBL (No Bus Latency), and ZeroSB (Zero Speed Bump) SRAMs, and NtRAM (No turnaround RAM) are equivalent SRAM devices that have been designed to eliminate dead clock cycles between read-to-write and write-to-read cycle transitions. ZBT™ is a trademark of Integrated Device Technology, Inc. (IDT), and IDT has cross-licensed the ZBT name to both Micron and Motorola. NoBL SRAM is offered by Cypress, where NoBL and ZBT SRAM are equivalent in form, fit, and function. ZeroSB SRAM is offered by NEC and NtRAM is offered by Samsung and Toshiba. The major application for this type of memory is network equipment.

These devices are synchronous static RAMs fabricated with advanced CMOS technology using full six-transistor memory cells. The devices are powered from either 3.3 V, 3 V, or 2.5 V and have either 3.3 V or 2.5 V LVTTL-compatible inputs and outputs. Many devices come in both pipelined and flow-through versions and have densities ranging from 2 Mb to 18 Mb, organized as 64 Kb × 32, 128 Kb × 36, 256 Kb × 18, 256 Kb × 36, 512 Kb × 18, 512 Kb × 36, or 1 Mb × 18. Clock frequencies range from 83 MHz to 200 MHz and higher, with maximum clock access times from 8.5 ns down to 3.2 ns.

The main feature of these families of devices is that there is no waiting when switching between reads and writes or writes to reads as shown in Figure 15.1. Address and control signals are applied to the SRAM during one clock cycle, and two cycles later, the associated data cycle occurs, be it read or write. In the figure, the main signals that illustrate this feature are CLK (memory clock), R/W# (read/write#), ADDRESS, DATAIN, and DATAOUT. Notice that data always appears on the bus 2 clock cycles after a read cycle is initiated and that write data must appear on the bus 2 clock cycles after a write cycle is initiated. These timings are independent of whether the previous command was a read or a write.[5]

5. In Figure 15.1, CEN# is the synchronous clock enable input, and ADV/LD# is a synchronous input that is used to load the internal registers with new address and control when it is sampled low at the rising edge of the clock with the chip selected. CE1# and CE2# are synchronous active low chip enables. BW1#–BW4# are synchronous byte write enables and OE# is the asynchronous output enable. ("#" indicates active low.)

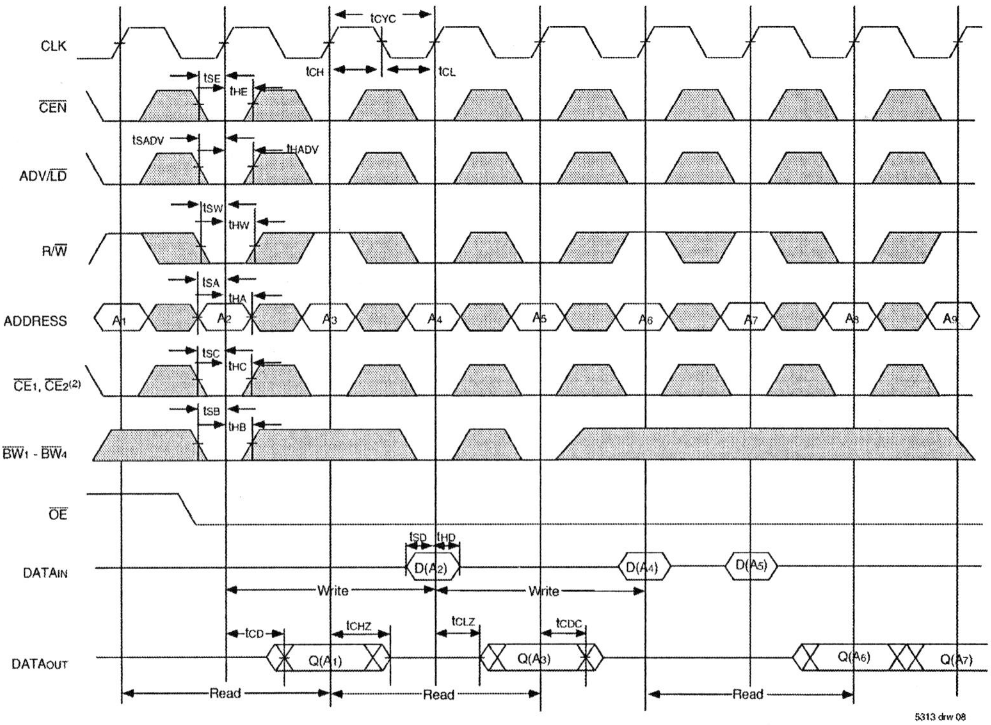

NOTES:

1. Q (A_1) represents the first output from the external address A_1. D (A_2) represents the input data to the SRAM corresponding to address A_2.
2. CE_2 timing transitions are identical but inverted to the $\overline{CE}_1$ and $\overline{CE}_2$ signals. For example, when $\overline{CE}_1$ and $\overline{CE}_2$ are LOW on this waveform, CE_2 is HIGH.
3. Individual Byte Write signals ($\overline{BW}x$) must be valid on all write and burst-write cycles. A write cycle is initiated when R/$\overline{W}$ signal is sampled LOW. The byte write information comes in two cycles before the actual data is presented to the SRAM.

Figure 15.1 Timing waveforms of combined read and write cycles for ZBT SRAM (IDT71T75802). [Idt03a] [Reprinted with permission of Integrated Device Technology, Inc.]

15.3 FCRAM (Fast Cycle RAM)

"FCRAM™ (Fast Cycle RAM)[6] is a new technology developed by the Fujitsu Corporation[7] that approaches the problem of DRAM/processor speed in a different way. With the improvements in CPU speeds over the last few years, designers have looked for a solution that would get them over the ever-present problem of relatively slower memory. Traditionally, various technologies such as EDO and SDRAM have attacked the problem with enhanced logic circuitry and peripherals that accessed the DRAM core. This resulted in advances, but these advances only targeted the interface to the DRAM, nothing else. While smaller process technologies allowed cores to go faster, the basic design stayed the same.

6. FCRAM is a trademark of Fujitsu Limited, Japan.
7. Toshiba Electronic Components is a second source for FCRAM devices.

FCRAM changed the DRAM core itself. The process included pipeline operation and core segmentation with the added benefit of power reduction. This new structure introduced advantages over traditional memory cores, which have remained virtually unchanged in the last two decades. One such advantage is the ability to send row and column information at the same time, as opposed to the standard sequential operation; in fact, a secondary command can be issued without needing to wait for the primary command to be completed. FCRAM cores are offered in memory chips with three different interfaces: SDR (single data rate), DDR (double data rate), and SRAM-like interfaces.[8] [Inn03a] FCRAM also uses a hidden precharge to reduce random access cycle time and uses a highly segmented memory core to reduce power consumption. In some devices, one segment of the memory core can remain powered and active while other segments of the memory core are turned off to save power." [Inn03a]

"The major features of FCRAM products are short t_{RC} (short row cycle time for high effective bandwidth), low power consumption, and flexible external interface. In recent years, the number of end products using DRAMs has increased rapidly in diverse market sectors. These range from computer products such as PCs, workstations, and servers, to consumer electronic products like digital televisions, digital still cameras, video disks, and set-top boxes, to telecommunications and networking-related products such as switches and routers, mobile/cellular phones, PDAs, and network servers.

Such a wide range of systems require various levels of performance and more diversified memory products, which has resulted in a segmentation of the DRAM market, based on applications and performance. However, conventional commodity DRAMs cannot satisfy all these varying requirements, so Application-Specific Memory (ASM) products are becoming essential, particularly for the rapidly evolving multimedia, networking, and graphics systems markets."[9] [Fuj03a]

FCRAM technically refers to just the memory core and is a type of DRAM since it is refreshed through a hidden precharge. The refresh is "hidden" as it is controlled by the FCRAM internally and does not need external control. "Precharge (reset) processing takes place automatically as operations are completed. From the outside, it appears as though precharge times have been eliminated." [Fuj01b]

"Address input can be made in parallel with the command input. With this pipelining scheme, it's possible to start a command operation while a data read or write operation is still in process, resulting in an excellent improvement of the cycle time." [Fuj03a]

FCRAM cores are used in three types of FCRAM memory devices:

- Network FCRAM — FCRAM with a DDR synchronous interface used in networking, graphics, and multimedia.
- Mobile FCRAM — FCRAM with an asynchronous SRAM interface for mobile phone applications.

8. Reprinted with permission of Innoventions Inc.
9. Reprinted with permission of Fujitsu Limited, "FIND," Vol.21 No.4, 2003, Technical Analysis, "Latest Trends for Mobile Phone Applications."

- Consumer FCRAM — FCRAM with a synchronous SDR SDRAM interface for use in consumer applications such as cameras, battery-driven devices, and car navigation systems.

Network FCRAM is also referred to as Network RAM and DDR FCRAM — Network FCRAM, Network RAM, and DDR FCRAM are the same thing. Network FCRAM tends to be high-density and high-speed (e.g., 512 Mb, 333 MHz/666 Mbps per pin) whereas Mobile FCRAM and Consumer FCRAM tend to be lower in speed and density (e.g., 16 Mb, 81 MHz/81 Mbps per pin). Mobile FCRAM and Consumer FCRAM are promoted by vendors as being high-speed and high-density, but this is in comparison to competitive memory devices used in mobile and consumer applications.

15.3.1 Network FCRAM

Network FCRAM provides fully synchronous operation using a DDR I/O where all data input and output is synchronized with both edges of the DQS data strobe. The clock is differential (CLK, CLK#) where output data (DQs and DQS) are aligned to the crossings of CLK and CLK#.

Network FCRAMs are addressed by row (upper address), column (lower address), and bank (typical FCRAMs have four banks). A memory access (a read or a write operation) is burst oriented, meaning that a memory access starts at a selected bank and address and continues for a set number in a programmed sequence.

15.3.2 Mobile FCRAM

"Before Mobile FCRAM, two types of memory were used in cellular phones: flash memory and low-power SRAM. Low-power SRAM was available and reasonably priced, but was limited to low densities. Mobile FCRAM was introduced as a competing technology to offer greater density, lower power, and faster cycle times. The Mobile FCRAM series adopted an asynchronous SRAM interface (the standard RAM in cellular phones) and refresh-free operation so that the cellular phone designer could upgrade easily to the memory without making major system architecture changes."[10] [Fuj01a]

Because the memory in the FCRAM is segmented, one memory segment may be fully powered and active while other segments are powered down. "The powered-down modes include SLEEP, NAP, and PARTIAL modes:

- SLEEP mode — Conventional power-down mode (1μs to 350μs exit time)
- NAP mode — Capable of high-speed recovery from power-down mode (1 μs + exit time)
- PARTIAL mode — Data in 8 Mb of 32 Mb devices or data in 16 Mb of 64 Mb devices can be retained

10. Reprinted with permission of Fujitsu Limited, "FIND," Vol.19 No.3, 2001, New Products.

In the conventional power-down mode (SLEEP mode) supported by a 16 Mb product, the exit time (t_{CHH}) from the power-down mode is 1 μs to 350 μs. In the NAP mode, the exit time (t_{CHHN}) is as short as 1 μs. In the conventional SLEEP mode, no data is retained in any area of the memory. In the PARTIAL mode, data retention is allowed in some areas and prohibited in others. Data can be retained in an 8 Mb area in a 32 Mb product and in a 16 Mb area in a 64 Mb product."[11] [Fuj02c]

"16 Mb, 32 Mb, and 64 Mb Mobile FCRAMs have been developed as ASMs for mobile phones. These ASMs simultaneously achieve a high-speed access time of 80/70/70 ns, a low power consumption with a standby current of 70/100/150 μA, and a power-down current of 10 μA.

Since mobile FCRAM is a pseudo-SRAM which adopts an asynchronous SRAM interface, it costs less than pure SRAM and can be easily introduced as a replacement for pure SRAM in existing systems."[11] [Fuj02b]

15.3.2.1 Stacked MCP (Multi-Chip Package)

"In the mobile phone industry, the largest demands by far are for miniaturization and increased functionality. However, high functionality can lead to product enlargement due to the increased number of parts. This dilemma must be solved by miniaturization of the parts themselves, and memory is no exception to this challenge.

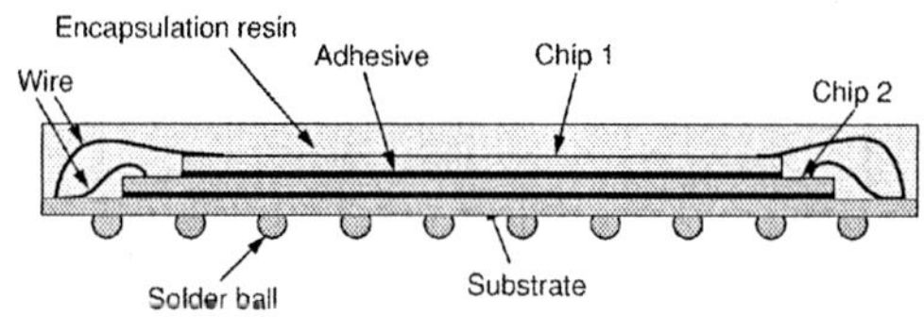

Figure 15.2 Structural diagram of stacked MCP. [Reprinted with permission of Fujitsu.]

Therefore, a commercialized, 2-chip, stacked MCP (see Figure 15.2) 9 mm × 9 mm series has been developed in which the substrate mounting area has been reduced to a maximum of approximately 61% of the area in existing MCPs with similar configurations. The two chips are usually an FCRAM core and a NOR-type, dual-operation flash memory. The 64 Mb NOR-type dual-operation flash memory mounted in all these products achieves a drastic improvement in chip miniaturization and electric features by means of new process technology. In turn, these improvements mean greater miniaturization and higher performance of the mobile devices." [Fuj02b]

15.3.3 Consumer FCRAM

"Consumer FCRAM is a new ASM (Application-Specific Memory) for consumer products and consists of two types of devices. The first device type is a packaged FCRAM which transfers data at faster speeds than the second product, which is a consumer FCRAM for SiP (System-in-a-Package). The SiP consumes very little power and occupies a smaller substrate area.

The consumer FCRAM family adopts an SDR (Single Data Rate) SDRAM interface that allows the system designer to easily proceed with system design using a conventional system configuration." [Fuj02a]

11. Reprinted with permission of Fujitsu Limited, "FIND," Vol.20 No.2, 2002, New Products.

15.3.3.1 The Low Latency of Consumer FCRAM

"Some Consumer FCRAM devices provide existing SDRAM capabilities with a CAS Latency = 2 (low latency), enabling one random access operation per two machine cycles. CAS (Column Address Strobe) Latency, or CL, is the delay time between instructions expressed as a number of clock cycles. CL is usually a number such as 1, 2, or 2.5. For CL = 2 devices, the maximum random access speed is 30 ns (for two 15 ns cycles at 66 MHz). In addition, these devices are designed to be easy to use with the widely used SDRAM interface to replace memories in existing systems.

Figure 15.3 shows a comparison of write operation timing between a general-purpose SDRAM and the FCRAM. In the general-purpose SDRAM (top of Figure 15.3), an active command (ACT) and row address are input at time t = 0, then a write command (WT) and write data are input at t = 2. At t = 4, a precharge command (PRE) is input, and the chip can move to the next random access command at t = 6.

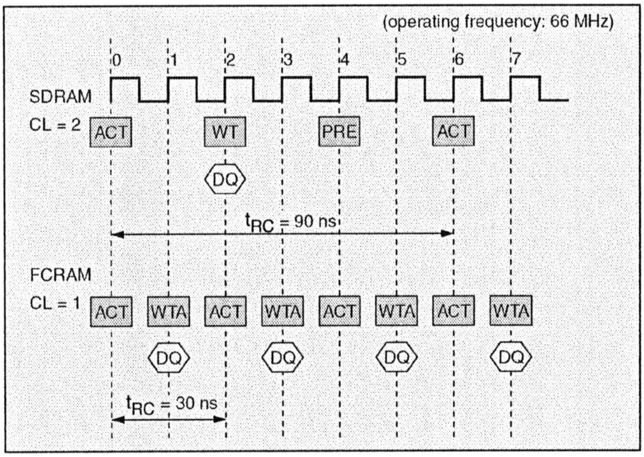

Figure 15.3 Write operation timing considerations: SDRAM vs. FCRAM. [Reprinted with permission of Fujitsu Microelectronics.]

By comparison, in the FCRAM, after the active command (ACT) and row address are input at t = 0, the write command (WTA) and write data are input on the next cycle at t = 1. There is no empty t_{RCD} latency (first access time) as in the general-purpose SDRAM, so that instructions can be packed together at maximum density with no wasted time. Because precharge operations are performed automatically within the chip, it is ready to go to the next random access operation at t = 2. The FCRAM can thus be said to have random access capability approximately three times that of a general-purpose SDRAM. In other words, the FCRAM can be said to have higher bus efficiency than the SDRAM at the same frequency.

Figure 15.4 shows a timing comparison of read operations (the read command is RDA). Here also, the FCRAM is able to pack instructions with no empty cycles in its read operations, achieving approximately three times the random access capability of a general-purpose SDRAM.

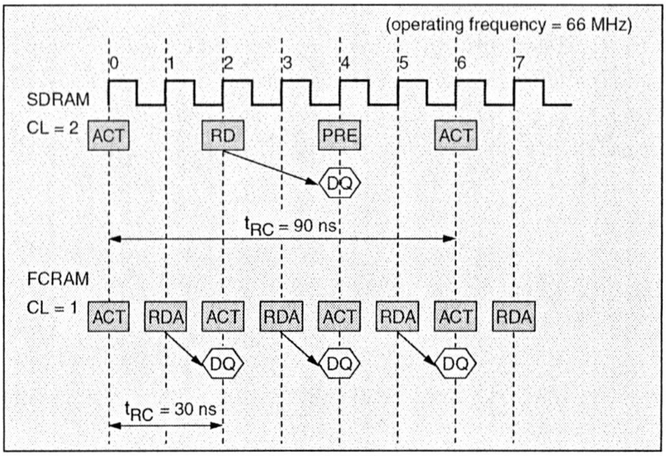

Figure 15.4 Read operation timing considerations: SDRAM vs. FCRAM. [Reprinted with permission of Fujitsu Microelectronics.]

Figures 15.5 and 15.6 show a comparison of first access time and cycle time between various devices of FCRAM, EDO, and SDRAM. In Figure 15.5, the top line, FC (16 M/CL = 1),[12] shows first access time and cycle time of an FCRAM operating at 66 MHz (15 ns cycle) with a latency of CL = 1. In comparison with a PC133/2-2-2 operation (133 MHz, 7.5 ns cycle, CL = 2), the first access time for the PC133/2-2-2 of t_{RAC} = 28.5 ns is longer than the FCRAM t_{RAC} of 25 ns, so that the FCRAM provides a faster operating speed. The comparison of cycle times shows that the PC133/2-2-2 product has a cycle time of t_{RC} = 52.5 ns, and the FCRAM is capable of operating as fast as t_{RC} = 30 ns, some 1.75 times faster than the PC133/2-2-2. Thus, even when operating at a frequency as low as 66 MHz, the FCRAM can provide high-speed performance surpassing a 133 MHz product.

Figure 15.6 shows the data from Figure 15.5 in terms of clock cycles. Because systems operate on the basis of clock cycles, this clearly illustrates the effective improvement in system performance. Here we see that the FCRAM's first access time is 1.5 times faster, and cycle time is 2.0 times faster, than the PC133/2-2-2 system.[13]

There are two additional considerations with respect to Consumer FCRAM worth mentioning. First, some memory controllers today do not support a CL = 1 operation. Switching to Consumer FCRAM may require modification to the memory controller. Second, in consideration of small-scale systems, the output buffer drive is limited to approximately half that of a PC memory."[14] [Fuj01b]

12. 16 M/CL = 1 means 16 Mb density device and a CAS Latency of 1 cycle (Column Address Strobe Latency).
13. See Chapter 13, Section 13.2.9.1, Speed Bin, on page 336 for the definition of "2-2-2."
14. Reprinted with permission of Fujitsu Limited, "FIND," Vol.19 No.1, 2001, New Products.

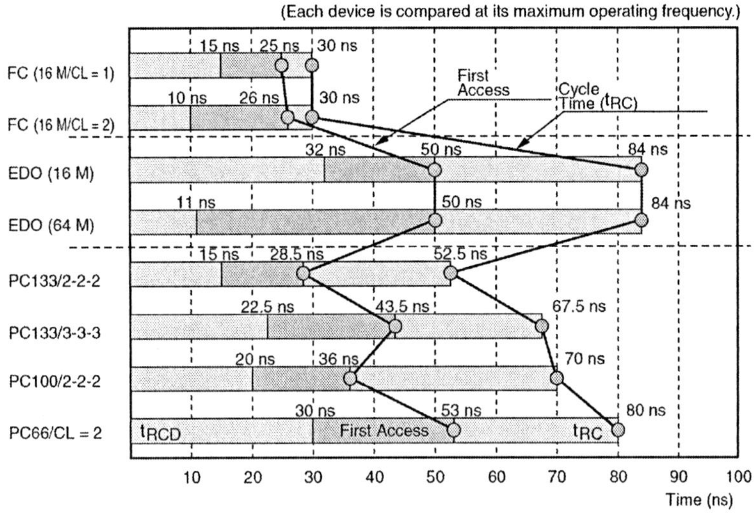

Figure 15.5 First access time (t_{RCD}) and cycle time (t_{RC}) simultaneous comparison. [Reprinted with permission of Fujitsu Microelectronics.]

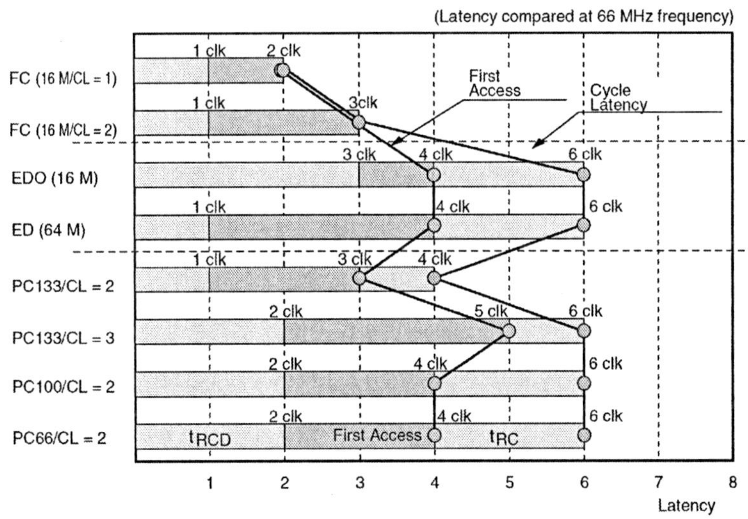

Figure 15.6 Latency comparison of first access time and cycle time. [Reprinted with permission of Fujitsu Microelectronics.]

15.3.3.2 Consumer FCRAM for SiP (System-in-a-Package)

"In the SiP (System-in-a-Package) technology, a logic chip and an FCRAM core are mounted in a single package to construct optimal systems and products. SiP makes it possible to consolidate products and technologies more cheaply than with SoC (System-on-a-Chip) configurations."[15] [Fuj02a]

The SiP results in reduced power supply current and voltage from the use of the FCRAM core, faster data transfer via the multi-bit output configuration, and optimal pad positioning for multi-chip mounting. Plane mounting and stack mounting are possible by shifting from the conventional center pad to edge pad. A SiP implementation can be mounted in a MCP such as that shown in Figure 15.2 on page 400.

"A direct connection between a logic chip and FCRAM in a single package reduces the number of external I/O pins, allowing these two components to function as parts of the system."[16] [Fuj02d]

As an example, "the MB811L646449 wafer has an input/output bus configuration with a 64-bit bus. This wide-bus configuration makes it possible to enclose the SiP-configured data bus wiring within the package without influencing the substrate's mounting area. The resulting data transfer rate, 5.18 Gbps at an operating frequency of 81 MHz, equals that obtained when a 32-bit SDRAM product is operated at 160 MHz or more." [Fuj02a]

15.3.4 Summary of FCRAM Features

Table 15.3 compares the applications and features of Network, Mobile, and Consumer FCRAM. The values in the table are taken from a sampling of devices to provide an indication of values for the different features. Because of this, no one device will conform to all of the values listed in any one column of this table.

Table 15.3 Representative values of Network, Mobile, and Consumer FCRAM[a]

Feature	Network FCRAM	Mobile FCRAM	Consumer FCRAM
Applications	Networking, Graphics, Multimedia	Mobile Phone, Cellular, PDAs	Cameras, Battery-Driven Devices, Car Navigation Systems
Interface	DDR (Synchronous) SSTL_2[b]	SRAM (Synchronous)	SDR SDRAM (Synchronous)
Packaging	60-ball BGA	48-ball FBGA 81-ball BGA MCP[c]	54-pin TSOP SiP MCP Wafer
V_{DD}	2.5 V ± 0.15 V	+2.3 V to 2.7 V +2.7 V to +3.1 V	+3.3 V ± 0.3 V +2.5 V ± 0.2 V 1.8 V

15. Reprinted with permission of Fujitsu Limited, "FIND," Vol.20 No.2, 2002, Technical Analysis.
16. Reprinted with permission of Fujitsu Limited, "FIND," Vol.20 No.2, 2002, New Products.

Table 15.3 Representative values of Network, Mobile, and Consumer FCRAM[a] (Continued)

Feature	Network FCRAM	Mobile FCRAM	Consumer FCRAM
$V_{DD}Q$ I/O Voltage	2.5 V ± 0.15 V	1.8 V	
Density	64, 256, 512 Mb	8, 16, 32, 64 Mb	16, 64 Mb
Clock Frequency	333 MHz	25 MHz [d]	66 MHz, 81 MHz
Bandwidth per Pin	666 Mbps	25 MHz	66 Mbps, 81 Mbps
Bus Width[e]	x8, x16, x18	x16	x64
Bandwidth per Device	To 11.988 Gbps+	400 Mbps	3.46 Gbps, 5.18 Gbps[f]
Power Modes	Power-Down	Sleep, Nap, Partial Power-Down	Power-Down
ICC1	840–1000 mA	25 mA	120–240 mA
t_{RC}	20–30 ns	70–80 ns	30–108 ns

a. Values in this table are taken from a sampling of devices and do not cover all FCRAM devices.

b. The SSTL_2 driver may be half-strength.

c. MCP is Multi-Chip Package; SiP is System-In-A-Package; FBGA is Fine-pitch Ball Grid Array.

d. Frequency and bandwidth are normally not specified for these types of devices, only latencies and cycle times.

e. Bus width is the number of data (DQ) pins per device.

f. This is Gbps (gigabits per second), not GB/s (Gigabytes per second), which was used in other tables in this chapter.

15.4 SigmaRAM (ΣRAM)

"SigmaRAM is a family of SRAM products jointly defined by the SigmaRAM Consortium. The consortium is a collaborative group of leading SRAM manufacturers who have joined together in establishing the industry's first open standard for networking SRAMs. Consisting of GSI Technology, Integrated Silicon Solutions Inc., and Sony Electronics Inc., the organization was formed with the goal of creating an open, JEDEC-standard memory product family targeting the high-speed networking and telecommunication markets.

Today's high-performance networking systems require a new kind of memory support. This includes very fast access time, high cycle rates, no bus turnaround penalty, low power, and high density. Meeting the demands of third-generation systems requires memory to be able to execute fast, random, multiple reads, change from reads to writes in one clock cycle, and sustain high data bandwidth. This describes the SigmaRAM family of synchronous SRAMs. Designed specifically for networking applications, the family of devices features higher speed, new packaging, an improved clocking scheme, multimode operation, and low power.

Clock speed is one of the key features of the SigmaRAM family. At a clock rate of 333 MHz, a 72-bit output can deliver up to 24 Gbps throughput. Compare this with 5–7 Gbps, the typical output of SRAMs designed for burst-cache operation.[17] The ability to change from reads

17. Cache SRAMs are typically narrower and slower. Metrics assume a x36 burst SRAM at 200 MHz with or without burst mode.

to writes in one clock cycle is another key feature supported in the SigmaRAM family. This is a necessity when performing fast lookup at OC192 packet speeds.

SigmaRAM packaging is also designed for networking. While legacy burst RAMs use advanced BGA packaging, the pinout favors the use of multilayer circuit boards requiring buried vias to reach pads deep in the array. SigmaRAMs were designed to minimize this problem by placing all signal pads within the first three outer rows. This maximizes the escape space and eases routing hassles. In addition, the omission of buried vias allows networking boards to utilize less expensive technology."[18] [Sig01a]

15.4.1 The SigmaRAM Family

"The SigmaRAM family consists of devices featuring common I/O buses. Each architecture optimizes a specific sequence of read and write commands. In addition, the family includes single and double data rate operation and supports late and double late write protocols in various combinations. The combination of wide bus widths, from x36 to x72, and high clock rates (up to 333 MHz) gives the SigmaRAM family very high performance." [Sig01a] See Figure 15.7.

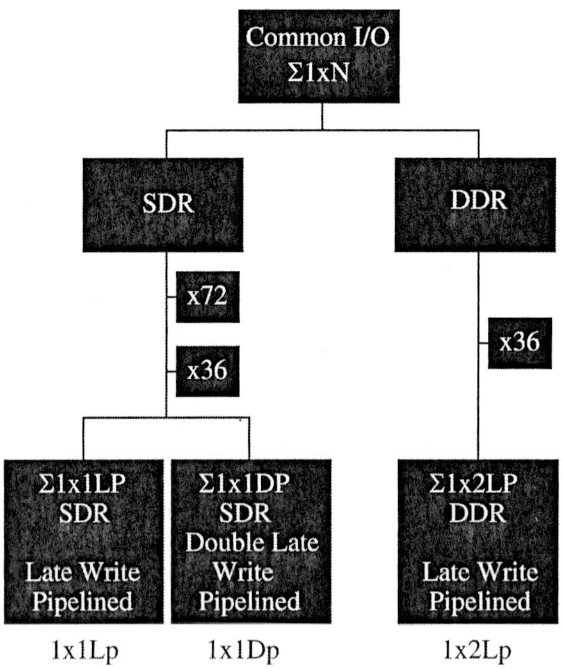

Figure 15.7 The SigmaRAM family includes common I/O SDR late write and double late write pipelined devices and common I/O DDR late write pipelined devices. [Sig01a] [Reprinted with permission of the SigmaRAM Consortium.]

18. Reprinted with permission of the SigmaRAM Consortium.

15.4.2 SigmaRAM Features

SigmaRAM comes with an LVCMOS or HSTL interface on common I/O devices. Modes of operation include late- and double-late-write modes, and pipelined read mode. SigmaRAM devices have echo clocks, programmable output drive strength, byte-write operation, and programmable chip enables.

"SigmaRAM supports the wide bus structures and fast random-address cycle times that traditional SRAMs cannot handle effectively. SigmaRAM also consolidates many different SRAM architectures and operation modes into a single product family for performance and design flexibility. These devices offer the widest data bus of 72 bits in the common I/O single data rate (SDR) architecture. SigmaRAM devices operating at 333 MHz are capable of sustaining up to 24 Gbps data throughput with entirely random addressing. Common I/O versions share one data bus, yet incur no penalty when alternating between reads and writes. These versions also offer double data rate (DDR) architectures that can sustain high throughput with fewer pins.

SigmaRAM devices use a 209-pin, ball grid array (BGA) package that provides twin benefits of scalability from 18 Mb to 144 Mb densities and ease in routing over traditional SRAM packaging for workstation and server applications.

Additional features of SigmaRAM devices include low power consumption resulting from a 1.8 V supply voltage and JEDEC-standard 1.5 and 1.8 V interfaces, a source synchronous interface (via echo clocks), programmable chip enables for easy depth expansion, programmable output drive strength, late and double late-write modes, pipeline read mode, coherent reads and writes, byte-write operation, IEEE 1149.1 JTAG boundary-scan testability, clamshell pinouts, and burst mode.

SigmaRAM device product roadmaps and datasheets are available at the SigmaRAM Consortium Web site (*www.sigmaram.com*)." [Sig01b]

15.5 RLDRAM (Reduced Latency DRAM)

"RLDRAM (Reduced Latency DRAM) was codeveloped by Micron Technology and Infineon Technologies. RLDRAM offers high density, high bandwidth, reduced cycle time, and SRAM-like access. These four features are the defining characteristics of RLDRAM. The unique internal memory architecture allows for an ultra-fast random access and closes the gap between DRAM and the faster, but more expensive SRAM. RLDRAMs are ideal for switch and router applications, and also for other high-bandwidth, high-speed, and latency-sensitive applications, such as L3 cache in high-end PCs or servers." [Xil03b] The participation of both Infineon and Micron ensure multiple sources for RLDRAM by developing pin- and function-compatible products. RLDRAM tends to compete with Network FCRAM (Network Fast Cycle RAM) in networking applications.

> Technical information on RLDRAM (Reduced Latency DRAM) is available at these Web sites: *www.rldram.com*, *www.micron.com/rldram*, and *www.infineon.com/memory/rldram*.

15.5.1 RLDRAM as a New Memory Standard for High-Speed Applications

"Data traffic in today's networks is steadily increasing. Growing user numbers, faster access systems, and emerging applications (such as virtual private networks and video-on-demand) call for sophisticated processing of large data packets at high speed.

The key building blocks of networks — high performance routers and switches — have to contend with spiraling data volumes. To meet this challenge, they need very fast memory chips that offer high density — ideally a fast SRAM chip with DRAM-like density of 256 Mb or more. Up until now, however, these haven't existed.

Fast random access does not just mean pushing the frequency up to pump in or out more data per second. The memory requirements of routers and switches differ from those of PCs and notebooks (where burst access mode generally suffices). Fast random access means that the memory chip allows the interleaved storage of small data packets at random addresses, enabling these packets to be read out at high speeds. Fast random access is ideally suited to the needs of routers and switches and cannot be achieved with standard burst-oriented memories that are optimized to move large chunks of data. The RLDRAM is based on a new internal memory architecture, which allows ultrafast random access that is twice the speed of standard DRAM memories." [Inf02a] [Source: Infineon Technologies AG, "A New Memory Standard for High-Speed Applications," Feb 2002.]

15.5.1.1 Networking Requirements

"The network infrastructure industry needs DRAM devices which are designed to achieve fast random access and high density. SDR (single data rate), DDR (double data rate) and other conventional memories have a rather long row cycle time (t_{RC}) of more than 60 ns. Row cycle time is the delay before you may access the same region (the bank) of a chip again. (By comparison, CAS latency (column address strobe latency) is the delay in clock cycles between registration of a read command and availability of the first output data.)

A t_{RC} of 60 ns is too slow for networking data processing, as a significant percentage of data packets in today's networks are small (around 40 bytes in size). Assuming a data rate of 40 Gbps, one of these packets is received and one is transmitted every 10 ns. This is possible today using fast SRAMs with access times of around 5 ns (and less). However, this is an extremely expensive solution for packet buffer memories between 128 MB and 512 MB.

The density requirements for OC-768 are even higher (up to 4 GB). In short, the DRAM used in next-generation high-end routers and switches must support fast random access, high densities, and high data throughput.

RLDRAM II supports higher data rates in next-generation routers and switches and offers a high density of 576 Mb with a short t_{RC} of 20 ns. In addition, RLDRAM is divided into 8 banks instead of the 4 banks commonly used in other memory solutions. By managing the data flow in the banks, that is, by cycling through banks in sequence, a very high effective data throughput rate is achieved.

Low t_{RC} (20 ns) and 8 banks allow fast random access with less bank conflicts than other memory architectures. SRAM-like access means that the whole read or write command together with the total address can be applied in one cycle. Many other memory architectures need two cycles for this. The fact that RLDRAM can do it in one cycle enables patterns with more read and write accesses, resulting in higher data throughput.

For RLDRAM II, three bit widths are possible: x9, x18, and x36. The highest possible operating frequency is 400 MHz using a DDR interface (i.e., DDR 800), which means that a maximum sustained bandwidth of 3.6 Gbps can be achieved for one device. Lower speed grades of 200 and 300 MHz are also available." [Inf02a]

Xilinx (e.g., Virtex-II) and other FPGAs, as well as ASICs, can be used as memory controllers for RLDRAM I and II devices.

15.5.1.2 Programmable Impedance Output Buffer

"The RLDRAM is equipped with programmable impedance output buffers. This allows a user to match the driver impedance to the system data bus impedance. To adjust the impedance, an external precision resistance (RQ) is connected between the ZQ pin and V_{SS}. The value of the resistor must be five times the desired impedance. For example, a 250 Ω resistor is required for an output impedance of 50 Ω. To ensure that the output impedance is one fifth of the value of RQ (within 15%), the range of RQ is 125 Ω to 300 Ω." [Inf03a] [Source: Infineon Technologies AG, HYB18RL28818/36AC Data Sheet, Version 1.60, July 2003.]

15.5.2 RLDRAM I and RLDRAM II

There are two generations of RLDRAM currently in the marketplace — RLDRAM I and RLDRAM II — with a roadmap for a third, RLDRAM III. Table 15.4 highlights some of the differences between the two current generations. RLDRAM III is roadmapped for operation at clock frequencies of 500 MHz to 1 GHz.

Table 15.4 Feature Comparison Between RLDRAM I and RLDRAM II

Feature	RLDRAM I	RLDRAM II
Density	256 Mb	288 Mb, 576 Mb
Clock Speed	200, 250, 300 MHz	200, 300, 400 MHz
Data Rate per Pin	400, 500, 600 Mbps	400, 600, 800 Mbps
Bandwidth per Chip, Maximum	2.4 GB/s	3.6 GB/s
Bit Width	x16, x32	x9, x18, x36
Interface	DDR (double data rate)	DDR (double data rate)
Error Checking Support	No	Yes
Organization	256 Mb (1 Mb × 8 banks × 32) 256 Mb (2 Mb × 8 banks × 16)	288 Mb (1 Mb × 8 banks × 36) 288 Mb (2 Mb × 8 banks × 18)
Row Cycle Time (t_{RC})	25 ns	20 ns
SRAM Type Interface	Yes	Yes

Table 15.4 Feature Comparison Between RLDRAM I and RLDRAM II (Continued)

Feature	RLDRAM I	RLDRAM II
Burst Length	2, 4	2, 4, 8
On-Die Termination (ODT)	No	Yes (switchable)
Output Driver Adjustment	No (50 Ω)	Yes (25–50 Ω programmable)
Write Clock per 18 bit	No	Yes
Balanced Read Latency, Write Latency	No	Yes
Optional Muxed Address Mode	No	Yes
Separate I/O	No	Yes (x9, x18)
Signaling Standard	HSTL	HSTL
V_{DD}	2.5 V, 1.8 V	1.8 V
I/O voltage $V_{DD}Q$	1.8 V	1.5 V, 1.8 V
Delay Locked Loop (DLL)	No	Yes
Banks	8	8
Package	144-ball T-FBGA 11 × 18.5 mm	144-ball P-TFBGA 11 × 18.5 mm
Differential Echo Clocks Synced with Data Out	Yes	Yes
IEEE 1149.1 JTAG	Yes	Yes
Technology	0.17 μm	0.11 μm

15.6 DDR SRAM (Double Data Rate SRAM — DDR, DDRII SRAM)

SRAM technologies include ZBT SRAM, NoBL SRAM, ZeroSB SRAM, NtRAM (these four are equivalents with different names given by different vendors), SigmaRAM, and QDR SRAM. These were discussed in detail previously in this chapter, except for QDR SRAM, which is discussed in Chapter 16 starting on page 429. ZBT SRAM and its three equivalents come in only SDR versions with either 3.3 V or 2.5 V LVCMOS I/Os. SigmaRAM vendors provide devices in both SDR and DDR versions, where they may be offered with 2.5 V low-voltage CMOS, 1.8 V very-low-voltage CMOS ("Expanded HSTL"), or 1.5 V HSTL I/Os. Figure 15.8 illustrates the evolution of SRAM over time as new technologies have led to higher and higher clock frequencies, shorter cycle times, and higher densities.

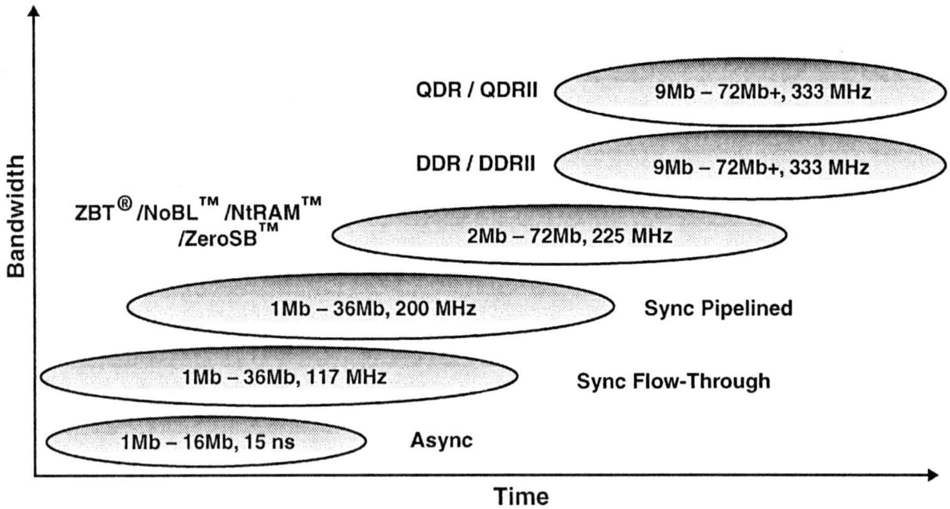

Figure 15.8 SRAM has evolved over time to exhibit higher clock frequencies, shorter cycle times, and higher densities. [Used with permission by Micron Technology, Inc. © 2004 Micron Technology, Inc. All rights reserved.]

15.6.1 DDRII SRAM Specified by QDR Consortium

"DDRII (Double Data Rate II) SRAMs[19] are a family of synchronous products defined and developed by the members of the QDR (Quad Data Rate) Consortium[20] which comprises Cypress, Hitachi, IDT, Micron, NEC, Renesas, and Samsung."[21] [Cyp02c] The QDR Consortium has defined DDR and DDR-II products, in addition to QDR SRAM products, where the DDR and DDR-II products are considered to be part of the QDR family. "The QDR product family includes DDR and DDR-II common I/O Burst 2, DDR common I/O Burst 4, and DDR separate I/O Burst 2 SRAM devices. Common I/O devices use the same bus for writing and reading whereas separate I/O devices have one bus for writing and a second bus for reading. 'Burst 4' refers to a burst size of 4 words. Each architecture is available in x8, x18, and x36 configurations."[22] [Mic01b] Please see Chapter 16, Quad Data Rate (QDR, QDRII) SRAM, starting on page 429 for more information about QDR and the QDR Consortium.

The QDR Consortium is the only entity which has provided a DDRII SRAM (2nd generation DDR SRAM) standard whereas both the QDR Consortium and individual vendors have defined DDR SRAM (1st generation DDR SRAM). The specifications among vendors for DDR SRAM and the DDR SRAM specification provided by the QDR Consortium can vary consider-

19. Industry documentation conventions leave out the dash between the "DDR" and the "II" when referring to SRAM (e.g., DDRII SRAM) whereas it includes a dash when referring to SDRAM, (e.g., DDR-II SDRAM).
20. Information on QDR technologies is available at the QDR Consortium Web site at *www.qdrsram.com*.
21. Reprinted with permission of Cypress Semiconductor Corporation.
22. 8,- 18-, and 36-bit widths.

ably. For example, DDR SRAM vendors may provide I/Os using either 2.5 V LVCMOS signaling technology or HSTL whereas the QDR Consortium uses only HSTL. Table 15.5 provides a feature comparison of DDR and DDRII SRAM as specified by the QDR Consortium.

Table 15.5 Feature Comparison of DDR and DDRII SRAM as Specified by the QDR Consortium [Mic03d]

Feature	DDRI	DDRII
Clock Frequency[a]		
2-Word Burst	N/A	333 MHz
4-Word Burst	200 MHz	333 MHz
Data Valid Window	1.4 ns @ 166 MHz	1.9 ns @ 166 MHz 0.98 ns @ 333 MHz
Latency	2 Cycles	2.5 Cycles
Density	9/18/36 Mb	18/36/72 Mb+
V_{CC}	2.5 V	1.8 V
I/O	HSTL (1.5 V or 1.8 V)[b]	HSTL (1.5 V or 1.8 V)
On-Die Termination	No	No
Programmable OCD[c]	No	Yes

a. These values are for the clock frequency. Data rates in Mbps per pin are twice these values.

b. HSTL 1.8 V is referred to as "Expanded HSTL" for which there are only vendor standards, but not a JEDEC standard.

c. OCD is off-chip driver.

15.6.2 Comparison of DDRII SRAM and DDR-II SDRAM I/Os

In Chapter 14, we saw that Double Data Rate II (DDR-II, DDR2) for SDRAM is specified by JEDEC standard JESD79-2, "DDR2 SDRAM Specification." This only specifies DDR-II (same as DDR2) for SDRAM, but not for SRAM. DDRII for SRAM is not specified by any JEDEC standard but only by the QDR Consortium. The fact that the two standards for DDR-II SDRAM and DDRII SRAM are different is easily realized when it's shown that DDR-II SDRAM uses SSTL_18 signaling whereas DDRII SRAM uses HSTL signaling for their I/Os.

15.7 Flash Memory

Flash memory is similar to EEPROM in that it is electrically erasable and programmable and retains its memory contents when power is removed (i.e., flash is nonvolatile memory). "The main difference between flash memory and traditional EEPROM is that flash memory requires data to be written or erased in blocks and EEPROM requires data to be written or erased one byte at a time."[23] [Web03] Because flash memory performs read and write operations a block at a time, it is generally faster than EEPROM and intended for different applications than EEPROM.

Flash memory is used in digital cellular phones, digital cameras, LAN switches, PC cards for notebook computers, digital set-top boxes, and embedded controllers. Flash is also used in digital camcorders, digital music players, handheld PCs, and global positioning devices. Flash can be implemented as embedded memory or in packaged, removable memory cards such as CompactFlash, MultiMediaCards, SmartMedia or SD (Secure Digital) Cards. At new, high density levels of 8 Gb, flash memory is now a viable choice for solid-state data storage, replacing magnetic tapes and low-density hard disk drives. NAND flash is also used in mobile applications, such as notebook PCs, tablet PCs, mobile handsets, MP3 players, and PDAs, that have space, weight, and power constraints.

Flash chips replaced earlier ROM BIOS chips in PCs so that the BIOS could be updated electronically in place instead of being replaced. As a result, flash memory is sometimes called "Flash ROM." Similarly, as flash memory replaced RAM devices, it was also sometimes referred to as "Flash RAM." Flash chips generally have lifespans from 100,000 to 300,000 write cycles.

Unlike DRAM and SRAM memory chips in which a single byte can be written, flash memory must be erased and written in fixed blocks, as mentioned before. These blocks typically range in size from 512 bytes up to 256 KB. Evolving out of EEPROM chip technology, which can be erased in place, flash memory is less expensive and more dense than EEPROM. The term was coined by Toshiba for its ability to be erased "in a flash."

15.7.1 Flash Densities, Access Times, and Other Features

Flash memories come in densities ranging from 1 Mb to 8 Gb and operate with random access times ranging from 45–120 ns. When flash is used in burst mode, subsequent accesses (following the initial random access) have access times as low as 8 ns. Most applications use flash devices with random access times of from 50–80 ns.

Flash operates at somewhat slower speeds than other types of memory. "Rarely updated, code-storage applications usually don't need superfast writes. They primarily require low-cost memories with simple interfaces that deliver high read throughput and low-cost packaging. NexFlash's four-signal SPI interfaces run at 33 MHz. NexFlash's latest generation SPI flash memories are available in 1, 2, and 4 Mb densities."[24] [Dip03a]

Some burst-mode flash devices may perform a typical 8-word sequential read at 40 MHz. For the same burst-mode device, but now operated in asynchronous mode, the read access time increases to 65 ns per byte. Other, faster, burst-mode flash devices have an initial access time of 54 ns followed by sequential byte transfers of 8 ns each (requiring a 125 MHz clock).

23. Reprinted with permission of Jupitermedia Corporation.

24. © Reed Business Information, a division of Reed Elsevier Inc.

Samsung has 4 Mb to 8 Gb NAND flash devices, including a 512 Mb NAND device with 50 ns serial page access, and a 70 nm 4 Gb NAND flash. Fujitsu's MirrorFlash ranges in density from 64 Mb to 256 Mb with a roadmap to 1 Gb and with random access times of 90 ns and subsequent burst access times of 25 ns.

Flash devices operate off of one of a variety of voltages, from 1.8 V to 2.2 V, 2.5 V, 3.3 V, and 5 V. NOR-type flash memories range from 8 Mb up to 64 Mb in 1.8 V families, from 2 Mb up to 128 Mb in 3 V families, and from 2 Mb up to 32 Mb in 5 V families. Flash devices powered at 1.8 V have access times ranging from 70 ns (13.5 ns subsequent burst-mode accesses) to 120 ns.

"Some flash memories include a Hidden-ROM (Hi-ROM) feature which provides a flash memory region that the system may access through a new command sequence. This is primarily intended for product designers who wish to use an electronic serial number (ESN) in the device with the ESN protected against modification. Once the Hi-ROM region is programmed, any further modification of that region is impossible. This ensures the security of the ESN once the product is shipped to the field."[25] [Fuj03b]

VHDL, Verilog, IBIS, and Denali simulation models are available for almost all flash parts.

15.7.2 Cell Phone Applications

"Cell phones generally have two processing elements in a handset, a DSP[26] and a microprocessor (e.g., an ARM[27] family processor). These processing elements handle different operations including the automatic hand-off from one cell to the next between carriers when moving, "roaming" features, and the correct calculation of billing for time used. The applications on the phone, such as µbrowser, J2ME virtual machine, or native applications such as games, may also be handled, even simultaneously, with cellular functions.

The memory used for the solution must be fast enough to keep up with processor requirements, yet be priced attractively enough to fit comfortably into the bill-of-materials for the total system. NOR flash meets this requirement with access times in the sub-70 ns range, with page-mode, or with 66–100 MHz synchronous (clocked) burst mode. Because of its speed, NOR flash allows the processor to execute code directly out of the flash, a feature called 'execute-in-place.' This feature negates the need for downloading a redundant copy into power hungry RAM for execution, and thus reduces parts count, power consumption, and system cost.

It is highly likely the cell phone will emulate the experience of the computer industry regarding the growth of memory content over time. In 2002, the 'sweet spot' for flash content in cell phones averaged slightly above 50 Mb across all phones shipped. By early 2004, that number had increased to over 200 Mb. Once again, Japan is laying down a potential path for others to follow with flash content in 3G phones approaching 300 Mb."[28] [Waw02]

25. Reprinted with permission of Fujitsu Limited, "FIND," Vol.21 No.4, 2003, Technical Analysis, "Latest Trends for Mobile Phone Applications."
26. DSP is digital signal processor.
27. ARM is Advanced RISC (Reduced Instruction Set Computer) Machines, Ltsd.
28. Reprinted with permission of the Semico Research Corporation.

NAND flash has started to play an important role in camera phones as the memory of choice in the external memory cards needed for storage of image files.

15.7.3 NOR vs. NAND Flash

NAND and NOR are the two major flash architectures. "NOR flash is characterized by fast random access to bytes or pages of information. NAND flash is characterized by slow initial access to pages of information with a serial readout from a buffer. NOR flash is best used in applications where code storage is required and data requirements are moderate (files are small but numerous, and accessed frequently). NAND flash is geared toward applications where files are large and infrequently accessed.

A comparison of page sizes of NOR and NAND flash shows that NOR flash has a 16 to 512 byte page while NAND flash has a 512 to 2K byte page. The byte-size granularity of NOR flash is helpful to designers who need to access small amounts of data quickly. This example also points out the usefulness of NOR parts that have the Read-While-Write (RWW) feature (see Section 15.7.5 on page 417). Effectively, a single NOR part can be internally partitioned so the processor can be writing to one array while simultaneously reading from the other array. This improves system performance and also removes the need for additional nonvolatile memory devices. Currently, NAND flash parts do not have this feature. Table 15.6 shows the comparison of page sizes between a 128 Mb NOR and NAND part along with some additional device parameters.

Table 15.6 Performance Comparison of 128 Mb NOR and NAND Flash Memory[a]

Parameter	Initial Access t_R	t_{RC}	t_{WC}	t_{PROG}	Page Size
NOR Flash	70 ns	80 ns	90 ns	218µs	32 Words (512 bits)
NAND Flash	35,000 ns (35µs)	50 ns	50 ns	200µs	528 Bytes (422 bits)

a. For NOR flash, initial access is the first access to data. For NAND flash, t_R is the memory cell array to starting access, t_{RC} is the read cycle time, t_{WC} is the write cycle time, and t_{PROG} is the programming time.

[Reprinted with permission of Semico Research Corporation.]

NOR flash will continue to be the memory type of choice on board the cell phone. NAND has entered the market as a removable memory resource and as an aftermarket storage add-on card. Even with the larger flash densities available today, the best solution seems to be the placement of the additional memory in external flash cards. The Sony Memory Stick and Secure Digital (SD) are good examples of solutions likely to be chosen by designers. Within these cards, it is also likely NAND flash will be the memory type used because it is designed to be used as a solid-state hard drive, reading sectors of information at a time versus reading words at a time (i.e., NOR). After the long first initial access, NAND can output large amounts of data quickly, which would fit applications like playing music or video files more closely than NOR.

NAND flash has already outstripped the densities required for the onboard nonvolatile memory in the phone. The growth of NAND flash density over time is more suited to meeting the needs of memory card vendors that will require very dense parts at very low cost per MB, but based on a minimum floor price.

In summary, each flash memory architecture has areas where it is best utilized:

- The better block granularity of NOR flash tends to be favored in applications for code storage where small amounts of data are written or accessed at frequent intervals.
- The true 'random access' nature of NOR fits these types of applications better than does NAND with its slower access times, allowing for direct execution from NOR flash.
- The ability of NAND to output data in the form of serial pages makes it a favorable solution for large file applications like playing MP3 files or viewing movies.
- The large block size of NAND best fits applications accessing large files that contain music or video data, functioning like a solid-state hard disk drive." [Waw02]

15.7.4 Significance of Block Size

"Flash is available in a variety of block sizes. The importance of this feature relates directly to the average size of each data sample or file your design stores in flash memory. For example, heavily compressed, black-and-white digital images take tens of kilobytes, whereas an uncompressed, high-resolution, 24-bit, color bitmap file may take tens of megabytes. Today's state-of-the-art audio-compression technologies typically achieve 20-to-1 compression ratios. A system capturing one-channel, 8-bit voice and sampling at an 8 KHz Nyquist frequency can store a 30 second compressed sample in a 12 KB file. On the other hand, a system capturing two-channel, 16-bit, high-fidelity sound at a 44 KHz sampling rate requires 264 KB to store the same 30 second sample and 5.28 MB to store it uncompressed.

The larger the block relative to the average sample size, the less efficiently you are able to use the media. This situation doesn't matter — up to a point — thanks to some clever engineering. As you selectively delete files, the flash-file system marks them invalid — not erasing them from the media — by programming file-header bits and using additional available space within the flash array to store replacement or additional samples. However, the array eventually becomes almost full of a combination of valid and 'deleted' files. The file system then must initialize a cleanup management function. The smaller the average file relative to block size, the more likely that a mix of valid and 'deleted' files resides in any block and the more media management the file system must do to create block-sized free space. Even if the file system performs this cleanup during periods when the system isn't accessing the flash (acting as a background task and minimizing performance impacts), the additional program and erase requirements impact power consumption.

Large-block-memory advocates point out that smaller blocks require more on-chip periphery circuits to decode and effectively isolate one block from all others, impacting die size and cost. However, this characteristic is technology-dependent as well; some approaches naturally lend themselves to smaller blocks. Cost is often the most important criterion for any memory, but a combination of secondary benefits may sometimes sway your decision to an alternative approach. A block that is significantly smaller than the average data sample or file may also complicate the design, burdening the flash-file system with multiple block erases per file operation and increasing power consumption.

Whenever you erase a block, you cycle it. When you write additional data to a partially programmed block but don't erase it, you don't cycle it. (Many flash-file systems exploit this important characteristic.) With increasing usage and subsequently higher cycle counts, a block's program and erase performance degrades because of the trapped charge in transistor oxides and other stress phenomena. Eventually, you will be unable to program or erase and thus will have to retire locations within the block — or even the entire block. All flash memories for data and file storage applications specify at least a minimum of 100,000 cycles. In most cases, this count exceeds your system's needs during its operating lifetime." [Dip97]

15.7.5 Read-While-Write (RWW) Flash

"The highest-volume applications for RWW devices are cellular phones and set-top boxes. However, customers are finding value utilizing simultaneous RWW flash memory in such varied applications as networking, PDAs, medical data recording devices, and handheld scanners."[29] [Amd03a]

Fujitsu offers a FLEXBank™ RWW architecture where the device is organized into four physical banks. This architecture has the capability of allowing a normal non-delayed read access from a non-busy bank of the array while an embedded write (either a program or an erase) operation is simultaneously taking place on any of the other three banks.

"Two banks in a RWW device cannot be written to at the same time. When using the RWW functionality, one of the simultaneous operations must be a read. The operation on another bank can be an erase or write." [Amd03a]

15.7.6 Page-Mode Flash Memory

"Just as flash memory overall performance increases, the internal architecture also becomes more complex. By comparing the three current flash architectures, standard, page mode, and burst mode, you can see how the complexity increases in order to achieve faster access times. The ellipses in Figure 15.9 show a relative performance versus complexity relationship. [Amd00a] The term 'performance' refers to the speed with which several read and write operations can be made. The major benefit of operating in page mode versus the standard mode is the greater speed.

29. © 2000, 2004 FASL LLC Reprinted with permission.

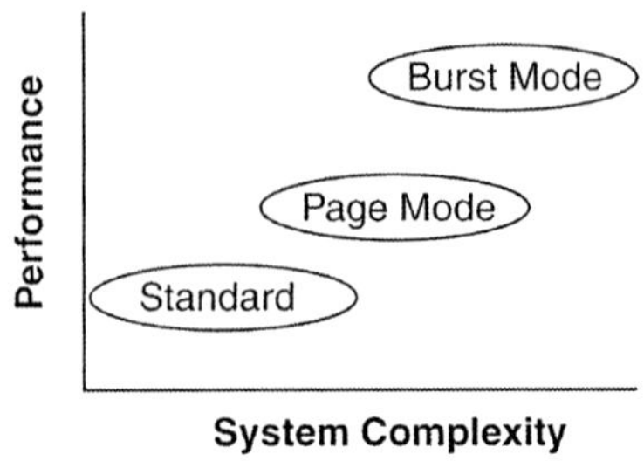

Figure 15.9 Flash memory performance increases with greater complexity. [Amd00a] [© 2000, 2004 FASL LLC. Reprinted with permission.]

A page-mode device is pin compatible with a standard flash device. This ensures an easy migration for a better performance part, with little added complexity to the overall system. If more performance is still desired, burst-mode devices can offer that performance. But the system complexity becomes much greater." [Amd00a]

15.7.6.1 Page Definition

"A 'page' is a small group of memory words that are accessed, internal to the memory, in parallel rather than one at a time. The time to reach the first word in the group is called the initial access time and is analogous to standard architecture flash memory access times. However, since all the words in the group are stored in an internal buffer following the initial access time, other words in the group can be delivered with a much reduced access time.

All pages in a memory device are the same size, but the size of the pages varies depending on the device. As an example, the page size of the Spansion™ S29GL512N device is 8 words, or 16 bytes, with the appropriate page being selected by the higher address bits A3–A19. The LSB bits A0–A2 (in word mode) and A-1[30] to A2 (in byte mode) select the specific word/byte within a page. The pages are therefore always aligned on an 8-word address boundary." [Amd00a]

15.7.6.2 How a Page-Mode Read Works

"The first read from the page-mode device is identical to a read from a standard flash device: CE# and OE# must go to V_{IL}, and a valid address must be placed on A19:A0. This first read has an access time t_{CE} or t_{ACC}, which is typical of a standard flash device. However, a subsequent page read access to a location anywhere within the same page is much faster. This access time is denoted as t_{PACC}. Fast page-mode accesses are obtained by keeping A3–A19 constant and changing A0–A2 to select the specific word, or changing A-1 to A2 to select the specific byte, within that page.

In our example using the Spansion S29GL512N device, t_{ACC} varies from 60 ns to 120 ns, but we will look at a device in which t_{ACC} = 90 ns, and t_{PACC} = 25 ns. Figure 15.10 clearly

30. In word mode, the DQ15/A-1 pin functions as DQ15 (MSB data input/output). In byte mode, this pin functions as A-1 (LSB address input).

depicts the initial access time (t_{ACC}) for the first read from a given page and the shorter, subsequent read access time from the page (t_{PACC}). Each read from within the page is triggered by changing the four address inputs, A-1, A0, A1, and A2." [Amd00a]

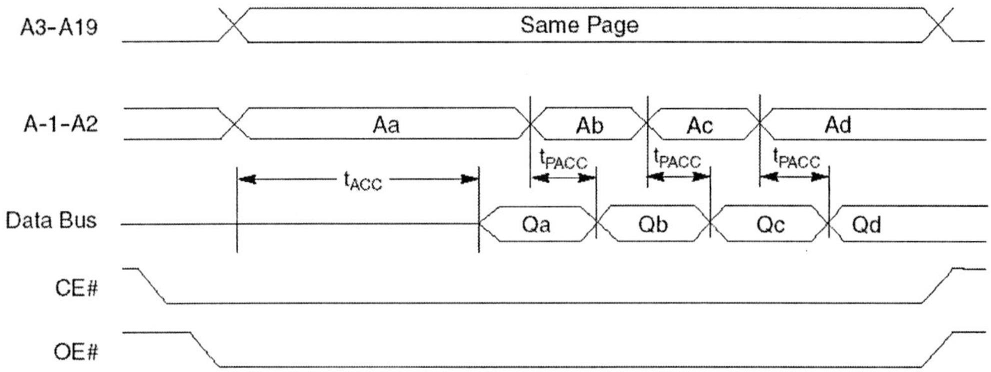

Figure 15.10 Timing diagram for the S29GL512N page-mode read (byte mode). [© 2000, 2004 FASL LLC. Reprinted with permission.]

15.7.6.3 Comparing a Page-Mode Device to a Standard Device

"By comparing a page-mode device and a standard device, the true advantage in speed can clearly be seen. Let's say that System A uses a page-mode flash memory device and System B uses a comparable standard flash memory device. Both systems are using the same processor and both flash devices have the same random access time. Table 15.7 shows how much time is required by both systems to read eight consecutive bytes from memory.

Table 15.7 Comparison of Read Times Between Page and Standard Devices

	System A: S29GL512N	System B: Comparable Standard Device
Byte 0	90 ns (t_{ACC})	90 ns (t_{ACC})
Byte 1	25 ns (t_{PACC})	90 ns
Byte 2	25 ns	90 ns
Byte 3	25 ns	90 ns
Byte 4	25 ns	90 ns
Byte 5	25 ns	90 ns
Byte 6	25 ns	90 ns
Byte 7	25 ns	90 ns
Total Time	**265 ns**	**720 ns**

[© 2000, 2004 FASL LLC Reprinted with permission.]

Both systems require a t_{ACC} of 90 ns to do the first random access read. However, System A can now do subsequent reads in 25 ns, whereas System B still requires 90 ns to complete each and every read command. By adding all the access times together, you see that System A, using a page-mode memory device, can read almost 2.5 times faster than System B." [Amd00a]

15.7.6.4 Requirements on a System Using Page-Mode Flash

"The system must provide memory interface logic that is aware of the page address boundaries and the difference in access times between an initial access and a subsequent access within the same page so that the access time (number of wait states) can be adjusted dynamically. Fortunately, many microprocessors have such interface logic already integrated into their memory interface." [Amd00a]

15.7.7 Burst-Mode Flash Memory

"Burst-mode devices offer improvements in system speed and performance by reducing sequential read access times. The burst-read capability offers an average access time reduction of more than 65% for an 8-word sequential read at 40 MHz. In the normal asynchronous mode, the read access time is 65 ns per byte. For a burst-mode operation, the initial access time is also 65 ns, followed by sequential byte transfers of 18 ns each.

Many burst-mode devices require four extra control pins to operate: Load Burst Address (LBA#), Burst Address Advance (BAA#), End of Burst Indicator (IND#), and a clock (CLK). LBA# indicates that the valid address is present on the address inputs. BAA# increments the address during the burst-mode operation. IND# Low indicates when the last word in the burst sequence is at the data outputs. Finally, the clock input can be tied to the system or microprocessor clock and provides the fundamental timing and internal operating frequency.

Many burst-mode devices have two different read modes: random read and burst-mode read. Random read is an asynchronous operation and is how data is normally read from a standard flash memory device. A valid address must be placed on the address lines and both CE# and OE# must be driven to V_{IL}. The valid data will be available on the data bus after a delay of t_{ACC}.

Burst-mode read is a synchronous operation that is tied to the rising edge of the clock. The microprocessor/microcontroller supplies only the initial address to the device. In the linear mode, the device delivers a continuous sequential word stream starting at the specified word and wraps around when the end of the internal 5-bit address counter is reached (11111). For example, if the initial address is xxxx0H,[31] the data is 0-1-1-3 . . . 28-29-30-31-0-1 . . .; if the initial address is xxxx2H, the data order is 2-3-4-5 . . . 28-29-30-31-0-1-2-3 . . . Data is repeated if more than 32 clocks are supplied.

31. H indicates a hex representation of the 19-bit address (A0:A18).

All subsequent addresses are automatically generated by the device at the rising edge of subsequent clock cycles by the assertion of the BAA# signal. Figure 15.11 shows the timing diagram of a burst-mode read operation.

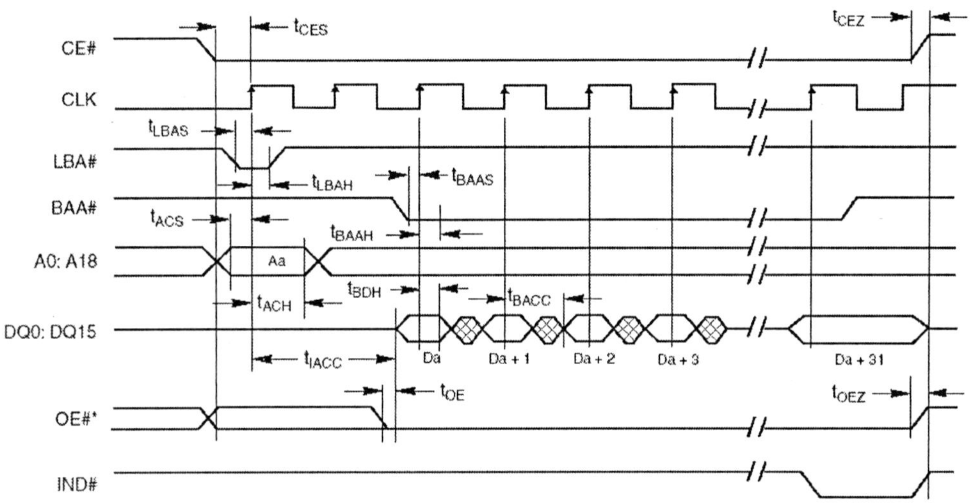

Figure 15.11 Timing diagram for burst-mode read operation. [Amd00b] [© 2000, 2004 FASL LLC Reprinted with permission.]

In Figure 15.11, notice how the various signals must act to achieve a burst-mode read. After the first address (labeled 'Aa') is stabilized, the LBA# signal is driven low at the first rising clock edge. This loads the first burst address into the device. The LBA# signal is then driven high and the address lines are 'don't care' for the remainder of the burst-mode read sequence.

After an initial access time of t_{IACC}, the data from the first address is available, and the device may now begin bursting the data out. The address is incremented by driving BAA# low (this is why the address lines are 'don't care'). The data is now available t_{BACC} after the previous read. At the end of the burst sequence, IND# goes low to indicate that the data is the last data in the sequence.

Table 15.8 compares the read times for burst and standard devices. In order to obtain optimum performance from a burst-mode device, the system's program should be structured such that the device can perform as many sequential reads as possible." [Amd00b]

15.7.8 MirrorFlash and Multi-Level Cell (MLC)

MirrorFlash™ and Multi-Level Cell (MLC) are two different architectures for increasing the density of flash memory devices. MirrorFlash is able to double the normal flash density by storing a full charge in each of two physically distinct locations within each cell. MLC, on the other hand, stores fractional levels of charge in one cell location.

Table 15.8 Comparison of Read Times Between Burst (System C) and Standard (System D) Devices

	System C: Am29BL162C	System D: Comparable Standard Device
Byte 0	65 ns (t_{ACC})	65 ns (t_{ACC})
Byte 1	18 ns (t_{PACC})	65 ns
Byte 2	18 ns	65 ns
Byte 3	18 ns	65 ns
Byte 4	18 ns	65 ns
Byte 5	18 ns	65 ns
Byte 6	18 ns	65 ns
Byte 7	18 ns	65 ns
Total time	**191 ns**	**520 ns**

15.7.9 Role of Software

"Cell phones have become rather sophisticated systems in their own right with the rise of continent-wide networks. The increasing demands of applications, cellular protocols, and resident operating systems now require cell phones to have a complex software mix resident in the device. The mix of code and data is further complicated by the addition of many different types of data in the phone. To this end, handset designers are increasingly turning to memory managers developed in conjunction with the hardware. For example, Intel Corporation has created software to handle the housekeeping function of code and data management in a phone. File systems and memory managers such as this allow cell phone OEMs to concentrate on evolving the features and functions of phones rather than writing software to solve issues that have already been resolved." [Waw02]

15.7.10 Flash Interfaces

"There are two types of flash interfaces. The first is the ATA interface, which has the same 512-byte block size as the standard hard disk sector. The second is the earlier linear flash, which is also used to execute a program directly from the chip (XIP). It requires Flash Translation Layer (FTL) or Flash File System (FFS) software to make it look like a disk drive."[32] [Com03]

32. Reproduced with permission from Computer Desktop Encyclopedia. Copyright © 1981–2003, The Computer Language Company Inc. All rights reserved.

15.7.11 Flash Cards

Flash memory chips are conveniently packaged as "flash cards" and come in several formats, including the full-size PC Card (ATA PC Card) and the smaller CompactFlash, SmartMedia, and similar formats:

"CompactFlash — A flash memory format invented by SanDisk Corporation. At 36.4 × 42.8 × 3.3 mm thick, it is about a third the size of a PC Card. It uses the same PC Card/ATA interface, but has 50 pins instead of 68, and cards support both 3.3 and 5 V operation. They can plug into a CompactFlash socket or into a standard Type II PC Card slot with an adapter. CompactFlash has become widely used for handheld digital devices.

A second-generation CompactFlash Type II card (CF2) increases the thickness from 3 to 5 mm, allowing for the inclusion of more electronics as well as a microminiaturized hard disk.

SmartMedia — An ultra-compact flash memory format developed by Toshiba. About the size of CompactFlash, but as thin as a credit card, SmartMedia cards are popular storage for digital cameras with capacities up to 128 MB. Available in 3.3 and 5 V variations, SmartMedia cards require no assembly in manufacture as they are actually flash memory chips in a unique chip package. The cards can be plugged into a SmartMedia socket or into a standard Type II PC Card slot with a PC Card adapter.

Memory Stick — A flash memory card from Sony designed for digital storage on cameras, camcorders, and other handheld devices. Introduced with 4 and 8 MB capacities, storage has increased to 128 MB. The tiny 2" long modules are about a tenth of an inch thick (0.85 × 1.97 × 0.11"). With digital cameras, the images are transferred to the computer by removing the card and inserting it into a PC Card adapter or Memory Stick drive. Cameras also use direct cable attachment to the computer's USB port.

Sony has also introduced Memory Sticks for use as I/O devices such as Memory Stick cameras, GPS systems, and Bluetooth modules; however, some units are only available in Japan.

Regular Memory Sticks have a dark lavender color, but there is a white version called 'MagicGate,' which came out when SDMI (Secure Digital Music Initiative) digital rights management was being introduced. The white MagicGate works like a regular Memory Stick, but also supports Sony's ATRAC3 audio recording format, which was used in concert with SDMI-compliant applications. Certain Sony music players such as Network Walkmans only accept the white MagicGate sticks.

Memory Stick Duo — In order to use Memory Sticks in cell phones and other small handheld devices, the Memory Stick Duo provides a smaller-footprint card that is inserted into an adapter that fits into a regular Memory Stick slot. The Duo card can then be transported from any Memory Stick device to a Duo device." [Com03]

15.8 FeRAM (Ferroelectric RAM) and MRAM (Magnetoresistive RAM)

"The contenders for imminent commercialization of next-generation nonvolatile memory appear to have been winnowed to two: ferroelectric RAM (FeRAM) and magnetoresistive RAM (MRAM)."[33] [Cle03]

"Hynix Semiconductor Inc. is sampling 4- and 8-Mb density FeRAMs that are manufactured in its 0.25-micron process. Texas Instruments Inc. has already built a 64-Mb FeRAM module on a 130-nanometer copper logic process, and will introduce embedded FeRAM in its 90-nm process. Japan's Oki Electric Industry Company has licensed a nondestructive readout (NDRO) FeRAM technology from Symetrix Corporation. Oki will serve as a foundry for this NDRO FeRAM using its 0.25-micron process, and the companies will collaborate on a 16-Mb device.

On the MRAM front, several companies are looking into MRAM's potential, including Cypress Semiconductor, IBM, Infineon Technologies, Motorola, and NEC. All of them see it as a universal replacement for SRAM, DRAM, and flash devices. Yet as promising as MRAM seems, it still lags behind FeRAM. Three high-profile companies — Hynix, Texas Instruments, and Oki — have announced advances on FeRAM products, which target such markets as cell phones, PDAs, and smart cards." [Cle03]

"MRAM is a nonvolatile memory technology that uses magnetic, thin film elements on a silicon substrate. Data is written and read by pulsing wires that are perpendicular to each other with one set above and the other below the magnetic bits. In most MRAM designs, the magnetic orientation is linear (north or south). In VMRAM (Vertical MRAM) designs, the magnetic elements are washer-shaped like the early magnetic cores, and the magnetic direction is clockwise or counterclockwise." [Com03]

"The basic attractions of MRAM are its speed and nonvolatility. MRAM can reach theoretical write times down to 2.3 nanoseconds. That's more than 1,000 times faster than the fastest nonvolatile flash and is 20 times faster than FeRAM. MRAM access times are as fast as 3 ns, or 20 times faster than DRAM. And since such reads require only 2 milliamps, MRAM consumes less than 1/100 the energy of DRAM.

MRAM's resistance to radiation is also one of the reasons it is considered a prime replacement for SRAM, which is expected to suffer more and more from density-induced, soft-error rates as it scales below 0.1 micron.

The first hurdle is manufacturability in a high-volume fab. Even extremely slight variations in the oxide, on the order of 0.1 angstrom, can change the magnetic tunneling junction resistance of MRAM by several percentage points, making the device unpredictable. But consistently keeping within the required oxide thickness in a real-world fab environment is tough, especially across an 8-inch wafer. The most serious problem is resistance variations, which can be handled with a self-reference sensing scheme.

33. Copyright © 2003 by CMP Media LLC, 600 Community Drive, Manhasset, NY 11030. Reprinted from EETIMES with permission.

Samsung recently entered the race to commercialize MRAM, a good sign since the company is known for executing well on high-volume, relatively high-margin products. Others have made progress, too. Motorola, which has researched MRAM for several years, is expected to sample devices soon as well as release a 4-Mb device for embedded systems. Toshiba and NEC are working together to develop core technology for a 256-Mb MRAM." [Cle03] However, some industry experts still feel that commercial use of MRAM is still years away.

"With all this progress in alternative technologies, one might think the companies sticking to traditional flash would be worried. They are — a little bit. But since these next-generation memories have been talked about for roughly a generation, nobody is losing sleep. Companies are continuing to focus on the technology in silicon now and to make it remain competitive even when the hopes and dreams of these universal memory types come true." [Cle03]

15.9 Memory Selection Guide

Please see the memory selection guide on the following two pages (Table 15.9). The selection guide compares the features of DDR SDRAM, RLDRAM, FCRAM, QDR SRAM, and DDR-II SRAM.

15.10 Exercises

15-1. (a) For GDDR3 technology, draw a diagram showing two data waveforms, one with a slew rate of 2 V/ns and the other with a slew rate of 4 V/ns. Add a differential clock to your drawing showing the true and complementary signals, and also add a reference voltage V_{REF} to the drawing. Have V_{REF} pass through the crossover voltages of the differential clock. (b) Show how power consumption with slew rate varies by use of this diagram. (c) Is the power consumption versus slew rate a linear function?

15-2. What transmission line impedances should be used for the PCB data traces connecting to GDDR3 SDRAM? Are these single-ended or differential characteristic impedances?

15-3. Among the following, what is the one key characteristic that differentiates each group of devices from other devices? (a) graphics memories, (b) SBT, NoBL, and ZeroSB SRAM and NtRAM, (c) FCRAM, (d) SigmaRAM, (e) RLDRAM, (f) flash, (g) FeRAM, and (h) MRAM.

15-4. Of the four extra control pins required for the burst-mode flash memory device, which pins are inputs and which are outputs? What does the signal on each pin indicate?

15-5. What is application-specific memory (ASM)?

15-6. What is the difference between an SiP and an SOC?

Table 15.9 Feature Comparison for DDR SDRAM, RLDRAM, FCRAM, QDR SRAM, and DDR-II SRAM

Xilinx Memory Selection Criteria										
Feature	DDR SDRAM		RLDRAM		FCRAM		QDR SRAM		DDR-II SRAM
	DDR-I	DDR-II	RL-I	RL-II	FC-I	FC-II	QDR-I	QDR-II	
Performance (MHz)	100-400	200-400	200,250,300	200,300,400	154-267	154-267	154-267	154-267	100-300
Xilinx Supported Performance (MHz)	200			200	154-267	333	66-200	100-300	
Density (Mb)	Components: 64Mb-512 Mb DIMMs: 32 Mb - 2G	256Mb-2Gb	256Mb	288Mb, 576Mb	256Mb, 512 Mb	288 Mb	0.5 – 2 Mb	8-36 Mb	9-36 Mb
I/O Standard (V)	SSTL-2.5V, Class I/II	SSTL 1.8V, Class I/II	HSTL 1.8V	HSTL 1.8V	SSTL 2.5, Class I/II	SSTL 1.8V, Class I/II	HSTL 2.5 V	HSTL 1.8V	HSTL 1.8V
Important Features	High density, high bandwidth	High density, has ODT (On Die termination), OCD (Off chip calibration), AL (Additive Latency)	High Bandwidth, High speed and reduced latency, DRAM with SRAM performance		High bandwidth, high speed, reduced latency, DRAM with SRAM performance		High bandwidth, high speed, reduced latency	High bandwidth, high speed, reduced latency, On Die termination	High bandwidth, high speed, reduced latency
Applications	Desk top PCs, servers, workstations, networking, graphics and mass storage		Switch and router applications and PC/L3 cache		Used in multimedia, high speed servers, mobile phones, networking applications like switch and routers and consumer applications like digital video camcorders and DVD recorders		Cache memory, buffering, FIFOs, LUTs, simultaneous read and write with equal ratio		Cache memory, LUTs, bursts of reads or writes
Relative Cost Comparison	Cheaper than other DRAMs and SRAMs		Higher than DDR SDRAMs		Higher than DDR SDRAMs		Higher than DRAMs		Higher than DRAMs
Burst Lengths	2,4,8	4,8	2,4	2,4,8	2,4	2,4	2,4	2,4	2,4
Component Data Widths	Components: 8,16,32 DIMMs: 32,64,72	Components: 4,8,16 DIMMs: 64, 72	16,32	9,18,36	8,16	9,18,36	8, 9, 18	8, 9, 18, 36, 72	8, 9, 18
No. of Banks	4	4	8	8	4	4	N/A	N/A	N/A

(continued)

Feature	DDR SDRAM		RLDRAM		FCRAM		QDR SRAM		DDR-II SRAM
	DDR-I	DDR-II	RL-I	RL-II	FC-I	FC-II	QDR-I	QDR-II	
Read Latency	Cas latency (2,2.5,3 clk cycles)	i) When AL=0, RL = Cas latency& RCD requirements similar to DDR-I. ii) When AL not zero, RL = CAS Latency + AL. No RCD requirements	Cas latency (5,6 clk cycles)		Cas latency (3, 4 clk cycles)	Cas latency (4,5,6,7clk cycles)	1.5 clock cycles		1.5 clock cycles
Read Command to Write Command Time	Burst length/2 + cas latency cycles	Burst length/2 +read latency cycles	1 clk cycle while addressing different banks, 8 clk cycles while addressing the same bank		Burst length/2 + Cas latency cycles		No latency		No latency
Write Command to Read Command Time	1 clk cycle after write burst	2 clk cycles	1 clk cycle while addressing different banks, 8 clk cycles while addressing the same bank		2.5 clk cycles		N/A	N/A	N/A
Row Cycle Time	12 clk cycles	13-15 clk cycles	5-8 clk cycles		5 clk cycles		N/A	N/A	N/A
Data Strobe	Non-free running	Differential strobe - Non-free running	Free running	Differential strobe -Free running	Non- free running	Unidirectional Free or non free running	Free running read and write clocks		Free running read and write clocks
Data Bits Per Strobe	8, 16, 32	8	8,16	9, 18 (for 36 bit wide device)	8	9, 18 (for 36 bit wide device)	One differential pair per device		One differential pair per
Read Data Validity	Data valid after cas latency cycles		Data valid signal from memory		Data valid after cas latency cycles		N/A	N/A	N/A
Data Access	Row activation required before column can be accessed		Row and column can be addressed together.		Row activation required before column can be accessed		N/A	N/A	N/A
Initialization Requirements	Initialization sequence required		Initialization sequence required		Initialization sequence required		No initialization sequence required		No initialization sequence required
Auto Refresh Requirements	Required every 15.6 us or up to 140.6 us during auto refresh	Refresh command required every 7.8 us	Required every 8 us.	Required every 3.9 us	Required every 7.8 us	Required every 3.9 us	No refresh required		No refresh required
External Precharge Requirements	Yes	Yes	No	No	No	No	No	No	No
Major Vendors	Samsung, Micron, Infineon		Micron, Infineon Page 2		Samsung, Toshiba, Fujitsu		Cypress, Samsung, IDT, NEC, Hitachi		Cypress, Samsung, IDT, NEC,

15-7. What does the term “SRAM-like access” mean?

15-8. (a) Which memory technology offers the widest data bus width in a single chip? (b) How many bits make up the bus?

15-9. When you select a flash memory chip based upon its block size, what do you base your block size selection upon?

15-10. Read-while-write is implemented in what kind of flash memory?

15-11. What is the major difference between a page-mode flash device and a burst-mode flash device for read operations? In both, reads following the initial access are much faster as can be seen by comparing Table 15.7 on page 419 with Table 15.8 on page 422.

15-12. What advantage does a read-while-write feature offer?

15-13. What are the three power-down modes available in Mobile FCRAM? Describe each mode.

15-14. What kind of interface does RLDRAM have? Why does this offer it an advantage?

15-15. For what types of applications would you use NOR flash memory and for what types of applications would you use NAND flash memory?

CHAPTER 16

Quad Data Rate (QDR, QDRII) SRAM

Chapter Objectives

- Introduce the QDR (quad data rate) Consortium and the QDR product family consisting of both QDR and DDR (double data rate) products.
- Show how QDR products have separate input and output ports that operate independently and simultaneously where each port is equipped with a DDR interface.
- Explain the types of applications for which QDR devices are best suited.
- Discuss the QDR clocking scheme, which takes into account the flight time delays of both the clock and data.
- Compare the features of QDR with the newer generation QDRII and explain their functional differences.
- Discuss echo clocks and echo clock timings.

16.1 Introduction to QDR

"QDR™ (quad data rate) SRAMs are a family of synchronous products defined and developed by the members of the QDR Consortium, which comprises Cypress, Hitachi, IDT, Micron, NEC, Renesas, and Samsung. Together, these members provide 70% of the synchronous SRAM volume worldwide. The QDR Consortium has defined two categories of QDR products: QDRI (also referred to as simply QDR) and QDRII."[1] [Cyp02c]

1. Reprinted with permission of Cypress Semiconductor Corporation.

> Information on QDR technologies is available at the QDR Consortium Web site at www.QDRSRAM.com.

The QDR Consortium has also defined DDR (double data rate) products that are considered to be part of the QDR family. "The QDR product family includes QDR Burst 2, QDR Burst 4, DDR common I/O Burst 2; DDR common I/O Burst 4, and DDR separate I/O Burst 2 SRAM devices. Each architecture is available in x8, x18, and x36 configurations.[2] In this chapter, we be discuss the QDRI and QDRII technologies.

The QDRII products operate at speeds of up to 333 MHz and are used in network switches, routers, and other high-bandwidth communication applications where they serve as the main memory for look-up tables, linked lists, statistics storage, and controller buffer memory. QDRII devices are used in OC-192 and OC-768 networking systems to achieve 36 Gbps or more of memory bandwidth. QDRII devices provide a data valid window that is 65% of the clock cycle, providing a 1.3 ns data valid window at 250 MHz. They come in a 165-pin, 13 mm × 15 mm FBGA package designed for clamshell applications, which provides a 40% space savings over traditional packages."[3] [Mic01b]

"All members of QDRI and QDRII families have separate input and output ports that operate independently and simultaneously. Each port is equipped with a DDR interface in order to maximize data throughput. Two ports operating in a DDR manner provide four pieces of data being transferred during each clock cycle, thereby leading to the name, quad data rate. The separate ports completely eliminate the possibility of data contention between the SRAM and memory controller, a concern with conventional common I/O devices. Separate port-selects allow for easy depth and width expansion. Variable impedance output buffers provide flexibility in drive strength in order to match the impedance seen by the device." [Cyp02c]

QDR standards are proprietary to the QDR Consortium. Only package and pinout information is offered as a JEDEC standard.

Controllers used with QDR devices include programmable devices such as the Xilinx Spartan-II and Altera APEX20KE FPGAs, as well as the Cypress' Delta39K CPLD. ASIC controllers also commonly drive QDR devices.

16.1.1 Choosing the Right SRAM

"QDR was specifically created for applications that have a nearly equal number of read and write cycles which must occur very close in time to each other. For example, applications which need to perform a sequence of reads and writes such as R-W-R-W . . ., R-R-W-R-R-W . . ., or R-W-W-R-W-W . . . precisely meet the conditions that favor QDR SRAMs over other SRAMs available today. Applications which average R/W (read/write) ratios greater than two or three (depending on the device burst length and clock frequency), or which employ data streaming,

2. 8-, 18-, and 36-bit widths.
3. Used with permission by Micron Technology, Inc., © 2004 Micron Technology, Inc. All rights reserved.

meet conditions which favor the DDR SRAM architecture." [Mic02a] Table 16.1 summarizes the optimal choices between QDR SRAM and DDR SRAM as a function of the burst length and ratio of reads to writes and writes to reads for the application.

Table 16.1 Optimal Choices Between QDR and DDR SRAM Devices[a]

Device	QDR	DDR SRAM
2-Word Burst	R/W < 3:1 2 Addresses / Clock	R/W > 3:1 1 Address / Clock
4-Word Burst	R/W < 2:1 1 Address / Clock	R/W > 2:1 1 Address / 2 Clocks

a. R/W means the ratio of read cycles to write cycles or the ratio of write cycles to read cycles.

"Most of the existing SRAM solutions are optimized for PC applications and have interfaces designed to move data efficiently for PC-type common I/O applications. In most networking applications, continuous movement of data through the SRAM is a necessity. In such applications, there are continuous transitions between read and writes through the memory. Common I/O devices like standard synchronous pipelined SRAMs do not perform well under these conditions.

No Bus Latency™ (NoBL) SRAMs have optimized the synchronous SRAM architecture to allow no latency in the read/write transitions and have 100% utilization of the I/O bus. In most networking applications, this improvement in throughput is not enough. Applications that need to have the read and write done simultaneously benefit from the QDR SRAM technology.

Applications such as ATM switches and routers also benefit from the fact that simultaneous reads and writes can be done on the QDR SRAM with the data through the QDR SRAM guaranteed even for simultaneous access to the same address location." [Cyp01a]

Figure 16.1 graphs the relative bandwidth per pin of all network SRAMs at a 250 MHz clock. It clearly illustrates that the optimal SRAM for an application is a function of the read-to-write ratio (and the write-to-read ratio) required for the application and the length of the burst.

SRAMs cost more than DRAMs when compared on a per-bit basis. However, in networking and other applications, SRAMs are more often selected over competing DRAM technologies because the performance (speed, latency, row cycle time) of the SRAMs is typically better. QDR SRAMs typically involve lower densities — for example, 9 Mb to 72 Mb with a roadmap to 288 Mb — than high-density SDRAMs, which can have densities as high as 512 Mb to 2 Gb.

"There are two implementations of the QDR architecture: 2-word burst and 4-word burst devices. Differences between them are shown in Table 16.2. There are two main design considerations in choosing between them other than the burst length: address rate and write data placement." [Mic02a]

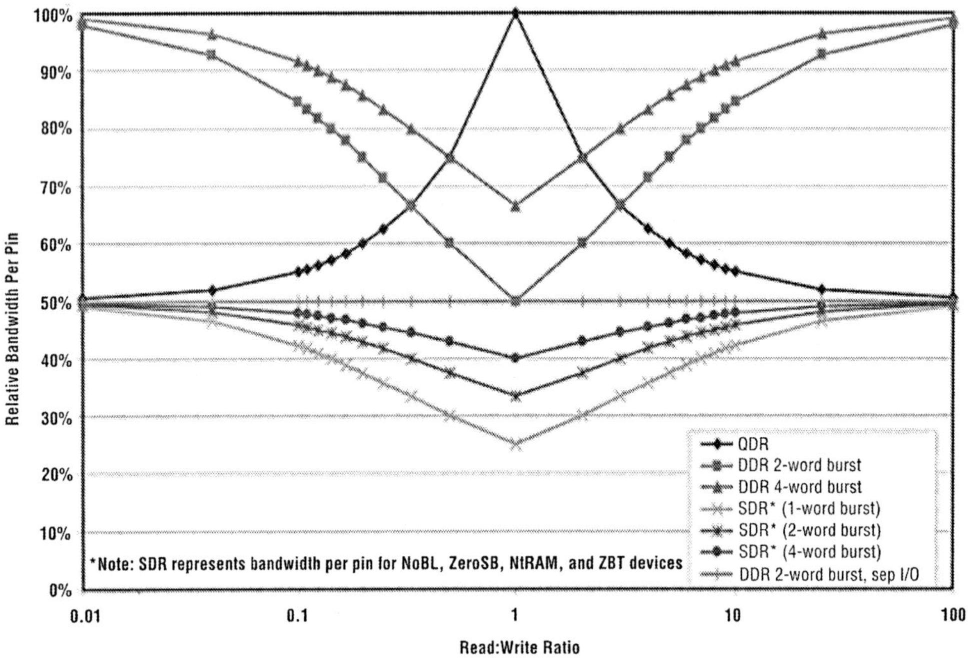

Figure 16.1 Relative bandwidth per pin of all network SRAMs at a 250 MHz clock. [Reprinted with permission of Micron Technology, Inc.]

Table 16.2 Differences Between 2-Word and 4-Word Burst QDR SRAMs[a]

Property	2-Word Burst	4-Word Burst
Address Rate	2 per Clock Cycle (DAR)	1 per Clock Cycle (SAR)
Supply First Input Data	At Time of Write Command	1 Clock Cycle after Write Command
Maximum Frequency First Generation Second Generation	 167 MHz 200 MHz	 200 MHz 333 MHz[b]
Bandwidth Parity with DDR	At R/W = 2 or W/R = 2	At R/W = 3 or W/R = 3
Write Address	Supplied at K# Rising Edge	Supplied at K Rising Edge
Data Transaction Size	2 Words	4 Words
Read Rate	2 Words per Clock Cycle	4 Words per 2 Clock Cycles
Write Rate	2 Words per Clock Cycle	4 Words per 2 Clock Cycles
Command Execution	Read from K↑ to K#↑ Write from K#↑ to K↑	Read from K↑ to K↑ Write from Next K↑ to K↑

a. "K#" is used to indicate the complement of clock signal K. "R/W" is used to indicate the average ratio of reads to writes or writes to reads, and "K↑" is used to indicate that the command execution takes place on the rising edge of clock K.

b. The first and second generations of devices referred to here are both for QDR (also referred to as QDRI), not for QDRII which is discussed later.

16.1.2 Address Rate

"The 2-word burst QDR SRAM can indefinitely sustain both a 2-word read and a 2-word write each clock cycle. Internally, the first half clock cycle is used to execute the read function, and the second half clock cycle is used to execute the write function. This requires accepting two addresses per clock cycle. The address bus is shared for the read and write data ports, necessitating double address rate (DAR) operation. The rising edge of master clock signal K (K↑) is used to register the read address. The rising edge of master clock signal K# (K#↑) is used to register the write address. K and K# are single-ended clock inputs to the SRAM, nominally 180° out of phase.

The 4-word burst QDR SRAM sustains both a 4-word read and a 4-word write every two clock cycles and internally utilizes the clock cycles differently than the 2-word burst device. The SRAM utilizes one clock cycle to perform an internal read. Data then exits the device for two consecutive clock cycles, resulting in four words of output data. The next clock cycle after a read, a write, can be then initiated, resulting in the need for four words of input data. This requires only one address per clock cycle. The rising edge of master clock signal K is used to register the read address. The next rising edge of master clock signal K is available to register the write address. If no write is required, that next rising edge of K cannot be used to initiate any new cycle. The 4-word burst cannot be terminated early; once it is initiated, it must finish. If the system can utilize the longer 4-word bursts of data, the design is unquestionably easier to implement since the effective address rate is one half as fast as the 2-word burst device. Also, given the same speed of silicon, the 4-word burst device can inherently be run at higher clock frequencies than the 2-word burst device. However, if the system architecture cannot operate on data bursts of that length, then the 2-word burst device is the obvious choice." [Mic02a]

16.1.3 Write Data Placement

"Write data placement is a relatively minor consideration. The 2-word burst and 4-word burst QDR SRAMs differ in the relative clock edges at which they capture input data for write cycles. Figure 16.2 illustrates this difference.

The 2-word burst device begins write data capture immediately upon write cycle indication: at the K rising edge, address, W#, and D1 are accepted; at the next K# rising edge, D2 are accepted and internally the write is executed. W# is the write enable, active low. D1 and D2 are the first and second data words of the 2-word burst. (D3 and D4 are the third and fourth data words of the 4-word burst below.)

The 4-word burst device begins write data capture one clock cycle after write cycle indication: at the first K rising edge, address and W# are accepted; at the following K rising edge, D1 are accepted; at the following K# rising edge, D2 are accepted; at the following K rising edge, D3 are accepted; at the following K# rising edge, D4 are accepted.

This difference in write placement has the minor effect of increasing the apparent write latency on the 4-word burst device. This difference is insignificant to system performance, although it must be correctly dealt with in the design." [Mic02a]

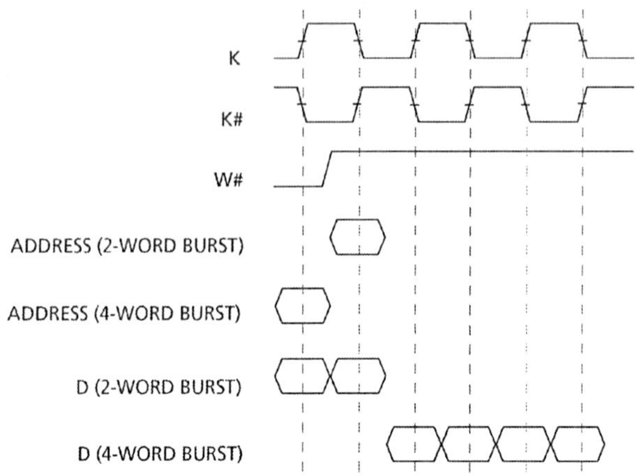

Figure 16.2 Write data placement. [Used with permission by Micron Technology, Inc., © 2004 Micron Technology, Inc., All rights reserved.]

16.1.4 Clocking Design

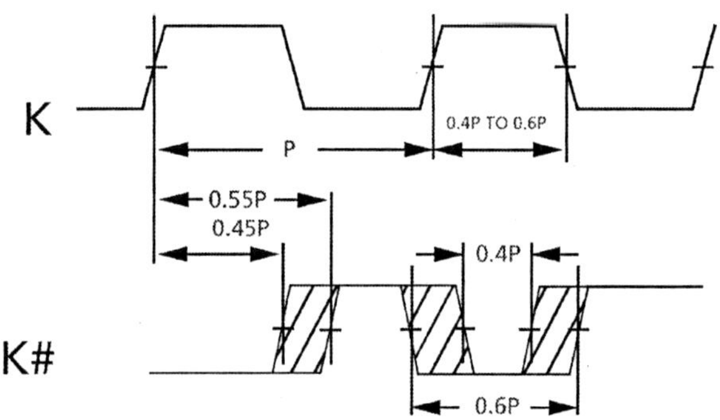

Figure 16.3 K to K# relationship. [Used with permission by Micron Technology, Inc., © 2004 Micron Technology, Inc., All rights reserved.]

"QDR SRAMs have two pairs of clock signals — K and K#, and C and C# — which are inputted to the device. K and K# comprise the master clock signals, nominally 180° out of phase. They may be generated directly by the controller chip (usually an ASIC with internal PLL or DLL), by a separate clock chip, or by an FPGA. Best system margins will be achieved when K and K# are exactly 180° out of phase (as in Figure 16.2), although the SRAMs will function correctly with the K# rising edge occurring at anywhere from 45% to 55% of the K clock period; that is, 162° to 198° out of phase. (See Figure 16.3.) Only rising edges of K and K# are used. The duty cycle of the individual K and K# clocks can range from 40/60% to 60/40%. This simplifies the task of generating the clocks. Figure 16.3 illustrates the allowable K–K# relationship for correct QDR SRAM operation. (Always consult the latest data sheets for specified timing relationships.)

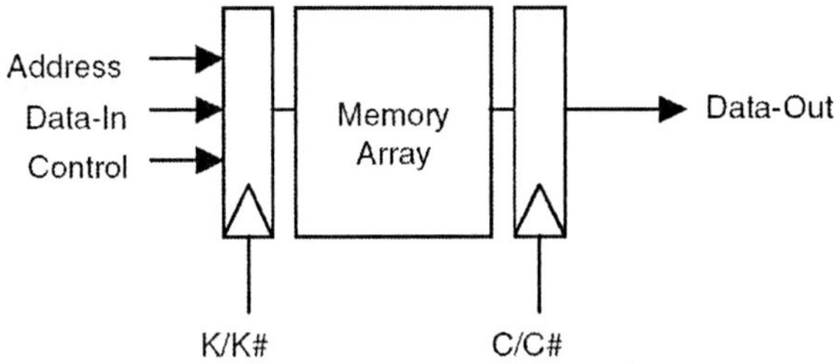

Figure 16.4 QDR SRAMs have both input and output clocks. [Reprinted with permission of Cypress Semiconductor Corporation.]

The second clock pair is C and C#, which comprise the output data clocks. This clock pair synchronizes the output timing of the output data bus (Q). Figure 16.4 illustrates the function of K/K# and C/C# in the simplified SRAM architecture. The time from K to C rising edges and K# to C# rising edges may be no sooner than coincident and no greater than 1/3 of the clock period at full device frequency operation. This limitation is necessary to ensure that output data is delivered before being overwritten by succeeding data. Also, C cannot precede K because data may not yet be internally available. For example, in the use of a 166 MHz clock, there may be from 0 to 2.0 ns skew from the K rising edge to the C rising edge, and similarly for K# to C#.

High frequencies bring in new complications and considerations. There are several ways of approaching the high-frequency design problem. The trick to making high-frequency systems work is to provide the maximum data windows to all devices in the system. Flight time delays of every individual signal must be considered." [Mic02a]

16.2 QDR SRAM Clocking Scheme

16.2.1 The QDR Device

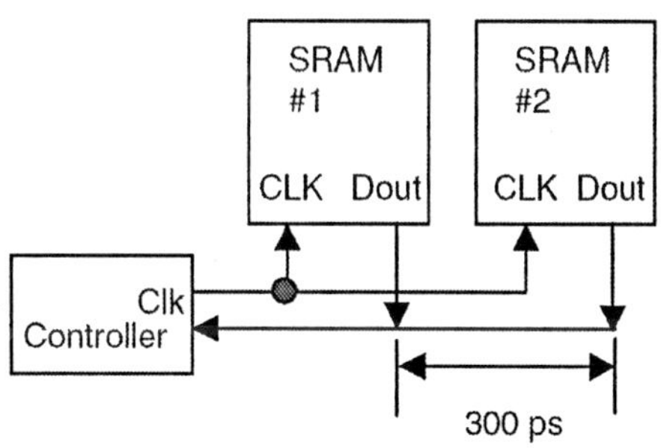

Figure 16.5 Flight time delays both the clock and output data. [Reprinted with permission of Cypress Semiconductor Corporation.]

"Most traditional synchronous SRAMs have a single clock input controlling both the input and output registers of the device. This approach works well in systems where the flight time from each SRAM to the receiving device is small when compared with the access time of the SRAM. When frequencies increase and access times shrink, flight times become significant. SRAMs farthest from the controller have the most difficult task during read operations. The first disadvantage is that the SRAM farthest from the controller receives the clock signal to trigger the output register during a read access a certain amount of time after each of the other SRAMs residing closer to the controller. As shown in Figure 16.5, it takes an additional 300 ps for the clock signal to reach SRAM #2 after reaching SRAM #1. The next disadvantage is that it will also take longer for the data generated by SRAM #2 to reach the controller than the data generated by SRAM #1. This, again, is due to the flight time difference between each SRAM and the controller. When the flight time of each SRAM becomes significant when compared with the access time of the SRAM, another approach is required.

The QDR SRAM family addresses this issue by providing a set of both input clocks (K/K#)[4] and output clocks (C/C#) as was shown in Figure 16.4. These independent clocks allow the user the ability to eliminate the flight time differences back to the controller, which is described in the next section." [Cyp00]

16.2.2 Using the Separate Input and Output Clocks

"One way of using the separate input and output clocks of the QDR SRAM is shown in Figure 16.6. It is important to note that the clock generated by the controller is fed to K/K# of the SRAM. This same clock signal is allowed to propagate past the farthest SRAM and is looped back to the controller. On the way back to the controller it is fed into the SRAM again and used as the output clock, C/C#. Of course, additional, separate clocks for each QDR SRAM may be needed to reduce loading and provide the signal integrity required. Each system designer will have to make his or her own trade-off.

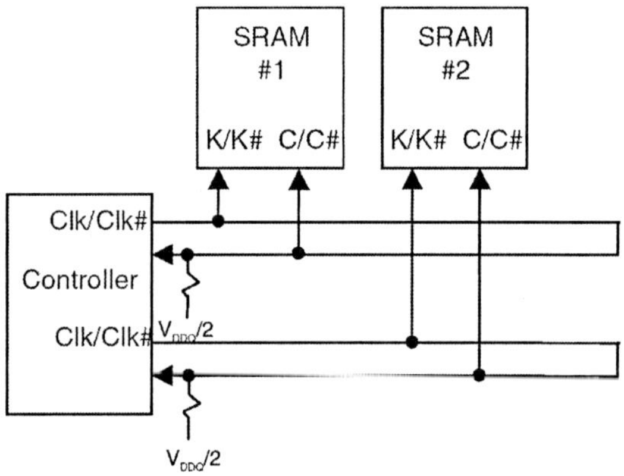

Figure 16.6 One approach to using separate input and output clocks of the QDR SRAMs. [Reprinted with permission of Cypress Semiconductor Corporation.]

First, let's evaluate the input timing to the SRAM. If care is taken to closely match the trace lengths of the signals from the controller to the SRAM (address, data, control, and clocks), then the only difference between the signals at the pins of each SRAM and those at the pins of the controller is the time skew caused by the flight time.

Let's take an example by assuming that an address is generated at t = 0 ns and a clock is generated at t = 1.0 ns. Both have a flight time to the SRAM of 0.2 ns. The address will arrive at the SRAM at t = 0.2 ns and the clock will arrive at 1.2 ns. The timing relationship between these two signals is maintained since both need to travel the same distance. This is true for all signals generated from the same controller if the trace lengths are tightly controlled. Therefore, generating the clock with the address, data, and control signals destined for the SRAM eliminates the flight time differences to each SRAM.

Now, lets take a look at the output timing in this configuration. First, notice that the clocks loop back to the controller after passing the last SRAM. On the way back to the controller, the out-

4. Both clocks are differential. The "#" symbol is used to indicate the complement of the clock signal, e.g., C# is the complement of C. Put differently, C is the true signal and C# is the complement signal of the differential pair.

put clock signals (C and C#) are delivered to the SRAM. By looping back the clock, the output clock is allowed to trigger the register of the SRAM farthest from the controller in advance of the other SRAM closer to the controller. By doing so, the device farthest away is given a 'head start' during read accesses when compared with the other SRAMs closer to the controller. In fact, the head start is the same as the flight time difference back to the controller. This technique will result in the data from all the SRAMs arriving at the controller at essentially the same time." [Cyp00]

16.2.3 Latching Data at the Controller

"The QDR family is able to provide an extremely high bandwidth of data. In order to take advantage of this, the user must be able to reliably latch the data at the receiving device. This section describes one method for latching the data provided by the SRAM.

In the implementation described in Section 16.2.2, the clock used to trigger the output register of the QDR SRAM is also fed back to the controller. The clock-to-data relationship at the pins of the SRAM is shown in Figure 16.7 at 167 MHz. Since both the data and clock have the same flight time back to the controller, the timing relationship between the two is maintained. As can be seen, clock #1 is responsible for clocking data 'A' out of the SRAM. Data 'A' is valid t_{CO} (clock to data output time) after the rising edge of the output clock C. Clock #2 (the rising edge of C# since the device only recognizes rising edges of the clocks) is responsible for clocking out data 'A+1.' In addition, the SRAM will 'hold' data 'A' for t_{DOH} (1.2 ns) from clock #2 (C#).

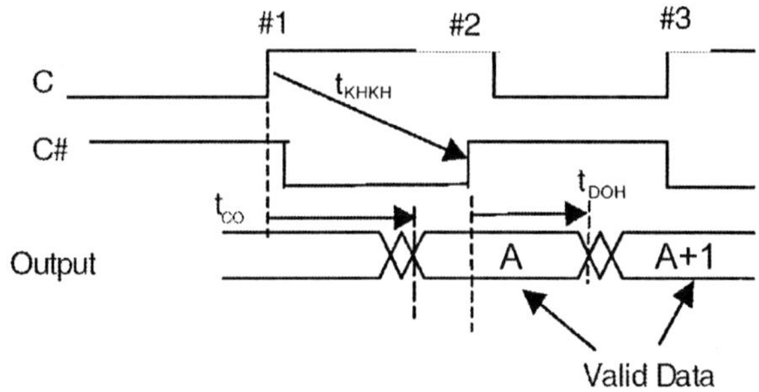

Figure 16.7 Timing diagram for QDR. [Reprinted with permission of Cypress Semiconductor Corporation.]

Let's assume that the QDR SRAM is being used at 167 MHz (t_{CYC} = 6.0 ns, the clock period). In addition, t_{KHKH} (rising edge of K to rising edge of K#) is a minimum of 2.7 ns. In addition, we will assume that t_{CO} = 2.5 ns and t_{DOH} = 1.2 ns. (t_{DOH} is the data hold time as shown in Figure 16.10 on page 442.) All of these are the worst-case spec limits for each parameter. Therefore, the worst case 'window' of valid data is:

$$\text{Window of valid data} = (t_{KHKH} - t_{CO}) + t_{DOH}$$

In this example, the window of valid data = (2.7 ns – 2.5 ns) + 1.2 ns = 1.4 ns.

As can be seen in Figure 16.7, the 1.4 ns 'window' of valid data is guaranteed in relation to the rising edge of C#. The data will be available 0.2 ns before the rising edge of C# (since t_{CO} is 0.2 ns less than the minimum t_{KHKH}) and held for 1.2 ns afterwards (the guaranteed value of

t_{DOH}). It is important to note that clock #1 is responsible for clocking out data 'A,' but in fact may not be the best clock to latch the data at the controller. In this example, C#, the clock on which data 'A' is held from, is the best clock with which to latch the data. Once the data is latched inside the controller, it can be synchronized with the controller clock." [Cyp00]

16.3 Comparison of QDR with QDRII

"While the QDR SRAM is able to satisfy bandwidth-hungry systems by providing four times the network performance of DDRII SRAM, a system designer has to overcome some difficulties in order to interface QDR to an FPGA or an ASIC. This is drawn from the fact that the data valid window is shorter on QDR, leaving little setup and hold time margins for the interfacing logic. The latest FPGAs have shorter setup and hold times, which eliminates this problem. But, nevertheless, there are many ASICs out in industry for which these timings will barely yield successful results. As the frequency of operation increases, latching of valid data from the QDR becomes difficult. This makes it unpredictable for a system to know when valid data from the QDR reaches the interfacing logic.

To address the above problems and to allow some flexibility to the users, QDRII was introduced by the QDR Consortium. This section covers the differences between QDR and QDRII and introduces the timings on QDRII. Lastly, it discusses some of the techniques which help in designing the memory portion of a system." [Cyp02c] A brief comparison of QDR with QDRII is provided in Table 16.3.

"QDRII is a family of synchronous SRAMs with an innovative architecture designed to help improve the bandwidth of QDR and also to increase the data valid window on the current QDR SRAM. QDRII SRAM is equipped with a pair of echo clocks (CQ, /CQ), which helps the

Table 16.3 QDR vs. QDRII Feature Comparison

Feature	QDR	QDRII
Frequency 2-Word Burst 4-Word Burst	 166 MHz 200 MHz	 250 MHz 333 MHz
Data Valid Window	1.4 ns @ 166 MHz	1.9 ns @ 166 MHz 0.98 ns @ 333 MHz
Data Hold Times	Positive	Negative
Read Latency	2 Cycles	2.5 Cycles
Clocks	No Echo Clocks	Echo Clocks
Clock to Data Output (t_{CO})	Longer	Shorter
Density	9/18/36 Mb	18/36/72 Mb+
I/O Signaling Standard	HSTL Class I	HSTL Class I
I/O Voltage	1.5 V	1.5 V
V_{CC} / Core Voltage	2.5 V	1.8 V
DLL Disable Option	No	Yes (DOFF Pin)

user in latching return data from the QDRII with known accuracy. Echo clocks are discussed at greater length later in this chapter. QDRII devices offered by all members of the QDR Consortium are compatible in terms of form, fit, and function.

Table 16.4 shows QDRII devices offered by Cypress Semiconductor Corporation. Speed grades ranging from 100 MHz to 333 MHz provide a wide variety of bandwidth options. QDRII SRAM operates with a 1.8 V core voltage and 1.5 V I/O supply. The I/O standard employed for QDR and QDRII is Class I HSTL.

Table 16.4 Devices in the Cypress Semiconductor Corporation QDRII Family

Device	Size	Density	Description
CY7C1310V18	2M × 8		
CY7C1312V18	1M × 18	18M	QDRII—Burst of 2
CY7C1314V18	512K × 36		
CY7C1311V18	2M × 8		
CY7C1313V18	1M × 18	18M	QDRII—Burst of 4
CY7C1315V18	512K × 36		

[Reprinted with permission of Cypress Semiconductor Corporation.]

Figure 16.8 shows the logic block diagram for the Cypress CY7C1315V18 18 Mb QDRII SRAM 4-word burst device. Note the presence of the input (K, /K), output (C, /C), and echo (CQ, /CQ) clocks in the figure." [Cyp02c]

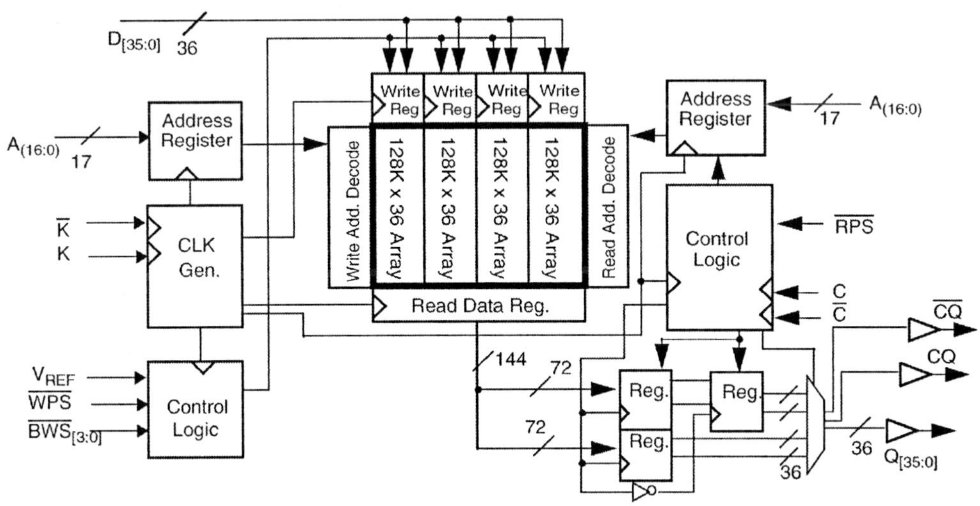

Figure 16.8 Logic block diagram for the Cypress CY7C1315V18 18 Mb QDRII SRAM 4-word burst architecture. [Reprinted with permission of Cypress Semiconductor Corporation.]

16.3.1 QDRII and QDR Functional Differences

"Although QDRII is designed to provide even higher performance than the standard QDR, and at the same time keep the functionality almost the same, some differences in functionality do exist.

There is no difference in write operations between QDRII and QDR. This includes write address placement and write data placement. This holds true for both burst of 2 and burst of 4 devices.

The only difference between QDR and QDRII exists in terms of read operations in the way data is outputted from the SRAM. There is no change in the way address and control signals are placed. The exception is that there is an additional half a clock cycle latency on QDRII on the first piece of data when performing a read. This means that during a read, instead of data being pumped out on the rising edge of C (or K in single clock mode) like on a standard QDR, the data is put out on the rising edge of /C (or /K in single clock domain) on QDRII. (As mentioned before, K and /K are the input source clocks and C and /C are the data output clocks.) This is the compensation that has to be paid for achieving a shorter t_{CO}, which are discussed shortly. Figure 16.9 shows the differences in read operation for both QDR and QDRII devices in a single clock domain for simplicity purposes. Only burst of 4 is shown here; burst of 2 follows similarly." [Cyp02c]

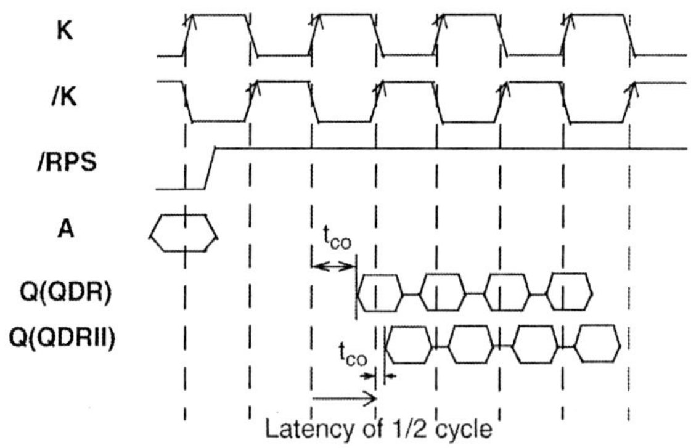

Figure 16.9 Read cycle on QDR and QDRII (burst of 4). [Reprinted with permission of Cypress Semiconductor Corporation.]

16.3.2 QDRII Features

"QDRII follows in the footsteps of QDR with some enhancements in terms of data valid window and echo clocks, besides an immense increase in frequency. Also, there is a difference in hold times, which are negative on QDRII." [Cyp02c]

16.3.2.1 Data Valid Window

"One of the biggest improvements QDRII provides over the standard QDR is the data valid window, which is 30% greater when compared between the SRAMs of the same frequencies. On a standard QDR, the clock to data out (t_{CO}) for the device running at 166 MHz is 2.5 ns and the data hold time (t_{doh}) is 1.2 ns, whereas on QDRII running at the same frequency, the t_{CO} is 0.4 ns and t_{doh} is –0.4 ns (more on the negative hold times shortly). Considering 50% duty cycle input clocks (3 ns pulse), the typical data valid window on the standard QDR running at 166 MHz is 1.4 ns, whereas the same is 1.9 ns on QDRII running at the same frequency. Figure 16.9 shows the t_{CO} parameter relative to the rising edges of the input clocks.

This improvement readily allows some of the latest ASICs and FPGAs with higher setup and hold times to interface to a QDRII SRAM." [Cyp02c]

16.3.2.2 Echo Clocks

"In a system, as the trace lengths between the SRAM controller and the QDR SRAM get longer and longer, it becomes difficult to predict and latch in the return data from the QDR SRAM. As a result, different schemes have to be adopted for different trace lengths in order to correctly capture the valid data. This is one of the problems that is overcome on QDRII with the help of the echo clocks.

Echo clocks are a pair of clocks that are outputs from the QDRII SRAM and are designated as CQ and /CQ. These clocks are nominally 180° out of phase with respect to each other. These clocks are always valid (free-running) and their frequency matches that of the input clocks (C and /C, or K and /K in the case of single clock domain). The rising edge of CQ is referenced to the rising edge of C (or K) and the falling edge of CQ is referenced to the rising edge of /C (or /K). The same is true for /CQ with its leading edge referenced to the leading edge of /C (or /K) and its trailing edge referenced to the leading edge of C (or K).

Echo clocks aid the user to correctly latch in the return data from QDRII SRAM at the controller. This is made possible by creating a constant relationship between the echo clocks and the data coming out of the QDRII SRAM. Every positive pulse of the echo clock is shaped in such a way that this pulse overlaps each data piece by 0.2 ns on both its leading and trailing edges. Figure 16.10 shows the relationship between the echo clocks and the output data of the QDRII SRAM." [Cyp02c]

16.3.3 Negative Hold Times

"Because of the DLL (delay-locked loop) built into QDRII devices in order to make the t_{CO} (clock to data output time) shorter and lengthen the data valid window, timings are significantly affected, especially the hold times. As the data has to be put out 0.4 ns after the rising edge of the clock (say C for 166 MHz), the minimum hold time has been made negative (to account for the slow process corner, low V_{CC}, and higher temperature) with respect to the next rising edge of the clock (/C) which will serve as the reference for the second piece of data. Figure 16.10 shows the negative hold times of both the echo clocks and the data (for a single clock domain).

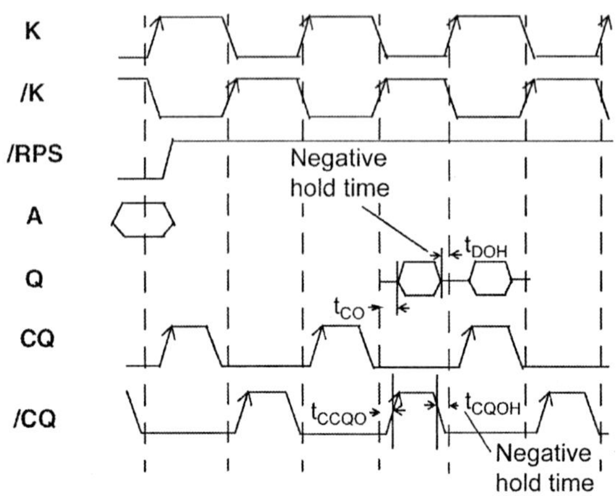

Figure 16.10 Echo clock timings on QDRII. [Reprinted with permission of Cypress Semiconductor Corporation.]

In Figure 16.10, t_{DOH} is the data output hold time after the rising edge of the K input clock, and t_{CQOH} is the echo clock hold time after the rise of the K input clock. t_{CQOH} can be referenced to either the rising edge of K or /K depending upon whether the hold time is being shown for CQ or for /CQ. Similarly, t_{DOH} can be referenced to either the rising edge of K or /K depending upon whether the hold time is being shown for the first data word or second data word. Both t_{DOH} and t_{CQOH} have negative values. t_{CCQO} is the time from the rising edge of the /K (or /C) clock to the echo clock being valid. As before, t_{CCQO} can be measured from either /K or K rising edges depending upon whether t_{CCQO} is being determined for CQ or /CQ." [Cyp02c]

16.3.4 Impact of DLL on the Operating Frequency Range

"The data valid window is considerably increased on the QDRII SRAM with the help of DLL circuitry, which resides in the SRAM. This helps in putting out the data earlier, specified by t_{CO} in the data sheet.

The DLL needs at least 1,024 clock cycles to lock up after power-up before any valid operations can be performed on the device. This includes the echo clocks as well; they will not be shaped and timed properly before 1,024 clock cycles.

The DLL is fine-tuned for a particular frequency range across every speed grade in which the device is offered. For example, from the data sheet, the minimum and maximum cycle times (t_{CYC} in the data sheet) for a 250 MHz device is specified as 4 and 5 ns respectively. This means that the frequency of the input clocks can only be between 250 MHz and 200 MHz, unlike any other synchronous SRAM, which can be operated all the way from 250 MHz down to 0 MHz.

Also, there is an option for turning off the DLL circuitry by tying the DOFF signal HIGH. It has to be noted under these conditions, the device no longer would operate at 250 MHz even if it was designated to, and that the maximum would be 200 MHz. Also, all the timings would be affected and would start to behave more or less like the standard QDR with degraded timings." [Cyp02c]

16.3.5 Choosing Between Burst of 2 and Burst of 4

"Like QDR, QDRII is offered in two kinds of implementation — burst of 2 and burst of 4. In terms of operation, QDRII implements the same kind of operation as QDR; that is, burst of 2 is associated with 2 pieces of data for every address location and burst of 4 is associated with 4 pieces of data for every address location. This is true for both reads and writes. The only difference to be noted is the additional half clock cycle latency that exists on QDRII during a read, as discussed before.

Both input and output data has to be always dealt with in a double data rate mode for both the devices. The exception is in the case of the address. On burst of 2, the address is toggled at double the frequency of the clock — the controller has to present two addresses, one for write and one for read every clock cycle if both reads and writes have to be performed every clock cycle. On burst of 4, read and write addresses are provided on different clock cycles, so the controller has to present only one address in every clock cycle or, in other words, needs to operate the address at just the clock frequency.

If a system can tolerate burst lengths of 4, then burst of 4 is easier to implement, as the address needs to be toggled only once per clock cycle. Also, burst of 4 can be operated at higher speeds than burst of 2. Moreover, for the same configuration, burst of 4 uses one address pin less than burst of 2, which might prove useful in certain systems." [Cyp02c]

16.3.6 Compatibility Considerations Between QDRII and QDR

"In terms of pinout, QDRII is completely compatible with QDR. The only extra pin that would be seen on QDRII is the DOFF pin used to enable or disable the internal DLL. The equivalent pin on QDR is a no-connect pin. So, effectively, the same foot pattern can be used for QDRII.

In both x8 and x16 configurations, pinout is defined in such a way that clamshelling (or mirror imaging) is possible, which results in easy routing of traces on the system board. In a x36 configuration, since all the no-connects that initially allowed for mirror imaging are taken by the extra data pins, clamshelling is not as easy. This requires some effort by the PCB designer if clamshelling is still preferred.

It is apparent that a x36 configuration can approximately replace two x16 SRAMs (no parity bits on x16), but at the same time, the user has to sacrifice 32 additional pins (without considering parity pins) on the controller side. It is a trade-off that has to be made by the system design engineer in choosing among the three configurations.

Since the initial latency for a read is different between QDR and QDRII and also, since the frequencies are different, the QDR controller should obviously be changed in order to accommodate QDRII. Lastly, the core power supply voltage at which QDRII operates is 1.8 V instead of the traditional 2.5 V utilized for the standard QDR." [Cyp02c]

16.3.7 Typical Implementation in a System

"It is preferable to have zero skew between K and /K and between C and/C. It is also equally recommended to have 50% duty cycle clocks. If a skew is introduced and the clocks are not symmetrical, then the echo clocks would not be shaped properly and also the data pieces would not be shaped symmetrically; as a result, setup and hold times may become an issue at the controller's side as one of the data pieces would be valid for a long time and the alternative piece would be shorter.

Let's look at some of the implementations as to how QDRII SRAMs sit on a system board and how the timings have to be set up from the QDRII controller's side to meet the setup and hold times on both the sides. There are two phases of operations to be discussed in terms of timings — writing into QDRII and reading back from QDRII." [Cyp02c]

16.3.7.1 Write Timings

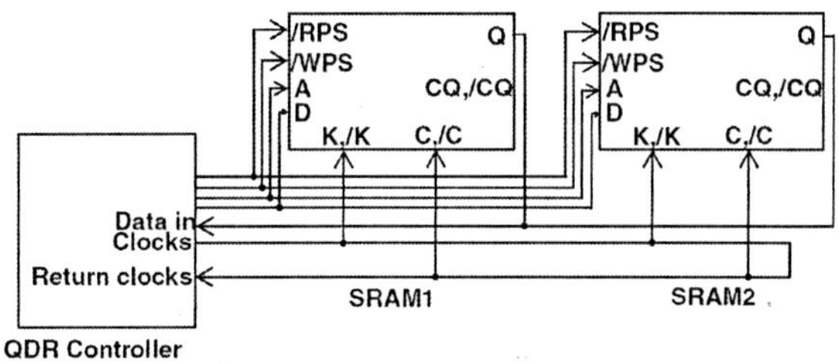

Figure 16.11 C and /C as the return clocks (width expansion). [Reprinted with permission of Cypress Semiconductor Corporation.]

"Some assumptions have to be made in terms of the delay between the two QDRII SRAMs and the delay between the controller and the QDRII. Although the easier case is to deal with two QDRII SRAMs sitting equidistant from the controller, we'll deal with the tougher case where the SRAMs are sitting side by side and the controller is sitting beside the first SRAM. Also, all the trace lengths are considered to be equidistant. Considering the 250 MHz burst of 2 SRAMs with 50% duty cycle and 180° phase difference clocks, let us assume that the trace length from the controller to the first SRAM is 3" and the trace length from the first SRAM to the second SRAM is 1". Figure 16.11 shows a scheme where a QDR controller (say, an ASIC) is interfaced to two QDR SRAMs.

Traditionally, one inch of trace corresponds to 0.2 ns of delay. Reckoning the time at 0 ns, /WPS (write port select) and A (address inputs) have to be generated at t = 0 ns. Since these have two loads, let's allow an additional flight delay of 0.1 ns to reach the desired levels before reaching the SRAMs. So, /WPS and A reach the SRAM1 at 0.7 ns (0.6 ns delay and 0.1 ns additional flight time) and SRAM2 at 0.9 ns. Depending upon whether K and /K are generated separately for SRAM1 and SRAM2 or together, K and /K have to be controlled at the controller. Let's take the case where the same clock routes to both the SRAMs and then returns back to the controller after routing to C and /C on the way. Allowing an additional flight delay of 0.3 ns for the clocks, K and /K have to be generated no more before than 0.3 ns. K and /K then reach SRAM1 at 1.2 ns (0.3 + 0.6 ns delay + 0.3 ns additional flight time) and SRAM2 at 1.4 ns. Since the setup times are 0.5 ns, this is met on both /WPS and Address signals. To provide more margins, K and /K can be generated in such a way that /WPS and Address can be valid around the rising edges of the clocks for more than 1 ns (setup and hold time window). Data is generated at 0.1 ns and these signals (flight delay is assumed to be zero as there is only one load for the data lines) would reach SRAM1 at 0.7 ns and SRAM2 at 0.9 ns, meeting the setup and hold times of 0.5 ns. Again the data could be made flexible by generating the clocks later, leaving enough margin for the setup time on the QDRII. As far as the hold time is concerned on the QDRII, Address A and data should not change at the controller until at least 1 ns and 1.1 ns, respectively, after which the next piece of address and data could be generated for the rising edge of /K. This ensures sufficient hold time of 0.5 ns. The above concept holds true for all trace lengths. Once you have set up all of the signals with respect to the clocks having enough margins, you don't have to worry about how far the SRAMs are. This assumes that all the signal traces are approximately equal and also that the drivers on the controller have enough signal strength to drive the longer traces.

Figure 16.12 summarizes the write timings. ASIC refers to the signals generated at the controller which, in this case, is an ASIC. SRAM1 and SRAM2 refer to the signals at the inputs of the respective SRAMs." [Cyp02c]

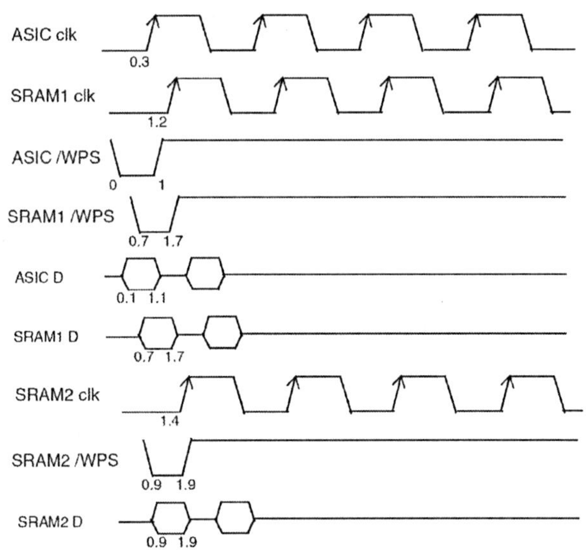

Figure 16.12 Write timings. [Reprinted with permission of Cypress Semiconductor Corporation.]

16.3.7.2 Read Timings

"There are several different implementations that a system designer could choose for the read timings. Some of these are discussed briefly next.

Using K and /K Alone While running at low frequencies (<133 MHz) and given shorter trace lengths, C and /C don't necessarily have to be used. The internal clock to the controller could be used to latch-in the return data as the data could be reasonably predicted at the controller's input. This means that only K and /K clocks could be used and C and /C clocks could be strapped to V_{DD}. For this scheme to work properly, the setup and hold times on the controller side have to be pretty small if the data has to be latched within a reasonable number of cycles. If this is not possible, then delay could be introduced on the data inputs into the controller, effectively increasing the setup or hold time. However, as the trace lengths become longer between the controller and the QDR SRAMs, it becomes highly difficult to accurately latch-in the data with the help of the controller's internal clock.

Using C and /C Clocks as the Return Clocks This implementation was introduced in Section 16.2.2 on page 436. While running at high frequencies, the C and /C clock pair can be used as the return clocks to the controller. This is done by running the K and /K clocks first to SRAM1 followed by SRAM2, then turning the clock back and routing to C on SRAM2 followed by C on SRAM1 and then returning to the controller as was shown earlier in Figure 16.11. If the loading on the K and /K clocks appears to be high, separate clocks could be routed to C and /C from the QDR controller in the same fashion.

An advantage with this kind of clocking is that the return data from both the SRAMs reach the controller at the same time. This follows from the fact that although K and /K clocks of SRAM1 are clocked first, C and /C are clocked the last after SRAM2. So, SRAM2 puts out the data first and then SRAM1, and both the pieces of data then reach the controller at the same time, again, given equal trace lengths for both the clock pairs and the data. Since there is a maximum clock to data valid timing, this should be taken as the timing between the rising edge of C and the first piece of data and the same is true for /C. Depending upon the setup and hold time on the controller side, C and /C have to be moved around with a delay element or by lengthening the trace with respect to the data trace such that there is enough setup and hold time available for the data when once both the return clock pair and the data reach the controller. Figure 16.13 employs this scheme. One important point to be noted here is that the round-trip time between SRAM1 and SRAM2 has to be less than the maximum skew allowed between K and C or /K and /C. In this way, the data at the controller side can be latched with the help of return clocks from the QDR. It is up to the system engineer to synchronize the return clocks with the internal master clocks of the controller.

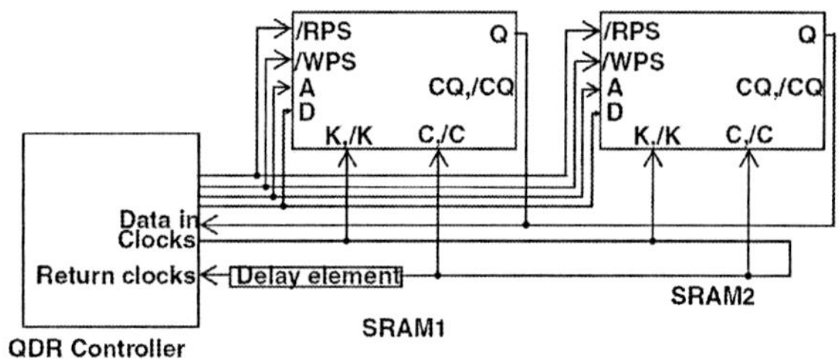

Figure 16.13 Delayed return clock approach (width expansion). [Reprinted with permission of Cypress Semiconductor Corporation.]

Using Echo Clocks as Return Clocks Previously what we have talked about is the use of C and /C as the return clocks. As said before, it is certainly possible to use C and /C as the return clocks and the data to be captured on the rising edges of these clocks. Since QDRII runs at 333 MHz and higher, it is recommended to use point-to-point links between the controller and the QDR SRAMs. Multiple loading points may distort the signal even if a slight noise creeps into the signal at these frequencies. Also, HSTL terminations have to be employed under multiple loads to ensure signal integrity. So, for point-to-point links, it is not necessary to use C and /C, instead echo clock pair CQ and /CQ may be used. As there is a definite relationship between the output data and the echo clocks, the data can always be latched into the controller with 100% accuracy. To be more precise, since the rising edge of the echo clocks always occur 0.2 ns before its associated data, echo clocks can be delayed such that they could be centered with respect to the data, assuming the setup and hold times are equal on the controller side.

Multiple SRAMs can also be interfaced with the QDR controller. Figure 16.14 shows four SRAMs connected to a QDR controller. SRAMs 1 and 2 and SRAMs 3 and 4 are equidistant from the controller, assuming a clamshell configuration. The idea here presents separate echo clocks running from each of the combinations so that the data could be latched at the controller using respective echo clocks.

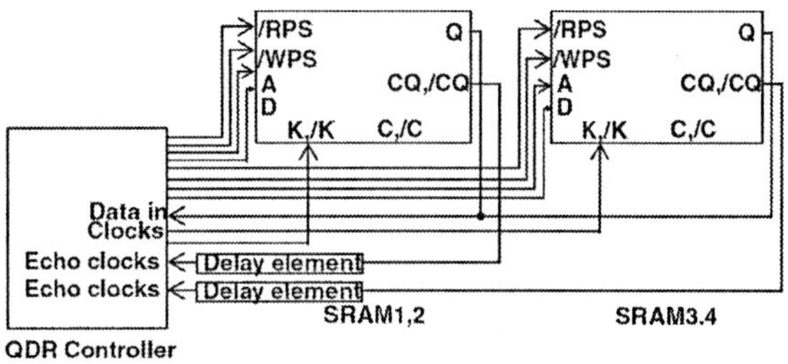

Figure 16.14 Echo clock approach. [Reprinted with permission of Cypress Semiconductor Corporation.]

The echo clocks from each of the combinations could be from the top or the bottom SRAM provided all the traces are equal. The delay elements aid in centering the data before being latched by the controller. These echo clocks can serve as data valid indicators.

When designing a multiple SRAM interface, care must be taken not to mix SRAMs from different vendors as this has the potential of varying timings and the relationships would no longer be guaranteed. For example, if SRAM 1 and 2 are from different vendors, the echo clock valid and the data valid timings may change from one SRAM to another. Since only one echo clock is being used from a combination, the data valid timing from the other SRAM whose echo clock is not used may change causing tighter setup and hold time restrictions on the controller." [Cyp02c]

16.3.8 Package

"The QDRII SRAM is offered in a 165 FBGA package. This is a 13 × 15 mm package, with a ball matrix of 11 × 15. The pitch of the balls is 1 mm. Moving to higher densities from here, the package is offered with the same ball count but in a 15 × 17 mm package to accommodate larger die. However, the footprint remains the same except that the sides stretch outward rendering it to be compatible with the 13 × 15 mm package." [Cyp02c]

16.3.9 Output Impedance Control

"The QDRII device, like the QDR, provides an option for the users to control the drive strength of the output drivers. Using the ZQ pin, the output impedance of the QDRII device can be varied all the way from 30 Ω and upwards to 70 Ω. This allows system designers to match the driver impedance to that of the system boards. To use this method of impedance control, one has to connect a resistor whose value is five times the required impedance at the ZQ pin. The ZQ circuit provides pull-up and pull-down tolerances of ±15%." [Cyp02c]

16.3.10 Data Line Terminations

"QDR uses Class I HSTL output buffers. One of the key benefits of the HSTL standard is that it allows low-swing, high-speed operation. The HSTL standard also calls for terminations on all inputs and outputs and also calls for an external reference voltage to be provided to all input buffers. The advantage of providing external reference voltage is that is provides a reference voltage to the input buffer that is dependent on the external supply voltage.

Adding terminations on all HSTL signals allows for low swing and low noise signaling. It is recommended that terminations be added on all HSTL signals. In the case of a single QDR SRAM on a bus connected to a controller, with proper design, terminations could be eliminated. In most cases with standard boards, it has been seen that HSTL buffers without terminations could be used for an interface with a single QDR device. Signal integrity simulations are necessary to make sure that this is possible. Figure 16.15 provides an example of a termination scheme used with QDRII." [Cyp02c]

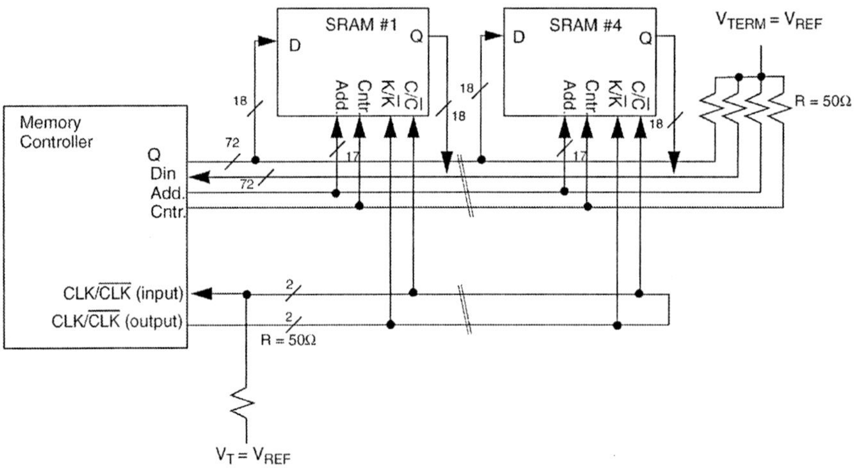

Figure 16.15 Termination scheme for application using four QDRII devices. [Reprinted with permission of Cypress Semiconductor Corporation.]

16.3.10.1 QDR IBIS and HSPICE Models

"Signal integrity simulations are a necessity in the case of designing more than one QDR device onto the same bus. QDR Consortium members provide both IBIS and HSPICE models to simplify the designer's signal integrity simulations." [Cyp02c]

16.3.11 Clamshell Design

"The QDR pin assignment allows devices to be located on both sides of a board and share signals, and is a mirror image pinout. This is referred to as device clamshelling (locating two devices on opposite sides of the PCB) and is supported for both width expansion and depth expansion. This was defined to simplify board layout and reduce the overall board area. Width and depth expansion require a few signals to be interrupted by vias and the signals are different for each.

The pinout on the QDR is defined such that if one side of the package has a data pin, the other side would be a 'no-connect' pin so that the trace running from the top to the bottom and vice versa (data to the no-connect pin) could actually come off on the other side of the PCB and get routed instead of drawing this trace from the inner layers. Since all these pins are closer, a common signal driving both the SRAMs must travel about the same distance, reducing reflections and increasing signal integrity." [Cyp02c]

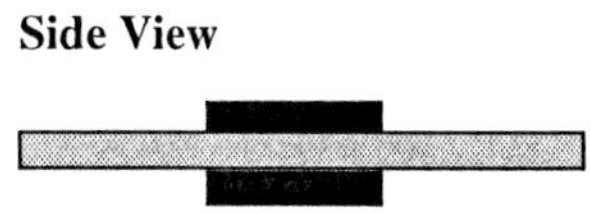

Figure 16.16 Clamshell configuration. [Reprinted with permission of Cypress Semiconductor Corporation.]

16.4 Exercises

16-1. Among the K, /K, C, /C, CQ, and /CQ clocks, which are input signals to the QDRII SRAM chip and which are signal outputs from the chip?

16-2. (a) Are echo clocks used during read or write operations, or both? (b) How are they used?

16-3. Can QDR or QDRII devices be operated at low clock frequencies such as 10 MHz? Why or why not?

16-4. What does the DOFF pin on QDRII devices allow you to do?

16-5. For high clock frequencies, why is it better to latch data into the controller with echo clocks rather than with the C and /C clocks?

16-6. (a) By what range can the output impedance of a QDRII device be varied? (b) How is this done?

16-7. (a) What types of applications are QDR and QDRII products used in? (b) Why are they so suitable for these types of applications?

16-8. (a) Is QDR technology a JEDEC standard? (b) Do QDR packages conform to a JEDEC standard? (c) Is QDR technology only applicable to SRAMs or is it also used with SDRAMs?

16-9. (a) What three types of controllers can be used with QDR SRAM memories? (b) What are the names of three manufacturers that provide one or more of these types of controllers?

16-10. Refer to Figure 16.1 on page 432, which shows the relative bandwidth per pin as a function of the read/write ratio for several different kinds of memory devices. What is the main concept that this figure illustrates?

16-11. With a 2-word burst QDR SRAM, is one address or are two addresses placed on the address bus every clock period when indefinitely sustaining both a 2-word read and a 2-word write each clock cycle?

16-12. What is the range of the allowable duty cycle for the K and K# clocks of QDR SRAMs for normal operation?

16-13. (a) For QDR SRAMs, what is the timing requirement limitation between the K and C clocks and between the K# and C# clocks. (b) What is the reason for this requirement?

16-14. (a) Name three additional features or feature improvements introduced with QDRII SRAM that QDR SRAM does not have. (b) Provide one trade-off, or negative characteristic of QDRII SRAM, that QDR SRAM does not have due to the enhancements contained in QDRII.

16-15. What causes the hold time of a QDRII device to be negative?

CHAPTER 17

Direct Rambus DRAM (DRDRAM)

Chapter Objectives

- Present the Rambus DRAM technology and Rambus' licensing of its intellectual property.
- Discuss the four main elements of a Rambus memory system.
- Compare the topology and architecture of a Rambus memory system with an SDRAM or DDR SDRAM memory system.
- Define peak and effective bandwidth and explain bank conflict and command bus bandwidth limitations of SDRAM.
- Compare the effective bandwidth and latency of RDRAM and SDRAM memory modules.
- Show how the reduced pincount of RDRAM devices can result in a much lower-cost design.
- Define long-channel and short-channel RDRAM configurations.
- Present a Rambus long-channel design example.

17.1 Direct Rambus DRAM™ (DRDRAM, RDRAM)[1]

"RDRAM, Direct RDRAM, and DRDRAM are three names for the same thing: Direct Rambus Dynamic Random Access Memory.[2] Rambus researches and develops memory technologies such as RDRAM and then licenses these technologies to chip manufacturers such as Samsung, Elpida, and Infineon, who manufacture the chips. Rambus does not market any RDRAM chips directly; it only markets intellectual property (IP) and licenses chip manufacturers.

RDRAM is used with Intel's 850, 850E, and 860 chipsets as shown in Figure 17.1 and with the SiS[3] R659 North Bridge chipset (uses PC1200 devices). SDRAM is used with other Intel chipsets. For Rambus devices, PCxxxx refers to a memory chip used in PCs or PC-like systems where xxxx is the data rate in Mbps per pin but where the memory clock rate is xxxx divided by 2. For example, for a PC1333 RDRAM, the RDRAM data rate is 1,333 Mbps and the RDRAM clock frequency is 667 MHz (rounded value).

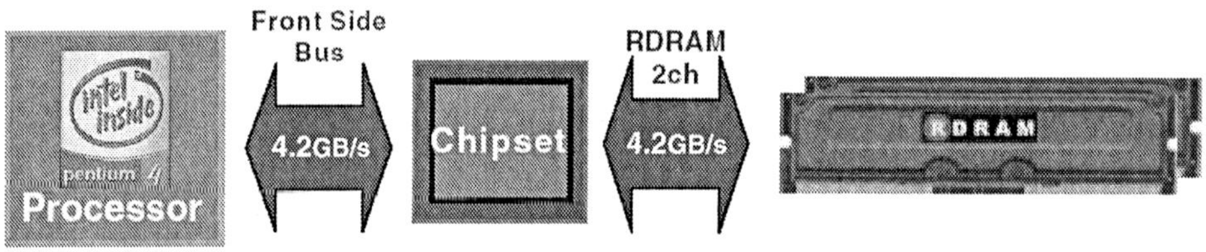

Figure 17.1 PC memory system architecture for RDRAM.

When working with modules of RDRAM devices, the modules are referred to as 'RIMMs™,' or Rambus In-line Memory Modules, rather than DIMMs. RIMMyyyy refers to a module having a total data rate for the entire RIMM of yyyy MB/s (Mega*Bytes*/s not Megabits/s). A RIMM4200 has a performance level of 4,200 MB/s or 4.2 GB/s. It uses one PC1066 chip for each of the two 16-bit channels, 2 chips total, thereby creating a 32-bit module. While the total data rate is precisely 4,264 MB/s, the module is referred to as a '4200.' Or, alternatively, 1,066 Mbps per pin, times 32 pins total for both chips, divided by 8 bits/Byte = 4.2 GB/s.

RIMMs are made by a variety of memory vendors. Rambus does not recommend any particular vendor's products and all components and modules are validated to be compatible with Rambus specifications by an independent third party.

A Rambus memory system has four main elements: a memory controller, RDRAM devices, the Rambus channel, and a clock generator. We discuss these in more detail in Section 17.2.1, which starts on page 461. Devices can be attached to the same PCB as the memory controller or put on RIMM modules. When dealing with RDRAM, we often refer to the

1. All text, tables, and figures in this chapter are reprinted with permission of Rambus Inc.
2. Direct RDRAM was a later generation of RDRAM, but the distinction has disappeared in today's marketplace.
3. SiS is Silicon Integrated Systems Corp. of Taiwan.

'RDRAM memory interface,' which is the interface between the memory controller (or chipset) and the RDRAM devices.

A representation of the balance in RDRAM architecture can be seen in Figure 17.2, which shows the placement of signals in an SDRAM and RDRAM memory system.

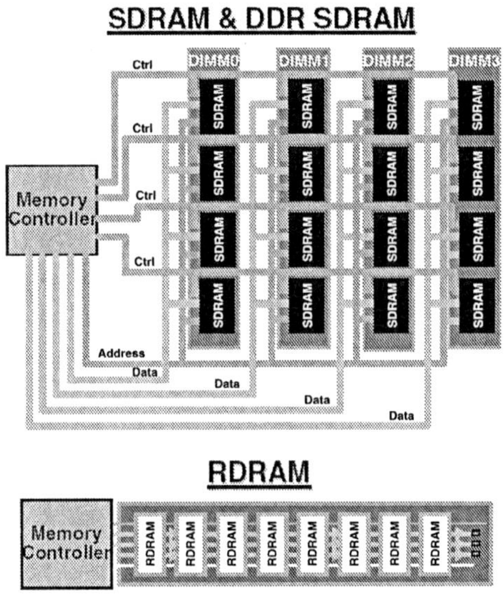

Figure 17.2 Electrical topology comparison.

The RDRAM memory interface is incorporated into both ASIC chips and systems. This interface features:

- The highest bandwidth per pin, reducing overall component count, controller pin-count and system board cost.
- Supported DRAM densities from 128–512 Mb with a roadmap to support up to 1 Gb
- Speed bins of 800, 1,066, 1,200, 1,333 Mbps with a roadmap to support up to 1,600 Mbps. (These speeds are the data rates. The memory clock rates are the above data rates divided by two.)
- Memory configurable into single or dual RIMM modules supporting bandwidths from 1.6 GB/s to 5.3 GB/s with a roadmap to support up to 12.8 GB/set quad-channel module.

The RDRAM interface achieves this performance through highly efficient protocol, low-voltage swing, separate control and address buses, and precise clocking to minimize skew between clock and data lines.

Some RDRAM devices are designed with 16-bit data buses and some are designed with 18-bit data buses. When RDRAM devices have 18-bit data buses, the two extra bits are for ECC (Error Correction Code). Not all RDRAM devices come in both 16-bit and 18-bit versions.

The RDRAM interface allows memory bandwidth (total data rate) to be scaled by very small amounts and allows smaller chip and board area by requiring significantly fewer I/Os for the same amount of bandwidth. RDRAM may also allow the implementation of a memory system with fewer chips as shown in Figure 17.3.

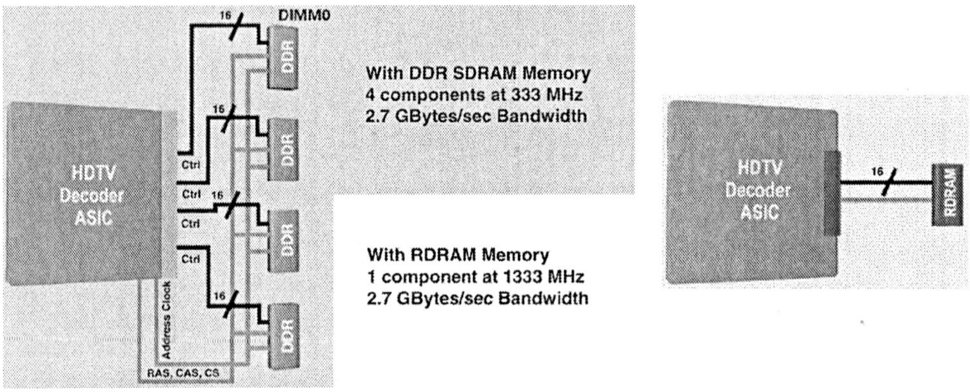

Figure 17.3 Comparison of consumer HDTV decoder designs showing reduced chip count with the use of RDRAM.

Rambus ASIC Cells (RACs) provide the physical layer for RDRAM devices and incorporate all of the required high-speed interface circuitry and logic. They are available in standard CMOS processes from TSMC and other leading suppliers." [Ram03a]

17.1.1 RDRAM Memory

"RDRAM memory uses a narrow, high-speed bus architecture to reduce pincount while maximizing bandwidth as shown in Figure 17.4. Command and data are sent along the RDRAM channel that is uniformly routed to each RDRAM device. This results in a system that has equal load and fanout on all signals, optimizing signal integrity and reducing performance loss due to signal stabilization time.

SDRAM and DDR SDRAM systems suffer from unequal line lengths and unequal loading (every SDRAM device must connect to the full command bus but only to a portion of the data bus). RDRAM uses matched line lengths and an equal loading topology to create an electrical 'racetrack' rather than a bird's nest of signal lines. The clean design and architecture of an RDRAM memory subsystem ensures reliable, high-frequency operation.

RDRAM memory uses separate buses to transmit row and column address information on both edges of the clock. This enables higher efficiency by allowing the issue of a column packet

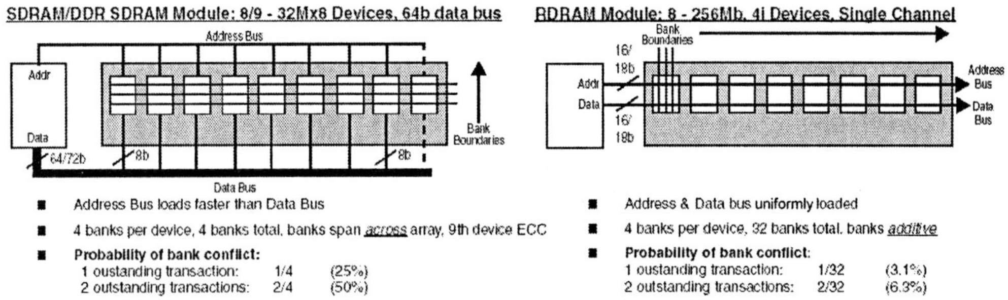

Figure 17.4 Memory system architecture comparison.

while a row packet is simultaneously being issued to another device, a feature not present in SDRAM or DDR. Bandwidth of the command and data buses is precisely matched to minimize delays introduced between subsequent operations. Equalization of write and read data packet lengths prevents added delay due to operation pipelining and misalignment, facilitates zero delay when transitioning from a read to a subsequent write operation and minimizes delay when transitioning from a write to a subsequent read operation. On-chip write posting enables an RDRAM to allow a pending read operation to go around a slower write operation to ensure delivery of critical read data first with minimal controller overhead.

A single RDRAM device is capable of serving an entire memory transaction. One of the benefits of this architecture is that it allows banks of individual devices to be additive, pooling them to decrease the probability of bank conflict and thus minimize bank conflict effects. In the 128 and 256 Mb RDRAM generation, 32 paired internal banks are supplied per RDRAM device (2(16d)). Future generations of the 256 Mb device and all devices at the 512/576 Mb generation and beyond will incorporate a reduced-cost, 4 independent bank core (4i). Since banks are additive, the effect of reduced device bankcount on performance is diminished and ranking is not needed to increase the number of banks.

RDRAM memory utilizes a 400 MHz clock signal that facilitates fast data transfer and allows for fine timing granularity for 'tiling' or pipelining of memory operations in 1.25 ns increments. RDRAM memory is capable of servicing a memory request every 10 ns.

These architectural characteristics result in a high-efficiency memory subsystem that delivers very high effective bandwidth in real world conditions: 1,301 MB/s effective bandwidth from 1,600 MB/s peak bandwidth at 81% efficiency. The efficiency and effective bandwidth of RDRAM memory increases rapidly as the transfer size increases beyond 32 bytes." [Ram01b]

17.1.2 The Memory Landscape

"RDRAM is often compared to SDRAM for PC memory performance issues. SDRAM and DDR SDRAM share many architectural and signaling features. Both use a parallel data bus, mainly available in component widths of x8 or x16, and both have a single addressing command bus that must be shared to transmit row and column addresses.

DDR SDRAM increases data bandwidth over conventional SDRAM by transmitting data on both edges of the synchronous clock signal using SSTL 2 signaling, thus in theory 'doubling' the data rate of the memory. It is important to note that DDR SDRAM does not double the address command bandwidth of the system by using both edges of the clock on the command bus, a factor that ultimately limits the real world performance gain from using DDR signaling on the data bus.

RDRAM memory combines a conventional DRAM core with a high-speed serial interface called the RDRAM channel. The channel uses 16 pins (2 bytes) for a data path operating at a high effective data rate per pin (1,333 MHz for a PC1333) by transmitting data on both edges of the clock. To facilitate maximum performance, the RDRAM channel utilizes double data rate signaling on non-multiplexed (i.e., not shared) row and column address command buses. Since each RDRAM device's datapath is as wide as the channel (16 bits), a single device can service an entire memory request, unlike SDRAM, which uses multiple devices in parallel to satisfy a request. Up to 32 RDRAM devices can be placed on each channel without a buffer. The channel is common to all devices and incorporates the command bus, the data bus and a serial control bus for initialization. The RDRAM channel is uniformly loaded as devices are added, facilitating maximum operating frequency by eliminating unequal signal paths. The RDRAM protocol supports many features that optimize the bandwidth and efficiency (real-world performance) of the overall system." [Ram01b]

17.1.3 Bandwidth and Latency

"Information on memory technologies tends to focus on peak system bandwidth and component latency. Peak system bandwidth can be defined as the maximum transfer rate in Mbps of a particular memory technology given ideal conditions. However, it is important to note that most applications do not create ideal conditions in their memory transaction profile: they are a mix of reads and writes, asynchronously dispatched to a wide variety of memory addresses. Effects particular to each memory system degrade this peak bandwidth to a lower number called effective bandwidth. Effective bandwidth can be expressed as the product of peak bandwidth and efficiency, a measure of the ability of the memory system to deliver on its peak bandwidth given real world memory transaction patterns.

In addition to examining the pitfalls of bandwidth measures, it is also important to look at latency. Devices are usually characterized by component latency, which is the time between row strobe (command reaches the DRAM's pins) and the delivery of the first bit of data on the output pin for an individual component. When evaluating system performance, it is useful to look at system latency, or the time from when a memory controller drives a requested address to when it samples the last of the read data." [Ram01b]

17.1.3.1 Comparing RDRAM with SDRAM and DDR SDRAM

"For the purposes of this discussion, five implementations of three general memory technologies are examined at a single module level: SDRAM (PC100 and PC133), DDR SDRAM (PC200 and PC266) and PC800 RDRAM memory. For the SDRAM and DDR SDRAM systems, a module construction of eight 32 Mbx8 devices is assumed for a 64-bit bus width. Four banks are assumed for each SDRAM/DDR device. An equivalent RDRAM system is constructed using eight 256 Mb 800 MHz RDRAM devices with 4 independent banks (4i)." [Ram01b]

17.1.3.2 Assumptions for Comparisons

"A simple memory transaction profile of 32 byte transfers consisting of 70% reads and 30% writes is used for this comparison, with 25% of all memory transactions requiring access to the same bank. Bank conflicts as well as read/writes are assumed to be randomly distributed. It is believed that this simple model reflects a broad spectrum of memory transaction and bank conflict profiles present in computing systems. It is also noted that ECC could be built into this system by the addition of another device to the SDRAM and DDR systems or by using ECC devices for the RDRAM system." [Ram01b]

17.1.3.3 Real-World Performance: Bandwidth

"Results of the bandwidth study for typical memories of different types, operating frequencies and core latency (t_{RAC}) timings can be referenced in Table 17.1.

Table 17.1 Effective Bandwidth of 256 MB Memory Modules — 32-Byte Transfers

Memory	BW_{PEAK}	Efficiency	BW_{EFF}	t_{RAC}	Notes
PC133 SDRAM	1066 MB/s	60%	638 MB/s	–40 ns	Typical SDRAM 133 MHz 1st bin
DDR200 SDRAM	1600 MB/s	45%	717 MB/s	–40 ns	Typical DDR SDRAM 200 MHz 2nd bin
PC800 RDRAM	1600 MB/s	81%	1301 MB/s	–45 ns	Typical RDRAM 800 MHz memory 2nd bin
DDR266 SDRAM	2133 MB/s	37%	798 MB/s	–39 ns	Typical DDR SDRAM 266 MHz 1st bin
PC1066 RDRAM	2132 MB/s	81%	1727 MB/s	–35 ns	RDRAM 1,066 MHz memory bin

SDRAM and DDR SDRAM share some architectural issues that impact their effective bandwidth. Each quotes peak bandwidth on a 64 bit bus, but these peak bandwidths decrease to substantially lower numbers under our study's approximation of real-world conditions as reflected in Table 17.1: PC100 decreases from 800 MB/s to 583 MB/s at 73% efficiency while PC133's peak bandwidth of 1,066 MB/s decreases to an effective bandwidth of 638 MB/s at 60% efficiency. Similarly DDR200 decreases from 1,600 MB/s peak to 717 MB/s at 45% efficiency while DDR266 decreases from 2,133 MB/s peak to 798 MB/s at 37% efficiency. There are two components largely responsible for this decrease in bandwidth for both memory types: bank conflict among adjacent transactions and constraints on the address command bus." [Ram01b]

17.1.3.4 SDRAM Bank Conflict and Command Bus Bandwidth Limitations

"Both SDRAM and DDR SDRAM suffer in delivering effective bandwidth due to their architecture. The losses can mainly be attributed to two factors: bank conflict and command bus bandwidth limitations.

The greatest loss in performance for both SDRAM and DDR SDRAM comes from bank conflict among adjacent transactions. Banks are defined as blocks of memory that share common sense amplifiers. Sense amplifiers are the components of DRAM that read the value stored in the DRAM cell for data retrieval and periodic refresh of a DRAM cell's contents. Implementing sense amplifiers takes a very large amount of silicon space relative to cells, so DRAM cells share sense amps to save cost by organizing into banks. Banking restricts the ability of the DRAM to service simultaneous requests to DRAM addresses/cells that lie in the same bank as a sense amp can only service one address/cell at a time. Thus, when two requests collide on a bank of memory that is already in use, delay must be introduced until the previous transaction completes. During this time, no new memory transactions can be initiated. Each SDRAM or DDR device is organized into 4 internal banks and these banks span across the module as each device only contributes part of the 64-bit interface. SDRAM/DDR controllers allow operations to pipeline to attempt to achieve the highest performance. Since service time is greater than issue time, the first memory operation will not complete before the second one is issued. Thus the probability of sequential memory accesses hitting the same bank is 25%, and it increases with each additional outstanding transaction. The performance degradation due to bank conflict is large: approximately 85% of SDRAM's efficiency loss and 80% of DDR SDRAM's efficiency loss can be attributed to bank conflict.

Ranking modules can be used in SDRAM and DDR SDRAM systems to provide more collective banks at the expense of more pins for chip enable signals to select which rank to activate as well as poorer upgrade granularity: modules must be added in pairs.

DDR SDRAM faces additional performance limitations due to the fact that the address command bus bandwidth has not been scaled to match the data bus bandwidth: the data rate is doubled to 200 or 266 MHz while the command rate is still at 100 or 133 MHz.

Both SDRAM and DDR SDRAM must connect the entire address command bus to every address pin on every device. Thus, the capacitive loading of the command bus increases at a faster rate than the data bus with the addition of each memory chip to the system. Because of this uneven loading, it may be required to transmit row/column address information for two consecutive cycles to allow sufficient bus settling time, commonly called the 'two-cycle addressing problem.' Two-cycle addressing seriously impedes performance by consuming command bandwidth. The limited bandwidth of the command bus restricts the ability of the memory system to concurrently process pipelined transactions. These constraints exhibit themselves in delays that may be introduced between sequential operations and when transitioning between read and write operations due to unequal lengths of these operations.

These signal integrity issues may also be solved by adding additional components such as address registers, clock recovery components, individual data line termination resistors, and bus switches. While improving the signal integrity of the system, addition of these components often introduces latency, which degrades system performance." [Ram01b]

17.1.3.5 Real-World Performance: Latency

"Table 17.2 documents component and system latency for the various memory architectures. Component latency is defined as the time between row strobe on the command pin and the delivery of the first bit of data on the output pin. Looking at the controller's perspective, system latency is the total time from when a controller drives an address to when it samples the last of the read data. The latency of the memory core (t_{RAC}) is the largest contributor to component latency. Despite relatively slow memory core offerings from memory vendors, (t_{RAC} = –45 ns for RDRAM memory, –39 ns for DDR266 SDRAM), RDRAM memory maintains very respectable system latency for transferring 16 and 32 bytes of data (t_{RD16} and t_{RD32})." [Ram01b]

Table 17.2 System Latency of 256 MB Memory Modules

Memory	t_{RD16}	t_{RD32}	t_{RD64}	t_{RD128}	t_{RAC}	Notes
PC133 SDRAM	67.5 ns	82.5 ns	112.5 ns	172.5 ns	–40 ns	Typical SDRAM 133 MHz 1st bin
DDR200 SDRAM	70 ns	80 ns	100 ns	140 ns	–40 ns	Typical DDR SDRAM 200 MHz 2nd bin
PC800 RDRAM	70 ns	80 ns	100 ns	140 ns	–45 ns	Typical RDRAM 800 MHz memory 2nd bin
DDR266 SDRAM	67.5 ns	75 ns	90 ns	120 ns	–39 ns	Typical DDR SDRAM 266 MHz 1st bin
PC1066 RDRAM	60 ns	67.5 ns	82.5 ns	112.5 ns	–35 ns	RDRAM 1,066 MHz memory bin

17.1.4 System Cost: Memory Granularity

"As computing systems evolve to meet changing market demands, memory footprints will continue to change to adapt to their needs. While PC memory applications continue to slowly increase their memory footprint over time, new fixed-function and Internet-oriented computing systems as well as embedded systems are making a push towards smaller memory footprints than PCs. Simultaneously, DRAM manufacturers continue to increase density of DRAM devices.

An RDRAM system has a very compelling memory granularity advantage over competing SDRAM or DDR based systems. Since a single RDRAM device can act as an entire memory subsystem, the granularity of the memory footprint is a single device: 32 MB for 256 Mb densities of RDRAM and 64 MB for 512 Mb generation. 64-bit DDR systems on the other hand, have a minimum granularity of 128 MB using x16 (16-bit wide) devices in 256 Mb technology. This granularity problem for SDRAM and DDR, illustrated in Tables 17.3 and 17.4, becomes more complex for designers of low-cost systems desiring to use the more widely used x8/16 SDRAM devices used in mainstream PCs. x32 specialty DDR SGRAM (graphics) devices have a granularity of two devices or a minimum footprint of 64 MB for 256 Mb technology." [Ram01b]

Table 17.3 Minimum Number of Devices

	SDRAM: 64-Bit Bus RDRAM: 1 Channel	SDRAM: 128-Bit Bus RDRAM: 2 Channels
SDRAM x8	8	16
SDRAM x16	4	8
SGRAM x32	2	4
RDRAM	1	2

Table 17.4 Minimum Memory Footprint

Memory Type	SDRAM: 64-Bit Bus RDRAM: 1 Channel		SDRAM: 128-Bit Bus RDRAM: 2 Channels	
	256 Mb Generic	512 Mb Generic	256 Mb Generic	512 Mb Generic
SDRAM x8	256 MB	512 MB	512 MB	1 GB
SDRAM x16	128 MB	256 MB	256 MB	512 MB
SGRAM x32	64 MB	128 MB	128 MB	256 MB
RDRAM	32 MB	64 MB	64 MB	128 MB

17.1.5 System Cost: Pincount and Bandwidth

"A single-channel RDRAM memory system delivers 1.3 GB/s effective bandwidth (1.6 GB/s peak bandwidth) in only 76 pins. Contrast this with a standard 64-bit SDRAM system that requires approximately 150 pins or a 64-bit DDR system that requires approximately 180 pins at approximately half of the effective bandwidth of RDRAM. 128-bit SDR or DDR systems can require 300 to 400 pins to match the effective bandwidth of a 76-pin RDRAM system.

Pincount impacts designs in many areas. The memory controller is usually most affected due to increase in pincount required to support SDRAM or DDR SDRAM systems. For systems targeted at lower-cost market segments, the ability to save 75 to 110 pins can significantly reduce controller cost by allowing a smaller package while delivering superior performance. In addition, for systems incorporating a memory interface integrated with a CPU, interface pincounts become even more critical.

Pincount also impacts die area and thus chip power due to increased I/O sites and the receiver, driver, and protection components necessary to support them. Other system considerations include issues created by simultaneous switching of I/Os (also referred to as delta–I effects), which can complicate designs." [Ram01b]

17.1.6 Motherboard Layers and Power Dissipation

"A single- or dual-channel RDRAM memory system can be routed in a cost-effective four-layer motherboard. Reference designs and documentation are available for RDRAM systems to ensure solid systems and reduce time to market. RDRAM devices and RIMM modules are designed, tested, and validated for cross-vendor compatibility.

System power budgets are shrinking due to reduced airflow, fanless systems, rack-mounted systems, and shrinking form factors. Simultaneously, performance requirements are increasing.

An RDRAM memory system consumes approximately equivalent power as a comparably built SDRAM or DDR SDRAM system." [Ram01b]

17.2 Long-Channel Design[4]

Rambus provides design guidelines for long and short channel designs. Short channel configurations are designs with up to 4 RDRAMs on the same PCB as the controller and tend to have bus lengths of no more than 4 inches. Long-channel systems are designs with from 8 to 32 RDRAMs on modules and connectors.

Sixteen-bit, long-channel systems typically consist of a motherboard, one or more memory modules, and a memory controller which resides on the motherboard. Such an arrangement allows the user to expand the memory size either by inserting more modules into the motherboard or by using higher-capacity modules.

A 16-bit, long-channel Rambus memory system is shown in Figure 17.5. Here, the RDRAM devices are placed on RIMM modules and these modules are inserted into connectors on the motherboard. Each of these modules can contain up to 16 RDRAMs. Each connector must contain either a RIMM module or a continuity module.

Multiple channels can be accommodated with additional 16-bit RIMMs. Additionally, in multiple-channel systems, two individual channels can be combined when using 32-bit RIMMs. This configuration allows for single slot expansion even with two channels.

17.2.1 Rambus Memory System

The Direct Rambus memory technology is designed to provide memory channel bandwidth of up to 3.2 GB/s using 1,600 MHz data transfer rates. It is intended for use in a broad range of applications, especially in the computer, consumer, and communications markets.

A Rambus memory system has four main elements: a memory controller, RDRAM devices, the Rambus channel, and a clock generator. The memory controller arbitrates memory requests, converts memory read and write requests to channel packets, and manages memory resources. The controller consists of an application logic core, a Rambus Memory Controller (RMC), and the Rambus ASIC Cell (RAC). The RAC acts as an interface between the controller logic and the Rambus channel. It handles the data width conversion between the controller logic and the channel as well as the signal level translation.

4. This section is reprinted from [Ram02].

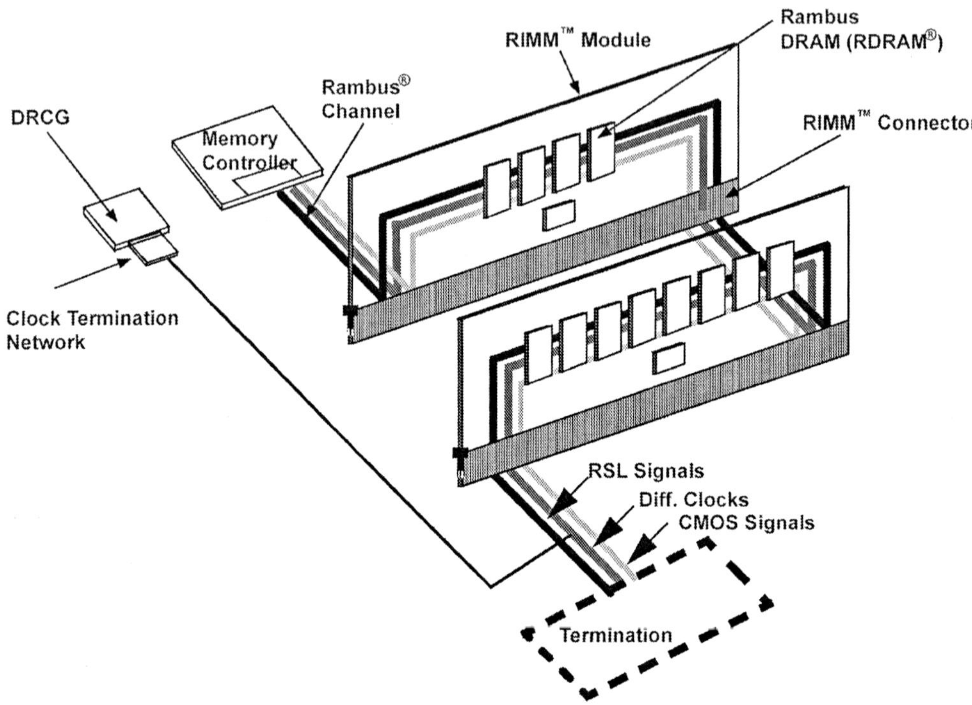

Figure 17.5 Overview of a 16-bit wide-channel system.

The RDRAMs are the memory resources in a Rambus system. They receive configuration and control instructions from the controller and send data to or receive data from the controller. The RDRAM consists of a standard DRAM core with an interface to the Rambus channel. The interface performs a similar function as the RAC. It handles the data width conversion between the DRAM core and the channel as well as the signal level translation.

The Rambus channel is the physical medium that connects the controller to the RDRAMs. It consists of 26 high-speed Rambus Signaling Level (RSL) signals, a high-speed differential clock pair, three CMOS signals, and associated power and ground lines. The RSL signals are low-voltage signals that carry timing, control information, addresses, and data between the RDRAMs and the controller. The CMOS signals carry initialization and operating mode information. The RSL signals, the differential clock, and two of the three CMOS signals require controlled impedance transmission lines that are terminated in their characteristic impedance.

The Direct Rambus Clock Generator (DRCG) provides the channel clock for the Rambus memory system. It is designed to receive a low-frequency clock from a stable clock source and multiply it to generate the channel clock. It also contains circuitry to align the phase of the channel clock with the internal clock of the memory controller in order to facilitate minimum latency transfers between the channel and the controller logic.

17.2.2 Schematics of a 16-Bit, Long-Channel Rambus System

A 16-bit, long-channel Rambus memory system consists of a memory controller, up to two RIMM connectors, a reference clock source, a Direct Rambus Clock Generator, power supplies, and discrete components for signal termination and power supply bypassing. Each RIMM connector can hold either a RIMM module or a continuity module, but the memory system must contain at least one RIMM module. Since each RIMM module may contain up to 16 RDRAMs, the total number of RDRAMs in the system will be less than or equal to 32. The schematic of a 16-bit, long-channel system is shown in Figure 17.6.

The physical interconnect between the memory controller and the RDRAMs is called the Rambus channel. This channel contains the following signals: two 9-bit data bytes, DQA[8:0] and DQB[8:0]; a 3-bit row control bus, ROW[2:0]; a 5-bit column control bus, COL[4:0] (ROW

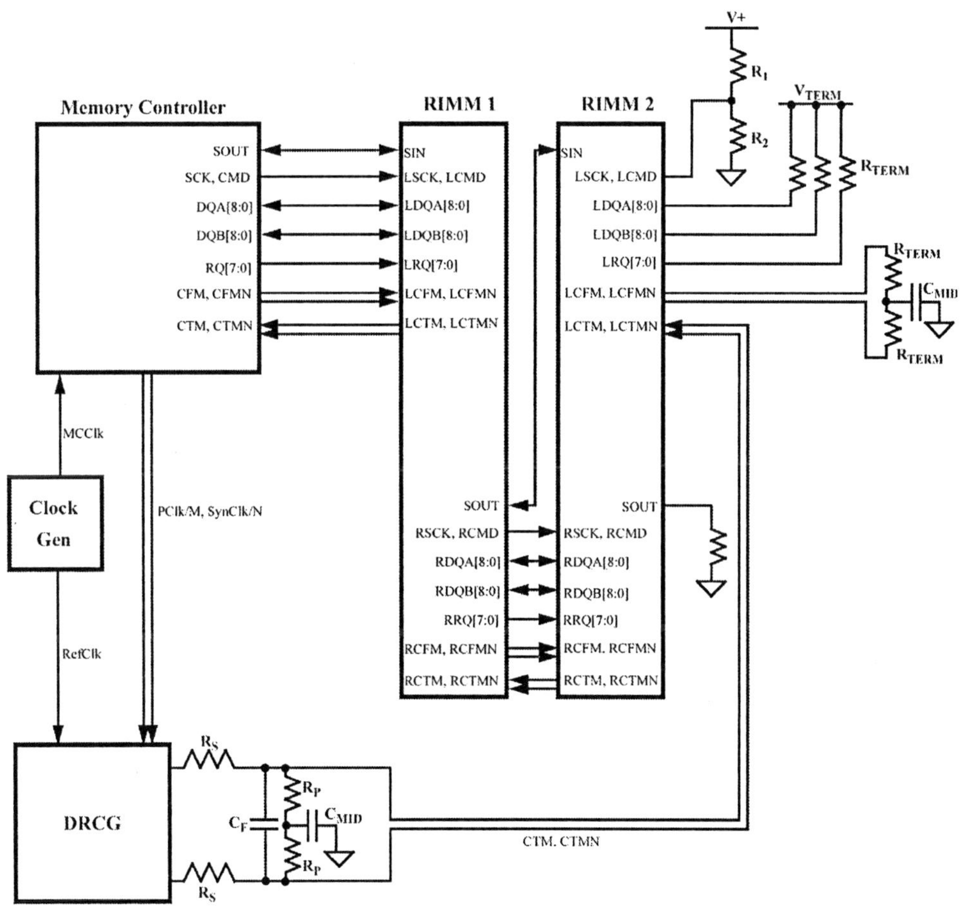

Figure 17.6 Overview of a Rambus 16-bit system motherboard.

and COL together referred to as RQ); a differential clock signal travelling toward the memory controller, CTM and CTMN (ClockToMaster); a differential clock signal travelling away from the memory controller, CFM and CFMN (ClockFromMaster); two high-speed CMOS signals, SCK and CMD; and a low-speed CMOS signal, SIO. The Rambus channel is a daisy chained interconnect, with the memory controller at one end of the bus, the termination network at the other end, and the 16-bit RIMM connectors and modules between the two ends.

The RIMM connector has two sets of pins located at opposite ends of the connector. Each set of pins constitutes a port. The two ports, the left port and the right port, are electrically identical and interchangeable. Correspondingly, each module also has a left port and a right port. The actual direction of the flow of signals on the modules is immaterial, and signals may either enter through the left port and exit through the right port or enter through the right port and exit through the left port. The port of entry is called the input port and the port of exit is called the output port. On a RIMM module, the signals enter through the input port, connect to all the devices on the module in a daisy chain, and then exit through the output port. On a continuity module, signals enter through the input port, traverse the length of the module and exit through the output port. The only exception to the bidirectional signal flow is the SIO signal. Each connector has an SIN[5] pin and an SOUT pin. The SIO signal should always enter a module through the SIN pin and exit through the SOUT pin.

On the motherboard, the Rambus channel should be routed from the controller to the input port of the first connector, then from the output port of the first connector to the input port of the second connector, and finally from the output port of the second connector to the termination network. Placing the two RIMM connectors parallel to each other on the motherboard is the most space-efficient scheme. In such a scheme, alternating the signal direction on neighboring modules results in the shortest routing possible between connectors on the motherboard. The termination network should always be placed after the second connector.

Figure 17.7 shows the clock generation in a Rambus system, which involves a main system clock source, the Direct Rambus Clock Generator (DRCG), and the controller that contains the Rambus ASIC Cell (RAC), the Rambus Memory Controller (RMC), and the gear ratio logic. The main clock source supplies a clock, MCClk, to the memory controller, and also supplies a reference clock, RefClk, to the DRCG. RefClk may or may not be the same frequency as the MCClk but should be harmonically related. Depending on the design, the controller may either internally multiply the MCClk or simply buffer it to obtain PClk, which is then used by the memory controller logic. Similarly, a PLL inside the DRCG multiplies RefClk to generate a differential clock at the desired frequency. This differential clock is routed through the channel twice: first from the termination end of the channel to the memory controller, and second from the memory controller to the termination end, where it is then terminated. The section of the clock net from the termination end to the memory controller is called ClockToMaster (CTM/

5. SIN and SOUT are serial I/O for reading from and writing to the control registers. SIN attaches to SIO0 of the first RDRAM on the module and SOUT attaches to SIO1 of the last RDRAM on the module.

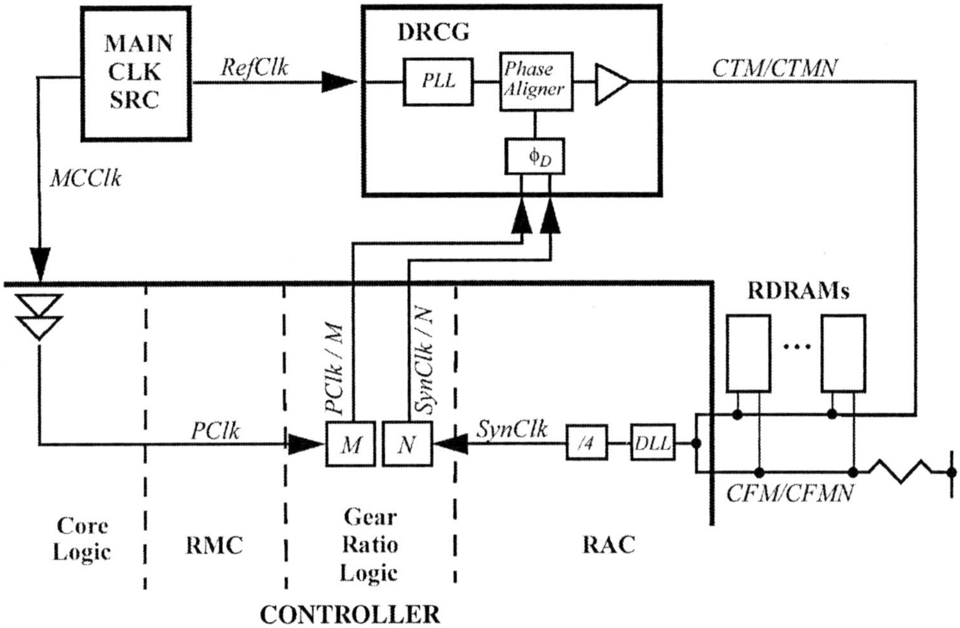

Figure 17.7 Clock generator architecture.

CTMN), while the section from the controller to the termination end is called ClockFromMaster (CFM/CFMN). Information transmitted by the RDRAMs to the controller is always aligned with ClockToMaster while information transmitted by the controller to the RDRAMs is always aligned with ClockFromMaster. The ClockToMaster and the ClockFromMaster nets are connected to each other on the die of the memory controller.

The RAC creates an internal version of CTM/CTMN, which is then divided by 4 to generate SynClk. The gear ratio logic divides PClk by M and SynClk by N to generate two output clocks at the same frequency. These output clocks are connected to the DRCG's phase detector inputs. The DRCG will slew the phase of CTM and CTMN until PClk/M and SynClk/N are phase aligned. This will ensure that PClk and SynClk are aligned inside the controller, thereby facilitating minimum latency data transfers between the RAC and the RMC.

The DQA, DQB, ROW, and COL signals (collectively known as RSL signals) are terminated at the opposite end of the channel from the memory controller by 27 Ω resistors pulled up to the termination voltage, V_{TERM}. Channel performance is improved by keeping the RSL termination an ohm below the channel impedance. The SCK and CMD[6] signals are also terminated at the far end of the channel with a Thevenin network, whose equivalent impedance equals 28 Ω

6. SCK is the serial input clock. It is the clock source used to read from and write to the RDRAM control registers. CMD is the serial command input used to read from and write to the control registers. CMD is also used for power management.

(the channel impedance), and whose Thevenin voltage is equal to the midpoint of the voltage swing of the CMOS signals (the CMOS signals swing between RAC V_{DD} and ground). In systems where the V_{CMOS}[7] supply is capable of sourcing sufficient current, the Thevenin termination network for the SCK and CMD lines may be connected to V_{CMOS} instead of the V_{DD} supply to the RDRAMs. In this case, the termination network will consist of two 56 Ω resistors connected between V_{CMOS} and GND.

The termination networks for all the signals are shown in Figure 17.8. While the SIO chain does not require a termination, the SOUT signal from the last RIMM connector must be pulled down to ground through a 10 KΩ resistor on the motherboard to prevent the line from floating.

The differential clock pair produced by the DRCG is terminated both at the DRCG and at the opposite end of the channel from the memory controller as shown in Figure 17.9. The termination network at the output of the DRCG not only matches the output impedance of the DRCG to the channel impedance, but also sets the voltage levels of the CTM/CTMN and CFM/CFMN

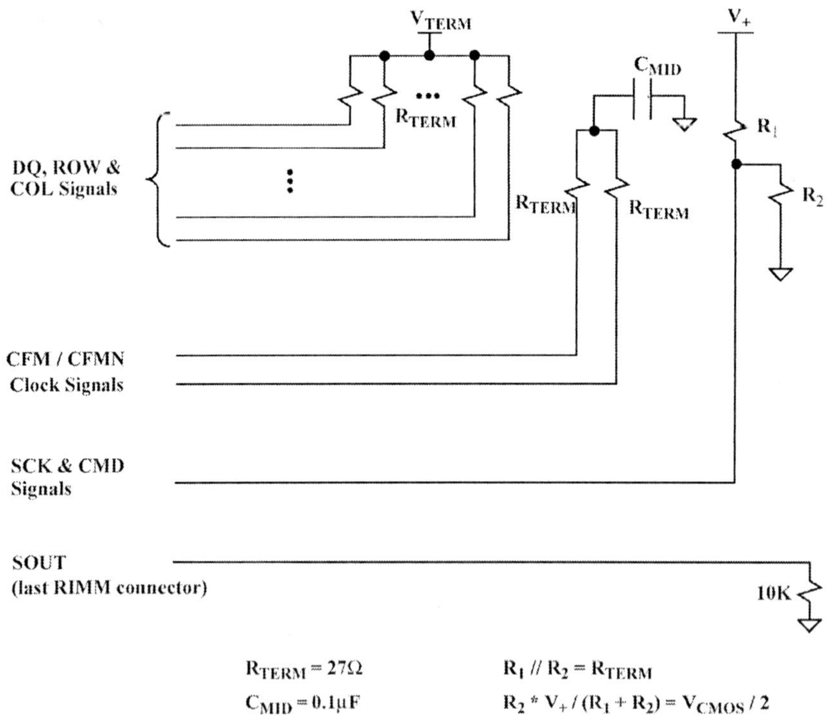

Figure 17.8 Channel termination.

7. V_{CMOS} is 1.8 V for V_{DD} = 1.8 V, V_{TERM} = 1.8 V, and V_{REF} = 1.4 V for 800 MHz operation or V_{REF} = 1.35 V for 1,066 MHz operation and beyond.

clocks. A small capacitor ($C_F \cong 5$ pF) between the two differential clock lines may help reduce the EMI generated by the motherboard but at the expense of increased clock jitter. Rambus recommends that C_F not be stuffed unless EMI considerations require it. Example values of these external components are shown in Figure 17.9 for a channel impedance of 28 Ω.

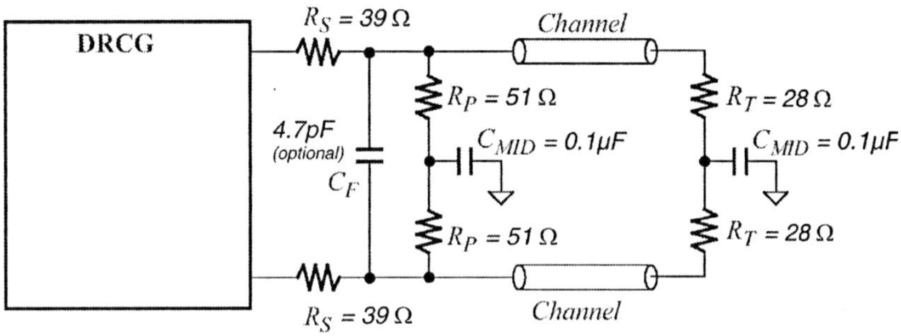

Figure 17.9 Clock output drivers with example component values shown for a channel impedance of 28 Ω.

The DQ, ROW, COL, SCK, and CMD signals and differential signals should all be terminated with discrete resistors. RPAKs (resistor packs) are not recommended since the close proximity of the resistors within the RPAKs causes significant coupling between neighboring signals.

Termination resistor values for a 28 Ω channel impedance are summarized in Table 17.5.

Table 17.5 Summary of Termination Resistor Values for 28 Ω Channel Impedance

16-Bit, Long-Channel Memory System Parameter	Specification/Value
DQA, DQB, ROW, and COL Signals	These signals are terminated at the opposite end of the channel from the memory controller by 27Ω termination resistors pulled up to the termination voltage, VTERM (Figure 17.6, Figure 17.8).
	Discrete termination resistors are recommended.
SCK and CMD Signals	These signals terminated at the opposite end of the channel from the memory controller with Thevenin network whose equivalent impedance equals 28 Ω and whose equivalent Thevenin voltage is equal to the midpoint of the CMOS voltage swing.
	In a system where V_{CMOS} can source sufficient amounts of current, the Thevenin network may be connected to VCMOS (Figure 17.6, Figure 17.8).
	Discrete termination resistors are recommended.

Table 17.5 Summary of Termination Resistor Values for 28 Ω Channel Impedance (Continued)

16-Bit, Long-Channel Memory System Parameter	Specification/Value
SIO Signal	Should always enter a module through the SIN pin and exit through the SOUT pin.
	Although the SIO chain doesn't require termination, the SOUT signal from the last RIMM connector must be pulled down to ground through a 10 KΩ resistor (Figure 17.6).
CTM/CTMN and CFM/CFMN Signals	The differential clock pair produced at the DRCG is terminated both at the DRCG and the opposite end of the channel from the memory controller (Figure 17.9).

17.3 Exercises

17-1. (a) What is a Rambus channel? (b) What signaling technology does RDRAM use for the 26 data, address, timing, and control lines running from the RDRAM to the controller? (c) What signaling technology does RDRAM use for three initialization and operating mode information lines? (d) What is carried on all other lines? (e) Which lines require controlled impedance transmission lines and termination in their characteristic impedance?

17-2. Refer to Figure 17.1 on page 452. For this 2-channel system at 4.2 GB/s, 16-bits per channel, what RDRAM chip is used? State your answer in terms of the format PCx . . . x.

17-3. (a) What is the maximum data rate in Mbps per pin that a PC1333 RDRAM device can transfer? (b) What is the maximum data rate in Mbps per RIMM module that a RIMM5200 can transfer? Specify your answer in peak data rates, not effective data rates. Also take into account that the "5200" in RIMM5200 is a rounded value.

17-4. What is the main difference between short-channel and long-channel configurations of memory chips and controllers?

17-5. What is the purpose of a Rambus continuity module?

17-6. What is the difference between peak data rates and effective data rates?

17-7. What are the differences between Direct RDRAM, Rambus DRAM, DRDRAM, and RDRAM in terms of the maximum possible data rate per pin?

17-8. Based upon information in this chapter, (a) what is the maximum data rate per pin for RDRAM devices currently available from vendors and (b) how is this likely to increase in the future?

17-9. What computer chipsets has RDRAM been used with in the past?

17-10. What is the difference between a RIMM and a DIMM?

17-11. (a) What two channel sizes do Rambus devices come in? (b) What types of applications are best suited for each?

17-12. (a) How many pins does a single-channel RDRAM memory channel require? (b) What benefits are there to the reduced pincount that Rambus DRAM devices offer?

17-13. How many layers are required to route a single- or dual-channel RDRAM memory system in a motherboard?

17-14. What termination resistor values are typically used to terminate RDRAM data, row address, and column address signals?

17-15. (a) Where does the last SOUT signal from the last RIMM connector go to? (b) Through what resistor value is it connected?

CHAPTER 18

Xtreme Data Rate (XDR) DRAM, FlexPhase and ODR

Chapter Objectives

- Present the Rambus XDR™ (Xtreme Data Rate) technology and provide an XDR system overview.
- Introduce FlexPhase™ technology and explain how its use by the controller compensates for bit-to-bit timing differences between signals.
- Introduce XDR's Octal Data Rate (ODR) technology, which transfers 8 bits of data per pin per clock.
- Show how ODR enables 3.2 Gbps data rates per pin using a 400 MHz memory clock.
- Discuss the three major blocks of the XDR I/O cell: DQ data blocks, control block, and RQ request block.
- Present the XDR DRAM block diagram and timing diagrams of the XDR logical layer.
- Discuss gaming console, HDTV/consumer, PC graphics, and PC main memory applications.

18.1 Introduction to XDR[1]

Rambus has developed a new DRAM interface with very-high-speed chip-to-chip data transfer technology — XDR (Xtreme Data Rate). This interface technology is implemented on standard CMOS DRAM memory cores and CMOS controller chips for applications such as high-performance main memory, PC graphics, game consoles, advanced digital consumer systems, high-

1. All text, tables, and figures in this chapter are reprinted with permission by Rambus Inc. [Ram03b]

performance networking systems, and other demanding applications requiring high-bandwidth memory subsystems. The Rambus XDR memory interface achieves an order of magnitude higher performance than today's memories while utilizing the fewest ICs. A single, 2-byte wide, 3.2 GHz XDR DRAM component provides up to 6.4 GB/s bandwidth over the XDR interconnect with a roadmap to 12.8 GB/s bandwidth for a 6.4 GHz component.

The Rambus XDR DRAM achieves an order of magnitude increase in DRAM bandwidth over today's best-of-class memory systems. It provides a total system solution to many of the complex issues and problems that engineers face in designing cost-effective, high-performance memory subsystems. The XDR interface uses a small number of very-high-speed signals to carry all address, data, and control information between the controller and DRAM components.

The XDR solution was engineered to be effective in small footprint, high-bandwidth consumer systems as well as in high-performance main memory applications.

Computer processors continue to place higher demands on the DRAM memory subsystem. XDR DRAM is ready to deliver on the increasing front-side bus performance requirements. As newer, higher performance processors with integrated memory interfaces become more commonplace, XDR memory will deliver the highest bandwidths necessary to meet the demands of these CPUs.

Consumer products such as game consoles and digital TV chipsets that demand high-performance memory bandwidth are able to use 6.4 to 12.8 GB/s bandwidth available from a single x16 XDR DRAM component. Intensive 3D graphics and real-time rendering of video animations continue to drive the need for high-bandwidth memories in gaming consoles. DTV chipsets are consolidating increasing numbers of high-definition MPEG2 image stream decodes with new advanced features such as hard-disk recording, video conferencing, and broadband gateways.

Desktop GPUs (graphics processing units) also have a strong need for high memory bandwidth. Next generation GPUs require increasing memory bandwidth for realistic real-time rendering of graphs and video sequences. These end-user experiences translate to upwards of 25–50 GB/s to supply the needed memory bandwidth.

18.1.1 XDR DRAM

The Rambus XDR DRAM is a general-purpose, high performance memory device suitable for use in a broad range of applications including computer main memory, graphics, video, high-performance network interfaces, and any other application where the memory subsystem must supply high bandwidth and low latency.

The memory is a CMOS DRAM available in varying bit capacities; it has native data bus widths from 1 to 32 bits. The DRAM data bus uses DRSL technology which operates from 3.2 to 6.4 GHz data rate per differential data pair while using conventional system and board design technologies. Typical x16 DRAM devices can sustain data bandwidths ranging from 6.4 to 12.8 GB/s.

The XDR architecture allows high sustained bandwidth for multiple, interleaved, randomly addressed memory transactions. The device's eight-bank architecture supports multiple interleaved transactions for maximum performance.

18.1.2 XDR DRAM Summary Table

Table 18.1 summarizes the features of XDR DRAM.

Table 18.1 Features of XDR DRAM

Highest Pin Bandwidth • 3.2 to 6.4 GHz data rate • Octal Data Rate (ODR) signaling • Bidirectional differential RSL (DRSL) • Programmable on-chip termination • Adaptive impedance matching **Highest Sustained Device Bandwidth** • 4.8 to 12.8 GB/s sustained data rate • Bank-interleaved transactions at full bandwidth • Dynamic request scheduling • Early-read-after-write support for maximum efficiency • Zero overhead refresh **Low Latency** • 1.25/2.0/2.5/3.33 ns request packets • Point-to-point data interconnect • Support for broad range of DRAM cores	**Low Power** • 1.8 V VDD • Programmable ultra-low-voltage I/O signaling (DRSL) • Low power PLL/DLL design • Power-down self-refresh support • Dynamic data width support with dynamic clock gating • Per pin I/O power-down • Sub-page activation support **Ease in System Design** • FlexPhase circuits minimize data line length matching • XDR interconnect uses minimum pin count • System cost and routing complexity are reduced

18.1.3 DRAM Block Diagram

A block diagram of the XDR DRAM device shows the major interconnects and internal blocks (see Figure 18.1).

The clock generation logic receives the CFM/CFMN clock reference and multiplies it by a factor of 4 to produce several internal clock signals.

The RQ11..0 pins receive the request packet. Two 12-bit words arrive in one t_{CYCLE} interval. The 1:2 Demux Block assembles the 24-bit request packet. These 24 bits are loaded into a register and decoded by the Decode Block.

The Decode Block produces several sets of control signals. These include the bank and row addresses for an activate (ACT) command, the bank and row addresses for a refresh activate (REFA) command, the bank address for a precharge (PRE) command, the bank address for a refresh precharge (REFP) command, and the bank and column addresses for a read (RD) or write (WR or WRM) command.

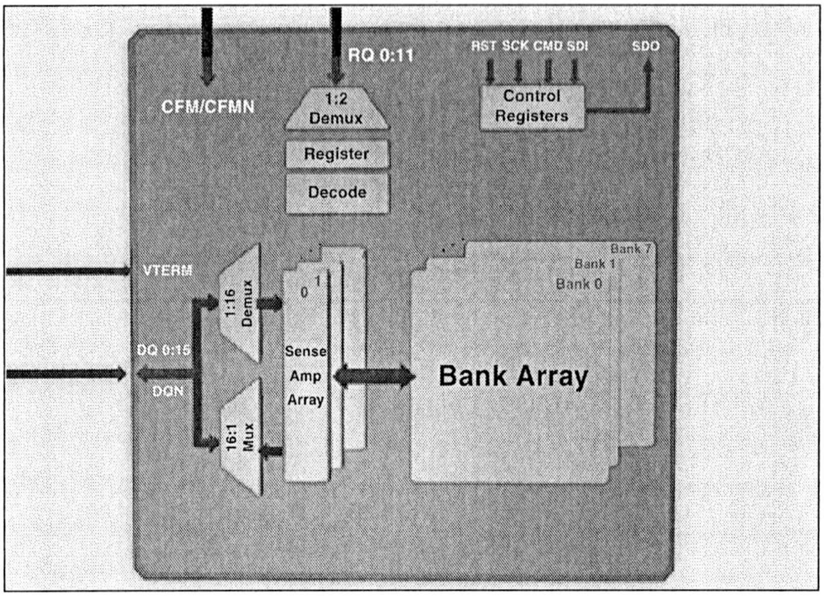

Figure 18.1 XDR DRAM block diagram.

A bank address is decoded for an ACT command. The indicated row of the selected bank is sensed and placed into the associated sense amp array for the bank. Sensing a row is also referred to as "opening a page" for the bank. Another bank address is decoded for a PRE command. The indicated bank and associated sense amp array are precharged to a state in which a subsequent ACT command can be applied. Precharging a bank is also called "closing the page" for the bank.

After a bank is given an ACT command and before it is given a PRE command, it may receive read (RD) and write (WR) column commands. These commands permit the data in the bank's associated sense amp array to be accessed. For a WR command, the indicated column of the associated sense amp array of the selected bank is written with the data received from the DQ15..0 pins. For a RD command, the indicated column of the selected bank's associated sense amp array is read and data is transmitted onto the DQ15..0 pins.

The DQ15..0 pins receive/transmit the write/read data packet (D) for a write/read transaction. Sixteen 16-bit words are received/transmitted in one t_{CC} interval.

The VTERM pin supplies a termination voltage for the DQ pins. The RST, SCK, and CMD pins connect to the Control Register block. These pins supply the data, address, and control needed to write the control registers. The read data for these registers is accessed through the SDO/SDI pins. These pins are also used to initialize the device.

The control registers are used to transition between power modes; they also are used for calibrating the high-speed transmit and receive circuits of the device. Additionally, they supply bank and row addresses for refresh operations.

18.1.4 XDR System Overview

The XDR memory solution was developed to support a broad range of applications and component implementations spanning multiple semiconductor and DRAM technology generations. The XDR system architecture consists of five primary elements (see Figure 18.2):

- DRAM
- Memory controller (XMC)
- I/O cell (XIO)
- Interconnect
- Clock generator (XCG)

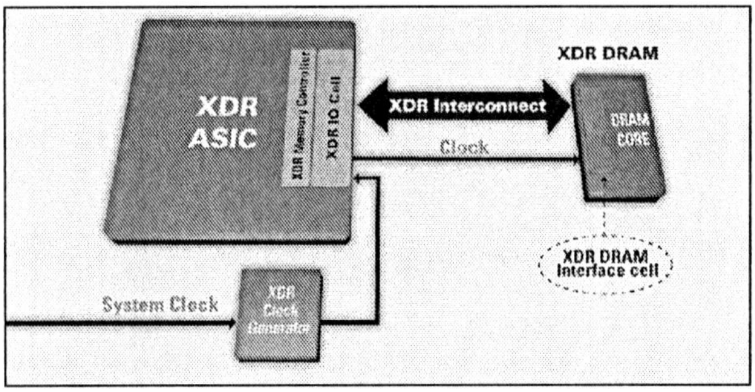

Figure 18.2 XDR system architecture.

The XDR DRAM is a memory component manufactured by Rambus' memory partners. The DRAM contains a compliant XDR DRAM interface specified by Rambus and a standard DRAM memory core. Compliant XDR DRAM components can vary in memory density, page size, core speed grade, and data bus width. Each vendor's XDR device is described by their individual vendor component datasheets. However, the XDR architecture ensures interoperability between components of similar architecture from different memory vendors.

The XMC is a configurable soft macro reference design provided by Rambus. Customers may integrate the XMC directly or use it as a reference design in developing and verifying their own memory controller.

The XIO cell residing on the ASIC controller consists of a collection of hard and soft macros, which provide the physical layer interface between the memory controller and the interconnect. The XIO cell blocks are flexible to embrace differing layout concerns. XDR reference memory controllers from Rambus or other providers offer flexibility to accommodate a wide variety of expected DRAM configurations and can be optimized to support customer specific environments.

XIO hard macros are available in a variety of foundry sources across multiple process technology generations.

As Figure 18.3 shows, the XDR interconnect is made up of a request bus, control signals, and a bidirectional data bus. A typical 32-bit XDR interconnect consists of 32 differential point-to-point data pairs, 12 request bus signals, 1 pair of differential clocks, and 4 serial control signals. A 32-bit XDR interface is capable of data bandwidth in excess of 25 GB/s.

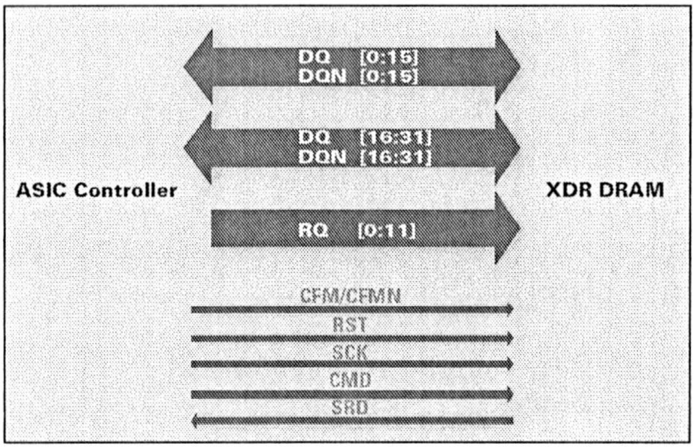

Figure 18.3 XDR interconnect.

The XCG produces low-noise, low-jitter differential clock signals for the memory system. This quad output differential clock chip provides differential clock signals for multiple XDR interconnects.

18.2 XDR Physical Layer

The XDR interconnect achieves its high speed with dense packaging, high-quality PCB transmission line design on standard PCBs, ultra-low-voltage differential signaling, and precise clocking. By employing these techniques on conventional CMOS IC and printed circuit board processes, Rambus' XDR system solution achieves high performance at low system cost.

18.2.1 Signaling

XDR solutions use Differential Rambus Signaling Levels (DRSL) for scalable speed point-to-point bidirectional data signals and Rambus Signaling Levels (RSL) for source synchronous bussed address and command signals to multiple DRAM devices. The point-to-point bus allows multi-GHz speeds while the bussed address allows scalable capacity of up to 36 DRAM devices per request bus.

The combination of DRSL and RSL signaling provides a high-performance interconnect for the most demanding of applications.

Ultra-low-voltage-swing differential signals minimize di/dt, thereby

- Reducing ground bounce
- Decreasing power consumption

- Reducing electromagnetic interference
- Increasing scalability to higher data rates
- Improving noise immunity via common-mode rejection

Although each DRSL signal pair is routed with controlled impedance, different data traces do not need to be line length matched. Each signaling differential bit pair is equipped with controller-based FlexPhase technology, which compensates for bit-to-bit timing differences between signals.

The request bus uses standard RSL signaling and transfers address and command information to the DRAMS. RSL is a single-ended, open-drain signaling standard. The bus switches at a frequency of 800 MHz and uses a 900 mV swing. These impedance controlled PCB traces must be length matched from the controller to the last DRAM on the request bus.

Figure 18.4 summarizes the XDR signal swing in comparison with other DRAM technologies.

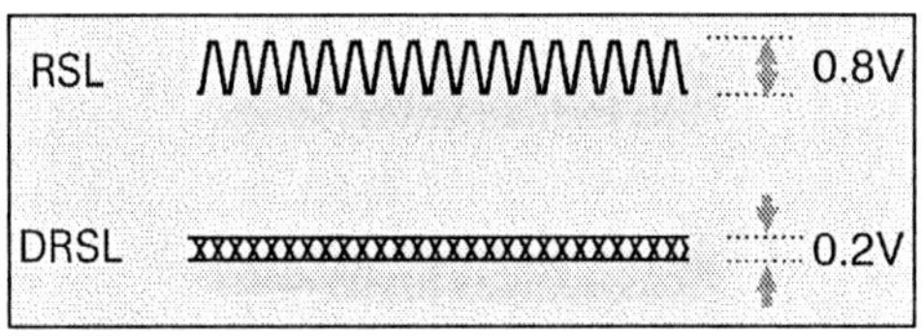

Figure 18.4 RSL and DRSL ultra-low-voltage signaling for XDR interconnect. [Reprinted with permission of Rambus, Inc.]

18.2.2 Clocking

XDR protocol uses a mesochronous interface, meaning that all data transfers are at a fixed but arbitrary phase with respect to clock edges. XDR FlexPhase circuits are internal to the XI0 cell and ensure maximum per-pin margin at high data rates.

Figure 18.5 is a block diagram of the clock distribution in an XDR system. The XDR Clock Generator (XCG) produces a Clock To Master (CTM) signal generated from a system clock input. The XCG differential clock output is wired to the ASIC CTM, and the XIO Cell PLL uses it to generate the internal XDR data clock. The CTM is routed through the ASIC package and is returned to the XDR interconnect as the CFM (clock from master) clock, which is delivered to each of the DRAM CFM input pins.

Each XDR DRAM generates internal data clocks from the external CFM reference to time data transfers to and from the DRAM's data pins. The DRAM also uses CFM to sample address and command information.

The CFM clock and the DRAM address/command information travel in parallel to minimize clock-to-data skew. Because the address/command transmission lines are matched, the clock and address/command signals remain synchronized as they propagate to their destinations.

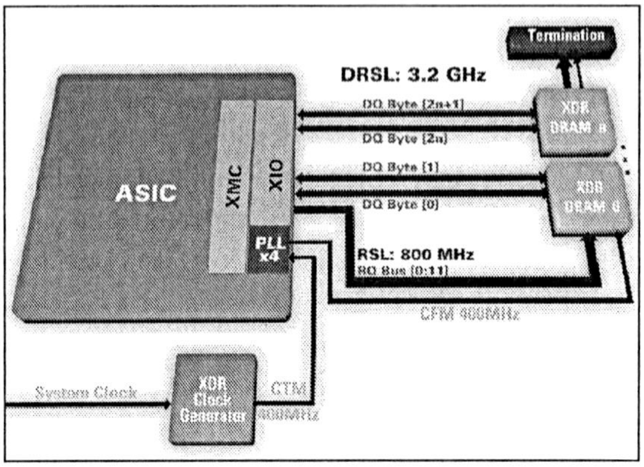

Figure 18.5 XDR system clocking and topology.

18.2.3 Data Transfers

Octal Data Rate (ODR) transfers are used between the XDR ASIC controller and the XDR DRAM. Figure 18.6 shows how the system clocking is generated by the XCG for address/command clocking and by the XIO internally for the corresponding data I/Os.

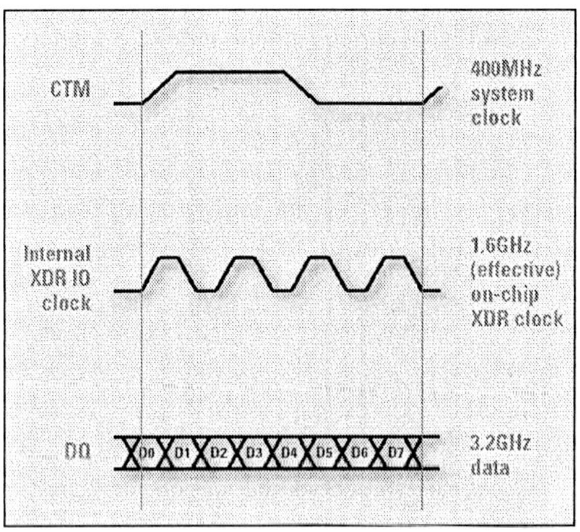

Figure 18.6 XDR's octal data rate (ODR).

XDR signaling operates at octal data rates, transferring 8 bits of data per clock. ODR enables 3.2 Gbps data rates with a 400 MHz clock and provides a scalable path to over 6.4 GHz as bandwidth needs increase.

The lower speed 400 MHz system clock is routed on the PCB between chips. On-chip, the 400 MHz clock is multiplied — up to 1.6 GHz with a PLL. This effective 1.6 GHz clock is subsequently used to transmit and receive data on both clock edges, resulting in 3.2 GHz data rates. The 1:8 relationship between clock and data rates results in ODR operation.

FlexPhase technology enables flexible phase, allowing precise on-chip alignment of data with the clock, thereby eliminating the need for PCB trace length matching and strict PCB timing constraints. Memory system designers no longer need to worry about matching adjacent data traces to an associated clock or strobe, allowing much simpler and more compact memory layouts. Unlike traditional serial link approaches, the technology does not use an embedded clock, allowing chip connections to avoid the added latency and 20% bandwidth penalty associated with 8- to 10-bit encoding.

FlexPhase circuits are contained in the ASIC's XIO cell. They allow the use of a CFM signal to generate an XIO data clock and to skew individual data bit output timing from the ASIC to the XDR DRAM to present a synchronized parallel data word to the addressed XDR DRAM. FlexPhase I/O circuits can compensate for systematic timing errors with a resolution as fine as 2.5 ps.

The aligned data word is clocked into the XDR DRAM using an internal DRAM clock generated from the external CFM clock. The same FlexPhase circuits are used to account for each bit's individual skew from the addressed DRAM to the ASIC to avoid burdening the XDR DRAM with extra circuitry. Figure 18.7 shows the FlexPhase circuits contained within each single data byte block in the XIO cell.

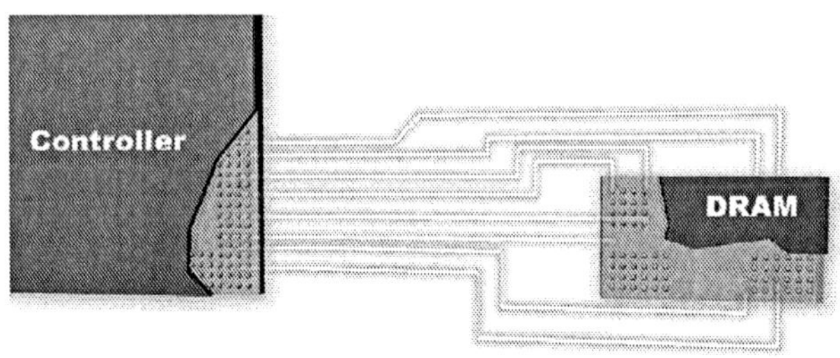

Figure 18.7 FlexPhase circuits on controller.

18.2.4 Address/Control Transfers

Address and command information is transferred to all XDR DRAMs via the XDR protocol. Figure 18.8 summarizes the signal swing characteristics and typical frequency of operation for address/command and data signals. Address/command information travels synchronously from the ASIC to the DRAM via the controlled impedance bus. CFM is routed in parallel and is length matched to the request bus to minimize skew and hence data clocking errors. Information is clocked on both edges of the CFM clock and is transferred at an 800 MHz data rate.

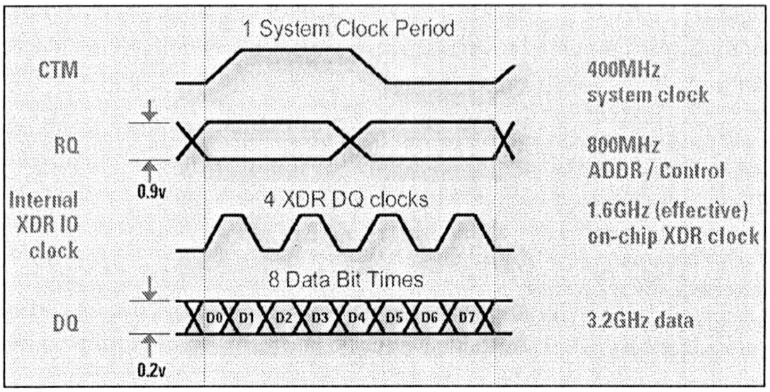

Figure 18.8 XDR's signal swing and frequencies.

18.2.5 XDR I/O Cell (XIO)

The XDR I0 (XI0) cell has the following major blocks:

- DQ data blocks
- Control block
- RQ request block

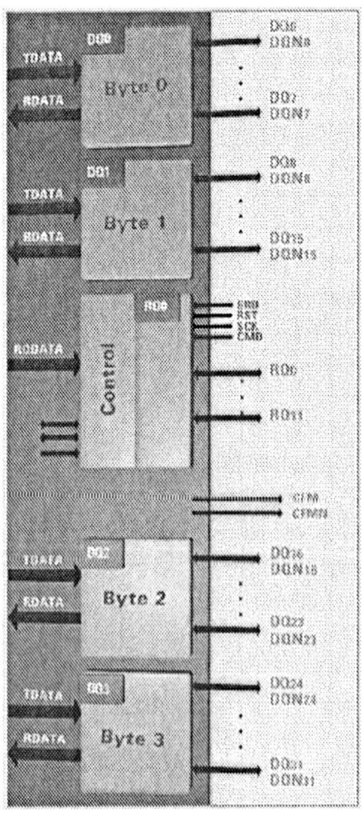

Figure 18.9 XDR I/O cell.

The XIO is a high-performance, low-latency controller interface to XDR DRAM memory subsystems. It is a versatile CMOS macro cell that can be seamlessly integrated into a wide variety of target processes. The general-purpose cell is independent of the logical memory controller design, enabling support for a wide variety of memory applications needing high bandwidth and low latency.

The XIO provides a wide, on-chip, CMOS-level signaling interface to the memory controller logic and a narrow, high-speed Differential Rambus Signaling Level (DRSL) interface to the external XDR memory system. The XIO performs the data serialization and deserialization functions associated with write and read operations, respectively. The interface supports a serialization ratio of 8 to 1. The internal interface uses an address bus and separate unidirectional read/write buses, which are synchronous to an internal clock supplied by the controller logic.

The XIO contains at least one 12-bit request bus block (RQ), one control block (CTL), and a variable number of 8-bit data blocks (DQs), typically four (see Figure 18.9). 9-bit XI0 cell bytes can be made available for applications requiring ECC-enabled DRAMS. The RQ block provides address and control information to the DRAMS. The CTL provides register access, initialization, maintenance, and testability functions. Each DQ block can transmit and receive data ranging from 3.2 to 6.4 GHz data rates.

The DQ block transmits and receives ultra-low-voltage DRSL (200–300 mV) differential signals with per bit FlexPhase circuits allowing arbitrary per-pin DQ data transmit and receive timing. The need for precise data bus trace length matching for the system data bus is eliminated.

The RQ block generates single-ended RSL (900 mV) signals and is bussed to DRAMs. A typical 32-bit data XIO cell block diagram appears in Figure 18.9. Note that the 32-bit XIO interface takes approximately 130 ASIC package pins to wire to the PCB and corresponding XDR DRAMs.

18.3 XDR Logical Layer

The timing diagrams in Figures 18.10 and 18.11 illustrate an XDR DRAM device's write and read transaction timing. Three sets of pins are used for normal memory access transactions: CFM clock pins, RQ[11:0] request pins, and DQ[15:0] data pins.

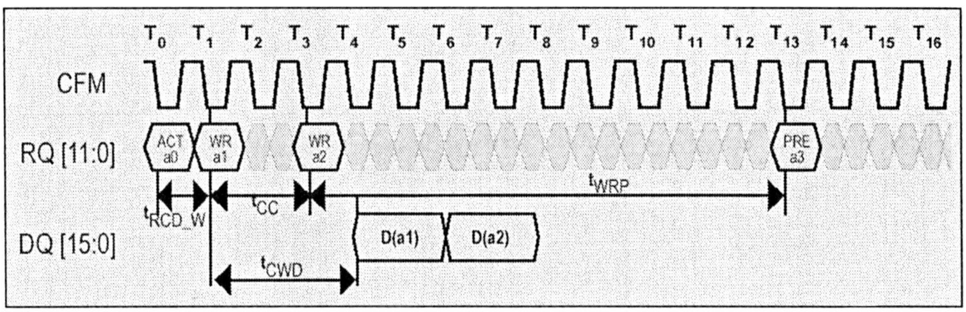

Figure 18.10 XDR write transaction.

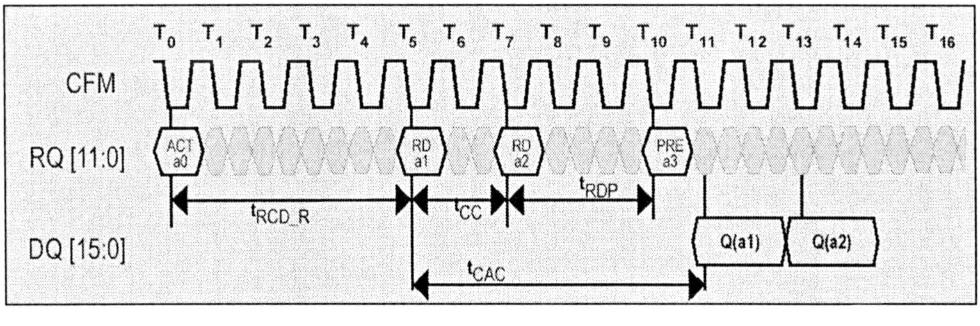

Figure 18.11 XDR read transaction.

A transaction is a collection of data packets (set of bit windows) needed to complete a memory access. Two buses carry packets: the RQ request bus and DQ data bus. Each packet on the RQ bus uses a set of 2-bit windows on each signal, while the DQ bus uses a set of 16-bit windows on each signal.

18.3.1 Write Transaction

In the write transaction diagram in Figure 18.10, a request packet (on the RQ bus) at clock edge T0 contains an activate (ACT) command. This command causes row Ra of bank Ba in the DRAM to be loaded into the sense amp array for the bank. A second request packet at clock edge T1 contains a write (WR) command. This causes the data packet D(a1) at edge T4 to be written to column Ca1 of the sense amp array for bank Ba. A third request packet at clock edge T3 contains another write (WR) command. This causes the data packet D(a2) at edge T6 also to be written to column Ca2. A final request packet at clock edge T13 contains a precharge (PRE) command to finish the sequence by closing the page.

The timing spaces between the write request packets are constrained by the following timing parameters in the diagram: $t_{RCD\text{-}W}$, t_{CC}, and t_{WRP}. In addition, the spacing between the request packets and data packets are constrained by the t_{CWD} parameter. The spacing of the CFM/CFMN clock edges is constrained by t_{CYCLE}.

18.3.2 Read Transaction

The read transaction in Figure 18.11 shows a request packet at clock edge T0 containing an ACT command. This causes row Ra of bank Ba of the memory component to load into the sense amp array for the bank. A second request packet at clock edge T5 contains a read (RD) command. This causes the data packet Q(a1) at edge T11 to be read from column Ca1 of the sense amp array for bank Ba. A third request packet at clock edge T7 contains another RD command. This causes the data packet Q(a2) at edge T13 also to be read from column Ca2. A final request packet at clock edge T10 contains a PRE command. The timing spaces between the read request packets are constrained by the following timing parameters in the diagram: $t_{RCD\text{-}R}$, t_{CC}, and t_{RDP}. In addition, the timing spaces between the request and data packets are constrained by the t_{CAC} parameter.

18.4 Applications

18.4.1 The Gaming Console Solution

Consumer gaming consoles are continuously improving in performance, targeting ever more realistic graphics, on-the-fly motion video rendering, and 3D worlds. The gaming consoles in the near future have very stringent performance and cost requirements. Built-in cost efficiencies must also be deployed to lower the product's cost as the product matures.

These platforms can be designed with architectures that require very few chips.

Using large amounts of processing horsepower requires that the processors be adequately fed with information (data) that is usually stored in the system's DRAM memory subsystem for quick access.

Competitive systems will require upwards of 25 GB/s of sustained bandwidth from the memory subsystem. Figure 18.12 shows a system that will meet these requirements.

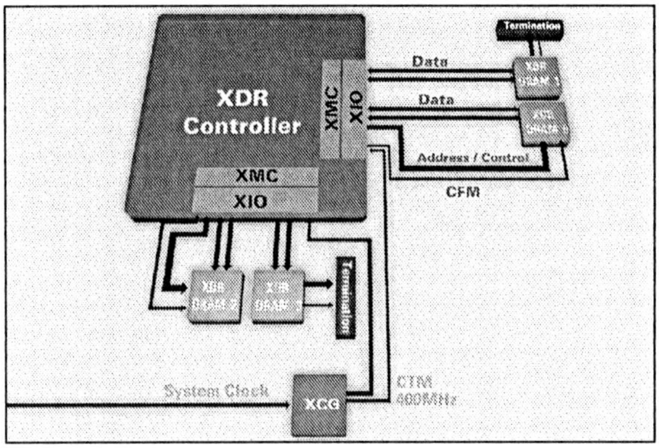

Figure 18.12 XDR consumer or graphics system topology.

XDR DRAM is the answer not only for its performance but also for its low-cost system attributes such as four-layer PCB design, compact chip placement, low EMR, and highest bandwidth per component for low chip-count systems.

18.4.2 The HDTV/Consumer Solution

Emerging consumer devices for High-Definition TV (HDTV) and other multimedia/entertainment applications are combining functions and feature sets of multiple devices into one system while simultaneously adding new features for users. Systems such as HD-ready cable or satellite set-tops are now dealing with multi-stream HD MPEG2 decode while also providing hard disk recording/program time-shifting features, graphical user interfaces, programming guides, and other advanced functions such as broadband Internet gateways and videoconferencing.

System cost pressures are driving a unification of formerly discrete CPU/media processors and their individual memory subsystems into a single chip with unified memory architecture. The XDR system solution provides up to 6.4 GB/s memory bandwidth from a single x16 XDR DRAM device. The 16-bit wide, 64 MB XDR memory solution uses only 90 CPU/media processor pins, including power and ground pins. For lower-cost systems, an x8 XDR DRAM provides 3.2 GB/s in 65 controller pins.

A single 16-bit XDR I0 cell on a HDTV chipset can enable a single chipset design to span from high-performance to cost-sensitive markets simply by varying the width of memory connected to it. In addition, XDR DRAM with FlexPhase ensures easy and cost effective four-layer PCB design with the fewest support components versus competing memory technologies.

18.4.3 The PC Graphics Solution

PC graphics subsystems have an insatiable need for dedicated memory bandwidth. Today's and tomorrow's graphics solutions require GPUs that constantly and consistently offer more compute horsepower to provide the most realistic, intensive 3D graphics rendering and complex video decoding to application developers and end users. Ultimately, the test is the end user's video and graphics experience.

Real-time rendering of lifelike video animations and the user experiences that these compute engines furnish form the battleground for graphics dominance. To support these goals, future GPUs will need upwards of 25 GB/s in memory bandwidth. A graphics provider must have not only the requisite performance but also a cost-competitive solution. A single-chip solution must be able to scale across a broad range of price/performance points.

The graphics system block diagram shown in Figure 18.12 will meet these memory bandwidth requirements. This extreme bandwidth solution uses a 64-bit memory bus with only four memory chips to yield between 25 and 30 GB/s bandwidth.

The controller is designed so that half the memory bus can be populated providing from 12–25 GB/s bandwidth to support other less bandwidth-demanding markets.

18.4.4 The PC Main Memory Solution

Many computer applications require a cost-effective, high-bandwidth, scalable, multi-density module solution with a small-form factor for their memory subsystems. As computer CPUs increase in raw MIPs/clock performance, so do their front-side buses, which are approaching GHz data rates.

The system designer is challenged by the increasing requirements of the memory subsystem to keep the CPU from stalling. Although many CPUs continue to increase their L1, L2, and L3 on-board caches, what counts is the bandwidth and latency performance of the off-board memory. This memory must be of adequate bandwidth to maximize data flow to and from the memory subsystem. It must also be of low enough latency to avoid pipeline stalls.

Increasing the performance of CPU buses constantly taxes the various subsystems that must provide a steady and quick stream of data from the typically slower DRAM subsystem. The XDR DRAM memory subsystem is a solution to this problem.

The modules that will populate these future workstations and PCs come in various densities and speeds. Table 18.2 shows various speeds and module densities supported by the XDR architecture.

Initial XDR modules will be 32 bits wide and support peak bandwidths of up to 16 GB/s, which rival other technologies needing many more modules with wider bus interfaces.

Higher-performance systems may use multiple module "channels" of XDR module memory to get to larger memory sizes and higher memory bandwidth subsystems.

Table 18.2 Speeds and Module Densities Supported by the XDR Architecture

DRAM	Module Capacity	Module Bandwidth		
		9.6 GB/s	12.8 GB/s	16 GB/s
	64 MB	✔	✔	✔
256 Mb	128 MB	✔	✔	✔
x2/x4/x8/x16	256 MB	✔	✔	✔
	512 MB	✔	✔	✔
	128 MB	✔	✔	✔
512 Mb	256 MB	✔	✔	✔
x2/x4/x8/x16	512 MB	✔	✔	✔
	1 GB	✔	✔	✔
	256 MB	✔	✔	✔
1 Gb	512 MB	✔	✔	✔
x2/x4/x8/x16	1 GB	✔	✔	✔
	2 GB	✔	✔	✔

Mainstream systems will focus on medium-size memory subsystems and may use a single channel of XDR module memory. These systems will ship with one module and have a connector for a field upgrade.

Figure 18.13 shows a mainstream memory subsystem supporting two modules connected either to a north bridge or directly to the processor.

The PCB for this system is easily routed in a four-layer design. Routing the data bus to the memory modules is simplified by FlexPhase technology, which compensates for any bit timing differences across the interface.

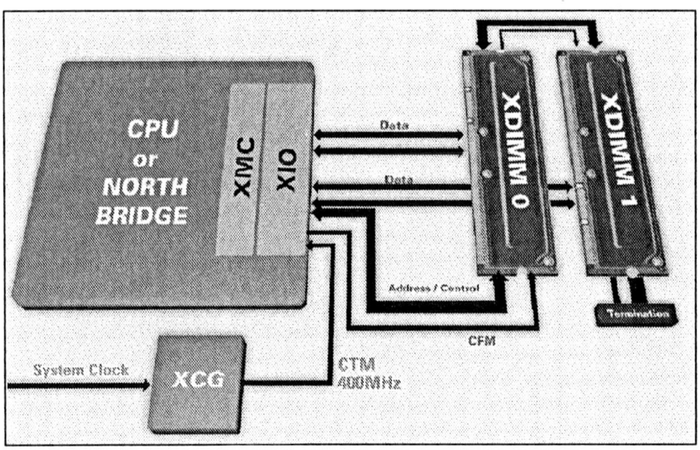

Figure 18.13 XDR module system topology.

Computer equipment operating at high clock speeds must take particular precautions to minimize effects of EMI emissions through extensive shielding or by using a technology that inherently lessens these effects. Using XDR modules, which employ ultra-low-voltage signaling and DRSL, minimizes the amount of EMR energy and its impact to system cost in containing such emissions.

XDR not only provides the needed bandwidth but does it cost-effectively by minimizing the number of modules needed, reducing the impact of EMR, and minimizing the PCB module placement and component count without sacrificing bandwidth or latency.

18.5 Exercises

18-1. (a) What is the highest external clock frequency that can be used in an XDR DRAM memory system? (b) At this maximum clock frequency, what is the maximum data rate per pin? (c) What is the total chip bandwidth that can be provided by a single 16-bit XDR DRAM chip?

18-2. What are the five primary elements of the XDR system architecture and what do they do?

18-3. What is the maximum number of devices that can be used in an XDR DRAM memory system per request bus?

18-4. To reach a data bandwidth of 25.6 GB/s for a 32-bit XDR interface, what would be the frequency of the required external system clock?

18-5. (a) In an XDR memory system, which signals use DRSL signaling and which signals use RSL signaling? (b) What is the difference between DRSL and RSL signaling?

18-6. What does the term "scalability" mean when we talk about "scalability to higher data rates" and "scalability to higher-density memory systems"?

18-7. Where are the FlexPhase circuits physically located with respect to the XDR memory system?

18-8. (a) What do the abbreviations CTM and CFM stand for? (b) What is the frequency of the CTM and CFM signals? (c) Which one of these signals is sent to the XDR DRAM memory chips? (d) What do the XDR DRAM memory chips do with the signal that is sent to them?

18-9. If the XDR memory system clock is 400 MHz, what is the frequency of the XDR I/O clock internal to the XDR DRAM chip?

18-10. Do FlexPhase circuits use an embedded clock within the data stream to synchronize data with the clock?

18-11. What are the three major blocks within the XDR I/O (XIO) cell and what do they do?

18-12. When data is sent from the XDR controller to the XDR DRAM, or from the XDR DRAM to the XDR controller, where is the data most likely to be a synchronized parallel data word (i.e., where all the bits of the data word are in perfect alignment with one another), at the controller or at the DRAM chip? Explain your answer.

18-13. Table 18.1 on page 473 summarizes the features of XDR DRAM. Under "Highest pin bandwidth," it lists data rates from 3.2 to 6.4 GHz. What range of XDR memory system clock frequencies corresponds to this range of data rates?

18-14. FlexPhase circuits can compensate for systematic timing errors to within what resolution? State your answer in ps.

18-15. (a) Is XDR DRAM implemented using a special DRAM memory core? (b) What are the special features of this memory core, if any? (c) Can XDR DRAM cores vary in memory density, page size, core speed grade, and data bus width and still be compliant with the XDR DRAM standard?

PART 4

Modeling, Simulation, and EDA Tools

CHAPTER 19

Differential and Mixed-Mode S-Parameters

Chapter Objectives

- Define S-parameters and explain how they are used in the design of high-speed systems.
- Discuss the advantages of using S-parameters over Y-, H-, and Z-parameters and over traditional circuit modeling techniques.
- Present the mathematics underlying S-parameters and two-port network theory.
- Discuss the mathematical relationship between single-ended, differential, and mixed-mode S-parameters.
- Compare calculated and measured differential return loss to single-ended return loss.
- Discuss the source of S-parameter measurement errors.
- Present direct-measurement and de-embedding processes to remove test fixture effects from S-parameter values.
- Discuss reference planes and how they change with calibration of an S-parameter test setup.
- Provide an actual example that calculates return loss and compares the results of a traditional calibration against that of a custom calibration that also removes the effects of the test fixture.

19.1 S-Parameters Bridge the Gap Between Chips and Systems

"With the tremendous growth in communications and networking, and the increase in demand for higher serial data bandwidth, there is a growing need to streamline the simulation of entire complex system environments — from chip, to package, to board, to connector, to backplane and back again to chip. The problem becomes even more challenging because the supply chain for multigigabit serial I/O connectivity requires tighter integration between many different parties — chip and board designers, system integrators, and backplane and connector suppliers.

Internet applications for data, audio, video, and graphics have strived to relieve the serial I/O bottleneck by introducing a myriad of standards such as 10-gigabit XAUI, Infiniband, and PCI Express. The explosion of multigigabit serial bandwidth in the past few years further challenges integration to these standards. The most sophisticated systems support 40 Gbps today, up from 1.125 Gbps to 10 Gbps in only the last couple of years. Only by bridging the circuit simulation gap between chips and systems can designers effectively address the design, simulation, and integration of multigigabit serial I/Os on multi-port networks.

At high frequencies, the scattering parameter (S-parameter) is the most accurate form of broadband frequency representation in off-chip passive networks such as packages, boards, and backplanes. S-parameters are preferred over other frequency-domain parameters such as Y and Z primarily because S-parameters range between the values of 0 and 1, and they are well behaved across a broad frequency range. In addition, S-parameters are the easiest parameters to model and measure using a vector network analyzer. By matching the signal impedance without reflection and saving the results to a standard file, designers can directly access S-parameter data for simulation.

However, most existing SPICE tools still require users to model S-parameters by converting S-parameter data from backplanes, boards and transmission lines into lumped RLGC models or SPICE subcircuits as shown in Figure 19.1. The figure depicts the process of measuring S-parameters for a transmission line and then extracting the RLGC parameters for use in SPICE simulations. This timeworn lumped RLGC approach is necessitated by the inability of existing SPICE tools to directly take in S-parameter data. The process also taxes the inherent capacity limitations within these tools.[1, 2]

For example, a 3.125 Gbps SerDes system consisting of a transmitter and PLLs (SPICE netlist), connectors, backplanes, line cards (S-parameter models), cables (transmission models), and a receiver (SPICE netlist) will require simulating more than 10,000 devices, including hundreds of thousands of RLC parasitics. The lumped model development time alone can take up to several weeks, not counting the SPICE simulation time and convergence issues. And at the end of this prolonged process, it is very likely that the system is only accurate within a narrow frequency range of a few GHz.

1. In Figure 19.1, R, L, G, C are the transmission lines per unit length resistance, inductance, conductance, and capacitance, respectively; Z_0 is the characteristic impedance; and γ is the transmission line's propagation constant.
2. The step labelled "De-embed S-parameters" is the process of eliminating test fixture effects from S-parameter values using models. De-embedding is discussed later in Section 19.5.2 on page 511.

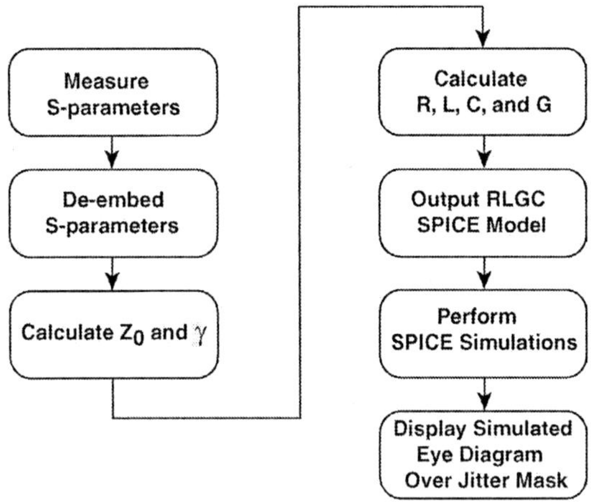

Figure 19.1 Transmission line parameter extraction process vs. streamlined process using next-generation SPICE tools.

So how can designers and integrators simulate S-parameters in time-domain netlists in huge systems down to picosecond accuracy? The linchpin technology lies with next-generation SPICE tools that perform true mixed-domain simulation with direct S-parameter data, and without restrictions in capacity and SPICE compatibility. The name of the game is to provide the fastest and most accurate path from measured S-parameters to eye diagrams.

Adapting to the ever-changing landscape of networking bandwidth requires the use of technology that bridges the time-domain and frequency-domain aspects of large systems. A true mixed-domain simulation solution with actual S-parameter data benefits the entire supply chain for multigigabit serial connectivity — from chip vendors, to board and system integrators of vendor IP, to the packaging, connector, and backplane manufacturers — to ensure significant system cost savings, easier integration, and most importantly, accurate and consistent operation across a broadband frequency range."[3] [Liu03]

19.2 The Rationale for Using S-Parameters

"Scattering parameters are a set of parameters describing the scattering and reflection of traveling waves when a network is inserted into a transmission line. S-parameters are normally used to characterize high-frequency networks where simple models valid at lower frequencies cannot be applied. S-parameters are normally measured as a function of frequency, so when looking at the formulas for S-parameters, it is important to realize that frequency is implied, and that complex gain (i.e., gain and phase) is also assumed. For this reason, S-parameters are often called complex scattering parameters." [Nat03]

3. Written by Yu Liu, Director of Simulation Products at Apache Design Solutions. Copyright © 2003 by CMP Media LLC, 600 Community Drive, Manhasset, NY 11030. Reprinted from EEDESIGN with permission.

"S-parameters are used because they

- Are relatively easy to obtain at high frequencies
  - — a vector network analyzer is used to measure voltage traveling waves
  - — circuits under test don't need to be shorted or open-circuited which could otherwise cause active devices to oscillate or self-destruct
- Relate to familiar measurements (gain, loss, reflection coefficient)
- Can be cascaded to predict system performance for multiple device interconnections
- Can be used to compute H, Y, or Z parameters if desired
- Can be easily imported and used in EDA (electronic design automation) tools"[4] [Bal98]

"The units of high-frequency measurements are different from those for lower-frequency measurements. At low frequencies, arbitrary circuit networks are often characterized by their H, Y, or Z parameters. The measurement of these parameters requires the placement of short and open circuits on the network ports. The use of shorts and opens at high frequencies presents several problems. At high frequencies, accurate shorts and opens are difficult to obtain because of parasitic electrical circuit elements. Also, active devices are often not stable with opens or shorts because of internal device parasitic electrical elements. To get around these problems, engineers measure S-parameters, which typically require 50 Ω terminations. It is much easier to get a good 50 Ω termination at high frequency than a good short or open, and active devices are usually stable with 50 Ω."[5] [Cas03]

19.2.1 Package Measurements

"Digital LSI and VLSI packages are now required to operate at clock rates of 50–250 MHz and beyond. This requires analog bandwidths of 500–2,500 MHz for good edge definition. At these frequencies, typical digital packages will exhibit electrical behavior completely unpredicted from low-frequency measurements and models. Complicating the issue, typical digital packages have metal pattern dimensions that resonate and/or couple to the outside environment-they can act as antennas and resonators. This is the microwave domain, where open lines may appear as short circuits and shorted lines as opens, depending on the frequency. Microwave designers have been successfully working with these issues for years, and now these techniques have been adapted to the high-speed digital design environment." [Cas03]

4. Reprinted with permission of Agilent Technologies.
5. Reprinted with permission of Cascade Microtech.

Sources of Information on S-Parameters

Information on S-parameters is readily available, but fragmented. Test equipment and chip vendors such as Agilent Technologies, Maxim Integrated Products, and Cascade Microtech provide considerable information through tutorials (Agilent) or application notes on their Web sites. S-parameter discussions are also found in microwave engineering textbooks such as references [Poz03], [Wai95], [Pen95], and [Ven92] and high-speed design books such as [Hal00]. Reference [Mav96] commits the entire book to the application of scattering parameters in network analysis. A very detailed discussion of mixed-mode S-parameters is found in Bockelman's Ph.D. dissertation [Boc97] and in the IEEE Transactions on Microwave Theory and Techniques [Boc95]. Web searches for technical literature can provide the most recent and up to date information on S-parameters, trends in their use, and new developments. The list of references at the end of S-parameter application notes and technical papers is also extremely useful. A Web site providing links to valuable S-parameter information can be found at *www.sss-mag.com/spara.html.*

19.3 Single-Ended S-Parameters[6]

19.3.1 Introduction

Linear networks, or nonlinear networks operating with signals sufficiently small to cause the networks to respond in a linear manner, can be completely characterized by parameters measured at the network terminals (ports) without regard to the contents of the networks. Once the parameters of a network have been determined, its behavior in any external environment can be predicted again without regard to the contents of the network.

S-parameters are important in digital design at gigahertz frequencies because they are easier to measure and work with at high frequencies than other kinds of parameters. They are conceptually simple, analytically convenient, and capable of providing a great insight into a measurement or design problem.

To show how S-parameters ease design at gigahertz frequencies, and how you can best take advantage of their abilities, this section describes S-parameters and relates them to more familiar concepts. Data obtained with a network analyzer is used to illustrate a design.

19.3.2 Two-Port Network Theory

Although a network may have any number of ports, network parameters can be explained most easily by considering a network with only two ports, an input port and an output port, like the network shown in Figure 19.2. To characterize the performance of such a network, any of several parameter sets can be used, each of which has certain advantages. Each parameter set is

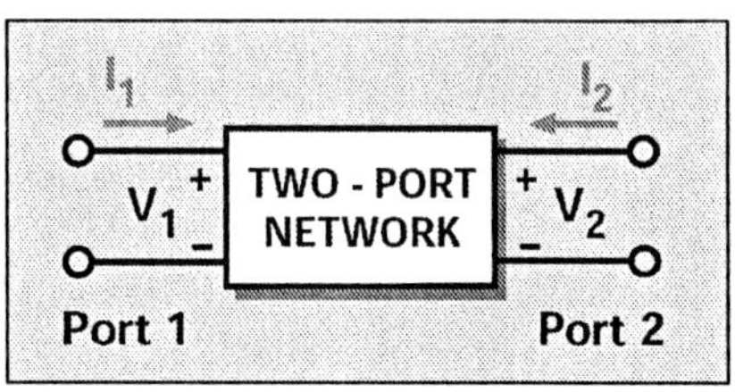

Figure 19.2 General two-port network.
[Reprinted with permission of Agilent Technologies, Inc.]

6. All text and figures in Section 19.3 and its subsections are taken from reference [Agi97].

related to a set of four variables associated with the two-port model. Two of these variables represent the excitation of the network (independent variables), and the remaining two represent the response of the network to the excitation (dependent variables). If the network of Figure 19.2 is excited by voltage sources V_1 and V_2, the network currents I_1 and I_2 will be related by the following equations (assuming the network behaves linearly):

$$I_1 = y_{11}V_1 + y_{12}V_2 \tag{19.1}$$

$$I_2 = y_{21}V_1 + y_{22}V_2 \tag{19.2}$$

In this case, with port voltages selected as independent variables and port currents taken as dependent variables, the relating parameters are called short-circuit admittance parameters, or y-parameters. In the absence of additional information, four measurements are required to determine the four parameters y_{11}, y_{12}, y_{21}, y_{22}. Each measurement is made with one port of the network excited by a voltage source while the other port is short circuited. For example, y_{21}, the forward transadmittance, is the ratio of the current at port 2 to the voltage at port 1 with port 2 short circuited, as shown in Equation (19.3).

$$y_{21} = \left.\frac{I_2}{V_1}\right|_{V_2 = 0\text{(output short circuited)}} \tag{19.3}$$

If other independent and dependent variables had been chosen, the network would have been described, as before, by two linear equations similar to Equations (19.1) and (19.2), except that the variables and the parameters describing their relationships would be different. However, all parameter sets contain the same information about a network, and it is always possible to calculate any set in terms of any other set.

"Scattering parameters," which are commonly referred to as S-parameters, are a parameter set that relates to the traveling waves that are scattered or reflected when an n-port network is inserted into a transmission line.

19.3.3 Using S-Parameters

The ease with which scattering parameters can be measured makes them especially well suited for describing transistors and other active devices. Measuring most other parameters calls for the input and output of the device to be successively opened and short circuited. This can be hard to do, especially at RF frequencies where lead inductance and capacitance make short and open circuits difficult to obtain. At higher frequencies these measurements typically require tuning stubs, separately adjusted at each measurement frequency, to reflect short or open circuit conditions to the device terminals. Not only is this inconvenient and tedious, but a tuning stub shunting the input or output may cause a transistor to oscillate, making the measurement invalid.

S-parameters, on the other hand, are usually measured with the device imbedded between a 50 Ω load and source, and there is very little chance for oscillations to occur.

Another important advantage of S-parameters stems from the fact that traveling waves, unlike terminal voltages and currents, do not vary in magnitude at points along a lossless transmission line. This means that scattering parameters can be measured on a device located at some distance from the measurement transducers, provided that the measuring device and the transducers are connected by low-loss transmission lines.

19.3.3.1 Derivation

Generalized scattering parameters were defined by K. Kurokawa in reference [Kur65]. These parameters described the interrelationships of a new set of variables (a_i, b_i). The variables a_i and b_i are normalized complex voltage waves incident on and reflected from the ith port of the network. They are defined in terms of the terminal voltage V_i, the terminal current I_i, and an arbitrary reference impedance Z_i, where the asterisk denotes the complex conjugate:

$$a_i = \frac{V_i + Z_i I_i}{2\sqrt{|\mathrm{Re}Z_i|}} \qquad b_i = \frac{V_i - Z_i^* I_i}{2\sqrt{|\mathrm{Re}Z_i|}} \tag{19.4}$$

For most measurements and calculations it is convenient to assume that the reference impedance Z_i is positive and real. For the remainder of this section, then, all variables and parameters will be referenced to a single positive real impedance, Z_0.

The wave functions used to define S-parameters for a two-port network are shown in Figure 19.3.

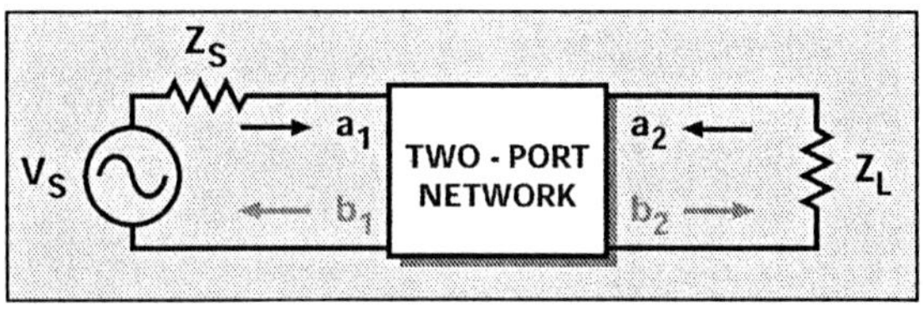

Figure 19.3 Two-port network showing incident waves (a1, a2) and reflected waves (b1, b2) used in S-parameter definitions. [Reprinted with permission of Agilent Technologies, Inc.]

The independent variables a_1 and a_2 are normalized incident voltages, as follows:

$$a_1 = \frac{V_1 + I_1 Z_0}{2\sqrt{Z_0}} = \frac{\text{voltage wave incident on port1}}{\sqrt{Z_0}} = \frac{V_{i1}}{\sqrt{Z_0}} \tag{19.5}$$

$$a_2 = \frac{V_2 + I_2 Z_0}{2\sqrt{Z_0}} = \frac{\text{voltage wave incident on port2}}{\sqrt{Z_0}} = \frac{V_{i2}}{\sqrt{Z_0}} \tag{19.6}$$

Dependent variables b_1 and b_2 are normalized reflected voltages:

$$b_1 = \frac{V_1 - I_1 Z_0}{2\sqrt{Z_0}} = \frac{\text{voltage wave reflected from port 1}}{\sqrt{Z_0}} = \frac{V_{r1}}{\sqrt{Z_0}} \tag{19.7}$$

$$b_2 = \frac{V_2 - I_2 Z_0}{2\sqrt{Z_0}} = \frac{\text{voltage wave reflected from port 2}}{\sqrt{Z_0}} = \frac{V_{r2}}{\sqrt{Z_0}} \tag{19.8}$$

The linear equations describing the two-port network are then:

$$b_1 = S_{11}a_1 + S_{12}a_2 \tag{19.9}$$

$$b_2 = S_{21}a_1 + S_{22}a_2 \tag{19.10}$$

and, when expressed as a matrix, becomes:

$$[b] = [S][a] \tag{19.11}$$

In Equation (19.11), [b] is an nx1 column matrix, [a] is an nx1 column matrix, and [S] is an nxn matrix, where n is the number of ports in the network (that is, n=2 in Figure 19.3).

The S-parameters S_{11}, S_{22}, S_{21}, and S_{12} are as follows:

$$S_{11} = \left.\frac{b_1}{a_1}\right|_{a_2 = 0} = \text{Input reflection coefficient with the output port terminated by a matched load } (Z_L{=}Z_0 \text{ sets } a_2{=}0) \tag{19.12}$$

$$S_{22} = \left.\frac{b_2}{a_2}\right|_{a_1 = 0} = \text{Output reflection coefficient with the input terminated by a matched load } (Z_S{=}Z_0 \text{ sets } a_1{=}0) \tag{19.13}$$

$$S_{21} = \left.\frac{b_2}{a_1}\right|_{a_2 = 0} = \text{Forward transmission (insertion) gain with the output port terminated in a matched load.} \tag{19.14}$$

$$S_{12} = \left.\frac{b_1}{a_2}\right|_{a_1 = 0} = \text{Reverse transmission (insertion) gain with the input port terminated in a matched load.} \tag{19.15}$$

Notice that

$$S_{11} = \frac{b_1}{a_1} = \frac{\frac{V_1}{I_1} - Z_0}{\frac{V_1}{I_1} + Z_0} = \frac{Z_1 - Z_0}{Z_1 + Z_0} \tag{19.16}$$

and

$$Z_1 = Z_0 \frac{1 + S_{11}}{1 - S_{11}}$$

where

$$Z_1 = \frac{V_1}{I_1}$$

is the input impedance at port 1.

This relationship between reflection coefficient and impedance is the basis of the Smith chart transmission-line calculator. Consequently, the reflection coefficients S_{11} and S_{22} can be plotted on Smith charts, converted directly to impedance, and easily manipulated to determine matching networks for optimizing a circuit design.

19.3.4 Smith Chart Transformation

The rectangular grid lines of the impedance plane are transformed to circles and arcs on the Smith chart as shown in Figure 19.4. The Smith chart is used to plot reflectances and the circular grid lines allow easy reading of the corresponding impedance.

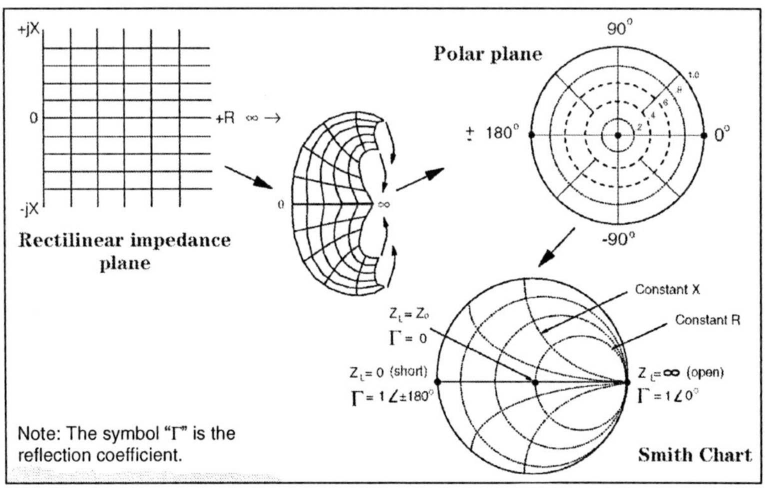

Figure 19.4 The Smith chart maps the rectilinear impedance plane onto the polar plane. [Reprinted with permission of Agilent Technologies, Inc.]

Vertical lines of constant resistance on the impedance plane are transformed into circles on the Smith chart. Horizontal lines of constant reactance on the impedance plane are transformed into arcs on the Smith chart. The transformation between the impedance plane and the Smith chart is nonlinear, causing normalized resistance and reactance values greater than unity to become compressed towards the right side of the Smith chart.

19.3.5 Advantages of S-Parameters

The previous equations show one of the important advantages of S-parameters, namely that they are simply gains and reflection coefficients, both familiar quantities to engineers. By comparison, some of the Y-parameters described earlier in this chapter are not so familiar. For example, the Y-parameter corresponding to insertion gain S_{21} is the "forward trans-admittance" y_{21} given by Equation (19.3). Clearly, insertion gain gives by far the greater insight into the operation of the network.

Another advantage of S-parameters springs from the simple relationship between the variables a_1, a_2, b_1, and b_2, and various power waves:

$|a_1|^2$ = Power incident on the input of the network
= Power available from a source impedance Z_0

$|a_2|^2$ = Power incident on the output of the network
= Power reflected from the load

$|b_1|^2$ = Power reflected from the input port of the network
= Power available from a Z_0 source minus the power delivered to the input of the network

$|b_2|^2$ = Power reflected from the output port of the network
= Power incident on the load
= Power that would be delivered to a Z_0 load

The previous four equations show that S-parameters are simply related to power gain and mismatch loss, quantities that are often of more interest than the corresponding voltage functions:

$$|S_{11}|^2 = \frac{\text{Power reflected from the network input}}{\text{Power incident on the network input}} \tag{19.17}$$

$$|S_{22}|^2 = \frac{\text{Power reflected from the network output}}{\text{Power incident on the network output}} \tag{19.18}$$

$$|S_{21}|^2 = \frac{\text{Power delivered to a } Z_0 \text{ load}}{\text{Power available from } Z_0 \text{ source}} \tag{19.19}$$

$$= \text{Transducer power gain with } Z_0 \text{ load and source}$$

$$|S_{12}|^2 = \text{Reverse transducer power gain with } Z_0 \text{ load and source ''} \tag{19.20}$$

19.4 Differential and Mixed-Mode S-Parameters[7,8]

"When S-parameters for single-ended networks are understood, one can extend this knowledge of S-parameters to characterize differential circuits. To arrive at an understanding of differential S-parameters, a differential amplifier will first be studied. An example of a differential amplifier is shown in Figure 19.5.

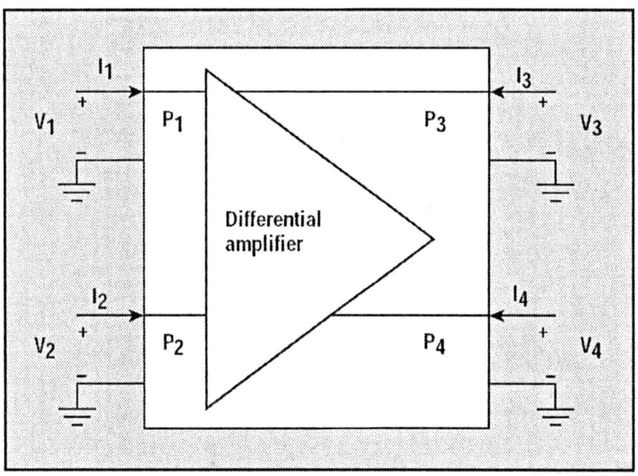

Figure 19.5 Example of a differential amplifier.

To characterize the differential amplifier in Figure 19.5, each lead can be identified as a port and the differential circuit labeled as a four-port network. This approach treats the amplifier as a single-ended device. To measure the S-parameters for this single-ended approach using a two-port vector network analyzer (VNA), terminate the two unused ports with 50 Ω and measure the two-port S-parameters for the two unterminated ports. Continue to terminate and measure

7. Except where noted, this section was written by Garth Sundberg of Maxim Integrated Products. Reprinted with permission of Maxim Integrated Products.
8. See also Appendix 23.A, "Generalized N-Port, Mixed-Mode S-Parameters," which starts on page 676.

the ports of the device in this fashion until enough information is gathered to construct the four-by-four S-parameter matrix. Assuming a good calibration for single-ended performance, the device is accurately characterized with the four-port S-parameters measured. These S-parameters do not provide much insight into the amplifier's differential (or common-mode) operation, because each port contains the differential and common-mode response. The amplifier is designed for differential operation.

A system similar to that used to describe the four transfer gains (A_{cc}, A_{dd}, A_{cd}, and A_{dc}, introduced by Middlebrook [Mid63]) is used to overcome this problem. (In these transfer gains, the subscript 'c' is used to indicate common-mode and the subscript 'd' is used to indicate differential-mode.) Bockelman and others [Boc95] introduced this system of S-parameters (called 'mixed-mode'). It begins by grouping ports one and two of the amplifier in Figure 19.5 together (to form a differential port one) and ports three and four of the amplifier in Figure 19.5 together (to form a differential port two). This grouping is shown in Figure 19.6.

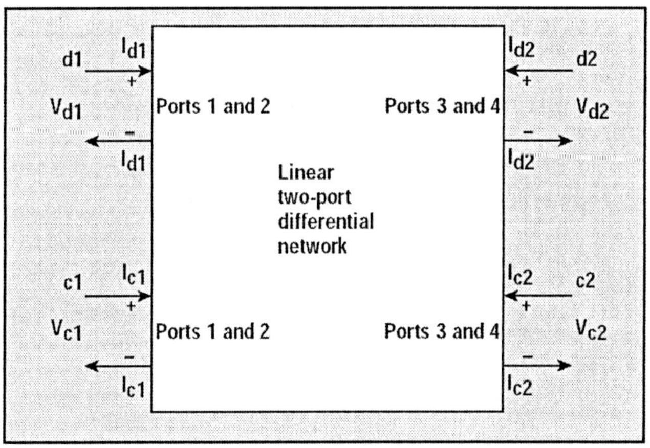

Figure 19.6 Example of a differential (mixed mode) circuit. [Reprinted with permission of Maxim Integrated Products.]

The network shown in Figure 19.6 is only a tool for visualizing the operation of a differential circuit. In reality, there are just two differential ports, with each port having both differential and common-mode signals.

Comparing the single-ended voltages and currents in Figure 19.5 with the differential voltages and currents in Figure 19.6, the differential and common-mode voltages and currents can be defined as

$$v_{d1} = v_1 - v_2 \qquad v_{c1} = \frac{v_1 + v_2}{2} \tag{19.21}$$

$$i_{d1} = \frac{i_1 - i_2}{2} \qquad i_{c1} = i_1 + i_2 \tag{19.22}$$

$$v_{d2} = v_3 - v_4 \qquad v_{c2} = \frac{v_3 + v_4}{2} \tag{19.23}$$

$$i_{d2} = \frac{i_3 - i_4}{2} \qquad i_{c2} = i_3 + i_4 \tag{19.24}$$

A way can be found to convert from single-ended S-parameters to mixed-mode S-parameters with this conversion from single-ended voltages and currents to differential and common-mode voltages and currents. Before the conversion is given, a review of what has been done in the past to measure circuits differentially will be presented.

In the past, if differential measurements were desired, a balun[9] (or hybrid coupler) would be needed, as seen in Figure 19.7. The problems associated with this method are (1) magnitude and phase imbalance of the baluns, (2) no way to measure mode conversion (that is, from differential to common-mode), (3) no rigorous definition of mixed-mode S-parameters, and (4) calibration of the system with the baluns is poorly defined.

Because of these problems, a carefully developed system is needed to describe the device differentially.

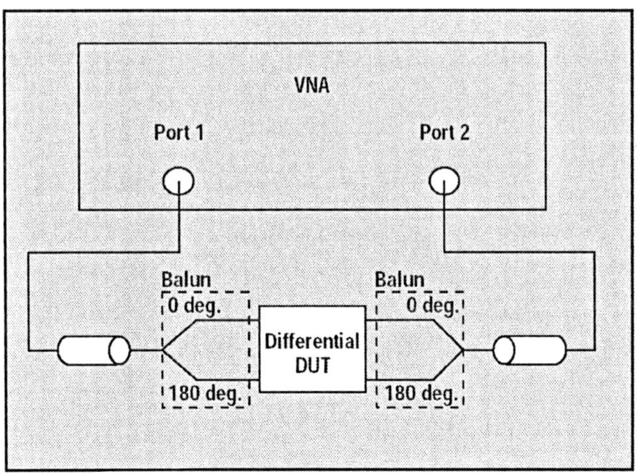

Figure 19.7 Sample setup to measure differential S-parameters with baluns. [Reprinted with permission of Maxim Integrated Products.]

9. The word "balun" means balanced-unbalanced: it's used to adapt a balanced device to an unbalanced one.

The need for mixed-mode S-parameters having been presented, the mixed-mode S-parameters can now be discussed. The definition for the incident and returning waves is given in Equations (19.25) to (19.28) for a differential and common-mode incident and returning power wave.

In Equations (19.25) to (19.28), v_{dn} is the differential voltage at port n, v_{cn} is the common-mode voltage at port n, i_{dn} is the differential current at port n, i_{cn} is the common-mode current at port n, Z_{dn} is the differential-mode characteristic impedance at port n ($2Z_{oo}$ for a coupled-line system), and Z_{cn} is the common-mode characteristic impedance at port n ($0.5*Z_{oe}$ for a coupled-line system).

$$a_{dn} = \frac{v_{dn} + i_{dn}Z_{dn}}{2\sqrt{Re(Z_{dn})}} \tag{19.25}$$

$$b_{dn} = \frac{v_{dn} - i_{dn}Z_{dn}}{2\sqrt{Re(Z_{dn})}} \tag{19.26}$$

$$a_{cn} = \frac{v_{cn} + i_{cn}Z_{cn}}{2\sqrt{Re(Z_{cn})}} \tag{19.27}$$

$$b_{cn} = \frac{v_{cn} - i_{cn}Z_{cn}}{2\sqrt{Re(Z_{cn})}} \tag{19.28}$$

To calculate Z_d, remember that Z = V/I; therefore, to get Z_d, take v_d (from Equation (19.21)) and divide it by i_d (from Equation (19.22)). A similar method is needed to calculate Z_c. Following these steps yields

$$Z_d = 2Z_{oo} \qquad Z_c = \frac{Z_{oe}}{2} \tag{19.29}$$

In Equation (19.29), Z_{oo} is the odd-mode impedance and Z_{oe} is the even-mode impedance of the system.

With the definition of the power waves in Equations (19.25) to (19.28), the mixed-mode S-parameters can be defined as

$$\begin{aligned}
b_{d1} &= S_{dd11}\, a_{d1} + S_{dd12}\, a_{d2} + S_{dc11}\, a_{c1} + S_{dc12}\, a_{c2} \\
b_{d2} &= S_{dd21}\, a_{d1} + S_{dd22}\, a_{d2} + S_{dc21}\, a_{c1} + S_{dc22}\, a_{c2} \\
b_{c1} &= S_{cd11}\, a_{d1} + S_{cd12}\, a_{d2} + S_{cc11}\, a_{c1} + S_{cc12}\, a_{c2} \\
b_{c2} &= S_{cd21}\, a_{d1} + S_{cd22}\, a_{d2} + S_{cc21}\, a_{c1} + S_{cc22}\, a_{c2}
\end{aligned} \tag{19.30}$$

where the following notation is used:

$$S_{ghij}=S_{(output\text{-}mode)(input\text{-}mode)(output\text{-}port)(input\text{-}port)} \tag{19.31}$$

which can be represented in the following format:

$$\begin{bmatrix} b_{d1} \\ b_{d2} \\ b_{c1} \\ b_{c2} \end{bmatrix} = \begin{bmatrix} \begin{bmatrix} S_{dd11} & S_{dd12} \\ S_{dd21} & S_{dd22} \end{bmatrix} & \begin{bmatrix} S_{dc11} & S_{dc12} \\ S_{dc21} & S_{dc22} \end{bmatrix} \\ \begin{bmatrix} S_{cd11} & S_{cd12} \\ S_{cd21} & S_{cd22} \end{bmatrix} & \begin{bmatrix} S_{cc11} & S_{cc12} \\ S_{cc21} & S_{cc22} \end{bmatrix} \end{bmatrix} \begin{bmatrix} a_{d1} \\ a_{d2} \\ a_{c1} \\ a_{c2} \end{bmatrix} \tag{19.32}$$

where S_{dd} are the differential S-parameters, S_{cc} are the common-mode S-parameters, S_{dc} are the S-parameters for the mode conversion that occurs when the device is excited with the common-mode signal and the differential signal is measured, and S_{cd} are the S-parameters for the mode conversion that occurs when the device is excited with a differential-mode signal and the common-mode response is measured. This mode conversion is unavoidable (whether intentional or not) because there is a common ground to the entire circuit and because of device mismatches and imbalances.

To convert from single-ended S-parameters to mixed-mode S-parameters, it is assumed that the device under test is being fed from differential input lines and that $Z_e = Z_o = Z_0$. The assumption of differential input lines is not limiting, as we can define the length of the lines to be arbitrarily small. The assumption of $Z_{oe} = Z_{oo} = Z_0$ implies that the differential input lines are not coupled. This is a valid assumption, as the lines of the VNA are coaxial and not coupled. See references [Boc95] and [Boc97].

Taking the definitions of v_d, v_c, i_d, and i_c from Equations (19.21) to (19.24) and using them in Equations (19.25) to (19.28), and taking Z_d to be $2Z_0$ (from Equation (19.29)), the following equations result:

$$a_{d1} = \frac{a_1 - a_2}{\sqrt{2}} \qquad a_{c1} = \frac{a_1 + a_2}{\sqrt{2}} \tag{19.33}$$

$$b_{d1} = \frac{b_1 - b_2}{\sqrt{2}} \qquad b_{c1} = \frac{b_1 + b_2}{\sqrt{2}} \tag{19.34}$$

$$a_{d2} = \frac{a_3 - a_4}{\sqrt{2}} \qquad a_{c2} = \frac{a_3 + a_4}{\sqrt{2}} \tag{19.35}$$

$$b_{d2} = \frac{b_3 - b_4}{\sqrt{2}} \qquad b_{c2} = \frac{b_3 + b_4}{\sqrt{2}} \tag{19.36}$$

A convenient matrix representation of Equations (19.33) to (19.36) is as follows:

$$\begin{bmatrix} a_{d1} \\ a_{d2} \\ a_{c1} \\ a_{c2} \end{bmatrix} = \frac{1}{\sqrt{2}} \begin{bmatrix} 1 & -1 & 0 & 0 \\ 0 & 0 & 1 & -1 \\ 1 & 1 & 0 & 0 \\ 0 & 0 & 1 & 1 \end{bmatrix} \begin{bmatrix} a_1 \\ a_2 \\ a_3 \\ a_4 \end{bmatrix} \tag{19.37}$$

$$\begin{bmatrix} b_{d1} \\ b_{d2} \\ b_{c1} \\ b_{c2} \end{bmatrix} = \frac{1}{\sqrt{2}} \begin{bmatrix} 1 & -1 & 0 & 0 \\ 0 & 0 & 1 & -1 \\ 1 & 1 & 0 & 0 \\ 0 & 0 & 1 & 1 \end{bmatrix} \begin{bmatrix} b_1 \\ b_2 \\ b_3 \\ b_4 \end{bmatrix} \tag{19.38}$$

More compactly, it can be represented as Equation (19.39):

$$a^{mm} = Ma^{se} \qquad b^{mm} = Mb^{se} \tag{19.39}$$

In Equation (19.39), the superscript ‘mm’ represents mixed-mode and the superscript ‘se’ represents the single-ended parameters; ‘M’ is given by Equation (19.40):

$$M = \frac{1}{\sqrt{2}} \begin{bmatrix} 1 & -1 & 0 & 0 \\ 0 & 0 & 1 & -1 \\ 1 & 1 & 0 & 0 \\ 0 & 0 & 1 & 1 \end{bmatrix} \tag{19.40}$$

Applying the conversion from a^{se} and b^{se} to a^{mm} and b^{mm} (given in Equation (19.39)) to the definition for single-ended S-parameters (given in Equation (19.11) on page 498) yields the following equation:

$$S^{mm} = MS^{se}M^{-1} \tag{19.41}$$

In the above equation, S^{mm} represents the mixed-mode S-parameters, S^{se} represents the single-ended S-parameters, and M^{-1} is the inverse of the M matrix. It is important to follow the port-number scheme given in Figure 19.5. If the four ports are not numbered in this fashion, then

Equation (19.40) for 'M' will not be correct. It will have to be arranged for Equation (19.41) to work correctly. Section 19.4 up to this point is referenced in [Boc95] and [Boc97]." [Max01b]

19.4.1 The Need for Differential and Mixed-Mode S-Parameters

"In dealing with differential devices and differential signals, what you really like to have are differential S-parameters rather than single-ended S-parameters. This is because differential S-parameters are more accurate for describing the differential behavior of a differential device or differential transmission line. The whole reason in performing the calculations described previously is that both the differential and common-mode response are compressed together and you cannot easily get either the differential or common-mode response from the single-ended S-parameters. This is why you go to the mixed-mode S-parameters. This conversion separates the differential-mode response from the common-mode response.

The technical discussion above on mixed-mode S-parameters is a mathematical treatment that tells you how to derive differential S-parameters from single-ended S-parameters. This allows you to go out in the lab and measure a differential device with single-ended S-parameter techniques, and then through matrix conversions and mathematics, calculate the differential S-parameters from the single-ended S-parameters. Your end goal is the differential S-parameters.

You may wonder if the mixed-mode S-parameters (e.g., S_{dc11}) are ever used for anything or whether they're just a mathematical by-product of calculating the differential S-parameters. It turns out that you do use these parameters. They are the conversion parameters from a common-mode signal to a differential-mode signal (S_{dc}) and the conversion from a differential-mode signal to a common-mode signal (S_{cd}). What these parameters represent is the non-ideal nature of devices. For example, in a differential amplifier any imbalance in the transistors (e.g., the transistors are not manufactured exactly the same, as there will be some small variation in each transistor from the manufacturing) will cause some of the differential input signal to come out of the amplifier as a common-mode signal. This is an undesirable result, but is unavoidable and needs to be characterized.

We care about the differential mode gain versus common-mode gain because designers use differential amplifiers to suppress common-mode noise (noise common to both of the input lines). If you have common-mode gain as well as differential-mode gain, you will actually be amplifying the common-mode noise, which is not desirable. In reality there is no way to stop common-mode gain, but you will want to minimize and characterize it so the common-mode response can also be modeled.

S-parameters are used instead of SPICE or IBIS models because the response at high-frequency is captured exactly. There will be no error from generating a model based on S-parameters. The drawback to this is that most S-parameters are not measured at dc and you will not be able to do a transient analysis with the measured S-parameters. This is why most people will make a SPICE or IBIS model. Also, the SPICE or IBIS model is usually faster to simulate than S-parameters. It used to be that it was difficult to measure the Y and Z parameters accurately at high frequencies, which is why S-parameters were invented. This is not as true any more as oscilloscopes can go to 40 GHz or higher these days. Still the S-parameters are an accurate way to capture high-frequency effects (e.g., when the size of the device becomes close to the size of your wavelength). You can still use Y and Z parameters, but these parameters typically assume a lumped-element approach which will not capture all high-frequency effects.

You use S_{dd11} (differential reflection loss) and S_{dd21} (differential insertion loss) as a way to see how good your backplane traces are. If you have a small S_{dd11} then you know your line impedance is close to your system impedance and almost all of your differential energy is making it onto the lines. The S_{dd21} will essentially tell you the loss of the line over frequency.

You may wonder why you would want to bother measuring S-parameters of a transmission line such as a PCB trace. You know what the parameters already are from the specifications of the board layout and traces. The answer is that you need to verify the impedance of the trace, as calculations may vary quite a bit from the actual board. Most times the PCB trace design equations will get close, but many simulators cannot adequately predict the effects of solder mask, loss tangent, and tolerance in the board. So the goal is to redo pre-layout simulations using the new, more accurate model or S-parameters to verify the design." [Max01b]

19.4.2 Return Loss Measurement Example

"The return loss of the MAX3950 10 Gbps deserializer was measured using an Agilent 8753D vector network analyzer to demonstrate this technique. This network analyzer goes only to 6 GHz, so just this data is presented. The measurement test setup is shown in Figure 19.8.

A standard SOLT (short-open-load-through) calibration was performed using the HP85033D 3.5 mm calibration kit to obtain the measured data. Once the calibration was performed, the measurement reference plane was moved to the end of the cables, as shown in

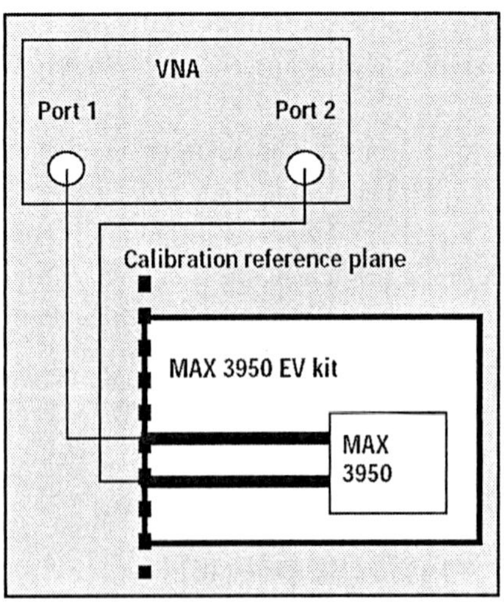

Figure 19.8 Measurement setup for MAX3950 return loss. [Reprinted with permission of Maxim Integrated Products.]

Figure 19.8. With no calibration kit built to measure the device on its circuit board, all S-parameters presented include the effects of the test fixture (SMA connectors and PCB traces which act as a transmission line) in addition to the characteristics of the MAX3950. That is, the calibration removes the distorting effects of the VNA and its test cables up to the calibration reference plane, but not the effects of the evaluation kit (test fixture). As a result, the actual return loss of the device will be better than that presented here. For future comparison, the single-ended return loss is presented in Figure 19.9.

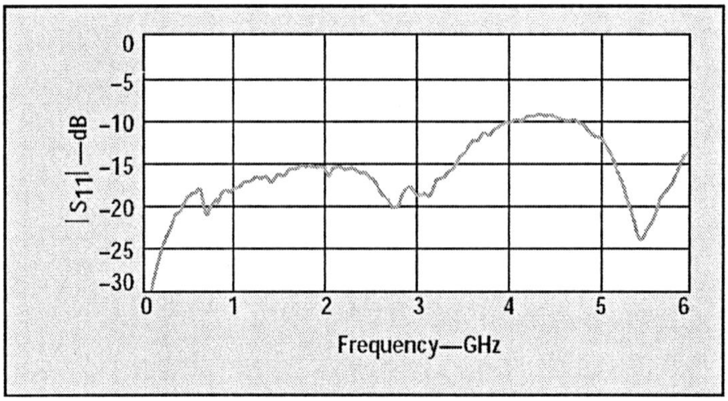

Figure 19.9 Single-ended return loss (S_{11}) for the MAX3950. [Reprinted with permission of Maxim Integrated Products.]

To get a true differential return loss, Equation (19.41) was applied to the measured data. This result is given in Figure 19.10. To validate the conversion, the differential return loss of the MAX3950 on the MAX3950 EV kit was measured using ATN Microwave's[10] ATN-4000 series differential network analyzer system. The results of the differential network analyzer are also presented in Figure 19.10.

It is clear that the actual differential return loss is different from the single-ended return loss. The difference between the measured data (i.e., measured with the ATN-4000 series network analyzer) and the calculated data (i.e., measured in a single-ended fashion and converted to differential) is most likely a result of two different VNAs with two different calibrations used for the measurement. The most accurate way to characterize the part is with the differential return loss, because the device is a differential part." [Max01b]

"Measurements of S-parameters and TDR response are fundamentally similar. S11 is normally expressed as a reflection magnitude and angle at a particular frequency. A TDR-derived ρ (reflection coefficient) is expressed as magnitude at a particular point in time or distance, assuming a specific propagation velocity. The biggest practical difference is that S-parameters can be used to develop a quantitative model, whereas TDR data generally provides qualitative data." [Cas03]

10. ATN Microwave was acquired by Agilent Technologies.

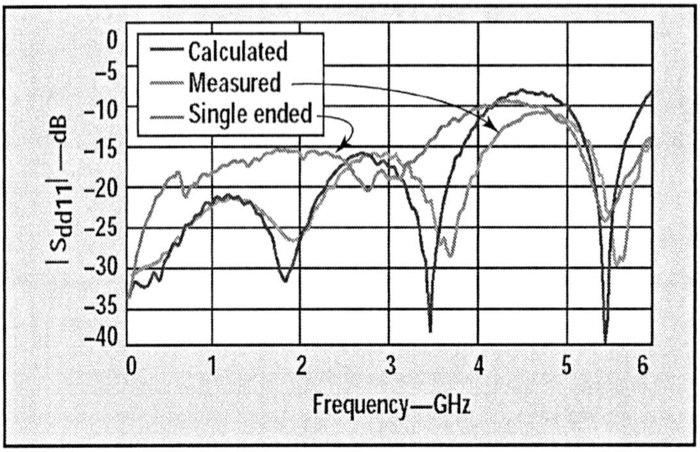

Figure 19.10 Single-ended and differential S-parameters of the MAX3950. [Reprinted with permission of Maxim Integrated Products.]

19.4.3 Touchstone Format

Table 19.1 lists example S-parameter data for Minicircuits' ERA-5SM device and illustrates the complex nature (magnitude and angle) of S-parameter values and their fluctuation with frequency. Figure 19.11 specifies a commonly used format for S-parameters which is Agilent Technologies' Touchstone format. This standard format allows for easy transfer of S-parameter values between equipment and simulation programs.

Table 19.1 S-Parameters for Minicircuits' ERA-5SM Device[a]

Freq	S11			S21			S12			S22		
MHz	db	Mag	Ang	db	Mag	Ang	db	Mag	Ang	db	Mag	Ang
100	–11.59	0.26	174.81	20.65	10.78	175.59	–23.27	0.07	0.46	–13.66	0.21	177.06
1000	–12.96	0.23	128.54	19.90	9.89	134.70	–22.86	0.07	5.51	–13.88	0.20	156.12
2000	–17.63	0.13	90.95	18.33	8.25	94.58	–21.94	0.08	7.65	–13.63	0.21	140.50
3000	–24.23	0.06	111.87	16.29	6.52	59.65	–20.97	0.09	6.40	–12.17	0.25	128.32
4000	–15.13	0.18	135.50	14.10	5.07	27.83	–19.97	0.10	1.18	–9.68	0.33	115.18
5000	–9.53	0.33	119.99	11.42	3.73	1.13	–19.49	0.11	–7.20	–7.70	0.41	102.84
6000	–9.43	0.34	125.06	10.46	3.33	–14.98	–19.90	0.10	–4.46	–9.41	0.34	107.96
7000	–7.79	0.41	114.10	9.09	2.85	–41.10	–18.06	0.13	–15.55	–9.11	0.35	104.02
8000	–7.25	0.43	92.06	7.99	2.51	–72.78	–16.87	0.14	–28.91	–9.18	0.35	89.03

a. The ERA-5SM device is not a digital device but an amplifier. The data still serves to exemplify the complex nature of S-parameters and their dependence upon frequency.

[Reprinted with permission of Minicircuits.]

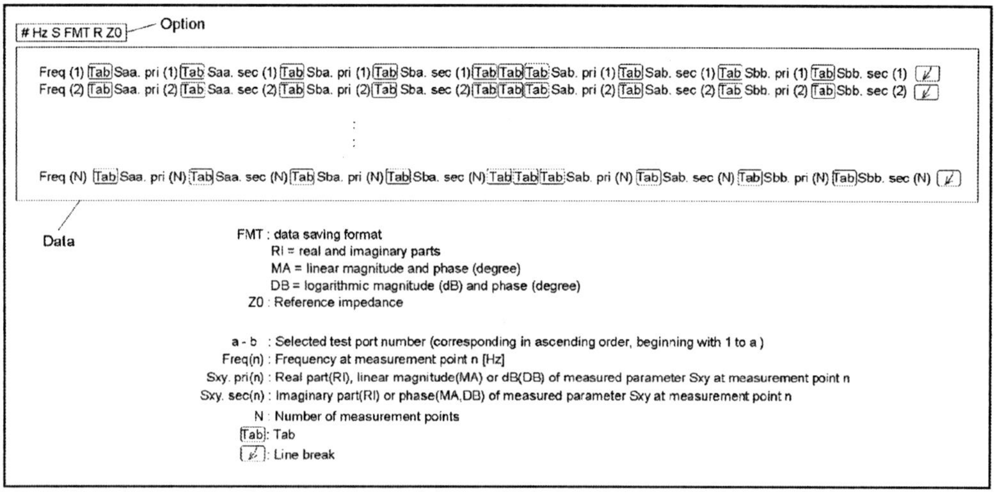

Figure 19.11 Agilent Technologies' format for a two-port Touchstone file. [Reprinted with permission of Agilent Technologies, Inc.]

19.5 Calibration

19.5.1 Measurement Errors

"Errors in vector network analyzer (VNA) measurements can be separated into three categories:

- Drift errors, which occur when the test system's performance changes after a calibration has been performed
- Random errors, which vary as a function of time
- Systematic errors, which include mismatch, leakage, and system frequency response

Calibration is the process for removing systematic errors from network analyzer measurements. To minimize drift errors, the test setup should be calibrated at appropriate frequencies, and maintained at a constant temperature. Random or noise-like errors are reduced by using a narrow IF bandwidth and by using averaging. Most VNAs include a trace averaging mode that will function over several sweeps to reduce the impact of random errors." [Max03]

19.5.2 Removing Test Fixture Effects: Direct Measurement and De-Embedding

"Accurate S-parameter characterization of a device under test (DUT) requires that test fixture characteristics be removed from the measured results. The test equipment typically used for characterizing a gigahertz component is the vector network analyzer (VNA), which uses standard 50 or 75 Ω coaxial interfaces at the test ports. The test equipment is calibrated at the coaxial interface (i.e., at the ends of coaxial test cables) defined as the 'measurement plane,' but the required measurements are at a different point, the point where the DUT attaches to the printed circuit board, or the 'device plane,' as shown in Figure 19.12. When the VNA is calibrated at the coaxial interface using any standard calibration kit, the DUT measurements include the unwanted test fixture effects.

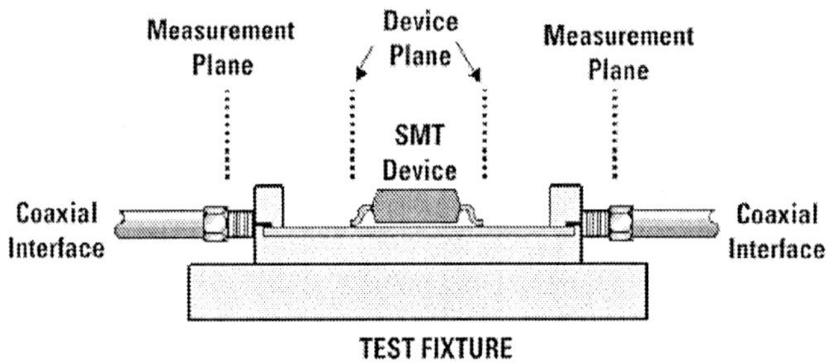

Figure 19.12 Test fixture configuration showing the measurement and device planes. [Agi01b] [Reprinted with permission of Agilent Technologies, Inc.]

Over the years, many different approaches have been developed for removing the effects of the test fixture from the measurement. These approaches fall into two fundamental categories: direct measurement and de-embedding. Direct measurement requires specialized calibration standards that are inserted into the test fixture and measured. The accuracy of the device measurement relies on the quality of these physical standards. De-embedding uses a *model* of the test fixture and mathematically removes the fixture characteristics from the overall measurement. This fixture de-embedding procedure can produce very accurate results for the DUT without complex calibration standards.

There are two main ways the de-embedding process can be implemented, the static approach and the real-time approach. The static approach uses measured data from a network analyzer and processes the data using T-parameter matrix calculations.[11] By comparison, the real-time approach uses the network analyzer to directly perform the de-embedding calculations, allowing the user to examine the de-embedding response in real-time. This technique is accomplished by modifying calibration error terms in the analyzer's memory.

The static approach does not allow for real-time feedback to the operator because the measured data needs to be captured and postprocessed in order to remove the effects of the test fixture. The static approach uses measured data from the test fixture and DUT gathered at the measurement plane. The data can be exported from the network analyzer or directly imported into a simulation tool, such as Agilent Technologies' ADS, over the GPIB. Using the fixture

11. For a detailed discussion of T-parameters and their use in de-embedding test fixtures, see reference [Agi01b].

model, the de-embedding process is performed using T-parameter matrix calculations or the negation model in ADS. Once the measurements are de-embedded, the data is displayed statically on a computer screen or can be downloaded into the analyzer's memory for display.

If real-time de-embedded measurements are required, the de-embedding calculation can be performed directly on the VNA using a different calibration model under the real-time approach. This different calibration model is the test fixture S-parameter model, and when it is incorporated into the calibration error terms in the VNA, real-time de-embedded measurements can be displayed directly on the VNA. This allows for real-time tuning of components without including the fixture as part of the measurement." [Max01b]

A detailed presentation of these techniques can be found in reference [Agi01b].

19.5.3 Calibration Example: Using Direct Measurement to Eliminate Test Fixture Effects

"The return loss S_{11} of the MAX390 was presented previously (in Section 19.4, which starts on page 501). But how accurate are those measurements? Like all measurements, they are only as accurate as the equipment, the methodology, and the operator are. Good test equipment and calibration standards are required to ensure the most accurate S-parameter measurements possible.

The calibration standards used depend on the measurement setup. Some possible calibration standards for the vector network analyzer are short, open, load, and thru (SOLT), which are provided in the Agilent 3.5 mm calibration kit or on an impedance standard substrate (ISS) like the one provided by Cascade Microtech with wafer probes. Other options include using through, reflect, line (TRL); line, reflect, match (LRM); line, reflect, reflect, match (LRRM); short, open, load, reciprocal (SOLR); or a custom-defined calibration kit. Whichever calibration scheme is used, the desired end result is the same: accurate and repeatable measurements.

An analogy of optics and light can be used to describe S-parameters. Imagine that the device under test being measured is an object behind a piece of dirty glass. To take an accurate picture of the object, the glass first needs to be cleaned. Think of calibration as cleaning or removing the glass to get the clearest possible picture.

In a typical VNA measurement system, there are cables which connect the VNA to the DUT (either through SMA connectors, wafer probes, or other connectors not mentioned here). This cable and connector have a significant effect in obscuring the measured data just like the dirty glass affecting a clear photo. Without a calibration, the response of the device and the connectors is being measured. A calibration is needed to remove the cable and connector effects.

One's position is defined with a calibration (for example, SOLT) on the Smith chart. The open standard defines the right edge of the Smith chart, the short standard defines the left edge of the Smith chart, and the load standard defines the center of the Smith chart (see the Smith chart in Figure 19.4 on page 499).

The through standard (needed only for two-port measurements) identifies the delay and the loss introduced by the cables and the connectors. Figure 19.13 shows a test setup with the reference planes defined before and after calibration. Returning to the analogy of optics, the reference plane is like the camera placement in relation to the object (that is, the side of the glass where the camera is placed to take the picture of the object).

In Figure 19.13, the transmission line elements are everything that is between the VNA and the DUT (cables, connectors, transmission lines on printed circuit boards, matching networks, and so on). The measurement results will not accurately report the DUT's performance if all the elements between the VNA and the DUT are not correctly identified and calibrated out.

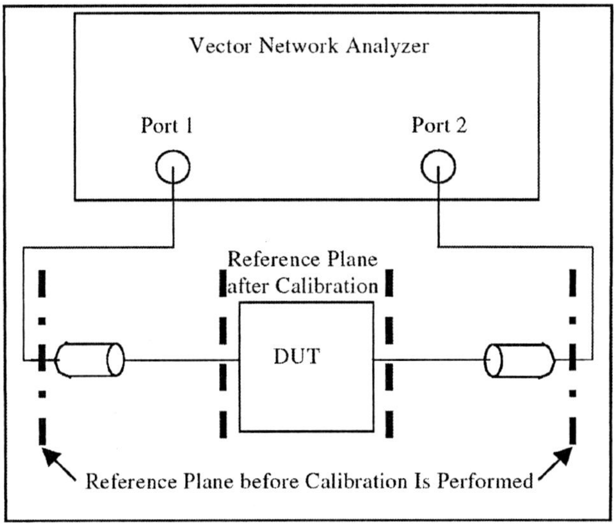

Figure 19.13 Sample measurement setup. [Reprinted with permission of Maxim Integrated Products.]

The importance of accurate characterization of the elements between the VNA and the DUT is illustrated by means of measurements of the return loss (S11) of the MAX3875 taken after two different calibrations were performed.

The first calibration was accomplished with the Agilent 3.5 mm D calibration kit. This procedure served to calibrate out the effects of the cable connecting the printed circuit board and the vector network analyzer. (See Figure 19.14.)

The second calibration was performed using custom-defined SOL (short, open, load) standards. This custom calibration kit was built with short, open, and load standards placed on the board at the point where the DUT is mounted (see Figure 19.15). One board is manufactured for each of the standards. This allows for the removal of transmission line effects of the PCB. The determination of these custom standards is outside the scope of this chapter, but the interested reader is referred to reference [God99]. It should be noted that, since 0402 series surface-mount components were used, this calibration kit loses accuracy above approximately 4 GHz.

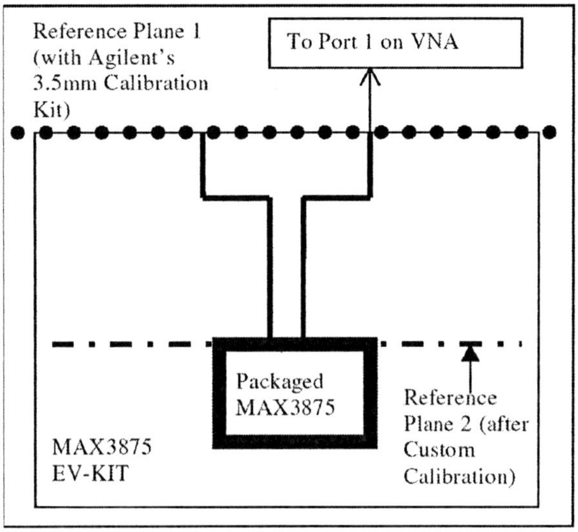

Figure 19.14 Different reference planes. [Reprinted with permission of Maxim Integrated Products.]

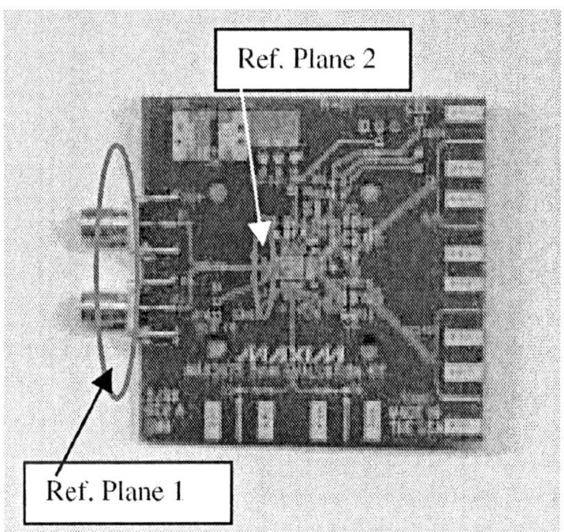

Figure 19.15 Calibration board. [Reprinted with permission of Maxim Integrated Products.]

One might think that calibrating the network analyzer to the end of the cables using a standard calibration kit would be sufficient. However, as illustrated in Figure 19.16, the measured return loss changes if the reference plane is moved from the edge of the MAX3875 EV kit (the evaluation kit for the MAX3875 device) to the input of the packaged part.

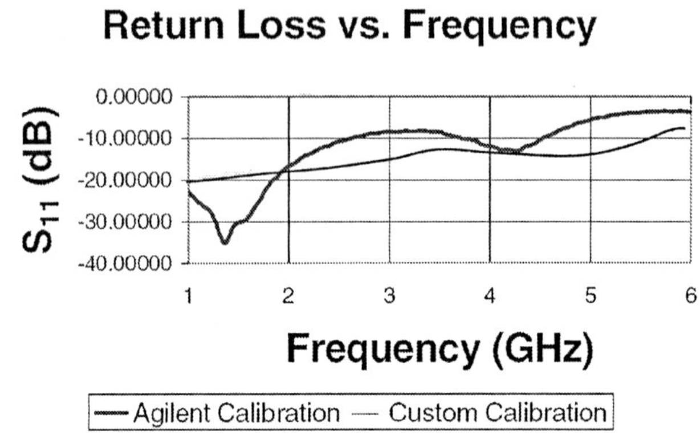

Figure 19.16 Return loss for the MAX3875. [Reprinted with permission of Maxim Integrated Products.]

The reason for this difference is apparent from the physical layout of the MAX3875 EV kit (evaluation kit, see Figure 19.16). Agilent's 3.5 mm D calibration kit only corrects for the cable from the network analyzer to the circuit board, not to the edge of the part. In fact, after a calibration is done with this calibration kit, there is still the SMA connector on the board, the coupling capacitor on the transmission line to the MAX3875, and the transmission line from the SMA connectors to the MAX3875. All three of these items on the circuit board corrupt the accurate measurement of the return loss." [Max01b]

19.6 Exercises

19-1. For a specific two-differential-port device or network such as a differential transmission line (differential trace pair on a PCB board) and for a single measurement of its S-parameters, are its single-ended S-parameters constants?

19-2. In Figure 19.10 on page 510, why aren't the calculated and measured differential return losses identical?

19-3. What is meant by "de-embedding"? What is the difference between calibration, de-embedding, and direct measurement of a test fixture?

19-4. What is the difference between the "measurement plane" of Figure 19.12 on page 512 and the "calibration reference plane" of Figure 19.8 on page 508?

19-5. Figure 19.10 on page 510 shows three curves (plots) of S-parameters of the MAX3950 versus frequency. The three curves are labelled "calculated," "measured," and "single-ended." Explain what each of the three curves represents, why they are different, and what the differences between them are.

19-6. What's the difference between "differential" and "mixed-mode" S-parameters?

19-7. What might be an easier and quicker way to determine differential S-parameters for a device than making single-ended S-parameter measurements using a VNA and subsequently converting the measured data to differential S-parameters through use of the equations contained in this chapter?

19-8. Two categories of approaches were discussed to remove the effects of the test fixture from S-parameter measurements, direct measurement and de-embedding. The process used to create the custom curve shown in Figure 19.16 on page 516 falls into which of these two categories? Why?

19-9. What is the difference between static and real-time de-embedding? Explain each term briefly in less than 30 words each.

19-10. Is calibration the same thing as de-embedding? Explain why or why not.

19-11. Put the data listed in Table 19.1 on page 510 in Touchstone format.

19-12. Explain the difference between drift errors, random errors, and systematic errors in making S-parameter measurements with a VNA. Provide an example of a cause for each type of error.

19-13. What advantages do S-parameters have over Y-, H-, or Z-parameters?

19-14. By taking a device with differential inputs and differential outputs and treating it as a 4-port, single-ended device, and subsequently determining the 4×4 matrix of single-ended S-parameters for it, why don't these S-parameters provide much insight into the device's differential operation?

19-15. For a differential circuit, why do we want to look at the differential return loss and exclude the common-mode return loss?

19-16. (a) Write out the full words meant by the following abbreviations: SOLT, ISS, TRL, LRM, LRRM, and SOLR. (b) What do these abbreviations all have in common?

CHAPTER 20

Time Domain Reflectometry (TDR), Time Domain Transmission (TDT), and VNAs

Chapter Objectives

- Explain the operational theory behind TDR and TDT.
- Discuss the Vector Network Analyzer (VNA), where it is used, and how it compares with TDR and TDR.
- Present the usefulness of TDR, TDT, and VNA test equipment in high-speed applications.
- Define differential impedance and common-mode, even-mode, odd-mode, and self impedances.
- Present the mathematics of time and frequency relationships as they apply to S-parameters and TDR and TDT impedances.
- Explain how multiple transmission line reflections distort the true impedance profile.
- Explain how IConnect® TDR software can improve accuracy and perform useful mathematical calculations.
- Discuss the features of a Tektronix TDR oscilloscope.

20.1 Time Domain Reflectometry (TDR)[1]

20.1.1 Introduction

"Time Domain Reflectometry (TDR) has traditionally been used for locating faults in cables. Currently, high-performance TDR instruments, coupled with add-on analysis tools, are commonly used as the tool of choice for failure analysis and signal integrity characterization of board, package, socket, connector, and cable interconnects at gigabit speeds.

1. Except where indicated, material in this chapter is reprinted with permission of TDA Systems.

Based on the TDR impedance measurements, the designer can perform signal integrity analysis of the system interconnect, and the digital system performance can be predicted accurately. A failure analyst can use TDR impedance measurements to locate a fault in the interconnect more accurately and quickly, allowing the analyst to focus on understanding the physics of the failure at this failure location. TDR measurement techniques for IC package characterization have been reported extensively and standardized by JEDEC." [Tda02a]

20.1.2 Single-Ended TDR Measurements

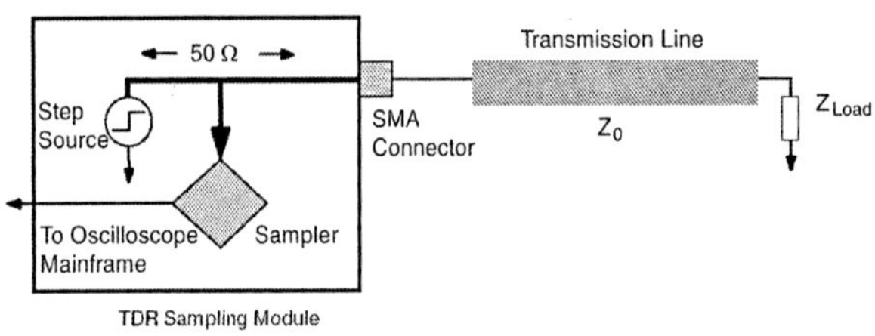

Figure 20.1 TDR oscilloscope is connected to the device under test (Z_{Load}) via cables (transmission lines), probes, and fixtures. [© Tektronix, Inc. Reprinted with permission. All rights reserved.]

"The TDR instrument is a very-wide-bandwidth equivalent sampling oscilloscope (18–20 GHz) with an internal step generator. It is connected to the device under test (DUT) via cables, probes, and fixtures (Figure 20.1). Because of the wide bandwidth of the oscilloscope, and to ensure that this bandwidth and fast rise time can be delivered to the DUT, you must use high-quality cables, probes, and fixtures, since these cables, probes, and fixtures can significantly degrade the rise time of the instrument, reduce the resolution, and decrease the impedance measurement accuracy. In a TDR probe, both a signal and a ground contact are normally required during the measurement.

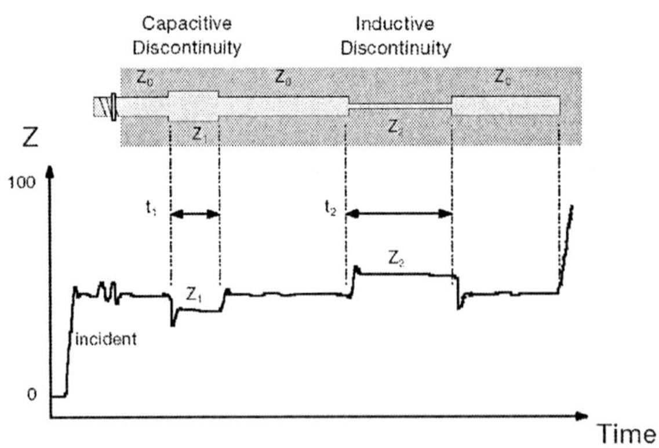

Figure 20.2 Reflected TDR signal waveform from a DUT providing impedance information about it. [© Tektronix, Inc. Reprinted with permission. All rights reserved.]

The extremely wide bandwidth and the internal source set the TDR in a class apart from any other oscilloscope, making it a high-frequency impedance and network[2] characterization tool. The oscilloscope step generator sends a step-like stimulus to the DUT. The signal is reflected from the DUT, and based on this reflection, we can look at the impedance, delay and other characteristics of the DUT as shown in Figure 20.2. (See [Til92], [Hew90]).

The equivalent resistance of the TDR source (R_{source}) defines the

2. The term "network" is used here to describe a potentially complicated mixture of cables, board traces, connectors, sockets, and IC packages.

characteristic impedance of the measurement system. Since R_{source} for the high-performance TDR instruments available today is 50 Ω, using non-50 Ω cables and probes can produce confusing results. Unlike with a regular oscilloscope, no active probes or resistor divider probes are allowed for use with TDR.

While the measurement system has to maintain the 50 Ω characteristic impedance, the DUT almost always differs from 50 Ω. Non-50 Ω impedances, such as 28 Ω in the case of Rambus or 75 Ω in the case of cable TV, can be measured quite accurately.

Because of the resistor divider effect between the 50 Ω resistance of the source and the 50 Ω characteristic impedance of the cable used to connect the instrument to the DUT, only half of the TDR source voltage V reaches the DUT initially (Figure 20.3). Z_0 is the characteristic impedance of the TDR measurement system (50 Ω), and Z_{DUT} is the impedance of the DUT.

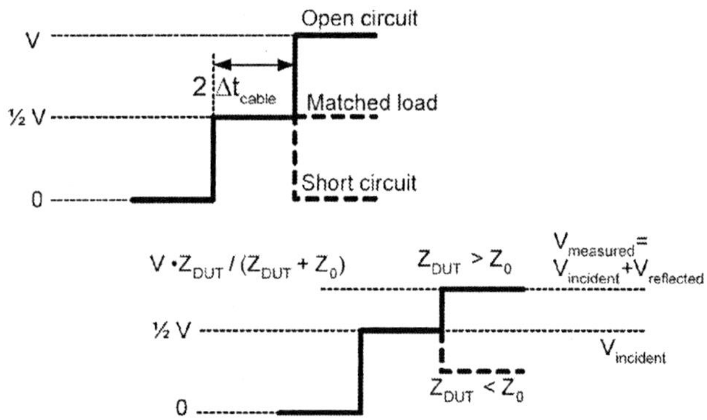

Figure 20.3 Typical shapes of the received TDR waveforms. [Reprinted with permission of TDA Systems.]

Then, if nothing is connected to the cable (open circuit condition), after the delay equal to the round-trip delay through the cable, the waveform will go up and reach the full incident voltage (Figure 20.3, top). Note that everything in TDR is round-trip delay. This applies not only to the cable interconnecting the TDR oscilloscope to the DUT, but also to all delay measurements on the DUT itself. In order to obtain accurate delay readout, the designer has to divide the measured delay by 2.

After the round-trip delay of the cable, the voltage reflected from the DUT arrives back to the oscilloscope and is added to the incident voltage on the oscilloscope to produce the measured voltage value $V_{measured}$. Then, the TDR oscilloscope uses the following equations to convert the voltage, which the oscilloscope measures, to impedance and reflection coefficients:

$$V_{reflected} = V_{incident} \frac{Z_{DUT} - Z_0}{Z_{DUT} + Z_0} \tag{20.1}$$

$$\rho = \frac{V_{reflected}}{V_{incident}} = \frac{Z_{DUT} - Z_0}{Z_{DUT} + Z_0} \qquad \tau = \frac{V_{transmitted}}{V_{incident}} = 1 + \rho = \frac{2Z_{DUT}}{Z_0 + Z_{DUT}} \tag{20.2}$$

$$Z_{DUT} = Z_0 \frac{1+\rho}{1-\rho} = Z_0 \frac{V_{incident} + V_{reflected}}{V_{incident} - V_{reflected}} = Z_0 \frac{V_{measured}}{2V_{incident} - V_{measured}} \tag{20.3}$$

where ρ is the reflection coefficient, τ is the transmission coefficient, and the other symbols used have been defined above. If you connect a shorting block to the cable or short the signal of the TDR probe to ground, creating a short circuit condition, the waveform will go to 0 volts. If you connect a 50 Ω termination ('matched load'), the waveform will stay at the $V_{incident}$ level, equal to V/2.

By observing Figure 20.3, you can see that there will be a better impedance measurement resolution between 0 and 50 Ω than between 50 Ω and infinity. Moreover, any impedance over 1,000 Ω is infinitely high, equivalent to an open circuit condition. Since the characteristic impedance of free space (which is what we are measuring with TDR) is only 377 Ω, this fact does not create any issues — no practical trace in a board, package, connector, or cable interconnect will have impedance which is much higher than the characteristic impedance of free space. It is worth noting that this limitation is not TDR-specific — there is no high-frequency measurement instrumentation that can measure impedance over 1,000 Ω at high frequency with any level of accuracy." [Tda02a]

20.1.2.1 TDR and Lumped Element Analysis

"Experienced TDR users can, without difficulty, recognize a 'dip' in a TDR waveform as a shunt capacitance and a 'spike' as a series inductance. Any L and C combination can also be represented as shown in Figure 20.4.

A series C or a shunt L, however, will represent a high-pass filter for the TDR signal, and the resulting reflection from the elements beyond such series C or shunt L can not be interpreted easily.

However, sometimes a dc blocking capacitor must be used in series with the TDR cable to protect the input of the TDR sampling head from the dc voltage on the line. This type of capacitor is a very-wide-band component (operating from dc to 15–25 GHz), and presents a high-pass filter with characteristics that are determined by the capacitor value, its parasitics, and the 50 Ω impedance of the cable on both sides of the capacitor. Since we know the characteristics of such filters well, we can still characterize the DUT accurately." [Tda02a]

20.1.2.2 TDR Resolution and Rise Time

"The issues of TDR resolution are often misunderstood or misrepresented because the TDR resolution is believed to be completely governed by the following rule of thumb: Two small discontinuities, such as two vias in a PCB, can still be resolved as two separate ones, as long as they are separated by at least half the TDR rise time.

If these two vias are not separated by half the TDR rise time as it reaches the vias, they will be shown by TDR as a single discontinuity. Assuming we use good cables, probes, and fixtures, and we can deliver the full 30 ps to 40 ps rise time of the instrument to the discontinuities in question, the minimal physical separation between these vias will be 15 ps to 20 ps. For FR4 board material with dielectric constant $\varepsilon_r = 4$, this results in 2.5–3 mm (0.1 in) resolution. Often this number (or other similar calculation) is quoted as the TDR resolution limit.

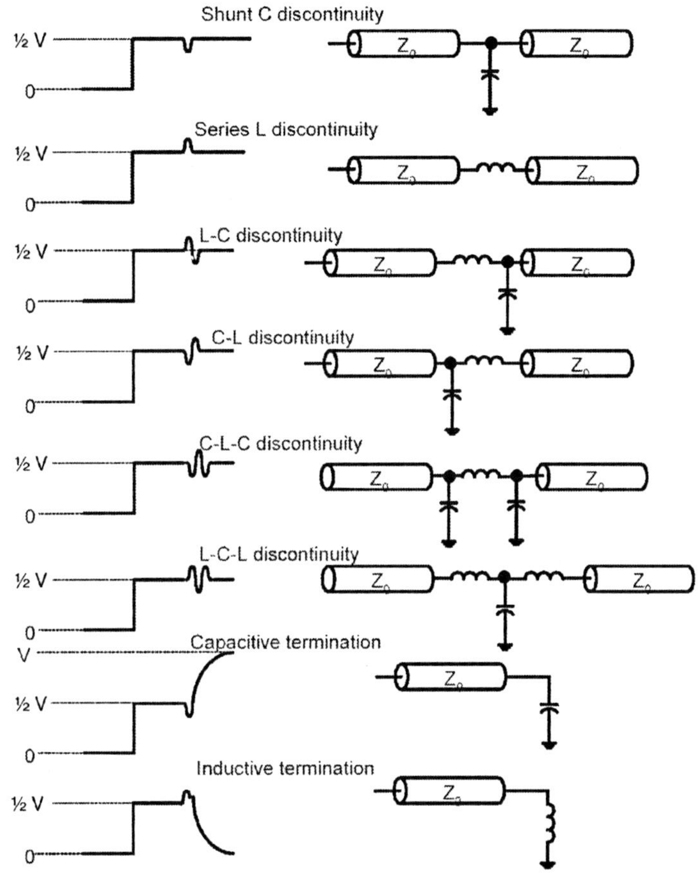

Figure 20.4 TDR waveforms for various inductive and capacitive lumped interconnect configurations. [Reprinted with permission of TDA Systems.]

The only time a designer may wish for a faster rise time is when the output drivers of the design becomes faster than the rise time of a TDR oscilloscope. Few of the rise time specifications or actual driver designs in the current digital signaling technologies approach the rise time of the TDR oscilloscopes currently on the market.

However, in real-life situations, the designer typically is looking to observe or characterize a single discontinuity, such as a single via, or a single bond wire in a package, rather than observe or characterize a group of vias or a group of bond wires. In this case, the above rule is totally irrelevant, and TDR can allow the designer to observe discontinuities of 1/10 to 1/5 of the TDR rise time, replacing the numbers above with transit times of 5 ps and lengths of less than 1 mm (25 milliinches) (see Figure 20.5).

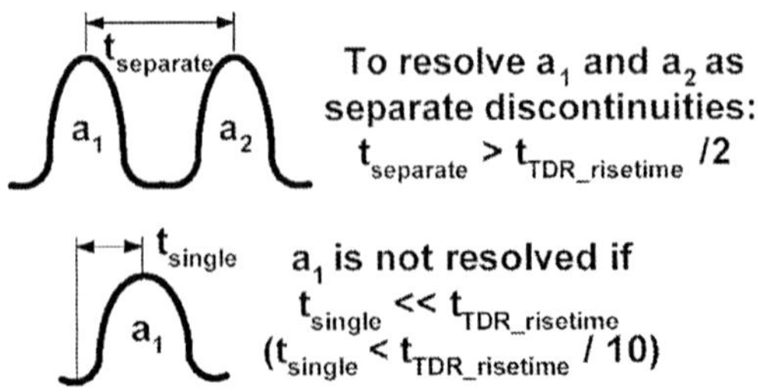

Figure 20.5 TDR resolution rules of thumb. [Reprinted with permission of TDA Systems.]

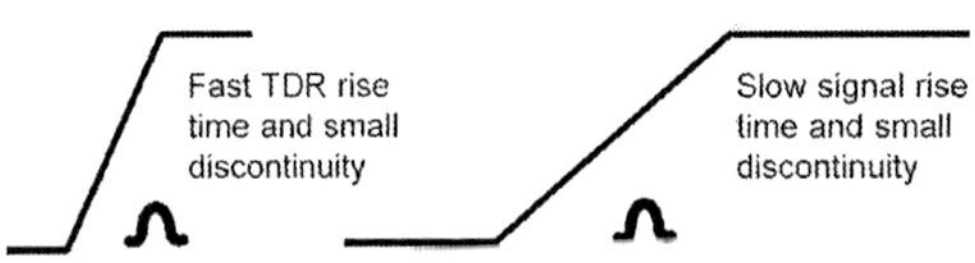

Figure 20.6 Fast TDR rise time and slow signal rise time. [Reprinted with permission of TDA Systems.]

Furthermore, there are well-developed relative TDR procedures for observing and characterizing even smaller discontinuities. For signal integrity modeling and lumped interconnect analysis, there are JEDEC standard procedures for package characterization, allowing the designer to measure sub-millimeter capacitive and inductive elements in 100 fF[3] and 200–300 pH range (see [Jed94], [Smo01]). For failure analysis applications, there are well-established procedures utilizing golden device comparisons (see [Ode99], [Smo00a]).

Taking this discussion further, practically all modern high-speed digital standards and applications — from Gigabit Ethernet to Infiniband — still have rise times which are slower than the 30–40 ps rise times of the TDR. Since TDR waveforms show the designer how a certain discontinuity would exhibit itself in the signal path, if the discontinuity does not cause a reflection of the fast TDR signal, it will have an even less effect on the slower real signal propagating through this discontinuity. Therefore, if the fast TDR rise time does not allow the designer to observe a certain discontinuity, then an even slower real signal will not show it either! (See Figure 20.6)." [Tda02a]

20.1.3 Differential TDR Measurements

"Differential TDR measurements are an important capability that the TDR instruments provide, since most of the modern signaling schemes and standards are differential — whether it is USB 2.0 or FireWire, Infiniband or Rapid I/O, SCSI or Fibre Channel, Gigabit Ethernet or Sonet — thus requiring differential impedance measurements. In addition, differential TDR is very useful for crosstalk characterization, whether it is crosstalk between two single-ended traces or crosstalk between differential pairs.

TDR instruments provide up to 8 single-ended or 4 differential channels allowing the designer to look at crosstalk between up to 4 differential pairs, making TDR the most capable high-frequency differential instrument currently available. (See Figure 20.7.)

3. fF is femtofarad.

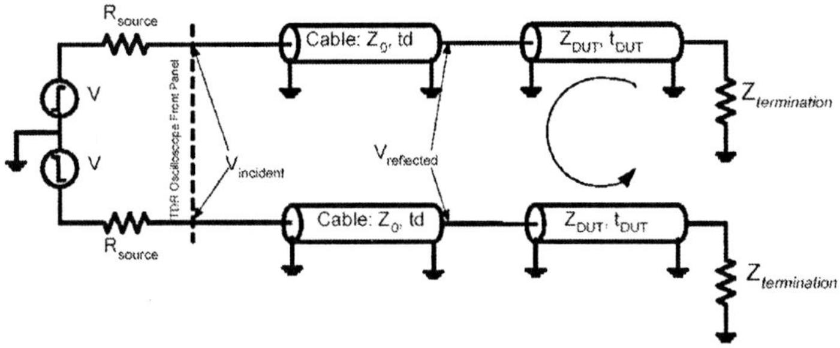

Figure 20.7 Differential TDR block diagram. [Reprinted with permission of TDA Systems.]

The TDR deskew capability ensures that both TDR signals in a differential pair reach the DUT traces at the same time, and corrects for delay differences between the cables, probes, and fixtures. If the TDR is not properly deskewed, the resulting skew can produce significant inaccuracies in differential impedance measurements and differential line modeling.[4]

Differential TDR measurements are necessary only if we are looking to obtain differential impedance, or to characterize coupling between the lines. If there is no coupling or other differential interaction between the lines, single ended TDR measurement will provide us with all the information that we need." [Tda02]

20.1.3.1 Differential and Odd-, Common- and Even-Mode Impedances

"Differential impedance is defined as the impedance between the two transmission lines when the two lines are driven differentially by opposite polarity signals, whereas the odd-mode impedance is the impedance of one line (referenced to ground) in the differential pair under the same drive conditions. Common-mode impedance is impedance between the two lines when the two lines are driven with a common-mode signal (same polarity signal), whereas the even-mode impedance is the impedance of a single line (referenced to ground) in the differential pair under the same drive conditions (see [Smo00b], [Tek03]). In many applications, it makes sense to use differential and common-mode impedances and delays to describe differential transmission line behavior.

Stating these definitions another way, differential impedance is the impedance measured between two conductors driven differentially, that is, with identical, but opposite-polarity signals — as opposed to the odd-mode impedance, which is the impedance of a single conductor (transmission line) in a differential pair referenced to ground when the two conductors are driven differentially. Even-mode impedance is the impedance (referenced to ground) of either conductor when the differential pair is driven with identical, same-polarity (even or common-mode) signals. Common-mode impedance is the impedance of the two conductors in a differential pair, when both conductors are driven with a single common-mode signal, relative to ground.

4. TDR oscilloscope manuals contain further details.

The odd-mode waveform can be obtained by applying a differential stimulus to a pair of pins under test, and acquiring only the positive switching channel on the TDR oscilloscope. The even-mode waveform can, in turn, be obtained by applying a common-mode stimulus to the same pair of pins under test, and acquiring the same channel as in the odd-mode measurement. [Tda02b]

"Tying together the two conductors in a differential line and driving them with a traditional single-ended TDR system, will yield a good measure of common-mode impedance. However, common-mode impedance is often less important than the differential impedance." [Tek01a]

In case of *mildly* non-symmetric differential pair, the following equations will apply:

$$Z_{differential} = Z_{oddline1} + Z_{oddline2} \quad (20.4)$$

$$Z_{common} = \frac{1}{2} \frac{(Z_{evenline1} + Z_{evenline2})}{2} \quad (20.5)$$

To measure the odd-mode impedances of line 1 and 2 ($Z_{oddline1}$ and $Z_{oddline2}$), the designer would connect the two outputs of the TDR sampling head to the two transmission lines under test (not forgetting to connect the ground plane if such ground plane is present) with the TDR signals switching in the opposite direction (i.e., differential stimulus), and measure the impedance of each line in the differential pair to ground. To measure the even-mode impedance ($Z_{evenline1}$ and $Z_{evenline2}$), the designer would switch the TDR sources in the same direction (i.e., common-mode stimulus), and measure the impedance of each line in the pair to ground. Different TDR oscilloscopes will have different procedures for measuring the differential and common-mode impedances, but the net results are the same. In case of a symmetrical differential pair, the following equations are true:

$$Z_{differential} = 2\ Z_{oddline1} \quad (20.6)$$

$$Z_{common} = \frac{Z_{evenline1}}{2} \quad (20.7)$$

In presence of coupling between the lines, the odd-mode impedance is always lower than the self-impedance of a single line, which in turn is lower than the even-mode impedance:

$$Z_{odd} = \sqrt{\frac{L_{self} - L_m}{C_{tot} + C_m}} \qquad Z_{self} = \sqrt{\frac{L_{self}}{C_{self}}} \qquad Z_{even} = \sqrt{\frac{L_{self} + L_m}{C_{total} - C_m}} \quad (20.8)$$

where L_{self}, C_{self}, L_m, and C_m are the self and mutual inductance and capacitance of the line pair per unit length, and $C_{total} = C_{self} + C_m$. Even though only the impedance of one of the lines in the line pair is measured to obtain odd, even and self impedance, the interaction between the lines will account for the difference between these impedance values." [Tda02]

Another commonly used symbol for the self-impedance of a transmission line is Z_0, the characteristic impedance of a single-ended transmission line to ground, where the transmission line has no mutual capacitive or mutual inductive coupling to other lines.

"Differential-mode and common-mode impedances are the most useful parameters from a practical design standpoint, whereas odd-mode and even-mode impedances are very useful from interconnect modeling and simulation point of view (see [Smo00b])." [Tda02]

Single line impedance clearly is not enough to characterize the differential transmission line pair, as you need to take line coupling into account. Differential impedance alone is not enough either, since you also need to take into account common-mode rejection and propagation.

20.1.4 Cables, Connectors, and Probes

"TDR cables and probes will degrade the rise time of the return signal measured on the TDR oscilloscope. The measured rise time displayed on the TDR oscilloscope screen is approximately

$$t_{measured} = \sqrt{t^2_{TDR} + 2\left(\frac{0.35}{f_{3dB}}\right)^2} \tag{20.9}$$

where t_{TDR} is the rise time measured on the TDR scope with no cable or probe connected. This is the rise time of the TDR as viewed on its screen with the TDR source pulse fed back directly to the TDR receiver. f_{3dB} is the 3 dB bandwidth of the cable and probe combination. The factor of 2 in this equation is due to the fact that the signal has to take a round trip through the cable before it is observed and measured on the oscilloscope. The $0.35/f_{3dB}$ term converts the 3 dB bandwidth to rise time. Using a cable and probe combination with a 3 dB bandwidth (f_{3dB}) of 10 GHz and a TDR scope with a rise time of 30 ps, the measured rise time displayed on the TDR oscilloscope will be about 58 ps. Using a 'faster' cable and probe combination with a 3 dB bandwidth of 17.5 GHz will result in a measured rise time which is lower, only 42 ps. In an application where the TDR cable length can be limited to under 2 ft, requesting a 'lowest-loss' flexible cable from your favorite high-quality, low-cost coaxial cable manufacturer is sufficient. If you are working with a 3 ft to 4 ft cable, however, or require full resolution and rise time that the oscilloscope can offer, you will have to work with a high-end microwave cable manufacturer. Semi-rigid cables can provide better performance than flexible ones, but are more difficult to use. SMA connectors are commonly used on the ends of TDR cables as they provide acceptable performance and can be mated directly to the 3.5 mm connector found on 20 GHz TDR sampling modules.[5]

5. 3.5 mm connector is specified to be a 26.5 GHz connector, whereas a typical SMA is rated to 12.5 or 18 GHz.

When using a probe for taking TDR measurement on a board, package, or connector, the designer has to define a ground location near the signal location. If such a ground location is not available, or if the spacing from signal to ground varies widely across the PCB, the designer may have to use a probe which has a long ground wire, or a variable-length wire. For a probe with a long ground wire, the parasitic inductance will be very large, and will not allow the designer to obtain a good quality TDR measurements. Variable-length ground wires, and variable-pitch (signal-to-ground spacing) probes do not provide sufficient measurement repeatability, and will not provide accurate impedance measurement results or signal integrity interconnect models.

In many cases, a simple, inexpensive, and convenient TDR probe can be fabricated by using a 3 inch length of semi-rigid coaxial cable with an SMA connector, exposing the center conductor of such cable, and either using the sleeve of the semi-rigid coax as the ground contact, or attaching a ground wire. Using a different diameter coax will result in a different probe pitch, and making the center and ground conductors shorter or longer can provide the right trade-off between convenience of use and performance. A differential TDR probe can be fabricated by using two single-ended coaxial cables of the same length, connecting their grounding sleeves together, and attaching the appropriate ground leads as necessary (Figure 20.8).

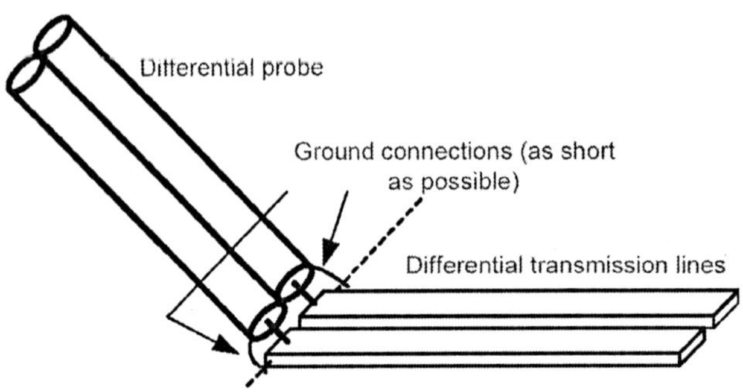

Figure 20.8 A differential probe connection to the differential DUT. [Reprinted with permission of TDA Systems.]

If your TDR measurements are sufficiently repeatable, you can sometimes de-embed your probe or cable parasitic inductance by either connecting your probe to a precision 50 Ω resistor on a calibration substrate, and subtracting this measurement from your DUT measurement, or by following the calibration and normalization procedures in your TDR oscilloscope. These procedures, however, do not replace the true impedance profile de-embedding described below, as they only de-embed the probe and the cable, but do nothing about the 'ghost' reflections inside the DUT itself, which are also discussed in the next section." [Tda02]

20.1.5 Multiple Reflections and the True Impedance Profile

"In case of single impedance interconnect, such as a test coupon on a PCB, or a controlled impedance cable, the TDR oscilloscope produces impedance readouts accurately. In other practical cases, however, such as a real trace on a board, which often has to travel through different layers with potentially different impedances, connected with vias, with connectors and packages mounted on the board, the situation is more complex, and we have to deal with an effect known as 'multiple reflections.' TDR oscilloscopes use Equations (17.1) through (17.3) to compute reflection coefficients and impedances of the DUT. These equations assume that the oscilloscope knows exactly what the incident voltage was at the impedance discontinuity in question. In reality (Figure 20.9), as the TDR signal propagates through multi-impedance interconnect, the incident voltage at each discontinuity changes, since at each discontinuity, a portion of the signal energy is reflected back to the oscilloscope, and only a portion of the signal energy continues to propagate. Moreover, the signal traveling back to the oscilloscope is re-reflected back into the DUT. As a result, the signal begins to bounce back and forth within the DUT, creating the effect of multiple (or 'ghost') reflections. The impedance measurement error in the oscilloscope quickly adds up, resulting in incorrect impedance readouts. The deeper into the DUT we are trying to measure impedance, the more multiple reflections we will accumulate, resulting in larger impedance measurement error. An impedance deconvolution algorithm, discussed in a number of publications (for example, see [Hay94], [Hsu97], [Smo99]), and implemented in IConnect TDR software, allows the designer to de-embed the multiple reflections and accurately compute the true impedance profile for each segment in the multi-impedance DUT.

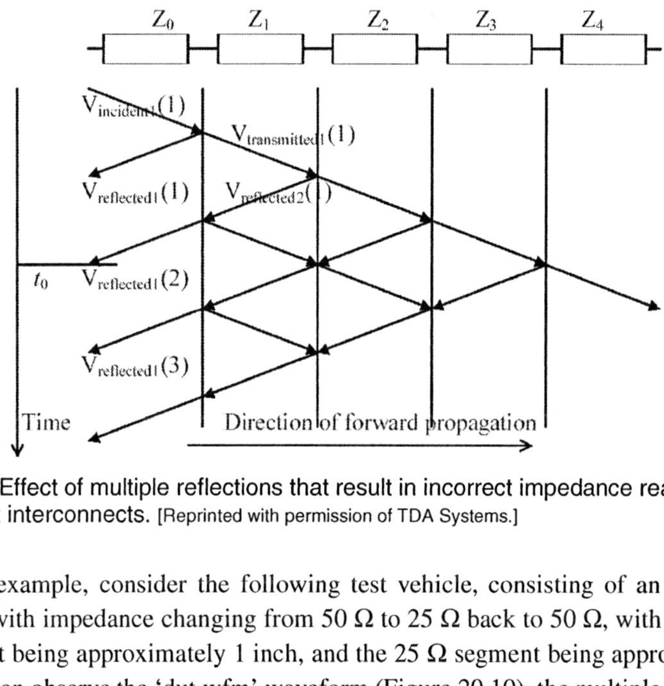

Figure 20.9 Effect of multiple reflections that result in incorrect impedance readouts for multisegment interconnects. [Reprinted with permission of TDA Systems.]

As an example, consider the following test vehicle, consisting of an outer board trace (microstrip) with impedance changing from 50 Ω to 25 Ω back to 50 Ω, with the length of each 50 Ω segment being approximately 1 inch, and the 25 Ω segment being approximately 5 inches long. As we can observe the 'dut.wfm' waveform (Figure 20.10), the multiple reflections exhibit themselves very strongly for this multisegment interconnect, not only giving an incorrect impedance reading (about 44 Ω) for the second 50 Ω segment, but also showing some ghost reflections in the area where the waveform should have shown the open circuit signature. (In Figure 20.10, the dut.wfm and step.wfm waveforms have been repositioned upwards for clearer viewing and the dut.wfm impedance values cannot be read directly from the vertical scale.) The true impedance profile, computed in IConnect, de-embeds the multiple reflection effects and provides the accurate true impedance profile for the DUT.

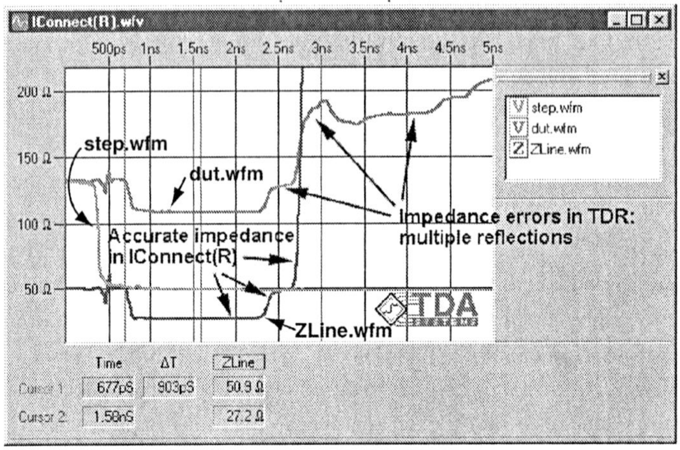

Figure 20.10 Accurate true impedance profile in IConnect vs. the reflection profile in the TDR oscilloscope. [Reprinted with permission of TDA Systems.]

In another example, we have looked at an open trace failure in a BGA package (Figure 20.11). In this example, the TDR waveform (reflection profile) provides a confusing indication about where the failure in the DUT occurred. Once the true impedance profile is computed, however, any confusion about the failure location is removed, and the failure location is clearly identified.

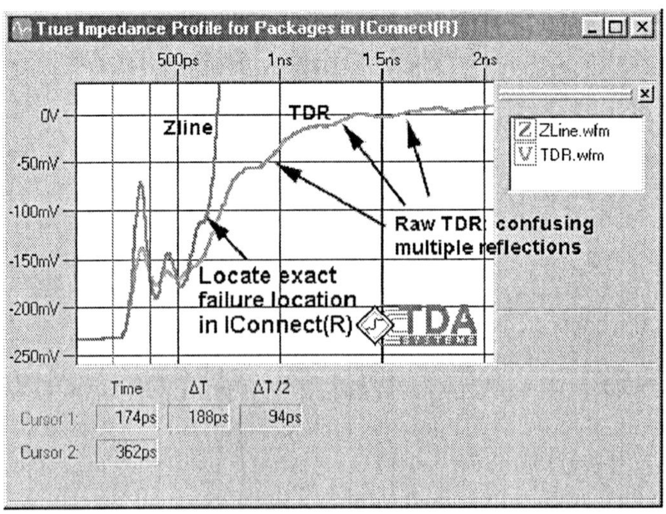

Figure 20.11 Exact failure location can be determined much more easily after the impedance profile has been computed in IConnect TDR software. [Reprinted with permission of TDA Systems.]

The TDR oscilloscope does not intentionally 'confuse' the instrument user. It shows the user the reflection profile of the DUT, pointing out how a real signal will be distorted by the DUT impedance mismatches and discontinuities. When a designer aims to merely observe the signal distortion, reflection profile is what is needed. If the designer's goal is to accurately measure the impedance in a multisegment interconnect, produce a SPICE or IBIS signal integrity model for the interconnect, or locate the failure, the impedance deconvolution algorithm is required to convert this reflection profile into the true impedance profile. In addition, when a designer attempts to zoom in on a portion of a multisegment interconnect, and window out the rest of the data, such windowing is only possible with the true impedance profile, not with the TDR reflection profile. Otherwise, the multiple reflections will remain in the windowed data and will distort the analysis results. Examples of such multisegment interconnects are a transmission line on a real circuit board, changing layers and going through connectors and packages, or a cable-connector assembly attached to a circuit board, or a modern BGA or flip chip package — as opposed to, let's say, a test coupon on a circuit board where the impedance can be read accurately without the use of the impedance deconvolution algorithm." [Tda02a]

20.1.6 Other TDR Measurement Issues

20.1.6.1 Using Good Measurement Practices

"To obtain good quality impedance, signal integrity modeling, and failure analysis data, it is important to follow general good measurement practices when using a TDR oscilloscope. The instrument should be turned on and its internal temperature should be allowed to stabilize for 20–30 minutes before performing any measurements. Calibration, compensation, and normalization for the instrument must be performed regularly, as specified by the instrument manufacturer. The internal instrument temperature must be within the specified range from the calibration points for the given instrument.

To maximize the resolution of the scope, particularly in the time axis, it is important to zoom in on the DUT — but at the same time to allow a window that is sufficiently long to include all the reflections related to the DUT. A window that is too short may prevent the designer from obtaining complete and accurate information about the DUT. When the designer intends to perform true impedance profile analysis, as implemented in IConnect TDR software, it is also important to window out the transition related to the sampling head to the cable interface, and focus on the DUT portion of the waveform, so as to ensure that the impedance deconvolution algorithm performs correctly (see Figure 20.12)." [Tda02a]

20.1.6.2 TDR Measurements of "Splits" and "Stubs"

"If the board trace under test splits into two or more directions, as is the case for address lines in a memory module, the TDR instrument shows the sum of all reflection from all the N legs in the split, but cannot separate which reflection came from which leg in the split. If the splits are of the same impedance and delay (as sometimes is the case in a 'star' interconnect topology), they can be simply represented by transmission lines running in parallel, and the impedance mea-

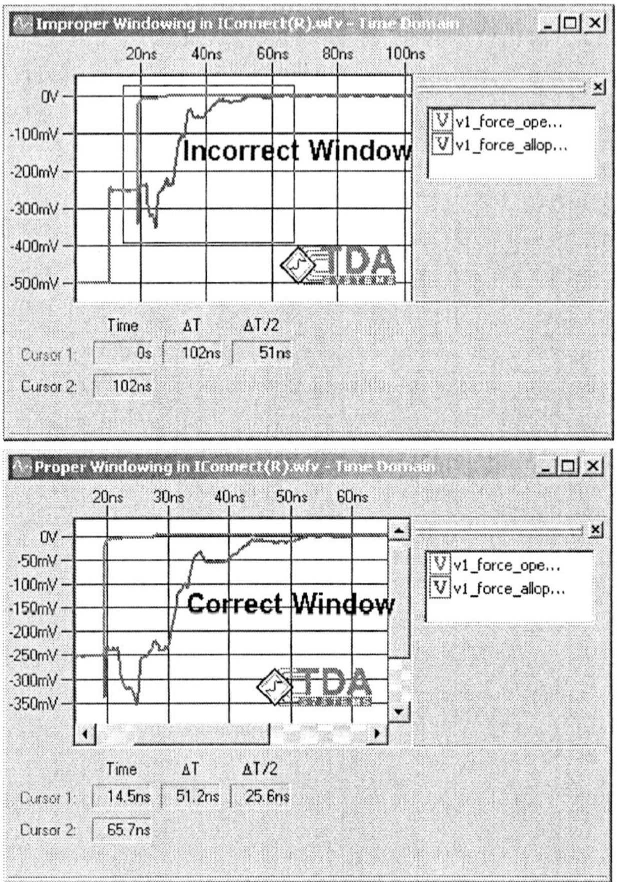

Figure 20.12 Improper windowing (top) and proper windowing (bottom) of a TDR waveform.
[Reprinted with permission of TDA Systems.]

sured by the TDR oscilloscope equals Z1 / N, with the delay of each trace being equal to the delay measured by TDR (Figure 20.13). In the case of a stub (which often takes place in a daisy chain configuration), if the length of each stub on the main bus is much shorter than the rise time of the signal propagating through the bus, the stubs can be treated as lumped capacitances loading the main bus, thus simplifying the measurement problem." [Tda02]

20.1.7 Frequency Domain and TDR

"The fast rise time of the TDR instrument allows the designer to obtain a wide-band characterization of the interconnect, since the incident TDR step contains all the harmonics up to approximately the 3 dB frequency of the TDR signal. The TDR measurement can be converted into the return loss (S11), and the TDT measurement into insertion loss (S21).

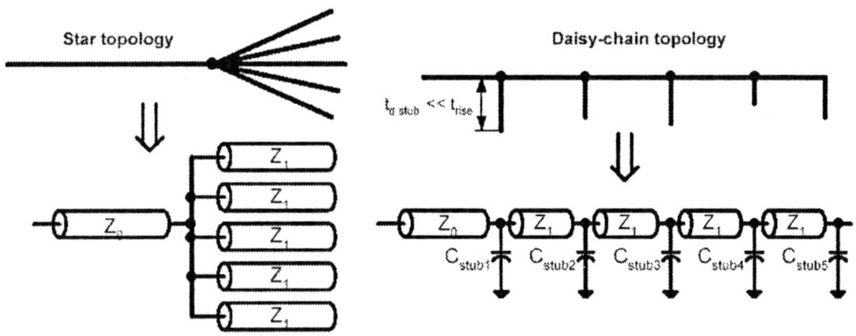

Figure 20.13 Taking TDR measurement of splits in a star topology and stubs in a daisy-chain topology. [Reprinted with permission of TDA Systems.]

Differential TDR measurements can also be converted into corresponding differential and mixed mode S-parameters as shown in Figure 20.14 (see [Smo02])." [Tda02a] In the top half of Figure 20.4, the first subscript of the S-parameters refers to the port number where the signal response is measured and the second subscript refers to the port number being stimulated. In the bottom half of Figure 20.4, the numerical subscripts are labeled the same way, but in addition, the first letter identifies where the response is being measured and the second letter identifies the stimulated port. For example, S_{12dc} is the S-parameter for a common-mode stimulus applied to

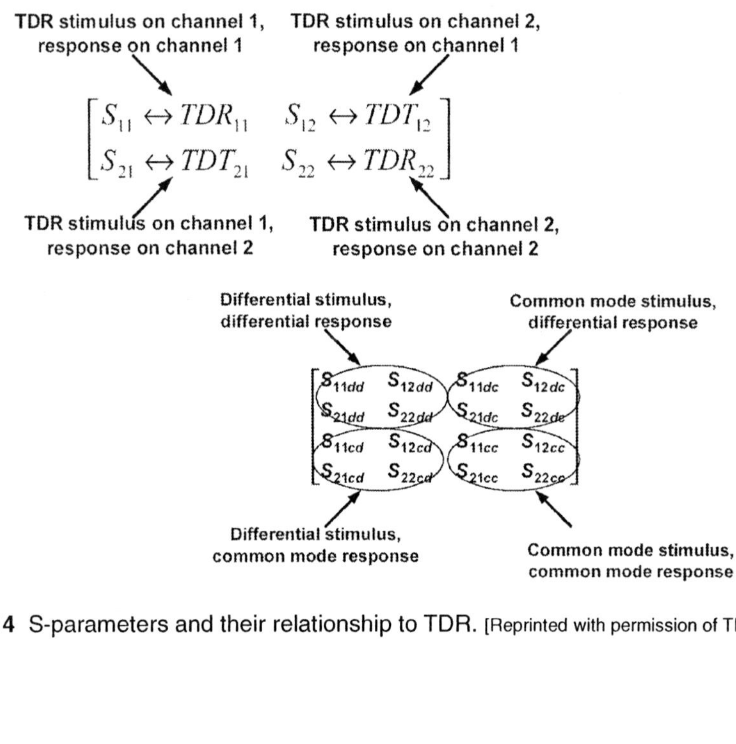

Figure 20.14 S-parameters and their relationship to TDR. [Reprinted with permission of TDA Systems.]

port 2 and a differential response measured on port 1. TDT stands for "Time Domain Transmission" and is discussed in Section 20.2 on page 536.

Software tools for extracting and validating interconnect models from measurements and even for computing S-parameters from TDR data, such as IConnect TDR software from TDA Systems, have been developed, expanding the applications for the TDR instruments. Furthermore, Time Domain Network Analysis (TDNA) calibration and measurement routines, designed to enhance the accuracy of TDR and TDT measurements in a manner similar to that of VNA calibration, yielding data in both time and frequency domain, have been reported and implemented in tools such as MultiCal software from United States National Institute of Standards and Technology (NIST). The calibration brought the accuracy of TDNA to a level similar to that of network analyzers.

20.1.8 Static Discharge Precautions

"High-bandwidth sampling heads are extremely static-sensitive. To ensure their continued performance in impedance and signal integrity applications, they require protection devices above and beyond those which suffice for conventional oscilloscopes. The following guidelines are essential when handling or using any high-bandwidth sampling module:

- Cap sampling inputs when not in use.
- Use grounded wrist straps of less than 10 mega ohms when in contact with the module.
- To prevent the possibility of probing a device that may have built up a charge, you should always discharge the device before probing or connecting it to the module. You can accomplish this by touching the probe ground lead or your finger (assuming you are using a wrist strap) to the point you want to TDR test. "Static isolation units" perform this function by using a microwave relay. With an isolation unit, the DUT cable is connected to ground until a foot switch is depressed and the relay switches, connecting the DUT to the TDR input."[6] [Tek01a]

20.1.9 Controlling Rise Time

"While in many cases using a TDR instrument with the fastest rise time available is desirable, very fast rise times can in some cases give misleading results. For example, testing the impedance of a microstrip on a circuit board with a 35 ps TDR step rise time system provides excellent resolution. However, most of the highest-speed logic families in use today cannot begin to match the 35 ps rise time of the TDR step. While SiGe GigaComm™ and SiGe GigaPro™ devices (both are ECL family devices) have edge rates of from 20–40 ns, most ECL is not this fast. The other ECL and high-speed logic families have output rise times in the 200 ps to 2 ns range. Reflections from small discontinuities such as stubs or sharp corners in the microstrip

6. Copyright Tektronix, Inc. Reprinted with permission. All rights reserved.

will be quite visible and may generate large reflections at 35 ps rise time. The same transmission line driven in actual operation by an ECL gate with a 375 ps rise time may produce negligible reflections.

This difference between measurement and reality could be the beginning of a signal integrity problem. In attempting to correct the misleading impedance readings, you might compromise the environment that the real operational signals need. It is often preferable to see the transmission line's TDR response to rise times similar to those of the actual circuit operation." [Tek01a]

20.1.10 Example of TDR Oscilloscope

"The TDS8000 digital sampling oscilloscope (see Figure 20.15) with its electrical bandwidth of 50 GHz provides the bandwidth that engineers need to accurately characterize fast designs. Equally important, the 80E04 electrical sampling module for TDR offers true differential TDR measurements. Two uncompromised channels in this single sampling module, driven by a common clock source, let engineers make impedance measurements that are much more accurate than measurements derived from a single-ended TDR acquisition. With most of today's TDR-capable instruments, the necessary ratio calculations are performed internally and a numerical result is displayed. With the TDS8000, TDR measurements can be displayed with units of volts (V), ohms (Ω), or ρ on the vertical magnitude scale and the horizontal axis represents units of time. Extended math functions are offered which not only calculate actual component values, but also the physical distance to points of interest along the transmission line.

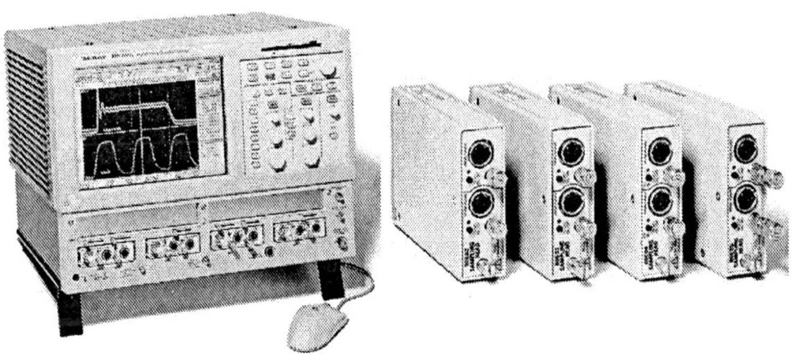

Figure 20.15 The Tektronix TDS8000 sampling oscilloscope with TDR 80E04 electrical sampling modules. [© Tektronics, Inc. Reprinted with permission. All rights reserved.]

Random noise can be a significant source of error when making measurements of small impedance variations. Fortunately, modern instruments can perform signal averaging to reduce the effects of random noise. The drawback in many instruments is that averaging can dramatically slow the processing speed, particularly if automatic measurements are being displayed. The TDS8000 oscilloscope addresses this problem with its built-in multiple processors, which can share the processing workload.

The TDS8000 sampling oscilloscope includes open access to the Windows® operating system. While the instrument remains a dedicated oscilloscope, the ability to access the Windows desktop creates a powerful new tool. Built-in applications such as WordPad, Paint, and a Web browser allow users to concurrently maintain lab notes while working with the instrument. This saves time and eliminates error-prone steps associated with transporting images for later report development and communicating measurement results. Other applications such as TDA Systems IConnect software can be installed in the instrument to accomplish TDR modeling and acquisition. Connection to a network enables Web-based information browsing, e-mail exchange, printing, and file sharing." [Tek01a]

20.2 Time Domain Transmission (TDT)

"TDT stands for Time Domain Transmission. We can stimulate the DUT on one side, and using the remaining channels on the TDR instrument, measure the transmission through the DUT, whether in single-ended or differential-mode (see Figure 20.16).

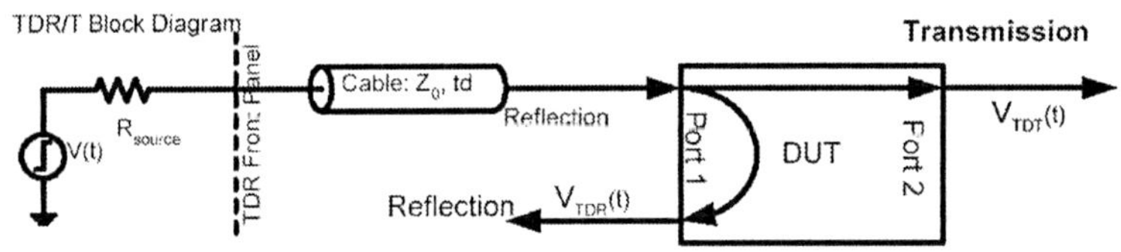

Figure 20.16 Time Domain Transmission (TDT). [Reprinted with permission of TDA Systems.]

TDT is useful for demonstrating the effect that the interconnect impedance discontinuities will produce at the receiver, as well as for characterization of lossy transmission line parameters, such as rise time degradation, return loss, and skin effect and dielectric loss (see [Bog01]). In addition, based on the TDT measurement, IConnect TDR software can predict the eye diagram degradation through the given interconnect. The same deskew capability, used to deskew the TDR channels during the differential measurement, can be used to deskew the TDR channels during the TDT measurements." [Tda02a]

20.3 TDR and TDT Simulations for a Loaded BLVDS Backplane

20.3.1 Loading Effects of Connectors, Stubs, and ICs on the Backplane

A typical BLVDS backplane configuration is depicted in Figure 20.17, shown with only a single plug-in card for illustrative purposes. Once plugged in, "the card adds a load to the bus which is mainly capacitive, resulting from the connector (2–3 pF), the PCB trace (2–3 pF), and the device (4–5 pF), for a total load of about 10 pF. Whether this load is a bulk load or distributed load depends upon the edge rate of the signals traversing the backplane and cards, and the physical distances involved.

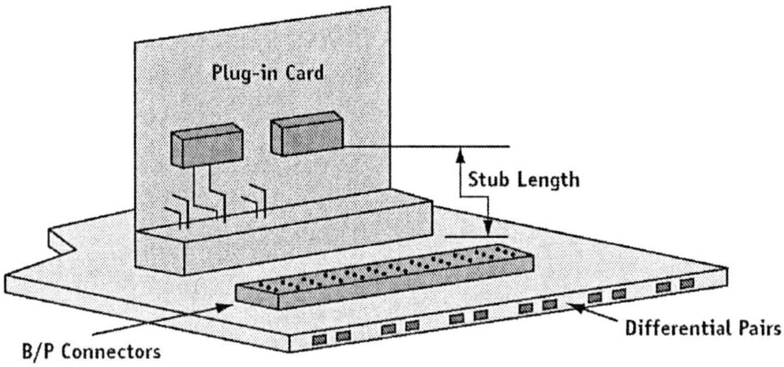

Figure 20.17 Backplane (B/P) configuration with plug-in card. [Reprinted with permission of North East Systems Associates, Inc. (NESA)]

For TDR and TDT simulations of the BLVDS backplane and plug-in cards described in this section and the next, a TDR signal rise time of 0.5 ns was modeled since typical BLVDS signal rise times are in the 0.5–1.0 ns range. The backplane was modeled with 20 slots and the simulations were performed using SPICE. Except where stated otherwise, the unloaded differential backplane trace impedances were 100 Ω.

Figure 20.18 displays a TDR simulation of backplane impedance versus loading. Case A shows the raw traces only (no via). Case B inserts the backplane connectors. Case C adds all the

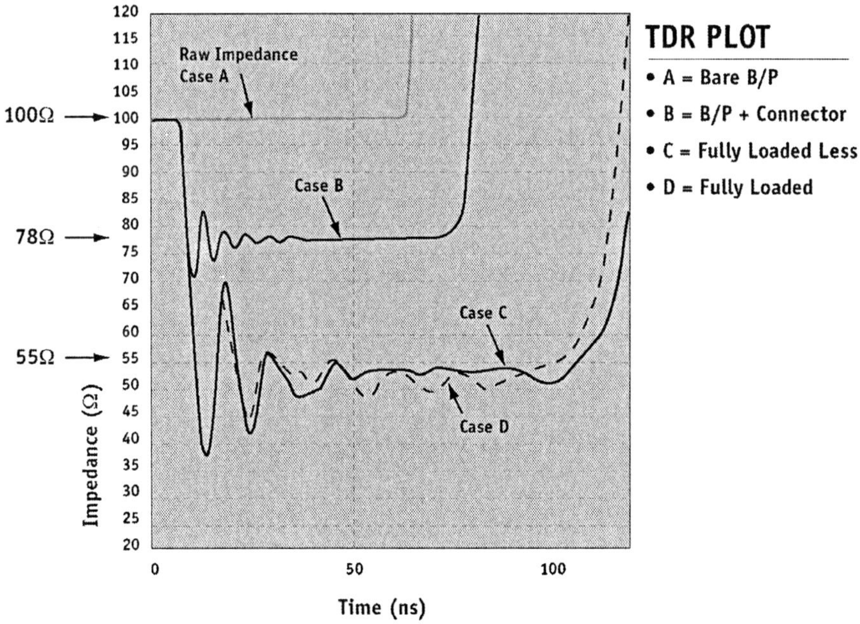

Figure 20.18 Time domain reflectometry (TDR) simulation for four situations. A: Bare backplane. B: Backplane with a connector. C: Fully loaded backplane without ICs. D: Fully loaded backplane with ICs. [Reprinted with permission of North East Systems Associates, Inc. (NESA)

cards into the bus, but with devices not yet installed on the cards. Lastly, Case D adds devices to all cards and is for a fully loaded system. Notice that about 50% of the loading is due only to the backplane connectors and their vias. Also notice how the velocity of the signal is slowed by the loading. This final value (Case D) is the impedance value that the two 56 Ω termination resistors are matched to."[7] [Nat00]

20.3.2 Effect of Plug-In Card Stub Lengths

"Stubs have one golden rule: The shorter the better! A long stub adds to the capacitive load, lowers the loaded backplane impedance even more, and tends to impact signal quality. For this reason, stub interconnects should be a microstrip and the number of vias should be limited (0 to 1 is best). Stub lengths should be typically 1 to 1.5 inches or less.

Figure 20.19 shows a TDR simulation comparing the effects of 0.5, 1, and 1.5 inch stub lengths. The figure also serves to define ZEFF – min and ZEFF – max, the minimum and maximum loaded-backplane impedance values (shown for the 1.5 inch stub length case). The figure shows that as the stub length increases, the value of ZEFF – min goes lower and lower. For com-

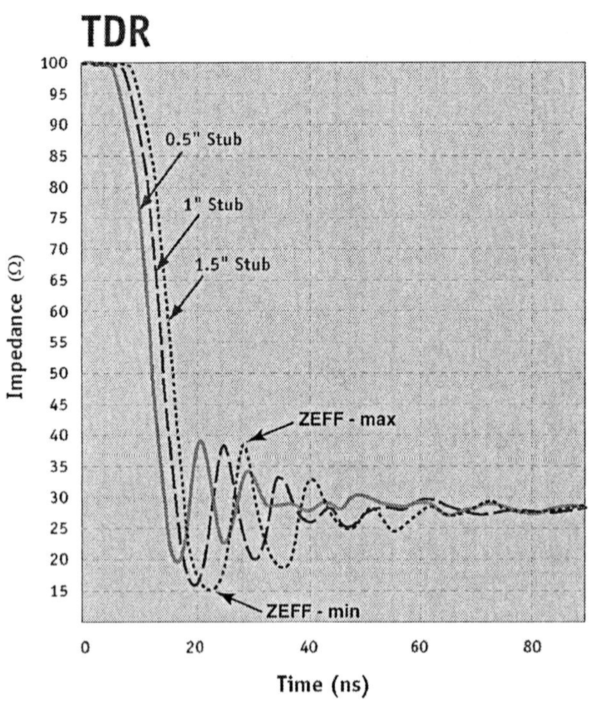

Figure 20.19 Time domain reflectometry (TDR) simulation comparing effects of 0.5, 1, and 1.5 inch stub lengths. [Reprinted with permission of North East Systems Associates, Inc. (NESA)]

7. Reprinted with permission of National Semiconductor Corporation.

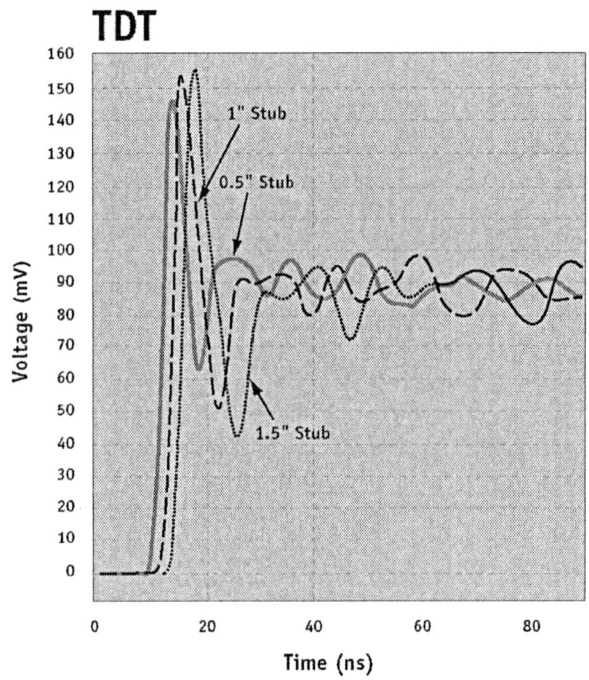

Figure 20.20 Time domain transmission (TDT) simulation comparing effects of 0.5, 1, and 1.5 inch stub lengths. [Reprinted with permission of North East Systems Associates, Inc. (NESA)]

parison, Figure 20.20 shows a TDT plot which also illustrates the effects of 0.5, 1, and 1.5 inch stub lengths. For this TDT plot, the vertical scale is in millivolts.

Figure 20.21 plots the results from other similar TDR simulations, but where the unloaded backplane differential impedance is 130 Ω rather than 100 Ω as was the case for the previous figures. The plot illustrates how increasing stub length lowers the backplane's loaded differential impedance (effective differential impedance), ZEFF. In the figure, ZEFF is the average of ZEFF – min and ZEFF – max. By designing the unloaded differential impedance to be 130 Ω rather than 100 Ω, the loaded backplane differential impedance can be kept to a higher value. This results in larger noise margins. For a more detailed discussion of this effect, see reference [Say98].

There are three layout concepts which help to maintain a high 'loaded' bus impedance and thereby increase noise margins. First, as mentioned above, design the unloaded differential trace impedance to be as high as practical, e.g., 130 Ω vs. 100 Ω. Second, limit the number of vias on the card's stub to minimize capacitance loading, and third, keep stub lengths as short as possible.

BLVDS is capable of operating in wildcard configurations (i.e., cards plugged into any slot). Depending on system noise margin goals, full card loading is not required for Bus LVDS devices to function properly. Termination selection is a bit tricky, but should be matched to a fully loaded case (or slightly higher). Signal wave shapes are good even for half-loaded or section-loaded busses. Single-ended technologies cannot support this feature due to noise margin violations." [Nat00]

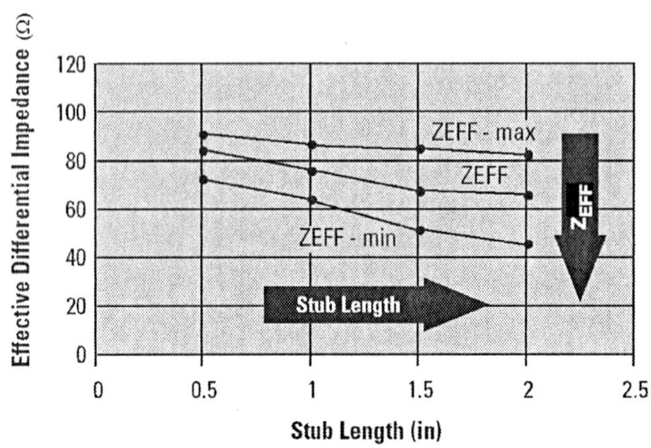

Figure 20.21 The TDR simulated differential impedance of a 20 slot, fully loaded BLVDS backplane with differential impedance in the unloaded state $Z_{DIFF} = 130\ \Omega$. [Reprinted with permission of North East Systems Associates, Inc. (NESA)]

20.4 Vector Network Analyzer (VNA)

20.4.1 VNA Instrumentation Background

VNA (Vector Network Analyzer) instruments were originally developed as a microwave design and measurement tool (see Figure 20.22). In microwave design, engineers are concerned primarily with narrowband and resonant systems such as mixers, filters, resonators, and power splitters/combiners. Engineers require exact data about the frequency band over which the circuit can operate, center frequency, and Q-factor. These requirements caused instrument manufacturers to continuously improve VNA accuracy and achieve very high dynamic range (signal-to-noise ratio) in the frequency domain, easily reaching 100 dB when careful measurement techniques are used. This high dynamic range and accuracy are achieved through narrowband filtering at each frequency point, as well as through calibration procedures that allow the instrument to correct for imperfections in the measurement path. Short-open-load-thru (SOLT) was the most commonly used calibration procedure, and advanced calibration procedures, such as thru-reflect line (TRL), were invented to improve the instrument accuracy and extend its frequency range. Powerful microwave design tools, such as Touchstone, were developed for analysis and synthesis of microwave electronic systems.

Differential VNA capabilities were not added to VNAs until very recently. Currently, a four-port single-ended or two-port differential capability is available. However, because high accuracy is required for narrowband microwave work, the focus of development for these instruments was on this accuracy rather than ease of use, making them less amenable to digital designers without microwave design background. The designer or test engineer needs to really understand and meticulously follow the calibration procedures in order to obtain the data from a network analyzer.

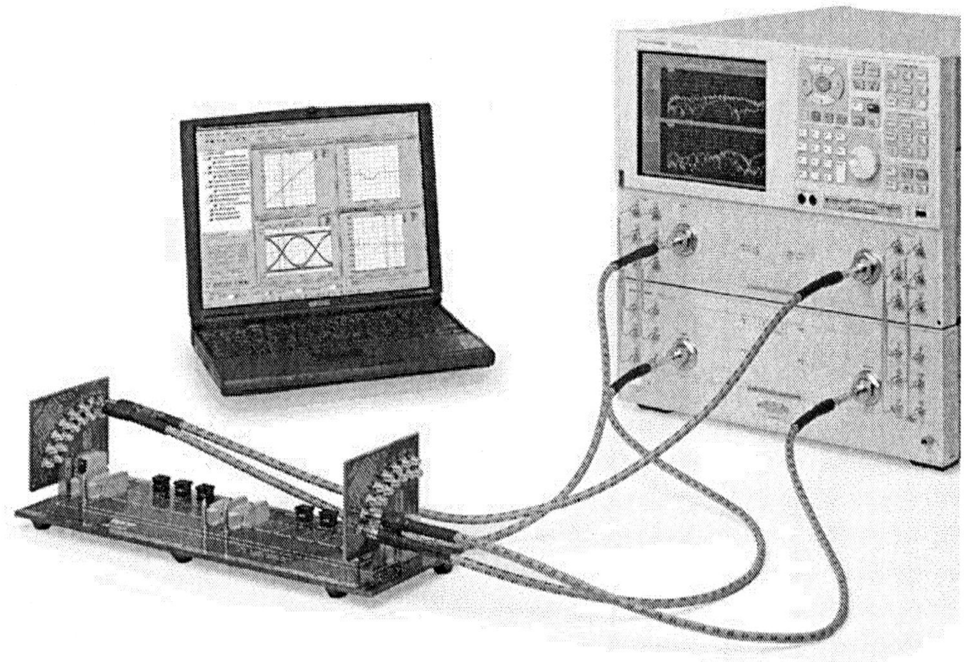

Figure 20.22 Agilent Technologies' E8364A PNA Series VNA with Physical Layer Test System (PLTS) options for backplane testing to 50 GHz. [Copyright © 2003 Agilent Technologies, Inc. Reproduced with permission of Agilent Technologies.]

To obtain accurate models for high-speed interconnects, a signal integrity engineer eventually needs to perform a measurement-based validation of the interconnect models. At that point, a choice needs to be made about the type of test equipment to be used to accomplish the job. Ease of use, accuracy, and cost are only a few of the attributes that will dictate the right instrumentation. The TDR oscilloscope and the VNA are the two most common instruments used today. The choice between these two sometimes moves to a more theoretical choice between time and frequency domain measurements. Occasionally, it becomes a choice of personal belief in one domain being superior over the other. In reality, there are appropriate times for the use of both instruments.

20.4.2 Network Analyzer Terminology: R, A, and B Channels vs. Port 1 and Port 2

On earlier network analyzers, "the incident wave is measured with the R (for reference) channel, the reflected wave is measured with the A channel, and the transmitted wave is measured with the B channel. With amplitude and phase information of these three waves, we can quantify the reflection and transmission characteristics of our device under test (DUT). Some of the common measured terms are scalar in nature (the phase part is ignored or not measured), while others are vector (both magnitude and phase are measured). For example, return loss is a scalar measurement of reflection, while impedance is a vector reflection measurement. Ratioed reflection is often shown as A/R and ratioed transmission is often shown as B/R, relating to the measurement channels used in the network analyzer." [Bal98] See Figure 20.23.

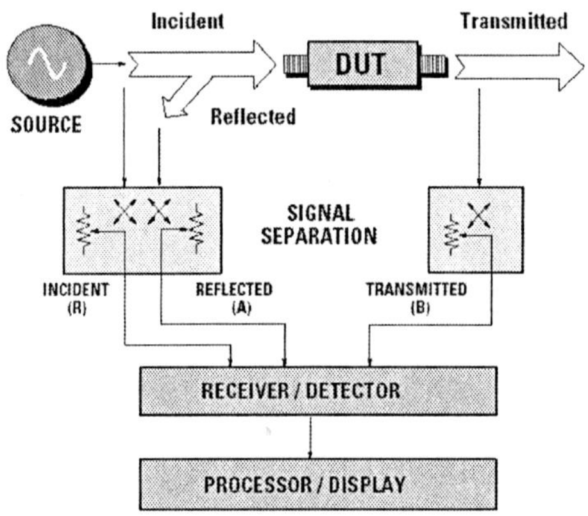

Figure 20.23 Vector network analyzer (VNA) block diagram with R, A, and B channels labeled. [Reprinted with permission of Agilent Technologies, Inc.]

New network analyzers today usually come with "Port 1" and "Port 2" rather than R, A, or B channels. The "Port 1, Port 2" terminology refers to network analyzer S-parameter test sets. The R, A, and B channels of older analyzers were inputs that were effectively direct sampler (mixer) access. These analyzers usually had outboard test sets, or users built their own. With the addition of directional bridges, splitters, and other components, VNA manufacturers now include this signal-separation circuitry in their network analyzers. It is this added circuitry that allows the analyzers to make S-parameter measurements, and it also creates ports 1 and 2. Many network analyzers have an option to add what is called "direct sampler access." This is essentially the R, A, and B inputs (channels). As an example, all Agilent Technologies' PNA analyzers have four input receivers (mixers not samplers) and have S-parameter test sets. This makes for the most versatility. Direct sampler access is frequently used for high power situations where the internal signal separation devices are rated too low in power dissipation.

20.4.3 The Differences Between VNAs, Scalar Network Analyzers (SNAs), and Spectrum Analyzers

The main difference between VNAs and scalar network analyzers (SNAs) is that VNAs can measure both magnitude and phase of a signal whereas scalar network analyzers can only measure signal magnitude, but not the phase. Vector means that the analyzer has the capability to perform vector (magnitude and phase) error corrections. A scalar network analyzer can make A/

R and B/R measurements without phase information. S-parameter measurements require the measurement of both magnitude and phase and require a VNA for the measurements. The term "network analyzer" is a broader term that can refer to either a VNA or scalar network analyzer.

"VNAs are the most powerful kind of network analyzer and can measure frequencies from 5 Hz up to 110 GHz. Designers and final test in manufacturing use VNAs because they measure and display the complete amplitude and phase characteristics of an electrical network. These characteristics include S-parameters, magnitude and phase, standing wave ratios (SWR), insertion loss or gain, attenuation, group delay, return loss, reflection coefficient, and gain compression.

VNA hardware consists of a sweeping signal source (usually internal), a test set to separate forward and reverse test signals, and a multichannel, phase-coherent, highly sensitive receiver (see Figure 20.23). In the RF and microwave bands, typical measured parameters are referred to as S-parameters, and are also commonly used in computer-aided design models.

An SNA measures only the amplitude portion of the S-parameters, resulting in measurements such as transmission gain and loss, return loss, and SWR. Once a passive or active component has been designed using the total measurement capability of a VNA, an SNA may be a more cost-effective measurement tool for the production line to reveal out-of-specification components. While SNAs require an external or internal sweeping signal source and signal separation hardware, they only need simple amplitude-only detectors, rather than complex (and more expensive) phase-coherent detectors." [Agi01a]

"Spectrum analyzers measure signal amplitude characteristics (carrier level, sidebands, harmonics) and can demodulate and measure complex signals. Spectrum analyzers are receivers only and can be used for scalar component testing (no phase measurements) when used with a tracking generator or external source." [Bal98]

"A network/spectrum analyzer eliminates the circuit duplication in a benchtest setup of a network and spectrum analyzer. Frequency coverage ranges from 10 Hz to 1.8 GHz. These combination instruments can be economical alternatives in design and test of active components like amplifiers and mixers, where analysis of signal performance is also needed." [Agi01a]

"Network analyzers are used to measure components, devices, circuits, and subassemblies. They contain both a source and multiple receivers, and generally display ratioed amplitude and phase information (frequency or power sweeps). A network analyzer is always looking at a *known* signal (in terms of frequency), since it is a stimulus-response system. With network analyzers, it is harder to get an accurate trace on the display, but very easy to interpret the results. With vector-error correction, network analyzers provide much higher measurement accuracy than spectrum analyzers.

Spectrum analyzers are most often used to measure signal characteristics such as carrier level, sidebands, harmonics, phase noise, and the like on *unknown* signals. They are most commonly configured as a single-channel receiver, without a source. Because of the flexibility needed to analyze signals, spectrum analyzers generally have a much wider range of IF bandwidths available than most network analyzers. Spectrum analyzers are often used with external sources for nonlinear stimulus-response testing. When combined with a tracking generator, spectrum analyzers can be used for scalar component testing (magnitude versus frequency, but no phase measurements). With spectrum analyzers, it is easy to get a trace on the display, but interpreting the results can be much more difficult than with a network analyzer." [Bal98]

20.4.4 Vector Error Correction of Systematic Measurement Errors

"Vector error correction is the process of characterizing systematic error terms by measuring known calibration standards and then removing the effects of these errors from subsequent measurements. Figure 20.24 shows the major systematic errors associated with network measurements. The errors relating to signal leakage are directivity and crosstalk. Errors related to signal reflections are source and load mismatch. The final class of errors is related to frequency response of the receivers, and are called reflection and transmission tracking. The full two-port error model includes all six of these terms for the forward direction and the same six (with different data) in the reverse direction, for a total of 12 error terms. This is why two-port calibration is often referred to as 12-term error correction.

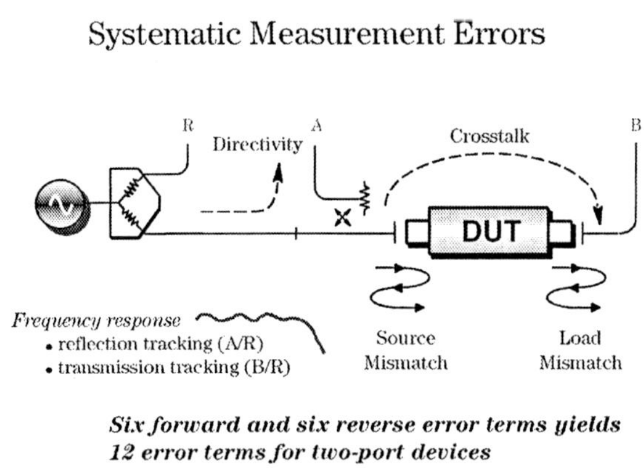

Figure 20.24 The major systematic errors associated with network measurements. [Reprinted with permission of Agilent Technologies, Inc.]

The two-port calibration usually requires 12 measurements on four known standards (short-open-load-thru or SOLT). Some standards are measured multiple times (e.g., the thru standard is usually measured four times). The standards themselves are defined in a cal-kit definition file which is stored in the network analyzer. For example, Agilent Technologies' network analyzers contain all of the cal-kit definitions for their standard calibration kits. In order to make accurate measurements, the cal-kit definition must match the actual calibration kit used." [Bal98]

20.4.5 The Mathematics of Time and Frequency

"VNAs measure power, but versus frequency, and transform it into scattering parameters (S-parameters). S-parameters are a ratio of the reflected or transmitted wave voltage to the incident wave voltage, and for the 2-port measurement shown in Figure 20.25, can be written in their simplified form as

$$S_{11}(f) = \frac{V_{reflected1}(f)}{V_{incident1}(f)} \qquad S_{21}(f) = \frac{V_{transmitted2}(f)}{V_{incident1}(f)}$$
$$S_{12}(f) = \frac{V_{transmitted1}(f)}{V_{incident2}(f)} \qquad S_{22}(f) = \frac{V_{reflected2}(f)}{V_{incident2}(f)} \tag{20.10}$$

VNAs can display S-parameter data versus frequency as real and imaginary values (magnitude and phase) or on a Smith chart. It is also worth noting that S_{11} and S_{21} parameters are frequently referred to as reflection and transmission coefficients and are similar to the reflection and transmission coefficients in TDR.

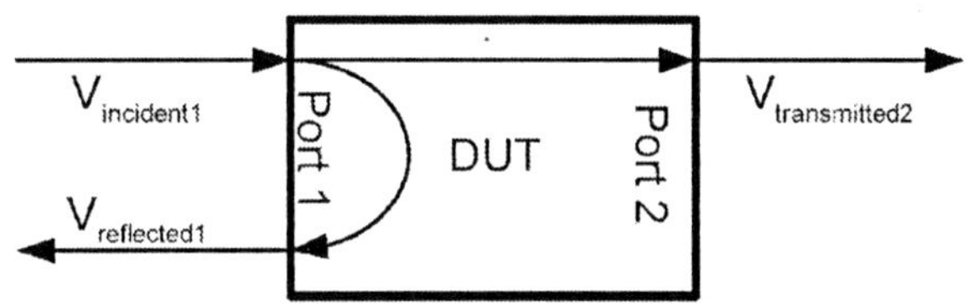

Figure 20.25 Two-port VNA measurement block diagram. [Reprinted with permission of TDA Systems.]

The basic block diagram of TDR and TDT, as compared with that of a 2-port VNA instrument, is shown in Figure 20.26 The close similarity between the two block diagrams of these two seemingly different measurement systems is obvious. The fundamental difference between the TDR and VNA instruments lies in the fact that TDR is measuring voltage versus time while VNA is measuring power versus. frequency, as well as in the way data is displayed. Moreover,

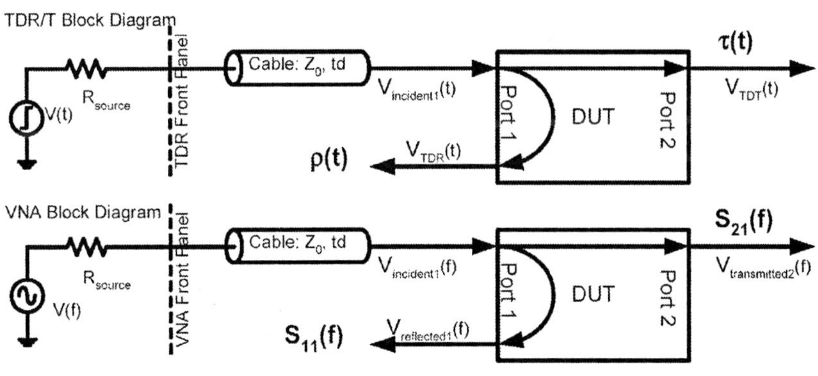

Figure 20.26 TDR/T measurement block diagram, compared with equivalent 2-port VNA block diagram. [Reprinted with permission of TDA Systems.]

we observe clear similarities between Equations (17.1) to (17.2) and (17.10). Since we know from basic math that voltage in the time and frequency domains are related by the Fast Fourier Transform (FFT)[8] and deconvolution in the time domain corresponds to division in the frequency domain, we can easily see that there is a relationship between the TDR reflection and transmission coefficients and VNA S-parameters:

$$\begin{aligned} S_{11}(f) &= FFT(\rho(t)) \qquad \rho(t) = \frac{1}{N}\,IFFT(S_{11}(f)) \\ S_{21}(f) &= FFT(\tau(t)) \qquad \tau(t) = \frac{1}{N}\,IFFT(S_{21}(f)) \end{aligned} \tag{20.11}$$

VNAs measure both the real and imaginary parts of the signal versus frequency, whereas TDRs measure voltage versus time. However, as the above equations clearly indicate, there is a clear one-to-one relationship between the real TDR data versus time and real/imaginary VNA data versus frequency (Figure 20.27). All the information about the frequency domain is contained in the TDR/TDT measurements. As these equations indicate, scattering-parameter computation is just as applicable to TDT (transmission) measurements as it is to TDR (reflection) measurements. TDT results in insertion loss (S21), TDR in return loss (S11)." [Tda02c]

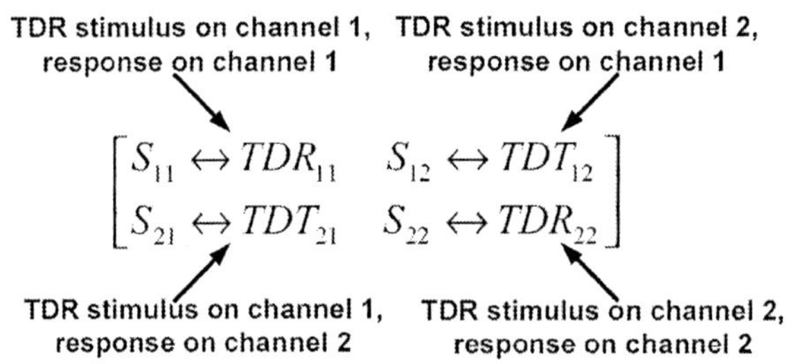

Figure 20.27 There is a clear relationship between TDR and TDT measurements in the time domain and network analyzer measurements in the frequency domain. [Reprinted with permission of TDA Systems.]

20.4.6 Practical Comparison of TDR and VNA Measurements

"Signal integrity interconnects are typically complex, multicomponent structures. For example, a path from a driver to receiver may begin in the driver package, go through several layers and vias on the printed circuit board (PCB), and go through a second package to reach the receiver.

8. To be more exact, it is the Fourier Transform (FT) that relates time and frequency domains. The Discrete Fourier Transform (DFT) is the computational method for handling sampled data, and FFT is just a computationally efficient implementation of DFT.

The signals that propagate through these interconnects are digital and therefore broadband. When we measure these signals to search for signal integrity issues such as crosstalk, we typically do not have a noise floor higher than about 10 mV out of the 1 V signal (1%). This corresponds to about –40 dB in the frequency domain (Figure 20.28). Such crosstalk typically is significantly below threshold and is ignored in digital system design. Resonance in digital interconnects may happen, but they are typically not very sharp and do not require the same measurement dynamic range as a resonance in a high-Q microwave resonator.

Digital designers live and design in the time domain and for them, TDR offers well-known advantages for the characterization of time domain interconnects. TDR is intuitive, allows visual operation, and has windowing capability. The visual nature of TDR measurements comes from the fact that TDR is a transient measurement and each change in the TDR waveform can be correlated to an appropriate lumped or distributed circuit component (Figure 20.4 on page 523). Multiple reflections inside the DUT can obscure the correlation, but the true impedance profile for the DUT can be computed in IConnect TDR software.

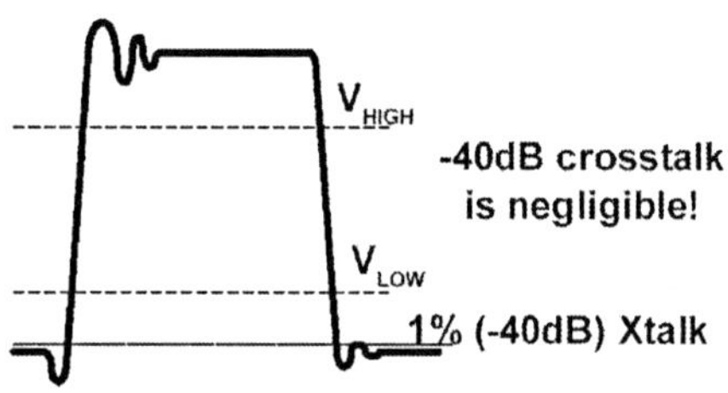

Figure 20.28 Crosstalk of –40 dB translates into 1% crosstalk in time domain. [Reprinted with permission of TDA Systems.]

For example, you can easily determine if a via in the signal path is capacitive or inductive by simply performing a TDR measurement and observing whether the TDR waveform goes up or down over the via region. The transmission lines surrounding the via can be easily windowed out and only the portion of the waveform related to the via can be analyzed. VNA measurements, on the other hand, are steady-state measurements. The same via, surrounded by two transmission lines, can not be easily isolated from the transmission lines with a VNA instrument in frequency domain. If the test structure allows the designer to isolate the via or de-embed the transmission lines, via inductance or capacitance can be obtained very accurately. If such de-embedding is possible, however, the via inductance can be obtained just as accurately with TDR using the technique for package characterization discussed in [Jed95], [Tda01].

For more complex interconnect structure characterization, the ability of a TDR instrument coupled with IConnect TDR software to obtain a SPICE or IBIS physical model for the interconnect exceeds that of VNA and VNA-based modeling tools. A physical model directly correlates its inductances, capacitances, and transmission lines to the interconnect geometrical features. Such a physical model provides valuable insight into which components — a package, board trace, or via — of the interconnect system may be causing the most signal integrity problems, and therefore, must be redesigned (see Figure 20.29). VNA-based modeling tools excel in creating a single element or a multi-element behavioral model — that is, a model that will reproduce the S-parameter behavior over the specified frequency range. This behavioral model can be quite accurate, but it does not provide a designer with an insight into which portions of the intercon-

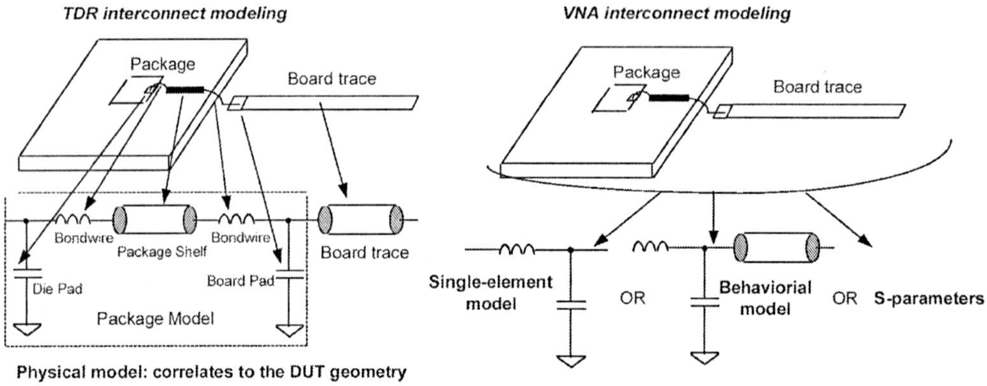

Figure 20.29 TDR models correlate well with the physical layout of the circuit. In contrast, VNA models are either single-element, behavioral, or S-parameter models. [Reprinted with permission of TDA Systems.]

nect (connectors? PCB vias? packages?) are the main causes of signal integrity problems and which must be replaced with an electrically better-performing component or material. In addition, for long transmission line structures, such behavioral models can be quite large and cumbersome to utilize.

TDR instruments have lower dynamic range than VNA instruments. This is a result of both narrowband noise filtering in the frequency domain and advanced frequency domain calibration techniques in VNAs. Typical TDR dynamic range is in the range of 25 dB to 40 dB whereas VNAs can provide 100 dB or even more of dynamic range. However, as noted above, high-speed interconnects do not normally require dynamic range higher than 40 dB, and the high dynamic range available from VNAs is not utilized on interconnect measurements. In addition, a TDR user can use digital averaging in a TDR oscilloscope, which has the same effect on the dynamic range as narrowband filtering in the VNA. As a matter of fact, every time you increase the averaging in the oscilloscope two times, you increase the dynamic range by 3 dB. Therefore, by using 128 averages instead of 1 (i.e., instead of no averages), you can increase dynamic range of the scope by 21 dB, easily reaching the 40 dB dynamic range, which is more than sufficient for accurate interconnect measurements. Without the frequency domain calibration, the accuracy of any frequency domain data is limited by the frequency at which the measurement fixturing or probing becomes non-ideal. In fact, uncalibrated S-parameter measurements are a valid way of qualitatively determining the maximum frequency to which your fixturing setup is accurate. If a VNA has a time domain option, it can be used much the same way as a regular TDR oscilloscope can be used. However, the same advantages that VNA instruments have — high dynamic range and accuracy, resulting from advanced calibration procedures — result in a more lengthy measurement process and higher learning curves for digital design engineers as compared to those of TDR oscilloscopes.

For a typical digital design engineer who has no microwave design background, dealing with frequency domain S-parameters and VNAs is a daunting task. Creating a SPICE or IBIS model for high-speed digital interconnects from frequency domain S-parameters is certainly not an intuitive process. The comfort level of this digital designer when dealing with differential and mixed-mode S-parameters may be even lower, whereas differential TDR measurements can be sufficiently intuitive for this designer because of past experience with differential signaling in digital designs. The designer needs SPICE or IBIS models in order to simulate signal integrity in the digital system interconnects, and a VNA-based solution may be unnecessarily complex and difficult to understand." [Tds02c]

20.4.6.1 Frequency Domain Measurements Using TDR

"As the digital signal speed increases dramatically into the gigahertz range, interconnect frequency-dependent losses become an important part of the interconnect model. Such losses result not only from signal attenuation and rise time degradation, but also, together with crosstalk, result in jitter and eye-diagram degradation. Frequency domain S-parameters are traditionally used to characterized losses, and such data can be very precise in describing the performance of the interconnect systems over the desired frequency range. However, S-parameter data can be difficult and sometimes impossible to use with a SPICE or IBIS simulator directly. S-parameters need to be converted into a SPICE or IBIS model, which, as noted above, typically produces a behavioral model that normally does not correlate to the physical layout of the interconnect and can be very complex for an electrically long, lossy interconnect transmission line.

IConnect TDR software offers a method to extract lossy line parameters directly from TDR and TDT measurements [Bog01], as well as, based on this lossy line data, to predict the eye diagram degradation in the interconnect. Since TDR is a more visual and easy-to-use instrument for a typical digital designer, this TDR-based loss extraction methodology may be more intuitive for the same designer as well. The lossy line models that IConnect extracts are designed to be used directly with off-the-shelf SPICE and IBIS simulators with lossy line simulation capabilities, thereby providing a very easy and intuitive way for a digital designer to include the transmission line loss in the digital system simulations. Additionally, the software can compute single-ended, differential, and mixed-mode S-parameters from TDR and TDT data acquired with the TDR oscilloscope. This is similar to many VNAs offering computational options to obtain TDR waveforms. Even though the dynamic range of this measurement is lower than the dynamic range of a VNA, it is more than sufficient to analyze losses and resonances that may affect the digital signal integrity.

Digital signal processing issues have to be taken into account when converting the TDR data into S-parameters, and VNA data into the time domain, beyond the straightforward FFT computation. First, appropriate windowing techniques must be applied in order to obtain correct representation of the spectrum of the TDR signals, as well as to go from VNA S-parameters to time domain data. Secondly, you have to keep in mind that there is a direct correlation between the frequency range f_{BW} of the data and its time step Δt, as well as between the frequency step Δf and the time range T:

$$f_{BW} = \frac{1}{2\Delta t} \qquad T = \frac{1}{\Delta f} \tag{20.12}$$

This can be easily remedied by effective and accurate interpolation techniques. Finally, because the TDR oscilloscope uses a step-like waveform, its available incident power decreases as 1/f with frequency, resulting in lower dynamic range at higher frequencies. On the other hand, VNAs do not measure the dc value, but rather extrapolate the data to obtain the value. This may lead to some errors when measuring a broadband structure such as the digital interconnect.

As an example, for a typical time domain window of 5 ns, with 10 ps time step and 500 points, the computed S-parameter data will have 50 GHz bandwidth and a 200 Mhz step. Clearly, for a 20 GHz TDR oscilloscope, the usable bandwidth is limited by that of the oscilloscope or the step generator, whereas the step of 200 Mhz may or may not be sufficiently fine, and interpolation of the TDR data may be necessary. A similar effect occurs when taking the VNA data into time domain. When measuring a 20 GHz frequency sweep with 500 points, the resulting frequency step is 40 Mhz, the time window will be 25 ns, and the time step 500 ps. This may not provide sufficient time resolution either, and interpolation of the VNA data may be necessary.

All these constraints lead to the fact that a 20 GHz TDR oscilloscope with 35 ps to 40 ps reflected rise time really could not provide 20 GHz S-parameters. A more reasonable range for S-parameters computed from TDR measurements is 4 GHz to 6 GHz without any calibration, or 8 GHz to 12 GHz with normalization, SOLT, or other calibration. Conversely, a 20 GHz network analyzer produces a much slower rise time than a 20 GHz TDR oscilloscope." [Tda02c]

20.4.7 VNAs with Time Domain Capability

Some VNAs have time domain capability, converting the data from frequency into time domain. Recently, differential test sets for 6 GHz and 20 GHz network analyzers have been developed. Such a time domain option in a VNA can be used in lieu of a TDR. However, a TDR instrument is still about half the price of a network analyzer with a comparable frequency range and rise time. As discussed above, these differential VNAs do provide higher dynamic range, but this improvement in dynamic range is not necessary and is not utilized by the digital interconnect characterization. The difficulty in using differential VNAs may outweigh the accuracy enhancements. Because of their intuitive nature and ease of use, TDR oscilloscopes continue to find wider application in differential interconnect characterization. In addition, IConnect TDR software is capable of producing differential and mixed-mode S-parameters from the TDR data, allowing the users to view the frequency domain response, and insertion and return losses.

20.4.8 Comparison, Summary, and Recommendations

"Because of its ease of use and intuitive nature, TDR is better suited for wideband and medium dynamic range structures such as digital interconnects. For these reasons, TDR remains the tool of choice for high-speed digital designers. With the ability of IConnect TDR software to produce 2-port and differential S-parameter results from the TDR oscilloscope measurements, a digital designer can obtain a complete characterization picture for interconnects. This includes the important extraction of lossy line SPICE and IBIS modeling information.

On the other hand, VNAs are definitely the preferred choice for analog and narrowband resonant system measurement. High-speed interconnects, however, are not strongly resonant, and the significantly higher dynamic range offered by VNAs is not put to good use on the interconnect measurements.

With certain TDR step generator accessories (see [And03]), faster risetime steps can be realized. These accessories allow even higher frequencies to be characterized with TDR, and enable TDR to resolve even smaller features. This extends even further the range of applications for which these instruments are suitable. If, even with these enhancements, the rise time of the TDR oscilloscope is not sufficient for the given high-speed digital application, or if a frequency range of higher than 20 GHz is required for loss characterization, or if the ultimate dynamic range of the VNA is required for a sharp resonance in a digital package, the 50–110 GHz VNA is the instrument of choice; however, these situations are rare in high-speed digital design today.

VNAs are very useful instruments in microwave and RF system design, wireless design, and other narrowband system design. With the wireless communication revolution well under way, there will be no shortage of need for these powerful instruments. Some signal integrity engineers come from microwave design background and will always prefer a VNA. For the vast majority of other signal integrity engineers who come from the digital design world, TDR will become the clear choice for most of their signal integrity measurements.

20.5 Exercises

20-1. What is the difference between a TDR oscilloscope and a traditional digital or analog oscilloscope?

20-2. (a) What is the characteristic impedance of cables and probes that you should use when making TDR measurements? (b) Is this characteristic impedance of the cables and probes a differential impedance, common-mode impedance, even-mode impedance, odd-mode impedance, or something else?

20-3. Why shouldn't you use active probes or resistor divider probes when making TDR measurements?

20-4. How would your TDR analysis be affected if in the lab you accidently used 75 Ω coaxial test cables instead of standard 50 Ω cables?

20-5. What is a deconvolution algorithm and what does it do?

20-6. What is meant by TDR deskew capability?

20-7. A single-ended TDR system has a 5 V internal signal source that transmits its TDR pulse into a 50 Ω cable, which in turn is connected to a DUT. The TDR source impedance is also 50 Ω. The DUT has sections with impedances ($Z_{section}$) of 60, 30, 75, and 100 Ω before ending in an open circuit. If the length of each section is 2 inches and the DUT is a circuit board with a propagation delay of 140 ps per inch, draw the impedance profile of what you'd expect to see on a TDR screen. In your drawing, also show the impedance profile of the 1 foot cable, which has a propagation delay of 100 ps per inch.

20-8. Is the characteristic impedance Z_0 of a transmission line equal to the even-mode, odd-mode, common-mode, differential-mode, or self-impedance of the transmission line?

20-9. In what situations might you want to use a VNA instead of a TDR system to verify the model of an interconnect?

20-10. Why do you need to determine the differential, common-mode, even-mode, and odd-mode impedances and where are they used?

20-11. Given the values for a TDR test setup of $Z_{DUT} = 67\ \Omega$, $Z_0 = 50\ \Omega$, and an internal TDR source voltage of 4 V, calculate the values of the reflection coefficient ρ, transmission coefficient τ, the incident voltage $V_{incident}$, the reflected voltage $V_{reflected}$, and the measured voltage $V_{measured}$. Determine $V_{measured}$ at a point where the front edge of the TDR pulse has traveled to the DUT and returned but where the TDR pulse is still being transmitted.

20-12. In Figure 20.4 on page 523, why do you suppose there aren't any examples that include lumped parasitic resistances?

20-13. Assume your TDR instrument transmits a TDR pulse capable of reaching the DUT with a rise time of 100 ps. Under what circumstances might you want to use a TDR that has an even faster rise time?

20-14. Refer to Figure 20.7 on page 525. What value for the termination impedance $Z_{termination}$ would you want to use when performing actual TDR testing?

20-15. Name two capabilities of IConnect software that a TDR system normally does not have by itself.

20-16. What are wildcard configurations?

20-17. What characteristic or feature of the TDR plots in Figure 20.18 on page 537 indicates that the velocity of the signal is slowed as the backplane loading increases?

20-18. In designing a BLVDS backplane, what beneficial effect is there to using a higher unloaded backplane impedance such as 130 Ω to that of using a lower unloaded backplane impedance such as 100 Ω?

20-19. In Figure 20.19 on page 538, the TDR plot for the three different stub lengths shows the values of ZEFF – min getting smaller and smaller as the stub length increases while ZEFF – max does not seem to change by very much. Also, as the plots settle down, they don't seem to differ in impedance hardly at all; at 80 ns out on the time scale, there's virtually no difference in the impedance values as a function of the stub length. How do you explain this?

CHAPTER 21

Modeling with IBIS

Chapter Objectives

- Explain how modeling devices using IBIS protects vendor proprietary information about the modeled circuit, in contrast to SPICE models.
- Discuss why IBIS models tend to run 25 times faster than SPICE models.
- Provide a behavioral diagram of IBIS and provide input and output IBIS structure models.
- Discuss the steps involved in generating an IBIS model.
- Discuss IBIS model validation on a simulator.
- Indicate the limitations of IBIS on its ability to model multigigabit FPGA I/Os.
- Provide an example of IBIS code for an SPI Serial EEPROM model.

21.1 An Introduction to IBIS (I/O Buffer Information Specification) Modeling[1]

21.1.1 Introduction

With time to market becoming shorter and shorter, system designers are struggling to release a product from concept to reality in a tightly budgeted time. The need to simulate before prototyping is very essential and the ability to simulate and simulate accurately has heightened even more. But in order to simulate a system-level board, all components on the board need to be modeled. Unfortunately many device models are not readily available from vendors.

1. This chapter is reprinted with permission of National Semiconductor Corporation. [Huq99]

IBIS (I/O Buffer Information Specification) is a behavioral modeling specification that is gaining worldwide popularity as a standard format to generate device models. IBIS solves many of the problems that prevented system designers from obtaining semiconductor vendors' SPICE models. The following sections discuss various aspects of IBIS including its history, advantages, compatibility, model generation flow, data requirements in modeling the input/output structures, and future trends.

21.1.2 About IBIS

IBIS is a fast and accurate behavioral method of modeling input/output buffers based on I-V (current/voltage) table data derived from measurement or full circuit simulation. It uses a standardized software parsable format in the form of an ASCII file to store the behavioral information needed to model device characteristics of integrated circuits. IBIS can be used by almost any simulation/EDA[2] tool in the industry. A wide range of industry leaders support the IBIS Open Forum.

21.1.3 History of IBIS

The originator of IBIS was Intel Corporation. At present the standard is being driven by the IBIS Forum, with over 28 members consisting of EDA vendors, computer manufacturers, semiconductor vendors, and universities.

IBIS v1.0 was released in April 1993. IBIS v1.0 was capable of modeling standard TTL or CMOS types of I/O structures. In June 1993, IBIS v1.1 was released with the major changes being the addition of more comments to the original specification. IBIS v2.0 was ratified in June 1994 and was a considerable improvement over IBIS v1.1. It added features such as multiple rail support (examples are the V+ and V– supplies for RS-232), ECL, terminator models, open drain, open collector, differential I/O, controlled slew rate, and definitions of complex package parameters, to name a few. IBIS v2.1 added more comments to clarify v2.0 and was ratified December 1994.

IBIS v3.0 was ratified in 1997 and had various advanced features. Some of the added features were driver selection, diode stored charge, package model extension, electrical board description, multistage drivers, series elements, and more. IBIS v3.1 and v3.2 clarified issues with v3.0.

Today IBIS is an approved standard within the EIA (Electronic Industry Alliance) and is also known under ANSI/EIA-656-A. The current version is IBIS v4.0, which was ratified on July 19, 2002. Version 4.1 was in review at the time this book was written.

> The latest version of the IBIS specification under development is available from the IBIS ANSI/EIA-656-A homepage Web site at *www.eigroup.org/ibis*. The official version (v3.2) is available from IHS Global as EIA-656, Revision A, or as IEC62014-1, for a modest fee at *http://global.ihs.com/*.

2. EDA is Electronic Design Automation.

21.1.4 Golden Parser

The software parser (known as the "Golden Parser") validates the IBIS model file. The Golden Parser checks the syntax of the IBIS model file to confirm that the data format meets the IBIS specification. The object code of the parser is available for free from the forum. Simulator vendors may also purchase the source code for a fee.

IBIS is backwards compatible. All models created today using the present version of the specification are guaranteed to work with future versions of IBIS. The IBIS forum is continually defining new and improved ways of modeling complex and unique I/O structures.

21.1.5 Advantages of IBIS

The IBIS model file protects proprietary information about the modeled circuit as no process or circuit design information is disclosed. A SPICE model, on the other hand, can disclose substantial information that semiconductor vendors consider to be confidential such as circuit nodal connections and process parameters. IBIS models are accurate, as nonlinear aspects of I/O structures as well as package parasitic and ESD structures are considered in the model parameters. Since IBIS is behavioral, the simulation time for an IBIS model can run up to 25 times faster than a structural model (SPICE). IBIS does not have non-convergence issues like SPICE and can practically run on any industry-wide platform as most EDA vendors support the IBIS specification. One of the most popular uses of IBIS is for signal integrity analysis on system boards. The models are very easy to create as they can be made from bench measurements or from simulation data.

Figure 21.1 shows a behavioral block diagram of IBIS and the components needed to create an input and an output model.

21.1.6 Input Structure Model

Information needed to model the input structure is shown in Figure 21.2. C_pkg, R_pkg, and L_pkg are the package parameters. Power_Clamp and GND_Clamp define the ESD (electrostatic discharge) structures on the inputs. The I-V table data defines these clamp structures. C_comp is the input capacitance of the die (pad) to ground.

21.1.7 Output Structure Model

Information needed to model the output structure is shown in Figure 21.3. Pull-up defines VOH/IOH (voltage output high/current output high). Pull-down defines the VOL/IOL (voltage output low/current output low) and ramp rate defines the dV/dT (derivative of voltage with respect to time) of the rising and falling waveforms.

The pull-up and pull-down data are created from the I-V tables. The remaining parameters are similar to the input structure except that they define the package parasitics of the output pin as well as the output capacitance of the output pin.

The pull-up and power_clamp data are "V_{CC} relative," meaning that the voltage values are referenced to V_{CC} and not to ground. The voltages in the tables are derived from the equation:

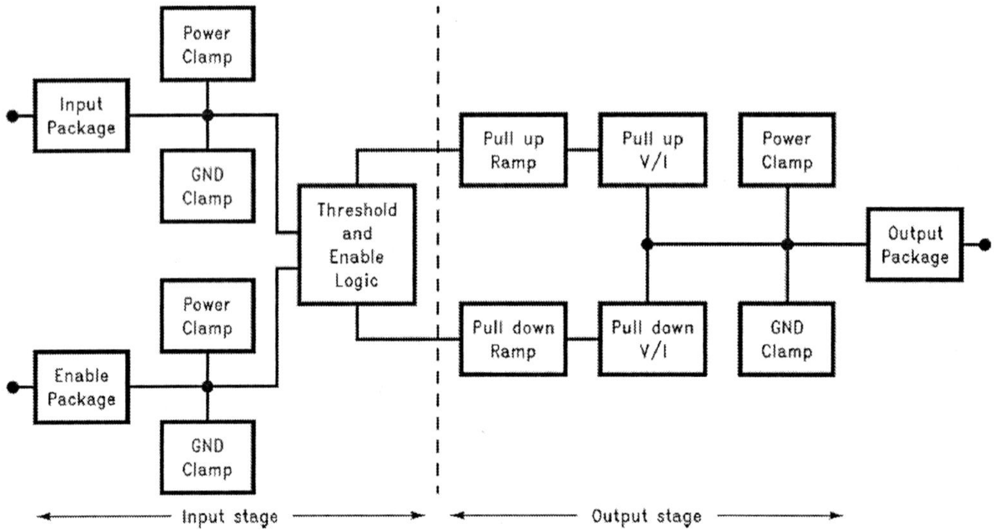

Figure 21.1 Behavioral diagram of IBIS.

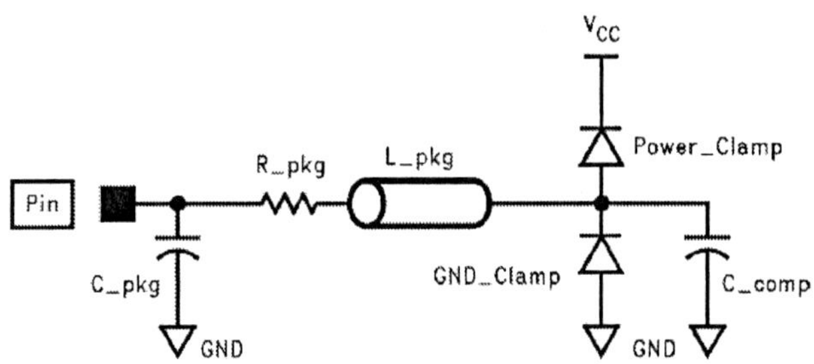

Figure 21.2 Input/enable structure model.

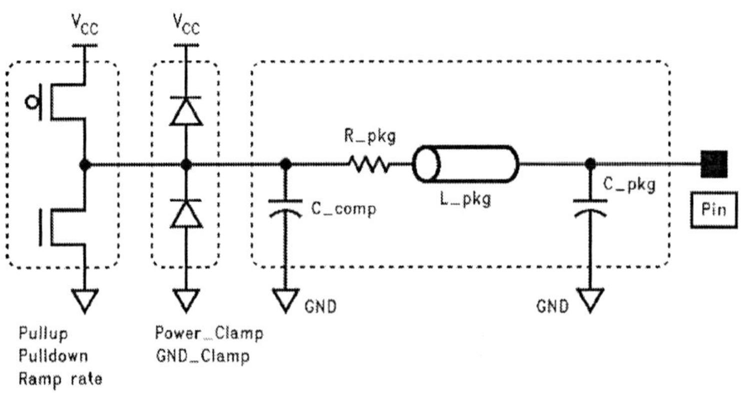

Figure 21.3 Output structure model.

$V_{table} = V_{CC} - V_{output}$. V_{CC} relative data is necessary for the simulator as the pull-up structure depends on the voltage between the output and V_{CC} and not the voltage between the output and ground pin.

An interconnect engineer can create a slow and a fast model using IBIS. The slow model is useful to determine flight time[3] and the fast model is useful to analyze overshoot, undershoot, crosstalk, and the like. By combining minimum IOH/IOL with maximum ramp time and maximum package parameters, a slow model is generated. To create a fast model, the maximum IOH/IOL, minimum ramp time, and minimum package information is used.

21.1.8 The IBIS Model Generation Process

The following steps are taken to create an IBIS model and are shown in Figure 21.4:

Step 1. All necessary I-V and other parameters are either measured on the bench, obtained from simulation, or provided by the semiconductor vendor.

Measurements of package parameters can be made through TDR (time domain reflectometry) techniques if they are not available from the semiconductor vendor. I-V data can be collected using a curve tracer or programmable supply with sinking and sourcing capabilities. Clamp curves can be generated by putting the device in TRI-STATE® and sweeping the I/O. For non TRI-STATE devices, the I-V tables are the summation of the clamp and pull-up/pull-down.

The required data is:

- R_pkg, L_pkg, C_pkg, and C_comp for all inputs, outputs, and enables.
- Power_Clamp and GND_Clamp (ESD structures if present) for I/O.
- Pull-up and pull-down (only if present in the model), and ramp rates for outputs.

Step 2. The interconnect engineer creates the IBIS ASCII file following the format defined in the IBIS standard. This can be done on a UNIX or DOS text editor.

Step 3. The ASCII model file is checked by the Golden Parser for possible syntax errors.

Step 4. If passed by the Golden Parser, the model is imported into a simulator and validated for accuracy. The model is now ready for use.

21.1.9 IBIS Model Validation on a Simulator

Model validation is the final and critical step to IBIS modeling. Visual inspection, s2iplt[4] checks, parser test, simulation of the model under known loads, comparison of IBIS simulation

3. Flight time is total propagation time down an interconnect.
4. There are two publicly available utilities for viewing the tables of IBIS models. North Carolina State University developed a Perl and GNUPLOT-based s2iplt utility suitable for UNIX-based applications. HyperLynx offers a second utility, winibis, at no charge, which is their Visual IBIS Editor for viewing, checking, and editing IBIS files. Many other companies provide their own IBIS development, viewing, and translation tools and products.

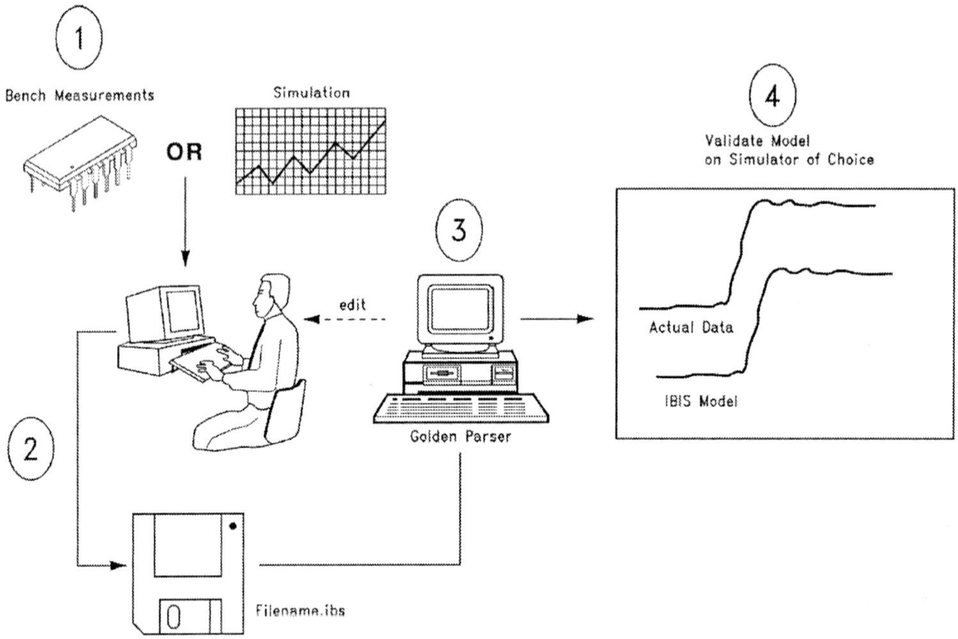

Figure 21.4 The IBIS model generation process.

data to bench observations or SPICE simulation data, edge rate, and signal swing-level verification are also done during this phase.

IBIS models are primarily used for SI (signal integrity) analysis. Analysis items are typically crosstalk, ringing, overshoot, undershoot, mismatched impedance, reflections, line termination analysis, topology scheme analysis, design rule generation, and multi-board simulation. IBIS is best suited for SI analysis on system boards and requires very little run time (IBIS is 25 times faster than SPICE).

21.1.10 Using SPICE to Model Multigigabit FPGA I/Os and Pre-Emphasis

Although IBIS models are sufficient for simulation of single-ended and low-speed differential I/O, the Xilinx RocketIO™ Multi-Gigabit Transceiver (MGT) and other FPGA vendor high-speed I/Os operate at rates that are too fast for existing IBIS simulators to process. Extended or advanced features of IBIS planned for the future are needed for such advanced buffers. For this reason, IBIS models for the RocketIO MGT are not made available by Xilinx, which instead provides SPICE models. Xilinx Virtex-II Pro SelectI/O-Ultra IBIS models are available from the Xilinx IBIS download page.

IBIS has limited capability to simulate transceivers and drivers that use pre-emphasis. However, SPICE is usually used to simulate these types of devices for full flexibility.

21.1.11 Future Trends of IBIS

IBIS continues to be referenced by many other standards and the IBIS Committee continues to work with other groups. Most integrated circuit vendors generate IBIS models today. IBIS modeling is accurate, easy to create, and compatible on a wide rage of simulation platforms. It solves the "no models available" problem in the industry. IBIS has created a common industry format for modeling using behavioral information and does not disclose any proprietary process parameters or circuit design details.

21.2 Example of an IBIS Model

Table 21.1 below shows the computer software code for a small IBIS model. This model was created by Teraspeed Consulting and is for an Atmel AT25640N SPI (serial peripheral interface) serial EEPROM in an 8-pin SOIC package. For this chip, the code is 356 lines long including comments. For larger or more dense chips, the code is much longer, for example, 2,457 lines for a 16-bit Bus LVDS serializer/deserializer chip. This model example was selected for its small size and does not exercise many of the IBIS features, but does give a good example of an input and an open-sink output. To use IBIS models such as this one, the typical procedure would be to download the IBIS model (IBIS code) from the chip manufacturer's Web site, save it to your local hard drive, and then use it as one of the many input files to your simulation tool. This model was developed from bench measurements of the device.

Table 21.1 Software Code for the IBIS Model of an Atmel AT25640N SPI Serial EEPROM in an 8-pin SOIC Package [Courtesy of Teraspeed Consulting.]

```
|***************************************************************
| COPYRIGHT(C) TERASPEED CONSULTING GROUP LLC
| 2003 ALL RIGHTS RESERVED. THIS WORK CONTAINS
| PROPRIETARY INFORMATION WHICH IS THE PROPERTY
| OF TERASPEED CONSULTING GROUP LLC OR ITS
| LICENSORS. USE OF THESE MATERIALS IS SUBJECT TO
| THE TERMS BELOW.
|
| USE: YOU MAY USE IBIS MODELS SOLELY FOR DEVELOPING
| PRINTED CIRCUIT BOARD DESIGNS. NO OTHER USE OR
| DISTRIBUTION OF THESE MATERIALS IS PERMITTED.
|
| DISCLAIMER OF WARRANTY:  UNLESS OTHERWISE AGREED
| IN WRITING, TERASPEED SOFTWARE AND ASSOCIATED
| FILES ARE PROVIDED "AS IS" AND WITHOUT WARRANTY.
| TERASPEED HAS NO OBLIGATION TO SUPPORT OR
| OTHERWISE MAINTAIN SOFTWARE.  TERASPEED MAKES
| NO WARRANTIES, EXPRESS OR IMPLIED WITH RESPECT TO
| SOFTWARE INCLUDING ANY WARRANTY OF MERCHANTABILITY
| OR FITNESS FOR A PARTICULAR PURPOSE.
|
```

```
| LIMITATION OF LIABILITY:  TERASPEED IS NOT LIABLE FOR
| ANY PROPERTY DAMAGE, PERSONAL INJURY, LOSS OF PROFITS,
| INTERRUPTION OF BUSINESS, OR FOR ANY OTHER SPECIAL,
| CONSEQUENTIAL OR INCIDENTAL DAMAGES, HOWEVER CAUSED,
| WHETHER FOR BREACH OF WARRANTY, CONTRACT, TORT
| (INCLUDING NEGLIGENCE), STRICT LIABILITY OR OTHERWISE.
| IN NO EVENT SHALL TERASPEED'S LIABILITY EXCEED THE
| AMOUNT PAID FOR THE PRODUCT GIVING RISE TO THE CLAIM.
|
| TRADEMARKS:  THE FOLLOWING ARE REGISTERED TRADEMARKS
| OR SERVICE MARKS OF TERASPEED CONSULTING GROUP LLC:
| THE FOLLOWING ARE TRADEMARKS OR SERVICE MARKS OF
| TERASPEED CONSULTING GROUP LLC:
| ALL OTHER BRANDS AND PRODUCT NAMES ARE TRADEMARKS OF
| THEIR RESPECTIVE OWNERS.
|
| COPIES:  ALL COPIES OF SOFTWARE, INCLUDING BUT NOT LIMITED
| TO ANY MODIFICATIONS, ENHANCEMENTS OR DERIVATIVE WORKS
| MUST CONTAIN ALL TERASPEED'S NOTICES AND LEGENDS,
| INCLUDING THOSE EMBEDDED IN SOFTWARE AND THOSE AFFIXED
| TO THE SOFTWARE MEDIUM AND CONTAINER, AS RECEIVED FROM
| TERASPEED.
|
| ****************************************************************
|
[IBIS Ver] 3.2
[File Name] at25640n.ibs
[File Rev] 3.0
[Source] Measured and created by Teraspeed Consulting Group
[Date]   03-24-2003
[Notes] Quality Check of this IBIS model: Model Passes IBIS Check.
Model
        parses into ICX IS.
        1. MIN TYP MAX information is based on Temprature and Voltage
           characteristics only.
|
[Copyright]
| ****************************************************************
| COPYRIGHT(C) TERASPEED CONSULTING GROUP LLC 2003
| ALL RIGHTS RESERVED. THIS WORK CONTAINS
| ROPRIETARY INFORMATION WHICH IS THE PROPERTY OF
| TERASPEED CONSULTING GROUP LLC OR ITS LICENSORS.
| USE OF THESE MATERIALS IS SUBJECT TO THE TERMS
| BELOW.
|
| USE: YOU MAY USE IBIS MODELS SOLELY FOR DEVELOPING
| PRINTED CIRCUIT BOARD DESIGNS. NO OTHER USE OR
| DISTRIBUTION OF THESE MATERIALS IS PERMITTED.
| ****************************************************************
|  Teraspeed Spectrum Model Builder
```

```
|
[Component] AT25640N
[Manufacturer] Atmel
|
[Package]
| variable    typ    min    max
R_pkg   0.188ohms 0.188ohms 0.188ohms
L_pkg   2.383nH   2.383nH   2.383nH
C_pkg   0.36pF    0.36pF    0.36pF
|
[Pin] signal_name  model_name  R_pin  L_pin C_pin
|
1    _CS       Input_1       0.154    2.924n    0.374p
2    SO        Output_1      0.15     2.504n    0.354p
3    _WP       Input_1       0.15     2.504n    0.354p
4    GND       GND           0.154    2.924n    0.374p
5    SI        Input_1       0.154    2.924n    0.374p
6    SCK       Input_1       0.15     2.504n    0.354p
7    _HOLD     Input_1       0.15     2.504n    0.354p
8    VCC       POWER         0.154    2.924n    0.374p
|
|
[Model] Output_1
Model_type Open_sink
|
Vref=3.1000V
Rref=500.0ohms
Cref=2.00000pF
Vmeas=1.65000V
C_comp 2.00000pF 2.00000pF 2.00000pF
|
| variable typ    min     max
[Voltage Range]          3.10000V  3.06300V  3.19900V
[POWER Clamp Reference]  3.10000V  3.06300V  3.19900V
[Pullup Reference]       3.10000V  3.06300V  3.19900V
[Pulldown Reference]     0.00000V  0.00000V  0.00000V
[GND Clamp Reference]    0.00000V  0.00000V  0.00000V
|
[GND Clamp]
|
-3.19900V    -348.921mA   -235.666mA   -232.330mA
-3.10000V    -334.438mA   -225.765mA   -222.450mA
-3.06300V    -329.025mA   -222.065mA   -218.758mA
-1.11860V    -44.5741mA   -27.6137mA   -24.7148mA
-1.04260V    -33.4558mA   -13.8229mA   -17.1304mA
-982.230mV   -24.6242mA   -9.91783mA   -11.1057mA
-982.160mV   -24.6160mA   -9.91330mA   -11.0987mA
-941.760mV   -19.9152mA   -1.97884mA   -7.06694mA
-860.790mV   -10.4936mA   -958.030uA   -1.10195mA
```

```
-860.780mV   -10.4929mA   -957.903uA   -1.10121mA
-800.080mV   -5.78021mA   -192.640uA   -635.285uA
-784.800mV   -5.03724mA   0.00000A     -517.997uA
-739.470mV   -2.83315mA   0.00000A     -170.047uA
-717.317mV   -2.29356mA   0.00000A     0.00000A
-678.890mV   -1.35760mA   0.00000A     0.00000A
-582.400mV   -558.979uA   0.00000A     0.00000A
-291.200mV   0.00000A     0.00000A     0.00000A
0.00000V     0.00000A     0.00000A     0.00000A
6.12600V     0.00000A     0.00000A     0.00000A
6.20000V     0.00000A     0.00000A     0.00000A
6.39800V     0.00000A     0.00000A     0.00000A
|
[Pulldown]
|
-3.19900V   -52.2227mA   -59.2071mA   -77.2635mA
-3.10000V   -50.6074mA   -57.3753mA   -74.8731mA
-3.06300V   -50.0037mA   -56.6907mA   -73.9798mA
0.00000V    -26.3064uA   -14.2426uA   -23.1022uA
20.6740mV    311.020uA    368.300uA    476.075uA
20.7620mV    312.456uA    369.199uA    478.200uA
101.640mV    1.63210mA    1.19547mA    1.71672mA
182.680mV    2.68406mA    2.02340mA    2.95773mA
279.090mV    3.93554mA    2.90671mA    4.43410mA
360.140mV    4.98763mA    3.64930mA    5.53070mA
360.230mV    4.98880mA    3.65003mA    5.53192mA
521.730mV    6.83335mA    4.96722mA    7.71700mA
582.410mV    7.52640mA    5.46213mA    8.41803mA
602.760mV    7.71786mA    5.62810mA    8.65313mA
643.310mV    8.09937mA    5.92648mA    9.12160mA
780.160mV    9.38691mA    6.93347mA    10.5830mA
800.120mV    9.57470mA    7.08035mA    10.7658mA
820.390mV    9.73071mA    7.22950mA    10.9515mA
941.750mV    10.6647mA    7.95252mA    12.0630mA
1.02250V    11.2862mA    8.43360mA    12.6922mA
1.02260V    11.2870mA    8.43408mA    12.6930mA
1.17890V    12.2009mA    9.18260mA    13.9110mA
1.19910V    12.3190mA    9.25594mA    14.0382mA
1.21900V    12.4141mA    9.32820mA    14.1635mA
1.25990V    12.6095mA    9.32820mA    14.4210mA
1.33060V    12.9473mA    9.32820mA    14.7527mA
1.34120V    12.9980mA    9.40521mA    14.8024mA
1.44220V    13.3956mA    10.1390mA    15.2762mA
1.48280V    13.5554mA    10.2160mA    15.4667mA
1.50290V    13.6345mA    10.3640mA    15.5610mA
1.52310V    13.7140mA    10.4368mA    15.6250mA
1.55810V    13.8157mA    10.5630mA    15.7359mA
1.65910V    14.1090mA    10.8107mA    16.0560mA
1.69920V    14.1800mA    10.9090mA    16.1359mA
1.71990V    14.2166mA    11.1290mA    16.1771mA
```

```
1.76020V    14.2879mA   11.2020mA   16.2574mA
1.82070V    14.3950mA   11.2826mA   16.3780mA
1.84080V    14.4212mA   11.3093mA   16.4180mA
1.98290V    14.6065mA   11.4986mA   16.6380mA
1.99860V    14.6270mA   11.5195mA   16.6408mA
2.01850V    14.6270mA   11.5460mA   16.6443mA
2.03910V    14.6270mA   11.5460mA   16.6480mA
2.04905V    14.6270mA   11.5460mA   16.6899mA
2.05930V    14.6473mA   11.5460mA   16.7330mA
2.09950V    14.7270mA   11.5460mA   16.7595mA
2.15005V    14.7559mA   11.5460mA   16.7929mA
2.24100V    14.8079mA   11.6781mA   16.8530mA
2.26110V    14.8194mA   11.7072mA   16.9380mA
2.28160V    14.8311mA   11.7370mA   16.9380mA
2.32185V    14.8541mA   11.7370mA   16.9380mA
2.36210V    14.8771mA   11.7790mA   16.9380mA
2.37945V    14.8870mA   11.7790mA   16.9380mA
2.49780V    14.9546mA   11.7790mA   17.1080mA
2.63025V    15.0303mA   11.7790mA   17.1080mA
2.63980V    15.0358mA   11.7886mA   17.1080mA
2.78180V    15.1169mA   11.9316mA   17.2950mA
2.89840V    15.1836mA   12.0490mA   17.3494mA
2.97925V    15.2298mA   12.0490mA   17.3871mA
3.06010V    15.2760mA   12.0970mA   17.4249mA
6.12600V    17.0281mA   12.0970mA   18.8556mA
6.20000V    17.0704mA   12.0970mA   18.8902mA
6.39800V    17.1835mA   12.0970mA   18.9826mA
|
[POWER Clamp]
|
6.29900V     0.00000A    0.00000A    0.00000A
6.20000V     0.00000A    0.00000A    0.00000A
6.16300V     0.00000A    0.00000A    0.00000A
37.0000mV    0.00000A    0.00000A    0.00000A
0.00000V     0.00000A    0.00000A    0.00000A
-99.0000mV   0.00000A    0.00000A    0.00000A
-211.600mV   0.00000A    0.00000A    0.00000A
-321.000mV   0.00000A    3.20936uA   0.00000A
-428.850mV   51.8486uA   7.99016uA   5.20436uA
-460.200mV   66.9200uA   9.37985uA   6.85241uA
-500.700mV   86.3903uA   29.6578uA   8.98146uA
-561.200mV   115.476uA   567.857uA   12.1619uA
-642.000mV   154.320uA   2.47089mA   16.4095uA
-677.700mV   502.430uA   3.58417mA   177.468uA
-717.900mV   1.38753mA   5.58537mA   535.089uA
-758.700mV   3.14976mA   7.64181mA   1.77566mA
-819.500mV   5.77583mA   11.5632mA   4.03740mA
-880.100mV   9.28553mA   16.7995mA   7.05361mA
-900.400mV   10.4612mA   18.5536mA   8.06401mA
-900.500mV   10.4670mA   18.5623mA   8.06899mA
```

```
-1.00160V     18.6949mA  31.1059mA    14.2696mA
-3.02600V     183.448mA  282.276mA    138.926mA
-3.10000V     189.470mA  291.479mA    143.483mA
-3.29800V     205.584mA  316.104mA    155.675mA
|
|
|
[Falling Waveform]
R_fixture = 50.0000
V_fixture = 3.10000
V_fixture_min = 3.06300
V_fixture_max = 3.19900
0.00000s   3.09355V  3.05678V  3.19287V
3.50000ns  3.09308V  3.05745V  3.20108V
3.90000ns  3.09303V  3.05752V  3.18369V
4.40000ns  3.09296V  3.05762V  3.10206V
6.00000ns  3.09275V  3.05792V  2.72916V
6.50000ns  3.09268V  3.05802V  2.64745V
6.70000ns  3.06973V  3.05805V  2.61477V
7.20000ns  3.01237V  3.05815V  2.55227V
8.00000ns  2.86347V  3.05830V  2.48349V
8.90000ns  2.69597V  3.05847V  2.44334V
9.70000ns  2.59387V  3.05862V  2.40765V
10.0000ns  2.55559V  3.05868V  2.40179V
11.3000ns  2.46171V  3.05893V  2.37643V
11.7000ns  2.44728V  3.05900V  2.36862V
12.1000ns  2.43285V  3.05908V  2.36527V
12.5000ns  2.41841V  3.04395V  2.36192V
12.9000ns  2.40398V  3.00902V  2.35857V
13.1000ns  2.40154V  2.99155V  2.35690V
13.8000ns  2.39300V  2.90557V  2.35104V
14.7000ns  2.38203V  2.79501V  2.35110V
16.0000ns  2.36618V  2.66609V  2.35118V
17.0000ns  2.35399V  2.59847V  2.35125V
18.3000ns  2.35422V  2.54598V  2.35133V
20.1000ns  2.35455V  2.50441V  2.35145V
22.4000ns  2.35496V  2.48152V  2.35160V
29.9000ns  2.35631V  2.47405V  2.35210V
|
[Rising Waveform]
R_fixture = 50.0000
V_fixture = 3.10000
V_fixture_min = 3.06300
V_fixture_max = 3.19900
0.00000s  2.35631V 2.47405V 2.35210V
3.60000ns 2.35715V 2.47503V 2.34744V
4.70000ns 2.35741V 2.47532V 2.31062V
5.20000ns 2.35753V 2.47546V 2.33141V
5.60000ns 2.35762V 2.47557V 2.40699V
5.80000ns 2.35767V 2.47562V 2.47839V
```

```
6.20000ns 2.34937V 2.47573V 2.69375V
6.40000ns 2.34522V 2.47579V 2.82965V
6.80000ns 2.33692V 2.47589V 3.03375V
7.20000ns 2.32862V 2.47600V 3.12973V
7.60000ns 2.32032V 2.47611V 3.15275V
8.00000ns 2.32806V 2.47622V 3.17577V
8.10000ns 2.33000V 2.47625V 3.17781V
8.50000ns 2.37895V 2.47636V 3.18597V
8.90000ns 2.49321V 2.47646V 3.19413V
9.00000ns 2.53987V 2.47649V 3.19617V
9.70000ns 2.86648V 2.47668V 3.18874V
10.0000ns 2.97010V 2.47676V 3.18555V
10.4000ns 3.03492V 2.47687V 3.18130V
10.5000ns 3.04057V 2.47690V 3.18024V
11.1000ns 3.07449V 2.47706V 3.18169V
11.9000ns 3.08711V 2.47728V 3.18361V
12.1000ns 3.09026V 2.47496V 3.18409V
14.4000ns 3.09069V 2.44832V 3.18963V
15.1000ns 3.09082V 2.48003V 3.19131V
15.5000ns 3.09089V 2.52930V 3.19227V
15.8000ns 3.09095V 2.59434V 3.19300V
16.7000ns 3.09111V 2.87915V 3.19516V
17.0000ns 3.09117V 2.95909V 3.19588V
17.2000ns 3.09120V 2.99061V 3.19637V
17.7000ns 3.09130V 3.03058V 3.19757V
19.3000ns 3.09159V 3.05735V 3.20142V
29.9000ns 3.09355V 3.05678V 3.19287V
|
[Ramp]
| variable typ min max
dV/dt_r 442.344m/1.00011n 349.638m/1.13583n 504.462m/894.730p
dV/dt_f 442.344m/3.06240n 349.638m/3.65170n 504.462m/2.80389n
R_load = 50.0000Ohms
|
|
[Model] Input_1
Model_type Input
Vinl = 800.000mV
Vinh = 2.00000V
C_comp 2.00000pF 2.00000pF 2.00000pF
|
| variable typ    min    max
[Voltage Range]          3.10000V  3.06300V  3.19900V
[POWER Clamp Reference] 3.10000V  3.06300V  3.19900V
[GND Clamp Reference]    0.00000V  0.00000V  0.00000V
|
[GND Clamp]
|
  -3.10000V    -265.620mA    -251.319mA    -272.073mA
  -1.04280V    -21.9107mA    -25.3950mA    -20.7000mA
```

```
  -1.04260V    -21.8870mA    -25.3730mA    -20.6818mA
-921.480mV    -11.2894mA    -15.2042mA    -9.65510mA
-901.020mV    -9.49920mA    -13.4864mA    -8.31819mA
-880.850mV    -8.42902mA    -11.7930mA    -7.00023mA
-861.060mV    -7.37901mA    -10.6187mA    -5.70710mA
-800.300mV    -4.15522mA    -7.01346mA    -2.91890mA
-800.090mV    -4.14407mA    -7.00100mA    -2.91236mA
-780.150mV    -3.08610mA    -6.23557mA    -2.29126mA
-759.890mV    -2.55142mA    -5.45786mA    -1.66020mA
-733.416mV    -1.85274mA    -4.44161mA    -1.24643mA
-719.110mV    -1.47520mA    -4.35366mA    -1.02283mA
-679.000mV    -948.216uA    -4.10707mA    -395.940uA
-643.580mV    -482.851uA    -3.88932mA    -172.350uA
-643.290mV    -479.040uA    -3.88753mA    -171.488uA
-603.020mV    -402.009uA    -3.63996mA    -51.8460uA
-523.024mV    -248.987uA    -3.14817mA    11.6144uA
-479.163mV    -229.485uA    -2.87852mA    46.4090uA
-328.408mV    -162.454uA    -1.95171mA    46.4090uA
-272.001mV    -137.374uA    -1.60494mA    46.8037uA
-177.653mV    -64.5346uA    -1.02491mA    47.4640uA
 -20.9791mV    56.4215uA    -61.7156uA    47.4640uA
  24.3070uV    65.7602uA     67.4080uA    47.4640uA
  20.7220mV    74.9631uA     96.0070uA    47.4640uA
  91.4560mV    106.414uA     96.0070uA    47.4640uA
 121.900mV    119.950uA     106.514uA    47.4640uA
 162.190mV    119.950uA     120.420uA    47.4640uA
 234.588mV    119.950uA     120.420uA    47.4640uA
 258.720mV    119.950uA     134.120uA    47.4640uA
   6.20000V    119.950uA     134.120uA    47.4640uA
|
[POWER Clamp]
|
6.20000V    0.00000A   0.00000A   0.00000A
-3.10000V   0.00000A   0.00000A   0.00000A

|
[End]
```

21.3 Exercises

21-1. What does the acronym IBIS stand for?

21-2. From an integrated circuit vendor's point of view, what major advantage does an IBIS model offer over a SPICE model?

21-3. List and explain the four steps in the IBIS model generation process.

21-4. Assume you are in the process of creating an IBIS model for a new ASIC (application-specific integrated circuit) chip that has just been developed by your company. The chip is now in your lab and you are getting ready to measure the parameters needed to define your IBIS model. What are all the parameters you need to measure in the lab in order to create your IBIS model? Assume none of these parameters has been obtained from simulation or other sources and that a SPICE model from the ASIC design is not available.

21-5. Access Web sites of two semiconductor manufacturers. Search their sites until you find an IBIS model for one or more of their digital ICs. Open up and scan through the IBIS model and get a feel for what the model includes and how long it is. Print out and turn in to your instructor the cover page for each of the two models to show what integrated circuit the model was for. Two things that are helpful to observe are that (1) the models are provided for free by the IC vendors and (2) the more complex the IC, the longer and more complicated is the IBIS model. Why do you think the models are provided for free?

21-6. What is the Golden Parser and what does it do?

21-7. Can models developed under IBIS v2.1 be run on an IBIS v4.0-compliant simulator?

CHAPTER 22

Mentor Graphics — EDA Tools for High-Speed Design, Simulation, Verification, and Layout

Edited by Matthew Hogan of Mentor Graphics Corporation.

Chapter Objectives

- Introduce Mentor Graphics' two major software tool suites used for high-speed design and identify the individual tools that comprise the suites.
- Discuss the high-speed design flow, which tools support each phase of the flow, and what the tools do.
- Explain why ICX, Tau™, and HyperLynx tools are required for designing high-speed circuits.
- Discuss system verification and interconnect synthesis.
- Describe field solvers and how they are integrated into the Mentor Graphics tool suites.
- Present circuit timing methods and symbolic timing.
- Discuss signal integrity and timing models, IBIS, SPICE, VHDL-AMS, TDML, and STAMP.

22.1 Overview of Mentor Graphics High-Speed Tools[1]

Mentor Graphics Corporation provides electronic design automation (EDA) software tools that support engineering needs in all market segments. These markets are typically referred to as the workgroup, enterprise, and ready-to-use markets and are supported by their Expedition Series, Board Station, and PADS tool suites.

1. This chapter is reprinted with permission of Mentor Graphics Corporation.

A typical flow consists of the core design creation process (DxDesigner and Expedition PCB) being augmented by additional tools to look at the specific problems inherent in high-speed design (HyperLynx and ICX) as shown in Figure 22.1.

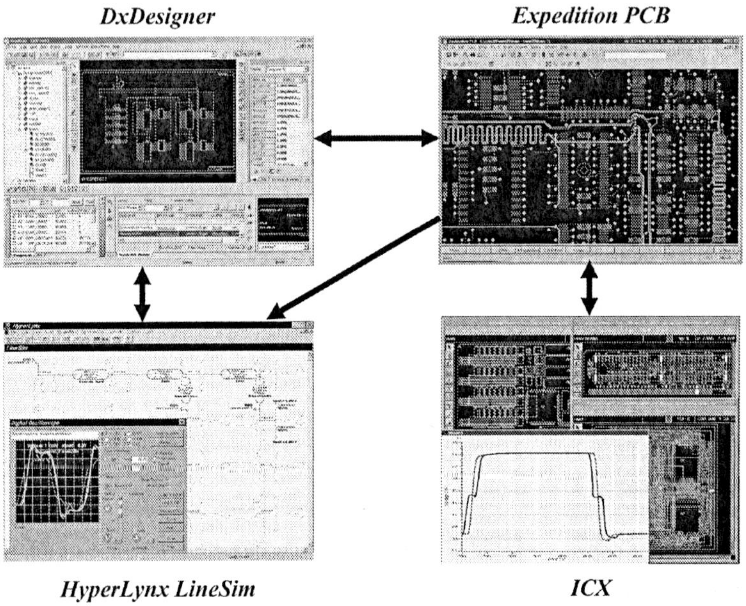

Figure 22.1 A typical integrated high-speed design flow using Mentor Graphics' tools.

Defining net connectivity for a printed circuit board (PCB) is most often done using a schematic capture tool. These tools allow customers to describe the connections between components on a PCB. DxDesigner is a powerful, but easy-to-use schematic capture tool that enables designers to be very productive in describing the design, creating design constraints, and managing device selection in the design process. During the design of a PCB, it is often advantageous to explore different design solutions and trade-offs. This is particularly true for designs that use components with fast switching speeds. Such components may pose design difficulties, and require analysis and verification using high-speed design tools. Before the PCB has been routed, signal integrity analysis on proposed design elements can be performed using HyperLynx LineSim™. This tool allows the designer to do "what-if" analysis on the design to make sure that ideas and configurations going into the design will work.

The logical design contained in the schematic can be physically realized by designing the actual physical placement of components, actual connections in copper on the PCB, and any additional elements that are going to be part of the design. These elements are put together using a PCB design layout tool, like Expedition PCB. Expedition PCB takes the netlist that was created in the schematic, and allows the board designer to place components, traces, vias, and other physical design elements in the PCB design. It is the physical elements created in Expedition PCB that will eventually go off to be manufactured. As this is an accurate representation of what will be built, it is also an important place for us to do a more rigorous signal integrity analysis. We can do this analysis with signal integrity tools, such as ICX. ICX provides an excellent platform to perform complex analysis and verification of a PCB. It is able to verify the entire design, comparing simulation results against a known set of constraints.

Ensuring that the PCB will function correctly not only requires that the net connectivity be correct, and that signal integrity analysis is performed, but also may require different types of analysis and verification, like EMC analysis (Quiet Expert), timing analysis (with Tau), or functional verification (with ModelSim), or it may even require that software be written for programmable components on the board. FPGAs (field programmable gate arrays) are a very powerful design element that are increasingly being used in PCB design. They provide exceptional design flexibility (the FGPA logic can be changed by reprogramming), but also provide another dimension to PCB design that must be managed: The large number of pins on FPGAs make it difficult for schematic capture (DxDesigner) and the board layout (Expedition PCB) tools to maintain the correct pin placement and physicals component description for the FPGA. Fortunately, there is a tool — FPGA BoardLink — that manages this data, and provides a seamless connection between the FGPA design tools, the schematic capture tool (DxDesigner) and the PCB layout tool (Expedition PCB). Without tools such as FPGA BoardLink, many days, if not weeks are spent redesigning the circuit, and re-entering the schematic to ensure that the FPGA design, the schematic, and the board design are in synchronization.

22.2 ICX

In 1996, Mentor Graphics purchased a startup company called Interconnectix, which developed a unique design solution — ICX — for high-speed board design that continues to this day. The solution that they pioneered in their product's 1994 release was the management of electrical constraints. Constraint management has been a growing necessity for the past 10 years. Not only is it necessary for designs to be simulated, but a methodical process of validating those simulation results against known acceptable limits needs to be established. This comparison is the foundation of design constraint management.

ICX is a powerful signal integrity verification tool (see Figure 22.2). ICX also includes industry-leading simulation model support to increase the accuracy and ease of verification through mixed-signal, multilingual, ADMS[2] simulation technology. Rather than going through a traditional design cycle of alternating routing and analysis, ICX's analysis is integrated in real time within the route process, delivering correct-by-construction designs (see Figure 22.3). In addition, routing is accomplished according to electrical constraints (see Figure 22.4) to improve

2. ADVance MS (ADMS) is the industry's first mixed-signal simulator that brings together digital and mixed-signal standard HDLs with SPICE in a single environment. ADMS is intended for demanding applications such as verification of mixed-signal systems-on-a-chip.

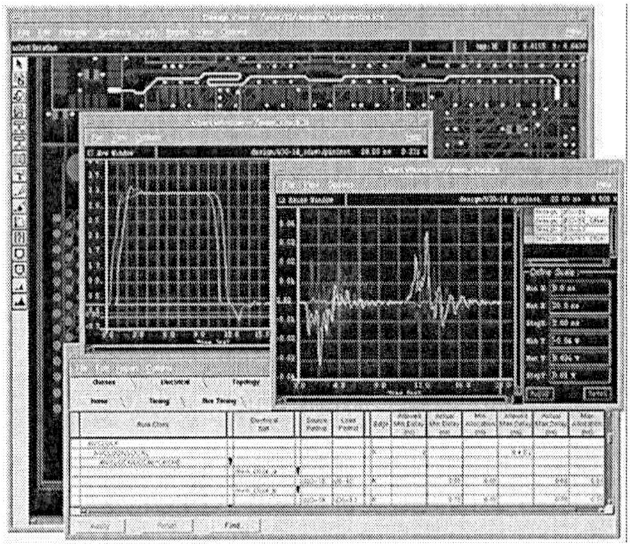

Figure 22.2 ICX is a powerful signal integrity verification tool.

design quality, rather than with rules representing converted physical constraints. By integrating physical design with electrical verification, the tools provide a single environment for rules entry, what-if design exploration, timing-driven placement, and electrically driven routing and system verification.

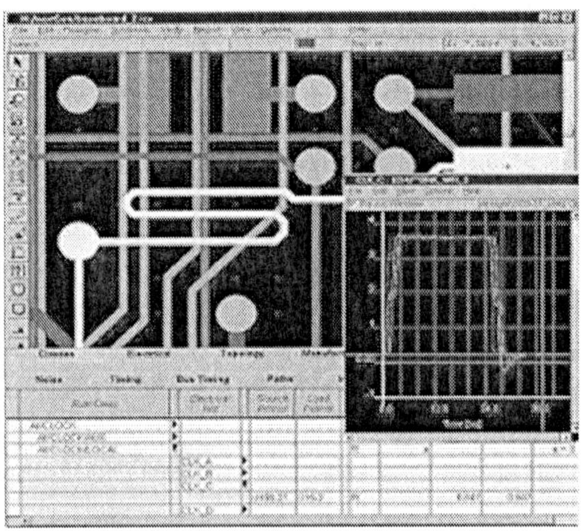

Figure 22.3 ICX's analysis is integrated in real time within the route process, delivering correct-by-construction designs.

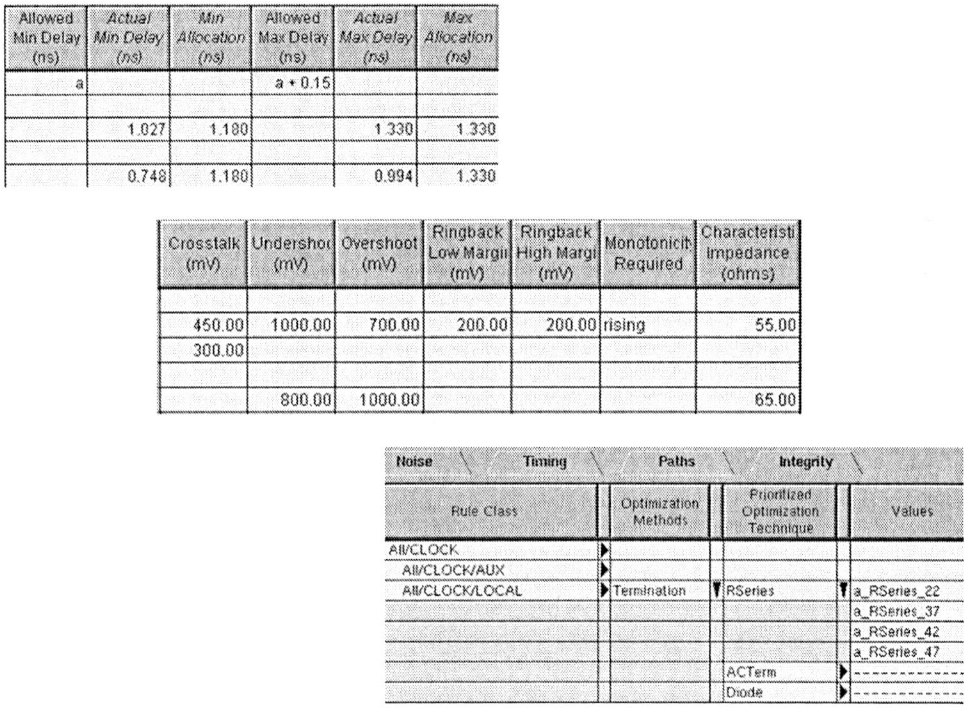

Allowed Min Delay (ns)	Actual Min Delay (ns)	Min Allocation (ns)	Allowed Max Delay (ns)	Actual Max Delay (ns)	Max Allocation (ns)
a			a + 0.15		
	1.027	1.180		1.330	1.330
	0.748	1.180		0.994	1.330

Crosstalk (mV)	Undershoo (mV)	Overshoot (mV)	Ringback Low Margi (mV)	Ringback High Margi (mV)	Monotonicit Required	Characteristi Impedance (ohms)
450.00	1000.00	700.00	200.00	200.00	rising	55.00
300.00						
	800.00	1000.00				65.00

Noise | Timing | Paths | Integrity

Rule Class	Optimization Methods	Prioritized Optimization Technique	Values
All/CLOCK			
All/CLOCK/AUX			
All/CLOCK/LOCAL	Termination	RSeries	a_RSeries_22
			a_RSeries_37
			a_RSeries_42
			a_RSeries_47
		ACTerm	------------
		Diode	------------

Figure 22.4 ICX provides a comprehensive constraint management system for both physical and electrical constraints.

22.2.1 IS_Analyzer

22.2.1.1 Hierarchical Rules Entry and Management

The IS_Analyzer™ tool leverages the power of a true hierarchical rules entry and management system. An interactive spreadsheet interface provides easy and effective management of both physical and electrical rules, which can be applied at both the class and net level in your design. Electrical requirements are specified in their native electrical terms without requiring translation into less-intuitive physical constraints. The integration of this sophisticated rules system enables true concurrent constraint management. Signal quality constraints include overshoot, undershoot, noise margin, impedance, monotonicity, and crosstalk. Delay constraints can be applied at the net, bus, and pin-to-pin levels. Path-level delays spanning multiple interconnects are also supported.

22.2.1.2 Analysis and Verification

IS_Analyzer features an integrated parameter extraction module and a multi-conductor transmission-line simulator, which works in both pre- and post-route design phases, and includes full-board and system verification. Components can be modeled using industry-standard IBIS,

SPICE, or VHDL-AMS models, yielding the most flexible simulation environment on the market. Electronic Board Description (EBD), coupled connector, and device package models are also supported. Corner cases are defined in order to evaluate best- and worst-case scenarios. Accurate delay results are produced with simulation, including push-out effects of crosstalk. Frequency-dependent losses, such as dielectric loss and skin effects, are also modeled. The simulation resolution is controllable at the class and net level to within one picosecond.

22.2.1.3 System-Level Definition and Analysis

IS_Analyzer allows quick and easy system-level definition. Use the spreadsheet interface to define system interconnects by simply dragging and dropping modules onto connectors. Signal quality and delay analysis can be performed on nets that span multiple boards through cables and connectors. System-wide reports enable design verification across all corner cases.

22.2.1.4 IS_Analyzer Major Benefits

The major benefits if the IS_Analyzer include the following:

- A single, integrated environment for analysis and verification of electronic systems
- Easy to use, hierarchical rules management system
- Critical timing and signal quality verification throughout the design process, including pre- and post-route
- Full-board or system-level design verification across all corner cases
- Multilingual environment allows simultaneous simulation of IBIS, SPICE, VHDL-AMS, and S-parameter models
- Interfaces with most popular CAD systems

The IS_Analyzer and the IS_Floorplanner™ tools generate accurate interconnect delay results based on transmission line simulations across multiple corners. Tau is able to import simulated interconnect delays for analysis of the complete timing path with interconnect considerations. Interconnect delays from the ICX tools can be imported at any stage of the design process: during placement, initial routing, or final routing. The combination of ICX and Tau provide a complete timing and signal integrity verification solution.

22.2.2 IS_Floorplanner

22.2.2.1 Hierarchical Rules Management

The IS_Floorplanner tool employs a powerful hierarchical constraint management system for both physical and electrical constraints, which can be applied at both the class and net level. Electrical requirements may be specified in their native electrical terms with no translation requirement into arbitrary physical constraints, enabling true, concurrent constraint management.

22.2.2.2 Hierarchical Floorplanning

Hierarchical floorplanning — derived from the original schematic hierarchy — greatly simplifies the placement process and allows graphical manipulation of logic blocks at any hierarchical level. Standard placement features such as move, copy, and rotate work on hierarchical blocks as well as individual components. Background analysis provides continuous feedback during floorplanning, indicating the design's ability to meet the electrical requirements.

22.2.2.3 A Complete What-If Analysis Environment

IS_Floorplanner enables pre-schematic what-if analysis, allowing the exploration of topology, termination, driver selection, placement, board stackup, characteristic impedance, and single-net routing to determine the optimum implementation (see Figure 22.5). Common topologies, including daisy chain, star, and H-tree are available from a predefined list. Pin-level model assignment is also available to optimize buffer strength for drivers and receivers, and terminators can be added and analyzed on the fly.

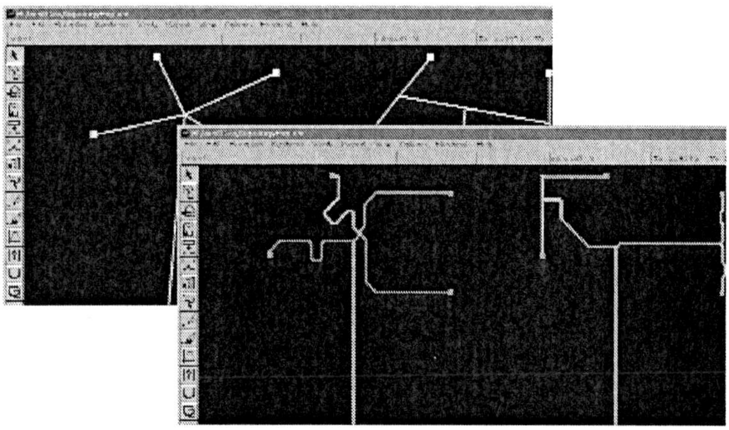

Figure 22.5 Perform what-if analysis on predefined topologies to determine optimal routing and analyze trade-offs prior to completing the schematic.

Multiple routing strategies can be analyzed using IS_Floorplanner's powerful, interactive router, or using the built-in autorouter. After manual routing, graphical feedback is provided, flagging violations of both physical and electrical constraints.

22.2.2.4 Analysis and Verification

IS_Floorplanner features an integrated parameter extraction module and a multi-conductor transmission line simulator. Components are modeled using industry-standard IBIS models, including EBD, connector and device package models, as well as SPICE and VHDL-AMS. Results from simulation are used to drive all aspects of the design process, from early what-if analysis, to placement and routing, as well as full-board verification. Corner cases are defined in order to evaluate best- and worst-case scenarios. Accurate delay results are produced with simulation including crosstalk push-out effects. Frequency-dependent losses, such as dielectric loss and skin effects, are also modeled, and simulation resolution is controllable at the class and net level to within one picosecond of accuracy.

22.2.2.5 Major IS_Floorplanner Benefits

The major benefits of the IS_Floorplanner include the following:

- A single integrated environment for design and verification of electronic systems
- Hierarchical constraint management system
- Timing-driven, hierarchical placement
- What-if capability for developing rules, validating models, and exploring design trade-offs
- Accepts timing constraints directly from the Tau tool; back annotates interconnect delay information to Tau
- Electrical constraints directly drive the routing process through integration with the IS_Synthesizer™ tool
- Full-board verification across all corner cases
- Multilingual environment, allows simultaneous simulation of IBIS, SPICE, VHDL-AMS, and S-parameter models
- Interfaces with most popular CAD systems

22.2.3 IS_Multiboard

22.2.3.1 System-Level PCB Design

The IS_MultiBoard™ tool extends the powerful analysis and verification capabilities of the ICX family of products beyond the single board to the system level (see Figure 22.6).

22.2.3.2 System-Level Definition

IS_MultiBoard enables seamless system-level definition. Use the spreadsheet interface to define system interconnects by simply dragging and dropping modules onto connectors. SPICE-like abstract circuit definitions are available for subsystems that have yet to be physically realized.

22.2.3.3 Analysis and Verification

IS_MultiBoard features an integrated parameter extraction module and a multi-conductor transmission line simulator, which works in both pre- and post-route design phases, and includes full-board and system verification. Components can be modeled using industry-standard IBIS, SPICE or VHDL-AMS models, yielding the most flexible simulation environment on the market. Electronic Board Description (EBD), coupled connector, and device package models are also supported. Corner cases are defined in order to evaluate best- and worst-case scenarios. Accurate delay results are produced with simulation, including push-out effects of

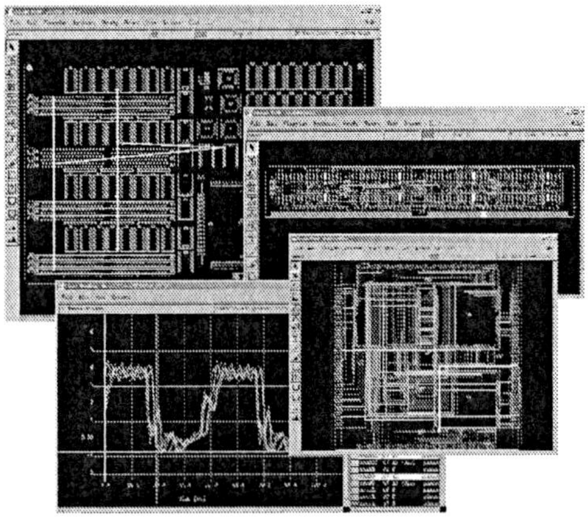

Figure 22.6 IS_MultiBoard integrates multiple PCB designs and provides simulation across the entire system.

crosstalk. Frequency-dependent losses, such as dielectric loss and skin effects, are also modeled. The simulation resolution is controllable at the class and net level to within one picosecond.

Signal quality and delay analysis can be performed on nets that span multiple boards through cables and connectors, and system-wide reports enable design verification across all corner cases. With IS_MultiBoard, parameter extraction for each subsystem is automatic and transparent. Automatic signal integrity optimization considers the effects of off-board segments of the net. Accuracy is further enhanced with IS_MultiBoard's coupled connector model support.

Major product benefits include the following:

- A single, integrated environment for analysis and verification of electronic systems
- Explore and optimize physical design trade-offs on multi-board signals
- Easy-to-use interface implements what-if strategies on any board within the system
- Easy-to-use hierarchical rules management system
- System-level design verification across all corner cases
- Drag and drop modules onto connectors for painless system definition
- Automatic pin mapping
- SPICE-like description language defines subsystems when physical databases are not available
- Multilingual environment allows simultaneous simulation of IBIS, SPICE, and VHDL-AMS models

22.2.4 IS_Optimizer

22.2.4.1 Electrically Driven Interconnect Design

The IS_Optimizer™ tool provides the capability to route up to 128 nets at a time to true electrical requirements without approximation to geometric interpretations. Simultaneous routing and simulation of delay and noise is carried out to produce a design that is both electrically and physically correct (see Figure 22.7).

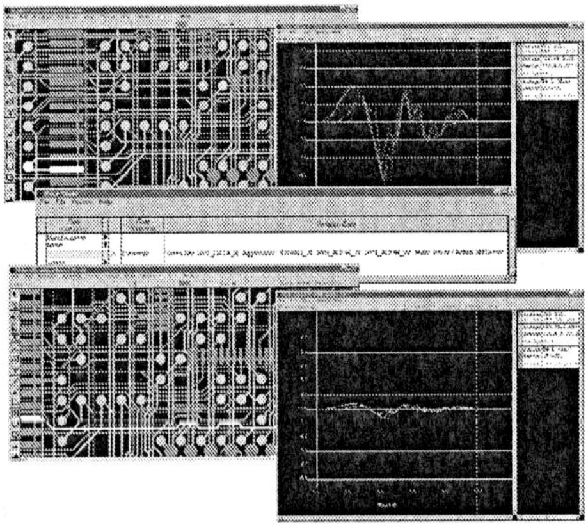

Figure 22.7 IS_Optimizer uses powerful synthesis technology to automatically update designs to meet true electrical constraints.

22.2.4.2 Hierarchical Rules Entry and Management

IS_Optimizer leverages the power of a true hierarchical rules entry and management system. An interactive spreadsheet interface provides easy and effective management of both physical and electrical rules, which can be applied at both the class and net level. Electrical requirements are specified in their native electrical terms and there is no translation into arbitrary physical constraints. The integration of this sophisticated rules entry system enables true concurrent constraint management.

22.2.4.3 Interconnect Optimization

IS_Optimizer is able to perform automated rerouting when electrical errors are detected, including corner case verification and optimization. Optimization also enables the correction of purely physically constrained designs to meet the true electrical constraint. IS_Optimizer maintains all of the electrical constraints for signal quality and delay, including constraints for topology, impedance, and differential pairs during a full-board optimization pass, and there are no restrictions on the number of nets that may be optimized at one time. Multiple optimization passes may be run to further reduce the impact of crosstalk and improve delay margins.

22.2.4.4 Interactive Synthesis

IS_Optimizer provides an interactive and semiautomatic routing environment, as well as real-time graphical feedback from the concurrent analysis system that allows the user to correctly route or reroute while maintaining electrical and mechanical constraints. Push-and-shove technology provides superior trace editing capabilities in both interactive and automatic modes.

Design reuse is also supported, enabling rapid reuse of layouts that range from simple fanout structures to complex sections of a design, saving hours of work on each design.

22.2.4.5 Underlying Architecture

IS_Optimizer's underlying architecture supports true 45-degree gridless routing using multipass rip up and retry strategies to produce electrically and mechanically correct designs. IS_Optimizer supports staggered or spiral blind and buried build-up technology, microvias, buried components, and fine-pitch devices such as BGAs. Diagonal layer biasing may be used to dramatically reduce interconnect delay and ensure signal integrity.

Easy-to-assign predefined and user-defined net topologies, including complex comb and H-tree structures, allow necessary routed structures to be created for proper load balancing. IS_Optimizer maintains its electrically correct environment by managing stubs electrically rather than physically.

Major product benefits include the following:

- A single, integrated environment for design and verification of electronic systems
- True hierarchical signal quality and delay rules for superior electrical performance
- Intelligently routes up to 128 critical nets at a time, based on electrical requirements
- Concurrent analysis of both electrical and physical parameters
- Integrated interactive and semiautomatic routing capability
- Multilingual environment allows simultaneous simulation of IBIS, SPICE, and VHDL-AMS models
- Interfaces with most popular CAD systems

22.2.5 IS_Synthesizer

22.2.5.1 Electrically Driven Interconnect Design

The IS_Synthesizer tool provides the capability to route a design to true electrical requirements without approximation to geometric interpretations. Simultaneous routing and simulation of delay and noise enable production of designs that are both electrically and physically correct (see Figure 22.8).

22.2.5.2 Hierarchical Rules Entry and Management

IS_Synthesizer leverages the power of a true hierarchical rules entry and management system. An interactive spreadsheet interface provides easy and effective management of both physical and electrical rules, which can be applied at both the class and net level. Electrical requirements are specified in their native electrical terms and there is no translation into arbitrary physical

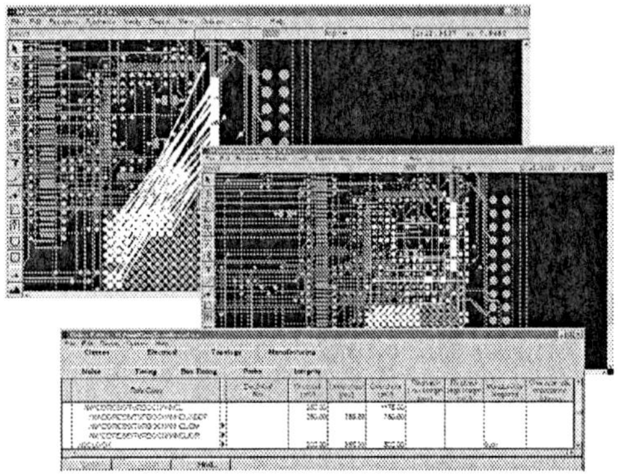

Figure 22.8 IS_Synthesizer offers a unique combination of real-time analysis and routing that automatically generates net topologies that satisfy the design's true electrical constraints.

constraints. The integration of this sophisticated rules entry system enables true concurrent constraint management.

22.2.5.3 Interconnect Optimization

IS_Synthesizer is able to perform automated rerouting when electrical errors are detected, including corner-case verification and optimization. Optimization also enables the correction of purely physically constrained designs to meet the true electrical constraint.

22.2.5.4 Interactive Synthesis

IS_Synthesizer provides both an interactive and semiautomatic routing environment. Real-time graphical feedback from the concurrent analysis system allows the user to correctly route or re-route while maintaining electrical and mechanical constraints. Push-and-shove technology provides superior trace editing capabilities in both interactive and automatic modes. Design reuse is also supported, enabling rapid reuse of layouts that range from simple fanout structures to complex sections of a design, saving hours of work on each design.

22.2.5.5 Powerful Underlying Technology

The underlying architecture supports true 45-degree gridless routing using multipass rip-up and retry strategies to produce electrically and mechanically correct designs. IS_Synthesizer supports staggered (spiral) blind and buried build-up technology, microvias, buried components, and fine-pitch devices such as BGAs (ball grid arrays). Diagonal layer biasing may be used to dramatically reduce interconnect delay and ensure signal integrity. Easy to assign predefined and user-defined net topologies, including complex comb and H-tree structures, allow necessary routed structures to be created for proper load balancing. IS_Synthesizer maintains its electrically correct environment by managing stubs electrically rather than physically.

IS_Synthesizer supports leading edge differential signal routing such as LVDS. Balanced-pair topologies, including symmetrical fanout and controlled-differential impedance, ensure correct skew between the differential pair, thus minimizing common-mode noise.

Major product benefits include the following:

- A single, integrated environment for design and verification of electronic systems
- True hierarchical signal quality and delay rules for superior electrical performance
- Intelligently routes high-speed designs based on electrical requirements
- Automatic load balancing based on receiver characteristics
- Concurrent analysis of both electrical and physical parameters
- Support of advanced topologies to manage complex electrical requirements
- Multilingual environment, allows simultaneous simulation of IBIS, SPICE, and VHDL-AMS models
- Interfaces with most popular CAD systems

22.3 Tau

The Tau board-level symbolic timing analysis tool performs comprehensive worst-case timing analysis and verification on designs using an advanced symbolic timing methodology, eliminating false paths that are typically reported in standard static timing tools. Tau employs models that contain both timing and functional information. By exploiting this combination, Tau restricts analysis to timing paths that are truly relevant, thus facilitating analysis and verification without the need for test vectors. Skew and phase shift between clocks is automatically computed within Tau by traversing the clock tree. Delay correlation within a single component eliminates overly conservative, worst-case (minimum/maximum) timing analysis.

Tau models are those that are created from the timing and functional information contained in manufacturers' data sheets, or from a number of electronic formats, including TDML and STAMP. The model creation tool included with Tau, TauLib, allows users to create Tau models from data sheets alone, or by reading these electronic formats. Tau models are generally created to be reused in subsequent designs to provide a repeatable and robust mechanism within your organization for circuit timing analysis. Tau employs symbolic timing analysis to provide the Expedition Series with the premier solution for the timing verification needs of board-level circuits. Symbolic timing analysis addresses the limitations of traditional static timing tools, such as the identification of a large number of false paths, and also eliminates many of the modeling issues found when using static timing tools.

22.3.1 Circuit Timing Methods and Symbolic Timing[3]

François Rabelais once said, "Nothing is so dear and precious as time." Yet, somehow, time is remarkably almost forgotten in the design of electronic circuits. Some designers enjoy the value and advantages of performing their own timing analysis. Others have had the perception that timing analysis and verification is an important part of a successful circuit design process largely

3. This section was written by Matthew Hogan and taken from his paper titled, "Circuit Timing Analysis: Mastering A Lost Art." Reference [Hog01].

eroded through the use of chipsets and other matched components. Conservative estimations used in manufacturers' design guides often sacrifice circuit performance for generalized solutions of timing, including effects from signal integrity. Independent validation of circuit timing is required to ensure correct and optimal timing for each design.

Electronic circuits contain either clocked (synchronous) or non-clocked (asynchronous) components. Clocked components must satisfy setup and hold constraints that are relative to that clock, before an output value may be propagated through the device. Non-clocked components have inputs that are constrained (via setup and hold times) relative to other non-clocked inputs. In both cases, these constraints and the propagation through the devices need to be considered to ensure the proper operation of the circuit.

22.3.1.1 Timing Methods

Designers have at their disposal a number of methods to ensure correct circuit timing. Static timing verification and functional simulation are two of the more traditional timing methods.

Static Timing Verification Static timing verification has been used successfully for many years in the ASIC space. In fact, it has been used so successfully that it often forms part of the sign-off process for an ASIC (application-specific integrated circuit) design. The almost universal adoption of static timing analysis in the ASIC world has been largely due to the availability of static timing models that were provided by ASIC vendors.

The models used for static timing verification generally do not contain any functional information. All of the timing links within a specific model are assumed to be valid all of the time. In all but the simplest of components this is not the case, especially when performing timing analysis at the board or system level. Analyzing these timing links all of the time produces a large number of false timing errors. Each static timing tool has a different way in dealing with these errors. The reporting of these false timing errors is undesirable, as the circuit designer would typically need to sort through a large number of timing errors reported in the design, and then tag (or label) the timing errors that are known to be false. The static timing tool does not present these tagged violations to the designer in subsequent reports.

Functional Simulation Functional simulation offers designers a very detailed view of circuit behavior. However, this detail comes at a price. Extensive runtimes and complete functional device models are required to perform the functional simulation. Exhaustive test vectors are required to form the list of input patterns that completely exercise all design paths. The coverage of the functional simulation is entirely dependent on the quality of these input test vectors.

Symbolic Timing Analysis Symbolic timing analysis provides designers with a more appealing solution for circuit timing analysis than both static timing verification and functional simulation. Symbolic timing analysis offers analysis speeds comparable with static timing, without reporting false timing paths. Much like static timing, symbolic timing uses simple timing models. However, the functional information used to differentiate timing is unique to symbolic timing analysis.

22.3.1.2 Symbolic Timing Analysis with Tau

Tau is the Mentor Graphics symbolic timing analysis and verification tool that empowers circuit designers with efficient circuit timing analysis. Tau automatically recognizes circuit functionality that alters or differentiates timing, and uses a netlist and bus timing models to provide an accurate representation of component timing variations in the design. Using a spreadsheet-based user interface, Tau presents the designer with a concise list of relevant timing equations. This saves the designer from being inundated with volumes of false timing errors that require individual assessment and elimination. Using Tau for symbolic timing analysis also improves the confidence in board-level timing, since the complete spread of device timing variations are considered. (See Figure 22.9.)

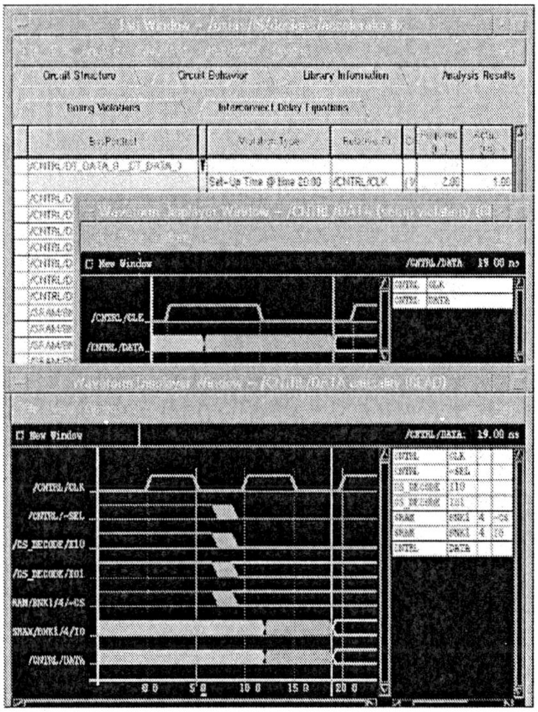

Figure 22.9 Tau provides designers with accurate circuit timing analysis.

22.3.1.3 Timing Models

Circuit designers often use off-the-shelf components in their designs. Timing models are generally not provided with the components. Component vendors provide product data sheets, but they are generally in paper or PDF formats. The timing information contained in the data sheets must either be transcribed into a timing model, or imported from an electronic version of this information (usually from timing waveforms). An open waveform standard holds promise in providing an open format for the exchange of these necessary waveforms. Timing Diagram Markup Language (TDML) is an industry-backed effort that establishes an open, industry standard language for the exchange of interactive timing diagrams.

The Tau modeling tool, TauLib, is among the growing number of tools that read the TDML open format. Waveform editing tools, such as Timing Designer from Forte Design Systems, Timing Diagrammer Pro from SynaptiCAD, Inc., and TimingTool (an Internet-based, free-to-use timing diagram editor, *www.timingtool.com*) also use this format. TauLib imports many electronic timing data formats, including STAMP, and is able to use the timing relationships within these files to create models, which are then consumed by Tau. The spreadsheet-based environment of TauLib allows easy editing of timing and functional relationships that are imported. These symbolic timing models need only describe bus functional information, as the complexities of the actual device behavior are treated as a black box.

While versatile at solving many design problems, FPGAs pose a unique dilemma for timing verification. Each turn of an FPGA yields different timing. Verifying this timing with confidence and in a reasonable time can be a daunting task. Timing data contained in STAMP files generated from FPGA vendor tools offer integration into the design process by being able to easily incorporate FPGA timing updates into the symbolic timing analysis. The STAMP format facilitates this by allowing multiple timing data files to be associated with a single model file. Validation of timing for different place and route runs from an FPGA design within the board design context may be performed. Tau models allow different timing variations to be quickly inserted into the timing analysis, providing an easy to use what-if environment for timing validation.

Tau models do not contain design specific information; rather they are more general purpose, and are therefore reusable in other designs. The reuse of symbolic timing models promotes greater efficiencies in the design process, since validated models are now available to the designer (ultimately reducing the time spent on model creation).

22.3.1.4 Using Tau in a Workflow

A tight integration to schematic- and physical-based flows allows timing information to be passed between Tau and the Mentor Graphics Interconnect Synthesis high-speed design and verification environment (also known as ICX). Schematic-based interfaces, including EDIF, enable extraction of netlist data for timing verification. PCB interfaces permit the physical PCB design to be used within the ICX environment. The Interconnect Synthesis tools allow simulation of the analog effects seen by these digital signals. Interconnect delays, which may contain physical phenomenon such as timing pull-in or push-out due to crosstalk with other nets, or additional delay caused by excessive ringing, are communicated between these tools. This establishes a complete view of both timing and signal integrity. This tight integration provides a powerful solution to both timing and signal integrity. Figure 22.10 shows this integration.

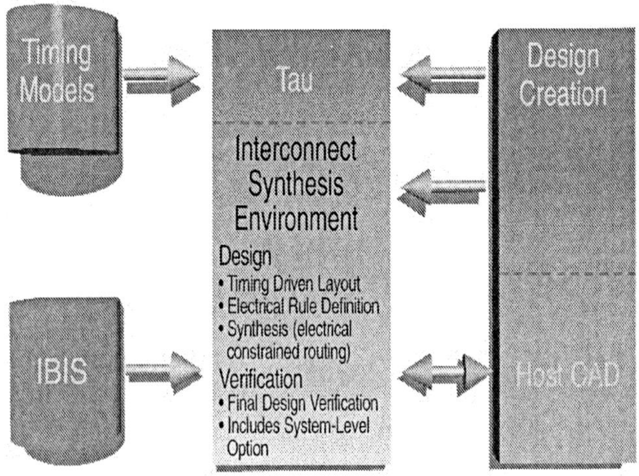

Figure 22.10 The Mentor Graphics interconnect synthesis environment.

22.3.1.5 Conclusion

Board-level designs do not need to have high-frequency clocks to suffer from timing problems; all that is required is aggressive timing and fast edge rates common to current driver technologies. FPGA technologies offer engineers many new design and timing alternatives. The exploration of these timing alternatives can be done in a controlled and repeatable manner by examining timing variations through employing complete board timing verification solutions. This is available today through symbolic timing analysis with Tau.

22.4 HyperLynx

HyperLynx is the industry standard for quick, accurate, easy-to-use signal integrity (SI) analysis. HyperLynx enables you to quickly and easily perform SI analysis on signals ranging from megabit to gigabit speeds. You no longer need to be a signal integrity expert or spend weeks in the lab to get results.

HyperLynx can help you do the following:

- Analyze driver strength, net topology, and termination, as well as their effects on signal integrity and EMC.
- Test nets across multiple boards or cables.
- Plan and analyze differential signals.
- Model lossy lines and frequency-dependent vias.
- Use eye diagrams (see Figure 22.11) to view effects of multi-bit sequences.
- Perform simultaneous simulation with HSPICE and IBIS models.

HyperLynx comes in two versions, HyperLynx GHz and HyperLynx EXT.

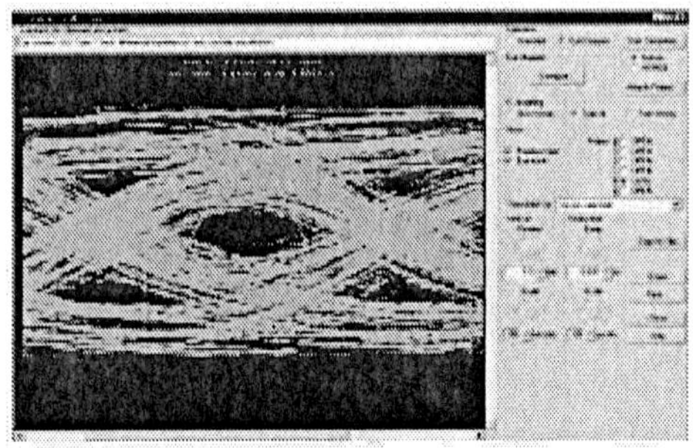

Figure 22.11 HyperLynx includes a complete suite of tools for pre/post-layout signal integrity, crosstalk, and EMC analysis.

22.4.1 HyperLynx GHz

Increasingly fast edge rates in today's integrated circuits (ICs) cause detrimental high-speed effects, even in PCB designs running at low operating frequencies. As driver ICs switch faster, a growing volume of boards suffer from signal degradation, including over/undershoot, ringing, glitching, crosstalk, and timing problems. When degradation becomes serious enough, the logic on a board can fail — making otherwise excellent designs stare blankly into space. HyperLynx EXT — including LineSim for pre-layout analysis and BoardSim™ for post-route analysis — targets designs at frequencies up to about 500 MHz (see Figure 22.12). At higher speeds, lossy effects (including skin effect and dielectric loss) and via modeling take on increasing importance; the HyperLynx GHz bundle models all of these effects and addresses the special needs of multigigabit design.

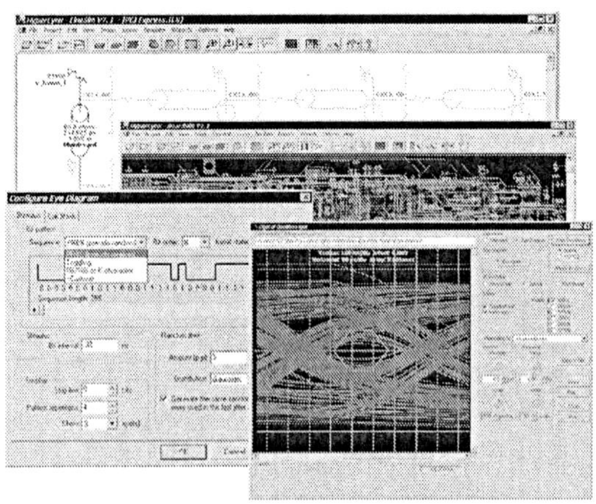

Figure 22.12 HyperLynx GHz includes a complete suite of tools for pre- and post-layout signal integrity, crosstalk, and EMC analysis.

22.4.1.1 Complete SI and EMC Analysis Suite

With HyperLynx, you can address high-speed PCB problems throughout your design cycle, beginning at the earliest architectural stages and moving through post-layout verification. The emphasis is on solving problems early — where it is less costly and more efficient; and on getting results without spending weeks in training; and on getting it right the first time, saving recurrent layout, prototype and test cycles in the lab. Hardware engineers, PCB designers, and signal integrity specialists alike can use HyperLynx EXT or GHz as a team — it's even easier than using an oscilloscope or spectrum analyzer in the lab.

22.4.1.2 LineSim GHz

Pre-layout simulation with LineSim GHz, part of HyperLynx GHz, lets you predict and eliminate signal integrity problems early, allowing you to proactively constrain routing, plan stack-ups, and optimize clock, critical signal topologies, and terminations prior to board layout. LineSim's intuitive point-and-click transmission-line modeling approach is an ideal way to get it right the first time:

- Simulate with IBIS or HSPICE models easier than ever.
- Accurate modeling of lossy transmission-line effects — including skin effect and dielectric loss.
- Full analysis of differential pairs to achieve desired differential impedance, and to optimize differential termination.
- Quickly enter complex interconnect scenarios, including ICs, transmission lines, cables, connectors, and passive components.
- Model any combination of transmission line: microstrip, buried microstrip, or symmetric/asymmetric stripline.
- Simulate immediately, using industry-standard IBIS models, HyperLynx's 7,000 model IC library, generic models, or build your own models from data book information.
- Visual IBIS Editor allows you to test and edit IBIS models, including a powerful V/I-V/t auto-correct option and drag-and-drop curve correction.

22.4.2 HyperLynx EXT

HyperLynx EXT — including LineSim for pre-layout analysis and BoardSim for post-route analysis — targets designs at frequencies up to about 500 MHz (see Figure 22.13). At higher speeds, lossy effects (including skin effect and dielectric loss) and via modeling take on gradually increasing importance; the HyperLynx GHz bundle models all of these effects and includes features that address the special needs of gigabit-per-second, SerDes-based designs.

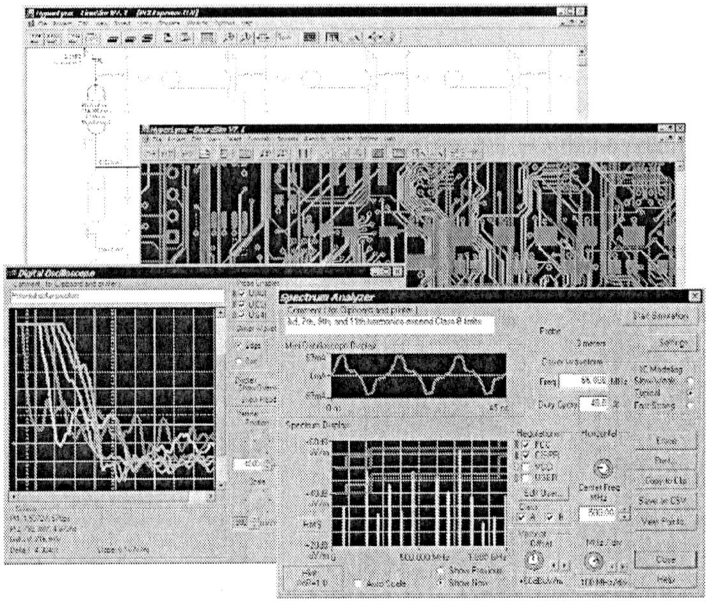

Figure 22.13 HyperLynx EXT includes a complete suite of tools for pre- and post-layout signal integrity, crosstalk, and EMC analysis.

22.4.2.1 LineSim EXT

LineSim EXT, part of HyperLynx EXT, lets you perform pre-layout simulation, allowing you to predict and eliminate signal integrity problems early to constrain routing, plan stackups, and optimize clock, critical signal topologies, and terminations prior to board layout (see Figure 22.14). Using LineSim's intuitive point-and-click transmission-line modeling approach allows you to do the following:

- Quickly enter complex interconnect scenarios, including ICs, transmission lines, cables, connectors, and passive components.
- Simulate immediately, using industry-standard IBIS models, HyperLynx's 7,000 model IC library, generic models, or build your own models from data book information.
- Experiment with stackup; impedance calculation is automatic.
- Model any combination of transmission lines: microstrip, buried microstrip, or symmetric/asymmetric stripline.
- Full analysis of differential pairs to achieve desired differential impedance, and to optimize differential termination.
- Visual IBIS Editor allows you to check and edit IBIS models, including a powerful V/I-V/t auto-correct option and drag-and-drop curve correction.
- Antenna Current Probe provides invaluable insight into the root cause of most EMC problems — excessive energy in critical nets on the PCB.

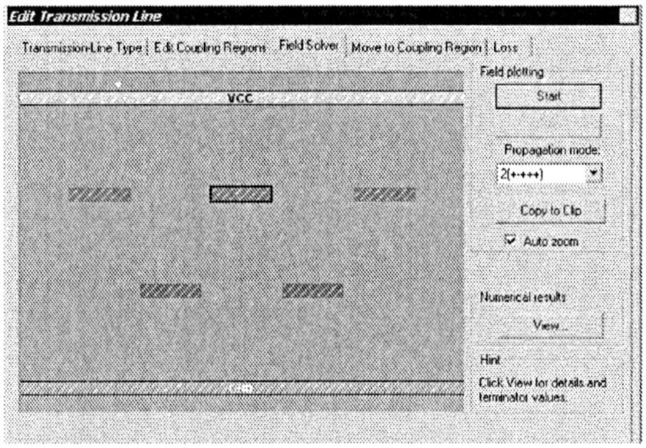

Figure 22.14 Pre-layout crosstalk analysis allows you to optimize spacing, stackup, and termination.

22.5 Mentor Graphics Field Solvers Used in ICX and HyperLynx[4]

22.5.1 Introduction

In signal integrity analysis for circuit boards, transmission line models must be generated for circuit simulators to use. One key ingredient of a transmission line model is the so-called characteristic impedance. This is simply the ratio of the voltage to current anywhere along the transmission line. It is an intrinsic property of a transmission line with a specific geometry. The characteristic impedance is obtained using an electromagnetic field solver. This is referred to simply as a "field solver."

Two field solvers are used in the Mentor Graphics ICX and HyperLynx family of tools. The ICX field solver and the HyperLynx field solver both use the so-called "method of moments" solution technique. More on this solution technique will be presented below. Other popular solution techniques include the finite element method (used by many Ansoft products) and the finite difference method.

Each of these solution techniques have their own advantages and disadvantages. For instance, method of moment techniques can handle "open boundary"[5] problems better than finite element techniques, but finite elements excel at problems with non-linear materials.[6] Other major distinctions between solvers can also be noted. One of these is a dimensional aspect, the solvers can be 2D, 2.5D, or 3D (dimensions).

4. The author of this section on field solvers is Dr. Robert G. Kaires of Mentor Graphics Corporation.
5. Open boundary problems are those that have fields that extend to infinity. An example is a microstrip transmission line.
6. An example of a non-linear material is ferrite. It isn't common to have this type of material in a PCB environment.

Another distinction is the order of the frequency approximation used. If the frequency is assumed to be zero, the solver is called "static" or "quasi-static." If full-frequency effects are included, the solver is called "full-wave." It is also possible to include some, but not all, frequency effects. An example of this would be a solver that takes into account so-called skin effect conductor losses[7] but not radiation losses.[8]

22.5.2 Geometry Problem Entry and Discretization

A large part of the problem solution technique is the geometry description and discretization. Geometry entry can be from a graphical user interface (GUI) or text-based file entry. Discretization refers to the fact that any numerical solution technique requires the problem geometry to be broken into pieces or "elements." More elements are needed where field strengths are greater. Element generation can be completely automatic and transparent to the user (as they are in Mentor Graphics tools) or, at the opposite extreme, can be completely (and tediously) specified manually by the user. Very often element generation is done in an automatic step but then the user is allowed to examine the element distribution and make changes (improvements). More elements can be added near corners or bends in the geometry where field strengths will be greater. Standalone field solvers with GUIs often take this approach. Users familiar with Ansoft products may be familiar with the process of generating a finite element "mesh" (an element mesh) and possibly refining the mesh. Some field solvers have the capability of iterating until an excellent element distribution is obtained. This is done by picking a crude element distribution to begin with, then solving the problem and examining the error in the field distribution. More elements are added where needed until a small field error is obtained.

There are two problems with this approach for signal integrity. First, it is too slow: iterations take time. Second, it is difficult to pick appropriate error criteria for stopping the iteration. In general, one must "baby-sit" the problem and stop it when necessary. Another approach is to use a fixed-element distribution. The challenge with this approach is to find an element distribution that yields accurate results with minimal elements. If too many elements are used, speed will be impacted.

Both the ICX and HyperLynx field solvers handle the problem of geometry generation and discretization in a fashion completely transparent to the user. Two-dimensional cross-sections representing transmission lines to be simulated are extracted from the circuit board database. The geometry is generated automatically and sent to the field solvers. The ICX and HyperLynx field solvers use a fixed, proprietary element distribution to yield fast, accurate results.

7. Skin effect losses are resistive losses that are generated in the metal of the traces. These losses are due to the tendency of current to move to the outside edges of a conductor as the frequency of the signal being transmitted by the conductor increases. This effect is due to the conductor's self-inductance. The greater the signal's frequency, the more the current crowds to the outside edge and the greater the resistive loss.
8. Radiation losses are resistive losses due to energy escaping to the atmosphere due to electromagnetic radiation. This effect generally becomes significant at a much higher frequency then the aforementioned skin effect losses.

22.5.3 Speed and Accuracy

There is always a trade-off (inverse relationship) between speed and accuracy in numerical field solvers. In signal integrity analysis for PCBs, speed is always of the utmost importance because hundreds, if not thousands of problems may need to be solved. Accuracy needs to be good, but overkill in this area is not a good idea. All field solvers (and all numerical algorithms) have limitations in the accuracy they achieve. Also keep in mind that overall accuracy[9] suffers for other reasons as well. Another factor affecting overall accuracy is the manufacturing tolerance of the PCB, which can result in geometry errors. Line widths, thickness, and length all have uncertainties. As well, material property errors, such as dielectric constant, thickness, and loss tangent have uncertainties. In addition they can vary with frequency and position on the circuit board. These effects are generally not taken into account. Also keep in mind that accuracy can be affected when the assumptions (physics) about the model break down. For instance, signal integrity programs model transmission lines using the so-called quasi-static approximation. This is sometimes referred to as a TEM[10] approximation. This approximation will break down at high enough frequencies.[11] So the model would be no longer valid, even if the numerical implementation and the geometric and material properties were perfect.

22.5.4 The Method of Moment Technique

The method of moment technique is used by both the ICX and HyperLynx field solvers. Some authors use the term "boundary element method." In actuality these methods are the same.[12] The essential feature that distinguishes this method from the finite element method is that only boundaries of the problem are "gridded" (divided into elements). In the finite element method, all of the problem space must be "meshed" into elements. Since elements cannot be placed in space all the way out to infinity, artificial boundaries must be introduced into the finite element technique. This is not necessary in the method of moments. The method of moments solution can also result in a smaller overall problem to solve since elements only exist on problem boundaries resulting in a smaller number of elements. This can translate to a faster solution time for a given problem.

9. "Overall accuracy" can be defined as the accuracy achieved when taking into account all factors such as numerical error, geometry error, material property error, etc. The best way to determine this accuracy is to compare simulated results with very carefully measured results. This is easier said than done, however, since errors caused by driver and receiver models would have to be de-embedded from the results.
10. TEM stands for Transverse Electro Magnetic. It means that the electric and magnetic fields are transverse to the direction of propagation. TEM propagation occurs on transmission lines with at least two conductors at low enough frequencies and with low enough losses.
11. This breakdown is referred to as "moding." It should not cause undue concern that we do not model this effect. If moding were to occur, one would have to switch to a better transmission line technology, such as coaxial cable. Moding would cause unacceptable signal integrity problems. If this occurs, you have reached the useful frequency limit of your technology. Therefore it would not make sense to model this effect.
12. Source is reference [Har83].

22.6 The Expedition Series and Design Flow

22.6.1 DxDesigner

DxDesigner is a multifaceted product for design definition. Its robust environment encompasses much more than just schematic capture. It can even be expanded into

- A gateway between design definition and the entire enterprise
- A methodology for design creation that enables team-based design

By redefining the entire front-end engineering process, DxDesigner becomes much more than a point-tool for schematic capture, it becomes a solution for design definition. By addressing all aspects of circuit design and simulation, component selection, library management, constraint definition, project management, and team-based design within a single integrated solution, DxDesigner enables engineers to better define the total product development process (see Figure 22.15).

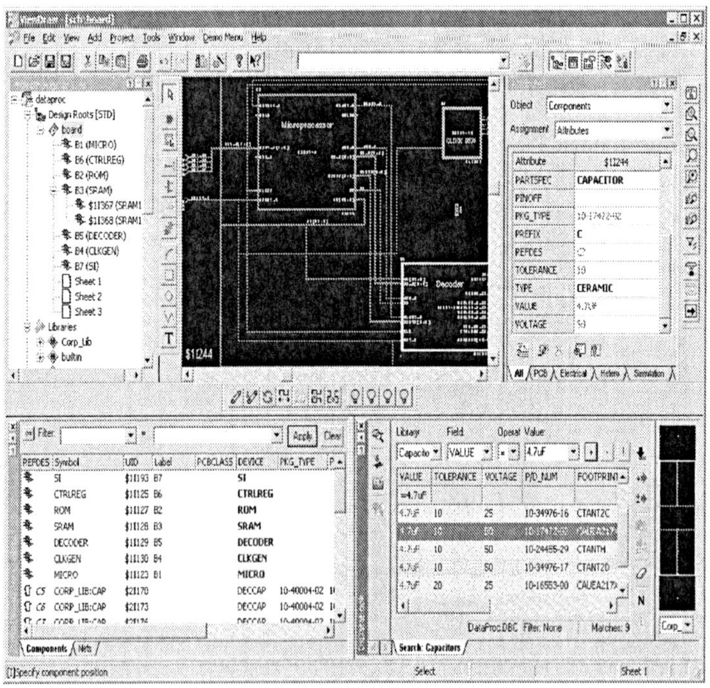

Figure 22.15 DxDesigner is the complete solution for design definition.

DxDesigner enables geographically dispersed teams to coordinate their efforts by using a centralized parts library that also services purchasing, component engineering, manufacturing, and design engineering teams. The use of advanced hierarchical design techniques enables the reuse of subsystems in future designs, speeding delivery of next-generation products.

DxDesigner also provides electronic engineers with a comprehensive solution to create high-integrity designs using the latest technology. DxDesigner can be used to find, test, and eliminate signal integrity (SI) problems in the first PCB spin, and then develop analysis-driven physical design constraints to ensure a high-quality layout for a product that works the first time.

Creating competitive products is about more than capturing a schematic and passing the design to layout. Libraries must be managed so that the optimal parts with the lowest cost and shortest lead-time are used in the design. Sections of designs must be reused to reduce development time and improve reliability. Simulation and signal integrity analysis must be performed to ensure that the design will function properly and can be manufactured. Constraints must be applied early in the process so that fewer design iterations are required. Finally, the design data must be distributed and integrated with the purchasing and manufacturing operations so that a complete product can be produced with the fewest errors in the least amount of time. The DxDesigner tool provides a complete solution for design creation, definition, and reuse. It provides everything needed for circuit design and simulation, component selection, library management, signal integrity planning, project management, and team-based design.

An open object model provides complete access to DxDesigner data and commands through standard VBScript or JScript languages, allowing for extensive customization.

22.6.1.1 Design Reuse

Design reuse is a powerful methodology for reducing development cost and time while increasing product functionality and reliability. DxDesigner has many features that were created with design reuse in mind. For instance, the library browser can be used to search for IP reuse blocks that may be available from other design projects. Sophisticated hierarchy support allows for multiple instantiations of a reuse block, and supports the ability to import the instance-specific data to a new top-level design. A complete cross-referencing tool can cross reference nets across hierarchical blocks, and can even identify the sheet number in the drawing package where a particular block is defined. Specialized attributes can identify a block for reuse, limiting the repackaging of components and adding a unique prefix to all the components in the block. An intuitive design navigator simplifies the search for parts and nets in a hierarchical design, making this design methodology practical and effective.

Design variants can also be defined in DxDesigner. A component (or block) can be uninstalled or substituted for another component while maintaining the integrity of a user-specified characteristic such as the PCB footprint. This flexibility allows multiple variants to be produced from a single base design. If changes are made, all the variants inherit the changes, thus reducing the required rework and maintenance of design data.

Finally, reusable design data can be stored in a repository where it is managed and available to other designers. A user can check a design out from the repository over the Internet, then use that design as is or modify it (the original design stored in the repository remains intact for future designers). All of these features combine to make design reuse accessible to any project team.

22.6.1.2 Integration with the Enterprise

In order to optimize a product development cycle, design data must be integrated with the rest of the corporate enterprise. DxDesigner has the ability to publish schematics, library data, and other design data through a web server, so any user with the appropriate permissions can gain access to this information using a Web browser. DxDesigner is also integrated with several product life cycle management systems, making design data, PDF schematics, and BOMs (bills of material) available throughout the company. It also supports a centralized, Internet-based library so that only one version of the corporate library needs to be maintained.

22.6.1.3 Constraint-Based Design

To build successful products the first time, designs must be constrained early in the design process. This prevents the production of designs using parts that cannot be sourced or are cost-prohibitive, or the production of boards that do not meet speed or manufacturing requirements. DxDesigner provides designers with the status of all component data so that they can choose the right part. It can also verify an entire design against the corporate database to ensure that all data is consistent and correct. The DxDesigner integrated constraint editor specifies physical and electrical constraints that can be read directly by Mentor Graphics signal integrity tools. Once analysis is complete, these constraints are automatically passed to the target layout system.

22.6.1.4 Variant Management

The Variant Manager tool manages the creation of multiple product configurations from a single design database. Variant Manager's single-point ECO (engineering change order) management minimizes errors, reduces costs, improves design quality, and enhances production efficiency.

22.6.2 Expedition PCB

Expedition PCB, powered by AutoActive technology (discussed below), is the PCB layout and autorouter tool and is an integral part of the tightly integrated Expedition Series. By combining ease of use with advanced functionality, Expedition PCB offers designers the leading technology for the creation of today's most complex designs. Expedition PCB includes interactive and customizable multipass autorouting controls for design challenges such as differential pair routing, net tuning, manufacturing optimization, and microvia and build-up technology. Figure 22.16 illustrates that Expedition PCB is designed for the layout of dense, difficult, high-technology PCB designs.

Designers today are increasingly challenged by the need to manage signal quality in order to achieve system performance and reduce prototype iterations. High-speed design with Expedition PCB is an integrated part of the AutoActive design environment.

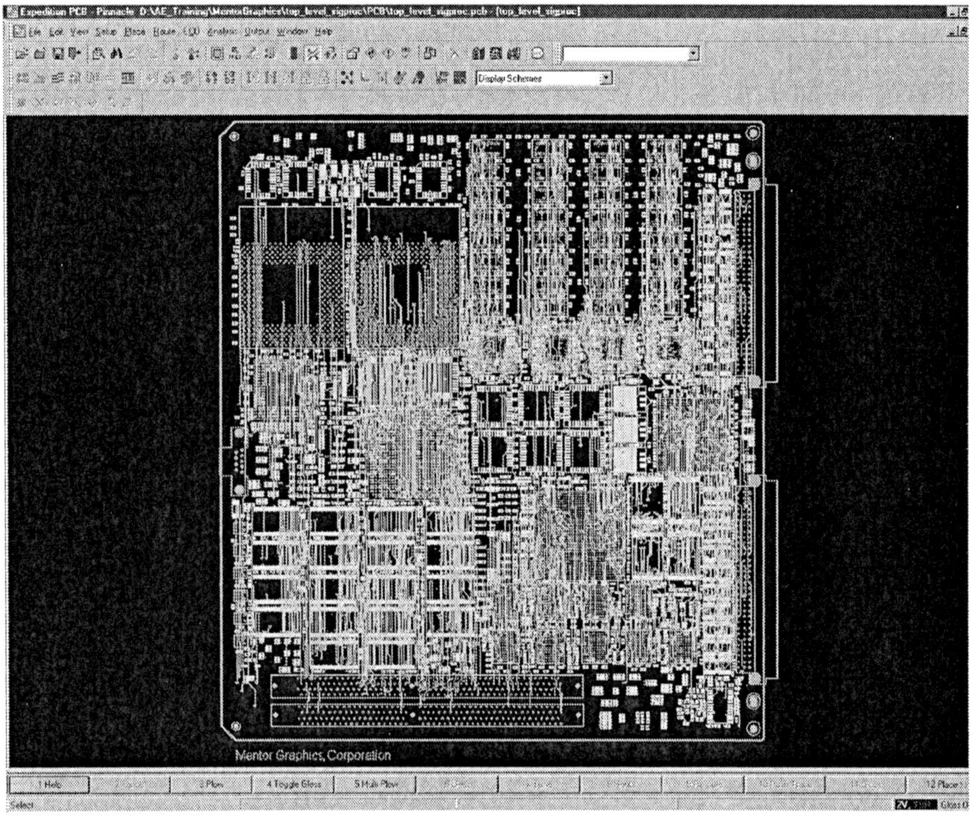

Figure 22.16 The Expedition Series is the most productive solution available for the creation of dense, difficult, high-technology PCB designs.

22.6.2.1 AutoActive Technology

AutoActive technology represents a revolutionary step forward for PCB design. The power of industry-leading autorouting technology is combined with interactive editing capabilities to produce a single, powerful, and easy-to-use design environment. This environment eliminates the burdens of jumping between tools to get your job done and managing differences between the constraints on the autorouter and on interactive editing.

AutoActive provides designers with greater control than ever before, with the ability to easily switch between automatic and manual editing. From simple tasks, such as defining board areas, to complex procedures that involve maintaining high-speed signal conditions, all objectives are accomplished with the system and the designer working together in real time. The net result of AutoActive technology is reduced design times, increased productivity, and unmatched design quality.

Features of AutoActive technology include the following:

- A single, integrated place and route editing environment that reduces total design time and increases productivity
- All physical rules and high-speed rules are maintained
- Correct-by-construction design that produces high-quality results with cleanup time eliminated
- Shape-based, true 45-degree routing
- The most advanced autorouting technology: Stop and start the autorouter at any time and all results will be correct-by-construction
- Dynamic cleanup of traces through the reduction of segments, prevention of acute angles, and application of pad entry rules

22.6.2.2 Dynamic Area Fills

Expedition PCB automatically clears area fills[13] around traces, vias and pads as the board is edited. Dynamic area fills are so fast that Expedition PCB allows you to keep your area fills turned on while you are doing all your edits. Moving a via pushes and shoves other vias, traces, and area fills, and connectivity is automatically maintained.

22.6.2.3 Rules by Area

The rules by area functionality greatly improves routing around BGAs and other fine-pitched parts. Rule areas represent complete rule sets that are obeyed by online and batch DRC and in interactive and automatic routing. Rule areas may be defined by layer and can be assigned to any polygon, rectangle, or circle. Trace widths and clearances automatically change when traversing into or out of the rule area. You may also change via sizes and spans[14] in a rule area to maximize route completion.

22.6.2.4 Multiplow with Variable Via Patterns

Expedition PCB's multiplow functionality allows you to simultaneously route multiple nets, including differential pairs, with true 45-degree routing. It can even handle routing through areas of staggered pins. Traces being routed will push and shove the other vias and traces out of the way and automatically clear area fills as needed. Changes can be easily made to a variety of selectable via patterns at the touch of a button, allowing enhanced flexibility for routing into dense areas of a design.

22.6.2.5 Dynamic Hazard Review

Design hazards are dynamically displayed and may be individually selected and colored for easy identification. When a hazard is fixed, it is removed dynamically from the hazard list.

13. Area fills are expanses of copper that have been specifically placed on the board. They are used for many purposes, including signal shielding, improved power distribution, increased current carrying capability, and others.
14. A via span describes the start and stop layers that a specific via has as it goes through the PCB. An example of via spans is shown in Figure 22.18 on page 600 (bottom left panel in the figure).

22.6.2.6 ECOs with Expedition PCB and the Expedition Series

ECOs (engineering change orders) can cause delays and introduce errors in the design process. The Expedition Series makes ECOs less painful and more accurate than ever before. Expedition PCB's powerful automation and tight system integration drastically reduces ECO completion times and eliminates synchronization errors. You can change rules, replace parts, and reroute automatically with no rule violations—all in real time.

22.6.2.7 Constraint Definition

Expedition PCB handles an extensive set of constraints to meet high-speed performance requirements whether you're routing interactively or automatically. A common constraint definition environment is shared between schematic capture and layout, allowing the evaluation of critical signals at any design stage. Constraints include same-layer and adjacent-layer differential pairs, controlled impedance, net scheduling, and delay.

22.6.2.8 Net Tuning

While routing interactively, graphic tuning aids are displayed for guidance. By graphic tuning we mean tuning for length and crosstalk. To tune for length, we add meander (also known as tromboning or serpentining). To tune for crosstalk, we adjust the trace separation. Nets modified out-of-tune during edits are automatically retuned. The Hazards dialog box dynamically updates as you edit nets, providing instant feedback relative to your constraints. Nets can also be tuned automatically within an autoroute pass. Tuned nets are automatically maintained as you complete the design.

22.6.2.9 Differential Pair Routing

Routing and editing differential pairs with Expedition PCB is accomplished with speed and ease that will change your view of high-speed design. Pair spacing rules can be established by both layer and net class. If one trace is edited, the other trace in the pair automatically moves with it. Adjacent layer differential pair routing capabilities add another valuable option for routing critical signals on a dense PCB (see Figure 22.17).

22.6.2.10 Advanced Interconnect Routing

The challenges of advanced interconnect are prevalent today with BGA, CSP, COB, and DCA[15] packages increasing board density. Build-up and microvia structures used in these board designs further complicate routing. Expedition PCB, powered by AutoActive technology, offers the leading technology for advanced interconnect designs (see Figure 22.18).

15. BGA is ball grid array, CSP is chip scale package, COB is chip-on-a-board, and DCA is direct chip attach.

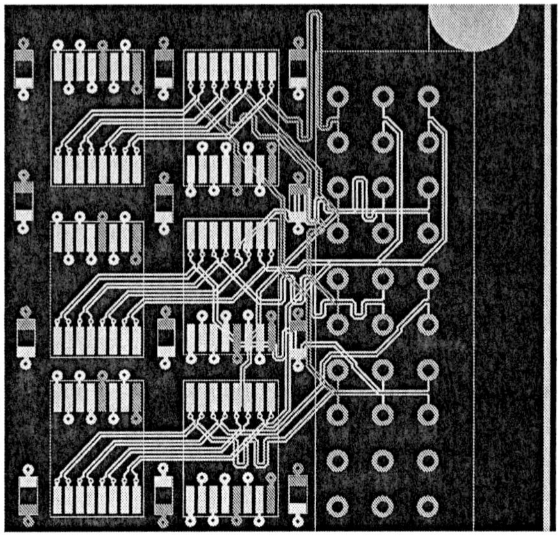

Figure 22.17 Routing and editing differential pairs with Expedition PCB is accomplished with speed and ease.

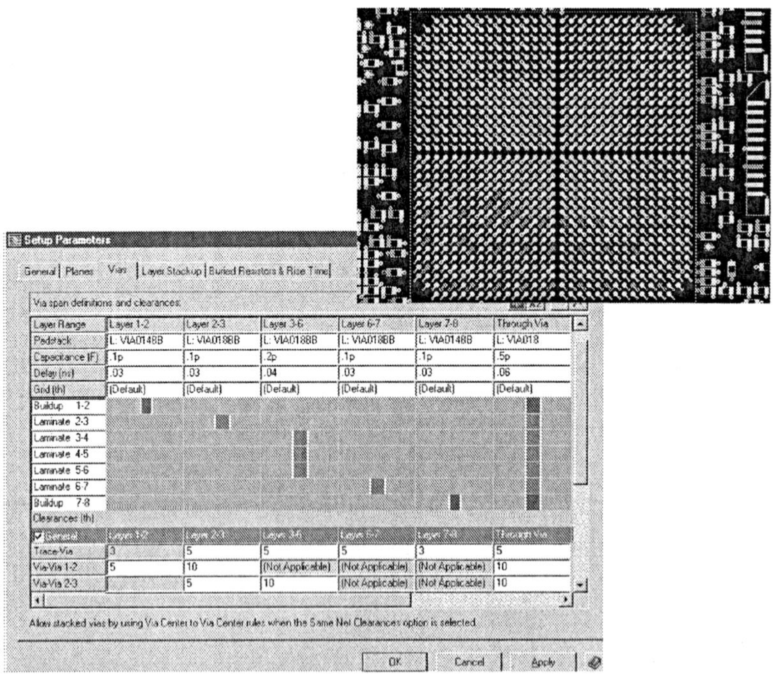

Figure 22.18 Expedition PCB offers the leading technology for the creation of advanced interconnect designs.

Expedition PCB supports the definition of complex via structure rules and the routing of microvia geometries, including comprehensive via-in-pad rules. Via spans between any two layers are possible. By moving beyond traditional laminate layer pairing, Expedition PCB facilitates the design of build-up structures on laminate to enable escape patterns from dense, high-pin-count devices. Build-up areas typically have a smaller clearance than the laminate beneath them. Expedition PCB can establish delay values and clearances per via span to address these issues. Additionally, Expedition PCB features true 45-degree routing for BGA fanout and staggered connectors, enabling localized rule definition to facilitate escape paths from dense areas.

22.6.2.11 TeamPCB

TeamPCB allows multiple designers to simultaneously work on the same PCB layout design. For PCB designs that are too large or complex to complete in the desired timeframe, TeamPCB enables collaboration across geographically dispersed or functionally organized PCB design groups, to maximize productivity and dramatically shorten design times. It is designed for immediate deployment and does not require time-consuming training or changes to a company's infrastructure or library to implement.

22.6.2.12 Design Reuse

The Design Reuse module creates and stores reusable blocks of circuitry, including schematic and PCB placement and routing data, in a central library. These blocks can then be placed and modified within the same design and across multiple designs. Design reuse automates this process and manages the design data to ensure error-free databases and reduce the overall PCB design cycle time.

22.6.2.13 DMS — Design Data Management

DMS integrates the electronic design process with the supply chain on the designer's desktop. It ensures complete data consistency, accuracy, and availability throughout the design enterprise. Additionally, DMS consolidates multiple data systems, enabling collaboration and life cycle management across multiple vendors, disciplines, and sites.

22.6.3 FPGA BoardLink

FPGA BoardLink automates the placement and wiring of the FPGA symbol within the board schematic, reducing this process from days to seconds. FPGA BoardLink maintains complete connectivity between the system-level design and FPGA design, facilitating route-to-simulation — either with the original FPGA hierarchy or with Synopsys LMC models programmed from FPGA layout files (see Figure 22.19).

Furthermore, FPGA BoardLink retains the correct PCB connectivity during revisions of the FPGA design. FPGA BoardLink not only creates the schematic connectivity needed for verification and layout of the PCB, but also documents which signal connections are made to which device pins and how these map to the original board-level bus structures.

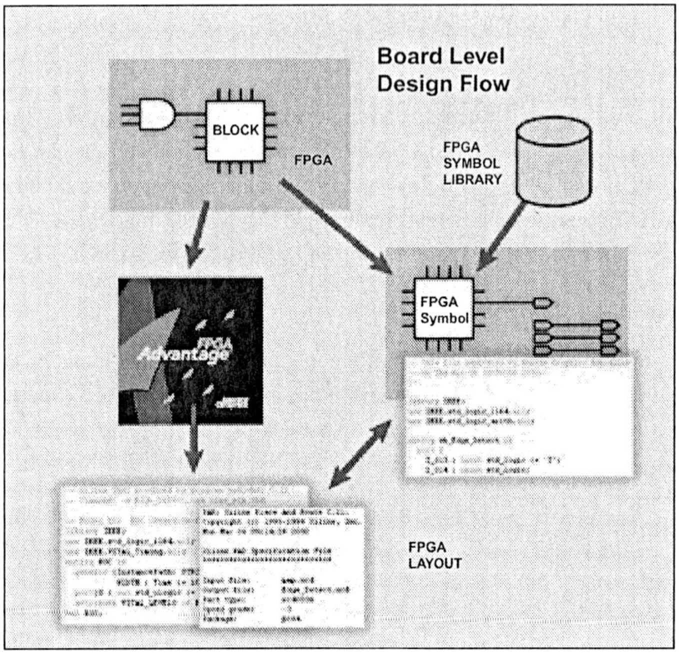

Figure 22.19 Board-level design flow.

22.6.3.1 FPGA On-Board

Today's FPGA devices offer I/O counts that exceed 500 pins. These large pin counts create a productivity obstacle when the FPGA design is incorporated into the board schematic. The effort required to manually place and connect several hundred pins is time consuming and prone to errors. Plus, this process may have to be repeated on subsequent FPGA design iterations.

Without FPGA BoardLink, the integration of an FPGA within a PCB design can take many hours each time an FPGA design is modified. Because FPGA BoardLink automates this process and maintains connectivity across design iterations, what used to take several hours or even days, now requires only seconds.

22.6.3.2 Reducing Design Times for FPGA/PCB Integration

FPGA BoardLink "slides" in an extra layer of hierarchy, mapping the user's pin names on an FPGA block to the actual pins on an FPGA symbol. Individual wire and original bused connections remain fully intact. FPGA BoardLink generates connectivity data directly from the FPGA vendor report files. The appropriate FPGA symbol is automatically retrieved from the FPGA package libraries based on the user's device and packages selections. The results can then be applied in a number of ways, depending on the overall design strategy, to complete the FPGA/PCB link. Users of DesignView normally complete a design starting from a top-level hierarchical block before using FPGA BoardLink. FPGA BoardLink can provide the starting-point for the system-level schematic, including a known good symbol. FPGA Advantage™ users design the FPGA using Renoir, ModelSim, and Leonardo Spectrum design flows. In this case, FPGA BoardLink builds a schematic to use within the PCB layout of the overall, complete system design.

22.6.3.3 FPGA Device Support

FPGA BoardLink supports all major FPGA vendors and devices, including large Virtex and Apex devices from Xilinx and Altera. These devices provide hundreds of programmable I/O pins that FPGA BoardLink manages and maintains for FPGA on-board integration. FPGA BoardLink is completely integrated within the DesignView environment and is available through a separate vendor configuration browser. This design-centric browser is used to manage all vendor-specific design tasks.

Output files from the FPGA vendor implementation tools are managed by the vendor configuration browser for the downstream tasks of board-level verification and PCB layout.

22.6.4 HDL Designer Series (FPGA Advantage)

HDL Designer Series includes customizable data management and analysis capabilities enabling designers to selectively view, search, organize, and correlate design data — dramatically improving the productivity of each engineer on your team. A flexible environment, HDL Designer Series integrates with the leading version management systems and supports your team's choice of simulation, synthesis, and other design tools. Robust data management and version control mechanisms help avoid costly mistakes and design iterations, and protect your company's design investment against the impact of staff turnover. The HDL Designer Series maintains a homegrown design environment and lowers the risks of distributed design projects reaching across extended teams.

22.6.4.1 The Design Manager Provides Complete Design Visibility

Large, complex designs require new design analysis and management techniques. The flexible and powerful Design Manager provides easy design navigation and the ability to search and organize data to aid in understanding design content. Sort, group, and filter your data based on any design unit attribute and instantly view any portion of the design hierarchy.

22.6.4.2 Tasks Support Repeatable Design Process

The use of Tasks enables predictable, repeatable design flow execution. Tasks allow common operations to be performed with a single click, saving invaluable design time. Tasks can range from individual "tools" to complex decision-based "flows." They are designed to serve the needs of designers trying to quickly capture a repeated operation and CAD support specialists responsible for maintaining complete design flows. Built-in tasks are provided to support all leading industry simulation and synthesis tools.

22.6.4.3 Mix Text, Tabular, and Graphical Editors for Varying Design Requirements

Design creation methodology must keep pace with larger, block-based designs. HDL Designer Series offers teams a choice of data entry mechanisms to match both individual preferences and specific design roles. The powerful DesignPad design-aware text editor and the unique Interface-Based Design™ (IBD™) connectivity table editor offer the fastest design iteration for experienced designers. These choices along with the configurable State Machine (see Figure 22.20), Block Diagram, Flow Chart, and Truth Table editors ensure that your team is the most efficient at each point in the design process. Accelerated HDL code entry DesignPad is a design-aware, fully customizable HDL code editor. DesignPad's power lies in its inherent understanding of HDL design. A code browser lets designers rapidly navigate HDL code and provides immediate access to any code blocks. DesignPad lets designers easily cross-reference error/warning messages and a powerful file comparison feature highlights differences and lets designers selectively apply changes to either file. Verilog and VHDL templates are included for faster and more accurate HDL entry. Speed-enhancing features include multi-tab, multi-window editing, and the capability to drag and drop components.

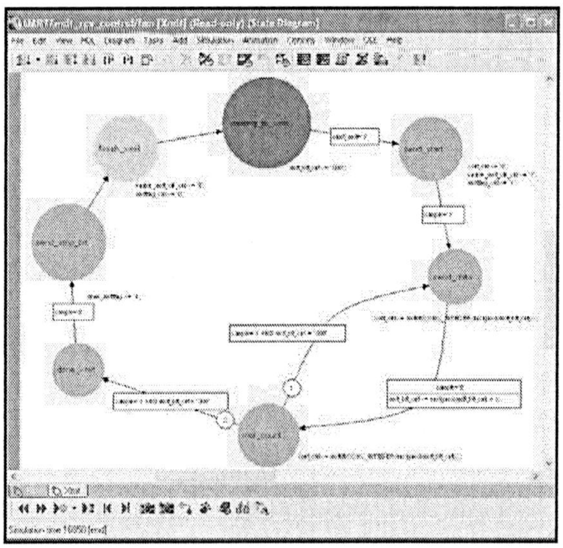

Figure 22.20 State Machine animation provides instant verification of state coverage.

22.6.4.4 Interfaced-Based Design Methodology Rapidly Defines Design Structure

The interface-based design editor is a unique tabular design methodology for rapidly editing structural HDL for large and complex designs. It allows you to rapidly describe design hierarchy and interfaces for blocks, reusable components, IP, and interconnects between design units. IBD works equally well as a standalone editor or as a complement to the Block Diagram editor with changes in either view instantly reflected in the other. (See Figure 22.21.)

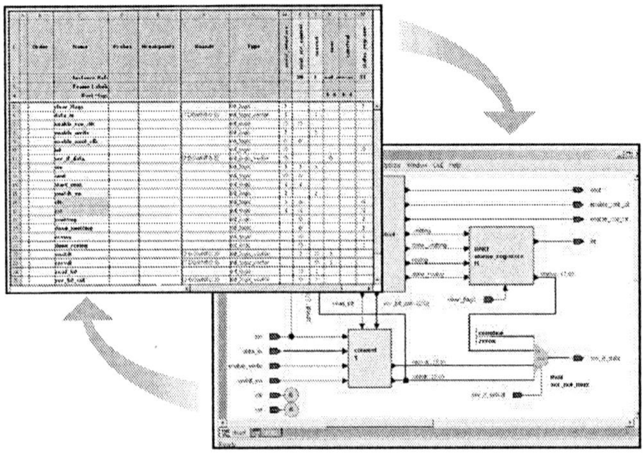

Figure 22.21 Interface-based design tables support compact structural HDL descriptions and can be viewed instantly as equivalent block diagrams.

22.6.4.5 Intuitive Graphical Editors

HDL Designer Series includes Block Diagram, State Machine, Flow Chart, Truth Table, and Tabular I/O editors for creating a design and, simultaneously, its corresponding documentation. Use of these graphical editors ensures a consistent coding style; further facilitating design reuse and maintenance. These editors use an intuitive, graphical methodology that is ideally suited to designers migrating to HDL methodologies or switching their primary design language, making them quickly productive.

22.6.4.6 Flexible ModuleWare Logic Generator

Flexible and easy to use, HDL Designer Series ModuleWare generates silicon-vendor-independent logic for a variety of common logic functions. Each function's parameters can be quickly tailored to fit your particular design problem. ModuleWare components are available for easy drag and drop into all three of the structural design editors: DesignPad, IBD, and Block Diagram. Most importantly, the code generated from ModuleWare is optimized for synthesis so you know you are always getting the most efficient implementation.

22.6.4.7 Practical IP and Design Reuse

Today, 45–50 percent of designs include IP reuse with 90 percent predicted by 2005. Design teams simply cannot complete large projects on time without the efficiency gained from reusing existing designs as building blocks.

22.6.5 ModelSim (FPGA Advantage)[16]

ModelSim SE is Mentor Graphic's UNIX-, Linux-, and Windows-based verification environment, combining high performance with advanced debugging capabilities (see Figure 22.22). ModelSim PE, by comparison, is Mentor's entry-level simulator and offers VHDL, Verilog, or mixed-language simulation. The information below describes ModelSim SE.

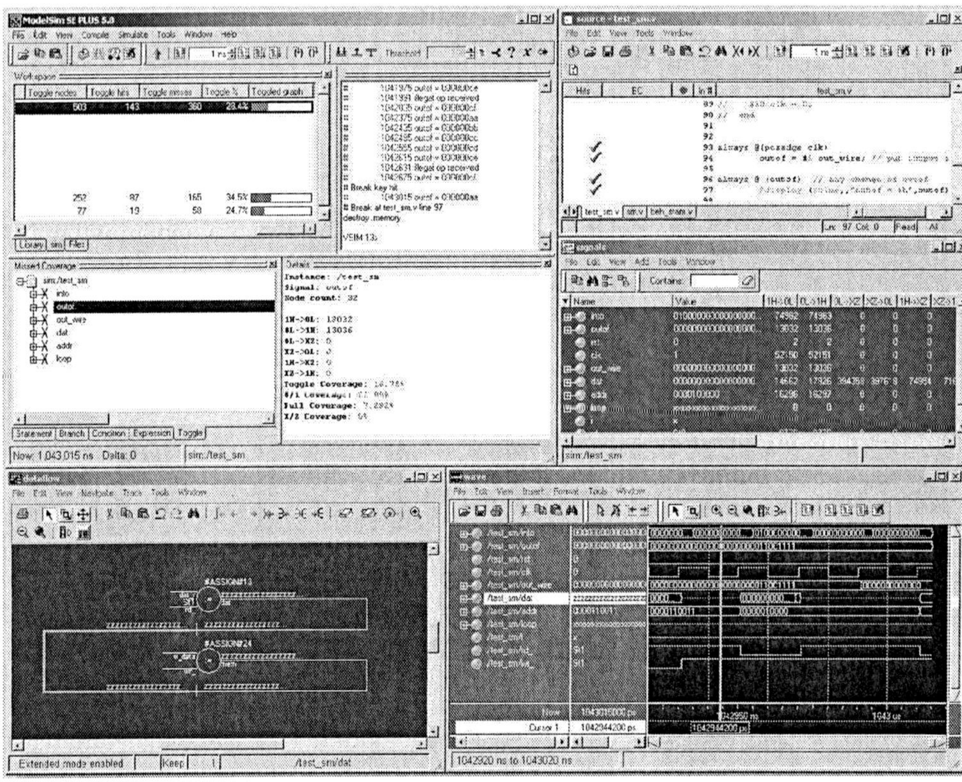

Figure 22.22 ModelSim SE is Mentor Graphic's UNIX-, Linux-, and Windows-based verification environment.

22.6.5.1 High-Performance, Trilingual Verification Environment

ModelSim SE is used on today's largest multimillion-gate designs. It provides advanced verification and debugging capabilities and supports 32- and 64-bit UNIX- and Linux-, and 32-bit Windows-based platforms.

16. ModelSim is the industry's leading mixed-language HDL simulator from Mentor Graphics. It simulates, verifies, and debugs all industry-standard HDL design languages, including VHDL, Verilog, Verilog 2001, SystemVerilog, PSL, and SystemC.

ModelSim is a trilingual simulator, meaning that it can accept inputs in any combination of VHDL, Verilog, and native SystemC[17] and has an integrated built-in C debugger. ModelSim provides a unified kernel for Verilog, VHDL, and SystemC for a true mixed-language environment. There is no learning curve with ModelSim's SystemC implementation since debugging SystemC code is done the same way as HDL code.

22.6.5.2 Verilog 2001/SystemVerilog

ModelSim SE fully supports IEEE 1364-2001 and has phase 1 implementation for SystemVerilog. SystemVerilog is an extension to IEEE 1364-2001 and provides additional constructs for modeling at higher levels of abstraction as well as additional design verification and test capabilities. SystemVerilog is an Accellera standard. For a complete list of supported constructs for Verilog 2001, please go to *www.ieee.org*. For a complete SystemVerilog LRM (language reference manual), please go to *www.accellera.org*. Advanced Code Coverage ModelSim has advanced coverage metrics integrated right into the tool.

22.6.6 Quiet Expert

A complete EMC methodology, Quiet Expert provides board designers and layout engineers with a powerful way to check their PCBs for incorrect or dangerous routing structures that may cause EMI problems. The tool will not only locate the problems by highlighting them in the integrated board viewer, but also report the cause of the problem and advice on possible solutions. It can be used throughout the design cycle — even on a partially routed board — so the design teams can have immediate feedback on the layout before it is complete. The ease-of-use model allows for all teams in the design phase to benefit from the technology. This method of analyzing the board early in the design cycle is the single most cost-effective way to analyze boards for potential EMI problems (see Figure 22.23).

22.6.6.1 EMC-Based Design Rule Checker

At the heart of the Quiet Expert analysis is the advanced design rule checker (DRC). The DRC analysis uses both physical and electrical parameters to verify the design and determine problems areas. The combination of using both allows the tools to provide a comprehensive EMC-based analysis. A powerful API enables creation of complex design rule checks that can include altering physical based rules parameters according to frequency. This is powerful in cases where higher frequency nets must be routed according to stricter guidelines than slower or less-critical nets. Rules can also be parameterized, facilitating adjustment based on design type or product line. Quiet Expert also includes built-in algorithms to predict radiation for any given net. These algorithms were pioneered by the University of Missouri-Rolla and provided to the UMR EMC

17. SystemC is a modeling platform consisting of both C++ class libraries and a simulation kernel for designing at the system-behavioral and RTL levels. It is suitable for creating cycle-level models of interfaces and hardware architectures. SystemC is emerging as the leading C/C++ design tool with support from all of the major EDA companies. SystemC was developed by Synopsys, CoWare, and Frontier Design (now Adelante Technologies) and is driven by the Open SystemC Initiative.

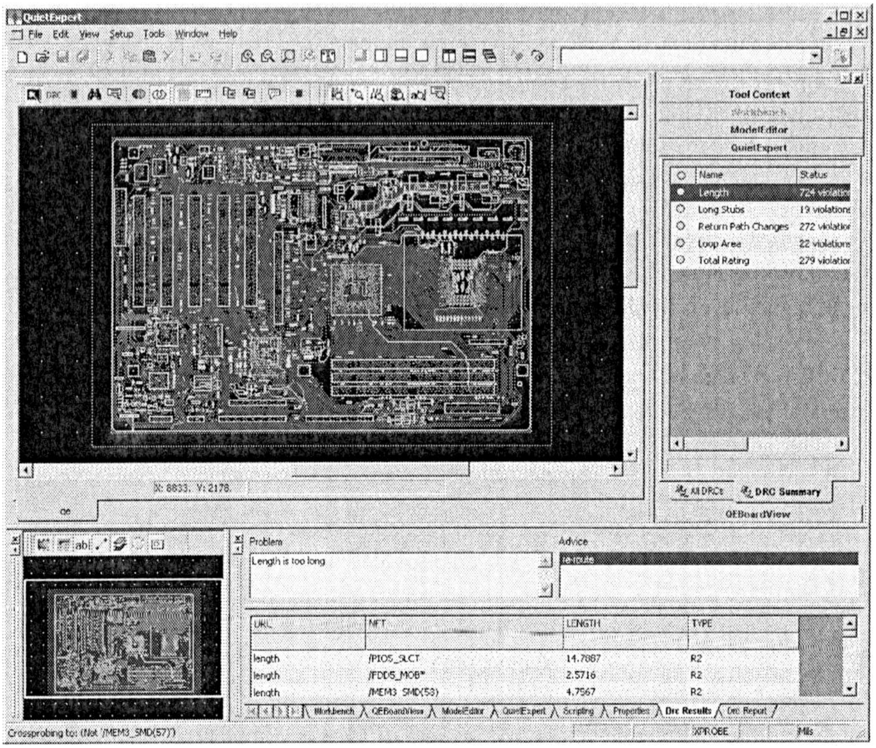

Figure 22.23 Quiet Expert is an integrated environment that easily identifies EMC problem areas on a PCB design.

Consortium, one of the biggest consortiums focused on EMC analysis in the United States. The results produced in this analysis allow the designers to compare relative EMC levels predicted from different nets on the board so violating nets can be prioritized.

22.6.6.2 Easy Viewing of Results

Results are easily viewed in several different ways. They are output in a spreadsheet-like window that conveniently displays all pertinent information of any particular DRC. These results can be sorted to any parameter, including net name and type of violations on that net. This facilitates finding areas of concern quickly and easily. Alongside the results are messages for problem and advice for that particular net. The messages can be customizable by the user, and can even be programmed for any international language.

These results are also automatically cross probed to the integrated board viewer so the user can immediately see where the problem exists on the board. Several highlighting options are available to the user to emphasize the violating area specifically. In addition, the viewer contains full layer display control to assist in examining the exact structure in question (see Figure 22.24).

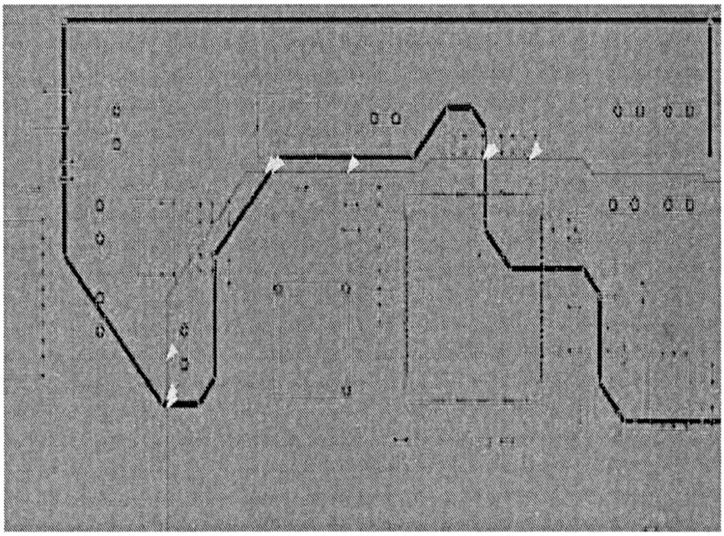

Figure 22.24 Complete highlighting of potential problems is supported. Here, a clock net crossing multiple gaps is shown with markers signifying impedance changes on the net.

22.6.6.3 Customizable Flow Control

A customizable flow control provides the flexibility to adjust to any corporate design flow. The result is a fully automated transition into the software, making it powerful yet easy to use. Proprietary tasks such as reading in different modeling formats, linking into schematic capture tools, or using databases managed by PDM[18] systems are all possible with the use of the flow control. Customizable flow control also supports automatic notification of the completion of analysis, database packing, and the team-based approach to PCB layout and analysis.

Major product features include the following:

- Complete EMC analysis tool for PCB layouts
- Scans design and flags areas of potential EMC problems
- Provides quantitative analysis on predicted EMI levels using algorithms pioneered at the University of Missouri-Rolla
- Advice on fixing potential problems aids in creating correct design to meet strict EMI regulations
- Full integration into most major layout tool flows
- Powerful DRC rule API allows for implementation of complex DRC rules for both EMI and signal integrity

18. PDM is program data management.

22.7 Signal Integrity and Timing Models

22.7.1 Signal Integrity Models

22.7.1.1 IBIS

An introduction to IBIS was presented in the previous chapter, Chapter 21, "Modeling with IBIS," starting on page 555. Also, Chapter 24, "IBIS Modeling and Simulation of High-Speed Fiber-Optic Transceivers," starting on page 681, provides a simulation example that uses IBIS models and compares the simulation results with those obtained with SPICE models. Section 22.8, "Use the Right Models for Simulation of Multigigabit Channels," beginning on page 613 in this chapter, discusses several considerations in the application of IBIS models. Other discussions on IBIS can be found in the index under two headings, "IBIS" and "models, IBIS."

22.7.1.2 SPICE

SPICE is a general-purpose circuit simulation program that enables engineers to model and simulate electronic circuits. An acronym for Simulation Program with Integrated Circuit Emphasis, it is very capable at solving a wide variety of circuit problems. This generalization, although an excellent feature for a general-purpose circuit simulator, is not ideal for solving problems in the signal integrity or PCB design spaces. Many optimizations can be — and are — made by other simulation techniques and modeling solutions. SPICE is often a convenient format to use as it is often the language used to describe and create many of the ICs used in PCB design. The input and output buffer models of these devices can be reduced (to improve simulation speed) and encrypted (although the SPICE model is now vendor specific, for a particular brand of SPICE) to preserve the intellectual property of the original design.

22.7.1.3 VHDL-AMS

VHDL-AMS is a modeling and hardware description language that can describe arbitrarily complex I/O buffer behaviors, while enabling very fast simulation speeds. VHDL-AMS uses the same base language constructs as digital designers use for VHDL in system, FPGA, and IC design. While simulation is not quite as fast as IBIS, VHDL-AMS has the advantage that it can model adaptive buffers and dynamic behaviors, such as is required when modeling transmitter pre-emphasis and receiver equalization. IBIS models[19] that do not call out specific SPICE or VHDL-AMS buffer models are unable to describe this behavior. There are no specific limitations on the types of logic that can be described, so single-ended signals are modeled as easily as differential signals. Transmitter pre-emphasis is when a signal switching at high frequency is amplified, or transmitted at a higher voltage, to compensate for the attenuation the signal will experience at its high switching rate on the PCB. Receiver equalization is when these

19. With IBIS v4.1, IBIS models now include the ability to call out, externally, buffer and circuit models that are described in SPICE, VHDL-AMS, and other languages. IBIS v4.1 models are a convenient way to encapsulate this adaptive model information as it uses a familiar framework to describe the rest of the component.

high-frequency signals are compensated for at the receiver. Often both pre-emphasis and equalization are required for the successful transmission of signals in a SerDes-type bit stream.

The advantage of both VHDL-AMS and static IBIS over SPICE is that their simulations run up to 200 times faster than SPICE. A system simulation that takes only one minute to run using VHDL-AMS models may take over three hours to run when modeled in SPICE. Because of the shorter runtimes, several simulation iterations can more easily be performed to examine the trade-offs of different signal drive levels used by pre-emphasis circuits, different amounts of equalization at the receiver, different internal termination resistances at the receiver, and other parameters selectable through programming or external control.

VHDL-AMS accurately describes analog, digital, and mixed-signal characteristics of integrated circuit I/O buffers, but does so without revealing circuit design details that are proprietary to the vendor. IBIS does this as well, but SPICE models contain proprietary vendor design information. To overcome this, SPICE models may be encrypted. Encryption of buffer models effectively hides the proprietary nature of the model, but does so at the cost of forcing the use of a specific SPICE simulator on all users of the models. Vendors prefer the use of VHDL-AMS and IBIS models to protect their intellectual property (IP) as it does not force the engineer using the model to specifically purchase another SPICE simulator, just for that model.

VHDL-AMS has tremendous flexibility in that VHDL-AMS code can be written only for those buffer characteristics of interest for which the simulation is being run. The code does not need to describe all features or functions of the buffer I/Os.

VHDL-AMS models are as accurate as SPICE models and are fully capable of modeling serial data links at today's 3.125 Gbps, 10 Gbps, and higher data rates.

22.7.2 Timing Models

Where IBIS, SPICE, and VHDL-AMS are considered signal integrity models, TDML and STAMP models are timing models used in timing simulations.

22.7.2.1 TDML

Timing Diagram Markup Language (TDML) is an electronic waveform standard that is based on extensible markup language (XML). TDML models expedite timing analysis, as component information is readily available in a machine-readable format (see Figure 22.25). In essence, TDML contains data book information in electronic format, as opposed to print or PDF format.

TDML has a growing industry momentum with support from many timing and waveform tools, including

- Tau (Mentor Graphics)
- TimingDesigner (Forte Design Systems)
- Timing Diagrammer (SynaptiCAD)
- TimingTool (available on the Web at *www.timingtool.com*)

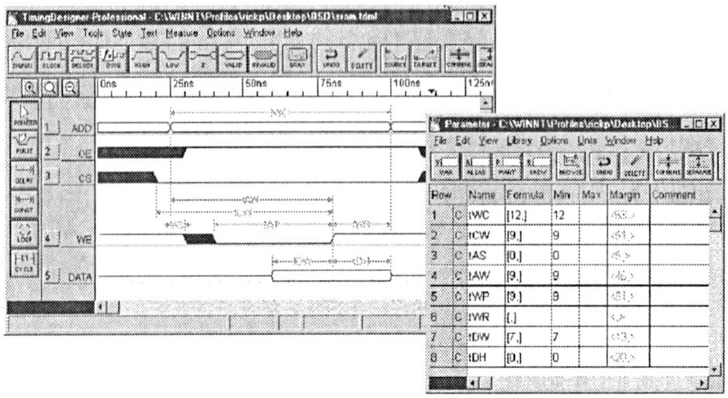

Figure 22.25 Manufacturer supplied timing — TDML.

22.7.2.2 STAMP

STAMP is another electronic format that is generally used by FPGA and ASIC design tools (see Figure 22.26) for timing analysis. Created by Synopsys, STAMP provides a mechanism for describing timing paths (arcs) and associates timing numbers to each of these timing paths. Although the STAMP format is not an open standard, it is considered a stable and viable format for expressing component timing. Models are described in two parts: model file (timing arcs) and data file (timing numbers). Functional model information is not provided in the STAMP format.

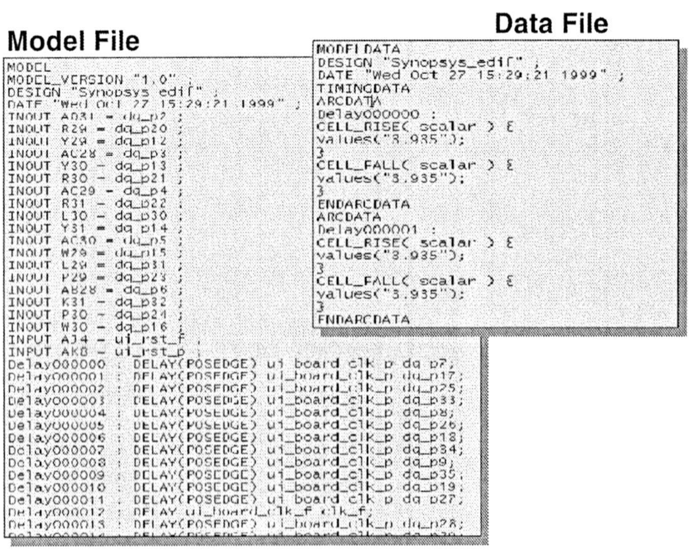

Figure 22.26 Manufacturer supplied timing — STAMP.

Further, multiple data files may be overlaid onto a single model file. This is beneficial when using custom or programmable devices (such as FPGAs and ASICs) that have one functional description and may have multiple timing versions from different place and routes.

22.7.2.3 TDML vs. STAMP

TDML models are usually created by device component vendors for devices that do not contain re-configurable or custom hardware, such as combinatorial logic, memory, and microprocessors. These models use graphical waveforms, and the timing relationships between the waveforms are easy to view and edit.

STAMP is used predominantly for FPGA and ASIC designs. Since a single logical design may have more than one set of overlaid timing numbers, different FPGA place-and-routes may also be overlaid to determine the most effective component placement. Critical paths may then be identified and then fed back into the FPGA routing to produce a design that maximizes timing requirements and is more tolerant of timing variations.

22.8 Use the Right Models for Simulation of Multigigabit Channels[20]

22.8.1 Introduction

Models drive simulations. Simulation of electrical circuits in modern systems is becoming more important as different engineers, teams, or companies design separate components and modules. Multiple design teams need the ability to validate the combination of all sections of the system design. Waiting to assemble actual prototypes is no longer feasible in today's market, so simulation tools provide the information needed in the earlier stages of the system design.

It used to be acceptable to divide digital functional simulation from analog simulation in system analysis and verification. Now, the digital function and analog effects are becoming more tightly mixed in IC I/O buffers. For a long time, designers used only SPICE for the few circuits that they simulated. IBIS was created to protect IP and speed up simulation time, thus allowing for the simulation of more circuits because of available time and models. This worked well until the I/O buffer technology complexity outpaced the IBIS specification development. Realizing that simulation can save time and money in the design cycle, but needing more flexibility in simulation tools, many designers have reverted to SPICE simulations for nets with complex drivers. With the flexibility of running a SPICE simulation comes its associated complexities and difficulties. The optimal situation is to have just the model information that is required to validate the design constraints without extra information that extends the simulation time.

This section studies an example of using VHDL-AMS, SPICE, S-parameter, IBIS, and transmission line models in a single simulation environment of multigigabit circuits to find an acceptable balance of model detail and simulation speed. VHDL-AMS, being a superset of VHDL, provides purely digital simulation for sections that don't need to be simulated as analog

20. This section was written by Weston Beal of Mentor Graphics Corporation. [Bea04]

circuits. These sections might generate complex control signals for a dynamically programmed line driver or multibit stimulus based on a specific bus protocol. VHDL-AMS also provides the digital-to-analog and analog-to-digital converter models to allow mixed signal simulation. SPICE and S-parameter models are used on some sections where those are the only models available from a third party. IBIS models are used when available to speed the simulation of some of the traditional I/O buffers in the circuit. Doing this simulation of multiple model types in one simulator increases overall usability and accuracy.

For years the signal integrity (SI) community has been pushing to do signal simulation earlier in the design cycle. Engineers started by mostly doing simulation of PCBs after the layout process was complete. Maybe the simulation was done after an error was found with a prototype in the lab. This simulation was useful as a verification step, but the method of fixing the error at that late stage in the design was not usually the most efficient. It is generally well understood that SI engineers can save time and design resources by doing SI simulation as early as possible in the design cycle. The type of simulation depends greatly on what types of models and simulators are available. Often, engineers just work with what is given to them and make the best of it. After all, that's what makes them good engineers, right?

For this case, simulating a multigigabit serial bus before or during the schematic design entry will be studied. The design depends on knowing the results of various scenarios. The purpose of our simulation is to determine the programmable configurations that should be used for the drivers and receivers of an Altera Stratix GX FPGA high-speed serial I/O (HSSI). It will be assumed that some mechanical constraints, such as component placement and connectors, are already established.

22.8.2 The Circuit under Study

Let's start by looking at what model is available for the FPGA. In this case the only model that was found is an encrypted SPICE model[21] of the transceiver I/O buffer. It's not really clear how much of the buffer is included in the simulation circuit, but at least the sections connected to the device pins are there. Looking into the model it is found that the whole circuit is designed as a top-level SPICE circuit. There are some subcircuits that are encrypted, but the buffer is not a separate subcircuit. The model as created by the vendor includes the I/O transceiver pair, a fairly complex package model, three inches of lossy transmission line, and various voltage sources to control the behavior of the driver and receiver. Hereafter, the circuit will be referred to as circuit "A."

21. HSPICE and Mentor's Eldo simulator provide encrypting utilities that make files unintelligible except by their simulator. This makes it possible for vendors to encrypt model files and pass them to customers, allowing the customers to use the model in simulation, but does not give away any proprietary information. The negative effect of encryption is that the user cannot change things that might be causing errors and must use the model in the targeted simulator and no other.

Figure 22.27 shows the level of detail included for the transceiver model. The main subcircuit for the transmitter is the block highlighted in the left section of Figure 22.27. The main subcircuit for the receiver is the block just below the transmitter block in the figure. Each of these subcircuits needs to be isolated with just the minimum number of elements to support them and get the correct behavior.

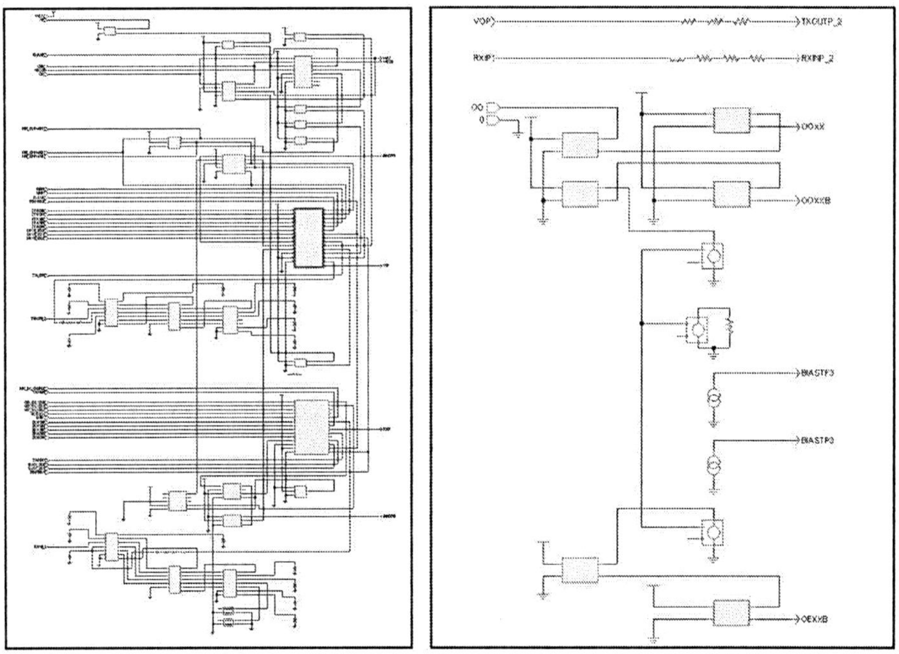

Figure 22.27 The schematic of the full HSSI transceiver is fairly complex.

Another potential problem with the original SPICE circuit from the vendor was that it used global voltage nodes. Global nodes are fine if you can be certain that your circuit will never be used in another model environment. But what if a model creator from another company uses the same global node name for another purpose and the end user needs to simulate a circuit that uses models from two different sources? The end user must then remove the global node declaration and edit any subcircuit that used the global nodes to declare the nodes in the subcircuit definition. Furthermore, what if one or both of the models is encrypted? Now the end user cannot edit the subcircuit. This could easily be a dead-end situation rendering the models useless. In the end it was found that all the subcircuits needed for the driver and receiver had voltage supply nodes on the subcircuit declaration so it wasn't a problem.

The HSSI model also implements the scale option (OPTION SCALE = 1u). This is also a potential danger. It seems that most IC model creators use the scale option to make their transistor models easier to create. Again, that's fine if they can ensure that their model will never be used in a different environment. Imagine what would happen if vendor A defines transistor models assuming a scale option of 1u and vendor B assumes no scaling option. Then the SI engineer at company C gets these two device models and needs to simulate them in the same circuit. In order for this to work the end user must understand all the implications of the scale option and change the scaling in one set of models or the other. This task can be so extensive that it overshadows the benefit of running the simulation. This study used only the HSSI model so the scale option did not cause a problem in the simulations described here.

Simulation of the original circuit from the vendor took about 21 minutes in real time or 14:27 of CPU time. However, most engineers really don't want to wait 20 minutes for each simulation. If the engineer is only going to do one simulation and look at the pretty waveforms, then this is fine. But in reality, this simulation will probably be run over and over during a couple of weeks to investigate different configurations of pre-emphasis, voltage swing, trace length, connectors, and termination.

22.8.3 Model Reduction

Engineers would really like to have separate subcircuits for the driver and the receiver. This makes circuit creation and modification easier. Also, with a circuit that contains so many transistor models it saves a lot of simulation time to remove as many transistors as possible.

The first step in reducing the simulation time is to separate the driver and receiver models and put them into individual subcircuits. Also, there was an issue of creating appropriate input stimulus to get the desired waveforms out of the driver. The model used some unknown circuits to derive multiple signals to control data and pre-emphasis sections of the driver. Some of the transistor-level subcircuits were replaced with simple logic functions that produced the correct control signals for the output stage and reduced the number of transistors in the simulation. This, in turn, decreased the time to simulate the circuit. The resulting subcircuits for the driver and receiver are shown in Figures 22.28 and 22.29, respectively.

Also, with the driver and receiver defined in subcircuits, the models can be instantiated on parts in a simple ICX electrical editor design. Running the simulation from ICX allows the easy modification of the interconnect and changing models in the circuit. It also allows the use of a prototype of a design kit that makes the programming of the driver and receiver behavior very quick, easy, and less error prone. The design kit includes a small GUI to set configuration parameters from drop-down lists. This utility writes SPICE include files that the driver and receiver models read to control their behaviors. With the reduced I/O buffer models used in circuit "A," simulation of the circuit took about 12 minutes real time or 7:38 CPU time to run in ICX. This is about 57 percent of the original simulation time.

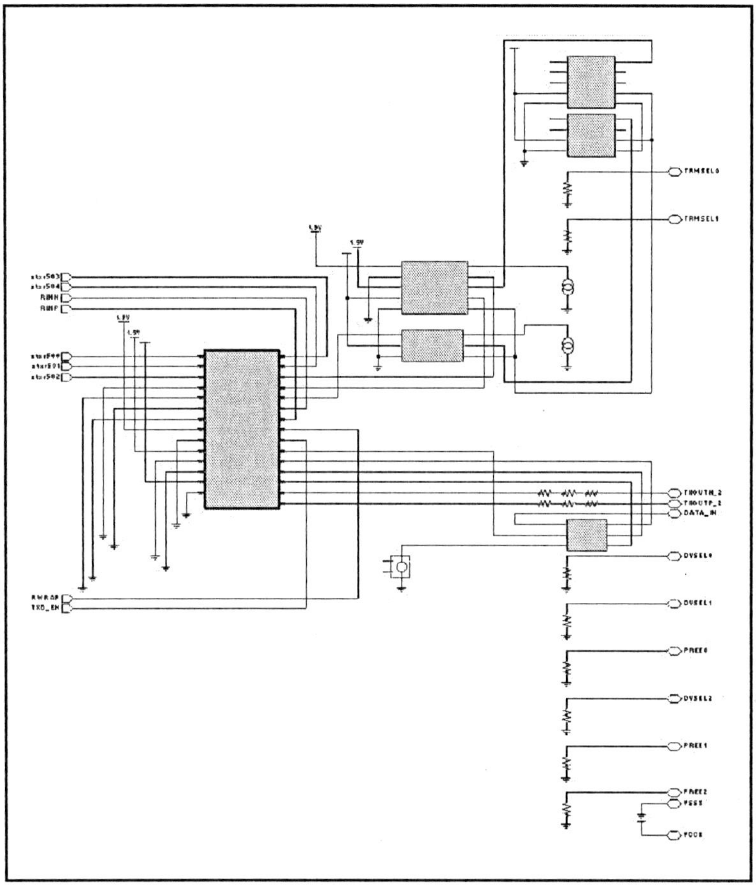

Figure 22.28 The driver subcircuit is reduced to speed up simulation time.

22.8.4 Package Model Simplification

The next step in reducing simulation time was to work with the package models. The package models from the vendor included three fully coupled sections of six pins as indicated in Figure 22.30.

The unused pins were terminated with 50 Ω resistors to ground at both ends. The model of the middle section used a 6-conductor frequency dependent coupled lossy line model. The first and last sections used discrete elements for each pin path with mutual inductance and capacitance values. This seemed like overkill for such a short path of about 1 cm.

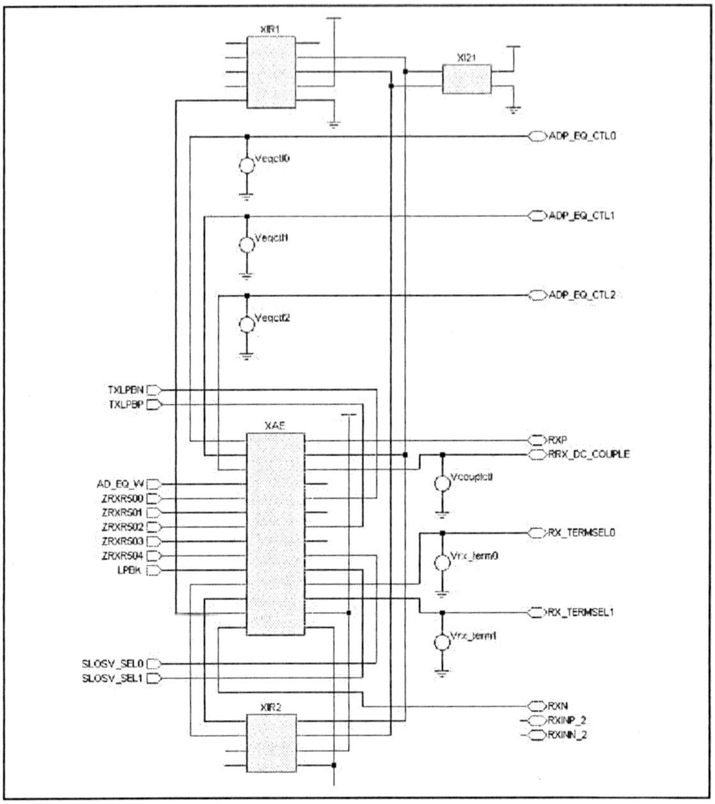

Figure 22.29 The receiver subcircuit is simplified significantly.

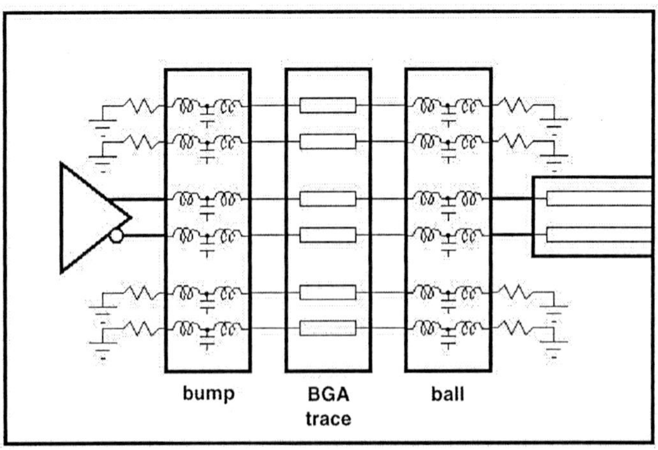

Figure 22.30 The package model includes six pins.

22.8.4.1 Package Element Elimination

By removing the elements representing the unused pins, it was thought a significant amount of simulation time might be saved. This was done by simply removing the extra elements without changing the values of the remaining component values to compensate for the removal of some elements. The simulation of circuit "A" with the reduced package model took 7:05 minutes of CPU time. The simulation time was still 93 percent of the time it took to simulate with the full package model and actually changed the resulting waveform beyond acceptable error. Considerable time could be spent correcting the 2-pin model to correctly represent the two pins used from the 6-pin model, but the slight reduction of simulation time did not warrant the time it would take to construct the accurate reduced package model.

22.8.4.2 IBIS RLC Package

Another approach to reduce the package model was to use the simple RLC parameters. Since the SPICE subcircuits of the driver and receiver are referenced from an IBIS model when used in ICX, this approach is trivial to set up. Using equations $L = Z \times T$ and $C = T \div Z$, where T is 83 ps (measured from die and pin waveforms) and Z is 50 Ω, gives L = 4.15 nH and C = 1.66 pF. The simulation of circuit "A" using the simple RLC package model ran in only 5:45 minutes of CPU time. This is 75 percent of the simulation time when using the full SPICE package model, so it would be advantageous to use this method. Also, the simple RLC model is not adequate to represent the package for the fast edge rates of the HSSI driver. The reduced overshoot in the simulated waveforms indicate that this RLC model matches the desired 50 Ω better than the full SPICE package model. After many iterations of changing L_pkg, C_pkg, and R_pkg in the IBIS model it was determined that a single section of RLC elements does not represent the package closely enough to match the simulated results at the driver pins.

The more important issue is whether the waveforms match at the receiver. The differential waveform at the receiver pins after 10 inches of lossy line are not close enough to continue down this path. The simple package model in conjunction with the VHDL-AMS buffer models (discussed later in this section) is overly optimistic. It doesn't show as much noise as the full SPICE model with extensive package model.

22.8.4.3 S-Parameter Package Model

Another method of modeling components, such as the package, at high frequencies is with S-parameters. It is fairly easy to isolate the package model and have the SPICE simulator generate an S-parameter set that represents the subcircuit. A 4-port S-parameter set was created representing the 6-pin network. The four ports are at both ends of the two pins used and the unused pins are still terminated with 50 Ω to ground. The S-parameters were computed at increments of 100 MHz up to 10 GHz. Using this S-parameter set to model the package in circuit "A" the simulation time was reduced to 5:39 minutes of CPU time. This is a good improvement at 74 percent of simulation time over the full package model, but the results were slightly different from the original simulation. The S-parameter model will be reused in later simulations to try to further qualify if the difference in simulation results is acceptable.

22.8.5 Behavioral Modeling

Since the most improvement in simulation time seems to come from reducing the I/O buffer model rather than the package model, the possibility of changing the model to a behavioral model will be investigated.

The first thought might be to create an IBIS model of the buffer. This idea is fleeting as soon as it is realized that IBIS does not model drivers with pre-emphasis well. Another approach would be to try macro models in SPICE, but a more appropriate method is VHDL-AMS (IEEE 1076.1). VHDL-AMS provides the syntax to make purely analog models, mathematical functions, and strictly digital signals mixed in the same environment. Applying this language to model the HSSI driver and receiver is a logical choice because the digital state machine syntax of VHDL can be used to control the pre-emphasis signals and the analog and mathematical functions used to model the input and output stages. The information required to make such a model is usually supplied in a device data sheet. It is not necessary to know the internal implementation because all that is needed is to match the behavior of the device at the pins. Nevertheless, with more information, the model might be made more accurate or versatile. The necessary level of model detail depends on the application.

Parameterized VHDL-AMS models were made of the HSSI driver and receiver. With some experimentation, the model parameters were adjusted such that the simulation results matched the results of SPICE simulations very closely. The circuit used for correlation was simply a 50 Ω resistor from each pin to 0.75 V (the common voltage = $V_{DD}/2$). The test circuit excluded the package model in order to more accurately match the behavior of just the driver.

The results of the VHDL-AMS model compare very well with the original SPICE simulation of the driver as shown in Figure 22.31. The Stratix GX HSSI driver has programmable output swing, pre-emphasis level, and output impedance. In the figure, the three different pairs of curves show different output voltage and pre-emphasis levels. The major benefit is that the simulation is much faster. Simulation time for VHDL-AMS model in the test circuit is reported as 0.07 seconds of CPU time. It seems that this time included only the matrix solution time and not the simulation setup so the real time was measured from simulation command until waveforms appear in real time. This time was only 9 seconds. This is extremely fast compared with the simulation of the full SPICE buffer model, but remember that this simulation included only the driver and a couple of resistors. The real test is to compare the simulation of circuit "A" with the VHDL-AMS buffer models versus SPICE models.

Now with a good behavioral model written in VHDL-AMS, the SI engineer can work faster to determine an appropriate package model. We'll go back to the full 6-pin package model as a baseline. The simulation of circuit "A" with the VHDL-AMS buffer models and the full SPICE package model took 27 seconds in real time. Compare this time with the simulation time of circuit "A" with the separated driver and receiver SPICE models. The VHDL-AMS model simulation took only 8 percent of the time for the SPICE model simulation. Using the S-parameter model of the package and keeping all other elements of the circuit equal reduced the simulation time to 12 seconds. The results are comparable for the signal transitions, even though the S-parameter results in Figure 22.32 show a bit more noise at the high and low levels.

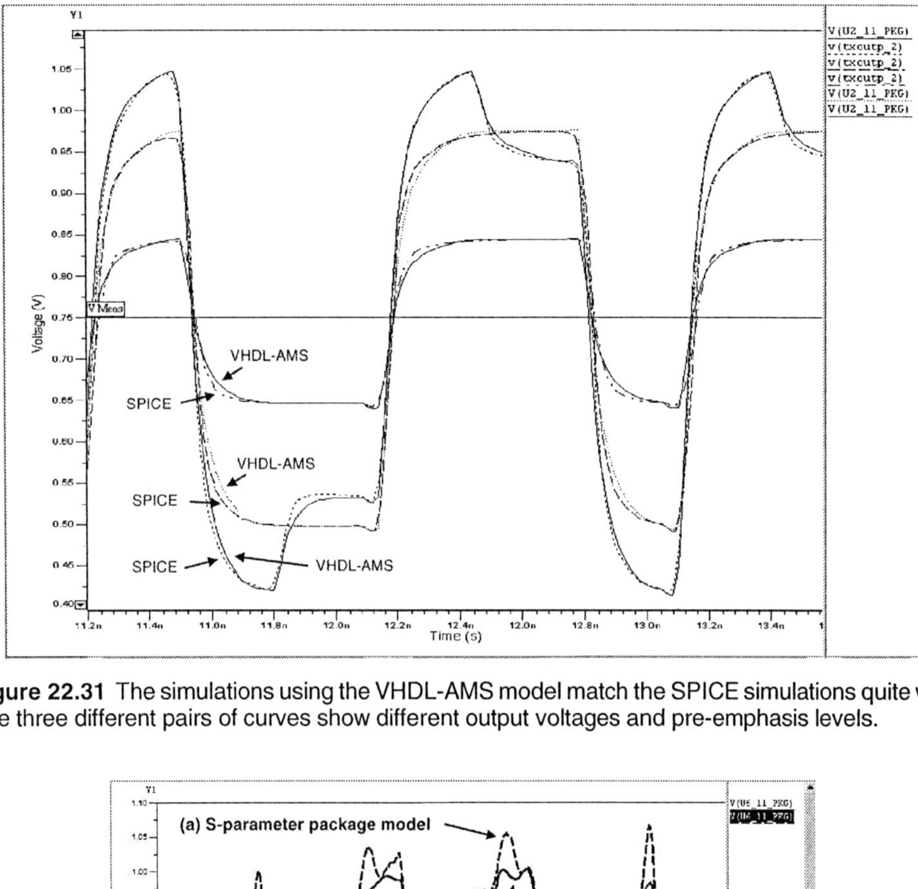

Figure 22.31 The simulations using the VHDL-AMS model match the SPICE simulations quite well. The three different pairs of curves show different output voltages and pre-emphasis levels.

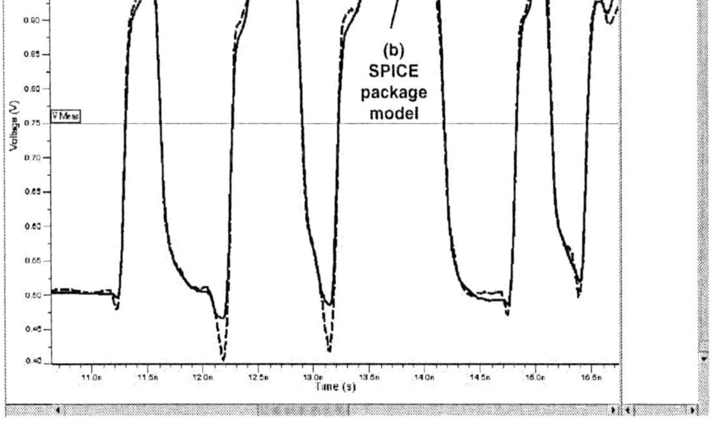

Figure 22.32 Simulated waveforms at the driver pins (after propagating through the driver package) for (a) S-parameter package model and (b) SPICE package model. The VHDL-AMS driver buffer model was used for both simulations.

Even though the resulting waveforms at the driver pins are a bit different, the more important comparison is at the receiver. Circuit "A" was modified by extending the three inch trace between the driver and receiver to 15 inches of trace on FR-4 and modeled with a lossy transmission line model. The results are shown in Figure 22.33 for the cases of full SPICE and S-parameter package models.

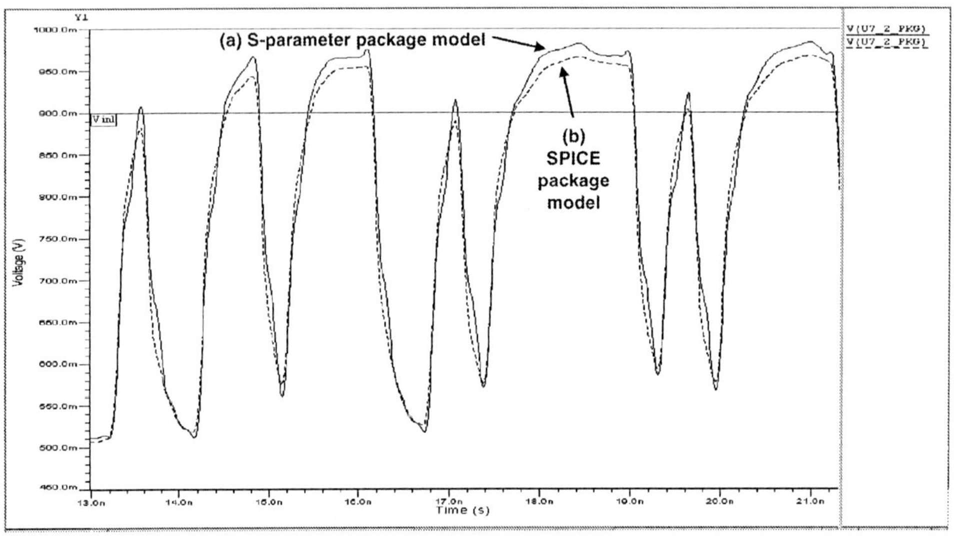

Figure 22.33 Simulated waveforms at the receiver (using VHDL-AMS driver buffer model) after 15 inches of FR-4 lossy transmission line for (a) S-parameter package model and (b) SPICE package model.

The simulation with the S-parameter package model still has a higher dc level. Figure 22.34 shows the eye diagrams of the differential signals from the two simulations.

From the eye diagrams it is observed that the S-parameter model is noisier and the eye closes just a little bit on the left. If this result leaves some margin around the eye mask then the error might be tolerated in order to achieve the decrease in simulation time.

22.8.6 Conclusion

The results of many simulations of the Altera high-speed serial I/O transceiver used in the Stratix GX FPGA have been shown. Different buffer models have been investigated in an attempt to decrease circuit simulation time. It is clear that the VHDL-AMS behavioral model runs much faster (about 3 percent simulation time) than a full transistor level model with very comparable results. On the other hand, trying to reduce the package model to a model format that saves significant time in simulation was not so straightforward. The results varied dramati-

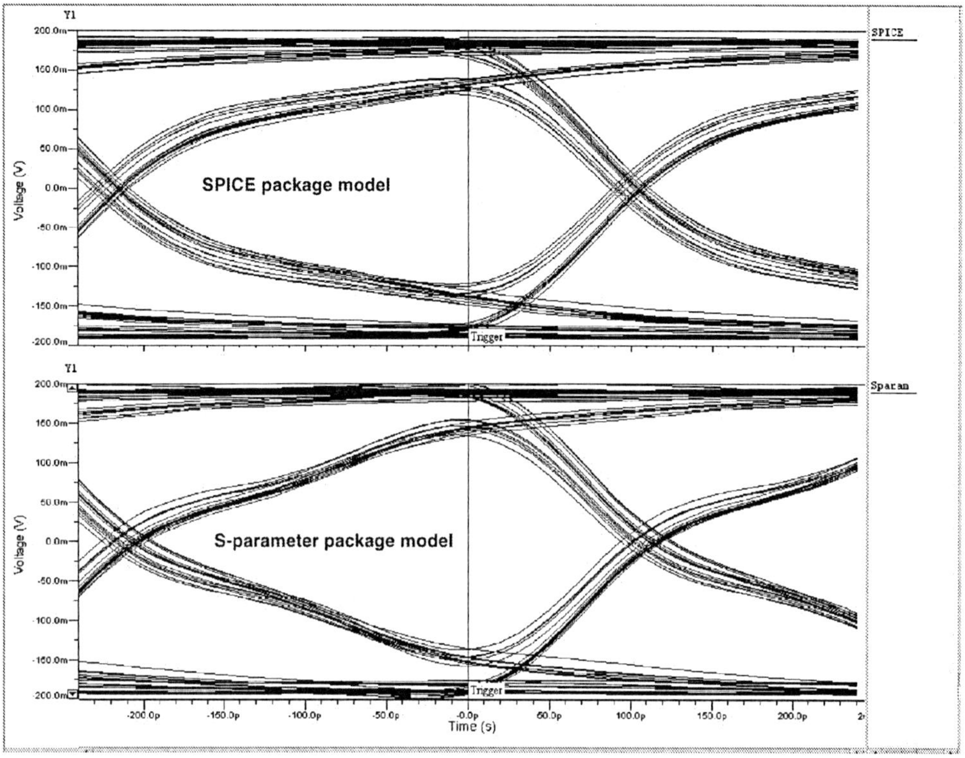

Figure 22.34 Simulated eye diagrams are still different at the receiver end of 15 inches of lossy transmission line. (VHDL-AMS driver buffer model used.)

cally in some cases and didn't reduce the simulation time enough to justify the model creation work nor to accept the errors that the quicker models introduced. In order to reduce simulation times in general, the focus should probably be on reducing or simplifying buffer models. For IC packages, traces, and connectors, the best models that can be obtained should be used and let the simulation time be a secondary concern.

22.9 Exercises

22-1. What is a "multiplow" capability?

22-2. What does "rules by area" mean?

22-3. Define the following terms: (a) via span, (b) area fill, (c) DRC, and (d) symbolic timing analysis.

22-4. What is a variant manager and what does it do?

22-5. What does a field solver do?

22-6. In what year did Mentor Graphics first have constraint management in their high-speed design tools?

22-7. (a) Name two model formats that can be used to describe an adaptive device buffer. (b) Can these be used with IBIS models?

22-8. What is a TDML model?

22-9. How does symbolic timing analysis differ from static timing analysis?

22-10. What problem does Quiet Expert help design engineers solve?

22-11. Name two timing model formats. How do they differ?

22-12. In the section titled, "Use the Right Models for Simulation of Multigigabit Channels," several approaches were taken in an attempt to make the simulation run faster, i.e., to save simulation time. (a) List three of these approaches, (b) indicate whether each of these three approaches worked without incurring unreasonable distortion of the modeled signal, and (c) explain why each of these three approaches did or did not work without incurring unreasonable distortion of the modeled signal.

22-13. Name at least nine benefits of using a floorplanning tool.

22-14. What is the difference between LineSim GHz and LineSim EXT?

22-15. What is the benefit to being able to assign design rules to nets in a hierarchical fashion?

PART 5

Design Concepts and Examples

CHAPTER 23

Advances in Design, Modeling, Simulation, and Measurement Validation of High-Performance Board-to-Board 5-to-10 Gbps Interconnects

Written by Brian Vicich of Samtec Inc., Scott McMorrow, Tom Dagostino, Jim Bell, and Bob Ross of Teraspeed[SM] Consulting Group LLC, and Rob Hinz of Cider Designs

Chapter Objectives

- Examine modeling methodology and its application to PCB traces, connector design, the breakout regions of connectors, and vias.
- Discuss passivity in frequency domain network measurement and simulation and how to correct error-induced nonpassivity in S-parameter network models.
- Discuss measurement accuracy issues with SMA launches and traces.
- Discuss how to measure and validate printed circuit board material parameters.
- Demonstrate the high correlation between eye patterns generated from measured S-parameters using a VNA and eye patterns created through trace and component modeling.
- Discuss the development of n-port mixed-mode S-parameters (in chapter appendix).

23.1 Introduction[1]

As an ever-increasing number of compact digital systems move to multigigabit data transmission speeds, designers are faced with the challenge of bringing robust, high-performance products to market at ever-increasing risk, due in part by outdated or incomplete modeling and

1. Material in this chapter is reprinted with permission of Samtec Inc., Teraspeed Consulting Group LLC, and the IEC (International Engineering Consortium). This material was first presented at DesignCon West 2004. Web sites for Samtec, Teraspeed Consulting, and the IEC can be found at *www.samtec.com*, *www.teraspeed.com*, and *www.iec.org*, respectively.

simulation techniques that do not take into account all sources of high-speed signal loss. In this chapter, we will present hybrid design, modeling, simulation, and measurement techniques that reduce risk in high-performance designs of 10 Gbps and above.

We will discuss the importance of modeling that is appropriate to the system elements being modeled, and then how to combine all of these modeling elements into a hybrid, unified time-domain simulation environment utilizing Synopsys HSPICE.

We will next describe the design of transparent measurement and validation boards, and ultimately show a stunning correlation between measured and simulated performance in the time and frequency domain for busses designed with high-volume, low-cost, board-to-board connector systems. S-parameter data from these test boards will be shown to match solutions generated from electromagnetic field solvers and simulation. In addition, time domain measurements will show near-exact correlation between simulation and measurement.

Finally, real eye pattern measurements of an operating 10 Gbps board-to-board system will be demonstrated to show the power of this hybrid methodology in characterizing and optimizing design performance.

Many multigigabit per second designs will utilize some form of board-to-board mezzanine interconnect, facilitated by connectors, cables, and flex assemblies.[2] These interconnections are often being asked to perform at exceptionally high data rates that were unthinkable just a few years ago. Yet, as the speeds of systems are increasing, cost is being driven out of products at accelerating rates. Simple board-to-board connector systems, such as Samtec's QStrip™, QPairs™, SamArray™, and other future low- and high-density connectors, along with flex circuit and high data rate cable assemblies, are now being asked to perform within custom interfaces in a wide variety of applications such as PCI Express, HyperTransport, OC-192 mezzanine interconnections, and the like. Connector and interconnect systems that were not explicitly designed to operate at these rates are nonetheless being designed into advanced systems.

In an effort to push existing and future products to their limits, ever-increasing challenges drive the need for precise modeling and simulation. At high data rates and high frequencies, exact modeling is not only important but also is absolutely necessary. Every effect must be accounted for in the modeling of connectors and all other elements in the interconnect path. Simple transmission line or lumped element models are no longer appropriate in all areas. Lossy material properties must also be accounted for and properly extracted, modeled, and simulated in the frequency domain. Three-dimensional (3D) electromagnetic effects can become a design "deal breaker" if not considered, modeled, and simulated. Even the boundary conditions for each modeled section need to be carefully weighed.

Unfortunately, no one tool can be relied on to handle the complete job at this time. Two-dimensional (2D) electromagnetic field solvers, such as Ansoft Maxwell 2D, produce accurate and reliable modeling results for uniform 2D trace geometries, if used correctly. Frequency-dependent models that include the effects of field penetration into conductors and dielectric

2. A flex circuit or assembly is a flexible circuit board used for high-speed interconnection and consisting of DuPont Pyralux® FR™ dielectric board adhesive material that corresponds with Kapton dielectric material.

losses can be accurately extracted. We will show that, with care, W-element models[3] can be created for Synopsys HSPICE, which have stunning accuracy and correlation to measured data.

The breakout region (BOR) is the PCB area surrounding the connector used for pads, escapes, via placement, and routing. In other words, it is the area where the traces are "broken out" from the connector pads and footprint to the area of uniform traces. Vias at breakout regions, connector interfaces, trace crossings or trace near-crossings, antipads,[4] and non-uniform transmission line sections (such as the "Y" to a differential via[5]) are problems that lend themselves to 3D full-wave modeling[6] and simulation techniques. Microwave tools such as CST MicroWave Studio, or Flomerics Micro-Stripes™ can be used to provide highly accurate frequency domain modeling of complex structures that cannot be adequately modeled with the current industry-standard 2D tools. But these 3D microwave tools generally produce "natural" S-parameters as their output, which must somehow be integrated into a time domain SPICE simulation environment, with HSPICE often being the de facto standard.

To facilitate the translation of frequency domain S-parameters into time domain SPICE models, a new class of highly accurate black box model fitters like Sigrity BroadBand SPICE is now available. With amazing accuracy, tools such as these can provide the "missing link" between the frequency domain and the time domain. Yet another path exists, utilizing Fast Fourier Transform, Inverse Transform, and frequency domain convolution techniques to perform fast and accurate simulations of S-parameters in the time domain. However, these approaches are not without potential pitfalls. Errors occur at all stages of modeling and simulation. Approximations, numerical errors, truncation effects, subtle undamped resonance, and other numeric or algorithmic limitations can render the time domain simulation of frequency domain extracted elements unstable or nonpassive. To be useful in a wide variety of simulations, extracted (or measured) S-parameter models must be guaranteed to be passive throughout the simulation process. Based upon our experience, Teraspeed Consulting has developed various techniques to "passivity correct" these models and ensure stable time domain simulation results under a wide variety of operating conditions.

Working as a team, Samtec and Teraspeed Consulting have developed a robust process that utilizes advanced 2D, 3D, frequency domain and time domain modeling, and measurement techniques in the development of an advanced modeling and simulation environment for current and future high speed connector systems. This process can be seen today in the Samtec Final Inch™ signal integrity offering for the QStrip, QPairs, and SamArray connector families, which can be used for the modeling and simulation of advanced interconnect systems.

3. A W-element model is a Synopsys HSPICE model for a coupled lossy line element. It assumes a square-root dependence of the skin effect with frequency and linear dependence of the dielectric loss with frequency.
4. An antipad is the insulating gap between the copper via and the copper plane surrounding the via.
5. A differential via is a pair of regular vias located close together and fed by a differential signal.
6. Full-wave modeling refers to using models or modeling techniques that make no substantial simplifying assumptions to Maxwell's equations.

In this chapter, we present hybrid design, modeling, simulation, and measurement techniques that reduce risk in high-performance designs of 10 Gbps and above.

We discuss the importance of modeling that is appropriate to the system elements being modeled, including

- High-accuracy stripline trace modeling with the Ansoft Maxwell 2D electromagnetic field solver and Synopsys HSPICE W-element table models
- Full-wave three-dimensional modeling utilizing CST MicroWave Studio for vias, connector transitions, and SMA launches[7]
- Lumped-element versus S-parameter modeling for connector cross-sections
- Full-wave simulation for the optimization of connector breakout and transition structures

We also discuss how to combine all of these modeling elements into a hybrid, unified time domain simulation environment utilizing HSPICE.

23.2 Modeling Methodology[8]

Figure 23.1 shows the complete Final Inch modeling methodology. Measurement, modeling, simulation and fabrication of test boards are combined into a complete design and evaluation process. Measurement of fundamental material properties at the front end, along with measurement validation at the back end, close the loop on the complete design, modeling, and simulation process. Modeling methods appropriate to the detail, bandwidth, and data rates required are used for each individual system element, from connector to pad, via, trace, and even measurement connectors. Measurements in the time and frequency domain refine the process during successive cycles.

23.2.1 Major System Elements

There are seven major system elements that make up a board-to-board interconnect design, as follows:

- Drivers
- Receivers
- Packages
- PCB traces
- Connectors
- Breakout region
- Vias

7. An SMA launch is the point of entry of a signal into a circuit board through an SMA connector, at which point the signal is assumed to be undistorted for signal analysis and modeling purposes.
8. The author acknowledges the Application Engineering and R&D teams at Agilent Technologies and also Tom Dagostino (now with Teraspeed Consulting Group LLC) for his advice in the model development process.

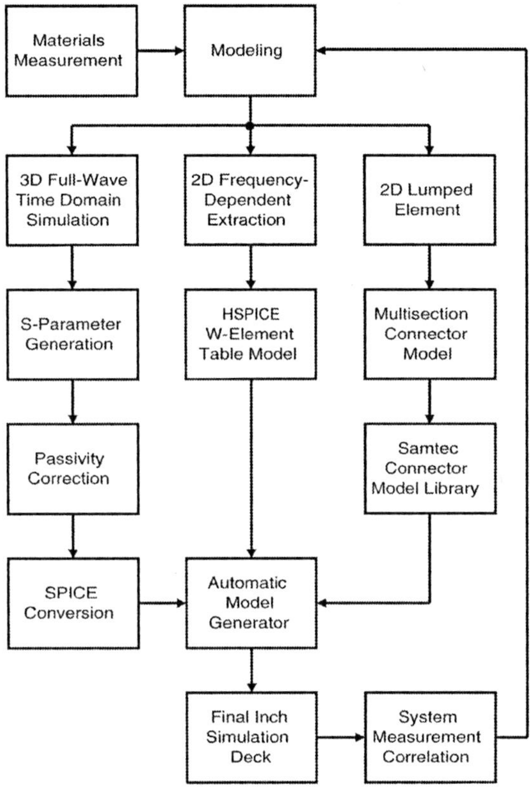

Figure 23.1 The Final Inch modeling and evaluation process.

Driver and receiver modeling are beyond the scope of this chapter. In place of these, high-quality SMA RF connectors are used to "launch" signals into the system for measurement of performance. For purposes of this chapter, drivers, receivers, and packages are treated as ideal elements. In reality, they are not. Each contributes its own impedance mismatch and losses to the entire interconnect equation, primarily due to input and output capacitance and normal device process variations. Package modeling can easily be shown to be an extension of the same modeling processes used for connector breakout regions. Final system design combines the methodology described here with accurate models for the drivers, receivers, and packages.

We focus on the following primary areas for the modeling of board-to-board interconnects:

1. PCB traces
2. Connector
3. Breakout region (BOR)
4. Vias

23.2.1.1 PCB Trace Modeling

PCB traces are controlled impedance transmission lines that provide the electrical connection between driver and receiver packages. Two conductors are required for a complete circuit, one to carry the signal forward and one for the return path. In multilayer PCBs, the ground/power plane(s) provide the signal return path, causing one of two naturally occurring transmission line types to be formed, either microstrip or stripline.

The ideal trace is a perfect conductor, causing no signal distortion or attenuation, and therefore having no length constraints. Real traces, however, are lossy, attenuating, and distort the signal as it propagates down the line until, at some finite length, the signal logic levels become unrecognizable to the receiver. Trace attenuation is both non-linear and a function of frequency. At gigabit frequencies one major contributor to attenuation is increased series resistance caused by skin effect. Skin effect can be defined as the tendency for current density to fall with increasing distance from the surface of a conductor. The current flows through an outer shell, or skin that shrinks with signal frequency. As a first approximation, the effective conductor resistance increases proportionally to the square root of frequency (see Figure 23.2).

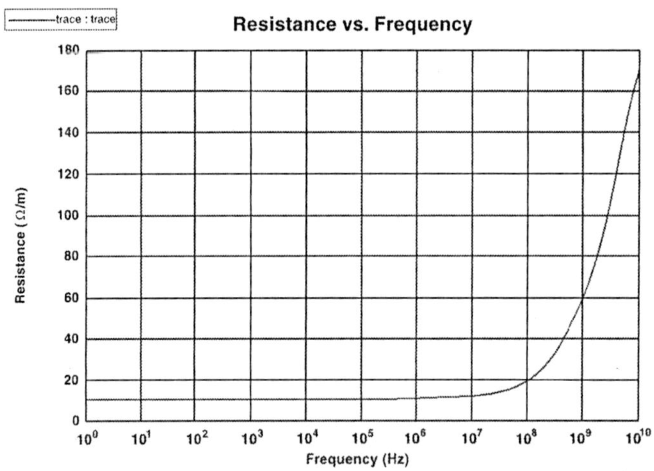

Figure 23.2 Frequency-dependent resistance of a 5-mil, 50 Ω trace.

In addition, at high frequencies energy loss in the dipole moment of dielectric materials causes dielectric loss, typically characterized by loss tangent. For epoxy resin materials such as the many varieties of "standard" FR-4, and practical trace dimensions, dielectric losses tend to dominate in the gigabit frequency range, which, as a first approximation, are proportional to frequency (see Figure 23.3).

The loss of a PCB trace can be described formally as follows:

$$\alpha_T = \alpha_C + \alpha_D + \alpha_R \tag{23.1}$$

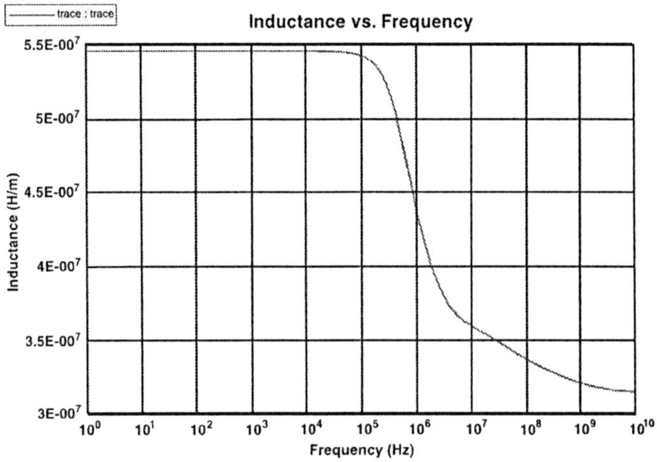

Figure 23.3 Frequency-dependent inductance of a 5-mil, 50 Ω trace.

where

α_T is attenuation loss
α_C is conductance loss (dominated by skin effect with square root f dependence)
α_D is dielectric loss (characterized by loss tangent with f dependence)
α_R is radiation loss

For stripline conductors with negligible radiation loss, this generally simplifies to

$$\alpha_T = \alpha_C + \alpha_D \tag{23.2}$$

For accurate high-frequency trace extraction, both lossy conductor and lossy dielectric properties must be included in the modeling process. In addition, losses are not isolated quantities. Electromagnetic causality requires that whenever dielectric loss varies, so also must permittivity (ε_r). In the lumped element circuit formulation of a transmission line, this means that the admittance matrix (lumped capacitance C and lumped conductance G) must track and be consistent across all frequencies. Loss tangent and ε_r cannot remain constant with frequency, but must both vary simultaneously. Errors in the modeling of loss tangent and dielectric constant can lead to causality errors, as has been reported in a wide variety of sources.

However, another much larger source of causality errors commonly occurs, but is generally unreported. Variation of electromagnetic field penetration through lossy conductors with frequency is well known to give rise to what is commonly known as skin effect attenuation. This means that, at higher frequencies, the net height of the induction loop shrinks from the center-to-center distance to the surface distance, reducing the net inductance. This frequency dependence reduces phase delay for higher frequency signal components. In the lumped element circuit formulation of a transmission line, this means that the impedance matrix (inductance L and resistance R) must track and be consistent across all frequencies.

The inductance of a PCB trace, a flex trace, or a cable center conductor can be decomposed into what has been described as the external inductance, that is, the inductance of the transmission line, assuming that the conductor is perfect and that all fields terminate on its surface and the internal inductance, which is formed by the average path of the fields which penetrate into the internal region of the conductors. It is this so-called internal inductance that gives rise to yet another potential causality issue.

At mid to lower frequencies, skin effect and field penetration below the conductor surface cause increased transmission line inductance with no subsequent change in capacitance. As a result, both the delay and impedance of the trace is increased. If lossy conductor resistance is modeled but lossy conductor inductance is not modeled and does not track the resistance change, then frequency-dependent dispersion (propagation delay changes with frequency) will not be properly modeled, which gives rise to an incorrect pulse response in the time domain. First order modeling, which considers only the attenuation caused by lossy conductors and lossy dielectric, produces incorrect simulation results that are often apparent in the time-domain pulse and eye patterns.

Most 2D electromagnetic transmission line RLGC parameter extractors do not correctly solve for the internal inductance. Most material property data sheets for commonly used PCB dielectrics do not specify the variation in properties from low to high frequencies, and if they do, often do not de-embed or remove the effect of internal inductance from the specification of the relative dielectric constant. These factors lead to uncertainty in the modeling and simulation results for a majority of high-speed multi-Gbps systems, and as a result, an over-reliance on actual measurements of interconnect channel behavior with TDR and VNA[9] approaches.

Unfortunately, measurements can only be performed on a small subset of the actual systems that will be produced. It is unlikely that the statistical outliers that limit worst-case performance will actually be measured. As a result, the lack of sufficient post-fabrication measurements prevents designers from having confidence that their multi-Gbps designs are robust.

However, with a robust and accurate interconnect modeling and simulation process that is matched by measurement correlation, we regain accurate characterization over process. We become able to design systems to their full capacities without increasing cost or sacrificing reliability. As a result, an engineer can design board-to-board interconnect systems that meet all design targets and system cost goals utilizing cost effective and readily available high-speed connectors.

Each Final Inch trace design is extracted utilizing Ansoft Maxwell 2D with complete dielectric and conductor loss modeling, producing tabular RLGC models at 31 different frequencies ranging from 100 Hz to 25 GHz. The results are then converted into a Synopsys HSPICE

9. TDR is time domain reflectometry and VNA is vector network analyzer. Please see Chapter 20 starting on page 519 for a detailed discussion of both TDR and VNAs.

W-element frequency-dependent, lossy-line model for highly accurate simulation results that have been correlated to measured results in both the time and frequency domains.

23.2.1.2 Connector Design

2D quasi-static and 3D full-wave electromagnetic field solvers are utilized extensively in the design and development of high-speed connector systems. Figure 23.4 shows an example of a connector breakout region (BOR) that has been designed synergistically in conjunction with the connector design process. 3D full-wave simulations of the internal connector structures, the connector surface-mount structure, the PCB pads, vias, and breakout traces were all used for the design and trade-off process. Insertion loss, return loss, crosstalk, and connector impedance

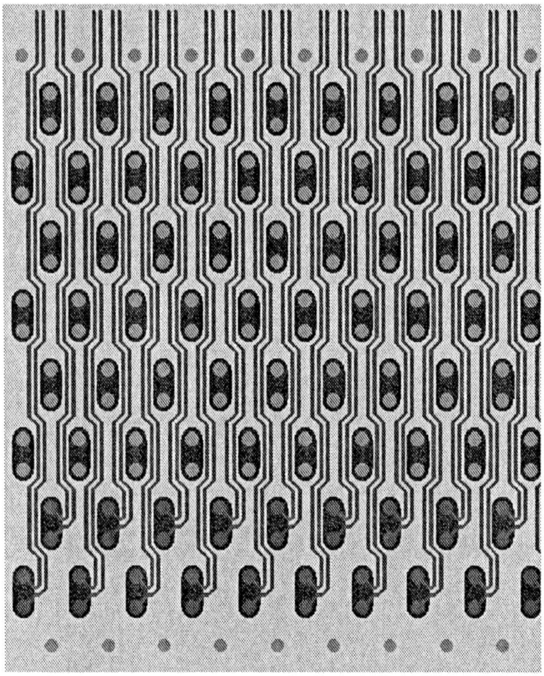

Figure 23.4 Final Inch BOR for an advanced, next-generation differential connector.

were all evaluated during the design process of the connector system and the PCB BOR patterns. Because the influence of the connector does not end at the printed circuit board, other connector systems have failed to live up to claims of high performance. In Final Inch philosophy, the printed circuit board, breakouts, vias, and traces are all considered an extension of the connector. As such, they are considered and designed in conjunction with the connector.

For electrically long connectors, frequency-dependent loss modeling is a must. Samtec connector models are currently delivered as coupled, multisection RLC HSPICE and PSPICE models with section sizes appropriate for edge rates up to 50 ps and bandwidth to 7 GHz. For many board-to-board stacking connectors and many surface-mount connectors, this is sufficiently accurate. Samtec's SPICE models have been validated to lab measurements for bit rates up to and including 12.5 Gbps with the capability of handing all the latest serial communications interface data rates including OC-192, XAUI, PCI Express, Serial ATA, OC-48, and many more.

23.2.1.3 Breakout Region (BOR) Modeling

The BOR is one of the most critical areas in any high-speed PCB design, and because of myriad conflicting tradeoffs that must be considered, it is often the most frustrating. Three of the most significant concerns are signal integrity, routability, and manufacturability.

Signal Integrity (SI) — If great care is not taken, it is quite possible that signal degradation in the BOR will be far more significant than that which occurs in the connector itself. With fast edge rate, high-bandwidth signals, the SI performance of a good connector can be ruined by poor design of the breakout region. At high frequencies, via-to-via crosstalk can become extremely high, causing noise, jitter, and eye closure problems that are often improperly attributed to the connector. With careful consideration of via placement to control the signal return path and inter-signal coupling, suitable SI performance can be maintained.

There are two more areas of a BOR design that require special attention from a signal integrity perspective. These are the fanout and via stub length. Figure 23.5 shows the via fanout area for a surface-mount connector.

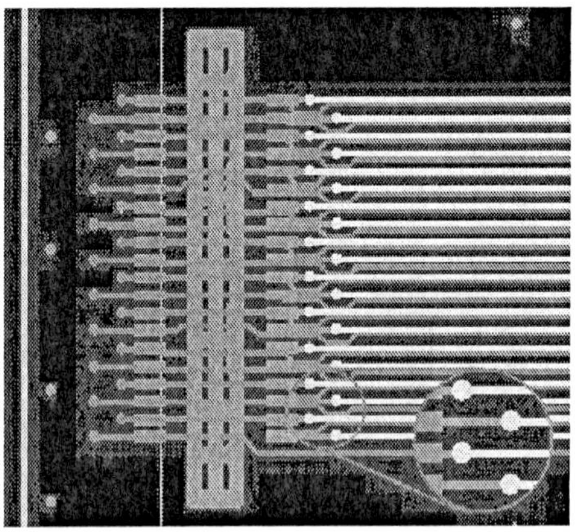

Figure 23.5 Typical BOR around surface-mount connector with staggered fanout vias.

Fanout — In most cases, vias are required to transfer signals from the connector pads to inner PCB layers. When these vias are located near a high-density connector, they must be staggered to create two or more rows of lower density via fields. This allows traces to pass between shorter fanout via row(s) on their way to longer fanout via rows. For breakout regions with only two rows of vias, we call these short fanout and long fanout. The farther away via fields are from the connector, the longer the breakout trace needs to be. The longer the breakout traces, the shorter the remaining non-breakout traces need to be in order for all traces to the connector to be equal in length.

Via Stub Length — Nets with short via stub lengths have significantly better signal integrity performance than those with long via stubs. For relatively thin boards, 0.100" and less in thickness, signal degradation due to via stubs is generally insignificant for data rates up to approximately 6 Gbps. However, for faster data rates and larger board thickness, stub size does matter. Longer via stubs cause increased insertion loss, impedance discontinuities, resonance, and jitter in the received signal.

As examples, refer to Figure 23.6. The via on the left side of the connector ties a signal from PCB top layer 1 to PCB internal signal layer 6; the via on the right side of the connector ties a signal from PCB top layer 1 to PCB internal signal layer 3. Layer 6 vias have a shorter via stub than the layer 3 vias, and therefore will have better signal quality as shown in Figure 23.7.

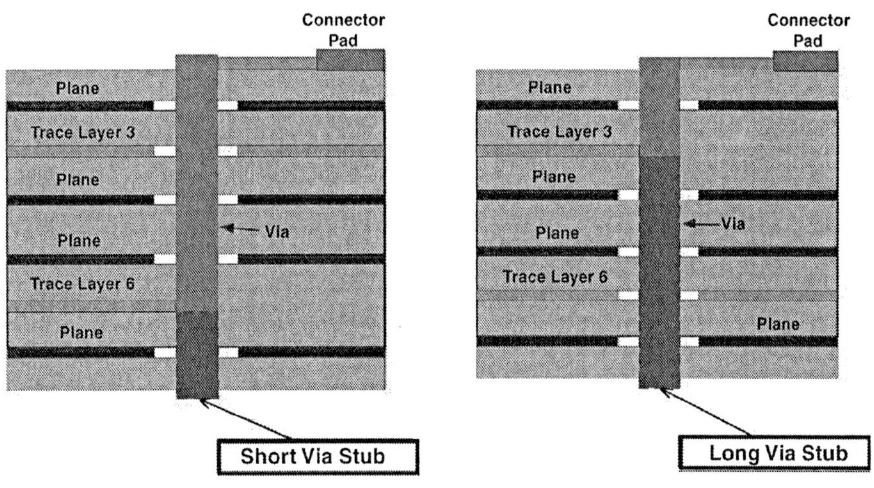

Figure 23.6 Via stubs.

Vias are used for signal escape in the BOR. For Final Inch designs, signal escape via placement was evaluated with CST MicroWave Studio, a 3D full-wave electromagnetic modeler and simulator, to determine via placement, spacing, and antipad sizing for high-speed interconnect paths. The performance-limiting effect of long via stubs was also taken into consideration in the layout, routing, and modeling process to guarantee the bounds of performance.

Final Inch via designs have been modeled, measured, and validated with PRBS sequences at bit rates from 1 Gbps to 12.5 Gbps.

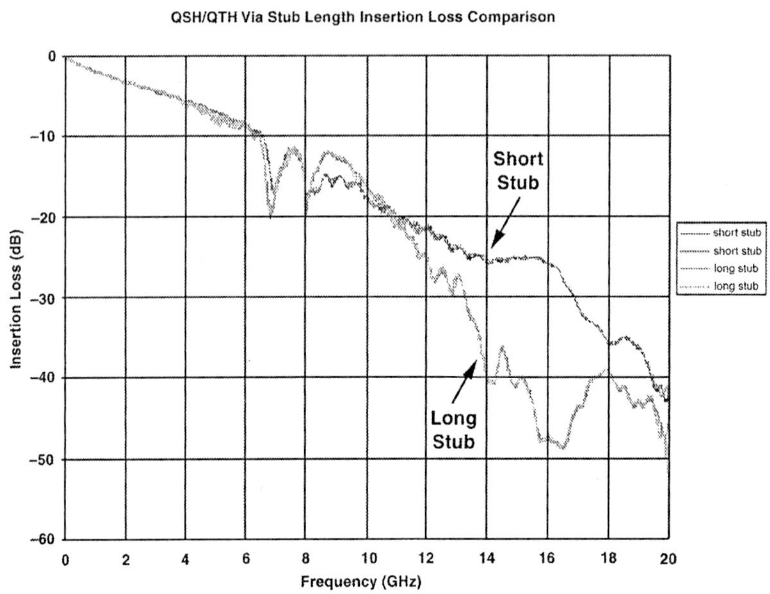

Figure 23.7 Via stub length insertion loss.

Routability — Currently, a large portion of the BOR design routing cannot be successfully automated. When automated routing is employed, the results will likely be suboptimal from a SI perspective. Therefore, a significant portion of the total PCB design effort must be focused on the BOR.

Connectors require solder through-holes or surface-mount pads for attachment to the board. Since signal traces are often routed on internal board layers, vias are usually required in the BOR to establish connection between the connector surface-mount pads and internal signal traces. Similarly, the through-holes required for connectors can be problematic. Vias and through-holes cause significant routing conflicts and possible signal degradation. Their placement must be carefully considered for high-performance interconnects.

Manufacturability — Breakout and routing solutions must be designed for manufacturability as well as SI performance. In-house or outside fabrication and assembly teams typically have their own sets of CAD rules designed to keep manufacturing defect rates as low as possible. Final Inch PCB designs are optimized to use the least aggressive manufacturing processes consistent with the SI performance and signal density required. Ample spacing is provided between vias to provide large routing channels. Trace-to-trace, via-to-via, via-to-plane, and other manufacturing and routing details are carefully considered. As a result, Final Inch designs are robust, simple to implement, and easy to manufacture.

23.2.1.4 Vias

Vias are metallized holes in the PCB that can be used for routing traces between layers, for mounting components onto PCBs, and for tying power planes to other power planes of the same voltage or tying ground planes together (called "via stitching"). We'll restrict our discussion to signal routing vias.

At Gbps data rates, the lumped approximation of the via is no longer valid, making the design process much more complex. The entire structure of the via must be understood and modeled. Vias are not uniform transmission line structures like stripline trace. Their very 3D nature requires special consideration in design and modeling. Single-ended vias are problematic creatures, requiring special care and feeding. Without a well-defined return path (usually created using neighboring ground vias), via performance suffers.

It's a misnomer to say that a via can be "impedance-controlled." A trace is a uniform impedance-controlled structure. A via, on the other hand, is not uniform, and can only be constrained within a reasonable region of operation. This is not easily performed but is absolutely necessary for high-performance systems.

First, a good return path is not always available. When this is the case, crosstalk increases, mismatches and reflections increase, and maximum speed suffers. Even under these difficult circumstances, with accurate modeling, these issues can be simulated and understood to yield high-performance results. Samtec's Final Inch models and test boards are designed specifically to allow the engineer to evaluate a multitude of connector and PCB signal-to-ground ratios for singled-ended designs. These models, extracted with 3D full-wave techniques, show the "hidden" effects of BOR discontinuities and crosstalk that are normally not included in either connector models or board-level simulation tools.

23.3 Simulation

A hallmark of the Final Inch modeling and simulation process is in the ability to combine models created with multiple creation and extraction methodologies and have them work correctly within the HSPICE simulation environment. 2D multisection lumped element SPICE models for connectors, W-element table models for uniform traces, and 3D S-parameter models that have been translated into time domain "black box" subcircuits all contribute to accurate, high-performance solutions.

To get to this point, several issues were solved:

1. Interoperability of Synopsys HSPICE W-element models with Sigrity BroadBand SPICE pole-zero fitted G-element models[10]
2. W-element simulation fidelity with slow risetime signals

10. The G-element is a voltage-controlled current source. One of the many options in HSPICE allows the transfer function to be described as a Laplace pole/zero model. Several commercial off the shelf (COTS) software packages, including Sigrity Broadband SPICE, will perform a pole/zero fit to frequency dependent S-parameters, and generate G-element Laplace behavioral models.

3. Simulation stability issues caused by nonpassive S-parameter extraction
4. Simulation stability issues caused by fitter and simulation numerical errors

23.3.1 HSPICE W-Element Issues

In the process of developing working Final Inch simulation decks,[11] several issues concerning the HSPICE W-element were uncovered: interoperability with G-element behavioral models and simulation fidelity with slow risetime signals. These issues were discovered with Synopsys HSPICE 2003.03-SP1 and prior versions.

In a nutshell, the algorithm for the W-element does not always automatically set the internal time step and bandwidth of the algorithm correctly. This results in incorrect lossy transmission line modeling and convergence issues when simulating with other behavioral G-elements in the circuit. The following additional HSPICE code will "fix" the problem and "trick" the simulator into operating correctly by forcing the algorithm to believe that an extremely fast risetime signal is present in the system. This is accomplished by adding a "Trojan" pulse source, not connected to the actual circuit, that fools the W-element algorithm into believing that a high-bandwidth, small step size simulation is required.

```
*******************************************
* code to force HSPICE W-element time step and bandwidth algorithm
* to work correctly for slow edge rate signals, and play well with
* other Laplacian and lumped element models
*
vfrog frog 0 pulse (1 0 0 25p 25p 75p 200p)
rfrog frog 0 50
*******************************************
```

Simple, isn't it? This "fix" resolved issues one and two on our list quite nicely. Issues three and four, which are essentially frequency-domain to time-domain conversion circuit passivity issues, will take a bit more creativity in the following section.

23.3.1.1 Passivity in Frequency Domain Network Measurement and Simulation

In the previous section we discussed the motivation for using frequency domain S-parameters to describe the electrical networks used for high-speed signaling. While there are advantages to using frequency domain network descriptions, there are also potential problems that must be carefully managed.

11. A simulation deck is the totality of software modules and files needed to run a computer simulation where the modules include models of circuit traces, connectors, vias, breakout regions, and other components. The modules also include control files and data input and output files.

The signals and performance criteria we are interested in are normally characterized in the time domain. The signals are generally random assemblages of bits represented by voltage levels and their associated rising and falling transitions. Performance criteria normally use metrics such as jitter, eye-closure, and timing push-out. All of these are expressed as functions of time.

This presents a fundamental difficulty because the network description is not expressed as a function of time, but rather as a function of frequency. To solve this dilemma, we must either (1) transform the time domain representation of our signal of interest into the frequency domain, apply it to the network transfer function, and then convert the response back to time domain, or (2) we can convert the network description to time domain and perform all simulations in time.

The network theory to accomplish these transformations is straightforward and uncomplicated. Unfortunately, the practical application of these principles is not. For example, the conversion from the time to the frequency domain is computed using the Fourier transform $F(j\omega)$:

$$F(j\omega) = \int_{-\infty}^{\infty} f(t)e^{-j\omega t}dt \tag{23.3}$$

and the conversion from the frequency domain to the time domain is computed using the inverse transform $f(t)$ as

$$f(t) = \frac{1}{2\pi}\int_{-\infty}^{\infty} F(j\omega)e^{j\omega t}d\omega \tag{23.4}$$

where t is time and ω is radian frequency.

While in practice we may use a different form of this transformation, it serves to illustrate an important point: that is, the value of $f(t)$ at each instant in time has implications for *all* frequencies. Conversely, the value of $F(j\omega)$ at each frequency has implications for *all* points in time.

Let's consider for a moment an S-parameter matrix that we constructed through careful measurement of an interconnect system. Likely, we measured the system using a VNA that gave us magnitude and phase for each port of interest across a range of frequencies. This range of frequencies is finite because of the limitations of the instrument performing the measurements.

This artificially truncated frequency domain response has the potential to create a problem. Since the time response of the system at any instant is dependent on the frequency response at *all* frequencies, truncating a system's frequency response has the potential to produce artifacts in time, including non-causality. Fortunately, over the last few years, most of the simulation tool vendors have done good work at minimizing this problem. Still, other problems remain.

23.3.1.2 Managing Error in Frequency Domain Network Parameters

Another significant difficulty when simulating systems using frequency domain parameters is dealing with errors. Error enters at a number of points in the modeling and simulation process.

When modeling a system using electromagnetic simulation, there are numerous potential sources for error. Certain simplifications in geometry and material properties are always required. Also, the simulation itself is subject to all the normal numerical issues found in iterative solvers: matrix condition, convergence, and finite precision. All these factors add to error. Modeling a system with measurement will also introduce errors. Measurement errors come from sources such as noise, calibration, and test fixture de-embedding.

Likewise the conversion process, from frequency to time domain, is also a source of errors. Many techniques have been used for this conversion. Most of the modern methods involve fitting an arbitrarily complex function to the data to be converted. The choice of function is driven by the desire to ultimately transform the function from frequency to time domain. This fitting procedure, by its nature, is only approximate and will induce errors in the resulting time domain function.

Time domain simulation in SPICE or a similar tool likewise has its own problems. Our main concern is ensuring integrity of the system model through its construction and conversion so that it arrives at the time domain simulation as a faithful representation of the actual network.

Some error is unavoidable. Generally we can tolerate errors in our simulations that result in a few percentage points variation from the actual physical result. What we cannot tolerate are errors in network parameters that, while small in their domain, produce very large errors under simulation. Fortunately, most errors are of the former type.

Unfortunately, however, some network parameter errors do cause very large errors during simulation. In particular, surprisingly small errors that result in nonpassivity of the network parameters of a passive network can cause time domain simulations to oscillate. This is a very disappointing result if significant time and money have been spent developing the network model.

23.3.1.3 Error-Induced Nonpassivity of Network Parameters

It is not uncommon to model a very large passive system, such as a 48-port, coupled-line system that might be found in a high-density connector breakout region. A model extraction such as this often requires days, if not weeks, of run time. The opportunity cost for this is very high if the resulting model result is unusable because it is nonpassive. We have developed several techniques for correcting modeling errors resulting in nonpassive network models. These techniques have allowed us to recover significant investments of time and money that were otherwise lost because of nonpassive network parameters.

In presenting these techniques, we must first examine what it means for a network to be passive. Developing the basis behind a formal test for passivity is quite involved and beyond the scope of this chapter, but we can develop a basic understanding of what makes a network nonpassive and then accept the formal test at face value.

Simply put, passive networks are networks that have positive real parts in the elements of the associated Z and Y parameter matrices. The imaginary part of the elements of the Z matrix represents the reactance of the element. The real part of the elements of the Z matrix is the resistance associated with each element. Negative resistances and conductances act as sources in the network. This fact has been used to great effect in the development of Gunn diode amplifiers and oscillators, which utilize negative incremental resistance to produce necessary gain.

Let's consider a seemingly trivial example. Assume we have a network that, for all frequencies, when 1 watt is incident on the network, 0.5 watts is reflected, and 0.5 watts is transmitted. This network can be described in S-parameter form by

$$\boldsymbol{S} = \begin{bmatrix} \frac{\sqrt{2}}{2} & \frac{\sqrt{2}}{2} \\ \frac{\sqrt{2}}{2} & \frac{\sqrt{2}}{2} \end{bmatrix} \tag{23.5}$$

On inspection, one might decide this network is reciprocal, passive, and lossless. One watt in results in exactly one watt out when applied to either port. While the network is reciprocal, it is neither lossless nor passive.

Assuming a 1 Ω system impedance, converting the $\boldsymbol{S}$ matrix to the associated $\boldsymbol{Z}$ matrix results in

$$\boldsymbol{Z} = \begin{bmatrix} \frac{-1}{\sqrt{2}-1} & \frac{-\sqrt{2}}{\sqrt{2}-1} \\ \frac{-\sqrt{2}}{\sqrt{2}-1} & \frac{-1}{\sqrt{2}-1} \end{bmatrix} = \begin{bmatrix} -2.414 & -3.414 \\ -3.414 & -2.414 \end{bmatrix} \tag{23.6}$$

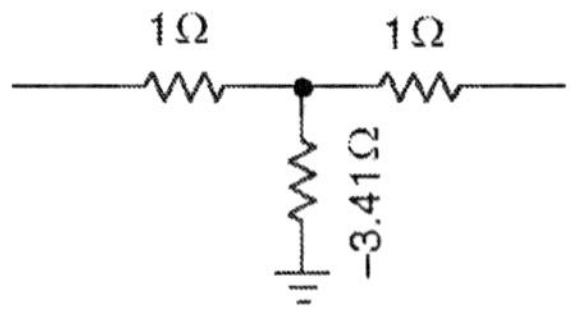

Figure 23.8 Circuit representation.

This network can be represented by the circuit shown in Figure 23.8. From Figure 23.8, it is obvious that the network represented by $\boldsymbol{S}$ is not passive. This is a non-intuitive result. It is clear we need a better tool than intuition. Now that we have developed a feel for network passivity, let's examine how small errors can result in nonpassive network parameters.

Consider the circuit of Figure 23.9 where the resistance and capacitive reactance are each 1 Ω. The Z-parameter matrix for this low pass network is

$$\boldsymbol{Z} = \begin{bmatrix} 1+j & -j \\ -j & j \end{bmatrix} \tag{23.7}$$

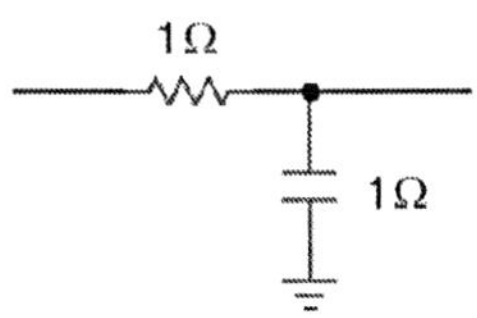

Figure 23.9 Circuit where resistance and capacitive reactance is 1.

The corresponding S-parameter matrix is:

$$\boldsymbol{S} = \frac{1}{13}\begin{bmatrix} 3+2j & -6-4j \\ -6-4j & -1+8j \end{bmatrix} \tag{23.8}$$

Now let's assume that we derived this network description from measurements and let's modify the data in small ways as follows:

$$S_{error} = \frac{1}{13}\begin{bmatrix} 2.9+1.8j & -5.9-3.6j \\ -5.9-3.6j & -1.2+7.3j \end{bmatrix} \tag{23.9}$$

which results in a Z matrix:

$$\boldsymbol{Z} = \begin{bmatrix} 1.1+j & -0.1-j \\ -0.1-j & 0.1+j \end{bmatrix} \tag{23.10}$$

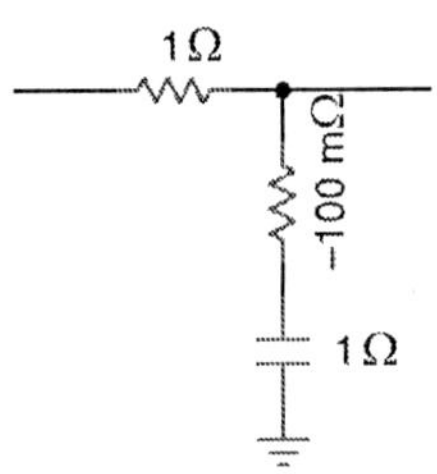

Figure 23.10 Network circuit.

The circuit for this network is shown in Figure 23.10. This network is nonpassive and is a trivial case, but illustrates how small errors can result in non-passive network parameters.

The sufficient case for passivity presented earlier, while adequate, is a bit cumbersome because it involves converting the S-parameter matrix to Z or Y parameters and then examining each element for positive real parts. There is a somewhat more direct method using the S-parameter matrix itself. The formal justification for it is very involved. If the reader is interested, we refer you to the literature (see [Roh68]). For a scattering matrix to be passive, it is sufficient condition that the scattering matrix satisfies the following condition:

$\boldsymbol{I} - \mathbf{S} \times \mathbf{S}'$ must have eigenvalues that have non-negative real parts

where $\boldsymbol{I}$ is the identity matrix and $\boldsymbol{S}'$ is the transpose (flip along diagonal) of the scattering matrix $\boldsymbol{S}$. We use this test in our tools for passivity correction.

23.3.1.4 Correcting Error Induced Nonpassivity in S-Parameter Network Models

As an example of correcting, we will examine an actual model derived from electromagnetic simulation. This model is nonpassive and oscillates under simulation. We can think of oscillation in terms of a basic feedback amplifier model (see Figure 23.11).

Oscillation occurs when the loop gain of positive feedback exceeds unity. The only way to avoid oscillation at all frequencies is to reduce the loop gain. If we test our example network for passivity, modeled as an 8-port S-parameter matrix, we see that the network has two regions of instability (see Figure 23.12).

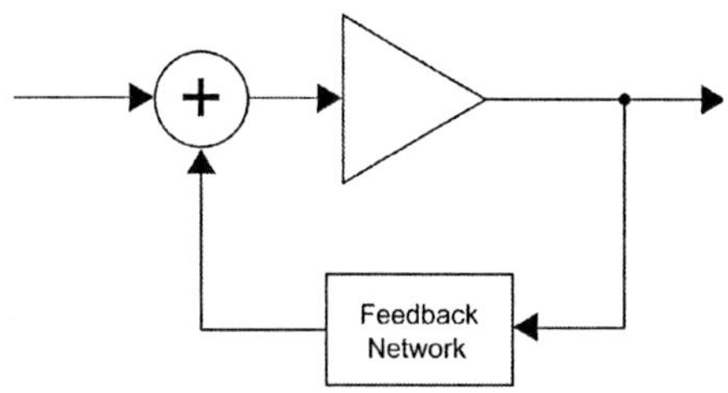

Figure 23.11 Basic feedback amplifier model.

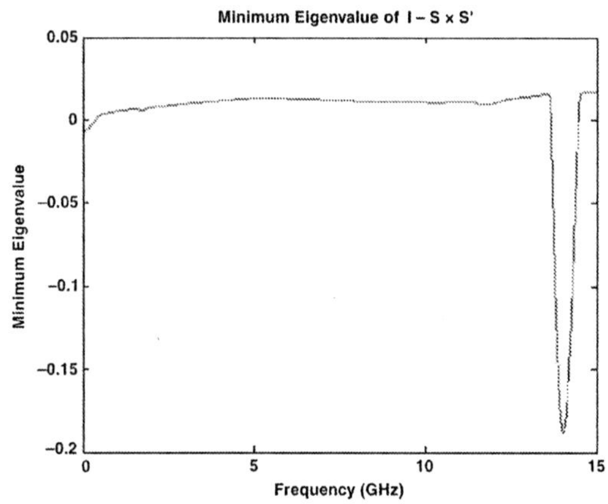

Figure 23.12 Eigenvalue display shows nonpassive S-parameter frequencies.

Recalling our test for passivity, the real part of all eigenvalues must be greater than or equal to zero for the network to be passive. From the plot of minimum eigenvalues, we can see that this network has two regions of nonpassivity, just above 0 Hz and also in the region of 14 GHz.

We wanted to find a simple way to fix networks like the one above. We did not want to look deeply into the mechanisms involved with the instability of a particular network, but understanding the basic mechanism of gain and feedback, we approached correction as a problem of reducing loop gain.

Our first attempt at a solution was to first identify the regions of nonpassivity. Once identified, we scale just those regions by multiplying the S-parameter matrix for those frequency points by a factor that causes the matrix to become passive. Our algorithm chooses the scale factor iteratively so that it is as close to 1 as possible (within a set threshold) but still results in a passive network. The scale factors for the example above are shown in Figure 23.13.

When compared with the minimum eigenvalue plot, it can be seen that the points of maximum deviation from passivity required the most correction. With this correction in place, the network converted to time domain and simulated without instability. This technique alone works quite well and gives us very reasonable simulation results. The adjustment to the network parameters is the minimum necessary to obtain passivity.

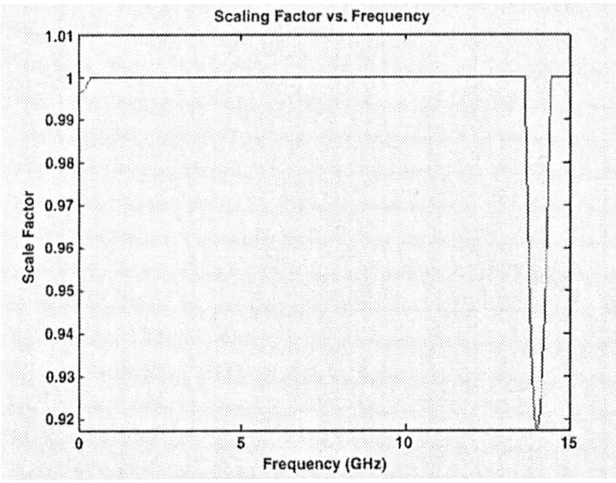

Figure 23.13 Computed S-parameter passivity scaling factor.

We continued to have some networks, however, that showed nonpassive behavior when simulated, even after they were passed through our process. When we carefully examined why this happened, we discovered that because our process made networks only just barely passive, the additional error introduced when the networks were converted to time domain caused them to become nonpassive again. To correct this difficulty, we applied an adjustable global scale factor that makes a very small adjustment to every frequency point, giving us a passivity margin. We call this scale factor a nudge, to differentiate it from the frequency-dependent scale factor we applied previously. Selection of a nudge factor less than 1 is only necessary if the network shows continued instability.

Figure 23.14 illustrates a passivity-corrected model that still shows some instability (oscillations) versus the same passivity-corrected model that has been nudged and which shows no oscillation. Figure 23.15 shows the close correlation of the passivity-corrected model versus the nudged passivity model before the onset of instability. The two curves are so well correlated that they appear to be one curve. Figure 23.16 shows a close-up of the nudge. A very slight decrease in gain is applied to compensate for algorithmic error terms in the transform from the frequency domain to the time domain. Figures 23.15 and 23.16 illustrate that nudging does not introduce significant error into the model.

As one additional correction, we also set a maximum dynamic range, as measured from the largest value in the network matrix, beyond which all values are set to zero. For example, if the maximum value is 1 and the maximum dynamic range is 60 dB, the minimum value we allow is –60 dB from 1, or 0.001. In this way, we hope to clean up a little of the numerical noise.

We have created a tool set that automates the correction processes we have described. These tools have all been coded in MATLAB by MathWorks. This process has saved us and our customers significant time and money. While it is not a rigorous mathematical solution to error-induced nonpassivity, it does produce good results from otherwise unusable models.

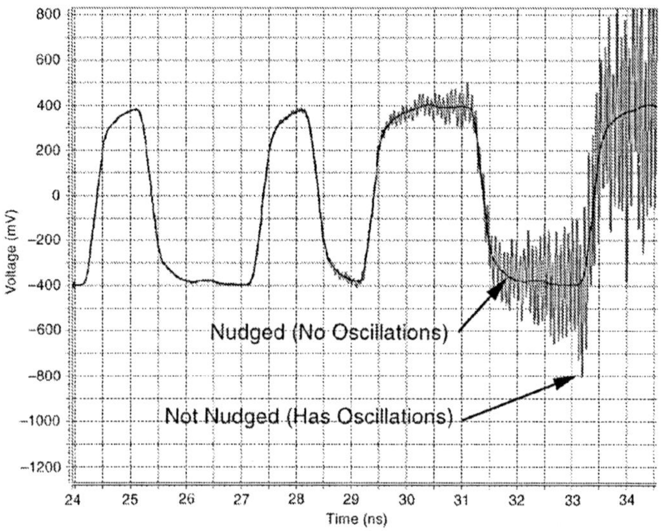

Figure 23.14 Passivity corrected model vs. nudged passivity corrected model, at the onset of instability.

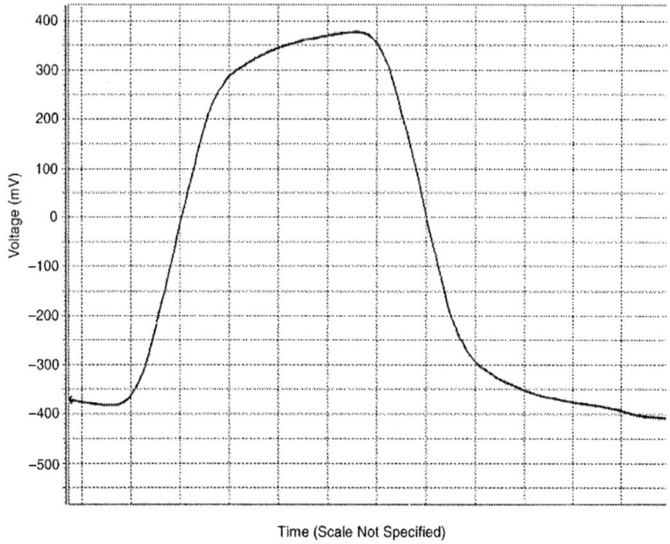

Figure 23.15 Passivity corrected model vs. nudged passivity corrected model. The two curves superimpose, indicating the extremely close correlation between the curves before the onset of instability.

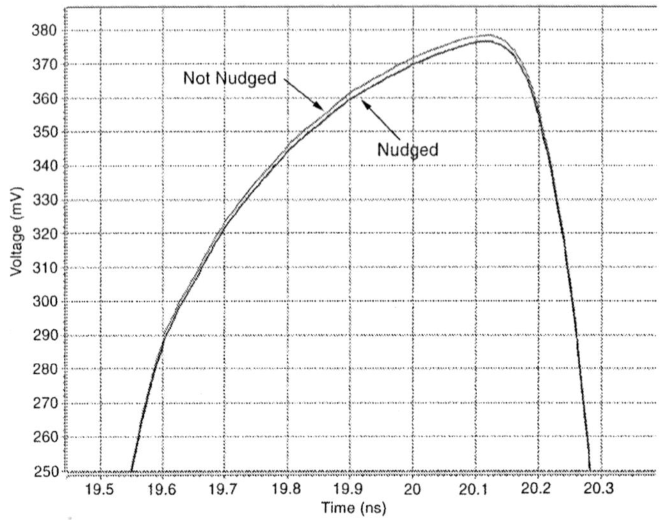

Figure 23.16 Close-up of the nudge.

23.4 Measurement

So far we have shown that when developing models for multi-Gbps operation, real-world measurements are the only way to close the loop between modeling and reality, and the only way to accurately measure "reality" is with high-accuracy test boards. But what about the test board itself?

23.4.1 The Design and Need for High-Accuracy Test Boards

Final Inch test and evaluation boards are based on the recommended layout documentation for the connector family. These boards contain the mounted connector, BOR, and matched-length transmission lines. These lines terminate into industry-standard coaxial connectors that provide easy connection to lab instruments. Probe points are also provided on the boards so that industry-standard micro probes can be used. These boards can be used to evaluate the connectors' performance in the customer's own laboratory using whatever signaling schemes they prefer. A test trace board is also included along with a user's guide.

There are several features designed into Final Inch test and evaluation boards that make them exceptionally accurate when it comes to measuring signal quality in the connector breakout regions. These features are discussed next.

23.4.1.1 High-Bandwidth SMA Launch

For accurate measurement of connectors, the BOR, and PCB traces, a vertical (top-mount) SMA array has been designed into each of Samtec's Final Inch test and evaluation boards.

To facilitate highly accurate broadband measurements, the SMA is placed on either the top or bottom side of the PCB. Depending upon which internal signal layer carries the signal under test, the launch is made from whichever side of the board leads to the shortest BOR via stub length.

23.4.1.2 Isolation of Traces Prior to Final Inch

Traces on the test and evaluation boards are routed on internal layers in stripline or offset stripline geometries. The traces have enough separation to be considered uncoupled; therefore, they make little or no contribution to crosstalk measurements. Trace separation has been maximized to provide the lowest coupled noise possible within the physical design constraints.

The electrical performance of each trace design has been characterized and optimized using Ansoft Maxwell 2D. Tabular RLGC models at 31 different frequencies ranging from 100 Hz to 100 GHz have been created and processed into a Synopsys HSPICE W-element frequency-dependent lossy line model. These models permit highly accurate simulation results in both time and frequency domain.

23.4.1.3 Reference Structures

The third board in each Final Inch kit contains reference structures designed as an aid to Final Inch evaluation. The test traces can be used to

- Calibrate simulation models
- De-embed the insertion loss performance of the connector with via breakouts
- Correlate characteristic impedance and dielectric loss for simulation modeling

23.5 Measurement Accuracy Issues

In any measurement system, there are three issues that need to be addressed. They are

- Test equipment accuracy
- The effects of fixturing for measurement
- The characteristics of the device under test (DUT)

Oscilloscopes, VNAs, current meters, and the like each have inherent errors associated with them. Each has a finite bandwidth, noise floor, linearity, drift, and so on that affects the final accuracy of the measurement. Some of these characteristics can be calibrated out, some cannot. Each instrument will measure to within a certain percentage of the actual input signal. Improvements in the accuracy of measurement can be achieved by calibration and/or applying correction factors to the measurements.

Other inherent characteristics of the equipment cannot be eliminated. Random noise, drifts over time, temperature, humidity, non-repeatable connections, and so on cannot be eliminated but can be reduced by good operating procedures. There are also limitations in the practical accuracy of the calibration standards applied.

Test fixtures add their own set of bandwidth filtering, noise, impedance discontinuities, resonances and other errors to the overall uncertainty of the measurements. A test board used to connect to the connectors has dielectric losses, imperfect Z_0, skin effect losses, crosstalk, and so on. The connection between the DUT and the test equipment is almost never perfect, resulting in reduced bandwidth, added aberrations, and/or additional noise.

The DUT has its own set of issues. How do you connect to it? Will the characteristic of interest be swamped out by limitations in the test equipment or fixturing? Will other characteristics of the DUT/fixture/equipment affect the parameter(s) of interest? Whenever you measure something, you are disturbing it.

23.5.1 SMA Launch

One of the most important aspects of this characterization project was the design of the system that took the test cables and connected them to the DUT. We have seen many different approaches to the task of connecting the test equipment to the DUT, including using semi-rigid coax soldered directly to the connectors, poorly designed connector launches to test boards, poorly defined traces with many discontinuities or worse, and the total ignorance of a controlled-impedance environment by use of spaghetti wiring between coax and the DUT.

Ideally, the test system should have a much wider bandwidth than the expected bandwidth of the DUT. A rule of thumb is to have three to five times measurement bandwidth versus DUT bandwidth. To accomplish this for Samtec connectors, a test vehicle was developed that contained a test cable to DUT bandwidth of 20 GHz when correction factors are applied.

Molex 73251-1850 SMA connectors were selected and a special board launch was designed. The quality of that design is demonstrated in Figures 23.17 and 23.18 illustrating the

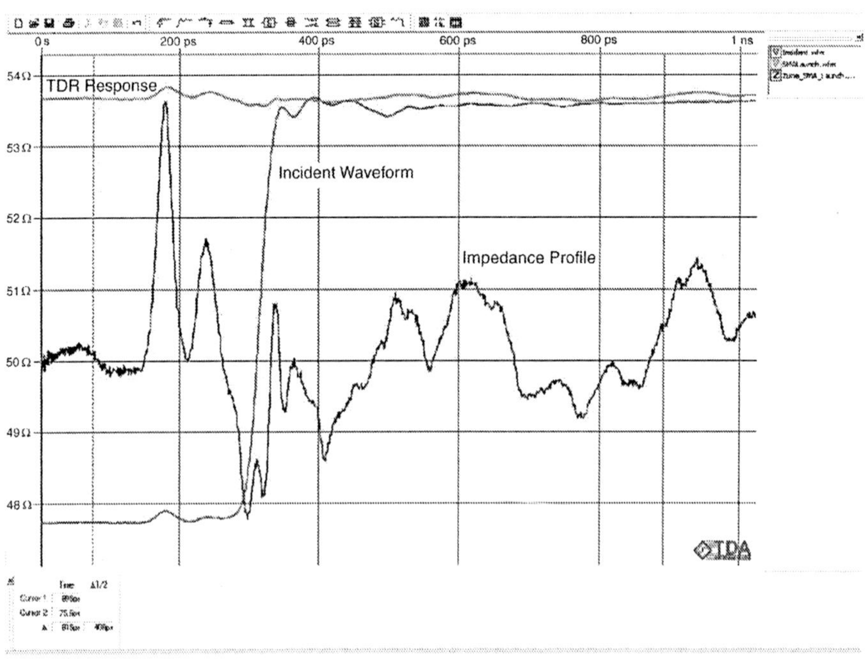

Figure 23.17 Plot of the impedance profile of the SMA connector launch to the test board.

TDR response of the SMA connector/board interface. Here we see how a 28 ps pulse from the Tektronix TDS8000/80E04 transitions the connector to board interface. Figure 23.17 shows the TDR profile of this launch. The upper trace is the reflection off the connector. It starts with the 50 Ω cable. At about 180 ps into the waveform, the connector's behavior starts. By 300 ps into the plot, the board's characteristics are shown. The transition is the incident pulse from the TDR head. The third trace shows the impedance vs. time of the cable, connector, launch, and board. Note the maximum deviation from the ideal 50 Ω is about 53.5 Ω, and it only lasts for a few ps.

It should be noted that the 53.5 Ω impedance discontinuity seen is internal to the SMA connector itself, at the male/female interface of the mating connectors between the cable and the board. Worst-case impedance for the SMA to PCB transition is seen to be 51.5 Ω to 48 Ω.

Note the lack of significant inductive (increase in impedance) or capacitive (decrease in impedance) discontinuity at the interface. This contrasted with other SMA launches we have seen as shown in Figure 23.18. The Teraspeed Consulting launch is about 50 Ω ± 3 Ω. Others had variations up to over 30 Ω. Also note some of the launches produce ringing that can swamp critical details for the first few hundred ps. These measurements of the SMA launch are in excellent agreement with the predicted results. This clean launch ensures wide bandwidth test signals can be applied to the DUT.

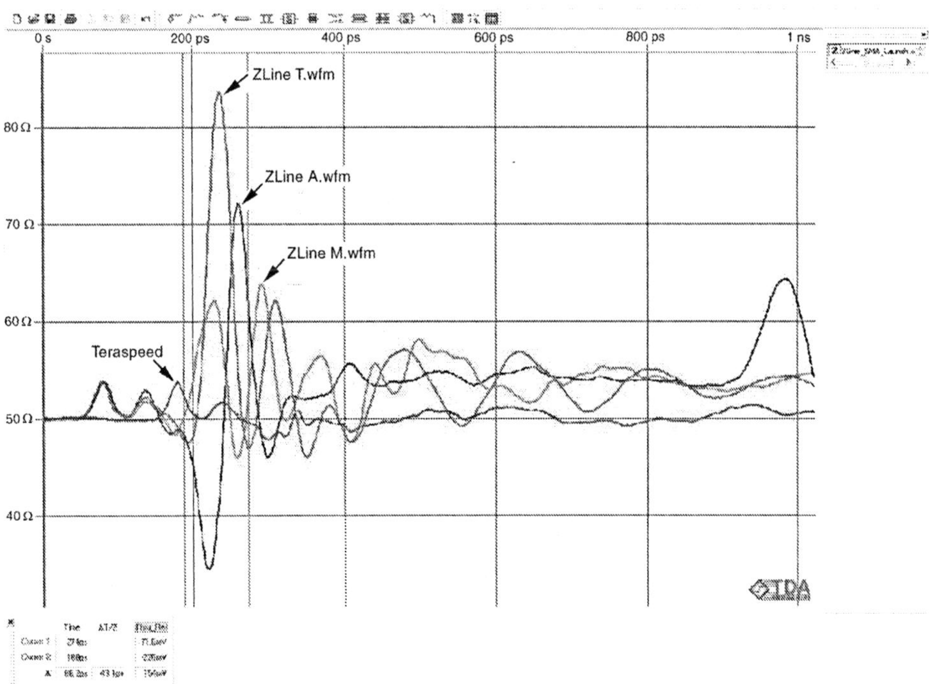

Figure 23.18 Comparison of various launches seen in other test boards using both end-launched and top-launched SMA.

23.5.2 Traces

Traces between the SMA connector and the DUT must be controlled. Teraspeed Consulting selected Photocircuits to fabricate the boards because of past experience with their ability to control trace impedance. Working with Photocircuits, we developed a stack-up and geometry to produce traces of 50 Ω ± 2 Ω, as seen in Figure 23.17.

In addition, the transition between the traces and the connector had to preserve the signal fidelity. Special attention was paid to this transition and this became part of the Final Inch project.

23.5.3 Measurements

Teraspeed Consulting used two complementary measurement techniques to characterize the connector, TDR/TDT for time domain and VNA for the frequency domain.

23.6 Frequency Domain Measurement

23.6.1 Calibration

Teraspeed Consulting used an Agilent 8720ES VNA to perform the frequency domain characterization of the connectors. Before measurements of the connectors could be made, a full two-port calibration of the VNA, which included the test fixture, was performed. Test traces on the test fixture were included in the calibration along with the test cables and SMA connectors. The test traces on the calibration board were of the same length, impedance, and on the same layers as the test traces on the connector test boards. The same SMA launch was used. Thus, the calibration system included the same set of characteristics as the connector test board but without the connectors under test.

The calibration constants were stored in the 8720ES and were applied to all measurements of the connectors. Thus, any deviation from an ideal set of measurements using this calibration would result solely from the characteristics of the connector.

23.6.2 Measurements

Full N-port characterization was made of the connectors. For the differential connectors, four-port measurements were made. For the single-ended connectors, crosstalk was measured out to two connections away.

23.6.3 Time Domain Measurements

Teraspeed Consulting used a Tektronix TDS8000 with 80E03 20 GHz sampling head and 80E04 20 GHz sampling head/TDR to characterize the connectors in the time domain. The 80E04 is used for TDR (time domain reflectometry) characterization and in conjunction with the 80E03 for TDT (time domain transmission) characterization. Both common-mode and differential-mode analysis were performed for differential connectors. Figure 23.19 shows the TDT characterization of a calibration board.

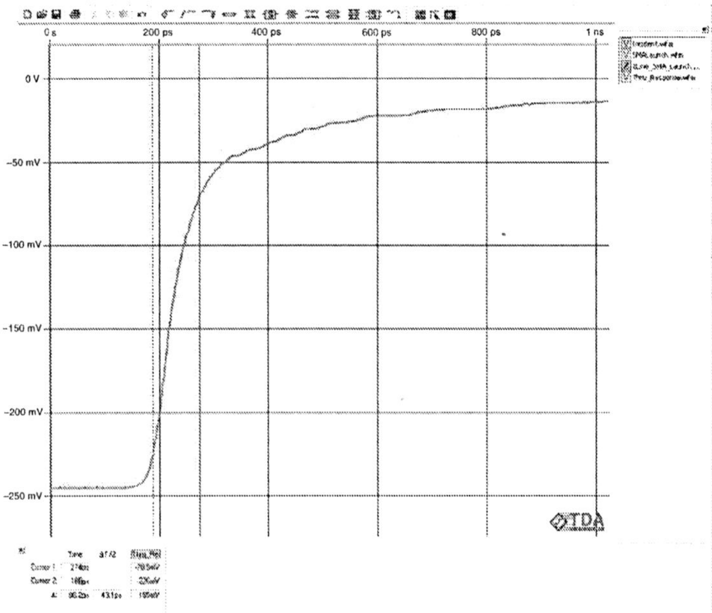

Figure 23.19 TDT of a calibration board showing the fast transition from the board plus the effects of skin loss over the 7-in. trace on the test calibration board.

23.6.4 Application Proof

Teraspeed Consulting connected a pair of Accelerant Networks AN6425 SerDes to the test system. Bit error rates as reported by the Accelerant Networks test software were monitored. Once the system self-calibrated, the receiver was disconnected and the output of the test board was connected to the scope to show the eye diagram of the transmitted data at 6.22 GB/s PAM-4. As you can see in Figure 23.20, the eyes are wide open and match the predicted response.

23.7 Validation of Material Parameters

This section discusses validating important material parameters. For high-speed design, the dielectric constant and loss tangent (or dissipation factor) of the PCB material itself are related to the signal propagation characteristics and loss degradation observed in microstrip and stripline designs. Vendors often give "worst-case" values measured at lower frequencies (for example, 1 MHz) where the limiting characteristics of interest are different at higher frequencies. The discussion revolves around characterizing a flex circuit board used for high-speed interconnection and consisting of DuPont Pyralux FR dielectric board adhesive material that corresponds with Kapton dielectric material.

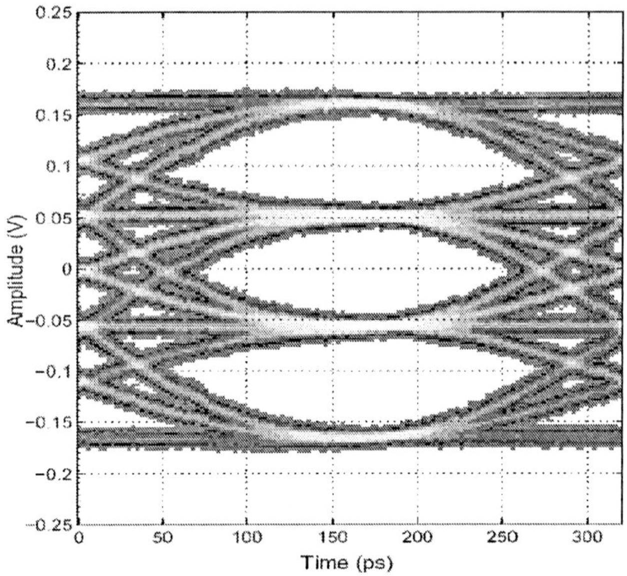

Figure 23.20 PAM-4 eye pattern for Accelerant Networks AN6425 at 6.22 Gbps with Samtec QSE/QTE Final Inch test board and a 1-meter long, 38 AWG micro coax assembly showing excellent eye opening.

Several promising stripline and resonator test structures, centered between two copper planes, were fabricated on a test board. Other methods, and the focus here, is on the most promising structure consisting of a stripline centered between two outside copper planes. SMA connectors were used to interface with the external test equipment, and the launch structures involved design techniques discussed elsewhere. A number of randomly distributed vias were used to connect the outside planes and de-Q planar resonance. Stripline structures (as opposed to surface traces) were used so that the electrical measurements would be related primarily to the material characteristic.

23.7.1 Terms

Some terms used in this discussion are discussed in detail in two recent, excellent references (see [Joh03] and [Bog03]). However, some distinctions are presented here that show up in actual characterization measurements.

In general, the permittivity of a material is a complex value. The reference value is the permittivity of free space (a perfect vacuum), $\varepsilon_0 = 8.854 \times 10^{-12}$ F/m, which is close to the permittivity of air. In practice, the relative permittivity ε_r is used with respect to air. The relative dielectric constant used in this presentation is defined as the real part of the relative permittivity of the material and also will use the symbol ε_r. The ε_r for air is 1, and the ε_r for typical FR4 material at 1 GHz is 4.2. The flex board material in this example has ε_r in the order of 3.0 to 3.5.

The loss tangent (or dissipation factor) designated here as tan δ of the material is the ratio of the imaginary-to-real part of the (relative) permittivity. It is also correlated with the conductivity of the dielectric material and dielectric loss. Typical values of tan δ are 0.035 or less for FR4-type printed circuit board material.

The term "effective" relative dielectric constant is used in this chapter. It is often defined in the context of non-homogeneous dielectric such as when electrical parameters are extracted from microstrip lines. In this case, the electrical parameters are a function of the dielectric material below and the air above as if it were surrounded entirely by a homogenous material with an effective ε_r.

In fact, the formula for calculating the effective dielectric constant ε_r for a line on a circuit board is based on an overall measurement:

$$\varepsilon_r = \frac{c^2}{v^2} \tag{23.11}$$

where

ε_r is effective relative dielectric constant
c is the speed of light
v is the velocity of propagation (from length and delay)

However, even if the material is homogeneous, the "effective" designator is generic in the sense that "homogeneous" material may still be composed of layers of adhesive and dielectric. Furthermore, the electric field does penetrate the conductive material in a manner that impacts the velocity of propagation at different frequencies. In this sense, "effective" is used here for ε_r.

23.7.2 Some Extracted Data

Some extraction methods related to these and other parameters are discussed later based on working with the flex board with these characteristics:

- Effective relative dielectric constant, ε_r 3.05 to 3.5
- Attenuation loss α_T versus frequency, around 1.6 dB/in at 20 GHz
- Loss tangent (tan δ), ranging from 0.0127 to 0.0163 with an average of 0.0145
- DuPont Pyralux FR Spec. at 1 MHz, $\varepsilon_r = 3.5$, tan $\delta = 0.02$

23.7.3 Overview of Characterization Methods

A large number of methods have been reported over the years and are still being reported in the technical literature for the characterization of dielectric constant and loss tangent. These are loosely classified as direct capacitive measurements, resonator measurements, and direct transmission line measurements. In addition, there are several hybrid approaches combining direct measurement and theoretical predictions based on physical dimensions and known formulas. This is actually a variation of using the theoretical formulas or calculated field solver results to iteratively bring the parameters in alignment with measured results. Such an approach remains most promising and useful. Several of these approaches have been standardized by IPC[12] and are publicly accessible for downloading at *www.ipc.org* in IPC-TM-650 Test Methods Manual. Number 2.5.5.3, "Permittivity (Dielectric Constant and Loss Tangent (Dissipation Factor) of Materials (Two Fluid Cell Method)," uses a capacitance bridge at 1 MHz to extract the capacitance and conductance at 1 MHz. Number 2.5.5.5.1, "Stripline Test for Complex Relative Permittivity of Circuit Board Materials to 14 GHz," describes a test fixture and a sequence of resonator measurements based on a disconnected length of stripline. Number 2.5.5.7, "Characteristic Impedance and Time Delay of Line on Printed Boards by TDR," describes a time domain measurement technique on a long microstrip. All of these methods describe physical fixtures and document processes to relate the measurements to physical board parameters.

Many other methods involving microstrips and stripline configurations are documented in books and the technical literature (e.g., see [Gup96], [Yue98], and [Pon98]). Some of these involve combining measurements and actual known formulas based on known physical dimensions (e.g., see [Yue98]). An extension of this is to use EDA tools ranging from free transmission line utilities to full 3D field solvers directly in the loop. To a large extent, EDA tools were used to validate and confirm some of the data in the stripline extraction discussed next.

23.8 Stripline Measurements

23.8.1 Stackup

Figure 23.21 shows a test board cross section and associated stackup used for characterizing the flex board. Differential traces are shown. Note that there is some trapezoidal etching. Also note that the material itself consists of a composite of dielectric and adhesive.

23.8.2 DC Resistance

In correlating measurements with theory, the actual effective width of the lines should be used. These widths are adjusted based on the etching factor. Table 23.1 summarizes the measured dc resistance of a set of traces designed to be 5 mils, 10 mils, and 15 mils wide. These results were used later in analysis based on multimeter measurements.

The dc resistances are directly proportional to length. However, they do not scale linearly with width. The actual widths appeared smaller than the design targets of 5 mils, 10 mils, and 15 mils. As shown in the cross-section drawing of Figure 23.21, the etching reduction appears trapezoidal. If x is the average etching reduction and is assumed constant for each of the specified widths, then x can be calculated for each pair of widths and the corresponding resistance for a given length. The calculation is based on the low-frequency inverse relationship between resistance and width expressed as

$$(w_a - x) R_1 = (w_b - x) R_2 \tag{23.12}$$

12. In 1999, IPC changed its name from the Institute of Interconnecting and Packaging Electronic Circuits to IPC (*www.ipc.org*).

or

$$x = (w_a R_1 - w_b R_2) / (R_1 - R_2) \quad (23.13)$$

where

R_1 and R_2 = measured resistances for different widths
w_a and w_b = the corresponding specified widths[13]

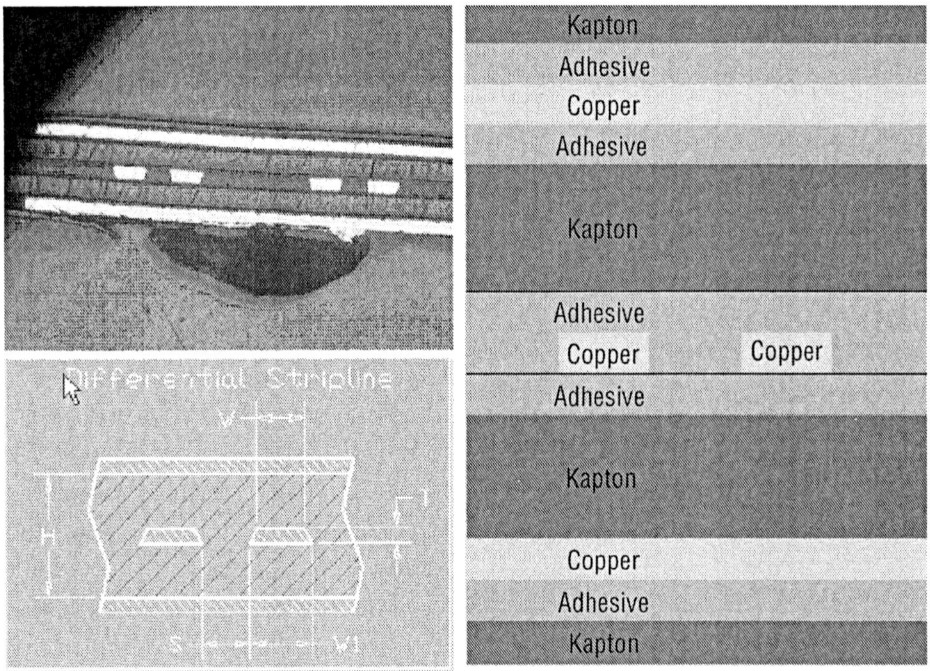

Figure 23.21 Cross-section and associated diagrams.

Table 23.1 Measured dc Resistance of Test Traces

Length	Design Width		
	5 Mils	**10 Mils**	**15 Mils**
1 in.	0.15 Ω	0.06 Ω	0.04 Ω (Adjusted to 0.0375 Ω)
2 in.	0.30 Ω	0.12 Ω	0.08 Ω (Adjusted to 0.075 Ω)
4 in.	0.60 Ω	0.24 Ω	0.15 Ω
Calculated Etching Adjusted Widths	3.33 mils	8.33 mils	13.33 mils

13. Uppercase symbols W_1 and W define real dimensions of actual traces as illustrated in Figure 23.21, whereas the lowercase symbols w_a and w_b used in the equations are two different values of *specified* widths.

For the 4-in. lines, each calculation produces $x = 1.67$ mils. The adjusted widths were used for some analytical comparisons later. The 2-in. and 1-in. lines also produce the same value if the 15-mil-wide lower resistance values are presumed to be the values in parenthesis if the meter had more resolution.

23.8.3 Characteristic Impedance and Delay

With the adjustments described in the previous section, the traces also had measured parameters that corresponded to closed form formulas where the following dimensions and copper thickness as shown in Figure 23.21 were assumed:

H = spacing between planes = 10 mils
$(W + W_1) / 2$ = etching adjusted width from table = 3.33 mils for 50 Ω design
T = copper thickness (1 oz.) = 1.4 mils

The measured impedance and delay of test traces is given in Table 23.2.

Table 23.2 Measured Impedance and Delay of Test Traces

Length	Design Width					
	5 Mils		10 Mils		15 Mils	
	Impedance	Delay	Impedance	Delay	Impedance	Delay
1 in.	53.5 Ω	165 ps	36.6 Ω	160 ps	28.9 Ω	155 ps
2 in.	53.4 Ω	305 ps	36.7 Ω	305 ps	29.2 Ω	305 ps
4 in.	53.1 Ω	615 ps	36.9 Ω	615 ps	29.3 Ω	610 ps

23.9 Stripline Results

23.9.1 Effective Relative Dielectric Constant

Figure 23.22 shows the effective dielectric constant up to 20 GHz based on averaged (because of symmetry) S_{12} and S_{21} delay measurements. The delay for the 1-in. trace is subtracted from the delay for the 4-in. trace in order to cancel common physical launch effects. The resulting delay for 3 in. of trace is inverted to convert it to a velocity and then scaled to the same units as the speed of light. The effective relative dielectric constant is proportional to the square of the reciprocal of this velocity per the formula given previously in Equation (23.11) on page 655.

Figure 23.22 shows some extraction results along with an overlaying power fit. Even though the test fixture was designed for a 50 Ω environment, some discontinuities did occur that contributed to the higher frequency ring. In this figure, y is the effective relative dielectric constant (the regression equation) as a function of x, and R^2 (R-square) is the fraction of the variance (not variation) in the data explained by the regression. R^2 approaches unity as the regression approaches a perfect fit.

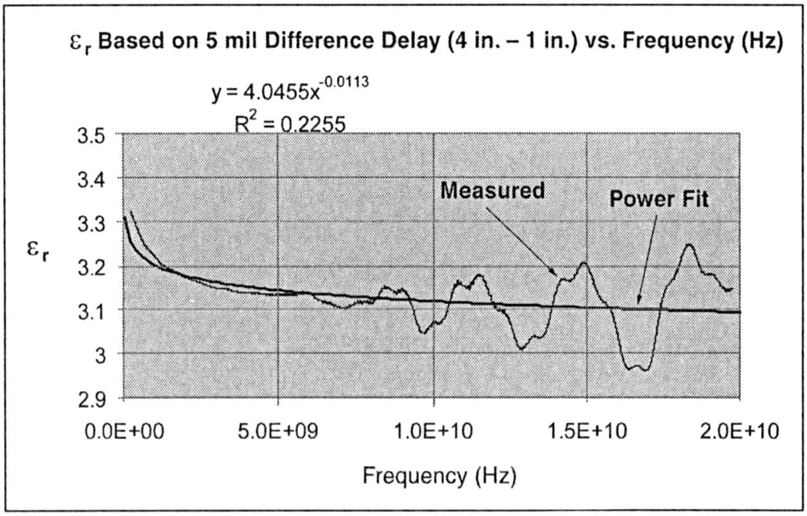

Figure 23.22 Effective relative dielectric constant versus frequency, based upon line difference delay measurements, with arbitrary power fit.

23.9.2 Total Losses for Stripline

The attenuation loss α_T was calculated from differential loss measurements. The common definition of attenuation loss versus frequency for stripline is given in reference [Yue98]:

$$\alpha_T(\omega) = 10 \log \{ |S_{11}(\omega)|^2 + |S_{21}(\omega)|^2 \} \tag{23.14}$$

where $\alpha_T(\omega)$ is the attenuation loss (per unit length) versus angular frequency, and ω is the angular frequency.

Because the 5-mil-wide line (nearly 50 Ω) was used, the loss was assumed to be just the S_{21} magnitudes. Calculations based on differences between the two lengths were assumed to cancel out to a first order any non-negligible common S_{11} impedance mismatch contributions to the attenuation loss. Since all structures were symmetrical, the actual calculations always involved averaging the S_{21} and S_{12} terms, and the S_{11} and S_{22} terms.

Figure 23.23 shows the difference in attenuation loss α_T for several 5-mil-wide striplines scaled to dB/in. in all cases. The scaled losses for the striplines are similar and predict about 1.6 dB/in. at 20 GHz. This extraction serves as the basis for loss tangent estimations made in the next section and also skin effect parameters for modeling.

The attenuation magnitudes of each pair of lines were subtracted to cancel common launch effects. The attenuation for the 1-in. line was subtracted from the attenuation for the 4-in. line, the attenuation for the 2-in. line was subtracted from the attenuation for the 4-in. line, and the attenuation for the 1-in. line was subtracted from the attenuation of the 2-in. line. The results were scaled to 1-in. lengths. The 5-mil-wide, 2-in. line extracted data showed higher oscillations

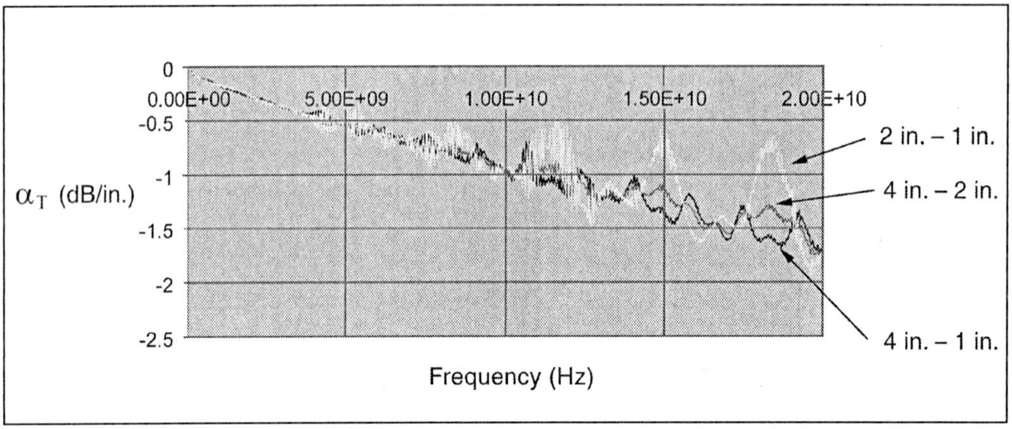

Figure 23.23 Scaled loss difference data for 5-mil lines and successive differences between 4-, 2-, and 1-in. lines.

and was possibly represented by a bad extraction or test structure. As a result, only the data based on 4-in. and 1-in. striplines was used for further analysis.

Data from 10-mil and 15-mil-wide stripline could also have been used. However, this data produces more oscillations over the frequency range because of significant line impedance differences from the reference test impedance of 50 Ω. Some preliminary processing was done, but detailed processing was done on the 5-mil-wide stripline.

23.9.3 Loss Tangent Estimation

The components contributing to the total attenuation loss α_T are caused by conductive loss and dielectric loss, linearly summed in terms of dB/unit length or nepers/unit length (see [Yue98]):

$$\alpha_T = \alpha_C + \alpha_D \tag{23.15}$$

where

α_T is the attenuation loss
α_C is the conductance loss (dominated by skin effect)
α_D is the dielectric loss from which loss tangent is extracted

The conductance loss component is broken into the series skin effect loss and a dc loss mismatch based on the 50 Ω characteristic impedance and measured dc resistance. This is done to put a measured dc constraint on subsequent curve fitting. There is some sensitivity in the results to dc constraint or to the fact that the fitting could be done without a dc constraint. The dc loss is calculated based on a voltage divider with the load:

$$\alpha_C = \alpha_{\text{SKIN}} + \alpha_{\text{dc}} \qquad (23.16)$$

where these approximate relationships hold over the frequencies of interest (in nepers/unit length), (see [Bog03]):

$$\alpha_{\text{dc}} = 20 \log (Z_L / (R_{\text{dc}} + Z_L)) / 8.6859 \qquad (23.17)$$

$$\alpha_{\text{SKIN}} = (R_0 / 2\, Z_0)\ \text{sqrt}\,(\omega / \omega_0) \qquad (23.18)$$

$$\alpha_D = (\delta\, \omega / 2\, v_0)\, (\omega / \omega_0)^{-\delta/\pi} = \delta\, \omega / 2\, v_0 \text{ (where } (\omega / \omega_0)^{-\delta/\pi} \text{ is nearly 1)} \qquad (23.19)$$

The equation factors are defined as follows:

R_{dc} is dc resistance (see Section 23.8.2 on page 656)
Z_L is load reference impedance (50 Ω)
Z_0 is the characteristic impedance (see Section 23.8.2)
$R_0(\omega_0)$ is ac resistance at ω_0 (where skin = conductance loss)
ω is angular frequency
$\delta = \tan\delta$, the loss tangent (for small values of δ)
v_0 is the velocity of propagation = 1 / delay per unit length

As illustrated by the equations, conductance loss is almost linearly related to the loss tangent parameter for the dielectric material. The process to estimate loss tangent is then to fit a quadratic equation to the measured loss data expressed as a function of the square root of frequency. This fitting approach was suggested in a similar extraction of loss components (see [Pon98]).

$$y = a_2\, x^2 + a_1\, x + a_0 \qquad (23.20)$$

The constant term a_0 is added as a dc constraint loss. The a_1 and a_2 extracted multipliers are then used for estimating the skin effect and conductance loss coefficients. This process is illustrated in Figures 23.24 and 23.25. A"scaled" (1×10^{10}) frequency is used in the process so that the extracted coefficients have four-digit resolution. The fitting could have been done without the a_0 constraint (sometimes with a resulting dc gain), or by setting it to 0. However, the results are numerically sensitive to its value. The a_0 value of –0.0260 dB/in. is illustrated for the 5-mil trace.

The quadratic factors leading to the loss tangent estimation are sensitive to the data range and initial dc resistance constraint. For example, Figure 23.25 (bottom figure, top trace) illustrates doing the estimation based only on a range through 5 GHz. The resulting quadratic factor is –0.6676 (versus –0.5389), producing a loss tangent value of 0.0147.

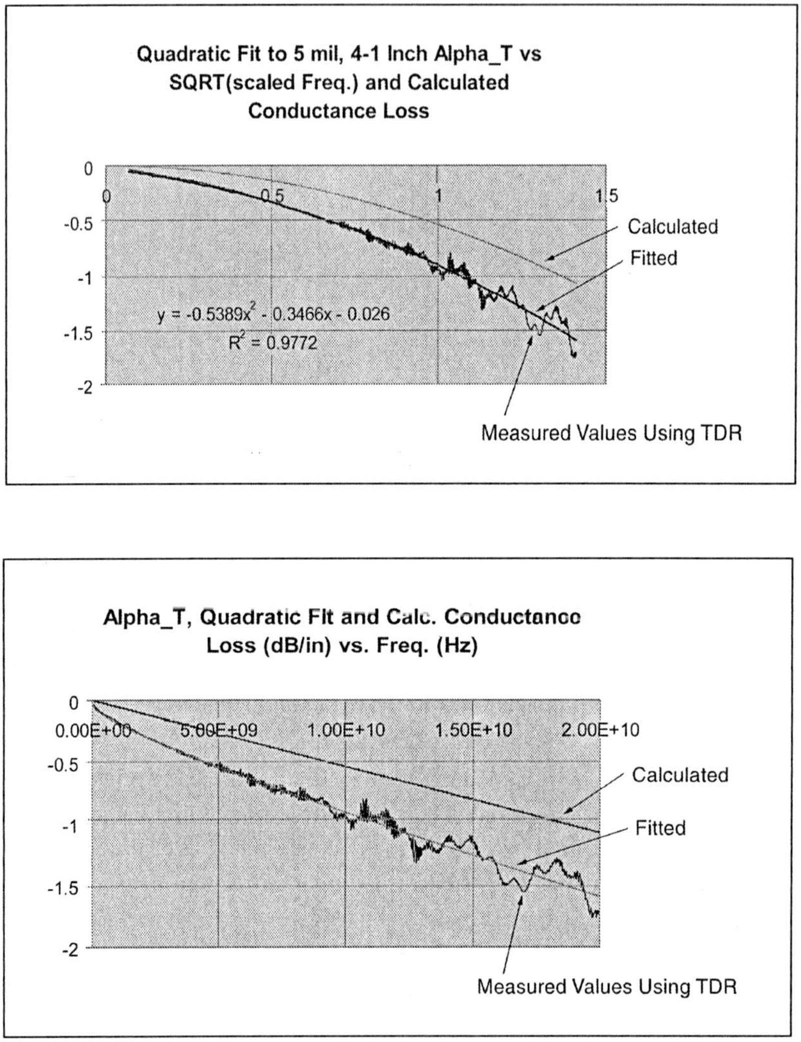

Figure 23.24 Quadratic fit over the entire 20 GHz range producing the scaled 0.5389 coefficient that is used to estimate the conductance loss and displayed to show the linear conductance loss contribution.

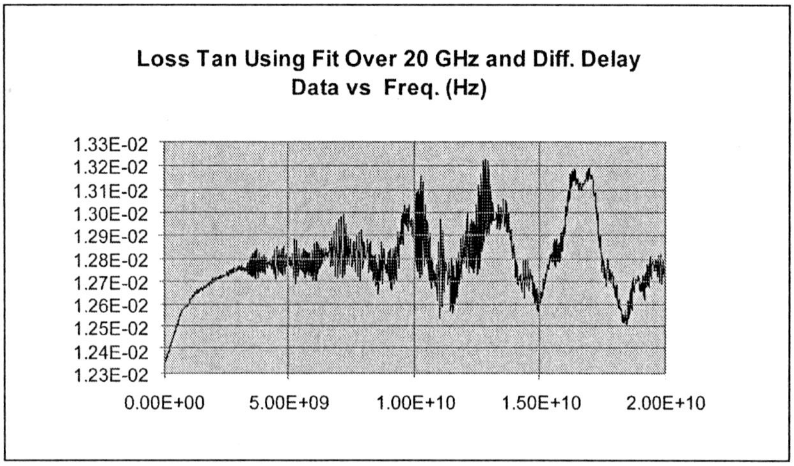

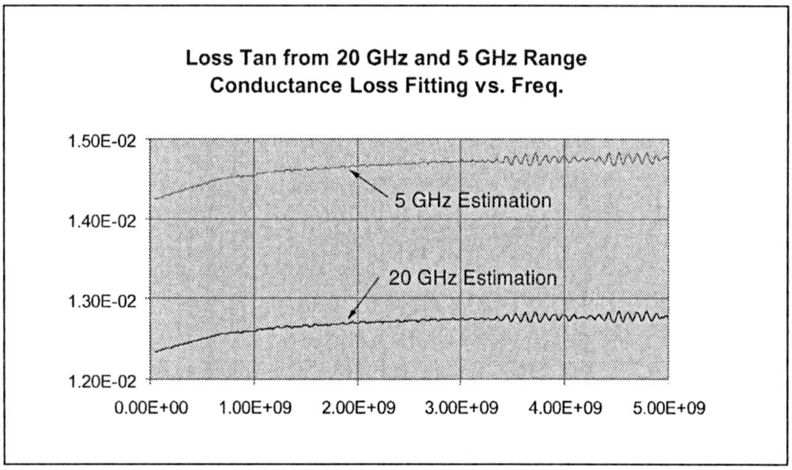

Figure 23.25 Corresponding loss tangent estimation for the 5-mil, 4-in. to 1-in. extraction and equation fit, and using the corresponding (unsmoothed) delay differences. Loss tangent estimation based on equation fitting over two different ranges with both results displayed to 5 GHz, average value about 0.0147.

23.10 Calculation Methods and Validation

Transmission line tools are useful for testing the results against theoretical predictions and testing the assumptions. Data for the 5-mil trace (effective width 3.33 mil) was entered into Ansoft Serenade SV 8.5, along with extracted dielectric constant and loss tangent values.

The assumptions were as follows:

Stripline width = 3.33 mils (taking into account the assumed actual width with etching)
Copper thickness = 1.4 mils
Spacing between plans = 10 mils
Length = 1 in. for per in. scaling
Relative dielectric constant = 3.1
Loss tangent = 0.014

With these assumed values, the resulting calculated characteristic impedance was 53.07 Ω, matching the TDR measured value. Table 23.3 shows the calculated trace losses, based upon these assumptions.

Table 23.3 Calculated Trace Losses

Frequency (GHz)	Dielectric Loss (dB/in.)	Conductor Loss (dB/in.)	Total Loss (dB/in.)
0.0	0.000	0.0063	0.0063
0.5	0.0285	0.0768	0.1053
1.0	0.0570	0.1086	0.1656
2.0	0.1140	0.1536	0.2676
5.0	0.2849	0.2428	0.5278
10.0	0.5699	0.3434	0.9133
15.0	0.8548	0.4206	1.2754
20.0	1.1398	0.4857	1.6254

The calculated dc loss of 0.0063 dB/in. differs from the assumed loss of 0.026 dB/in. (from measured 0.15 Ω per 1 in., 5-mil stripline) used as the dc loss constraint. It may be unrelated to dc loss. However, lower dc constraints did produce better fits. Constraint handling is still under investigation.

The data in Figure 23.25 compares very well with these calculated losses. A process based on interactive tuning of parameters to match measurements seems to work. Some literature (such as [Yue98]) propose processes using both actual measurements and physical structure information to estimate parameters. Consequently, a mathematical utility (which already embeds the equations) can be used by interactive tuning to estimate the parameters.

23.10.1 Comparison of Connector Simulations with and without Final Inch and Discussion of Advantages

It is always good to add a bit of science to our conjectures about these things. To illustrate our point, let's run two SPICE simulations on the system shown in Figure 23.26, consisting of two PCBs each with SMAs for injecting or measuring test signals, 50 Ω traces of known physical and electrical characteristics, a breakout region around each mating connector, and a QPairs QTH-DP 5 mm connector to connect the two PCBs together. Further, let us suppose the model elements for these components were created according to Table 23.4.

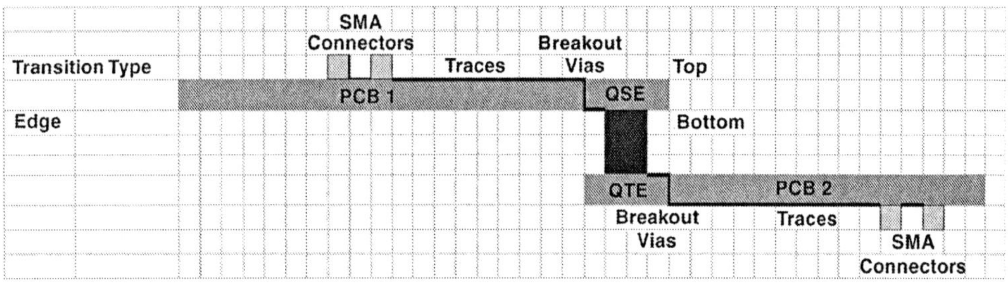

Figure 23.26 Final Inch test system.

Table 23.4 Model Elements Used in Each SPICE Simulation

Models	Simulation without Final Inch	Simulations with Final Inch
Connector	Samtec Lossy 3D Full-Wave Model	Samtec Lossy 3D Full-Wave Model
SMAs	None	3D Full-Wave Model
Traces	None	Lossy 2D Frequency-Dependent Table Model
BOR	None	Coupled 3D Full-Wave Model

For this particular experiment let's assume FR4 with an $\varepsilon_r = 4.2$ and a loss tangent of 0.02, which agree with correlation measurements taken of Final Inch test and evaluation designs. For stimulus we'll use a $2^5\!-\!1$ PRBS sequence for the victim net and a 1010 pattern for aggressors, 65 ps edge rates, and frequencies of both 1 Gbps and 10 Gbps.

The results of the simulations are shown in the following Figures 23.27 through 23.35.

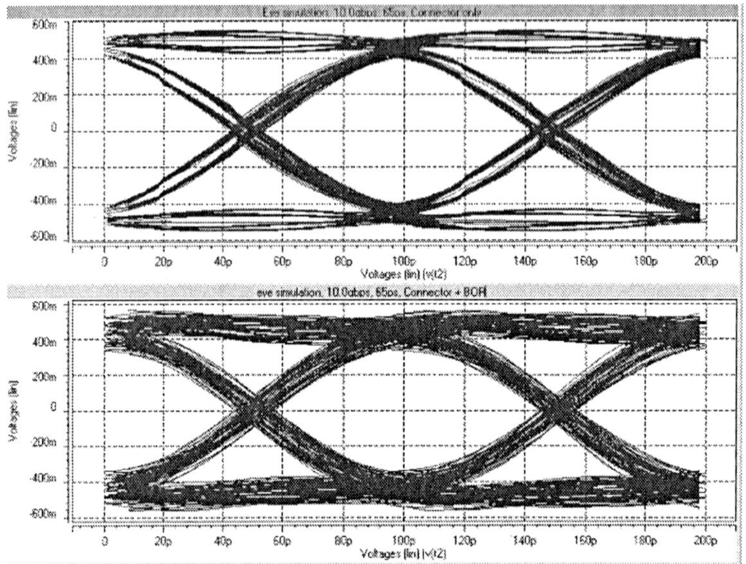

Figure 23.27 Eye comparisons at 10 Gbps, 65 ps edge rates. Connector only (top) vs. connector + BOR (bottom).

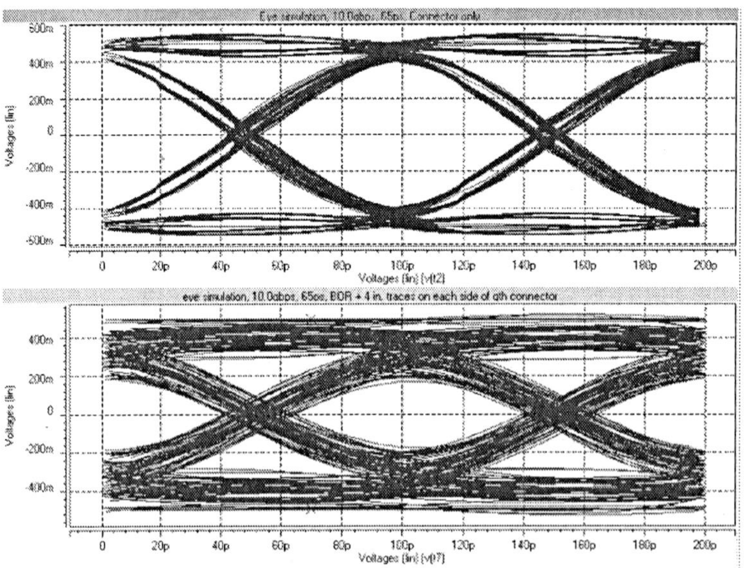

Figure 23.28 Eye comparisons at 10 Gbps, 65 ps edge rates. Connector only (top) vs. connector + BOR + 4 in. of trace (bottom).

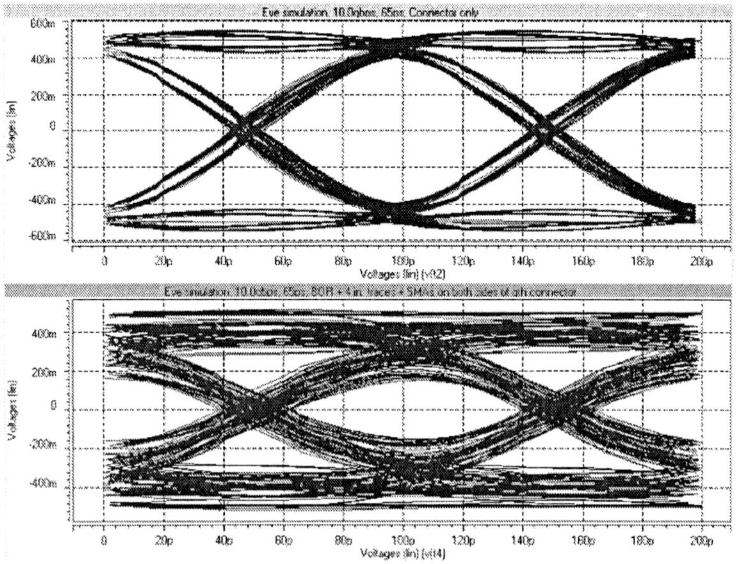

Figure 23.29 Eye comparisons at 10 Gbps, 65 ps edge rates. Connector only (top) vs. connector + BOR + 4 in. of trace + SMAs (bottom).

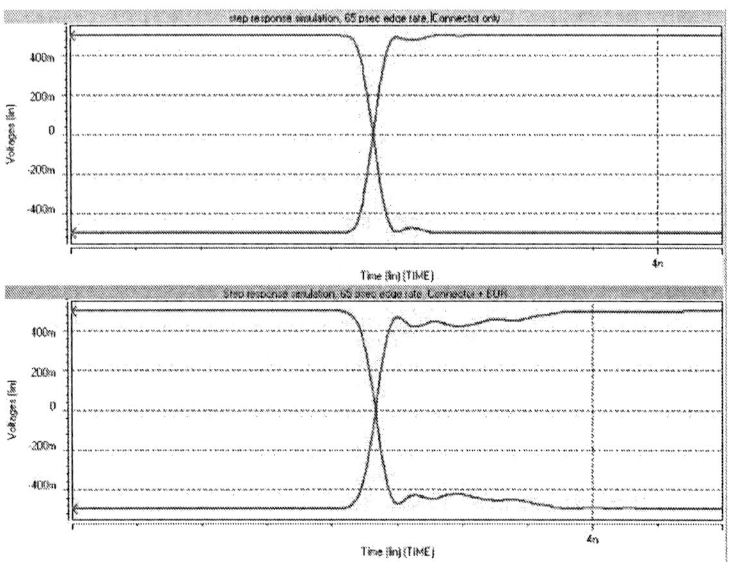

Figure 23.30 Step response comparison, 65 ps edge rates. Connector only (top) vs. connector + BOR (bottom).

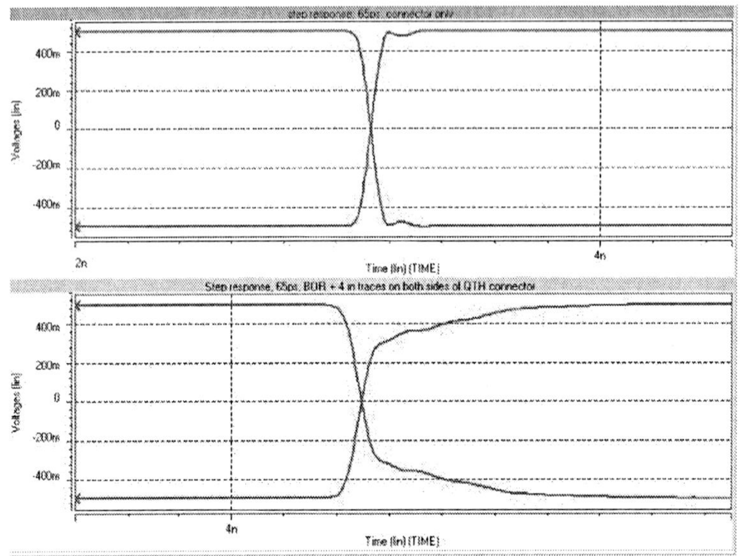

Figure 23.31 Step response comparison, 65 ps edge rates. Connector only (top) vs. connector + BOR + 4 in. of trace on both sides of the QTH connector (bottom).

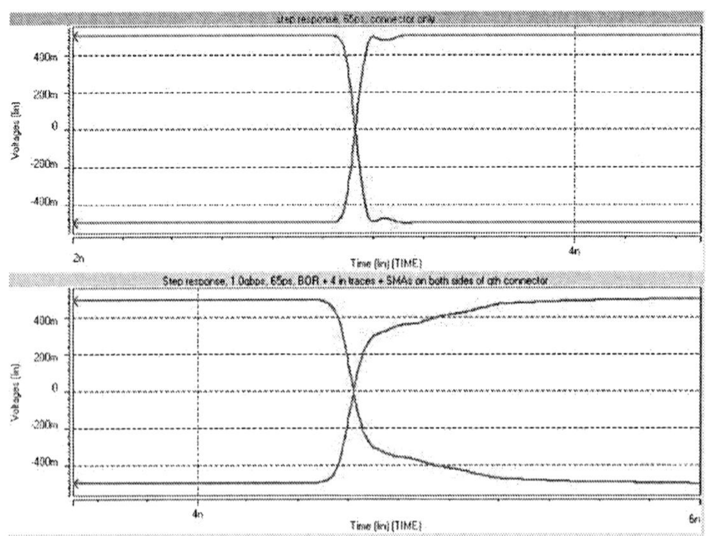

Figure 23.32 Step response comparison, 65 ps edge rates. Connector only (top) vs. connector + BOR + 4 in. of trace on both sides of the QTH connector + SMAs (bottom).

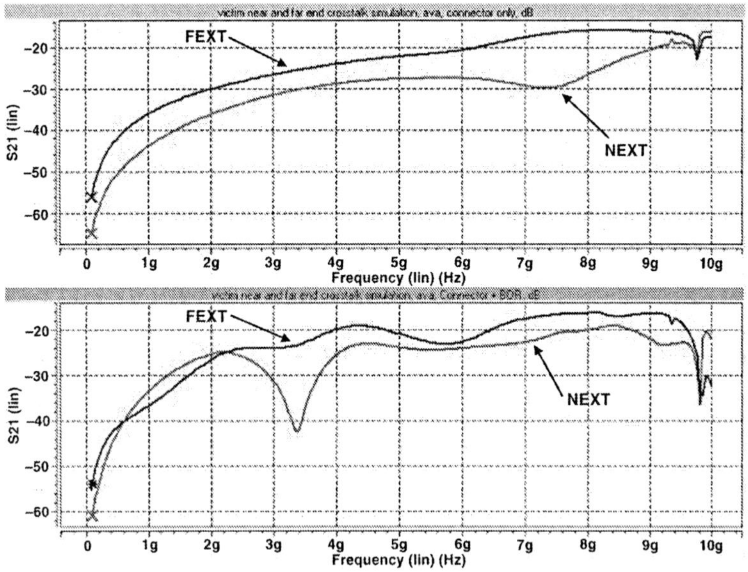

Figure 23.33 Victim crosstalk comparison in dB, Connector only (top) vs. connector + BOR (bottom). NEXT is near-end crosstalk and FEXT is far-end crosstalk.

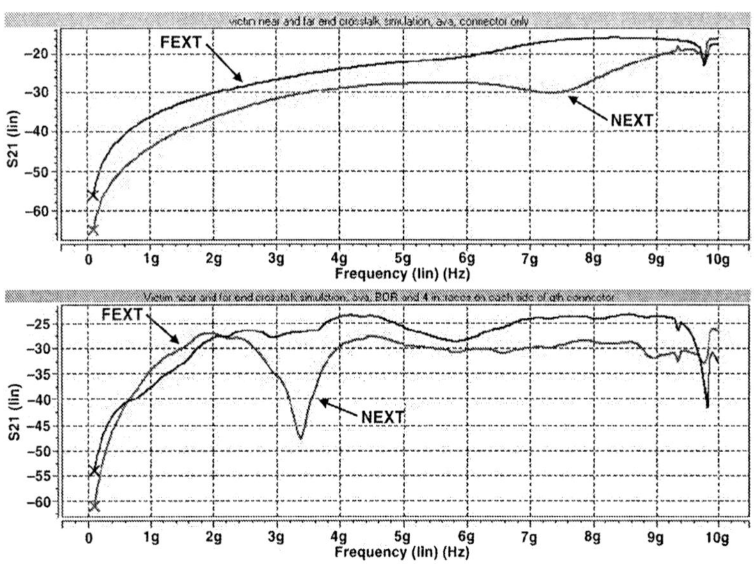

Figure 23.34 Victim near-end and far-end crosstalk comparison in dB. Connector only (top) vs. connector + BOR + 4 in. of trace (bottom).

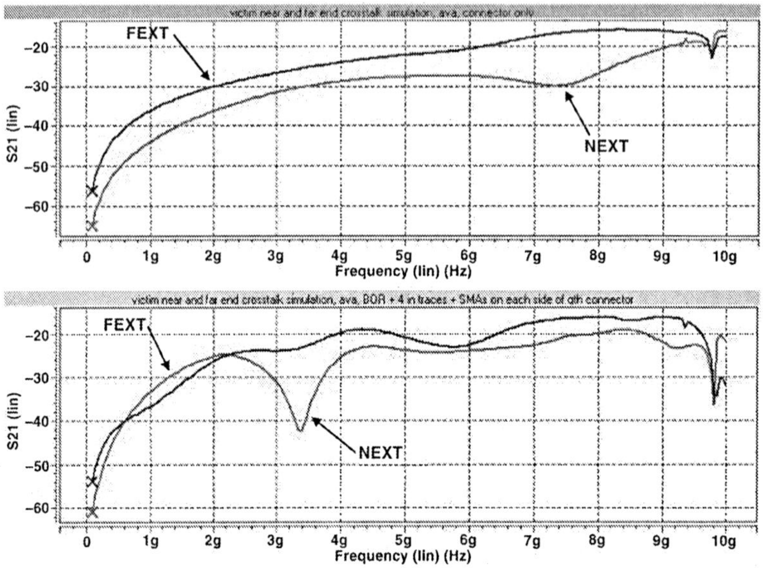

Figure 23.35 Victim near-end and far-end crosstalk comparison in dB; connector only (top) vs. connector + BOR + 4 in. of trace + SMAs (bottom).

23.10.2 SMA Test Trace Comparisons, Modeled vs. Measured

Model comparisons were performed for test circuits comprising an SMA connector tied to the internal trace of a PCB which in turn flowed to another SMA connector, as shown in Figure 23.36. Two types of models were created and simulated for comparison: (1) models generated from actual S-parameter measurements of the test circuit by a VNA, and (2) models generated from the circuit's physical dimensions using physical modeling and material parameter validation techniques discussed earlier in this chapter.

Figure 23.36 Test trace and test trace board.

Test traces were measured with an Agilent 8720, 20 GHz VNA for all two-port S-parameters. Measurements were made on stripline trace layer 3 and layer 4 with 20 GHz of measured bandwidth (as shown in Figure 23.36). Two-port S-parameters were converted to HSPICE G-element Laplace "black box" models with Sigrity BroadBand SPICE for comparison with the models derived from VNA measurements. 20 GHz bandwidth models were then extracted and used for subsequent simulations in Synopsis HSPICE. We refer to these simulations as those for the VNA extracted model.

For comparison, a second set of simulations of these same test traces was also performed in Synopsys HSPICE, but based upon the physical modeling and material parameter validation techniques discussed earlier in this chapter. We refer to these simulations as those for the modeled test trace.

The simulation and measurement parameters used were the same for all simulations:

> Victim input stimulus — 1volt ac, linearly swept from 100 MHz to 10 GHz
> Victim signals terminated at measurement point, 50 Ω to ground
> Passive signal terminated at both end points, 50 Ω to ground

Figure 23.37 compares the simulated eye diagrams for the modeled test trace versus the VNA extracted model for both signal layers 3 and 4. The eye diagrams show a very high level of correlation.

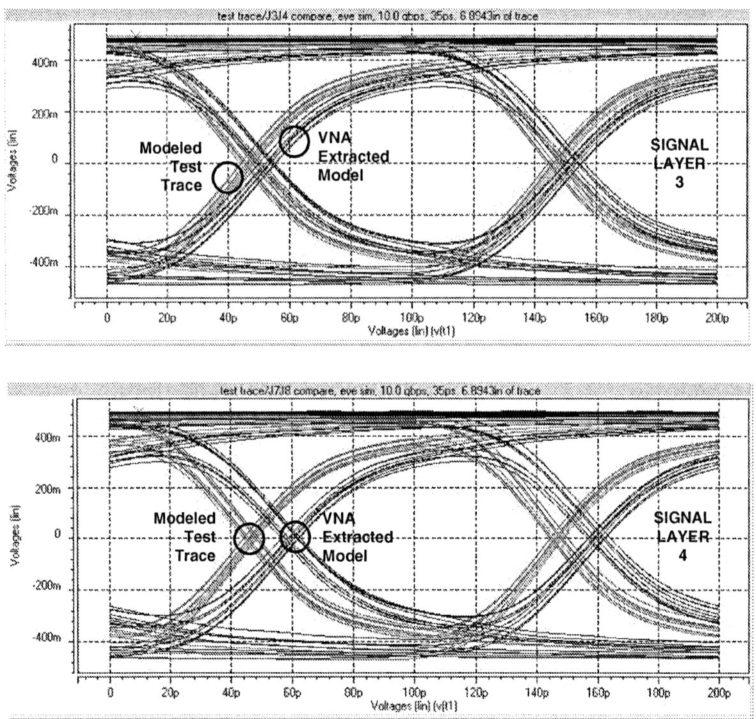

Figure 23.37 Eye comparisons for the modeled test trace vs. the VNA extracted model at 10 Gbps and 35 ps edge rate.

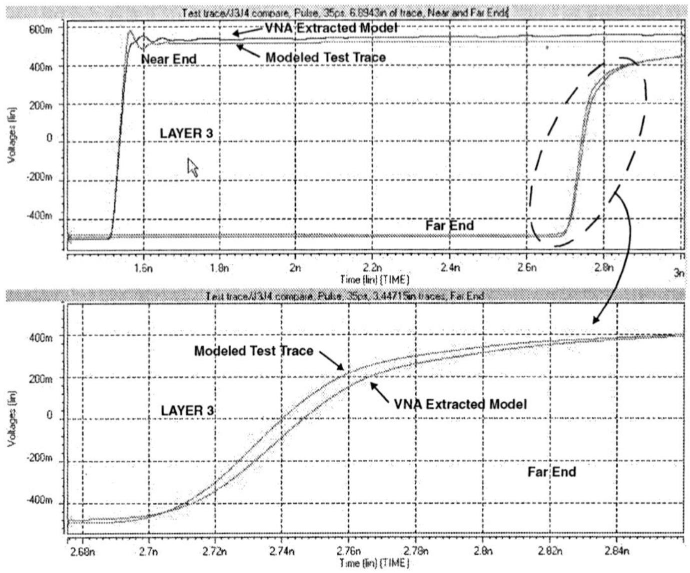

Figure 23.38 Step response comparison for the layer 3 modeled test trace vs. the VNA extracted model for (top) both the near and far end of the trace, and (bottom) expanded view of the step response at the far end.

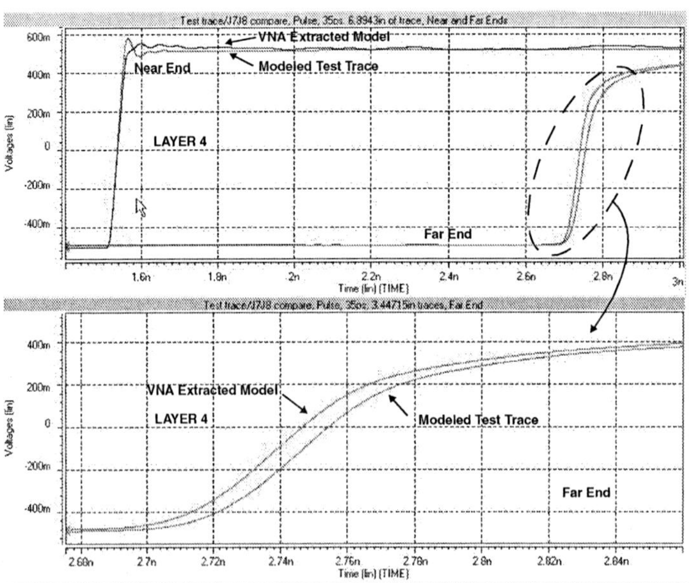

Figure 23.39 Step response comparison for the layer 4 modeled test trace vs. the VNA extracted model for (top) both the near and far end of the trace, and (bottom) expanded view of the step response at the far end.

The simulated pulse responses of the modeled test traces versus the VNA extracted models as shown in Figures 23.38 and 23.39 also illustrate very good correlation:

- The signal propagation time correlated to within 7.7 ps across two SMA connectors and a total trace length of 6.8943 in.
  - With an estimated delay of 172 ps/in., total delay error is 0.6%.
- Differences between the impedance of layer 3 and layer 4 can be easily seen in the slight differences of their step response.
- Slight differences between modeling and measurement are due to the lossless 3D modeling of the SMA connectors.
  - These differences are minimal when considering the full range of PCB design variables and their tolerances.

23.11 Conclusions

Samtec and Teraspeed Consulting are pushing advances in design, modeling, simulation, and measurement validation of high-performance, board-to-board 5-to-10 Gbps interconnects. At the same time, they are helping Samtec customers cut their design costs by providing design aids based on real connector/BOR physical and modeling results. The new design and verification process that makes this possible, called Final Inch, is a coherent design, simulation, and verification methodology. This is summarized in the following steps:

Step 1 — Select appropriate modeling for each element.
Step 2 — Design best performance BOR (in conjunction with connector design).
Step 3 — Integrate models into a coherent, hybrid simulation environment.
Step 4 — Deliver the design and simulation environment through Final Inch design kits.

The complete Final Inch package makes designing a high-speed interconnect system faster and easier then ever before.

Ease of Design — The BOR for each connector has been designed by signal integrity and CAD engineers who are experts in the field of high-frequency board design, and Samtec provides a complete set of design aids to help their customers add it to their own designs. They can either follow the recommended layout documentation, or import the CAD data directly into their design. Test and evaluation boards are also provided, which are based on the recommended layout documentation for the connector family. These boards can be used to evaluate the BOR and connectors' performance in the customer's own laboratory using whatever signaling schemes they prefer.

Ease of Simulation — The element models included in Final Inch packages originate from the best-of-the-best in modeling techniques: 2D multi-sectioned RLC models created using quasi-static field solvers for electrically short structures, 2D quasi-static lossy models created for electrically long structures, and 3D full-wave solvers used to model three dimensional structures within the transition regions of connectors and vias. Few engineers have access to all of these specialized modeling tools, each with its own language, abstractions, and data formats, so once the models are validated against the connector's test and validation board set lab measurements, they are converted into models compatible with Synopsys HSPICE and become the model library for the electrical design package's test and evaluation board HSPICE simulation deck, a simulation environment instantly usable by most design engineers.

Virtual Prototyping — Since the HSPICE deck has been fully correlated to measurements from the test and evaluation board, the user can create a virtual prototyping tool by replacing the SMA element models with the user's serial driver and receiver, saving additional development time.

And the Big Payback — Predictable performance at 10 Gbps rates and beyond.[14]

23.12 Exercises

23-1. Provide a one- or two-sentence definition for each of the following terms: (a) SMA launch, (b) W-element model, (c) G-element model, (d) differential via, (e) antipad, (f) breakout region, (g) simulation deck, (h) flex assembly, and (i) via stitching.

23-2. What happens when a good signal return path is not available?

23-3. (a) If a long breakout trace is used, is the location of its connector via closer to or farther away from the connector relative to a short breakout trace? (b) For a long breakout trace, is its corresponding non-breakout trace longer or shorter than the non-breakout trace corresponding to a short breakout trace?

23-4. In dealing with the BOR portions of a multi-GHz design, what are the most significant design concerns?

23-5. (a) In what format are Samtec connector models delivered? (b) What maximum edge rates and bandwidths do the model section sizes support?

23-6. What are two main effects that occur in PCBs that cause propagating GHz signals to become distorted?

23-7. At what stage of the high-speed design process are fundamental material properties measured?

14. For more information on Final Inch, visit Samtec's Web site at *www.samtec.com/FI* or contact the Final Inch Group at *finalinch@samtec.com*. For additional information on Teraspeed Consulting Group LLC, visit the Teraspeed Consulting Web site at *www.teraspeed.com*.

23-8. What is meant by "full-wave" modeling?

23-9. This chapter talks about hybrid design, modeling, simulation, and measurement techniques that reduce the risk in high-performance designs of 10 Gbps and above. What two entities does the term "hybrid" refer to when used in this context?

23-10. What is one tool used to facilitate the translation of frequency-domain S-parameters into time-domain SPICE models?

23-11. In validating material parameters, why were stripline traces used instead of microstrip traces?

23-12. What three correction factors were applied to S-parameter network models to correct the model for nonpassivity?

23-13. For an S-parameter matrix to be passive, what requirement must be met? Express your answer in terms of the identity matrix, the scattering matrix, and the transpose of the scattering matrix.

23-14. If small errors in the measurement of S-parameters in the frequency domain result in non-passive network parameters for a passive network, what effect can this have on time-domain simulations of that network?

23-15. What was the motivation for using frequency-domain S-parameters to describe the electrical networks used for high-speed signaling?

23-16. How were the two simulation issues of (a) W-element simulation fidelity with slow rise-time signals, and (b) the interoperability of Synopsys HSPICE W-element models with Sigrity BroadBand SPICE pole-zero fitted G-element models resolved?

23-17. Are Samtec's Final Inch models and test boards intended for both single-ended and differential designs?

Appendix 23.A Generalized N-Port, Mixed-Mode S-Parameters

23.A.1 Why Do We Care?

As system performance increases, there is an increasing demand placed on the design engineer's ability to accurately predict system behavior. This, in turn, has forced an increased reliance on sophisticated system simulations and high frequency, wide bandwidth measurements. Many, if not most, of these systems produce data in the form of frequency domain S-parameters.

S-parameters are network parameters that were developed principally to provide a method of specifying electrical characteristics by direct measurement of a network in a normal operating environment. These are usually single-ended measurements on each port of the network. In RF systems, this is normally a one-port or two-port measurement made using a VNA.

One-port and two-port measurements are not very useful for high-speed digital system design where differential signaling predominates. Characterization of a differential signaling path requires, at a minimum, a four-port measurement — one measurement set for each end of each member of the pair. This too has certain deficiencies. In high-speed systems, the reason differential signaling is preferred is because it is relatively insensitive to symmetric discontinuities; that is, discontinuities that affect both members of the pair equally. Many of these discontinuities create unavoidable problems for single-ended signaling. Something as simple as a via transition from one metal layer to another can cause serious problems with a single-ended signal operating at high edge rates. The same transition on a symmetric coupled differential pair will have a much smaller effect. Because of this, single-ended measurements can significantly underpredict actual system performance.

To address this issue, in the last few years there has been an increased use of mixed-mode S-parameters. Briefly, mixed-mode S-parameters are a mathematical manipulation of the *natural* S-parameters that describe the performance of differential network differentially rather than single-ended. Mixed-mode S-parameters describe differential networks in terms of pure modes, differential, common, and mixed modes, common to differential, and differential to common. See Chapter 19 starting on page 491 for a detailed discussion of differential and mixed-mode S-parameters.

While mixed-mode S-parameters provide a distinct improvement in our ability to visualize and predict performance of differential signaling paths, they still have a limitation in that they describe only a single differential pair. Real systems, at least the interesting ones, contain multiple differential pairs. Often these pairs are closely spaced with significant inter-pair coupling. Understanding the effects of this coupling is necessary for a clear understanding of system operation. As a result, there is a need to extend the mixed-mode S-parameter technique from two ports to multiple ports. To address the need for multiport, mixed-mode S-parameters, we will show an extension of the two-port, mixed-mode form to N-ports.

23.A.2 Development of N-Port, Mixed-Mode S-Parameters[15]

The extension of two-port, mixed-mode S-parameters should really reside in the realm of the obvious. The key to understanding this extension is simply to understand the formulation of the mixed-mode S-parameters.

Excitation of a differential port can be completely described by linear combinations of even- and odd-mode excitation of the pair elements. Formally, this means that the excitation is a linear combination of the excitation vectors:

$$\begin{bmatrix} 1 & -1 \\ 1 & 1 \end{bmatrix} \times \begin{bmatrix} v1_p & v2_p \\ v1_m & v2_m \end{bmatrix} = \begin{bmatrix} v1_d & v2_d \\ v1_c & v2_c \end{bmatrix} \tag{A.23.1}$$

where v represents voltage and where subscripts p and m are plus and minus terminal voltages for the differential pair, and subscripts d and c are differential and common-mode voltages.

The formulation of mixed-mode S-parameters involves a linear transformation of the natural S-parameters:

$$S_{mm} = MS_{nat}M^{-1} \tag{A.23.2}$$

$\boldsymbol{S_{mm}}$ and $\boldsymbol{S_{nat}}$ are the mixed-mode and natural S-parameter matrices, respectively. The work of the linear transformation is done with the matrix $\boldsymbol{M}$. Let's consider for a moment a four-port, single-ended network:

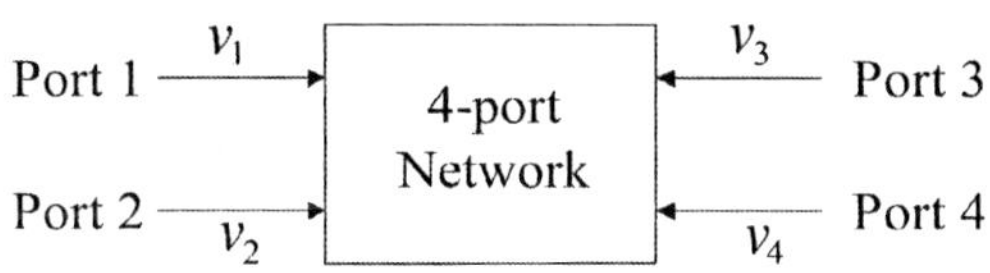

Figure 23.A.1 Four-port network.

To convert the natural S-parameters of this network to mixed-mode, matrix $\boldsymbol{M}$ is constructed as follows:

$$\boldsymbol{M} = \frac{1}{\sqrt{2}} \times \begin{bmatrix} 1 & -1 & 0 & 0 \\ 0 & 0 & 1 & -1 \\ 1 & 1 & 0 & 0 \\ 0 & 0 & 1 & 1 \end{bmatrix} \tag{A.23.3}$$

15. This appendix uses slightly different notation from the S-parameter notation used in Chapter 19. Here, the single-ended S-parameters S^{se} are referred to as natural S-parameters S_{nat}. Also, the notation such as S_{dc12} used in Chapter 19 is given by S_{12}^{dc} in this appendix.

Recalling our even- and odd-mode excitation matrix above, the important thing to recognize here is that the first row of the matrix $\boldsymbol{M}$ is the formation of a differential (odd-mode) excitation of ports 1 and 2 of the network, and row three is the formation of the common (even-mode) excitation of ports 1 and 2. It can be seen from this that $\boldsymbol{M}$ not only performs the linear transformation between $\boldsymbol{S_{nat}}$ and $\boldsymbol{S_{mm}}$, it also determines which single-ended ports will combine to form differential ports. For completeness, we should note that the $\boldsymbol{M}$ matrix not only reforms the excitation but also the response into differential and common modes.

$\boldsymbol{S_{mm}}$ then becomes

$$\boldsymbol{S_{mm}} = MS_{nat}M^{-1} = \begin{bmatrix} S^{dd} & S^{dc} \\ S^{cd} & S^{cc} \end{bmatrix} \tag{A.23.4}$$

where S^{dd} and S^{cc} are the pure differential- and common-mode S-parameters, and S^{dc} and S^{cd} are the mixed-mode differential-to-common and common-to-differential S-parameters, respectively.

$$\boldsymbol{S_{dd}} = \begin{bmatrix} S_{11}^{dd} & S_{12}^{dd} \\ S_{21}^{dd} & S_{22}^{dd} \end{bmatrix} \quad \boldsymbol{S_{dc}} = \begin{bmatrix} S_{11}^{dc} & S_{12}^{dc} \\ S_{21}^{dc} & S_{22}^{dc} \end{bmatrix} \tag{A.23.5}$$

$$\boldsymbol{S_{cd}} = \begin{bmatrix} S_{11}^{cd} & S_{12}^{cd} \\ S_{21}^{cd} & S_{22}^{cd} \end{bmatrix} \quad \boldsymbol{S_{cc}} = \begin{bmatrix} S_{11}^{cc} & S_{12}^{cc} \\ S_{21}^{cc} & S_{22}^{cc} \end{bmatrix} \tag{A.23.6}$$

Extending the $\boldsymbol{M}$ matrix to N ports is simply a matter of adding the necessary even- and odd-mode excitations. For example, consider the 8-port network shown in Figure 23.A.2:

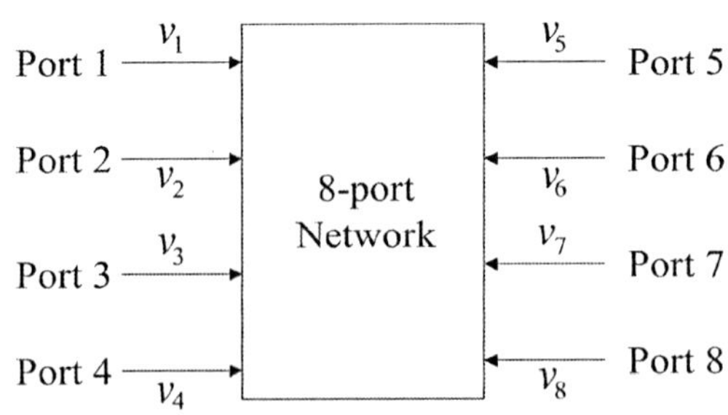

Figure 23.A.2 Eight-port network.

The M matrix to combine ports 1 and 2, 3 and 4, 5 and 6, and 7 and 8 into differential pairs will be

$$M = \frac{1}{\sqrt{2}} \times \begin{bmatrix} 1 & -1 & 0 & 0 & 0 & 0 & 0 & 0 \\ 0 & 0 & 1 & -1 & 0 & 0 & 0 & 0 \\ 0 & 0 & 0 & 0 & 1 & -1 & 0 & 0 \\ 0 & 0 & 0 & 0 & 0 & 0 & 1 & -1 \\ 1 & 1 & 0 & 0 & 0 & 0 & 0 & 0 \\ 0 & 0 & 1 & 1 & 0 & 0 & 0 & 0 \\ 0 & 0 & 0 & 0 & 1 & 1 & 0 & 0 \\ 0 & 0 & 0 & 0 & 0 & 0 & 1 & 1 \end{bmatrix} \quad \text{(A.23.7)}$$

Like the four-port example, S_{mm} is

$$S_{mm} = MS_{nat}M^{-1} = \begin{bmatrix} S^{dd} & S^{dc} \\ S^{cd} & S^{cc} \end{bmatrix} \quad \text{(A.23.8)}$$

but now

$$S_{dd} = \begin{bmatrix} S_{11}^{dd} & S_{12}^{dd} & S_{13}^{dd} & S_{14}^{dd} \\ S_{21}^{dd} & S_{22}^{dd} & S_{23}^{dd} & S_{24}^{dd} \\ S_{31}^{dd} & S_{32}^{dd} & S_{33}^{dd} & S_{34}^{dd} \\ S_{41}^{dd} & S_{42}^{dd} & S_{43}^{dd} & S_{44}^{dd} \end{bmatrix} \quad S_{dc} = \begin{bmatrix} S_{11}^{dc} & S_{12}^{dc} & S_{13}^{dc} & S_{14}^{dc} \\ S_{21}^{dc} & S_{22}^{dc} & S_{23}^{dc} & S_{24}^{dc} \\ S_{31}^{dc} & S_{32}^{dc} & S_{33}^{dc} & S_{34}^{dc} \\ S_{41}^{dc} & S_{42}^{dc} & S_{43}^{dc} & S_{44}^{dc} \end{bmatrix} \quad \text{(A.23.9)}$$

$$S_{cd} = \begin{bmatrix} S_{11}^{cd} & S_{12}^{cd} & S_{13}^{cd} & S_{14}^{cd} \\ S_{21}^{cd} & S_{22}^{cd} & S_{23}^{cd} & S_{24}^{cd} \\ S_{31}^{cd} & S_{32}^{cd} & S_{33}^{cd} & S_{34}^{cd} \\ S_{41}^{cd} & S_{42}^{cd} & S_{43}^{cd} & S_{44}^{cd} \end{bmatrix} \quad S_{cc} = \begin{bmatrix} S_{11}^{cc} & S_{12}^{cc} & S_{13}^{cc} & S_{14}^{cc} \\ S_{21}^{cc} & S_{22}^{cc} & S_{23}^{cc} & S_{24}^{cc} \\ S_{31}^{cc} & S_{32}^{cc} & S_{33}^{cc} & S_{34}^{cc} \\ S_{41}^{cc} & S_{42}^{cc} & S_{43}^{cc} & S_{44}^{cc} \end{bmatrix} \quad \text{(A.23.10)}$$

Expanding:

$$S_{dd} = \frac{1}{2} \begin{bmatrix} S_{11} - S_{21} - S_{12} + S_{22} & S_{13} - S_{23} - S_{14} + S_{24} & S_{15} - S_{25} - S_{16} + S_{26} & S_{17} - S_{27} - S_{18} + S_{28} \\ S_{31} - S_{41} - S_{32} + S_{42} & S_{33} - S_{43} - S_{34} + S_{44} & S_{35} - S_{45} - S_{36} + S_{46} & S_{37} - S_{47} - S_{38} + S_{48} \\ S_{51} - S_{61} - S_{52} + S_{62} & S_{53} - S_{63} - S_{54} + S_{64} & S_{55} - S_{65} - S_{56} + S_{66} & S_{57} - S_{67} - S_{58} + S_{68} \\ S_{71} - S_{81} - S_{72} + S_{82} & S_{73} - S_{83} - S_{74} + S_{84} & S_{75} - S_{85} - S_{76} + S_{86} & S_{77} - S_{87} - S_{78} + S_{88} \end{bmatrix} \quad \text{(A.23.11)}$$

$$S_{cc} = \frac{1}{2}\begin{bmatrix} S_{11}+S_{21}+S_{12}+S_{22} & S_{13}+S_{23}+S_{14}+S_{24} & S_{15}+S_{25}+S_{16}+S_{26} & S_{17}+S_{27}+S_{18}+S_{28} \\ S_{31}+S_{41}+S_{32}+S_{42} & S_{33}+S_{43}+S_{34}+S_{44} & S_{35}+S_{45}+S_{36}+S_{46} & S_{37}+S_{47}+S_{38}+S_{48} \\ S_{51}+S_{61}+S_{52}+S_{62} & S_{53}+S_{63}+S_{54}+S_{64} & S_{55}+S_{65}+S_{56}+S_{66} & S_{57}+S_{67}+S_{58}+S_{68} \\ S_{71}+S_{81}+S_{72}+S_{82} & S_{73}+S_{83}+S_{74}+S_{84} & S_{75}+S_{85}+S_{76}+S_{86} & S_{77}+S_{87}+S_{78}+S_{88} \end{bmatrix} \quad \text{(A.23.12)}$$

$$S_{dc} = \frac{1}{2}\begin{bmatrix} S_{11}-S_{21}+S_{12}-S_{22} & S_{13}-S_{23}+S_{14}-S_{24} & S_{15}-S_{25}+S_{16}-S_{26} & S_{17}-S_{27}+S_{18}-S_{28} \\ S_{31}-S_{41}+S_{32}-S_{42} & S_{33}-S_{43}+S_{34}-S_{44} & S_{35}-S_{45}+S_{36}-S_{46} & S_{37}-S_{47}+S_{38}-S_{48} \\ S_{51}-S_{61}+S_{52}-S_{62} & S_{53}-S_{63}+S_{54}-S_{64} & S_{55}-S_{65}+S_{56}-S_{66} & S_{57}-S_{67}+S_{58}-S_{68} \\ S_{71}-S_{81}+S_{72}-S_{82} & S_{73}-S_{83}+S_{74}-S_{84} & S_{75}-S_{85}+S_{76}-S_{86} & S_{77}-S_{87}+S_{78}-S_{88} \end{bmatrix} \quad \text{(A.23.13)}$$

$$S_{cd} = \frac{1}{2}\begin{bmatrix} S_{11}+S_{21}-S_{12}-S_{22} & S_{13}+S_{23}-S_{14}-S_{24} & S_{15}+S_{25}-S_{16}-S_{26} & S_{17}+S_{27}-S_{18}-S_{28} \\ S_{31}+S_{41}-S_{32}-S_{42} & S_{33}+S_{43}-S_{34}-S_{44} & S_{35}+S_{45}-S_{36}-S_{46} & S_{37}+S_{47}-S_{38}-S_{48} \\ S_{51}+S_{61}-S_{52}-S_{62} & S_{53}+S_{63}-S_{54}-S_{64} & S_{55}+S_{65}-S_{56}-S_{66} & S_{57}+S_{67}-S_{58}-S_{68} \\ S_{71}+S_{81}-S_{72}-S_{82} & S_{73}+S_{83}-S_{74}-S_{84} & S_{75}+S_{85}-S_{76}-S_{86} & S_{77}+S_{87}-S_{78}-S_{88} \end{bmatrix} \quad \text{(A.23.14)}$$

This formulation can be extended to any number of ports, limited only by memory and processing power.

We have used this technique in our custom programs to examine the operation of 48-port networks with tens of thousands of frequency points. Our tool reads a Touchstone file containing the natural S-parameters, and then computes the differential S-parameters for each frequency point. For networks this large, visualizing the data becomes difficult. We have also developed a custom, programmable plotting program. We also plan to extend the plotting tool to display maximum crosstalk by summing mixed mode crosstalk amplitudes for differential modes, as well as crosstalk in impure differential environments where there is some common mode excitation. All the tool development has been done using MATLAB by MathWorks. In the future we plan to convert the tools to stand-alone programs.

CHAPTER 24

IBIS Modeling and Simulation of High-Speed Fiber-Optic Transceivers

Chapter Objectives

- Compare the results of simulations performed using IBIS models with those performed using SPICE models.
- Introduce electrical board description (EBD) files and discuss how they are used in modeling and simulation.
- Discuss how both IBIS and EBD files are used together in simulating fiber-optic transceivers.
- Provide a procedure for the verification of IBIS models for multigigabit applications.
- Provide an example of using differential time-domain reflectometry (TDR) to verify simulation results.
- Compare signal shapes from simulations of "non-ideal" circuits utilizing IBIS and SPICE model representations of the same buffers.

24.1 Introduction[1]

The I/O Buffer Interface Specification (IBIS) in its current version v4.0 allows the use of electrical board description (EBD) files which make the accurate modeling of very-high-speed, internally ac-coupled transceivers possible. At data rates of 2 Gb and higher the internal printed circuit board (PCB) layout of the transceiver has an influence on the signal shape at the component pin. This chapter reviews some of the techniques that are used within Agilent Technologies to develop and verify models for the latest generation of 2.5 Gb transceiver components. Signal integrity simulations for typical applications, which employ these models, is discussed.

1. This chapter was written by Herbert Lage [Lag02]. Reprinted with permission of Agilent Technologies.

The wide adoption of the Gigabit Ethernet networking standard has increased system data rates well into the 1,000 MHz range in recent years. This boundary has been pushed even further by the latest transceiver components. Fibre Channel transceivers operating at 2 Gbps data rates are available as well as 10 Gbps Ethernet transceiver components conforming to the Xenpak multisource agreement (MSA) with its four 3.125 Gbps XAUI interfaces. Digital system design has therefore moved deep into the multigigahertz range, with signal rise and fall times of the order of 100 ps or less.

It is a well-known fact that signal integrity (SI) simulations become necessary when system designs break the 50–100 MHz barrier,[2] and are virtually mandatory in the GHz range. SI simulations are used to ensure the quality and accurate timing of electrical signals (see [Gre99]). The benefit of SI analysis early in the design cycle is well established, as it allows the identification and resolution of SI problems such as overshoot, ringing, crosstalk, delay mismatches, and the like, before the first prototype is built.

Good component models are a prerequisite to carry out meaningful and realistic SI simulations. Component models can be created in SPICE, IBIS, and a number of proprietary formats.[3] Due to its standardization, IBIS has increasingly gained support with component manufacturers and acceptance within the SI community in recent years.[4]

24.2 IBIS Models for High-Speed Fiber-Optic Transceivers

Agilent Technologies has recently released a range of small-form factor (SFF) high-speed transceiver products for the Ethernet, Fibre Channel, and the metropolitan telecom markets. Pluggable SFF Fibre Channel transceivers like the HFBR-5720L operate at a data rate of 2.125 Gbps, while the SONET-compliant family of the HFCT-5942L operates at the OC-48 data rate of 2.488 Gbps.

The construction of IBIS models for components operating in the 2–3 Gbps range poses additional challenges. The rise and fall times of these high-speed components are on the order of 100 ps, which equates to an electrical length l_e of about 1.5 cm in FR-4. In order to treat a transceiver as one single lumped component, the physical length l_p needs to satisfy the following condition (see [Joh93]):

$$l_p < l_e / 6 \qquad (24.1)$$

2. Signal integrity is introduced in Chapter 3, starting on page 31, and is discussed throughout this book.
3. Modeling is discussed in several places throughout this book. Please refer to the index under "models" for several other places in this text that discuss all aspects of modeling.
4. IBIS is discussed extensively in this textbook. Chapter 21, "Modeling with IBIS," starting on page 555, introduces IBIS and provides the code listing for an IBIS example. Please refer to the index under "IBIS" for several additional discussions on IBIS.

The physical length l_p of SFF transceivers without the optical connector is about 1 to 2 cm and therefore clearly too large to meet the above condition. This implies that the structure of the internal printed circuit board (PCB) of the transceiver component needs to be accounted for and treated as a distributed element.

The short effective length of signals with such fast edge rates makes it furthermore necessary to provide termination of the inputs and outputs as close to the laser driver and receiver postamplifier integrated circuits (ICs) as possible. Termination resistors and ac coupling capacitors for the high-speed signals are therefore included on the internal PCB of the transceiver and need to be accounted for in accurate electrical models of the components. Figure 24.1 shows a circuit diagram of the input and output connections of fiber-optic transceiver HFCT-5942L as a typical example.

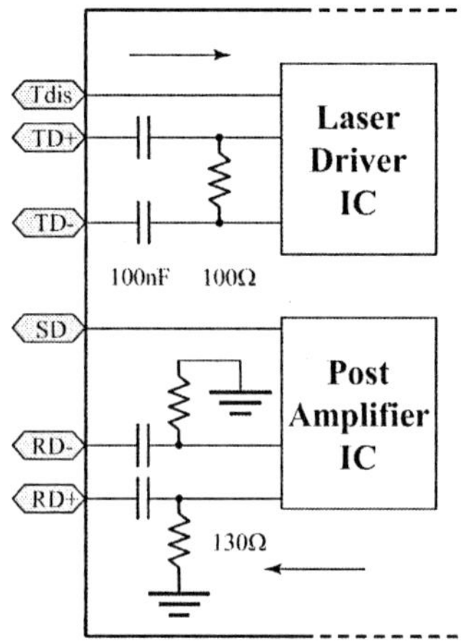

Figure 24.1 Ac coupling, termination and bias resistors for the high-speed data inputs (TD+, TD–) and outputs (RD+, RD–) of transceiver component HFCT-5942L.

The IBIS standard version v4.0, which was ratified in July 2002, allows for the description of component internal PCBs. The IBIS v4.0 standard document refers to components with such a level of complexity also as "board-level components." The connections between the active and passive elements on the PCB are defined in an electrical board description (EBD) file, which model developers will supply in addition to the IBIS files of a component. The IBIS files themselves contain the electrical characteristics of the input and output buffers of ICs inside the board-level component, and also the parameters of any additional passive elements like termination resistors or series capacitors. For a complete description of a board-level component both the EBD and IBIS files are required.

This chapter will describe the techniques that are used within Agilent Technologies to develop and validate complete IBIS models; that is, the combined IBIS and EBD models. The emphasis will be on the effect of the EBD model file on the combined models.[5]

In order to check the validity of using behavioral IBIS models at 2 Gbps data rates and higher, the results of simulations using a SPICE model will be compared with IBIS simulations. For this we will go beyond standard test load verification checks and we will analyze the effect of real-world topologies on the signal shape launched by the outputs of a high-speed transceiver. This should indeed provide a stringent test of the IBIS model quality and enable us to check out the limits within which IBIS modeling can provide accurate results. After the validity of the IBIS modeling approach has been established, a typical signal integrity case study utilizing IBIS models will be carried out.

5. Details of the development of input and output buffer models for the IBIS file itself can be found in [Cha01].

24.3 The Electrical Board Description (EBD) File

The EBD file describes how the user-accessible pins are connected with the internal structure of a component, for example, a fiber-optic transceiver. EBD files are structured very similar to IBIS files. The syntax specification can be found in Section 3 of the IBIS v4.0 standard document and will not be repeated here in depth; however, some key aspects that are important for the understanding of the models for fiber-optic transceivers shall be highlighted.

An EBD file must contain certain keywords like [IBIS Ver], [File Name], [Manufacturer], and so on, and a complete list of the user-accessible pins of the component under the heading [Pin List]. The actual description of the signal paths follows after the keyword [path description]. A path description is required for each pin whose signal name is not "GND," "POWER," or "NC" (not connected).

```
[Path Description] RD+
Pin 10
|
Len = 1 L=1.50n C=0.60p R=0.01 /
Len = 1 L=0.78n C=0.18p R=0.01 /
Node c02.1
Node c02.2
Len = 1 L=0.57n C=0.06p R=0.01 /
Node r131.1
Len = 1 L=1.85n C=0.09p R=0.01 /
|
Node hfct_5942L.10
```

Figure 24.2 EBD File: Path description for pin 10 of transceiver HFCT-5942L.

An example for a path description is shown in Figure 24.2 for pin 10 (Receiver Data Out, RD+) of the HFCT-5942L. The corresponding section of the internal PCB is shown in Figure 24.3. It is evident from Figure 24.2 that all path descriptions either start or end in an external pin. They are always read from the top, i.e., the example begins at external pin 10, which is followed by two transmission line sections, which then connect to node "c02.1" This node links the EBD file with the IBIS file containing the component parameters for the serial 100 nF capacitor. A table at the end of the EBD file following the keyword [Reference Designator Map] must resolve the node names and map them onto IBIS file names that have to reside in the same directory as the EBD file. The path description shown in Figure 24.2 finally ends in node "hfct_5942L.10," which is a reference to the IBIS file containing the properties of the high-speed output buffer of the receiver's postamplifier IC.

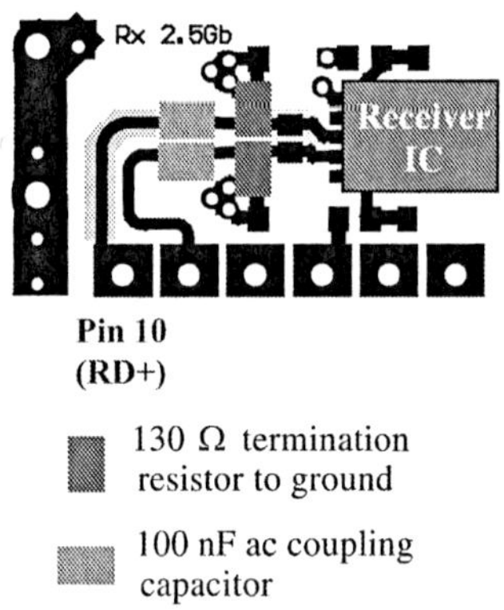

Figure 24.3 Receiver PCB of the HFCT-5942L showing the signal path from external pin to the postamplifier IC.

It is also worth noting that transmission lines are described in terms of their total series inductance L, total series capacitance C, and total resistance R. The statement "Len = 1" tells the simulation software to interpret this line as a distributed element (with the length of one "unit"). The transmission line parameters L and C (in values per unit length) can be easily converted into the perhaps more familiar characteristic impedance Z_0 and time delay t_d parametric description for lossless lines (see [Das96]):

$$Z_0 = \sqrt{\frac{L}{C}} \tag{24.2}$$

$$t_d = \sqrt{LC} \tag{24.3}$$

24.4 Verification of IBIS Models

Before IBIS models are released it is important to verify that the model accurately reflects the component's performance characteristics. Within Agilent Technologies we have adopted the following approach for the verification of the high-speed signal lines:

1. The simulated eye diagram of the component's data output pins into a typical load needs to closely resemble measured receiver eye diagrams.
2. If available, differential time domain reflectometry (TDR) measurements of the high-speed inputs will be fitted in order to confirm parasitic capacitances and inductances of the input signal paths.
3. The differential input signal must be reproduced with little deterministic jitter at the input buffers of the laser driver die. This is a necessary condition for the generation of jitter-free optical eye diagrams.

We will explain the merits of this approach in the following discussion. Figure 24.4(a) shows a measured receiver eye diagram in comparison to the simulated output of the unpackaged postamplifier die in Figure 24.4(b). The data in Figure 24.4(b) therefore excludes any contribution from the IC package parasitics and the EBD file.

Adding IC package parasitics and the electrical description of the internal PCB leads to the eye

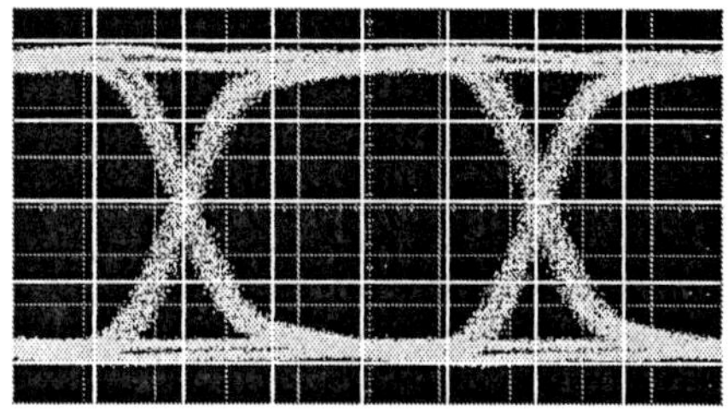

(a)

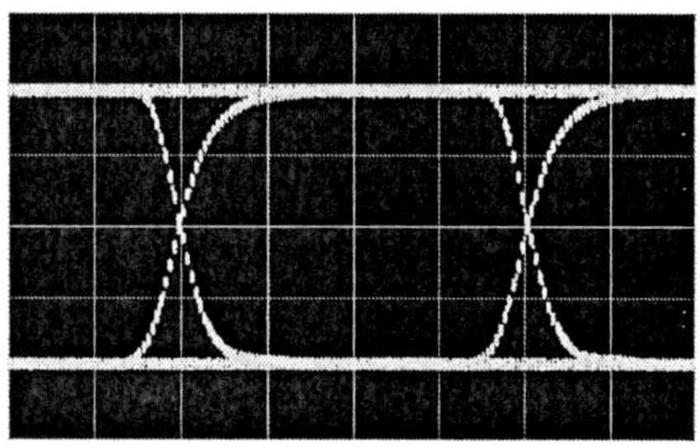

(b)

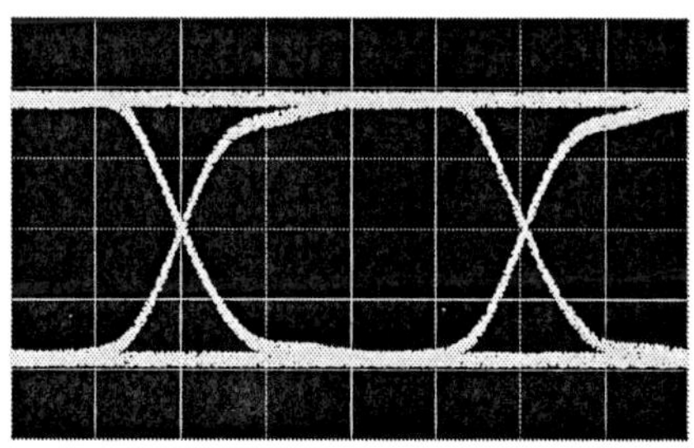

(c)

Figure 24.4 (a) Single-ended receiver eye diagram of the HFCT-5942L, (b) Output signal of the unpackaged receiver postamplifier die, and (c) Simulated single-ended receiver eye diagram using the full component model including the EBD file.

diagram in Figure 24.4(c). A low level of noise has been added to facilitate the comparison with the measured eye in Figure 24.4(a). It is clear from Figures 24.4(a) and (b) that the edge rates and the overall signal shape of the postamplifier output are altered under the influence of the package parasitics and the layout of the internal PCB. It is also apparent that by using an accurate IBIS v4.0 model a good overall fit between the simulation and the measurement could be achieved, where the model also reproduces the finer details in the ones and zeros.

It is worth noting, however, that such a good agreement between component model, simulation, and measurement was accomplished by increasing the impedance Z_0 of the short transmission line between the termination resistor and the postamplifier IC pin above its nominal value of 50 Ω. The increase that is required for a good match is beyond typical manufacturing tolerances. Even though it is not entirely clear at this point why this is necessary, we believe that the excess distributed inductance provided by this increase reflects either IC package parasitics or discontinuities between the IC and the PCB that could not be captured in the package model parameters of the IBIS file. This effectively blurs the line between the PCB description which is provided by the EBD file and the IC package model parameters normally accounted for in the IBIS file. It should be emphasized that this level of detail is only resolved due to the extremely fast rise and fall times of the HFCT-5942L's postamplifier IC of about 50 ps (at the die level). Here the EBD file provides the model developer with the means to expand the model's complexity to match the requirements of today's high-speed fiber-optic components.

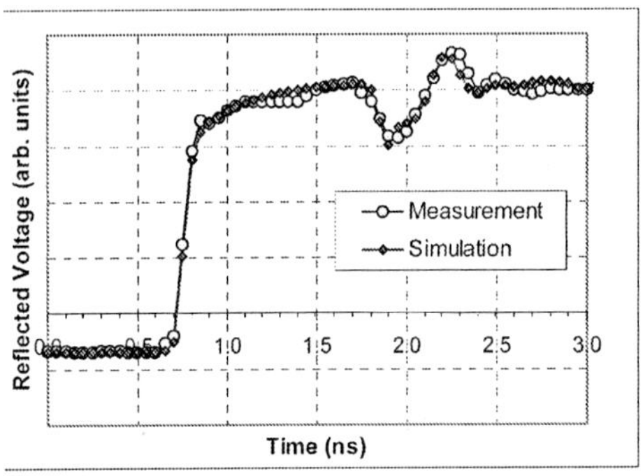

Figure 24.5 Differential TDR measurements and simulation taking into account the internal PCB and the IC package parasitics.

The high-speed inputs of the HFCT-5942L have been characterized with differential TDR measurements.[6] Differential TDR proves to be particularly useful for the characterization of differentially terminated input signals where single-ended measurements are difficult to interpret or would require modifications to the transceiver circuit board. The measured TDR signal is shown in Figure 24.5 together with a fit to the data using a detailed representation of the component's internal PCB. Again, the overall fit is satisfactory. Similar to the case of the high-speed outputs, the impedance of the transmission lines between the 100 Ω differential termination resistor and the laser driver IC pins (see also Figure 24.1 on page 683) had to be increased above their nominal value of 50 Ω to achieve a good match. As in the case for the outputs, this is more likely to be a contribution of the IC package parasitics that are rolled into the transmission line rather than an actual deviation of the transmission line from 50 Ω. Unfortunately, a more detailed analysis is not possible at this

6. A description of this technique can be found in references [Bog01b], [Bog00], [Smo01].

point, as it would require a higher measurement resolution. Higher resolution measurements or an independent characterization of the IC package should help us to further clarify the details of the transceiver IBIS model at the interface of the internal PCB and the IC package in the future.

The EBD model and IC package parasitics that have been derived from the fit to the TDR measurements are checked in a simulation where a pseudo-random signal is supplied to the high-speed inputs TD+ and TD– of the transceiver. In order for the model parameters to be consistent, the jitter of the differential input voltage at the input buffer of the laser driver die needs to be less than the jitter of the optical output signal. Furthermore, the eye diagram needs to be open with little infringement of the inner section. The results of the simulation are shown together with the eye of the driving pulse sequence in Figure 24.6. The upper eye diagram shows the externally generated signal. The lower eye diagram shows the differential voltage at the input buffer of the laser driver die.

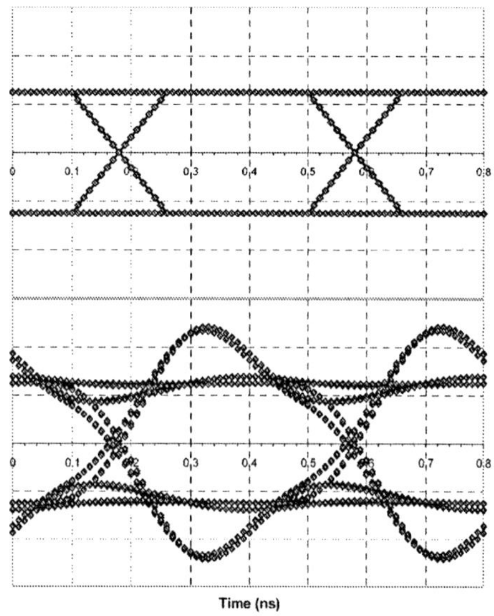

Figure 24.6 Differential input signals.

It is evident from Figure 24.6 that even though there is some overshoot in the differential signal at the leading edges of the eye diagram, there is very little infringement of the inner section. In addition, the deterministic jitter that is created by the parasitic inductances and capacitances of the input stage are of the order of 20 ps (equivalent to 50 mUI[7]) peak-to-peak, which is below the measured maximum jitter generation of 70 mUI [Agi03b]. This gives us confidence that the developed IBIS model accurately reflects the electrical performance characteristics of the transceiver inputs.

24.5 IBIS Models in "Real-World" Circuits

In the previous section we have seen that simulations utilizing IBIS models can accurately reproduce the behavior of components in "ideal" environments, that is, when the buffers drive the same test loads that were used to generate the IBIS models. However, in order to gain confidence in using IBIS models at multigigabit data rates, it will be very interesting to compare the signal shapes from simulations of "non-ideal" circuits utilizing IBIS as well as SPICE model representations of the same buffers. This will be a much more stringent test of the quality of the IBIS model and of the validity of using a behavioral modeling approach for transceiver components at these elevated bit rates.

7. mUI is milli unit interval where the full unit interval is taken as the width of one eye.

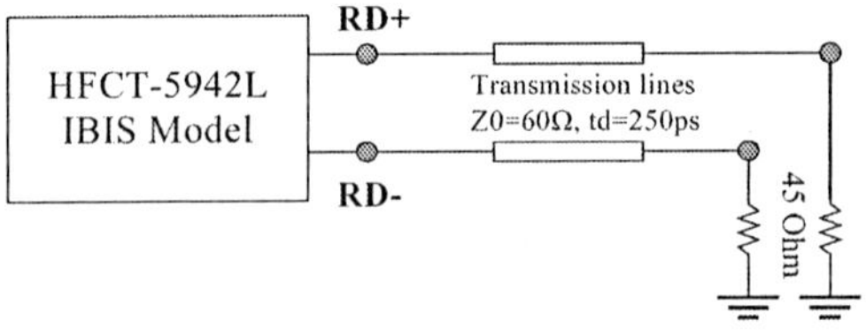

Figure 24.7 Circuit schematic for the simulation test comparing SPICE and IBIS component models.

Let us consider the following situation: A transceiver of the HFCT-5942L family drives a 250 ps long transmission line (nominal $Z_0 = 50\ \Omega$) which is terminated by a 50 Ω resistor. However, due to manufacturing tolerances the actual line impedance is 60 Ω, and resistors with a 10% tolerance have been chosen for the termination. The worst reflections will therefore occur for a resistor value of 45 Ω. Figure 24.7 shows this situation schematically.

The results of a full SPICE simulation are compared with a simulation utilizing the transceiver's IBIS model in Figure 24.8. A pseudo-random 2.5 Gbps pattern was used in both cases. The signal displayed in Figure 24.8 is the voltage drop at the termination resistor. Both simulations were carried out in Avant! HSPICE, which allows the simultaneous use of IBIS and SPICE models of the same component. For the full SPICE simulation, an HSPICE model of the PECL output buffers of the postamplifier IC of the HFCT-5942L was used. In both cases the EBD file was translated into a SPICE netlist using the T-Line element for the description of sections with distributed elements.

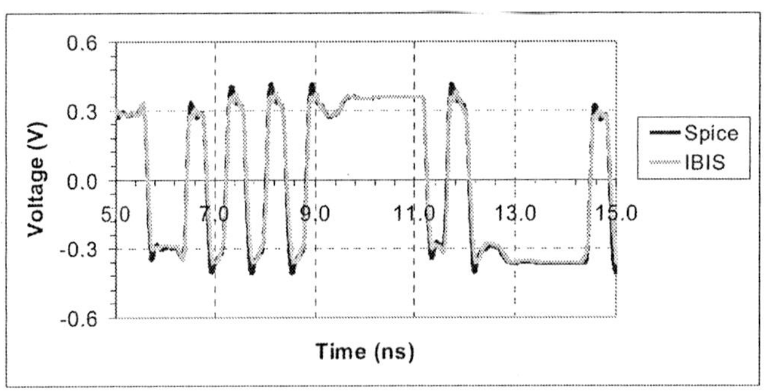

Figure 24.8 Simulation results for the setup shown in Figure 24.7. Displayed is the voltage at the termination resistor.

The comparison is very encouraging. The overall signal shapes are very similar; the edge rates and dc voltage levels are accurately reproduced by the IBIS model. There is a little difference in the fine detail on a sub-100 ps timescale that will require further investigation in the future. However, this difference is only 5% in terms of the total signal swing and we believe that this is acceptable given the differences in the modeling approach between transistor-level SPICE models and the behavioral IBIS description.

In order to test the IBIS model under more extreme conditions we removed the termination resistors and left the ends of the transmission lines unterminated. The open ends of the transmission lines fully reflect the output signal from the component. Since the component's output buffers cannot fully absorb the reflected signals due to the low impedance of the emitter followers, a fraction of the signal will be bouncing back and forth. The precise signal shape at any point in time depends on the previous bit sequence. This is a stringent test for the IBIS model as it is possible in this setup for small differences to accumulate over time. The same pseudo-random sequence as for the simulations in Figure 24.8 was used.

The result of this extreme-case simulation is displayed in Figure 24.9. The IBIS model simulations trace the HSPICE model simulations with a high degree of accuracy. As expected, some differences get more pronounced, as can be seen in the finer details between the main peaks. However, the main peaks are accurately reproduced, both in terms of edge rate and maximum peak level, even though the reflected voltage levels can reach up to three times their steady state value. This is a very encouraging result proving that IBIS models can be used with confidence in signal integrity simulations. It furthermore indicates that the results of simulations using IBIS models can be trusted even when extreme cases are analyzed.

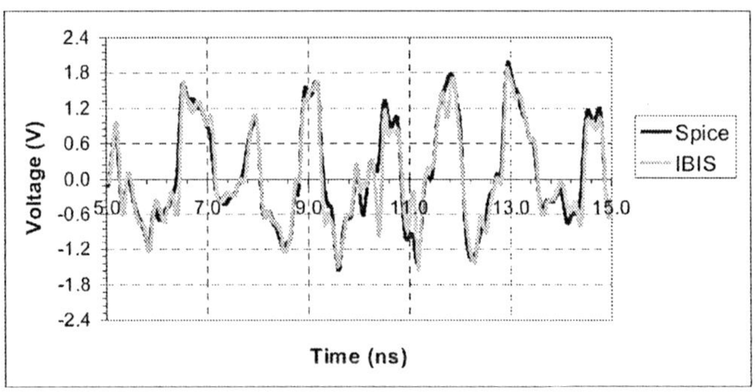

Figure 24.9 Same bit sequence as in Figure 24.8, but the termination resistors in Figure 24.7 have been removed so that the transmission lines are now open ended.

The quality of IBIS models critically depends on the quality of the SPICE model from which it is derived, or, if the IBIS model is based on measurements, on the accuracy of the measurements. The high bit rate makes tough demands on both, and it is imperative to verify the model's performance against either SPICE simulations or additional characterization data of a large population of parts as can be found, for example, in the characterization report of transceiver components.

24.6 Signal Integrity Case Study

In the previous section we have seen that simulations utilizing IBIS models can be very useful tools for the analysis of the signal integrity of a relatively simple high-speed electrical circuit. We will take this one step further in this section and take a look at one of Agilent Technologies' recent reference designs comprising a Fibre Channel transceiver, HFBR-5720L, and a serializer/deserializer IC (SerDes), HDMP-2630.[8]

One characteristic feature of the reference design's PCB is that, due to the pin layout of the two components, the high-speed differential pairs connecting the transceiver outputs (inputs) with the inputs (outputs) of the SerDes need to cross each other. This is achieved by re-routing the differential pair starting at the RD+/– pins of the HFBR-5720L from layer 3 of the PCB to layer 6 halfway between the transceiver and the SerDes.

This situation is schematically shown in Figure 24.10 where the via has been modeled as a single lumped inductive element. This is a very simple model neglecting the via landing pad capacitances, but we still expect it to be a reasonable first-order approximation (see [Joh93]). For the simulation of the two components, an IBIS model was used for the HFBR-5720L transceiver, and an HSPICE model for the HDMP-2630 SerDes. The HFBR-5720L IBIS model is available for download from Agilent Technologies Web site; the HSPICE model for the HDMP-2630 SerDes can be obtained from Agilent Technologies upon request.

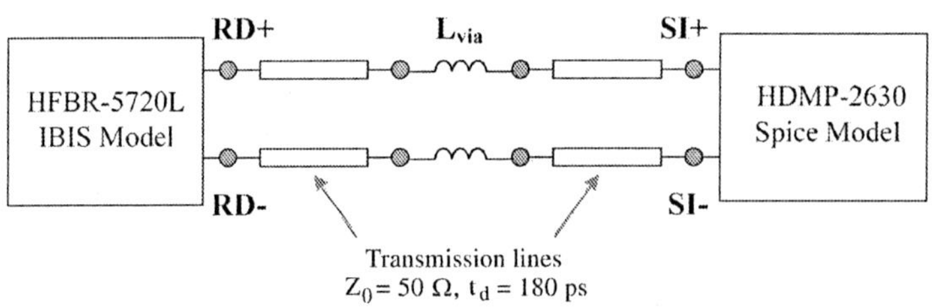

Figure 24.10 Schematic circuit diagram for the signal integrity study of the 2.125 Gbps Fibre Channel physical layer reference design for Agilent Technologies' HFBR-5720L and HDMP-2630.

The single-ended eye diagrams of the electrical signal at the pins of the SerDes and at the die are shown in Figures 24.11 and 24.12, respectively. L_{via} has been set to 0.5 nH in this simulation to provide a lower limit reference for the following case study where we will systematically increase L_{via} to investigate its effect on the signal propagation. The signal will be probed at the SerDes IC pin.

8. The reference design is described in detail in references [Agi03c] and [Wu01].

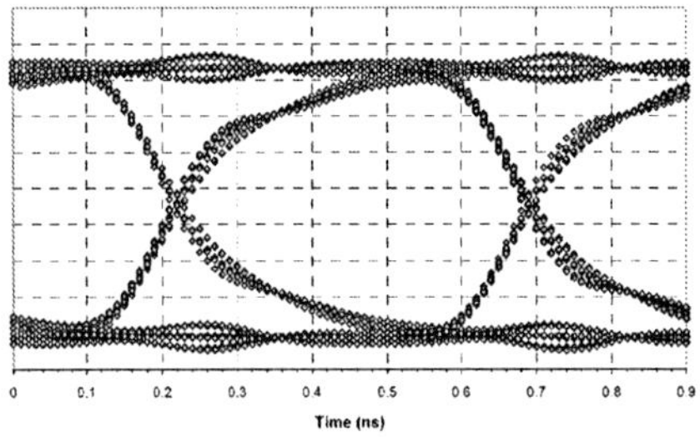

Figure 24.11 Signal at pin of SerDes HDMP-2630.

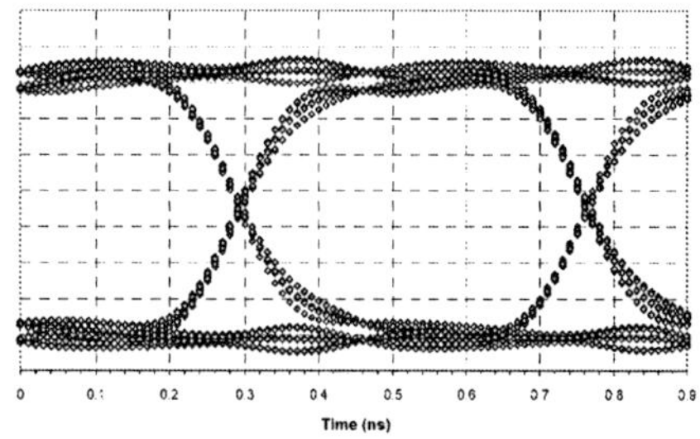

Figure 24.12 Signal at die of SerDes HDMP-2630.

The effect of increasing L_{via} in steps to 1.0 nH, 2.0 nH, and 4.0 nH is displayed in Figure 24.13. The succession of graphs nicely shows that the increase of L_{via} not only creates jitter on the rising and falling edges, but also leads to a closure of the eye as it takes more time for the signal to reach its steady-state level. Increasing the via inductance from 1 nH to 4 nH almost doubles the deterministic jitter from 14 ps to 27 ps, and leads to a 10% reduction of the eye opening.

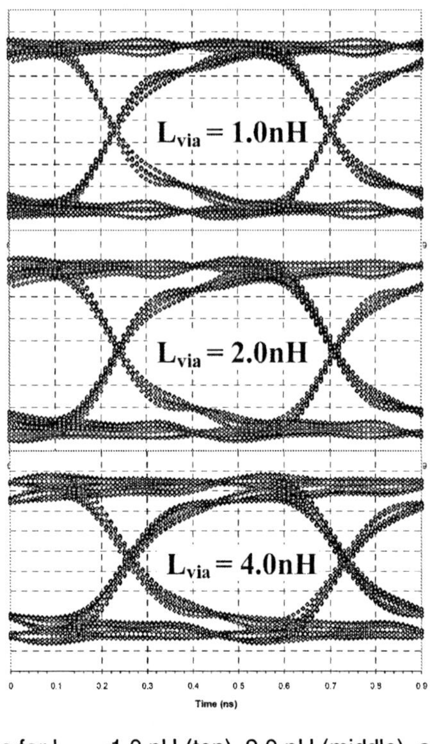

Figure 24.13 Eye diagrams for L_{via} =1.0 nH (top), 2.0 nH (middle), and 4.0 nH (bottom). Shown is the single-ended signal at the pins of the HDMP-2630.

It is also clear from Figure 24.13 that the via only has a small effect on the signal shape as long as its inductance does not exceed approximately 2 nH. Care has been taken in the reference design to provide as good a connection between the transceiver and the SerDes as possible. The most likely value for the via inductance will be between 1 and 2 nH, which is in line with typical lumped parameter values for multilayer PCBs (see [Alf03]). This is also evident in the measurements shown in reference [Agi03c], which are reproduced below in Figure 24.14. The diagram at the bottom shows measured data at the test points of the reference board where clear and open eye diagrams are a good indicator for the low inductance of the via and thus the high quality of the PCB design. The upper eye diagram has been simulated by adding the required circuit connections for the test points to the schematic in Figure 24.10 on page 690. The via inductance was set to 1.5 nH. We believe that the agreement between measurement and simulation is very satisfactory.

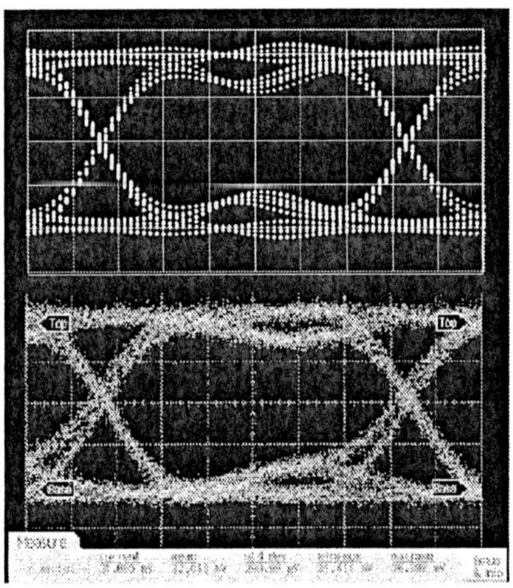

Figure 24.14 The diagram at the bottom shows the signal at the test point RX(+) of the reference design board. The diagram at the top shows the simulated signal at the test points.

24.7 Summary

In this chapter we have reviewed the IBIS v4.0 standard and highlighted the importance of the EBD file for the description of fiber-optic transceivers at 2 Gbps and higher data rates. We have shown that accurate models can be developed and we have reviewed some of the verification techniques that are used within Agilent Technologies to ensure the accuracy of the IBIS models at these high data rates.

The validity of the behavioral modeling approach has been verified by comparing SPICE and IBIS models of the same output buffers at OC-48 data rates for a "real-world" circuit simulation where good agreement was achieved. This then allowed us to carry out a signal integrity study for one of Agilent Technologies' reference designs where the effect of vias on the signal propagation along differential pairs was investigated. This showed that up to about 2 nH inductance of the via can be tolerated. Higher inductance causes the received signal to deteriorate visibly.

24.8 Exercises

24-1. What is a "reference design"?

24-2. If a differential periodic signal has an eye width of 0.8 ns, how much jitter does it have, as measured in picoseconds, if the jitter is equal to 66 mUI?

24-3. What was the purpose behind doing both SPICE and IBIS simulations for the same component?

24-4. (a) What is an EBD? (b) What information does it contain? (c) Who supplies it?

24-5. For the simulations, what two files were required to fully describe a board-level component?

24-6. Does an EBD file have to contain path descriptions for power and ground pins?

24-7. (a) In the signal integrity case study using the Fibre Channel transceiver and SerDes, what was the range of inductance values used to model vias? (b) What was the effect as the inductance of the via was increased?

24-8. Where is the receiver portion of a fiber-optic transceiver usually terminated?

24-9. (a) At what frequencies do signal integrity simulations start to become necessary? (b) At what frequencies are they mandatory?

24-10. What are signal integrity simulations used to ensure?

24-11. Before a transceiver can be considered a lumped element, what requirement has to be met?

24-12. Name five elements that the EBD file had to contain in order to be used in the simulations described in this chapter.

24-13. Is the jitter of the differential input voltage at the input buffer of the laser driver die less than or greater than the jitter of the optical output signal?

24-14. (a) What major element of a fiber-optic component was described with an IBIS file? (b) What major element of a fiber-optic component was described by an EBD file?

24-15. Before an IBIS model of a fiber-optic transceiver is released to users, what three criteria are used to verify the accuracy of the model?

CHAPTER 25

Designing with LVDS

Chapter Objectives

- Provide low-voltage differential signaling (LVDS) design concepts for printed circuit boards and backplanes.
- Compare the layer stack-up for LVDS boards against the layer stack-up for traditional high-speed boards and provide layout rules for high-speed designs.
- Show how differential impedances are calculated and why they are so important.
- Discuss failsafe biasing of LVDS and techniques to lower EMI.
- Present laboratory results for trace coupling experiments and show how trace coupling impacts board design.
- Discuss microstrips, broadside and edge-coupled striplines, and calculate their differential impedances.
- Set up an LVDS signal quality/eye pattern test circuit and a bit error rate test circuit and determine maximum data rates for each.

25.1 Layer Stack-Up and PCB Design[1]

Due to the high-speed switching rates obtainable by LVDS, a minimum of a four-layer PCB construction and FR-4 material is recommended. This allows for two signal layers and full power and ground planes. The stack is: signal (LVDS), ground, power, signal. [Nat00]

1. All figures, tables, and text in this chapter are reprinted with permission of National Semiconductor Corporation except where noted otherwise.

Differential traces are highly recommended for LVDS driver outputs and receiver inputs. Employing differential traces ensures a low-emission design and maximum common-mode rejection of any coupled noise. Differential traces require that the spacing between the differential pair be controlled. This distance should be held as small as possible to ensure that any noise coupled onto the lines will primarily be common-mode. Also, by keeping the pair close together the maximum canceling of fields is obtained. The differential impedance of the trace pair should be matched to the selected interconnect media (the cable's differential characteristic impedance).

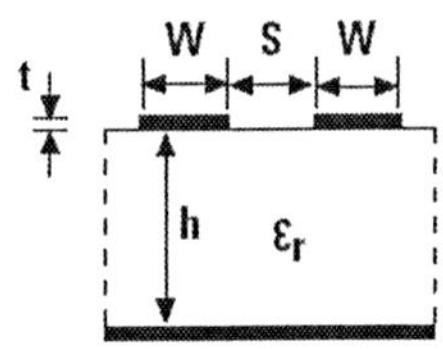

Figure 25.1 Parameters of the PCB and microstrip traces.

Microstrip differential traces lie on the surface of the PCB whereas stripline traces are buried within the PCB. When striplines are located side-by-side, they are referred to as "edge-coupled." When striplines are located one above the other, they are referred to as "broadside" striplines. This is discussed in more detail later on. (See "Differential Traces" on page 700 and Figure 25.3 on page 700).

Figure 25.1 shows differential microstrip traces on a printed circuit board and their dimensional parameters. W, S, t, and h represent trace width, spacing between traces, trace thickness, and spacing of the traces above the nearest ground or power plane, respectively. ε_r is the dielectric constant of the PCB material. For the microstrip line, the differential impedance, Z_{DIFF}, is given by

$$Z_{DIFF} \cong 2 Z_0 (1 - 0.48\ e^{-0.96 S/h})\ \Omega \tag{25.1}$$

where Z_0 is the single-ended characteristic impedance of each trace considered separately with respect to ground (or with respect to the solid plane in Figure 25.1). At first glance, it may appear that Z_{DIFF} does not depend on W, t, or ε_r. However, Z_0 does depend on these which in turn makes Z_{DIFF} dependent on them as well. Z_0 is the single-ended characteristic impedance of each trace individually.

As an example, let h = 24 mils, S = 11 mils and $Z_0 = 70\ \Omega$. Calculating the differential impedance, Z_{DIFF}, is

$$\begin{aligned} Z_{DIFF} &\cong 2 Z_0 (1 - 0.48\ e^{-0.96(11/24)})\ \Omega \\ &= 2(70)(0.69086)\ \Omega \\ &= 96.72\ \Omega \end{aligned} \tag{25.2}$$

Termination of LVDS lines is required to complete the current loop and for the drivers to properly operate. This termination in its simplest form is a single surface-mount resistor (surface-mount resistor minimizes parasitic elements) connected across the differential pair as close to the receiver inputs as possible (should be within 0.5 inch (13 mm) of input pins). Its value should be selected to match the interconnect differential characteristic impedance. The closer the match, the higher the signal fidelity and the less common-mode reflection will occur (lower emissions too). A typical value is 100 Ω ± 1%.

LVDS signals should be kept away from TTL and CMOS logic signals to minimize noise coupling from the large-swing TTL and CMOS signals. This has been accomplished on the PCB by routing TTL or CMOS signals on a different signal layer (bottom) than the LVDS signals (top) wherever possible. If both CMOS (or TTL) and LVDS signals are required on the same layer, a CMOS (or TTL) signal should never be routed within three times (3S) the distance between the differential pair (S) as shown in Figure 25.2. Adjacent LVDS differential pairs should be at least 2S away also.

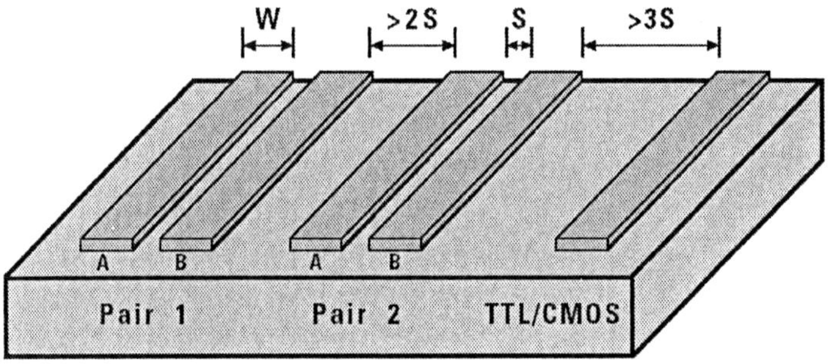

Figure 25.2 Pair spacing for differential lines.

These design guidelines are referred to as the "S Rules" as other separations are defined using the edge-to-edge "S" distance between the traces of the LVDS pairs. The "S Rules" can be summarized as follows:

- The distance between two pairs should be >2S.
- The distance between a pair and a TTL/CMOS signal should be >3S at a minimum. (Even better, locate the TTL/CMOS signals on a different plane isolated by a ground plane.)
- If a guard ground trace or ground fill is used, it should be >2S away.

Bypassing capacitors are recommended for each IC package. A 0.1 μF is sufficient on a quad driver or receiver device but additional smaller value capacitors may be added (e.g., 0.001 μF) if desired. Traces connecting V_{CC} and ground should be wide (low impedance, not 50 Ω dimensions) and employ multiple vias to reduce inductance. Bulk bypassing should be provided at the main power connection to the PCB as well.

25.2 PCB Board Layout Tips

Now that we have seen that LVDS has super speed, and very low power, noise, and cost, many people might assume that switching to LVDS (or any differential technology) will solve all of their noise problems. It will not, but it can help a lot! LVDS has low-swing, differential, ~3.5 mA current-mode outputs that can help reduce noise/EMI significantly, but these outputs switch (rise and fall) in less than a nanosecond. This means that every interconnect will act as a transmission line except for the very shortest ones. Therefore, knowledge of ultra-high-speed board design and differential signal theory is required. Designing high-speed differential boards is not difficult or expensive, but familiarizing yourself with these techniques before you begin your design will be helpful.

The fast edge rate of an LVDS driver means that differential impedance matching is very important even for short runs. Discontinuities in differential impedance will create reflections which will degrade the signal and also show up as common-mode noise. Common-mode noise on the line will not benefit from the canceling magnetic field effect of differential lines and will be radiated as EMI. You should use controlled differential impedance traces as soon as you can after the signal leaves the IC. Try to keep stubs and uncontrolled impedance runs to <12 mm (0.5 in). Also, avoid 90° turns since this causes impedance discontinuities; use 45° turns, radius or bevel PCB traces. Minimize skew between conductors within a differential pair. Having one signal of the pair arrive before the other creates a phase difference between voltages along the signal pairs which looks like and radiates as common-mode noise. Use bypass capacitors at each package and make sure each power or ground trace is wide and short (do not use 50 Ω dimensions) with multiple vias to minimize inductance to the power planes. A detailed list of suggestions for designing with LVDS is provided throughout this chapter. The suggestions are inexpensive and easy to implement. By using these suggestions as guidelines, your LVDS-based systems should provide maximum performance and be quick and easy to develop.

25.2.1 PC Board

There are guidelines which should be considered when designing or laying out a high-speed printed circuit board containing LVDS devices. These are:

a. Use at least four PCB board layers (top to bottom): LVDS signals, ground, power, TTL signals. Dedicating planes for V_{CC} and ground are typically required for high-speed design. The solid ground plane is required to establish a controlled (known) impedance for the transmission line interconnects. A narrow spacing between power and ground can also create an excellent high frequency bypass capacitance.

b. Isolate fast edge rate CMOS/TTL signals from LVDS signals. Otherwise the noisy single-ended CMOS/TTL signals may couple crosstalk onto the LVDS lines. It is best to put TTL and LVDS signals on different layers which should be isolated by the power and ground planes.

c. Keep the LVDS port side of drivers and receivers as close to connectors as possible. This helps to ensure that noise from the board is not picked up onto the differential lines and escapes the board as EMI from the cable interconnect. This recommendation also helps to minimize skew between the lines. Skew tends to be proportional to length and so limiting length also limits skew.

d. Bypass each LVDS device and also use distributed bulk capacitance. Surface mount capacitors placed close to the power and ground pins of each device work best. At the power supply input to a PCB, a 4.7 μF or 10 μF 35 V tantalum capacitor works well when placed between the power voltage bus and ground. Choosing a capacitor value which best filters the largest power/ground frequency components (usually 100 to 300 MHz) is best. This can be determined by checking the noise spectrum of V_{CC} across bypass capacitors. The voltage rating of tantalum capacitors is critical and must not be less than five times V_{CC}. Some electrolytic capacitors also work well. For bypassing power pins, one or two multilayer ceramic (MLC) surface mount capacitors (0.1 μF and 0.01 μF) should be used in parallel between each V_{CC} pin and ground if possible. For best results, the capacitors should be placed as close as possible to the V_{CC} pins to minimize parasitic effects that defeat the frequency response of the capacitance. Wide (>4 bits) and PLL-equipped (e.g., Channel Link and FPD Link) LVDS devices should have at least two capacitors per power type, while other LVDS devices are usually fine with a 0.1 μF capacitor. The bottom line is to use good bypassing practices. EMI problems many times start with power and ground distribution problems. EMI can be greatly reduced by keeping power and ground planes quiet.

e. Power and ground pins should be connected to power and ground planes, respectively, with wide low-impedance traces. Do not use 50 Ω design rules on power and ground traces. Their job is to provide a low-impedance connection.

f. Keep ground PCB return paths short and keep ground traces wide. Provide a return path that creates the smallest loop for image currents to return along.

g. Cables should employ a ground return wire connecting the grounds of the two systems. This provides for common-mode currents to return on a short known path. See 12.4.4, "Cable Ground and Shield Connections," on page 310.

h. Use two vias to connect bypass capacitor pads to power and ground in order to minimize inductance effects. Surface mount capacitors should be used as they are compact and can be located close to device pins.

25.2.2 Traces

There are additional guidelines for LVDS trace design when using microstrip or stripline as follows:

a. Edge-coupled microstrip, edge-coupled stripline, or broad-side stripline all work well for differential lines.

b. Traces for LVDS signals should be closely coupled and designed for 100 Ω differential impedance. See Section 25.2.3.

c. Edge-coupled microstrip lines offer the advantage that a higher differential impedance is possible (100 to 150 Ω) than for striplines. Also it may be possible to route from a connector pad to the device pad without any via. This provides a "cleaner" interconnect. A limitation of microstrip lines is that they can only be routed on the two outside layers of the PCB which in turn limits routing channel density.

d. Stripline may be either edge-coupled or broad-side lines. Since they are embedded in the board stack and typically sandwiched between ground planes, they provide additional shielding. This limits radiation and also coupling of noise onto the lines. They also require the use of via to connect to them.

25.2.3 Differential Traces

When designing for differential traces, controlled impedance PCB traces which match the differential impedance of the transmission medium (i.e., cable) and termination resistor should be used. The differential pair traces should be routed as close together as possible as soon as they leave the IC. This helps to eliminate reflections and ensures that noise is coupled as common-mode. In fact, we will show in "EMI Test Results" on page 713 that differential signals which are 1 mm apart radiate far less noise than traces 3 mm apart since magnetic field cancellation is much better with the closer traces. Plus, noise induced on the differential lines is much more likely to appear as common mode which is rejected by the receiver.

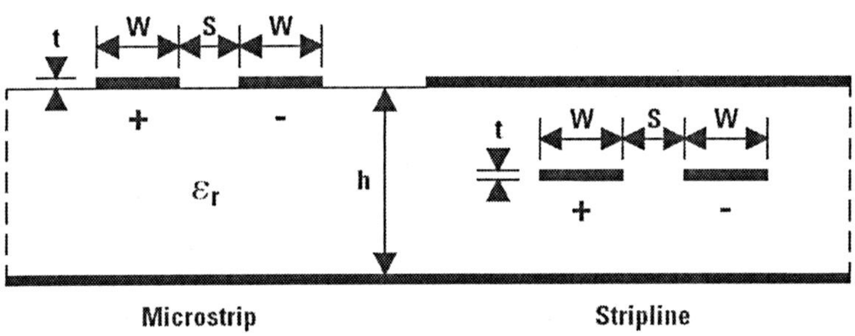

Figure 25.3 Cross-sectional view of microstrip and edge-coupled stripline PCB configurations.

Figure 25.3 shows the geometry for both a microstrip and edge-coupled stripline. Figure 25.4 shows the geometry for a broadside stripline. When designing for a specific differential impedance Z_{DIFF} for edge-coupled lines, it is recommended that you adjust trace width "W" to alter Z_{DIFF}. It is recommended to not adjust "S" which should be the minimum spacing specified by your PCB vendor for line-to-line spacing. The following equations can be used to calculate differential impedance for edge-coupled differential lines:

$$Z_{DIFF} \cong 2 Z_0 \left(1 - 0.48\, e^{-0.96(S/h)}\right)\ \Omega \quad \text{Microstrip}^2 \tag{25.3}$$

2. Equation (25.3) is the same as Equation (25.1) on page 696; is repeated here for convenience and comparison.

$$Z_{DIFF} \cong 2 Z_0 \left(1 - 0.374\, e^{-2.9(S/h)}\right)\ \Omega \quad \text{Edge-Coupled Stripline}^3 \tag{25.4}$$

where

$$Z_0 = \frac{60}{\sqrt{0.475\varepsilon_r + 0.67}} \ln\left(\frac{4h}{0.67(0.8W + t)}\right)\ \Omega \quad \text{Microstrip} \tag{25.5}$$

$$Z_0 = \frac{60}{\sqrt{\varepsilon_r}} \ln\left(\frac{4h}{0.67\pi(0.8W + t)}\right)\ \Omega \quad \text{Edge-Coupled Stripline} \tag{25.6}$$

and where

$$W = \frac{7.463h}{\exp\left(\frac{Z_0\sqrt{0.475\varepsilon_r + 0.67}}{60}\right)} - 1.25t \quad \text{Microstrip} \tag{25.7}$$

$$W = \frac{2.375h}{\exp\left(\frac{Z_0\sqrt{\varepsilon_r}}{60}\right)} - 1.25t \quad \text{Edge-Coupled Stripline} \tag{25.8}$$

In the above equations, all geometric variables must be in the same dimensional units. As a practical matter, the above equations may be used by practicing engineers for rough planning or checks of PCB layout data, but they are not used in the main engineering design processes. Today, EDA (electronic design automation) tools assist the engineer by doing most of the hard mathematical work. The engineer will specify the desired differential impedances, critical routing paths, use of microstrip versus stripline, and other higher-level parameters, but the tools will take it from there. As a practical matter, the tools will likely be more accurate than the above equations as well. The good news is that you shouldn't have to spend any time memorizing these equations!

A broadside stripline structure can also be used. The dimensions for this type of line are shown in Figure 25.4. Note that dimensions "h" and "S" are defined differently in this figure. Broadside-coupled striplines can be useful in backplane design as these use only one routing channel and may be easier to route through the connector pin field. The equation for the single-ended characteristic impedance of a broadside stripline is

$$Z_0 = \frac{80}{\sqrt{\varepsilon_r}} \ln\left[\frac{1.9(2h + t)}{0.8W + t}\right]\left[1 - \frac{h}{4(h + S + t)}\right] \quad \text{Ohms Broadside Stripline}^4 \tag{25.9}$$

3. For edge-coupled striplines, the term "0.374" may be replaced with "0.748" for lines which are closely coupled ($S < 12$ mils).
4. Always use consistent dimensions (e.g., all dimensions in mils, centimeters, or millimeters) for S, h, W, and t. See reference [AN-905].

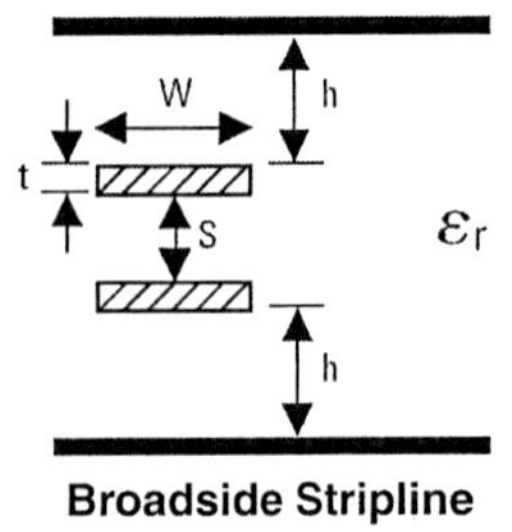

Figure 25.4 Cross-sectional view of broadside stripline.

The accuracy of the above equation is about the same as the accuracy for the previous equations for the microstrip and edge-coupled stripline.[5]

Common values of dielectric constant ε_r for various printed circuit board (PCB) materials is given below in Table 25.1. Consult your PCB manufacturer for actual numbers for the specific material that you plan to use. Note that in most LVDS applications, the widely used FR-4 PCB material is acceptable. GETEK is about 1.5 times as expensive as FR-4, but can be considered for 1000+ MHz designs. Also note that ε_r will vary within a single board. It is not uncommon for FR-4 PCBs to vary by 10% across one board, affecting skew. This is another good reason to keep differential lines close together.

The loss tangents listed in the table indicate the lossiness of PCB materials (dielectrics) in an ac electric field. The distributed conductance (G, mhos per unit length) of PCB traces at high frequencies is not caused by the flow of free charges but is a measure of internal dielectric losses in the insulating PCB material. These losses result from repeated reversals of the dielectric polarization by the signal's ac electric field. Since the loss is thus on a per-cycle basis, it tends to

Table 25.1 Common Values of Dielectric Constant (ε_r) and Loss Tangent (tan δ) for Various Printed Circuit Board (PCB) Materials

PCB Material	Dielectric Constant (ε_r)	Loss Tangent (tan δ)
Air	1.0	0
PTFE (Teflon)	2.1–2.5	0.0002–0.002
BT Resin	2.9–3.9	0.003–0.012
Polyimide	2.8–3.5	0.004–0.02
Silica (Quartz)	3.8–4.2	0.0006–0.005
Polyimide/Glass	3.8–4.5	0.003–0.01
Epoxy/Glass (FR-4)	4.1–5.3	0.002–0.02
GETEK	3.8–3.9	0.010–0.015 (1 MHz)
ROGERS4350 Core	8 ± 0.05	0.004 @ 10 GHz, 23° C
ROGERS4403 Prepreg	3.17 ± 0.05	0.005 @ 10 GHz, 23° C

5. These equations were derived from empirical data (experiment or observation). The actual accuracy has not been determined experimentally. Various sources cite possible errors for Z_{DIFF} of up to ±10%. The practical ranges for Z_0 and Z_{DIFF} are from 20 Ω to about 150 Ω. For Z_0 of a microstrip line, the maximum error is less than 3% if $w \le 2h$. For stripline Z_0, best accuracy results if parameters are kept within these guiding ratios: $b - t > 2w$ and $b > 4t$. See [AN-905].

be directly proportional to frequency over wide ranges of frequency. The greater the loss tangent of the PCB material, the greater is the signal attenuation as the signal propagates down its differential PCB traces. For most commonly used PCB insulating materials, the loss factor tan δ lies between 10^{-3} and 10^{-5}. [Chi68]

For single-ended coaxial transmission lines, the conductance per unit length, G, referenced to ground is equal to[6]

$$G = \omega C \tan\delta \tag{25.10}$$

where C is the capacitance per unit length referenced to ground, ω is the radian frequency (2πf) of the signal, and tan δ is the loss tangent.[7] For coaxial lines, C is constant within 0.5% over 60 Hz to 10 GHz and assuming that tan δ is somewhat constant, the equation reveals that G is directly proportional to frequency. Only at gigahertz frequencies does G becomes significant and attenuate the signal as it propagates down the line.

Skew can be minimized by matching electrical lengths between traces of a pair. Skew between the signals of a pair is equivalent to a phase difference between signals which destroys the magnetic field cancellation benefits of differential signals. EMI will result! (Note that the velocity of propagation is $v = c/(\sqrt{\varepsilon_r})$, where c (the speed of light) = 0.2997 mm/ps or 0.0118 in/ps). A general rule is to match lengths of the pair to within 100 mils. Do not rely solely on the auto-route function for differential traces. Carefully review dimensions to match trace length and to insure isolation between pairs of the differential lines. Minimize the number of via and other discontinuities on the line. Avoid 90° turns (these cause impedance discontinuities). Use arcs or 45° bevels instead.

Within a pair of traces, the distance between the two traces should be minimized to maintain common-mode rejection of the receivers. On the printed circuit board, this distance should remain constant to avoid discontinuities in differential impedance. Minor violations at connection points are allowable. The key to "imbalances" is to make as few of them as possible and make them as small as possible. Differential transmission works best on a balanced interconnect. Both lines of the pair should be as identical as possible for the best results.

25.2.4 Termination

The following are some guidelines for consideration with respect to termination:

a. Use a termination resistor which best matches the differential impedance of your transmission line. It should be between 90 Ω and 130 Ω for point-to-point cable applications. Remember that the current-mode outputs need the termination resistor to generate the proper differential voltage. LVDS is not intended to work without a resistor termination.

6. This equation is calculated in reference [Chi68].
7. Tan $\delta = \varepsilon''/\varepsilon'$ where the absolute permittivity of the dielectric material is represented by the complex number ε with $\varepsilon = \varepsilon' - j\varepsilon''$ and $j = \sqrt{-1}$. ε' is the real part of ε and ε'' is the negative imaginary part of ε. [Chi68]

b. Typically a single resistor across the pair at the receiver end suffices.

c. Surface-mount resistors are best. PCB stubs, component leads, and the distance from the termination to the receiver inputs should be minimized. The distance between the termination resistor and the receiver should be <7 mm (12 mm maximum).

d. Resistor tolerance of 1% or 2% is recommended. Note that from a reflection point of view, a 10% mismatch in impedance causes a 5% reflection. The closer the match the better. Match to the nominal differential impedance of the interconnect.

e. Center tap capacitive termination, as shown in Figure 25.5, may also be used in conjunction with two 50 Ω resistors to filter common-mode noise at the expense of extra components if desired. This termination is not commonly used or required. In Figure 25.5, $R = Z_{DIFF}$ (between 100 and 120 Ω) and C should be approximately 50 pF. Components should be surface mount components, placed close to the receiver. Use 1–2% resistors.

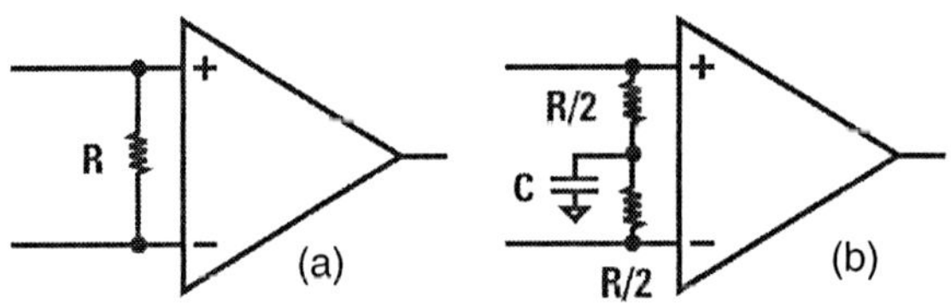

Figure 25.5 Common differential termination schemes. (a) Shunt termination, and (b) center tap capacitive termination.

25.2.5 Unused Pins

The following are recommendations for unused pins:

- LVDS inputs — Leave unused LVDS receiver inputs open (floating) unless directed differently by the specific component's datasheet. Their internal failsafe feature will lock the outputs high. These unused receiver inputs should not be connected to noise sources like cables or long PCB traces — float them near the pin. LVDS receivers are high-speed, high-gain devices, and only a small amount of noise, if picked up differentially, will cause the receiver to respond. This causes false transitions on the output and increased power consumption.
- LVDS and TTL outputs — Leave all unused LVDS and TTL outputs open (floating) to conserve power. Do not tie them to ground.
- TTL inputs — Tie unused TTL transmitter/driver inputs and control/enable signals to power or ground. In certain cases they may be left open if the datasheet supports this condition. Some devices provide internal pull-down (or pull-up) devices to bias the pins. Again, consult the datasheet for information regarding the device's features. This type of information is typically included in the pin description table.

25.2.6 Probing LVDS Transmission Lines

Always use a high-impedance (>100 kΩ), low capacitance (<0.5 pF) probe/scope with a wide bandwidth (>1 GHz). Improper probing will give deceiving results. LVDS is not intended to be loaded with a 50 Ω load to ground. This will distort the differential and offset voltages of the driver. Differential probes are recommended over two standard scope probes due to match and balance concerns. Bandwidth of the probe/scope combination should be at least 1 or 2 GHz. Tektronix and Agilent (HP) both make probes that are well suited for measuring LVDS signals.

25.2.7 Loading LVDS I/O — Preserving Balance

Avoid placing any devices which heavily load the low, ~3.5 mA LVDS output drive. If additional ESD protection devices are desired, use components which do not add a significant load to the LVDS output. Some of the connectors with integrated polymer ESD protection are a good option. Try not to disturb the differential balance. Treat both members of a pair equally.

25.2.8 Results of Good vs. Bad Design Practices

25.2.8.1 Impedance Mismatches

It is very common for designers to automatically use any off-the-shelf cables and connectors and 50 Ω autorouting when doing new designs. While this may work for some LVDS designs, it can lead to noise problems. Remember that LVDS is differential and does have low swing, current-mode outputs to reduce noise, but that its transition times are quite fast. This means impedance matching (especially differential impedance matching) is very important. Those off-the-shelf connectors and that cheap blue ribbon cable are not meant for high-speed signals (especially differential signals) and do not always have controlled impedance. Figure 25.6 shows a time-domain reflectometer (TDR) impedance trace of such a system. (See Chapter 20 starting on page

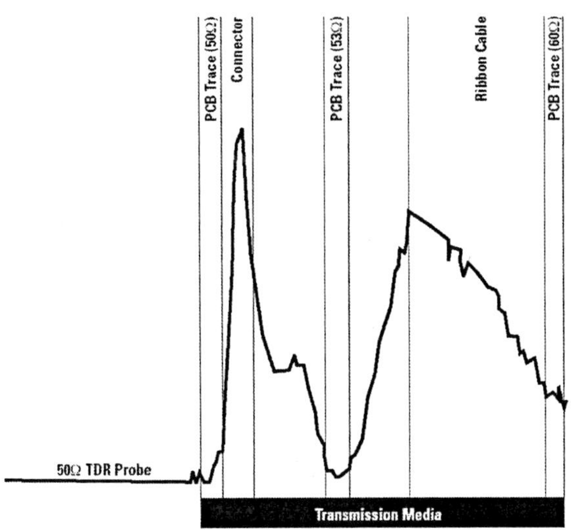

Figure 25.6 TDR plot of transmission media with mismatched impedance.

519 for a detailed explanation of TDR.) As one can plainly see, impedances are neither matched nor controlled. Beware, this example is not worst case — it is a typical example reflecting common TTL design practices. The reflections caused by impedance mismatching will generate a lot of noise and EMI.

Figure 25.7 a much improved design which follows most of the high-speed differential design practices discussed in this chapter. The TDR differential impedance plot is much flatter and noise is dramatically reduced.

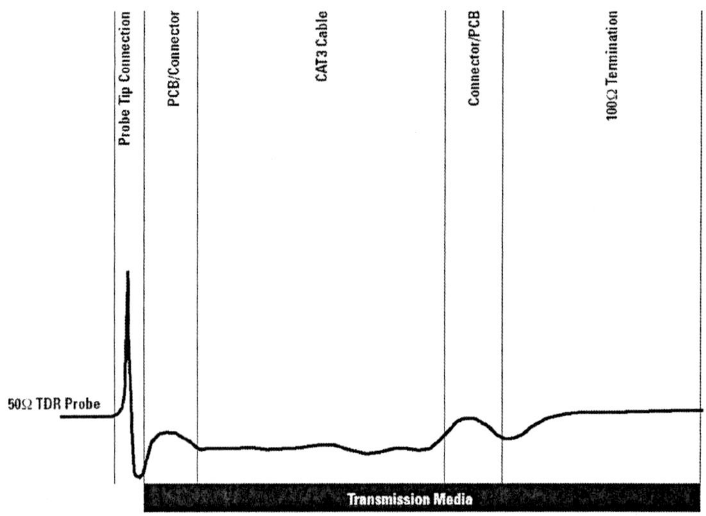

Figure 25.7 TDR plot with minimized impedance variations for best performance.

25.2.9 Crosstalk Between TTL and LVDS Signals

The next two figures, Figures 25.8 and 25.9, show the effects of TTL coupling onto LVDS lines. Figure 25.8 shows the LVDS waveforms before coupling, while Figure 25.9 shows the effects of a 25 MHz, 0 V to 3 V TTL signal upon the LVDS signals running adjacent for 4 inches. The result is an LVDS waveform modulated by the TTL signal. Note that the LVDS pair is not affected exactly equally — the signal which runs closest to the TTL trace is affected more than the other. This difference will not be rejected by the receiver as common-mode noise and though it will not falsely trigger the receiver, it does degrade the signal quality of the LVDS signal reducing noise margin. The common-mode noise will be rejected by the receiver, but can radiate as EMI.

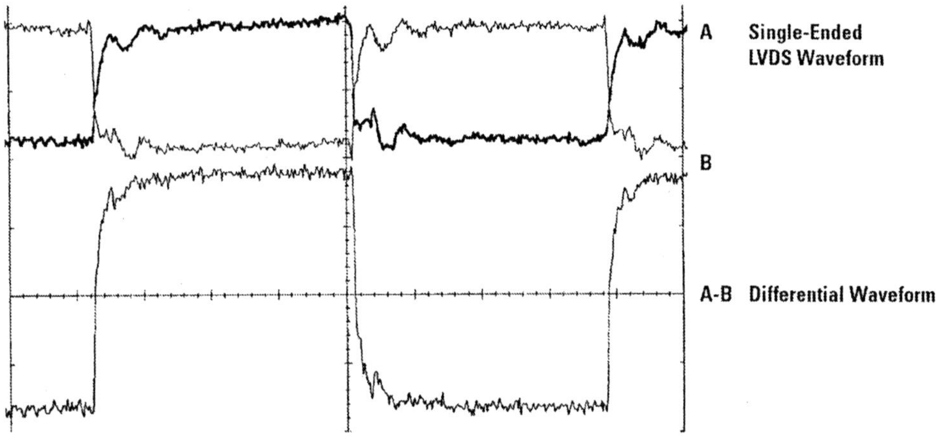

Figure 25.8 LVDS signals before crosstalk.

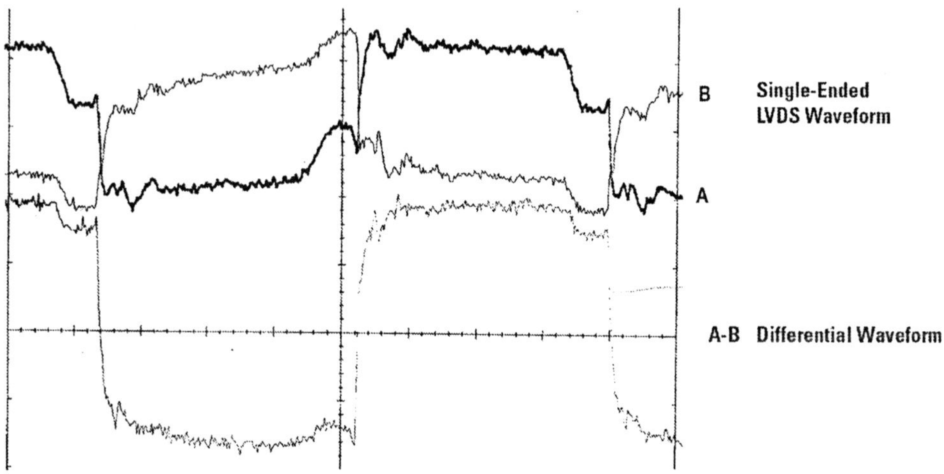

Figure 25.9 LVDS signals affected by TTL crosstalk.

25.2.10 Lowering Electromagnetic Interference (EMI)

25.2.10.1 LVDS and Lower EMI

High-speed data transmission usually means fast edge rates and high EMI. LVDS, however, has many positive attributes that help lower EMI:

1. The low output voltage swing (~350 mV)
2. Relatively slow edge rates, dV/dt ~ 0.350 V / 0.350 ns = 1 V/ns
3. Differential (odd-mode operation) so magnetic fields tend to cancel

4. "Soft" output corner transitions

5. Minimum I_{CC} spikes due to low current-mode operation and internal circuit design

To realize these advantages, however, designers must take care to ensure the close proximity of the pair conductors and to avoid creating impedance imbalances within a pair. The following sections describe these EMI-friendly design practices.

25.2.10.2 Electromagnetic Radiation of Differential Signals

Today's increasing data rates and tougher electromagnetic compatibility (EMC) standards are making electromagnetic radiation an increasing concern. System designers are usually most concerned with far field electromagnetic radiation, propagated through transverse electromagnetic (TEM) waves which can escape through shielding, causing a system to fail EMC tests. Fields around a conductor are proportional to voltage or current, which are small in the case of LVDS. The fields are distorted by and interact with their environment, which is why EMI is so hard to predict. The fields can be distorted to advantage, however, and such is the case with tightly coupled differential lines ("+" and "–" signals in close proximity with one another). In single-ended lines like CMOS or TTL shown in Figure 25.10, almost all the electric field lines are free to radiate away from the conductor. These fields may be intercepted by other objects, but some can travel as TEM waves which may escape the system causing EMI problems.

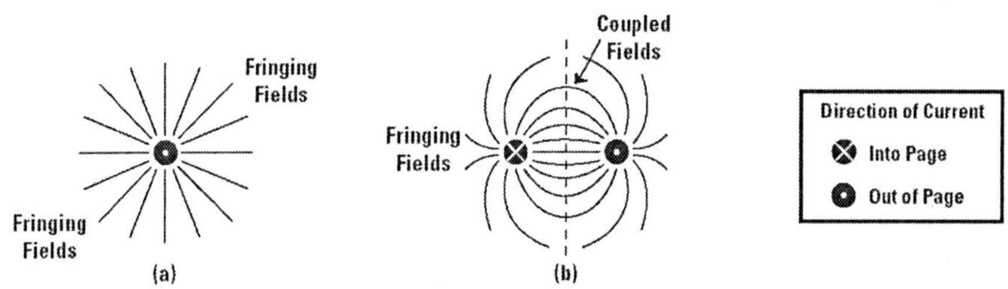

Figure 25.10 Electromagnetic field radiation in a single-ended signal (a) vs. cancellation in differential signals through coupling (b).

Balanced differential lines, however, have equal but opposite ("odd" mode) signals. This means that the concentric magnetic fields lines tend to cancel and the electric fields (shown in Figure 25.10) tend to couple. These coupled electric fields are "tied up" and cannot escape to propagate as TEM waves beyond the immediate vicinity of the conductors. Only the stray fringing fields are allowed to escape to the far field. Therefore, for coupled differential signals, much less field energy is available to propagate as TEM waves versus single-ended lines. The closer the "+" and "–" signals, the tighter or better the coupling.

Clearly, the voltages and currents of the two ("+" and "–") conductors are not always equal and opposite. For LVDS, the dc currents should never flow in the same direction as shown in Figure 25.11(a), but factors can cause an imbalance in currents, shown in Figure 25.11(c), ver-

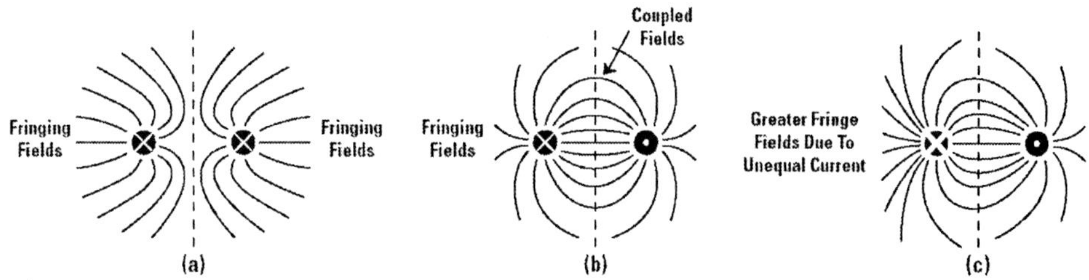

Figure 25.11 Even or common-mode signals (a), ideal equal and opposite odd-mode signals (b), and unbalanced signals (c) on differential lines.

sus the ideal case, shown in Figure 25.11(b). When this imbalance happens, an excess of field fringing occurs since the field strength of the two conductors is unequal. The extra fringe fields can escape as TEM waves and lead to more EMI.

Similar effects can be seen in microstrip and stripline PCB traces shown in Figure 25.12. The ideal cases for microstrip and stripline are represented by (a) and (b). Here we see that the microstrip ground plane helps couple additional field lines from below, tying up more field lines and reducing EMI. Stripline almost completely shields the conductors and therefore can significantly decrease EMI, but has the penalty of slower propagation velocity (about 40% slower than microstrip), more PCB layers, additional vias, and difficulty in achieving 100 Ω Z_{DIFF}. More

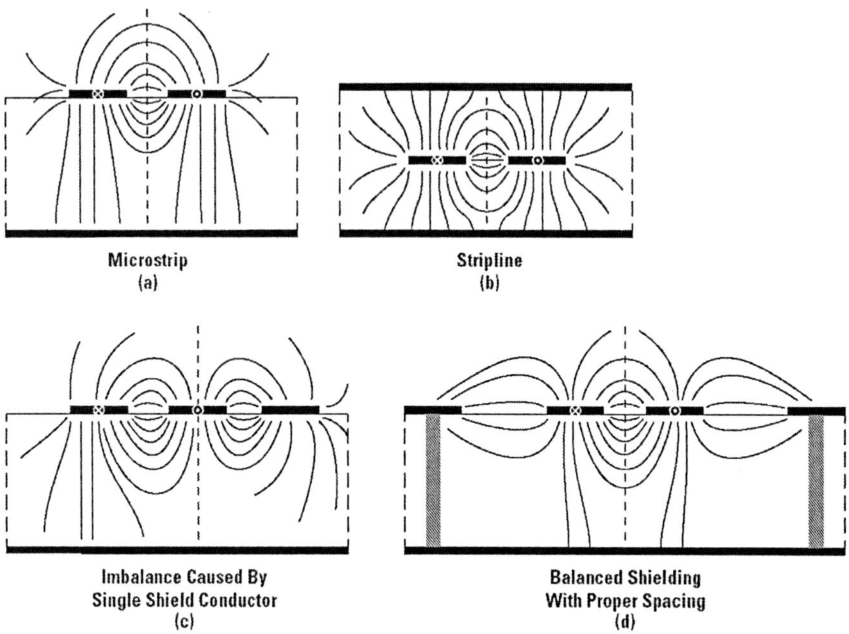

Figure 25.12 Ideal differential signals on microstrip (a) and stripline (b), negative effects of unbalanced shielding (c), and positive effects of balanced shielding (d).

shielding can be achieved using microstrip without significantly impacting propagation velocity using shield traces, as shown in Figure 25.12(d), but be careful to add the shield trace (preferably ground) on both sides of the pair, as in Figure 25.12(d). Running the shield trace — or any trace — on one side (c) creates an imbalance which can increase EMI. Ground trace shields should have frequent vias to the underlying ground plane at regular (<¼ wavelength) intervals, and should be placed at least 2S from the pair. (Ground trace shields are of questionable value. (See [Joh93].)

25.2.10.3 Design Practices for Low EMI

As discussed in the preceding paragraphs, the two most important factors to consider when designing differential signals for low EMI are close coupling between the conductors of each pair and minimizing the imbalances between the conductors of each pair. Let us discuss close coupling first.

In order for sufficient coupling to occur, the space between the conductors of an LVDS pair should be kept to a minimum (see Figure 25.13). Matched transmission impedance must also be maintained. Stripline power and ground planes/traces should not be closer than the distance between conductors to preserve closer coupling between the conductors versus the power and ground planes. A good rule is to keep S < W, S < h, and x greater or equal to the larger of 2S or 2W. The best practice is to use the closest spacing (S) allowed by your PCB vendor and then adjust trace widths (W) to control differential impedance. These EMI reduction rules are similar to the S Rules discussed in Section 25.1, "Layer Stack-Up and PCB Design," on page 695.

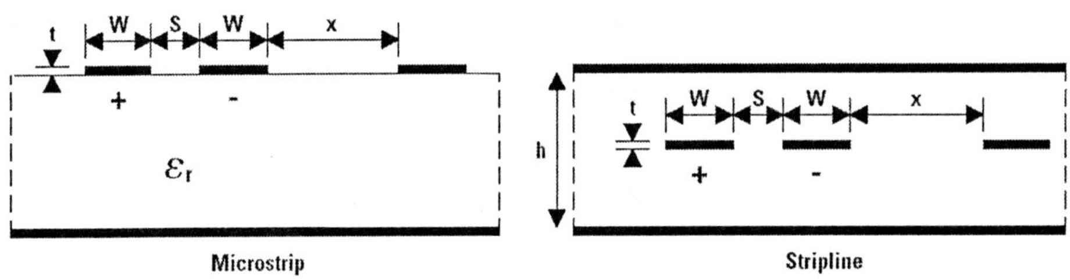

Figure 25.13 For sufficient coupling (canceling) of electromagnetic fields, the distance between the + and – signal should be minimized. Make S < W, S < h, and x ≥ 2W and 2S for good coupling.

Close coupling between conductors of a pair not only reduces EMI, but it helps to ensure noise coupled onto the conductors will do so equally as common-mode noise which will then be rejected by the receiver (see Figure 25.14). Since the differential pair is a current loop, minimizing conductor spacing also reduces the antenna loop.

Imbalance minimization is the other important factor in reducing EMI. Although fields result from the complex interaction between objects of a system and are difficult to predict (especially in the dynamic case), certain generalizations can be made. The impedance of your signal traces should be well-controlled. If the impedance of one trace changes versus another, the voltage and fields of one signal will be different from its partner. This will tend to create more fringing fields and therefore more EMI as we have seen.

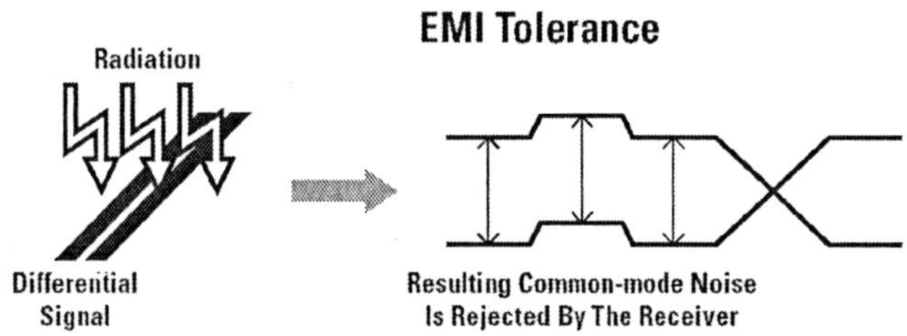

Figure 25.14 Close coupling not only reduces EMI but improves EMI tolerance too.

The basic rule to follow is: if any discontinuity must be introduced in proximity to differential lines, it should introduced equally to both members of the pair. Examples of discontinuities include components, vias, power and ground planes, PCB traces, and the like. Remember that the key word is balance. Figure 25.15 provides several examples of differential trace imbalances in a layout where the basic rule of balance has been violated.

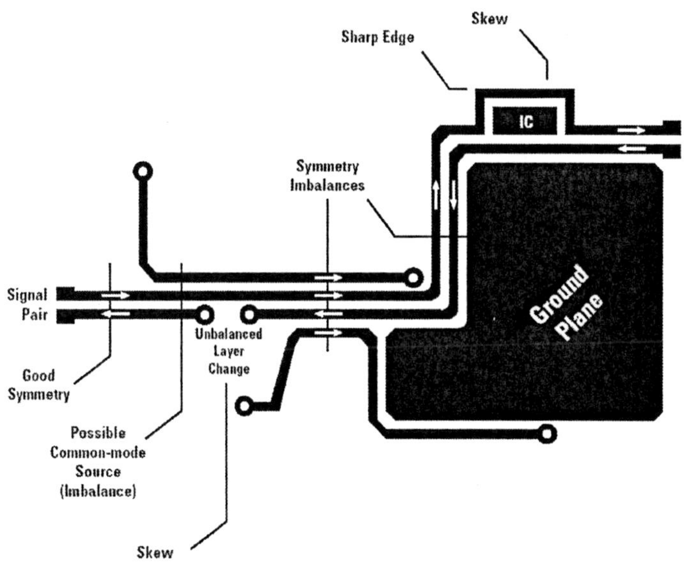

Figure 25.15 This PCB layout contains many sources of differential signal imbalance that will tend to increase electromagnetic radiation.

Unfortunately, unless you have an elaborate EMI lab, fields resulting from imbalances cannot easily be measured. Waveforms, however, are easy to measure. Since fields are proportional to voltage/current amplitude at any given point in time, any factors affecting the time (delay, velocity, etc.) and/or amplitude (attenuation, etc.) properties of the signals can increase EMI and can be seen on a scope. Figure 25.16 illustrates how waveforms — easily seen on a scope — can help predict far field EMI. First, the beneficial field canceling effects of ideal differential signals (b) versus single-ended signals (a) are compared.

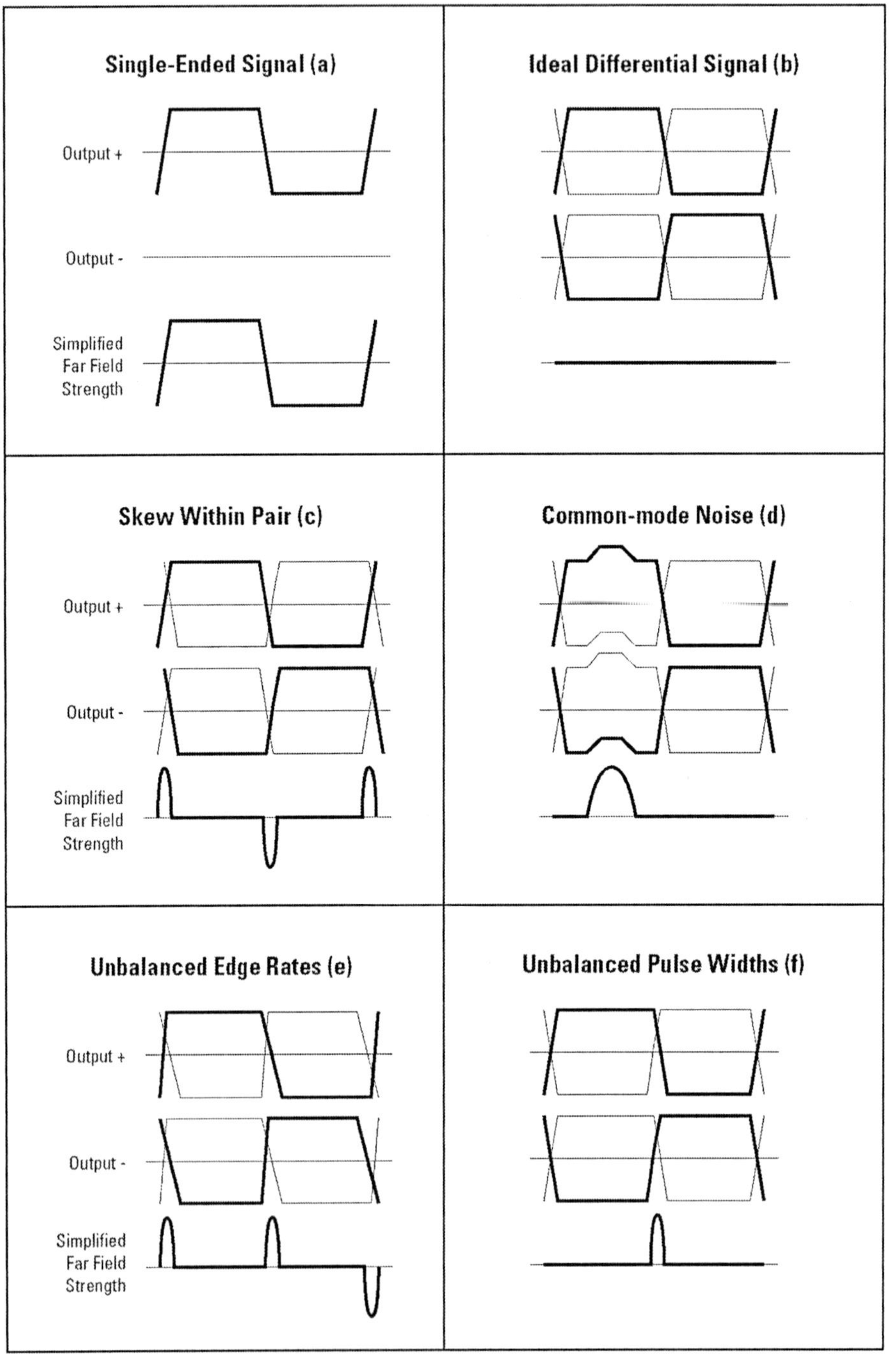

Figure 25.16 Diagram showing simplified far field radiation under various situations.

A real differential signal, however, is non-ideal and contains skew, unbalanced pulse widths and edge rates, common-mode noise, unbalanced attenuation, and so forth, as shown in (c) through (f) of Figure 25.16. These affect the relative amplitudes of the fields at any given moment, reducing the canceling effects of the differential signals, and potentially increasing EMI. Thus, the waveforms of one conductor of a pair should balance or mirror the other to minimize EMI.

25.2.10.4 EMI Test Results

The PCB setup shown in Figure 25.17 was used to examine the effects on EMI of closely coupled differential signals versus uncoupled signals. The setup compares two sets of LVDS signals: one set in which pair spacing is equal to trace width (S = W) and another set in which S >> W so that the pair members are no longer closely coupled (although the differential impedance of the transmission line is still 100 Ω). In the figure, each pair of differential traces is terminated with a 100 Ω resistor (depicted by a vertical rectangle) and each end of the resistor is run to an SMB test connector. The SMB connectors are used to verify proper circuit operation and an HP (Hewlett-Packard) close field E-field probe is used to pick up (measure) the resulting electric fields.

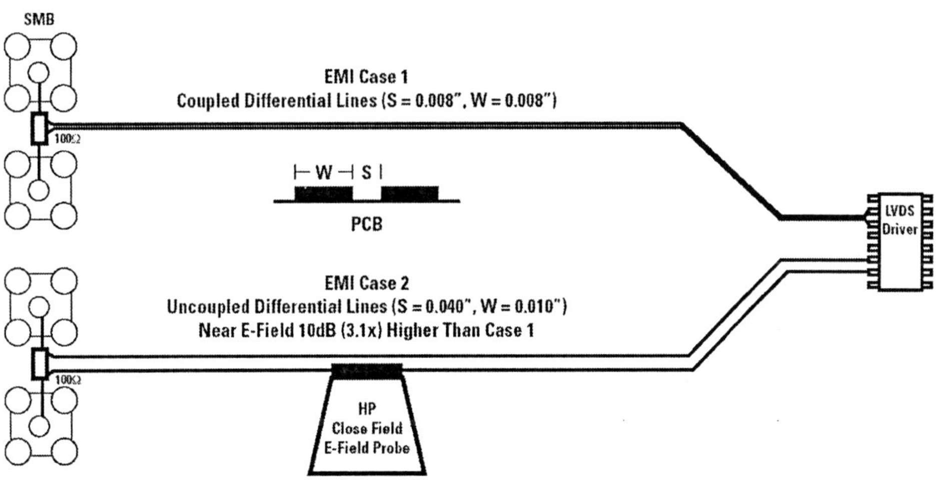

Figure 25.17 EMI test setup.

Near (close) field electric field measurements were made for both cases while using a 32.5 MHz 50% duty cycle clock as the source. The plots shown in Figures 25.18 and 25.19 depict the E-field strength (electric field strength) results for case 2, the uncoupled case. The first plot shows the E-field strength over 200 MHz to 1 GHz. The second plot looks more closely at the frequencies between 30 MHz and 300 MHz. The electric field noise shows up as "spikes" which occur at harmonics of the input frequency.

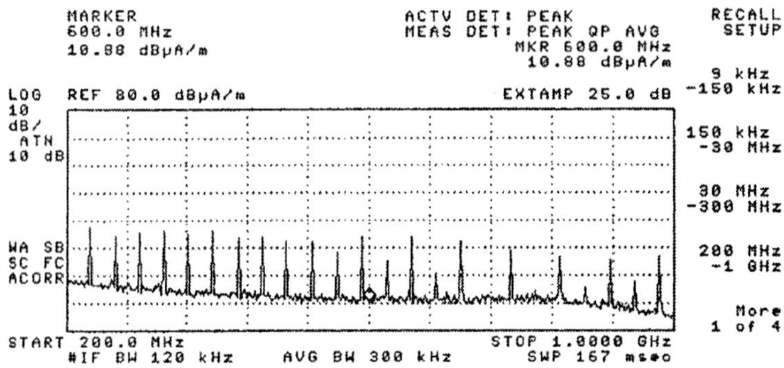

Figure 25.18 Near E-field strength for uncoupled signals (200 MHz to 1 GHz).

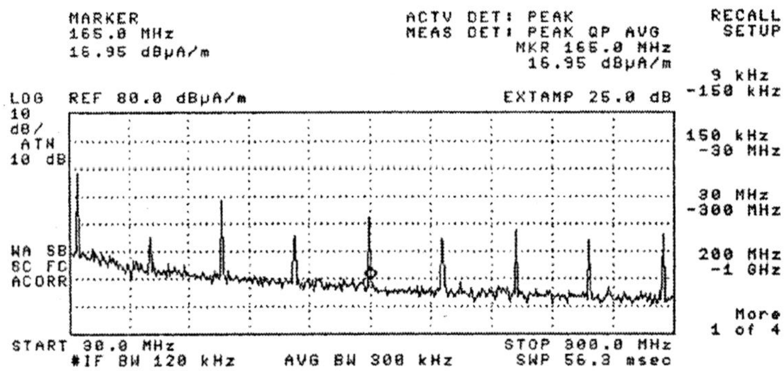

Figure 25.19 Near E-field strength for uncoupled signals (30–300 MHz).

The next two plots (see Figures 25.20 and 25.21) show the E-field strength for case 1 in which the differential pair is closely coupled. Notice that the harmonics are significantly reduced. When comparing Figure 25.20 (closely coupled) with Figure 25.18 (uncoupled) for the 200 MHz to 1 GHz frequency range, the E-field spikes in Figure 25.20 are roughly 10 dB lower (shorter) for the closely coupled case. Similarly, comparing Figure 25.21 (closely coupled) with Figure 25.19 (uncoupled) for the 30–300 MHz frequency range, the E-field spikes in Figure 25.21 are again roughly 10 dB lower for the closely coupled case. In the figures, each vertical division is 10 dB μA/m (microamp per meter).

In the far field, the EMI of the closely coupled pair should radiate much less due to the coupling of the electric fields. Even in the near field, however, the closely coupled pair generated much weaker electric fields. The closely coupled pair showed about 10 dB (>3 times) lower electric field strength than the uncoupled pair. This test illustrates two things:

1. Use of differential signals versus single-ended signals can be used effectively to reduce emissions.

2. The EMI advantages of differential signs will be lost or greatly diminished unless the signals are closely coupled.

This test used uncoupled LVDS signals to represent single-ended signals. Most single-ended signals such as TTL or GTL have a much greater swing and involve much greater currents, so their EMI would be expected to be even greater than what is seen here.

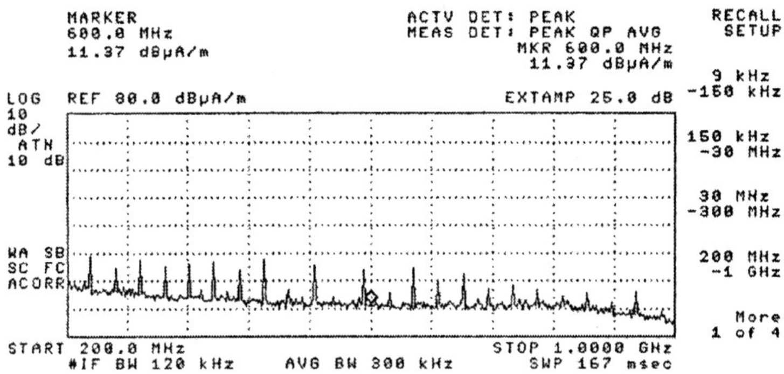

Figure 25.20 Near E-field strength for closely coupled signals (200 MHz to 1 GHz).

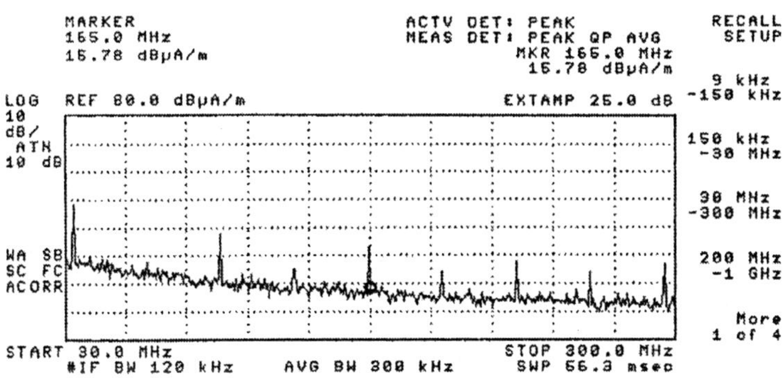

Figure 25.21 Near E-field strength for closely coupled signals (30–300 MHz).

25.2.11 Ground Return Paths

A conductor that carries current requires an opposite mirror current to return through some part of the system. This return current path will be the path of least impedance (path of least resistance at low frequencies and path of least inductance at high frequencies).

Since LVDS is differential, the signal current that flows in one conductor of a pair will flow back through the other conductor, completing the current loop. This is ideal, because the current return antenna loop area is minimized since the traces of a pair are closely spaced. Real signals, however, will have some common-mode noise current which must return also. This common-mode current will be capacitively coupled to ground and return to the driver through the path of least impedance. Therefore, a short ground current return path is needed between the driver and receiver in differential systems.

On PCBs, the best current return path is a uniform, unbroken ground plane beneath the LVDS signals. The ground plane will allow the common-mode (even-mode) current to return directly under the LVDS signals. This closely coupled path is the path of least impedance and means that the current loop area is minimized.

Similarly, in cables, a ground return wire or wires should be used between driver and receiver. This allows the return path to be in close proximity to the signal pairs reducing the current loop area. (See Chapter 12, “High-Speed Interconnects and Cabling,” starting on page 295).

25.2.12 Cable Shielding

Shielding is an effective way to reduce EMI. Shielding should be connected directly to both driver enclosure and receiver enclosure when possible. Shields are not designed to handle significant ground return currents, so it may be necessary to construct a filter network which isolates the shield from ground at one end (see Section 12.4.4, “Cable Ground and Shield Connections,” on page 310).

25.2.13 Common-Mode Noise Rejection

A test was conducted to demonstrate the common-mode noise rejection ability of LVDS receivers. Concern over the noise immunity of LVDS has been expressed because of the low voltage swing of LVDS (±350 mV swing with less than ±100 mV thresholds). Provided that the differential signals run close together through controlled impedance media, most of the noise on LVDS lines will be common-mode. In other words, EMI, crosstalk, power/ground shifts, and so on will appear equally on each pair and this common-mode noise will be rejected by the receiver.

The test setup for common-mode noise rejection is shown in Figure 25.22. The test setup is as follows: driver: DS90C031 (one channel), receiver: DS90C032 (one channel), $V_{CC} = 5$ V, $T_a = 25°$ C.

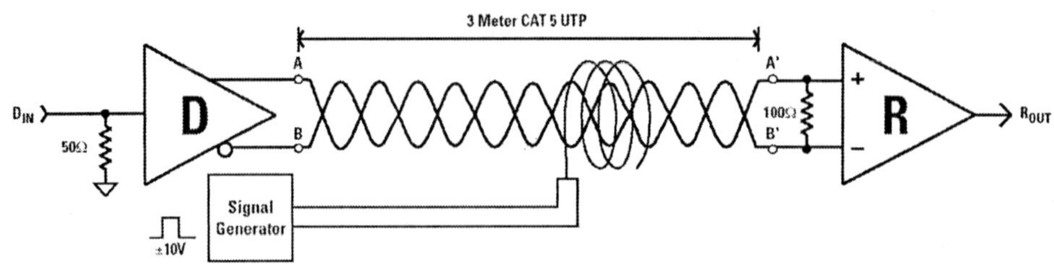

Figure 25.22 Common-mode noise rejection test setup.

An LVDS driver and receiver are connected by 3 meters of CAT5 UTP cable. While LVDS signals are being transmitted, a signal generator is radiating common-mode noise onto both conductors of the cable interconnect.

Figure 25.23 shows the reference LVDS single-ended waveforms obtained at the inputs to the receiver with no noise coupled onto the cable. The signal generator is off and the receiver output is clean and available for comparison with other test cases. Figure 25.24 shows the single-ended waveforms with +0.5 V to +1.75 V peak-to-peak common-mode noise (test case #1). The resulting receiver output is seen to still be clean. Figure 25.25 shows an expanded view of a portion of the same waveforms in Figure 25.24 obtained by expanding the horizontal time scale on the oscilloscope.

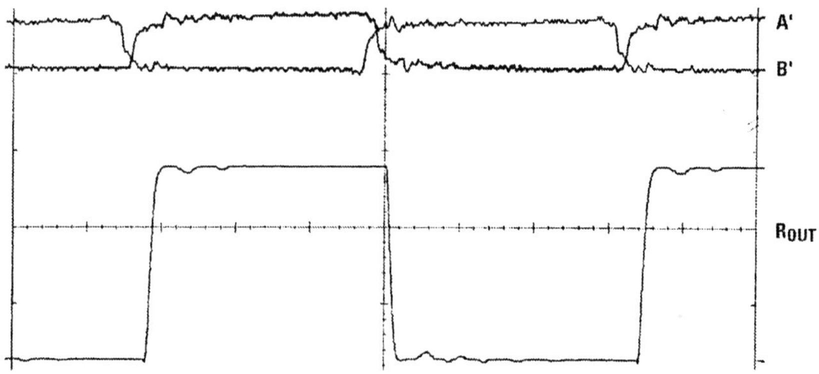

Figure 25.23 Reference waveforms showing LVDS signal and receiver outputs.

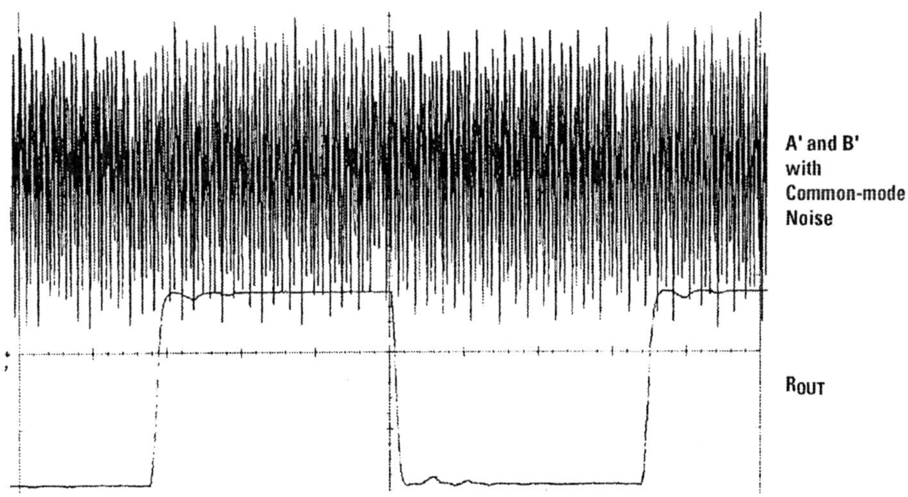

Figure 25.24 Coupled common-mode noise of +0.5 V to +1.75 V peak-to-peak and the resulting clean receiver output.

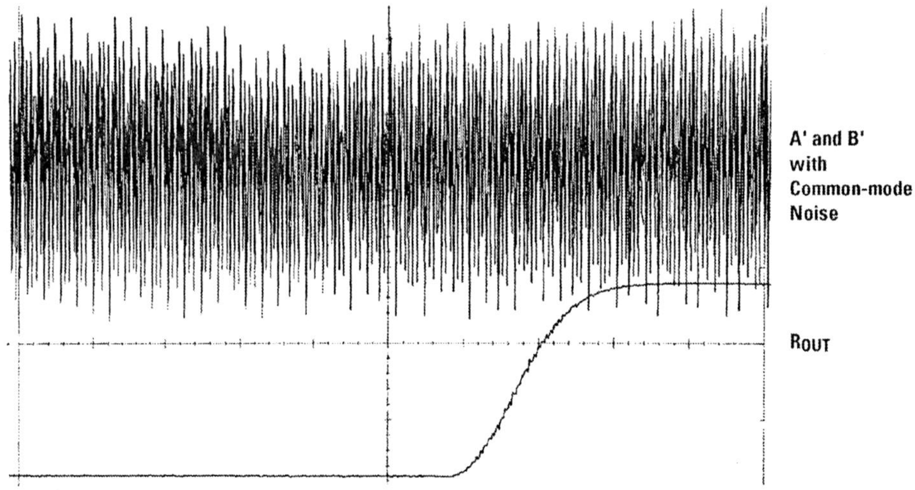

Figure 25.25 Expanded view of the same coupled common-mode noise waveforms and clean receiver output that were shown in Figure 25.24.

Figure 25.26 shows that a clean receiver output is still obtained with coupled common-mode noise of up to –0.5 V to +3.25 V (test case #2). That is, common-mode noise rejection occurs with V_{CM} noise (common-mode noise voltage) of up to –0.5 V to +3.25 V peak-to-peak. As mentioned earlier, the devices used for these tests were 5 volt devices.

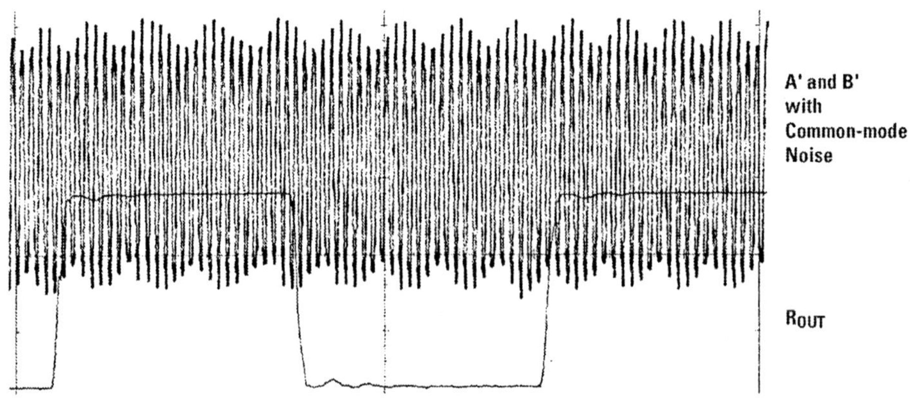

Figure 25.26 Clean receiver output despite –0.5 V to +3.25 V peak-to-peak common-mode noise.

25.3 LVDS Configurations

Although LVDS drivers and receivers are typically used in a point-to-point arrangement as shown in Figure 25.27, other topologies are possible. The point-to-point configuration does provide the best signal path and should be used for very-high-speed interconnect links. Point-to-point links are commonly used in conjunction with crosspoint switches.

Figure 25.27 Point-to-point configuration.

The configuration shown in Figure 25.28 allows bidirectional communication over a single twisted-pair. Data can flow in only one direction at a time. The requirement for two terminating resistors reduces the signal (and thus the differential noise margin) if using standard LVDS drivers. A better solution would be to employ Bus LVDS (BLVDS) drivers which are designed for double termination loads. They provide levels compatible with LVDS and do not trade off noise margin. Common-mode range for LVDS and BLVDS is ±1 V (typical), so cable lengths tend to be as long as 15 meters.

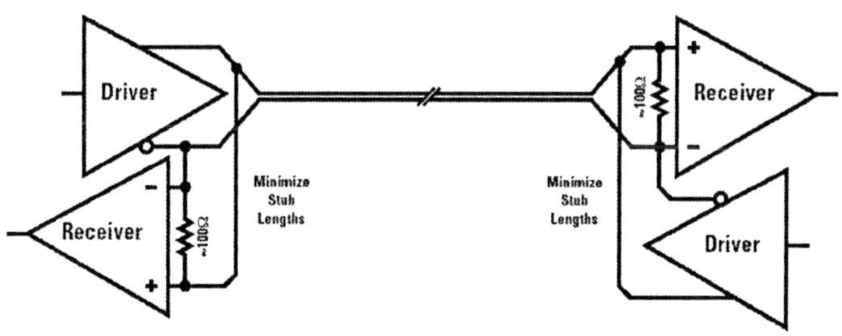

Figure 25.28 Bidirectional half-duplex configuration.

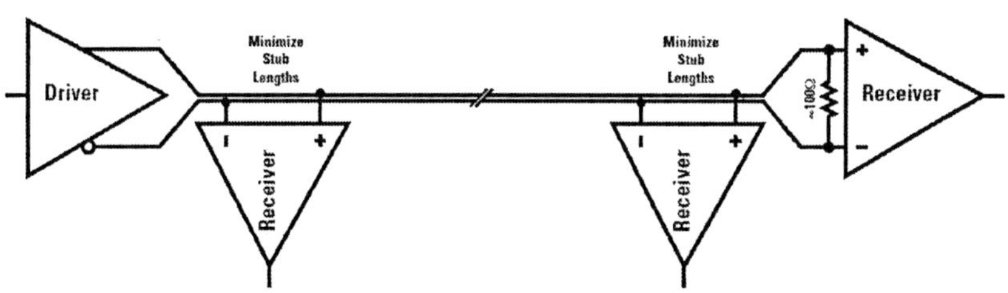

Figure 25.29 Multidrop configuration.

Since LVDS receivers have high impedance inputs, a multidrop configuration (see Figure 25.29) can also be used if transmission distance is short and stub lengths are less than ~12 mm (as short as possible). Use receivers with power-off high impedance if the network needs to remain active when one or more nodes are powered down. This application is good when the same set of data needs to be distributed to multiple locations.

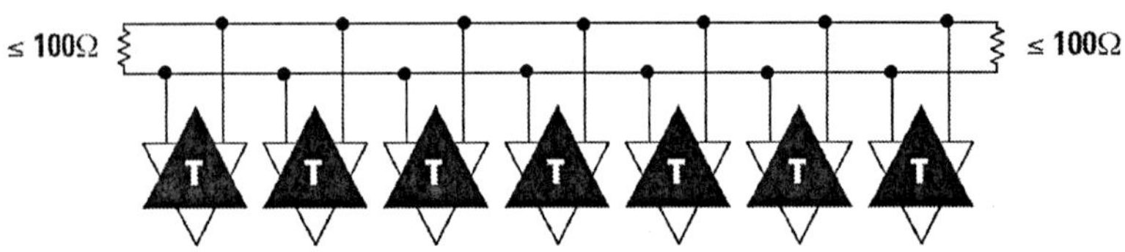

Figure 25.30 Multipoint configuration.

A multipoint bus as shown in Figure 25.30 supports multiple drivers, but only one is allowed to be active at any given time. With Bus LVDS devices, double terminated busses can be used without trading off signal swing and noise margin. Termination should be located at both ends of the bus. Failsafe biasing should be considered if a known state on the bus is required when all drivers are in TRI-STATE®. As with the multidrop bus, stubs off the mainline should be kept as short as possible to minimize transmission line problems.

25.4 Failsafe Biasing of LVDS

25.4.1 LVDS Failsafe Conditions

Most LVDS receivers have internal failsafe circuitry that forces the output to be in a known logic state (high) under certain fault conditions. These conditions include open, shorted, and terminated receiver inputs. Always consult the component's datasheet to determine which type of failsafe protection is supported. Here is a summary of LVDS failsafe conditions:

- Open input pins — Unused receiver inputs should be left open. Do not tie unused receiver inputs to ground or other voltages. The internal failsafe bias resistors will pull the "+" input high, and the "–" input low, thus guaranteeing a high, stable output state. This minimizes power dissipation and switching noise.
- Terminated input pins — If the cable is removed and the inputs to the receiver have a termination resistor across them, then the output will be stable (high). Noise picked up at the input, if differential in nature, can cause the device to respond. If this is the case, see Section 25.4.2 just below.
- Terminated input pins in noisy environments — See Section 25.4.2 just below if failsafe must be guaranteed in noisy environments when the cable is disconnected from the driver's end or if the driver is in TRI-STATE.

- Shorted inputs — The receiver output will remain in a high state when the inputs are shorted. This is considered fault condition protection only. It is not specified across the input voltage range of the receiver.

25.4.2 Boosting Failsafe in Noisy Environments

The internal failsafe circuitry is designed to source/sink a very small amount of current, providing failsafe protection for floating receiver inputs, shorted receiver inputs, and terminated receiver inputs as described above and in the component's datasheet. It is not designed to provide failsafe in noisy environments when the cable is disconnected from the driver's end or if the driver is in TRI-STATE. When this happens, the cable becomes a floating antenna which can pick up noise. If the cable picks up more differential noise than the internal failsafe circuitry can overcome, the receiver may switch or oscillate. If this condition can happen in your application, it is recommended that you choose a balanced and/or shielded cable which will reduce the amount of differential noise on the cable. In addition, you may wish to add external failsafe resistors to create a larger noise margin. However, adding more failsafe current will tend to unbalance the symmetrical LVDS output drive (loop) current and degrade signal quality somewhat. Therefore, a compromise should be the ultimate goal. External failsafe can be added, but must be small enough not to significantly affect driver current.

25.4.3 Choosing External Failsafe Resistors

The diagram in Figure 25.31 shows that many present LVDS devices typically have an internal failsafe voltage of about –10 to –30 mV. If the receiver will not always be driven by the driver in your application and the cable is expected to pick up more than 10 mV of differential noise you may need to add additional failsafe resistors as shown in Figure 25.32. The resistors are chosen by first measuring/predicting the amount of differential-mode noise you will need to overcome. V_{FSB} (failsafe bias voltage) is the offset voltage generated across the termination resistor (100 Ω). Note that you do not need to provide a bias (V_{FSB}) which is greater than the receiver threshold

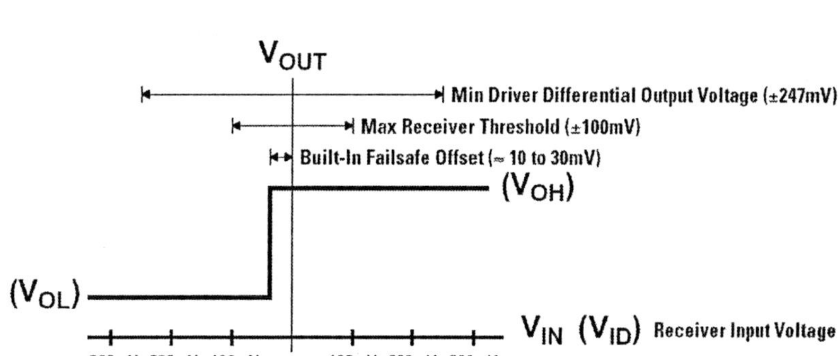

Figure 25.31 Typical differential input voltage (V_{ID}) vs. receiver logic state.

(100 mV), typically +15 mV or +20 mV is sufficient. You only need enough to overcome the differential noise, since the internal failsafe circuitry will always guarantee a positive offset. In fact, making VFSB too large will contend with the driver output causing the driven signal to become imbalanced and reduce signal quality.

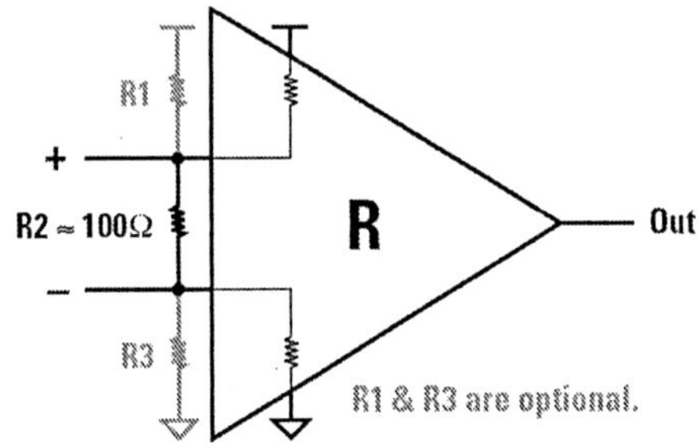

Figure 25.32 Diagram showing simplified internal failsafe circuitry and optional external "helper" failsafe resistors.

For best results, these procedures should be followed when choosing external failsafe resistors:

1. First ask the question, "Do I need external failsafe?" If your LVDS driver is always active, you will not need external failsafe. If the cable is never disconnected from the driver end while the system is active and/or your cable will not pick up much differential-mode noise, you may not need to boost failsafe.
2. Measure/predict the amount of differential-mode noise at the receiver end of the cable in worst-case conditions. If you have a lot of noise, use a balanced cable like twisted-pair which tends to pick up mostly common-mode noise, not differential-mode noise. Do not use simple ribbon cables which can pick up differential-mode noise due to fixed positions of the conductors. Use a shielded cable if possible. Using a balanced and/or shielded cable is the best way to prevent noise problems in noisy environments.
3. Once you have chosen the appropriate cable, measure the amount of differential voltage at the receiver under worst-case conditions. Set this equal to V_{FSB} in Equation (25.11) and solve for the external failsafe resistors R1 and R3.

$$V_{FSB} = \frac{R2}{R1 + R2 + R3} V_{CC} \tag{25.11}$$

$$I_{BIAS} = \frac{V_{CC}}{R1 + R2 + R3} << I_{LOOP} \text{ (Use } I_{BIAS} \le 0.1(I_{LOOP})\text{)} \tag{25.12}$$

$$V_{CM} = \frac{R3 + (R2)/2}{R1 + R2 + R3} V_{CC} = 1.2V \Rightarrow R1 \approx R3\left(\frac{V_{CC}}{1.2V} - 1\right) \quad (25.13)$$

$$R_{TEQ} = \frac{R2(R1 + R3)}{R1 + R2 + R3} = \text{match transmission line } Z_{DIFF} \quad (25.14)$$

4. You now have an equation relating R1 to R3. Choose R1 and R3 so that: (a) they approximately satisfy the third equation for $V_{CM} = 1.2$ V, and (b) they are large enough that they do not create a bias which will contend with the driver current ($I_{BIAS} << I_{LOOP}$, equation two). In general, R1 and R3 should be greater than 20 kΩ for $V_{CC} = 5$ V and greater than 12 kΩ for $V_{CC} = 3.3$ V. Remember that you want just enough I_{BIAS} to overcome the differential noise, but not enough to significantly affect signal quality.
5. The external failsafe resistors may change your equivalent termination resistance, R_{TEQ}. Fine tune the value of R2 to match R_{TEQ} to within about 10% of your differential transmission line impedance (for less than a 5% reflection), or better yet, to almost equal the differential impedance.

25.4.4 Power-Off High Impedance Bus Pins

Power-off high impedance is a useful feature and most second- and third-generation LVDS receivers provide this feature. Power-off high impedance is typically listed as a condition and feature of the I_{IN} (input current) parameter. This feature is useful in applications that employ more than one receiver and they are powered from local power supplies. If the power is turned off to one node, it should not load down the line and prevent communication between other powered-up nodes.

25.5 Eye Pattern Test Circuit

LVDS drivers and receivers are typically used in an uncomplicated point-to-point configuration as shown in Figure 25.33. This figure details a test circuit that was used to acquire eye pattern measurements. The components that were used to create the test circuit are listed in Table 25.2.

25.5.1 Eye Pattern Test Procedure

A pseudo-random bit sequence (PRBS) generator was connected to the driver input, and the resulting eye pattern (measured differentially across test points TP-A' and TP-B' in Figure 25.33) was observed on the oscilloscope. Different cable lengths were tested, and the frequency of the input signal was increased until the measured jitter equaled 20% with respect to the unit interval for the particular cable length. The coding scheme used was NRZ (nonreturn to zero). Jitter was measured twice at two different voltage points. Jitter was first measured at the 0 V differential voltage (optimal receiver threshold point) for minimum jitter, and measured a second time at the maximum receiver threshold points (±100 mV) to obtain the worst-case or maximum jitter at the receiver thresholds. Occasionally jitter is measured at the crossing point alone and, although this will result in a much lower jitter point, it ignores the fact that the receivers may not switch at that very point. For this reason, this signal quality test measured jitter at both points.

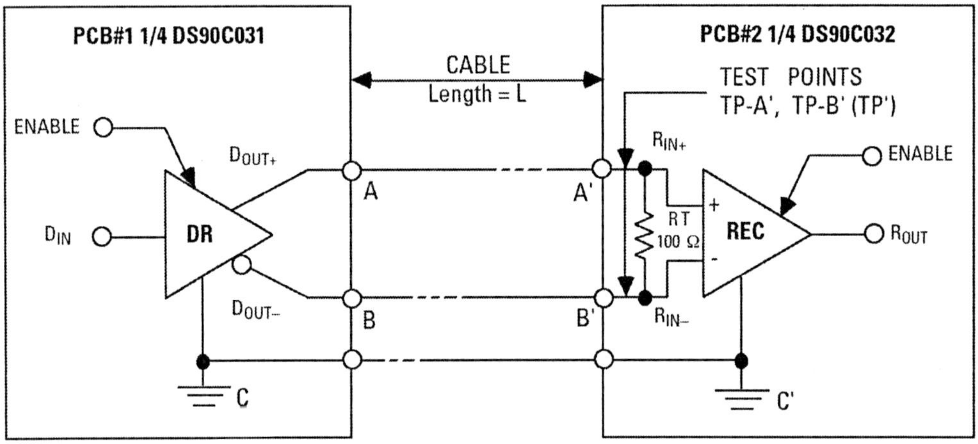

Figure 25.33 LVDS signal quality test circuit.

Table 25.2 Components Used in Eye Pattern Test Circuit

Component	Description
PCB#1	DS90C031 LVDS quad driver soldered to the PCB with matched PCB traces between the device (located near the edge of the PCB) and the connector. The connector is an AMP amplite 50 series connector. The DS90C031 requires +5 V power.
Cable	The cable used for this testing was Berk-Tek part number 271211. This is a 105 Ω (differential mode) 28 AWG stranded twisted-pair cable (25-pair with overall shield) commonly used on SCSI applications. This cable represents a common data interface cable. For this test, cable lengths tested were 1, 2, 3, 5, and 10 meter(s). Cables longer than 10 meters were not tested, but may be employed at lower data rates.
PCB#2	DS90C032 LVDS quad receiver soldered to the PCB with matched PCB traces between the device (located near the edge of the PCB) to the connector. The connector is an AMP amplite 50 series connector. A 100 Ω surface-mount resistor was used to terminate the cable at the receiver input pins.

25.5.2 Eye Pattern Test Results and Data Points

As described in the test procedure just above, the first of the two jitter measurements was made at the 0 V differential point. For the case with the 1 meter cable, 490 ps of jitter at 400 Mbps were measured, and with the 10 meter cable, 1.160 ns of jitter at 180 Mbps were measured. The full results are shown in Table 25.3.

Table 25.3 20% Jitter Table at 0 V Differential (Minimum Jitter)

Cable Length (m)	Data Rate (Mbps)	Unit Interval (ns)	Jitter (ns)
1	400	2.500	0.490
2	391	2.555	0.520
3	370	2.703	0.524
5	295	3.390	0.680
10	180	5.550	1.160

The second of the two jitter measurements was made at ±100 mV levels. For the 1 meter cable, 1 ns of jitter was measured at 200 Mbps, and for the 10 meter cable, 2 ns of jitter occurred at 100 Mbps. See Table 25.4.

Table 25.4 20% Jitter Table at ±100 mV (Maximum Jitter)

Cable Length (m)	Data Rate (Mbps)	Unit Interval (ns)	Jitter (ns)
1	200	5.000	1.000
2	190	5.263	1.053
3	170	5.882	1.176
5	155.5	6.431	1.286
10	100	10.000	2.000

Care should be taken in long cable applications using LVDS. When directly coupled, LVDS provides up to ±1 V common-mode rejection. Long cable applications may require larger common-mode support. If this is the case, transformer coupling or alternate technologies (such as RS-485) should be considered.

Figures 25.34 and 25.35 are a graphical representation of the relationship between data rate and cable length for the application under test. Both curves assume a maximum allotment of 20% jitter with respect to the unit interval. Basically, data rates between 200–400 Mbps are possible at shorter lengths, and rates of 100–200 Mbps are possible at 10 meters. Note that employing a different coding scheme, cable, wire gauge (AWG), etc. will create a different relationship between maximum data rate versus cable length. Designers are greatly encouraged to experiment with the specific components and parameters for the design they plan to implement.

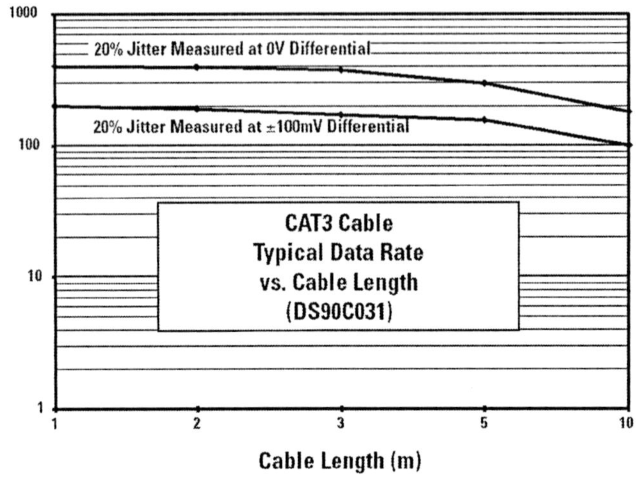

Figure 25.34 Typical data rate (MHz) vs. cable length for 0–10 meter CAT3 cable.

Eye patterns provide a useful tool to analyze jitter and the resulting signal quality as it captures the effects of a random data pattern. They provide a method to determine the maximum cable length for a given data rate or vice versa. Different systems, however, can tolerate different levels of jitter. Commonly 5%, 10%, or 20% is acceptable with 20% jitter usually being an upper practical limit. More than 20% jitter tends to close down the eye opening, making error-free recovery of NRZ data more difficult.

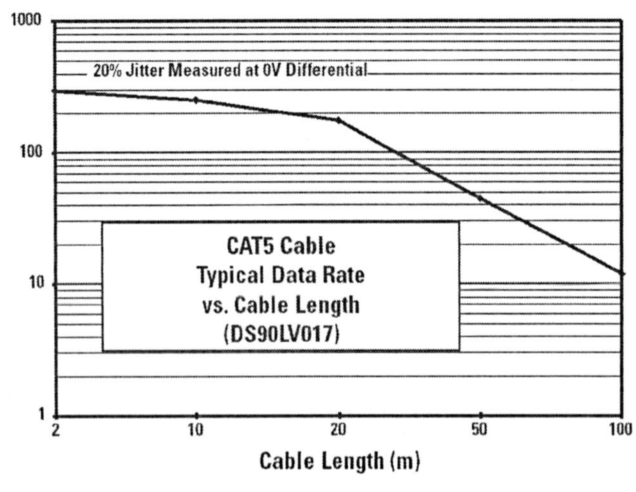

Figure 25.35 Typical data rate (MHz) vs. cable length for 2–100 meter CAT5 cable (see [AN-1088]).

25.6 BER Test Circuit

LVDS drivers and receivers are typically used in an uncomplicated point-to-point configuration as shown in Figure 25.36. Four drivers were used to drive 1 to 5 meters of standard twisted-pair cables at selected data rates. Four receivers were used to recover the data at the load end of the cable. Figure 25.36 details the test circuit that was used; is the same one used for the eye pattern tests discussed previously in Section 25.5, "Eye Pattern Test Circuit," on page 723. The test circuit components were listed before in Table 25.2, with the difference being that, for these BER tests, only cable lengths of 1 and 5 meters were used.

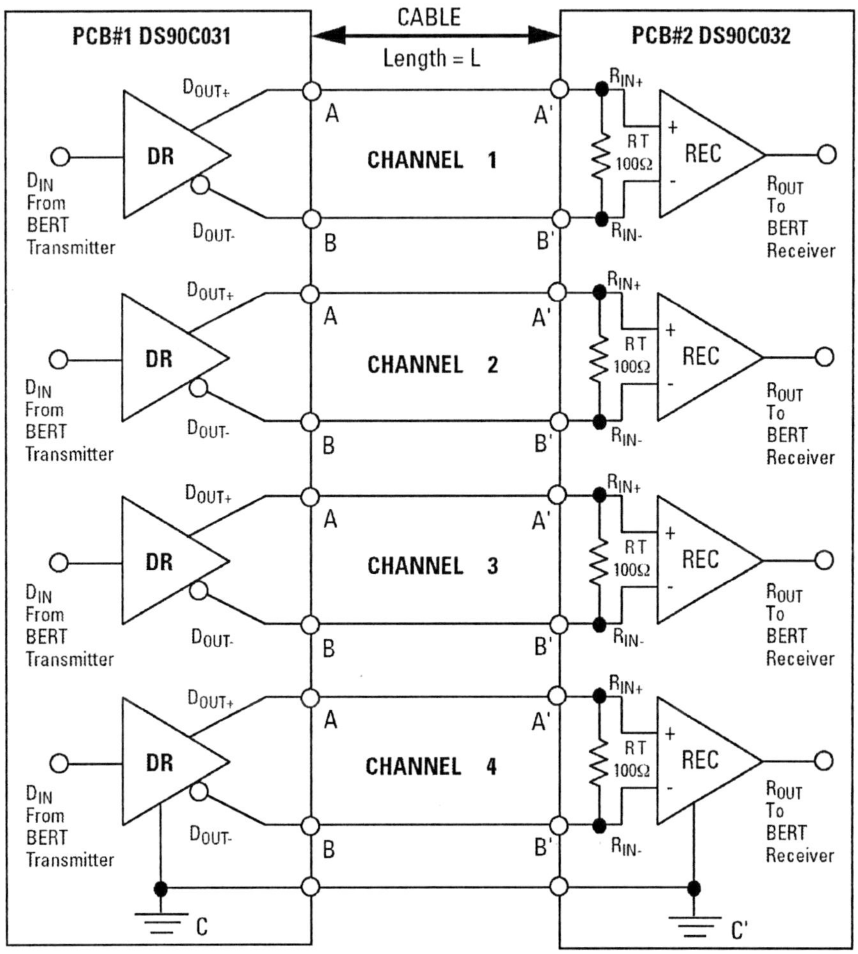

Figure 25.36 LVDS BER test circuit.

25.6.1 BER Test Procedure

A parallel high-speed BER transmitter/receiver set (Tektronix MultiBERT-100) was employed for the tests. The BERT transmitter was connected to the driver inputs, and the receiver outputs were connected to the BERT receiver inputs. Different cable lengths and data rates were tested. The BER tester was configured to provide a PRBS (pseudo random bit sequence) of $2^{15} - 1$ (32,767 bit long sequence). In the first test, the same input signal was applied to all four of the LVDS channels under test. For the other tests, the PRBS was offset by 4-bits providing a random sequence between channels. The coding scheme used was NRZ. Upon system test configuration, the test was allowed to run uninterrupted for a set amount of time. At completion of the time block, the results were recorded which included elapsed seconds, total bits transmitted, and number of bit errors recorded. Three tests were documented at room temperature using a power supply voltage of +5.0 V.

25.6.2 BER Tests and Results

The goal of the BER tests was to demonstrate that error rates of less than 1×10^{-12} are obtainable. The test conditions and results are shown above in Table 25.5. All three of the tests ran error free and demonstrated extremely low bit error rates using LVDS technology. The tests concluded that error rates of less than 1×10^{-12} can be obtained at 100 Mbps operation across 5 meters of twisted-pair cable. BER tests only provide a "Go – No Go" data point if zero errors are detected.

Table 25.5 BER Test Conditions and Results

Test #	Test Conditions	Test Results
1	Data Rate = 50 Mbps Cable Length = 1 meter PRBS Code = $2^{15} - 1$ NRZ For this test, the PRBS code applied to the four driver inputs was identical. This created a "simultaneous output switching" condition on the device.	Total Seconds: 87,085 (1 day) Total Bits: $1{,}723 \times 10^{13}$ Errors = 0 Error Rate = $< 1 \times 10^{-12}$
2	Data Rate = 100 Mbps Cable Length = 1 meter PRBS Code = $2^{15} - 1$ NRZ For this test, the PRBS code applied to the four driver inputs was offset by four bits. This creates a random pattern between channels.	Total Seconds: 10,717 (~3 hr.) Total Bits: 4.38×10^{12} Errors = 0 Error Rate = $< 1 \times 10^{-12}$
3	Data Rate = 100 Mbps Cable Length = 5 meter PRBS Code = $2^{15} - 1$ NRZ For this test, the PRBS code applied to the four driver inputs was offset by four bits. This creates a random pattern between channels.	Total Seconds: 10,050 (~2.8 hr.) Total Bits: 4×10^{12} Errors = 0 Error Rate = $< 1 \times 10^{-12}$

It is recommended to conduct further tests to determine the point of failure (data errors). This will yield important data that indicates the amount of margin in the system. This was done in the tests conducted by increasing the cable length from 1 meter to 5 meters, and also adjusting the data rate from 50 Mbps to 100 Mbps. Additionally, bench checks were made while adjusting the power supply voltage from 5.0 V to 4.5 V and 5.5 V, adjusting clock frequency, and by applying heat and cold to the device under test. No errors were detected during these checks (tests were checks only and were not conducted over time, e.g., 24 hours). BER tests demonstrated that the PRBS patterns were transmitted error free across the link.

25.7 Exercises

25-1. A differential microstrip pair has both traces on the surface of the PCB and a stripline has both traces buried inside the PCB. Why do you think normally it would not be a good idea to have one trace of a differential pair on the surface of the PCB and the other trace inside the PCB?

25-2. LVDS differential traces are usually terminated with a 100 Ω surface-mount resistor. (a) How much power is consumed by a current of 3.5 mA through the 100 Ω termination resistor? (Just calcualte the worst-case dc steady-state condition for this answer.) (b) Why don't chip manufacturers just put the termination resistor inside the chip?

25-3. Explain the difference between balanced cables and unbalanced cables. Give at least two examples of each.

25-4. What's the difference between a broadside stripline and an edge-coupled stripline?

25-5. What are two benefits to using differential connections between LVDS drivers and receivers?

25-6. Calculate the differential impedance Z_{DIFF} where Z0 is 65W for (a) h = 10 mils and S = 10 mils, (b) h = 5 mils and S = 10 mils, and (c) h = 10 mils and S = 5 mils. (d) What is the effect of decreasing the trace height above the ground plane from 10 to 5 mils? (e) What is the effect of decreasing the spacing between the traces from 10 to 5 mils?

25-7. What are the S Rules? List and explain each rule.

25-8. List 12 PCB layout and design tips for LVDS signals, devices, and cabling in addition to the S Rules.

25-9. The differential impedance for an edge-coupled stripline was given by Equation (25.4) on page 701 as

$$Z_{DIFF} \cong 2Z_0 \left(1 - 0.374\, e^{-2.9(S/h)}\right)\ \Omega$$

where S is the spacing between traces, h is the distance between the ground planes that enclose the stripline traces, and Z_0 is the characteristic impedance of a single trace. The footnote for that equation states that, for edge-coupled striplines, the term "0.374" may be replaced with "0.748" for lines that are closely coupled ($S < 12$ mils). (a) Let $S = 12$ mils, $h = 5$ mils, and $Z_0 = 50\ \Omega$. Calculate the differential impedance using the "0.374" term. (b) Now let $S = 12$ mils, $h = 5$ mils, and $Z_0 = 50\ \Omega$ as before, but calculate the differential impedance using the "0.748" term. (c) Compare your two answers from (a) and (b). By what percentage do the two answers differ? (Use the smaller of the two values as the reference for computing your percentage.) Does this suggest anything about the accuracy of this equation? (d) Repeat (a), (b), and (c) above, but now with $h =$ 25 mils.

25-10. Assume you have a differential sinusoidal signal propagating down a differential trace pair on a PCB. (a) At a signal frequency of 2.5 GHz, what minimum delay in either the true or complementary signal would cause that signal to be shifted by half of one wavelength so that the two signals are perfectly reinforcing and radiating EMI? (b) What trace routing condition could cause a delay difference between the two traces by this amount as the two traces go from the differential driver to the differential receiver? (c) If the traces were surrounded by epoxy/glass (FR-4) material, what amount of skew, measured in inches, would it take to create this condition in the differential signal? (For your calculations in (c) and (d), use minimum values from Table 25.1 on page 702.) (d) If the traces were surrounded entirely by Rogers 4350 core material, what amount of skew measured in inches would it take to create this condition in the differential signal? (e) Would the signals have to be perfectly reinforcing to generate EMI?

25-11. (a) Calculate the loss tangent at a frequency of 10 GHz for a coaxial cable where the capacitance per unit length is 1 pF/inch and the conductance per unit length at 10 GHz is 2.83 millimhos/inch. (b) Does the conductance per inch vary with frequency? (c) Does the loss tangent vary with frequency? (d) What units does the loss tangent have?

25-12. What are four guidelines to consider when terminating an LVDS receiver with a shunt resistor?

25-13. Should unused LVDS inputs, TTL inputs, LVDS outputs, and TTL outputs be tied to ground, tied to V_{CC}, or left disconnected with regard to LVDS devices?

25-14. What are the two most important factors to consider when designing differential signals for low EMI?

25-15. During lab checkout and evaluation of a prototype board containing differential LVDS signals, how could you tell the difference between EMI being generated due to unbalanced pulse widths on the true and complementary signals versus EMI being generated due to common-mode noise? Both conditions result in a positive going EMI pulse. Assume you can measure the radiated EMI with a large loop of wire connected to a sensitive oscilloscope probe.

25-16. (a) When analyzing the eye pattern generated while using the test circuit shown in Figure 25.33 on page 724, in what two ways was jitter measured? (b) What is the advantage to measuring jitter in each way? (c) Is jitter measured at the driver or at the receiver? (d) In what two units is jitter specified?

25-17. Table 25.2, "Components Used in Eye Pattern Test Circuit," on page 724, lists the test setup for making the eye pattern measurements reported in this chapter. What other equipment was required in order to perform these eye pattern tests? Provide a specification for each additional piece of equipment that it would have to meet in order to complete these measurements.

25-18. (a) When working with NRZ data and LVDS transmitters and receivers, what is the maximum amount of jitter that can be present at the driver for error-free recovery of the data? (b) What's the maximum amount of jitter that can be present at the receiver for error-free recovery of the data? (c) Should you measure the jitter at the 0 V crossing point or at the ±100 mV threshold points?

25-19. (a) When making jitter measurements using the eye pattern test circuit, by what percentage did the maximum data rate drop when going from a 1 meter cable to a 10 meter cable? (b) Would you conclude that the transmission distance has a major impact on the maximum data rate or only a minor inpact? (c) By what percentage did the data rate drop when measuring jitter at the ±100 mV threshold levels versus the 0 V differential voltage point? (d) Does your jitter measurement technique also have a major impact on the maximum data rate?

25-20. (a) What is the difference between the eye pattern test circuit of Figure 25.33 on page 724 and the BER test circuit of Figure 25.36 on page 727? (b) What is different about the testing approach taken with one test circuit from the testing approach of the other?

25-21. Table 25.5, "BER Test Conditions and Results," on page 728, indicated that data data rates of 50 Mbps and 100 Mbps could run for up to several days without a bit error occuring. What BER was determined under this test?

25-22. CAT5 cable is generally considered to be higher quality, have larger bandwidth, and cost more than CAT3 cable. In comparing Figure 25.34 on page 726 which uses CAT3 cable with Figure 25.35 on page 726 which uses CAT5 cable, there seems to be a discrepancy. At 2 MHz, the CAT3 cable actually seems to have a higher data rate than the CAT5 cable, but at 10 MHz, the CAT5 cable seems to have the higher data rate. How can you account for this apparent discrepancy?

25-23. Which do you feel is a better test for determining signal integrity: the eye pattern test or the BER test? Explain your answer.

CHAPTER 26

Designing to 10 Gbps Using SerDes Transceivers, Serializers, and Deserializers

Chapter Objectives

- Provide examples of Bus LVDS SerDes, serializer, and deserializer architectures.
- Provide overview of bus topologies, backplanes, and PCB recommendations.
- Discuss cable and connector selection and power and ground design concepts for SerDes applications.
- Discuss SerDes receiver reference and output clocks, and the transmit clock jitter tolerance.
- Discuss probing Bus LVDS signals.
- Discuss loopback testing with SerDes transceivers.
- Compare deserializer locking to random data versus sync patterns.
- Discuss the procedure to construct a jitter eye mask and then quantify the interconnect jitter margin.
- Discuss troubleshooting methodology for SerDes designs.

26.1 Introduction and the DS92LV16 (2.56 Gbps)[1]

This chapter provides an overview on the use of high-speed SerDes (serializer/deserializer) transceivers, serializers, and deserializers. Important engineering design concepts are presented through discussions of three actual devices. These devices are as follows:

1. Reprinted with permission of National Semiconductor Corporation, except where otherwise noted.

- DS92LV16, a 16-bit Bus LVDS SerDes that has 2.56 Gbps full-duplex throughput.
- DS25C400, a quad 2.5 Gbps SerDes transceiver with total throughput of 10 Gbps.
- DS92LV8028, an 8-channel 10:1 serializer with total throughput of all channels of 5.28 Gbps.

The first device to be discussed is the DS92LV16, shown in Figure 26.1 in a block diagram representation. The DS92LV16 is a member of National's robust and easy-to-use Bus LVDS serializer/deserializer (SerDes) family already popular in a wide variety of telecom, datacom, industrial, and commercial backplane/cable interconnect applications. The DS92LV16 is similar to the original 10-bit Bus LVDS SerDes products, but provides a wider, 16-bit data bus payload. The DS92LV16 is very flexible and performs over a wide, 25–80 MHz frequency range. The DS92LV16 costs about $14 per chip for the less expensive grade in quantities of 1,000, although prices vary over time not only with the purchase quantity, but with chip market demand, package styles, and the financial clout of customers.

The data throughput of the transceiver is maximized when the transmit clock TCLK is set to the highest specified rate of 80 MHz. At a TCLK of 80 MHz, 1.28 Gbps of data can be transfered. This is calculated as (80 MHz)(16-bit bus) = 1.28 Gbps and represents the data rate

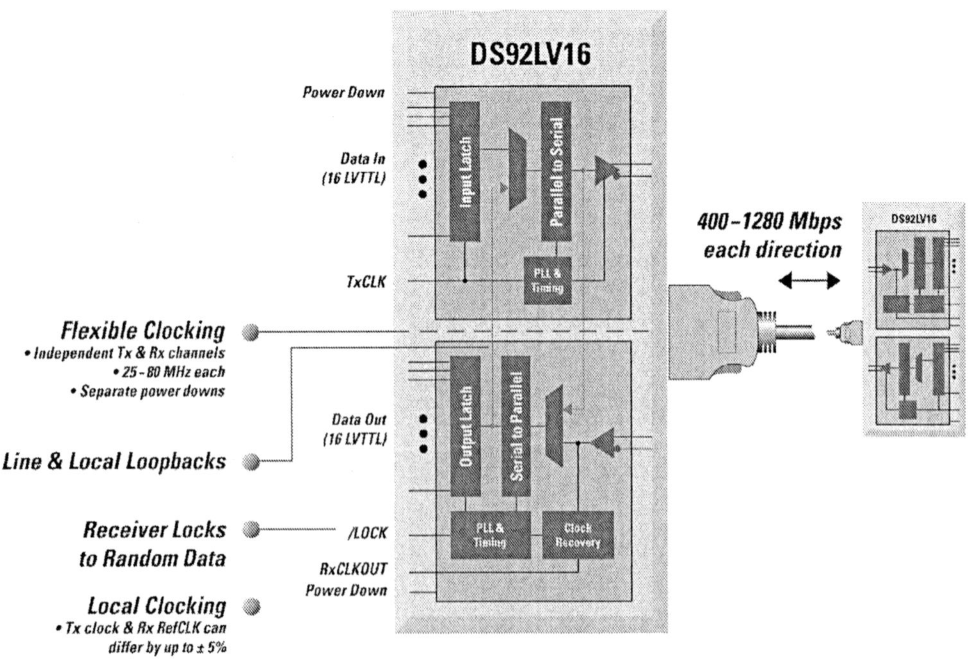

Figure 26.1 Block diagram of the DS92LV16, a 16-bit BLVDS SerDes chip with full-duplex throughput of 2.56 Gbps.

when transmitting or receiving a single channel of serialized data. When the device is operated in full-duplex mode where data is simultaneously transmitted and received, its total throughput is 2.56 Gbps.

Both the transmit clock and receiver reference clock have high jitter tolerance, allowing the use of low-cost clock sources. The serializer and deserializer sections are fully independent and can be operated at different frequencies. This is useful when upstream and downstream rates are not balanced. In addition, the serializer and deserializer blocks can be powered down independently if only one transmission direction is needed. The receiver locks to random data, eliminating the need to interrupt normal traffic with PLL (phase-locked loop) training patterns after hot plug events. The usual loss-of-lock feedback path from receiver to transmitter is also not required. Line and local loopback test modes allow the designer to segregate portions of the system to facilitate system diagnostics.

26.2 Bus LVDS SerDes Architecture

The Bus LVDS SerDes architecture was developed in conjunction with major telecommunications customers who desired features not provided by common 8B/10B datacom SerDes chipsets. Instead of scrambling each byte into a new 10-bit code as with 8B/10B coding, the Bus LVDS SerDes coding scheme frames each data word with two embedded clock bits, one high and one low, as shown in Figure 26.2. These embedded clock bits create a rising edge transition which provides precise timing and framing information to the receiver on every clock cycle. (See Figure 26.3.)

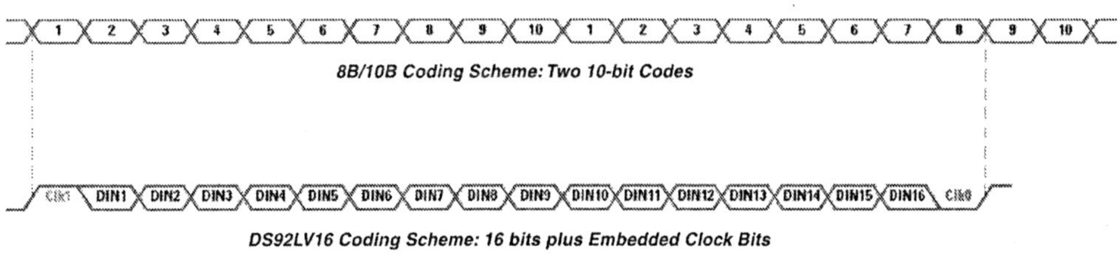

Figure 26.2 DS92LV16 embedded clock coding scheme showing 16 data bits plus clock 0, 1 bits.

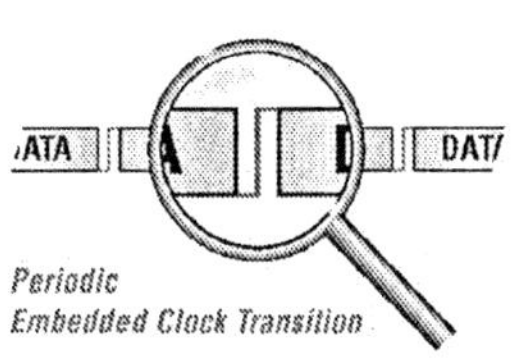

Figure 26.3 Low-to-high transition of embedded clock bits.

The resulting benefits of the Bus LVDS SerDes architecture are highlighted in Figure 26.4 and include the following:

- Lock to random data for true hot plug capability — lock can be achieved without interrupting traffic with PLL training patterns and without a loss-of-lock feedback path from receiver to transmitter.
- The relaxed ±5% RefCLK frequency specification (±5% tolerance relative to the transmit clock TCLK) means the receiver does not require a precise reference clock source to recover data reliably.
- Jitter-tolerant TCLK transmit clock input eliminates the need to use a high-grade telecom/datacom transmit clock source and also simplifies external loopback operations.
- Non-byte-oriented bus widths such as 10 bits are available.

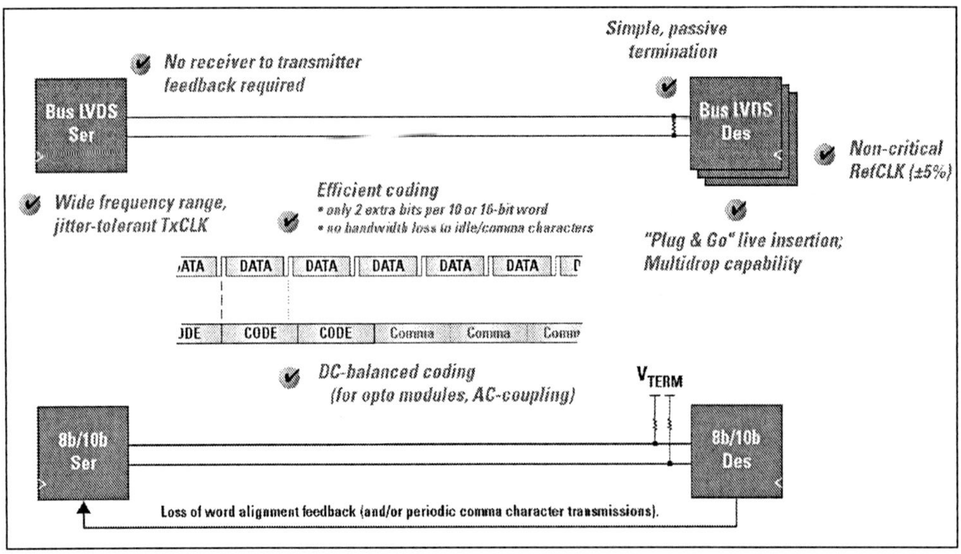

Figure 26.4 Comparison of BLVDS SerDes chipset and 8B/10B datacom SerDes chipset technologies.

26.3 Bus Topologies/Applications

The DS92LV16 is designed mainly for point-to-point applications, but you may also use it for multidrop interconnects under certain circumstances which are described below. Figure 26.5 illustrates several bus topologies, with their performance and applicability described in Table 26.1. In implementing a design, the differential characteristic impedances of cables and backplane traces and termination resistors should be matched to within ±10% for less than a 5% incident wave reflection. Stub lengths should be minimized, and connector pin assignments and signal routing should be done in a manner to ensure that matched differential traces of a pair have identical lengths. This is illustrated in Figure 26.6.

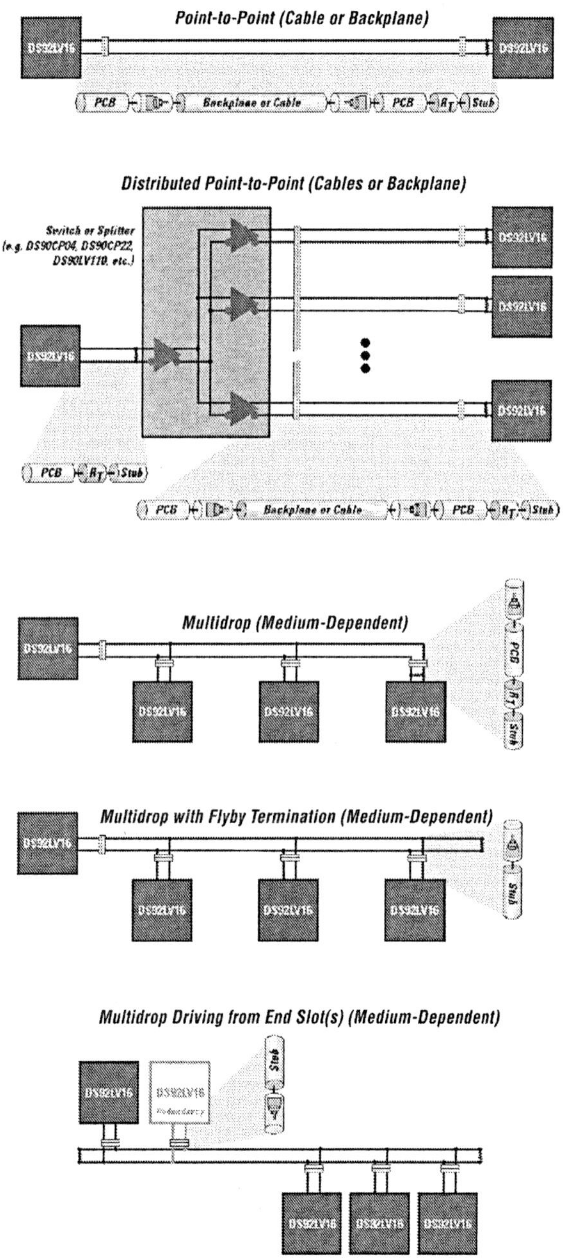

Figure 26.5 Illustration of several different bus topologies (from top to bottom): point-to-point, distributed point-to-point, multidrop, multidrop with flyby termination, and multidrop driving from the end slot.

Table 26.1 Typical Bus Performance Using the DS92LV16

Backplane Topology	Configuration	Performance	DS92LV16
Point-to-Point	1 Tx → 1 Rx	Chipset Maximum	Recommended
Distributed Point-to-Point	1 Tx → Switch → Multiple Rx's	Chipset Maximum	Recommended
Limited Multidrop	1 Tx → 2–4 Rx's	Up to 500–600 Mbps	Limited Application
Multidrop	1 Tx → 10–20 Rx's	Up to 300–400 Mbps	Not Recommended
Multipoint	10–20 Transceivers	Up to 200–300 Mbps	Not Recommended

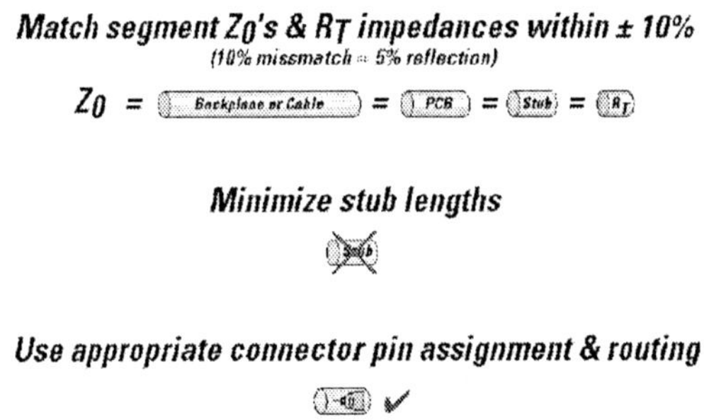

Figure 26.6 Rules to follow when routing signals: match impedances, keep stub lengths short, and keep differential traces balanced.

26.3.1 Point-to-Point

The DS92LV16 can deliver a full 1.28 Gbps payload in each direction over point-to-point backplanes and cables (2.56 Gbps when simultaneously serializing and deserializing in both directions). The distance depends on cable quality, transmission speed, and other factors. This performance is also possible when driving multiple receivers through a low jitter LVDS-to-LVDS switch/distribution chip in a distributed point-to-point configuration. In all point-to-point applications, you must place a termination resistor close to each receiver input.

26.3.2 Multidrop

The DS92LV16 may be used in limited multidrop (also known as point-to-multipoint) backplane applications driving a few receiver loads. The output structure is a modified Bus LVDS-type circuit that regulates the V_{OD} (output differential voltage) over a narrow range of loads. The driver expects to see a 100 Ω load, but may also operate with loads as low as 50 Ω with reduced voltage swing. The number of receivers allowed depends upon the bandwidth of the interconnecting path, the spacing of the nodes, and also the length of the resulting stubs. The DS92LV16's fast Bus LVDS output edge rates are optimized for the high throughput of the part, thus stub length must be minimized mechanically or electrically using "stub hider" buffer devices.

26.4 Backplanes

26.4.1 Point-to-Point

The DS92LV16 achieves the maximum 1.28 Gbps performance in controlled 100 Ω differential impedance point-to-point or distributed point-to-point FR-4 backplanes, though other backplane materials such as GETEK, Rogers, or hybrids may also be employed if desired. The backplane should be designed with sufficient bandwidth to support the desired data rate. For example, for 1 Gbps (500 MHz), a media bandwidth of 1.5 GHz is typically required.

26.4.2 Multidrop

In some instances, the DS92LV16 may be used in multidrop backplanes driving a 2–4 receiver load. Actual performance depends on the bandwidth of the interconnect. Pay close attention to minimizing stub lengths (connector plus trace) to less than 2 cm. Depending on data rate, "stub hider" chips should be used to reduce stub lengths seen by the signal on the backplane (see Figure 26.7). Receiver loads should be widely spaced to reduce the transmission segment to stub length ratio and/or placed far away from the transmitter to allow the edge rate to be slowed by the interconnect before reaching the first stub. Due to fast edge rates, the DS92LV16 is typically not recommended for driving more than a few multidrop loads. If the DS92LV16 were to be used with more than a few multidrop loads, there would be considerably more signal reflections due to the larger number of stubs and loads. The fast edge rates of this device would generate transmission line effects at each of these. These in turn would close down the eye opening of the signal at the receiver requiring a slower bus frequency. This device is optimized for high data throughput and when operated at much slower frequencies, could generate additional ringing and distortion of the signal.

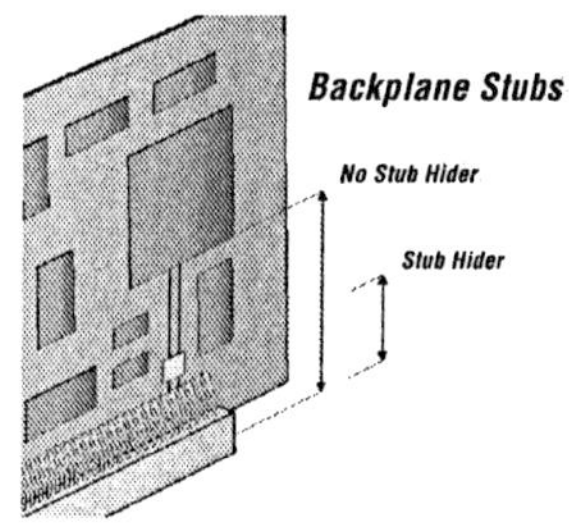

Figure 26.7 Use of stub hider chips can reduce stub lengths seen by backplane signals.

26.4.3 Termination

Termination is required and should be chosen to match the loaded differential impedance of the transmission line. In a point-to-point backplane, the termination value is typically 100 Ω. If a smaller output signal swing is desired, the line may be terminated at both the transmitter output as well as the receiver input, reducing the effective load seen by the transmitter by half. In a multidrop backplane, the added capacitive loading of the extra receivers lowers the differential impedance, requiring a lower value termination resistor. Again, the termination resistor(s) value chosen should match the loaded backplane impedance. Note that loads less than 50 Ω reduce the output signal swing below ±350 mV, compromising noise margins as shown in Figure 26.8. The typical capacitance of the DS92LV16 serial Bus LVDS inputs and outputs is about 5 pF.

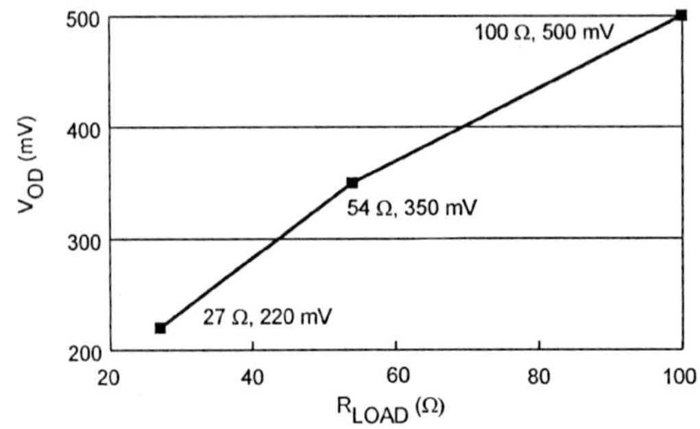

Figure 26.8 Typical V_{OD} vs. differential impedance.

26.5 PCB Recommendations

The following are PCB design recommendations, some of which are depicted in Figure 26.9:

- Use at least 4 PCB board layers (Bus LVDS signals, ground, power, and TTL signals).
- Minimize any resulting stub lengths. In a point-to-point bus, the resulting stub is typically the distance between the termination resistor and the receiver input pins of the DS92LV16. Place the small surface-mount resistor as close to the receiver input pins as possible. In a limited multidrop bus, the termination is typically at the far end of the bus only. The resulting stubs are from the main line to the receiver inputs and these distances should be minimized.

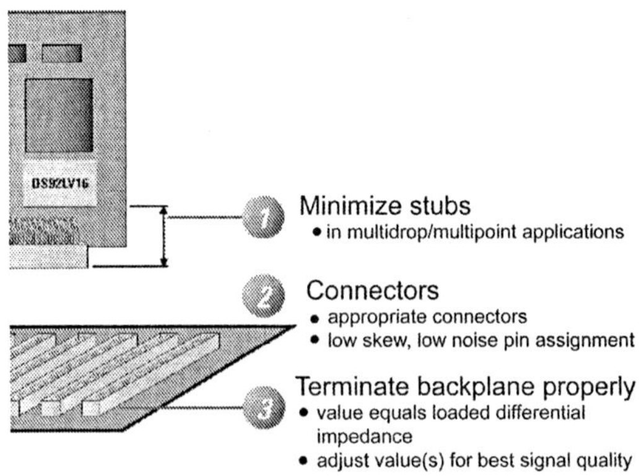

Figure 26.9 Backplane with plug-in card.

- Segment impedance matching — in a backplane point-to-point bus, the location of the serializer and deserializer is less critical. More important is matching the impedance of all transmission line segments. These impedances should be matched within 10% of the nominal value.

- Controlled impedance differential traces of 100 Ω are typically recommended for Bus LVDS signals. Edge-coupled microstrip, edge-coupled stripline, or broadside-coupled stripline differential traces may be used as shown in Figure 26.10. These traces should be closely-coupled (i.e., "S" in the figure should be minimized) to ensure coupled noise will appear as common-mode and be rejected by the receiver. This has the added benefit that closely coupled lines that are excited with odd-mode transmission tend to radiate less electromagnetic energy.

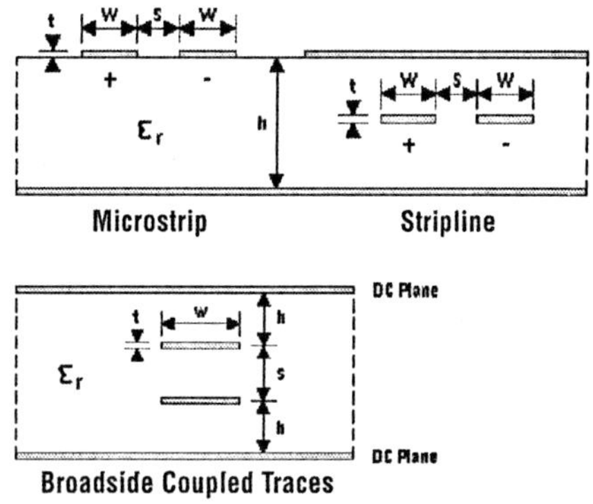

Figure 26.10 Geometries and parameters for microstrip, edge-coupled stripline, and broadside-coupled stripline differential traces.

- Treat the LVDS pair as one signal, avoiding influences that cause imbalances within the pair. Minimize skew within the pair. Maintain balance.
- Leave unused Bus LVDS receiver inputs open (floating) — the internal failsafe in the DS92LV16 will pull the input to a valid state.
- Termination of the Bus LVDS signals is required. The termination resistor value should match the differential impedance of the transmission line. 100 Ω is a typical value for point-to-point applications. It is better to err with too large a termination resistor (for a positive incident wave reflection) than too small.
- Isolate TTL and CMOS signals from Bus LVDS signals, placing them at least "3S" or "2W" away — whichever is larger. This will help to prevent them from coupling onto the LVDS lines.

There are just a few common practices that should be followed when designing PCBs for LVDS signaling. Designing with LVDS is actually quite easy if you remember that LVDS is a high speed, differential signaling technology. LVDS pairs should be considered transmission lines and any traces over a few centimeters should have controlled impedance. LVDS termination is always required. Each LVDS pair should be treated as one signal and both wires of each pair should see the same electrical path. Some good and bad trace routing practices are shown in Figure 26.11.

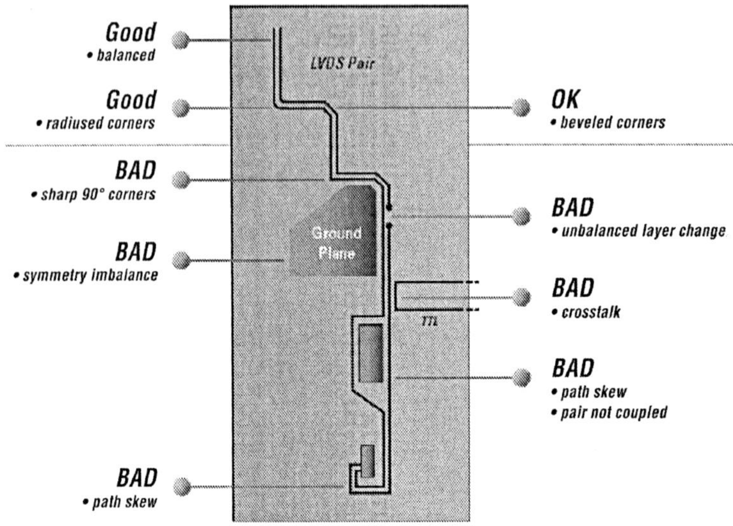

Figure 26.11 Some good and some bad trace routing practices.

26.6 Cables and Connectors

26.6.1 Cables

The DS92LV16 can be used over a wide variety of balanced cables depending on distance and signal quality requirements. In general, twinax or twisted-pair cables are recommended.

- Typical connections can be made using category 5 (CAT5) twisted pair ethernet cable with RJ-45 connectors. See Figure 26.12 for typical data rates versus cable distance, and Figure 26.13 for cable assemblies.
- Higher speed links: AMP Z –PACK® cables or other twinax cables can be used.
- Higher pincount: SCSI-type cables and connectors or LVDS-type cable assemblies such as the 3M™ MDR and Amphenol SKEWCLEAR® systems are commonly used.
- Very short distances (less than 0.3 meter): flex circuit (following correct PCB layout) or unbalanced ribbon cable may even be used.

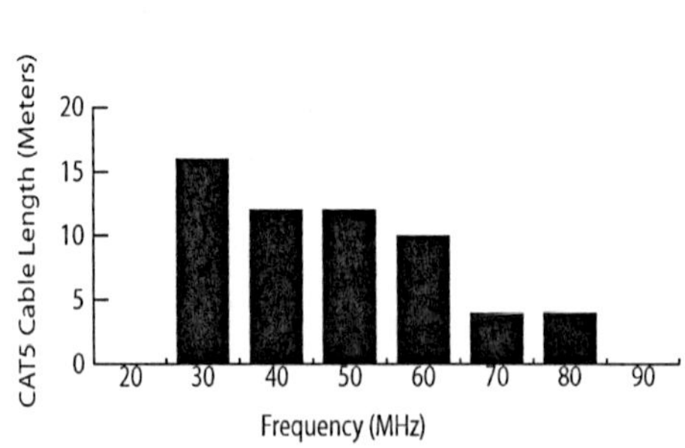

Figure 26.12 Typical CAT5 ethernet cable length versus frequency.

If multiple Bus LVDS pairs are run in the same cable assembly, use individually shielded pairs to minimize crosstalk between the pairs. At least one ground conductor should be used in the cable (connecting to the DS92LV16 analog grounds (AGNDs)) to provide a known, low-impedance return path for common-mode currents.

A termination resistor, placed as close as possible to the receiver inputs in point-to-point applications, is required. The resistor value should be chosen to match the differential impedance of the cable.

26.6.2 Connectors

Connectors with differential conductors offer the highest performance, but are not necessary in typical, lower speed or short distance applications. For single-ended connectors, keep LVDS pins away from other signals, particularly TTL/CMOS/LVTTL/LVCMOS signals. The "+" and "–" signals of a pair should be routed on the same row in multi-row connectors to help minimize skew. For high-speed applications (greater than 800 Mbps), using differential-optimized connectors is recommended.

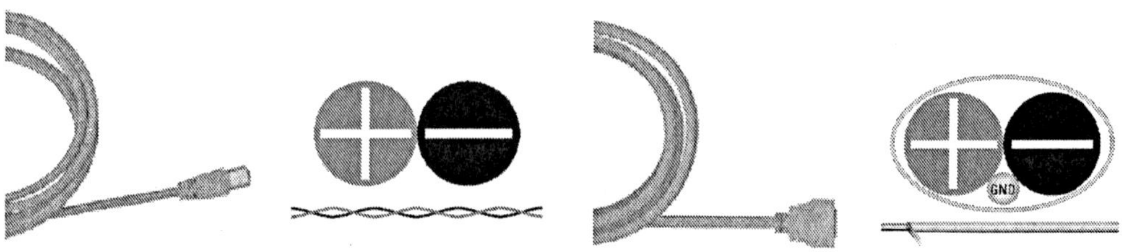

CAT5 ethernet cable with RJ-45 connectors. *Twinax-type cable assembly.*

Figure 26.13 CAT5 Ethernet cable and twinax-type cable can be used for slower speed LVDS applications.

26.7 Power and Ground

26.7.1 General Recommendations

A solid power/ground system is the foundation on which a reliable interconnect system is built. Design circuit board layout and stackup for the system to provide noise-free power to the device. Good layout practice will separate high-frequency and high-voltage-level inputs and outputs to minimize unwanted stray noise pickup, feedback, and interference. Power system performance may be greatly improved by using thin dielectrics (4 to 10 mils) for power/ground sandwiches. This increases the intrinsic capacitance of the PCB power system, which improves power supply filtering — especially at high frequencies — making the value and placement of external bypass capacitors less critical. External bypass capacitors should include both RF[2] ceramic and tanta-

2. RF is radio frequency.

lum electrolytic types. Use RF capacitors in the range of 0.001 μF to 0.1 μF. Use tantalum capacitors in the range of 2.2 μF to 10 μF. The voltage rating of tantalum capacitors should be at least three (five preferred) times the power supply voltage being used.

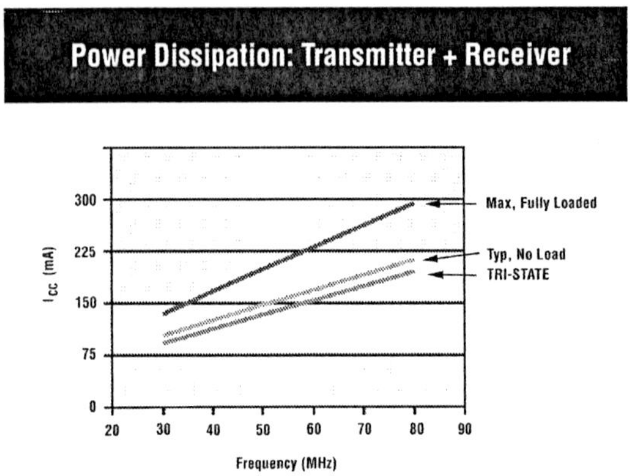

Figure 26.14 Supply current (mA) plotted versus frequency (MHz) for a DS92LV16 transmitter and receiver pair.

It is recommended practice to use two vias at each power/ground pin as well as all RF bypass capacitor terminals. Dual vias reduce the interconnect inductance by up to half, thereby extending the effective range of bypass components. Locate RF capacitors as close as possible to the supply pins and use wide, low-impedance traces, not 50 Ω traces. Surface-mount capacitors are recommended due to lower parasitics. When using multiple capacitors per supply pin, locate the smallest value closest to the supply pin. A bulk capacitor is recommended at the point of power entry. This is typically in the range of 50 μF to 100 μF and will smooth low-frequency noise.

Some devices have separate power/ground pins for different portions of the circuit blocks. Connecting these to separate power/ground planes on the PCB is typically not required. Pin description tables typically describe which blocks are connected to which power/ground pins. In some cases, an external filter may be used to provide clean power to sensitive circuits such as PLLs.

26.7.2 DS92LV16 Bypassing Recommendations

General device-specific bypassing recommendations are given below and are illustrated in Figure 26.15. Actual best practice depends on other board and system-level criteria including board density, power rail voltage and power supply type, and the supply needs of other integrated circuits on the board.

26.7.2.1 PVDD/PGND PLL Supply

The PVDD pins (the DS92LV16 has 2 PVDD pins) supply the PLL circuits. PLLs require a clean supply — less than 100 mV noise peak-to-peak — for the minimization of jitter. PVDD noise in the frequency range 200 kHz to 5 MHz can increase jitter and reduce noise margins. Certain power supplies may have switching frequencies or high harmonic content in this range. If this is the case, filtering of this noise spectrum may be required. A notch filter response is best to provide stable VDD, suppression of the noise band, and good high-frequency response (clock fundamental). This may be accomplished with a CRC or CLC pie filter. If employed, a separate pie filter is recommended for each PLL to minimize the voltage drop due to series resistance. Separate board power planes for each PVDD are typically not required.

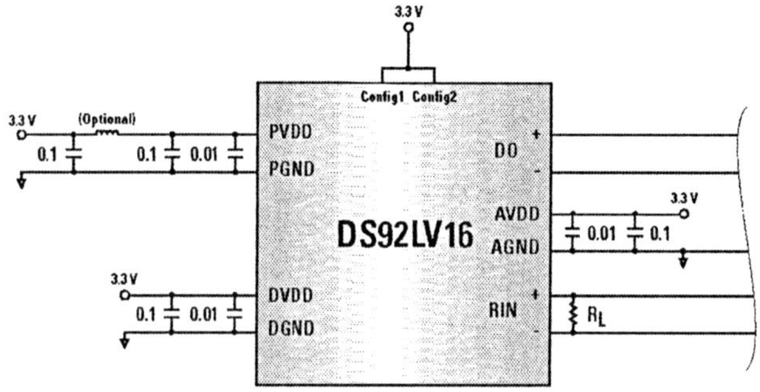

Figure 26.15 Typical connection diagram showing power supply bypassing. Employ sufficient bypassing to ensure less than 100 mV peak-to-peak noise on supply pins, especially PVDD. Each DVDD pin should have separate bypass capacitance.

26.7.2.2 AVDD/AGND LVDS Supply

The AVDD pins (the DS92LV16 has 4 AVDD pins) supply the LVDS circuits. Due to the nature of the DS92LV16 design, large currents are not drawn through these pins and a 0.1 μF capacitor is sufficient for these pins. If space is available, a 0.01 μF capacitor may be used in parallel with the 0.1 μF capacitor for additional high-frequency filtering. Connect the AGND ground to the cable ground to provide a return path for any common-mode (even) currents. Most of the LVDS current is odd-mode and returns within the pair, though a small amount of even-mode current may be present due to coupled noise. The cable ground should offer a low-impedance path to AGND.

26.7.2.3 Comparing Power Consumption

There is no industry-standard method for specifying power consumption in SerDes datasheets. Power consumption varies significantly with output load, pattern, clock frequency, and the number of channels tested. Even so-called "worst-case" power consumption conditions vary greatly from vendor to vendor. For instance, many vendors do not include the receiver parallel output bus load current.

The DS92LV16 maximum datasheet power consumption is specified at worst-case voltage (3.6 V), temperature, and pattern (checkerboard, i.e., alternating LVTTL ones and zeros) with 15 pF load on all CMOS outputs and both transmit and receive channels running at maximum clock frequency (80 MHz). In the datasheet, the total supply current I_{CCT} (including the load current) is 225 mA typical and 320 mA maximum. At the worst-case voltage of 3.6 V, the maximum power consumption is 1.152 watts.

26.7.2.4 DVDD/DGND Digital Supply

The DVDD pins supply the digital portion of the device and also the receiver output drivers. The receiver DVDD is more critical than the transmitter DVDD because it must power the receiver outputs during multiple output switching conditions. Four DS92LV16 receiver outputs are powered by each DVDD pin, suggesting that a minimum local bypass capacitance of 28 nF is required. This is calculated by taking 4 times the maximum short circuit current (4 × 85 mA = 340 mA), multiplying by the output rise time (4 ns), and then dividing by the maximum allowed VDD droop (assume 50 mV), which yields 27.2 nF.[3] Rounding up, a standard 0.1 μF is selected for each DVDD pin.

26.7.3 Power-Up Sequencing

The DS92LV16 does not require a specific power-up sequence. Best practice is: power up all VDD pins together, apply clocks (RefCLK & TCLK), and then bring /PWDN power-down pins high to enable the transmit and/or receive channels.

The chip has a power-on reset (POR) circuitry which holds the outputs in TRI-STATE® until the supply is above approximately 2.5 V. Since all VDD pins (PVDD, AVDD, and DVDD) are related internally, they must be powered up and down together. Employ sufficient bypassing to ensure less than 100 mV peak-to-peak noise on supply pins, especially PVDD. Each DVDD pin should have separate bypass capacitance.

26.8 Clocking

26.8.1 Transmit Clock (TCLK)

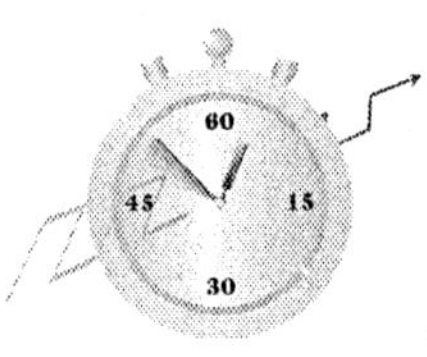

The transmit clock (TCLK) rising edge is used by the transmit block to latch in the DIN[0:15] parallel bus and run the transmitter's PLL. The TCLK input is LVTTL/LVCMOS compatible.

The TCLK clock input tolerates jitter of 80 ps (maximum root-mean-square, RMS) allowing the use of non-telecom/datacom-grade clock sources. The receiver output clock RCLK can also be fed back to TCLK and data retransmitted as an external loopback or repeater, though only two "hops" total are recommended before filtering out accumulated clock jitter.

Figure 26.16 shows the DS92LV16 transmitter's maximum TCLK input jitter tolerance versus jitter frequency for three different TCLK frequencies, 38.8 MHz, 52 MHz, and 78.8 MHz. The vertical axis shows the maximum allowed magnitude of jitter in picoseconds, that is, the amount of root-mean-square (RMS) variation of the bit end around the ideal bit position. It is measured in time (ps). The horizontal axis is the jitter frequency, measured in hertz, and is the rate of frequency at which the bit end is oscillating around the ideal bit position. From the figure, you can see that the least amount of jitter is tolerated around the frequency range of 200 KHz to

3. Current $i = C\,(dV/dt) \approx C\,(\Delta V/\Delta t)$, or $C \approx i\,(\Delta t/\Delta V)$. Multiply result by 4 for extra safety margin.

2 MHz where all three curves make a major dip. Excessive jitter, especially in this range of 200 kHz to 2 MHz, can appear on the serial outputs reducing link noise margins. If TCLK is interrupted, the chipset will go through the resynchronization procedure upon reapplication of valid TCLK.

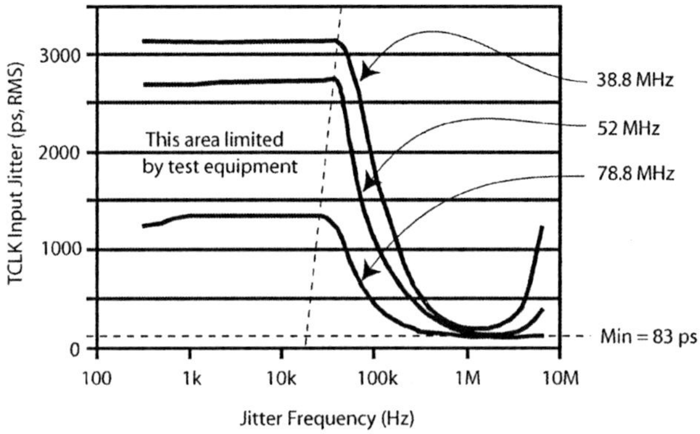

Figure 26.16 Maximum TCLK input jitter tolerance vs. jitter frequency under typical conditions.

26.8.2 Receiver Reference Clock (RefCLK)

The receiver reference clock is an LVTTL/LVCMOS compatible input and is used by the receiver to establish receiver PLL lock and prevent locking to a false harmonic. Therefore, the RefCLK characteristics are not very critical and the RefCLK frequency needs to be only within ±5% of the TCLK frequency of the transmitter from which the chip is receiving data. Unless a loopback test is being performed, this RefCLK frequency need not be the same frequency as the chip's own TCLK signal since the transmit and receive blocks are fully independent (i.e., upstream and downstream data rates are independent).

The relaxed RefCLK specifications mean that the RefCLK signal can be generated locally using a common, low-cost clock source. In fact, the RefCLK signal could even be removed once the receiver is locked (though this is not recommended).

26.8.3 Receiver Output Clock (RCLK)

The receiver output clock is an LVCMOS output whose rising edge is aligned to the middle of the receiver output data ROUT[0:15]. The subsequent ASIC/FPGA device should latch in ROUT[0:15] data meeting the ASIC's/FPGA's input setup and hold times. Output jitter for RCLK is not specified on the datasheet since it depends on the jitter of the incoming data and the TCLK clock source.

26.8.3.1 Clock Jitter

The transmitter will pass a certain amount of jitter contained on the TCLK clock to the serial data. This can affect noise margins. Similarly, some of the jitter on the clock 0,1 rising edge will be passed by the receiver to the RCLK output. Although the DS92LV16 is very tolerant to jitter, minimizing cycle-to-cycle jitter in the range of 200 kHz to 2 MHz on both the TCLK (DVDD digital) power bus and the PLL power bus will maximize noise margins.

26.9 Inputs and Outputs

26.9.1 Unused LVTTL Inputs

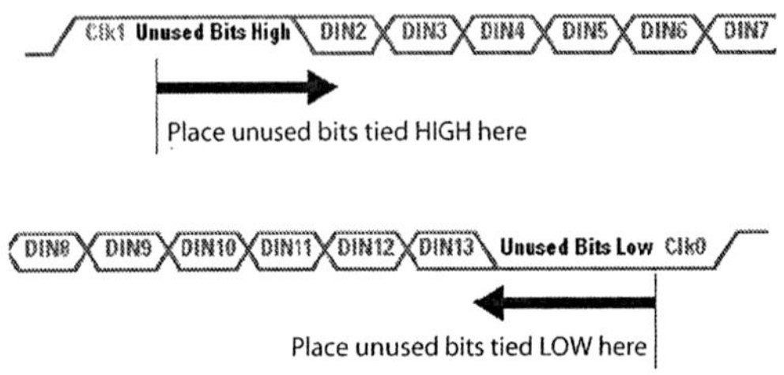

Figure 26.17 Unused least-significant bits (LSBs) may be tied high and unused most-significant bits (MSBs) may be tied low.

Unused LVTTL inputs may be left floating or tied high or low. The device datasheet indicates the presence of pull-down devices to bias unused pins. These internal impedances tend to be in the 200 kΩ range and may be overridden with lower value pull-up resistors if desired. If not using the full 16 bits of the payload, careful attention needs to be taken to avoid disrupting the random lock feature. These unused bits will still be transmitted, so in order for the receiver to maintain random data lock capability, care must be taken not to create a situation where the unused bits cause a repetitive multiple low-to-high transitions (RMTs). Avoid this by tying bits high at the LSB inputs (starting with DIN0) or by tying bits low at the MSB inputs (inputs ending with DIN16) as shown in Figure 26.17.

26.9.2 Floating Bus LVDS Receiver Inputs and Failsafe

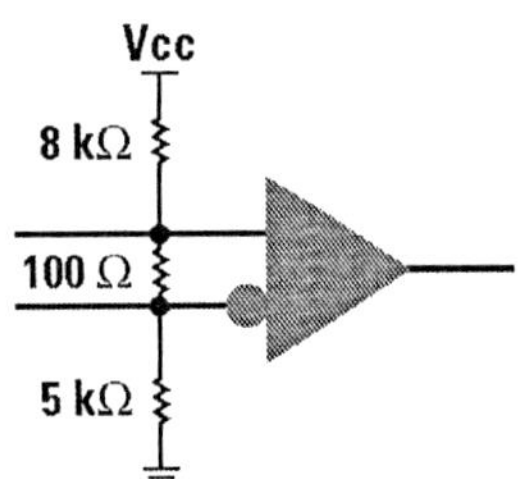

Figure 26.18 An example of external failsafe biasing of the DS92LV16's receiver.

In the event that the DS92LV16 receiver is disconnected from the backplane/cable, the internal failsafe circuitry is designed to reject a certain amount of differential noise (about 10 mV) from being interpreted as data or clock. This seems like a very small threshold, but balanced, closely coupled LVDS lines tend to pick up noise as common mode, not differential. The CMOS ROUT[0:15] outputs of the DS92LV16 deserializer will be TRI-STATE when the incoming serial data is removed and the receiver loses lock. Figure 26.18 shows how additional failsafe biasing can be implemented externally at the expense of two additional resistors.

26.9.3 Receiver CMOS Output Drive

The receiver CMOS outputs are specified at ±9 mA. This is typically sufficient to drive 2–3 LVTTL/LVCMOS loads. If more loads will be driven, especially at higher clock speeds, a logic buffer is recommended. Note that, depending on actual configuration (number of loads, stub lengths, segment distances, and so on), the receiver output bus may need to be treated as a transmission line and proper LVTTL/LVCMOS termination techniques employed.

26.10 Evaluating the DS92LV16

26.10.1 Evaluation Board

The DS92LV16 evaluation kit provides the ability to evaluate the DS92LV16 over AMP Z-PACK cables (1 meter cable included) or any user-supplied interconnect (modified to connect to the boards).

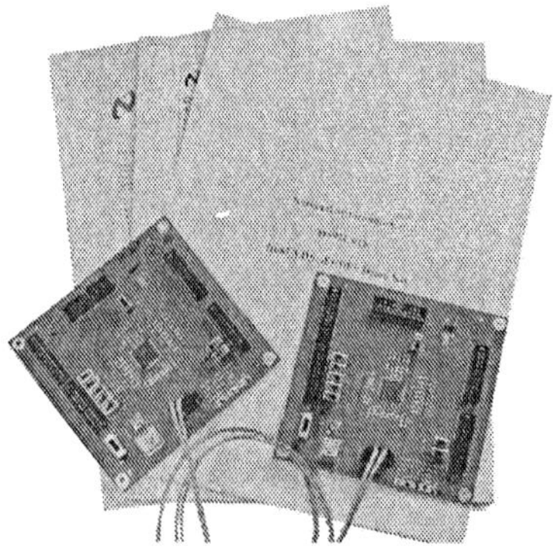

Figure 26.19 National's DS92LV16 evaluation kit.

26.10.2 Probing Bus LVDS Signals

LVDS signals are high-speed, low-swing signals. Improper probing can result in deceiving results since the probe and/or scope can filter high-speed components of the signal. Using an oscilloscope with a bandwidth greater than 1 GHz (such as the Agilent 86100 or Tektronix 694C) and a high-speed differential probe (such as the Tektronix P6247/8 or P6330) is highly recommended. LVDS drivers are not compatible with 50 Ω probes. Figure 26.20 shows the Tektronix P6247 1 GHz differential probe with the detail of the differential probe contacts. Figure 26.21 shows the PRBS (pseudo-random bit sequence) pattern at the serializer output. Note the 16 data bits plus the two clock bits. The corresponding sync pattern output is also

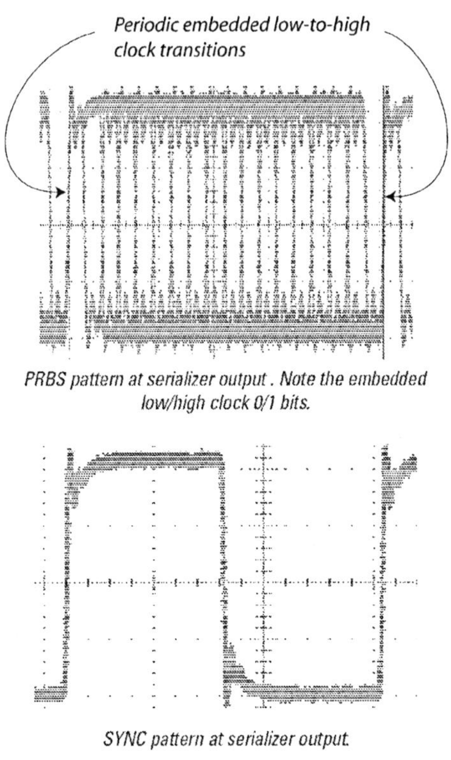

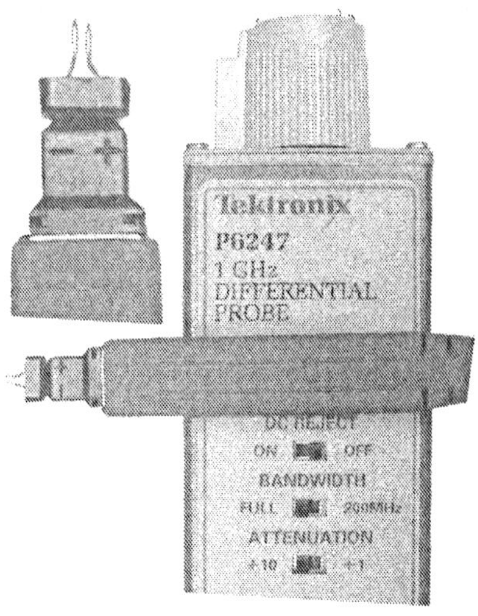

Figure 26.20 Tektronix P6247 1 GHz differential probe (above). Note two probe contacts and polarity indication in the probe tip detail.

Figure 26.21 PRBS and SYNC patterns at serializer output (left).

shown. In the PRBS serializer output pattern, the low-to-high clock transitions are identified by the two arrows and are the transitions from the low voltage level of the clock stop bit (clock0) to the high voltage level of the clock1 start bit.

26.11 Loopback Testing

Two testing modes allow the designer to isolate and test selected portions of the system. These modes are the (1) local loopback mode and (2) the line loopback mode. The local loopback mode is shown in Figure 26.22(a) and the line loopback mode is shown in Figure 26.22(b).

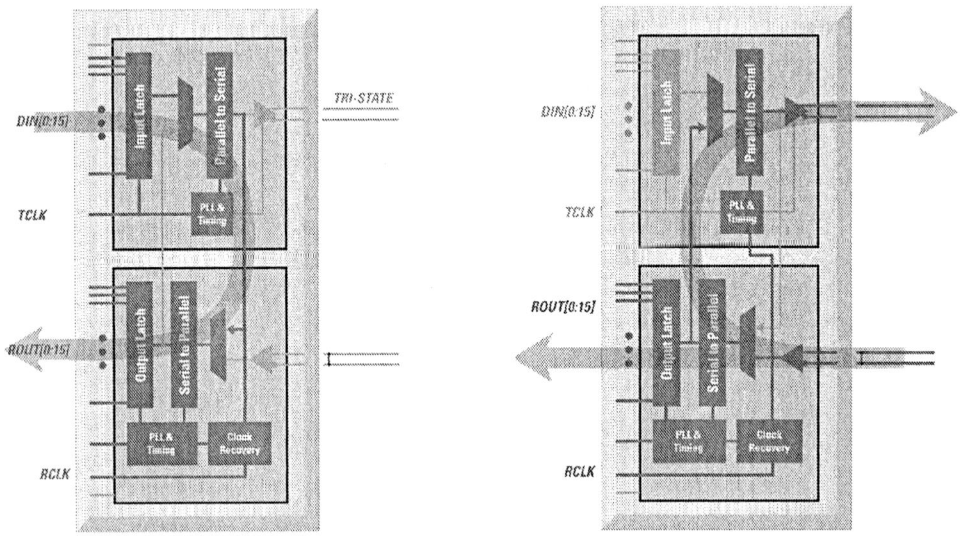

(a) (b)

Figure 26.22 (a) Local loopback mode and (b) line loopback mode.

26.11.1 Local Loopback

Asserting the LOCAL_LE pin (local loopback enable pin) high connects the parallel LVTTL data bus DIN[0:15] internally to the receiver output ROUT[0:15]. This mode includes all the functional blocks of the SerDes pair except the Bus LVDS input and output structures (the transmitter Bus LVDS outputs are in TRI-STATE). Switching the DS92LV16 to local loopback mode, therefore, means the receiver will relock to the loopback data and RefCLK must be within ±5% of the local TCLK frequency. The local loopback mode allows the local card to verify it is sending, receiving, and processing data properly.

26.11.2 Line Loopback

The DS92LV16 will enter the line loopback mode when the LINE_LE pin (line loopback enable pin) is brought high. This connects the Bus LVDS transmitter output (DO) pins internally to the receiver input (RIN) pins and to the parallel LVTTL data bus ROUT[0:15]. This loopback test mode includes the serializer/deserializer blocks. The line loopback mode allows the system to verify that the board-to-board links are operating properly.

Due to the relaxed clocking requirements of the DS92LV16, external line loopback within an FPGA or ASIC can also easily be performed to test portions of the system logic in addition to the DS92LV16 and interconnect. This is shown in Figure 26.23.

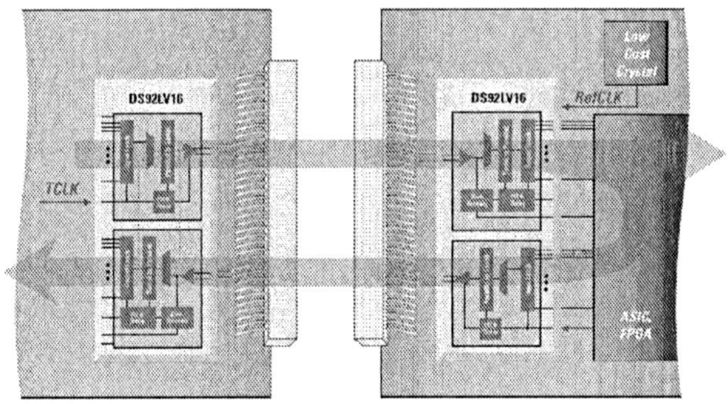

Figure 26.23 External line loopback within an FPGA or ASIC can also be performed to test portions of the system logic in addition to the DS92LV16 and interconnect.

26.12 Lock to Random Data vs. SYNC Patterns

26.12.1 SYNC Patterns

SYNC patterns, "H1111111100000000L" where "H" is the clock1 bit (high) and "L" is the clock0 bit (low), are sent automatically by the transmitter when the SYNC pin is held high. The receiver will lock to this pattern within a guaranteed time (see t_{DSR} specified in the datasheet). When the deserializer's /LOCK pin goes high, the receiver has locked to the incoming data stream and valid data appears at the receiver outputs.

26.12.2 Lock to Random Data

The receiver will also lock to random data without external system intervention, providing a powerful "plug and go" system capability not provided by most SerDes chipsets. In order to achieve this, the receiver looks for a consistent low-high transition caused by the clock 0,1 bits. If there are two data bits consistently "stuck" low-high over multiple clock cycles (known as a repetitive multiple transition or RMT), the receiver will not lock until these bits change and the receiver recognizes the unique embedded clock low-to-high transition. This is illustrated in Figure 26.24.

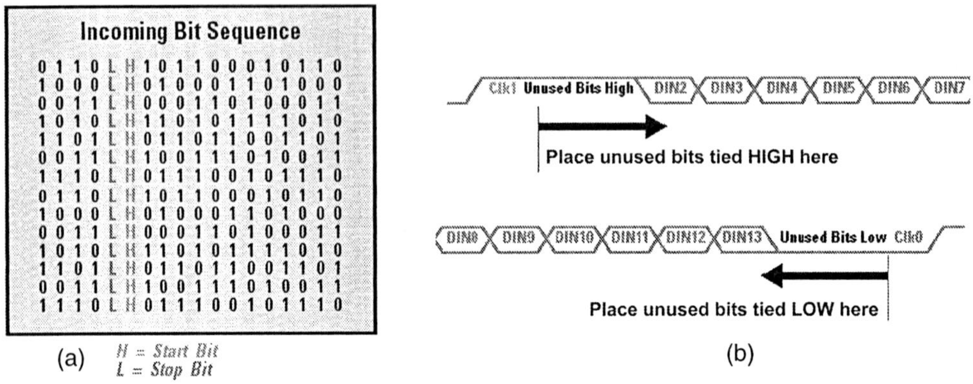

Figure 26.24 Incoming data steam arranged in 18-bit words. (a) Showing no RMTs (receiver will acquire lock) and (b) with RMTs. RMTs must disappear before the receiver can acquire lock.

26.12.3 Sending SYNC Patterns vs. Lock to Random Data

Sending SYNC patterns has the advantage that the receiver lock time is guaranteed to be less than t_{DSR} max. This can be useful in some systems and the sending of SYNC patterns can be achieved by controlling the SYNC pin through system logic or by feeding back the /LOCK pin to the SYNC pin. The receiver will lock to random data without system intervention, though the lock time will depend on the data pattern. This feature is useful in systems in which there is no quick or easy feedback path from receiver to transmitter or in systems that can't afford to interrupt data flow by sending special training patterns. Both lock modes, sending SYNC patterns and the locking to random data, are viable synchronization methods. Either may be employed and in fact, some systems may choose to employ both methods.

26.12.4 Once Lock Is Achieved

After the receiver is locked, it will stay locked regardless of data pattern: fixed (with or without RMTs), random, periodic, SYNC patterns, or a mix of these patterns. The receiver loses lock and enters the "acquire lock" state only after it detects two consecutive clock 0,1 bit errors. Once locked, the data between the clock bits do not affect the receiver's PLL.

26.13 Interconnect Jitter Margin

26.13.1 Interconnect Jitter Mask

When you look at data waveforms arriving at LVDS receiver inputs, the time interval for each data bit of the waveform can be thought of as having two associated factors, jitter and some spare jitter margin. Where jitter becomes so great that the margin disappears, the LVDS chipset (a serializer and deserializer) will no longer operate error free. Where margin does exist, it can be expressed as the amount of additional jitter that would be required to introduce bit reading errors at the receiver. Where jitter margin is expressed as a percentage, it is the percentage of the bit time interval taken up by the margin. Jitter is expressed in units of time, usually in picoseconds (ps), as is jitter margin (except when jitter margin is expressed as a percentage). The term "jitter budget" specifies the amount of allowed jitter assigned to the system designer that he or she should not exceed in order to meet the system design specification.[4]

Interconnect jitter margin also indicates the amount of interconnect loading that can occur on the link before bit errors occur. Jitter margin is determined from the ideal receiver noise margin (RNMI) t_{RNMI} specification and the received data waveform. Datasheets specify both left and right t_{RNMI}, indicating how much jitter the receiver can tolerate on either side of the ideal bit edge before bit errors occur. These two t_{RNMI} specifications also allow the construction of an eye mask (described in the next paragraph) to validate the application's signal quality. Jitter sources include:

- External noise including excessive power supply noise, crosstalk, etc.
- Interconnect effects such as inter-symbol interference (ISI), attenuation, etc.
- Clock jitter, i.e., excessive jitter on the serializer transmit clock TCLK.

Figure 26.25 shows a single bit interval of the ideal waveform appearing at the SerDes receiver's inputs. The bit interval is measured from the start edge (left-most black vertical line) to the stop edge (right-most black vertical line). The parameters $t_{RNMI\text{-left}}$, $t_{RNMI\text{-right}}$, $t_{DJIT\text{-left}}$,

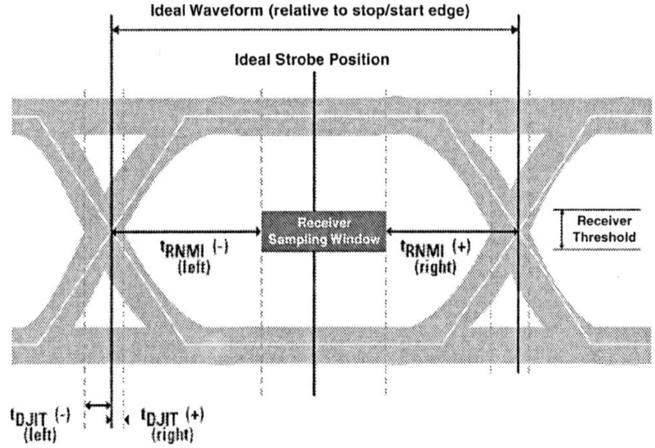

Figure 26.25 Ideal waveform appearing at the SerDes receiver's inputs. (This eye pattern is computer-generated for clarity.)

4. Some application notes use the terms "jitter budget" and "jitter margin" interchangeably. In this text, we make a distinction between the two terms, with jitter budget being the maximum allowed jitter listed in the design specification, and jitter margin as the amount of additional jitter that would be required to cause the system to fail.

and $t_{DJIT\text{-}right}$ are shown in the figure. These four parameters are specified in the device datasheet and, for the DS92LV16, are –230 ps (minimum value at a TCLK of 80 MHz), +230 ps (maximum value at 80 MHz), –75 ps (minimum value at 80 MHz), and +100 ps (maximum value at 80 MHz), respectively. Parameter t_{DJIT} (right or left) is the deterministic jitter described more fully below. The receiver thresholds shown in the right side of the figure indicate the receiver voltage input levels which, once reached, will cause the receiver output to change state. The shaded box labeled "Receiver Sampling Window" is the jitter mask. The ideal data strobe would occur centered on this sampling window or, at least, within the sampling window. The jitter "mask" is so called because it "masks" or blocks out a portion of the "eye" created by the waveform. Margins must be adequate enough to ensure that no portion of the waveform crosses or covers up the jitter mask. The purpose behind determining the jitter mask is to establish the amount of jitter margin that exists, that is, how close you are to having receiver bit errors.

Table 26.2 lists several common transmit clock (TCLK) frequencies and the corresponding calculated bit widths after serialization by an LVDS serializer. The ideal bit width is 1/18th of the TCLK period after serialization. For proper accuracy, calculations should be made to the nearest picosecond. You need to know precisely when the bit starts and stops so you can accurately determine the amount of margin or "eye opening" around the receiver jitter mask.

Table 26.2 Ideal Bit Widths (ns) Calculated at Common Frequencies After Serialization by an LVDS SerDes Device at Frequency TCLK (Transmit Clock)

MHz	ns	MHz	ns
30	1.852	50	1.111
30.72	1.808	52	1.068
33	1.684	61.44	0.904
35	1.587	62.5	0.889
38.8	1.432	66	0.842
39.44	1.409	78.88	0.704
40	1.389	80	0.694

26.13.2 Validating Signal Quality

In conjunction with bit error rate test (BERT) measurements, eye pattern measurements are a good way to validate the robustness of a high-speed link. While BER tests generally produce a pass/fail result, viewing the eye pattern at the receiver inputs gives a more qualitative indication of how much margin a given link has. The amount of eye opening around the receiver input jitter mask is a qualitative measure of the link's noise margin.

26.13.3 Steps to Construct a Jitter Mask

The receiver input jitter mask is constructed by taking the following steps:

1. Make sure the link is operating and sending the desired pattern (normally a PRBS pattern).
2. Capture the entire eye pattern at the receiver inputs using a high-bandwidth oscilloscope in variable persistence mode with a high-speed (>1 GHz), high-impedance (>100 kΩ) differential probe. Using variable persistence mode is important, as you won't get an eye pattern without it. The TCLK signal should be used as the trigger. Ideal bit edge positions are always multiples of T/18 away from the ideal clock0 – clock1 low-to-high transition (where the period of TCLK is T = 1/TCLK). The receiver jitter margin for any bit on the eye pattern can be calculated based on its ideal bit edge positions. The serial eye pattern needs to be measured at the receiver inputs after the interconnect's loading effects (backplane and/or cable) have occurred. A PRBS pattern or real system data pattern should be used to construct the eye pattern. When calculating ideal bit edge positions, do not round up or down to the nearest 10 picoseconds. Instead, calculate to the nearest picosecond.

    Figure 26.26 shows the full eye pattern captured by the scope set to variable persistence mode. Note the 16 data bits positioned between the start and stop bits for a total of 18 bits. For purposes of this discussion, we'll refer to the 16 data bits as a "data word" and the 18 bits consisting of the data word plus start and stop bits as a "data frame" of serialized data.

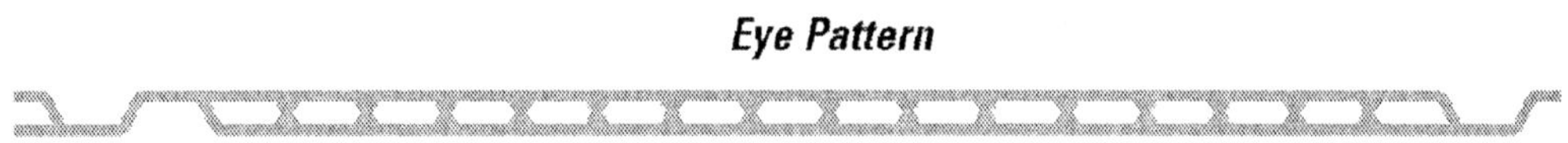

Figure 26.26 Full eye pattern captured by the scope set to variable persistence mode.

3. Use the vertical cursor function of the scope to place a vertical cursor over the average 0 V differential crossing point of the clock0 bit – clock1 bit low-to-high transition. The 0 V differential crossing point can be thought of as being about halfway between the top and bottom edges of the eye pattern waveform at the offset voltage V_{OS} (typically 1.2 V referenced to ground). This cursor position is now the ideal time point to which all other positions will be referenced. The vertical cursor position is used to determine the exact start and stop positions of each bit interval. Figure 26.27 shows a complete ideal data frame of 18 bits where each bit interval is 1/18 of the TCLK period T where T is 1/TCLK. The start clock bit is clock1 and is labeled as C1 in the figure and the stop clock bit is clock0 and is labeled as C0. These are typically referred to as the start and stop bits C1 and C0, respectively. Figure 26.28 shows the data eye pattern aligned to the ideal waveform. The receiver jitter margin for any bit of the eye pattern can be measured based on its ideal calculated bit edge positions.

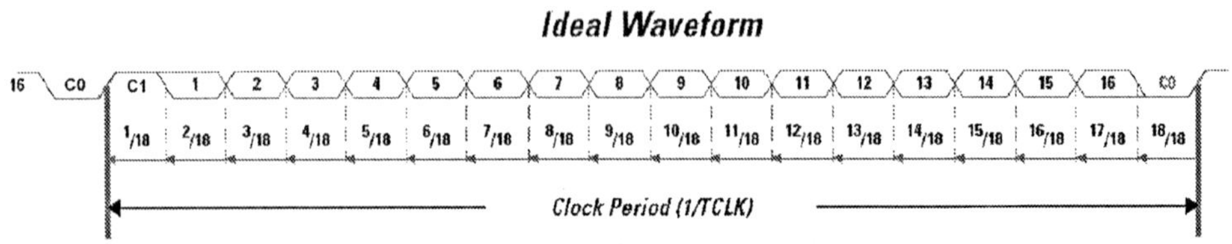

Figure 26.27 Ideal bit edge positions are always multiples of T/18 away from the ideal clock0 – clock1 low-to-high transition where T is the period of frequency TCLK (T = 1/TCLK).

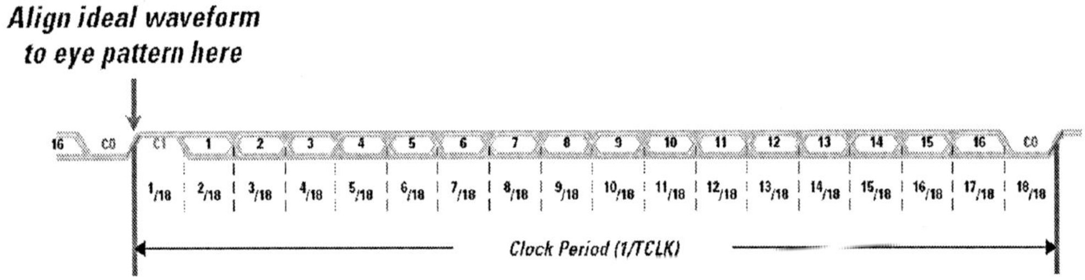

Figure 26.28 The receiver jitter margin for any bit on the eye pattern can be measured based on its ideal calculated bit edge positions.

4. Select the bit eye pattern in which to place the receiver input jitter mask. The worst-case bit, that is, the bit with the most “closed” eye, is recommended. The ideal crossing points on either side of the nth data bit will be n*T/18 and (n+1)*T/18 away from the ideal stop-start crossing point set in step 3, where T is period of the TCLK frequency (16 data bits plus 2 clock bits = 18 bits). Bits are numbered from 1 to 16 so that the first data bit is n = 1.
5. Calculate the ideal bit edges, being careful to avoid rounding errors, and place a cursor at both ideal bit edges of the selected bit.
6. Add the absolute value of $t_{RNMI\text{–}left}$ ($t_{RNMI\text{–}left}$ is usually a negative number) to the ideal left bit edge and subtract $t_{RNMI\text{–}right}$ from the ideal right bit edge to get the positions of the receiver input jitter mask. The values of $t_{RNMI\text{–}left}$ and $t_{RNMI\text{–}right}$ were stated earlier and are obtained from the device’s datasheet.
7. The height of the jitter mask is the receiver’s differential input threshold $V_{TH} - V_{TL}$. V_{TH} is the differential input high voltage threshold and V_{TL} is the differential input low-voltage threshold. Both are found in the device’s datasheet. $V_{TH} - V_{TL}$ is typically ±50 mV (100 mV total), though the datasheet value of ±100 mV (200 mV total) can be used if a more conservative estimate is desired. Look back at Figure 26.25 on page 753 to see the receiver thresholds.
8. The receiver input jitter mask should now look like a box inside the eye of the bit as shown in Figure 26.29. The eye opening (shown in the middle shade of gray in the fig-

ure) around the box is the relative noise margin of the link. As long as the captured eye opening stays out of this noise margin "box" (the jitter mask), the chipset should operate error-free.

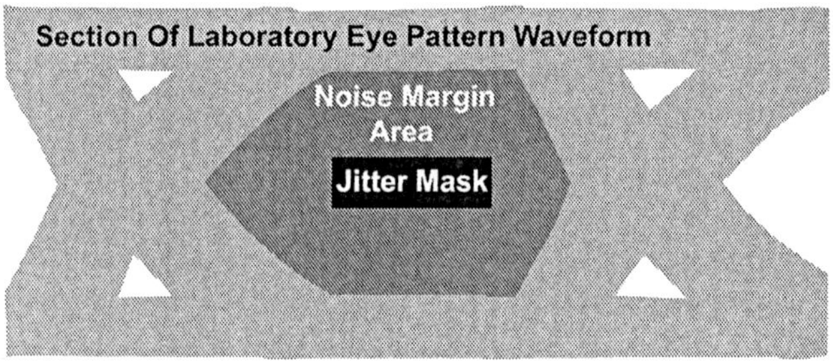

Figure 26.29 Laboratory eye pattern waveform (light gray), noise margin area (middle gray), and jitter mask (receiver sampling window, black) for a transmitter-receiver link.

The amount of jitter margin can be observed visually as the amount of eye opening around the jitter mask. The jitter margin can also be specified quantitatively as a time value in picoseconds (ps) or as a percentage. For example, referring back to Table 26.2 on page 754, we see that for a TCLK of 80 MHz, the serialized ideal data bit interval will be 0.694 ns (694 ps) in width. If the jitter mask can be shifted left or right by 140 ps before it bumps into the eye diagram (laboratory waveform), we would say that the jitter margin is 140 ps. Expressed as a percentage, the jitter margin would be (140 ps / 694 ps) × 100% or 20.2%.

The above eight steps to create a jitter mask are summarized below in Table 26.3:

Table 26.3 Summary of Steps to Construct a Jitter Mask

Step 1:	Get the transmitter-receiver link up and running.
Step 2:	Capture an eye pattern at the receiver inputs on your scope set to variable persistence mode.
Step 3:	Mark the start of a data frame in the eye pattern by placing a vertical scope cursor over a C0 to C1 clock transition at the average 0 V differential crossing point.
Step 4:	Identify the worst-case data or clock bit, the bit with the most "closed" eye.
Step 5:	Calculate the ideal bit edges of the worst-case bit and place cursors at both of these ideal bit edges by measuring over from the cursor at the start of the data frame.
Step 6:	Determine the left and right borders of the jitter mask (receiver sampling window) by adding the absolute value of $t_{RNMI-left}$ to the ideal left bit edge and subtracting $t_{RNMI-right}$ from the ideal right bit edge.
Step 7:	Determine the upper and lower borders of the jitter mask (V_{TH} and V_{TL} referenced to the zero-crossing point). The jitter mask is now fully determined.
Step 8:	Determine the amount of "eye opening" around the jitter mask. This is the jitter margin. Express jitter margin in ps or as a percentage of the bit width.

26.13.4 Alternative Jitter Estimates Using Only the Device Datasheet

Another approach to estimating the amount of jitter margin is to identify the amount of jitter margin that a transmitter-receiver chipset has before inclusion in any particular design. The available margin associated with only the chipset can then be allocated to, or split among, the planned interconnect and final margin. The planned interconnect would be the cable connection or backplane traces connecting the transmitter and receiver. Such an interconnect introduces signal delay and distortion thereby increasing the jitter. The final jitter margin is the margin the system designer would like to see exist as a safety buffer once the full link is designed and operating.

The chipset jitter margin can be determined solely from the device datasheets and needs no laboratory measurements. The $t_{RNMI\text{–}left/right}$ and $t_{DJIT\text{–}right/left}$ specs are used in these calculations. The quantity $(t_{RNMI}-t_{DJIT})/t_{RNMI}$ times 100% of the worst-case side (see Figure 26.30) is the percentage jitter margin left before bit errors are expected to occur.[5]

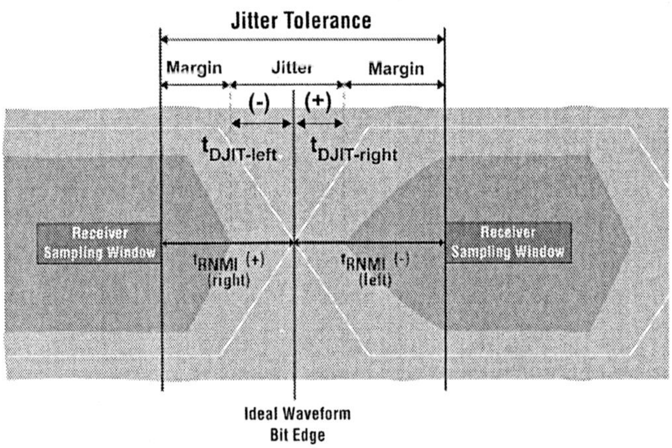

Figure 26.30 The t_{RNMI} specs can also be used around the ideal bit position to gauge the percentage of jitter margin left before bit errors could occur.

The total amount of jitter margin available for the interconnect (i.e., how much jitter margin is allowed for the cable or backplane) and final margin is calculated similarly by subtracting maximum transmitter deterministic jitter t_{DJIT} from receiver ideal noise margin t_{RNMI}:

Jitter margin left side: $-t_{RNMI\text{–}left} - t_{DJIT\text{–}right}$ {$t_{RNMI\text{–}left}$ is a negative number}

Jitter margin right side: $t_{RNMI\text{–}right} + t_{DJIT\text{–}left}$ {$t_{DJIT\text{–}left}$ is a negative number}

5. In the percentage equation, only the absolute values (magnitudes) of t_{RNMI} and t_{DJIT} should be used.

The smaller value from the above two calculations is the more conservative value and should be used for the remaining jitter margin available for the interconnect and final margin. The advantage of these calculations is that a jitter estimate which excludes the interconnect can be made directly from datasheet parameters without any laboratory data measurements. This remaining jitter margin can then be allocated to system-specific noise sources or distortions caused by system transmission media such as cables or backplanes.

26.14 Troubleshooting

The DS92LV16 is a robust (high-quality and reliable), easy-to-use SerDes device and application issues should be rare. If application problems do occur, the causes can normally be traced to a few simple, easy-to-check areas listed in order relevance and depicted in Figure 26.31:

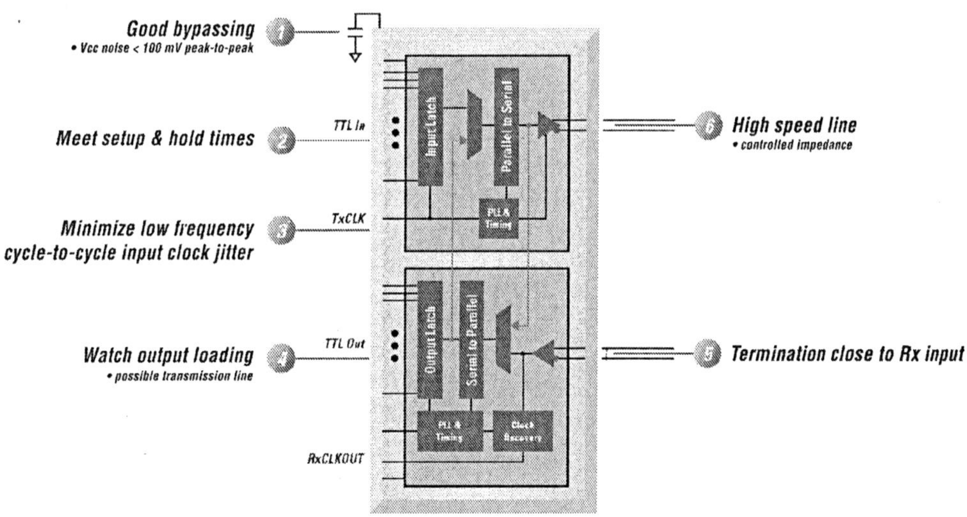

Figure 26.31 Areas to check when troubleshooting a SerDes device.

- Power supply noise: Excessive supply noise, especially on the PLL supply (PVCC), can add jitter to the transmitter serial output and affect the receiver data sampling. Keep supply noise under 100 mV peak-to-peak on the PVCC pins.
- Transmit clock: Minimize excessive cycle-to-cycle jitter on the transmit clock TCLK in the range of 200 kHz to 2 MHz which can add jitter to the transmitter serial output.
- Serial Bus: Should follow LVDS PCB layout and backplane recommendations, using proper termination, avoiding long stubs, and ensuring that the termination resistor is close to the receiver input.
- Parallel bus: The parallel LVTTL signals should not violate any setup and hold times and should be free from excessive overshoot and undershoot. [Nat02a]

26.15 Quad 2.5 Gbps (10 Gbps) Serializer/Deserializer (SerDes)

26.15.1 General Description

An example of a high-throughput device is the DS25C400, which is diagramed in Figure 26.32. The DS25C400 is a four-channel serializer/deserializer (SerDes) for high-speed serial data transmission over controlled-impedance transmission media such as a printed circuit board backplane or twin-axial cable. It is capable of transmitting and receiving serial data of 2.125–2.5 Gbps in "high data rate mode" or 1.0625–1.25 Gbps in "low data rate mode," per channel. The low or high data rate mode is determined by the EN_HDR (high data rate enable) pin.

Each transmit section of the DS25C400 contains a low-jitter clock synthesizer, an 8-bit or 10-bit parallel-to-serial converter with built-in 8B/10B encoder, and a CML (current-mode logic) output driver with selectable pre-emphasis optimized for backplane applications. (Pre-emphasis is discussed in a couple of pages in Section 26.15.3 on page 761. Its receive section contains an input limiting amplifier with on-chip terminations and selectable equalization levels, a clock/data recovery PLL, a comma detector, and a serial-to-parallel converter with built-in 8B/10B decoder. The DS25C400 has built-in local loopback test mode, pseudo-random pattern generator, and error detector to support self-testing.

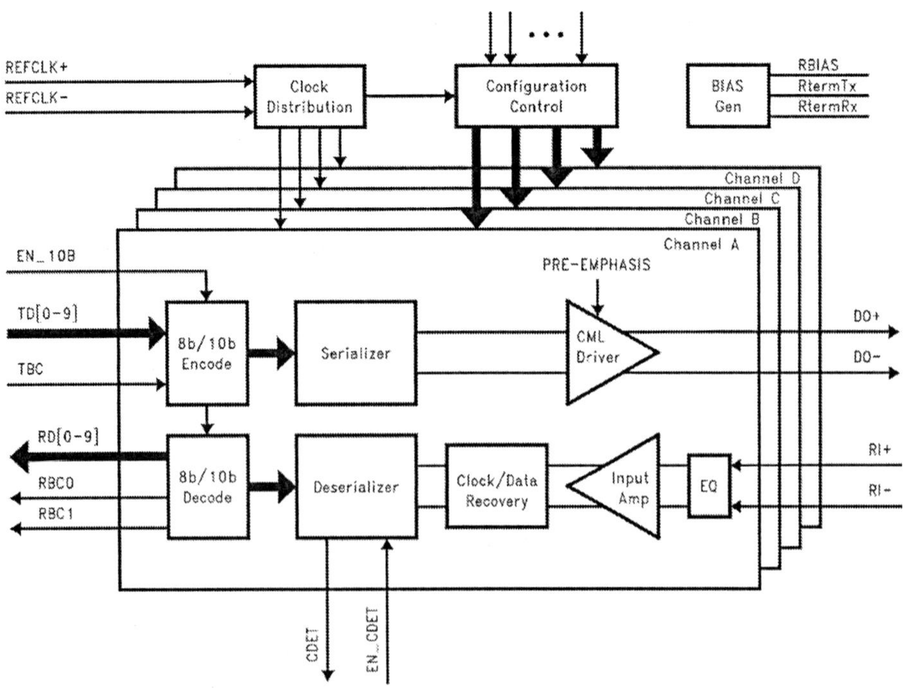

Figure 26.32 General functional diagram of the DS25C400 quad 2.5 Gbps serializer/deserializer.

The DS25C400 requires no external components for its clock synthesizers and clock recovery PLLs. Three external resistors are needed to set the proper bias currents and compensate for process variations to achieve tight tolerance on-chip terminations.

26.15.2 DS25C400 Features

A summary of DS25C400 features is as follows:

- Quad serializer/deserializer
- Data rate per channel: 2.125–2.5 Gbps in high data rate mode or 1.0625–1.25 Gbps in low data rate mode, modes selectable by a pin
- Supports 106.25–125 MHz differential reference input clock
- Low jitter clock synthesizers for clock distribution
- 8-bit or 10-bit parallel I/O Interface conforms to SSTL_18 Class 1 (also interfaces to 1.8 V HSTL or 1.8 V LVCMOS)
- On-chip 8B/10B encoder and decoder
- High-speed serial CML drivers
- High-speed serial CML on-chip terminations
- Selectable pre-emphasis and equalization
- On-chip comma detect for character alignment
- On-chip local loopback test mode
- On-chip pattern generator and error checker to support BIST (built-in self-test)
- Hot plug protection
- Low power, 420 mW (typical) per channel
- 324-ball TE-PBGA package
- Operating temperature –40°C to +85° C

26.15.3 Selectable Pre-Emphasis to Improve Signal Quality

"Pre-emphasis is magnitude boosting of high-frequency spectral components before launching the signal (wave) onto the transmission line. Transmission lines embedded in most standard PCB materials (FR4, Rogers 43xx, Nelco and Rogers) suffer varying degrees of dispersion and loss in the 1 GHz spectrum. This is mostly due to conductance losses (leakage from the copper trace to any other conducting structure) and skin effect. Dispersion is a phenomenon whereby spectral components travel at different velocities. The waveform looks smeared when it arrives at the receiver.

Both of these characteristics (dispersion and loss) play into a diminished and poorly received signal. By boosting the high-frequency spectral components, the composite signal arrives at the receiver looking the way it was intended. The magnitude of the high-frequency components is diminished as the wave travels through the transmission line, but since high-frequency components start out larger than lower frequency ones when pre-emphasis is used, they arrive with a much less distorted magnitude.

Pre-emphasis is done by simply increasing the maximum amplitude of the signal for one bit period. If the signal is 1 bit in duration, the amplitude is allowed to rise to a value which is some percentage greater in magnitude. At this point, if the signal is to stay at the same logic state, the driver sends a decreased magnitude signal, or nominal logic level. Every time a transition occurs, the greater magnitude level is used. For all times after this that the same level is to be transmitted, the nominal magnitude is used." [Xil03b]

Much of the jitter added by backplanes and cables is caused by inter-symbol interference (ISI) distortion. The DS25C400's configurable pre-emphasis compensates for ISI effects and is especially effective driving longer cables (up to 10–20 meters). To compensate against edge degradation due to bandwidth-limited transmission medium, the driver has three selectable steps of pre-emphasis, as shown in Table 26.4, providing additional current drive for one bit period following a data level transition. The pre-emphasis improves signal quality at the receiving end of the transmission medium and enhances error rate performance of the downstream receiver.

Table 26.4 Selectable Pre-Emphasis Currents by Pins PSEL1 and PSEL0

PSEL1	PSEL0	Descriptions
0	0	Pre-emphasis disabled. Drive current = 24 mA.
0	1	Pre-emphasis enabled. Drive current = 29.3 mA at first bit after data transition.
1	0	Pre-emphasis enabled. Drive current = 34.7 mA at first bit after data transition.
1	1	Pre-emphasis enabled. Drive current = 39.7 mA at first bit after data transition.

The serialized data bit stream is output at DO±, driven by a differential current-mode logic (CML) driver. Both DO+ and DO– are terminated with on-chip resistors to V_{DDHS} (2.5 V). The values of the internal termination resistors are tightly controlled to 50 Ω ± 10%. With an external load of 50 Ω to V_{DDHS} or ac coupled to GND, the driver provides single-ended voltage swing of 533 mV nominal, with a common-mode voltage of about (V_{DDHS} – 0.27 V).

The pre-emphasis current is only applied to the load termination for the first bit after a data transition. However, the pre-emphasis current increases the power supply dc current by the same amount. For power and thermal calculations, consideration must be paid to on-chip versus off-chip power dissipation.

26.15.4 Equalization Filtering at Receiver

The purpose of the equalization is to overcome ISI distortion occurring on limited bandwidth backplanes and connectors. The receiver front-end provides three steps of equalization filter (see Table 26.5) to improve the eye opening of the input data at the receiver inputs, RI±. The equalization filter is a first-order one, designed to equalize transmission loss and reduce ISI for long board traces in a backplane. [Nat02b]

Table 26.5 Three Steps of Equalization Filtering to Improve the Eye Opening of Receiver Input Data

EQ1	EQ0	Descriptions
0	0	Equalization disabled.
0	1	Equalization filter's zero location set at about 800 MHz.
1	0	Equalization filter's zero location set at about 500 MHz.
1	1	Equalization filter's zero location set at about 400 MHz.

26.16 Eight-Channel 10:1 Serializer for 5.28 Mbps

26.16.1 Serializer General Description

Another example of a high-throughput device is the DS92LV8028, which is diagramed in Figure 26.33. The DS92LV8028 integrates eight serializer devices into a single chip. The DS92LV8028 can simultaneously serialize up to eight 10-bit data streams. The 10-bit parallel inputs are LVTTL signal levels. At the maximum clock rate of 66 MHz, the throughput of the chip is (66 MHz)(8 channels)(10 bits/channel) or 5.28 Gbps. The serialized outputs are LVDS signals with extra drive current for point-to-point and lightly loaded multidrop applications. Each serializer block in the DS92LV8028 operates independently by using strobes from a single shared PLL. The DS92LV8028 uses a single +3.3 V power supply with a typical power dissipation of 740 mW.[6] Each serializer channel has a unique power down control to further conserve power consumption. For high-speed LVDS serial data transmission, line quality is essential, thus the DS92LV8028 includes an @SPEED TEST function. Each serializer channel has the ability to internally generate a PRBS data pattern. This pattern is received by specific deserializers (SCAN921224) that have the complement PRBS verification circuit. The deserializer checks the data pattern for bit errors and reports any errors on the test verification pins on the deserializer.

26.16.2 Serializer Features

The following is a summary of the features of the DS92LV8028 8-channel 10:1 serializer:

- All eight channels synchronous to one parallel clock rate, from 25–66 MHz
- Duplicates function of multiple DS92LV1021 and '1023 10-bit serializer devices
- Serializes from one to eight 10-bit parallel inputs into data streams with embedded clock
- Eight 5 mA modified Bus LVDS outputs capable of driving double terminations
- @Speed Test — PRBS generation to check LVDS transmission path to SCAN921224 or SCAN921260

6. Test conditions for this power consumption level: +3.3 V supply, PRBS data, and TCLK of 66 MHz.

- On-chip filtering for PLL
- 740 mW typical power dissipation (loaded, PRBS, 66 MHz, 3.3 V)
- High impedance inputs and outputs on power off
- Single power supply at +3.3 V (+/–10%)
- 196-pin LBGA package
- JTAG pins reserved for next version of device
- Industrial temperature range operation: –40 to +85°C

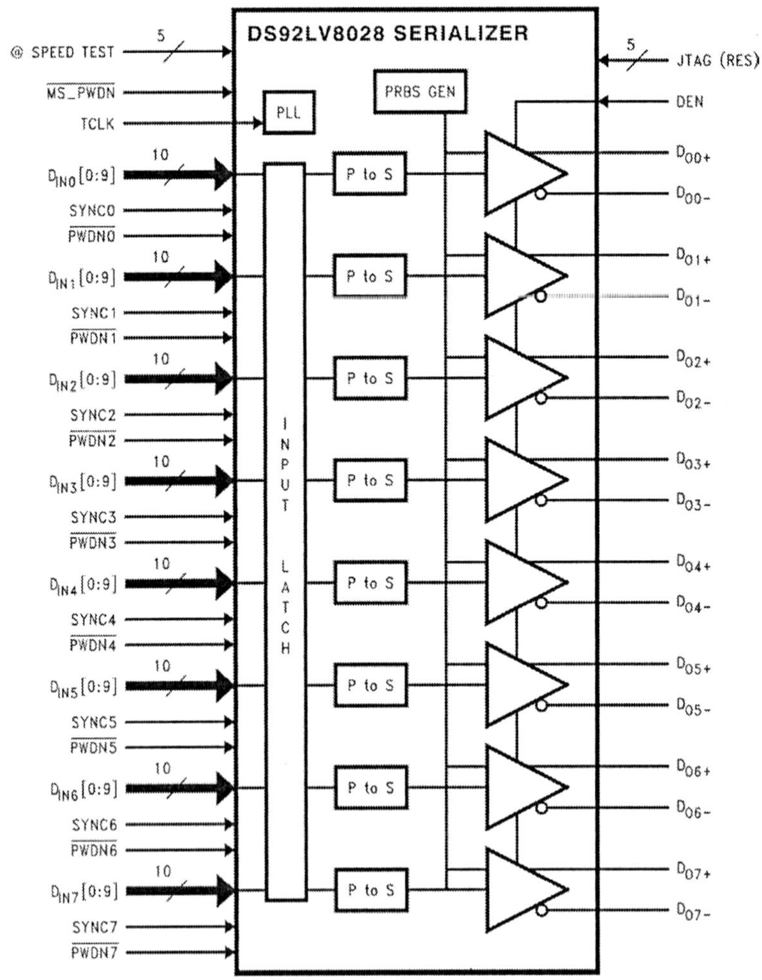

Figure 26.33 Block diagram of the DS92LV8028 8-channel 10:1 serializer with maximum data rate of 5.28 Gbps.

26.16.3 Serializer Functional Description

The DS92LV8028 combines eight 10:1 serializers into a single chip. Each of the eight serializers accepts 10 or less data bits. The serializers then multiplex the data into a serial stream with embedded clock bits and route to the LVDS output. The LVDS output is a 5 mA current loop driver. It provides enough drive for point-to-point and lightly loaded multidrop applications. The serialized data stream is compatible with the DS92LV1210, DS92LV1212A, DS92LV1224, and DS92LV1260 10-bit deserializers from National Semiconductor.

Each of the eight channels on the DS92LV8028 has its own serializer function but shares a single PLL. There is a single Transmit Clock (TCLK) for all eight channels. The data on all eight 10-bit interfaces is latched into the device with the rising edge of TCLK. Each of the serialized data streams is independent of the others and includes the embedded clock information. The skew between the serializer outputs is minimal.

There is a master power-down signal (MS_PWDN) to put the entire device into a low power consumption state. In addition, there is a power-down control signal for each of the eight channels. This allows the device to efficiently operate as one to eight 10-bit serializers. The @SPEED TEST signal initiates the sending of a random data pattern over the LVDS links. This allows for testing the links for bit error rates at the frequency they will be carrying data. In addition, JTAG boundary scan circuits will be added to the device at a later date. The JTAG signal pins are reserved on this version.

The DS92LV8028 has four operating modes. They are the Initialization, Data Transfer, Resynchronization, and @Speed Test states. In addition, there are two passive states: power-down and TRI-STATE. The following sections describe each operating mode and passive state.

26.16.3.1 Initialization

Before the '8028 serializes and transmits data, it and the receiving deserializer device(s) must initialize the link. Initialization refers to synchronizing the serializer's and the deserializer's PLLs to local clocks. The local clocks should be the same frequency, or within the specified range if from different sources. After all devices synchronize to local clocks, the deserializers synchronize to the serializers as the second and final initialization step.

Step 1: After applying power to the serializer, the outputs are held in TRI-STATE and the on-chip power-sequencing circuitry disables the internal circuits. When V_{CC} reaches V_{CC}OK (2.1 V), the PLL in the serializer begins locking to the local clock (TCLK). A local on-board data source or other source provides the specified clock input to the TCLK pin. After locking to TCLK, the serializer is now ready to send data or SYNC patterns, depending on the level of the SYNC input or a data stream at the data inputs. The SYNC pattern sent by the serializer consists of six ones and six zeros switching at the input clock rate.

Step 2: The deserializer PLL must synchronize to the serializer to complete the initialization. (Refer to the deserializer datasheet for operation details during this step of the Initialization State.) The deserializer identifies the rising clock edge in a synchronization pattern or non-repetitive data pattern. Depending on the data pattern that is being transmitted, the deserializer will synchronize to the data stream from the serializer after some delay. At the point where the deserializer's PLL locks to the embedded clock, the LOCK pin goes low and valid data appears on the output. The user's application determines control of the SYNC signal input. One recommendation is a direct feedback loop from the LOCK pin on the deserializer. The serializer stops sending SYNC patterns when the SYNC input returns to a low state.

26.16.3.2 Data Transfer

After initialization, the serializer accepts data from the inputs DINn0 to DINn9 where n is the channel number. The serializer uses the rising edge of the TCLK input to latch incoming data. If the SYNCn input is high for 4 TCLK cycles, the data on DINn0-DINn9 is ignored and SYNC pulses are transferred.

The serial data stream includes a start bit and stop bit appended by the serializer, which frame the 10 data bits. The start bit is always high and the stop bit is always low. The start and stop bits also function as clock bits embedded in the serial stream.

The serializer transmits the data and clock bits (10 + 2 bits) at 12 times the TCLK frequency. For example, if TCLK is 40 MHz, the serial rate is 40 × 12 = 480 Mbps. Since only 10 bits are from input data, the serial "payload" rate is 10 times the TCLK frequency. For instance, if TCLK = 40 MHz, the payload data rate is 40 × 10 = 400 Mbps. TCLK is provided by the data source and must be in the range 25 MHz to 66 MHz nominal.

The serializer outputs (DO0± to DO7±) can drive a point-to-point connection or lightly loaded multidrop connections. The outputs transmit data when the driver enable pin (DEN) is high, MS_PWDN and PWDNn are high, and SYNCn is low. When DEN is driven low, all the serializer output pins will enter TRI-STATE.

When any one of eight attached deserializer channels synchronizes to the input from the serializer, it drives its LOCK pin low and synchronously delivers valid data on the output. The deserializer locks to the embedded clock, uses it to generate multiple internal data strobes, and drives the embedded clock on the RCLK pin. The RCLK is synchronous to the data on the ROUT pins. While LOCK is low, data on ROUT is valid. Otherwise, ROUT is invalid.

26.16.3.3 Resynchronization

Whenever one of the connected DS92LV1212, '1212A, '1224, or '1260 deserializers loses lock, it will automatically try to resynchronize to the data stream from the serializer. If the data stream is not a repetitive pattern, then the deserializer will automatically lock.

For example, if the deserializer's received embedded clock edge is not detected two times in succession, the PLL loses lock and the LOCK pin is driven high. The '1212, '1212A, '1224, or '1260 deserializers will automatically begin searching for the embedded clock edge. If it is a random data pattern, the deserializer will lock to that stream. If the data pattern is repetitive, the deserializer's PLL will not lock in order to prevent the deserializer from locking to the data pattern rather than the clock. We refer to such patterns as repetitive-multiple-transition, RMT.

Therefore, if the data stream is not random data or the deserializer is the DS92LV1210, there needs to be a feedback path from the deserializer to the serializer. This feedback path can be as simple as connecting the deserializer's LOCK pin to the serializer's SYNC pin. This will automatically signal the serializers to send SYNC patterns whenever the deserializer loses lock.

The user has the choice of allowing the deserializer to resynchronize to the data stream, or to force synchronization by pulsing the serializer SYNC pin. This scheme is left up to user discretion.

26.16.3.4 @Speed Test

Since the high-speed LVDS serial data transmission line quality is essential to the chipset operation, a means of checking this signal integrity is built into the DS92LV8028 serializer. Each serializer channel has the ability to transfer an internally generated PRBS data pattern. This pattern traverses the transmission line to the deserializer. Specific deserializers (SCAN921224, for example) have the complement PRBS pattern verification circuit. The deserializer checks the data pattern for bit errors and reports any errors on the test verification pins on the deserializer.

The @Speed Test feature uses 5 signal pins. The BIST_SEL[0:3][7] and BIST_ACT pins together determine the functions of the BIST mode. The BIST_ACT signal activates the test feature. BIST_SEL[0:2] selects one of eight channels as the output for the BIST pattern. All channels perform BIST when BIST_ ACT = H and BIST_SEL[0:3] = 08H (8 hex). See Table 26.6. No BIST function is performed when BIST_SEL[0:3] are set from 9H (9 hex) to FH (F hex) even when BIST_ACT is set at high. BIST_SEL pins are pull-up internally.

The JTAG pins are reserved on this version of the serializer. They will be JTAG-compliant functionality on the next version. The @Speed Test will also be available through a JTAG command when available.

Table 26.6 Truth Table for BIST (Built-In Self-Test) Mode

BIST_ACT	BIST_SEL <3>	BIST_SEL <2>	BIST_SEL <1>	BIST_SEL <0>	MODE
H	L	L	L	L	BIST on channel 0
H	L	L	L	H	BIST on channel 1
H	L	L	H	L	BIST on channel 2
H	L	L	H	H	BIST on channel 3
H	L	H	L	L	BIST on channel 4
H	L	H	L	H	BIST on channel 5
H	L	H	H	L	BIST on channel 6
H	L	H	H	H	BIST on channel 7
H	H	L	L	L	BIST on all channels
L	X	X	X	X	No BIST
L	H	H	H	H	Default — No BIST

7. BIST is built-in self-test.

26.16.3.5 Power-Down

The power-down state is a low-power sleep mode that the serializer and deserializer typically occupy while waiting for initialization, or to reduce power when there are no pending data transfers. The DS92LV8028 serializers enter power-down when MS_PWDN is driven low. In power-down, the PLL stops and the outputs go into TRI-STATE. To exit power-down, the system drives MS_PWDN high.

Each of the serializers in the '8028 also has an individual power-down, PWDNn control pin. This control enables the deactivation of individual serializers while allowing others to operate normally. The benefit is that spare serializers can be allocated for backup operation, but not consuming power until employed for data transfers.

Upon exiting power-down, the serializer enters the initialization state. The system must then allow time to initialize before data transfer can begin.

26.16.3.6 TRI-STATE

When the system drives the DEN pin low, the serializer outputs enter TRI-STATE. This will TRI-STATE the output pins (DO0± to DO7±). When the system drives DEN high, the serializers will return to the previous state as long as all other control pins remain static (PWDNn, TCLK, SYNCn, and DINn[0:9]).

26.16.4 Serializer Application Information

26.16.4.1 Using the DS92LV8028

The DS92LV8028 is an easy-to-use serializer that combines eight 10:1 serializers into a single chip with a maximum payload of 5.28 Gbps. Each of the eight serializers accepts 10 or less data bits. The serializers then multiplex the data into a serial data stream with embedded clock bits and route to the LVDS output at up to 660 Mbps per channel. The LVDS output is a 5 mA current loop driver that can be used for point-to-point and lightly loaded multidrop applications. Each of the eight channels has their own serializer function but share a single Transmit Clock (TCLK) with a single PLL for the entire chip. The data on all eight channels is latched into the device with the rising edge of TCLK and the data stream is compatible with the DS92LV1210, DS92LV1212A, DS92LV1224, and DS92LV1260.[8]

If using less than 10 bits of data, it is recommended to tie off adjacent bits to the embedded clock bits to prevent causing an RMT in the data payload. For example, if only using 8 bits, tie D0 high and D9 low.

8. From National Semiconductor.

26.16.4.2 Power Considerations

The all-CMOS design of the serializer and deserializer makes them inherently low-power devices. Additionally, the constant current source nature of the LVDS outputs minimizes the slope of the speed versus I_{CC} curve of CMOS designs.

26.16.4.3 PCB Layout and Power System Considerations

Circuit board layout and stack-up for the BLVDS devices should be designed to provide low-noise power feed to the device. Good layout practice will also separate high-frequency or high-level inputs and outputs to minimize unwanted stray noise pickup, feedback, and interference. Power system performance may be greatly improved by using thin dielectrics (2 to 4 mils) for power/ground sandwiches. This arrangement provides plane capacitance for the PCB power system with low-inductance parasitic, especially proven effective at high frequencies above approximately 50 MHz, and makes the value and placement of external bypass capacitors less critical. External bypass capacitors should include both RF ceramic and tantalum electrolytic types. RF capacitors may use values in the range of 0.01 μF to 0.1 μF. Tantalum capacitors may be in the 2.2–10 μF range. The voltage rating of the tantalum capacitors should be at least five times the power supply voltage being used.

It is a recommended practice to use two vias at each power pin as well as at all RF bypass capacitor terminals. Dual vias reduce the interconnect inductance by up to half, thereby reducing interconnect inductance and extending the effective frequency range of the bypass components. Locate RF capacitors as close as possible to the supply pins, and use wide, low-impedance traces (not 50 Ω traces). Surface-mount capacitors are recommended due to their smaller parasitics. When using multiple capacitors per supply pin, locate the smaller value closer to the pin. A large bulk capacitor is recommend at the point of power entry. This is typically in the 50–100 μF range and will smooth low frequency switching noise. It is recommended to connect power and ground pins straight to the power and ground plane, with the bypass capacitors connected to the plane with via on both ends of the capacitor. Connecting a power or ground pin to an external bypass capacitor will increase the inductance of the path.

A small body size X7R chip capacitor, such as a 0603, is recommended for external bypass. Its small body size reduces the parasitic inductance of the capacitor. User must pay attention to the resonance frequency of these external bypass capacitors, usually in the range of 20–30 MHz range. To provide effective bypassing, very often, multiple capacitors are used to achieve low impedance between the supply rails over the frequency of interest. At high frequency, it is also a common practice to use two vias from power and ground pins to the planes, reducing the impedance at high frequency.

Pin description tables typically provide guidance on which circuit blocks are connected to which power pin pairs. In some cases, an external filter many be used to provide clean power to sensitive circuits such as PLLs.

Use at least a four-layer board with a power and ground plane. Locate CMOS (TTL) swings away from the LVDS lines to prevent coupling from the CMOS lines to the LVDS lines. Closely coupled differential lines of 100 Ω are typically recommended for LVDS interconnect.

The closely coupled lines help to ensure that coupled noise will appear as common-mode and thus is rejected by the receivers. Also, the tight coupling of the lines will radiate less.

26.16.4.4 Transmission Media

The DS92LV8028 serializers can be used in point-to-point configuration of a backplane across PCB traces or through cable interconnects. In point-to-point configurations, the transmission media needs only to be terminated at the receiver end. The DS92LV8028 may also be used with double terminations for a total load or 50 Ω for use in certain limited multidrop applications. Termination impedances lower than 50 Ω is not recommended.

26.16.4.5 Termination

Termination of the LVDS interconnect is required. For point-to-point applications, termination should be located at the load end. The nominal value is 100 Ω to match the line's differential impedance. Place the resistor as close to the receiver inputs as possible to minimize the resulting stub between the termination resistor and receiver.

26.16.4.6 DS92LV8028 BLVDS Serializer Bypass Recommendations

General device-specific guidance is given below. Exact guidance cannot be given as it is dictated by other board-level /system-level criteria. This includes the density of the board, power rails, power supply, and other integrated circuit power supply needs.

The DVDD pins supply the digital portion of the device. A 0.1 μF capacitor is sufficient for these pins.

The PVDD pin is the PLL (phase-locked loop) section power supply. The PVDD pin supplies the PLL circuit. The PLL(s) require clean power for the minimization of jitter. A supply noise frequency in the 300 kHz to 1 MHz range can cause increased output jitter. Certain power supplies may have switching frequencies or high harmonic content in this range. If this is the case, filtering of this noise spectrum may be required. A notch filter response is best to provide a stable VDD, suppression of the noise band, and good high-frequency response (clock fundamental). This may be accomplished with a pie filter (CRC or CLC). The pie filter should be located close to the PVDD power pin. Separate power planes for the PVDD pins are typically not required.

The AVDD pin supplies the LVDS portion of the circuit. The DS92LV8028 has nine AVDD pins. Due to the nature of the design, current draw is not excessive on these pins. A 0.1 μF capacitor is sufficient for these pins. If space is available, a 0.01 μF may be used in parallel with the 0.1 μF capacitor for additional high-frequency filtering.

26.16.4.7 Grounds

The AGND pin should be connected to the signal common in the cable for the return path of any common-mode current. Most of the LVDS current will be odd-mode and return within the interconnect pair. A small amount of current may be even-mode due to coupled noise, and driver imbalances. This current should return via a low-impedance, known path. Separate solid ground planes are recommended for DVDD, PVDD, and AVDD. Using a split plane may have a potential problem of ground loops, or difference in ground potential at various ground pins of the device. [Nat02c]

26.17 Exercises

26-1. Answer the following questions regarding the DS92LV16 BLVDS SerDes device: (a) What is the maximum data rate (exclude clock bits) at which it can transmit? (b) What is the maximum data rate at which it can receive? (c) If it is both transmitting and receiving simultaneously, what is its maximum full-duplex data rate? (d) By how much can the transmit and receive clocks of two connected DS92LV16s differ and still allow proper data transmission between them? (e) What is the range of frequencies over which the transmit clock and receive clock can properly operate? (f) How many bits of parallel data are converted into the serial bit stream? (g) What is the mathematical relationship among the number of bits in the parallel data, the transmit clock frequency, and the transmitted serial data rate? (h) If two embedded clock bits are added to every 16 data bits, what is the maximum full-duplex signaling rate of bits being driven onto serial lines?

26-2. Why isn't the first data bit in a data frame of an eye pattern at location (n – 1)T/18 to (n)T/18 where n = 1 as measured from the start of the data frame? ("T" is the length or period of the data frame in time and "n" is the bit number within the data frame where n takes on values from 1 to 18.)

26-3. What is t_{DSR}? What does it tell you?

26-4. How do you determine the left, right, top, and bottom edges of the jitter mask?

26-5. Refer to Figure 26.25 on page 753. (a) Assuming an 80 MHz TCLK and that the figure is drawn to scale (i.e., ignore actual values for $t_{RNMI\text{-}left}$, $t_{RNMI\text{-}right}$, $t_{DJIT\text{-}left}$, and $t_{DJIT\text{-}right}$), what is the maximum jitter in the transmitter-receiver system link represented by this figure? Express your answer in both ps and as a percentage. (b) What is the amount of jitter margin? (c) Now assume that $t_{RNMI\text{-}left}$, $t_{RNMI\text{-}right}$, $t_{DJIT\text{-}left}$, and $t_{DJIT\text{-}right}$ are –230 ps (minimum value at a TCLK of 80 MHz), +230 ps (maximum value at 80 MHz), –75 ps (minimum value at 80 MHz), and +100 ps (maximum value at 80 MHz), respectively. What is the maximum jitter in ps and as a percentage, and what is the jitter margin? State any assumptions you make.

26-6. The following statement was made about the DS92LV16 device: "The receiver locks to random data, eliminating the need to interrupt normal traffic with PLL (phase-locked loop) training patterns after hot plug events." What key feature or capability of this device allows the receiver to lock to random data? (Hint: What gets added to the data bits?)

26-7. What is the difference between a point-to-point bus topology and a distributed point-to-point bus topology?

26-8. What are one advantage and one disadvantage to using a "stub hider" buffer device?

26-9. What are three rules of thumb for routing BLVDS differential traces?

26-10. The DS92LV16 is capable of driving from 10 to 20 receivers in a multidrop configuration at reduced data rates ranging from 300–400 Mbps. The ability to do this depends upon the bandwidth of the interconnecting paths, the spacing of the nodes, and the length of the resulting stubs. However, this heavy loading, even at reduced data rates, is not recommended by the device manufacturer. (a) What is the maximum number of receivers and corresponding data rates recommended for this device in a multidrop configuration? (b) What types of adverse effects could happen if you were to disregard this recommendation?

26-11. When connecting a BLVDS driver and BLVDS receiver through a cable, given that each transmitted BLVDS signal has both true (signal) and inverted (return path) traces, and that most BLVDS can tolerate as much as one full volt between the driver and receiver grounds, is there any need to include a ground connection in the cable? If not, why not, and if there is, why?

26-12. For the DS92LV16, what is the maximum amount of peak-to-peak noise that can be present on any of its power supply pins?

26-13. Why is it recommended to use two vias to feed power to each power and ground pin as well as for connecting all RF bypass capacitor terminals when laying out a circuit card using the DS92LV16?

26-14. What is the difference between the DS92LV16's receiver's RefCLK and RCLK? What is the purpose of each?

26-15. Unused LVTTL/LVCMOS input pins on the DS92LV16 may be left floating, tied low, or tied high. If you want to use only 10 of the 16 input pins and leave the 6 others floating, which 10 input pins should you assign signals to and which 6 input pins should you leave floating?

26-16. Explain the difference between local and line loopback modes. Give an example of where you might use each mode.

26-17. What are two ways in which a DS92LV16 receiver can lock to incoming data or synchronize to the transmitter?

26-18. Assuming the jitter parameters for the DS92LV16 are as follows:

$$t_{RNMI\text{-}left} = -230 \text{ ps},$$
$$t_{RNMI\text{-}right} = 230 \text{ ps},$$

$$t_{DJIT-left} = -75 \text{ ps, and}$$
$$t_{DJIT-right} = 100 \text{ ps,}$$

how much jitter margin can be split between a cable or backplane interconnect and final jitter margin?

26-19. (a) What is the difference between the data coding schemes used by the DS25C400 and the DS92LV16 SerDes devices? (b) What is the difference between their output drivers? (c) Why is it that the DS25C400 requires a transmit clock of 125 MHz to create a data rate of 1.25 Gbps but the DS92LV16 only requires a transmit clock of 80 MHz to create a data rate of 1.28 Gbps?

26-20. The DS25C400 uses a differential input transmit clock whereas the DS92LV16 uses a LVCMOS/LVTTL transmit clock input. Is there an advantage to using a differential transmit clock? Is there a disadvantage?

26-21. LVDS has an output drive capability of about 3.5 mA per differential output. BLVDS, on the other hand, provides about 10–11 mA of output drive capability per output. What is the current drive capability of the DS92LV8028 8-channel 10:1 serializer per output and how does it compare with the drive capability for LVDS and BLVDS? Can the DS92LV8028 drive multidrop, multipoint, or full BLVDS loads?

CHAPTER 27

WarpLink SerDes System Design Example

Chapter Objectives

- Provide a practical design example using Motorola's MC92610 WarpLink 2.5 Gbps quad serializer-deserializer (SerDes) data interface device.
- Discuss backplane design rules and layer, line card, and test card layer stackups.
- Present backplane and chassis design considerations.
- Discuss backplane mesh interconnects and backplane fabric interconnects.
- Present WarpLink signal integrity HSPICE simulations and simulation matrix constants.
- Discuss and present simulation matrix eye pattern parameters.
- Present eye patterns generated with and without pre-emphasis to an XAUI mask overlay.
- Perform TDR simulations and compare the results with actual TDR and eye pattern measurements.

27.1 WarpLink Design Overview[1]

This example describes the technical design process used in the development of a WarpLink Reference Design Platform using Motorola's MC92610 WarpLink 2.5 Gbps quad serializer-deserializer (SerDes) data interface device [Mot02a]. The reference design platform is a system designed with backplane, daughter cards (line, switch, and test cards), and connectors that make 3.125 Gbaud (gigabaud) channels feasible. The reference design platform utilizes a combination

1. Material in this chapter is the copyright of Motorola. Used by permission.

of good design practices to make it feasible to use an FR-4 backplane with standard components and fabrication processes to realize 3.125 Gbaud XAUI[2]-compliant channels.

The MC92610 WarpLink SerDes device is not a reduced-swing ECL or CML device, but has proprietary controlled-impedance drivers for its serial outputs. Each true and complementary output has a 50 Ω source impedance which is the sum of series resistance plus its controlled channel resistance. The source impedance remains 50 Ω whether pre-emphasis is used or not. The device is typically powered from a 1.8 V power supply and referenced to ground. It has an offset voltage of about 0.9 V and at 3.125 Gbaud, has an output signal which swings approximatley ±400 mV around the offset, from about 0.45–1.3 V. The MC92610 has 100 Ω of termination resistance built into its receiver. This is in the form of two 50 Ω resistors connected in series with their common point bypassed to ground with an internal capacitor. The terminated signal is then ac-coupled internally to the actual receiver which has an input impedance of 5 KΩ to 10 KΩ due to internal bias resistors.

The MC92610 is primarily intended for point-to-point applications. While the reference design platform design discussed in this chapter consists of a backplane with several daughter cards, all MC92610 data connections are either point-to-point or routed to switches.

A brief description of critical component selection is followed by a discussion of design considerations related to high-speed signal integrity performance. The technical "Right-by-Design" process utilized by North East Systems Associates Inc. (NESA) included simulations that were verified through measurements of the fabricated hardware. The simulation and experimental results highlight the special features of the WarpLink 2.5 device making it possible to have 3.125 Gbaud wire-speed transmissions through FR-4 PWB material.

27.2 Introduction

27.2.1 WarpLink 2.5 Quad Device

The WarpLink 2.5 Quad device is a SerDes interface that transfers data between chips across a board, a backplane, or cables. It handles four full-duplex redundant data links. Each serial transceiver transmits and receives 8B/10B coded data at a nominal rate of 2.5 gigabits per second (Gbps) through 3.125 Gbaud links. The transmitter section accepts parallel SSTL_2/HSTL-compatible data, clock, and control signals and serializes the data into 4-differential pairs of CML high-speed data. The serial transceivers drive 50 Ω media (100 Ω differential). On-chip link termination allows for no external terminating resistors being needed. The block diagram for one of the four identical transceivers contained in the WarpLink MC92610 device is shown in Figure 27.1. Note that the transmitter block has two transmitters, a primary and a redundant one. Each quad chip then has a total of eight transmitters, four primary and four redundant. In a simi-

2. XAUI is a 10 Gigabit Attachment Unit Interface proposed by IEEE Standard 802.3ae. Applications inlcude 10 Gigabit Ethernet LAN, MAN, and WAN switches and routers, next-generation network interface cards, fiber headends, advanced test equipment (ATE), and high-speed serial copper connections.

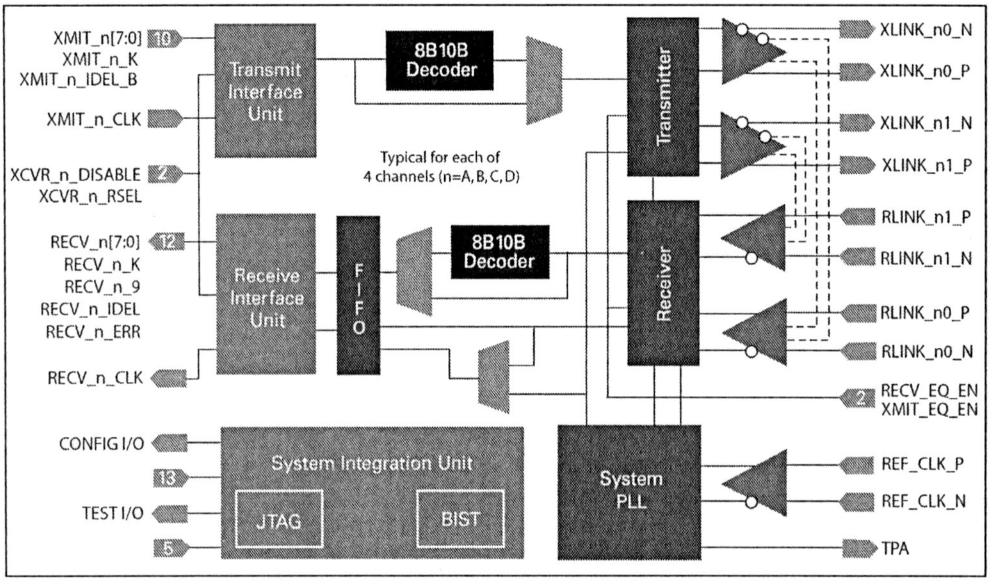

Figure 27.1 MC92610 quad SerDes transceiver block diagram for one of four identical channels.

lar fashion, each quad chip has a total of eight receivers, four primary and four redeundant. Each of the four channels per device has its own parallel input bus XMIT_n[7:0] and its own parallel output bus RECV_n[7:0]. Each channel has only one input bus and only one output bus. The figure's signal names use the letter "n" as a place holder for the link (channel) identifier letter "A" through "D." The MC92610 comes in a 324 pin MAPBGA[3] package (19 × 19 mm body size, 1.0 mm ball pitch).

Each bit of coded data is transmitted differentially out of the enabled driver. The primary driver outputs, XLINK_n0_P/XLINK_n0_N, are enabled by asserting XCVR_n_RSEL low. The redundant driver outputs, XLINK_n1_P/XLINK_n1_N, are enabled by asserting XCVR_n_RSEL high. For the receiver block, a low on XCVR_n_RSEL also causes the primary link inputs RLINK_n0_[P/N] to be received and redundant link inputs RLINK_n1_[P/N] to be ignored.

Broadcast mode is enabled by asserting XMIT_EN_ALL (not shown in Figure 27.1) high. In broadcast mode, all primary and redundant link outputs of all four transmitters are enabled, independent of the asserting the XCVR_n_RSEL signals. When XMIT_EN_ALL is low, only the selected link driver is active as described previously. Both the primary and redundant link outputs are disabled when the transceiver is disabled by asserting XCVR_n_DISABLE high.

The transmit driver is a 50 Ω controlled-impedance driver. The media over which the signals are transmitted has high-frequency loss that contributes significantly to a distortion known as inter-symbol interference (ISI). In order to offset, or equalize, the loss at high frequency, the

3. MAPBGA is Mold Array Process Ball Grid Array, a JEDEC standard.

MC92610's transmit drivers provide additional gain at high frequencies. This is termed transmit equalization (or pre-emphasis). Transmit equalization has the greatest benefit when driving longer lengths of coax or when traversing across a large backplane. Transmit equalization is of less benefit for short links. Transmit equalization is enabled by asserting XMIT_EQ_EN high.

The media through which the signals are received also has high-frequency loss that contributes significantly to ISI. In order to offset, or equalize, the loss at high frequency, MC92610's input amplifiers provide additional gain at high frequencies. This is termed receive equalization. Receive equalization has the greatest benefit when receiving signals through longer lengths of coax or when traversing across a large backplane. Receive equalization is also of less benefit for short links. Receive equalization is enabled by asserting RECV_EQ_EN high.

In the transmit direction, the near-end WarpLink device receives data on its four parallel transmit interfaces (parallel data inputs) which accommodate 8-bit uncoded or 10-bit precoded data bytes. When configured for the 8-bit mode, the device performs 8B/10B encoding on the uncoded data. It then serializes the coded data and sends it onto the four corresponding primary serial differential transmitters. Data can be sent out on the four redundant transmitters or on both primary and redundant transmitters simultaneously. Coded serial data comes out of each transmitter at 3.125 Gbaud wire-speed carrying 2.5 Gbps of user data through channels across a board, a backplane, or cables to far-end WarpLink devices' serial receivers.

In the receive direction, serial coded data coming from far-end WarpLink devices' serial transmitters are received by the near-end device on one of four primary or redundant serial receivers. The near-end device deserializes the data and, if configured for the 8-bit mode, performs 8B/10B decoding. The device then sends the data out on the four corresponding parallel receive interfaces.

The WarpLink 2.5 Quad is packaged in a 324-pin PBGA, with a 19 mm × 19 mm (0.75 × 0.75 in) body and a 1 mm ball-to-ball pitch. The device typically uses 1.8 watts. Its core and link power supply inputs require 1.8 V. The HSTL I/O power supply inputs (for the parallel and digital I/Os) use either 1.5 V or 1.8 V.

For further detail on the WarpLink 2.5 Quad's rich feature set, including selectable speed range, double data rate, 8B/10B, link synchronization, and recovered clock mode, refer to the WarpLink 2.5 Quad User's Manual.

27.2.2 WarpLink Reference Design Platform Goals

With the introduction of the WarpLink SerDes family, consisting of the WarpLink Quad (1.0 Gbps), WarpLink Quad Double Data Rate (DDR) (1.0 Gbps), and WarpLink 2.5 Quad (2.5 Gbps) devices, designing boards and backplanes capable of handling up to 3.125 Gbaud channels could prove challenging. The WarpLink Reference Design Platform was developed to assist designers in laying out the interconnection system as well as to provide an example of a system designed with a backplane, daughter cards, and connectors that make 3.125 Gbaud channels feasible. The development work served as a learning experience for Motorola, with the aim of sharing the final outcome with WarpLink customers.

A critical portion of the development effort centered on the simulation. In addition to simulations performed at 3.125 Gbaud, simulations at 5 Gbps were also carried out to give a better understanding of the current capabilities and limitations of the technology.

The platform was designed to accommodate WarpLink 2.5, as it is the fastest device in the family. FR-4 was chosen for the backplane and boards for three reasons: (1) material availability, (2) relative lower cost when compared with other fabrication materials, and (3) most PWB fabrication houses can process the material. Several channels were designed to be long enough to demonstrate WarpLink 2.5's XAUI compliance (50 cm or 19.68 in).

Standard available components were selected, specifically the ERNI HMZD connector, a high-density connector intended for high-speed signaling. Motorola's Timing Solutions devices, the MC100ES6111 (Low Voltage 2.5/3.3 V Differential ECL/PECL/HSTL Fanout Buffer), the MC100ES6222 (Low Voltage 1:15 Differential ECL/PECL Clock Divider and Fanout Buffer), and the MPC9456 (Low Voltage Single or Dual Supply 2.5 V and 3.3 V LVCMOS Clock Distribution Buffer), were examples of currently available clock network distribution devices that are intended for use in such a system.

27.2.3 WarpLink Reference Design Platform Overview

27.2.3.1 Architectural Overview

Mesh and Fabric The WarpLink Reference Design Platform consists of a 9U[4] 19 inch Compact PCI chassis with eight double-height slots. The slots accept three types of cards: the line card, the switch card, and the test card. WarpLink devices on the cards are connected to one another through the backplane interconnect, which supports both mesh and fabric switching schemes (described below). In this design example, both mesh and fabric switching nets are designed into the same backplane.

For the mesh-switching scheme, each slot has eight mesh ports. The mesh channels in the backplane connect seven mesh ports per slot to a mesh port on each of the other seven slots. The one remaining unconnected port is looped back to itself. The connection is done such that slot-*m*, port-*n* is connected to slot-*n*, port-*m*. Figures 27.2 and 27.3 illustrate the mesh interconnects of the backplane for slot 1 and 2, respectively. In the figures, each arrow represents a full-duplex transmit and receive connection (two differential pairs of traces). In this scheme, 2.5 gigabits of link data can go directly from any slot to any other slot.

In the fabric-switching scheme, each slot has eight line ports: four primary, four redundant. Slots 7 and 8 each have 32 fabric ports in addition to the line ports. The fabric channels of the backplane connect the four primary line ports of every slot to four of the 32 fabric ports on slot 8. The four redundant line ports of every slot are connected to four of the 32 fabric ports on slot 7. Figure 27.4 shows the fabric interconnects of the backplane. In this scheme, 10 gigabits of

4. 9U refers to the size of a printed wiring board, specifically the VME/VXI D size which is 13.386 by 14.437 inches (340 × 366.7 mm).

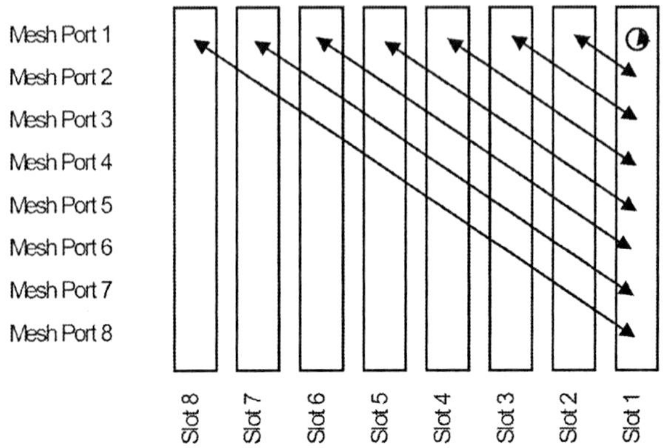

Figure 27.2 Backplane mesh interconnects from slot 1.

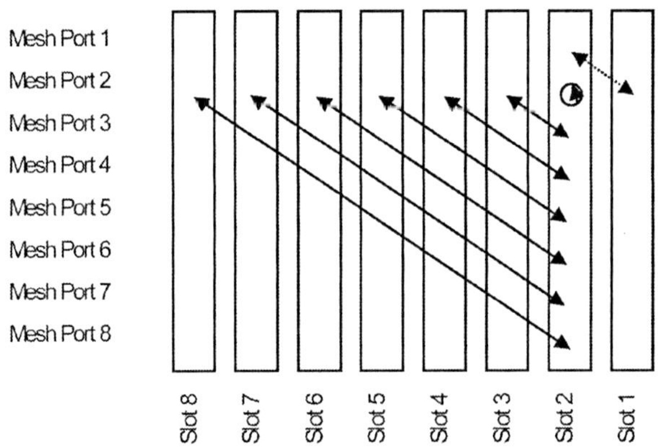

Figure 27.3 Backplane mesh interconnects from slot 2.

link data can go from any slot to any other slot through slots 7 or 8. Data links going through slot 8 are primary and those through slot 7 are redundant. The line ports are additional and different ports from the mesh ports.

Clock Distribution The system has two point-to-point distributed clock networks. The primary network originates from slot 8 and is distributed to all slots. The redundant one is from slot 7, which is also distributed to all slots. The card in each slot selects its own on-board clock, or the clock distributed by either the primary network or the secondary network. Motorola's MC100ES6111 device (on the card in slot 7 or 8) is used to drive the clock distribution network. An MC100ES6222 clock receiver on each line card is used to receive the distributed clock, and the MC100EP8111 HSTL clock buffer on the line card brings the clock signals to HSTL levels to be used by WarpLink.

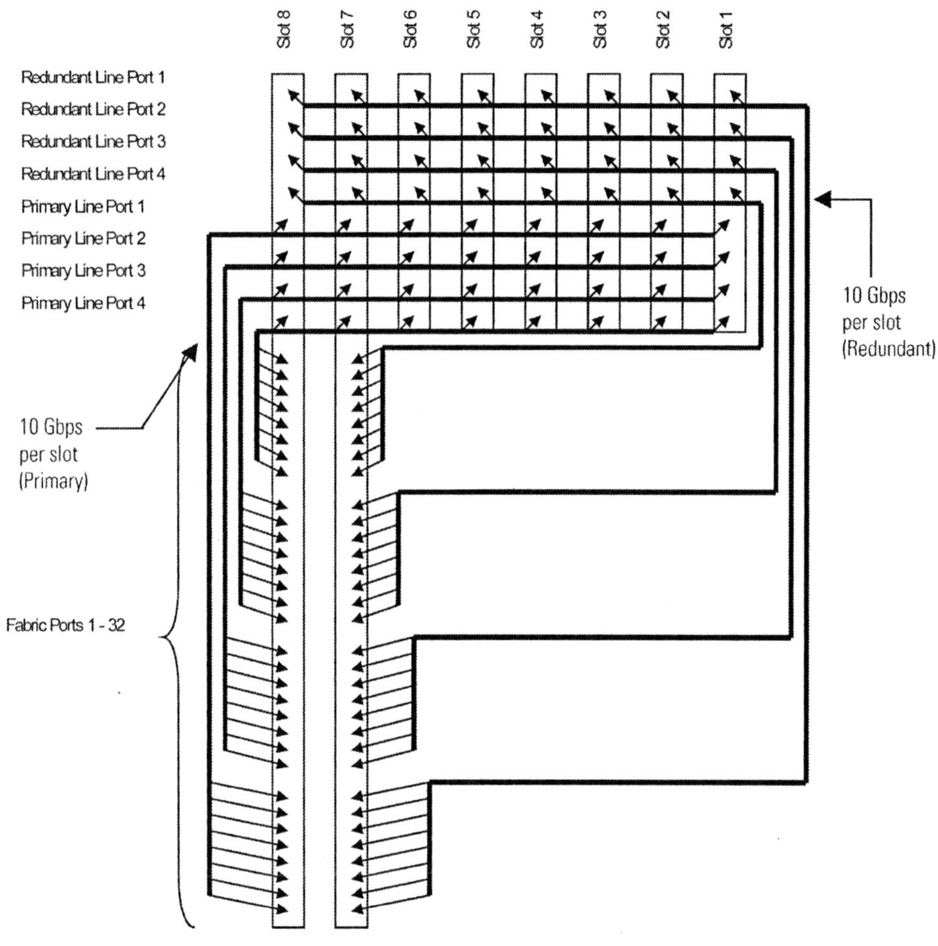

Figure 27.4 Backplane fabric interconnects.

Processor Bus and Switch Cards The system has a multidrop 32-bit bus going to all slots that can be used as a processor bus. A host card in slot 7 or 8 with a processor can send and receive control and status messages to all slots. Slot 8 houses the primary control card, and slot 7 the redundant control card. The multidrop 32-bit bus is designed to terminate Gunning Transceiver Logic Plus (GTLP) drivers. In this discussion, the terms "host card," "control card," and "switch card" will be used interchangeably.

27.2.3.2 Backplane

The key subassembly of the WarpLink Reference Design Platform is the backplane. It is a 14.96-inch × 15.57-inch FR-4, 28-layer printed circuit board (PCB). Eight columns of connectors form the eight slots into which the cards are inserted. The majority of the backplane traces form the 3.125 Gbaud channels that interconnect the WarpLink 2.5 devices on the daughter cards through the mesh and fabric topologies described previously. Other traces form the clock distribution network and the processor bus.

An important aspect of the backplane design was the selection of the high-speed differential connector from ERNI. The ERmet® ZD 8 × 10 connector is used as a differential board-to-backplane connector rated to support electrical signaling up to 5 Gbps. The ERmet ZD is compatible with the mechanical characteristics of the ERmet 2 mm-HM specifications as defined in IEC 61076-4-101.[5] (2 mm is the pin-to-pin spacing. HM stands for "Hard Metric.")

27.2.3.3 Line Card

One of the cards that reside in the WarpLink Reference Design Platform is the line card. It can go into any of the eight slots. In the mesh topology, all eight slots can be populated with line cards. In the fabric topology, slots 1–6 can be populated with line cards, while slots 7 and 8 must be populated with the switch cards. Each line card has two WarpLink 2.5 Quad devices on it.

Each WarpLink 2.5 Quad device has four SERDES devices, each capable of supporting one full-duplex data link. The parallel side of each SERDES device is not redundant. (Each device has four parallel input buses, one for each of the four channels.) Therefore, each SERDES device has only one parallel input port. This parallel port of the full-duplex data link supported by a SERDES device consists of 8 or 10 pins for the parallel transmit interface and 8 or 10 pins for the parallel receive interface. The serial side, however, is redundant. Therefore, each SERDES device has two serial ports: a primary one and a redundant one. Each of the serial ports consists of 4 pins: positive-transmit, negative-transmit, positive-receive, and negative-receive.

A line card with two WarpLink 2.5 Quad devices, therefore, can support eight full-duplex data links, to be used in either the mesh or fabric configuration.

The eight serial primary ports provided by the two devices are connected to the eight mesh ports of the slot when the card is inserted. The mesh ports are then connected to other mesh ports as previously discussed. In the mesh configuration, the primary serial ports are selected, and data always passes through the two WarpLink devices via their parallel ports and primary serial ports.

The eight serial redundant ports provided by the two devices are connected to the eight fabric-topology line ports of the slot when the card is inserted. The line ports are then connected to the fabric ports on slot 7 and 8 as previously discussed. In the fabric configuration, the redundant serial ports are selected, and data always passes through the two WarpLink devices via their parallel ports and redundant serial ports.

On the parallel side of the WarpLink devices, the HSTL parallel interfaces can be connected to ASICs, FPGAs, or headers to be controlled and monitored by pattern generators and logic analyzers.

5. The title to IEC 61076-4-101 is Connectors for Electronic Equipment — Part 4-101: Printed Board Connectors with Assessed Quality — Detail Specification for Two-Part Connector Modules, Having a Basic Grid of 2.0 mm for Printed Boards and Backplanes in Accordance with IEC 6091.

27.2.3.4 Switch Card

The switch card has the mesh and line ports of a line card, but also contains the 32 fabric ports. This card can only go into slots 7 and 8. There are 10 WarpLink devices on the switch card. Two of the 10 devices are used to form the line card portion: the devices' serial primary ports are connected to the mesh ports and the redundant ports to the line ports. The 32 serial redundant ports of the remaining eight devices are connected to 32 fabric ports of the slot into which the card is inserted. The 32 primary ports of these eight devices are unused. On the parallel side of the WarpLink devices, the interfaces can be connected to ASICs or FPGAs.

27.2.3.5 Test Card

The test card is a passive card with SMA connectors and 2x10 headers. It does not have any WarpLink or active devices on it. The card is used to characterize channels from slot to slot. For example, Time Domain Reflectometry (TDR) measurements can be performed with this test card to show via, connector, and backplane trace impedance. The card can also be used to independently characterize transmitters and receivers operating in the system. When the card is in a slot, controlled test signals from a signal generator can be fed into the near-end test card's transmit port of a link to characterize the corresponding receiver of a far-end WarpLink device. Also, signals from a far-end WarpLink transmitter can be measured off the near-end receive port, through a pair of SMA connectors on the test card.

27.3 Detailed Design Descriptions

27.3.1 WarpLink Reference Backplane

The WarpLink Reference Backplane is specifically designed for the Motorola WarpLink 2.5 Quad component. The overall design is representative of a *typical* backplane used in a present-day communications-system chassis. Figure 27.5 details the overall backplane as well as component selection and placement on the WarpLink backplane. Figure 27.6 details the layer stackup of the backplane.

27.3.1.1 Backplane/Chassis Design Considerations

Before we discuss the details of the Motorola WarpLink backplane, it would be helpful to give an overview of the process and constraints under which a backplane and chassis are designed. The design of the backplane must consider the design elements of the system as a whole. The major parameters that shape a chassis/backplane system must be established early on in the design, once the intended market is established. These include (1) the total power dissipated by the system, (2) the system logical architecture, (3) the approximate dimensions of the chassis, and (4) the subsequent regulatory requirements for the intended market. One of the outcomes of coupling the system architecture with the total power dissipation is the determination of the number of daughter card slots within the chassis dimension limitations. Knowing the maximum power dissipation of the various types of daughter cards and the number desired in a system helps to establish the constraints that limit the possible system configurations. At times, thermal

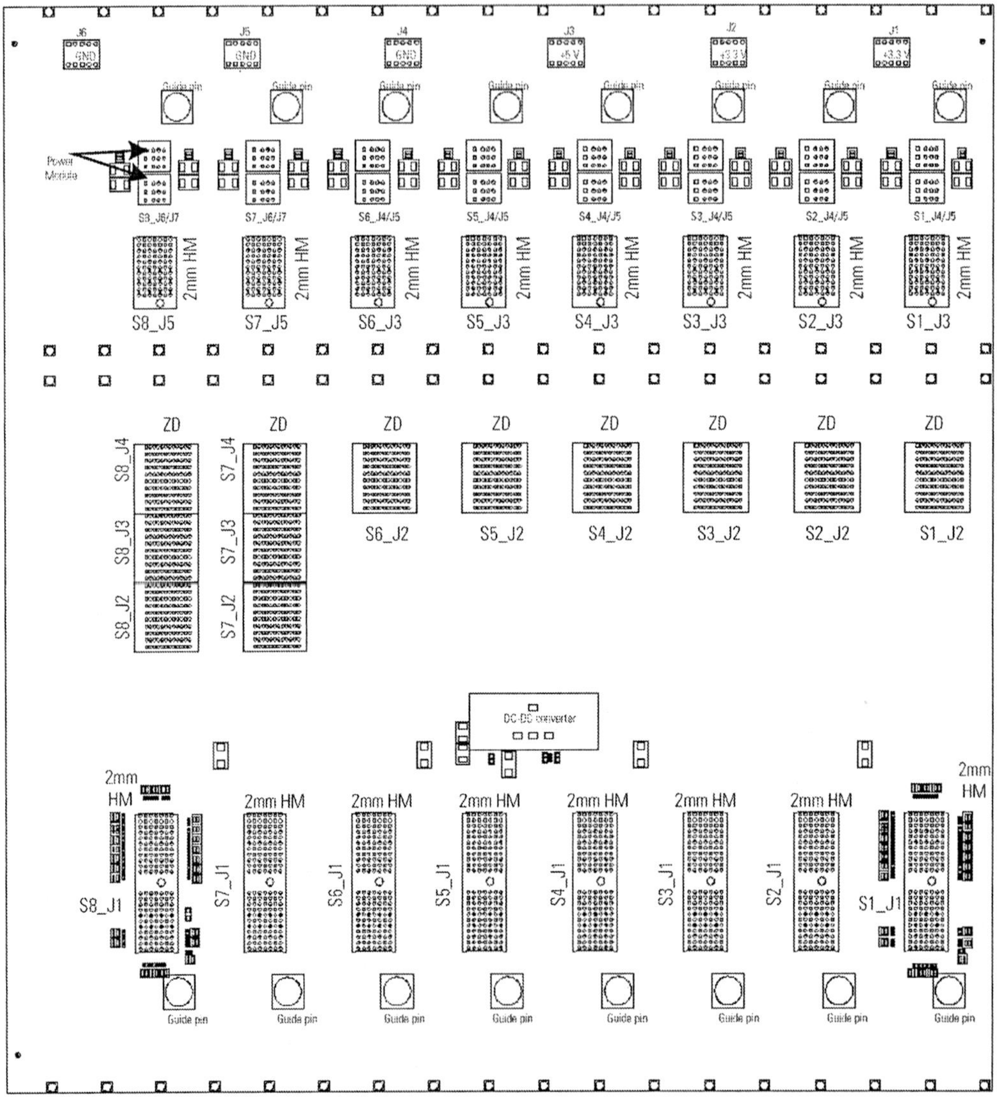

Figure 27.5 WarpLink reference backplane (front view).

constraints can directly impact the electrical architecture of a system. This is especially true if the chassis is sized so that it places limits on the total power that can be dissipated. A first pass formulation of a system takes place once a balance is reached between the requirements set by the system architecture coupled with the intended market and the total power that can be dissipated within the chassis enclosure. The thermal analysis and subsequent physical constraints on the slot pitch are required to determine the total number of daughter cards possible in the system.

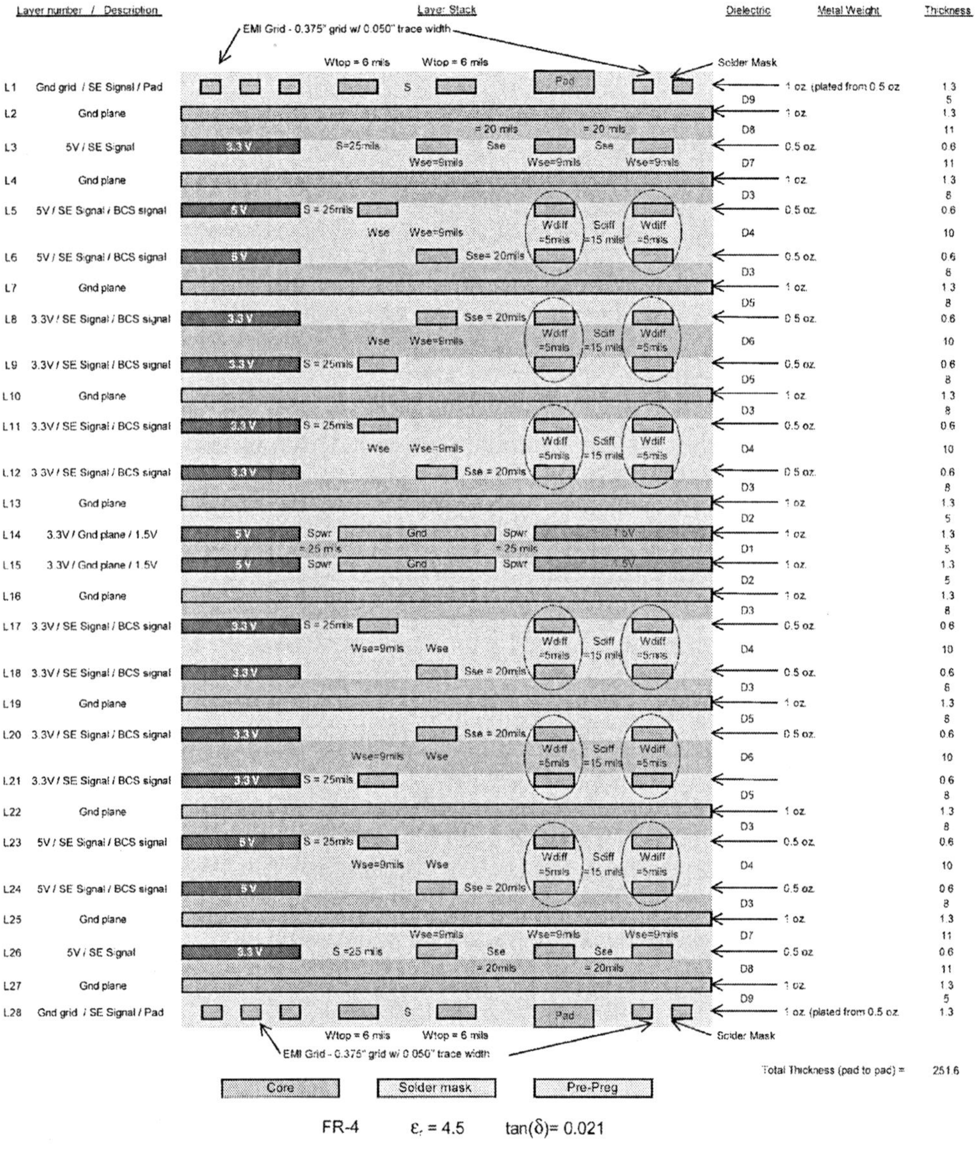

Figure 27.6 WarpLink reference backplane layer stackup.

Thus, the ability of a chassis to cool itself can limit the number and possibly the type of components used, including logic devices.

The design of the backplane begins in parallel to the thermal evaluation of the system, but is not set until the number of daughter card slots and overall system configuration are established. Once the number of slots is determined, the next major step is the identification of the type and number of signals per slot. A definition of the signaling needed in the WarpLink Reference Backplane was based on the architecture defined by the Motorola design team. The general architecture of the WarpLink Reference Design Platform is discussed in Section 27.2, "Introduction," on page 776 of this chapter.

27.3.1.2 WarpLink Backplane Physical Description

The physical configuration of the Motorola WarpLink backplane used in the reference system contains eight slots on a 1.6-inch card-to-card pitch. The Motorola WarpLink Reference Design Platform was designed to suit both switch fabric and mesh architectures. The logical configuration of the backplane defines slots 1 through 6 as pin-compatible line card slots, and slots 7 and 8 as pin-compatible switch/line card slots. When a switch fabric is inserted into slot 7, 8, or both, the system can function in either the fabric or mesh architecture. If a line card is inserted into slot 7, 8, both, or neither, the system will operate in the mesh architecture. The ability of the WarpLink to do this is due to the primary and redundant serial I/O characteristics of the device.

The signal allocations for each slot type are detailed below in Table 27.1.

Table 27.1 Signal Allocations

Signal Types	Slot 1–6 (Line Card)	Slot 7–8 (Switch/Line Card)	Signal Speed
Differential Gigabit Serial Links (Mesh/Fabric)	32 pairs	96 pairs	3.125 Gbd
Differential Clocks	2 pairs	8 pairs	156 MHz
Single-Ended Address/Data/Control Buses	36 pins	36 pins	<50 MHz

The signal speed identification is critical to the design rules formulated for the layout of a gigabit backplane. For example, the addition of unnecessary vias in the gigabit lines would adversely affect their signal transmission characteristics. Each via presents an impedance discontinuity that will contribute to the distortion of the opening of the receiver input eye pattern. This can reduce available voltage margins as well as time margins through an increase in signal jitter. Therefore, all differential routes of the WarpLink Reference Design Platform backplane had vias placed only where absolutely needed — at the device pads and at the connectors (plated-through holes).

A number of issues were assessed in arriving at a final connector selection for the Motorola WarpLink backplane. The ERmet ZD high-speed backplane connector was chosen for the gigabit nets for its high-frequency signal transmission characteristics and because it is side stackable with a standard 2 mm hard metric connector (2 mm pin-to-pin spacing). The ERmet 2 mm hard metric connector was selected for the slower speed single-ended and differential clock nets.

The backplane not only connects the eight slots of the Motorola WarpLink system; it also has been designed with features that help exhibit its gigabit transmission capabilities. The longest gigabit traces required to connect slots 1 to 8 are about 12 to 15 inches. The WarpLink backplane also contains four differential nets that are approximately 0.5 meter long. These nets demonstrate WarpLink's ability to operate over a long lossy channel. They are also used to test WarpLink's compliance to XAUI specifications. The backplane also contains four traces designed to inject crosstalk onto gigabit links. This serves to demonstrate WarpLink's noise tolerance. These noise-generating nets have been placed 5 and 10 mils away from the victim gigabit nets. A recommended design rule spacing of 15 mils (minimum) was implemented for the remaining gigabit nets in order to minimize any crosstalk.

The gigabit links are confined to the middle of the backplane (using buried traces) and are connected to the daughter cards through the ERNI Components ERmet ZD high-speed backplane connector. The slow-speed (single-ended) buses are positioned in the lower connector section of the backplane and are connected to the daughter cards via ERNI Components ERmet 2 mm hard metric (HM) connectors. The clock nets run across the upper connector section of the backplane and are connected to the daughter cards via with the ERNI 2 mm HM connectors as well. The backplane provides 3.3 V and 5 V power to the daughter cards by way of the ERmet power modules at the top of each slot.

27.3.1.3 Backplane Design Rules and Layer Stackup

The stackup cross-section of the WarpLink Reference Backplane board is shown in Figure 27.6. Both broadside-coupled differential (labeled as BCS) and single-ended (SE) routes are utilized in the layout. In the figure, "Sse" is single-ended spacing and "Wse" is the width of a single-ended trace. The characteristic impedance of the single-ended nets was designed to be 50 Ω and the broadside-coupled differential nets 100 Ω (both ±10%). All differential pairs were routed direct without vias. Vias for single-ended, multidrop buses were used only within the connector regions and at the trace ends for termination access. The WarpLink backplane was designed as a 28-layer, 0.250-inch thick PCB. The backplane contains six broadside-coupled routing channels (12 layers) and 14 single-ended routing layers (12 of which are shared with the differential pairs). Broadside-coupled differential routing was chosen over edge-coupled or uncoupled differential interconnects for its ability to maintain the routing channel within the gigabit connector footprints — board thickness was not a critical parameter. This is a typical routing scheme when using high-speed differential connectors on backplanes of substantial board thickness. One-ounce copper is 1.35 mils thick.[6]

The WarpLink Reference Design Platform has been designed for three types of daughter cards. Two of the cards, the switch card and the line card, contain active components in addition to the WarpLink SerDes. The third card is the passive test card. The design intent of the test card is to provide access to the gigabit nets for direct sampling of the WarpLink signals as well as providing access to the backplane for passive signal integrity measurements (via SMA test connectors).

6. Plating weight is the number of ounces of copper uniformly plated onto a flat surface per square foot.

27.3.2 Line, Switch, and Test Cards

27.3.2.1 Daughter Card Design Considerations

An important design consideration for the daughter cards should be the pin assignment of the high-speed connector. This pin assignment, in a way, is affected by the number of high-speed routing layers in a card. The number of layers should be minimized so that the card would be thin enough to fit the guiding rails of the chassis. From a signal integrity viewpoint, this should also be done so that the vias' stubs would be as short as possible, especially for the 3.125 Gbaud traces that are routed in the upper layers. For the chosen gigabit connector, four was the minimum number of high-speed layers we could have. Once the number of layers is determined, a preliminary hand-routing exercise should be carried out to determine the pin assignment, or to determine whether the existing pin assignment is or is not routable given these four layers. Care should be exercised when assigning pins to make both the line card and the backplane routable. Often, a couple of iterations would be required. This hand-routing exercise should be done early in the design cycle before committing to a pin assignment.

Another trivial design consideration for the daughter cards should be the placement of the WarpLink device. The device should be placed as close to the connector as possible. For the line card, one WarpLink was placed close to the connector, and the other one further away to emulate a typical layout in which many WarpLink devices might have been used.

27.3.2.2 Test Card and Line Card Layer Stackups

The stackups of the test and line cards are presented here to document the standard design practice used in the card layout and fabrication. Figures 27.7 and 27.8 detail the test card and line card stackups, respectively. The gigabit differential nets of the test card and line card were designed to be the same. The length of the traces in the test card emulated the lengths seen on the line card.

The major difference between the test and line card stackups is the addition of power planes to the line card. The test card has thicker dielectric material between layers 6 and 7 to account for the missing plane layers and maintain the same overall card thickness.

27.4 WarpLink Signal Integrity HSPICE Simulations

27.4.1 WarpLink Gigabit Simulations

As part of the Motorola WarpLink Reference Design Platform effort, NESA completed an extensive HSPICE signal integrity study for the gigabit links and differential clocks. Some of the simulation model features include: coupled lossy-line models; worst-case card via effects due to stubs; WarpLink driver pre-emphasis control; and data rates running at 2.5 Gbaud and 3.125 Gbaud. Although the majority of the results focus on the differential interconnects for the active WarpLink device, simulation cases were completed for the clock interconnects as well. The data in this summary includes receiver input eye patterns as well as TDR impedance profiles of the various differential interconnects.

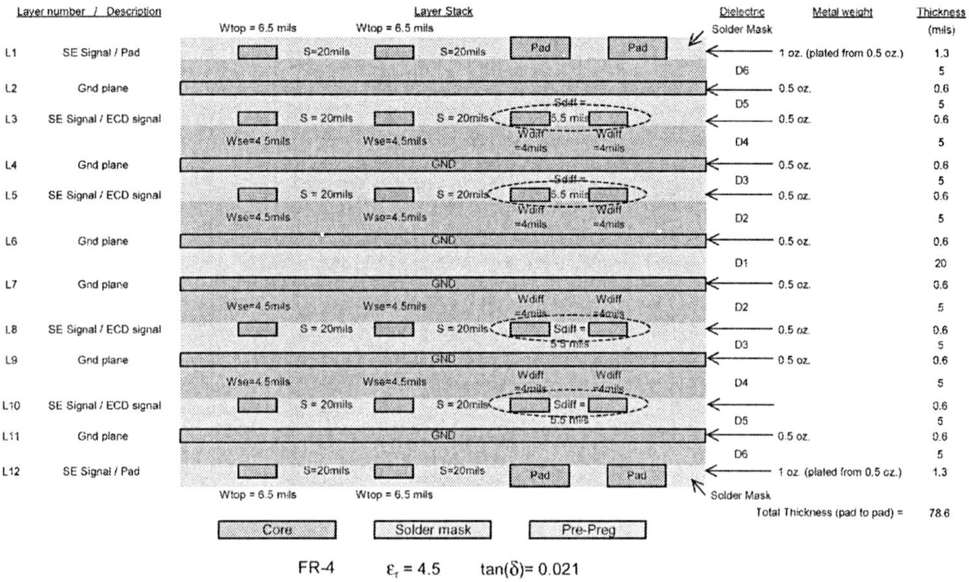

Figure 27.7 WarpLink test card layer stackup.

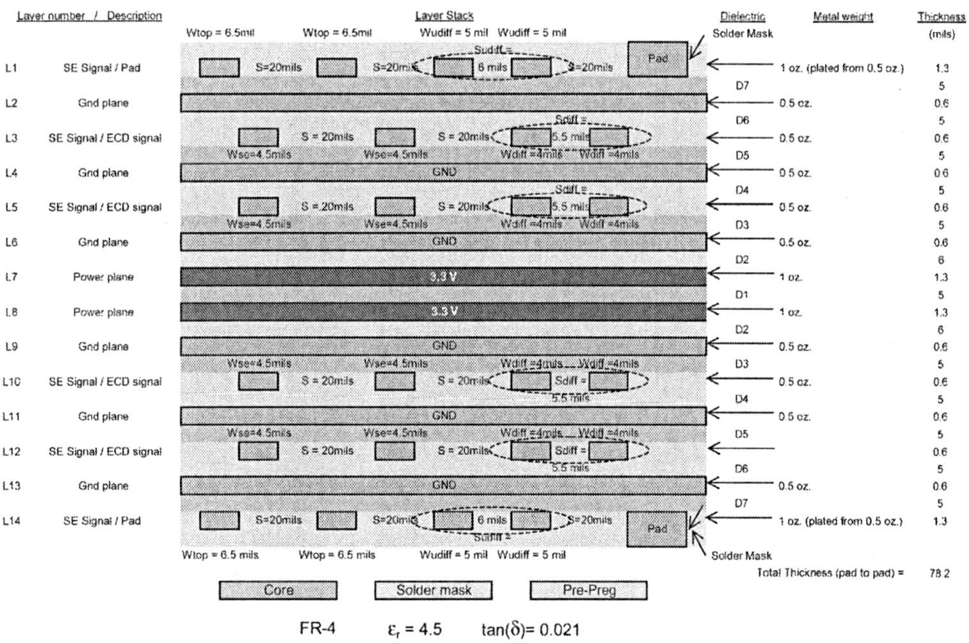

Figure 27.8 WarpLink line card layer stackup.

The Motorola WarpLink single-threaded HSPICE model was inserted as the driver and receiver into the interconnect environment constructed by NESA. Figure 27.9 details the interconnect topology for the gigabit links. Motorola provided the WarpLink device and package HSPICE models. Models for the ERmet ZD connector were supplied by ERNI Components. The coupled, lossy transmission line and via models were created from the PCB parameters by NESA. In Figure 27.9, the small square symbol labeled "via" is used to indicate impedance effects due to a via.

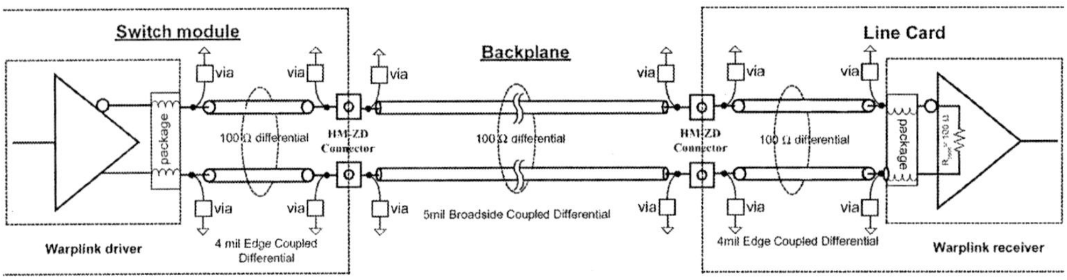

Figure 27.9 WarpLink gigabit interconnect.

The construction of the HSPICE models was followed by the identification of the range of data rates and transmission line lengths needed to prove out the gigabit capabilities of WarpLink. Table 27.2 identifies the variables and ranges of the values for the simulation matrix. Table 27.3 lists the constants, which are fixed parameters within the HSPICE simulation matrix. Table 27.4 identifies the simulation matrix and the resultant eye pattern parameters. Note that simulations were run at 65°C to better emulate the case temperature (versus the typical 25°C).

Table 27.2 Simulation Matrix Value

Variables	Range
WarpLink Device	
Temperature	0°C, 25°C, and 100°C
Supply Voltage (1.8 V ± 10%)	1.6 V, 1.8 V, and 2.0 V
Driver Pre-Emphasis (Transmit Equalization)	off, on
Process	SS, TT, FF
Interconnect	
Backplane Trace Length	2, 10, and 20 inches
Line and Switch Card Vias	One-Layer Transition with Stub and Full-Board Transition

The eye parameters measured at the receiver input include eye height, eye width, peak-to-peak jitter, and XAUI mask. The precise measurement point of the eye pattern is between the package and the silicon at the receiver input. A comparison of the odd-numbered cases (pre-emphasis off) in Table 27.4 to the even-numbered cases (pre-emphasis on) shows a great improvement for all eye metrics when pre-emphasis is engaged. The XAUI channel eye mask has been overlaid on the receiver input eye to illustrate compliance (refer to the eye pattern plots below).

Cases 9 and 10 and 13 and 14 contain the *stub* via model. In a typical backplane system, parasitic stubs are seen for vias where a signal traverses a via for less than the thickness of the backplane. For example, a differential pair routed on layers 5 and 6 will only pass through approximately 40 mil of the 250-mil backplane. The remaining 210 mils becomes a parasitic capacitive stub, reducing the trace impedance, and contributing to the degradation of the gigabit signal.

Table 27.3 HSPICE Simulation Matrix Constants

Constants	Value
Device	
Gigabit Tx/Rx Device	WarpLink 2 / WarpLink 2
Package	
324 PBGA (WL)	19 mm × 19 mm Body, 1 mm Ball Pitch
Data	
Data Rate	3.125 Gbps
Data Pattern	K28.5
Etch	
Switch/Mesh Card Trace Length	6 in
Line Card Trace Length	4 in
Board Parameters	
Driver Location	Switch/Mesh
Receiver Location	Line
Backplane Material	FR-4
Switch/Uplink Module Material	FR-4
Line Card Material	FR-4
Backplane Thickness	0.250 in
Switch/Mesh Card Thickness	0.078 in
Line Card Thickness	0.078 in
Etch Width (Backplane) BC[a] Diff	5 / 5 / 5 mils (w/s/w)
Etch Width (Line and Switch) EC[b] Diff	4 / 5.5 / 4 mils (w/s/w)
Switch/Mesh Card Via Count	2 — 1 @ package pad, 1 @ connector
Backplane Via Count	2 — 1 @ each connector
Line Card Via Count	2 — 1 @ connector pad, 1 @ package
Backplane Connectors	
Gigabit nets — Switch/Mesh Modules	8 - Row HM-ZD - GH Row
Gigabit Nets — Line Card	8 - Row HM-ZD - GH Row

a. BC is broadside-coupled.

b. EC is edge-coupled.

Table 27.4 Simulation Matrix Eye Pattern Parameters

Case	Temperature (°C) /65/100	Supply Voltage ±10%	Process Corner	Pre-Emphasis On/Off	Data Rate (Gbps)	Backplane Length (in.)	Via Type	Eye Height (mV)	Eye Width (ps)	Jitter (p-p) (ps)	Within XAUI Mask
1	0	2.0	FF	Off	3.125	2	Full	650	275	45	Yes
2	0	2.0	FF	On	3.125	2	Full	740	280	40	Yes
3	0	2.0	FF	Off	3.125	20	Full	300	220	100	Yes
4	0	2.0	FF	On	3.125	20	Full	450	260	60	Yes
5	100	1.6	SS	Off	3.125	2	Full	360	255	65	Yes
6	100	1.6	SS	On	3.125	2	Full	450	270	50	Yes
7	100	1.6	SS	Off	3.125	20	Full	170	220	100	No
8	100	1.6	SS	On	3.125	20	Full	220	240	80	Yes
9	100	1.6	SS	Off	3.125	20	Stub	120	190	130	No
10	100	1.6	SS	On	3.125	20	Stub	190	240	80	No
11	65	1.8	TT	Off	3.125	10	Full	370	260	60	Yes
12	65	1.8	TT	On	3.125	10	Full	480	280	40	Yes
13	65	1.8	TT	Off	3.125	10	Stub	280	235	85	Yes
14	65	1.8	TT	On	3.125	10	Stub	440	275	45	Yes

Figures 27.10 and 27.11 show examples of the simulated eye pattern for the WarpLink interconnect for cases 13 and 14. The measurement of the eye width and peak-to-peak jitter are made at the zero crossing of the pattern and the eye height is made at the center of the eye opening. Since this is a *zero average value* pattern, there should be no dc shift of the receiver input.

Figures 27.10 and 27.11 are typical of the improvement seen when the WarpLink transmit pre-emphasis feature is utilized. As shown in the simulation results of Table 27.4, the eye height was improved by an average of 26%, the eye width improved by 10%, and the peak-to-peak jitter improved by an average of 30%.

27.4.2 WarpLink Interconnect Impedance Profile

The Time Domain Reflectometry (TDR) interconnect simulation shown in Figures 27.12 and 27.13 illustrates the impact of a via stub. A differential TDR source (NESA Inc. proprietary simulation model) replaces the WarpLink driver while leaving the WarpLink package in the model. The receiver model is also removed and the interconnect is left open-circuited, which offers a clear indication of the end of the path.

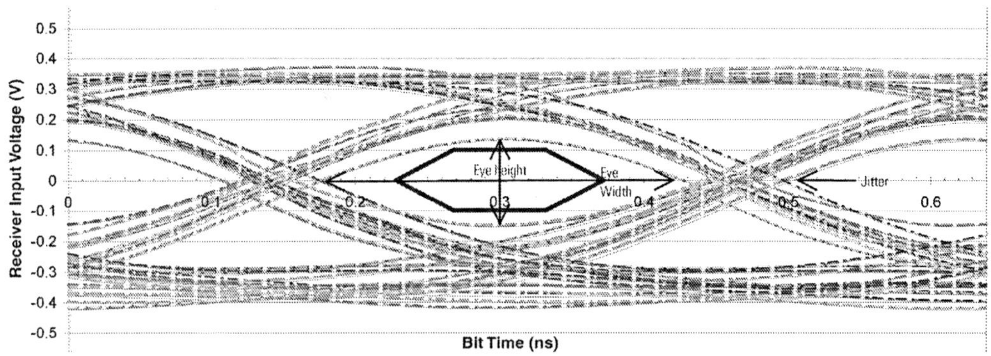

Figure 27.10 Case 13 with XAUI mask overlay. No pre-emphasis. 65°C, nominal supply voltage, TT process corner, 10-inch long backplane etch, via stubs, 3.125 Gbps data rate.

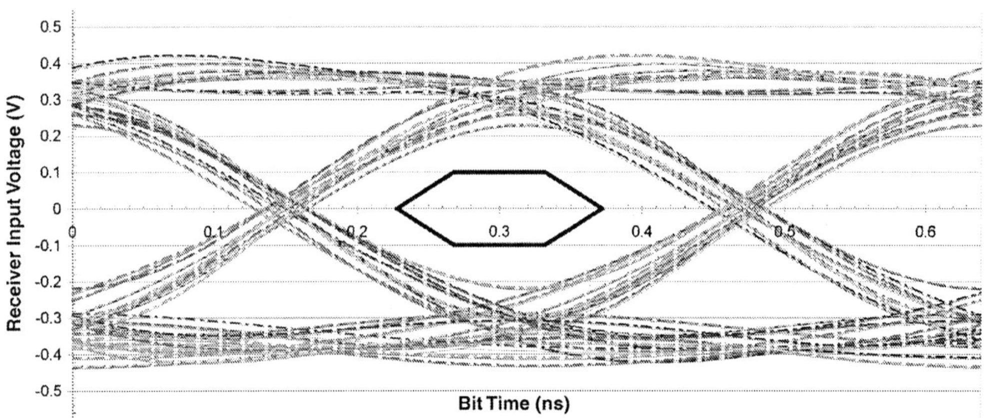

Figure 27.11 Case 14 with XAUI mask overlay. With pre-emphasis. 65°C, nominal supply voltage, TT process corner, 10-inch long backplane etch, via stubs, 3.125 Gbps data rate.

Table 27.5 details the six variations of the WarpLink gigabit interconnect. Figures 27.12 and 27.13 illustrate the difference of the impedance profile for a *stub* via and a *through* via, respectively.

Table 27.5 WarpLink Gigabit Interconnect Variations

Case	Backplane Length (in.)	Via Type	Case	Backplane Length (in.)	Via Type
1	2	Full	5	20	Full
2	2	Stub	6	20	Stub
3	10	Full	7	40	Full
4	10	Stub	8	40	Stub

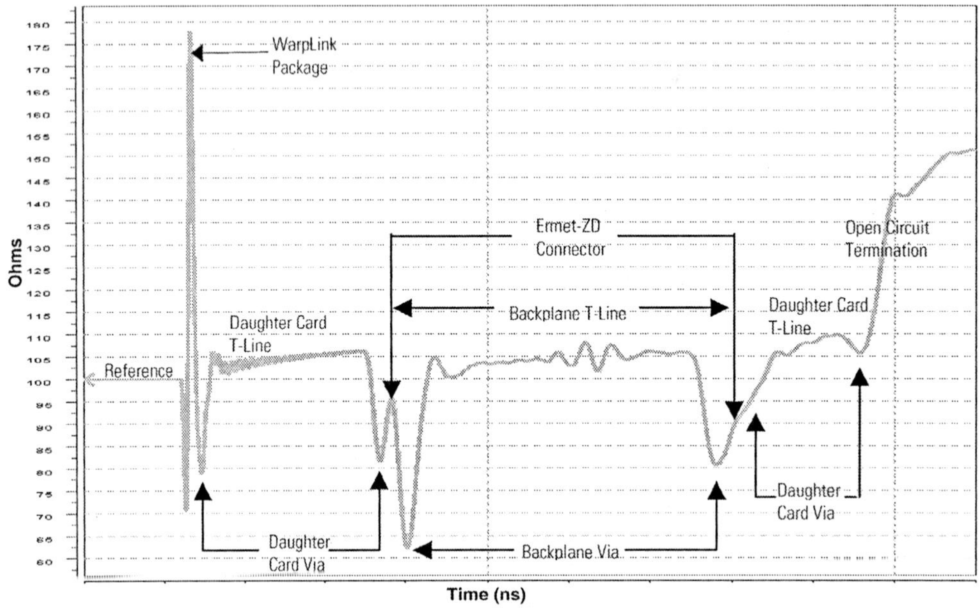

Figure 27.12 TDR of 10-inch backplane with via stubs.

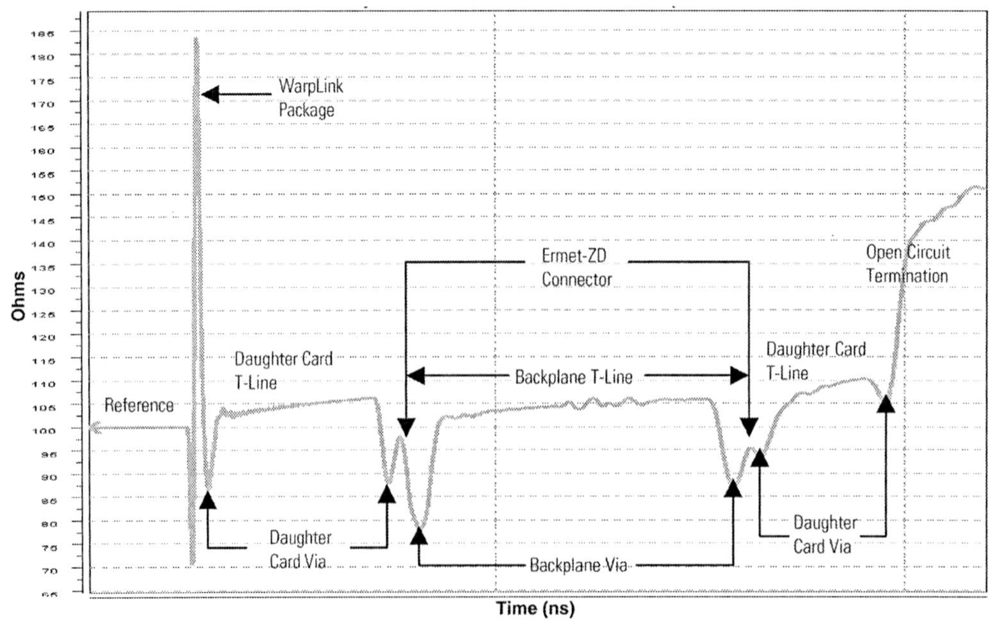

Figure 27.13 TDR of 10-inch backplane with through vias.

The TDR highlights a number of issues for the interconnect. One of the goals of a high-frequency digital interconnect is to maintain as constant a characteristic impedance as possible. The vias and package impedance profiles of Figures 27.12 and 27.13 show significant excursions from the reference impedance of 100 Ω.

A comparison of the impedance profiles show that the *through* via (full board transition) produce smaller impedance discontinuity, Figure 27.13, than the stub via case, Figure 27.12. The package model presents a significant discontinuity in the simulation of the WarpLink interconnect, for both via cases. The large inductive spike is likely due to the lumped-element package model used in the simulations.

The discontinuity of the ERmet ZD connector reflected its ability to transmit 5 Gbps signals. The peak impedance level within the connector region was approximately 95–98 Ω. The effects of the vias on both the daughter card and backplane influence this level due to risetime roll-off of the TDR pulse. The lossy characteristic of the transmission line is evident as seen in the increasing impedance along the length of the TDR profile.

27.4.3 WarpLink Reference System Clock Simulations

The clock architecture for the WarpLink Reference Design Platform is point-to-point. The clock operates at a digital frequency of 156.25 MHz. This data rate allowed the clock to be connected to the daughter card thought a standard 2 mm hard metric connector, the ERmet 2 mm-HM. The interconnect path models used for the clock simulations are almost identical to that of the gigabit WarpLink models. The main differences were the replacement of the ERmet ZD with the ERmet 2 mm-HM connectors and the inclusion of the clock driver and receivers in place of the WarpLink devices.

The variables and range of the simulation matrix for the clock are shown in Table 27.6. The results of the simulations, exemplified by the waveform in Figure 27.14, show little distortion and a well-controlled clock signal.

Table 27.6 Simulation Matrix Values

Variables	Range
Clock Buffer Device	
Temperature	0°C, 25°C, and 100°C
Process	SS, TT, FF
Supply Voltage (3.3 V ± 10%)	3.0 V, 3.3 V, and 3.6 V
Interconnect	
Backplane Trace Length	2, 10, and 20 inches
Line Card and Switch Card Vias	One-Layer Transition with Stub and Full-Board Transition

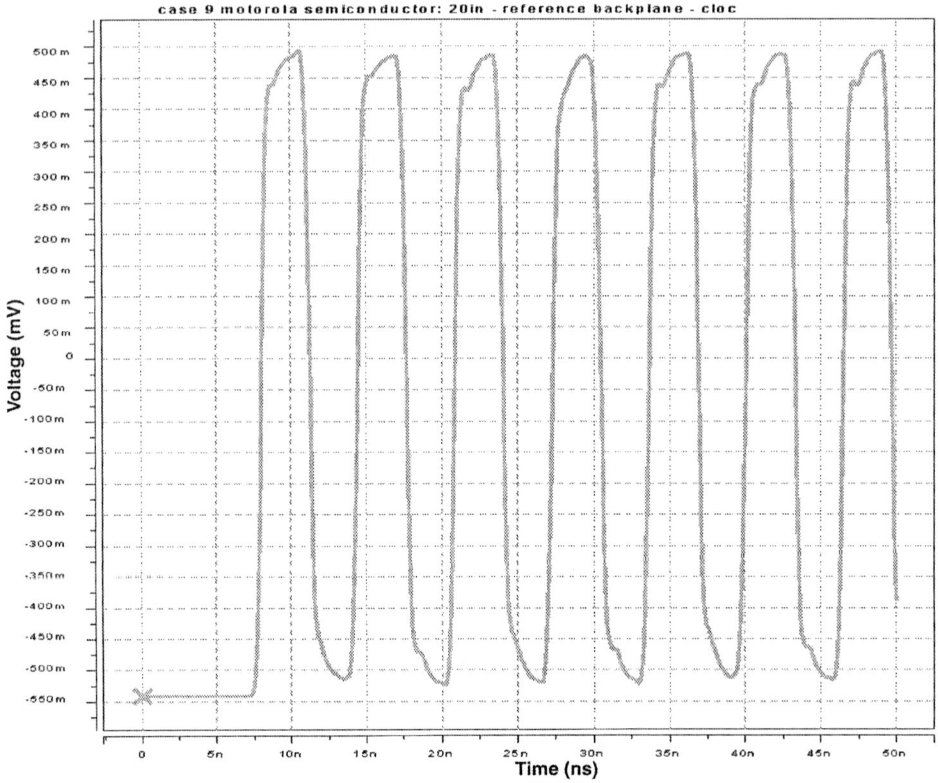

Figure 27.14 Typical clock receiver input simulation.

27.5 Descriptions of Passive Signal Integrity Measurements

27.5.1 Time Domain Reflectometry (TDR)

TDR was used to measure the impedance profile of the WarpLink Reference Design Platform's backplane interconnects. The impedance of an interconnect is important to determine since impedance mismatches due to surrounding structures can cause reflections, which in turn cause aberrations in signal fidelity. To measure the differential impedance, two signals of a pair are driven simultaneously with step sources of equal amplitude and opposite polarity. The far ends of the lines are either open circuit or match terminated and the reflections are sampled at the TDR heads. The impedance is then determined using both the incident and reflected wave components as an automatic function of TDR and DSO mainframe (digital storage oscilloscope mainframe).

27.5.2 Differential Time Domain Crosstalk

Crosstalk measurements were completed using the TDR step generator to perturb an aggressor line while monitoring a victim line via the differential input channels. Crosstalk measurements are basically a form of transmission measurement, where instead of measuring at the output ports of a given excited pair, one measures at the near- or far-end ports of a neighboring pair. Any remaining open ports are terminated, ideally in their characteristic impedance. Such measurements will illustrate undesired signal coupling occurring between neighboring lines or differential pairs. The crosstalk can be expressed both in terms of the voltage amplitude as well as in percentage of the driver or receiver voltage.

Coupling the stimulus and measurement techniques of the crosstalk measurement with the signals used to generate eye patterns can produce some interesting effects. By using a clock aggressor signal and a standard PRBS signal through the victim line, distortion can be observed when the added signal crosstalk is present.

27.5.3 Eye Diagrams

Eye diagrams can demonstrate the *goodness* of a component or connection path by passing a random bit stream through and monitoring the output over a span of time. Where the *hits* take place in time and voltage, and the distribution of the hits, can determine whether the device under test (DUT) is transmitting data that can be properly interpreted when compared with a *mask* or other criteria set in advance.

To measure an eye pattern across the WarpLink backplane, the HP8133A pulse generator was used in the PRBS23 data pattern mode, with the differential pulse voltage characteristics set to ±500 mV level at 2.5 Gbps and 3.125 Gbps. Once the equipment was set up, the procedure consisted of introducing the bit pattern into a line or differential signal pair and monitoring the output at the far end on the HP54750A.

27.5.4 Time Domain Test Equipment

The equipment used for the passive measurements included NESA's in-house Time Domain Reflectometer/Digital Storage Oscilloscope (TDR/DSO) and 3 GHz pulse generator with PRBS capabilities. A list of the hardware used to complete the relevant tasks follows:

Test Equipment:

1. Hewlett-Packard (HP) 54750A DSO mainframe
2. HP54754A 18 GHz differential TDR/2 channel plug-in module
3. HP54751A 20 GHz 2-channel plug-in module
4. HP8133A (w/option 2) 3 GHz pulse generator fixturing
5. W. L. Gore Phaseflex Cables with APC 3.5 mm connectors (BW: dc –26.5 GHz)
6. SMA, and APC loads, adapters, probes, and miscellaneous fixturing

The TDR/DSO equipment required for these investigations had to be very capable in terms of bandwidth, accuracy, and software support. NESA's in-house HP54750A TDR/DSO is capable of completing differential TDR and crosstalk, and can be used as a four-channel sampler for measurements such as bit-pattern data rate tests.

The HP8133A 3 GHz pulse generator was selected for its high bandwidth and many high-end features including two sets of differential outputs. The HP8133A can supply standard pulses, 32-bit pattern data sequences for data rate testing, and pseudo-random bit sequences (PRBS), useful for such measurements as eye diagrams. This unit allowed for testing up to 3.125 Gbps. These measurements were collected and stored electronically using Hewlett-Packard's Benchlink Lightwave™ software.

27.6 Passive Measurement Results

A passive signal integrity study was performed as part of the design of the WarpLink Reference Design Platform. This study was performed in support of the active measurement and demonstration testing. The evaluation of the backplane establishes the quality of the impedance control of the PCB construction as well as setting a baseline for the eye pattern data before any equalization is used to recover the gigabit signal.

27.6.1 TDR Results

Figures 27.15 and 27.16 show the simulated and measured TDR impedance profiles for the WarpLink gigabit interconnect. The measured TDR profile was taken using the test card designed as part of this project, to access the channels on the backplane. The far end of the backplane interconnect was left open-circuited to ease detection. Although the impedance profiles are on different time scales the features of the interconnect do match. Using this type of evaluation, we can confirm that the simulation result shown in the section, "WarpLink Signal Integrity HSPICE Simulations" on page 788, uses an interconnect model that is representative of the physical interconnect. Note that the actual measurement does not include the package or the second daughter card in the impedance profile.

The TDR measurement was also used to confirm the quality of the PCB fabrication. The broadside couple differential interconnects routes were specified to be 100 Ω ± 10%. This is the case for the WarpLink Reference Backplane and test card. The model and the measurement also similarly model the increasing impedance along the length of the interconnect due to accumulated losses.

27.6.2 Eye Diagram Measurement Results

As part of the passive signal-integrity study, eye diagrams were taken for a variety of route lengths on the WarpLink backplane. Table 27.7 contains a full set of data measurements for both 2.5 Gbaud as well as 3.125 Gbaud. As expected, the longer the interconnect got and the higher the data rate got, the smaller the eye opening got. These results proved the need for equalization in gigabit signaling over long interconnects.

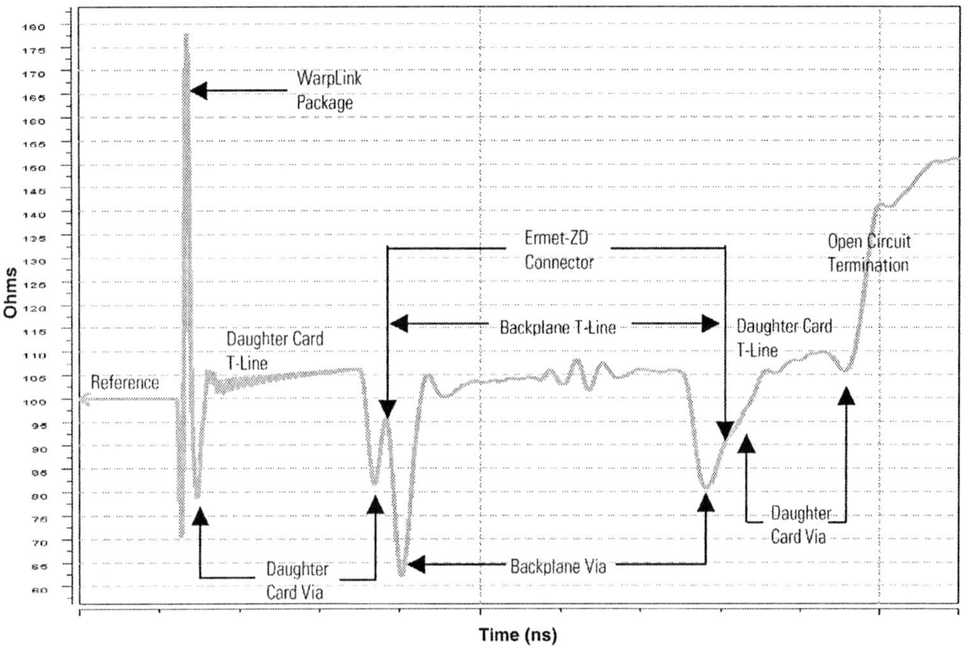

Figure 27.15 Simulated TDR of 10-inch backplane with via stubs.

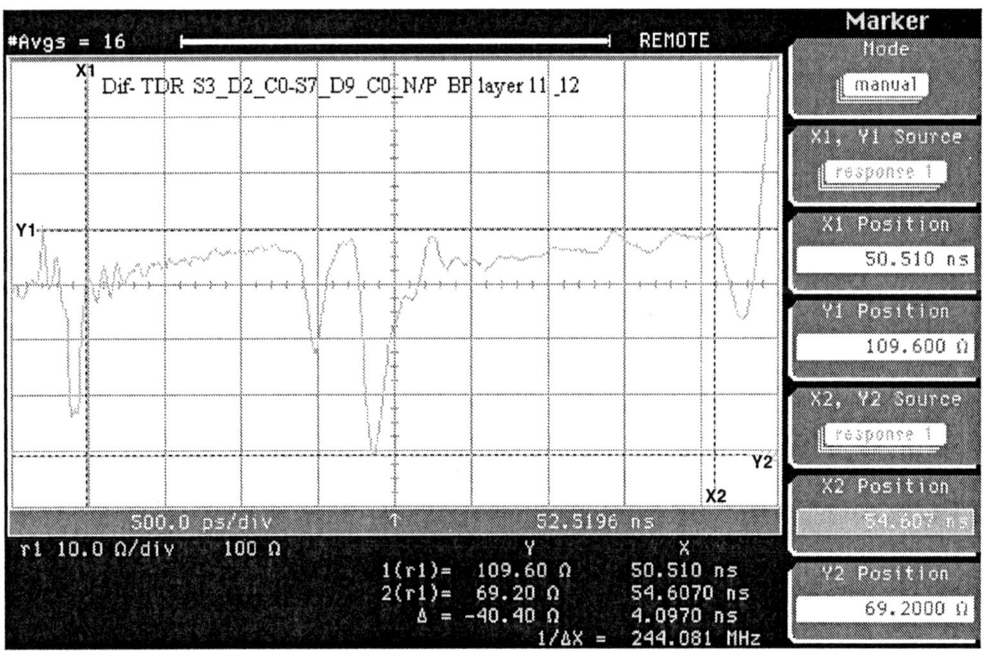

Figure 27.16 Measured TDR of 4.6-inch test card and 6.9-inch backplane.

Table 27.7 2.5/3.125 Gbaud Data Measurements

Net Name	Total Length (in.)	Data Rate (Gbps)	Eye Height (mV)	Eye Width (ps)	Jitter (p-p) (ps)
Reference Signal	Cables Only	2.5	850.0	358.0	40.7
Reference Signal	Cables Only	3.125	840.0	255.5	61.1
S4_D1_D0-S4_D1_D0_[N.P]	8.1	2.5	537.8	295.8	101.5
S4_D1_D0-S4_D1_D0_[N.P]	8.1	3.125	480.2	183.1	133.5
S7_D10_D0-S8_D10_C0_[N.P]	11.3	2.5	501.3	291.6	107.1
S7_D10_D0-S8_D10_C0_[N.P]	11.3	3.125	342.3	215.0	105.0
S2_D1_C0-S3_D1_B0_[N.P]	18	2.5	376.9	252.3	145.9
S2_D1_C0-S3_D1_B0_[N.P]	18	3.125	163.2	170.0	150.0
S1_D2_C0-S7_D9_A0_[N.P]	29.0	2.5	253.0	185.5	212.7
S1_D2_C0-S7_D9_A0_[N.P]	29.0	3.125	39.8	83.0	237.0

Figure 27.17(b) demonstrates the distortion seen for the differential pair S4_D1_D0-S4_D1_D0_P|N, a moderately long interconnect, 8.1 inches in total length, when compared to a reference path measurement shown in Figure 27.17(a). The direct comparison allows us to attribute the PCB interconnect of the WarpLink backplane and test card as the cause of the distortion in the eye opening.

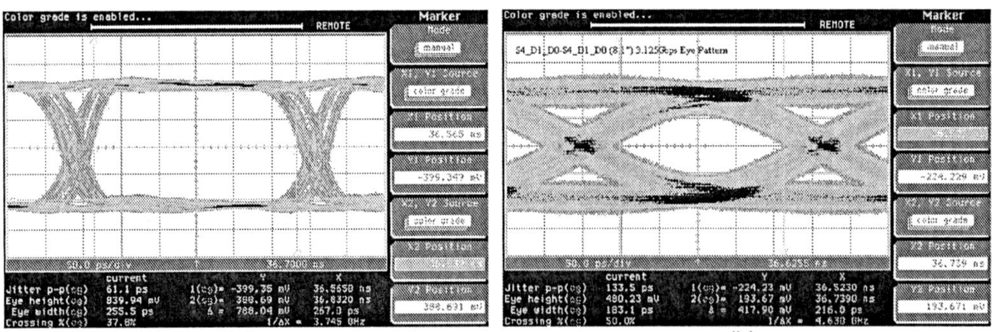

(a) (b)

Figure 27.17 Passive eye patterns. (a) Reference path, and (b) S4_D1_D0-S4_D1_D0_[P|N].

27.7 Active Measurement Results

While the passive measurements quantify the quality of the passive transmission channels, the active measurements show how WarpLink 2.5 Quad performs over the same transmission channels. Active eye patterns driven by a WarpLink transmitter through a variety of transmission channels of the system's backplane were obtained. XAUI compliance was confirmed by overlaying the measured eye patterns to the eye mask defined in the XAUI specification. The following sections describe the test setup and the various channels chosen for the active measurements presented here.

27.7.1 Test Setup

Eye diagrams from two different transmission channels are presented. The transmission channel connecting slot 8 and slot 1 was chosen as an example of the longest, and thus worst case, backplane channel that was not artificially lengthened. The second transmission channel chosen connects slot 7 and slot 1. This channel was artificially lengthened to present a condition that exceeds an interconnect length defined by the XAUI standard.

A line card was installed in either slot 7 or 8. A test card from which the eye diagrams were obtained was in slot 1. The transmitter was configured for the built-in self-test (BIST) mode, configured for PN Equation 2 of the 23-bit PN generator (please refer to the WarpLink 2.5 Quad Product Brief for more details of the BIST mode). The full data path from a WarpLink transmitter on a line card in slots 7 or 8 to the test card in slot 1 consisted of:

1. WarpLink device
2. Package via
3. Line card FR-4 trace
4. Line card connector via
5. Line card — backplane ERmet ZD connector
6. Backplane connector via
7. Backplane FR-4 trace
8. Backplane connector via
9. Backplane connector
10. Test card — backplane ERmet ZD connector
11. Test card FR-4 trace
12. SMA via
13. SMT — SMA connector
14. SMA right-angle adapter
15. 36 inches of Rosenberger rf-coaxial cable
16. ac-coupling SMA adapter
17. TDS-8000 scope

The WarpLink device is clocked via the clock distribution networks originating from slot 7. (Please refer to "Clock Distribution" on page 780 for more details). The clock trigger of the TDS-8000 is also clocked with the same clock distribution network, by way of the test card in slot 1.

27.7.2 Eye Diagrams from Slot 8 to Slot 1

The interconnect between slots 8 and 1 are the longest channel in the 19" rack WarpLink Design Reference system that is not artificially lengthened. The total length of the transmission channel in FR-4 is 18.7". This includes 1.8" of FR-4 on the line card, 12.2" on the backplane, and 4.7" on the test card. Without pre-emphasis/equalization, WarpLink exceeds the XAUI mask with a 5% margin. With preemphasis/ equalization, the WarpLink gigabit link satisfies the XAUI mask with a 9% margin. The comparison of Figure 27.18(a) and (b) shows a clear benefit of pre-emphasis/equalization to the signal fidelity of the eye diagram.

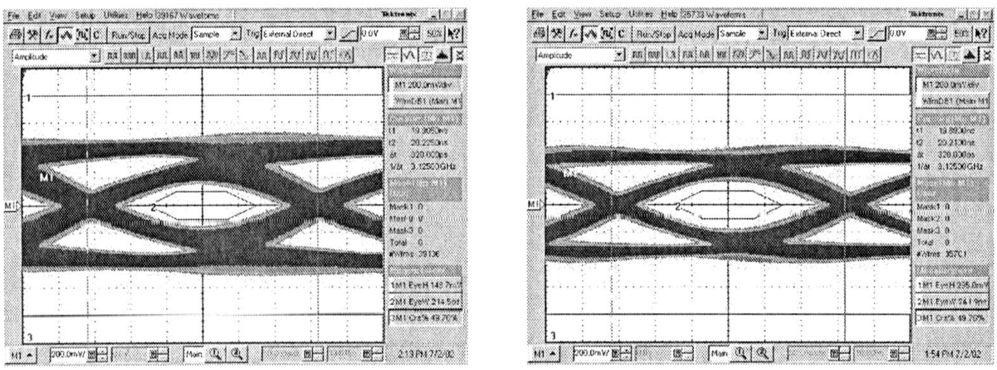

(a) (b)

Figure 27.18 Active eye patterns Tx – Slot 8 to Rx – Slot 1. (a) Without equalization, and (b) with equalization.

27.7.3 Eye Diagrams from Slot 7 to Slot 1

The artificially lengthened interconnect between slots 7 and 1 presents a condition that is worse than that defined by the XAUI standard. This is a worst-case routing condition that exemplifies a routing implementation on a large backplane. The total length of the transmission channel in FR-4 is 26.2". This includes 1.8" of FR-4 on the line card, 19.7" on the backplane, and 4.7" on the test card. With preemphasis/equalization on, WarpLink passes the XAUI mask with a 3% margin. As expected via simulation the resultant eye does not conform to the XAUI mask with pre-emphasis/equalization disabled. Again, the comparison of Figure 27.19 (a) and (b) shows a clear benefit of pre-emphasis/equalization to the signal fidelity of the eye diagram.

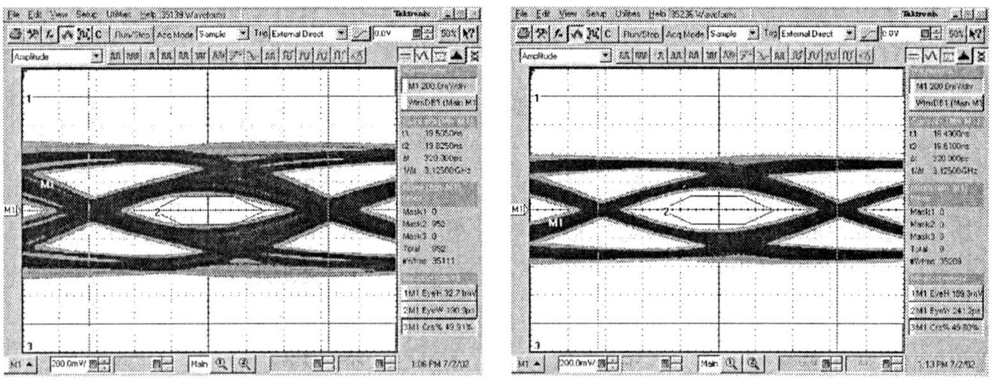

(a) (b)

Figure 27.19 Active eye patterns Tx – Slot 7 to Rx – Slot 1. (a) Without equalization, and (b) with equalization.

27.8 Summary and Conclusions[7]

The Motorola WarpLink Reference Design Platform demonstrates the capabilities of the WarpLink 2.5 Quad SERDES device. The device was operated at wire speed of 3.125 Gbaud to provide an aggregate speed of 10 Gbps per device. The device's pre-emphasis/equalization feature proved indispensable for data transfer at gigabit wire speeds over long interconnects. In particular, WarpLink 2.5 Quad has demonstrated the ability to operate at, and beyond the XAUI specification of 3.125 Gbaud over 50cm long channels.

The WarpLink Reference Design Platform is built with common design practices, components, and PCB material (FR-4). Unrealistic and expensive gigabit design requirements — such as blind vias, via drill-back, bottom-layer-only routing, and low dielectric constant PCB materials — were not required.

Using a fully matured simulation and design methodology, NESA implemented a full end-to-end gigabit simulation matrix of cases. The currently available SerDes model for the WarpLink 2.5 Quad device was verified to emulate actual measured results. Using this modeling strategy, areas of risk for the gigabit links were identified and properly addressed while still in the design phase.

7. This design study was prepared as a joint effort by Quang Xuan Nguyen of the Motorola WarpLink design team and Dr. Edward Sayer 3rd from North East Systems Associates Inc. (NESA, *www.nesa.com*). NESA was contracted by Motorola to do the design, perform simulations, and define the layout rules for the reference backplane. Their expertise lies in the area of the practical design and manufacturing techniques required for high-performance systems and backplanes. Additional thanks go to Michael Baxter of NESA and Thecla Chomicz of Motorola for their technical expertise and assistance in performing the extensive passive and active measurements presented here.

27.9 Exercises

27-1. How is it that the WarpLink MC92610 SerDes device, which is designed for point-to-point applications (single-load applications), can be used in the Reference Design Platform when the platform has 8 plug-in cards (8 potential loads)?

27-2. What logic family are the WarpLink device serial drivers in? ECL? CML?

27-3. Refer to the backplane mesh interconnects shown in Figures 27.2 and 27.3 on page 780. (a) For the mesh architecture, what is the relationship between the number of slots and the number of mesh ports of each slot? (b) For an 8-slot mesh backplane, what is the number of full-duplex connections that the backplane must provide? Include in this number the connections of ports that loop back on themselves. Exclude clock networks from this number. (c) For an 8-slot mesh backplane, what is the total number of traces needed to implement these mesh interconnects? (d) For a mesh backplane with "s" slots, what is the total number of full-duplex connections that the backplane must provide? (e) For a mesh backplane with "s" slots, what is the total number of traces required to implement these interconnects? (f) As the number of slots increases, how does the number of backplane interconnects grow? (g) Why do you want to include interconnects in the backplane for unconnected ports that loop back on themselves?

27-4. Refer to the backplane fabric architecture shown in Figure 27.4 on page 781. (a) For the 8-slot backplane shown in the figure, how many full-duplex line port to fabric port interconnects are required? (b) For this same 8-slot backplane, how many actual traces are required to be routed in the backplane? (Exclude clock traces from your number.) (c) For a backplane with "s" slots, how many actual traces are required to be routed in the backplane? (d) Which backplane architecture requires more traces, the mesh-switching scheme or the fabric-switching scheme?

27-5. Why was the Reference Design Platform built and tested?

27-6. What is the difference between signaling rates at 3.125 Gbaud (gigabaud) and data rates at 2.5 Gbps (gigabits per second)?

27-7. (a) Does the MC92610 WarpLink device provide transmit equalization, receiver equalization, or both? (b) What is the difference between the two?

27-8. Refer to Table 27.4, "Simulation Matrix Eye Pattern Parameters," on page 792 and the corresponding eye patterns shown in Figures 27.10 and 27.11 on page 793. Describe the improvement seen in the eye patterns when the WarpLink transmit pre-emphasis feature is utilized.

27-9. For the Reference Design Platform, how many layers are in the backplane? Line card? Test card?

27-10. The Reference Design Platform has a 32-bit bus going to all slots that can be used as a processor bus. This bus is designed to terminate Gunning Transceiver Logic Plus (GTLP) drivers. What are two features of this bus that make it well suited for GTLP drivers?

27-11. (a) Which ports of the MC92610 WarpLink 2.5 Gbps Quad SerDes device are redundant, the serial ports or the parallel ports? (b) Which ports are not redundant? (c) For this device, what do we mean when we say a port is redundant? (d) When this device is used on a line card and the card is inserted into the Reference Design Platform backplane, which device ports become connected to the mesh ports of the slot, the primary or redundant ports? (e) When this device is used on a line card and the card is inserted into the same backplane, which device ports become connected to the fabric-topology line ports of the slot, the primary or redundant ports?

27-12. What are the major parameters that shape a chassis/backplane system that must be established early on in the design once the intended product market is established?

27-13. To demonstrate WarpLink's noise tolerance, the backplane of the Reference Design Platform includes some noise-generating nets close to victim gigabit nets. (a) How close were the noise-generating nets placed to the victim gigabit nets to assure coupling of the noise onto the victim nets? (b) What is the recommended design rule for the spacing of gigabit nets in order to minimize any crosstalk? (c) When we talk about spacing, are we talking about the spacing between the differential pairs of the gigabit signals, or are we talking about the spacing between the true and inverted traces within the differential pair?

27-14. The backplane of the Reference Design Platform contains six broadside-coupled routing channels using 12 layers of the board to route the serial differential signals. Why were broadside-coupled differential traces used instead of edge-coupled differential traces?

27-15. (a) In the TDR analysis of the 10-inch backplane, which kind of via located in the backplane at the connectors produced the most impedance discontinuity, the "through via" or the "stub via"? (b) In the TDR plots of Figure 27.12 on page 794 and Figure 27.13 on page 794, what features in the plots indicate that this is the case?

PART 6

Emerging Protocols and Technologies

CHAPTER 28

Electrical Optical Circuit Board (EOCB)

Chapter Objectives

- Explain that OEM manufacturers in datacom, telecom, and high-end computing are setting the development agenda for printed circuit boards (PCBs) with performance specifications that can only be met using optical waveguides.
- Identify the major companies and development teams that are bringing electrical optical circuit boards into development and manufacture.
- Discuss optical waveguide materials and the three-dimensional fabrication techniques being used to manufacture printed circuit boards with waveguides.
- Present performance advantages to "all-optical" processing and routing of signals and the advantages of optical interconnects over their electrical counterparts.
- Present an optical backplane market forecast and an optical packaging roadmap to the future.

28.1 The Photonic PCB Industry and Development Programs[1]

The challenge of optoelectronics and integration of photonics into the printed circuit has started to take off. In the next seven years, expectations are that photonic PCBs will grow to a $700 million industry. This chapter looks at the issues, materials and current processes being researched to create this integrated optoelectronic circuit board by European, Japanese, and North Ameri-

1. Except where otherwise noted, this chapter was written by Happy Holden of Westwood Associates. Reprinted with permission of Westwood Associates.

can organizations. In addition to reviewing the global players in polymer photonics, this chapter will review the current programs of four of the following eight groups globally [Hol02a]:

- Terahertz Photonics (Truemode Backplane™)
- University of Ulm, Fraunhofer Institute, DaimlerChrysler, Siemens (EOCB)
- NTT[2]
- Dupont, Hewlett-Packard (PolyGuide™)
- University of Texas
- NIST,[3] 3M, Goodyear (TOPCat™)
- JIEP[4]
- 3M

28.2 Optoelectronic Printed Circuits Based on HDI-Microvia Technology

The widespread use of new electronic components employing ball-grid array (BGA), chip scale packaging (CSP), and other evolving technology form-factors means new fabrication techniques must be used to create printed circuit boards (PCBs) that will accommodate parts with extremely tight lead pitches and small geometries. In addition, extremely fast clock speeds and signal bandwidths challenge systems designers to find better ways of overcoming the negative effects that radio frequency interference (RFI) and electromagnetic interference (EMI) have on their product's performance. Optoelectronics printed circuits are based on HDI-microvia technology. High-density interconnect (HDI) printed circuits are now being designed in ever-increasing quantities for very high speed applications. In addition to the extremely high-frequency electrical signals, the need now includes the requirement for integrated optical waveguides. Finally, increasingly restrictive cost targets are compounding problems associated with today's smaller, denser, lighter, and faster systems. In the area of optoelectronics, the need for integrated waveguides is being met with major programs.

One of the new challenges is to create and implement MOEMS (micro optoelectromechanical systems), which are the miniaturized combination of optics, electronics, and mechanical functions. A related technology is MEMS (micro electromechanical systems), the miniaturized combination of electronic and mechanical functions. [Hol03]

Traditionally, signals on fiber-optic cables have been converted to electrical signals at the location where the fiber-optic cable connects to a printed circuit board or backplane. The signals are then processed and routed as electrical signals, and then often converted back into optical signals for outgoing transmission over fiber-optic cables once again. This process is referred to as the O-E-O solution, or optical-electrical-optical solution.

2. NTT Corporation is Nippon Telegraph and Telephone Corporation.
3. NIST is the National Institute of Standards and Technology.
4. JIEP is the Japan Institute of Electronics Packaging.

New technology which offers an alternative to the O-E-O process routes optical signals right on the PCBs and backplanes thereby eliminating the optical-to-electrical and electrical-to-optical conversions. The advantages to the "all-optical" solution — that is, retaining a signal in its optical form during processing and routing — are ease of scalability (the ability to vary the number of devices served or vary the loading on the equipment), improved signal integrity, elimination of a bottleneck of high-speed signal transmission, minimized board real estate, and reduced crosstalk and EMI leakage. It also results in low skew, high-density signal routing, reduced printed circuit board thickness and size, reduced need for high-dielectric constant materials, increased design flexibility, provides high reliability, and is cost-effective.

Similarly, there are advantages of optical interconnects over traditional electrical ones. These advantages include low crosstalk and EMI and higher interconnect density. Interconnects can be placed on a 80 to 100 μm pitch and there is no need for signal-ground pairs or the careful design of transmission line geometries. There is greater voltage isolation and reduced timing jitter, especially for low-voltage supplies and long lines. Designs become easier, with fewer board layers, bit-rate independence, and an absence of electromagnetic wave considerations such as impedance matching, inductive and capacitive coupling, backreflection, and grounding issues. [Hol03]

28.2.1 Benefits of HDI

Staying competitive and delivering the products people want means seeking out and embracing the best available technologies and design methodologies. The use of PCBs incorporating microvia circuit interconnects is currently one of the most viable solutions on the market. Adopting microvia technology means products can utilize the newest, smallest, and fastest devices, meet stringent RFI/EMI requirements, and keep pace with downward-spiraling cost targets. Integrated photonics including waveguides are built on this HDI-microvia foundation.

28.2.2 Microvia Technologies

Microvias, as the name implies, are vias of less than or equal to six mils (150 microns) in diameter. Their most typical use today is in blind and buried vias used to create interconnections through one dielectric layer within a PCB. (A blind via connects a surface layer to one or more internal layers but does not go through the board to the opposite outside layer. A buried via connects internal layers [usually two signal layers adjacent to one another] but does not connect with either of the two surface layers.) Microvias are commonly used in blind via constructions where the outer layers of a multilayer PCB are connected to the next adjacent signal layer. Used in all forms of electronic products, they effectively allow for the cost effective fabrication of high-density assemblies. IPC[5] has selected high-density interconnection structures (HDIS) as a term to refer to all of these various microvia technologies.

5. IPC is the Association Connecting Electronics Industries.

28.2.3 Use of Microvias in PCBs

28.2.3.1 Better Electrical Performance/Signal Integrity

Due to the physical structure of microvias, there is a reduction in switching noise. This is attributable to the decreased inductance and capacitance of the via as its physical size becomes smaller and shorter. A microvia will have nearly one-tenth the electrical parasitics of a throughhole. (A typical via, smaller than a throughhole and larger than a microvia, will have parasitic values on the order of magnitude of 0.5 pF for capacitance and 1 nH for inductance. For high-speed design, vias should be modeled and parasitic values determined precisely.) Another advantage of using microvia technology for creating interconnects is a reduction in signal reflections and crosstalk between traces. The corresponding increase in routability area also allows designers to place traces further apart when necessary to reduce crosstalk.

28.2.3.2 Improved RFI/EMI/ESD

In the realm of RFI and EMI, increased routability area combined with the microvia's physical via-in-pad implementation allows designers to place more ground plane around components. By doing this, the size of ground return loops decreases and improved RFI and EMI performance is realized. These benefits are highlighted in Figure 28.1. [Hol02b]

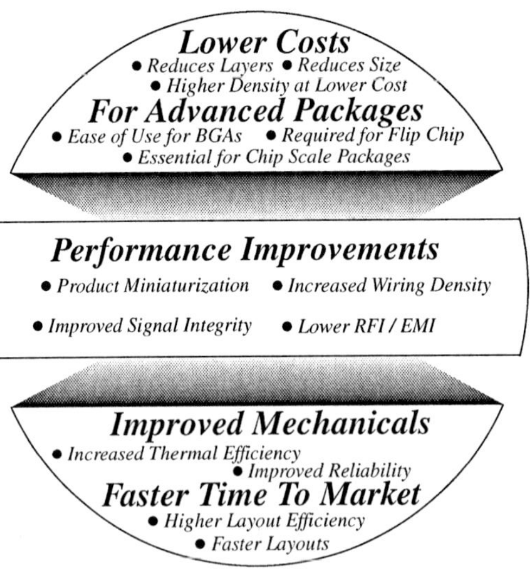

Figure 28.1 Benefits of HDI come from five major areas. [Reprinted with permission of Westwood Associates.]

28.2.4 Photonics and Electrical Performance

The performance of conventional electrical interconnection technology is limited through underlying physical properties. The most important disadvantages and problems respectively are as follows:

- Emission of and susceptibility to electromagnetic radiation
- Limited data-rate-length product caused by attenuation and dispersion, mainly affected by the high-frequency skin effect and the frequency dependent loss factor tan δ of the dielectric materials
- High number of single interconnects (pin-count) at component level and connector level caused by the limited data-rate per single interconnect
- High effort to guarantee a high level of signal integrity

HDI's contribution comes mainly from the adage, "Smaller and closer is better!" That is, HDI's main contribution is *miniaturization*! Optoelectronics requires this precision and miniaturization. The signal integrity improvements for photonics come from three phenomena:

1. Reduction of noise
2. EMI radiation reduction
3. Improved signal propagation and lower attenuation noise

28.3 Photonics and Waveguides

While an electrical signal can be multiplexed on a single wire or trace, many laser frequencies can be multiplexed on a single waveguide. This results in a 1,000X increase in information handling while not being influenced by magnetic and electrical fields like electronic signals.

The integrated optical board and backplanes are predicted to have steady growth over the years. Figure 28.2 shows the predicted growth forecast by ElectroniCast. By 2008, the value of rigid and flex photonic boards is predicted to be $565 million.

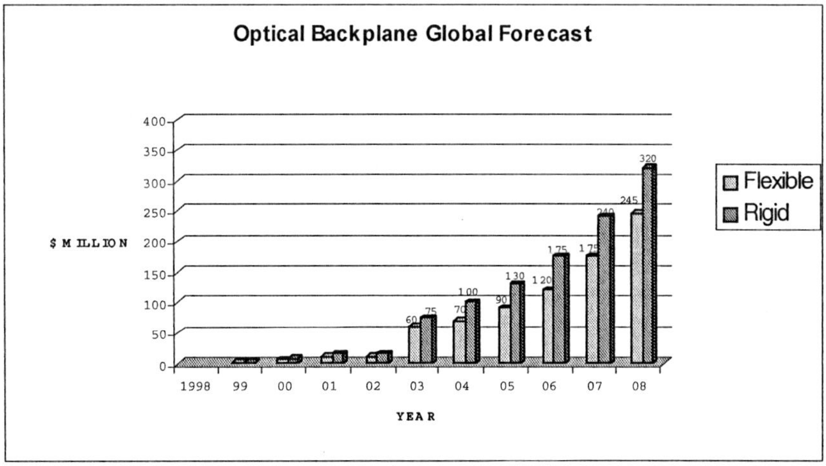

Figure 28.2 Optical backplane market forecast by ElectroniCast. [Courtesy of ElectroniCast.]

The need to route "long-line optical signals" in an efficient manner and at lower costs has created the interest in integrated optical waveguides and methods of optical routing. The current method of signal routing is to convert the optical signal to an electronic signal, interrogate it, switch it to its next destination, and then convert it back to an optical signal. All of this forms what the industry calls "terabit routers" and can require up to 42 large, complex multilayers. With integrated optics, this can likely be done by one or two unique PCB assemblies. The "board-to-board" optical circuits may look like the illustration in Figure 28.3.[6] A major challenge to implementing this technology is determining how to read packet header information encoded in a light beam so you know where to route the light beam next.

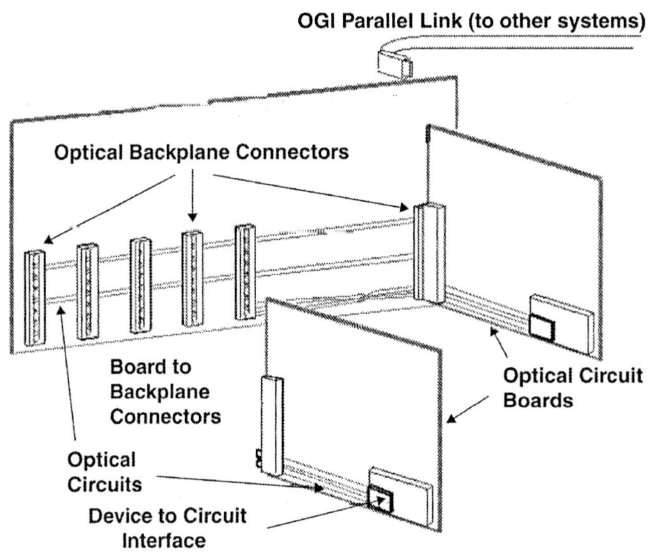

Figure 28.3 Board-to-board optical interconnections. [Wat01] [Reprinted with permission of 3M Corporation.]

It has to be considered that the design process is of the same importance within the value supply chain as the manufacturing process. The most important components and manufacturing technologies to be developed are as follows:

- Optical waveguides and techniques for integrating them into printed circuit boards
- Electrical-optical and optical-electrical converters for realizing optical transmitters and receivers

6. In Figure 28.3, "OGI" refers to Optical Graphics Interface.

- Robust coupling concepts to realize optical
  - component-to-board connectors
  - board-to-board connectors
  - board-to-fiber connectors

The following tools, models, algorithms and methodologies are necessary to provide an efficient design of electrical-optical printed circuit boards:

- Simulation models and algorithms to enable time domain simulation of optical interconnects
- Routing and general design rules for optical interconnects and entire electrical-optical printed circuit boards
- Methodologies for the analysis and optimization of electrical-optical printed circuit boards in respect of functional, technological, and cost requirements

Numerous techniques are under development for multimode waveguide lamination, through-hole light coupling, surface finishes, and soldering connectors. [Erb01]

Integrated optoassemblies will have a number of factors new to PCBs:

- Optical waveguide materials
- 3D fabrication techniques
- New components
- 3D assembly techniques

28.3.1 Optical Waveguide Materials

The optical materials being considered for integrated waveguides are currently polymers. Polymers have a number of advantages:

- Stability — High thermal stability and long-term photostability. Tested for compliance to Telecordia (previously Bellcore) international standards such as 1209 and 1221 for > 600 hours at 85°C and 85% RH (relative humidity), solder temperature >230°C and degradation temperature >350°C.
- Well established — A huge body of data was obtained over the last 100 years on polymers including all the popular photoresists.
- Useful — Unique properties you cannot get anywhere else (such as bend radius, modulators, and index tuning) and unique processing options (such as photolithographic, RIE (reactive ion etching), direct laser writing, molding, and printing).

Polymers also have a number of disadvantages:

- Unstable — Many have low thermal stability (point of failure below 80°C) with photodegration (laser dyes < days) and are sensitive to delamination, moisture, and chemicals.
- Unknown — New materials require new processes, equipment, and experience.
- Useless — Some polymers have losses of 20 dB/km (decibels/kilometer) while optical glass is <0.1 dB/km. The packaging cost of polymers is 80% of the device cost.

Candidate materials are acrylates, halogenated acrylates, cyclobutenes, polyimides, and polysiloxane. Some of the most popular are listed in Table 28.1.

Table 28.1 Candidate Polymers for Optical Waveguides

Manufacturer	Polymer Type	Patterning Techniques	Waveguide Optical Losses[a] 840 nm	1300 nm	1550 nm
Allied Signal	Halogenated Acrylate	Lithographic, RIE Laser	0.01	0.03	0.07
	Acrylate	Lithographic, RIE Laser	0.02	0.2	0.5
Dow Chemical	Benzocyclobutene	RIE	0.8	1.5	—
	Perfluorocyclobutene	RIE	0.01	0.02	0.03
DuPont	Acrylate (PolyGuide)	Lithographic, RIE	0.2	0.6	—
	Teflon AF	RIE	—	—	—
Amoco	Fluorinated Polyimide	Lithographic	—	0.4	1.0
BF Goodrich	Polynorbornenes	Lithographic	0.18	—	—
General Electric	Polyetherimide	RIE, Laser	0.24	—	—
JDSU	Acrylate	RIE	—	—	0.6
NTT	Halogenated Acrylate	RIE	0.02	0.07	1.7
	Polysiloxane	RIE	0.17	0.43	—
Asahi	Cytop	RIE	—	0.3	—

a. dB/cm

[Courtesy of Lucent Technologies Inc.]

Table 28.2 shows transmission losses of some polymers (acrylate and fluorinated acrylate). For comparison, Figure 28.4 illustrates transmission loss versus wavelength for PMMA (polymethyl methacrylate), D-PMMA (deuterated PMMA, also abbreviated as dPMMA), Cytop (made by Asahi), PCS (polycarbonate silica), and silica. Tables 28.1 and 28.2 show losses in dB/cm whereas Figure 28.4 shows losses in dB/km. [Erb01]

Table 28.2 Wavelength-Dependent Loss (dB/cm) of Conventional and Optimized Polymer Waveguides

λ(nm)	Acrylate	Fluorinated Acrylate
840	0.02	0.001
1300	0.15	0.03
1550	0.45	0.06

[Courtesy of Lucent Technologies Inc.

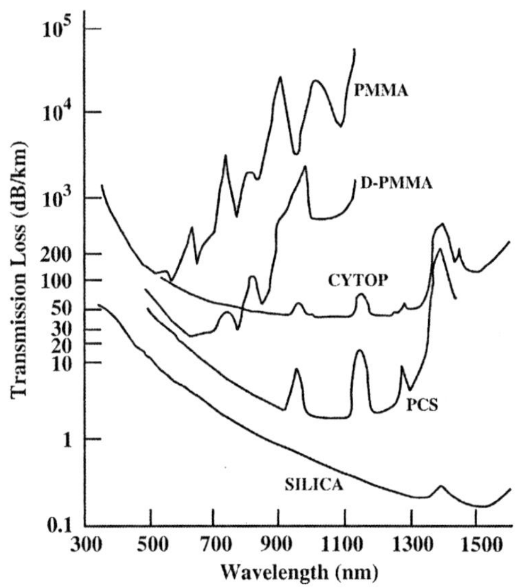

Figure 28.4 Transmission loss (dB/km) as a function of wavelength (nm) for PMMA, D-PMMA, Cytop, PCS, and silica. [Courtesy of Lucent Technologies Inc.

28.3.2 3D Fabrication Techniques

Eight integrated waveguide development programs and their three-dimensional fabrication techniques have been described in company and organization literature. These programs were listed in Section 28.1 on page 809. Four of these fabrication techniques will be discussed in this chapter presently:

- Truemode Backplane
- EOCB
- PolyGuide
- TOPCat

28.3.2.1 Terahertz Photonics' Truemode Backplane[7]

"Vital to all embedded optical backplanes is a material and process for the deposition of low-loss multimode waveguides. The main selection criteria for the material are that it exhibit low loss, is photo writable for ease of manufacture, and is able to withstand the temperature and pressure regimen of lamination (180°C at 350 psi) and solder reflow (up to 250°C for 20 s) during PCB manufacture.

Terahertz Photonics has developed a polymer called Truemode Backplane specifically to meet these needs. As with all Truemode polymers, the new polymer is a highly cross-linked multifunctional acrylate. Optimization of the composition ensures good adhesion to FR4, enhanced thermal stability, humidity resistance and ultra-low loss of <0.03 dB/cm at a wavelength of 850 nm." [Ter03] Figure 28.5 shows a typical layer stackup for embedding a Truemode Backplane polymer waveguide in a PCB.

Figure 28.5 Lamination stackup for implementation of Truemode Backplane waveguides. [Courtesy of Terahertz Photonics.]

A typical link budget for a Truemode Backplane looks something like this:

VCSEL[8] output	0 dB	(10 Gbps w/10^{-12} BER)
Propagation loss	5 dB	
Bends	1 dB	
Splitter	3 dB	
Coupling in/out	3 dB	(worst case)
Total	12 dB	
Margin	5 dB[9]	

7. Courtesy of Terahertz Photonics
8. VCSEL is vertical cavity surface emitting laser.
9. Assumes 17 dB of loss was budgeted for the system.

"At 0.05 dB/cm, the propagation loss is essentially irrelevant. Only manufacturability and cost are . . ." challenges.[10]

"In addition to providing material for optical backplanes, Terahertz Photonics has intellectual property in the critical area of interconnects to the optical backplane." [Ter03]

For Telecom Components — Solica™ Silica and Truemode Polymer "Silica is the dominant material from which passive optical components are manufactured because silica is environmentally stable and has low loss when coupled to silica optical fiber. Silica is used in a variety of planar components including splitters and couplers and interferometric components such as interleavers and arrayed waveguide gratings (AWGs).

The strong position that planar components have is set to increase as multiple functions such as combined attenuators and AWGs are implemented on single chips to reduce costs. In addition, there is activity among leading component companies to use silica as the platform for hybrid devices with indium phosphide source lasers and detectors.

For certain applications, the limitations of silica require that an additional material be used either of itself or in conjunction with silica. When ultra-low cost is required, for example in PON splitters,[11] there are few materials that can compete with polymer which is spun onto the substrate and UV (ultraviolet light) cured. The performance of the polymer remains critical as it must exhibit low loss and good environmental properties.

For components which must be dynamic, such as variable optical attenuators, optical polymers with a high thermo-optic effect are also of benefit when combined with silica. For components which are required to be athermalized,[12] such as AWGs, the negative thermo-optic effect of polymer is also of benefit." [Ter03]

28.3.2.2 Electrical Optical Circuit Board

The second waveguide fabrication technique takes into consideration new requirements resulting in a new concept for printed circuit boards that are characterized by a hybrid layer-stack consisting of electrical and optical layers (see Figure 28.6). This concept provides the required compatibility concerning the design and manufacturing processes, and it enables the required free optical routing. The optical waveguides are part of separate optical layers resulting in an electrical optical circuit board (EOCB).[13]

The EOCB process uses an embossing mold (shim) to hot-emboss the foil structure of the waveguide. Topas, the optical material, fills the deformed core. After curing, the core is lami-

10. Dr. Frank Tooley, Chief Technical Officer, Terahertz Photonics.
11. "PON is Passive Optical Network. A PON splitter is located in the distribution network and is used for delivering broadband network access service to business and residential customers. It distributes an optical signal from a common transmitter to multiple subscribers, merges signals from all subscribers, and connects them to the common OLT receiver (optical line terminal receiver located in the central office). The PON splitter passes light in both directions. The same amount of optical loss is observed in both the splitting and merging directions." [Lee03]
12. "Athermalize" means to reduce or eliminate the effects of temperature changes.
13. See references [Gri02], [Gri01a], [Tag01], [GRI01b], [Kra99], and [Kra00].

nated into the EOCB. This hot-embossing process, which is one possible way of fabrication, is depicted in Figure 28.7(a) and is based upon first manufacturing rib waveguides and then laminating them in a second step on a polymer substrate. Finally, the optical layer is obtained after filling with the overcladding. (Cladding is the process of laminating or attaching another material over a subject. Overcladding is the process of laminating or attaching another material over the subject in such a way as to envelop the three-dimensional features of the subject.) In order to meet the high-temperature requirements already mentioned, high-Tg (high glass transition temperature) materials such as COC[14] or special polycarbonate can be used.

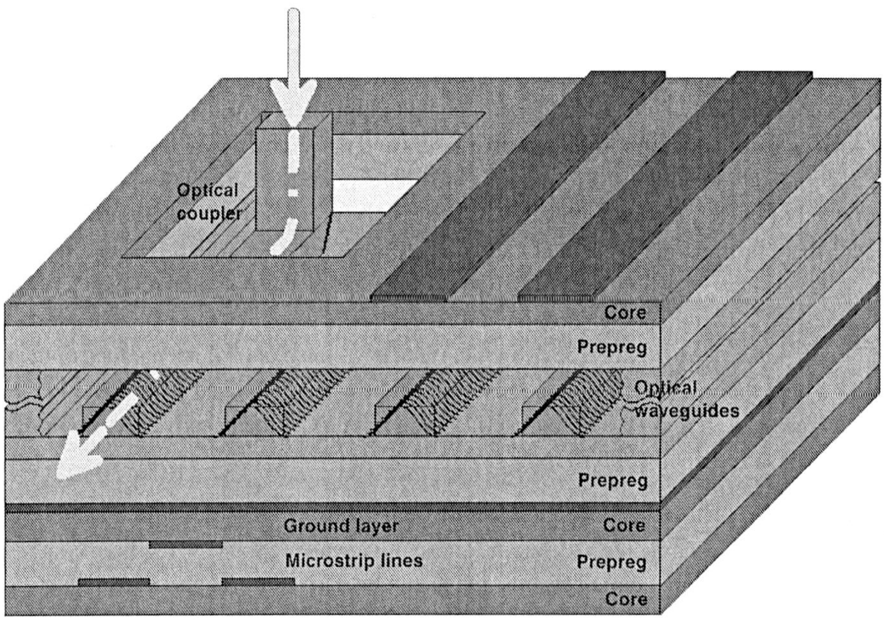

Figure 28.6 Concept of a printed circuit board with an integrated optical layer containing optical multimode waveguides. [Gri02] [© C-LAB (Joint venture of University of Paderborn and Siemens Business Services), Paderborn, Germany.]

In another hot-embossing process using polymers for waveguide core and cladding, so-called hot embossing processes can be applied in order to manufacture the optical layers (refer to Figure 28.7(b)). This technology enables the required degrees of freedom as well as a precise fabrication of arbitrary passive structures like power splitters and starcouplers. The entire fabrication process of the optical layer is widely compatible to standard lamination processes. It is based on the structuring of a polymer foil, filling the resulting grooves with fluid pre-polymeric core material, and curing. Finally, a second polymer foil is laminated on the first one. [Gri01b], [Kra99], [Kra00]

14. COC is Cyclo Olefin Copolymer, a high-temperature thermoplastic from Europe.

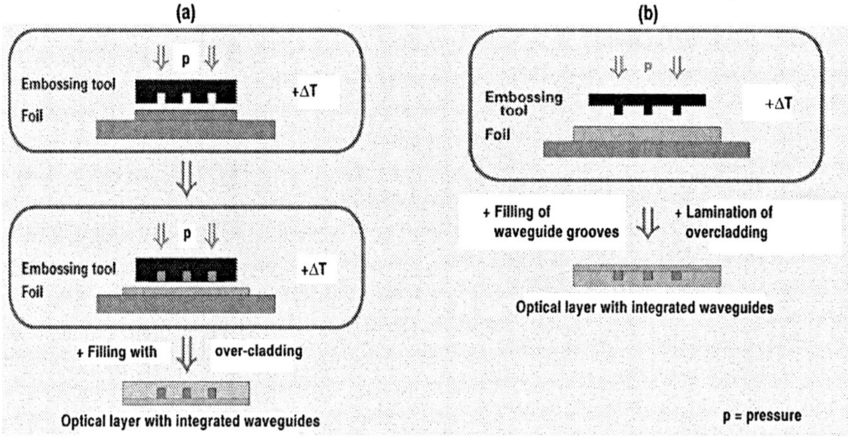

Figure 28.7 Manufacturing optical layers by hot embossing, (a) rib-cladding process and (b) groove-filling process. [Gri02] [Courtesy of *PC FAB* magazine, UP Media Group.]

About four years ago, electrical optical printed circuit boards that enabled onboard datarates of 3 Gbps could be developed, manufactured, and demonstrated. This work was part of the German projects "Electrical/Optical Circuit Board" and "OptoSys," which were and still are supported by the German government, Department of Education and Research (BMBF) under grants 16SV802/6 and 01BP801/01. (Dortmund, Ulm, and the Fraunhofer Institute for Applied Solid-State-Physics in Freiburg are participating on a sub-order basis.) An example of the European program can be seen in Figures 28.8 and 28.9. Figure 28.8 shows the Siemens C-LAB test vehicle, called OptoSys, fabricated by a collaboration of Siemens AG, Infineon Technologies AG, ILFA Feinstleitertechnik Gmbh, DiamlerChristler AG, University of Paderborn, Dortmund and Ulm, and the Fraunhofer Institute for Applied Physics in Freiburg, funded by the German government. On one side is the transmission electronics and VCSEL that transmits the optical signal; on the left side is the photoreceptor and associated circuitry; connecting the two is the integrated optical waveguides. Figure 28.9 shows a close-up of the VCSEL with cooling.

The EOCB technology has been developed under a consortium of Andus, FhG IZM Berlin, ILFA, Mikropacks, OECA, Robert Bosch, Siemens, and the University of Paderborn.

28.3.2.3 PolyGuide

PolyGuide is a photosensitive polymer developed nearly 10 years ago. The fabrication process requires vacuum lamination, adhesion promotion, and then a photo mask to expose the material to UV light as shown in Figure 28.10. The material is then post-cured and will perform as a waveguide.

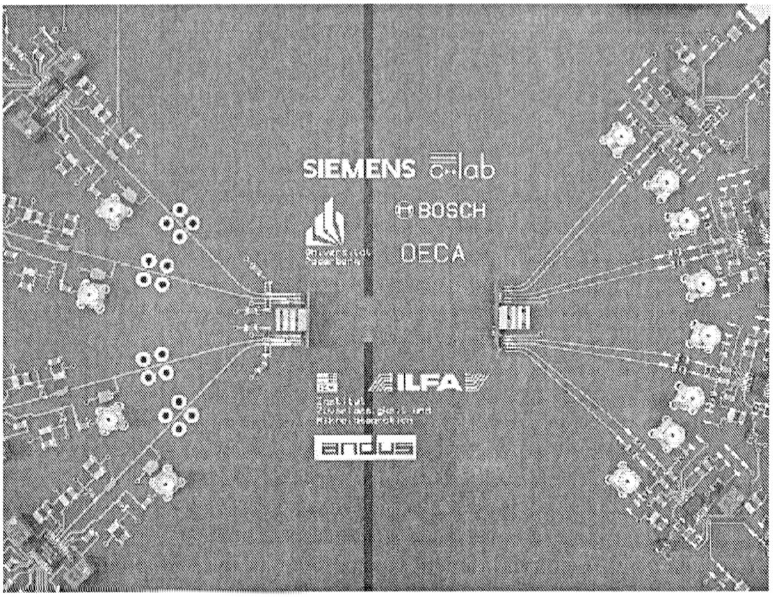

Figure 28.8 C-LAB test vehicle for optical waveguides. [Gri02] [© C-LAB (Joint venture of University of Paderborn and Siemens Business Services), Paderborn, Germany.]

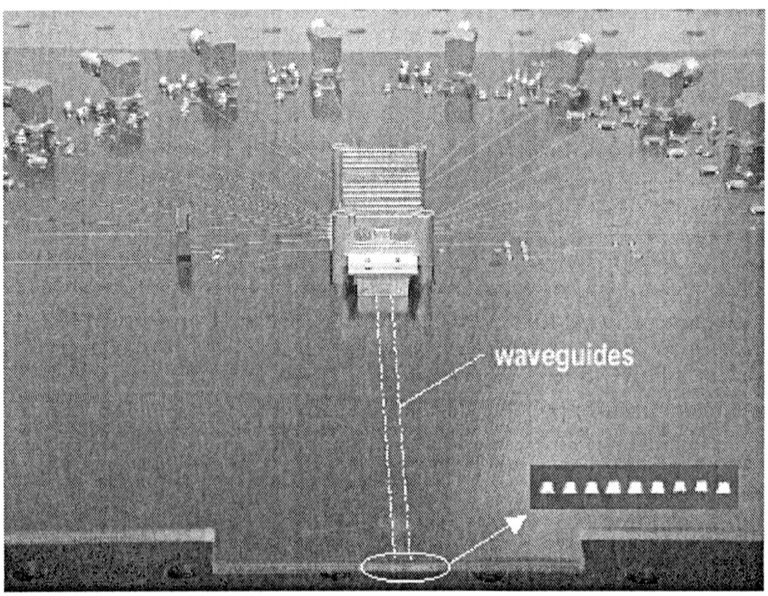

Figure 28.9 Close-up of the VCSEL laser transmitter for the C-LAB test vehicle. [Gri02] [© C-LAB (Joint venture of University of Paderborn and Siemens Business Services), Paderborn, Germany.]

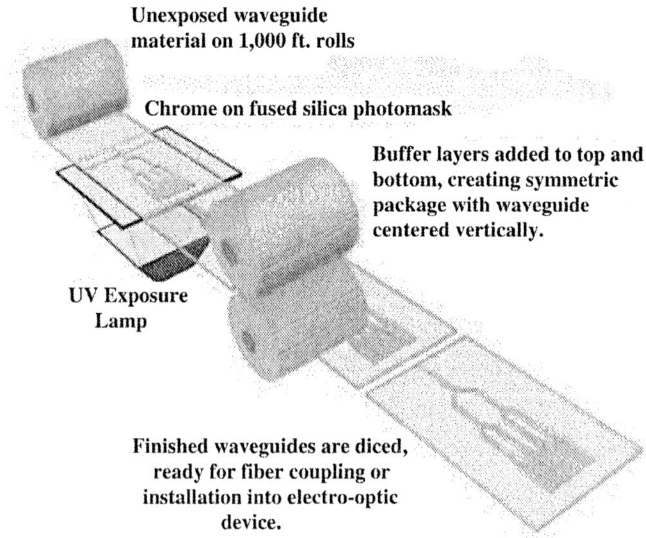

Figure 28.10 PolyGuide optical waveguide manufacturing process. [Wat01] [Courtesy Bruce L. Booth of Optical CrossLinks.]

28.3.2.4 TOPCat

TOPCat is the result of an NIST-sponsored ATP project[15] and is a planar polymer waveguide. The material is a polynorbornene where the manufacturing process is a "core-first" microreplication. A laser is used to create an embossing tool and the tool is coated with the pre-polymer material. The excess is then struck off and the materials cured and finally overcoated with a cladding. The waveguide core can be lifted out and an overcladding applied. Refer to Figure 28.11 for an example of the waveguide fabrication process. Figure 28.12 is one of the finished planar polymer waveguides. The lab results with the BF Goodrich polynorborene polymer proved to be very encouraging. The waveguide showed these characteristics:

- 0.3 dB/cm losses
- Survived 2,000 hours at 125°C (centigrade)
- Survived 1,000 hours at 85°C/85% RH (relative humidity)

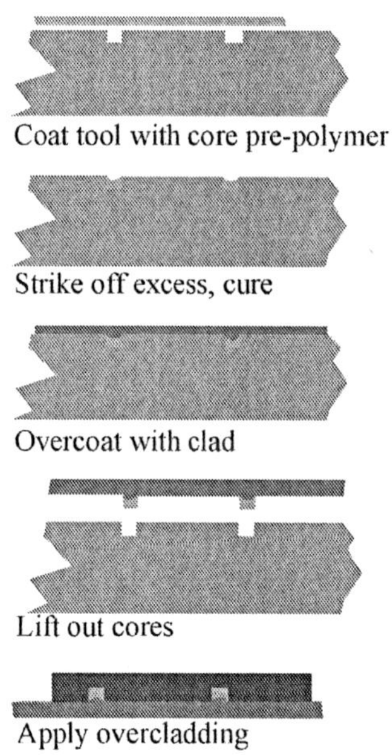

Figure 28.11 TOPCat planar optical waveguide fabrication process. [Wat01] [Reprinted with permission of 3M Corporation.]

15. ATP is the Advanced Technology Program, sponsored by the National Institute of Standards and Technology.

Figure 28.12 TOPCat planar optical waveguide cores with overcladding. [Wat01] [Reprinted with permission of 3M Corporation.]

28.3.3 New Components

New key components are required to use an integrated waveguide. The three components are listed below, and shown in Figure 28.13.

1. VCSEL (vertical cavity surface emitting laser) — A VCSEL emits laser light normal to the surface of the wafer in a beam with small angular divergence.
2. Micromirrors — small mirrors that can reflect the laser signals and be integrated into the polymer waveguide channels.
3. Optical receiver — Photo diodes/arrays — Silicon devices with optical sensitivity in the wavelengths of the VCSEL lasers.

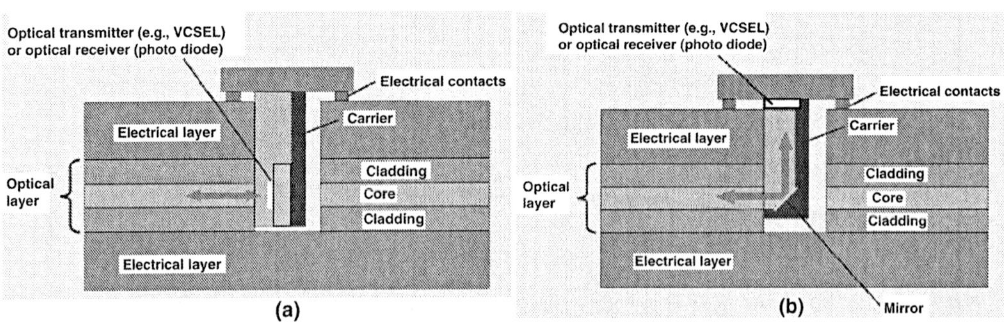

Figure 28.13 Optical components required to support integrated waveguides: (a) direct coupling of optical transmitters and receivers to board-integrated optical waveguides, and (b) indirect coupling through a 90° beam deflection with micromirrors. [Gri02] [Courtesy PC FAB magazine, UP Media Group.]

28.3.4 3D Assembly Techniques

With the new optical components listed above, new 3D assembly techniques are going to be required. Many optical components require micron-level tolerances to create the proper optical alignment. Key to these assembly capabilities will be the Z-axis plateau tolerances of the printed circuit. X-Y-angular alignment will require new kinematic mounts that provide micron-precise alignments. A future optoelectronics assembly might look like the one portrayed in Figure 28.14.

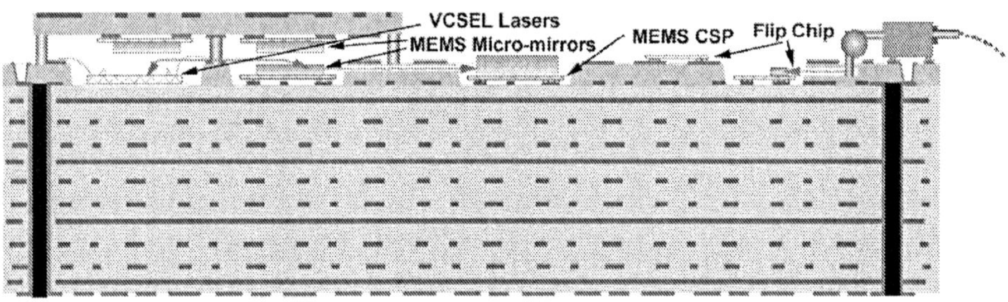

Figure 28.14 Integrated 3D assembly with thin-film, flip chip, MEMS, and PCB VCSEL lasers. [Reprinted with permission of Westwood Associates.]

28.3.5 NTT, University of Texas, and JIEP

Other similar programs are being conducted by NTT in Japan. Their Telecommunications Energy Laboratory-Network Hardware Integration Laboratory has the goal of "research and development on optical devices and modules and their integration for innovative photonic networks." Specifically, they are working in five major areas:

- Telecommunications Mechatronics — Optical switches and tunable optical devices based on optomechanical and precise microprocessing technology, and their applications for future optical telecommunications.
- Photonic Interconnection Technology — Innovative photonic interconnection technologies for high-speed and high-throughput photonic network systems. Figure 28.15[16] shows the intended chip-to-chip optical interconnections: optical input or output interface packages for compact and cost-effective photonic devices for CWDM (Course Wave Division Multiplex) systems, as well as advanced polishing technologies for photonic devices. Figure 28.16 shows an 84-pin, plastic BGA with integrated VCSEL devices and microlens (ϕ400 μm).

16. In the figure, "LSI" is large scale integration and "PD" is photodiode.

- Opto-Micromachine Systems — Micro-optomechanical components and modules for future photonic networks utilizing optical MEMS with optical signal processing technologies.
- Optical Thin-Film Technology — High-quality thin-film devices for optical communications based on cyclotron resonance (ECR) plasma deposition technology.
- Micro Optical Circuit Technology — Design and fabrication of micro-optical integrated circuits and devices with high functionality.

The University of Texas, like the University of Arkansas (HiDEC), has been conducting research with thin-film deposited circuits to make multichip module optical interconnects. [Erb01]

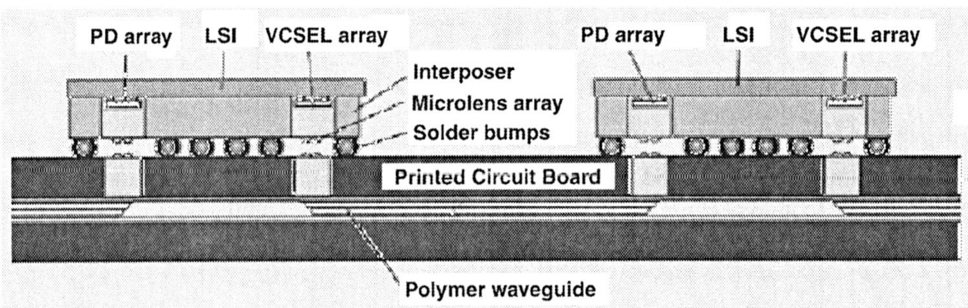

Figure 28.15 NTT optical waveguide components required to support chip-to-chip interconnections. [Ish01] [Courtesy NTT.]

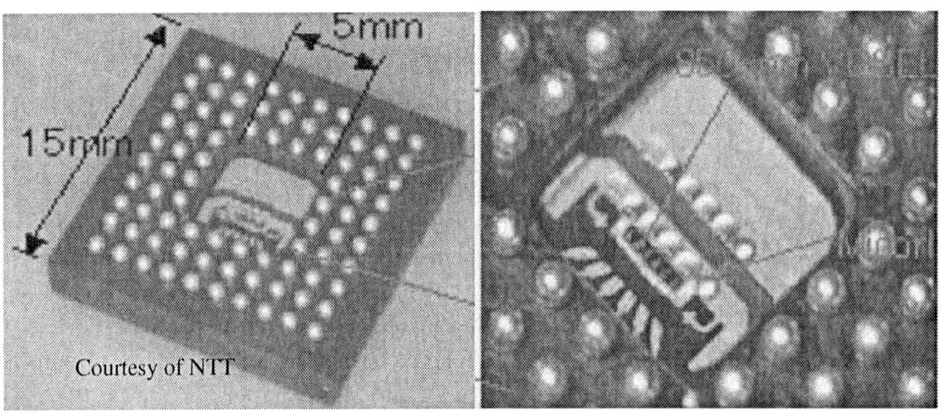

Figure 28.16 Closeup of the VCSEL and other optical components. The solder bumps are used to align the various components. [Ish01] [Courtesy NTT.]

The Japan Institute of Electronics Packaging (JIEP) was established in April 1998 by merging two organizations, the Society for Hybrid Microelectronics and the JIPC. Together with SEMI of Japan, they have put together a very complete "optical packaging" roadmap (refer to Figure 28.17). It provides details for the module, board, and subsystem packaging requirements for optical waveguides out to the year 2010.

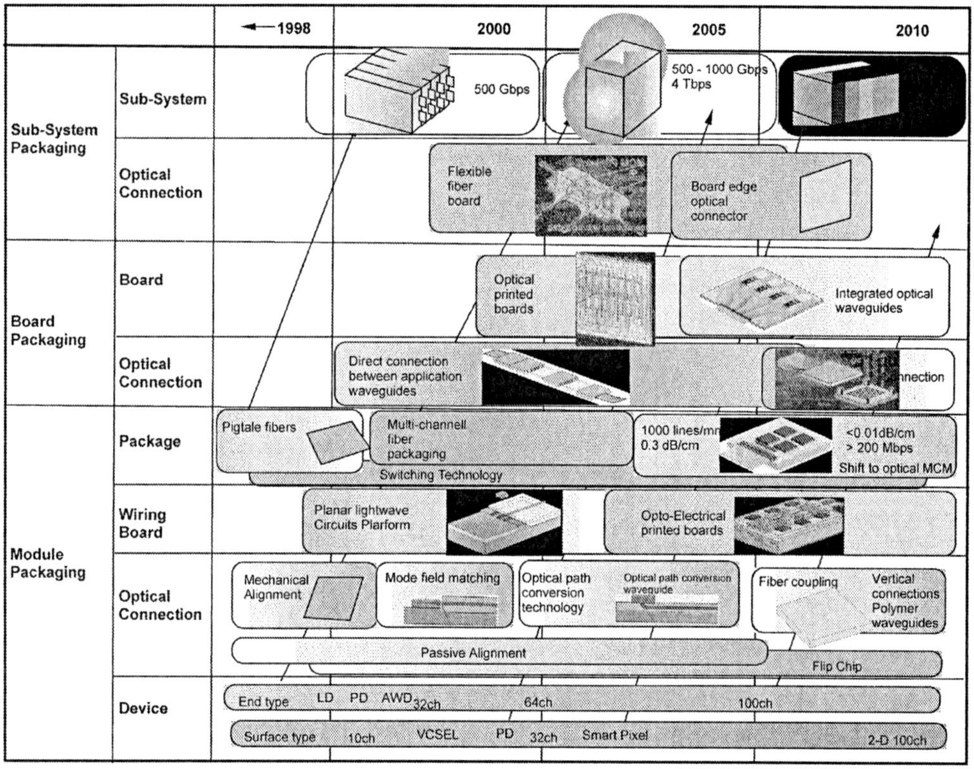

Figure 28.17 JIEP optical packaging roadmap. [Reprinted with permission of Westwood Associates.]

28.4 Conclusion

Telecommunication rates continue to increase. Longline optical communications is the standard. The challenge now is to bring faster optical communication rates into the metro area and to the individual or commercial user. The performance limitations of electrical interconnections and wires (traces) due to their physical field limitations can be overcome with optical interconnections. The basic technologies are optical waveguide materials, waveguide fabrication, low-cost optical components (transmitters and receivers), and optical assembly. These technologies can be made compatible with existing printed circuit board design and manufacturing processes and tolerances. Both electrical and optical interconnections are now being fabricated into a single board with significant improvement in end-product performance.

28.5 Exercises

28-1. What do the abbreviations EOCB, RIE, O-E-O, MOEMS, MEMS, PCS, VCSEL, PMMA, D-PMMA, dPMMA, JIEP, and CSP stand for? Provide both the words that the abbreviation represents and a short definition (35 words or less) for each item.

28-2. Go to the Web sites for two of the manufacturers of electrical optical circuit boards, Truemode Backplane, PolyGuide, or TOPCat listed in this chapter (or other photonic printed circuit board manufacturers that you can identify). For each of the two manufacturers, describe that manufacturer's present capabilities to manufacture photonic printed circuit boards and the products that they offer, as discussed on their Web sites. Limit each of the two discussions to 100 to 250 words each. You may have to do a Web search on key optical materials in order to locate Web sites.

28-3. Explain the difference between the two contrasting concepts of "O-E-O" and "all-optical." What are several advantages of the all-optical approach over the O-E-O concept?

28-4. What fabrication technology is most integrated waveguide technology built upon?

28-5. What is the maximum diameter of a typical microvia?

28-6. How do the parasitic inductance and capacitance of a microvia compare with those of a throughhole?

28-7. What is the major difference between the Truemode Backplane and an EOCB?

28-8. Name and briefly describe the four fabrication techniques for integrated waveguides discussed in this chapter. Limit your descriptions to 100 words or less (each) while still presenting the main concepts of the three-dimensional fabrication process.

28-9. What are the five major areas being researched in Japan by NTT on the development and integration of optical devices and modules?

28-10. What is driving the research and product development supporting the integration of waveguides into PCBs?

28-11. What are three names for technologies that combine both mechanical and optical properties of devices?

28-12. What is "cladding"? What is "overcladding"?

CHAPTER 29

RapidIO

Chapter Objectives

- Explain where RapidIO™ is used and the need for interconnect standards for high-performance embedded systems.
- Present that RapidIO is now an ISO/IEC international standard and is being mapped into Advanced TCA, CompactPCI, VME Switched Serial VXS, Switched Mezzanine XMC, and next-generation VME form factors.
- Explain that higher system performance requires point-to-point switched interconnects.
- Discuss how RapidIO is targeted toward control plane applications in network equipment.
- Detail RapidIO packet formats and the protocol for packets and control symbols.
- Discuss the RapidIO 8/16 LP-LVDS physical layer.
- Discuss the wide range of performance levels in terms of peak and sustained data rates.

29.1 RapidIO: The Interconnect Architecture for High-Performance Embedded Systems[1]

RapidIO. This chapter describes RapidIO, a high-performance, low pin count, packet switched system level interconnect architecture. The interconnect architecture is an open standard that addresses the needs of a wide variety of embedded infrastructure applications. Applications include interconnecting microprocessors, memory, and memory-mapped

1. The author of this chapter is Dan Bouvier. [Bou03] © RapidIO Trade Association. Reprinted with permission.

I/O devices in networking equipment, storage subsystems, and general-purpose computing platforms. This interconnect is intended primarily as an intra-system interface, allowing chip-to-chip and board-to-board communications with performance levels ranging from 1 gigabit per second to 60 gigabits per second. Two families of RapidIO interconnects are defined: a parallel interface for high-performance microprocessor and system connectivity and a serial interface for serial backplane, DSP, and associated serial control plane applications. The serial and parallel forms of RapidIO share the same programming models, transactions, and addressing mechanisms. Supported programming models include basic memory-mapped I/O transactions, port-based message passing, and globally shared distributed memory with hardware-based coherency. RapidIO also offers a very high degree of error management and provides a well-defined architecture for recovering from and reporting transmission errors. The RapidIO interconnect is defined as a layered architecture which allows scalability and future enhancements while maintaining backward compatibility.

29.2 RapidIO Is Now an International Standard

"The International Standards Organization (ISO) and the International Electrotechnical Committee (IEC) have ratified the RapidIO interconnect specification as ISO/IEC DIS 18372. RapidIO continues to gain support as both a chip and board interconnect. Task groups have also been mapping RapidIO to a wide range of form factors, including Advanced TCA (PICMG 3.5), CompactPCI (PICMG 2.18), VME Switched Serial VXS (VITA 41.2), Switched Mezzanine XMC (VITA 42), and next-generation VME (VITA 46)." [Ele03]

29.3 Embedded System Development

Computer and embedded system development continues to be burdened by divergent requirements. On one hand the system performance must continue to increase at a nearly exponential rate, while on the other hand the system cost must remain constant or even decrease. Several applications, such as those found in networking and telecommunications infrastructure equipment, are also burdened with increasing capability and reliability requirements.

The connections between microprocessors and peripherals have traditionally been composed of a hierarchy of buses (see Figure 29.1). Devices are placed at the appropriate level in the hierarchy according to the performance level they require. Low-performance devices are placed on lower-performance buses, which are bridged to the higher-performance buses so as to not burden the higher-performance devices. Bridging may also be used to address legacy interfaces.

The need for higher levels of bus performance is driven by two key factors: First, the need for higher raw data bandwidth to support higher peripheral device performance requirements; and second, the need for more system concurrency. The overall system bandwidth requirements have also increased because of the increasing use of DMA (direct-memory access), smart processor-based peripherals, and multiprocessing in systems.

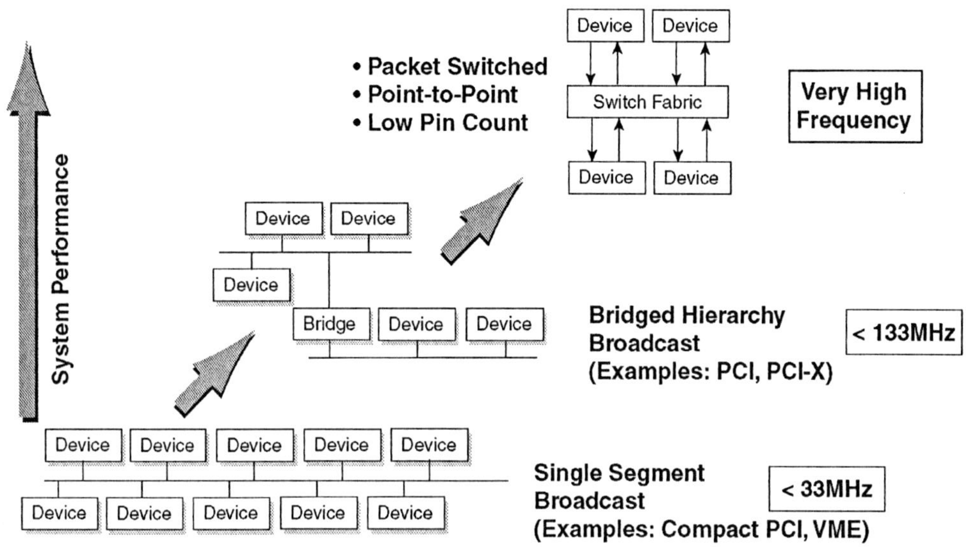

Figure 29.1 Higher system performance levels require adoption of point-to-point switched interconnects. [©RapidIO Trade Association. Reprinted with permission.]

29.3.1 Why RapidIO?

Over the past several years the shared multidrop bus has been exploited to its full potential. Many techniques have been applied, such as increasing frequency, widening the interface, pipelining transactions, splitting transactions, and allowing out-of-order completion. Continuing to work with a bus in this manner creates several design issues. Increasing bus width, for example, reduces the maximum achievable frequency due to skew between signals. More signals will also result in more pins on a device, traces on boards, and larger connectors, resulting in a higher product cost and a reduction in the number of interfaces a system or device can provide.

Worsening the situation is the desire to increase the number of devices that can communicate directly with each other. As frequency and width increase, the ability to have more than a few devices attached to a shared bus becomes a difficult design challenge. In many cases, system designers have inserted a hierarchy of bridges to reduce the number of loads on a single bus.

Developed as an open standard, RapidIO is designed to addresses the needs of present and future high-performance embedded systems. In embedded system applications, RapidIO has just a limited-to-no impact on the software infrastructure that operates over it. It can be implemented with relatively few transistors and it offers low operation latency and high bandwidths.

High-performance embedded equipment often contains separate control plane and data-forwarding planes as shown in Figure 29.2. The data-forwarding plane is responsible for moving data through the system, while the control plane manages the data movement. RapidIO provides all of the necessary attributes to be useful in the control plane. RapidIO, in some applications, may also be useful as the basis for data-forwarding plane implementations.

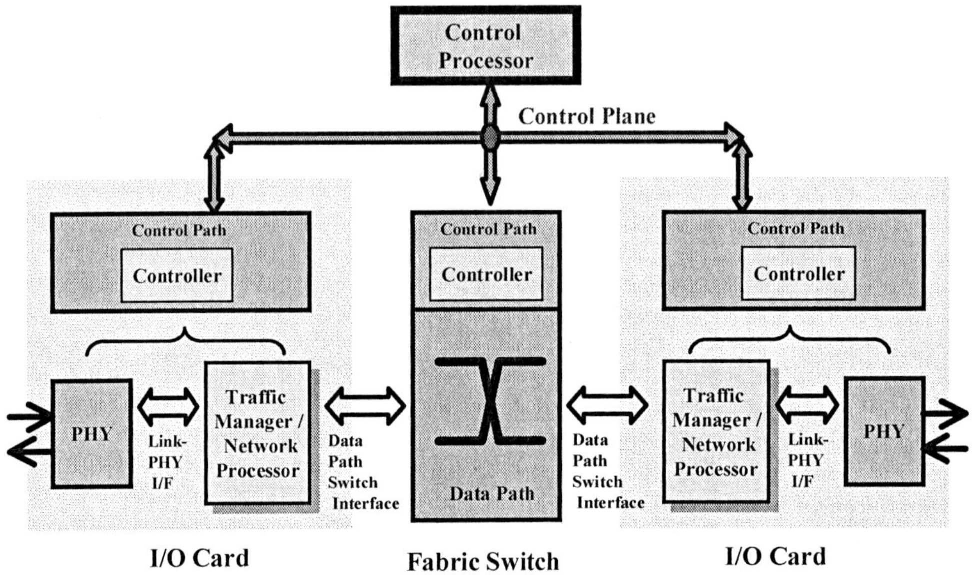

Figure 29.2 RapidIO is targeted toward control plane applications in network equipment.

29.3.2 Interconnect Landscape

RapidIO is categorized as an intra-system interconnect, as shown in Figure 29.3. Specifically, RapidIO is targeted at the intra-system interconnect applications in the high-performance embedded equipment market. This market has distinctly different requirements from the desktop and server computer spaces. The embedded market has historically been served by a number of different vendors. The openness of the RapidIO Trade Association is well-suited to this environment. The RapidIO Trade Association counts among its current members nearly two dozen leading vendors of microprocessors, DSPs, FPGAs, ASICs, and embedded memories.

InfiniBand is targeted as a System-Area Network (SAN) interconnect. A SAN is used to cluster systems together to form larger, highly available systems. SANs usually connect whole computers together within distances of up to 30 meters. Operations through a SAN are typically handled through software drivers using message channels or remote direct memory access (RDMA). InfiniBand competes more directly with Fibre Channel and Ethernet-based system area networking technologies such as iSCSI.

HyperTransport and PCI Express share some common characteristics with RapidIO but are more appropriately described as point-to-point versions of PCI. While they maintain compatibility with the PCI interconnect architecture from a software viewpoint, which is very important for desktop computer markets, they do not offer the scalability, robustness, and efficiency required by embedded systems developers.

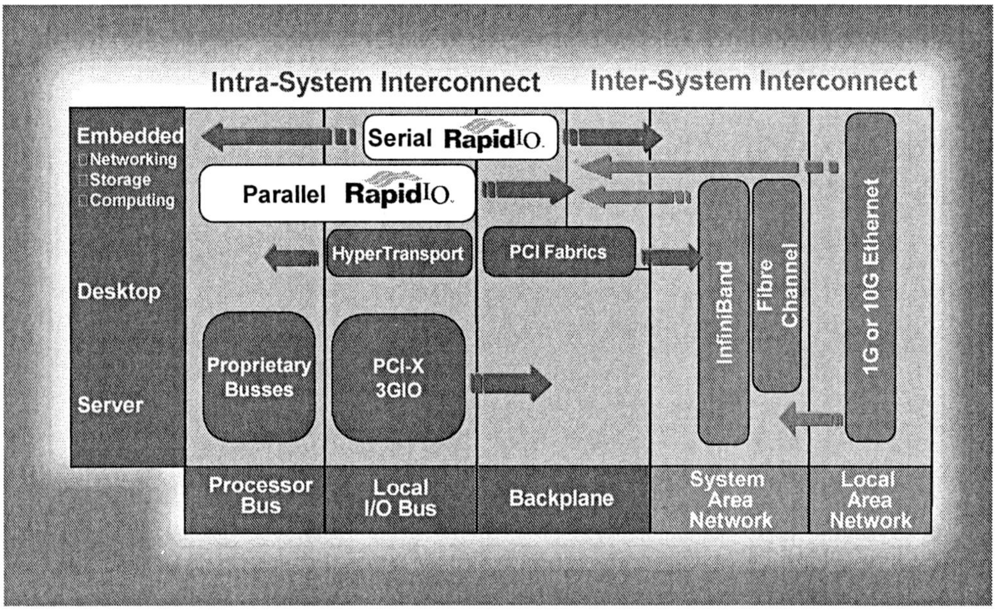

Figure 29.3 RapidIO connects processors, memory, and peripherals within a subsystem or across a backplane.

29.3.3 Where Will It Be Used?

The RapidIO interconnect is targeted for use in environments where multiple devices must work in a tightly coupled architecture. Figure 29.4 illustrates a generic system containing memory controllers, processors, and I/O bridges connected using RapidIO switches.

In computing applications the PCI bus is frequently used. Enterprise storage applications, for example, use PCI to connect multiple disk channels to a system. As disk throughput has increased so has the need for higher system throughput. To meet the electrical requirements of higher bus frequencies, the number of supportable devices per bus segment must be decreased. In order to connect the same number of devices, more bus segments are required. These applications require higher bus performance, more device fan-out, and greater device separation. PCI-to-PCI bridge devices could be used to solve this problem but only within a tree-shaped hierarchy that increases system latency and cost as more PCI devices are added to the system.

RapidIO can be used for transparent PCI-to-PCI bridging allowing for a flattened architecture utilizing fewer pins and offering greater transmission distances. Figure 29.5 shows one such bridged system. In this example, several PCI-X bridges are connected together using RapidIO switches. An InfiniBand or Ethernet Channel Adapter can be provided if the system is to become part of a wider SAN.

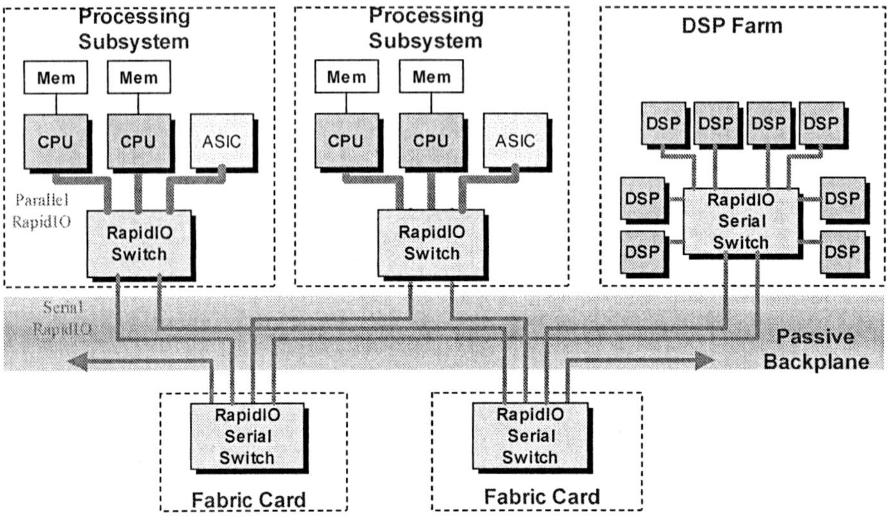

Figure 29.4 RapidIO allows a variety of devices to be interconnected.

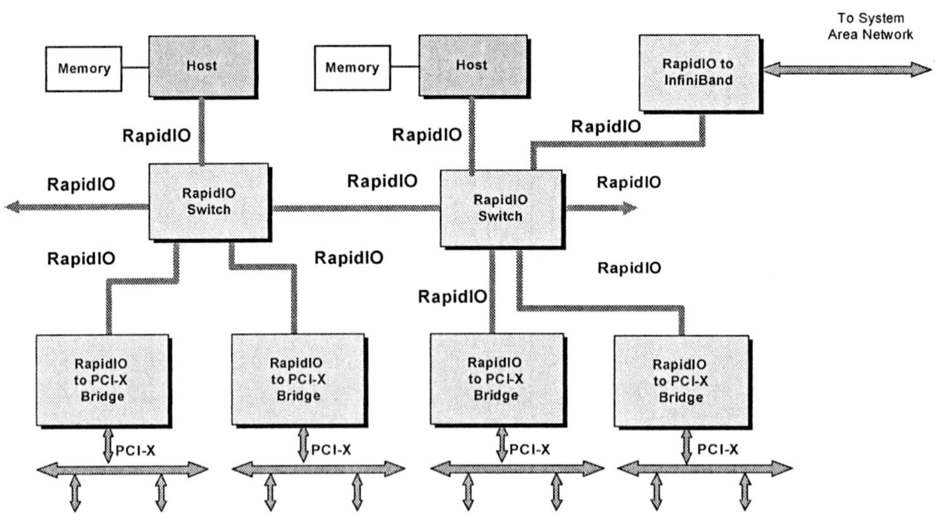

Figure 29.5 RapidIO as a PCI bridging fabric reduces device pin count.

Many systems require the partitioning of functions into field-replaceable units. These printed circuit boards have traditionally been interconnected using multidrop interfaces like VME or PCI. Higher system-level performance can be achieved by using RapidIO. RapidIO is also well-suited for hot-swap applications because RapidIO's point-to-point topology enables the removal of devices with little or no electrical impact to neighboring devices or subsystems.

29.3.4 Philosophy

The architecture of RapidIO was driven by certain principal objectives:

- *Focus on applications that connect devices which are within the box or chassis:*

  As an example, since devices are not intended to be connected with several meters of cable, the interface can use a simple signaling and flow control scheme.
- *Limit the impact on software:*

  Many applications are built on a huge legacy of memory-mapped I/O. In such applications the interconnect must not be visible to software. Offering more abstracted interfaces requires a large software re-engineering effort.
- *Confine protocol overhead:*

  Because of the typical application dependence on latency and bandwidth, it is important that the protocol use no more transaction overhead than is absolutely necessary.
- *Partition the specifications to limit the necessary function set to only those needed for the application:*

  This limits design complexity and enables future enhancements without impacting the top-to-bottom specification.
- *Manage errors in hardware:*

  In meeting the high standards of availability, yet limiting the impact to performance and software infrastructure, it is important that the interconnect be able to detect errors. The transmission media should have the capability of detecting multiple-bit errors. Going a step further, the interconnect must utilize hardware mechanisms to survive single-bit errors and many multiple-bit errors.
- *Limit silicon footprint:*

  Many applications require the use of a high-performance interconnect, but have a limited transistor budget. It is important that the interface be able to fit within a small transistor count. As an example, it is important that it fit within commonly available FPGA technology.
- *Leverage common process technology I/O:*

  Since this interface is targeted for intra-system communications, it should not require separate discrete physical layer hardware. This means that the I/O technology should be power efficient and limit the use of exotic process technology. Further, the interface should leverage existing industry-standard I/O technology and the support infrastructure associated with such standards.
- *RapidIO is specified in a three-layer architectural hierarchy (Figure 29.6):*

  The logical layer specification defines the overall protocol and packet formats. This is the information necessary for end points to initiate and complete a transaction. The transport layer specification provides the necessary route information for a packet to move from end point to end point. The physical layer specification describes the device level interface specifics such as packet transport mechanisms, flow control, electrical characteristics, and low-level error management.

This partitioning provides the flexibility to add new transaction types to the logical specification without requiring modification to the transport or physical layer specifications. RapidIO as a PCI bridging fabric reduces device pin count.

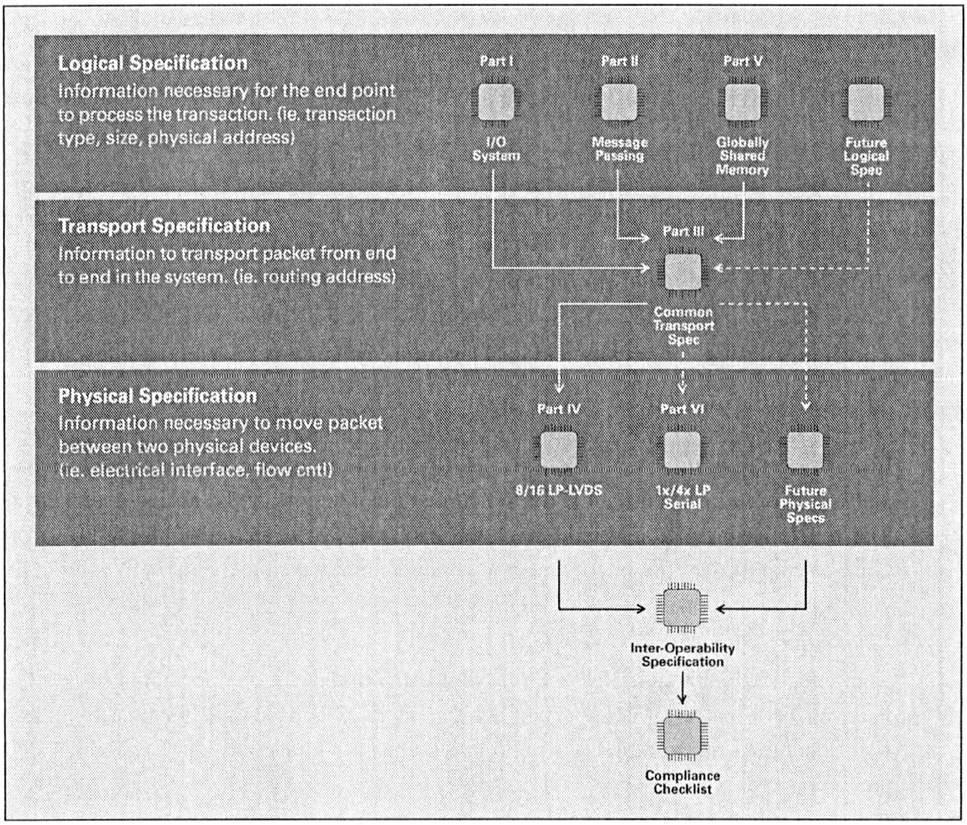

Figure 29.6 The RapidIO specifications hierarchy allows flexibility in usage and future specification development.

29.4 RapidIO Protocol Overview

29.4.1 Packets and Control Symbols

RapidIO operations are based on request and response transactions. Packets are the communication element between end point devices in the system. A master or initiator generates a request transaction, which is transmitted to a target. The target then generates a response transaction back to the initiator to complete the operation. The RapidIO transactions are encapsulated into packets, which include all of the necessary bit fields to ensure reliable delivery to the targeted end point. RapidIO end points are typically not connected directly to each other but instead have intervening connection fabric devices. Control symbols are used to manage the flow of transactions in the RapidIO physical interconnect.

Control symbols are used for packet acknowledgement, flow control information, and maintenance functions. Figure 29.7 shows how transactions progress through the system.

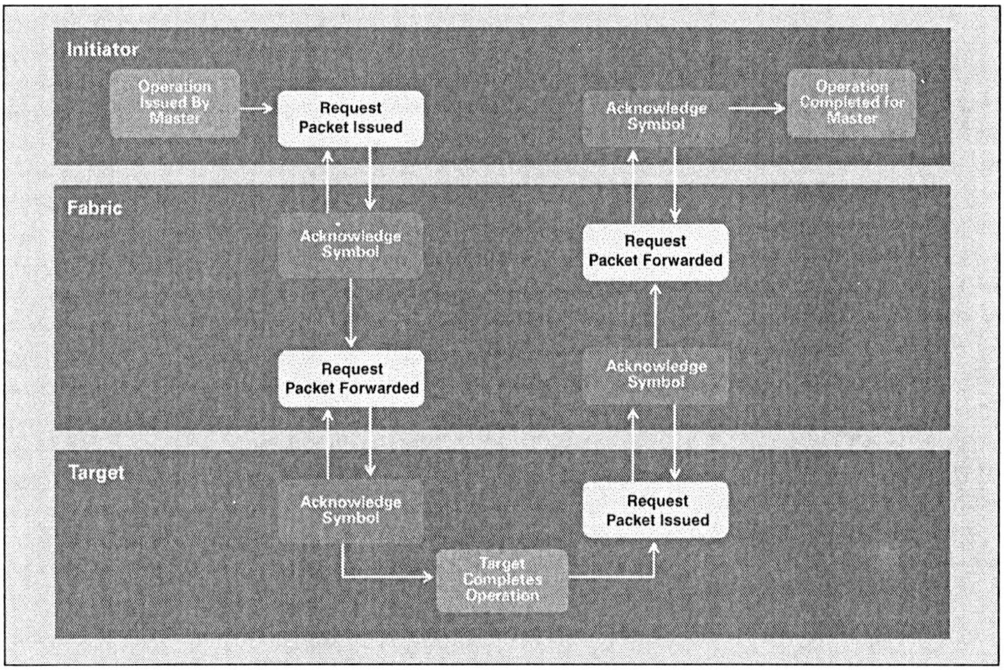

Figure 29.7 Transactions are constructed with request and response packet pairs.

In this example, the initiator begins an operation in the system by generating a request transaction. The request packet is sent to a fabric device, which is acknowledged with a control symbol. The packet is forwarded to the target through the fabric device. The target completes the request transaction and generates a response transaction. The response packet returns through the fabric device using control symbols to acknowledge each hop. Once the packet reaches the initiator and is acknowledged, the operation is considered complete.

29.4.2 Packet Format

The RapidIO packet is comprised of fields from the three-level specification hierarchy. Figure 29.8 shows typical request and response packets. Certain fields are context-dependent and may not appear in all packets.

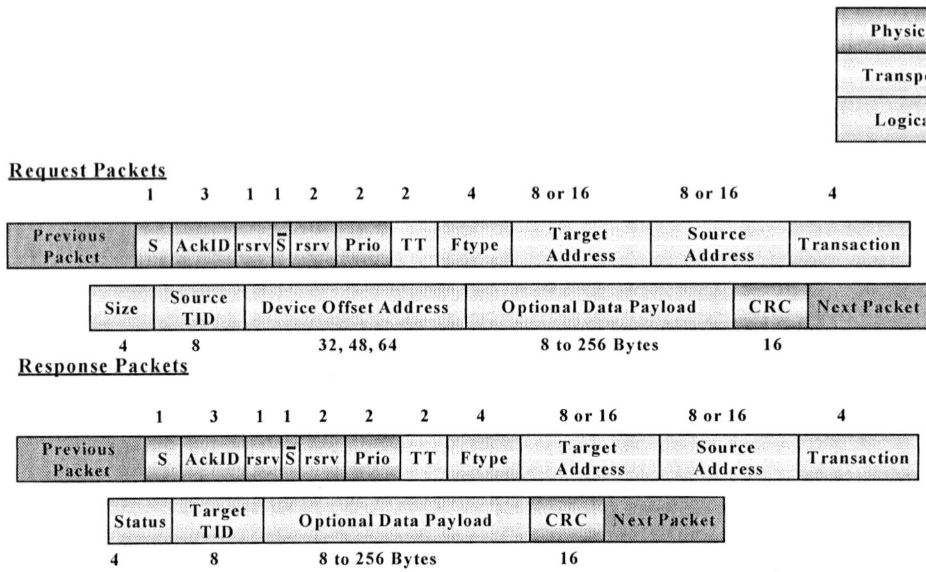

Figure 29.8 The RapidIO packet contains fields from the specification hierarchy.

The request packet begins with physical layer fields. The "S" bit indicates whether this is a packet or control symbol. The "AckID" indicates which packet the fabric device should acknowledge with a control symbol. The "Prio" field indicates the packet priority used for flow control. The "TT," "Target Address," and "Source Address" fields indicate the type of transport address mechanism used, the device address where the packet should be delivered, and where the packet originated. The "Ftype" and "Transaction" indicate the transaction that is being requested. The "Size" is an encoded transaction size. RapidIO transaction data payloads range from 1 byte to 256 bytes in size. The "source TID" indicates the transaction ID. RapidIO devices may have up to 256 outstanding transactions between two end points. For memory mapped transactions the "Device Offset Address" follows. For write transactions a "Data Payload" completes the transaction followed by a 16-bit CRC.

Response packets are very similar to request packets. The "Status" field indicates whether the transaction was successfully completed. The "TargetTID" corresponds to the request packet source transaction ID.

The Serial RapidIO physical layer bits are slightly different than those used in Parallel RapidIO packet headers.

29.4.3 Transaction Formats and Types

One of the attributes of a software-transparent interconnect is the requirement for a rich set of transaction functions. The RapidIO transaction is described through two fields: the packet format type "Ftype," and the "Transaction." To ease the burden of transaction deciphering, transactions are grouped by format as shown in Table 29.1.

Table 29.1 RapidIO Transaction Formats

Functions	Transaction Types
I/O Non-Coherent Functions	NREAD (read non-sharable memory)
	NWRITE, NWRITE_R, SWRITE (write non-sharable memory)
	ATOMIC (read-modify-write to non-sharable memory)
Port-Based Functions	DOORBELL (generate an interrupt)
	MESSAGE (write to port)
System Support Functions	MAINTENANCE (read or write configuration, control, and status registers)
User-Defined Functions	Open for application specific transactions
Cache Coherence Functions	READ (read globally shared cache line)
	READ_TO_OWN (write globally shared cache line)
	CASTOUT (surrender ownership of globally shared cache line)
	IKILL (instruction cache invalidate)
	DKILL (data cache invalidate)
	FLUSH (return globally shared cache line to memory)
	IO_READ (read non-cacheable copy of globally shared cache line)
OS Support Functions	TLBIE (TLB invalidate)
	TLBSYNC (force completion of TLB invalidates)

29.4.4 Message Passing

When data must be shared amongst multiple processing elements in a system, a protocol must be put in place to maintain ownership. In many embedded systems this protocol is managed through software mechanisms. If the memory space is accessible to multiple devices, then locks or semaphores are used to grant access to one device. In other cases, processing elements may only have access to locally owned memory space. In these "shared nothing" machines, a mechanism is needed to pass data from the memory of one processing element to another. This can be done using software visible mailbox hardware.

RapidIO provides a useful message passing facility. The RapidIO message passing protocol extensions describe transactions that enable mailbox and doorbell communications. A RapidIO mailbox is a port through which one device may send a message to another device. The receiving device determines what to do with the message after it arrives. A RapidIO message can range in length from zero to 4,096 bytes. A receiver can have one to four addressable message queues to capture inbound messages.

The RapidIO doorbell is a lightweight port-based transaction, which can be used for in-band interrupts. A doorbell message has a 16-bit software definable field, which can be used for a variety of messaging purposes between two devices.

29.4.5 Globally Shared Memory

One of the protocol extensions offered in RapidIO is support for a globally shared distributed memory system. This means that memory may be physically located in different places in the machine yet may be cached amongst different processing elements.

For RapidIO, a directory-based coherency scheme is specified to support this. For this method each memory controller is responsible for tracking where the most current copy of each data element resides in the system. A directory entry is maintained for each device participating in the coherency domain. Simple coherency states of Modified, Shared, or Local (MSL) are tracked for each element. Figure 29.9 shows an example of a "read with intent of modification" request to a memory controller. For this example other devices in the system have shared copies of the data. The memory controller indexes the requested data in its directory and subsequently issues appropriate invalidation transactions to the sharers. At the completion of this set of operations the memory controller forwards the latest copy of the data to the requestor to complete the transaction.

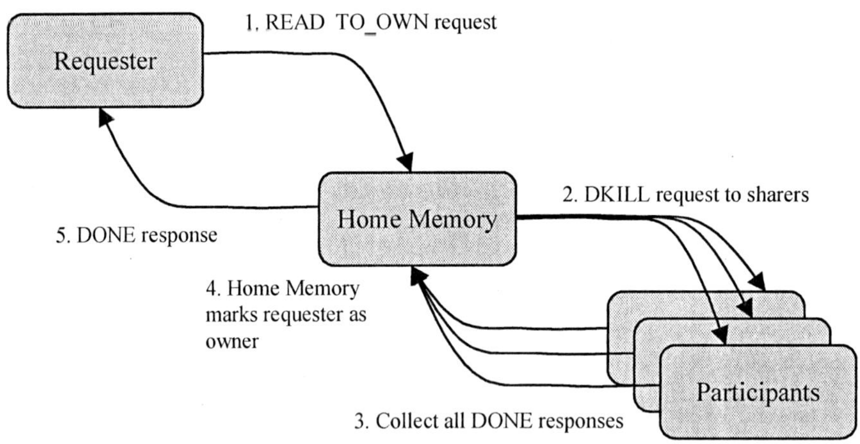

Figure 29.9 The memory controller is responsible for managing coherency in the directory-based scheme.

29.4.6 Future Extensions

RapidIO is partitioned to support future protocol extensions. This can be done at the user level through the application of user-definable transaction format types, or through future use of reserved fields. RapidIO is architected so that switch fabric devices do not need to interpret packets as they flow through, making future forward compatibility much easier to achieve.

29.4.7 Flow Control

Flow control is an important aspect of any interconnect. The objective is for devices to be able to complete transactions in the system without being blocked by other transactions. Bus-based interconnects use arbitration algorithms to be sure that devices make forward progress and that higher-priority transactions take precedence over lower-priority ones. With switch-based interconnects, transactions enter at different points in the system and there is no opportunity to employ a centralized arbitration mechanism. This creates the need for alternate methods to manage transaction flow in the system.

One of the objectives of RapidIO is to limit overhead and complexity as much as possible, especially in the area of flow control. For RapidIO, flow control is specified as part of the physical specification. This is because transaction flow is largely dependent on the physical interconnect and system partitioning. Each RapidIO packet carries with it a transaction priority. Each transaction priority is associated with a transaction request flow. There currently are three transaction request flows defined. Transaction request flows allow higher-priority requests to move ahead of lower-priority requests. Transaction ordering rules are managed within a transaction request flow but not between flows.

Transaction request flows make it possible for system implementers to structure traffic through the machine to guarantee deterministic behavior. For example, an application may require that certain time-critical operations be delivered with a guaranteed latency. This data can be identified as a higher-priority request flow. To handle critical data, the switch fabric may contain queuing structures that allow the movement of higher priority traffic ahead of lower priority traffic. These queuing structures may include interval timers to guarantee that flows from different source ports targeted to a common destination port are given opportunity to move critical data through — resulting in isochronous delivery behavior for certain data classes.

RapidIO also describes three types of flow control mechanisms: retry, throttle, and credits. The retry mechanism is the simplest mechanism and is required not only for flow control but also as a component of hardware error recovery. A receiver, unable to accept a packet because of a lack of resources or because the received packet was corrupt, may respond with a retry control symbol. The sender will retransmit the packet.

The throttle mechanism makes use of the idle control symbol. Idle control symbols may be inserted into a packet during packet transmission. This allows a device to insert wait-states in the middle of packets. A receiving device can also send a throttle control symbol to the sending device requesting that it slow down by inserting idle control symbols.

The credit-based mechanism is useful for devices that implement transaction buffer pools, especially fabric devices. In this scheme certain control symbols contain a buffer status field that represents the current status of the receiver's buffer pool for each transaction flow. A sender only sends packets when it knows the receiver has available buffer space. This information enables switch fabric devices to discover downstream blocking conditions and to make better packet output selections.

29.5 Physical Interface

The RapidIO logical packet description is defined to be physical layer independent. This means he RapidIO protocol could be transmitted over anything from serial to parallel interfaces, from copper to fiber media. The first physical interface considered and defined is known as the 8- or 16-bit link protocol end point specification (8/16 LP-LVDS). This specification is defined as having 8 or 16 data bits in each direction along with clock and frame signals in each direction (Figure 29.10).

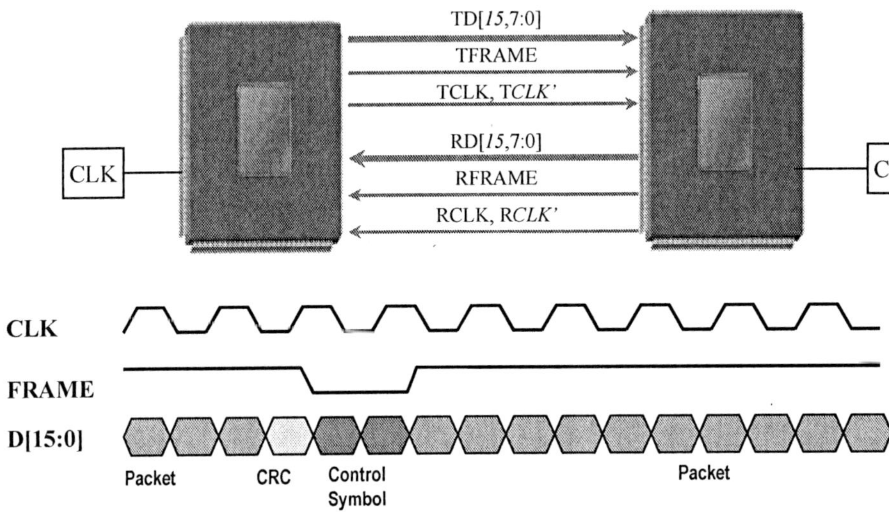

Figure 29.10 The RapidIO 8/16 LP-LVDS physical layer provides for full-duplex communications between devices.

The 8/16 LP-LVDS interface is a source synchronous interface. This means that a clock is transmitted along with the associated data. Source synchronous clocking allows longer transmission distances at higher frequencies. Two clock pairs are provided for the 16-bit interface to help control skew. The receiving logic is able to use the receive clock for re-synchronization of the data into its local clock domain.

The FRAME signal is used to delineate the start of a packet or control symbol. It operates as a non-return to zero (NRZ) signal where any transition marks an event.

29.5.1 Parallel Electrical Interface

RapidIO adopts the IEEE 1596.3 Low Voltage Differential Signals (LVDS) standard as the basis for the electrical interface.[2] LVDS is a low-swing (250 to 400 mV), constant-current differential

2. While RapidIO uses the IEEE 1596.3 LVDS standard as the basis for its electrical interface, the IEEE 1596.3 standard has been withdrawn by IEEE. LVDS is still standardized by TIA/EIA-644A and TIA/EIA-899.

signaling technology-targeted toward short-distance printed circuit board applications. LVDS is technology independent and can be implemented in CMOS. Differential signals provide improved noise margin, immunity to externally generated noise, lower EMI, and reduced numbers of power and ground signal pins. LVDS has a simple receiver-based termination of 100 Ω. At higher frequencies the RapidIO specification recommends additional source termination to reduce signal reflection effects.

The target frequencies of operation are from 250 MHz to more than 1 GHz. Data is sampled on both edges of the clock. The resulting bandwidth scales to 4 GB/s for the 8-bit and 8 GB/s for the 16-bit interfaces.

29.5.2 The Serial RapidIO Controller

Since the Serial RapidIO specification is only defined in the physical layer (RapidIO technology defines the physical layer as the electrical interface and device-to-device link protocol), most of the controller remains the same. As a result, much of the design knowledge and verification infrastructure are preserved. This eases system-level switching between parallel and serial links.

During the initial development stages of the Serial RapidIO specifications the designers decided to preserve as many of the concepts found in the RapidIO parallel specification as feasible. The parallel specification includes the concept of packets and in-band control symbols. These were delineated and differentiated by both a separate frame signal and an "S" bit in the header. In the serial link specification this delineation is accomplished using spare characters ("K-codes") found in the 8B/10B encoding technique. In this way, the sending device indicates to the receiving link partner the start of a packet, end of packet, or embedded control symbol using these codes.

29.5.3 Link Protocol

A unique feature of parallel RapidIO technology is that packet transmission is managed on a link-by-link basis. In the past, with synchronous buses, a mastering device had to exchange handshake signals with the target device. These signals indicated whether a transaction was acknowledged and accepted by the target device.

With a source synchronous interface like the RapidIO specification defines, it is not practical to rely on a synchronous handshake since the receive port of a link is decoupled from the sending port. Therefore, many interconnects have ignored this issue and rely on an end-to-end handshake to guarantee delivery. However, this has the disadvantage of not allowing precise detection and recovery of errors and forces far longer feedback loops for flow control.

To address this issue RapidIO technology uses embedded control symbols for link-level communication between devices. Packets are explicitly tagged between each link with a sequence number otherwise known as AckID. The AckID is independent of the end-to-end transaction ID. Using control symbols, the receiving device indicates for each packet whether it has been received along with additional buffer status information. Receiving devices can immediately detect a lost packet, and through control symbols, can re-synchronize with the sender and recover it without software intervention. The receiving device then forwards the packet to the next switch in the fabric, and so on, until the final target of the packet is reached.

29.5.4 Enhanced Flow Control

Serial RapidIO technology allows for the longer transmission distances and thus longer loop latencies in providing feedback between the receiver and transmitter on a link. Consequently the Serial physical layer specification increases the number of "AckID" values from 8 to 32. Additionally, the Serial RapidIO specification now defines a transmitter-controlled flow control scheme whereby the receiving port provides information to its link partner about the amount of buffer space it has available. With this information, the sending port can allocate the use of the receive buffers of the receiving port. The sending port does not have to be concerned that one or more of the packets shall be forced to retry.

29.5.5 PCS and PMA Layers

The Serial RapidIO specification uses a Physical Coding Sublayer (PCS) and Physical Media Attachment (PMA) sublayer to organize packets into a serial bit stream at the sending side and to extract the bit stream at the receiving side (this terminology is adopted from IEEE 802.3). (See Figure 29.11.)

Besides encoding for transmission and decoding for reception, the PCS function is also responsible for idle sequence generation, lane striping, lane alignment, and de-striping on reception. The PCS uses 8B/10B encoding for transmission over the link.

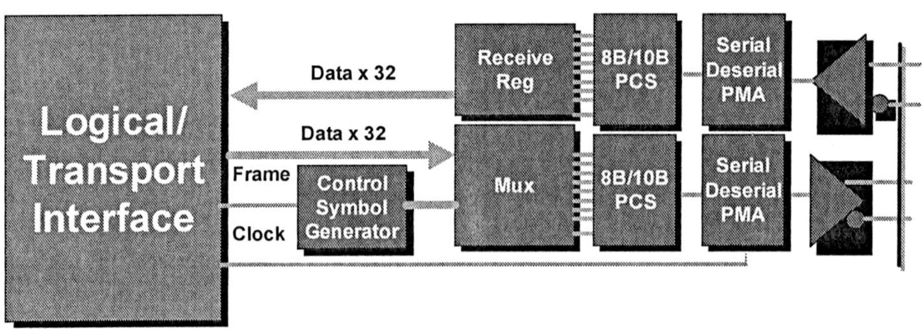

Figure 29.11 The RapidIO Serializer/Deserializer function integrates directly with the logic and transport layers.

The PCS layer also provides the mechanisms for automatically determining the operational mode of the port as either 1-lane or 4-lane, and provides for clock difference tolerance between the sender and receiver without requiring flow control.

The PMA function is responsible for serializing 10-bit parallel code-groups to/from a serial bit stream on a lane-by-lane basis. Upon receiving data, the PMA function provides alignment of the received bit stream to 10-bit code-group boundaries, independently on a lane-by-lane basis. It then provides a continuous stream of 10-bit code-groups to the PCS, one stream for each lane. The 10-bit code groups are not observable by layers higher than the PCS.

29.5.6 Electrical Interface

Serial RapidIO technology uses differential current steering drivers based on those defined in the 802.3 XAUI specifications. This signaling technology was developed to drive long distances over backplanes.

For the Serial RapidIO technology, two transmitter specifications were designated: a short-run transmitter and a long-run transmitter. The short-run transmitter is used mainly for chip-to-chip connections either on the same printed circuit board or across a single connector such as that for a mezzanine card. The minimum swings of the short run specification reduce the overall power used by the transceivers. A user can further reduce the power by lowering the termination voltages.

The long-run transmitter uses larger "voltage swings" that are capable of driving across backplanes. This allows a user to drive signals across two connectors and common printed circuit board material.

To ensure interoperability between drivers and receivers of different vendors and technologies, ac coupling must be used at the receiver input.

29.6 Maintenance and Error Management

RapidIO steps beyond traditional bus-based interfaces by providing a rich set of maintenance and error-management functions. These enable the initial system discovery, configuration, error detection, and recovery methods. A discovery mechanism with certain similarities to PCI is used to find devices in the system and configure them.

29.6.1 Maintenance

A maintenance operation is supported by all RapidIO devices and is used to access predefined maintenance registers within each device. The registers contain information about the device, including its capabilities and memory-mapped requirements. Also included are error detection and status registers such as watchdog timer settings. For more complex multi-bit errors, software may make use of these registers to recover or shut down a RapidIO device.

29.6.2 System Discovery

Through the use of the maintenance operation, a RapidIO host may traverse a system and configure each device in the network. Since RapidIO allows complex system topologies such as meshes, a mechanism is provided so that the host can determine if it has been to a node before. For high-reliability applications, RapidIO allows more than one host to configure the machine.

29.6.3 Error Coverage

The mean time between failure (MTBF) and mean time to repair (MTTR) of a system are often critical considerations in embedded infrastructure applications. RapidIO applications require reliable packet delivery. It is intolerable for an error to go undetected or for a packet to be dropped. Further, once detected, it is desirable to recover from these errors with minimal system interruption. Because of these factors and since RapidIO operates at very high frequencies, it is necessary to provide strong error coverage. Much of the RapidIO error coverage is handled through the physical layer specification. This allows different coverage strategies depending on the physical environment without affecting the transport and logical specifications.

Several error detection schemes are deployed to provide coverage. Packets are covered using a CCITT16 cyclic redundancy check (CRC). Packets are transmitted and acknowledged following a strict sequence using the "AckID" field.

29.6.4 Error Recovery

The control symbol is at the heart of the hardware error recovery mechanism. Should a packet be detected with a bad checksum, control symbols are sent to verify that both sender and receiver are still synchronized and a packet retry is issued. Should a transaction be lost, a watchdog time-out occurs and the auto recovery state machines attempt to re-synchronize and retransmit.

If an interface fails severely, it may not be able to recover gracefully in hardware. For this extreme case, RapidIO hardware can generate interrupts so that system software can invoke a higher-level error recovery protocol. Software may query the maintenance registers to reconstruct the current status of the interface and potentially restart it. An in-band device reset control symbol is provided.

29.7 Performance

RapidIO is intended as a processor and memory interface where both latency and bandwidth must be considered. RapidIO may also be used to carry data streams in which deterministic delivery behavior is required. Separate clock and frame signals are provided to eliminate encoding and decoding latency. Source routing and transaction priority tagging limit the blocking of packets, especially those of a critical nature. Large data payloads of up to 256 bytes and responseless stream write operations move larger volumes of data with less transaction overhead.

29.7.1 Packet Structures

The RapidIO packet is structured to promote simplified construction and parsing of packets in a wider on-chip parallel interface, limiting the amount of logic operating on the narrower high-frequency interface. Packets are organized in byte granularities and further in 32-bit word alignments. In this way fields land consistently in specific byte lanes on the receiving device, limiting the need for complex assembly and parsing logic.

29.7.2 Source Routing and Concurrency

Traditional bus-based systems, such as those using PCI, have relied on address broadcasting to alert targets of an operation. This is effective since all devices monitor a common address bus and respond when they recognize a transaction to their address domain. Unfortunately, only one master can be broadcasting an address at a time.

Switch-based systems can employ two methods to route transactions: broadcast or source routing. For the broadcast scheme, a packet is sent to all connected devices. It is expected that one and only one device actually will respond to the packet. The advantage of the broadcast scheme is that the master does not need to know where the receiver resides. The disadvantage is that there is a large amount of system bandwidth wasted on the paths for which the transaction was not targeted.

To take full advantage of available system bandwidth, RapidIO employs source routing. This means that each packet has a source-specified destination address that instructs the fabric specifically where the transaction is to be routed. With this technique only the path between the sender and receiver is burdened with the transaction. The switch devices make decisions on a much smaller number of bits resulting in small route tables and lower latency. This method leaves open bandwidth on other paths for other devices in the system to communicate with each other concurrently. This scheme does not preclude the use of broadcast or multicast routing.

29.7.3 Packet Overhead

A performance concern in any tightly coupled intra-system interconnect is the transaction overhead required for the interface. Such overhead includes all bytes sent to complete a transaction such as arbitration, addresses, acknowledgements, error coverage, and so on. Figure 29.12 shows some typical RapidIO operations and the number of bytes required to complete each operation. It is important to remember that RapidIO is a full-duplex interface and therefore the interface can be fully pipelined with several outstanding operations at various stages of completion. Reply overhead does not contend with sending overhead. This is different from traditional buses, which require turnaround cycles and arbitration phases that add to the overhead.

29.7.4 Bandwidth

The RapidIO 8/16LP-LVDS specification presently defines four operating frequencies and two widths. The Serial RapidIO specification defines three operating frequencies and two widths (x1, x4). Together RapidIO offers sustained data payload bandwidths ranging from just under a gigabit per second for the 1.25 GHz 1x serial interface up to 60 gigabits per second for the 1 GHz x16 parallel interface. Table 29.2 shows the operating points for both serial and parallel RapidIO with peak and sustained full-duplex data bandwidth. In the table, sustained operations include all transaction overhead and assume fully loaded full-duplex communications. 32-byte operation is 50% efficient and 256-byte operation is greater than 90% efficient. For unidirectional traffic, 32-byte is greater than 70% efficient and 256-byte is greater than 95% efficient. The table indicates that the 8/16LP-LVDS physical layer can scale to the needs of the application.

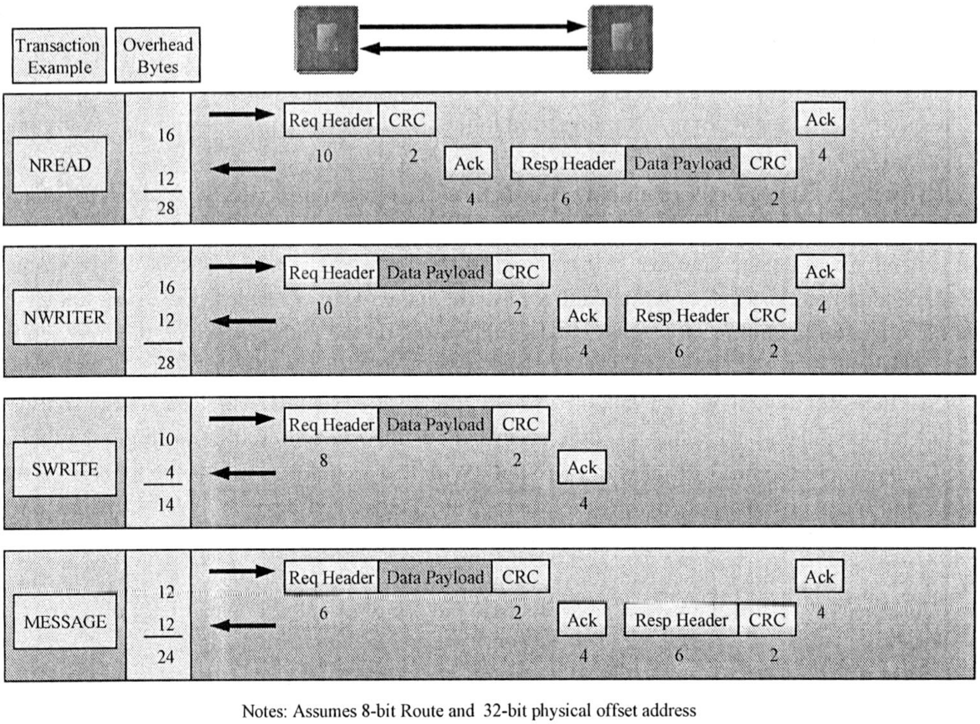

Figure 29.12 Operation overhead includes the total bytes sent in each direction to complete an operation.

Table 29.2 RapidIO Physical Layers Offer a Wide Range of Performance Levels to Meet the Needs of Applications

Parallel RapidIO						
Clock Rate	8-bit Mode			16-bit Mode		
	PEAK	Sustained 32-byte Op	Sustained 256-byte Op	PEAK	Sustained 32-byte Op	Sustained 256-byte Op
250 MHz	8 Gbps	4 Gbps	7.5 Gbps	16 Gbps	8 Gbps	15 Gbps
500 MHz	16 Gbps	8 Gbps	15 Gbps	32 Gbps	16 Gbps	30 Gbps
750 MHz	24 Gbps	12 Gbps	22.5 Gbps	48 Gbps	24 Gbps	45 Gbps
1 GHz	32 Gbps	16 Gbps	30 Gbps	64 Gbps	32 Gbps	60 Gbps

Serial RapidIO						
Clock Rate	1-bit Wide			4-bit Wide		
	PEAK	Sustained 32-byte Op	Sustained 256-byte Op	PEAK	Sustained 32-byte Op	Sustained 256-byte Op
1.25 GHz	2 Gbps	1 Gbps	1.8 Gbps	8 Gbps	4 Gbps	7.2 Gbps
2.5 GHz	4 Gbps	2 Gbps	3.6 Gbps	16 Gbps	8 Gbps	14.4 Gbps
3.125 GHz	5 Gbps	2.5 Gbps	4.5 Gbps	20 Gbps	10 Gbps	18 Gbps

29.7.5 Operation Latency

In load/store environments, where processors or other devices are dependent on the results of an operation before proceeding, latency becomes a key factor to system performance. It is assumed that the operations of interest here are small byte or cache-line oriented operations. Since RapidIO is narrower than traditional parallel buses, an operation requires more clock cycles for data transmission; also RapidIO has extra overhead associated with routing and error management. However, RapidIO, being a point-to-point technology with separate send and receive paths, has virtually no contention or arbitration overhead, can operate at higher frequencies and can send and receive data simultaneously. The combination of these factors leads to operation latency comparable to existing parallel buses.

29.8 Summary

The RapidIO interconnect provides a robust packet-switched system-level interconnect. It provides a partitioned architecture that can be enhanced in the future. It enables higher levels of system performance while maintaining or reducing implementation costs. A RapidIO end point can be implemented in a small silicon footprint. Proven industry-standard signaling schemes (LVDS, XAUI) are used for the physical interfaces. Error management includes the ability to detect multi-bit errors and survive most multi-bit and all single-bit errors. Even with all these capabilities, RapidIO's protocol overhead and latency are comparable to current bus technologies and significantly better than local area network-based fabric technologies such as Ethernet.

29.9 Exercises

29-1. (a) What range of data rates in Gbps is covered by RapidIO? (b) For the minimum value in this range, what's the architecture (serial or parallel), number of bytes in the sustained operation, clock rate, and bit width? (c) For the maximum value in this range, what's the architecture (serial or parallel), number of bytes in the sustained operation, clock rate, and bit width?

29-2. What types of applications is RapidIO intended for?

29-3. (a) How does a bridged hierarchy bus structure work and what are its advantages over a single hierarchy bus (single bus structure)? (b) What bus structure is being used to achieve even higher system performance? Answer each question with just two or three sentences.

29-4. What negative effect does increasing the bus width have as an approach to increasing the bus data throughput rate and performance?

29-5. What is the function of control planes and data forwarding planes?

29-6. (a) What is a SAN and what is it used for? (b) Which technologies among the following are typically used for SAN interconnects: PCI Express, HyperTransport, InfiniBand, Fibre Channel, and iSCSI?

29-7. To meet the challenge of higher bus frequencies, what advantages does RapidIO offer over bus structures that reduce the number of supportable devices per bus segment to obtain the higher frequencies and then increase the number of bus segments?

29-8. Can RapidIO interconnects be used in live insertion applications?

29-9. List the eight principal objectives that drive the RapidIO architecture.

29-10. (a) What is the RapidIO 8/16LP-LVDS specification? (b) What do the "8," "16," and "LP" stand for in the name? (c) What signaling technology does it use?

CHAPTER 30

PCI Express and ExpressCard

Chapter Objectives

- Explain how PCI Express is a technology for next-generation interconnect that will serve as a standard local I/O bus for a wide variety of future computing platforms.
- Illustrate how the PCI Express bus is a fully serial interface that is easily scalable by adding lanes.
- Reveal why the multidrop, parallel bus implementation of PCI 2.2 and PCI-X is close to its practical limits of performance.
- Summarize the requirements for a third-generation local I/O bus.
- Illustrate the layered architecture of PCI Express and describe its physical, link, transaction, software, and configuration/operating system layers.
- Introduce the ExpressCard and describe its funtionality and small-form factor.

30.1 PCI Express as Next-Generation I/O[1]

This chapter looks at the success of the widely adopted PCI bus and describes a higher-performance, next-generation I/O interconnect, called PCI Express architecture, that will serve as a standard local I/O bus for a wide variety of future computing platforms. Key PCI attributes such as its usage model and software interfaces are maintained, whereas its bandwidth-limiting, parallel bus implementation is replaced by a long-life, fully serial interface. A split-transaction protocol is implemented with attributed packets that are prioritized and optimally delivered to their

1. Reprinted by permission of Intel Corporation, Copyright Intel Corporation 2004 [Bha03], except where noted.

target. The new PCI Express architecture comprehends a variety of form factors to support smooth integration with PCI and to enable new system form factors. PCI Express architecture will provide industry leading performance and price/performance.

The PCI bus has served us well for the last 10 years and it will play a major role in the next few years. However, today's and tomorrow's processors and I/O devices are demanding much higher I/O bandwidth than PCI 2.2 or PCI-X can deliver and it is time to engineer a new generation of PCI to serve as a standard I/O bus for future generation platforms. There have been several efforts to create higher-bandwidth buses and this has resulted in the PC platform supporting a variety of application-specific buses alongside the PCI I/O expansion bus as shown in Figure 30.1.

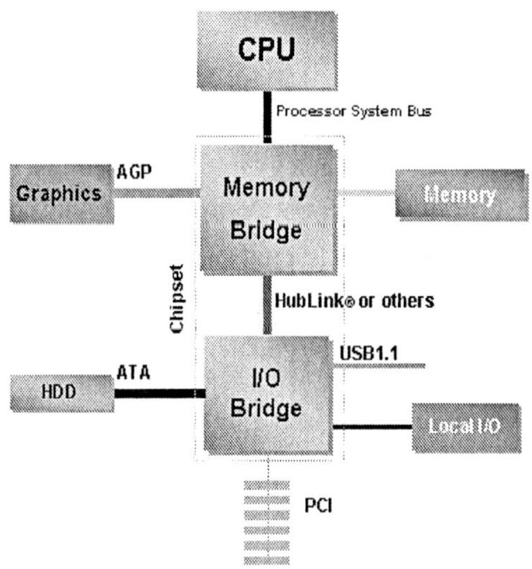

Figure 30.1 Today's PC has multiple local buses with different requirements.

The processor system bus continues to scale in both frequency and voltage at a rate that will continue for the foreseeable future. Memory bandwidths have increased to keep pace with the processor. Indeed, as shown in Figure 30.1, the chipset is typically partitioned as a memory hub and an I/O hub since the memory bus often changes with each processor generation. One of the major functions of the chipset is to isolate these ever-changing buses from the stable I/O bus.

Close investigation of the 1990's PCI signaling technology reveals a multidrop, parallel bus implementation that is close to its practical limits of performance: It cannot be easily scaled up in frequency or down in voltage, its synchronously clocked data transfer is signal skew limited, and the signal routing rules are at the limit for cost-effective FR4 technology. All approaches to pushing these limits to create a higher bandwidth, general-purpose I/O bus result in large cost increases for little performance gain. The desktop solution of Figure 30.1 is only part of the problem of diverging local I/O bus standards. To PCI's credit, it has been used in applications not envisaged by the original specification writers, and variants and extensions of PCI can be found in desktop, mobile, server, and embedded communications market segments.

Today's software applications are more demanding of the platform hardware, particularly the I/O subsystems. Streaming data from various video and audio sources are now commonplace on the desktop and mobile machines and there is no baseline support for this time–dependant data within the PCI 2.2 or PCI-X specifications. Applications such as video-on-demand and audio redistribution are putting real-time constraints on servers too. Many communications applications and embedded-PC control systems also process data in real-time. Today's platforms — an example desktop PC is shown in Figure 30.2 — must also deal with multiple concurrent transfers at ever-increasing data rates. It is no longer acceptable to treat all data as equal — it is more important, for example, to process streaming data first since late real-time data is as useless as no data. Data needs to be "tagged" so that an I/O system can prioritize its flow throughout the platform.

Applications, such as Gigabit Ethernet and InfiniBand, require higher bandwidth I/O. A third-generation I/O bus must include additional features alongside increased bandwidth.

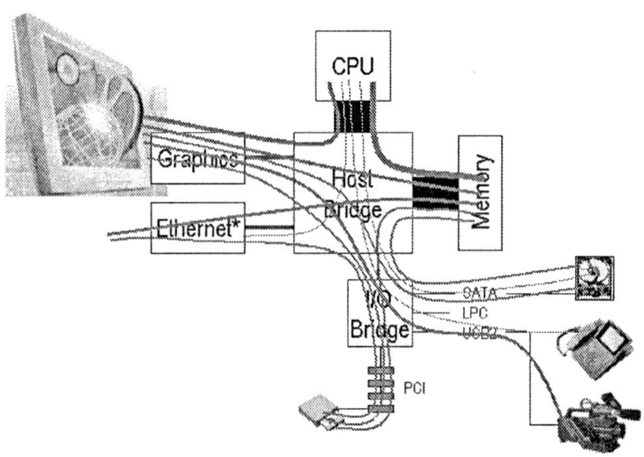

Figure 30.2 Multiple concurrent data transfers.

The following list summarizes the requirements for a third-generation local I/O bus:

- Supports multiple market segments and emerging applications:
  Unifying I/O architecture for desktop, mobile, server, communications platforms, workstations, and embedded devices.
- Low cost and high volume:
  Cost at or below PCI cost structure at the system level.

- PCI-compatible software model:

  Boot existing operating systems without any change. PCI-compatible configuration and device driver interfaces.

- Performance:

  Scalable performance via frequency and additional lanes. High bandwidth per pin. Low overhead. Low latency.

- Support multiple platform connection types:

  Chip-to-chip, board-to-board via connector and docking station, and enable new form factors.

- Advanced features:

  Comprehend different data types. Power management. Quality of service. Hot-plug and hot-swap support. Data integrity and error handling. Extensible. Base mechanisms to enable embedded and communications applications.

- Non-goals:

  Coherent interconnect for processors, memory interconnect, cable interconnect for cluster solutions.

30.2 PCI Express Architecture Overview

Recent advances in high-speed, low-pin-count, point-to-point technologies offer an attractive alternative for major bandwidth improvements. A PCI Express multidrop, parallel bus topology contains a host bridge and several endpoints (the I/O devices) as shown in Figure 30.3. Multiple point-to-point connections introduce a new element, the switch, into the I/O system topology also shown in Figure 30.3. The switch replaces the multidrop bus and is used to provide fan-out

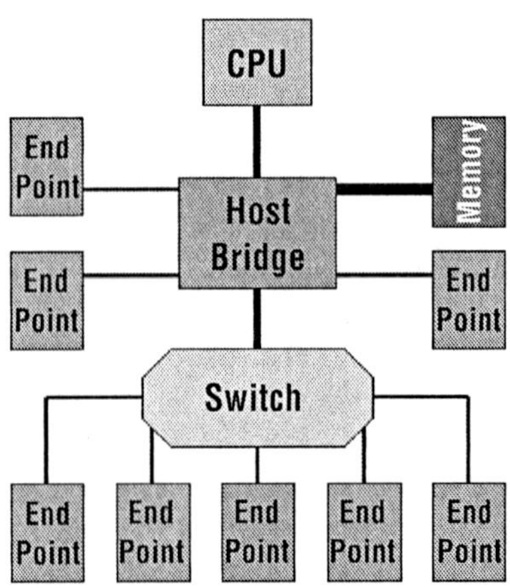

Figure 30.3 A switch is added to the system topology.

for the I/O bus. A switch may provide peer-to-peer communication between different endpoints and this traffic, if it does not involve cache-coherent memory transfers, need not be forwarded to the host bridge. The switch is shown as a separate logical element but it could be integrated into a host bridge component.

The low-signal-count, point-to-point connections may be constructed with connectors and cables. The PCI Express mechanicals will enable new classes of system partitioning (the "boring beige box" is no longer required!).

Figures 30.4 through 30.6 show typical platforms using the PCI Express architecture. The multiple, similar parallel buses of today's platform are replaced with PCI Express links with one or more lanes. Each link is individually scalable by adding more lanes so that additional bandwidth may be applied to those links where it is required — such as graphics in the desktop platform and bus bridges (e.g., PCI Express-to-PCI-X) in the server platform.

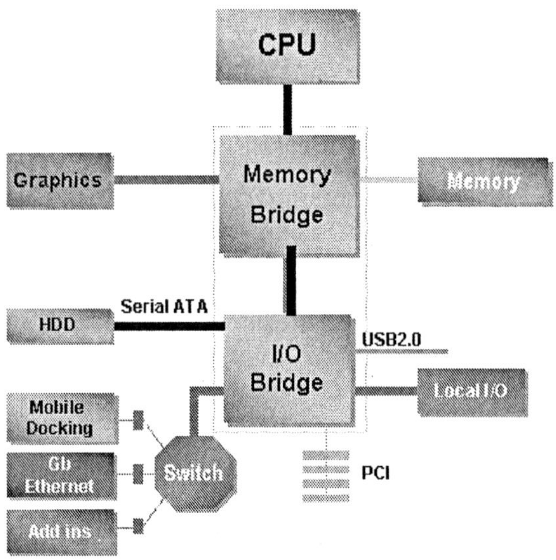

Figure 30.4 General-purpose desktop/mobile I/O interconnect for future years.

A PCI Express switch provides fanout capability and enables a series of connectors for add-in, high performance I/O. The switch is a logical element that may be implemented within a component that also contains a host bridge, or it may be implemented as a separate component.

It is expected that PCI will coexist in many platforms to support today's lower bandwidth applications until a compelling need, such as a new form factor, causes a full migration to a fully PCI Express based platform.

The server platform requires more I/O performance and connectivity including high-bandwidth PCI Express links to PCI-X slots, Gigabit Ethernet, and an InfiniBand fabric. Figure 30.5

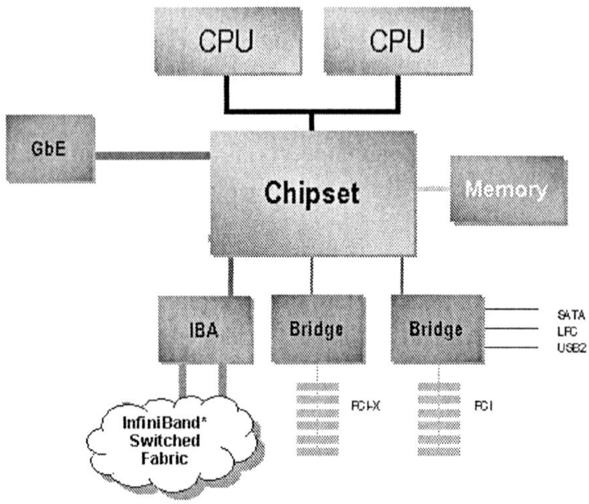

Figure 30.5 PCI Express-based server or workstation system.

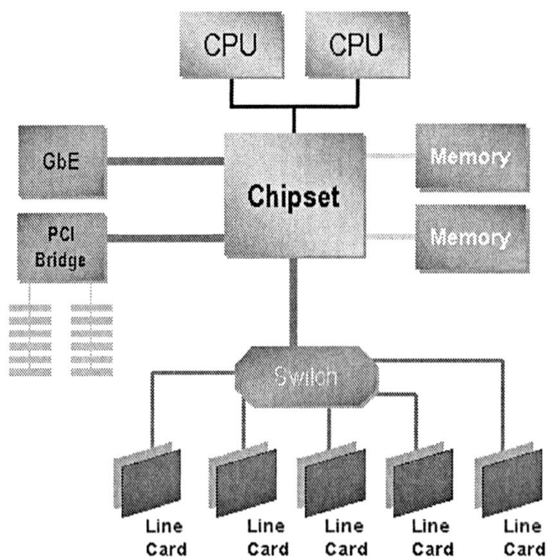

Figure 30.6 PCI Express-based networking communications system.

shows how PCI Express provides many of the same advantages for servers, as it does for desktop systems. The combination of PCI Express for "inside the box" I/O, and InfiniBand fabrics for "outside the box" I/O and cluster interconnect, allows servers to transition from "parallel shared buses" to high-speed serial interconnects.

The networking communications platform could use multiple switches for increased connectivity and quality of service for differentiation of different traffic types. It too would benefit from a multiple PCI Express links that could be constructed as a modular I/O system.

30.3 PCI Express Architecture

The PCI Express architecture is specified in layers as shown in Figure 30.7. Compatibility with the PCI addressing model (a load-store architecture with a flat address space) is maintained to ensure that all existing applications and drivers operate unchanged. PCI Express configuration uses standard mechanisms as defined in the PCI Plug-and-Play specification. The software layers will generate read and write requests that are transported by the transaction layer to the I/O devices using a packet-based, split-transaction protocol. The link layer adds sequence numbers and a cyclical redundancy check (CRC) to these packets to create a highly reliable data transfer mechanism. The basic physical layer consists of a dual-simplex channel that is implemented as a transmit pair and a receive pair. The initial speed of 2.5 gigatransfers/second/direction provides a 200 MB/s communications channel that is close to twice the classic PCI data rate.

The remainder of this section will look deeper into each layer starting at the bottom of the stack.

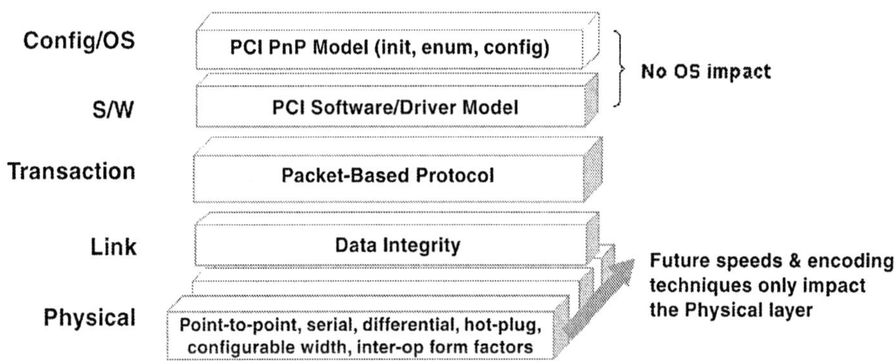

Figure 30.7 The PCI Express architecture is specified in layers.

30.3.1 Physical Layer

The fundamental PCI Express link consists of two, low-voltage, differentially driven pairs of signals: a transmit pair and a receive pair, as shown in Figure 30.8. A data clock is embedded using the 8B/10B encoding scheme to achieve very high data rates. The initial frequency is 2.5 gigatransfers/second/direction and this is possible to increase with silicon technology advances of up to 10 gigatransfers/second/direction (the theoretical maximum for signals in copper). The physical layer transports packets between the link layers of two PCI Express agents.

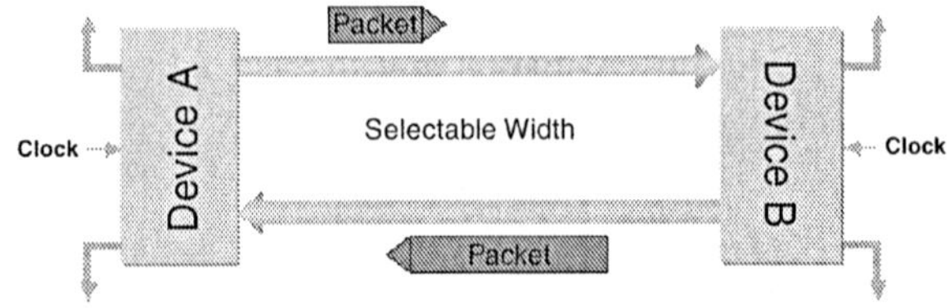

Figure 30.8 A PCI Express link uses transmit and receive signal pairs.

The bandwidth of a PCI Express link may be linearly scaled by adding signal pairs to form multiple lanes. The physical layer supports x1, x2, x4, x8, x12, x16, and x32 lane widths and splits the byte data as shown in Figure 30.9. Each byte is transmitted, with 8B/10B encoding, across the lane(s). This data disassembly and reassembly is transparent to other layers.

During initialization, each PCI Express link is set up following a negotiation of lane widths and frequency of operation by the two agents at each end of the link. No firmware or operating system software is involved.

The PCI Express architecture comprehends future performance enhancements via speed upgrades and advanced encoding techniques. The future speeds, encoding techniques, or media would only impact the physical layer.

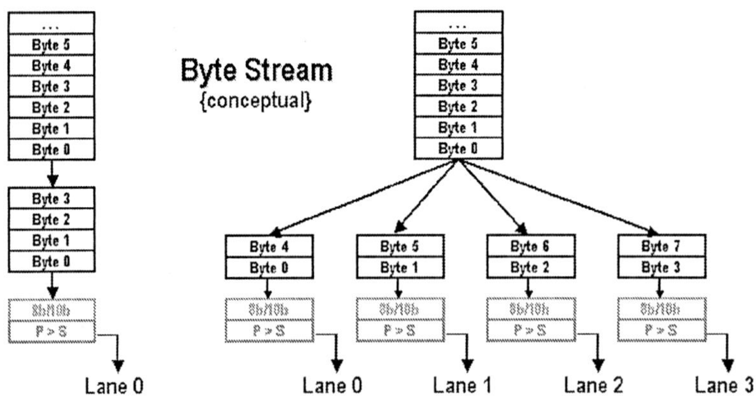

Figure 30.9 A PCI Express link consists of one or more lanes.

30.3.2 Link Layer

The primary role of a link layer is to ensure reliable delivery of the packet across the PCI Express link. The link layer is responsible for data integrity and adds a sequence number and a CRC to the transaction layer packet as shown in Figure 30.10.

Most packets are initiated at the transaction layer (see "Transaction Layer" on page 859). A credit-based, flow control protocol ensures that packets are only transmitted when it is known

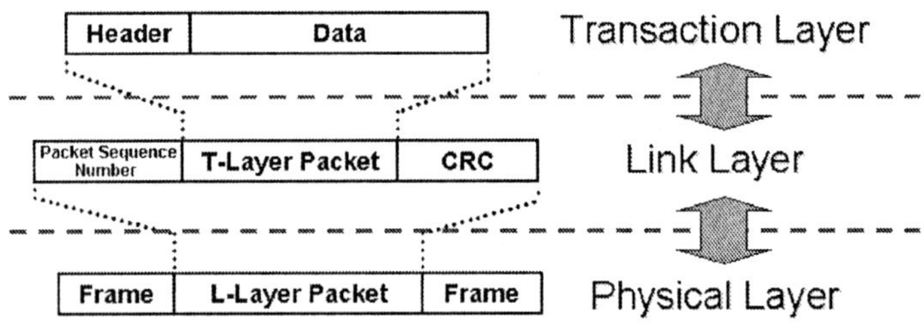

Figure 30.10 The link layer adds data integrity features.

that a buffer is available to receive this packet at the other end. This eliminates any packet retries, and their associated waste of bus bandwidth due to resource constraints. The link layer will automatically retry a packet that was signaled as corrupted.

30.3.3 Transaction Layer

The transaction layer receives read and write requests from the software layer and creates request packets for transmission to the link layer. All requests are implemented as split transactions and some of the request packets will need a response packet. The transaction layer also receives response packets from the link layer and matches these with the original software requests. Each packet has a unique identifier that enables response packets to be directed to the correct originator. The packet format supports 32-bit memory addressing and extended 64-bit memory addressing. Packets also have attributes such as "no-snoop," "relaxed-ordering," and "priority," which may be used to optimally route these packets through the I/O subsystem.

The transaction layer supports four address spaces: it includes the three PCI address spaces (memory, I/O, and configuration) and adds a message space. PCI 2.2 introduced an alternate method of propagating system interrupts called Message Signaled Interrupt (MSI). Here a special-format memory write transaction was used instead of a hard-wired sideband signal. This was an optional capability in a PCI 2.2 system. The PCI Express specification reuses the MSI concept as a primary method for interrupt processing and uses message space to support all prior side-band signals, such as interrupts, power-management requests, resets, and so on, as in-band messages. Other "special cycles" within the PCI 2.2 specification, such as "interrupt acknowledge," are also implemented as in-band messages. You could think of PCI Express messages as "virtual wires" since their effect is to eliminate the wide array of sideband signals currently used in a platform implementation.

30.3.4 Software Layers

Software compatibility is of paramount importance for a third-generation local I/O bus. There are two facets of software compatibility: initialization (enumeration) and runtime. PCI has a robust initialization model wherein the operating system can discover all of the add-in hardware devices present and then allocate system resources, such as memory, I/O space, and interrupts, to create an optimal system environment. The PCI configuration space and the programmability of I/O devices are key concepts that are unchanged within the PCI Express architecture; in fact, all operating systems will be able to boot without modification on a PCI Express-based platform.

The runtime software model supported by PCI is a load-store, shared memory model — this is maintained within the PCI Express architecture which will enable all existing software to execute unchanged. New software may use new capabilities.

30.3.5 Mechanical Form Factors

The low signal count of a PCI Express link will enable both an evolutionary approach to I/O subsystem design and a new modular approach that will encourage new system partitioning.

30.3.5.1 ExpressCard

Initial implementations of PCI Express-based add-in cards will coexist alongside the current PCI-form-factor boards (both full-size and "half-height"). For example, a higher-bandwidth link, such as a 16-lane connection to a graphics card, could use a new connector placed alongside the existing PCI or AGP connector in the area previously occupied by those types of connectors, as shown in Figure 30.11.

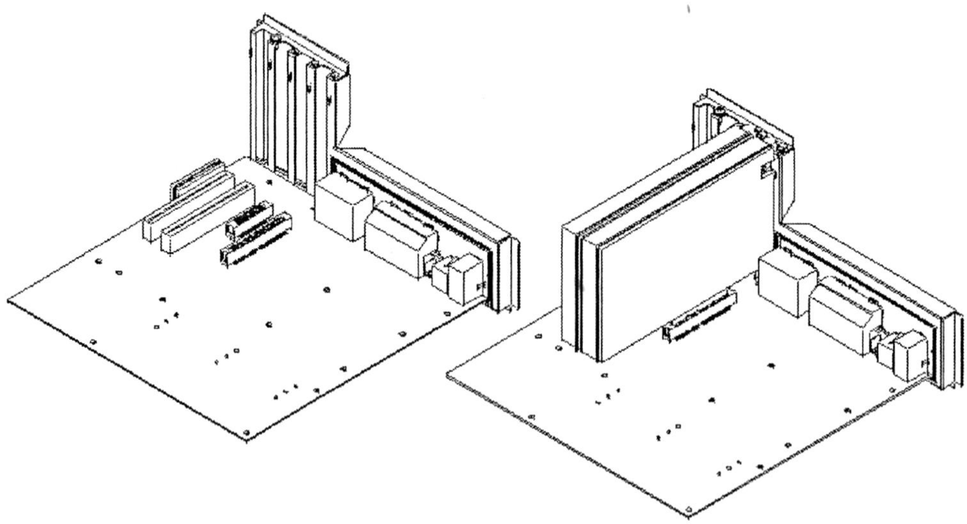

Figure 30.11 Using an additional connector alongside the PCI connector.

In addition to the base implementations of PCI Express-based add-in cards that replace AGP and PCI cards, other form factors are in the process of being developed. The PCMCIA industry group (*www.pcmcia.org*) has announced the development of a new specification, code-named ExpressCard, that takes the next step in PC card evolution. This ExpressCard will offer integrated desktop and mobile external expansion into a single standard. The ExpressCard standard is targeted for OEMs seeking small footprint for thinner mobile designs and sealed systems for desktop designs. ExpressCard is the next evolution of the PC Card, combining a smaller form factor, faster performance with the PC cards, reliability, and ease of use. ExpressCard supports hot-swappable device sharing between mobile and desktop devices, similar to how USB devices can be shared. Desktop users will benefit from ExpressCards plug-and-play ease of use, which eliminates the need to open the chassis to add new features. The new standard is operating-system independent.

"Formerly known as NEWCARD, ExpressCard is the hot-swappable expansion module for PCs and laptops. It was developed through a coalition of system OEMs, card manufacturers, and connector and component manufacturers including Dell, HP, IBM, Intel, Lexar Media, Microsoft, SCM Microsystems, and Texas Instruments.

Based on PCI Express architecture and Universal Serial Bus (USB) 2.0 interfaces, ExpressCard directly connects to chipsets, removing the need for a bridge component. PCMCIA specifies that card manufacturers can also use the standard silicon in building their products.

The standard modules come in two form factors identified primarily by their width in millimeters — ExpressCard 34 and ExpressCard 54, as shown in Figure 30.12. Both modules are 75 mm long and 5 mm high and both have 26 pins compared with the 68-pin, bus-controller card that it replaces. Both also put out less than 1.3 watts of dissipation.

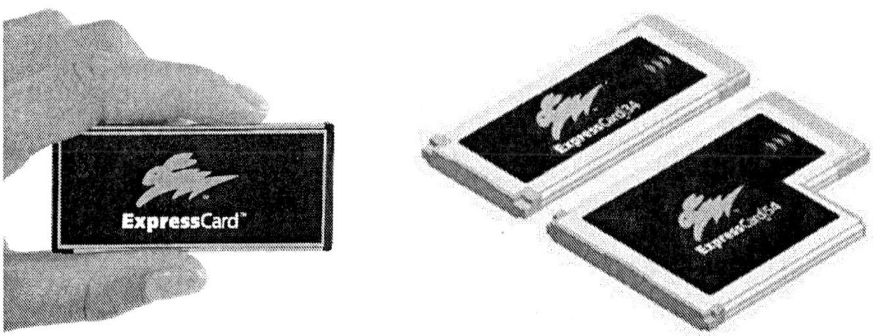

Figure 30.12 ExpressCard modules are available in two widths, 34 mm and 54 mm. [Reprinted with permission of PCMCIA.]

ExpressCard is a major milestone because it can be used for a wide range of applications including wired and wireless LANs, broadband modems, audio/video streaming, TV tuners/decoders, I/O adapters (e.g., 1394a/b), and magnetic disk drives — and it can be used without having to pop the casing every time you do an upgrade.

ExpressCard is seen as a major leap forward in delivering thinner, lighter, and faster systems to consumers. By drawing upon the PCI Express and USB standards, it is anticipated that it will be easier for desktop and notebook users to add hardware capabilities to their systems than before. Despite the module's mobile and hot-swappable benefits, ExpressCard is not being considered for traditional PDAs because it would be too thick for most handheld devices. While there are a lot of embedded possibilities for ExpressCard, the focus is currently on mainstream PCs and laptops.

In addition to vendors, ExpressCard has a strong collaboration with PCMCIA, the USB Implementers Forum (USB IF), and the Peripheral Component Interconnect-Special Interest Group (PCI-SIG). The three are working together on a compliance program. Companies that develop ExpressCard-compliant products will be able to use the ExpressCard name and rabbit logo on their products to indicate plug and play interoperability."[2] [Sin03]

30.4 Development Timeline

The PCI-SIG board of directors approved the PCI Express Specification Rev 1.0 in July 2002. The specification is available to PCI-SIG members and is currently at Rev 1.0A. The PCI-SIG Web site is located at *www.pci-sig.com*. Intel's developer network site, established to help the industry, is located at *www.express-lane.org/*.

30.5 Summary

PCI Express's differential, point-to-point connection provides a very-high-speed interconnect using few signals. Its message space eliminates all prior sideband signals resulting in minimal implementation signals. PCI Express's 200 MB/s/pin will translate to the lowest-cost implementation for any required bandwidth.

The PCI Express architecture meets all of the requirements of a third-generation I/O bus. Its advanced features and scalable performance will enable it to become a unifying I/O solution across a broad range of platforms — desktop, mobile, server, communications, workstations, and embedded devices. A PCI Express link is implemented using multiple, point-to-point connections called lanes and multiple lanes can be used to create an I/O interconnect whose bandwidth is linearly scalable. This interconnect will enable flexible system partitioning paradigms at or below the current PCI cost structure. PCI Express is software compatible with all existing PCI-based software to enable smooth integration within future systems.

30.6 Exercises

30-1. Is PCI Express expected to quickly replace PCI 2.2 and PCI-X?

30-2. (a) Why is PCI Express described as being easily scalable? (b) What does easily scalable mean?

2. Copyright 2004 Jupitermedia Corporation. All rights reserved. *www.Wi-fiplanet.com.*

30-3. (a) What are the two sizes for ExpressCards? State all three dimenstions for each card. (b) How many pins does each module have? (c) How many pins do the older bus-controller cards have?

30-4. (a) How are bytes to be transmitted over a PCI Express link split out among each of the lanes in a four-lane application? (b) Conceptually, does the 8B/10B encoding occur before or after the bytes are split out?

30-5. What two things does the link layer add to the transaction layer packet?

30-6. Do both the PCI Express interconnect bus and ExpressCard allow hot-swapping?

30-7. Describe the effect that the disassembly and reassembly of bytes occuring at the physical layer has upon the link layer.

30-8. (a) What is the initial data rate in Gbps for one direction of a single lane? (b) What data rate is this expected to increase to as a result of advances in silicon technology?

30-9. What encoding scheme is used to achieve very high data rates?

30-10. List the five layers of the PCI Express architecture and explain, in a single sentence for each layer, what that layer does.

PART 7

Lab and Test Instrumentation

CHAPTER 31

Electrical and Optical Test Equipment

Written by Greg Le Cheminant of Agilent Technologies.

Chapter Objectives

- Discuss the differences between real-time and equivalent-time oscilloscopes.
- Explain how equivalent-time oscilloscopes can work, given that their sampling rate is much lower than their bandwidth.
- Discuss bit error ratio testers (BERTs) and the process to make a BER measurement.
- Define jitter and explain the three tools available for jitter analysis.
- Explain how BERT is also used to characterize optical systems.
- Introduce optical spectrum analyzers (OSAs) and their ability to display power as a function of optical wavelength.
- Discuss photodetection and define optical modulation amplitude and extinction ratio.
- Explain how to characterize fiber properties and measure optical receiver sensitivity.

There are a variety of tools available for the high-speed designer, each designed to provide a specific class or type of information. In this chapter, we will provide an overview of several of the more common tools ranging from the oscilloscope to the bit error ratio tester (BERT). Features and measurement issues associated with these instruments will be discussed.

31.1 Oscilloscopes[1]

31.1.1 Classes of Oscilloscopes — Real-Time vs. Equivalent-Time

Between the two main classes of oscilloscopes, analog and digital, digital technology has largely overtaken analog technology and the majority of oscilloscopes sold today are digital. Within the classification of "digital oscilloscopes," there are two basic sub-classes, real-time oscilloscopes and equivalent-time oscilloscopes. A special class of the equivalent-time oscilloscope is the time-domain reflectometer (TDR). Real-time oscilloscopes are the most common while equivalent-time oscilloscopes are used for measurements of the fastest signals. Real-time oscilloscopes have bandwidths of 6 GHz and higher and can sample data at rates of up to 20 Gsa/sec (gigasamples/second) per channel. By comparison, equivalent-time oscilloscopes can make waveform measurements at data rates of 40 Gbps and have a bandwidth of up to 80 GHz or higher. In any form of a digital oscilloscope, the signal amplitude is converted from analog signal levels to digitized values (sampling), processed, and displayed. However, the sampling process for the equivalent-time oscilloscope is much different from that of the real-time oscilloscope.

31.1.2 Real-Time Oscilloscopes

The real-time oscilloscope is based upon an extremely fast analog-to-digital converter (ADC). The bandwidth of this type of oscilloscope is determined by the speed or sample rate of the ADC. In the sampling theorem limit, the oscilloscope measurement bandwidth is half the sample rate. Practically it is less than that, as reconstruction of a signal from samples requires some form of interpolation. Thus today's 20 GSa/s oscilloscopes specify their bandwidths to be 6 GHz.

The sampling process is initiated by a trigger event. In that the oscilloscope displays amplitude as a function of time, and time is anchored to the trigger event, triggering represents one of the most important aspects of a high-speed waveform measurement. A trigger is usually some signal provided to the oscilloscope such that when a defined amplitude or signal condition is achieved, the sampling process is initiated. There are a variety of triggering scenarios and signals. The trigger may come directly from the signal to be measured or come from a separate signal such as a system clock that is synchronous to the observed signal. Examples of trigger events can be as simple as a rising or falling signal edge crossing a specific amplitude threshold or as sophisticated as an event occurring as a result of complex logic.

The ADC determines the amplitude resolution of the oscilloscope and converts the amplitude to a binary number. The larger the number of bits output by the ADC, the higher the vertical resolution of the oscilloscope. Eight-bit resolution is common. A higher-resolution ADC provides a more accurate amplitude measurement, but at the expense of sampling rate.

Getting the signal to the oscilloscope is a potential source for measurement error and needs to be carefully considered. The oscilloscope will typically have both a high-impedance as

1. All text and figures in this chapter are reproduced with permission of Agilent Technologies Inc. Copyright 2004 Agilent Technologies Inc.

well as a 50 Ω input setting. The high-impedance setting is intended to avoid loading down the circuit. A simple model of the oscilloscope (not the probe) input would be a 1 MΩ resistor with a parallel capacitance, both shunted to ground. (The capacitance value is dependent upon the particular oscilloscope model, but usually is in the low pF range.) The capacitance causes the overall impedance to decrease as frequency is increased. Thus very-high-speed oscilloscopes need to have minimal shunt capacitance. The 50 Ω setting is used for measurement of systems that have an inherent 50 Ω impedance, common in high-speed transmission systems. Although the oscilloscope impedance is low, and in some situations can load down circuitry, the impedance is flat over a wide frequency range and can provide a good termination for a high-speed transmission system.

High-performance probes come in differential, single-ended, and solder-in configurations. In high-performance systems, the probe and a transmission line cable are connected to an amplifier which is external to the scope and is built into the end of a second cable. This second cable has the external amplifier in one end and the other end of the cable is plugged into the scope input channel. (See Figure 31.1.)

It is a common requirement to examine signals where there is no convenient connector to attach a cable. Some form of probe is required to acquire the signal and bring it to the oscilloscope. It is important that the circuitry of the probe does not significantly change the overall circuitry seen by the signal to be measured. An altered circuit can change the signal being

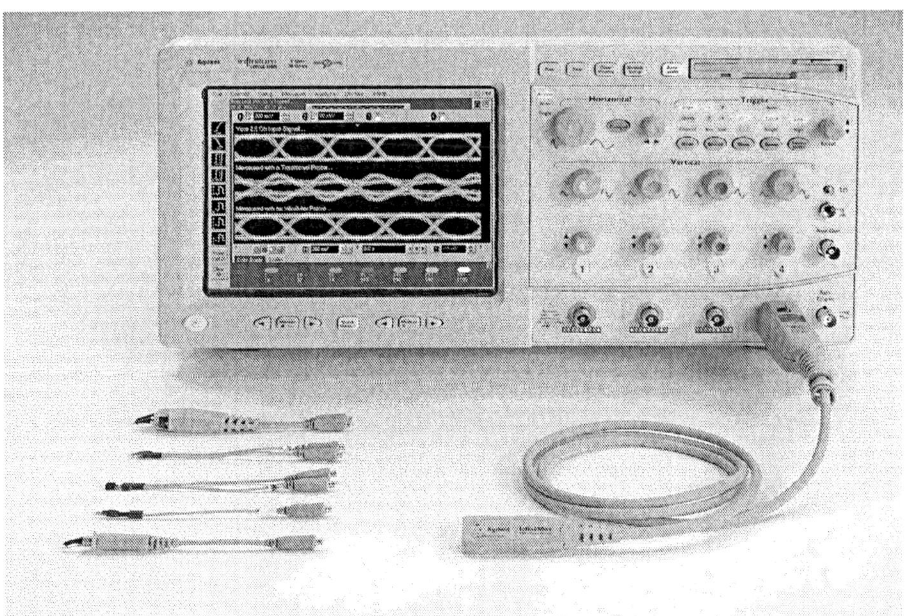

Figure 31.1 Agilent Infiniium 54850 real-time oscilloscope can sample at 20 GSa/s per channel on all four channels simultaneously.

examined and lead to erroneous waveform displays. As signal speeds reach the Gbps rate, the parasitic inductance and capacitance of basic probes makes them virtually unusable. By contrast, the design of high-speed probes minimizes the physical dimensions and places the probe amplification as close to the probe tip as possible. The most modern designs use a transmission line system between the probe tip and the amplifier. This yields bandwidths in excess of 7 GHz with minimal circuit loading.

There is a wide range of oscilloscope performance and pricing. Key performance parameters include sample rate/bandwidth and memory depth. Memory depth is used to describe the number of sample points that can be acquired in a single waveform, thus allowing the signal to be viewed over an extended length of time. Other important parameters include probe solutions and cost. In general, the wider the bandwidth and the faster the sampling rate, the more expensive are the scope and probes.

A 2-channel, 500 MHz oscilloscope with 8 MB of memory such as the Agilent 54642A sells for approximately $8,000, while a 6 GHz 4-channel oscilloscope with 32 MB of memory and precision 7 GHz probing (Agilent 54855A) will cost over $60,000.

31.1.3 Equivalent-Time Oscilloscopes

The equivalent-time oscilloscope is also known as the wide-bandwidth oscilloscope and, as its name suggests, has bandwidths well above the 6 GHz available on the real-time oscilloscope (see Figure 31.2). (The term "wide bandwidth" was coined when real-time scopes had band-

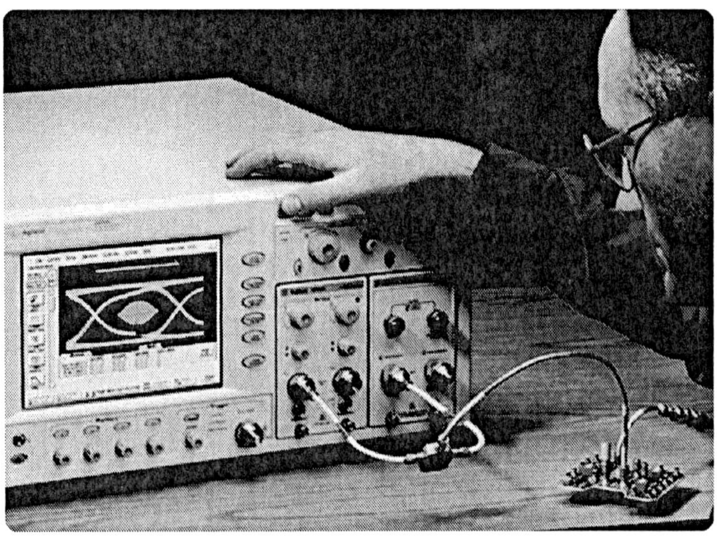

Figure 31.2 Agilent 86100C Infiniium DCA-J equivalent-time oscilloscope with bandwidth in excess of 80 GHz.

widths below 1 GHz and the equivalent-time oscilloscope could operate over 20 GHz. Today's real-time scopes are also wide bandwidth, but the "wide-bandwidth" name typically implies an equivalent-time scope.) Depending upon the plug-in configuration, bandwidths of 20, 50, or even 80 GHz can be achieved. Thus these instruments provide the highest precision in characterizing the shape of the waveform.

The bandwidth of an equivalent-time scope does come at a cost. These instruments are not capable of a direct analog-to-digital conversion of a waveform. The rate at which samples of the waveform are taken is under 1 Msa/sec. This would seem to contradict the high-bandwidth performance of such an instrument. The key to understanding how the high bandwidth can be achieved at a low sample rate is in understanding how these oscilloscopes are triggered as well as the types of signals that can be measured. To view a pulse train similar to what is seen with the real-time oscilloscope, the pulse train viewed on an equivalent-time scope must be one that systematically repeats. There must also be a trigger signal or timing reference that provides a synchronizing event (trigger) for the oscilloscope. This signal must provide a timing edge at the beginning of every repetition of the data sequence.

When the trigger event occurs, a single-amplitude sample of the waveform is taken. (This is in contrast to the complete waveform record that is obtained by the real-time oscilloscope for a single-trigger event.) With the equivalent-time scope, and when the sample is completed and the waveform repeats, the trigger event occurs at the exact relative time to the pattern as for the first pass. The oscilloscope is triggered again, but rather than sample at the same time as when the first sample was taken, the sampling point is incremented or delayed a small amount in time. The sample is taken at a slightly later point in the pattern. The process is repeated until enough samples are taken to fill the time span of the oscilloscope display and the waveform is completely constructed. Even though the rate at which samples are taken is relatively slow, the instrument bandwidth can be very high.

The bandwidth of the equivalent-time scope is set by the sampling electronics. Rather than having the bandwidth dictated by sampling rate or speed of the ADC, it is determined by how accurately the amplitude of the sample is as a function of the signal frequency. Eventually as frequencies become extremely fast, amplitude inaccuracies will occur. This rolloff effect then defines the instrument bandwidth.

Sometimes digital signals are not repetitive and the process discussed above breaks down. However, very accurate measurements using the equivalent-time scope can still be made. Once again, the key lies in how the oscilloscope is triggered. Rather than a trigger occurring once per repetition of the pattern, a clock trigger is used. The clock trigger provides a trigger event for every bit in the waveform. When the oscilloscope receives the first clock edge a sample is taken. The sample may be of a logic 1 or a logic 0, depending upon when the oscilloscope triggers. The low speed of the sampling process does not allow the next clock edge to trigger the oscilloscope. In fact, hundreds or even thousands of bits may be transmitted by the signal source before the oscilloscope can respond to another edge. They will not be viewed. However, when the oscilloscope does finally respond, the clock trigger ensures that the oscilloscope is synchronized to the data. The sampling location occurs at a very precise location relative to the bit position of the signal. To reconstruct the waveform, the second sample is taken at a time slightly delayed from the trigger (compared with the first sample).

Like the first sample, the second sample is as likely to be taken on a logic 1 as a logic 0 as the trigger event occurred at an essentially random bit location in the pattern, but at a precise location in the bit period. Keep in mind that the second sample is taken many bits later in the data stream and the level of the first sample has no bearing on the level of the second. As the process is repeated for subsequent samples, the waveform displayed consists of samples from both logic 1's and logic 0's. The waveform effectively has two levels and is referred to as the eye diagram. Triggering the equivalent-time oscilloscope with a clock signal synchronous to the data then allows the equivalent-time oscilloscope to also be used on signals that are not repetitive. As digital communications signals are transmitted at faster and faster rates, an equivalent-time oscilloscope may be the only tool capable of accurate waveform analysis. The eye diagram is the most convenient display mode. The equivalent-time oscilloscope is also referred to as a digital communications analyzer (DCA), as it will have built-in analysis features to measure the eye diagram.

In a research and development environment, many engineers and scientists have learned how to quickly gauge the quality of a signal through a quick visual inspection of the eye diagram. In the most basic sense, an "open" eye diagram is indicative of a quality signal, while a "closed" eye is indicative of signal impairments.

Recall that it is desirable to have an open eye diagram in both the vertical and horizontal sense. It is difficult to measure a simple numerical parameter that can describe the openness of the eye diagram. Instead, a process called "eye-mask" testing is used. An eye mask is a closed shape created by grouping together a constellation of solid polygons, usually rectangular or hexagonal, that represent where the eye diagram waveform may *not* exist. A typical eye mask consists of a polygon located in the center of the eye diagram, as well as one polygon above and one polygon below. The eye diagram is not allowed to intersect or "violate" any of the mask polygons (i.e., the closed shape referred to as the "mask"). The minimum acceptable opening for the eye diagram is then defined by the size and shape of the central polygon. The XAUI jitter template on the cover of this book is a good example of an eye mask.

Most time-domain reflectometers (TDR is discussed in Chapter 20 starting on page 519) are built around an equivalent-time oscilloscope. The edge speeds required to achieve high two-event resolution dictate a very wide bandwidth receiver for the TDR. Thus the equivalent-time oscilloscope actually functions as several instruments in one — a general-purpose oscilloscope, a digital communications analyzer, and a TDR. Recent developments have added a fourth dimension to the scope, jitter analysis. This topic was discussed in Section 3.4 on page 43 and in Section 26.13 on page 752.

There are several key parameters for the equivalent-time oscilloscope. Bandwidth is usually the most important. The instrument bandwidth is determined by the high frequency samplers, which are interchangeable pieces to the instrument mainframe. Thus the scope can be configured by the user to measure a range of signal rates. Sampling rate tends not to be an important parameter, as the measurement bandwidth is independent of the sampling rate. The oscilloscope will have some uncertainty as to when samples are acquired. Thus a signal edge will have some "width" or spread in its display, even if the signal is perfect. This instrument imperfection is referred to as timebase jitter. Typical values are on the order of 1 ps RMS. This value can be insignificant as long as the jitter of the signal being measured is at least double (2 ps RMS). However, at 10 and 40 Gbps signaling rates, the instrument instability may exceed that of the signal. Special instrument configurations are available to reduce the scope jitter to below 0.2 ps.

Equivalent-time oscilloscopes are generally lower in price than the real-time oscilloscopes. A dual, 20 GHz channel equivalent-time scope will sell for $25,000 (Agilent 86100 and 86112A). A dual 75 GHz, equivalent-time scope with 0.2 ps jitter will sell for approximately $75,000 (Agilent 86118A and 86107A).

31.2 Bit Error Ratio Testers (BERTs)

Whether a digital signal is traveling between chips on a printed circuit board, over a few meters of cable between equipment racks, or across the ocean on an optical fiber, the most important measurement parameter of all is bit error ratio. This is used to indicate if any of the transmitted bits were received in error. Modern communications systems perform extremely well. BERs on the order of 1 bit received in error per trillion bits transmitted are common. This is often expressed as a BER of 1 E–12. It is common for this parameter to be incorrectly referred to as the bit error rate instead of bit error ratio. However, the 1E–12 number mentioned above does not indicate the rate as a function of time at which bits are received in error. This would require knowing both the transmission rate in bits per second and the BER (ratio of errored bits to transmitted bits).

Measuring BER is typically achieved using a BER tester or "BERT" (see Figure 31.3). A BERT consists of two major building blocks. First there is a pattern generator. The pattern generator performs two major functions. It determines the specific data sequence that the system under test will transmit. This is commonly referred to as the test pattern. It is essential that specific, known patterns be used so that it can be determined if the system under test correctly transported the data. The pattern generator also can provide control over characteristics of the test signal such as waveform amplitude and to some degree the pulse shape. The second block of the BERT is the error detector. The error detector is used to verify that each bit sent through the system under test has been received correctly.

The process of performing a BER measurement is as follows:

1. The pattern generator is configured to produce a specific data sequence which is fed to the system or circuitry under test.

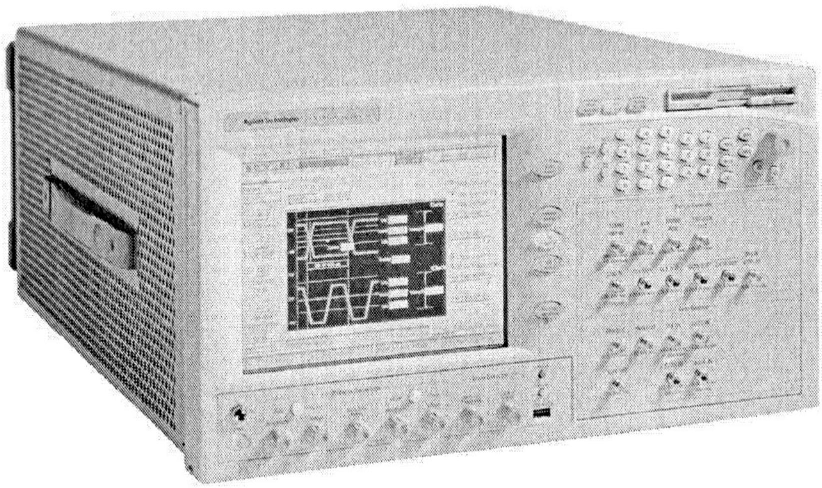

Figure 31.3 Agilent N4901A 13.5 Gbps serial bit error ratio tester (BERT).

2. The system transmits the pattern to its receiver. The receiver must determine the logic level for each incoming bit. In most cases, the signal at the receiver decision circuit will have been degraded due to attenuation and dispersion in the transmission path. The system receiver effectively produces a regenerated output signal/pattern according to what it thought the incoming bits were. A perfect receiver would never make a mistake in assessing the logic level of the input signal.
3. A separate pattern generator within the error detector section of the BERT produces a pattern identical to that originally fed to the system under test.
4. The error detector pattern and the signal/pattern from the system receiver are aligned in time and compared bit for bit. Whenever there is disagreement, an error is counted.
5. The BER is determined from the number of bits found in error divided by the total number of bits checked.

An error detector can be viewed as a decision circuit followed by an "exclusive-or" (EXOR) logic circuit. Data from the system/circuit under test is fed to the error detector. A clock signal is also fed to the error detector. This clock signal is often a clock signal that the receiver of the system/circuit under test has derived from the data being transmitted. The clock is used to time the decision circuit and EXOR gate.

There are three steps to setting up a BERT. First, the timing of the decision circuit must be optimized within one clock cycle to sample the data signal at the ideal point in time (usually the middle of the bit period). Thus a "clock-to-data" alignment must be performed. This is usually an automatic adjustment available in the BERT. Second, there is also an optimum amplitude threshold the error detector decision circuit will use to determine if incoming data is at the 1 or 0 level. A BERT will also have an alignment procedure to determine this optimum level. The third basic step to setting up a BER test is to align the internal pattern generator of the error detector with the pattern from the system/circuit under test so that the EXOR function is being performed on identical locations in each pattern. This too is an automatic procedure available in all modern BERTs.

Although a pattern generator can produce virtually any pattern sequence desired, most BER measurements are performed using pseudo-random binary sequences (PRBS). A PRBS is a data pattern that attempts to replicate truly random data yet is completely deterministic (a requirement for performing a BER measurement). PRBS patterns have lengths of 2^N-1. For example, a 2^7-1 pattern is 127 bits in length and will include all sequences which are 7 bits in length with the exception of 7 sequential 0s. A $2^{10}-1$ PRBS is 1,023 bits in length and has all sequences which are 10 bits in length with the exception of 10 sequential 0s. Pattern generators commonly produce pattern lengths up to $2^{31}-1$. Long PRBS patterns tend to provide higher levels of "stress" on circuitry. Consider how a circuit trying to derive a clock signal from data might perform when the pattern can have 31 consecutive 1s ($2^{31}-1$ PRBS) versus a run of only 7 (2^7-1 PRBS).

PRBS patterns are generated using a series of shift registers with feedback taps. One of the unique properties of a PRBS is that when compared with itself, the BER will be 50 percent except when the patterns from the system under test and the error detector are exactly aligned. This allows the alignment process to be performed quickly and efficiently

There are several beneficial properties of PRBS patterns. The spectral content of a PRBS is quite broad; thus it can expose resonances in test devices. PRBS patterns can be decimated to produce shorter PRBS patterns. This is useful when testing multiplexing and demultiplexing circuits. (For example, the shorter patterns produced by a 1 to 4 demultiplexer are still PRBSs and can thus be easily tested for BER.) PRBS patterns can also be generated at extremely high speeds, limited only by the hardware generating the pattern.

In addition to PRBS patterns, specialized patterns can also be created and generated with a BERT. A special pattern might be created that causes a transmitter to produce more jitter than if a PRBS were used. Communications protocols often require specific bit sequences to "frame" the data payload. Thus a pattern may be built to include the framing bits. These patterns are not produced by a specific circuit topology, but instead are loaded into the memory of the pattern generator. The pattern generator then produces a data sequence according to the loaded pattern. (The internal pattern generator of the error detector works in a similar manner.)

With BERs of 1×10^{-12} and even lower being common in today's high-speed systems, what should be the expected time required to verify that a system is operating at a low BER? It is ideal to collect 100 or even 1,000 errors to be confident that the desired error ratio performance is being achieved. However, at an error ratio of 1×10^{-12} and a data rate on the order of 1 Gbps, it will take 11 and one half days to collect 1,000 errors! If the data rate is increased to 10 Gbps and the number of errors collected is reduced to only 100, it will still take several hours to complete the measurement.

There are techniques that can be used to try to determine BER in a reduced time frame. Both techniques to be discussed rely on making BER measurements in non-ideal conditions, thus increasing the BER and making an extrapolation to determine the ideal BER. One technique is to stress the signal being tested. The other technique is to adjust the sampling threshold of the error detector to non-ideal levels. A common technique to stress a signal is to simply attenuate it prior to reaching the system receiver. As the signal level is reduced, the likelihood is increased that a receiver will incorrectly interpret 1s as 0s and 0s as 1s. The BER will be artificially increased and then take less time to measure. By measuring the BER at several levels of attenuation, a BER versus received power curve can be plotted. Extrapolation of the curve can then be used to determine the BER with no external signal attenuation. Extrapolation requires an assumption that the dominant error mechanisms are the same at both high and low signal levels. This usually implies random noise. However, it is possible that a deterministic, but extremely low-probability, mechanism exists and is not seen except at high power levels where noise is less likely to cause errors. Thus final system qualification may require a lengthy BER measurement at full signal power.

The BER will also be artificially increased if the error detector sampling threshold is increased or decreased from the ideal. However, this type of BER analysis is typically performed on the signal at the test receiver input rather than its output. The BER analysis is not made on the complete system, but instead is an assessment of the quality of the signal being presented to the receiver. This is called a Q-factor measurement. Q-factor is essentially the signal to noise ratio of the signal. If the dominant mechanism for errors in a system is random noise, then BER can be estimated by the Q-factor parameter. The procedure is beyond the scope of this text, but is a common measurement available in many BERTs and is documented within the test equipment literature.

The key parameters for the BERT include the range of data rates it can both produce and receive, the patterns it can generate, and the quality of the waveforms (amplitudes, shape, jitter etc.). High-performance BERTs operating in the 50 Mbps to 3.2 Gbps range cost approximately $80,000 (Agilent N4906), while a BERT that can operate to 13.5 Gbps will cost approximately $130,000 (Agilent N4901).

31.3 Pulse Generators

Pulse generators (see Figure 31.4) are used to produce signals to drive electrical circuits. The signals may be high-quality square waves, precision pulse signals, or may be as sophisticated as the pattern generator section of a BERT. A common use of a pulse generator is to inject a pristine signal into a circuit and then observe the circuit output response with an oscilloscope. Pulse generators are often used to replace digital signals with a known, adjustable signal to perform functions such as clock, data, or control

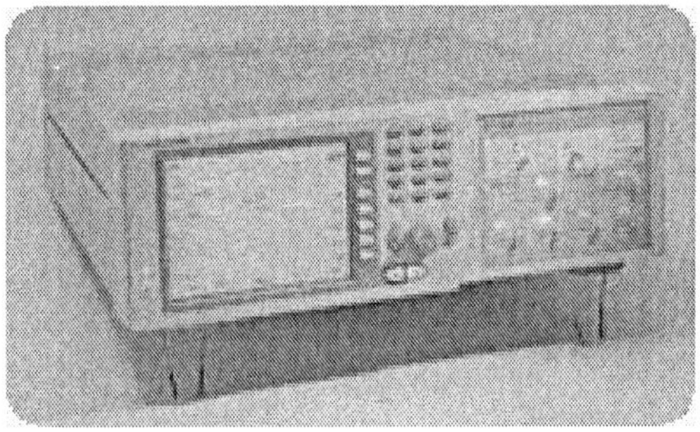

Figure 31.4 Agilent 81134A 3.35 GHz dual-channel pulse generator.

31.4 Jitter Analyzers

Jitter has been referred to as the deviation of signal position from its ideal location. A basic example would be if a signal edge were early or late from its ideal location. In digital communication circuits, if jitter becomes excessive, decision circuits may be making logic decisions at the edges of a bit period rather than the center of the bit. Mistakes can be made and the BER degraded.

Basic jitter can be observed with the oscilloscope. For example, a digital data stream can be observed relative to the system clock. The clock can be considered ideal. If the data is jitter-free, the edges will always appear at the same relative location to a clock edge. Jitter can also be observed on the eye diagram. If the data is jitter free, all the rising edges will lay on top of each other as will all the falling edges. If there is jitter, the crossing points of the eye diagram will spread out and the center of the eye will begin to close.

As signal rates increase, jitter becomes more difficult to control and the analysis of jitter becomes more complex. Rather than simply knowing how large the jitter is, the causes of jitter must also be determined and the jitter is further analyzed to determine what types of jitter are present. Jitter may be caused by mechanisms such as clock oscillator phase noise. In this case, the jitter is random in nature. Jitter may be due to certain data sequences stressing circuit performance. This jitter is deterministic, as it will always be observed whenever the specific data sequences are encountered. An external clock signal may radiate onto the data clock, causing a periodic jitter. This is also deterministic, but is uncorrelated to the data sequence.

Three tools are available for jitter analysis: the real-time oscilloscope, the time-interval analyzer, and a specialized version of the equivalent-time oscilloscope. The process of a complete jitter analysis involves sending a known data pattern through the system being tested. The location of each edge of that pattern is then determined and compared with the ideal. The instruments then analyze the edge locations to extract the random, data-dependent, and periodic components of the jitter.

In the case of the real-time oscilloscope, the instrument is configured to capture the entire pattern as a single waveform. Thus a very deep memory scope may be required if the pattern length is long. From this long waveform record, the edge positions are extracted and analyzed.

The time interval analyzer is a threshold detection instrument. Rather than acquire a waveform, it simply detects the precise time at which the signal crosses a set amplitude and the signal's edge location can be precisely determined. By locating the threshold crossing of the edges of the pattern, the information necessary to perform a complete jitter analysis is available. Time interval analyzers have similar ranges of operation as the real-time oscilloscope (~3 Gbps).

The situation becomes more complex for the equivalent-time oscilloscope. Recall that to see the individual edges of a signal, a pattern trigger is required. To observe all the edges in a data pattern, the pattern trigger position in the data pattern must be sequentially incremented, or the scope time position must be incremented. This coupled with the relatively low sampling rate results in a very time consuming measurement, perhaps taking several hours to complete. Specialized versions of equivalent-time oscilloscopes are now available that automatically sequence through patterns and focus the data sampling only on the signal edges. Measurement times are generally under one minute, even for long patterns. Given the wide bandwidth of this scope, comprehensive jitter analysis is now achieved at virtually any data rate (see Figure 31.5).

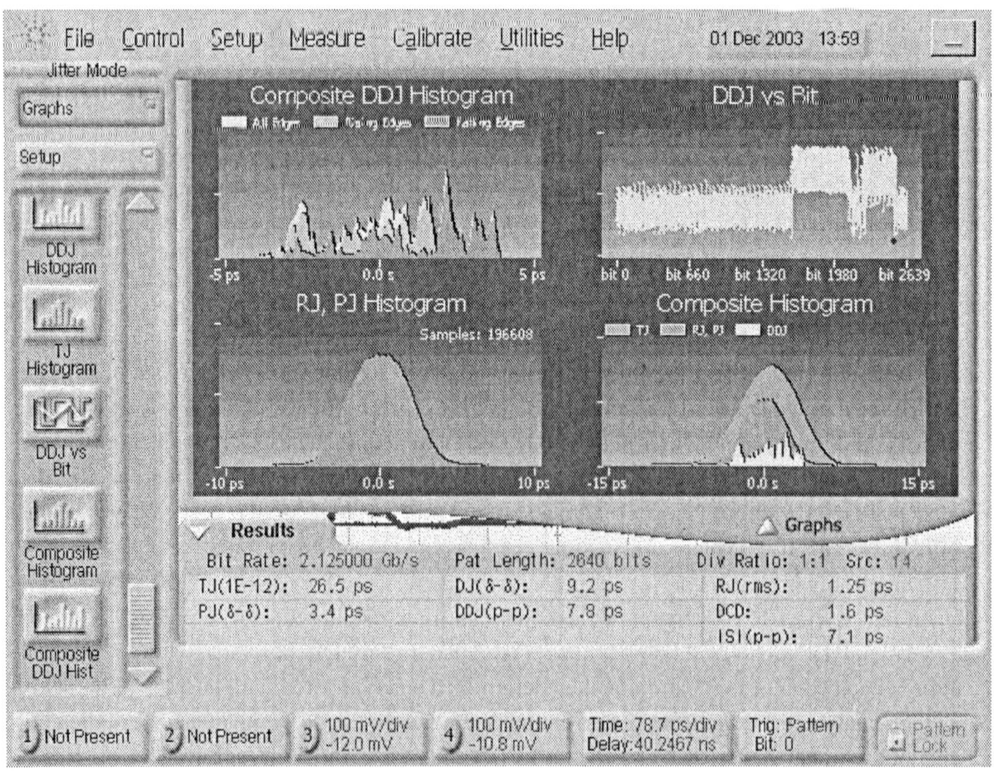

Figure 31.5 Jitter analysis test results generated by an equivalent-time scope.

31.5 Logic Analyzers

In many ways, the logic analyzer (see Figure 31.6) is similar to a multi-channel oscilloscope. Signal levels are recorded as a function of time. But rather than displaying the exact signal levels and waveform shapes, the basic logic level is recorded and displayed. The logic analyzer can be thought of as both a timing analyzer and a state analyzer. For state analysis, an external clock is used. The clock is usually from the circuit under test and is often the master clock of the system. When the clock signal transitions, the logic state of each input is latched and stored into memory. Thus a listing of the logic state for each clock state is displayed. To monitor logic states on a continuous basis (timing analysis), the logic analyzer uses its own timing to continuously latch the logic states.

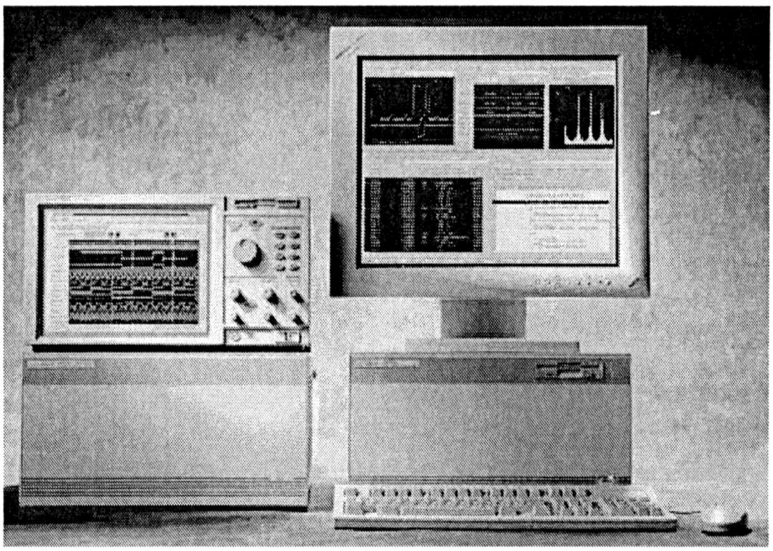

Figure 31.6 Agilent 16702B logic analysis system with a maximum data rate of 2.5 Gbps with no requirement for a header to be designed into the target.

As digital systems can run very fast, the instrument memory can quickly be swamped. A clock qualifier input is available to cause the logic analyzer to ignore certain defined clock cycles. The storage can be controlled to only record data that meets a specific criteria such as a specific logical sequence. Another technique for saving instrument memory is to only record when logical states change as opposed to the state at every clock cycle.

Logic analyzers can also provide a capability to capture glitches. A glitch is a signal that crosses the logic threshold level multiple times between samples. This provides a valuable troubleshooting tool, as the instrument can be configured specifically to locate and display these troublesome events.

It is often useful to have true oscilloscope capability as well as logic analysis. It can be useful to observe analog waveforms on the same display as logic levels. This is achieved with a mixed-signal oscilloscope (MSO). For example, if false states are due to an edge risetime problem, it would be useful to observe both the logic state as well as the analog signal performance.

31.6 Characterizing Optical Systems

The basic fiber-based communications system consists of a laser transmitter that converts electrical data to intensity-modulated light, a fiber-optic cable as short as a few meters or as long as 100+ kilometers, and a photodetector receiver that converts the information on the light back into an electrical signal.

Just as with a purely electrical system, an indicator of how well the optical communications system performs is the BER. Acceptable BERs are in the range of one error per one billion to one trillion bits transmitted. It is rare for a newly designed system to operate at these performance levels at initial turn on. Prior to system construction, the individual elements must be characterized to determine their potential for proper system-level performance.

Most transmitters for high-speed digital communications are laser-based, although shorter, slower data-rate systems may use light-emitting diodes. Several parameters are used to specify and qualify the performance of a laser transmitter, for example, the spectrum of the transmitter signal, the strength of the signal, and how well the signal carries information.

The propagation velocity of light is determined by the index of refraction of the fiber. Unfortunately, this parameter is wavelength-dependent. If an optical pulse or "bit" composed of energy at a range of wavelengths propagates along a fiber, the signal elements will travel at different velocities and reach the end of the fiber at different times. This will cause the pulse to be spread out in time. The effect is known as chromatic dispersion. If the fiber is long enough, the spreading of the pulse can be significant enough to severely distort the pulse causing it to interfere with adjacent bits. This will degrade the BER. Thus, transmitters used in long-haul systems must have carefully controlled spectra. On the other hand, short-haul systems have a relaxed spectral tolerance.

31.6.1 Optical Spectrum Analyzer (OSA)

Spectral measurements determine a laser's output wavelength(s). The tool of choice is the optical spectrum analyzer (OSA, see Figure 31.7), which directly displays laser power as a function of wavelength. Fundamentally, an OSA is a precision optical filter followed by a power meter. The filter scans across the wavelength range of interest while the power meter measures signal strength during the scan. The OSA then displays power versus wavelength. Many short-haul systems employ inexpensive vertical cavity surface emitting lasers (VCSEL). A VCSEL will emit power at a multitude of wavelengths. To minimize the effect of chromatic dispersion, a long-haul system laser must emit energy at predominantly one narrow range of wavelengths.

When the amplitude of a laser is turned on and off to represent digital information, the wavelength of the transmitter may shift. For example, as the current through a laser diode is

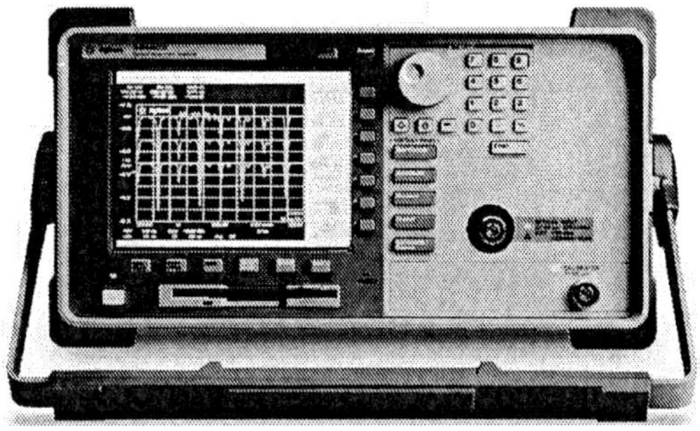

Figure 31.7 Agilent 86145B portable optical spectrum analyzer (OSA) that is best suited for 50 GHz and wider WDM applications.

increased, the physical mechanisms that control wavelength are modified. This is often referred to as "chirp" and can degrade system performance through the chromatic dispersion effect. Measuring chirp can be difficult because the signal power is changing over a wide range ("off to on" and "on to off").

Test systems are available to perform chirp measurements and consist of the tunable filter of an OSA, a digital communications analyzer (DCA), and analysis software. The tunable filter of an OSA converts the change in wavelength to a change in power. The digital communications analyzer acquires the resulting waveform out of the OSA and the analysis software extracts the chirp parameters.

31.6.2 Photodetection Using Equivalent-Time Oscilloscope

To determine how precisely an optical transmitter converts an electrical input signal into an optical output signal, the wide-bandwidth (equivalent-time) oscilloscope discussed above is used. The wide-bandwidth scope has a high-speed photodetector or "optical-to-electrical" (O/E) detector built into it. The photodetector is essentially an envelope detector, where the optical power level is converted linearly to a current level. In an optical digital communications system, the laser turns on to transmit a logic 1 and turns almost off to transmit a logic 0. When the laser signal is passed through the optical detector, an electrical output signal is created which is a close approximation to the original electrical signal that modulated the laser. The waveform out of the photodetector for a communication signal will look like a pulse train showing a pattern of "ones" and "zeroes." As with electrical pulses, digitally modulated optical signals are viewed as pulse trains, but more commonly as eye diagrams. For error-free communications, the amplitude separation between the logic one signal level and the logic zero signal level should be large. Thus the mask testing approach discussed above is the preferred method of characterizing the laser waveshape.

31.6.3 Optical Modulation Amplitude (OMA) and Extinction Ratio

In addition to the relative opening of the eye, the absolute opening needs to be quantified. A parameter that is used to indicate the separation between the mean one level and the mean zero level is optical modulation amplitude (OMA). On the other hand, extinction ratio is the ratio of the one level to the zero level (whereas OMA is the difference between the one and zero level). Extinction is an indicator of how efficiently the electrical signal modulating the laser is converted to optical signal power. In the ideal case, very little power is consumed transmitting a zero. Low zero powers are indicated by high extinction ratios. Typical values of the extinction ratio are 10 or higher. Note that as the extinction ratio becomes high, the one level and the OMA converge to the same value, and the average power is approximately half of the one level.

31.6.4 Power Meters

The power of a laser is distributed across a range of wavelengths and will quickly vary in intensity when carrying information. A simple measurement of the overall strength of a laser is the average power measurement. This measurement is most accurately performed with an average power meter. Like the OSA, a typical power meter has a photodetector whose electrical output signal is proportional to the incoming optical power. However, the electrical signal will be proportional to the aggregate power from the entire signal spectrum. Unlike the photodetector used in the DCA, the power meter detector is far too slow to respond to the changes in power due to modulation. Instead, it will yield the average strength of the signal.

Modern power meters are extremely accurate, with uncertainties as low as ±0.1%. To achieve these low levels of uncertainties, the instrumentation and user must minimize potential measurement errors caused by sensitivities to changes in wavelength, linearity, connector loss, reflections, and polarization.

31.6.5 Characterizing Fiber Properties

Several important properties of the fiber must be characterized. Fiber loss and dispersion will degrade BER. Loss reduces the separation between one and zero levels. Fiber also has dispersive properties that can cause narrow communication pulses to temporarily broaden and decrease in peak amplitude. As pulses become too wide, they can interfere with adjacent pulses and confuse the receiver. Two primary dispersion mechanisms are chromatic and polarization mode dispersion (PMD). Characterization of both PMD and chromatic dispersion can be achieved with a single test system.

31.6.6 Optical Receiver Sensitivity

The system receiver's principal task is to correctly determine if received bits are logic ones or zeros using some form of a photodetector or decision circuit. The main performance metric for a receiver is sensitivity. Sensitivity is determined by characterizing the BER as a function of received power.

In a typical test setup, an optical data signal is sent to the receiver under test, with a precision variable attenuator between the optical signal source and the optical receiver. The optical receiver's electrical output is then passed to a BER tester which compares each bit (1 or 0) produced with the known data pattern sent to the receiver. Whenever there is disagreement, a bit error is logged. After an extremely large number of bits are checked, a bit error ratio can be determined.

To determine sensitivity, the power level being sent to the receiver is attenuated to the point that a nonzero error ratio is measured. The power level is noted. The attenuation is increased, perhaps 1 dB, and a new error ratio is measured. The process is repeated until a ratio versus received power curve can be generated.

31.6.7 Optical Amplifiers

A key element of long-haul communications systems is the optical amplifier. In older optical communication systems, when the optical signal strength became low, the optical signal was reconverted to an electrical signal, regenerated, and then retransmitted by another laser.

By comparison, optical amplifiers boost an optical signal without electrical regeneration. The amplifier essentially increases the number of photons in an optical pulse, counteracting the effects of fiber attenuation. Noise, gain, and gain flatness versus wavelength are critical parameters. Testing is performed with precision tunable optical sources, power meters, and optical spectrum analyzers. Optical amplifiers typically operate over a wavelength range of about 40 nm, although work is being done to extend the range.

To increase the transmission capacity of the system, many different wavelengths are transmitted simultaneously on the fiber through dense wavelength division multiplexing (DWDM). Optical spectrum analyzers become valuable tools for analyzing and managing the many signals used in a DWDM system.

31.6.8 Multi-Wavelength Meter

Another tool designed for use in the DWDM realm is the multi-wavelength meter. This instrument acts like a conventional wavelength meter, which provides superior accuracy in measuring wavelength. However, unlike the common wavelength meter, it has the added capability of being able to measure multiple signals simultaneously. Thus the many channels of a DWDM system can be precisely characterized and monitored.

The Agilent 86120B (see Figure 31.8), 81620C, and 81622A multi-wavelength meters are Michelson interferometer-based wavelength meters that allow you to measure the average wavelength of the input signal. In addition, the multi-wavelength meters have advanced digital processing and accurately and easily differentiate and measure up to 1,000 (200 and 100 for the 86120C and 86120B, respectively) discrete wavelengths. They offer the following measurement capabilities and specifications:

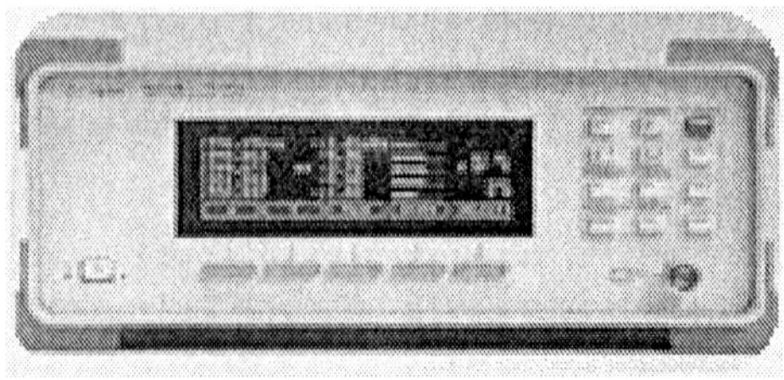

Figure 31.8 Agilent 86120B multi-wavelength meter.

- Measurement of from 1 to 1,000 discrete wavelengths and power
- Measurement of average wavelength and total power
- Up to ±0.2 ppm wavelength accuracy
- Up to 5 GHz wavelength resolution
- Calibration for evolution in air or vacuum
- Wavelength units in nm, THz, or wave number (1/cm)
- Amplitude units in dBm, mW, or μW
- OSNR and averaged OSNR for WDM — SONET/SDH systems
- Rugged design to withstand strong shocks and vibration

31.6.9 Reference Receiver

Eye-mask testing is a key element of most industry standards that specify high-speed, optically-based communications systems. To achieve consistent results across the industry, it is essential to specify the frequency response of the measurement channel. Thus, in addition to defining the shape of the eye mask, the measurement system is also defined through the concept of a reference receiver. A reference receiver usually consists of a photodetector followed by a low-pass filter and is most commonly used in front of an oscilloscope for a waveform measurement. That is, the optically-based communications system is connected to the reference receiver, which in turn is connected to the oscilloscope.

The combined response of the photodetector and low-pass filter typically follows a fourth order Bessel-Thomson frequency response. This response is chosen since it closely approximates the response of a Gaussian filter. A Gaussian response yields minimal distortion of the waveform.

It is interesting to note the bandwidth that is normally specified for a reference receiver. In most communications standards, the –3 dB bandwidth is set to be 75 percent of the optical bit rate. For example, a reference receiver for a 10 Gbps system would have a bandwidth of 7.5 GHz. Initially this seems counterintuitive. Intentionally reducing the measurement bandwidth is likely to change the shape of the eye diagram. Effects like overshoot and ringing mentioned above can literally disappear when the bandwidth is reduced. The reduced bandwidth of the reference receiver can actually yield a waveform that will pass the eye-mask test that would otherwise fail with a wider-bandwidth receiver.

To understand the logic behind this, consider the main intent of the test. The usability of the transmitter in a real communications system needs to be verified. This is a much different objective than trying to produce the most accurate image of the waveform shape. Most communications systems have receivers with bandwidths just wide enough to allow accurate determination of input signal levels. If the receiver bandwidth is wide, internal noise will increase, and likely degrade system BER. If the receiver bandwidth is too low, a signal making a 0 to 1 transition will be sluggish and not reach full amplitude within the bit period. Somewhere in between these extremes is an ideal receiver bandwidth. Although communication system receivers are not normally specified to have a "75 percent of bit rate" bandwidth, they do not normally have relatively wide bandwidths. Thus the reference receiver measurement approach has proven to be an effective way to verify transmitter performance.

31.7 Test Equipment Specifications

Test and measurement tools are used to learn important information about high-speed digital circuits. Measurement results are generally very accurate, yet every instrument will provide data with some measurement uncertainty. Depending upon the tool and the measurement, assessing the precision of a test result can vary from being very simple to being a complex analysis composed of many different instrument parameters. Understanding instrument specifications can be enhanced by understanding some basic terminology. Specification is a term usually used to describe guaranteed instrument performance. This then tells you what the test equipment manufacturer is able to confirm with 100 percent confidence. The manufacturer is usually capable of verifying this performance through some form of reference standard procured from a national standards laboratory. It is common for instruments to perform at a higher level than the manufacturer is capable or willing to guarantee. The term "typical performance" may be used to describe an expected, but not guaranteed level of operation. For example, 80 percent of instruments produced may achieve this higher level of performance.

Whether to use specified or typical measurement performance depends on what level of measurement uncertainty is required. When performance must be unquestionably guaranteed, using the more conservative "specification" parameter is probably the best choice. When typical results are acceptable for the devices being tested, typical test equipment performance, and hence measurement uncertainty should be acceptable.

31.8 Exercises

31-1. What is the difference between bit error rate and bit error ratio?

31-2. What are the three kinds of test equipment available that can perform jitter analysis?

31-3. Why must the laser for a long-haul system emit energy at predominantly one narrow range of wavelengths?

31-4. How is it that an electrical instrument, the equivalent-time oscilloscope, is able to precisely measure an optical transmitter's conversion of an electrical input signal into into an optical output signal?

31-5. What is the difference between how optical signals are boosted in power for long-haul applications today compared with how this was done years ago?

31-6. What is the maximum sampling rate for a real-time oscilloscope as stated in this chapter? Is this the sampling rate for one channel, or the equivalent rate for all of the oscilloscope's channels?

31-7. What is the difference between equipment "specifications" and equipment "typical performance"?

31-8. (a) What is the "extinction ratio"? (b) What is the "optical modulation amplitude (OMA)"? (c) What is the difference between the two?

31-9. What is a reference receiver and how is it used?

31-10. What is usually the longest PRBS (bit pattern) that a BERT can generate?

31-11. In high-performance probe systems, where does the probed signal first get amplifed by the test equipment?

31-12. What is an OSA and what does it do?

31-13. In designing a real-time oscilloscope, what is the trade-off when going to a higher-resolution ADC — that is, what do you give up for the benefit of accuracy?

31-14. What is usually an acceptible range for the BER for an optical system?

31-15. What are the basic elements of a fiber-based communication system?

Acronyms

ABT
Advanced BiCMOS technology

ADM
Add drop mux

AN
Application note

ANSI
American National Standards Institute

ASIC
Application-specific integrated circuit

ASM
Application-specific memory

ASSP
Application-specific standard product

ATA
Advanced Technology Attachment

ATAPI
Advanced Technology Attachment Packet Interface

B/P
Backplane

BBUL
Bumpless build-up layer

BER
Bit error rate, bit error ratio

BERT
Bit error ratio testing

BIST
Built-in self-test

BLVDS
Bus LVDS (Bus Low Voltage Differential Signaling)

BSDL
Boundary Scan Description Language

BTL
Backplane transceiver logic

CAN
Control area network

CAT3
Category 3 (cable classification)

CAT5
Category 5 (cable classification)

CBTL
CMOS backplane transceiver logic

CDR
Clock and data recovery

CD-ROM
Compact Disk - Read Only Media

CD-RW
Compact Disk - Read/Write

CISPR
International Special Committee on Radio Interference

CLB
Configurable logic block

CML
Current-mode logic

CMOS
Complementary metal-oxide semiconductor

CPLD
Complex programmable logic device

CRC
Cyclic redundancy check

CSP
Chip scale package

D
Driver

DCC
Digital cross connect

DCI
Digitally controlled impedance

DCM
Digital clock manager

DCR
dc resistance, direct current resistance

DDWG
Digital Display Working Group

DES
Data Encryption Standard

DLL
Delay-locked loop

DMA
Direct memory access

DRAM
Dynamic random access memory

DIMM
Dual inline memory module

DDR
Double data rate

DRSL
Differential Rambus signaling level

DTDR
Differential time domain reflection

DTDT
Differential time domain transmission

DUT
Device under test

DVD
Digital Versatile Disk

DVI
Digital visual interface

DWDM
Dense-wavelength-division multiplexing

EBGA
Enhanced ball grid array

ECC
Error correction code

ECL
Emitter coupled logic

EDA
Electronic design automation

eDRAM
Embedded DRAM

EIA
Electronic Industries Association

EMC
Electromagnetic compliance

EMI
Electromagnetic interference

EN
Enable

ERC
Edge rate control

ESD
Electrostatic discharge

eSRAM
Embedded SRAM

ETL
Enhanced transceiver logic

EVK
Evaluation kit

FACT
Fairchild Advanced CMOS Technology

FAST
Fairchild Advanced Schottky TTL

FCC
Federal Communications Commission

FPD
Flat panel display

FPD-LINK
Flat panel display link

FPGA
Field-programmable gate array

FPSC
Field-programmable system-on-a-chip

Gbaud
Gigabaud

Gbps
Gigabits per second

GB/s
Gigabytes per second

GTL
Gunning Transceiver Logic

GTLP
Gunning Transceiver Logic Plus

GTO
Gradual turn on/off

HDMI
High-definition multimedia interface

Hi-Z
High impedance

HSTL
High-speed transceiver logic

IC
Integrated circuit

I/O
Input/output

IBIS
I/O Buffer Information Specification

IDC
Insulation displacement connector

IEEE
Institute of Electrical and Electronics Engineers

IOB
Input output block

IP
Intellectual property

ISI
Inter symbol interference

JEDEC
Joint Electron Device Engineering Council (former name, now known simply as JEDEC)

JTAG
Joint Test Action Group

kbps
kilobits per second

LAN
Local area network

LCX
Low-voltage CMOS

LDI
LVDS display interface

LFBGA
Low-profile, fine-pitch ball-grid array

LLP
Leadless leadframe package

LUT
Look-up table

LVC
Low-voltage CMOS

LVCMOS
Low-voltage complementary metal-oxide semiconductor

LVDS
Low Voltage Differential Signaling

LVPECL
Low voltage positive emitter-coupled logic

LVT
Low-voltage technology

LVTTL
Low-voltage transistor-transistor logic

MAPBGA
Mold array process ball grid array

Mbps
Megabits per second

MB/s
Megabytes per second

MCP
Multichip package

MDR
Mini delta ribbon

MIPS
Million instructions per second

MLC
Multilayer ceramic

M-LVDS
Multipoint low voltage differential signaling

MSA
Multisource agreement

MSPS
Megasamples per second

MT
Mega transfers

MTBF
Mean time between failures

MUX
Multiplexer

NECL
Negative emitter coupled logic

NRE
Non-recurring engineering

NRZ
Non return to zero

NSMD
Non solder mask defined

OCD
Off-chip driver

OCM
On-chip memory

ODT
On-die termination

OEA
Operating environment architecture

OOB
Out of band

PAL
Programmable array logic

PBGA
Plastic ball grid array

PC
Personal computer

PCB
Printed circuit board

PCI
Peripheral component interconnect

PECL
Positive emitter-coupled logic

PHDL
Philips Hardware Description Language

PHY
Physical layer device

PLL
Phase lock loop

POR
Power-on reset

PRBS
Pseudorandom binary sequence

PTH
Plated through-hole

PWB
Printed wire board

QDR
Quad data rate

R
Receiver

RAID
Redundant array of inexpensive disks

RF
Radio frequency

RFI
Radio frequency interference

RIMM
Rambus inline memory module

RLGC
Resistance, inductance, conductance, capacitance (usually per unit length)

RMT
Repetitive multiple transition

RS
Recommended standard

RSDS
Reduced swing differential signal

RSKM
Receiver input skew margin

RSL
Rambus signaling level

RWW
Read-while-write

SATA
Serial Advanced Technology Attachment

SCI
Scalable Coherent Interface

SCSI
Small Computer System Interface

SDH
Synchronous Digital Hierarchy

SDI
Serial Digital Interface

SDR
Single data rate

SDV
Serial Digital Video

Ser/Des
Serializer/deserializer

SI
Signal integrity

SLL
Standard Linear and Logic

SLVS
Scalable low-voltage signaling

SMD
Solder mask defined, surface-mount device

SMPTE
Society of Motion Picture and Television Engineers

SOS
Simultaneous output switching

SPICE
Simulation program with integrated circuit emphasis

SPLD
Simple programmable logic device

SSTL
Stub-series terminated logic

STN
Super twist neumatic

STN-DD
Super twist neumatic-dual drive

SUT
System under test

TDR
Time domain reflectometry

TDT
Time domain transmission

TEM
Transverse electromagnetic

TFT
Thin film transistor

TIA
Telecommunications Industry Association

TMDS
Transition-Minimized Differential Signaling

TP
Test point

TTL
Transistor transistor logic

TVSOP
Thin, very small-outline package

TWP
Twisted pair

TX
Transmitter

UBT
Universal bus transceiver

UISA
User instruction set architecture

UTP
Unshielded twisted pair

VBT
Virtual backplane transceiver

VCM
Voltage common-mode

VEA
Virtual environment architecture

WCSP
Wafer chip scale packaging

WDM
Wavelength Division Multiplexing

XAUI
10-Gigabit Attachment Unit Interface

REFERENCES

[Adu03] XGSX4 3.1875 GHz Transceiver, *www.aduroinc.com*, Aduro Inc., 2003.

[Agi01a] 10 Hints for Making Better Network Analyzer Measurements, Application Note 1291-1B, Agilent Technologies, Oct 25, 2001.

[Agi01b] De-embedding and Embedding S-Parameter Networks Using a Vector Network Analyzer, Application Note 1364-1, Agilent Technologies, Mar 22, 2001.

[Agi02] Save Trace Data in Touchstone Format, Agilent E5070B/E5071B ENA Series RF Network Analyzers, Agilent Technologies, no. 16000-95001, Second Edition, August 2002.

[Agi02b] X2 MSA Q&A, Agilent Technologies, July 22, 2002.

[Agi03a] N1957B Physical Layer Test System, 10 MHz to 50 GHz, *www.agilent.com*, Agilent Technologies, 2003.

[Agi03b] Agilent Technologies HFCT-5942xx Single Mode Laser Small Form Factor Transceivers for ATM, SONET OC-48/SDH STM-16, Data Sheet, Agilent Technologies.

[Agi03c] 2.125/1.0625GBd Fibre Channel Physical Layer Reference Design for Agilent HFBR-5720L and HDMP-2630/2631, Application Note 1229, Agilent Technologies.

[Agi97] S-Parameter Techniques for Faster, More Accurate Network Design, Application Note 95-1, Agilent Technologies, 1997.

[Alf03] Peter Alfke, Printed Circuit Board Design Considerations, Web site of Xilinx, Inc., *www.xilinx.com/xcell/xl28/xl28_22.pdf.*

[Ali00] F. Alicke, F. Bartholdy, and others, Comparing Bus Solutions, Application Report SLLA067, Texas Instruments, Mar 2000.

[All03] Roger Allan, "Coax Backplane Connector Handles Rates To 40 Gbits/s," *Electronic Design*, Oct 13, 2003.

[Alt03a] Part 1. Introduction to HardCopy Stratix Devices, HardCopy Device Handbook, Altera Corporation, *www.altera.com*, vol. 1, June 2003.

[Alt03b] HardCopy Stratix Overview, Altera Corporation, *www.altera.com*, 2003.

[Alt03c] HardCopy: ASIC Gain without the Pain, Altera Corporation, *www.altera.com*, June 2003.

[Alt03d] Component Selector Guide, Altera Corporation, *www.altera.com*, Sept 2003.

[Amd00a] Understanding Page Mode Flash Memory Devices, Application Note, Publication# 23711, Rev. A, Advanced Micro Devices, May 25, 2000.

[Amd00b] Understanding Burst Mode Flash Memory Devices, Application Note, Publication# 23710, Rev. A, Advanced Micro Devices, Mar 23, 2000.

[Amd03a] Ask AMD Flash Memory Knowledge Base, *www.amd.com*, Advanced Micro Devices, 2003.

[AN-5017] LVDS Fundamentals, Application Note AN-5017, Fairchild Semiconductor, Dec 2000.

[Ana96] High Speed Design Techniques, Analog Devices, Norwood, MA, 1996.

[And03] J. Andrews, 10ps Risetime Network TDT & TDR Measurements using the PSPL 15ps Pulse Generator & HP 50 GHz oscilloscope, PicoSecond Pulse Labs Application Note.

[Ans03a] SIwave™: Full-Wave PCB and Package Analysis for Signal Integrity and Power Delivery, SIwave™ Product Overview, *www.ansoft.com*, Ansoft Corporation, 2003.

[Ans03b] Maxwell 2D: Powerful, Accurate Software for Two-Dimensional, Electromagnetic, Electromechanical, and Thermal Analysis, Maxwell 2D Product Overview, *www.ansoft.com*, Ansoft Corporation, 2003.

[Ans03c] Maxwell 3D: Powerful, Accurate Software for Three-Dimensional, Electromagnetic, Electromechanical, and Thermal Analysis, Maxwell 3D Product Overview, *www.ansoft.com*, Ansoft Corporation, 2003.

[Ans03d] Spicelink™ Electromagnetically Charged EDA for Signal-Integrity Analysis, Spicelink™ Product Overview, *www.ansoft.com*, Ansoft Corporation, 2003.

[Bal98] David Ballo, "Network Analyzer Basics: Back to Basics Seminar," Agilent Technologies, 1998.

[Bau03] Jason Baumbach, Julian Jenkins, and Jon Withrington, "Common I/O Design Strategies for High-Speed Interfaces," *Electrical Engineering Times*, Jan 27, 2003.

[Bea04] Weston Beal, "Use the Right Models for the Simulation of Multi-Gigabit Channels," Mentor Graphics Corporation, Feb 2004.

[Bha03] Ajay V. Bhatt, Creating a Third Generation I/O Interconnect, Desktop Architecture Labs, Intel Corporation, 2003.

[Bir01] Randy Bird, Fiber Optics 101: A Primer, *www.certcities.com*, Irvine, CA, Apr 6, 2001.

[Boc95] D. E. Bockelman, W. R. Eisenstadt, "Combined Differential and Common-Mode Scattering Parameters: Theory and Simulation," *IEEE Transactions on Microwave Theory and Techniques*, vol. MTT-43, July 1995.

[Boc97] D. E. Bockelman, Ph.D. dissertation, University of Florida, 1997.

[Bog99] Eric Bogatin, "Signal Integrity and HDI Substrates," *The Board Authority*, vol. 1(2), June 1999, pp. 22–26. (A PDF copy is available for download at *www.Megatest.com*.)

[Bog00] Eric Bogatin and Mike Resso, "Differential Impedance Measurements with Time Domain Reflectometry," *Insight*, vol. 5, no. 4, 2000.

[Bog01] E. Bogatin, S. Corey, and M. Resso, "Practical Characterization, Analysis and Simulation of Lossy Lines," DesignCon 2001, Santa Clara, CA, Jan 2001 (TDA Systems Application Note LOSS-0601).

[Bog01b] Eric Bogatin and Mike Resso, "Using TDR for Differential Impedance Design and Verification," *Insight*, vol. 6, no. 1, 2001.

[Bog03] Eric Bogatin, *Signal Integrity — Simplified*, Upper Saddle River, NJ: Prentice Hall, 2003.

[Bou03] Dan Bouvier, RapidIO™: The Interconnect Architecture for High Performance Embedded Systems, white paper, RapidIO Trade Association, 2003.

[Bro00] Shona L. Brown and Kathleen M Eisenhardt, "Patching: Restitching Business Portfolios in Dynamic Markets," *Harvard Business Review*, OnPoint Enhanced Edition, Feb 1, 2000.

[Cad01a] Dennis Nagle, High-Speed Constraint Management For Complex High-Speed PCB Designs, Technical Paper, Cadence Design Systems, Inc., 2001.

[Cad02a] Lynne Green, Getting On Board with Optics, Cadence Community Educational Series, Cadence Design Systems, Inc., Feb 2002.

[Cad02b] Todd Westerhoff, Solution Space Analysis For High-Speed Design, Technical Paper, Cadence Design Systems, Inc., 2002.

[Cad02c] Ron Matthews, Modern High-Speed Design Methodology Integrating EMC and Signal Integrity, Technical Paper, Cadence Design Systems, Inc., Nov 2002.

[Cad02d] Heiko Dudek, Differential Signals: A Design Technology for Today, Technical Paper, Cadence Design Systems, Inc.

[Cad02e] Vadim Heyfitch, Modeling Traces For Multi-Gigabit Data Transmission in PCBs, Technical Paper, Cadence Design Systems, Inc., Oct 2002.

[Cad02f] Lynne Green, Modeling and Simulation for Signal Integrity, Technical Paper, Cadence Design Systems, Inc., Nov 2002.

[Cad03a] SPECCTRAQuest High-Speed System Level Design and Analysis, Data Sheet, Cadence Design Systems, Inc., Sep 2003.

[Cad03b] Zhen Mu, Effective Use of IBIS Models in Gigabit-Per-Second PCB Simulations, Technical Paper, Cadence Design Systems, Inc., 2003.

[Cad03c] High-Speed PCB Design and Analysis, *www.cadence.com/products/si_pk_bd/highspeed_pcb_dt.aspx*, Cadence Design Systems, Inc., 2003.

[Cas03] High Speed Digital Microprobing: Principles and Applications, Application Note, Cascade Microtech, 2003.

[Cha01] Mark Chang, "Introduction to IBIS Modeling of Fiber Optic Transceivers," *ONIDS* 2001.

[Chi68] Robert A. Chipman, *Theory and Problems of Transmission Lines, Schaum's Outline Series in Engineering*, New York: McGraw-Hill, 1968.

[Cle03] Mike Clendenin, "MRAM, FeRAM are at the Finish Line," *EE Times*, Mar 27, 2003.

[Com03] Technical Encyclopedia, *EDN* Web site, *www.reed-electronics.com/ednmag*.

[Coo01] Clyde F. Coombs, Jr., *Coombs' Printed Circuits Handbook*, Fifth Edition, New York: McGraw-Hill, 2001.

[Cra02] Robert Cravotta, "29th Annual Microprocessor Directory," *EDN*, Oct 3, 2002.

[Cyp99] Serializing High-Speed Parallel Buses to Extend Their Operational Length, Application Note, Cypress Semiconductor Corporation, Mar 12, 1999.

[Cyp00] Quad Data Rate (QDR) SRAM Clocking Scheme, Cypress Semiconductor Corp., Feb 16, 2000.

[Cyp01a] Interfacing the QDR with Altera APEX20KE, Cypress Semiconductor Corporation, Mar 12, 2001.

[Cyp01b] ADR SRAM: Assembly Considerations, Cypress Semiconductor Corporation, June 19, 2001.

[Cyp02a] A Comparison of CML and LVDS for High-speed Serial Links, Application Note, Rev. 0, Cypress Semiconductor Corporation, Dec 20, 2002.

[Cyp02c] QDRII SRAM: A Design Guide, Cypress Semiconductor Corporation, June 12, 2002.

[Cyp02d] Data Sheet, 18-Mb QDR-II SRAM 4 Word Burst Architecture (CY7C1315V18), Document no. 38-05181, Rev. A, Cypress Semiconductor Corporation, July 25, 2002.

[Das96] David J. Dasher, "Measure Parasitic Capacitance and Inductance Using TDR," *Hewlett-Packard Journal*, Aug 1996.

[Deh02] Frank Dehmelt, Transmission at 200 Mbps in VME Card Cage Using LVDM, Application Report SLLA088A, Texas Instruments, Jan 2002.

[Den03a] Denali Software Inc., *www. denali.com*, 2003.

[Dip97] Brian Dipert, "Data Storage In A Flash," *EDN Access*, July 3, 1997.

[Dip02a] Brian Dipert, "Third Annual Programmable-Logic Directory," *EDN*, Sep 5, 2002.

[Dip03a] Brian Dipert, "Flash-Memory Transformation Responds to Application and Competition Evolutions," *EDN*, Sep 22, 2003.

[Ele00] Fifty Years of Fiber Optics, ElectroniCast, 2003.

[Ele02a] "Flexible Transceiver Protects Backplane Investments," *EE Times*, July 3, 2002.

[Ele03] "RapidIO Now an International Standard," *Electronic Design Update* e-Newsletter, Dec 29, 2003.

[Elp01a] User's Manual, Synchronous DRAM, Document no. E0124N10 (Ver. 1.0), Elpida Memory, May 2001.

[Elp02a] How to Use DDR SDRAM: User's Manual, Document no. E0234E30 (Ver. 3.0), Elpida Memory, Apr 2002.

[Erb01] Christoph Erben, "New Materials in Optoelectronics — Advantages and Challenges for Polymers," Second Optoelectronics Packaging Workshop, Austin, TX, Aug 21–22, 2001, pp. B1–B22.

[Fai99] Designing with Bushold, Application Note AN-5006, Fairchild Semiconductor, Sep 1999.

[Fai00] DDR SDRAM Signaling Design Notes, Preliminary, Rev 1.0, Fairchild Semiconductor, Oct 27, 2000.

[Fai01a] LVDS Description and Technology Characteristics, Interface Products Databook, Fairchild Semiconductor, Feb 2001.

[Fai01b] GTLP in BTL Applications, Application Note AN-5013, Fairchild Semiconductor, Feb 2001.

[Fai02a] Section 7 — Backplane Transceiver Technologies, Backplane Designer's Guide, Fairchild Semiconductor, Mar 2002.

[Fai02b] GTLP Description and Technology Characteriestics, Interface Products Databook, Fairchild Semiconductor, Dec 2000.

[Fol03] "Content Addressable Memory," Free On-Line Dictionary of Computing, *foldoc.doc.ic.ac.uk/foldoc*, 2003.

[Fre03] Louis E. Frenzel, "Optical Communications Battling Back From Adversity," *EDN*, Nov 24, 2003.

[Fuj01a] 16 M-Bit Mobile FCRAM: MB82D01171A-90/90L/90LL, *www.fme.fujitsu.com*, Fujitsu Microelectronics Europe, 2001. Also in *FIND*, New Products, vol. 19, no. 3, 2001.

[Fuj01b] 16 M-Bit SDRAM-Type FCRAM: MB81E161622, *www.fme.fujitsu.com*, Fujitsu Microelectronics Europe, 2001. Also in *FIND*, New Products, vol. 19, no. 1, 2001.

[Fuj02a] FCRAM's Solution for Consumer Products — Consumer FCRAM Family, *www.fme.fujitsu.com*, Fujitsu Microelectronics Europe, 2002. Also in *FIND*, Technical Analysis, vol. 20, no. 2, 2002.

[Fuj02b] A Combination of Flash Memory and Mobile FCRAM/SRAM — 2 Chip Stacked MCP, *www.fme.fujitsu.com*, Fujitsu Microelectronics Europe, 2002. Also in *FIND*, New Products, vol. 20, no. 2, 2002.

[Fuj02c] 32M-bit/64M-bit Mobile FCRAM MB82D02172A/MB82D04172, *www.fme.fujitsu.com*, Fujitsu Microelectronics Europe, 2002. Also in *FIND*, New Products, vol. 20, no. 2, 2002.

[Fuj02d] 16M-bit Consumer FCRAM for SiP, *www.fme.fujitsu.com*, Fujitsu Microelectronics Europe, 2002. Also in *FIND*, New Products, vol. 20, no. 2, 2002.

[Fuj03a] NOR-FLASH, MirrorFLASH, FCRAM, Multi-Chip Packages & Packaging Technology, *www.fme.fujitsu.com*, Fujitsu Microelectronics Europe, 2003. Also in *FIND*, Technical Analysis: "Latest Trends for Mobile Phone Applications," vol. 21, no. 4, 2003.

[Fuj03b] NOR-Flash, MirrorFlash, FCRAM, Multi-Chip Packages & Packaging Technology: Product Overview, *www.fme.fujitsu.com*, Fujitsu Microelectronics Europe, 2003. Also in *FIND,* Technical Analysis: "Latest Trends for Mobile Phone Applications," vol. 21, no. 4, 2003.

[Gii02] High-Speed Optical Data Link Modules: Market Review and Forecast 2002 (1Gb/s and Above), *www.gii.co.jp*, Global Information, Inc., 2002.

[Gin02] Kevin Gingerich and John Goldie, "New TIA Standard Enables Multipoint LVDS," *EDN*, Feb 21, 2002.

[God99] E. Godshalk and G. Sundberg, "Characterization of Surface Mount Components at Microwave Frequencies Using Wafer Probes," 53rd Automatic Radio Frequency Techniques Group Conference Digest, Anaheim, California, June 1999, pp. 131–138.

[Goe94] Ashok K. Goel, *High-Speed VLSI Interconnections*, New York: John Wiley & Sons, Inc., 1994.

[Gol02] John Goldie, Sorting Out Backplane Driver Alphabet Soup, Application Note AN-1123, National Semiconductor, 2002. Also found in *Electronic Design*, vol. 46, no. 25, Nov 2, 1998, Penton Publishing.

[Gol03a] John Goldie, LVDS, CML, ECL-Differential Interfaces with Odd Voltages, Planet Analog (*www.PlanetAnalog.com*), Jan 21, 2003.

[Gre99] Lynne Green, "Signal Integrity," *IEEE Circuits and Devices*, Nov 1999.

[Gri01a] E. Griese, A. Himmler, and others, "Self-Aligned Coupling of Optical Transmitter and Receiver Modules to Board-Integrated Multimode Waveguides," *Proceedings of SPIE*, vol. 4455, 2001, pp. 243–246.

[Gri01b] Elmar Griese, "A High-Performance Hybrid Electrical-Optical Interconnection Technology for High-Speed Electronic Systems," *IEEE Transactions Advanced Packaging*, vol. 24, no. 3, Aug 2001, pp. 375–383.

[Gri02] Elmar Griese, "Optical Interconnection Technology for PCB Applications," *PCB Fab*, June 2002, pp. 20–36.

[Gup96] K. C. Gupta, Ramesh Garg, Inder Bahl, and Prakash Bhartia, *Microstrip Lines and Slotlines*, Second Edition, Boston, MA: Artech House, 1996.

[Hal00] Stephen H. Hall, Garrett W. Hall, and James A. McCall, *High-Speed Digital System Design: A Handbook of Interconnect Theory and Design Practices*, New York: John Wiley & Sons, Inc., 2000.

[Har00] Charles A. Harper, *High Performance Printed Circuit Boards*, New York: McGraw-Hill, 2000.

[Har83] R. F. Harrington, T. K. Sarkar, "Boundary Elements and the Method of Moments," Boundary Elements: Proceedings of the Fifth International Conference, Hiroshima, Japan, Nov 1983.

[Hay74] William H. Hayt, Jr., *Engineering Electro-Magnetics*, Third Edition, New York: McGraw-Hill, 1974.

[Hay94] L. A. Hayden and V. K. Tripathi, "Characterization and Modeling of Multiple Line Interconnections from TDR Measurements," *IEEE Transactions on Microwave Theory and Techniques*, vol. 42, Sep 1994, pp. 1737–1743.

[Hew90] Advanced TDR Techniques, Hewlett-Packard Application Note 62-3, May 1990.

[Hog01] Matthew Hogan, "Circuit Timing Analysis: Mastering a Lost Art," Mentor Graphics, Sept 2001.

[Hol02a] Happy Holden, *The Developing Technologies of Integrated Optical Waveguides in Printed Circuits*, West Haven, CT: Westwood Associates, 2002.

[Hol02b] Happy Holden, HDI's Beneficial Influence On High-Frequency Signal Integrity, Part I & II, New Technical Publications on Mentor Graphics' Web site: *www.mentor.com/pcb*, 2002.

[Hol03] Happy Holden, *Next Generation: Opto-Electronics, Photonics and Embedded Waveguides*, West Haven, CT: Westwood Associates, 2003.

[Hor89] Paul Horowitz and Winfield Hill, *The Art of Electronics*, Second Edition, Cambridge University Press, 1989.

[Hsu97] C. W. Hsue and T. W. Pan, "Reconstruction of Nonuniform Transmission Lines from Time-Domain Reflectometry," *IEEE Transactions on Microwave Theory and Techniques*, vol. 45, no. 1, Jan 1997, pp. 32–38.

[Huq99] Syed B. Huq, An Introduction to IBIS (I/O Buffer Information Specification) Modeling, Application Note AN-1111, Interface Databook, pp. 13-3 to 13-10, National Semiconductor, 1999.

[Ibm01] Overview of DDR SDRAM for the IBM PPC440GP, Application Note, IBM Corporation, Sep 5, 2001.

[Idt03a] Data Sheet, IDT71T75602/802, 512x36, 1Mx18, 2.5V Synchronous ZBT SRAM, Integrated Device Technology, Inc., May 2003.

[Iec02] Safety of Laser Products, Part 1: Equipment Classification, Requirements and User's Guide, IEC, 2002.

[Inf02a] Ulrich Backhausen, A New Memory Standard for High-Speed Applications, Infineon Technologies AG, Feb 2002.

[Inf03a] Data Sheet, 288 Mb RLDRAM, HYB18RL28818/36AC, Infineon Technologies AG, 2003.

[Inf03b] PAROLI 2 Parallel Optical Link, brochure, *www.infineon.com*, Infineon Technologies AG, 2003.

[Inf03c] PAROLI Backplane Solution, Slide Presentation, *www.infineon.com*, Infineon Technologies AG, 2003.

[Inn03a] What Is FCRAM (Fast Cycle RAM) Memory?, *www.memoryinformation.com*, Innoventions Inc., 2003.

[Inn03b] What Is SRAM (Static RAM) Memory?, *www.memoryinformation.com*, Innoventions Inc., 2003.

[Inn03c] What Is EDO and FPM?, *www.memoryinformation.com*, Innoventions Inc., 2003.

[Inq03] Toshiba and Infineon Announce 32Mbit FeRAM, The Inquirer, *www.theinquirer.net*, Feb 14, 2003.

[Ins02] XGMII Extender Sublayer (XGXS) and 10 Gigabit Attachment Unit Interface (XAUI), IEEE Std. 802.3ae, Clause 47, Institute of Electrical and Electronic Engineers, Apr 2002.

[Int97] Pentium II Processor Developer's Manual, Document no. 243502-001, Intel Corporation, Oct 1997.

[Int00] Pentium III Xeon™ Processor at 500 and 550 MHz, Document no. 245094-002, Intel Corporation, Feb 2000.

[Int01] Evolution of Gigabit Technology: From the Backbone to the Desktop, Intel Corporation, 2001.

[Int02a] Expanding Moore's Law, The Exponential Opportunity, *www.intel.com*, Intel Corporation, Fall 2002 Update.

[Int02b] Intel Pentium 4 Processor on 0.13 Micron Process in the 478-pin Package Data Sheet, Intel Corporation, 2002.

[Int03a] Intel Pentium 4 Processor with 512-KB L2 Cache on 0.13 Micron Process at 2 GHz–3.06 GHz, with Support for Hyper-Threading Technology at 3.06 GHz, Data Sheet, Document no. 298643-007, Intel Corporation, Jan 2003.

[Int03b] Research, Silicon, Moore's Law, Overview, *www.intel.com*, Intel Corporation, 2003.

[Irw02] Scott Irwin, Usage Models for Multi-Gigabit Serial Transceivers, Xilinx, white paper: Virtex-II Series, WP157 (v1.0), Mar 15, 2002.

[Ish01] Y. Ishii, S. Koike, Y. Arai, and Y. Ando, "Short Wave SFF Small Form Factor Transceivers," *Proceedings of ECTC 2001 Conference*, Orlando, FL, May 30, 2001.

[Isr02] Joshua Israelsohn, "Fiber Lights the Short Haul," *EDN*, Mar 21, 2002.

[Isr03] Joshua Israelsohn, "The X-Modules," *EDN*, Feb 6, 2003.

[Jed94] Guidelines for Measurement of Electronic Package Inductance and Capacitance Model Parameters, JEDEC Publications JEP-123, 1994.

[Jed95] Guideline for Measurement of Electronic Package Inductance and Capacitance Model Parameters, JEDEC Publication #123, JC-15 Committee, Oct 1995.

[Jed03b] JEDEC Standard JESD79-2, DDR2 SDRAM Specification, Sep 2003.

[Jes95] EIA/JESD8-6, High Speed Transceiver Logic (HSTL) A 1.5 V Output Buffer Supply Voltage Based Interface Standard for Digital Integrated Circuits, Electronic Industries Association, Aug 1995.

[Jes96] EIA/JESD8-8, Stub Series Terminated Logic for 3.3 Volts (SSTL_3), Electronic Industries Association, Aug 1996.

[Jes02a] JESD8-15, Stub Series Terminated Logic for 1.8 V (SSTL_18), JEDEC Solid State Technology Association, Oct 2002.

[Jes02b] JESD8-9B, Addendum No. 9B to JESD8 — Stub Series Terminated Logic For 2.5 Volts (SSTL_2), JEDEC Solid State Technology Association, Oct 2002.

[Joh93] Howard Johnson and Martin Graham, *High-Speed Design: A Handbook of Black Magic*, Upper Saddle River, NJ: Prentice Hall PTR, 1993.

[Joh97] Howard Johnson, "The I/O Buffer Information Specification," *Printed Circuit Design*, May 1997, p. 17.

[Joh03] Howard Johnson and Martin Graham, *High-Speed Signal Propogation, Advanced Black Magic*, Upper Saddle River, NJ: Prentice Hall, 2003.

[Jup02a] ATI Promotes New GDDR3 Graphics Memory Specification, *www.earthweb.com*, Oct 7, 2002.

[Ken02] Clark S. Kent and Lois R. Lane, "Wave Dynamics of Capes in Sub-Stratospheric Flight," *Proceedings of the American Superhero Society*, 65145-72, Daily Planet, Metropolis, NY, 2002.

[Kol02] Paul Kolesar and Jane Ehrgott, "Laser Safety Standards Update and Impact on 850 nm Serial PMD," Lucent Technologies, presented at *IEEE 802ae* Albuquerque, Mar 2000.

[Kra99] D. Krabe and W. Scheel, "Optical Interconnects by Hot Embossing for Module and PCB Technology — The EOCB Approach, *Proceedings of 49th Electronics Components & Technology Conference*, San Diego, CA, June 1999, pp. 1164–1166.

[Kra00] D. Krabe, F. Ebling, N. Arndt-Staufenbiel, G. Lang, and W. Scheel, "New Technology for Electrical/Optical Systems on Module and Board Level: The EOCB approach," *Proceedings of 50th Electronics Components & Technology Conference*, Las Vegas, NV, May 2000, pp. 970–974.

[Kur65] K. Kurokawa, "Power Waves and the Scattering Matrix," *IEEE Transactions on Microwave Theory and Techniques*, vol. MTT-13, no. 2, Mar 1965.

[Lag02] Herbert Lage, "IBIS Modeling and Simulation of High-Speed Fiber-Optic Transceivers," *ONIDS (Optical Network Interface Design Symposium) 2002*, Agilent Technologies, Apr 26, 2002.

[Lee03] Peter Lee, PON Q&A: What Is a Passive Optical Network?, APA Optics, Inc. (Blaine, MN) 2003.

[Lip99] Jim Lipman, "Models Make the Difference in High-Speed PC-Board Design," *Electronic Design News*, Apr 15, 1999.

[Liu03] Yu Liu, "S-Parameters Bridge Gap Between Chips and Systems," *EEdesign*, Feb 10, 2003.

[Mat95] Richard E. Matick, *Transmission Lines For Digital And Communication Networks: An Introduction to Transmission Lines, High-Frequency, and High-Speed Pulse Characteristics and Applications*, Piscataway, NJ: IEEE Press, 1995.

[Mav96] R. Mavaddat, *Network Scattering Parameters*, World Scientific Publishing Co., 1996.

[Max00] Introduction to LVDS, PECL, and CML, Application Note HFAN-01.0, Rev 0, Maxim Integrated Products, Sep 2000.

[Max01a] MAX3800 3.2Gbps Adaptive Equalizer and Cable Driver Data Sheet, Maxim Integrated Products, Rev 1, Nov 2001.

[Max01b] Single-Ended and Differential S-Parameters, Application Note: HFAN-5.1.0, Rev 0, Maxim Integrated Products, Mar 2001.

[Max02] MAX9157 Quad Bus LVDS Transceiver, Data Sheet, Maxim Integrated Products, Jan 2002.

[Max03] Improve Device S-Parameter Measurement with Fixture Design and Calibration, Application Note 1913, Maxim Integrated Products, Mar 6, 2003.

[Mer03] Rick Merritt, "Board Design Revamped," *EE Times*, Oct 13, 2003.

[Mic97] Standard DC Specifications, 10K And 100K Series DC Electrical Characteristics, ECLinPS and ECLinPS Lite, Rev. E, Micrel Semiconductor, June 1997.

[Mic98a] ECLinPS Lite Family Specifications, Rev. F, Micrel, Dec 1998.

[Mic00a] ECLinPS Lite Ordering Information and Marking Instructions, Rev. F, Micrel, Aug 2000.

[Mic01a] High Bandwidth Solutions, High-Speed PECL and LVPECL Termination, Newsletter, vol. 1, Micrel Semiconductor, Q4, 2001.

[Mic01b] QDRII and DDRII SRAM Specifications Now Public, Micron Web site: *www.micron.com*, Micron Technology, Inc., Oct 15, 2001.

[Mic02a] QDR SRAM Design Guide, Technical Note TN-54-01, Rev. 10/02, Micron Technology, Inc.

[Mic03a] 2-Input XOR/SNOR, SY10EL07, SY100EL07 Final, Rev. F, Data Sheet, Micrel, Feb 2003.

[Mic03b] Data Sheet, SuperLite™ SY55852U D Flip-Flop, Rev. B, Micrel Inc., Mar 2003.

[Mic03c] Jeff Janzen, "DDR2 Offers New Features and Functionality," *Designline*, vol. 12, no. 2, Micron Technology, Inc. (Boise, Idaho), 3Q, 2003.

[Mic03d] Next-Generation Network SRAM, slide presentation, *www. qdrsram.com*, Micron Technology, Inc., 2003.

[Mic03e] Graphics DDR3 On-Die Termination And Thermal Considertions, Technical Note TN-44-01, Rev. A, Micron Technology, Inc., May 2003.

[Mic03f] The Future of Memory: GDDR3 SDRAM Functionality, Technical Note TN-44-01, Rev. A, Micron Technology, Inc., Jan 2003.

[Mid63] R. D. Middlebrook, *Differential Amplifiers*, New York: Wiley, 1963.

[Min03] S-Parameter Data, Model ERA-5SM Amplifier, Mini-Circuits, 2003.

[Mon00] Mark I. Montrose, *Printed Circuit Board Design Techniques for EMC Compliance*, Second Edition, Piscataway, NJ: IEEE Press, 2000.

[Mot01a] With Strong Commitment to Roadmap, Motorola to Introduce Silicon Germanium Family of High-End Timing Solutions, Smart Networks Developer Forum, New Orleans, LA, press release, May 21, 2001.

[Mot02a] WarpLink™ Reference Design Platform, white paper, BR1570/D, Rev. 0, Motorola Inc., 2002.

[Mot02b] MPC603EUM: MPC603e RISC Microprocessor Users Manual, *http://e-www.motorola.com*, Rev. 3, Q2, 2002, p. xxvii.

[Mot03] Overview, Motorola SerDes Transceivers, *http://e-www.motorola.com*, 2003.

[Mu02] Zhen Mu, Combining Multiple Simulations to 12.5 Gbps to Guide CAD Tool Use, Sycamore Networks, Inc., 2002.

[Nat98] DS36276 FAILSAFE Multipoint Transceiver, data sheet, National Semiconductor, July 1998.

[Nat00] *LVDS Owner's Manual*, National Semiconductor, Second Edition, Santa Clara, CA, Summer 2000.

[Nat02a] DS92LV16 Design Guide, Rev. 1.01, National Semiconductor Corporation, Feb 2002.

[Nat02b] Data Sheet, DS25C400, preliminary, National Semiconductor Corporation, Sep 2002.

[Nat02c] Data Sheet, DS92LV8028, National Semiconductor Corporation, Sep 2002.

[Nat02d] IBIS Code, DS90C031TM, LVDS Quad Driver, *www.national.com*, National Semiconductor Corporation, 2002.

[Nat03] Definition of S-Parameters, Web site, NI Developer Zone, Measurement Encyclopedia, National Instruments.

[Nik03] Ronald Nikel, Designing 5-Gbit/s Serdes: Steering Through a Road Filled with Potholes, *www.CommsDesign.com/story/OEG20030327S0010*, Mar 27, 2003.

[Ode99] C. Odegard and C. Lambert, "Comparative TDR Analysis as a Packaging FA Tool," *Proceedings from the 25th International Symposium for Testing and Failure Analysis*, Santa Clara, CA, Nov 14–18, 1999.

[Ons88] MECL System Design Handbook, Rev. 1A, HB205/D, ON Semiconductor, May 1988.

[Ons02a] Clock and Data Management Solutions: Designing at High Speed Using ECL and SiGe (High Performance Data Management, Advanced Clock Management and Communication ASSPs), Application Note BRD8017/D, Rev. 4, ON Semiconductor, Apr 2002.

[Ons02c] Data Sheet, MC100EP16VS, 3.3V/5VECL Differential Receiver/Driver with Variable Output Swing, ON Semiconductor, Rev. 2, Jan 2002.

[Ons03a] GigaComm Device Data, Application Note DLD604/D, ON Semiconductor, *www.onsemi.com*, 2003.

[Ons03b] NBSG11 (2.5V/3.3V SiGe 1:2 Differential Clock Driver with RSECL Outputs), Data Sheet, Rev. 6, ON Semiconductor, *http://onsemi.com*, Apr 2003.

[Ons03c] Mark Barrera, ECLinPSE, ECLinPS LiteE, ECLinPS PlusE and GigaCommE Marking and Ordering Information Guide, Application Note, AND8002/D, Rev. 2, ON Semiconductor, Apr 2003.

[Pcd02a] Jamie Metcalfe, Jack Woida, Robert Bielby, and others, "Emerging Technologies and Industry Trends," *Printed Circuit Design*, Sep 2002.

[Pcg03] The PC Guide, *www.PCGuide.com*, 2003.

[Pen95] Raymond S. Pengelly, *Microwave Field-Effect Transistors: Theory, Design, and Applications*, Hardcover Third Edition, Noble Publishing, Dec 1995.

[Phi02a] GTLPH16612/GTL1655 Data Sheet, Philips Semiconductor, Nov 2002.

[Phi03a] GTLP explained, Philips Semiconductor, 2003.

[Pic03] Small Form Factor Pluggable (SFP) Longwave Transceiver, *www.picolight.com*, PicoLight, 2003.

[Pon98] George Ponchak, Alan Downey, "Characterization of Thin Film microstrip Lines on Polyimide," *IEEE Transactions on Components, Packaging, and Manufacturing Technology Part B*, vol. 21, no. 2, May 1998, pp. 171–176.

[Pon01] Peter Pondillo, Optical Fiber Local Area Networks: Bandwidth, Data Rate, and Link Length — What Does It All Mean?, white paper WP410, Corning Incorporated, Corning, NY, Oct 2001.

[Pow98] Jon Powell and Don Mazur, "IBIS Evolves: Keeping Pace with Signal Integrity Issues," *Printed Circuit Design*, Oct 1998, p. 34.

[Poz03] David M. Pozar, *Microwave Engineering*, Second Edition, New York: John Wiley & Sons, 2003.

[Ram01a] Gigahertz Rambus Signaling Technologies: RSL, QRSL, and SerDes Technology Overview, Rambus Inc., Los Altos, CA, June 2001.

[Ram01b] RDRAM Memory: Leading Performance and Value over SDRAM and DDR, Rambus, Inc., Document WP0001-R, Ver. 1.2, Aug 2001.

[Ram01c] RDRAM: Maximizing the Value of PCs and Workstations, Rambus, Inc., Document WP0003-R, Ver. 1.0, Aug 2001.

[Ram01d] RDRAM Performance Anaylsis: 2x16d versus 4i Bank Architecture, Rambus, Inc., Los Altos, CA, Document WP0002, Ver. 1.0, Apr 2001.

[Ram02] Direct Rambus Long Channel Design Guide, DL-0102-1.4, Rambus, Inc., 2002.

[Ram03a] RDRAM Summary, High Performance Memory Interface Solution, Rambus, 2003.

[Ram03b] XDR DRAM: System Design Overview, Rambus, Inc., *www.rambus.com*, 2003.

[Ram03c] Differential Rambus Signaling Levels, Rambus, Inc., *www.rambus.com*, 2003.

[Reb02] Jagdish Rebello, "10 Gigabit Ethernet in LANs, MANs, and WANs: Real or Just Plain Hype?" *i-Suppli*, Apr 2002.

[Roh68] R. A. Rohrer, "Lumped Network Passivity Criteria," *IEEE Trans. Circuit Theory*, vol. CT-15, no. 1, Mar 1968, pp. 24–29.

[Sam03a] 512Mb M-die DDR2 SDRAM Specification, Preliminary, Rev. 0.92, Samsung Electronics, June 2003.

[Sam03b] Data Sheet, 256M GDDR SDRAM, K4D551638D-TC, Rev 1.3, Samsung Electronics, Apr 2003.

[Sam03c] Success in Graphics Is All About Partnerships, Position Paper CG2040-A, Samsung Semiconductor, Inc., Mar 7, 2003.

[Say98] Edward P. Sayre, Jinhua Chen, and Michael Baxter, Signal Integrity and Validation of National's Bus LVDS (BLVDS) Technology in Heavily Loaded Backplanes, Technical white paper, North East Systems Associates, Inc. (*www.nesa.com*), Westford, MA, May 1998.

[Say02a] Edward P. Sayre, Michael Baxter, and others, Signal Integrity and Validation of National's Bus LVDS (BLVDS) Technology in Heavily Loaded Backplanes, North East System Associates (NESA) (*www.nesa.com*), Westford, MA.

[Say03] Edward P. Sayre, Edward P. Sayre, III, and Michael A. Baxter, DDR-II SDRAM Technology and Design Trends, North East Systems Associates, Inc. (NESA) (*www.nesa.com*), Westford, MA, 2003.

[Sch01] Henning Schroder, "Photonic Advantages and Challenges for Polymers," *Second Optoelectronics Packaging Workshop*, Austin, TX, Aug 21–22, 2001, pp. B1–B22.

[Sho02a] Paul Shockman, Interfacing with ECLinPS, AND8066/D, Rev. 2, ON Semiconductor, May 2002.

[Sho02b] Paul Shockman, Termination of ECL Logic Devices, Publication Number AND8020/D, Rev. 4, ON Semiconductor, Aug 2002.

[Sho96] Masakazu Shoji, *High-Speed Digital Circuits*, Reading, MA: Addison-Wesley, 1996.

[Sig01a] SigmaRAM Targets High Speed Networking Applications, white paper, SigmaRAM Consortium, *www.SigmaRAM.com*, June 2001.

[Sig01b] SigmaRAM Consortium Announces 24-Gb/s, 18-Mbit, 333-MHz Synchronous SRAM Product Family for Networking and Telecom Markets, press relese, SigmaRAM Consortium, *www.SigmaRAM.com*, June 18, 2001.

[Sil03] SiliconPipe Technology Makes 40Gbps Copper-Backplane a Reality, press release, SiliconPipe, Nov 11, 2003.

[Sil04] Silicon Pipe Web site, *www.siliconpipe.com*, 2004.

[Sin03] Michael Singer, NEWCARD Reborn As ExpressCard, *www.Wi-fiplanet.com*, Sep 16, 2003.

[Smo99] D. A. Smolyansky and S. D. Corey, "Printed Circuit Board Interconnect Characterization from TDR Measurements," *Printed Circuit Design Magazine*, May 1999, pp. 18–26, (TDA Systems Application Note PCBD-0699).

[Smo00a] D.A. Smolyansky, "Electronic Package Failure Analysis Using TDR," *Proceedings from the 26th International Symposium for Testing and Failure Analysis*, Bellevue, WA, 2000.

[Smo00b] D. A. Smolyansky and S. D. Corey, "Characterization of Differential Interconnects from Time Domain Reflectometry Measurements," *Microwave Journal*, vol. 43, no. 3, Mar 2000, pp. 68–80, (TDA Systems application note DIFF-1199).

[Smo01] D. A. Smolyansky, "TDR Techniques for Characterization and Modeling of Electronic Packaging," *High Density Interconnect Magazine*, Mar 2001 (Part 1) and Apr 2001 (Part 2), (TDA Systems application note PKGM-0101).

[Smo02] D. Smolyansky and S. Corey, M. Resso, "Choosing The Right Signal Integrity Tools for InfiniBand Measurements," *DesignCon 2002*, Santa Clara, CA, Jan 2002, (TDA Systems Application Note TDFD-0202).

[Sta00] William Stallings, *Data & Computer Communications*, Sixth Edition, Upper Saddle River, NJ: Prentice-Hall, Inc., 2000.

[Tag01] M. R. Taghizadeh, H. Thienpont, and G. E. Jabbour (Eds.), "Micro and Nano-optics for Optical Interconnection and Information Processing," *Proceedings of SPIE*, 2001, vol. 4455, pp. 247–250.

[Tda00] Characterization of Differential Interconnects from TDR Measurements, Application Note DIFF-1199, TDA Systems, Mar 2000.

[Tda01] TDR Techniques for Characterization and Modeling of Electronic Packaging, TDA Systems application note PKGM-0101; also in *HDI Magazine*, Mar and Apr 2001 (published in two parts).

[Tda02a] TDR Primer, Application Note TDRP-0402, TDA Systems, Apr 2002. Also printed in *Circuit Design Magazine*, Apr 2002.

[Tda02b] Computing Self and Mutual Capacitance and Inductance Using Even and Odd TDR Measurements, Application Note CECO-0502, TDA Systems, 2002.

[Tda02c] Choosing Signal Integrity Measurement Tools: Time or Frequency Domain?, Application Note TDFD-0202, TDA Systems, Jan 2002.

[Tek01a] TDR Impedance Measurements: A Foundation for Signal Integrity, Application Note, Tektronix, Apr 2001.

[Tek03] "Differential Ohms Measurement with the 11800-Series Oscilloscope," Tektronix Technical Brief 47W- 7520.

[Ter03] Materials & Processes, for Optical Backplanes, for Telecom Components, Terahertz Photonics Web site, *www.thzonline.com*, 2003.

[Tex01] TLK3101 2.5 Gbps to 3.125 Gbps Transceiver Data Sheet, Texas Instruments, Revised Feb 2001.

[Tex02a] SN65CML100 1.5-Gbps LVDS/LVPECL/CML-To-CML Translator/Repeater, Data Sheet, Texas Instruments, Nov 2002.

[Tex02b] TLK2501 1.5 to 2.5 Gbps Transceiver, Data Sheet, Texas Instruments, Feb 2002.

[Tex02c] Comparing Bus Solutions, Application Note SLLA067, Texas Instruments, Mar 2002.

[Tex02d] SLL Advanced Bus Interface Logic Products, Selection Guide and Reference, Bus Interface Solutions Overview, Texas Instruments, 2002.

[Tex03a] SN65LVDM179 Full-Duplex LVDM Transceiver Data Sheet, Texas Instruments, Mar 2003.

[Tha00] Jonathan Thatcher, "10Gigabit Ethernet: Brief Introduction to the IEEE 802.3ae project," *Networld + Interop* Presentation, May 2000.

[Tia01] TIA/EIA-644-A, Electrical Characteristics of Low Voltage Differential Signaling (LVDS) Interface Circuits, Telecommunications Industry Association, Feb 2001.

[Tia02] TIA/EIA-899, Electrical Characteristics of Multipoint-Low-Voltage Differential Signaling (M-LVDS) Interface Circuits for Multipoint Data Interchange, Telecommunications Industry Association, Feb 26, 2002.

[Til92] M.D. Tilden, "Measuring Controlled-Impedance Boards with TDR," *Printed Circuit Fabrication*, Feb 1992, Tektronix Application Note 85W-8531-0.

[Tos03a] Data Sheet, Network FCRAM, TC59LM913/05AMB-50, -55, -60, *www.toshiba.com*, Toshiba, Apr 21, 2003.

[Tos03b] Read-While-Write (RWW) Flash: It's Not Just for Cellular Phones Anymore, Toshiba, 2003.

[Ven92] George Vendelin, Anthony M. Pavio, and Ulrich L. Rohde, *Microwave Circuit Design Using Linear and Nonlinear Techniques*, John Wiley & Sons, Reprint Edition, May 1992.

[Vic04] Brian Vicich, Scott McMorrow, and others, "Advances in Design, Modeling, Simulation and Measurement Validation of High Performance Board-to-Board 5-to-10 Gbps Interconnects," *DesignCon West*, 2004.

[Wai95] James R. Wait, *Electromagnetic Waves in Stratified Media* (IEEE/OUP Series on Electromagnetic Wave Theory), Reprint Edition, July 1995.

[Wak01] John F. Wakerly, *Digital Design: Principles & Practices*, Third Edition Updated, Upper Saddle River, NJ: Prentice-Hall, Inc., 2001.

[Wat01] Jim Watson, "Chip-to-Chip Optical Interconnect—Optical Pipedream?," *OIDA Integration and Packaging Meeting*, Boulder, CO, July 24–25, 2001.

[Waw02] Richard Wawrzyniak, Key Trends in the Cellular Phone Market for Flash Technology, Semico Research Corporation, Spring 2002.

[Web03] Webopedia, *www.webopedia.com*, Jupitermedia Corporation, 2003.

[Wei03] Curt Weinstein, Selecting the Right Transmission Medium, *www.corning.com*, Corning, Inc., 2003.

[Wha03] IT-Specific Encyclopedia, Whatis.com, *http://whatis.techtarget.com*, 2003.

[Wil03] Ron Wilson, "Startup Takes On Memory/Core-Logic Link," *EE Times*, Nov 10, 2003.

[Wu01] Francis Wu, "Designing with Agilent's Small Factor Hot-Pluggable Optical-Transceiver," *ONIDS*, 2001.

[X2m02a] A Cooperation Agreement for a Small Versatile 10 Gigabit Transceiver Package, X2 MSA group, Issue 0.9, July 31, 2002.

[X2m02b] Industry Leaders Announce "X2" Multi-Source Agreement for 10 Gigabit Pluggable Optical Transceivers, X2 MSA group, July 22, 2002.

[Xil01] High Speed Transceiver Logic (HSTL), Virtex Tech Topic, VTT008 (v2.0), Xilinx Corporation, May 5, 2001.

[Xil02a] Synthesizable FCRAM Controller, Application Note XAPP266 (1.0), Xilinx, Inc., Feb 27, 2002.

[Xil03a] Austin Lesea, Signal Integrity: Tips and Tricks, Xilinx Corporation, 2003.

[Xil03b] Memory Corner — RLDRAM, Overview, *www.xilinx.com*, Xilinx, Inc. 2003.

[Xil03c] Xilinx Memory Selection Criteria, Xilinx Web site: *www.xilinx.com*, Xilinx, Inc. 2003.

[Xpa02a] A Cooperation Agreement for a Small Form Factor Pluggable 10 Gbit/s Transceiver Package, Xpak MSA Group, Revision 2.2, Dec 2002.

[Xpa02b] Xpak MSA Group Announces Build-to-Specification Availability, Enhancements to Small-Form-Factor 10-Gigabit Ethernet Standard, Xpak MSA Group press release, Sep 3, 2002.

[You01] Brian Young, *Digital Signal Integrity: Modeling and Simulation with Interconnects and Packages*, Upper Saddle River, NJ: Prentice Hall PTR, 2001.

[Yue98] Heping Yue, Kathleen L. Virga, and John L. Prince, "Dielectric Constant and Loss Tangent Measurement Using a Stripline Fixture," *IEEE Transactions on Components, Packaging, and Manufacturing Technology Part B*, vol. 21, no. 4, Nov 1998, pp. 441–446.

About the Author

Tom Granberg has earned several technical degrees — a B.S. in physics from Washington State University, an M.S. and a Ph.D. in electrical engineering from the University of Missouri-Columbia. He also holds an M.B.A. from the University of Colorado at Denver. Tom has worked for dominant networking companies Cisco Systems and SkyStream Networks and in ASIC design emulation at Quickturn Design Systems (a Cadence Company). He has also worked in digital signal processing, digital imaging systems, and flat panel sensors, and at companies including Condor Systems, Martin Marietta (now Lockheed Martin), Storage Technology, and Honeywell Test Instruments. He lives in Santa Clara, California — the heart of Silicon Valley — and enjoys access to many of the world's newest technologies.

INDEX

C

D

F

M

N

O

P

Q

R

W

X

Y